EIGHTH [illegible]

Biolog

VOLUME 1

Jonathan B. Losos
Harvard University

Kenneth A. Mason
Purdue University

Susan R. Singer
Carleton College

based on the work of

Peter H. Raven
Director, Missouri Botanical Gardens;
Engelmann Professor of Botany
Washington University

George B. Johnson
Professor Emeritus of Biology
Washington University

Boston Burr Ridge, IL Dubuque, IA New York San Francisco St. Louis
Bangkok Bogotá Caracas Lisbon London Madrid
Mexico City Milan New Delhi Seoul Singapore Sydney Taipei Toronto

Biology, EIGHTH EDITION
VOLUME 1

Copyright © 2008 by The McGraw-Hill Companies, Inc. All rights reserved. Printed in the United States of America. Except as permitted under the United States Copyright Act of 1976, no part of this publication may be reproduced or distributed in any form or by any means, or stored in a data base retrieval system, without prior written permission of the publisher.

This book is a McGraw-Hill Learning Solutions textbook and contains select material from *Biology*, Eighth Edition by Jonathan B. Losos, Kenneth A. Mason, Susan R. Singer, based on the work of Peter H. Raven and George B. Johnson. Copyright © 2008 by The McGraw-Hill Companies, Inc. Reprinted with permission of the publisher. Many custom published texts are modified versions or adaptations of our best-selling textbooks. Some adaptations are printed in black and white to keep prices at a minimum, while others are in color.

4 5 6 7 8 9 0 QSR QSR 0 9 8

ISBN 13: 978-0-07-333748-7
ISBN-10: 0-07-333748-X

Editor: Shirley Grall
Production Editor: Jessica Portz
Printer: Digital Impressions
Binder: Quebecor World

Multilevel figures

take students from a macro to micro view using "blow-out" arrows to help students put concepts into context.

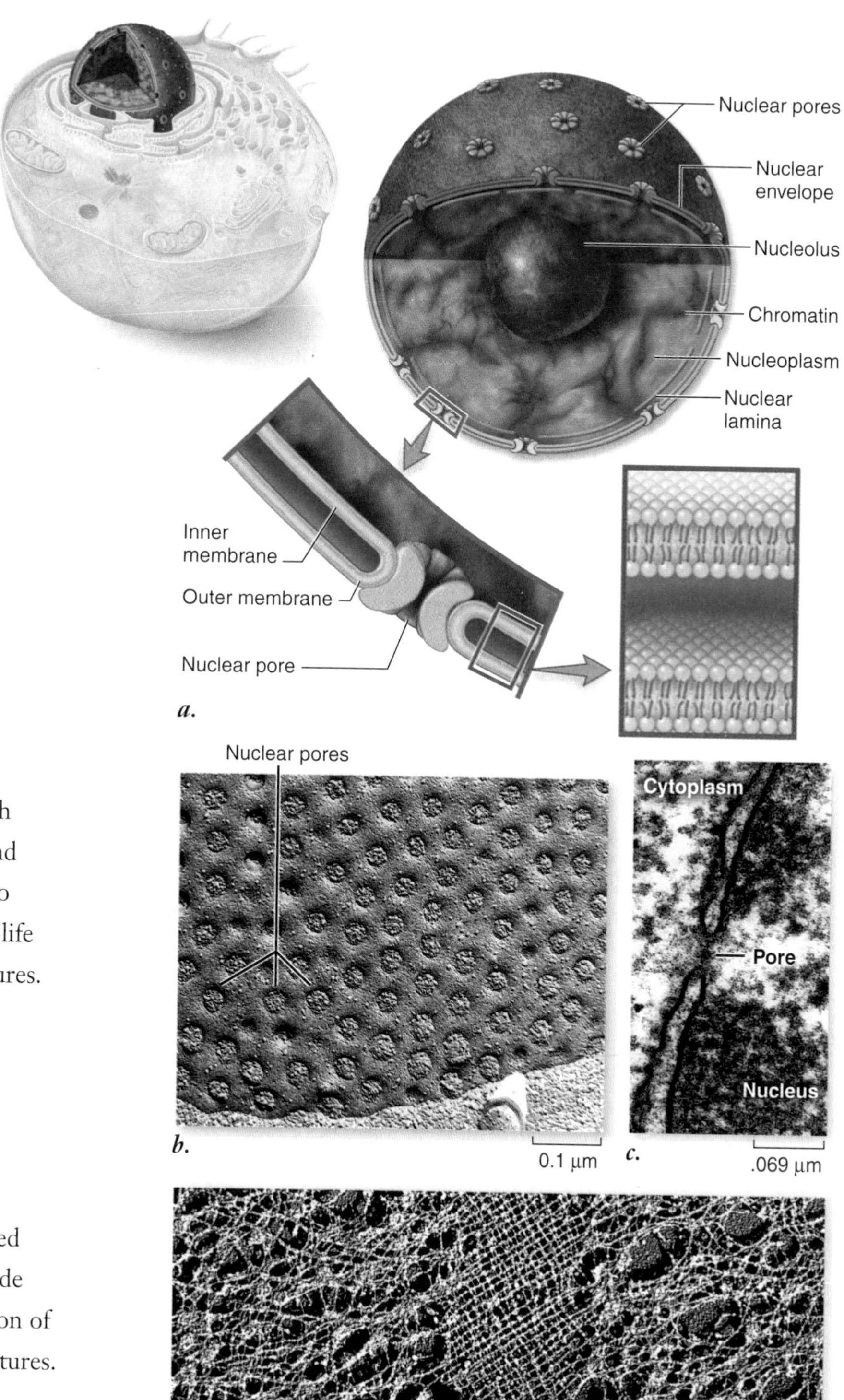

Illustrations are paired with high-quality LM, SEM, and TEM photomicrographs to provide students with real-life examples of cellular structures.

Whenever possible, a measurement bar is provided with a micrograph to provide students with an appreciation of the scale of biological structures.

Consistent color coding

means that students immediately recognize the biological structures used throughout the book. Their study time is spent learning concepts rather than orienting themselves to figure conventions. In some figures, color coding is also used to give the student visual cues to how information is related.

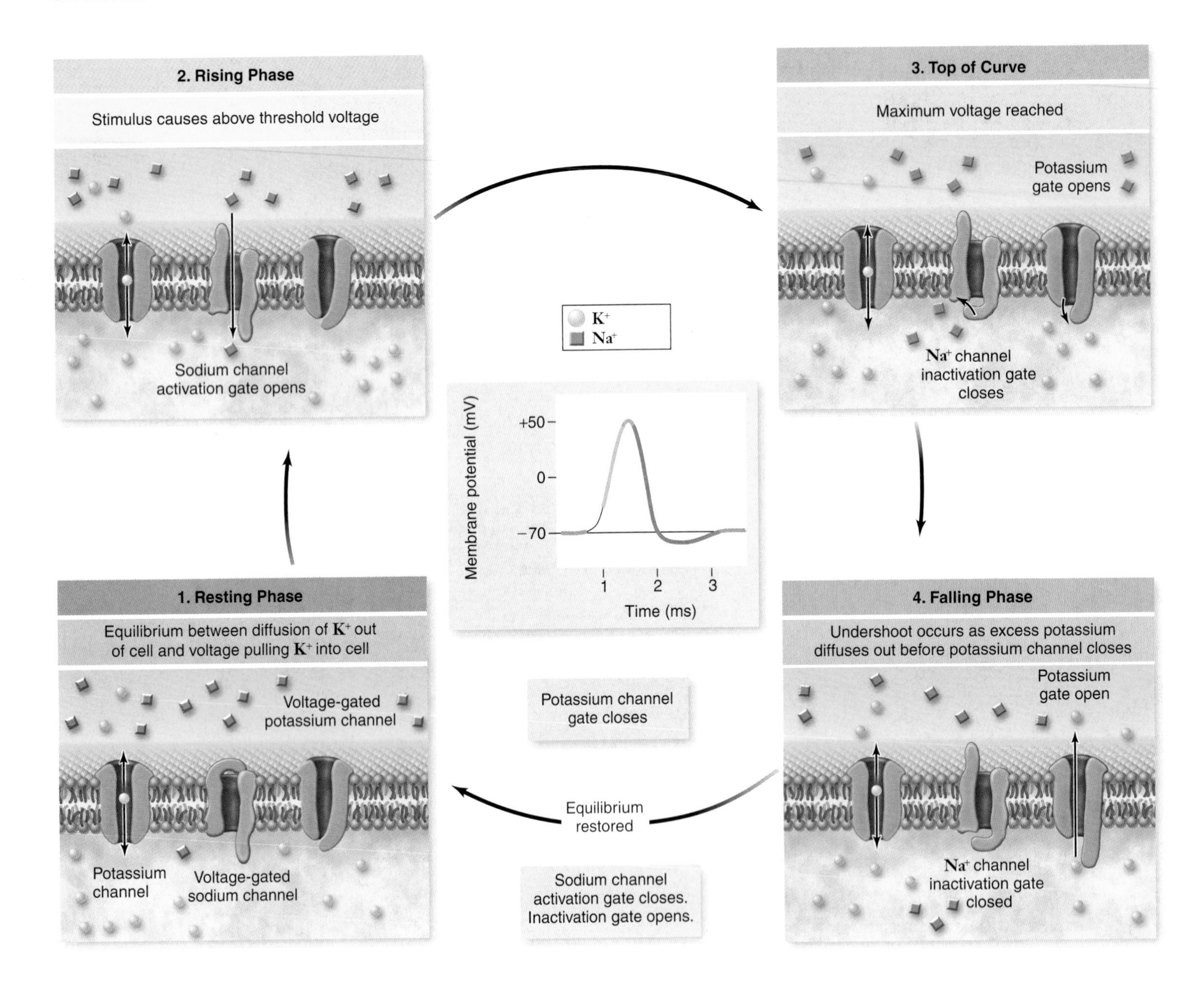

Consistent Pedagogical Aids to Promote Learning

Each chapter in the eighth edition is structured using the same set of pedagogical devices, which enables the student to develop a consistent learning strategy. These tools work together to provide a clear content hierarchy, break content into smaller, more accessible chunks, repeat important concepts, and provide students with opportunities for higher level thought.

chapter 24

Genome Evolution

introduction

GENOMES CONTAIN THE RAW MATERIAL for evolution, and many clues to evolution are hidden in the ever-changing nature of genomes. As more genomes have been sequenced, the new and exciting field of comparative genomics has emerged and has yielded some surprising results and many, many questions. Comparing whole genomes, not just individual genes, enhances our ability to understand the workings of evolution, to improve crops and to identify the genetic basis of disease so that we may develop more effective treatments with minimal side effects. The focus of this chapter is on how comparative genomics is enhancing our understanding of genome evolution and how this new knowledge can be applied to improve our lives.

concept outline

24.1 Comparative Genomics
- *Evolutionary differences accumulate over long periods*
- *Genomes evolve at different rates*
- *Plant, fungal, and animal genomes have unique and shared genes*

24.2 Evolution of Whole Genomes
- *Ancient and newly created polyploids guide studies of genome evolution*
- *Plant polyploidy is ubiquitous, with multiple common origins*
- *Polyploidy induces elimination of duplicated genes*
- *Polyploidy can alter gene expression*
- *Transposons jump around following polyploidization*

24.3 Evolution Within Genomes
- *Individual chromosomes may be duplicated*
- *DNA segments may be duplicated*
- *Genomes may become rearranged*
- *Gene inactivation results in psuedogenes*
- *Horizontal gene transfer complicates matters*

24.4 Gene Function and Expression Patterns
- *Chimp and human gene transcription patterns differ*
- *Speech is uniquely human: An example of complex expression*

24.5 Nonprotein-Coding DNA and Regulatory Function

24.6 Genome Size and Gene Number
- *Noncoding DNA inflates genome size*
- *Plants have widely varying genome size*

24.7 Genome Analysis and Disease Prevention and Treatment
- *Distantly related genomes offer clues for causes of disease*
- *Closely related organisms enhance medical research*
- *Pathogen–host genome differences reveal drug targets*

24.8 Crop Improvement Through Genome Analysis
- *Model plant genomes provide links to genetics of crop plants*
- *Beneficial bacterial genes can be located and utilized*

471

Chapter openers

include an outline comprised of the chapter headings, which provides a consistent framework for the student. Declarative, numbered main headings and sentence-style supporting headings result in a cogent overview of the content to be covered.

1. Interim summaries review key points from the section so students can easily identify the take-away message.

2. Numbered main headings clearly identify the start of a new concept section.

3. Inquiry questions challenge students to think about what they are reading at a more sophisticated level.

are shown in figure 55.11. Oysters produce vast numbers of offspring, only a few of which live to reproduce. However, once they become established and grow into reproductive individuals, their mortality rate is extremely low (type III survivorship curve). Note that in this type of curve, survival and mortality rates are inversely related. Thus, the rapid decrease in the proportion of oysters surviving indicates that few individuals survive, thus producing a high mortality rate. In contrast, the relatively flat line at older ages indicates high survival and low mortality.

In hydra, animals related to jellyfish, individuals are equally likely to die at any age. The result is a straight survivorship curve (type II).

Finally, mortality rates in humans, as in many other animals and in protists, rise steeply later in life (type I survivorship curve).

Of course, these descriptions are just generalizations, and many organisms show more complicated patterns. Examination of the data for *P. annua*, for example, reveals that it is most similar to a type II survivorship curve (figure 55.12).

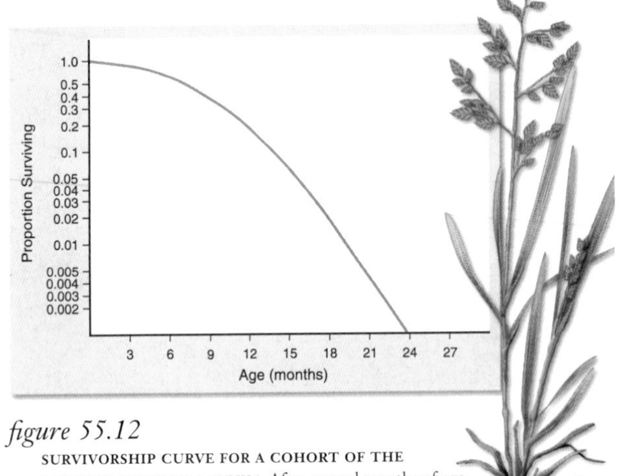

figure 55.12

SURVIVORSHIP CURVE FOR A COHORT OF THE MEADOW GRASS *POA ANNUA*. After several months of age, mortality increases at a constant rate through time.

1

The growth rate of a population is a sensitive function of its age structure. The age structure of a population and the manner in which mortality and birthrates vary among different age cohorts determine whether a population will increase or decrease in size.

inquiry

?

Suppose you wanted to keep meadow grass in your room as a houseplant. Suppose, too, that you wanted to buy an individual plant that was likely to live as long as possible. What age plant would you buy? How might the shape of the survivorship curve affect your answer?

3

55.4 Life History and the Cost of Reproduction

2

Natural selection favors traits that maximize the number of surviving offspring left in the next generation. Two factors affect this quantity: how long an individual lives, and how many young it produces each year.

Why doesn't every organism reproduce immediately after its own birth, produce large families of offspring, care for them intensively, and perform these functions repeatedly throughout a long life, while outcompeting others, escaping predators, and capturing food with ease? The answer is that no one organism can do all of this, simply because not enough resources are available. Consequently, organisms allocate resources either to current reproduction or to increasing their prospects of surviving and reproducing at later life stages.

The complete life cycle of an organism constitutes its **life history.** All life histories involve significant trade-offs. Because resources are limited, a change that increases reproduction may decrease survival and reduce future reproduction. As one example, a Douglas fir tree that produces more cones increases its current reproductive success—but it also grows more slowly. Because the number of cones produced is a function of how large a tree is, this diminished growth will decrease the number of cones it can produce in the fu-

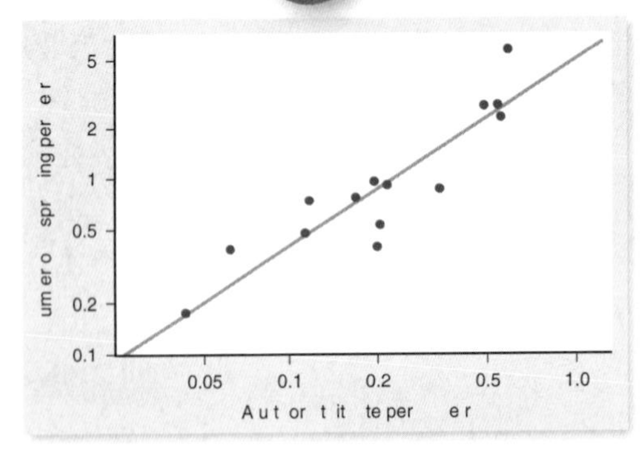

figure 55.13

REPRODUCTION HAS A PRICE. Data from many bird species indicate that increased fecundity in birds correlates with higher mortality, ranging from the albatross (lowest) to the sparrow (highest). Birds that raise more offspring per year have a higher probability of dying during that year.

1154 *part* VIII *ecology and behavior*

1. **Summary Tables** are used extensively to help students study and review the chapter content. Illustrations are added in some cases to further aid students in recall.

2. **Self Test Questions** are a mixture of knowledge and comprehension questions that test a student's basic understanding of the main concepts from the chapter.

3. **Challenge Questions** are application and analysis questions that measure a student's ability to use terms and concepts learned from the chapter in new situations.

4. **Concept Review** summarizes the main concepts from the chapter and their supporting ideas. Key figures are cited to alert students to particularly relevant illustrations.

1

TABLE 15.2 Differences Between Prokaryotic and Eukaryotic Gene Expression

Characteristic	Prokaryotes	Eukaryotes
Introns	No introns, although some archaeal genes possess them.	Most genes contain introns.
Number of genes in mRNA	Several genes may be transcribed into a single mRNA molecule. Often these have related functions and form an operon. This coordinates regulation of biochemical pathways.	Only one gene per mRNA molecule; regulation of pathways accomplished in other ways.
Site of transcription and translation	No membrane-bounded nucleus, transcription and translation are coupled.	Transcription in nucleus; mRNA moves out of nucleus for translation.
Initiation of translation	Begins at AUG codon preceded by special sequence that binds the ribosome.	Begins at AUG codon preceded by the 5′ cap (methylated GTP) that binds the ribosome.
Modification of mRNA after transcription	None; translation begins before transcription is completed.	A number of modifications while the mRNA is in the nucleus: Introns are removed and exons are spliced together; a 5′ cap is added; a poly-A tail is added.

review questions

2

SELF TEST

1. Which of the following statements is NOT part of … theory?
   a. All organisms are composed of one or more cells.
   b. Cells come from other cells by division.
   c. Cells are the smallest living things.
   d. Eukaryotic cells have evolved from prokaryotic cells.
2. The most important factor that limits the size of a cell is—
   a. the amount of proteins and organelles that can be made by a cell
   b. the rate of diffusion
   c. the surface-area-to-volume ratio of the cell
   d. the amount of DNA in the cell
3. What type of microscope would you use to examine the surface details of a cell?
   a. Compound light microscope
   b. Transmission electron microscope
   c. Scanning electron microscope
   d. Confocal microscope
4. All cells have all of the following except—
   a. Plasma membrane
   b. Genetic material
   c. Cytoplasm
   d. Cell wall
5. Eukaryotic cells are more complex than prokaryotic cells. Which of the following would you NOT find in a prokaryotic cell?
   a. Cell wall
   b. Plasma membrane
   c. Nucleus
   d. Ribosomes
6. The difference between a gram-positive and gram-negative bacteria is—
   a. the thickness of the peptidoglycan cell wall
   b. the type of polysaccharide present in the cell wall
   c. the type and amount of protein in the cell wall

11. Proteins can move from the Golgi apparatus to—
    a. the extracellular fluid
    b. transport vesicles
    c. lysosomes
    d. all of the above
12. Lysosomes function to—
    a. carry proteins to the surface of the cell
    b. add short-chain carbohydrates to make glycoproteins
    c. break down organelles, proteins, and nucleic acids
    d. remove electrons and hydrogen atoms from hydrogen peroxide
13. What do chloroplasts and mitochondria have in common?
    a. Both are present in animal cells.
    b. Both have an outer membrane and an elaborate inner membrane.
    c. Both are present in all eukaryotic cells.
    d. Both organelles function to produce glucose.
14. Eukaryotic cells are composed of three types of cytoskeletal filaments. How are these three filaments similar?
    a. They contribute to the shape of the cell.
    b. They are all made of the same type of protein.
    c. They are all the same size and shape.
    d. They are all equally dynamic and flexible.
15. Animal cells connect to the extracellular matrix through—
    a. glycoproteins
    b. fibronectins
    c. integrins
    d. collagen

3

CHALLENGE QUESTIONS

1. Eukaryotic cells are typically larger than prokaryotic cells (refer to figure 4.2). How might the difference in the cellular structure of eukaryotic versus a prokaryotic cell help to explain this observation?
2. The smooth endoplasmic reticulum is the site of synthesis of the phospholipids that make up all the membranes of a cell—especially the plasma membrane. Use the diagram of an animal cell (figure 4.6) to trace a pathway that would carry a phospholipid molecule from the SER to the plasma membrane. What endomembrane compartments would the phospholipids travel through? How can a phospholipid molecule move between membrane compartments?
3. Use the information provided in table 4.3 to develop a set of predictions about the properties of mitochondria and chloroplasts if these organelles were once free-living prokaryotic cells. How do your predictions match with the evidence for endosymbiosis?
4. In evolutionary theory, homologous traits are those with a similar structure and function derived from a common ancestor. Analogous traits represent adaptations to a similar environment, but from distantly related organisms. Consider the structure and function of the flagella found on eukaryotic and prokaryotic cells. Are the flagella an example of a homologous or analogous trait? Defend your answer.
5. The protist, *Giardia lamblia*, is the organism associated with water-borne diarrheal diseases. *Giardia* is an unusual eukaryote because it seems to lack mitochondria. Explain the existence of a mitochondria-less eukaryote in the context of the endosymbiotic theory.

concept review

4

4.1 Cell Theory

Modern cell theory states that organisms are composed of one or more cells. Cells are the smallest unit of life and arise from preexisting cells.

- Cell size is constrained by the effective distance of diffusion within a cell, from the surface to the interior of the cell.
- As a cell increases in size the surface area increases as a square function and the volume increases as a cubic function.
- Large cells deal with the diffusion problem by having more than one nucleus or by becoming flattened or elongated.
- The visualization of cells and their components is facilitated by microscopes and staining cell structures.
- All cells have DNA, a cytoplasm, a plasma membrane, and ribosomes.

4.2 Prokaryotic Cells (figure 4.3)

Prokaryotic cells do not have a nucleus or an internal membrane system, and they lack membrane-bounded organelles.

- The plasma membrane is surrounded by a rigid cell wall that maintains shape and helps maintain osmotic balance.
- The plasma membrane in some prokaryotes is infolded and provides similar functions to eukaryotic internal membranes.
- Bacteria have a cell wall made up of peptidoglycan. Archaeal cell walls have different architecture.
- Archaeal plasma membranes differ from bacteria and eukaryotes.
- The plasma membrane in some archaea is a monolayer composed of saturated lipids attached to glycerol at each end.
- Structurally Archaea resemble prokaryotes, but functionally they more closely resemble eukaryotes.
- Prokaryotic flagella rotate because of proton transfer.

4.3 Eukaryotic Cells (figures 4.6 and 4.7)

Eukaryotic cells have a membrane-bounded nucleus, an endomembrane system, and many different organelles.

- The nucleus contains genetic information.
- The nuclear envelope consists of two phospholipid bilayers; the outer layer is contiguous with the ER.
- The inside of the nuclear envelope is covered with nuclear lamins, which maintain the shape of the nucleus.
- Nuclear pores allow exchange of small molecules between the nucleoplasm and the cytoplasm.
- DNA is organized with proteins into chromatin.
- The nucleolus is a region of the nucleoplasm where rRNA is transcribed and ribosomes are assembled.
- Ribosomes are composed of RNA and protein and use information in mRNA to direct the synthesis of proteins.

4.4 The Endomembrane System

The endomembrane system forms compartments and vesicles and provides channels to carry molecules and surfaces for synthesis of macromolecules.

- The endoplasmic reticulum (ER) creates channels and passages within the cytoplasm (figure 4.11).
- The interior compartment of the ER is called the cisternal space, or lumen.
- The rough endoplasmic reticulum (RER) has ribosomes on the surface and is composed mainly of flattened sacs. RER is involved in protein synthesis and modification.
- The smooth endoplasmic reticulum (SER) lacks ribosomes and is composed more of tubules. SER is involved in synthesis of carbohydrates and lipids and in detoxification.
- The Golgi apparatus receives vesicles from the ER on the *cis* face, modifies and packages macromolecules, and transports them in vesicles formed on the *trans* face (figure 4.13).
- Lysosomes are vesicles containing enzymes that break down macromolecules located in food vacuoles and recycle the components of old organelles (figure 4.14).
- Microbodies contain enzymes and grow by incorporating lipids and proteins before they divide.
- Peroxisomes contain enzymes that catalyze oxidation reactions, resulting in the formation of hydrogen peroxide.
- Plants have many specialized vacuoles. The conspicuous central vacuole, surrounded by the tonoplast membrane, is used for storage, maintaining water balance, and growth.

4.5 Mitochondria and Chloroplasts: Cellular Generators

Mitochondria and chloroplasts have a double-membrane structure, contain their own DNA, can synthesize proteins, can divide, and are involved in energy metabolism.

- Mitochondria produce ATP using energy-contained macromolecules (figure 4.17).
  - *The inner membrane is extensively folded into layers called cristae.*
  - *The intermembrane space is a compartment between the inner and outer membrane.*
  - *The mitochondrial matrix is a compartment consisting of the fluid within the inner membrane.*
- Chloroplasts use light to generate ATP and sugars (figure 4.18).
  - *In addition to a double membrane, chloroplasts also have stacked membranes called grana that contain vesicles called thylakoids.*
  - *The fluid stroma surrounds the thylakoids.*
- Evidence indicates that mitochondria and chloroplasts arose via endosymbiosis.

4.6 The Cytoskeleton

The cytoskeleton is composed of three different fibers that support cell shape and anchors organelles and enzymes (figure 4.20).

- Actin filaments, or microfilaments, are long, thin polymers responsible for cell movement, cytoplasmic division, and formation of cellular extensions.
- Microtubules are hollow structures that are used in cell movement and movement of materials within a cell.
- Intermediate filaments are stable structures that serve a wide variety of functions.
- Paired centrioles, located in the centrosome, help assemble the nuclear division apparatus of animal cells (figure 4.21).
- Molecular motors move vesicles along microtubules.

4.7 Extracellular Structures and Cell Movement

Extracellular structures provide protection, support, strength, and cell recognition.

- Plants have cell walls composed of cellulose fibers. Fungi have cell walls composed of chitin.
- Animals have a complex extracellular matrix.
- Cell crawling occurs as actin polymerization forces the cell membrane forward while myosin pulls the cell forward.
- Eukaryotic flagella have a 9 + 2 structure and arise from a basal body.
- Cilia are shorter and more numerous flagella.

www.ravenbiology.com

chapter 4 cell structure 83

Designed to help students maximize their learning experience in biology—we offer the following options to students:

ARIS

ARIS (Assessment, Review, and Instruction System) is an electronic study system that offers students a digital portal of knowledge.

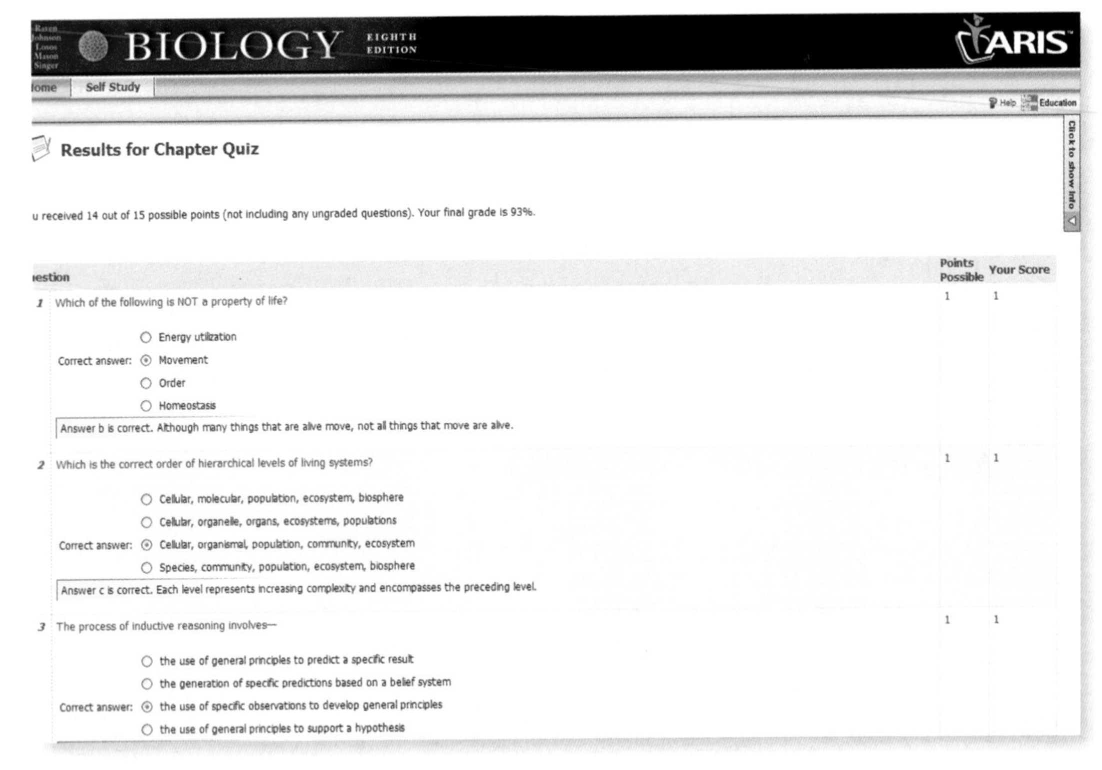

Students can readily access a variety of **digital learning objects,** which include

- chapter level quizzing
- pretests
- animations
- videos
- flashcards
- answers to Inquiry Questions
- answers to all end-of-chapter questions
- MP3 and MP4 downloads of selected content
- learning outcomes and assessment capability woven around key content

Student Study Guide

Helping students focus their time and energy on important concepts, the study guide offers students a variety of tools:

1. **Tips for Mastering Key Concepts**—provides an overview of the chapter, summarizes key points, and gives helpful hints on topics to consider.
2. **Map of Understanding**—illustrates relationships between the major concepts in the chapter.
3. **Key Terms**—listed by the section of the chapter in which they occur.
4. **Learning by Experience**—a variety of activities designed to help students actually use the knowledge acquired from the chapter.
5. **Exercising Your Knowledge**—approximately 30 multiple-choice questions per chapter.
6. **Assessing Your Knowledge**—provides answers to the chapter questions, including a guide to which questions test comprehension of the concepts and which test for knowledge of detail. Students are directed to the appropriate text sections for review.

Content Delivery Flexibility

Biology is available in many formats in addition to the traditional textbook to give instructors and students more choices when deciding on the format of their biology text. Choices include:

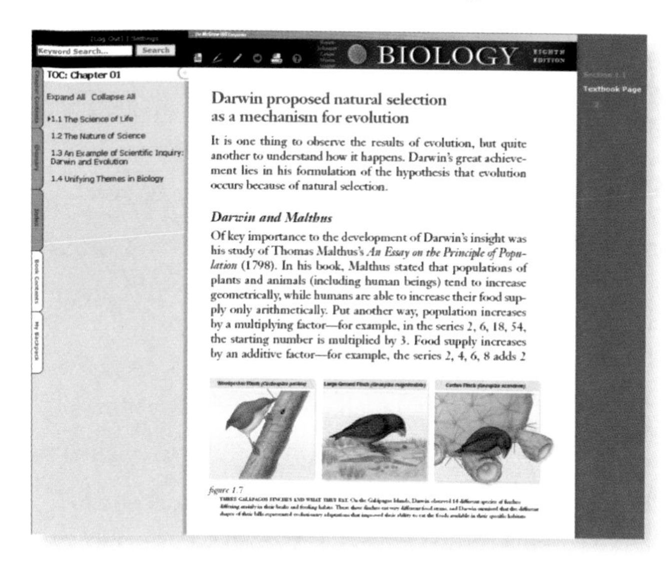

eBook

The entire text is available electronically through the ARIS website. This electronic text offers not only the text in a digital format but includes embedded links to figures, tables, animations, and videos to make full use of the digital tools available and further enhance student understanding.

Color Custom by Chapter

For even more flexibility, we offer the *Biology* text in a full-color, custom version that allows instructors to pick the chapters they want included. Students pay for only what the instructor chooses.

Volumes

The complete text has been split into three natural segments to allow instructors more flexibility and students more purchasing options.

Volume 1—Units 1 (Chemistry), **2** (Cell Biology), and **3** (Genetics)

Volume 2—Units 6 (Plant Biology) and **7** (Animal Biology)

Volume 3—Units 4 (Evolution), **5** (Diversity), and **8** (Ecology)

Dedicated to providing high-quality and effective supplements for instructors, the following Instructor supplements were developed for *Biology:*

ARIS with Presentation Center

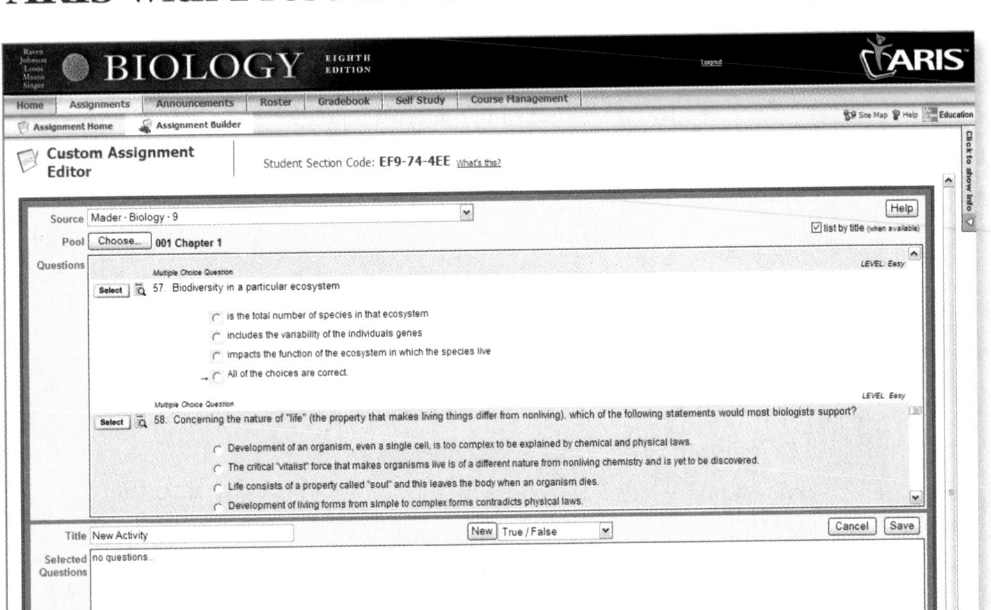

Assessment, Review, and Instruction System, also known as ARIS, is an electronic homework and course management system designed for greater flexibility, power, and ease of use than any other system. Whether you are looking for a preplanned course or one you can customize to fit your course needs, ARIS is your solution.

In addition to having access to all student digital learning objects, ARIS allows instructors to:

Build Assignments

- Choose from prebuilt assignments or create your own custom content by importing your own content or editing an existing assignment from the prebuilt assignment.
- Assignments can include quiz questions, animations, and videos—anything found on the website.
- Create announcements and utilize full-course or individual student communication tools
- Assign *unique multilevel tutorial questions* developed by content experts that provide intelligent feedback through a series of questions to help students truly understand a concept; not just repeat an answer.

Track Student Progress

- Assignments are automatically graded.
- Gradebook functionality allows full course management, including
  - Exporting your gradebook to Excel, WebCT, or BlackBoard
  - Dropping the lowest grades
  - Weighting grades and manually adjusting grades
  - Manipulating data, allowing you to track student progress through multiple reports

Offer More Flexibility

- **Sharing Course Materials with Colleagues**—Instructors can create and share course materials and assignments with colleagues with a few clicks of the mouse, allowing multiple-section courses with many instructors (and TAs) to continually be in synch if desired.
- **Integration with BlackBoard or WebCT**—Once a student is registered in the course, all student activity within McGraw-Hill's ARIS is automatically recorded and available to the instructor through a fully integrated gradebook that can be downloaded to Excel, WebCT, or BlackBoard.

Presentation Center

Build instructional materials where-ever, when-ever, and how-ever you want!

ARIS Presentation Center is an online digital library containing assets such as photos, artwork, animations, PowerPoint slides, and other media presentations that can be used to create customized lectures, visually enhanced tests and quizzes, compelling course websites, or attractive printed support materials.

Access to your book, access to all books!

The Presentation Center library includes thousands of assets from many McGraw-Hill titles. This ever-growing resource gives instructors the power to utilize assets specific to an adopted textbook as well as content from all other books in the library.

Nothing could be easier!

Accessed from the instructor side of your textbook's ARIS website, the ARIS Presentation Center's dynamic search engine allows you to explore by discipline, course, textbook chapter, asset type, or keyword. Simply browse, select, and download the files you need to build engaging course materials. All assets are copyrighted McGraw-Hill Higher Education but can be used by instructors for classroom purposes.

Instructor's Testing and Resource CD-ROM

This cross-platform CD-ROM provides these resources for instructors:

- **Instructor's Manual**—This manual contains chapter synopses, learning outcomes, concept maps, common student misconceptions, instructional strategies, conceptual demonstrations, ideas for laboratories, examples of higher level assessment, sources for additional web resources, and service learning ideas.
- **Test Bank**—The test bank offers questions that can be used for homework assignments or the preparation of exams.
- **Computerized Test Bank**—This software can be utilized to quickly create customized exams. The user-friendly program allows instructors to sort questions by format or level of difficulty; edit existing questions or add new ones; and scramble questions and answer keys for multiple versions of the same test.

Student Response System

Wireless technology brings interactivity into the classroom or lecture hall. Instructors and students receive immediate feedback through wireless response pads that are easy to use and engage students. This system can be used by instructors to:

- Take attendance
- Administer quizzes and tests
- Create a lecture with intermittent questions
- Manage lectures and student comprehension through the use of the gradebook
- Integrate interactivity into their PowerPoint presentations

Transparencies

This boxed set of overhead transparencies includes every piece of line art in the textbook plus every table. The images have been modified to ensure maximum readability in both small and large classroom settings.

Biology Laboratory Manual, Eighth Edition
Darrell S. Vodopich, *Baylor University*
Randy Moore, *University of Minnesota-Minneapolis*
0-07-299522-X / 2008

This laboratory manual is designed for an introductory majors' biology course with a broad survey of basic laboratory techniques. The experiments and procedures are simple, safe, easy to perform, and especially appropriate for large classes. Few experiments require a second class meeting to complete the procedure. Each exercise includes many photographs, traditional topics, and experiments that help students learn about life. Procedures within each exercise are numerous and discrete so that an exercise can be tailored to the needs of the students, the style of the instructor, and the facilities available.

New to this edition:

- **Clearer Information on Safety**—A new safety icon will be used throughout the text, replacing the current "CAUTION" that appears in red font.
- ***Laboratory Safety Rules*** table has been added to the *Welcome* chapter.
- **Process of Science Exercises**—A new exercise on *Process of Science* now appears in chapter 1.
- **New Figures and Tables**—More than 70 tables and figures have been extended with additions, revisions, or replacements.

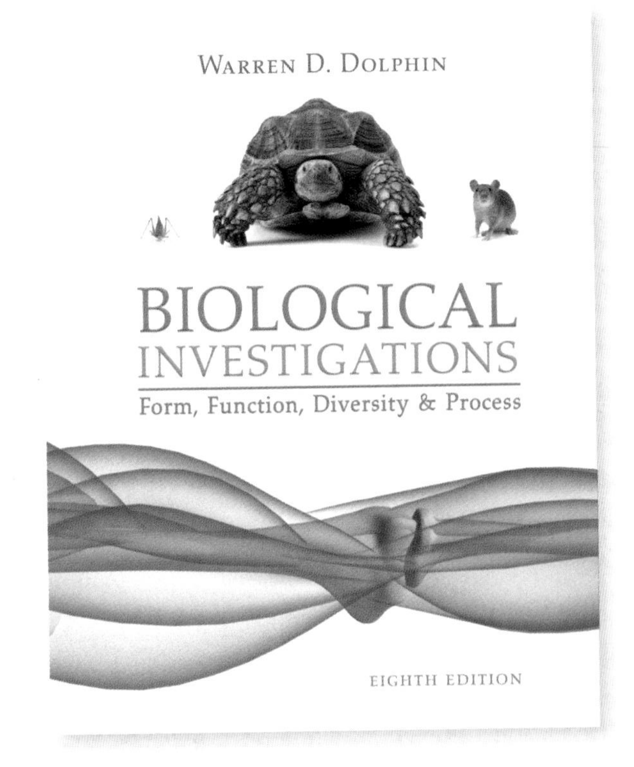

Biological Investigations Lab Manual, Eighth Edition
Warren D. Dolphin, *Iowa State University*
0-07-299287-5 / 2008

This independent lab manual can be used for a one- or two-semester majors' level general biology lab and can be used with any majors' level general biology textbook. The labs are investigative and ask students to use more critical thinking and hands-on learning. The author emphasizes investigative, quantitative, and comparative approaches to studying the life sciences.

New to this edition:

- **Revised Art Program**—More than 90 figures have been revised or replaced throughout the textbook.
- **How to Evaluate Web Material**—Information on evaluating material gathered on the web has been added to Lab Topic 1.

contents

Part I The Molecular Basis of Life

Part II Biology of the Cell

Biology

chapter 1

The Science of Biology

introduction

YOU ARE ABOUT TO EMBARK ON A JOURNEY—a journey of discovery about the nature of life. Nearly 180 years ago, a young English naturalist named Charles Darwin set sail on a similar journey on board H.M.S. *Beagle*; a replica of this ship is pictured here. What Darwin learned on his five-year voyage led directly to his development of the theory of evolution by natural selection, a theory that has become the core of the science of biology. Darwin's voyage seems a fitting place to begin our exploration of biology: the scientific study of living organisms and how they have evolved. Before we begin, however, let's take a moment to think about what biology is and why it's important.

concept outline

1.1 The Science of Life

This is the most exciting time to be studying biology in the history of the field. The amount of data available about the natural world has exploded in the last 25 years, and we are now in a position to ask and answer questions that previously were only dreamed of.

We have determined the entire sequence of the human genome, and are in the process of sequencing the genomes of other species at an ever increasing pace. We are closing in on a description of the molecular workings of the cell in unprecedented detail, and we are in the process of finally unveiling the mystery of how a single cell can give rise to the complex organization seen in multicellular organisms. With robotics, advanced imaging, and analytical techniques, we have tools available that were formerly the stuff of science fiction.

In this text, we attempt to provide a view of the science of biology as it is practiced now, while still retaining some sense of how we got to this exciting position. In this introductory chapter, we examine the nature of biology and the nature of science in general to begin to put into context the information presented in the rest of the text.

Biology unifies much of natural science

The study of biology does much to unify the information gained from all of the natural sciences. Biological systems are the most complex chemical systems that we know of on Earth, and their many functions are both determined and constrained by the principles of chemistry and physics. Put another way, there are no new laws of nature to be gleaned from the study of biology—but that study does illuminate and illustrate the workings of those natural laws.

The intricate chemical workings of cells are based on all that we have learned from the study of chemistry, and every level of biological organization is governed by the nature of energy transactions learned from the study of thermodynamics. Biological systems do not represent any new forms of matter, and yet they are the most complex organization of matter known. The complexity of living systems is made possible by a constant source of energy: the Sun. The conversion of this energy source into organic molecules by photosynthesis can be understood using the principles of chemistry and physics.

As modern scientists take on more difficult problems, the nature of how we do science is changing as well. Science is becoming more interdisciplinary, combining the expertise of a variety of scientists in exciting new fields such as nanotechnology. Biology is at the heart of this multidisciplinary approach because biological problems often require many different approaches to arrive at solutions.

Life defies simple definition

In its broadest sense, biology is the study of living things—*the science of life*. Living things come in an astounding variety of shapes and forms, and biologists study life in many different ways. They live with gorillas, collect fossils, and listen to whales. They read the messages encoded in the long molecules of heredity and count how many times a hummingbird's wings beat each second.

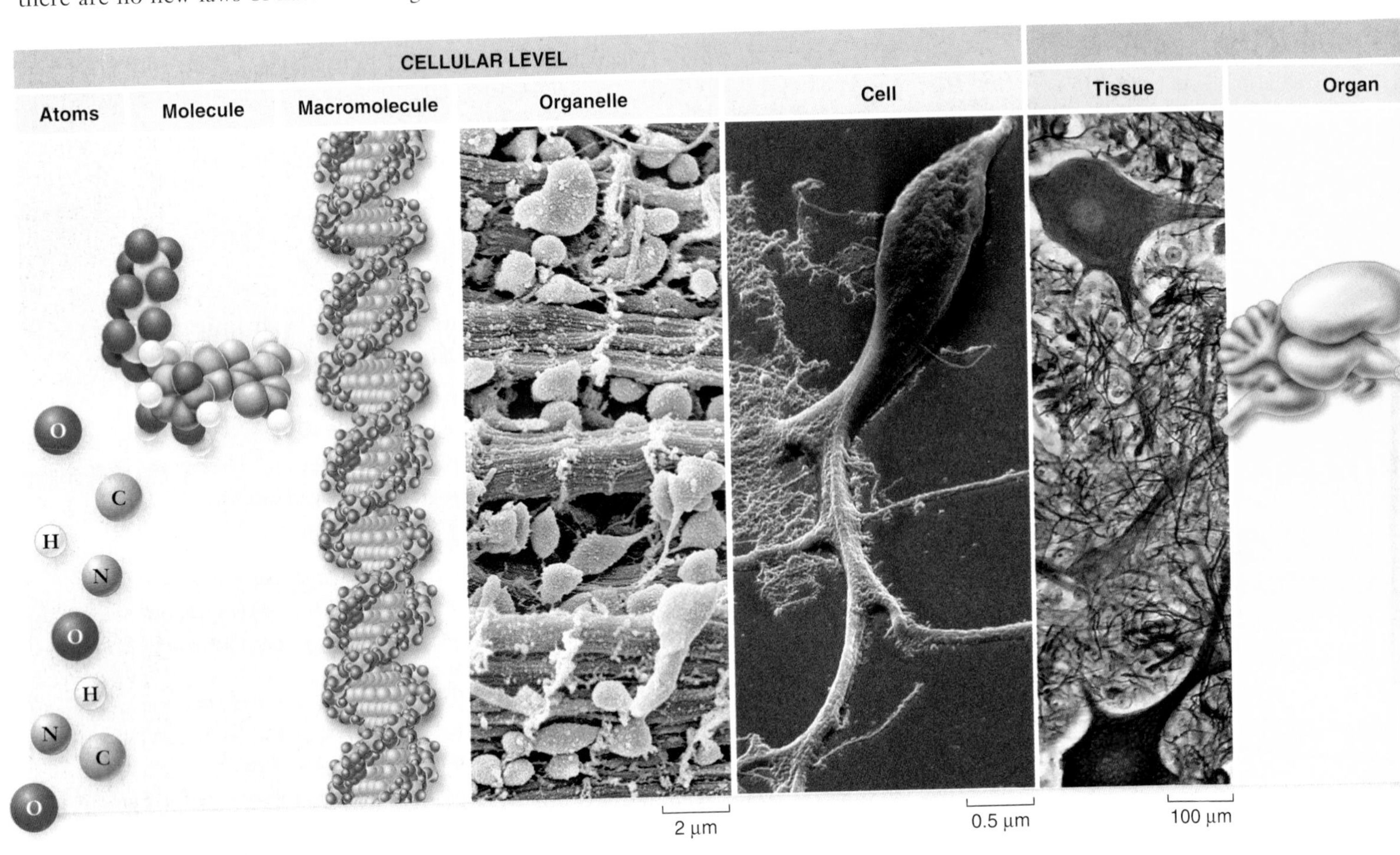

What makes something "alive"? Anyone could deduce that a galloping horse is alive and a car is not, but why? We cannot say, "If it moves, it's alive," because a car can move, and gelatin can wiggle in a bowl. They certainly are not alive. Although we cannot define life with a single simple sentence, we can come up with a series of seven characteristics shared by living systems:

- **Cellular organization.** All organisms consist of one or more cells. Often too tiny to see, cells carry out the basic activities of living. Each cell is bounded by a membrane that separates it from its surroundings.
- **Ordered complexity.** All living things are both complex and highly ordered. Your body is composed of many different kinds of cells, each containing many complex molecular structures. Many nonliving things may also be complex, but they do not exhibit this degree of ordered complexity.
- **Sensitivity.** All organisms respond to stimuli. Plants grow toward a source of light, and the pupils of your eyes dilate when you walk into a dark room.
- **Growth, development, and reproduction.** All organisms are capable of growing and reproducing, and they all possess hereditary molecules that are passed to their offspring, ensuring that the offspring are of the same species.
- **Energy utilization.** All organisms take in energy and use it to perform many kinds of work. Every muscle in your body is powered with energy you obtain from the food you eat.
- **Homeostasis.** All organisms maintain relatively constant internal conditions that are different from their environment, a process called **homeostasis.**
- **Evolutionary adaptation.** All organisms interact with other organisms and the nonliving environment in ways that influence their survival, and as a consequence, organisms evolve adaptations to their environments.

Living systems show hierarchical organization

The organization of the biological world is hierarchical—that is, each level builds on the level below it.

1. **The Cellular Level.** At the cellular level (figure 1.1), **atoms,** the fundamental elements of matter, are joined together into clusters called **molecules.** Complex biological molecules are assembled into tiny structures called **organelles** within membrane-bounded units we call **cells.** The cell is the basic unit of life. Many independent organisms are composed only of single cells. Bacteria are single cells, for example. All animals and plants, as well as most fungi and algae, are multicellular—composed of more than one cell.
2. **The Organismal Level.** Cells in complex multicellular organisms exhibit three levels of organization. The most basic level is that of **tissues,** which are groups of similar

figure 1.1

HIERARCHICAL ORGANIZATION OF LIVING SYSTEMS. Life is highly organized from the simplest atoms to complex multicellular organisms. Along this hierarchy of structure, atoms form molecules that are used to form organelles, which in turn form the functional subsystems within cells. Cells are organized into tissue, and then into organs and organ systems such as the nervous system pictured. This extends beyond individual organisms to populations, communities, ecosystems and finally the entire biosphere.

cells that act as a functional unit. Tissues, in turn, are grouped into **organs,** body structures composed of several different tissues that act as a structural and functional unit. Your brain is an organ composed of nerve cells and a variety of associated tissues that form protective coverings and contribute blood. At the third level of organization, organs are grouped into **organ systems.** The nervous system, for example, consists of sensory organs, the brain and spinal cord, and neurons that convey signals.

3. **The Populational Level.** Individual organisms can be categorized into several hierarchical levels within the living world. The most basic of these is the **population,** a group of organisms of the same species living in the same place. All populations of a particular kind of organism together form a **species,** its members similar in appearance and able to interbreed. At a higher level of biological organization, a **biological community** consists of all the populations of different species living together in one place.
4. **Ecosystem Level.** At the highest tier of biological organization, a biological community and the physical habitat within which it lives together constitute an ecological system, or **ecosystem.** For example, the soil, water, and atmosphere of a mountain ecosystem interact with the biological community of a mountain meadow in many important ways.
5. **Emergent Properties at Every Level.** At each higher level in the living hierarchy, novel properties emerge. These **emergent properties** result from the way in which components interact, and they often cannot be deduced just from looking at the parts themselves. Examining individual cells, for example, gives little clue of what a whole animal is like. You, as a human, have the same array of cell types as does a giraffe. It is because the living world exhibits many emergent properties that it is difficult to define "life."
6. **The Biosphere.** The entire planet can be thought of as an ecosystem that we call the biosphere.

The previous descriptions of the common features and organization of living systems begins to get at the nature of what it is to be alive. The rest of this book illustrates and expands on these basic ideas to try to provide a more complete account of living systems.

Biology is a unifying science that uses the knowledge gained from other natural sciences to study living systems. There is no simple definition of life, but living systems share a number of properties that together describe life. Living systems are also organized hierarchically, with new properties emerging that may be greater than the sum of the parts.

1.2 The Nature of Science

Much like life itself, the nature of science defies simple description. For many years scientists have written about the "scientific method" as though there is a single way of doing science. This oversimplification has contributed to confusion on the part of nonscientists about the nature of science.

At its core, science is concerned with understanding the nature of the world around us by using observation and reasoning. To begin with, we assume that natural forces acting now have always acted, that the fundamental nature of the universe has not changed since its inception, and that it is not changing now. A number of complementary approaches allow understanding of natural phenomena—there is no one "right way."

Scientists also attempt to be as objective as possible in the interpretation of the data and observations they have collected. Because scientists themselves are human, this is not completely possible; because science is a collective endeavor subject to scrutiny, however, it is self-correcting. Results from one person are verified by others, and if the results cannot be repeated, they are rejected.

Much of science is descriptive

The classic vision of the scientific method is that observations lead to hypotheses that in turn make experimentally testable predictions. In this way, we dispassionately evaluate new ideas to arrive at an increasingly accurate view of nature. We discuss this way of doing science later in this chapter, but it is important to understand that much of science is purely descriptive: In order to understand anything, the first step is to describe it completely. Much of biology is concerned with arriving at an increasingly accurate description of nature.

The study of biodiversity is an example of descriptive science that has implications for other aspects of biology in addition to social implications. Efforts are currently underway to classify all life on the Earth. This ambitious project is purely descriptive, but it will lead to a much greater understanding of biodiversity as well as the impact of our species on biodiversity.

One of the most important accomplishments of molecular biology at the dawn of the 21st century was the completion of the sequence of the human genome. Many new hypotheses about human biology will be generated by this knowledge, and many experiments will be needed to test these hypotheses, but the determination of the sequence itself was descriptive science.

Science uses both deductive and inductive reasoning

The study of logic recognizes two opposite ways of arriving at logical conclusions: deductive reasoning and inductive reasoning. Science makes use of both of these methods, although induction is the primary way of reasoning in hypothesis-driven science.

Deductive reasoning

Deductive reasoning applies general principles to predict specific results. Over 2200 years ago, the Greek scientist Eratosthenes used Euclidean geometry and deductive reasoning to accurately estimate the circumference of the Earth (figure 1.2). Deductive reasoning is the reasoning of mathematics and

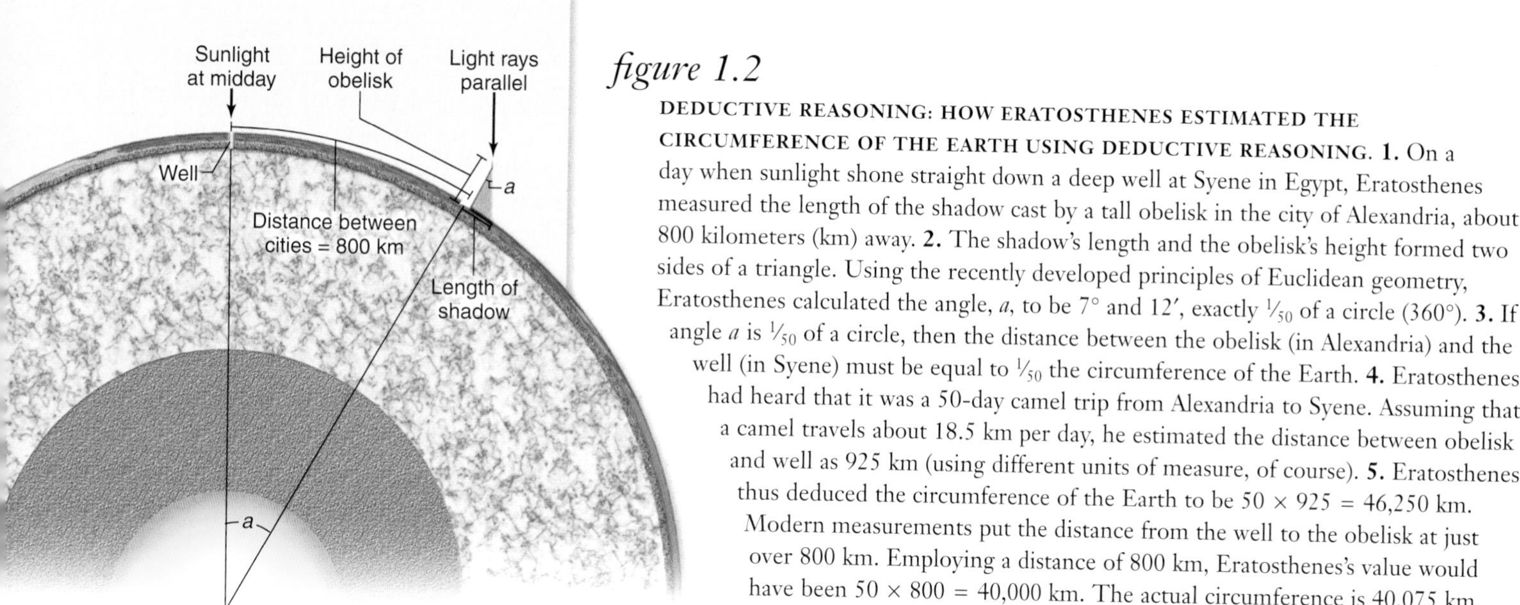

figure 1.2

DEDUCTIVE REASONING: HOW ERATOSTHENES ESTIMATED THE CIRCUMFERENCE OF THE EARTH USING DEDUCTIVE REASONING. 1. On a day when sunlight shone straight down a deep well at Syene in Egypt, Eratosthenes measured the length of the shadow cast by a tall obelisk in the city of Alexandria, about 800 kilometers (km) away. **2.** The shadow's length and the obelisk's height formed two sides of a triangle. Using the recently developed principles of Euclidean geometry, Eratosthenes calculated the angle, *a*, to be 7° and 12′, exactly $1/50$ of a circle (360°). **3.** If angle *a* is $1/50$ of a circle, then the distance between the obelisk (in Alexandria) and the well (in Syene) must be equal to $1/50$ the circumference of the Earth. **4.** Eratosthenes had heard that it was a 50-day camel trip from Alexandria to Syene. Assuming that a camel travels about 18.5 km per day, he estimated the distance between obelisk and well as 925 km (using different units of measure, of course). **5.** Eratosthenes thus deduced the circumference of the Earth to be $50 \times 925 = 46{,}250$ km. Modern measurements put the distance from the well to the obelisk at just over 800 km. Employing a distance of 800 km, Eratosthenes's value would have been $50 \times 800 = 40{,}000$ km. The actual circumference is 40,075 km.

philosophy, and it is used to test the validity of general ideas in all branches of knowledge. For example, if all mammals by definition have hair, and you find an animal that does not have hair, then you may conclude that this animal is not a mammal. A biologist uses deductive reasoning to infer the species of a specimen from its characteristics.

Inductive reasoning

In **inductive reasoning,** the logic flows in the opposite direction, from the specific to the general. Inductive reasoning uses specific observations to construct general scientific principles. For example, if poodles have hair, and terriers have hair, and every dog that you observe has hair, then you may conclude that all dogs have hair. Inductive reasoning leads to generalizations that can then be tested. Inductive reasoning first became important to science in the 1600s in Europe, when Francis Bacon, Isaac Newton, and others began to use the results of particular experiments to infer general principles about how the world operates.

An example from modern biology is the action of homeobox genes in development. Studies in the fruit fly, *Drosophila melanogaster,* identified genes that could cause dramatic changes in developmental fate, such as a leg appearing in the place of an antenna. When the genes themselves were isolated and their DNA sequence determined, it was found that similar genes were found in many animals, including humans. This led to the general idea that the homeobox genes act as switches to control developmental fate.

Hypothesis-driven science makes and tests predictions

Scientists establish which general principles are true from among the many that might be true by systematically testing alternative proposals. If these proposals prove inconsistent with experimental observations, they are rejected as untrue. Figure 1.3 illustrates the process.

After making careful observations, scientists construct a **hypothesis,** which is a suggested explanation that accounts for those observations. A hypothesis is a proposition that might be

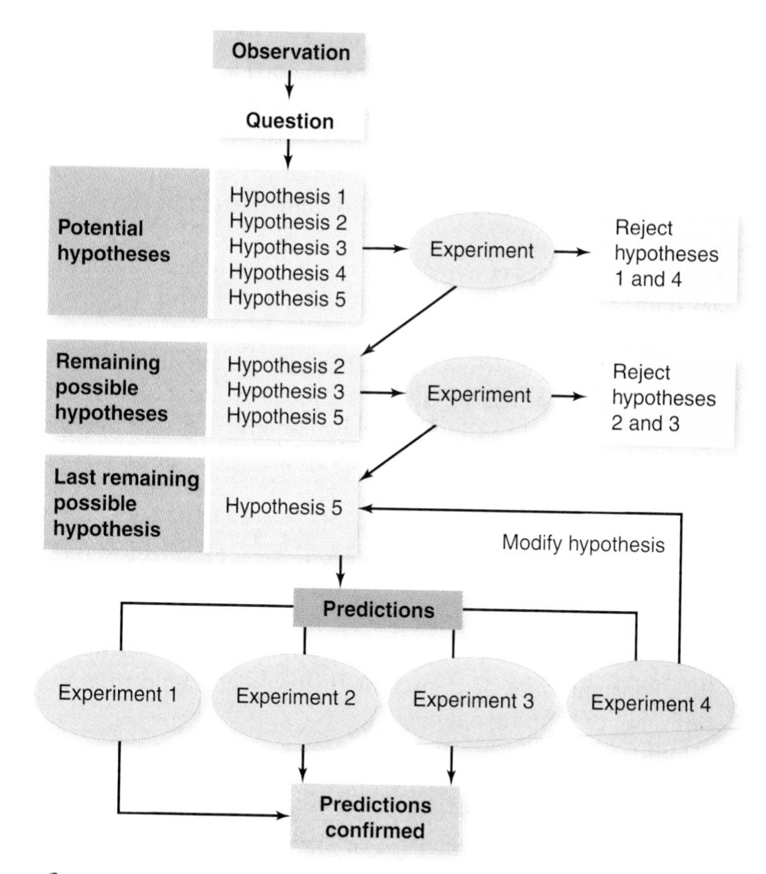

figure 1.3

HOW SCIENCE IS DONE. This diagram illustrates how scientific investigations proceed. First, scientists make observations that raise a particular question. They develop a number of potential explanations (hypotheses) to answer the question. Next, they carry out experiments in an attempt to eliminate one or more of these hypotheses. Then, predictions are made based on the remaining hypotheses, and further experiments are carried out to test these predictions. The process can also be iterative. As experimental results are performed, the information can be used to modify the original hypothesis to fit each new observation.

true. Those hypotheses that have not yet been disproved are retained. They are useful because they fit the known facts, but they are always subject to future rejection if, in the light of new information, they are found to be incorrect.

This process can also be *iterative*, that is, a hypothesis can be changed and refined with new data. For instance, geneticists Beadle and Tatum studied the nature of genetic information to arrive at their "one-gene/one-enzyme" hypothesis (chapter 15). This hypothesis states that a gene represents the genetic information necessary to make a single enzyme. As investigators learned more about the molecular nature of genetic information, the hypothesis was refined to "one-gene/one-polypeptide" because enzymes can be made up of more than one polypeptide. With still more information about the nature of genetic information, investigators found that a single gene can specify more than one polypeptide, and the hypothesis was refined again.

Testing hypotheses

We call the test of a hypothesis an **experiment.** Suppose that a room appears dark to you. To understand why it appears dark, you propose several hypotheses. The first might be, "There is no light in the room because the light switch is turned off." An alternative hypothesis might be, "There is no light in the room because the lightbulb is burned out." And yet another hypothesis might be, "I am going blind." To evaluate these hypotheses, you would conduct an experiment designed to eliminate one or more of the hypotheses.

For example, you might test your hypotheses by flipping the light switch. If you do so and the room is still dark, you have disproved the first hypothesis: Something other than the setting of the light switch must be the reason for the darkness. Note that a test such as this does not prove that any of the other hypotheses are true; it merely demonstrates that the one being tested is not. A successful experiment is one in which one or more of the alternative hypotheses is demonstrated to be inconsistent with the results and is thus rejected.

As you proceed through this text, you will encounter many hypotheses that have withstood the test of experiment. Many will continue to do so; others will be revised as new observations are made by biologists. Biology, like all science, is in a constant state of change, with new ideas appearing and replacing or refining old ones.

Establishing controls

Often scientists are interested in learning about processes that are influenced by many factors, or **variables.** To evaluate alternative hypotheses about one variable, all other variables must be kept constant. This is done by carrying out two experiments in parallel: a test experiment and a control experiment. In the **test experiment,** one variable is altered in a known way to test a particular hypothesis. In the **control experiment,** that variable is left unaltered. In all other respects the two experiments are identical, so any difference in the outcomes of the two experiments must result from the influence of the variable that was changed.

Much of the challenge of experimental science lies in designing control experiments that isolate a particular variable from other factors that might influence a process.

Using predictions

A successful scientific hypothesis needs to be not only valid but also useful—it needs to tell us something we want to know. A hypothesis is most useful when it makes predictions because those predictions provide a way to test the validity of the hypothesis. If an experiment produces results inconsistent with the predictions, the hypothesis must be rejected or modified. In contrast, if the predictions are supported by experimental testing, the hypothesis is supported. The more experimentally supported predictions a hypothesis makes, the more valid the hypothesis is.

As an example, in the early history of microbiology it was known that nutrient broth left sitting exposed to air becomes contaminated. There were two hypothesis proposed to explain this observation: spontaneous generation and the germ hypothesis. Spontaneous generation held that there was an inherent property in organic molecules that could lead to the spontaneous generation of life. The germ hypothesis proposed that preexisting microorganisms that were present in air could contaminate the nutrient broth.

These competing hypotheses were tested by a number of experiments that involved filtering air and boiling the broth to kill any contaminating germs. The definitive experiment was performed by Louis Pasteur, who constructed flasks with curved necks that could be exposed to air, but that would trap any contaminating germs. When such flasks were boiled to sterilize them, they remained sterile, but if the curved neck was broken off, they became contaminated (figure 1.4).

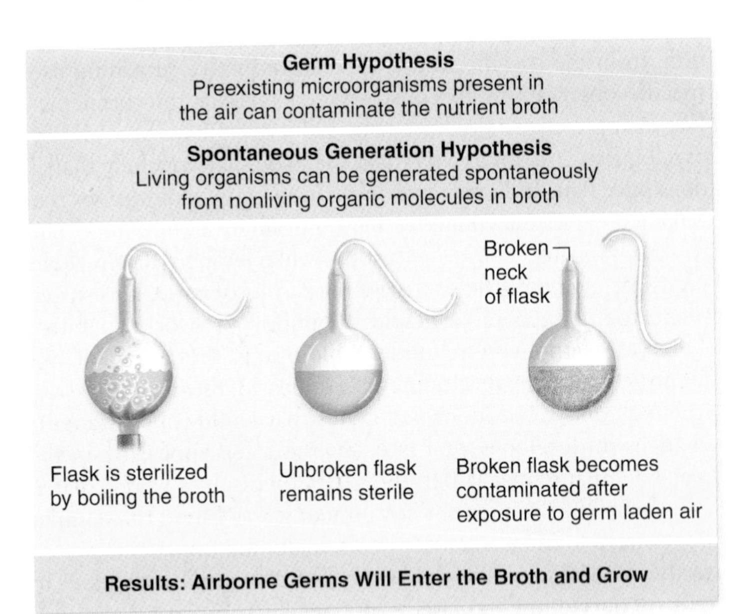

figure 1.4

EXPERIMENT TO TEST SPONTANEOUS GENERATION VS. GERM HYPOTHESIS. Pasteur built swan-necked flasks to prevent airborne contamination. When the flask is heated, it kills any germs in the flask. The flask will remain sterile unless the neck is broken, in which case it becomes contaminated. Spontaneous generation predicts growth should occur in either flask, while germ theory predicts growth only when the sterile flask is exposed to air.

This result was predicted by the germ hypothesis—that when the sterile flask is exposed to air, airborne germs will arrive in the broth and grow. The spontaneous generation hypothesis predicted no difference in results with exposure to air. This experiment disproved the hypothesis of spontaneous generation and supported the hypothesis of airborne germs under the conditions tested.

Reductionism breaks larger systems into their component parts

Scientists often use the philosophical approach of **reductionism** to understand a complex system by reducing it to its working parts. Reductionism has been the general approach of biochemistry, which has been enormously successful at unraveling the complexity of cellular metabolism by concentrating on individual pathways and specific enzymes. By analyzing all of the pathways and their components, scientists now have an overall picture of the metabolism of cells.

Reductionism has limits when applied to living systems, however—one of which is that enzymes do not always behave exactly the same in isolation as they do in their normal cellular context. A larger problem is that the complex interworking of many networked functions leads to emergent properties that cannot be predicted based on the workings of the parts. Biologists are just beginning to come to grips with this problem and to think about ways of dealing with the whole as well as the workings of the parts. The emerging field of systems biology is aimed toward this different approach.

Biologists construct models to explain living systems

Biologists construct models in many different ways for a variety of uses. Geneticists construct models of interacting networks of proteins that control gene expression, often even drawing cartoon figures to represent that which we cannot see. Population biologists build models of how evolutionary change occurs. Cell biologists build models of signal transduction pathways and the events leading from an external signal to internal events. Structural biologists build actual models of the structure of proteins and macromolecular complexes in cells.

Models provide a way to organize how we think about a problem. Models can also get us closer to the larger picture and away from the extreme reductionist approach. The working parts are provided by the reductionist analysis, but the model shows how they fit together. Often these models suggest other experiments that can be performed to refine or test the model.

As researchers gain more knowledge about the actual flow of molecules in living systems, more sophisticated kinetic models can be used to apply information about isolated enzymes to their cellular context. In systems biology, this modeling is being applied on a large scale to regulatory networks during development, and even to modeling an entire bacterial cell.

The nature of scientific theories

Scientists use the word **theory** in two main ways. The first meaning of *theory* is a proposed explanation for some natural phenomenon, often based on some general principle. Thus, we speak of the principle first proposed by Newton as the "theory of gravity." Such theories often bring together concepts that were previously thought to be unrelated.

The second meaning of *theory* is the body of interconnected concepts, supported by scientific reasoning and experimental evidence, that explains the facts in some area of study. Such a theory provides an indispensable framework for organizing a body of knowledge. For example, quantum theory in physics brings together a set of ideas about the nature of the universe, explains experimental facts, and serves as a guide to further questions and experiments.

To a scientist, theories are the solid ground of science, expressing ideas of which we are most certain. In contrast, to the general public, the word *theory* usually implies the opposite—a *lack* of knowledge, or a guess. Not surprisingly, this difference often results in confusion. In this text, *theory* will always be used in its scientific sense, in reference to an accepted general principle or body of knowledge.

Some critics outside of science attempt to discredit evolution by saying it is "just a theory." The hypothesis that evolution has occurred, however, is an accepted scientific fact—it is supported by overwhelming evidence. Modern evolutionary theory is a complex body of ideas, the importance of which spreads far beyond explaining evolution. Its ramifications permeate all areas of biology, and it provides the conceptual framework that unifies biology as a science. Again, the key is how well a hypothesis fits the observations. Evolutionary theory fits the observations very well.

Research can be basic or applied

In the past it was fashionable to speak of the "scientific method" as consisting of an orderly sequence of logical, either/or steps. Each step would reject one of two mutually incompatible alternatives, as though trial-and-error testing would inevitably lead a researcher through the maze of uncertainty that always impedes scientific progress. If this were the case, a computer would make a good scientist. But science is not done this way.

As the British philosopher Karl Popper has pointed out, successful scientists without exception design their experiments with a pretty fair idea of how the results are going to come out. They have what Popper calls an "imaginative preconception" of what the truth might be. Because insight and imagination play such a large role in scientific progress, some scientists are better at science than others—just as Bob Dylan stands out among songwriters or Claude Monet stands out among Impressionist painters.

Some scientists perform *basic research*, which is intended to extend the boundaries of what we know. These individuals typically work at universities, and their research is usually supported by grants from various agencies and foundations.

The information generated by basic research contributes to the growing body of scientific knowledge, and it provides the scientific foundation utilized by *applied research*. Scientists who conduct applied research are often employed in some kind of industry. Their work may involve the manufacture of food additives, the creation of new drugs, or the testing of environmental quality.

Research results are written up and submitted for publication in scientific journals, where the experiments and conclusions are reviewed by other scientists. This process of careful evaluation, called *peer review*, lies at the heart of modern science. It helps to ensure that faulty research or false claims are not given the authority of scientific fact. It also provides other scientists with a starting point for testing the reproducibility of experimental results. Results that cannot be reproduced are not taken seriously for long.

Science uses many methods to arrive at an understanding of the natural world. Science can be descriptive, amassing observations to gain an increasingly accurate account of the world. Both deductive reasoning and inductive reasoning are used in science. Hypothesis-driven science builds hypotheses based on observation. When a hypothesis has been extensively tested, it becomes an accepted theory. Theories are coherent explanations of observed data at present, but they may be modified to fit new data.

1.3 An Example of Scientific Inquiry: Darwin and Evolution

Darwin's theory of evolution explains and describes how organisms on Earth have changed over time and acquired a diversity of new forms. This famous theory provides a good example of how a scientist develops a hypothesis and how a scientific theory grows and wins acceptance.

Charles Robert Darwin (1809–1882; figure 1.5) was an English naturalist who, after 30 years of study and observation, wrote one of the most famous and influential books of all time. This book, *On the Origin of Species by Means of Natural Selection*, created a sensation when it was published, and the ideas Darwin expressed in it have played a central role in the development of human thought ever since.

The idea of evolution existed prior to Darwin

In Darwin's time, most people believed that the different kinds of organisms and their individual structures resulted from direct actions of a Creator (and to this day, many people still believe this). Species were thought to have been specially created and to be unchangeable, or immutable, over the course of time.

figure 1.5

CHARLES DARWIN. This newly rediscovered photograph taken in 1881, the year before Darwin died, appears to be the last ever taken of the great biologist.

In contrast to these ideas, a number of earlier naturalists and philosophers had presented the view that living things must have changed during the history of life on Earth. That is, **evolution** has occurred, and living things are now different from how they began. Darwin's contribution was a concept he called *natural selection*, which he proposed as a coherent, logical explanation for this process, and he brought his ideas to wide public attention.

Darwin observed differences in related organisms

The story of Darwin and his theory begins in 1831, when he was 22 years old. He was part of a five-year navigational mapping expedition around the coasts of South America (figure 1.6), aboard H.M.S. *Beagle*. During this long voyage, Darwin had the chance to study a wide variety of plants and animals on continents and islands and in distant seas. Darwin observed a number of phenomena that were of central importance to him in reaching his ultimate conclusion.

Repeatedly, Darwin saw that the characteristics of similar species varied somewhat from place to place. These geographical patterns suggested to him that lineages change gradually as species migrate from one area to another. On the Galápagos Islands, 960 km (600 miles) off the coast of Ecuador, Darwin encountered a variety of different finches on the various islands. The 14 species, although related, differed slightly in appearance, particularly in their beaks (figure 1.7).

Darwin thought it was reasonable to assume that all these birds had descended from a common ancestor arriving from the South American mainland several million years ago. Eating different foods on different islands, the finches' beaks had changed during their descent—"descent with modification," or evolution. (These finches are discussed in more detail in chapters 21 and 22.)

In a more general sense, Darwin was struck by the fact that the plants and animals on these relatively young volcanic islands resembled those on the nearby coast of South America. If each one of these plants and animals had been created independently and simply placed on the Galápagos Islands, why didn't they resemble the plants and animals of islands with similar climates—

figure 1.6

THE FIVE-YEAR VOYAGE OF H.M.S. *BEAGLE*. Most of the time was spent exploring the coasts and coastal islands of South America, such as the Galápagos Islands. Darwin's studies of the animals of the Galápagos Islands played a key role in his eventual development of the concept of evolution by means of natural selection.

such as those off the coast of Africa, for example? Why did they resemble those of the adjacent South American coast instead?

Darwin proposed natural selection as a mechanism for evolution

It is one thing to observe the results of evolution, but quite another to understand how it happens. Darwin's great achievement lies in his formulation of the hypothesis that evolution occurs because of natural selection.

Darwin and Malthus

Of key importance to the development of Darwin's insight was his study of Thomas Malthus's *An Essay on the Principle of Population* (1798). In his book, Malthus stated that populations of plants and animals (including human beings) tend to increase geometrically, while humans are able to increase their food supply only arithmetically. Put another way, population increases by a multiplying factor—for example, in the series 2, 6, 18, 54, the starting number is multiplied by 3. Food supply increases by an additive factor—for example, the series 2, 4, 6, 8 adds 2

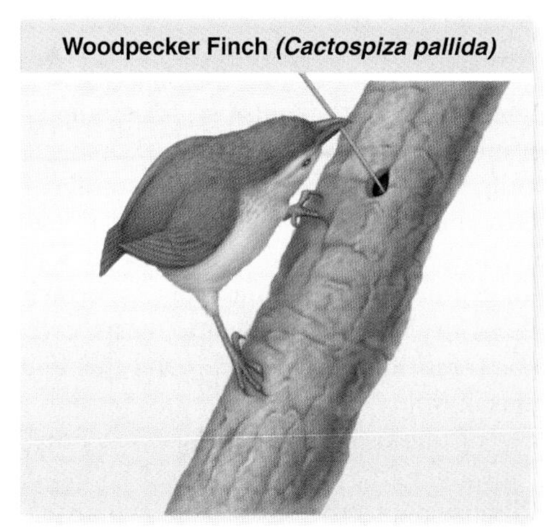

figure 1.7

THREE GALÁPAGOS FINCHES AND WHAT THEY EAT. On the Galápagos Islands, Darwin observed 14 different species of finches differing mainly in their beaks and feeding habits. These three finches eat very different food items, and Darwin surmised that the different shapes of their bills represented evolutionary adaptations that improved their ability to eat the foods available in their specific habitats.

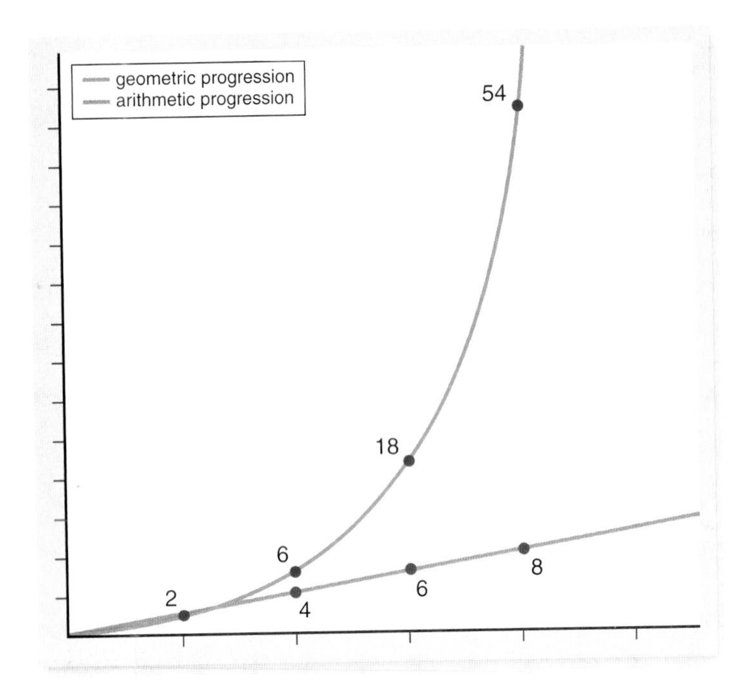

figure 1.8

GEOMETRIC AND ARITHMETIC PROGRESSIONS. A geometric progression increases by a constant factor (for example, the curve shown increases ×3 for each step), whereas an arithmetic progression increases by a constant difference (for example, the line shown increases +2 for each step). Malthus contended that the human growth curve was geometric, but the human food production curve was only arithmetic.

inquiry

What is the effect of reducing the constant factor by which the geometric progression increases? Might this effect be achieved with humans? How?

to each starting number. Figure 1.8 shows the difference that these two types of relationships produce over time.

Because populations increase geometrically, virtually any kind of animal or plant, if it could reproduce unchecked, would cover the entire surface of the world within a surprisingly short time. Instead, populations of species remain fairly constant year after year, because death limits population numbers.

Sparked by Malthus's ideas, Darwin saw that although every organism has the potential to produce more offspring than can survive, only a limited number actually do survive and produce further offspring. Combining this observation with what he had seen on the voyage of the *Beagle*, as well as with his own experiences in breeding domestic animals, Darwin made an important association: Individuals possessing physical, behavioral, or other attributes that give them an advantage in their environment are more likely to survive and reproduce than those with less advantageous traits. By surviving, these individuals gain the opportunity to pass on their favorable characteristics to their offspring. As the frequency of these characteristics increases in the population, the nature of the population as a whole will gradually change. Darwin called this process *selection.*

Natural selection

Darwin was thoroughly familiar with variation in domesticated animals, and he began *On the Origin of Species* with a detailed discussion of pigeon breeding. He knew that animal breeders selected certain varieties of pigeons and other animals, such as dogs, to produce certain characteristics, a process Darwin called **artificial selection.**

Artificial selection often produces a great variation in traits. Domestic pigeon breeds, for example, show much greater variety than all of the wild species found throughout the world. Darwin thought that this type of change could occur in nature, too. Surely if pigeon breeders could foster variation by artificial selection, nature could do the same—a process Darwin called **natural selection.**

Darwin drafts his argument

Darwin drafted the overall argument for evolution by natural selection in a preliminary manuscript in 1842. After showing the manuscript to a few of his closest scientific friends, however, Darwin put it in a drawer, and for 16 years turned to other research. No one knows for sure why Darwin did not publish his initial manuscript—it is very thorough and outlines his ideas in detail.

The stimulus that finally brought Darwin's hypothesis into print was an essay he received in 1858. A young English naturalist named Alfred Russel Wallace (1823–1913) sent the essay to Darwin from Indonesia; it concisely set forth the hypothesis of evolution by means of natural selection, a hypothesis Wallace had developed independently of Darwin. After receiving Wallace's essay, friends of Darwin arranged for a joint presentation of their ideas at a seminar in London. Darwin then completed his own book, expanding the 1842 manuscript he had written so long ago, and submitted it for publication.

Testing the predictions of natural selection

More than 120 years has elapsed since Darwin's death in 1882. During this period, the evidence supporting his theory has grown progressively stronger. We briefly explore some of this evidence here; in chapter 22, we will return to the theory of evolution by natural selection and examine the evidence in more detail.

The fossil record

Darwin predicted that the fossil record would yield intermediate links between the great groups of organisms—for example, between fishes and the amphibians thought to have arisen from them, and between reptiles and birds. Furthermore, natural selection predicts the relative positions in time of such transitional forms. We now know the fossil record to a degree that was unthinkable in the 19th century, and although truly "intermediate" organisms are hard to determine, paleontologists have found what appear to be transitional forms and found them in the appropriate positions in time.

Recent discoveries of microscopic fossils have extended the known history of life on Earth back to about 3.5 billion years ago (BYA). The discovery of other fossils has supported Darwin's predictions and has shed light on how organisms have, over this enormous time span, evolved from the simple to the complex. For vertebrate animals especially, the fossil record is rich and exhibits a graded series of changes in form, with the evolutionary sequence visible for all to see.

figure 1.9

HOMOLOGY AMONG VERTEBRATE LIMBS. The forelimbs of these five vertebrates show the ways in which the relative proportions of the forelimb bones have changed in relation to the particular way of life of each organism.

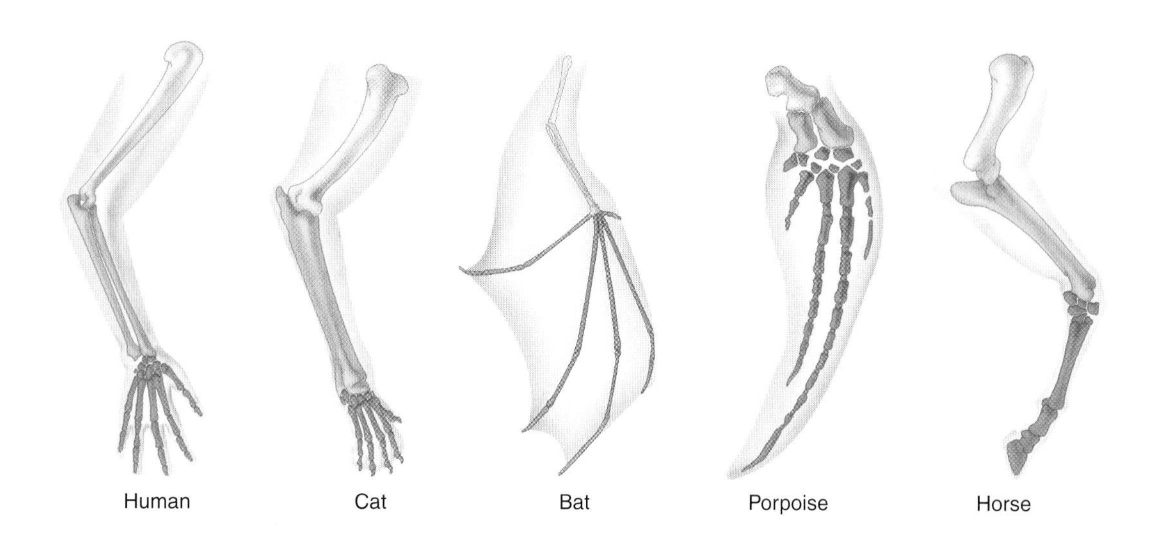

The age of the Earth

Darwin's theory predicted the Earth must be very old, but some physicists argued that the Earth was only a few thousand years old. This bothered Darwin, because the evolution of all living things from some single original ancestor would have required a great deal more time. Using evidence obtained by studying the rates of radioactive decay, we now know that the physicists of Darwin's time were very wrong: The Earth was formed about 4.5 BYA.

The mechanism of heredity

Darwin received some of his sharpest criticism in the area of heredity. At that time, no one had any concept of genes or how heredity works, so it was not possible for Darwin to explain completely how evolution occurs.

Even though Gregor Mendel was performing his experiments with pea plants in Brünn, Austria (now Brno, the Czech Republic), during roughly the same period, genetics was established as a science only at the start of the 20th century. When scientists began to understand the laws of inheritance (discussed in chapters 12 and 13), this problem with Darwin's theory vanished.

Comparative anatomy

Comparative studies of animals have provided strong evidence for Darwin's theory. In many different types of vertebrates, for example, the same bones are present, indicating their evolutionary past. Thus, the forelimbs shown in figure 1.9 are all constructed from the same basic array of bones, modified for different purposes.

These bones are said to be **homologous** in the different vertebrates; that is, they have the same evolutionary origin, but they now differ in structure and function. They are contrasted with **analogous** structures, such as the wings of birds and butterflies, which have similar function but different evolutionary origins.

Molecular evidence

Evolutionary patterns are also revealed at the molecular level. By comparing the genomes (that is, the sequences of all the genes) of different groups of animals or plants, we can more precisely specify the degree of relationship among the groups. A series of evolutionary changes over time should involve a continual accumulation of genetic changes in the DNA.

This difference can be seen clearly in the protein hemoglobin (figure 1.10). Rhesus monkeys, which like humans are primates, have fewer differences from humans in the 146-amino-acid hemoglobin β-chain than do more distantly related mammals, such as dogs. Nonmammalian vertebrates, such as birds and frogs, differ even more.

The sequences of some genes, such as the ones specifying the hemoglobin proteins, have been determined in many organisms, and the entire time course of their evolution can

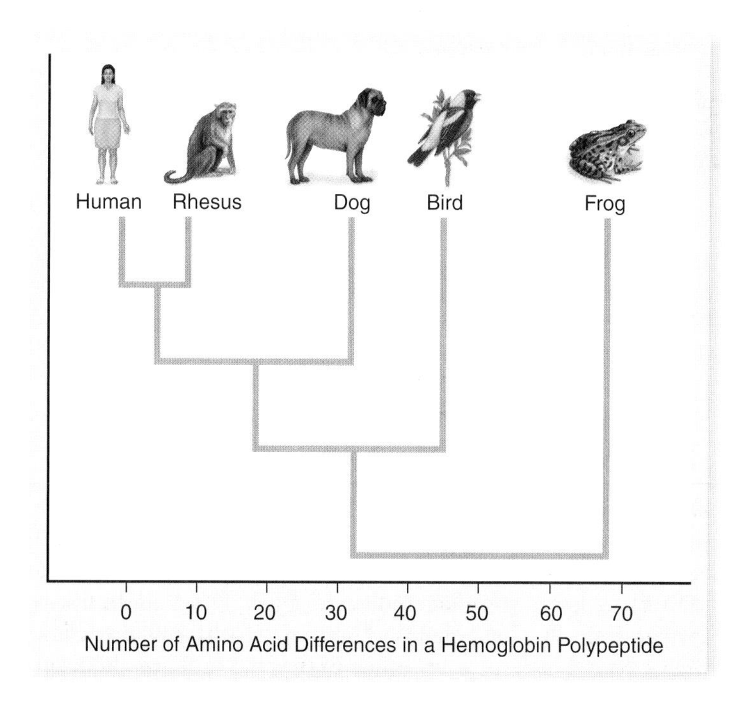

figure 1.10

MOLECULES REFLECT EVOLUTIONARY PATTERNS. Vertebrates that are more distantly related to humans have a greater number of amino acid differences in the hemoglobin polypeptide.

inquiry

Where do you imagine a snake might fall on the graph? Why?

be laid out with confidence by tracing the origins of particular nucleotide changes in the gene sequence. The pattern of descent obtained is called a **phylogenetic tree.** It represents the evolutionary history of the gene, its "family tree." Molecular phylogenetic trees agree well with those derived from the fossil record, which is strong direct evidence of evolution. The pattern of accumulating DNA changes represents, in a real sense, the footprints of evolutionary history.

Darwin's theory of evolution by natural selection presents an example of the process of science. Darwin observed differences in related organisms and proposed the hypothesis of natural selection to explain these differences. The predictions generated by natural selection have been tested and continue to be tested by analysis of the fossil record, genetics, comparative anatomy, and even the DNA of living organisms.

1.4 Unifying Themes in Biology

The study of biology encompasses a large number of different subdisciplines, ranging from biochemistry to ecology. In all of these, however, unifying themes can be identified. Among these are cell theory, the molecular basis of inheritance, the relationship between structure and function, evolution, and the emergence of novel properties.

Cell theory describes the organization of living systems

As was stated at the beginning of this chapter, all organisms are composed of cells, life's basic units (figure 1.11). Cells were discovered by Robert Hooke in England in 1665, using one of the first microscopes, one that magnified 30 times. Not long after that, the Dutch scientist Anton van Leeuwenhoek used microscopes capable of magnifying 300 times and discovered an amazing world of single-celled life in a drop of pond water.

In 1839, the German biologists Matthias Schleiden and Theodor Schwann, summarizing a large number of observations by themselves and others, concluded that all living organisms consist of cells. Their conclusion has come to be known as the **cell theory.** Later, biologists added the idea that all cells come from preexisting cells. The cell theory, one of the basic ideas in biology, is the foundation for understanding the reproduction and growth of all organisms.

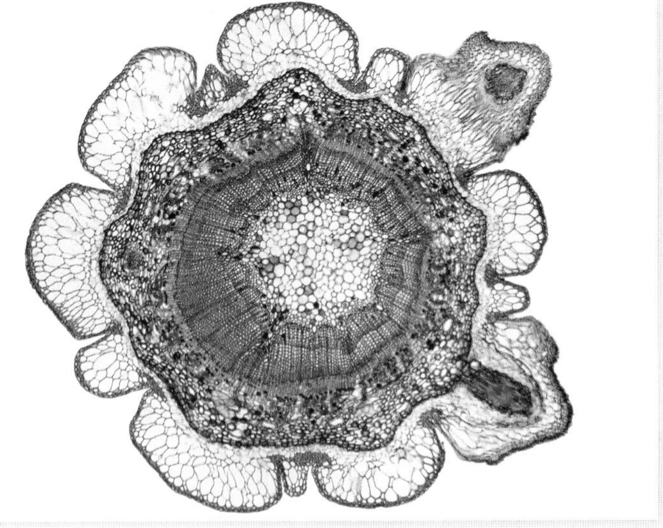

figure 1.11

LIFE IN A DROP OF POND WATER. All organisms are composed of cells. Some organisms, including the protists, shown in part *(a)* are single-celled. Others, such as the plant shown in cross section in part *(b)* consist of many cells.

The molecular basis of inheritance explains the continuity of life

Even the simplest cell is incredibly complex—more intricate than any computer. The information that specifies what a cell is like—its detailed plan—is encoded in **deoxyribonucleic acid (DNA),** a long, cablelike molecule. Each DNA molecule is formed from two long chains of building blocks, called nucleotides, wound around each other (figure 1.12). Four different nucleotides are found in DNA, and the sequence in which they occur encodes the cell's information. Specific sequences of several hundred to many thousand nucleotides make up a **gene,** a discrete unit of information.

The continuity of life from one generation to the next—heredity—depends on the faithful copying of a cell's DNA into daughter cells. The entire set of DNA instructions that specifies a cell is called its *genome.* The sequence of the human genome, 3 billion nucleotides long, was decoded in rough draft form in 2001, a triumph of scientific investigation.

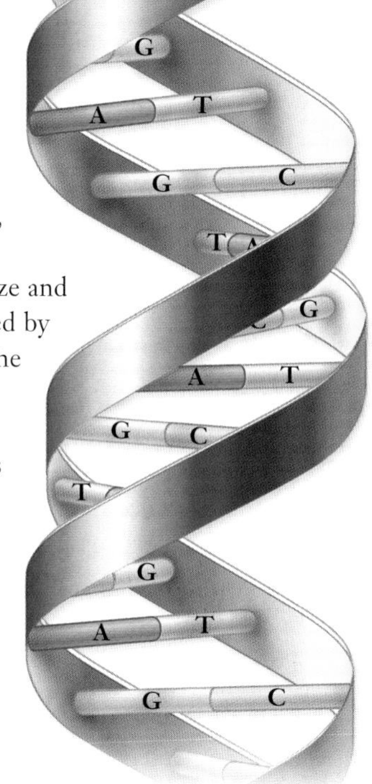

figure 1.12

GENES ARE MADE OF DNA. Winding around each other like the rails of a spiral staircase, the two strands of DNA make a double helix. Because of their size and shape, the nucleotide represented by the letter A can only pair with the nucleotide represented by the letter T, and likewise, the letter G with the letter C. This means that whatever the sequence is on one strand, the sequence on the other strand will be complementary. From each strand, the other can be easily assembled.

The relationship between structure and function underlies living systems

One of the unifying themes of molecular biology is the relationship between structure and function. Function in molecules, and larger macromolecular complexes, is dependent on their structure.

Although this observation may seem trivial, it has far-reaching implications. We study the structure of molecules and macromolecular complexes to learn about their function. When we know the function of a particular structure, we can infer the function of similar structures found in different contexts, such as in different organisms.

Biologists study both aspects, looking for the relationships between structure and function. On the one hand, this allows similar structures to be used to infer possible similar functions. On the other hand, this knowledge also gives clues as to what kinds of structures may be involved in a process if we know about the functionality.

For example, suppose that we know the structure of a human cell's surface receptor for insulin, the hormone that controls uptake of glucose. We then find a similar molecule in the membrane of a cell from a different species—perhaps even a very different organism, such as a worm. We might conclude that this membrane molecule acts as a receptor for an insulin-like molecule produced by the worm. In this way, we might be able to discern the evolutionary relationship between glucose uptake in worms and in humans.

The diversity of life arises by evolutionary change

The unity of life that we see in certain key characteristics shared by many related life-forms contrasts with the incredible diversity of living things in the varied environments of Earth. The underlying unity of biochemistry and genetics argues that all life has evolved from the same origin event. The diversity of life arises by evolutionary change leading to the present biodiversity we see.

Biologists divide life's great diversity into three great groups, called domains: Bacteria, Archaea, and Eukarya. The domains Bacteria and Archaea are composed of single-celled organisms with little internal structure (termed *prokaryotes*), and the domain Eukarya is made up of organisms composed of a complex, organized cell or multiple complex cells (termed *eukaryotes*).

Within Eukarya are four main groups called kingdoms (figure 1.13). Kingdom Protista consists of all the unicellular eukaryotes except yeasts (which are fungi), as well as the multicellular algae. Because of the great diversity among the protists, many biologists feel kingdom Protista should be split into several kingdoms.

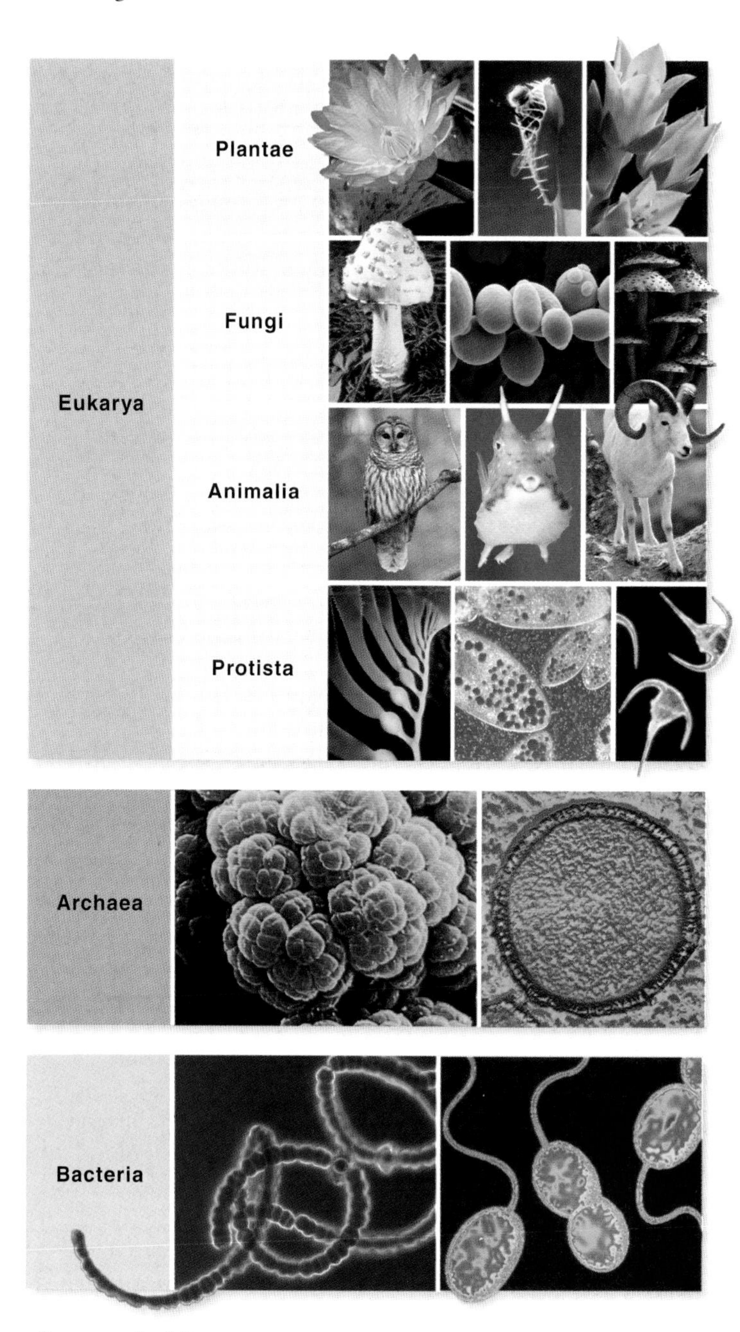

figure 1.13

THE DIVERSITY OF LIFE. Biologists categorize all living things into three overarching groups called domains: Bacteria, Archaea, and Eukarya. Domain Eukarya is composed of four kingdoms: Plantae, Fungi, Animalia, and Protista.

Kingdom Plantae consists of organisms that have cell walls of cellulose and obtain energy by photosynthesis. Organisms in the kingdom Fungi have cell walls of chitin and obtain energy by secreting digestive enzymes and then absorbing the products they release from the external environment. Kingdom Animalia contains organisms that lack cell walls and obtain energy by first ingesting other organisms and then digesting them internally.

Evolutionary conservation explains the unity of living systems

Biologists agree that all organisms alive today have descended from some simple cellular creature that arose about 3.5 BYA. Some of the characteristics of that earliest organism have been preserved. The storage of hereditary information in DNA, for example, is common to all living things.

The retention of these conserved characteristics in a long line of descent usually reflects that they have a fundamental role in the biology of the organism—one not easily changed once adopted. A good example is provided by the homeodomain proteins, which play a critical role in early development in eukaryotes. Conserved characteristics can be seen in approximately 1850 homeodomain proteins, distributed among three different kingdoms of organisms (figure 1.14). The homeodomain proteins are powerful developmental tools that evolved early, and for which no better alternative has arisen.

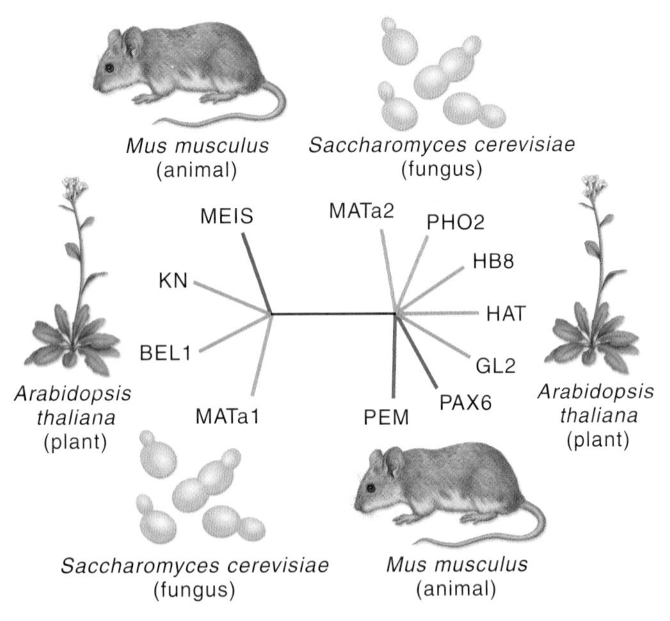

figure 1.14

TREE OF HOMEODOMAIN PROTEINS. Homeodomain proteins are found in fungi (*brown*), plants (*green*), and animals (*blue*). Based on their sequence similarities, these 11 different homeodomain proteins (uppercase letters at the ends of branches) fall into two groups, with representatives from each kingdom in each group. That means, for example, the mouse homeodomain protein PAX6 is more closely related to fungal and flowering plant proteins, such as PHO2 and GL2, than it is to the mouse protein MEIS.

Cells are information-processing systems

One way to think about cells is as highly complex nanomachines that process information. The information stored in DNA is used to direct the synthesis of cellular components, and the particular set of components can differ from cell to cell. The control of gene expression allows differentiation of cell types in time and space, leading to changes over developmental time into different tissue types—even though all cells in an organism carry the same genetic information.

Cells also process information that they receive about the environment. Cells sense their environment through proteins in their membranes, and this information is transmitted across the membrane to elaborate signal-transduction chemical pathways that can change the functioning of a cell.

This ability of cells to sense and respond to their environment is critical to the function of tissues and organs in multicellular organisms. A multicellular organism can regulate its internal environment, maintaining constant temperature, pH, and concentrations of vital ions. This homeostasis is possible because of elaborate signaling networks that coordinate the activities of different cells in different tissues.

Emergent properties arise from the organization of life

As mentioned earlier, the hierarchical organization of life leads to emergent properties. The idea that the whole is greater than the sum of its parts is true of biological systems. At present, these emergent properties are not predictable, but they are observable. As biologists gain a greater understanding of the organization of biological systems, the problem of emerging properties becomes one of the most interesting challenges they face. The new science of systems biology is aimed at solving this problem, and it represents one of the most exciting areas of future research.

Biology as a science is broad and complex, but some unifying themes help to organize this complexity. Cells are the basic unit of life, and they are information-processing machines. The structures of molecules, macromolecular complexes, and even higher levels of organization are related to their functions. The diversity of life can be classified and organized based on similar features, and evolutionary conservation indicates important functions. The organization of living systems leads to emergent properties that cannot be predicted at present.

concept review

1.1 The Science of Life

Biological systems are complex chemical systems, and their functions are both determined and constrained by the principles of chemistry and physics.

■ The study of biological systems is interdisciplinary because solutions require many different approaches to solve a problem.

■ Although life is difficult to define, living systems have seven characteristics in common. All organisms:

- *are composed of one or more cells.*
- *are complex and highly ordered.*
- *respond to stimuli.*
- *are capable of growth, reproduction, and transmission of genetic information to their offspring.*
- *need energy to accomplish many kinds of work.*
- *maintain relatively constant internal conditions independent of the environment by a process called homeostasis.*
- *evolve adaptations to their environment.*

■ The organization of living systems is hierarchical, progressing from atoms to the biosphere.

■ At each higher level of organization emergent properties arise that are greater than the sum of the parts.

1.2 The Nature of Science

At its core, science is concerned with understanding the nature of the world by using observation and reasoning.

■ Much of science is concerned with an increasingly accurate description of nature.

■ There are two ways to arrive at a logical conclusion.

- *Deductive reasoning applies general principles to predict specific results.*
- *Inductive reasoning uses specific observations to construct general scientific principles.*

■ Hypothesis-driven science makes and tests predictions.

- *A hypothesis is constructed from careful observations.*
- *Hypotheses are iteratively changed and refined as new data are generated.*
- *A scientific experiment is a test of a hypothesis.*
- *Experiments involve a test in which a variable is manipulated and a control in which the variable is not manipulated.*
- *Hypotheses are rejected if they make a prediction that cannot be verified experimentally.*
- *A hypothesis can be supported by experiments but never proved.*

■ Scientists use reductionism to study components of a larger system. This has limits because parts may act differently when they are isolated from the larger system.

■ Biologists use models to organize how we think about a problem.

■ Scientists use the word *theory* in two main ways: as a proposed explanation for some natural phenomenon and as a body of concepts that explains facts in an area of study.

■ Scientists engage in basic and applied research.

1.3 An Example of Scientific Inquiry

Darwin's theory of evolution is a good example of how a scientist develops a hypothesis and how a scientific theory grows and gains acceptance.

■ Darwin proposed natural selection as a coherent, logical explanation for how life changed during the history of the Earth.

■ Darwin observed variation in similar species from place to place.

■ Sparked by Malthus's ideas, Darwin noted that species produce many offspring, but only a limited number survive and reproduce.

■ Darwin observed that the traits of offspring can be changed by artificial selection.

■ Darwin proposed that individuals that possess traits that increase survival and reproductive success will become more numerous in populations.

■ This is called natural selection, or to use Darwin's language, descent with modification.

■ Wallace independently came to the same conclusions from his own studies.

■ Natural selection has been tested using data from many fields.

- *The fossil record shows intermediate links between major groups of organisms.*
- *The age of the Earth, argued to be young in Darwin's time, has been shown by studying rates of radioactive decay to be 4.5 billion years.*
- *The research of Mendel and others provided evidence that traits can be inherited as discrete units.*
- *Comparative anatomy provides evidence of evolution from the study of homologous structures.*
- *Molecular data from the study of DNA and proteins provide evidence for changes over time.*
- *Phylogenetic trees using molecular data support the organismal relationships observed in the fossil record.*

■ Taken together, the preceding facts strongly support evolution by natural selection.

■ No data to conclusively disprove evolution has been found since Darwin.

1.4 Unifying Themes in Biology

Unifying themes in biology embrace the many complex subdisciplines of biology. They are:

■ Cell theory describes the basic unit of life and is the foundation of understanding growth and reproduction in all organisms.

■ Hereditary information encoded in genes and found in the molecule DNA is passed on from one generation to the next.

■ The structure and function of organic molecules are interdependent.

■ The diversity of life along with the underlying similarities in biochemistry and genetics support the contention that all life evolved from a single source.

■ Evolution is conservative, and all living organisms share characteristics found in original life forms because they serve an important function.

■ Cells can sense and respond to environmental changes through proteins located on their cell membranes.

■ At each higher level of hierarchical organization new emergent properties arise that could not have been predicted from lower levels of organization.

review questions

SELF TEST

1. Which of the following is NOT a property of life?
   a. Energy utilization
   b. Movement
   c. Order
   d. Homeostasis
2. Which is the correct order of hierarchical levels of living systems?
   a. Cellular, molecular, population, ecosystem, biosphere
   b. Cellular, organelle, organs, ecosystems, populations
   c. Cellular, organismal, population, community, ecosystem
   d. Species, community, population, ecosystem, biosphere
3. The process of inductive reasoning involves—
   a. the use of general principles to predict a specific result
   b. the generation of specific predictions based on a belief system
   c. the use of specific observations to develop general principles
   d. the use of general principles to support a hypothesis
4. A hypothesis in biology is best described as—
   a. a possible explanation of an observation
   b. an observation that supports a theory
   c. a general principle that explains some aspect of life
   d. an unchanging statement that correctly predicts some aspect of life
5. What is the significance of Pasteur's experiment to test the germ hypothesis?
   a. It proved that heat can sterilize a broth.
   b. It demonstrated that cells can arise spontaneously.
   c. It demonstrated that some cells are germs.
   d. It demonstrated that cells can only arise from other cells.
6. Which of the following is NOT an example of reductionism?
   a. Analysis of an isolated enzyme's function in an experimental assay
   b. Investigation of the effect of a hormone on cell growth in a petri dish
   c. Observation of the change in gene expression in response to specific stimulus
   d. An evaluation of the overall behavior of a cell
7. A scientific theory is—
   a. a guess about how things work in the world
   b. a statement of how the world works that is supported by experimental data
   c. a belief held by many scientists
   d. both a and c
8. How is the process of natural selection different from that of artificial selection?
   a. Natural selection produces more variation.
   b. Natural selection makes an individual better adapted.
   c. Artificial selection is a result of human intervention.
   d. Artificial selection results in better adaptations.
9. How does the fossil record help support the theory of evolution by natural selection?
   a. It demonstrates that simple organisms predate more complex organisms.
   b. It provides evidence of change in the form of organisms over time.
   c. It shows that diversity existed millions of years ago.
   d. Both a and b.
10. The theory of evolution by natural selection is a good example of how science proceeds because—
    a. It rationalizes a large body of observations.
    b. It makes predictions that have been tested by a variety of approaches.
    c. It represents Darwin's belief of how life has changed over time.
    d. Both a and b.
11. How does the field of molecular genetics help support the concept of evolution?
    a. Comparisons of genes demonstrate a relationship between all living things.
    b. Different organisms have different genomes.
    c. Sequencing allows for the identification of unique genes.
    d. The number of genes in an organism increase with the complexity of the organism.
12. The cell theory states—
    a. Cells are small.
    b. Cells are highly organized.
    c. There is only one basic type of cell.
    d. All living things are made up of cells.
13. The molecule DNA is important to biological systems because—
    a. It can be replicated.
    b. It encodes the information for making a new individual.
    c. It forms a complex, double-helical structure.
    d. Nucleotides form genes.
14. In which domain of life would you find only single-celled organisms?
    a. Eukarya
    b. Bacteria
    c. Archaea
    d. Both b and c
15. Evolutionary conservation occurs when a characteristic is—
    a. important to the life of the organism
    b. not influenced by evolution
    c. reduced to its least complex form
    d. found in more primitive organisms

CHALLENGE QUESTIONS

1. Exobiology is the study of life on other planets. In recent years, scientists have sent various spacecraft out into the galaxy in search for extraterrestrial life. Assuming that all life shares common properties, what should exobiologists be looking for as they explore other worlds?
2. The classic experiment by Pasteur (see figure 1.4) tested the hypothesis that cells arise from other cells. In this experiment cell growth was measured following sterilization of broth in a swan-neck flask or in a flask with a broken neck.
   a. Which variables were kept the same in these two experiments?
   b. How does the shape of the flask affect the experiment?
   c. Predict the outcome of each experiment based on the two hypotheses.
   d. Some bacteria (germs) are capable of producing heat-resistant spores that protect the cell and allow it to continue to grow after the environment cools. How would the outcome of this experiment have been affected if spore-forming bacteria were present in the broth?

Do you need additional review? *Visit* www.ravenbiology.com *for practice quizzes, animations, videos, and activities designed to help you master the material in this chapter.*

chapter 2

The Nature of Molecules

introduction

ABOUT 12.5 BILLION YEARS AGO, an enormous explosion likely marked the beginning of the universe. With this explosion began a process of star building and planetary formation that eventually led to the formation of Earth, about 4.5 billion years ago. Around 3.5 billion years ago, life began on Earth and started to diversify. To understand the nature of life on Earth, we first need to understand the nature of matter that forms the building blocks of all life.

Starting with the earliest speculations about the world around us, the most basic question has always been, "What is it made of?" The ancient Greeks recognized that larger things may be built of smaller parts. This concept was not put on solid experimental ground until the early 20th century, when physicists began trying to break atoms apart. From those humble beginnings to the huge particle accelerators used today, the picture that emerges of the atomic world is fundamentally different from that of the macroscopic world around us.

To understand how living systems are assembled, we must first understand a little about atomic structure, about how atoms can be linked together by chemical bonds to make molecules, and about the ways in which these small molecules are joined together to make larger molecules, until finally we arrive at the structure of a cell. Our study of life on Earth therefore begins with physics and chemistry. For many of you, this chapter will be a review of material encountered in other courses.

concept outline

2.1 The Nature of Atoms

Any substance in the universe that has mass and occupies space is defined as **matter.** All matter is composed of extremely small particles called **atoms.** Because of their size, atoms are difficult to study. Not until early in the last century did scientists carry out the first experiments revealing the physical nature of atoms.

Atomic structure includes a central nucleus and orbiting electrons

Objects as small as atoms can be "seen" only indirectly, by using complex technology such as tunneling microscopy (figure 2.1). We now know a great deal about the complexities of atomic structure, but the simple view put forth in 1913 by the Danish physicist Niels Bohr provides a good starting point for understanding atomic theory. Bohr proposed that every atom possesses an orbiting cloud of tiny subatomic particles called **electrons** whizzing around a core, like the planets of a miniature solar system. At the center of each atom is a small, very dense nucleus formed of two other kinds of subatomic particles: **protons** and **neutrons** (figure 2.2).

Atomic number and the elements

Within the nucleus, the cluster of protons and neutrons is held together by a force that works only over short, subatomic distances. Each proton carries a positive (+) charge, and each neutron has no charge. Electrons carry a negative (–) charge. Typically, an atom has one electron for each proton and is, thus, electrically neutral. Different atoms are defined by the number of protons, a quantity called the **atomic number.** The chemical behavior of an atom is due to the number and configuration of electrons, as we will see later in this chapter. Atoms with the same atomic number (that is, the same number of protons) have the same chemical properties and are said to belong to the same element. Formally speaking, an **element** is any substance that cannot be broken down to any other substance by ordinary chemical means.

figure 2.1

SCANNING TUNNELING MICROSCOPE IMAGE. The scanning tunneling microscope is a nonoptical form of imaging. A probe ending in a single atom is passed over the surface of the material to be imaged. Quantum effects between probe and surface allow imaging. This image shows a lattice of oxygen atoms (shown in dark blue) on a rhodium crystal (shown in light blue).

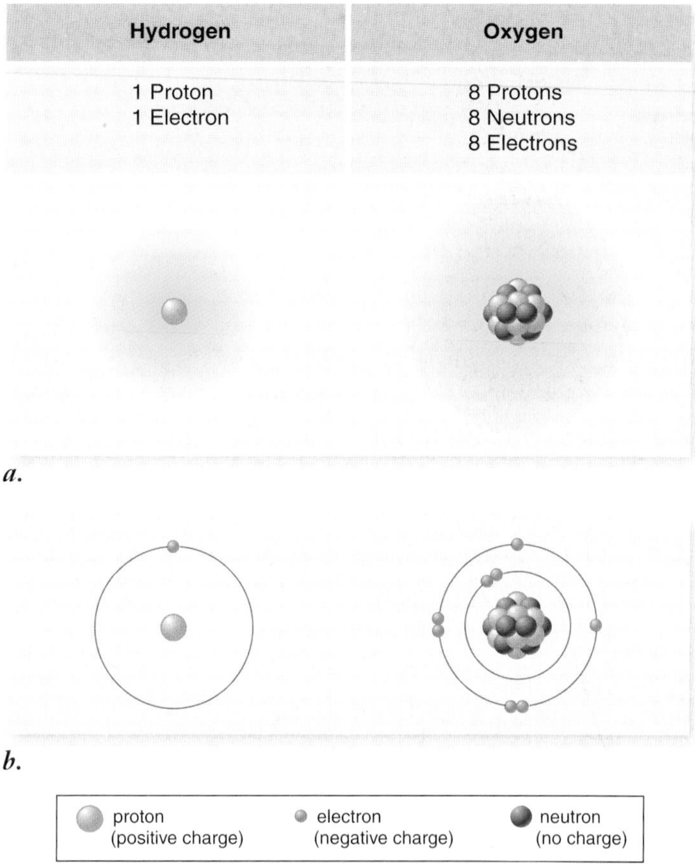

figure 2.2

BASIC STRUCTURE OF ATOMS. All atoms have a nucleus consisting of protons and neutrons, except hydrogen, the smallest atom, which usually has only one proton and no neutrons in its nucleus. Oxygen, for example, typically has eight protons and eight neutrons in its nucleus. In the simple "Bohr model" of atoms pictured here, electrons spin around the nucleus at a relatively far distance. ***a.*** Atoms are depicted as a nucleus with a cloud of electrons. The cloud of electrons is not shown to scale. ***b.*** The electrons are shown in discrete energy levels. These are described in greater detail in the text and the next two figures.

Atomic mass

The terms *mass* and *weight* are often used interchangeably, but they have slightly different meanings. *Mass* refers to the amount of a substance, but *weight* refers to the force gravity exerts on a substance. An object has the same mass whether it is on the Earth or the Moon, but its weight will be greater on the Earth because the Earth's gravitational force is greater than the Moon's. The **atomic mass** of an atom is equal to the sum of the masses of its protons and neutrons. Atoms that occur naturally on Earth contain from 1 to 92 protons and up to 146 neutrons.

The mass of atoms and subatomic particles is measured in units called *daltons*. To give you an idea of just how small these units are, note that it takes 602 million million billion (6.02×10^{23}) daltons to make 1 gram. A proton weighs approximately 1 dalton (actually 1.007 daltons), as does a neutron (1.009 daltons). In contrast, electrons weigh only 1/1840 of a dalton, so their contribution to the overall mass of an atom is negligible.

Electrons

The positive charges in the nucleus of an atom are neutralized or counterbalanced by negatively charged electrons, which are located in regions called **orbitals** that lie at varying distances around the nucleus. Atoms with the same number of protons and electrons are electrically neutral, having no net charge, and are called **neutral atoms.**

Electrons are maintained in their orbitals by their attraction to the positively charged nucleus. Sometimes other forces overcome this attraction, and an atom loses one or more electrons. In other cases, atoms gain additional electrons. Atoms in which the number of electrons does not equal the number of protons are known as **ions,** and they are charged particles. An atom having more protons than electrons has a net positive charge and is called a **cation.** For example, an atom of sodium (Na) that has lost one electron becomes a sodium ion (Na^+), with a charge of +1. An atom having fewer protons than electrons carries a net negative charge and is called an **anion.** A chlorine atom (Cl) that has gained one electron becomes a chloride ion (Cl^-), with a charge of −1.

Isotopes

Although all atoms of an element have the same number of protons, they may not all have the same number of neutrons. Atoms of a single element that possess different numbers of neutrons are called **isotopes** of that element.

Most elements in nature exist as mixtures of different isotopes. Carbon (C), for example, has three isotopes, all containing six protons (figure 2.3). Over 99% of the carbon found in nature exists as an isotope that also contains six neutrons. Because the total mass of this isotope is 12 daltons (6 from protons plus 6 from neutrons), it is referred to as carbon-12 and is symbolized ^{12}C. Most of the rest of the naturally occurring carbon is carbon-13, an isotope with seven neutrons. The rarest carbon isotope is carbon-14, with eight neutrons. Unlike the other two isotopes, carbon-14 is unstable: Its nucleus tends to break up into elements with lower atomic numbers. This nuclear breakup, which emits a significant amount of energy, is called *radioactive decay*, and isotopes that decay in this fashion are **radioactive isotopes.**

Some radioactive isotopes are more unstable than others, and therefore they decay more readily. For any given isotope, however, the rate of decay is constant. The decay time is usually expressed as the **half-life,** the time it takes for one-half of the atoms in a sample to decay. Carbon-14, for example, often used in carbon dating of fossils and other materials, has a half-life of 5730 years. A sample of carbon containing 1 gram of carbon-14 today would contain 0.5 gram of carbon-14 after 5730 years, 0.25 gram 11,460 years from now, 0.125 gram 17,190 years from now, and so on. By determining the ratios of the different isotopes of carbon and other elements in biological samples and in rocks, scientists are able to accurately determine when these materials formed.

Radioactivity has many useful applications in modern biology. Radioactive isotopes are one way to label, or "tag," a specific molecule and then follow its fate, either in a chemical reaction or in living cells and tissue. The downside, however, is that the energetic subatomic particles emitted by radioactive substances have the potential to severely damage living cells, producing genetic mutations and, at high doses, cell death. Consequently, exposure to radiation is carefully controlled and regulated. Scientists who work with radioactivity follow strict handling protocols and wear radiation-sensitive badges to monitor their exposure over time to help ensure a safe level of exposure.

Electrons determine the chemical behavior of atoms

As mentioned earlier, the key to the chemical behavior of an atom lies in the number and arrangement of its electrons in their orbitals. The Bohr model of the atom shows individual electrons as following discrete circular orbits around a central nucleus; however, such a simple picture does not reflect reality. Modern physics indicates that it is not possible to precisely locate the position of any

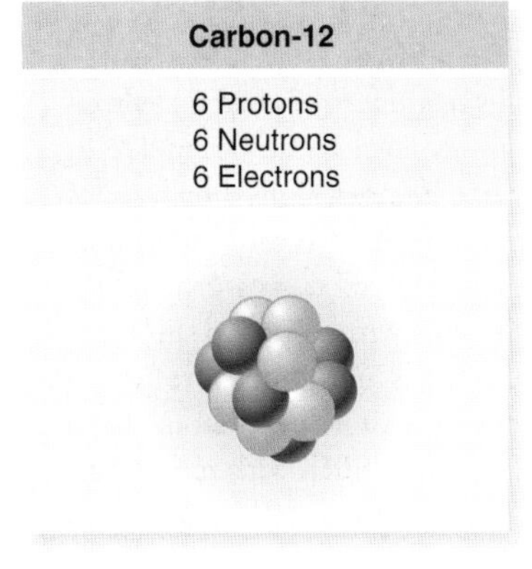

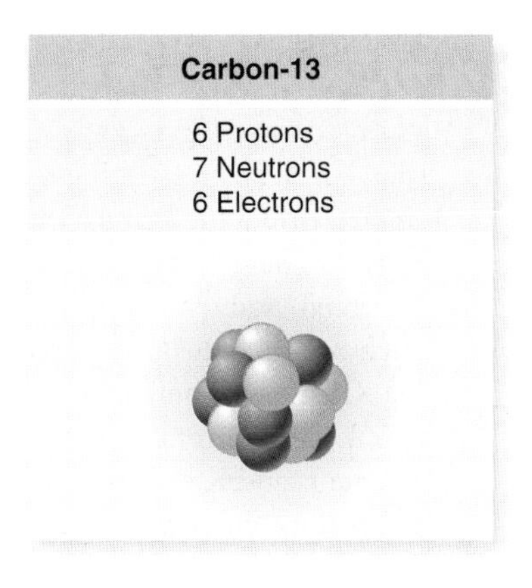

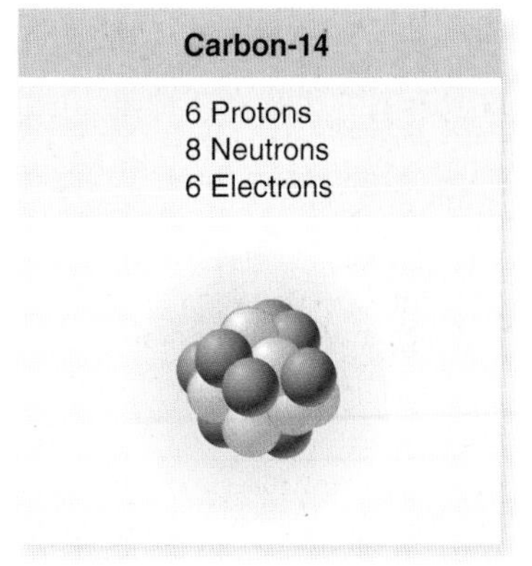

figure 2.3

THE THREE MOST ABUNDANT ISOTOPES OF CARBON. Isotopes of a particular element have different numbers of neutrons.

individual electron at any given time. In fact, an electron could be anywhere, from close to the nucleus to infinitely far away from it.

A particular electron, however, is more likely to be located in some areas than in others. An orbital is defined as the area around a nucleus where an electron is most likely to be found. These orbitals represent probability distributions for electrons, that is, regions with a high probability of containing an electron. Some electron orbitals near the nucleus are spherical (*s* orbitals), while others are dumbbell-shaped (*p* orbitals) (figure 2.4). Still other orbitals, more distant from the nucleus, may have different shapes. Regardless of its shape, no orbital may contain more than two electrons.

Almost all of the volume of an atom is empty space, because the electrons are on average very distant from the nucleus, relative to its size. If the nucleus of an atom were the size of a golf ball, the orbit of the nearest electron would be a mile away. Consequently, the nuclei of two atoms never come close enough in nature to interact with each other. It is for this reason that an atom's electrons, not its protons or neutrons, determine its chemical behavior, and it also explains why the isotopes of an element, all of which have the same arrangement of electrons, behave the same way chemically.

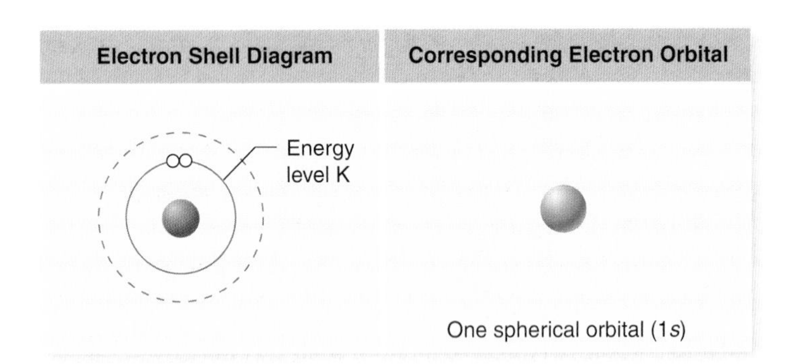

a.

Atoms contain discrete energy levels

Because electrons are attracted to the positively charged nucleus, it takes work to keep them in their orbitals, just as it takes work to hold a grapefruit in your hand against the pull of gravity. The formal definition of energy, as you probably recall, is the ability to do work.

The grapefruit held above the ground is said to possess *potential energy* because of its position; if you release it, the grapefruit falls, and its potential energy is reduced. Conversely, if you carried the grapefruit to the top of a building, you would increase its potential energy. Electrons also have a potential energy that is related to their position. To oppose the attraction of the nucleus and move the electron to a more distant orbital requires an input of energy, which results in an electron with greater potential energy. Chlorophyll captures energy from light during photosynthesis in this way, as you'll see in chapter 8—light energy excites electrons in the chlorophyll molecule. Moving an electron closer to the nucleus has the opposite effect: Energy is released, usually as radiant energy (heat or light), and the electron ends up with less potential energy (figure 2.5).

One of the initially surprising aspects of atomic structure is that electrons within the atom have discrete **energy levels.** These discrete levels correspond to quanta (sing., quantum), which means specific amount of energy. To use the grapefruit analogy again, it is as though a grapefruit could only be raised to particular floors of a building. Every atom exhibits a ladder of potential energy values, a discrete set of orbitals at particular energetic "distances" from the nucleus.

Because the amount of energy an electron possesses is related to its distance from the nucleus, electrons that are the same distance from the nucleus have the same energy, even if they occupy different orbitals. Such electrons are said to occupy the same energy level. The energy levels are denoted

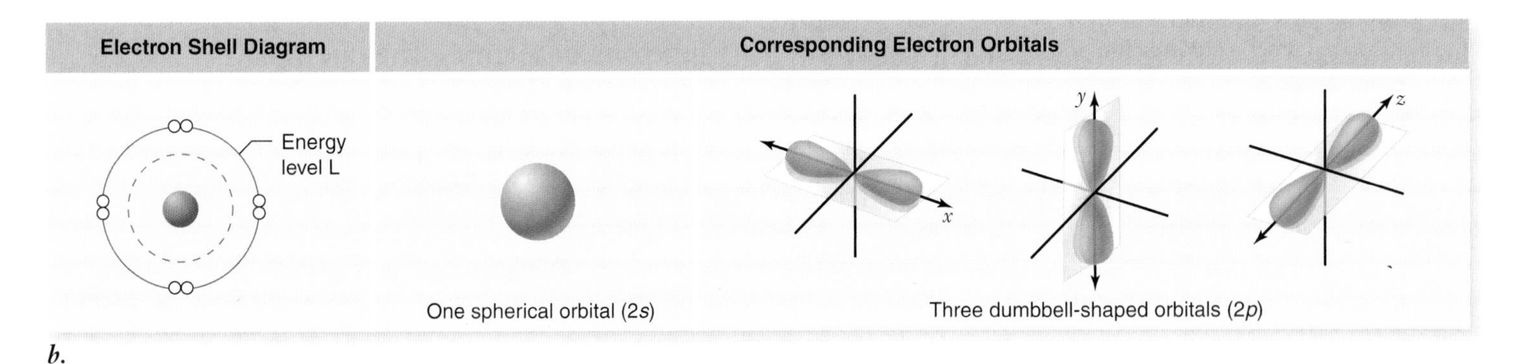

b.

Electron Shell Diagram

Electron Orbitals

Neon

c.

figure 2.4

ELECTRON ORBITALS. ***a.*** The lowest energy level or electron shell—the one nearest the nucleus—is level K. It is occupied by a single *s* orbital, referred to as 1*s*. ***b.*** The next highest energy level, L, is occupied by four orbitals: one *s* orbital (referred to as the 2*s* orbital) and three *p* orbitals (each referred to as a 2*p* orbital). Each orbital holds two paired electrons with opposite spin. Thus, the K level is populated by two electrons, and the L level is populated by a total of eight electrons. ***c.*** The neon atom shown has the L and K energy levels completely filled with electrons and is thus unreactive.

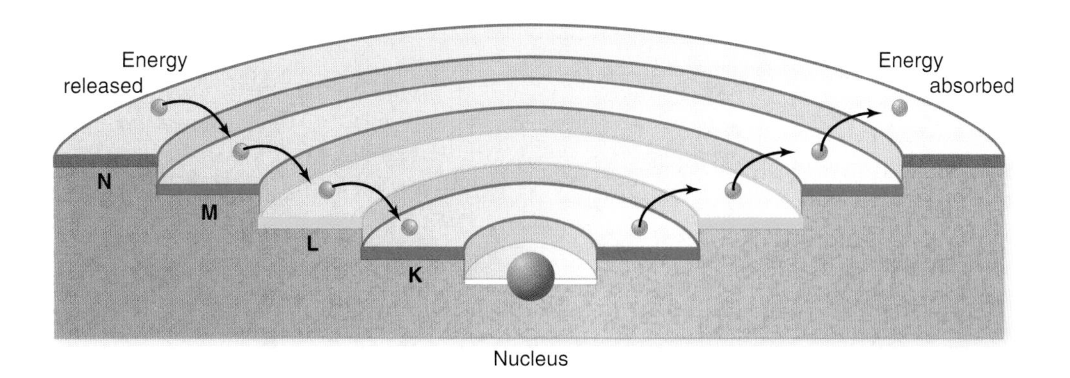

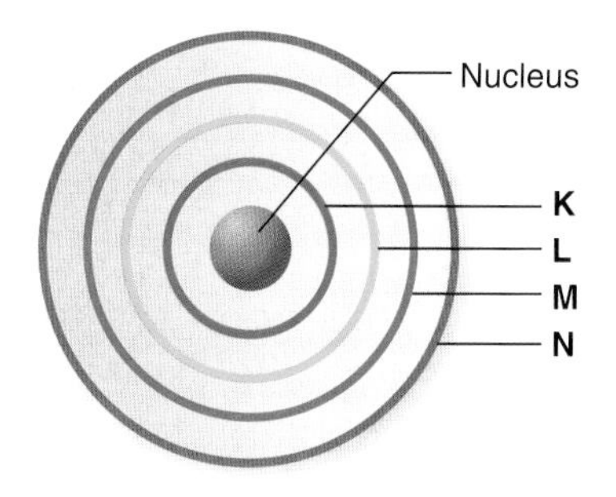

figure 2.5

ATOMIC ENERGY LEVELS. Electrons have energy of position. When an atom absorbs energy, an electron moves to a higher energy level, farther from the nucleus. When an electron falls to lower energy levels, closer to the nucleus, energy is released. The first two energy levels are the same as shown in the previous figure.

with letters K, L, M, and so on (figure 2.5). Be careful not to confuse energy levels, which are drawn as rings to indicate an electron's *energy*, with orbitals, which have a variety of three-dimensional shapes and indicate an electron's most likely *location*. Electron orbitals are arranged so that as they are filled, this successively fills each energy level. This filling of orbitals and energy levels is what is responsible for the chemical reactivity of elements.

During some chemical reactions, electrons are transferred from one atom to another. In such reactions, the loss of an electron is called **oxidation,** and the gain of an electron is called **reduction.**

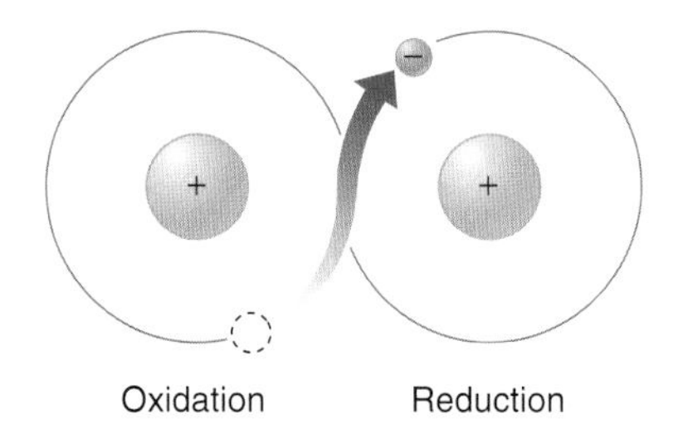

Notice that when an electron is transferred in this way, it keeps its energy of position. In organisms, chemical energy is stored in high-energy electrons that are transferred from one atom to another in reactions involving oxidation and reduction (described in chapter 7). Because oxidation and reduction can be coupled, such that one atom or molecule is oxidized while another is reduced in the same reaction, we call these *redox reactions*.

An atom consists of a nucleus of protons and neutrons surrounded by a cloud of electrons. The number of electrons largely determines the chemical behavior of an atom. Atoms that have the same number of protons but different numbers of neutrons are called isotopes. Isotopes of an atom differ in atomic mass but have similar chemical properties. Electrons are localized about a nucleus in regions called orbitals. No orbital can contain more than two electrons, but many orbitals may be the same distance from the nucleus and, thus, contain electrons of the same energy.

2.2 Elements Found in Living Systems

Ninety elements occur naturally, each with a different number of protons and a different arrangement of electrons. When the nineteenth-century Russian chemist Dmitri Mendeleev arranged the known elements in a table according to their atomic number, he discovered one of the great generalizations of science: The elements exhibited a pattern of chemical properties that repeated itself in groups of eight. This periodically repeating pattern lent the table its name: the periodic table of elements (figure 2.6).

The periodic table displays elements according to atomic number and properties

The eight-element periodicity that Mendeleev found is based on the interactions of the electrons in the outermost energy level of the different elements. These electrons are called **valence electrons,** and their interactions are the basis for the elements' differing chemical properties. For most of the atoms important to life, the outermost energy level can contain no more than eight electrons; the chemical behavior of an element reflects how many of the eight positions are filled. Elements possessing all eight electrons in their outer energy level (two for helium) are **inert,** or nonreactive. These elements, which include helium (He), neon (Ne), argon (Ar), and so on, are termed the *noble gases*. In sharp contrast, elements with seven electrons (one fewer than the maximum number of eight) in their outer energy level, such as fluorine (F), chlorine (Cl), and bromine (Br), are highly reactive. They tend to gain the extra electron needed to fill the energy level. Elements with only one electron in their outer energy level, such as lithium (Li), sodium (Na), and potassium (K), are also very reactive; they tend to lose the single electron in their outer level.

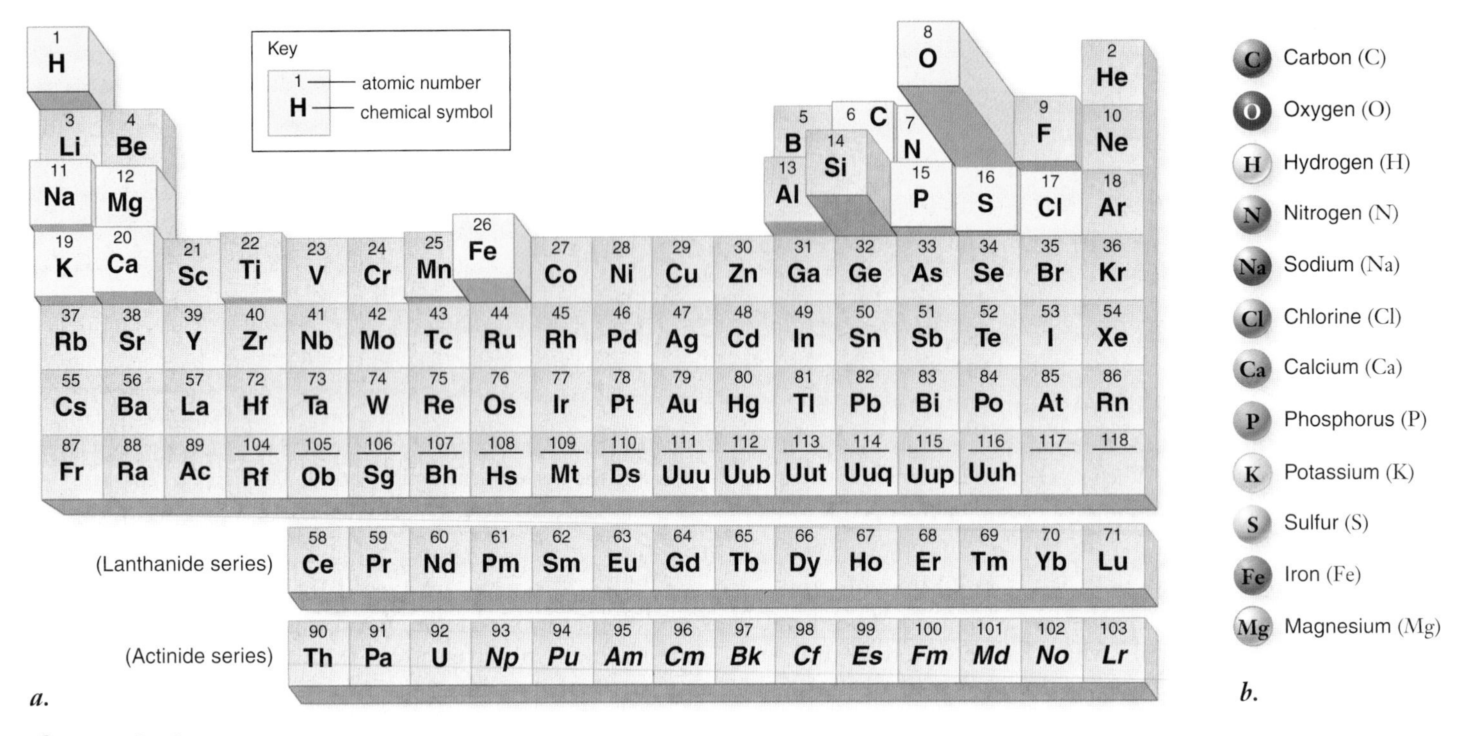

figure 2.6

PERIODIC TABLE OF THE ELEMENTS. ***a.*** In this representation, the frequency of elements that occur in the Earth's crust is indicated by the height of the block. Elements shaded in green are found in living systems in more than trace amounts. ***b.*** Common elements found in living systems are shown in colors that will be used throughout the text.

Mendeleev's periodic table leads to a useful generalization, the **octet rule,** or *rule of eight* (Latin *octo,* "eight"): Atoms tend to establish completely full outer energy levels. For the main group elements of the periodic table, the rule of eight is accomplished by one filled *s* orbital and three filled *p* orbitals (figure 2.7). The exception to this is He, in the first row, which needs only two electrons to fill the 1*s* orbital. Most chemical behavior of biological interest can be predicted quite accurately from this simple rule, combined with the tendency of atoms to balance positive and negative charges. For instance, you read earlier that sodium ion (Na^+) has lost an electron, and chloride ion (Cl^-) has gained an electron. In the following section, we describe how these ions react to form table salt.

Of the 90 naturally occurring elements on Earth, only 12 (C, H, O, N, P, S, Na, K, Ca, Mg, Fe, Cl) are found in living systems in more than trace amounts (0.01% or higher). These elements all have atomic numbers less than 21, and thus, have low atomic masses. Of these, four elements (carbon, hydrogen, oxygen, and nitrogen) constitute 96.3% of the weight of your body. The majority of molecules that make up your body are compounds of carbon, which we call **organic** compounds. These organic compounds contain primarily these four elements (CHON) explaining their prevalence in living systems. Some trace elements, such as zinc (Zn) and iodine (I), play crucial roles in living processes even though they are present in tiny amounts. Iodine deficiency, for example, can lead to enlargement of the thyroid, causing a bulge at the neck called a goiter.

Nonreactive	Reactive
2 protons 2 neutrons 2 electrons	7 protons 7 neutrons 7 electrons
K, 2+	K, L, 7+
Helium	Nitrogen

figure 2.7

ELECTRON ENERGY LEVELS FOR HELIUM AND NITROGEN. Green balls represent electrons, blue ball represents the nucleus with number of protons indicated by number of (+) charges. Note that the helium atom has a filled K shell and is thus unreactive, whereas the nitrogen atom has five electrons in the L shell, three of which are unpaired, making it reactive.

The periodic table shows the elements in terms of atomic number and repeating properties. Only 12 elements are found in significant amounts in living organisms.

2.3 The Nature of Chemical Bonds

A group of atoms held together by energy in a stable association is called a **molecule.** When a molecule contains atoms of more than one element, it is called a **compound.** The atoms in a molecule are joined by **chemical bonds;** these bonds can result when atoms with opposite charges attract (ionic bonds), when two atoms share one or more pairs of electrons (covalent bonds), or when atoms interact in other ways (table 2.1). We will start by examining **ionic bonds,** which form when atoms with opposite electrical charges (ions) attract.

TABLE 2.1 Bonds and Interactions

Name	Basis of interaction	Strength
Covalent bond	Sharing of electron pairs	Strong
Ionic bond	Attraction of opposite charges	↑
Hydrogen bond	Sharing of H atom	
Hydrophobic interaction	Forcing of hydrophobic portions of molecules together in presence of polar substances	↓
van der Waals attraction	Weak attractions between atoms due to oppositely polarized electron clouds	Weak

Ionic bonds form crystals

Common table salt, the molecule sodium chloride (NaCl), is a lattice of ions in which the atoms are held together by ionic bonds (figure 2.8). Sodium has 11 electrons: 2 in the inner energy level (K), 8 in the next level (L), and 1 in the outer (valence) level (M). The single, unpaired valence electron has a strong tendency to join with another unpaired electron in another atom. A stable configuration can be achieved if the valence electron is lost to another atom that also has an unpaired electron. The loss of this electron results in the formation of a positively charged sodium ion, Na^+.

The chlorine atom has 17 electrons: 2 in the K level, 8 in the L level, and 7 in the M level. As you can see, one of the orbitals in the outer energy level has an unpaired electron. The addition of another electron fills that level and causes a negatively charged chloride ion, Cl^-, to form.

When placed together, metallic sodium and gaseous chlorine react swiftly and explosively, as the sodium atoms donate electrons to chlorine to form Na^+ and Cl^- ions. Because opposite charges attract, the Na^+ and Cl^- remain associated in an **ionic compound,** NaCl, which is electrically neutral. The electrical attractive force holding NaCl together, however, is not directed specifically between individual Na^+ and Cl^- ions, and no discrete sodium chloride molecules form. Instead, the force exists between any one ion and all neighboring ions of the opposite charge. The ions aggregate in a crystal matrix with a precise geometry. Such aggregations are what we know as salt crystals. If a salt such as NaCl is placed in water, the electrical attraction of the water molecules, for reasons we will point out later in this chapter, disrupts the forces holding the ions in their crystal matrix, causing the salt to dissolve into a roughly equal mixture of free Na^+ and Cl^- ions.

As living systems always include water, ions are more important than ionic crystals. Important ions in biological systems include Ca^{2+}, which is involved in cell signaling, K^+ and Na^+, which are involved in the conduction of nerve impulses.

Covalent bonds build stable molecules

Covalent bonds form when two atoms share one or more pairs of valence electrons. Consider gaseous hydrogen (H_2) as an example. Each hydrogen atom has an unpaired electron

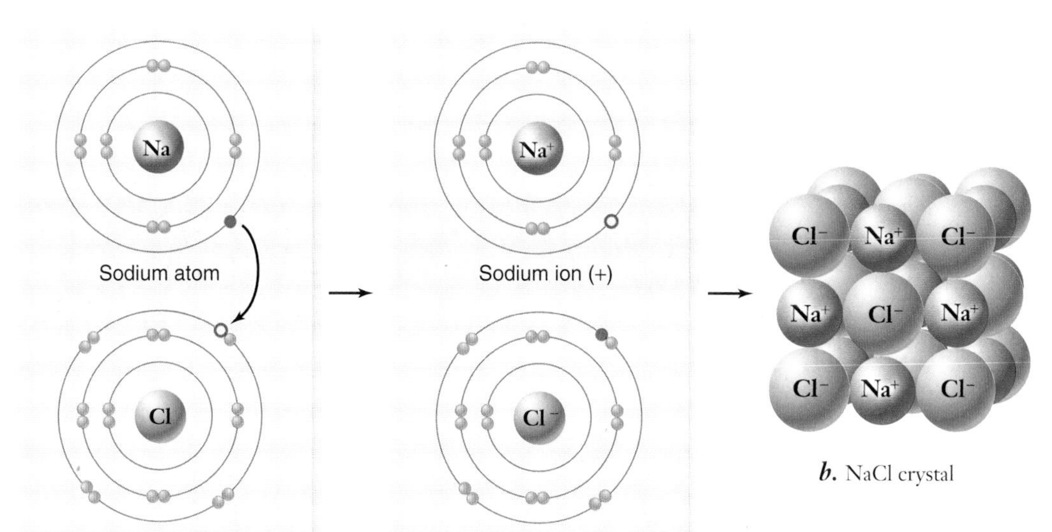

figure 2.8

THE FORMATION OF IONIC BONDS BY SODIUM CHLORIDE.

a. When a sodium atom donates an electron to a chlorine atom, the sodium atom becomes a positively charged sodium ion, and the chlorine atom becomes a negatively charged chloride ion. ***b.*** The electrostatic attraction of oppositely charged ions leads to the formation of a lattice of Na^+ and Cl^-.

and an unfilled outer energy level; for these reasons, the hydrogen atom is unstable. However, when two hydrogen atoms are in close association, each atom's electron is attracted to both nuclei. In effect, the nuclei are able to share their electrons. The result is a diatomic (two-atom) molecule of hydrogen gas.

The molecule formed by the two hydrogen atoms is stable for three reasons:

1. **It has no net charge.** The diatomic molecule formed as a result of this sharing of electrons is not charged because it still contains two protons and two electrons.
2. **The octet rule is satisfied.** Each of the two hydrogen atoms can be considered to have two orbiting electrons in its outer energy level. This state satisfies the octet rule, because each shared electron orbits both nuclei and is included in the outer energy level of *both* atoms.
3. **It has no unpaired electrons.** The bond between the two atoms also pairs the two free electrons.

Unlike ionic bonds, covalent bonds are formed between two individual atoms, giving rise to true, discrete molecules.

The strength of covalent bonds

The strength of a covalent bond depends on the number of shared electrons. Thus **double bonds,** which satisfy the octet rule by allowing two atoms to share two pairs of electrons, are stronger than **single bonds,** in which only one electron pair is shared. In practical terms, more energy is required to break a double bond than a single bond. The strongest covalent bonds are **triple bonds,** such as those that link the two nitrogen atoms of nitrogen gas molecules (N_2).

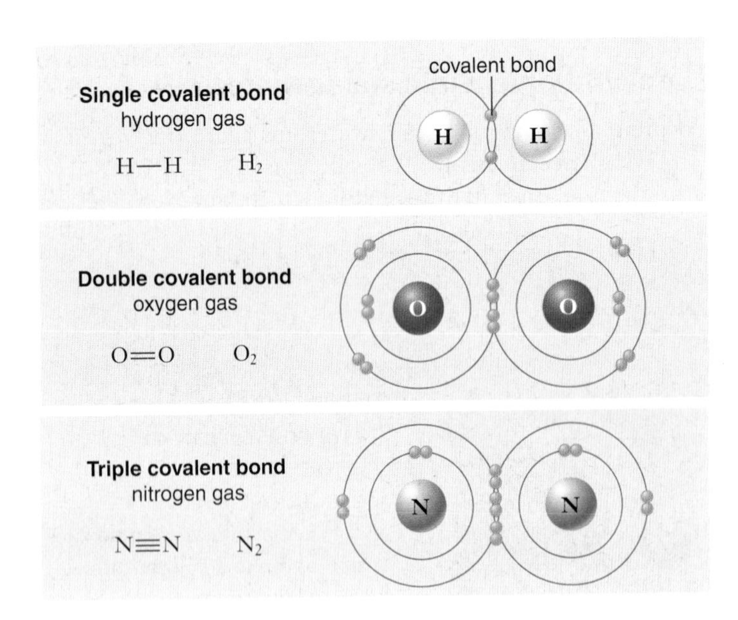

Covalent bonds are represented in chemical formulas as lines connecting atomic symbols, where each line between two bonded atoms represents the sharing of one pair of electrons. The *structural formulas* of hydrogen gas and oxygen gas are H—H and O=O, respectively, and their *molecular formulas* are H_2 and O_2. The structural formula for N_2 is N≡N.

Molecules with several covalent bonds

A vast number of biological compounds are composed of more than two atoms. An atom that requires two, three, or four additional electrons to fill its outer energy level completely may acquire them by sharing its electrons with two or more other atoms.

For example, the carbon atom (C) contains six electrons, four of which are in its outer energy level and are unpaired. To satisfy the octet rule, a carbon atom must form four covalent bonds. Because four covalent bonds may form in many ways, carbon atoms are found in many different kinds of molecules. CO_2 (carbon dioxide), CH_4 (methane), and C_2H_5OH (ethanol) are just a few examples.

Polar and nonpolar covalent bonds

Atoms differ in their affinity for electrons, a property called **electronegativity.** In general, electronegativity increases left to right in a row of the periodic table and decreases down the columns of the table. Thus the elements in the upper-right corner have the highest electronegativity.

For bonds between identical atoms, for example, H_2 or O_2, the affinity for electrons is obviously the same, and the electrons are equally shared. Such bonds are termed **nonpolar,** and the resulting compounds are also referred to as nonpolar.

For atoms that differ greatly in electronegativity, electrons are not equally shared. The shared electrons are more likely to be found near the atom with greater electronegativity, and less likely to be located near the atom of lower electronegativity. In this case, although the molecule is still electrically neutral (same number of protons as electrons), the distribution of charge is not uniform. This unequal distribution results in regions of partial negative charge near the more electronegative atom, and regions of partial positive charge near the less electronegative atom. Such bonds are termed **polar covalent bonds,** and the molecules polar molecules. When drawing polar molecules, these partial charges are usually symbolized by the lowercase Greek letter delta (δ). The partial charge seen in a polar covalent bond is relatively small—far less than the unit charge of an ion. For biological molecules, we can predict polarity of bonds by knowing the relative electronegativity of a small number of important atoms (table 2.2). Notice that although C and H differ slightly in electronegativity, this small difference is negligible, and C—H bonds are considered nonpolar.

The importance of polar and nonpolar molecules will be explored later as it is an important feature of the chemistry of water. Water (H_2O) is a polar molecule with electrons more concentrated around the oxygen atom.

TABLE 2.2 Relative Electronegativities of Some Important Atoms

Atom	Electronegativity
O	3.5
N	3.0
C	2.5
H	2.1

Chemical reactions alter bonds

The formation and breaking of chemical bonds, which is the essence of chemistry, is termed a *chemical reaction.* All chemical reactions involve the shifting of atoms from one molecule or ionic compound to another, without any change in the number or identity of the atoms. For convenience, we refer to the original molecules before the reaction starts as *reactants,* and the molecules resulting from the chemical reaction as *products.* For example:

$$6H_2O + 6CO_2 \longrightarrow C_6H_{12}O_6 + 6O_2$$

$$\textit{reactants} \longrightarrow \textit{products}$$

You may recognize this reaction as a simplified form of the photosynthesis reaction, in which water and carbon dioxide are combined to produce glucose and oxygen. Most animal life ultimately depends on this reaction, which takes place in plants. (Photosynthetic reactions will be discussed in detail in chapter 8.)

The extent to which chemical reactions occur is influenced by three important factors:

1. **Temperature.** Heating the reactants increases the rate of a reaction because the reactants collide with one another more often. (Care must be taken that the temperature is not so high that it destroys the molecules altogether.)
2. **Concentration of reactants and products.** Reactions proceed more quickly when more reactants are available, allowing more frequent collisions. An accumulation of products typically slows the reaction and, in reversible reactions, may speed the reaction in the reverse direction.
3. **Catalysts.** A catalyst is a substance that increases the rate of a reaction. It doesn't alter the reaction's equilibrium between reactants and products, but it does shorten the time needed to reach equilibrium, often dramatically. In living systems, proteins called enzymes catalyze almost every chemical reaction.

Many reactions in nature are reversible, meaning that the products may themselves be reactants, and the reaction proceeds in reverse. We can write the preceding reaction in the reverse order:

$$C_6H_{12}O_6 + 6O_2 \longrightarrow 6H_2O + 6CO_2$$

$$\textit{reactants} \longrightarrow \textit{products}$$

This reaction is a simplified version of the oxidation of glucose by cellular respiration, in which glucose is broken down into water and carbon dioxide in the presence of oxygen. Virtually all organisms carry out forms of glucose oxidation; details are covered later, in chapter 7.

An ionic bond is an attraction between ions of opposite charge in an ionic compound. Such bonds exist between an ion and all of the oppositely charged ions in its immediate vicinity. A covalent bond is a stable chemical bond formed when two atoms share one or more pairs of electrons. In polar covalent bonds, unequal electron sharing results in an imbalance of charge, called polarity, between the bonded atoms. Chemical reactions make and break bonds, combining reactants to form products.

2.4 Water: A Vital Compound

Of all the molecules that are common on Earth, only water exists as a liquid at the relatively low temperatures that prevail on the Earth's surface. Three-fourths of the Earth is covered by liquid water (figure 2.9). When life was beginning, water provided a medium in which other molecules could move around and interact, without being held in place by strong covalent or ionic bonds. Life evolved in water for 2 billion years before spreading to land. And even today, life is inextricably tied to water. About two-thirds of any organism's body is composed of water, and all organisms require a water-rich environment, either internal or external, for growth and reproduction. It is no accident that tropical rain forests are bursting with life, while dry deserts appear almost lifeless except when water becomes temporarily plentiful, such as after a rainstorm.

a. Solid

b. Liquid

c. Gas

figure 2.9

WATER TAKES MANY FORMS. ***a.*** When water cools below 0°C, it forms beautiful crystals, familiar to us as snow and ice. ***b.*** Ice turns to liquid when the temperature is above 0°C. ***c.*** Liquid water becomes steam when the temperature rises above 100°C, as seen in this hot spring at Yellowstone National Park.

Water's structure facilitates hydrogen bonding

Water has a simple molecular structure, consisting of an oxygen atom bound to two hydrogen atoms by two single covalent bonds (figure 2.10). The resulting molecule is stable: It satisfies the octet rule, has no unpaired electrons, and carries no net electrical charge.

The single most outstanding chemical property of water is its ability to form weak chemical associations, called **hydrogen bonds.** These bonds form between the partially negative O atoms and the partially positive H atoms of two water molecules. While these bonds have only 5 to 10% of the strength of covalent bonds, they are responsible for much of the chemical organization of living systems.

The electronegativity of O is much greater than that of H (see table 2.2), and so the bonds between these atoms are polar. In fact, water is a highly polar molecule, and *the polarity of water underlies water's chemistry and the chemistry of life.*

If we consider the shape of a water molecule, we can see that water's two covalent bonds have a partial charge at each end: δ^- at the oxygen end and δ^+ at the hydrogen end. The most stable arrangement of these charges is a *tetrahedron,* in which the two negative and two positive charges are approximately equidistant from one another. The oxygen atom lies at the center of the tetrahedron, the hydrogen atoms occupy two of the apexes, and the partial negative charges occupy the other two apexes (figure 2.10*b*). The bond angle between the two covalent oxygen–hydrogen bonds is 104.5°. This value is slightly less than the bond angle of a regular tetrahedron, which would be 109.5°. In water, the partial negative charges occupy more space than the partial positive regions, and therefore the oxygen–hydrogen bond angle is slightly compressed.

Water molecules are cohesive

The polarity of water allows water molecules to be attracted to one another: that is, water is **cohesive.** Each water molecule is attracted at its oxygen end, which is δ^-, to the hydrogen end, of other molecules, which is δ^+. The attraction produces hydrogen bonds among water molecules (figure 2.11). Each hydrogen bond is individually very weak and transient, lasting on average only a hundred-billionth (10^{-11}) of a second. The cumulative effects of large numbers of these bonds, however, can be enormous. Water forms an abundance of hydrogen bonds, which are responsible for many of its important physical properties (table 2.3).

Water's cohesion is responsible for its being a liquid, not a gas, at moderate temperatures. The cohesion of liquid water is also responsible for its **surface tension.** Small insects can walk on water (figure 2.12) because at the air–water interface, all the surface water molecules are hydrogen-bonded to molecules below them.

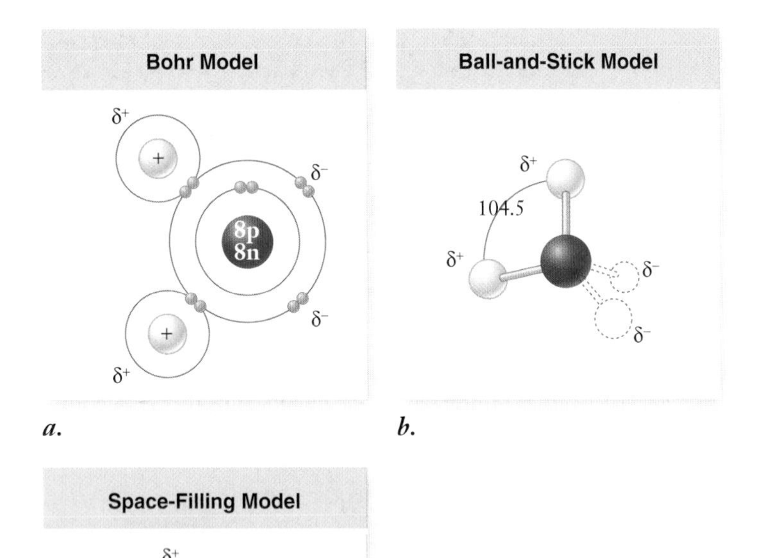

figure 2.10

WATER HAS A SIMPLE MOLECULAR STRUCTURE. ***a.*** Each water molecule is composed of one oxygen atom and two hydrogen atoms. The oxygen atom shares one electron with each hydrogen atom. ***b.*** The greater electronegativity of the oxygen atom makes the water molecule polar: Water carries two partial negative charges (δ^-) near the oxygen atom and two partial positive charges (δ^+), one on each hydrogen atom. ***c.*** Space-filling model shows what the molecule would look like if we could see it.

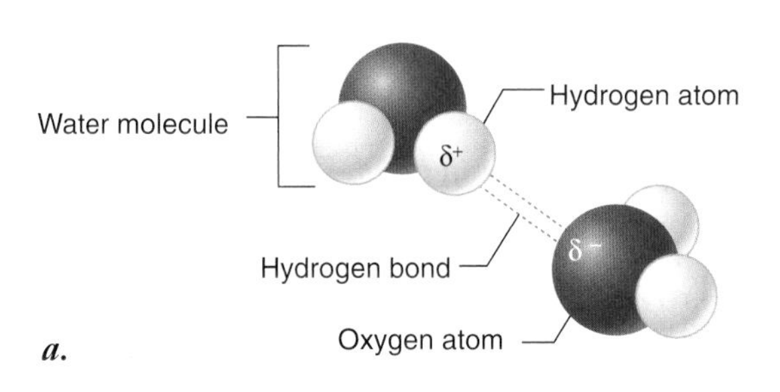

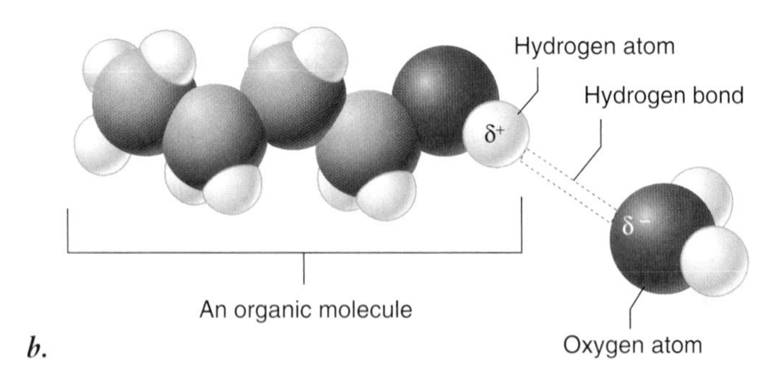

figure 2.11

STRUCTURE OF A HYDROGEN BOND. ***a.*** Hydrogen bond between two water molecules. ***b.*** Hydrogen bond between an organic molecule (*n*-butanol) and water. H in *n*-butanol forms a hydrogen bond with oxygen in water. This kind of hydrogen bond is possible any time H is bound to a more electronegative atom (see table 2.2).

TABLE 2.3	The Properties of Water	
Property	**Explanation**	**Example of Benefit to Life**
Cohesion	Hydrogen bonds hold water molecules together.	Leaves pull water upward from the roots; seeds swell and germinate.
High specific heat	Hydrogen bonds absorb heat when they break and release heat when they form, minimizing temperature changes.	Water stabilizes the temperature of organisms and the environment.
High heat of vaporization	Many hydrogen bonds must be broken for water to evaporate.	Evaporation of water cools body surfaces.
Lower density of ice	Water molecules in an ice crystal are spaced relatively far apart because of hydrogen bonding.	Because ice is less dense than water, lakes do not freeze solid, allowing fish and other life in lakes to survive the winter.
Solubility	Polar water molecules are attracted to ions and polar compounds, making them soluble.	Many kinds of molecules can move freely in cells, permitting a diverse array of chemical reactions.

Water molecules are adhesive

The polarity of water causes it to be attracted to other polar molecules as well. This attraction for other polar substances is called **adhesion.** Water is adhesive to any substance with which it can form hydrogen bonds. This property explains why substances containing polar molecules get "wet" when they are immersed in water, but those that are composed of nonpolar molecules (such as oils) do not.

The attraction of water to substances having surface electrical charges is responsible for capillary action: If a glass tube with a narrow diameter is lowered into a beaker of water, water will rise in the tube above the level of the water in the beaker, because the adhesion of water to the glass surface, drawing it upward, is stronger than the force of gravity, drawing it down. The narrower the tube, the greater the electrostatic forces between the water and the glass, and the higher the water rises (figure 2.13).

The chemistry of life is water chemistry. Water can form hydrogen bonds with itself and with other polar molecules because of its polar characteristics. Hydrogen bonding makes water cohesive: the molecules stick to each other. The cohesive nature of water is responsible for its high surface tension. Water molecules are also adhesive: they stick to other polar molecules. This property is responsible for the phenomenon of capillary action.

figure 2.12

COHESION. Some insects, such as this water strider, literally walk on water. In this photograph you can see how the insect's feet dimple the water as its weight bears down on the surface. Because the surface tension of the water is greater than the force that one foot brings to bear, the strider glides atop the surface of the water rather than sinking. The high surface tension of water is due to hydrogen bonding between water molecules.

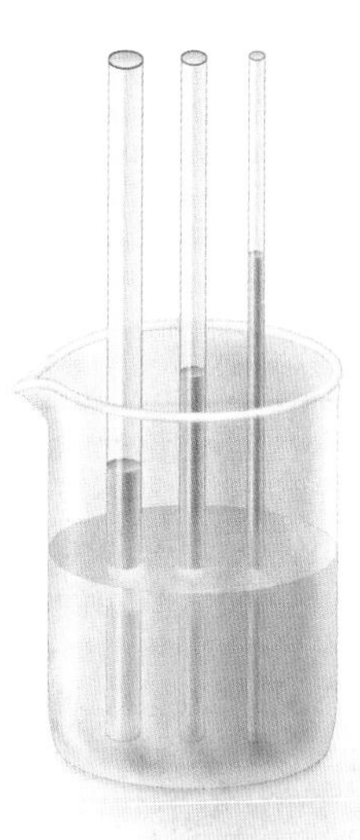

figure 2.13

ADHESION. Capillary action causes the water within a narrow tube to rise above the surrounding water; the adhesion of the water to the glass surface, which draws water upward, is stronger than the force of gravity, which tends to draw it down. The narrower the tube, the greater the surface area available for adhesion for a given volume of water, and the higher the water rises in the tube.

2.5 Properties of Water

Water moderates temperature through two properties: its high specific heat and its high heat of vaporization. Water also has the unusual property of being less dense in its solid form, ice, than as a liquid. In addition, water acts as a solvent for polar molecules and exerts an organizing effect on nonpolar molecules. Water can also dissociate to form ions. All these properties result from its polar nature.

Water's high specific heat helps maintain temperature

The temperature of any substance is a measure of how rapidly its individual molecules are moving. In the case of water, a large input of thermal energy is required to break the many hydrogen bonds that keep individual water molecules from moving about. Therefore, water is said to have a high **specific heat,** which is defined as the amount of heat that must be absorbed or lost by 1 gram of a substance to change its temperature by 1 degree Celsius (°C). Specific heat measures the extent to which a substance resists changing its temperature when it absorbs or loses heat. Because polar substances tend to form hydrogen bonds, the more polar a substance is, the higher is its specific heat. The specific heat of water (1 calorie/gram/°C) is twice that of most carbon compounds and nine times that of iron. Only ammonia, which is more polar than water and forms very strong hydrogen bonds, has a higher specific heat than water (1.23 calories/gram/°C). Still, only 20% of the hydrogen bonds are broken as water heats from 0° to 100°C.

Because of its high specific heat, water heats up more slowly than almost any other compound and holds its temperature longer when heat is no longer applied. This characteristic enables organisms, which have a high water content, to maintain a relatively constant internal temperature. The heat generated by the chemical reactions inside cells would destroy the cells if not for the absorption of this heat by the water within them.

Water's high heat of vaporization facilitates cooling

The **heat of vaporization** is defined as the amount of energy required to change 1 gram of a substance from a liquid to a gas. A considerable amount of heat energy (586 calories) is required to accomplish this change in water. Because the transition of water from a liquid to a gas requires the input of energy to break its many hydrogen bonds, the evaporation of water from a surface causes cooling of that surface. Many organisms dispose of excess body heat by evaporative cooling, for example, through sweating in humans and many other vertebrates.

Solid water is less dense than liquid water

At low temperatures, water molecules are locked into a crystal-like lattice of hydrogen bonds, forming solid ice (see figure 2.9). Interestingly, ice is less dense than liquid water because the hydrogen bonds in ice space the water molecules relatively far apart. This unusual feature enables icebergs to float. If water did not have this property, nearly all bodies of water would be ice, with only the shallow surface melting annually. The buoyancy of ice is important ecologically because it means bodies of water freeze from the top down and not the bottom up. Liquid water beneath the surface of ice that covers most lakes in the winter allows fish and other animals to overwinter without being frozen.

The solvent properties of water help move ions and polar molecules

Water molecules gather closely around any substance that bears an electrical charge, whether that substance carries a full charge (ion) or a charge separation (polar molecule). For example, sucrose (table sugar) is composed of molecules that contain polar hydroxyl (OH) groups. A sugar crystal dissolves rapidly in water because water molecules can form hydrogen bonds with individual hydroxyl groups of the sucrose molecules. Therefore, sucrose is said to be *soluble* in water. Water is termed the *solvent,* and sugar is called the *solute.* Every time a sucrose molecule dissociates, or breaks away, from a solid sugar crystal, water molecules surround it in a cloud, forming a **hydration shell** that prevents it from associating with other sucrose molecules. Hydration shells also form around ions such as Na^+ and Cl^- (figure 2.14).

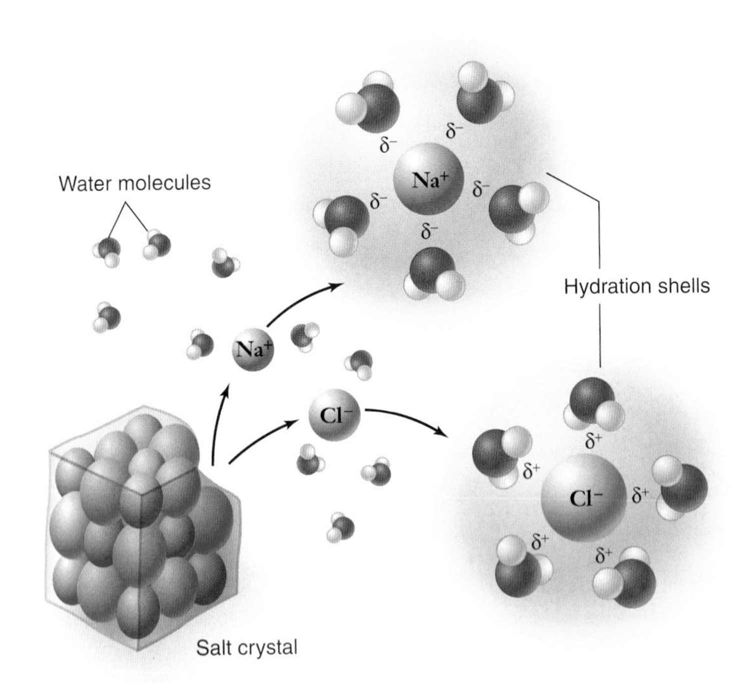

figure 2.14

WHY SALT DISSOLVES IN WATER. When a crystal of table salt dissolves in water, individual Na^+ and Cl^- ions break away from the salt lattice and become surrounded by water molecules. Water molecules orient around Cl^- ions so that their partial positive poles face toward the negative Cl^- ion; water molecules surrounding Na^+ ions orient in the opposite way, with their partial negative poles facing the positive Na^+ ion. Surrounded by hydration shells, Na^+ and Cl^- ions never reenter the salt lattice.

Water organizes nonpolar molecules

Water molecules always tend to form the maximum possible number of hydrogen bonds. When nonpolar molecules such as oils, which do not form hydrogen bonds, are placed in water, the water molecules act to exclude them. The nonpolar molecules are forced into association with one another, thus minimizing their disruption of the hydrogen bonding of water. In effect, they shrink from contact with water, and for this reason they are referred to as **hydrophobic** (Greek *hydros*, "water," and *phobos*, "fearing"). In contrast, polar molecules, which readily form hydrogen bonds with water, are said to be **hydrophilic** ("water-loving").

The tendency of nonpolar molecules to aggregate in water is known as **hydrophobic exclusion.** By forcing the hydrophobic portions of molecules together, water causes these molecules to assume particular shapes. This property can also affect the structure of proteins, DNA, and biological membranes. In fact, the interaction of nonpolar molecules and water is critical to living systems.

Water can form ions

The covalent bonds of a water molecule sometimes break spontaneously. In pure water at 25°C, only 1 out of every 550 million water molecules undergoes this process. When it happens, a proton (hydrogen atom nucleus) dissociates from the molecule. Because the dissociated proton lacks the negatively charged electron it was sharing, its positive charge is no longer counterbalanced, and it becomes a hydrogen ion, H^+. The rest of the dissociated water molecule, which has retained the shared electron from the covalent bond, is negatively charged and forms a hydroxide ion, OH^-. This process of spontaneous ion formation is called *ionization:*

$$\underset{\textit{water}}{H_2O} \longrightarrow \underset{\textit{hydroxide ion}}{OH^-} + \underset{\textit{hydrogen ion (proton)}}{H^+}$$

At 25°C, a liter of water contains one ten-millionth (or 10^{-7}) mole of H^+ ions. A **mole** is defined as the weight of a substance in grams that corresponds to the atomic masses of all of the atoms in a molecule of that substance. In the case of H^+, the atomic mass is 1, and a mole of H^+ ions would weigh 1 gram. One mole of any substance always contains 6.02×10^{23} molecules of the substance. Therefore, the **molar concentration** of hydrogen ions in pure water, represented as $[H^+]$, is 10^{-7} moles/liter. (In reality, the hydrogen ion usually associates with another water molecule to form a hydronium ion, H_3O^+.)

Water does not change temperature rapidly because of its high specific heat. In living systems, high water content maintains a near-constant temperature. Water's high heat of vaporization allows cooling by evaporation. Because polar water molecules cling to one another, it takes considerable energy to separate them. Water also clings to other polar molecules, causing them to be soluble in a water solution, but water tends to exclude nonpolar molecules. Water dissociates to form ions. The hydrogen ion concentration of pure water is 10^{-7} moles per liter.

2.6 Acids and Bases

The concentration of hydrogen ions, and concurrently of hydroxide ions, in a solution is described by the terms *acidity* and *basicity*. Pure water, having an $[H^+]$ of 10^{-7} mole/liter, is considered to be neutral, that is, neither acidic nor basic. Recall that for every H^+ ion formed when water dissociates, an OH^- ion is also formed, meaning that the dissociation of water produces H^+ and OH^- in equal amounts.

The pH scale measures hydrogen ion concentration

The **pH scale** (figure 2.15) is a more convenient way to express the hydrogen ion concentration of a solution. This scale defines **pH,** which stands for "partial hydrogen," as the negative logarithm of the hydrogen ion concentration in the solution:

$$pH = -\log [H^+]$$

Because the logarithm of the hydrogen ion concentration is simply the exponent of the molar concentration of H^+, the pH equals the exponent times −1. For water, therefore, an $[H^+]$ of 10^{-7} moles/liter corresponds to a pH value of 7. This is the neutral point—a balance between H^+ and OH^-—on the pH scale. This balance occurs because the dissociation of water leads to equal amounts of H^+ and OH^-.

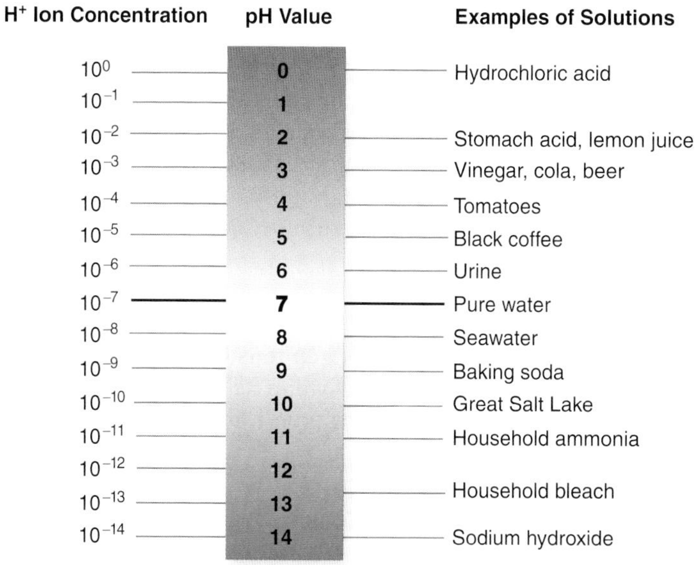

H^+ Ion Concentration	pH Value	Examples of Solutions
10^0	0	Hydrochloric acid
10^{-1}	1	
10^{-2}	2	Stomach acid, lemon juice
10^{-3}	3	Vinegar, cola, beer
10^{-4}	4	Tomatoes
10^{-5}	5	Black coffee
10^{-6}	6	Urine
10^{-7}	7	Pure water
10^{-8}	8	Seawater
10^{-9}	9	Baking soda
10^{-10}	10	Great Salt Lake
10^{-11}	11	Household ammonia
10^{-12}	12	
10^{-13}	13	Household bleach
10^{-14}	14	Sodium hydroxide

figure 2.15

THE pH SCALE. The pH value of a solution indicates its concentration of hydrogen ions. Solutions with a pH less than 7 are acidic, whereas those with a pH greater than 7 are basic. The scale is logarithmic, so that a pH change of 1 means a 10-fold change in the concentration of hydrogen ions. Thus, lemon juice is 100 times more acidic than tomato juice, and seawater is 10 times more basic than pure water, which has a pH of 7.

Note that, because the pH scale is *logarithmic*, a difference of 1 on the scale represents a 10-fold change in hydrogen ion concentration. A solution with a pH of 4 therefore has 10 times the $[H^+]$ of a solution with a pH of 5 and 100 times the $[H^+]$ concentration of a solution with a pH of 6.

Acids

Any substance that dissociates in water to increase the concentration of H^+ ions (and lower the pH) is called an **acid.** The stronger an acid is, the more H^+ ions it produces and the lower its pH. For example, hydrochloric acid (HCl), which is abundant in your stomach, ionizes completely in water. A dilution of 10^{-1} moles/liter of HCl dissociates to form 10^{-1} moles/liter of H^+ ions, giving the solution a pH of 1. The pH of champagne, which bubbles because of the carbonic acid dissolved in it, is about 2.

Bases

A substance that combines with H^+ ions when dissolved in water, and thus lowers the $[H^+]$, is called a **base.** Therefore, basic (or alkaline) solutions have pH values above 7. Very strong bases, such as sodium hydroxide (NaOH), have pH values of 12 or more. Many common cleaning substances, such as ammonia and bleach, accomplish their action because of their high pH.

Buffers help stabilize pH

The pH inside almost all living cells, and in the fluid surrounding cells in multicellular organisms, is fairly close to 7. Most of the enzymes in living systems are extremely sensitive to pH; often even a small change in pH will alter their shape, thereby disrupting their activities and rendering them useless. For this reason, it is important that a cell maintain a constant pH level.

But the chemical reactions of life constantly produce acids and bases within cells. Furthermore, many animals eat substances that are acidic or basic. Cola drinks, for example, are moderately strong (although dilute) acidic solutions. Despite such variations in the concentrations of H^+ and OH^-, the pH of an organism is kept at a relatively constant level by buffers (figure 2.16).

A **buffer** is a substance that resists changes in pH. Buffers act by releasing hydrogen ions when a base is added and absorbing hydrogen ions when acid is added, with the overall effect of keeping hydrogen ion concentration relatively constant.

Within organisms, most buffers consist of pairs of substances, one an acid and the other a base. The key buffer in human blood is an acid–base pair consisting of carbonic acid (acid) and bicarbonate (base). These two substances interact in a pair of reversible reactions. First, carbon dioxide (CO_2) and H_2O join to form carbonic acid (H_2CO_3), which in a second reaction dissociates to yield bicarbonate ion (HCO_3^-) and H^+.

If some acid or other substance adds H^+ ions to the blood, the HCO_3^- ions act as a base and remove the excess H^+ ions by forming H_2CO_3. Similarly, if a basic substance removes H^+ ions from the blood, H_2CO_3 dissociates, releasing more H^+ ions into the blood. The forward and reverse reactions that interconvert H_2CO_3 and HCO_3^- thus stabilize the blood's pH.

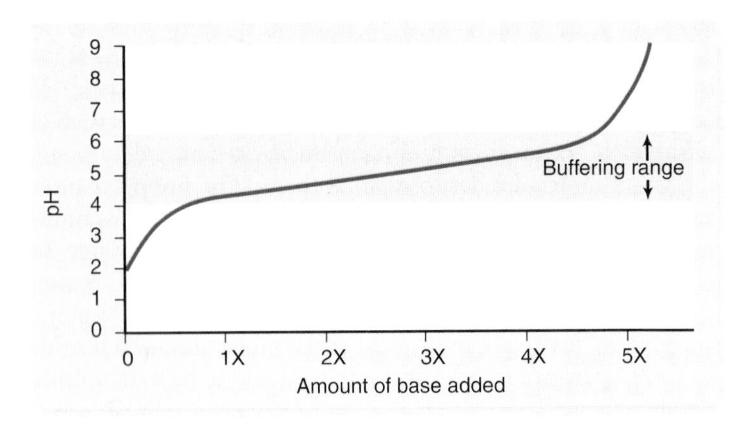

figure 2.16

BUFFERS MINIMIZE CHANGES IN pH. Adding a base to a solution neutralizes some of the acid present, and so raises the pH. Thus, as the curve moves to the right, reflecting more and more base, it also rises to higher pH values. A buffer makes the curve rise or fall very slowly over a portion of the pH scale, called the "buffering range" of that buffer.

inquiry

For this buffer, adding base raises pH more rapidly below pH 4 than above it. What might account for this behavior?

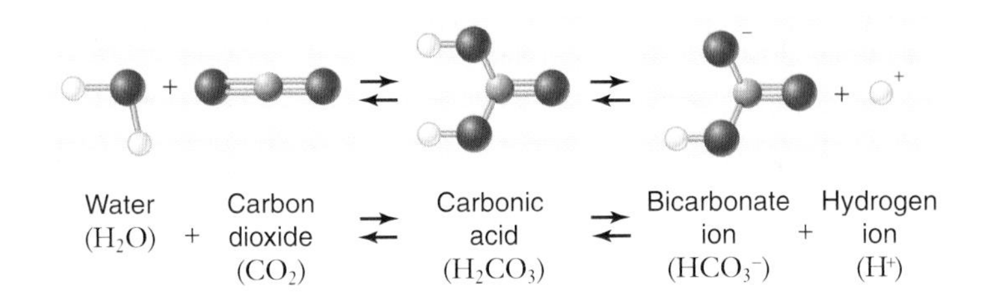

The reaction of carbon dioxide and water to form carbonic acid is a crucial one because it permits carbon, essential to life, to enter water from the air. The Earth's oceans are rich in carbon because of the reaction of carbon dioxide with water.

In a condition called blood acidosis, human blood, which normally has a pH of about 7.4, drops to a pH of about 7.1 or below. This condition is fatal if not treated immediately. The reverse condition, blood alkalosis, involves an increase in blood pH of a similar magnitude and is just as serious.

The pH of a solution is the negative logarithm of the H^+ ion concentration in the solution. Thus, low pH values indicate high H^+ concentrations (acidic solutions), and high pH values indicate low H^+ concentrations (basic solutions). Even small changes in pH can be harmful to life. Buffer systems in organisms, such as the carbon dioxide/bicarbonate system in humans, help to maintain pH within a narrow range.

2.1 The Nature of Atoms

All matter is composed of atoms (figure 2.2).

- Atoms are composed of a nucleus with positively charged protons and neutral neutrons surrounded by one or more orbitals containing negatively charged electrons.
- To be electrically neutral an atom must have the same number of protons as electrons.
- Atoms that gain or lose electrons are called ions.
- If an atom loses electrons it has a positive charge and is called a cation. If the atom gains an electron it has a negative charge and is called an anion.
- Each element is defined by its atomic number, which is the number of protons found in the nucleus.
- Atomic mass is the sum of the mass of protons and neutrons in an atom.

An atom is called an isotope if the number of neutrons exceeds the number of protons.

- Isotopes are different forms of the same element that have different numbers of neutrons and thus different atomic mass.
- Radioactive elements are unstable and break up into smaller number elements.
- The rate of decay for any radioactive element is constant.
- Decay is expressed as a half-life, the amount of time for 50% of atoms to decay.

Electrons determine the behavior of atoms.

- The potential energy of electrons increases as distance from the nucleus increases. Excited electrons can temporarily move to a higher energy level and increase their potential energy.
- The loss of electrons from an atom is called oxidation.
- The gain of electrons is called reduction.
- Electrons can be transferred from one atom to another in coupled redox reactions (see figure in left column, page 21).

2.2 Elements Found in Living Systems

The periodic table is based on the interactions of valence electrons.

- There are 90 naturally occurring elements in the Earth's crust.
- Twelve of these elements are found in living organisms in greater than trace amounts.
- Elements with filled outermost electron orbitals are inert. These are found in the last column of the periodic table.

2.3 The Nature of Chemical Bonds

Molecules contain two or more similar atoms joined by chemical bonds. Compounds contain two or more different elements.

- Ionic bonds occur when two different ions with opposite electrical charges are attracted to each other. Ionic bonds can be very strong, but not as strong as a covalent bond (figure 2.8*b*).
- Covalent bonds occur when one or more pairs of electrons are shared between two atoms. One atom may form covalent bonds with several other atoms.
- Covalent bonds are the strongest and they are responsible for the stability of organic molecules.
- Electronegativity is the affinity of an atom for electrons. It increases across rows and decreases down columns of the periodic table.
- Nonpolar covalent bonds involve equal sharing of electrons between atoms.
- Polar covalent bonds involve unequal sharing of electrons between atoms. This occurs between atoms with large differences in electronegativity.
- Chemical reactions make, break, or otherwise alter bonds. Temperature, pH, and catalysts will affect reaction rates.

2.4 Water: A Vital Compound

Life can be understood through the chemistry of water (figure 2.11*a*).

- Hydrogen bonds are weak interactions between a partially positive H in one molecule and a partially negative oxygen in another molecule.
- Cohesion is the tendency of water molecules to adhere to one another due to hydrogen bonding.
- Adhesion occurs when water molecules adhere to other polar molecules.

2.5 Properties of Water

Water has several properties because it is polar.

- Water has a high specific heat because it takes a considerable amount of energy to disrupt hydrogen bonds. The large amount of water in organisms helps them maintain body temperature.
- Water has a high heat of vaporization, which is used for cooling. It takes a considerable amount of heat to break enough hydrogen bonds for liquid water to become a gas.
- The hydrogen bonds between water molecules in the solid phase are spaced farther apart than in the liquid phase. As a result ice floats.
- Water is a good solvent for polar substances and ions. Water will tend to exclude nonpolar substances.
- Molecules or portions of molecules that are polar are hydrophilic. These substances will be attracted to water.
- Molecules that are nonpolar are hydrophobic molecules. These substances will be repelled by water.
- Because of hydrophobic exclusion, nonpolar molecules or their components will tend to aggregate or form particular shapes. This can influence the shape of biological molecules.

2.6 Acids and Bases (figure 2.15)

Water forms two ions when covalent bonds break.

- The hydrogen ion (H^+) is positively charged, and the hydroxide ion (OH^-) is negatively charged.
- The relationship between H^+ and OH^- is expressed as pH. This is defined as the negative logarithm of H^+ ion concentration.
- The pH scale is logarithmic and a difference of 1 on a pH scale means a 10-fold change in concentration of hydrogen ions.
- If the concentration of hydrogen ions is greater than that of hydroxide ions, the solution is acidic and the pH is below 7 units. If the concentration of hydrogen ions is less than that of the hydroxide ions, the solution is basic and the pH is above 7 units.

SELF TEST

1. The property that distinguishes one atom (carbon for example) from another atom (oxygen for example) is—
   a. The number of electrons
   b. The number of protons
   c. The number of neutrons
   d. The combined number of protons and neutrons
2. If an atom has one valence (outer energy level) electron, it will most likely form—
   a. One polar, covalent bond
   b. Two nonpolar, covalent bonds
   c. Two covalent bonds
   d. An ionic bond
3. An atom with a net positive charge must have—
   a. More protons than neutrons
   b. More protons than electrons
   c. More electrons than neutrons
   d. More electrons than protons
4. The isotopes C^{12} and C^{14} differ in—
   a. The number of neutrons
   b. The number of protons
   c. The number of electrons
   d. Both b and c
5. An atom with more electrons than protons is called—
   a. An element
   b. An isotope
   c. A cation
   d. An anion
6. Which of the following is NOT a property of the elements most commonly found in living organisms?
   a. The elements have a low atomic mass
   b. The elements have an atomic number less than 21
   c. The elements possess eight electrons in their outer energy level
   d. The elements are lacking one or more electrons from their outer energy level
7. Which of the following atoms would you predict could be a cation?
   a. Fluorine (F)
   b. Helium (He)
   c. Potassium (K)
   d. Boron (B)
8. Refer to the element pictured. How many covalent bonds could this atom form?

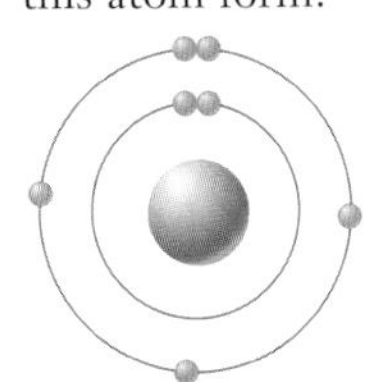

   a. Two
   b. Three
   c. Four
   d. None
9. Refer to the element pictured. How many covalent bonds could this atom form?

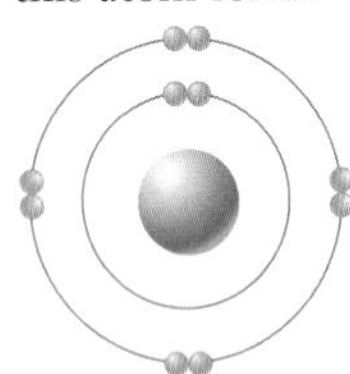

   a. Two
   b. Three
   c. Four
   d. None
10. An ionic bond is held together by
    a. shared valence electrons
    b. attractions between ions of the same charge
    c. charge attractions between valence electrons
    d. attractions between ions of opposite charge
11. How do polar covalent bonds differ from nonpolar covalent bonds?
    a. In a polar covalent bond the electrons are shared equally between the atoms.
    b. In a nonpolar covalent bond there is a charge attraction between the atomic nuclei.
    c. There is a large difference in electronegativity of the atoms in a nonpolar bond.
    d. There is a large difference in electronegativity of the atoms in a polar bond.
12. A hydrogen bond can form—
    a. when hydrogen is part of a polar covalent bond
    b. only in water
    c. between any large electronegative atoms like oxygen
    d. when two atoms of hydrogen share an electron
13. Which of the following properties of water is NOT a consequence of its ability to form hydrogen bonds?
    a. Cohesiveness
    b. High specific heat
    c. Ability to function as a solvent
    d. Neutral pH
14. A substance with a high concentration of hydrogen ions is—
    a. called a base
    b. called an acid
    c. has a high pH
    d. both b and c

CHALLENGE QUESTIONS

1. Elements that form ions are important for a range of biological processes. You have learned about the cations, sodium (Na^+), calcium (Ca^{2+}) and potassium (K^+) in this chapter. Use your knowledge of the definition of a cation to identify other examples from the periodic table.
2. A popular theme in science fiction literature has been the idea of silicon-based life-forms in contrast to our carbon-based life. Evaluate the possibility of silicon-based life based on the chemical structure and potential for chemical bonding of a silicon atom.
3. Recent efforts by NASA to search for signs of life on Mars have focused on the search for evidence of liquid water in the planet's history rather than looking directly for biological organisms (living or fossilized). Use your knowledge of the influence of water on life on Earth to construct an argument justifying this approach.

Do you need additional review? *Visit* www.ravenbiology.com *for practice quizzes, animations, videos, and activities designed to help you master the material in this chapter.*

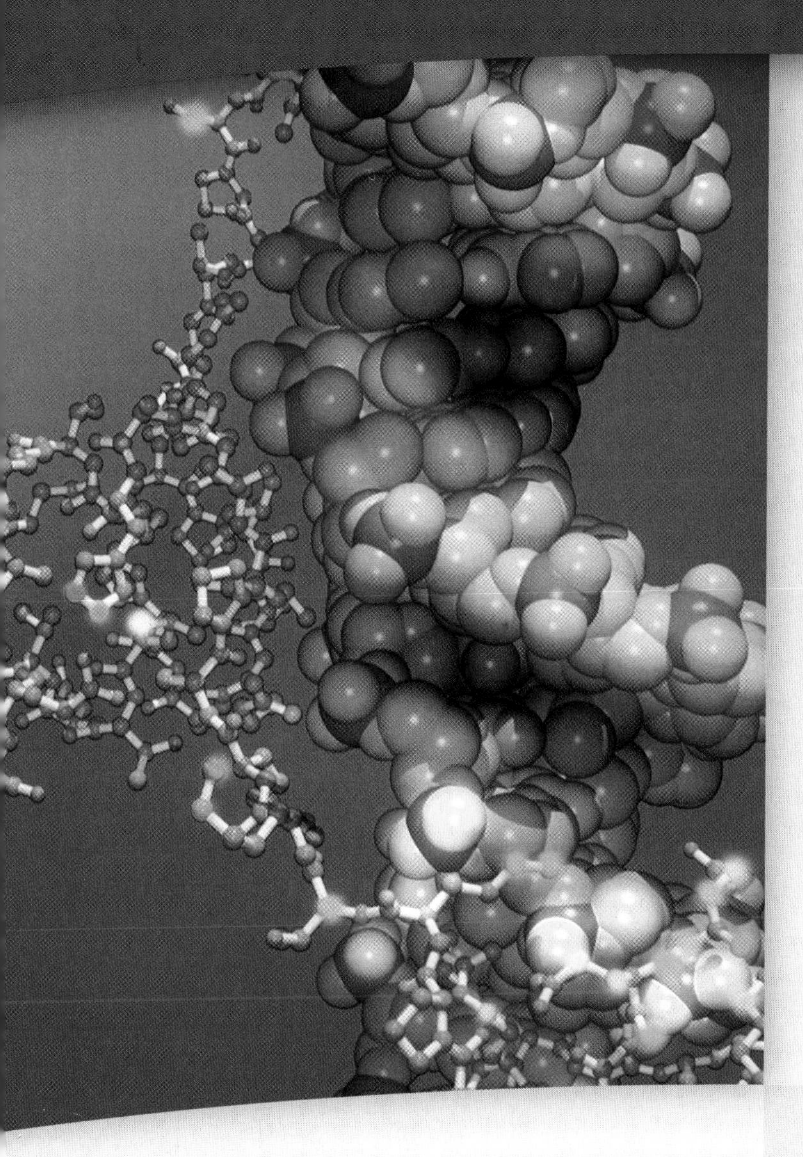

chapter 3

The Chemical Building Blocks of Life

introduction

A CUP OF WATER CONTAINS more molecules than there are stars in the sky. But many molecules are much larger than water molecules; they consist of thousands of atoms, forming hundreds of molecules that are linked together into long chains. These enormous assemblies, which are almost always synthesized by living things, are **macromolecules.** As you may know, biological macromolecules can be divided into four categories: *carbohydrates*, *nucleic acids*, *proteins*, and *lipids*, and they are the basic chemical building blocks from which all organisms are assembled.

Biological macromolecules all involve carbon-containing compounds, so we begin the discussion with a brief summary of carbon and its chemistry. The study of the chemistry of carbon, because of its biological significance, is known as *organic chemistry*.

concept outline

3.1 Carbon: The Framework of Biological Molecules

In chapter 2, we reviewed the basics of chemistry. No new laws of chemistry are found in biological systems, and biological systems do not violate the laws of chemistry. Thus, chemistry forms the basis of living systems.

The framework of biological molecules consists predominantly of carbon atoms bonded to other carbon atoms or to atoms of oxygen, nitrogen, sulfur, or hydrogen. Because carbon atoms can form up to four covalent bonds, molecules containing carbon can form straight chains, branches, or even rings, balls, and coils.

Molecules consisting only of carbon and hydrogen are called **hydrocarbons.** Because carbon–hydrogen covalent bonds store considerable energy, hydrocarbons make good fuels. Gasoline, for example, is rich in hydrocarbons, and propane gas, another hydrocarbon, consists of a chain of three carbon atoms, with eight hydrogen atoms bound to it. The chemical formula for propane is C_3H_8, and its structural formula is shown as

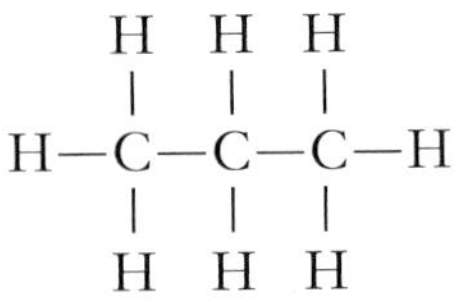

```
    H   H   H
    |   |   |
H — C — C — C — H      Propane structural formula
    |   |   |
    H   H   H
```

Theoretically speaking, there is no limit to the length of a chain of carbon atoms. As described in the rest of this chapter, the four main types of biological molecules often consist of huge chains of carbon-containing compounds.

Functional groups account for differences in molecular properties

Carbon and hydrogen atoms both have very similar electronegativities, so electrons in C—C and C—H bonds are evenly distributed, with no significant differences in charge over the molecular surface. For this reason, hydrocarbons are nonpolar. Most biological molecules that are produced by cells, however, also contain other atoms. Because these other atoms frequently have different electronegativities, molecules containing them exhibit regions of partial positive or negative charge, and so are polar. These molecules can be thought of as a C—H core to which specific molecular groups, called **functional groups,** are attached. For example, an attached —OH group is a functional group called a *hydroxyl group.*

Functional groups have definite chemical properties that they retain no matter where they occur. The hydroxyl and carbonyl (C=O) groups, for example, are both polar because of the electronegativity of the oxygen atoms (as described in chapter 2). Other functional groups include the acidic carboxyl (COOH) and phosphate (PO_4) groups, and the basic amino (NH_2) group. Many of these functional groups can also participate in hydrogen bonding. Hydrogen bond donors and acceptors can be predicted based on the electronegativities given previously in table 2.2. Figure 3.1 illustrates these biologically important functional groups and lists the macromolecules in which they are found.

Functional Group	Structural Formula	Example	Found In
Hydroxyl	—OH	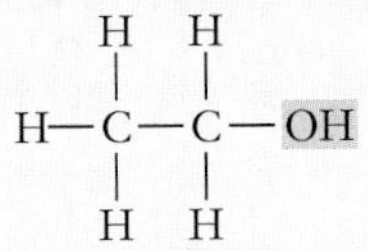 H—C(H)(H)—C(H)(H)—OH Ethanol	carbohydrates, proteins, nucleic acids, lipids
Carbonyl	—C(=O)—	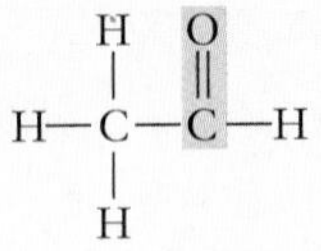 H—C(H)(H)—C(=O)—H Acetaldehyde	carbohydrates, nucleic acids
Carboxyl	—C(=O)OH	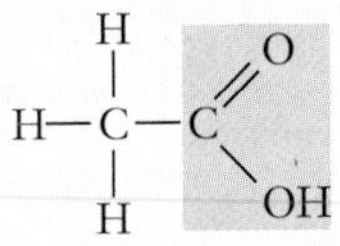 H—C(H)(H)—C(=O)OH Acetic acid	proteins, lipids
Amino	—N(H)H	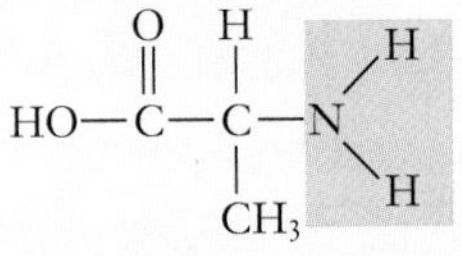 HO—C(=O)—C(H)(CH$_3$)—N(H)H Alanine	proteins, nucleic acids
Sulfhydryl	—S—H	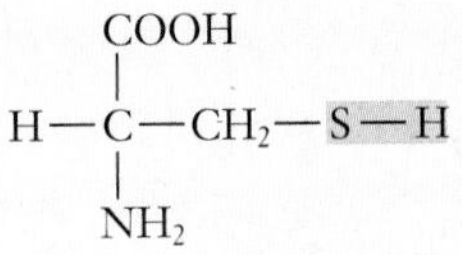 H—C(COOH)(NH$_2$)—CH$_2$—S—H Cysteine	proteins
Phosphate	—O—P(O$^-$)(=O)—O$^-$	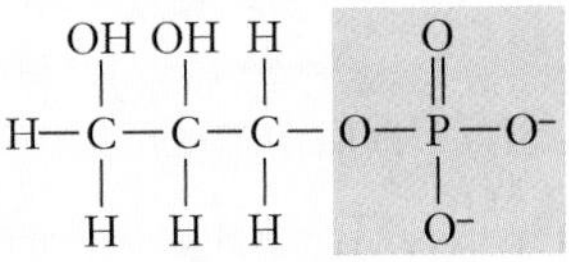 H—C(OH)(H)—C(OH)(H)—C(H)(H)—O—P(=O)(O$^-$)—O$^-$ Glycerol phosphate	nucleic acids
Methyl	—C(H)(H)—H	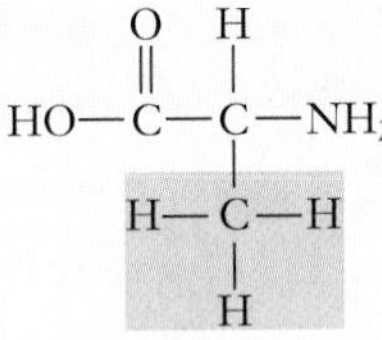 HO—C(=O)—C(H)(NH$_2$)—C(H)(H)—H Alanine	proteins

figure 3.1

THE PRIMARY FUNCTIONAL CHEMICAL GROUPS. These groups tend to act as units during chemical reactions and confer specific chemical properties on the molecules that possess them. Amino groups, for example, make a molecule more basic, whereas carboxyl groups make a molecule more acidic. These functional groups are also not limited to the examples in the "Found In" column but are widely distributed in biological molecules.

TABLE 3.1	Macromolecules		
Macromolecule	**Subunit**	**Function**	**Example**
CARBOHYDRATES			
Starch, glycogen	Glucose	Energy storage	Potatoes
Cellulose	Glucose	Plant cell walls	Paper; strings of celery
Chitin	Modified glucose	Structural support	Crab shells
NUCLEIC ACIDS			
DNA	Nucleotides	Encodes genes	Chromosomes
RNA	Nucleotides	Needed for gene expression	Messenger RNA
PROTEINS			
Functional	Amino acids	Catalysis; transport	Hemoglobin
Structural	Amino acids	Support	Hair; silk
LIPIDS			
Fats	Glycerol and three fatty acids	Energy storage	Butter; corn oil; soap
Phospholipids	Glycerol, two fatty acids, phosphate, and polar R groups	Cell membranes	Phosphatidylcholine
Prostaglandins	Five-carbon rings with two nonpolar tails	Chemical messengers	Prostaglandin E (PGE)
Steroids	Four fused carbon rings	Membranes; hormones	Cholesterol; estrogen
Terpenes	Long carbon chains	Pigments; structural support	Carotene; rubber

Isomers have the same molecular formulas but different structures

Organic molecules having the same structural formula can exist in different forms called **isomers.** If the differences exist in the actual structure of the carbon skeleton, we call them *structural isomers.* Later you will see that glucose and fructose are structural isomers of $C_6H_{12}O_6$. A more subtle form of isomer is called a *stereoisomer*, and these molecules have the same carbon skeleton but differ in how the groups attached to this skeleton are arranged.

Enzymes in biological systems usually recognize only a single, specific stereoisomer. A subcategory of stereoisomers, called *enantiomers*, are actually mirror images of each other. A molecule that has mirror-image versions is called a **chiral** molecule. When carbon is bound to four different molecules, this inherent asymmetry exists (figure 3.2).

Chiral compounds are characterized by their effect on polarized light. Polarized light has a single plane, and chiral molecules rotate this plane either to the left or to the right. We therefore call the two chiral forms *D* for *dextrorotatory* and *L* for *levorotatory*. Living systems tend to produce only a single enantiomer of the two possible forms; for example, in most organisms we find primarily D-sugars and L-amino acids.

Biological macromolecules include carbohydrates, nucleic acids, proteins, and lipids

As mentioned at the beginning, biological macromolecules are traditionally grouped into carbohydrates, nucleic acids, proteins, and lipids (table 3.1). In many cases, these macromolecules are polymers. A **polymer** is a long molecule built by linking together a large number of small, similar chemical subunits called **monomers,** like railroad cars coupled to form a train. The nature of a polymer is determined by the monomers used to build the polymer. For example, complex carbohydrates such as starch are polymers of simple ring-shaped sugars; nucleic acids (DNA and RNA) are polymers of nucleotides (figure 3.3); proteins are polymers of amino acids, and fats are polymers of fatty acids (see figure 3.3). These long chains are built via chemical reactions termed *dehydration reactions* and are broken down by *hydrolysis reactions.*

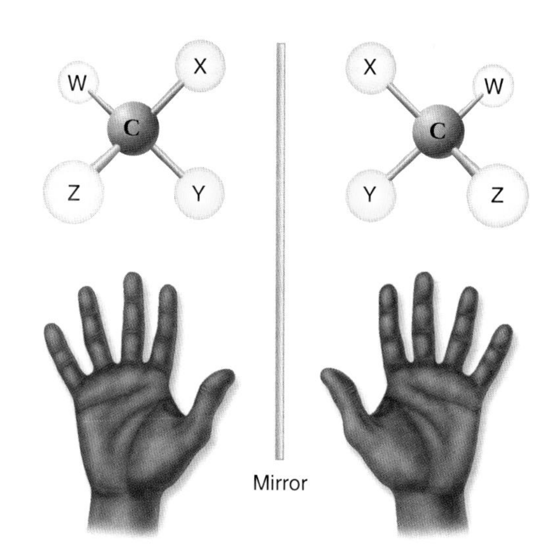

figure 3.2

CHIRAL MOLECULES. When carbon is bound to four different groups, the resulting molecule is said to be chiral. This kind of molecule can have stereoisomers that are mirror images. The two molecules shown have the same four groups but cannot be superimposed, much like your two hands. These types of stereoisomers are called *enantiomers.*

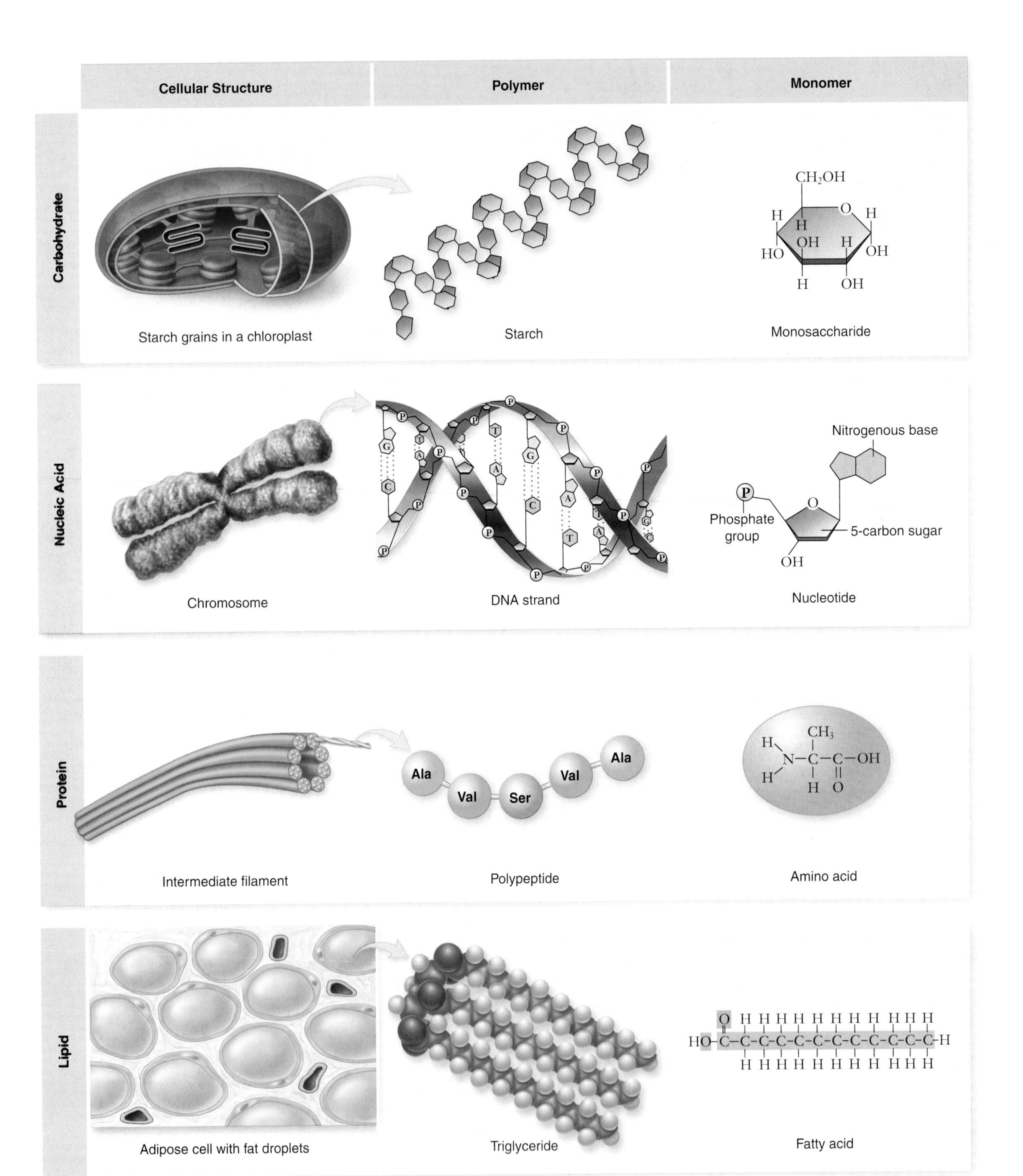

figure 3.3

POLYMER MACROMOLECULES. The four major biological macromolecules are shown. Carbohydrates, nucleic acids, and proteins all form polymers and are shown with the monomers used to make them. Lipids do not fit this simple monomer–polymer relationship, however, because they are constructed from glycerol and fatty acids. All four types of macromolecule are also shown in their cellular context.

The dehydration reaction

Despite the differences between monomers of the major polymers, the basic chemistry of their synthesis is similar: To form a covalent bond between two monomers, an —OH group is removed from one monomer, and a hydrogen atom (H) is removed from the other (figure 3.4*a*). For example, this simple chemistry is the same for linking amino acids together to make a protein or assembling glucose units together to make starch. This reaction is also used to link fatty acids to glycerol in lipids. This chemical reaction is called condensation or a **dehydration reaction,** because the removal of —OH and —H is the same as the removal of a molecule of water (H_2O). For every subunit that is added to a macromolecule, one water molecule is removed. These and other biochemical reactions require that the reacting substances are held close together and that the correct chemical bonds are stressed and broken. This process of positioning and stressing, termed *catalysis,* is carried out in cells by enzymes.

The hydrolysis reaction

Cells disassemble macromolecules into their constituent subunits by performing reactions that are essentially the reverse of dehydration—a molecule of water is added instead of removed (figure 3.4*b*). In this process, which is called **hydrolysis,** a hydrogen atom is attached to one subunit and a hydroxyl group to the other, breaking a specific covalent bond in the macromolecule.

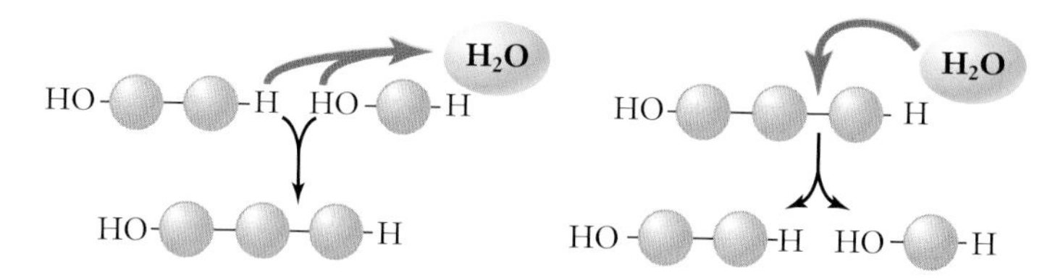

figure 3.4

MAKING AND BREAKING MACROMOLECULES. *a.* Biological macromolecules are polymers formed by linking monomers together by dehydration synthesis. This process releases a water molecule for every bond formed. *b.* Breaking the bond between subunits involves a process called hydrolysis, which reverses the loss of a water molecule by dehydration.

Living systems are made up of four main types of macromolecule. Macromolecules are polymers, which consist of long chains of similar subunits that are joined by dehydration reactions and are broken down by hydrolysis reactions.

3.2 Carbohydrates: Energy Storage and Structural Molecules

Carbohydrates are a loosely defined group of molecules that contains carbon, hydrogen, and oxygen in the molar ratio 1:2:1. Their empirical formula (which lists the number of atoms in the molecule with subscripts) is $(CH_2O)_n$, where *n* is the number of carbon atoms. Because they contain many carbon–hydrogen (C—H) bonds, which release energy when oxidation occurs, carbohydrates are well suited for energy storage. Sugars are among the most important energy-storage molecules, and they exist in several different forms.

Monosaccharides are simple sugars

The simplest of the carbohydrates are the **monosaccharides** (Greek *mono,* "single," and Latin *saccharum,* "sugar"). Simple sugars may contain as few as three carbon atoms, but those that play the central role in energy storage have six (figure 3.5). The empirical formula of six-carbon sugars is:

$$C_6H_{12}O_6 \quad \text{or} \quad (CH_2O)_6$$

Six-carbon sugars can exist in a straight-chain form, but in an aqueous environment they almost always form rings.

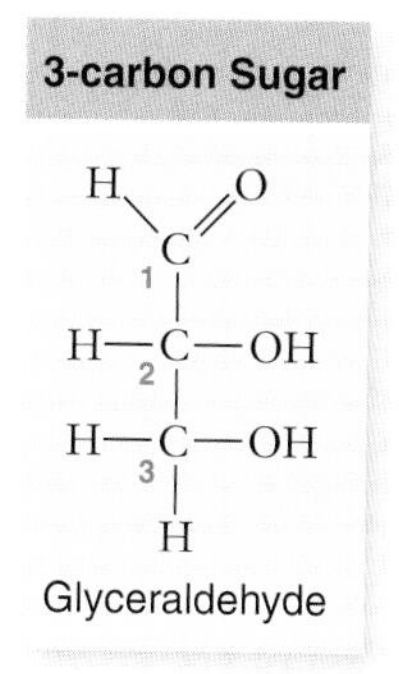

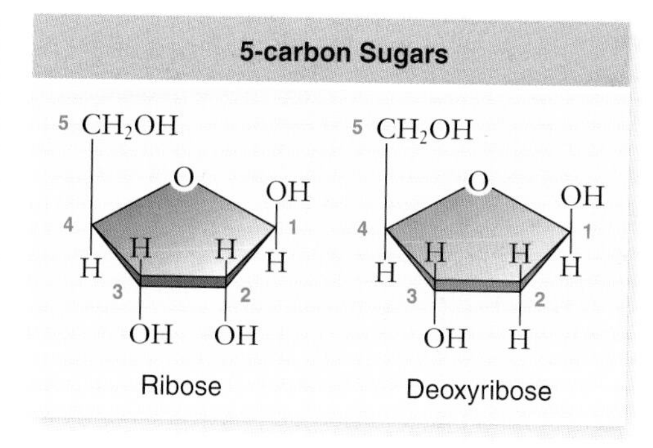

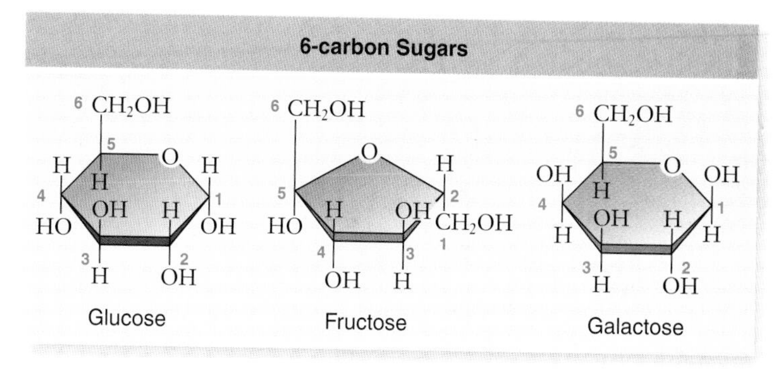

figure 3.5

MONOSACCHARIDES. Monosaccharides, or simple sugars, can contain as few as three carbon atoms and are often used as building blocks to form larger molecules. The five-carbon sugars ribose and deoxyribose are components of nucleic acids (see figure 3.14). The six-carbon sugar glucose is a component of large energy-storage molecules. The blue numbers refer to the carbon atoms; monosaccharides are conventionally numbered from the more oxidized end.

figure 3.6

STRUCTURE OF THE GLUCOSE MOLECULE. Glucose is a linear, six-carbon molecule that forms a six-membered ring in solution. Ring closure occurs such that two forms can result: α-glucose and β-glucose. These structures differ only in the position of the OH bound to carbon 1. The structure of the ring can be represented in many ways; the ones shown here are the most common, with the carbons conventionally numbered (in *blue*) so that the forms can be compared easily. The bold lines represent portions of the molecule that are projecting out of the page toward you.

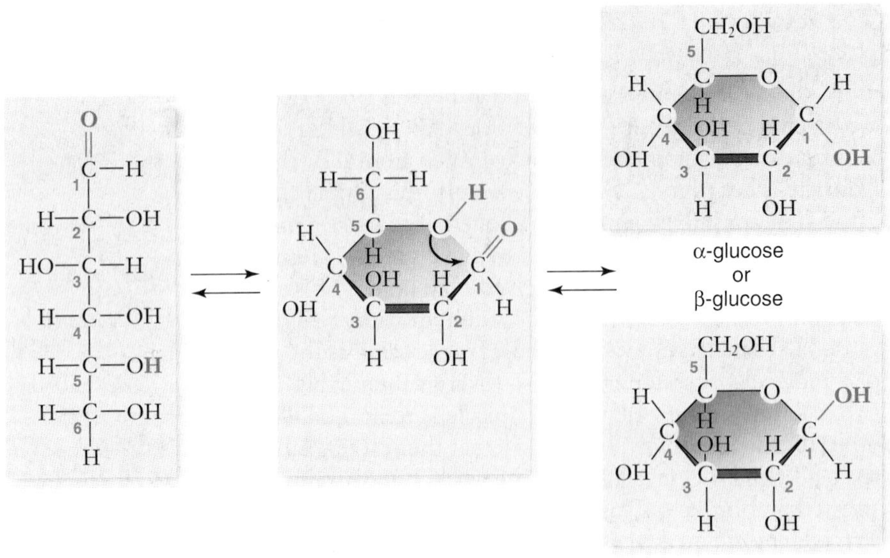

The most important of the six-carbon monosaccharides for energy storage is glucose, which you first encountered in the examples of chemical reactions in chapter 2. Glucose has seven energy-storing C—H bonds (figure 3.6). Depending on the orientation of the carbonyl group (C=O) when the ring is closed, glucose can exist in two different forms: alpha (α) or beta (β).

Sugar isomers have structural differences

Glucose is not the only sugar with the formula $C_6H_{12}O_6$. Both structural isomers and stereoisomers of this simple six-carbon skeleton exist in nature. Fructose is a structural isomer that differs in the positioning of the carbonyl carbon (C=O); galactose is a stereoisomer that differs in the position of OH and H groups relative to the ring (figure 3.7). These differences often account for substantial functional differences between the isomers. Your taste buds can discern them: Fructose tastes much sweeter than glucose, despite the fact that both sugars have identical chemical composition. Enzymes that act on different sugars can distinguish both structural and stereoisomers of this basic six-carbon skeleton. The different stereoisomers of glucose are also important in the polymers that can be made using glucose as a monomer, as you will see later in this chapter.

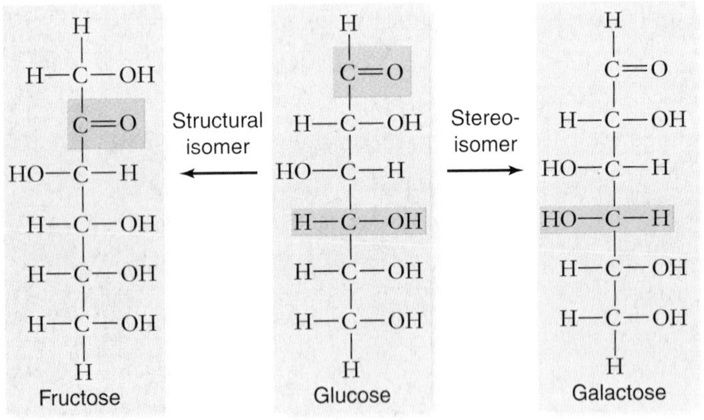

figure 3.7

ISOMERS AND STEREOISOMERS. Glucose, fructose, and galactose are isomers with the empirical formula $C_6H_{12}O_6$. A structural isomer of glucose, such as fructose, has identical chemical groups bonded to different carbon atoms. Notice that this results in a five-membered ring in solution (see figure 3.5). A stereoisomer of glucose, such as galactose, has identical chemical groups bonded to the same carbon atoms but in different orientations (the —OH at carbon 4).

Disaccharides serve as transport molecules in plants and provide nutrition in animals

Most organisms transport sugars within their bodies. In humans, the glucose that circulates in the blood does so as a simple monosaccharide. In plants and many other organisms, however, glucose is converted into a transport form before it is moved from place to place within the organism. In such a form, it is less readily metabolized during transport.

Transport forms of sugars are commonly made by linking two monosaccharides together to form a **disaccharide** (Greek *di*, "two"). Disaccharides serve as effective reservoirs of glucose because the normal glucose-utilizing enzymes of the organism cannot break the bond linking the two monosaccharide subunits. Enzymes that can do so are typically present only in the tissue where the glucose is to be used.

Transport forms differ depending on which monosaccharides are linked to form the disaccharide. Glucose forms transport disaccharides with itself and with many other monosaccharides, including fructose and galactose. When glucose forms a disaccharide with the structural isomer fructose, the resulting disaccharide is *sucrose*, or table sugar (figure 3.8*a*). Sucrose is the form in which most plants transport glucose and is the sugar that most humans and other animals eat. Sugarcane and sugar beets are rich in sucrose.

When glucose is linked to the stereoisomer galactose, the resulting disaccharide is *lactose*, or milk sugar. Many mammals supply energy to their young in the form of lactose. Adults have greatly reduced levels of lactase, the enzyme required to cleave lactose into its two monosaccharide components, and thus they cannot metabolize lactose efficiently. Most of the energy that is channeled into lactose production is therefore reserved for offspring. For this reason lactose as an energy source is primarily for offspring in mammals.

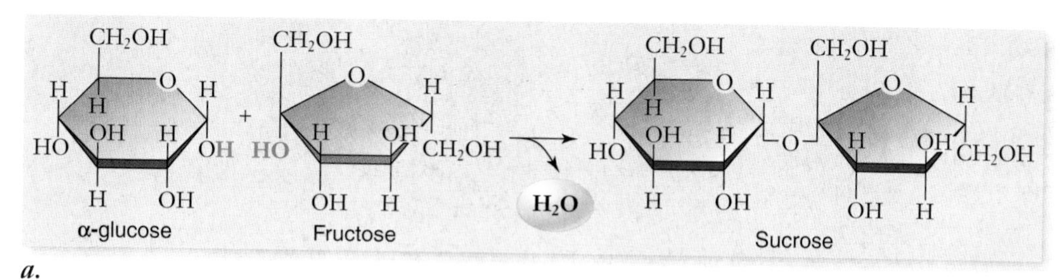

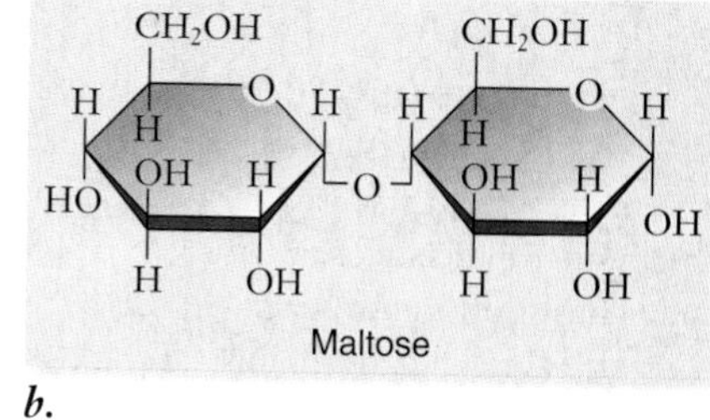

figure 3.8

HOW DISACCHARIDES FORM. Some disaccharides are used to transport glucose from one part of an organism's body to another; one example is sucrose ***(a)***, which is found in sugarcane. Other disaccharides, such as maltose ***(b)***, in grain are used for storage.

Polysaccharides provide energy storage and structural components

Polysaccharides are longer polymers made up of monosaccharides that have been joined through dehydration synthesis. **Starch,** a storage polysaccharide, consists entirely of α-glucose molecules linked in long chains. **Cellulose,** a structural polysaccharide, also consists of glucose molecules linked in chains, but these molecules are β-glucose. Because starch is built from α-glucose we call the linkages α linkages; cellulose has β linkages.

Starches and Glycogen

Organisms store the metabolic energy contained in monosaccharides by converting them into disaccharides, such as *maltose* (figure 3.8*b*), which are then linked together into the insoluble polysaccharides called *starches.* These polysaccharides differ mainly in the way in which the polymers branch.

The starch with the simplest structure is *amylose,* which is composed of many hundreds of α-glucose molecules linked together in long, unbranched chains. Each linkage occurs between the number 1 carbon of one glucose molecule and the number 4 carbon of another, making them α-1,4 linkages (figure 3.9*a*). The long chains of amylose tend to coil up in water, a property that renders amylose insoluble. Potato starch is about 20% amylose (figure 3.9*b*).

Most plant starch, including the remaining 80% of potato starch, is a somewhat more complicated variant of amylose called *amylopectin.* Pectins are branched polysaccharides with the branches occurring due to bonds between the 1 carbon of one molecule and the 6 carbon of another (α-1,6 linkages). These short amylose branches consist of 20 to 30 glucose subunits (figure 3.9*b*).

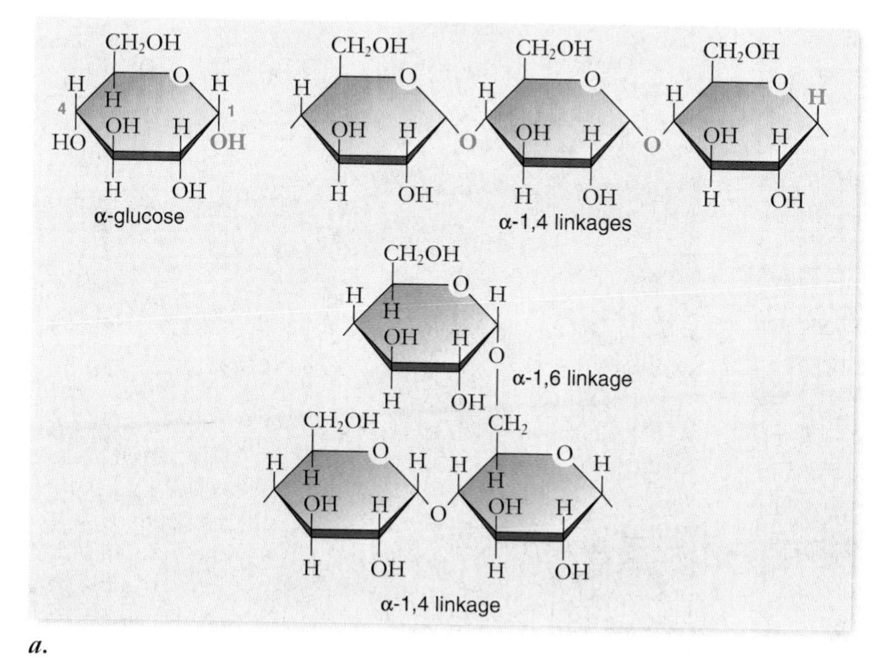

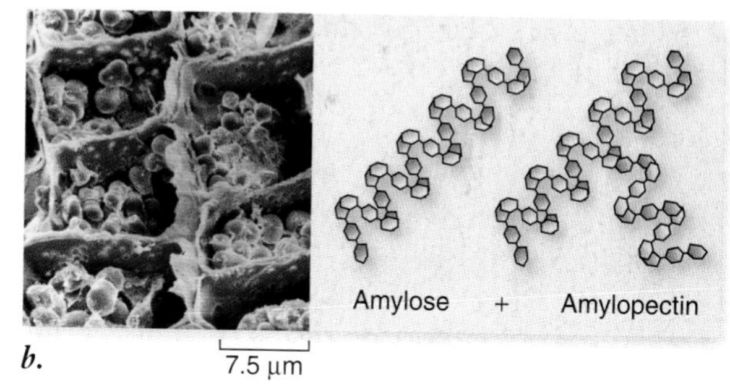

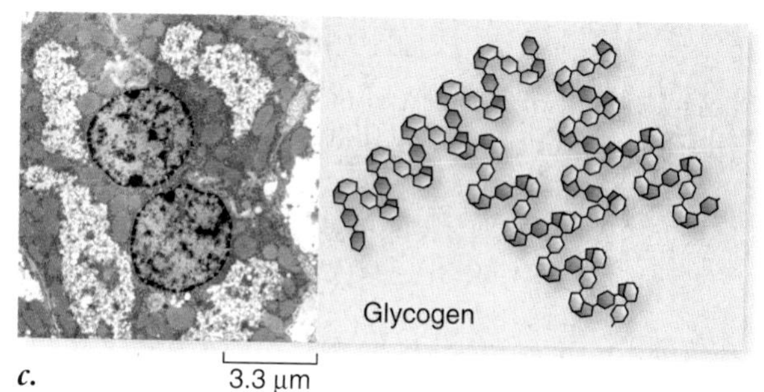

figure 3.9

POLYMERS OF GLUCOSE: STARCH AND GLYCOGEN.
a. Starch chains consist of polymers of α-glucose subunits joined by α-1,4 glycosidic linkages. These chains can be branched by forming similar α-1,6 glycosidic bonds. These storage polymers then differ primarily in their degree of branching. ***b.*** Starch is found in plants and is composed of amylose and amylopectin, which are unbranched and branched, respectively. The branched form is insoluble and forms starch granules in plant cells. ***c.*** Glycogen is found in animal cells and is highly branched and also insoluble, forming glycogen granules.

figure 3.10

POLYMERS OF GLUCOSE: CELLULOSE. Starch chains consist of α-glucose subunits, and cellulose chains consist of β-glucose subunits. ***a.*** Thus the bonds between adjacent glucose molecules in cellulose are β-1,4 glycosidic linkages. ***b.*** Cellulose is unbranched and forms long fibers. Cellulose fibers can be very strong and are quite resistant to metabolic breakdown, which is one reason wood is such a good building material.

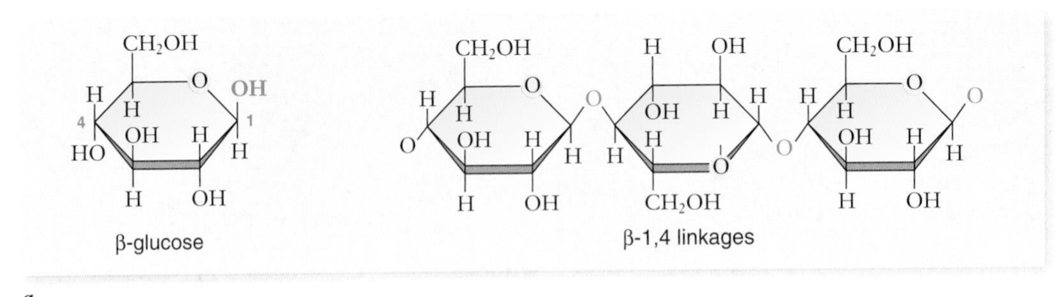

a.

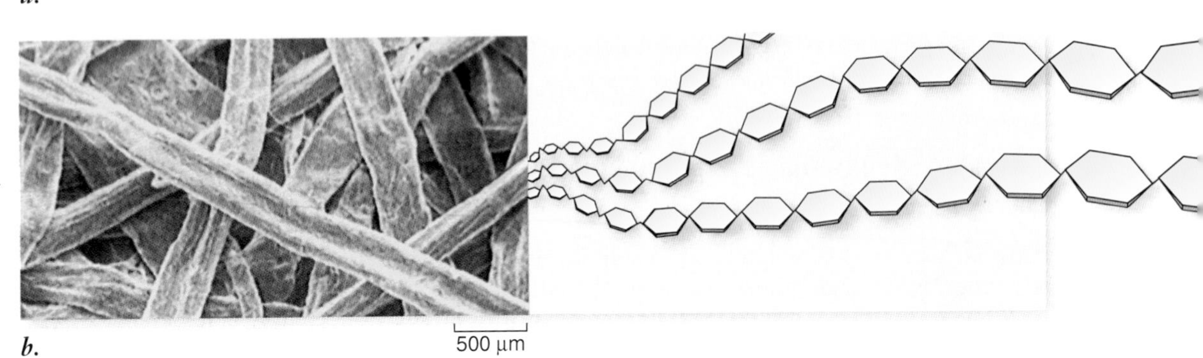

b.

The comparable molecule to starch in animals is **glycogen.** Like amylopectin, glycogen is an insoluble polysaccharide containing branched amylose chains. Glycogen has a much greater average chain length and more branches than plant starch (figure 3.9*c*).

Cellulose

Although some chains of sugars store energy, others serve as structural material for cells. For two glucose molecules to link together, the glucose subunits must be of the same form. *Cellulose* is a polymer of β-glucose (figure 3.10). The bonds between adjacent glucose molecules still exist between the 1 carbon of the first glucose and the 4 carbon of the next glucose, but these are β-1,4 linkages.

A chain of glucose molecules consisting of all β-glucose has properties that are very different from those of starch. These long, unbranched β-linked chains make tough fibers. Cellulose is the chief component of plant cell walls (figure 3.10). It is chemically similar to amylose, with one important difference: The starch-hydrolyzing enzymes that occur in most organisms cannot break the bond between two β-glucose units because they only recognize α linkages.

Because cellulose cannot be broken down readily by most creatures, it works well as a biological structural material. Those few animals able to break down cellulose find it a rich source of energy. Certain vertebrates, such as cows, can digest cellulose by means of bacteria and protists harbored in their digestive tracts that provide the necessary enzymes.

Chitin

Chitin, the structural material found in arthropods and many fungi, is a modified form of cellulose where an N-acetyl group replaces a hydroxyl in each glucose unit. When cross-linked by proteins, it forms a tough, resistant surface material that serves as the hard exoskeleton of insects and crustaceans (figure 3.11; see chapter 33). Few organisms are able to digest chitin and use it as a major source of nutrition, but most possess a chitinase enzyme, probably to protect against fungi.

Sugars are among the most important energy-storage molecules in organisms. Monosaccharides have three to six carbon atoms; the six-carbon monosaccharides typically have a ring form. The structural differences between sugar isomers can confer substantial functional differences upon the molecules. Disaccharides consist of two linked monosaccharides. Starches are polymers of α-glucose. Most starches are branched, rendering the polymer insoluble. Structural carbohydrates such as cellulose in plants are chains of sugars such as β-glucose that are not easily digested.

figure 3.11

CHITIN. Chitin is the principal structural element in the external skeletons of many invertebrates, such as this lobster.

3.3 Nucleic Acids: Information Molecules

The biochemical activity of a cell depends on production of a large number of proteins, each with a specific sequence. The information necessary to produce the correct proteins is passed through generations of organisms, even though the protein molecules themselves are not.

Nucleic acids are the information-carrying devices of cells, just as disks contain the information that computers use, blueprints carry the information that builders use, and road maps display the information that travelers use. Two main varieties of nucleic acids are **deoxyribonucleic acid (DNA;** figure 3.12) and **ribonucleic acid (RNA).**

The way DNA encodes the genetic information used to assemble proteins (as discussed in detail in chapter 14) is similar to the way the letters on a page encode information. Unique among macromolecules, nucleic acids are able to serve as templates to produce precise copies of themselves. This characteristic allows genetic information to be preserved during cell division and during the reproduction of organisms. DNA, found primarily in the nuclear region of cells, contains the genetic information necessary to build specific organisms.

Cells use RNA to read the cell's DNA-encoded information and to direct the synthesis of proteins. RNA is similar to DNA in structure and consists of transcripted copies of portions of the DNA. These transcripts serve as blueprints specifying the amino acid sequences of proteins. This process will be described in detail in chapter 15.

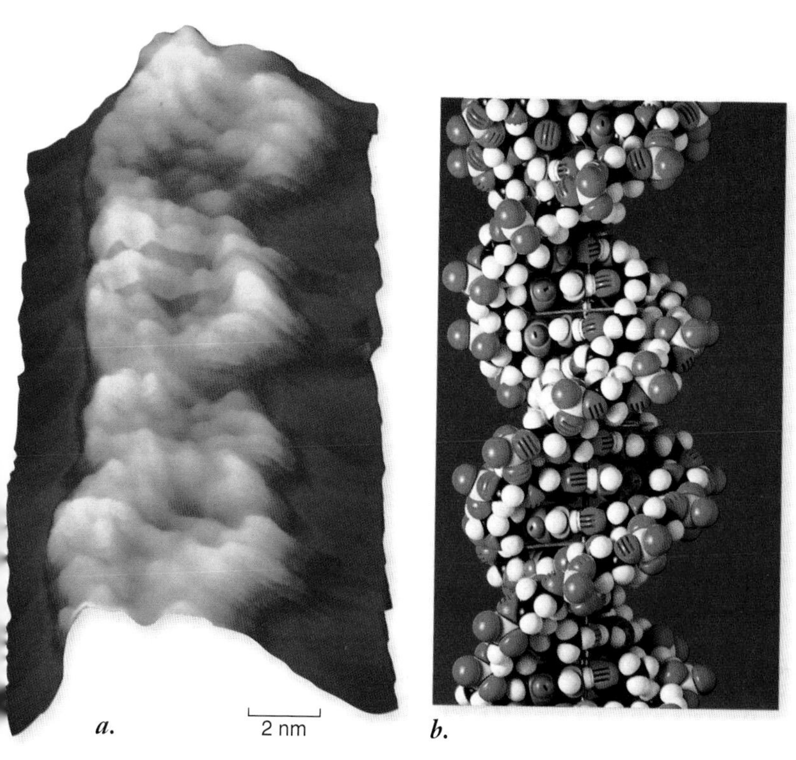

figure 3.12

IMAGES OF DNA. ***a.*** A scanning-tunneling micrograph of DNA (false color; 2,000,000) showing approximately three turns of the DNA double helix. ***b.*** A space-filling model for comparison to the image of actual DNA in ***(a)***.

Nucleic acids are nucleotide polymers

Nucleic acids are long polymers of repeating subunits called **nucleotides.** Each nucleotide consists of three components: a pentose or five-carbon sugar (ribose in RNA and deoxyribose in DNA); a phosphate ($-PO_4$) group; and an organic nitrogenous (nitrogen-containing) base (figure 3.13). When a nucleic acid polymer forms, the phosphate group of one nucleotide binds to the hydroxyl group from the pentose sugar of another, releasing water and forming a phosphodiester bond by a dehydration reaction. A **nucleic acid,** then, is simply a chain of five-carbon sugars linked together by phosphodiester bonds with a nitrogenous base protruding from each sugar (figure 3.14). These chains of nucleotides, *polynucleotides*, have different ends: a phosphate on one end and an $-OH$ from a sugar on the other end. We conventionally refer to these ends as 5′ ("five-prime," $-PO_4$) and 3′ ("three-prime," $-OH$) taken from the numbering on carbons of the sugar (figure 3.14).

Two types of nitrogenous bases occur in nucleotides. The first type, *purines*, are large, double-ring molecules found in both DNA and RNA; the two types of purines are adenine (A) and guanine (G). The second type, *pyrimidines*, are smaller, single-ring molecules; they include cytosine (C, in both DNA and RNA), thymine (T, in DNA only), and uracil (U, in RNA only).

DNA carries the genetic code

Organisms use sequences of nucleotides in DNA to encode the information specifying the amino acid sequences of their proteins. This method of encoding information is very similar to the way in which sequences of letters encode information in a

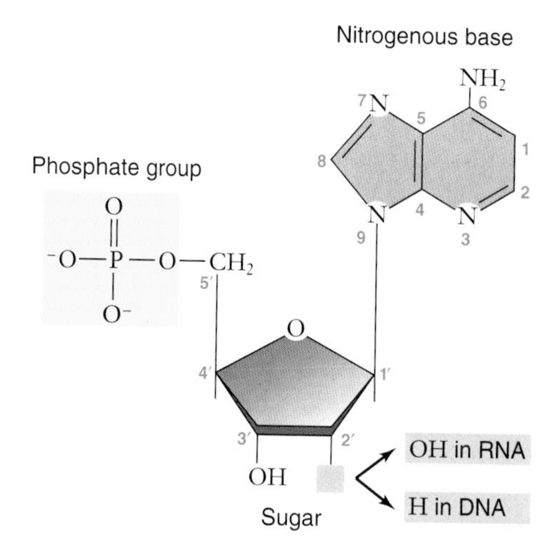

figure 3.13

STRUCTURE OF A NUCLEOTIDE. The nucleotide subunits of DNA and RNA are made up of three elements: a five-carbon sugar (ribose or deoxyribose), an organic nitrogenous base (adenine is shown here), and a phosphate group. Notice that the numbers on the sugar all are given as "primes" (1′, 2′, etc.) to distinguish them from the numbering on the rings of the bases.

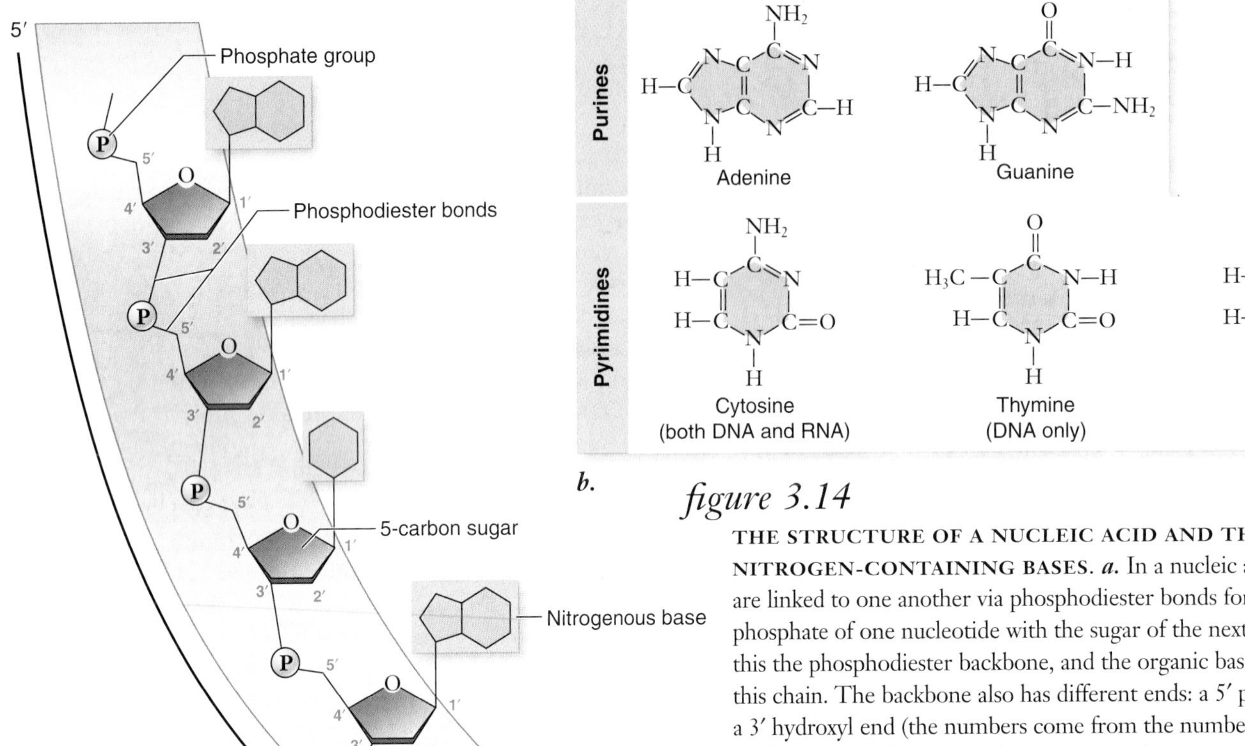

a. *b.*

figure 3.14

THE STRUCTURE OF A NUCLEIC ACID AND THE ORGANIC NITROGEN-CONTAINING BASES. ***a.*** In a nucleic acid, nucleotides are linked to one another via phosphodiester bonds formed between the phosphate of one nucleotide with the sugar of the next nucleotide. We call this the phosphodiester backbone, and the organic bases protrude from this chain. The backbone also has different ends: a 5′ phosphate end and a 3′ hydroxyl end (the numbers come from the numbers in the sugars). ***b.*** The organic nitrogenous bases can be either purines or pyrimidines. The base thymine is found in DNA, and the base uracil is found in RNA.

sentence. A sentence written in English consists of a combination of the 26 different letters of the alphabet in a certain order; the code of a DNA molecule consists of different combinations of the four types of nucleotides in specific sequences, such as CGCTTACG. The information encoded in DNA is used in the everyday metabolism of the organism and is passed on to the organism's descendants.

DNA molecules in organisms exist not as single chains folded into complex shapes, like proteins, but rather as two chains wrapped about each other in a long linear molecule. The two strands of a DNA polymer wind around each other like the outside and inside rails of a spiral staircase. Such a spiral shape is called a helix, and a helix composed of two chains is called a **double helix.** Each step of DNA's helical staircase is composed of a base-pair, consisting of a base in one chain attracted by hydrogen bonds to a base opposite it on the other chain (figure 3.15).

The base-pairing rules are rigid: Adenine can pair only with thymine (in DNA) or with uracil (in RNA), and cytosine can pair only with guanine. The bases that participate in base-pairing are said to be **complementary** to each other. Additional details of the structure of DNA and how it interacts with RNA in the production of proteins are presented in chapters 14 and 15.

RNA is a transcript of a DNA strand

RNA is similar to DNA, but with two major chemical differences. First, RNA molecules contain ribose sugars, in which the 2 carbon is bonded to a hydroxyl group. (In DNA, this hydroxyl group is replaced by a hydrogen atom.) Second, RNA molecules utilize uracil in place of thymine. Uracil has the same structure as thymine, except that one of its carbons lacks a methyl (—CH_3) group.

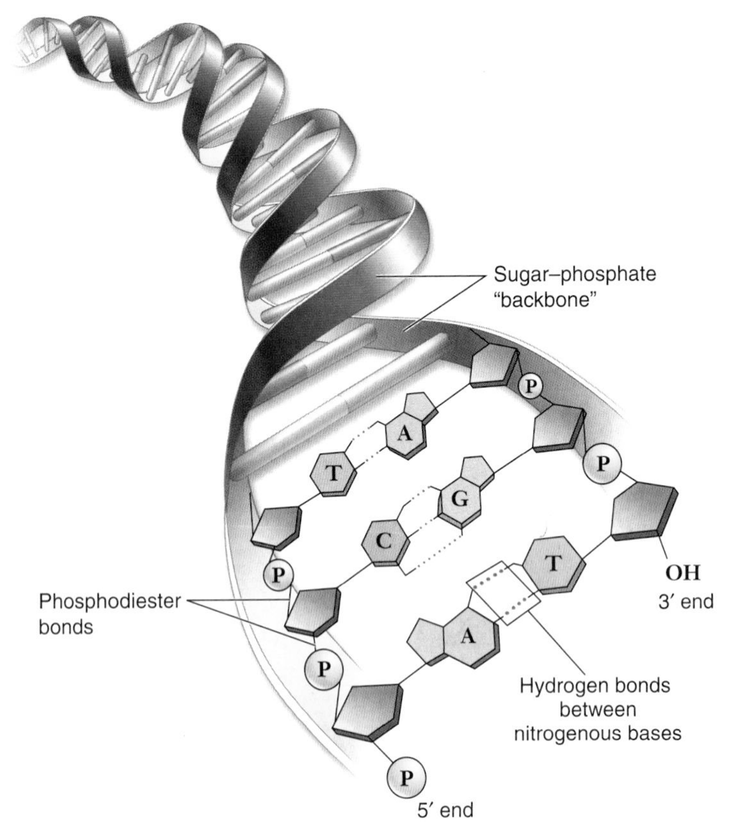

figure 3.15

THE STRUCTURE OF DNA. DNA consists of two polynucleotide chains running in opposite directions wrapped about a single helical axis. Hydrogen bond formation (dashed lines) between the organic bases, called base-pairing, causes the two chains of a DNA duplex to bind to each other and form a double helix.

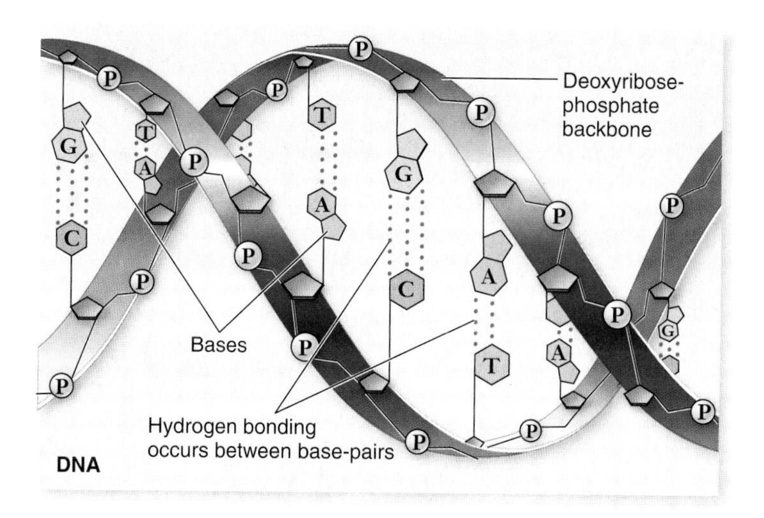

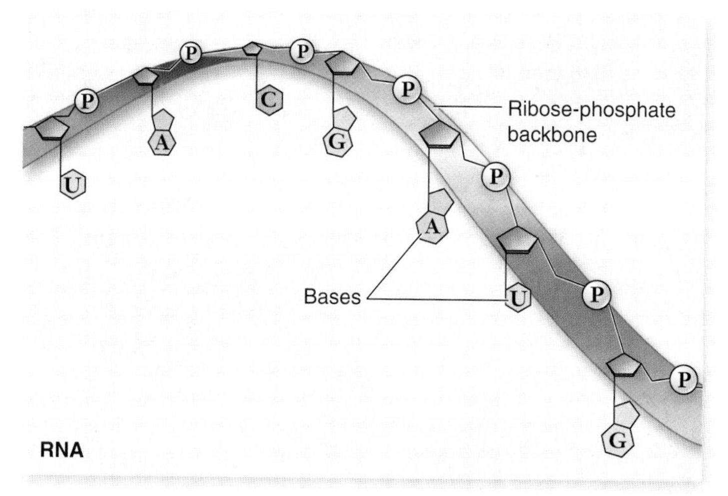

figure 3.16

DNA VERSUS RNA. DNA forms a double helix, uses deoxyribose as the sugar in its sugar–phosphate backbone, and utilizes thymine among its nitrogenous bases. RNA is usually single-stranded, uses ribose as the sugar in its sugar–phosphate backbone, and utilizes uracil in place of thymine.

Transcribing the DNA message into a chemically different molecule such as RNA allows the cell to tell which is the original information storage molecule and which is the transcript. DNA molecules are always double-stranded (except for a few single-stranded DNA viruses), whereas the RNA molecules transcribed from DNA are typically single-stranded (figure 3.16). These differences separate the role of DNA in storing hereditary information from the role of RNA in using this information to specify the sequence of amino acids in proteins.

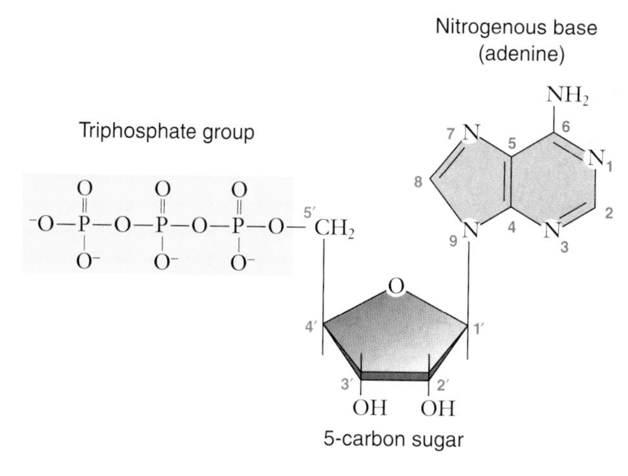

figure 3.17

ATP. Adenosine triphosphate (ATP) contains adenine, a five-carbon sugar, and three phosphate groups.

Other nucleotides are vital components of energy reactions

In addition to serving as subunits of DNA and RNA, nucleotide bases play other critical roles in the life of a cell. For example, adenine is a key component of the molecule **adenosine triphosphate (ATP;** figure 3.17), the energy currency of the cell. ATP is used to drive energetically unfavorable chemical reactions, to power transport across membranes, and to power the movement of cells; in short, it is the most common form of energy in a cell.

Two other important nucleotide-containing molecules are **nicotinamide adenine dinucleotide (NAD^+)** and **flavin adenine dinucleotide (FAD).** These molecules function as electron carriers in a variety of cellular processes. You will see the action of these molecules in detail when we discuss photosynthesis and respiration (chapters 6–8).

A nucleic acid is a long chain of five-carbon sugars with an organic base protruding from each sugar. DNA is a double-stranded helix that stores hereditary information as a specific sequence of nucleotide bases. RNA is a single-stranded molecule that transcribes this information to direct protein synthesis.

3.4 Proteins: Molecules with Diverse Structures and Functions

Proteins are the most diverse group of biological macromolecules, both chemically and functionally. The number of protein functions within cells is so large that we could not begin to list them all. We can, however, group these functions into seven categories, as follows (figure 3.18). This list is a summary only; details are covered in later chapters.

1. **Enzyme catalysis.** Enzymes are biological catalysts that facilitate specific chemical reactions. Because of

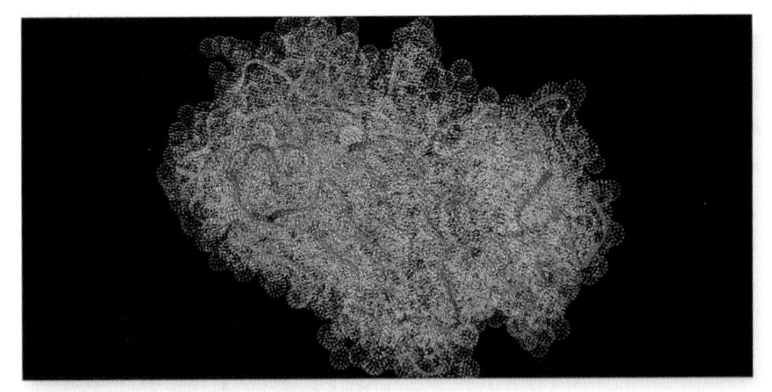

Enzyme catalysis: space-filling model of an enzyme

Defense: venom

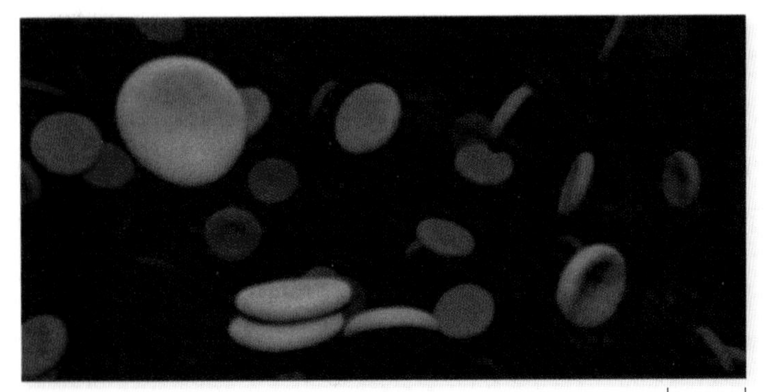

Transport: hemoglobin 3.3 μm

Support: keratin

Motion: actin and myosin

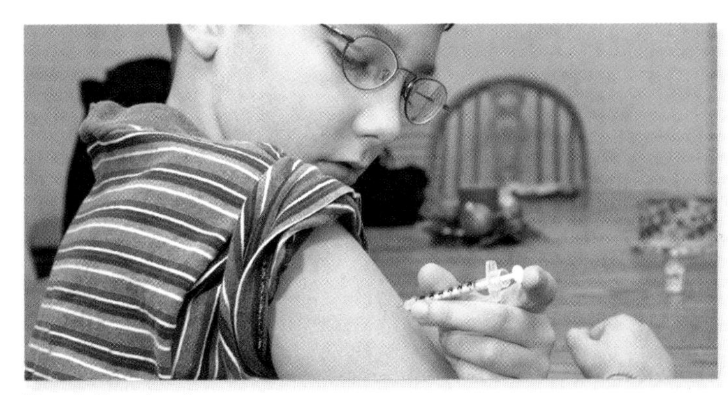

Regulation: insulin

Storage: calcium

figure 3.18

PROTEIN FUNCTIONS. Proteins perform a variety of roles in living systems. This includes active roles such as enzymes, defensive proteins such as venoms and actin and myosin in muscle. Proteins also have a structural role such as keratin, and act to transport O_2, or sequester important ions like calcium. Proteins also have regulatory roles as signaling molecules like insulin, or as receptors.

this property, the appearance of enzymes was one of the most important events in the evolution of life. Enzymes are globular proteins with a three-dimensional shape that fit snugly around the molecules they act on, facilitating chemical reactions by stressing particular chemical bonds.

2. **Defense.** Other globular proteins use their shapes to "recognize" foreign microbes and cancer cells. These cell-surface receptors form the core of the body's endocrine and immune systems.
3. **Transport.** A variety of globular proteins transport small molecules and ions. The transport protein

hemoglobin, for example, transports oxygen in the blood, and myoglobin, a similar protein, transports oxygen in muscle. Iron is transported in blood by the protein transferrin.

4. **Support.** Protein fibers play structural roles. These fibers include keratin in hair, fibrin in blood clots, and collagen, which forms the matrix of skin, ligaments, tendons, and bones, and is the most abundant protein in a vertebrate body.
5. **Motion.** Muscles contract through the sliding motion of two kinds of protein filaments: actin and myosin. Contractile proteins also play key roles in the cell's cytoskeleton and in moving materials in cells.
6. **Regulation.** Small proteins called hormones serve as intercellular messengers in animals. Proteins also play many regulatory roles within the cell—turning on and shutting off genes during development, for example. In addition, proteins also receive information, acting as cell-surface receptors.
7. **Storage.** Calcium and iron are stored by binding as ions to storage proteins.

Table 3.2 Summarizes these functions and includes examples of proteins in the human body that carry them out.

Proteins are polymers of amino acids

Proteins are linear polymers of 20 different amino acids. **Amino acids,** as their name suggests, contain an amino group (—NH_2) as well as an acidic carboxyl group (—COOH). The specific order of amino acids determines the protein's structure and function. Many scientists believe amino acids were among the first molecules formed on the early Earth. It seems highly likely that the oceans that existed early in the history of the Earth contained a wide variety of amino acids.

Amino acid structure

The generalized structure of an amino acid can be represented as amino and carboxyl groups bonded to a central carbon atom,

TABLE 3.2 The Many Functions of Protein

Function	Class of Protein	Examples	Examples of Use
Enzyme catalysis	Enzymes	Hydrolytic enzymes	Cleave polysaccharides
		Proteases	Break down proteins
		Polymerases	Synthesize nucleic acids
		Kinases	Phosphorylate sugars and proteins
Defense	Immunoglobulins	Antibodies	Mark foreign proteins for elimination
	Toxins	Snake venom	Blocks nerve function
	Cell surface antigens	MHC proteins	"Self" recognition
Transport	Circulating transporters	Hemoglobin	Carries O_2 and CO_2 in blood
		Myoglobin	Carries O_2 and CO_2 in muscle
		Cytochromes	Electron transport
	Membrane transporters	Sodium–potassium pump	Excitable membranes
		Proton pump	Chemiosmosis
		Glucose transporter	Transports glucose into cells
Support	Fibers	Collagen	Forms cartilage
		Keratin	Forms hair, nails
		Fibrin	Forms blood clots
Motion	Muscle	Actin	Contraction of muscle fibers
		Myosin	Contraction of muscle fibers
Regulation	Osmotic proteins	Serum albumin	Maintains osmotic concentration of blood
	Gene regulators	*lac* repressor	Regulates transcription
	Hormones	Insulin	Controls blood glucose levels
		Vasopressin	Increases water retention by kidneys
		Oxytocin	Regulates uterine contractions and milk production
Storage	Ion binding	Ferritin	Stores iron, especially in spleen
		Casein	Stores ions in milk
		Calmodulin	Binds calcium ions

with an additional hydrogen and a functional side group indicated by R. These components completely fill the bonds of the central carbon:

$$\begin{array}{c} R \\ | \\ H_2N-C-COOH \\ | \\ H \end{array}$$

The unique character of each amino acid is determined by the nature of the R group. Notice that unless the R group is an H atom, as in glycine, amino acids are chiral and can exist as two enantiomeric forms: D or L. In living systems, only the L-amino acids are found in proteins, and D-amino acids are rare.

The R group also determines the chemistry of amino acids. Serine, in which the R group is $-CH_2OH$, is a polar molecule. Alanine, which has $-CH_3$ as its R group, is nonpolar. The 20 common amino acids are grouped into five chemical classes, based on the R group:

1. Nonpolar amino acids, such as leucine, often have R groups that contain $-CH_2$ or $-CH_3$.
2. Polar uncharged amino acids, such as threonine, have R groups that contain oxygen (or $-OH$).
3. Charged amino acids, such as glutamic acid, have R groups that contain acids or bases that can ionize.
4. Aromatic amino acids, such as phenylalanine, have R groups that contain an organic (carbon) ring with alternating single and double bonds. These are also nonpolar.
5. Special function amino acids have unique individual properties. Methionine is often the first amino acid in a chain of amino acids; proline causes kinks in chains; and cysteine links chains together.

Each amino acid affects the shape of a protein differently, depending on the chemical nature of its side group. For example, portions of a protein chain with numerous nonpolar amino acids tend to fold into the interior of the protein by hydrophobic exclusion.

Peptide bonds

In addition to its R group, each amino acid, when ionized, has a positive amino (NH_3^+) group at one end and a negative carboxyl (COO^-) group at the other. The amino and carboxyl groups on a pair of amino acids can undergo a dehydration reaction to form a covalent bond. The covalent bond that links two amino acids is called a **peptide bond** (figure 3.19). The two amino acids linked by such a bond are not free to rotate around the N—C linkage because the peptide bond has a partial double-bond character, unlike the N—C and C—C bonds to the central carbon of the amino acid. This lack of rotation about the peptide bond is one factor that determines the structural character of the coils and other regular shapes formed by chains of amino acids.

A protein is composed of one or more long unbranched chains. Each chain is called a **polypeptide** and is composed of amino acids linked by peptide bonds. The terms *protein* and *polypeptide* tend to be used loosely and may be confusing. For proteins that include only a single polypeptide chain, the two terms are synonymous.

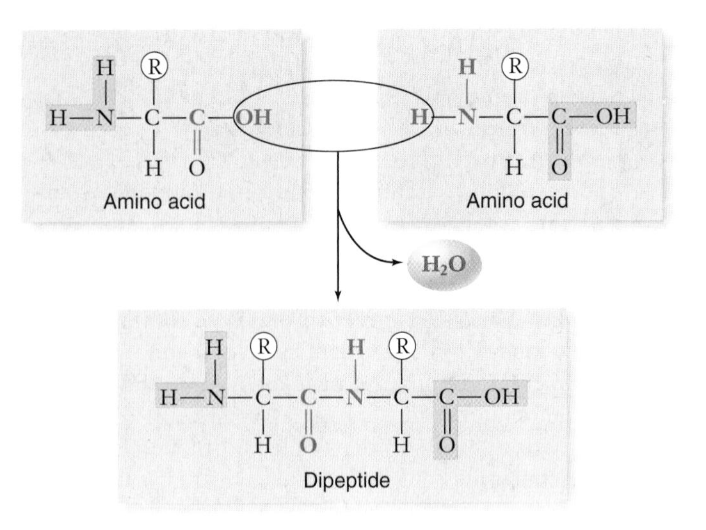

figure 3.19

THE PEPTIDE BOND. A peptide bond forms when the amino end of one amino acid joins to the carboxyl end of another. Reacting amino and carboxyl groups are shown in red and nonreacting groups are highlighted in green. Notice that the resulting dipeptide still has an amino end and a carboxyl end. Because of the partial double-bond nature of peptide bonds, the resulting peptide chain cannot rotate freely around these bonds.

The pioneering work of Frederick Sanger in the early 1950s provided the evidence that each kind of protein has a specific amino acid sequence. Using chemical methods to remove successive amino acids, then identify them, Sanger succeeded in determining the amino acid sequence of insulin. In so doing he demonstrated clearly that this protein had a defined sequence, which was the same for all insulin molecules in the solution. Although many different amino acids occur in nature, only 20 commonly occur in proteins. Figure 3.20 illustrates these 20 amino acids and their side groups.

Proteins have levels of structure

The shape of a protein determines the protein's function. One way to study the shape of something as small as a protein is to look at it with very short wavelength energy—in other words, with X-rays. X-rays can be passed through a crystal of protein to produce a diffraction pattern. This pattern can then be analyzed by a painstaking procedure that allows the investigator to build up a three-dimensional picture of the position of each atom. The first protein to be analyzed in this way was myoglobin, soon followed by the related protein hemoglobin.

As more and more proteins were studied, a general principle became evident: In every protein studied, essentially all the internal amino acids are nonpolar ones—amino acids such as leucine, valine, and phenylalanine. Water's tendency to hydrophobically exclude nonpolar molecules literally shoves the nonpolar portions of the amino acid chain

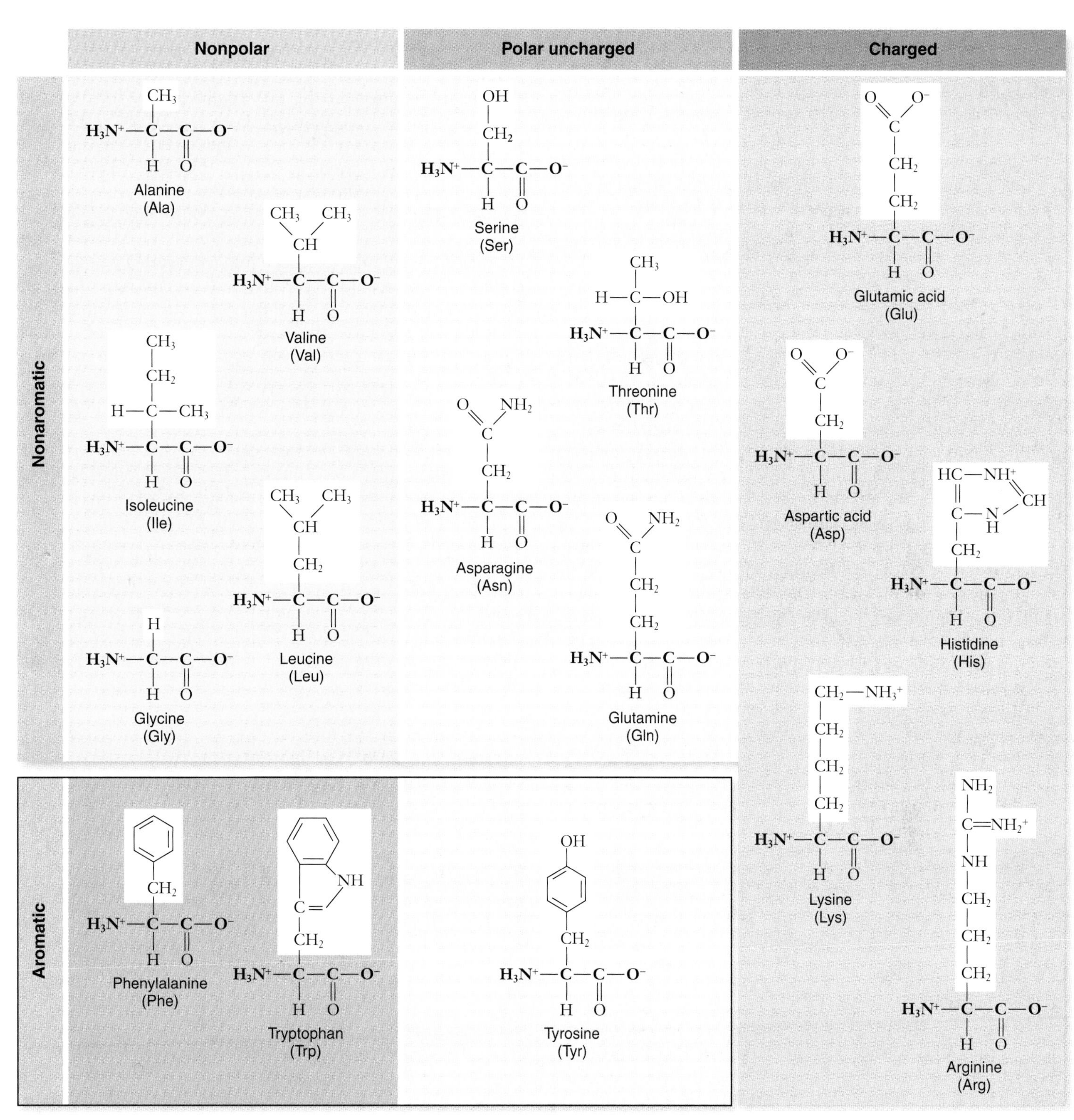

figure 3.20

THE 20 COMMON AMINO ACIDS. Each amino acid has the same chemical backbone, but differs in the side, or R, group. Six of the amino acids are nonpolar because they have $—CH_2$ or $—CH_3$ in their R groups. Two of the six contain ring structures with alternating double and single bonds, which classifies them also as aromatic. Another six are polar because they have oxygen or a hydroxyl group in their R groups. Five others are capable of ionizing to a charged form. The remaining three special function amino acids have chemical properties that allow them to help form links between protein chains or kinks in proteins.

into the protein's interior (figure 3.21). This effect positions the nonpolar amino acids in close contact with one another, leaving little empty space inside. Polar and charged amino acids are restricted to the surface of the protein, except for the few that play key functional roles.

The structure of proteins has traditionally been discussed in terms of a hierarchy with four levels of structure: *primary*, *secondary*, *tertiary*, and *quaternary* (figure 3.22). We will discuss this view and then integrate it with a more modern view arising from our increasing knowledge of protein structure.

Primary structure: amino acid sequence

The **primary structure** of a protein is its amino acid sequence. Because the R groups that distinguish the amino acids play no role in the peptide backbone of proteins, a protein can consist of any sequence of amino acids. Thus, because any of 20 different amino acids might appear at any position, a protein containing 100 amino acids could form any of 20^{100} different amino acid sequences (that's the same as 10^{130}, or 1 followed by 130 zeros—more than the number of atoms known in the universe). This important property of proteins permits great diversity.

Consider the protein hemoglobin, used in your blood to transport oxygen. Hemoglobin is composed of two α-globin peptide chains and two β-globin peptide chains. The α-globin chains differ from the β-globin chains in the sequence of amino acids. Furthermore, any alteration in the normal sequence of either of the types of globin proteins, even by a single amino acid, can have drastic effects on the function of the protein.

Secondary structure: hydrogen bonding patterns

The amino acid side groups are not the only portions of proteins that form hydrogen bonds. The peptide groups of the main chain can also do so. These hydrogen bonds can be with water, or with other peptide groups. If the peptide groups formed too many hydrogen bonds with water, it would lead to proteins that behave like a random coil and not produce the kinds of globular structures that are common in proteins. Linus Pauling suggested that the peptide groups could interact with one another if the peptide was coiled into a spiral that he called the **α helix.** We now call this sort of regular interaction of groups in the peptide backbone **secondary structure.** Another form of secondary structure can occur between regions of peptide aligned next to each other to form a planar structure called a **β sheet.** These can be either parallel or antiparallel depending on the orientation of the adjacent sections of peptide with respect to each other.

These two kinds of secondary structure create regions of the protein that are cylindrical (α helices) and planar (β sheets). A protein's final structure can include regions of each type of secondary structure. For example, DNA-binding proteins usually have regions of α helix that can lay across DNA and interact directly with the bases of DNA. Porin proteins that form holes in membranes are composed of β sheets arranged to form a pore in the membrane. Finally in hemoglobin, mentioned earlier, the α- and β-globin peptide chains that make up the final molecule each have characteristic regions of secondary structure.

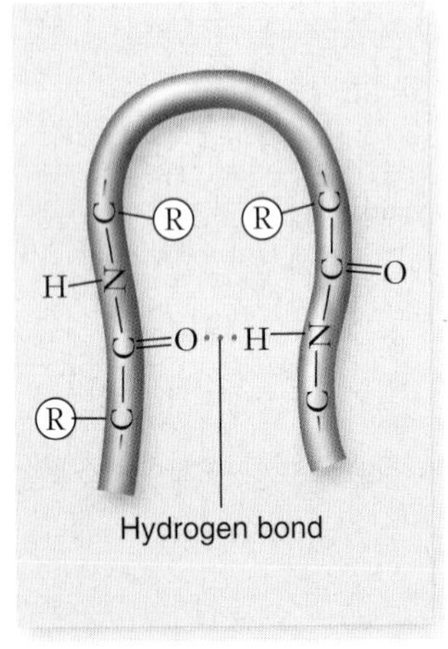

a.

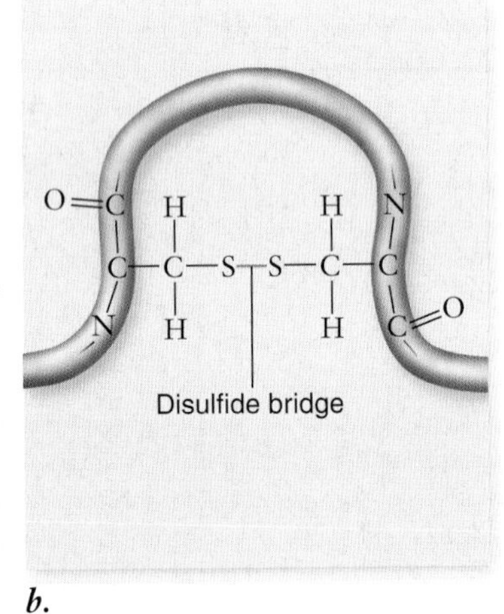

b.

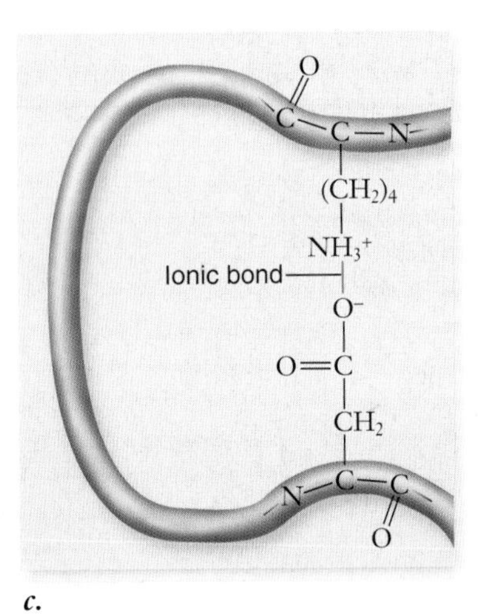

c.

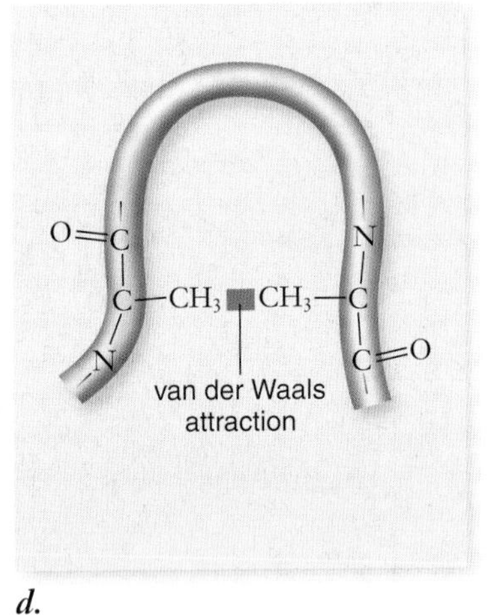

d.

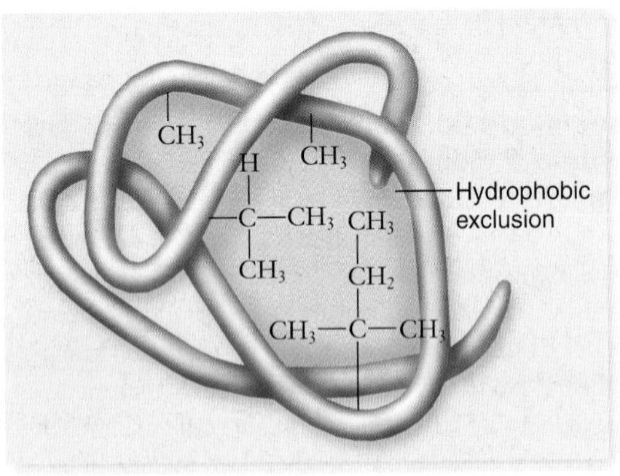

e.

figure 3.21

INTERACTIONS THAT CONTRIBUTE TO A PROTEIN'S SHAPE. Aside from the bonds that link the amino acids in a protein together, several other weaker forces and interactions determine how a protein will fold. ***a.*** Hydrogen bonds can form between the different amino acids. ***b.*** Covalent disulfide bridges can form between two cysteine side chains. ***c.*** Ionic bonds can form between groups with opposite charge. ***d.*** van der Waals attractions occur, which are weak attractions between atoms due to oppositely polarized electron clouds. ***e.*** Polar portions of the protein tend to gather on the outside of the protein and interact with water, whereas the hydrophobic portions of the protein, including nonpolar amino acid chains, are shoved toward the interior of the protein.

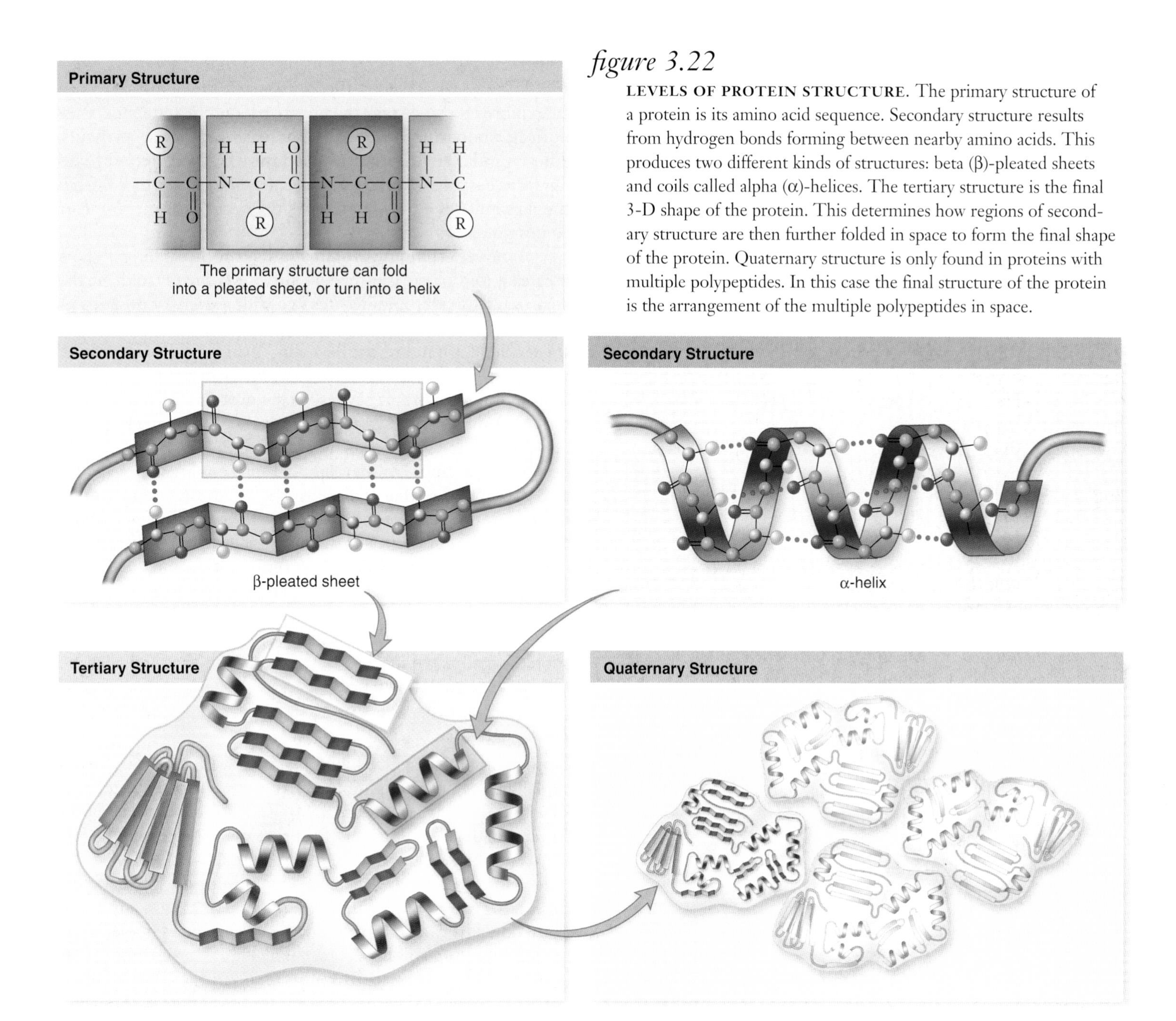

figure 3.22

LEVELS OF PROTEIN STRUCTURE. The primary structure of a protein is its amino acid sequence. Secondary structure results from hydrogen bonds forming between nearby amino acids. This produces two different kinds of structures: beta (β)-pleated sheets and coils called alpha (α)-helices. The tertiary structure is the final 3-D shape of the protein. This determines how regions of secondary structure are then further folded in space to form the final shape of the protein. Quaternary structure is only found in proteins with multiple polypeptides. In this case the final structure of the protein is the arrangement of the multiple polypeptides in space.

Tertiary structure: folds and links

The final folded shape of a globular protein is called a protein's **tertiary structure.** This tertiary structure contains regions that have secondary structure and determines how these are further arranged in space to produce the overall structure. A protein is initially driven into its tertiary structure by hydrophobic exclusion from water. Ionic bonds between oppositely charged R groups bring regions into close proximity, and disulfide bonds (covalent links between two cysteine R groups) lock particular regions together. The final folding of a protein is determined by its primary structure—the chemical nature of its side groups (see figure 3.22). Many small proteins can be fully unfolded ("denatured") and will spontaneously refold into their characteristic shape.

The tertiary structure is stabilized by a number of forces including hydrogen bonding between R groups of different amino acids, electrostatic attraction between R groups with opposite charge (also called salt bridges), hydrophobic exclusion of nonpolar R groups, and covalent bonds in the form of disufides. The stability of a protein, once it has folded into its tertiary shape, is strongly influenced by how well its interior fits together. When two nonpolar chains in the interior are in very close proximity, they experience a form of molecular attraction called van der Waals' forces. Individually quite weak, these forces can add up to a strong attraction when many of them come into play, like the combined strength of hundreds of hooks and loops on a strip of Velcro. These forces are effective only over short distances, however; no "holes" or cavities exist in the interior of proteins. The variety of different nonpolar amino acids, with a different-sized R group with its own distinctive shape, allows nonpolar chains to fit very precisely within the protein interior.

It is therefore not surprising that changing a single amino acid can drastically alter the structure, and thus the function of a protein. The sickle cell version of hemoglobin (HbS), for example, is a change of a single glutamic acid for a valine in the β-globin chain that causes the protein to aggregate into clumps. Note that this exchanges a charged amino acid for a nonpolar one on the surface of the protein, leading the protein to become sticky and form aggregates. Another variant of hemoglobin called HbE, actually the most common in human populations, causes a change from glutamic acid to lysine at a different site in the β-globin chain. In this case the structural change is not as dramatic, but it still impairs function, resulting in multiple types of anemia and thalassemia. More than 700 structural variants of hemoglobin are known, with up to 7% of the world's population being carriers for forms that are clinically relevant.

Quaternary structure: subunit arrangements

When two or more polypeptide chains associate to form a functional protein, the individual chains are referred to as subunits of the protein. The arrangement of these subunits is termed the **quaternary structure** of the protein. In proteins composed of subunits, the interfaces where the subunits contact one another are often nonpolar, and they play a key role in transmitting information between the subunits about individual subunit activities.

As mentioned earlier, the protein hemoglobin is composed of two α-chain subunits and two β-chain subunits. So each α- and β-globin chain has a primary structure consisting of a specific sequence of amino acids. This then assumes a characteristic secondary structure consisting of α helices and β sheets that are then arranged into a specific tertiary structure for each α- and β-globin subunit. Lastly, these subunits are then arranged into their final quaternary structure that, in this case, is the final structure of the protein. For proteins that consist of only a single peptide chain, the enzyme lysozyme for example, the tertiary structure is the final structure of the protein.

Motifs and domains are additional structural characteristics

To directly determine the sequence of amino acids in a protein is a laborious task. Although the process has been automated, it remains slow and difficult.

The ability to sequence DNA changed this situation rather suddenly. Sequencing DNA was a much simpler process, and even before it was automated, the number of known sequences rose quickly. With the advent of automation, the known sequences increased even more dramatically, and today the entire sequence of hundreds of bacterial genomes and more than a dozen animal genomes, including that of humans, has been determined. Because the DNA sequence is directly related to amino acid sequence in proteins, biologists now have a large database of protein sequences to compare and analyze. This new information has also stimulated thought about the logic of the genetic code and whether underlying patterns exist in protein structure. Our view of protein structure has evolved with this new information. Researchers still view the four-part hierarchical structure as being important, but two new terms have entered the biologist's vocabulary: motif and domain.

Motifs

As biologists discovered the 3-D structure of proteins (an even more laborious task than determining the sequence) they noticed similarities between otherwise dissimilar proteins. These similar structures are called **motifs,** or sometimes "supersecondary structure." The term *motif* is borrowed from the arts and refers to a recurring thematic element in music or design.

One very common protein motif is the β-α-β motif, which creates a fold or crease; the so-called "Rossmann fold" at the core of nucleotide-binding sites in a wide variety of proteins. A second motif that occurs in many proteins is the β barrel, a β sheet folded around to form a tube. A third type of motif, the helix-turn-helix, consistes of two α helices separated by a bend. This motif is important because many proteins use it to bind to the DNA double helix (figure 3.23; see also chapter 16).

Motifs indicate a logic to structure that investigators still do not understand. Do they simply represent a reuse by evolution of something that already works, or do they represent an optimal solution to a problem, such as how to bind a nucleotide? One way to think about it is that if amino acids are letters in the language of proteins, then motifs represent repeated words or phrases. Motifs have been very useful in the analysis of the proliferation of known proteins. Databases of protein motifs are now maintained, and these can be used to search new unknown proteins for known motifs. This process can shed light on the function of an unknown protein.

Domains

Domains of proteins are functional units within a larger structure. They can be thought of as substructure within the tertiary structure of a protein (figure 3.23). To continue the metaphor: Amino acids are letters in the protein language, motifs are words or phrases, and domains are paragraphs.

Most proteins are made up of multiple domains that perform different aspects of the protein's function. In many cases, these domains can be physically separated. For example, transcription factors (discussed in chapter 16) are enzymes that bind to DNA and initiate its transcription. If the DNA-binding region is exchanged with a different transcription factor, using molecular biology techniques, then the specificity of the factor for DNA can be changed without changing its ability to stimulate transcription. Such "domain-swapping" experiments have been performed with many transcription factors, and they indicate, among other things, that the DNA-binding and activation domains are functionally separate.

These functional domains of proteins may also help the protein to fold into its proper shape. As a polypeptide chain folds, the domains take their proper shape, each more or less independently of the others. This action can be demonstrated experimentally by artificially producing the fragment of a polypeptide that forms the domain in the intact protein, and showing that the fragment folds to form the same structure as it exhibits in the intact protein. A single polypeptide chain connects the domains of a protein, like a rope tied into several adjacent knots.

Domains can also correspond to the structure of the genes that encode them. Later, in chapter 15, you will see that genes

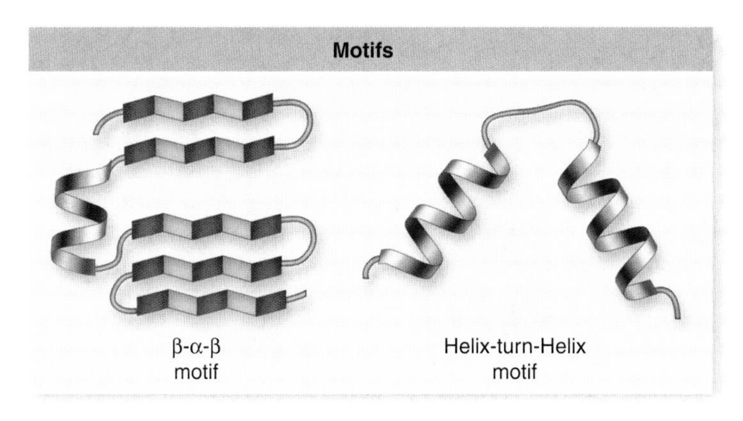

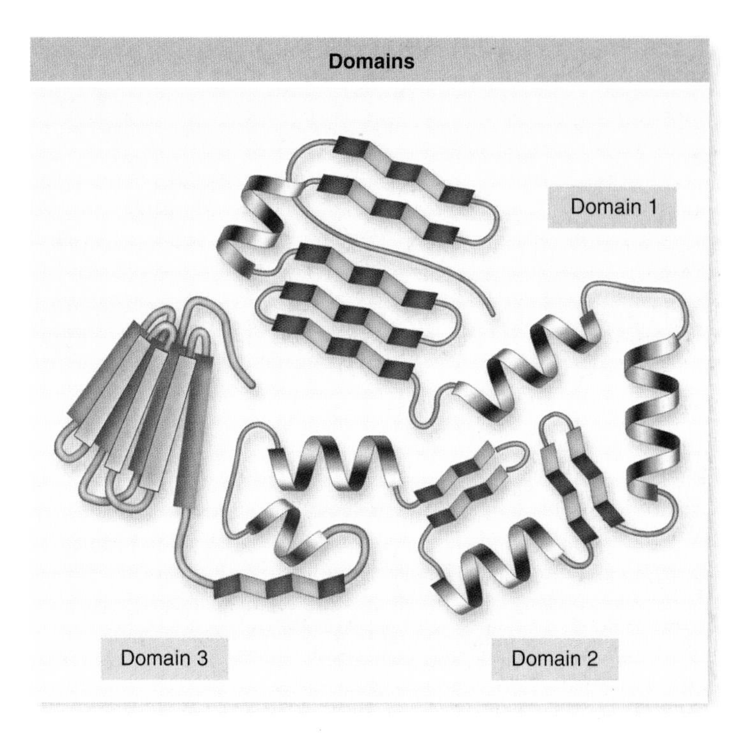

figure 3.23

MOTIFS AND DOMAINS. The elements of secondary structure can combine, fold, or crease to form motifs. These motifs are found in different proteins and can be used to predict function. Proteins also are made of larger domains, which are functionally distinct parts of a protein. The arrangement of these domains in space is the tertiary structure of a protein.

in eukaryotes are often in pieces within the genome, and these pieces, called *exons*, sometimes encode the functional domains of a protein. This finding led to the idea of evolution acting by shuffling protein-encoding domains.

The process of folding relies on chaperone proteins

Until recently, investigators thought that newly made proteins fold spontaneously as hydrophobic interactions with water shove nonpolar amino acids into the protein interior. We now know this view is too simple. Protein chains can fold in so many different ways that trial and error would simply take too long. In addition, as the open chain folds its way toward its final form, nonpolar "sticky" interior portions are exposed during intermediate stages. If these intermediate forms are placed in a test tube in an environment identical to that inside a cell, they stick to other, unwanted protein partners, forming a gluey mess.

How do cells avoid having their proteins clump into a mass? A vital clue came in studies of unusual mutations that prevent viruses from replicating in bacterial cells—it turns out that the virus proteins produced inside the cells could not fold properly. Further study revealed that normal cells contain **chaperone proteins,** which help other proteins to fold correctly.

Molecular biologists have now identified many proteins that act as molecular chaperones. This class of proteins has multiple subclasses, and representatives have been found in essentially every organism that has been examined. Furthermore, these proteins seem in all cases to be essential for viability as well, illustrating their fundamental importance. Many are heat shock proteins, produced in greatly increased amounts when cells are exposed to elevated temperature; high temperatures cause proteins to unfold, and heat shock chaperone proteins help the cell's proteins to refold properly.

One class of these proteins, called chaperonins, has been extensively studied. In the bacterium *Escherichia coli* (*E. coli*), one example is the GroE chaperonin, an essential protein. In mutants in which the GroE chaperonin is inactivated, fully 30% of the bacterial proteins fail to fold properly. Chaperonins associate to form a large macromolecular complex that resembles a cylindrical container. Proteins can move into the container, and the container itself can change its conformation considerably (figure 3.24). Experiments have shown that an improperly folded protein can enter the chaperonin and be refolded. The details of how this is accomplished are not clear, but seem to involve changes in the hydrophobicity of the interior of the chamber.

The flexibility of the structure of chaperonins is amazing. We tend to think of proteins as being fixed structures, but this is clearly not the case for chaperonins and this flexibility is necessary for their function. It also illustrates that even domains that may be very widely separated in a very large protein are still functionally connected. The folding process within a chaperonin utilizes the hydrolysis of ATP to power these changes in structure necessary for function. This entire process can occur in a cyclic manner until the appropriate structure is achieved. Cells use these chaperonins both to accomplish the original folding of some proteins and to restore the structure of incorrectly folded ones.

Some diseases may result from improper folding

Chaperone protein deficiencies may be implicated in certain diseases in which key proteins are improperly folded. Cystic fibrosis is a hereditary disorder in which a mutation disables a vital protein that moves ions across cell membranes. In at least some cases, this protein appears to have the correct amino acid sequence, but it fails to fold into its final, functional form. Researchers have also speculated that chaperone deficiency may be a cause of the protein clumping in brain cells that produces the amyloid plaques characteristic of Alzheimer disease.

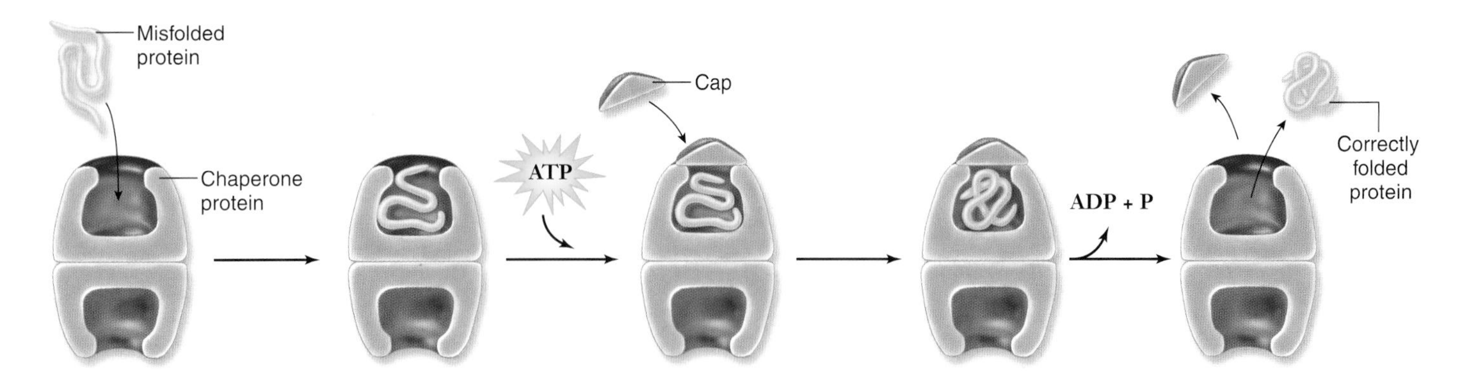

figure 3.24

HOW ONE TYPE OF CHAPERONE PROTEIN WORKS. This barrel-shaped chaperonin is from the GroE family of chaperone proteins. It is composed of two identical rings each with seven identical subunits, each of which has three distinct domains. An incorrectly folded protein enters one chamber of the barrel, and a cap seals the chamber. Energy from the hydrolysis of ATP allows for structural alterations to the chamber, changing it from hydrophobic to hydrophilic. This change allows the protein to refold. After a short time, the protein is ejected, either folded or unfolded, and the cycle can repeat itself.

Denaturation inactivates proteins

If a protein's environment is altered, the protein may change its shape or even unfold completely. This process is called **denaturation** (figure 3.25). Proteins can be denatured when the pH, temperature, or ionic concentration of the surrounding solution is changed.

When proteins are denatured, they are usually rendered biologically inactive. This action is particularly significant in the case of enzymes. Because practically every chemical reaction in a living organism is catalyzed by a specific enzyme, it is vital that a cell's enzymes remain functional.

Denaturation of proteins is involved in the traditional methods of salt curing and pickling: Prior to the general availability of refrigerators and freezers, the only practical way to keep microorganisms from growing in food was to keep the food in a solution containing a high concentration of salt or vinegar, which denatured the enzymes of most microorganisms and prevented them from growing on the food.

Most enzymes function within a very narrow range of physical parameters. Blood-borne enzymes that course through a human body at a pH of about 7.4 would rapidly become denatured in the highly acidic environment of the stomach. Conversely, the protein-degrading enzymes that function at a pH of 2 or less in the stomach would be denatured in the relatively basic pH of the blood. Similarly, organisms that live near oceanic hydrothermal vents have enzymes that work well at the temperature of this extreme environment (over 100°C). They cannot survive in cooler waters, because their enzymes do not function properly at lower temperatures. Any given organism usually has a tolerance range of pH, temperature, and salt concentration. Within that range, its enzymes maintain the proper shape to carry out their biological functions.

When a protein's normal environment is reestablished after denaturation, a small protein may spontaneously refold into its natural shape, driven by the interactions between its nonpolar amino acids and water (figure 3.26). This process is termed **renaturation,** and it was first established for the enzyme ribonuclease (RNase). The renaturation of RNase led to the doctrine that primary structure determines tertiary structure. Larger proteins can rarely refold spontaneously, however,

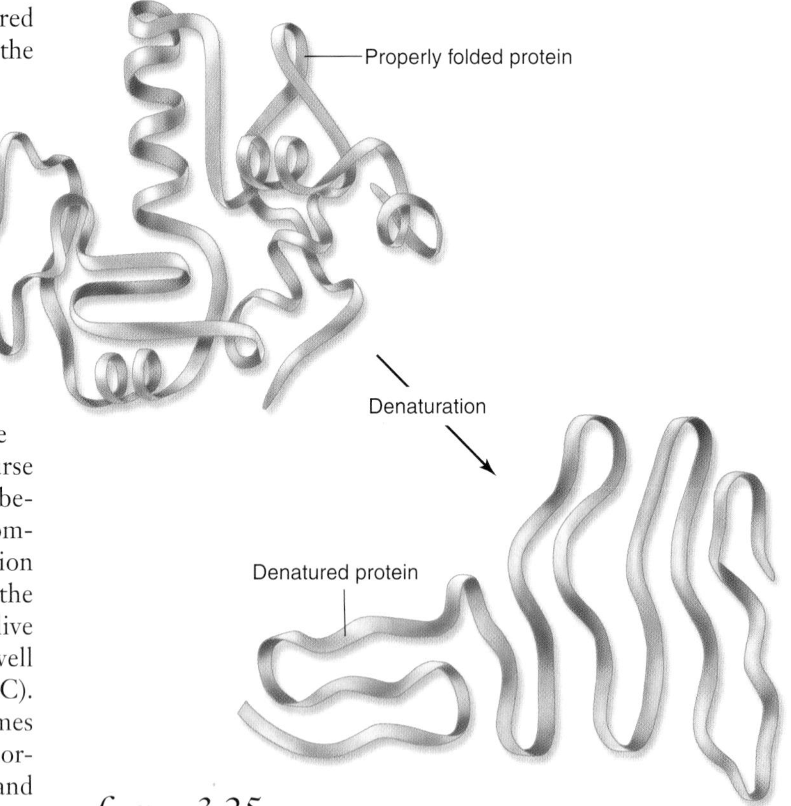

figure 3.25

PROTEIN DENATURATION. Changes in a protein's environment, such as variations in temperature or pH, can cause a protein to unfold and lose its shape in a process called denaturation. In this denatured state, proteins are biologically inactive.

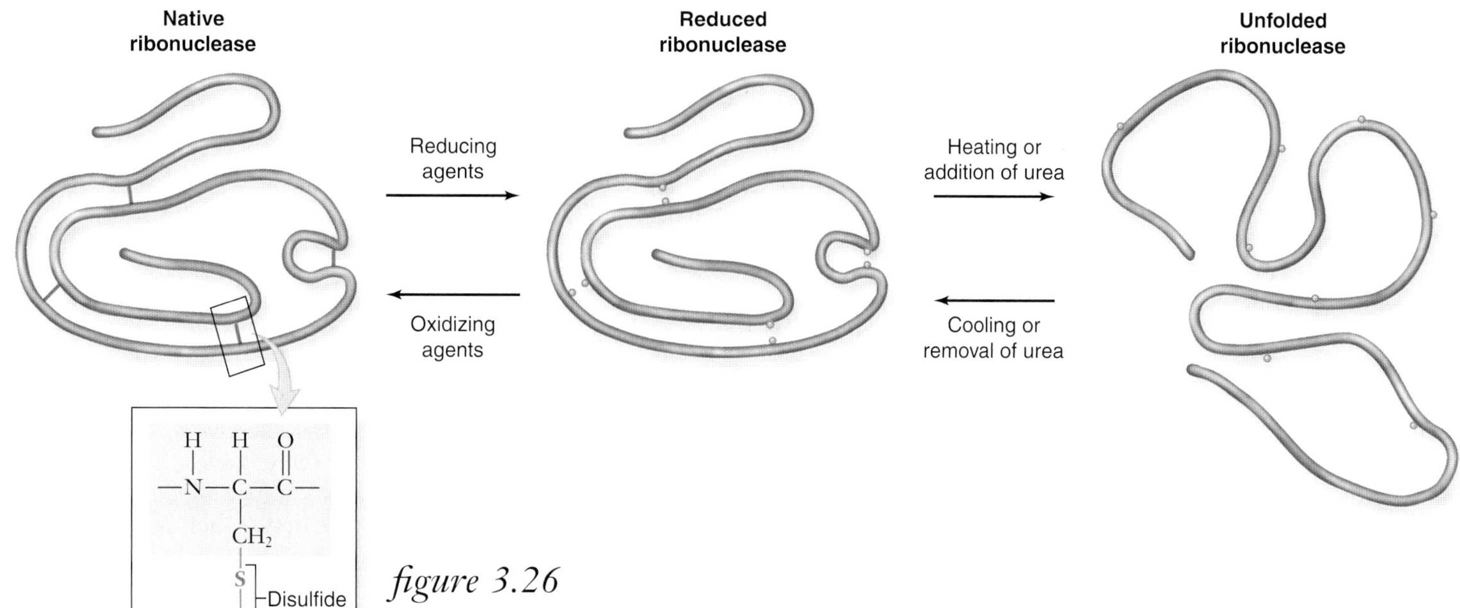

figure 3.26

PRIMARY STRUCTURE DETERMINES TERTIARY STRUCTURE. When the protein ribonuclease is treated with reducing agents to break the covalent disulfide bonds that cross-link its polypeptide chain, and then placed in urea or heated, the protein denatures (unfolds) and loses its enzymatic activity. Upon cooling or removal of urea and treatment with oxidizing agents, it refolds and regains its enzymatic activity, demonstrating that for this protein, no information except the amino acid sequence of the protein is required for proper folding. This is not true of all proteins.

because of the complex nature of their final shape, so this simple dictum needs to be qualified.

The fact that some proteins can spontaneously renature implies that tertiary structure is strongly influenced by primary structure. In an extreme example, the *E. coli* ribosome can be taken apart and put back together experimentally. Although this process requires temperature and ion concentration shifts, it indicates an amazing degree of self-assembly. That complex structures can arise by self-assembly is a key idea in the study of modern biology.

It is important to distinguish denaturation from **dissociation.** For proteins with quaternary structure, the subunits may be dissociated without losing their individual tertiary structure. For example, the four subunits of hemoglobin may dissociate into four individual molecules (two α-globin and two β-globin) without denaturation of the folded globin proteins, and they will readily reassume their four-subunit quaternary structure.

Proteins are a diverse class of macromolecule that perform many different functions. Proteins are made up of 20 different kinds of amino acids. The amino acids fall into five chemical classes, each with different properties that determine the nature of the resulting protein. Protein structure can be viewed at four levels: (1) the amino acid sequence, or primary structure; (2) coils and sheets, called secondary structure; (3) the three-dimensional shape, called tertiary structure; (4) individual polypeptide subunits associated in a quaternary structure. Different proteins often have similar substructures called motifs and can be broken down into functional domains. Proteins have a narrow range of conditions in which they fold properly; outside that range, proteins tend to unfold.

3.5 Lipids: Hydrophobic Molecules

Lipids are a somewhat loosely defined group of molecules with one main chemical characteristic: They are insoluble in water. Storage fats such as animal fat are one kind of lipid. Oils such as olive oil, corn oil, and coconut oil are also lipids, as are waxes such as beeswax and earwax.

Lipids have a very high proportion of nonpolar carbon–hydrogen (C—H) bonds, and so long-chain lipids cannot fold up like a protein to sequester their nonpolar portions away from the surrounding aqueous environment. Instead, when they are placed in water, many lipid molecules spontaneously cluster together and expose what polar (hydrophilic) groups they have to the surrounding water, while sequestering the nonpolar (hydrophobic) parts of the molecules together within the cluster. You may have noticed this effect when you add oil to a pan containing water, and the oil beads up into cohesive drops on the water's surface. This spontaneous assembly of lipids is of paramount importance to cells, as it underlies the structure of cellular membranes.

Fats consist of complex polymers of fatty acids attached to glycerol

Many lipids are built from a simple skeleton made up of two main kinds of molecules: fatty acids and glycerol. Fatty acids are long-chain hydrocarbons with a carboxylic acid (COOH) at one end. Glycerol is a three-carbon polyalcohol (three —OH groups). Many lipid molecules consists of a glycerol molecule with three fatty acids attached, one to each carbon of the glycerol backbone. Because it contains three fatty acids, a fat molecule is commonly called a **triglyceride** (the more accurate chemical name is *triacylglycerol*). This basic structure is depicted in figure 3.27. The three fatty acids of a triglyceride need not be identical, and often they differ markedly from one another. The hydrocarbon chains of fatty acids vary in length; the most common are even-numbered chains of 14 to 20 carbons. The many C—H bonds of fats serve as a form of long-term energy storage.

If all of the internal carbon atoms in the fatty acid chains are bonded to at least two hydrogen atoms, the fatty acid is said to be **saturated,** which refers to its having all the hydrogen atoms it can have (see figure 3.27). A fatty acid that has double bonds between one or more pairs of successive carbon atoms is said to be **unsaturated.** Those fatty acids with more than one double bond are termed **polyunsaturated.**

Having double bonds changes the behavior of the molecule because no free rotation can occur about a C=C double bond as it can with a C—C single bond. This characteristic mainly affects melting point: that is, whether the fatty acid is a solid fat or a liquid oil at room temperature. Fats containing polyunsaturated fatty acids have low melting points because their fatty acid chains bend at the double bonds, preventing

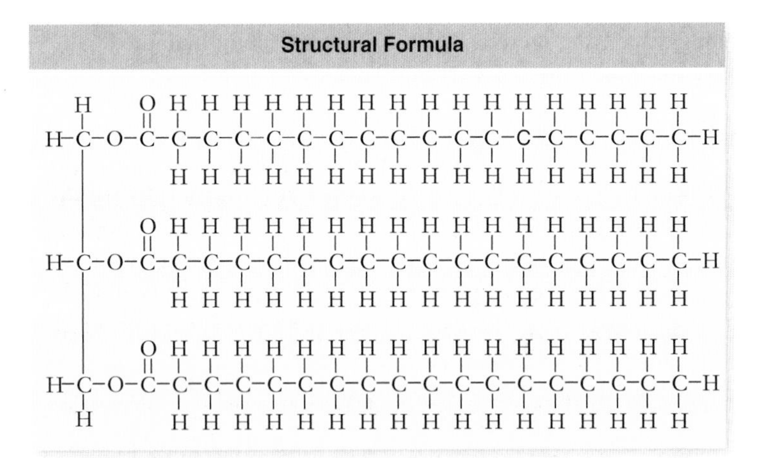

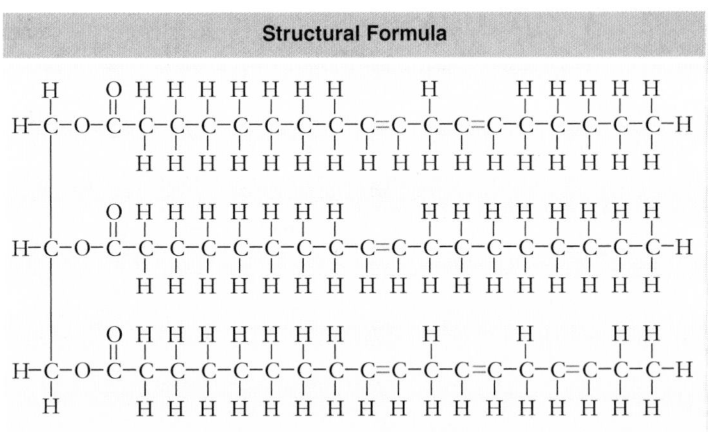

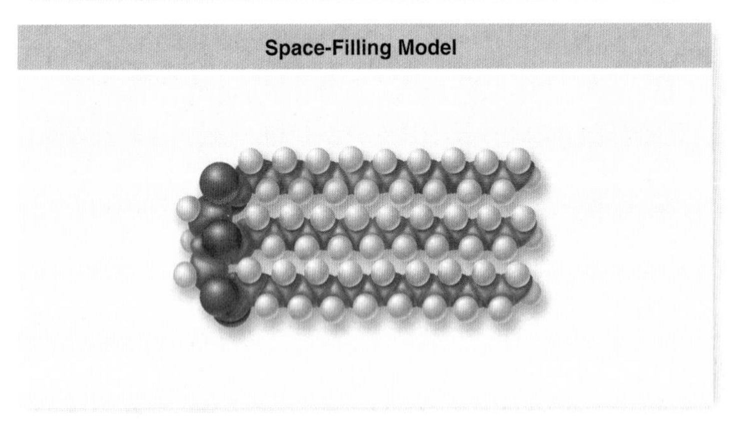

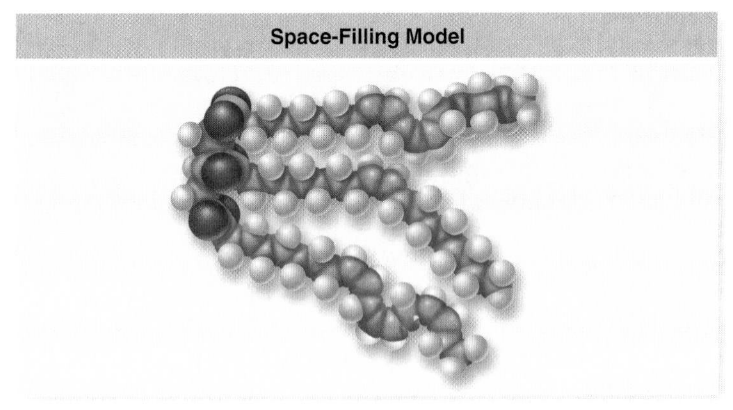

a. *b.*

figure 3.27

SATURATED AND UNSATURATED FATS. ***a.*** A saturated fat is composed of triglycerides that contain three saturated fatty acids, those with no double bonds and, thus, a maximum number of hydrogen atoms bonded to the carbon chain. Many animal fats are saturated. ***b.*** Unsaturated fat is composed of triglycerides, contains three unsaturated fatty acids, that is, they have one or more double bonds and, thus, fewer than the maximum number of hydrogen atoms bonded to the carbon chain. This example includes both a monounsaturated and two polyunsaturated fatty acids. Plant fats are typically unsaturated. The many kinks of the double bonds prevent the triglyceride from closely aligning, producing liquid oils at room temperature.

the fat molecules from aligning closely with one another. Most saturated fats, such as animal fat or those in butter, are solid at room temperature.

Placed in water, triglycerides spontaneously associate together, forming fat globules that can be very large relative to the size of the individual molecules. Because fats are insoluble in water, they can be deposited at specific locations within an organism, such as in vesicles of adipose tissue.

Organisms contain many other kinds of lipids besides fats (figure 3.28). *Terpenes* are long-chain lipids that are components of many biologically important pigments, such as chlorophyll and the visual pigment retinal. Rubber is also a terpene. *Steroids*, another class of lipid, are composed of four carbon rings. Most animal cell membranes contain the steroid cholesterol. Other steroids, such as testosterone and estrogen, function as hormones in multicellular animals. *Prostaglandins* are a group of about 20 lipids that are modified fatty acids, with two nonpolar "tails" attached to a five-carbon ring. Prostaglandins act as local chemical messengers in many vertebrate tissues. Later chapters explore the effects of some of these complex fatty acids.

Fats are excellent energy-storage molecules

Most fats contain over 40 carbon atoms. The ratio of energy-storing C—H bonds in fats is more than twice that of carbohydrates (see section 3.2), making fats much more efficient molecules for storing chemical energy. On average, fats yield about 9 kilocalories (kcal) of chemical energy per gram, as compared with about 4 kcal per gram for carbohydrates.

Most fats produced by animals are saturated (except some fish oils), whereas most plant fats are unsaturated (see figure 3.27). The exceptions are the tropical plant oils (palm oil and coconut oil), which are saturated despite their fluidity at room temperature. An oil may be converted into a solid fat by chemically adding hydrogen. Peanut butter sold in stores is usually artificially hydrogenated to make the peanut fats solidify, preventing them from separating out as oils while the jar sits on the store shelf. However, artificially hydrogenating unsaturated fats seems to eliminate the health advantage they have over saturated fats. The hydrogenation reaction produces *trans*-fatty acids that appear to increase the level of cholesterol carried in blood. Therefore, it is currently thought that margarine, made from hydrogenated corn oil, is no better for your health than butter.

When an organism consumes excess carbohydrate, it is converted into starch, glycogen, or fats reserved for future use. The reason that many humans in developed countries gain weight as they grow older is that the amount of energy they need decreases with age, but their intake of food does not. Thus, an increasing proportion of the carbohydrates they ingest is available to be converted into fat.

A diet rich in fats is one of several factors that are thought to contribute to heart disease, particularly atherosclerosis, a condition in which deposits of fatty substances called plaque adhere to the lining of blood vessels, blocking the flow of blood. Fragments of plaque, breaking off from a deposit and clogging arteries to the brain, are a major cause of strokes.

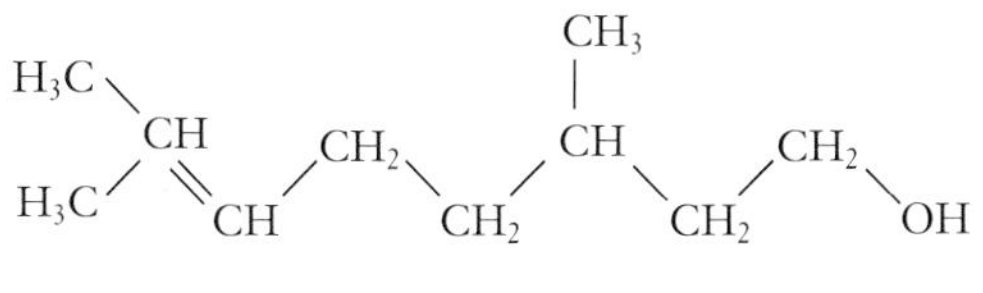

a. Terpene (citronellol)

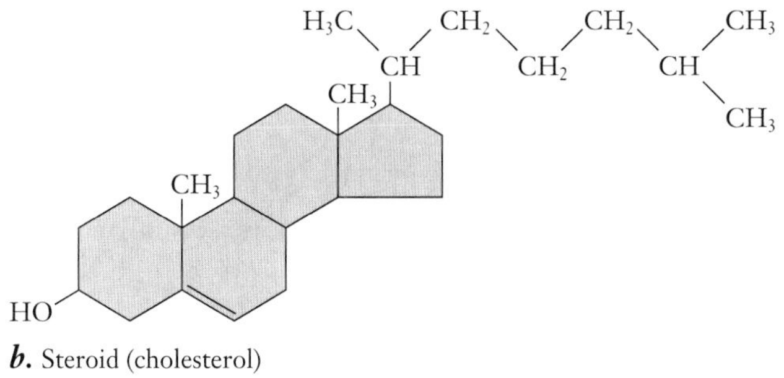

b. Steroid (cholesterol)

figure 3.28

OTHER KINDS OF LIPIDS. ***a.*** Terpenes are found in biological pigments, such as chlorophyll and retinal, and ***(b)*** steroids play important roles in membranes and as the basis for a class of hormones involved in chemical signaling.

Phospholipids form membranes

Complex lipid molecules called **phospholipids** are among the most important molecules of the cell because they form the core of all biological membranes. An individual phospholipid can be thought of as a substituted triglyceride, that is, a triglyceride with a phosphate replacing one of the fatty acids. The basic structure of a phospholipid includes three kinds of subunits:

1. *Glycerol*, a three-carbon alcohol, with each carbon bearing a hydroxyl group. Glycerol forms the backbone of the phospholipid molecule.
2. *Fatty acids*, long chains of $—CH_2$ groups (hydrocarbon chains) ending in a carboxyl (—COOH) group. Two fatty acids are attached to the glycerol backbone in a phospholipid molecule.
3. *A phosphate group* ($—PO_4^{2-}$) attached to one end of the glycerol. The charged phosphate group usually has a charged organic molecule linked to it, such as choline, ethanolamine, or the amino acid serine.

The phospholipid molecule can be thought of as having a polar "head" at one end (the phosphate group) and two long, very nonpolar "tails" at the other (figure 3.29). This structure is essential for these molecules' function, although it first appears paradoxical. Why would a molecule need to be soluble in water, but also not soluble in water? The formation of a membrane shows the unique properties of such a structure.

In water, the nonpolar tails of nearby lipid molecules aggregate away from the water, forming spherical *micelles*, with the tails facing inward (figure 3.30*a*). This is actually how detergent molecules work to make grease soluble in water. The grease is soluble within the nonpolar interior of the micelle and the polar surface of the micelle is soluble in water. With phospholipids, a

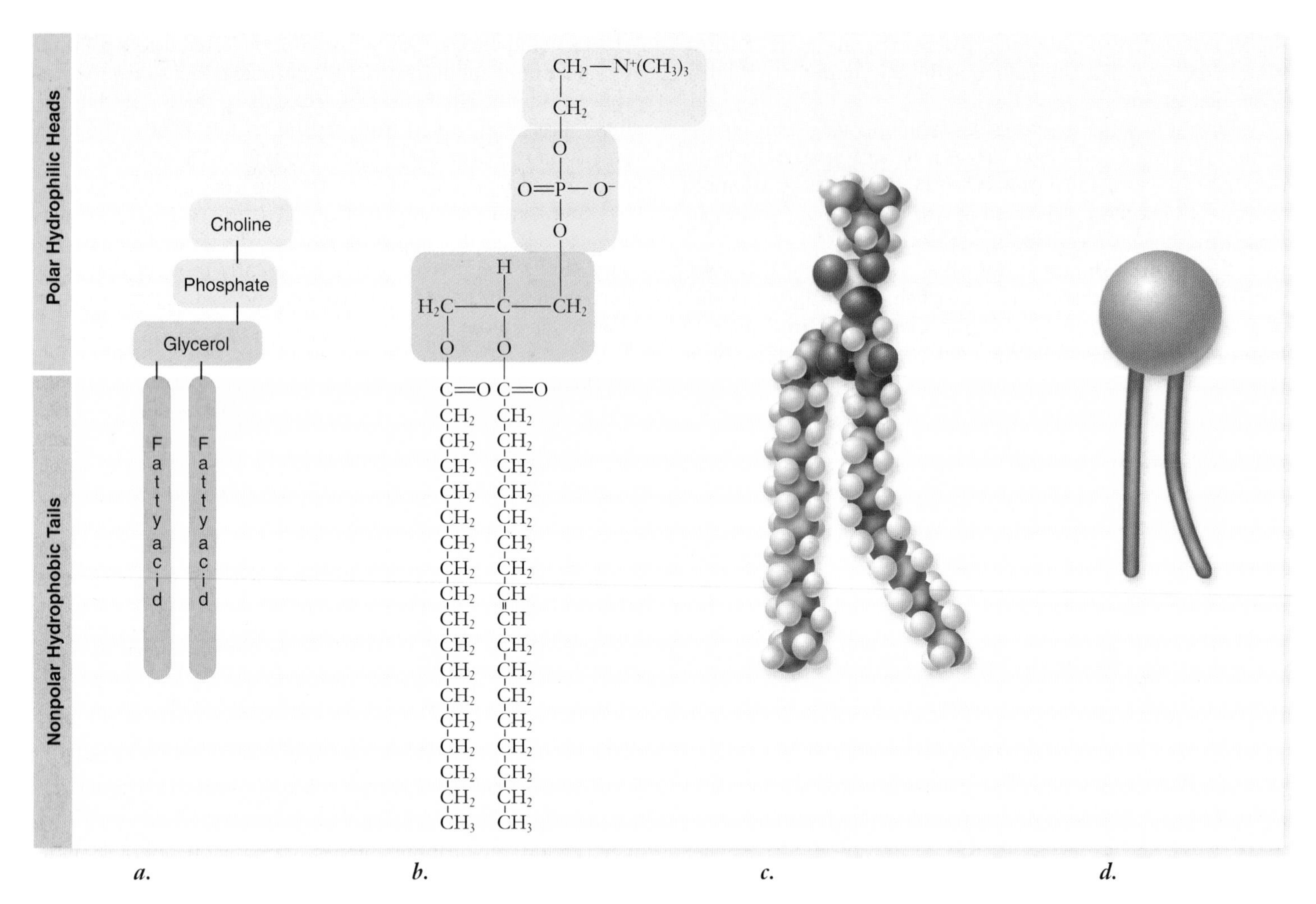

figure 3.29

PHOSPHOLIPIDS. The phospholipid phosphatidylcholine is shown as ***(a)*** a schematic, ***(b)*** a formula, ***(c)*** a space-filling model, and ***(d)*** an icon used in depictions of biological membranes.

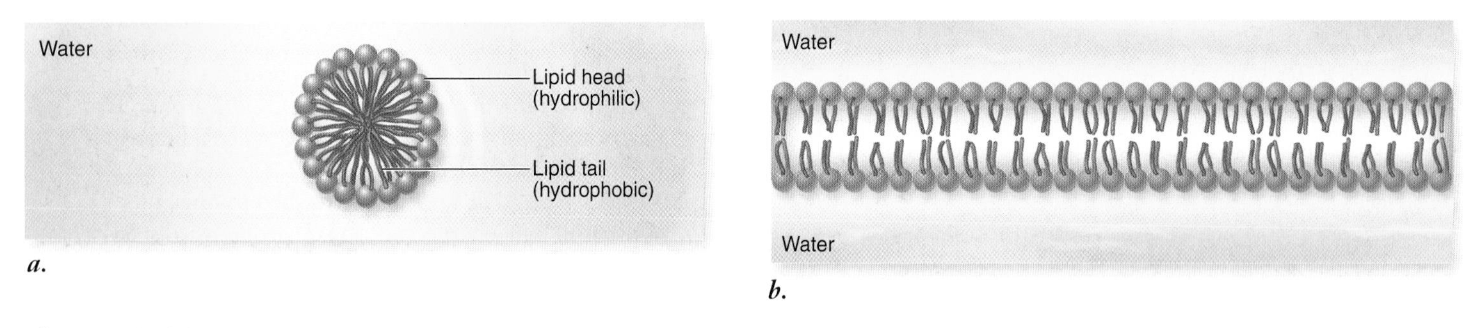

figure 3.30

LIPIDS SPONTANEOUSLY FORM MICELLES OR LIPID BILAYERS IN WATER. In an aqueous environment, lipid molecules orient so that their polar (hydrophilic) heads are in the polar medium, water, and their nonpolar (hydrophobic) tails are held away from the water. ***a.*** Droplets called micelles can form, or ***(b)*** phospholipid molecules can arrange themselves into two layers; in both structures, the hydrophilic heads extend outward and the hydrophobic tails inward. This second example is called a phospholipid bilayer.

more complex structure forms in which two layers of molecules line up, with the hydrophobic tails of each layer pointing toward one another, or inward, leaving the hydrophilic heads oriented outward, forming a bilayer (figure 3.30*b*). Lipid bilayers are the basic framework of biological membranes, discussed in detail in chapter 5.

Triglycerides are made of fatty acids and glycerol. Cells also contain a variety of different lipids that play important roles in cell metabolism. Because the C—H bonds in lipids are nonpolar, they are not water-soluble, and aggregate together in water. This kind of aggregation by phospholipids forms biological membranes.

concept review

3.1 Carbon: The Framework of Biological Molecules

Carbon forms the backbone of all biological molecules. Carbon can be arranged into chains and rings and is used in combination with other atoms to form biological molecules.

- Carbon can form four covalent bonds.
- Hydrocarbons consist of carbon and hydrogen, and their bonds store considerable energy.
- Biological molecules are built using functional groups that confer specific chemical properties.
- Carbon and hydrogen have similar electronegativity so C—H bonds are not polar.
- Oxygen and nitrogen have greater electronegativity than carbon and hydrogen leading to polar bonds.
- Isomers are molecules that have the same formula but different structure.
- Structural isomers differ in actual structure, whereas stereoisomers differ in how the structural groups are attached.
- Enantiomers are stereoisomers that are mirror images of each other and they rotate the plane of polarized light.
- Polymers are long chains of similar chemical subunits or monomers.
- Most important biological macromolecules are polymers.
- Biological polymers are formed by elimination of water, or a dehydration reaction.
- Biological polymers can be broken down by adding water, or a hydrolysis reaction.

3.2 Carbohydrates: Energy Storage and Structural Molecules

The empirical formula of a carbohydrate is $(CH_2O)_n$. Carbohydrates are used for energy storage and as structural molecules.

- Simple sugars contain three to six carbon atoms and exist as structural isomers and stereoisomers.
- Monosaccharides contain one subunit, disaccharides contain two and polysaccharides contain more than two.
- Glucose is used to make three important polymers: starch, glycogen, and cellulose.
- Starch and glycogen are branched polymers of α-glucose made by plants and animals, respectively, used for energy storage.
- Cellulose is an unbranched polymer of β-glucose made by plants for cell walls.

3.3 Nucleic Acids: Information Molecules

Nucleic acids are polymers formed by phosphodiester bonds between nucleotides. Nucleic acid molecules are used for information storage.

- DNA and RNA are polymers composed of nucleotide monomers.
- DNA uses the sugar deoxyribose, and RNA uses the sugar ribose.
- Nucleic acid polymers contain four different nucleotides. In DNA these include the bases adenine, guanine, cytosine, and thymine. In RNA, thymine is replaced by uracil.
- DNA exists as a double helix, but RNA is usually a folded single chain.
- DNA is stabilized by hydrogen bonds between bases. These form specific base pairs: adenine with thymine and guanine with cytosine.
- DNA encodes the information for amino acid sequences in proteins using four different nucleotides.
- RNA is made by copying DNA and is used to make proteins.
- Adenosine triphosphate (ATP) is a nucleotide that is also used to provide energy in cells.
- The nucleotides NAD^+ and FAD are used to transport electrons.

3.4 Proteins: Molecules with Diverse Structures and Functions

Proteins are a structurally diverse category of molecules. Proteins are made from amino acid monomers. Proteins carry out a diverse array of functions.

- Most enzymes are proteins that act as catalysts for metabolic reactions.
- Proteins defend our bodies, transport gases and ions, provide structure, contract and provide motion, receive information, regulate cell activities, and store bound ions.
- Proteins are linear polymers of 20 different kinds of amino acids.
- Amino acids are joined by peptide bonds to make polypeptides.
- The 20 common amino acids are characterized by R groups that may be polar, nonpolar, or charged.
- Protein structure is defined by the following hierarchy: primary, secondary, tertiary, and quaternary.
- Primary structure refers to the amino acid sequence. Secondary structure is based on hydrogen-bonding patterns that can create helices or planar sheets. Tertiary structure refers to the three-dimensional folded shape, and quaternary structure is formed by the association of two or more polypeptides.
- Motifs are similar structures found in dissimilar proteins. Domains are functional subunits within a tertiary structure.
- Chaperonins assist in the folding of proteins. Chaperone deficiencies may cause disease.
- Denaturation refers to an unfolding of tertiary structure. Disassociation refers to separation of quaternary subunits with no changes to their tertiary structure.

3.5 Lipids: Hydrophobic Molecules

Lipids are composed of fatty acids and glycerol and are insoluble in water. Lipids are long-term energy storage molecules. Phospholipids form the basis for biological membranes.

- Fatty acids can be saturated or unsaturated.
- Saturated fatty acids contain the maximum number of hydrogen atoms.
- Unsaturated fatty acids contain one or more double bonds between carbon atoms.
- Phospholipids contain two fatty acids and one phosphate attached to glycerol. The phosphate head is hydrophilic, and the lipid tail is hydrophobic.
- Membranes composed of phospholipids have the hydrophilic heads of each layer directed to the outside of the membrane, and the hydrophobic tails directed toward the center.

SELF TEST

1. How is a polymer formed from multiple monomers?
   a. From the growth of the chain of carbon atoms
   b. By the removal of an —OH group and a hydrogen atom
   c. By the addition of an —OH group and a hydrogen atom
   d. Through hydrogen bonding
2. Why are carbohydrates important molecules for energy storage?
   a. The C—H bonds found in carbohydrates store energy
   b. The double bonds between carbon and oxygen are very strong
   c. The electronegativity of the oxygen atoms means that a carbohydrate is made up of many polar bonds
   d. They can form ring structures in the aqueous environment of a cell
3. Plant cells store energy in the form of ________, and animal cells store energy in the form of __________
   a. fructose; glucose
   b. disaccharides; monosaccharides
   c. cellulose; chitin
   d. starch; glycogen
4. Which carbohydrate would you find as part of a molecule of RNA?
   a. Galactose
   b. Deoxyribose
   c. Ribose
   d. Glucose
5. What makes cellulose different from starch?
   a. Starch is produced by plant cells, and cellulose is produced by animal cells.
   b. Cellulose forms long filaments, and starch is highly branched.
   c. Starch is insoluble, and cellulose is soluble.
   d. All of the above.
6. A molecule of DNA or RNA is a polymer of—
   a. monosaccharides
   b. nucleotides
   c. amino acids
   d. fatty acids
7. What chemical bond is responsible for linking amino acids together to form a protein?
   a. Phosphodiester
   b. β-1,4 Linkage
   c. Peptide
   d. Hydrogen
8. The double helix structure of a molecule of DNA is stabilized by—
   a. phosphodiester bonds
   b. peptide bonds
   c. an α helix
   d. hydrogen bonds
9. Which of the following is NOT a difference between DNA and RNA?
   a. Deoxyribose sugar versus ribose sugar
   b. Thymine versus uracil
   c. Double-stranded versus single-stranded
   d. Phosphodiester versus hydrogen bonds
10. What monomers make up a protein?
    a. Monosaccharides
    b. Nucleotides
    c. Amino acids
    d. Fatty acids
11. Which part of an amino acid has the greatest influence on the overall structure of a protein?
    a. The (—NH_2) amino group
    b. The R-group
    c. The (—COOH) carboxyl group
    d. Both a and c
12. A mutation that alters a single amino acid within a protein can alter—
    a. the primary level of protein structure
    b. the secondary level of protein structure
    c. the tertiary level of protein structure
    d. all of the above
13. Which of these factors contributes to the diversity of protein form and function in the cell?
    a. Quaternary interactions between peptides
    b. Formation of α helices and β-pleated sheets
    c. The linear sequence of amino acids that makes up the polymer
    d. All of the above
14. What chemical property of lipids accounts for their insolubility in water?
    a. The length of the carbon chain
    b. The large number of nonpolar C—H bonds
    c. The branching of saturated fatty acids
    d. The C=C bonds found in unsaturated fatty acids
15. The spontaneous formation of a lipid bilayer in an aqueous environment occurs because—
    a. The polar head groups of the phospholipids can interact with water.
    b. The long fatty acid tails of the phospholipids can interact with water.
    c. The fatty acid tails of the phospholipids are hydrophobic.
    d. Both a and c.

CHALLENGE QUESTIONS

1. Spider webs are made of "silk," which is a long, fibrous protein. The threads you see in a web are actually composed of many individual proteins. One important structural motif within spider's silk protein is a "β-crystal". β-Crystals are regions where the β-pleated sheets from the multiple individual protein fibers stack one upon the other. What chemical bonds are required for the formation of the β-crystals? What level of protein structure is responsible for the formation of the silk you see in the web? Predict how the presence of the β-crystal motif would influence the physical properties of the silk protein.
2. How do the four biological macromolecules differ from one another? Refer to the diagram of the monomer structure in figure 3.3 and summarize what "clues" you use to distinguish between these important molecules.
3. Hydrogen bonds play an important role in stabilizing and organizing biological macromolecules. Consider the four macromolecules discussed in this chapter. Describe three examples where hydrogen bond formation affects the form or function of the macromolecule.
4. The cells of your body are distinct even though they all contain the same genetic information. Use the information in table 3.2 to develop an explanation for the diversity of specialized cellular structure and function found within your body.

Do you need additional review? *Visit* www.ravenbiology.com *for practice quizzes, animations, videos, and activities designed to help you master the material in this chapter.*

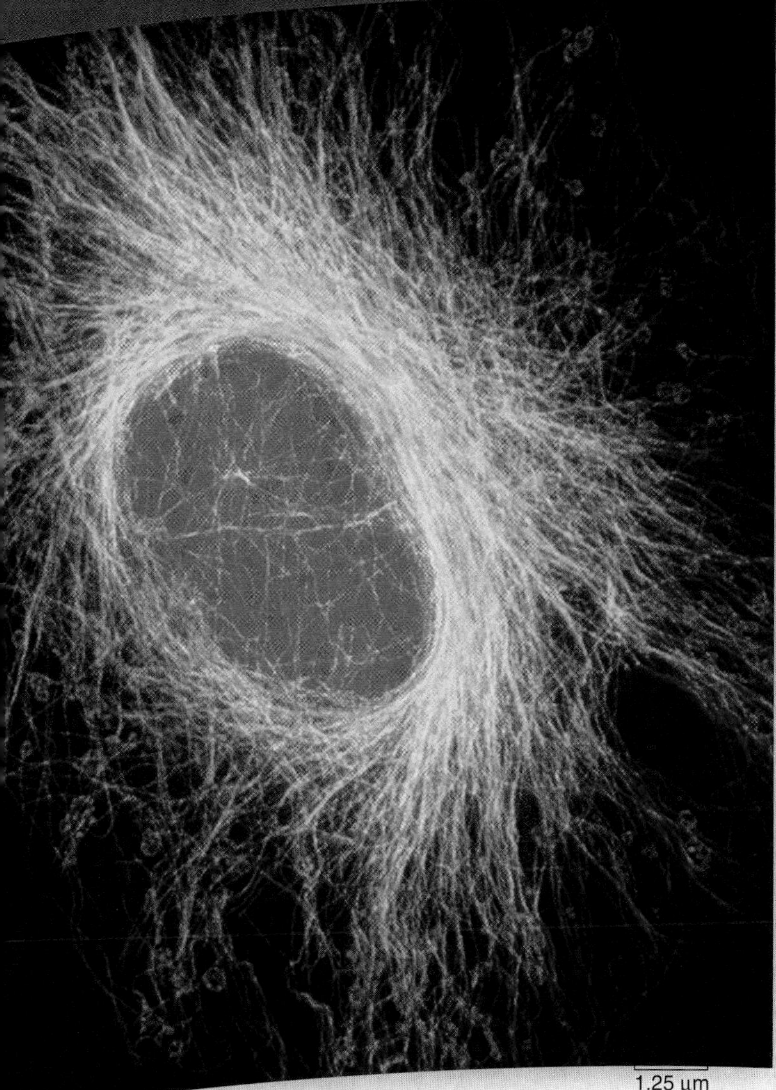

chapter 4

Cell Structure

introduction

ALL ORGANISMS ARE COMPOSED OF CELLS. The gossamer wing of a butterfly is a thin sheet of cells and so is the glistening outer layer of your eyes. The hamburger or tomato you eat is composed of cells, and its contents soon become part of your cells. Some organisms consist of a single cell too small to see with the unaided eye, while others, such as humans, are composed of many specialized cells, such as the fibroblast cell shown in the striking fluorescence micrograph on this page. Cells are so much a part of life that we cannot imagine an organism that is not cellular in nature. In this chapter, we take a close look at the internal structure of cells. In chapters 5 to 10, we will focus on cells in action—how they communicate with their environment, grow, and reproduce.

concept outline

4.1 Cell Theory

A general characteristic of cells is their microscopic size. Although there are exceptions, a typical eukaryotic cell is 10 to 100 micrometers (μm) (10 to 100 millionths of a meter) in diameter, most prokaryotic cells are only 1 to 10 μm in diameter.

Because cells are so small, their discovery did not occur until the invention of the microscope in the seventeenth century. Robert Hooke was the first to observe cells in 1665, naming the shapes he saw in cork *cellulae* (Latin, "small rooms"). This comes down to us as *cells.* Another early microscopist, Anton van Leeuwenhoek first observed live cells, which he termed tiny "animalcules." After these early efforts, a century and a half passed before biologists fully recognized the importance of cells. In 1838, botanist Matthias Schleiden stated that all plants "are aggregates of fully individualized, independent, separate beings, namely the cells themselves." In 1839, Theodor Schwann reported that all animal tissues also consist of individual cells. Thus was born cell theory.

Cell theory is the unifying foundation of cell biology

The cell theory was proposed to explain the observation that all organisms are composed of cells. While it sounds simple, it is a far-reaching statement about the organization of life.

In its modern form, **cell theory** includes the following three principles:

1. All organisms are composed of one or more cells, and the life processes of metabolism and heredity occur within these cells.
2. Cells are the smallest living things, the basic units of organization of all organisms.
3. Cells arise only by division of a previously existing cell.

Although life likely evolved spontaneously in the environment of early Earth, biologists have concluded that no additional cells are originating spontaneously at present. Rather, life on Earth represents a continuous line of descent from those early cells.

Cell size is limited

Most cells are relatively small for reasons related to the diffusion of substances into and out of cells. The rate of diffusion is affected by a number of variables, including surface area available for diffusion, temperature, concentration gradient of diffusing substance, and the distance over which diffusion must occur. As the size of a cell increases, the length of time for diffusion from the membrane to the interior of the cell increases as well. Larger cells need to synthesize more macromolecules, have correspondingly higher energy requirements, and produce a greater quantity of waste. Molecules used for energy and biosynthesis must be transported through the membrane. Any metabolic waste produced must be removed, also passing through the membrane. The rate at which this transport occurs depends on both the distance to the membrane, as well as the area of membrane available. For this reason, an organism made up of many relatively small cells has an advantage over one composed of fewer, larger cells.

The advantage of small cell size is readily apparent in terms of the **surface area-to-volume ratio.** As a cell's size increases, its volume increases much more rapidly than its surface area. For a spherical cell, the surface area is proportional to the square of the radius, whereas the volume is proportional to the cube of the radius. Thus, if two cells differ by a factor of 10 in radius, the larger cell will have 10^2, or 100 times, the surface area, but 10^3, or 1000 times, the volume of the smaller cell (figure 4.1).

The cell surface provides the only opportunity for interaction with the environment, because all substances enter and exit a cell via this surface. The membrane surrounding the cell plays a key role in controlling cell function, and because small cells have more surface area per unit of volume than large ones, the control is more effective when cells are relatively small.

Although most cells are small, some quite large cells do exist, and these have apparently overcome the surface-area-to-volume problem by one or more adaptive mechanisms. For example, some cells, such as skeletal muscle cells, have more than one nucleus, allowing genetic information to be spread around a large cell. Some other large cells, such as neurons, are long and skinny, so that any given point in the cytoplasm is close to the plasma membrane, and thus diffusion between the inside and outside of the cell can still be rapid.

Microscopes allow visualization of cells and components

Other than egg cells, not many cells are visible to the naked eye (figure 4.2). Most are less than 50 μm in diameter, far smaller than the period at the end of this sentence. So, to visualize cells we

Cell radius (r)	1 unit	10 unit
Surface area ($4\pi r^2$)	12.57 unit^2	1257 unit^2
Volume ($\frac{4}{3}\pi r^3$)	4.189 unit^3	4189 unit^3
Surface Area / Volume	3	0.3

figure 4.1

SURFACE-AREA-TO-VOLUME RATIO. As a cell gets larger, its volume increases at a faster rate than its surface area. If the cell radius increases by 10 times, the surface area increases by 100 times, but the volume increases by 1000 times. A cell's surface area must be large enough to meet the metabolic needs of its volume.

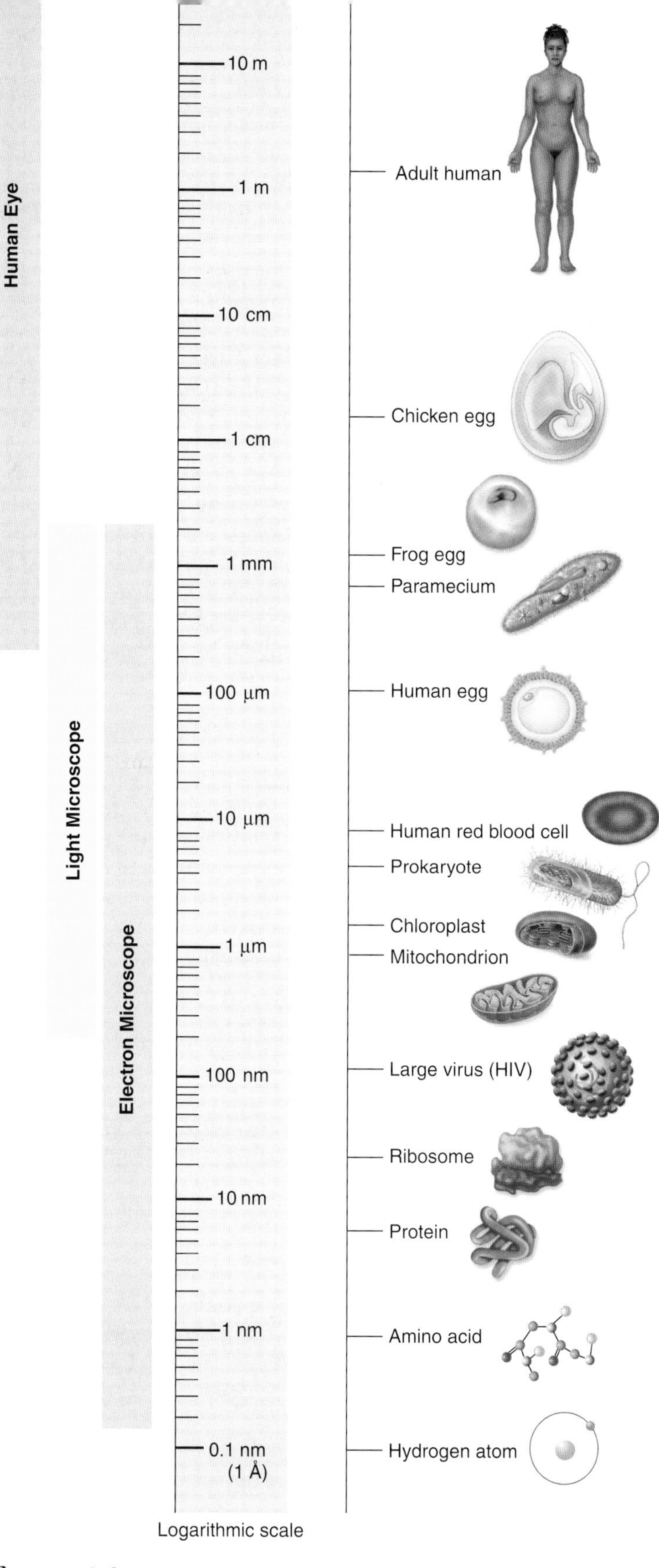

figure 4.2

THE SIZE OF CELLS AND THEIR CONTENTS. Most cells are microscopic in size, although vertebrate eggs are typically large enough to be seen with the unaided eye. Prokaryotic cells are generally 1 to 10 µm across.
1 m = 10^2 cm = 10^3 mm = 10^6 µm = 10^9 nm

need the aid of technology. The development of microscopes and their refinement has allowed us to explore cells.

The resolution problem

How do we study cells if they are too small to see? The key is to understand why we can't see them. The reason we can't see such small objects is the limited resolution of the human eye. **Resolution** is defined as the minimum distance two points can be apart and still be distinguished as two separated points. When two objects are closer together than about 100-µm, the light reflected from each strikes the same photoreceptor cell at the rear of the eye. Only when the objects are farther than 100 µm apart will the light from each strike different cells, allowing your eye to resolve them as two objects rather than one.

Types of microscopes

One way to increase resolution is to increase magnification so that small objects appear larger. The first microscopists used glass lenses to magnify small cells and cause them to appear larger than the 100-µm limit imposed by the human eye. The glass lens adds increased focusing power. Because the glass lens makes the object appear closer, the image on the back of the eye is bigger than it would be without the lens.

Modern **light microscopes,** which operate with visible light, use two magnifying lenses (and a variety of correcting lenses) to achieve very high magnification and clarity (table 4.1). The first lens focuses the image of the object on the second lens, which magnifies it again and focuses it on the back of the eye. Microscopes that magnify in stages using several lenses are called **compound microscopes.** They can resolve structures that are separated by at least 200 nanometers (nm).

Light microscopes, even compound ones, are not powerful enough to resolve many of the structures within cells. For example, a cell membrane is only 5 nm thick. Why not just add another magnifying stage to the microscope to increase its resolving power? The reason is that when two objects are closer than a few hundred nanometers, the light beams reflecting from the two images start to overlap each other. The only way two light beams can get closer together and still be resolved is if their wavelengths are shorter. One way to avoid overlap is by using a beam of electrons rather than a beam of light. Electrons have a much shorter wavelength, and an **electron microscope,** employing electron beams, has 1000 times the resolving power of a light microscope. **Transmission electron microscopes,** so called because the electrons used to visualize the specimens are transmitted through the material, are capable of resolving objects only 0.2 nm apart—just twice the diameter of a hydrogen atom!

A second kind of electron microscope, the **scanning electron microscope,** beams the electrons onto the surface of the specimen. The electrons reflected back from the surface, together with other electrons that the specimen itself emits as a result of the bombardment, are amplified and transmitted to a screen, where the image can be viewed and photographed. Scanning electron microscopy yields striking three-dimensional images, and it has improved our understanding of many biological and physical phenomena (see table 4.1).

TABLE 4.1 Microscopes

LIGHT MICROSCOPES

Bright-field microscope:
Light is simply transmitted through a specimen, giving little contrast. Staining specimens improves contrast but requires that cells be fixed (not alive), which can cause distortion or alteration of components.

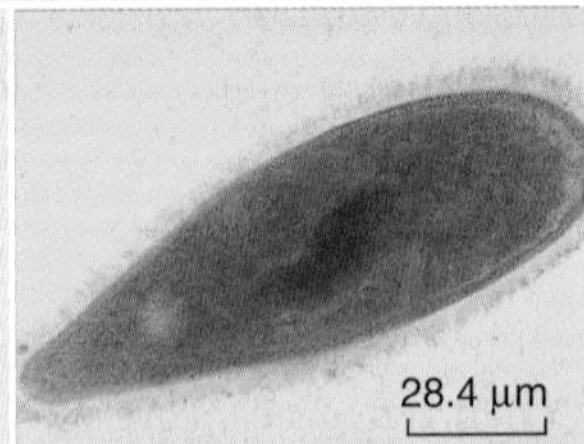

Dark-field microscope:
Light is directed at an angle toward the specimen; a condenser lens transmits only light reflected off the specimen. The field is dark, and the specimen is light against this dark background.

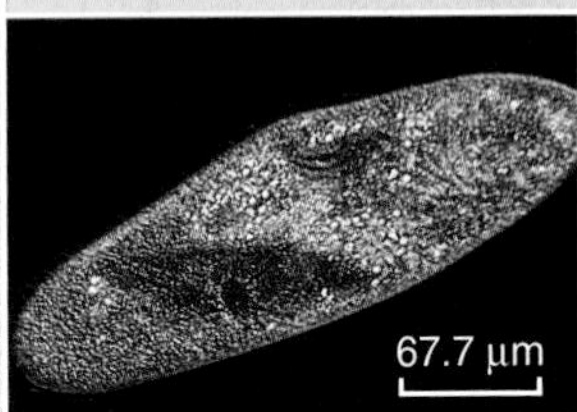

Phase-contrast microscope:
Components of the microscope bring light waves out of phase, which produces differences in contrast and brightness when the light waves recombine.

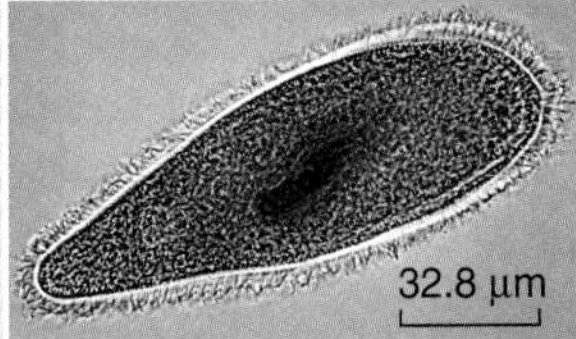

Differential-interference–contrast microscope:
Out-of-phase light waves to produce differences in contrast are combined with two beams of light traveling close together, which create even more contrast, especially at the edges of structures.

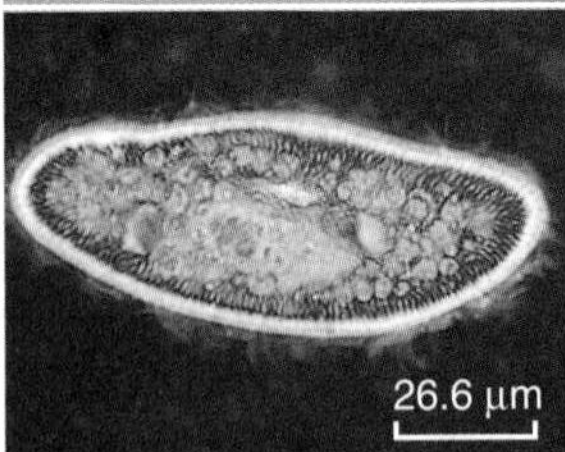

Fluorescence microscope:
Fluorescent stains absorb light at one wavelength, then emit it at another. Filters transmit only the emitted light.

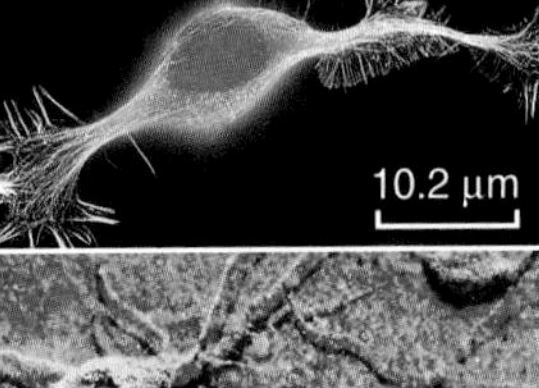

Confocal microscope:
Light from a laser is focused to a point and scanned across the fluorescently stained specimen in two directions. Clear images of one plane of the specimen are produced, while other planes of the specimen are excluded and do not blur the image. Multiple planes can be used to reconstruct a 3-D image.

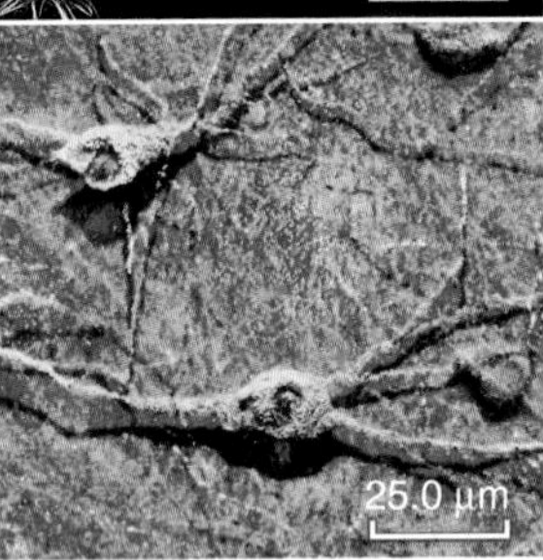

ELECTRON MICROSCOPES

Transmission electron microscope:
A beam of electrons is passed through the specimen. Electrons that pass through are used to expose film. Areas of the specimen that scatter electrons appear dark. False coloring enhances the image.

Scanning electron microscope:
An electron beam is scanned across the surface of the specimen, and electrons are knocked off the surface. Thus, the surface topography of the specimen determines the contrast and the content of the image. False coloring enhances the image.

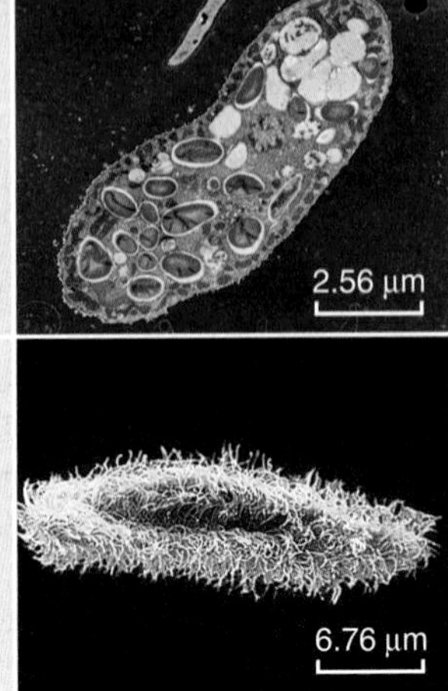

Using stains to view cell structure

Although resolution remains a physical limit, we can improve the images we see by altering the sample. Certain chemical stains increase the contrast between different cellular components. Structures within the cell will absorb or exclude the stain, producing contrast that aids resolution.

Staining techniques have become even more powerful with the use of stains that bind to specific types of molecules. This method employs antibodies that bind, for example, to a particular protein. This process, called *immunohistochemistry*, uses antibodies generated in animals such as rabbits or mice. When these animals are injected with specific proteins, they produce antibodies that bind to the injected protein, and the antibodies can be purified from their blood. These purified antibodies can then be chemically bonded to enzymes, to stains, or to fluorescent molecules. When cells are incubated in a solution containing the antibodies, the antibodies bind to cellular structures that contain the target molecule and can be seen with light microscopy. This approach has been used extensively in the analysis of cell structure and function.

All cells exhibit basic structural similarities

The general plan of cellular organization varies in the cells of different organisms, but despite these modifications, all cells resemble one another in certain fundamental ways. Before we begin a detailed examination of cell structure, let's first summarize four major features all cells have in common: a nucleoid or nucleus where genetic material is located, cytoplasm, *ribosomes* to synthesize proteins, and a plasma membrane.

Centrally located genetic material

Every cell contains DNA, the hereditary molecule. In **prokaryotes,** the simplest organisms, most of the genetic material lies in a single circular molecule of DNA. It typically resides near the center of the cell in an area called the **nucleoid,** but this area is not segregated from the rest of the cell's interior by membranes.

By contrast, the DNA of eukaryotes, which are more complex organisms, is contained in the nucleus, which is surrounded by a double membrane structure called the nuclear envelope. In both types of organisms, the DNA contains the genes that code for the proteins synthesized by the cell. (Details of nucleus structure are described later in the chapter.)

The cytoplasm

A semifluid matrix called the **cytoplasm** fills the interior of the cell. The cytoplasm contains all of the sugars, amino acids, and proteins the cell uses to carry out its everyday activities. Although it is an aqueous medium, cytoplasm is more like jello than water due to the high concentration of proteins and other macromolecules. In eukaryotic cells, in addition to the nucleus, the cytoplasm also contains specialized membrane-bounded compartments called **organelles.** The part of the cytoplasm that contains organic molecules and ions in solution is called the **cytosol** to distinguish it from the larger organelles suspended in this fluid.

The plasma membrane

The **plasma membrane** encloses a cell and separates its contents from its surroundings. The plasma membrane is a phospholipid bilayer about 5 to 10 nm (5 to 10 billionths of a meter) thick, with proteins embedded in it. Viewed in cross section with the electron microscope, such membranes appear as two dark lines separated by a lighter area. This distinctive appearance arises from the tail-to-tail packing of the phospholipid molecules that make up the membrane (see chapter 5).

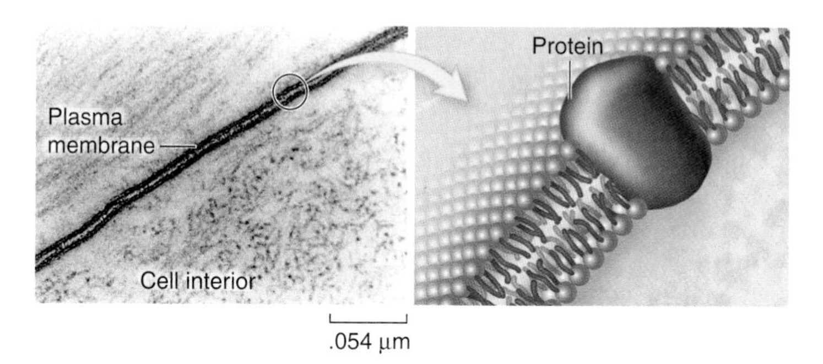

The proteins of the plasma membrane are in large part responsible for a cell's ability to interact with the environment. *Transport proteins* help molecules and ions move across the plasma membrane, either from the environment to the interior of the cell or vice versa. *Receptor proteins* induce changes within the cell when they come in contact with specific molecules in the environment, such as hormones, or with molecules on the surface of neighboring cells. These molecules can function as *markers* that identify the cell as a particular type. This interaction between cell surface molecules is especially important in multicellular organisms, whose cells must be able to recognize one another as they form tissues.

We'll examine the structure and function of cell membranes more thoroughly in chapter 5.

All organisms are cells or aggregates of cells. All cells arise from preexisting cells. Multicellular organisms usually consist of many small cells rather than a few large ones because of the limits of diffusion of small molecules into the cell.

Most cells and their components are so small they can only be viewed using microscopes that use lenses to focus beams of reflected light or electrons. Different kinds of microscopes and different staining procedures are used, depending on what subcellular component is being resolved.

All cells are bounded by a plasma membrane and filled with a semifluid substance called cytoplasm. The genetic material is found in the central portion of the cell and, in eukaryotic cells, is contained in a membrane-bounded structure called a nucleus.

4.2 Prokaryotic Cells

When cells were visualized with microscopes, two basic cellular architectures were recognized: eukaryotic and prokaryotic. These refer to the presence or absence, respectively, of a membrane-bounded nucleus that contains genetic material. We have already mentioned that in addition to lacking a nucleus, prokaryotic cells do not have an internal membrane system or numerous membrane-bounded organelles.

Prokaryote Cells Have Relatively Simple Organization

Prokaryotes are the simplest organisms. Prokaryotic cells are small, consisting of cytoplasm surrounded by a plasma membrane and encased within a rigid **cell wall,** with no distinct interior compartments (figure 4.3). A prokaryotic cell is like a one-room cabin in which eating, sleeping, and watching TV all occur.

Prokaryotes are very important in the ecology of living organisms. Some harvest light by photosynthesis, others break down dead organisms and recycle their components, cause disease, and have uses

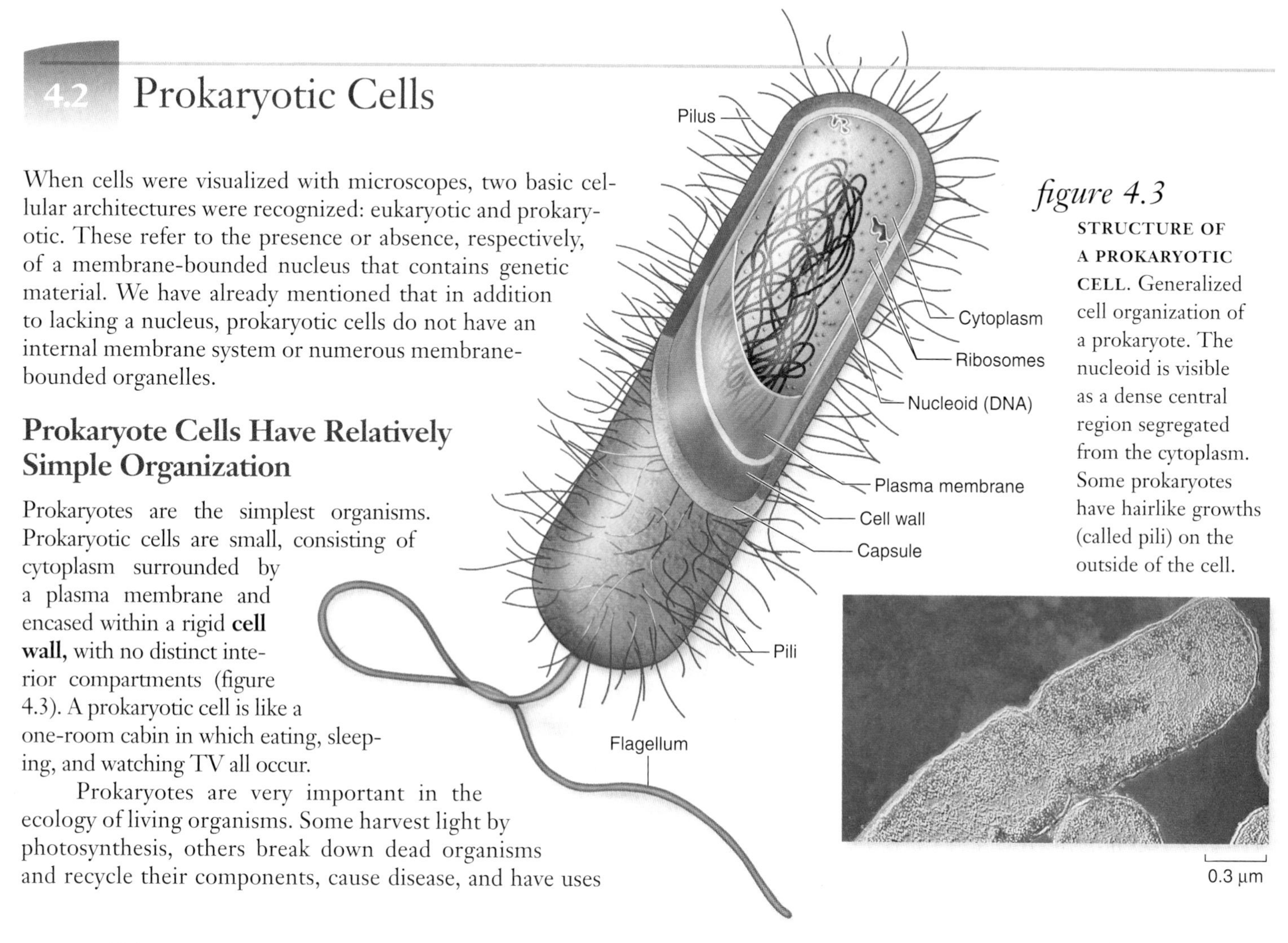

figure 4.3

STRUCTURE OF A PROKARYOTIC CELL. Generalized cell organization of a prokaryote. The nucleoid is visible as a dense central region segregated from the cytoplasm. Some prokaryotes have hairlike growths (called pili) on the outside of the cell.

in many important industrial processes. There are two main domains of prokaryotes: archaea and bacteria. Chapter 28 covers prokaryotic diversity in more detail.

Although prokaryotic cells do contain complex structures like **ribosomes,** which carry out protein synthesis, most have no membrane-bounded organelles characteristic of eukaryotic cells. Prokaryotes also lack the elaborate cytoskeleton found in eukaryotes, although they appear to have molecules related to actin, which is found in microfilaments (discussed later in the chapter). These actinlike proteins form supporting fibrils near the surface of the cell, but the cytoplasm of a prokaryotic cell appears to be one unit with no internal support structure. Consequently, the strength of the cell comes primarily from its rigid cell wall (see figure 4.3).

The plasma membrane of a prokaryotic cell carries out some of the functions organelles perform in eukaryotic cells. For example, some photosynthetic bacteria, such as the cyanobacterium *Prochloron* (figure 4.4), have an extensively folded plasma membrane, with the folds extending into the cell's interior. These membrane folds contain the bacterial pigments connected with photosynthesis. In eukaryotic plant cells, photosynthetic pigments are found in the inner membrane of the chloroplast.

Because a prokaryotic cell contains no membrane-bounded organelles, the DNA, enzymes, and other cytoplasmic constituents have access to all parts of the cell. Reactions are not compartmentalized as they are in eukaryotic cells, and the whole prokaryote operates as a single unit.

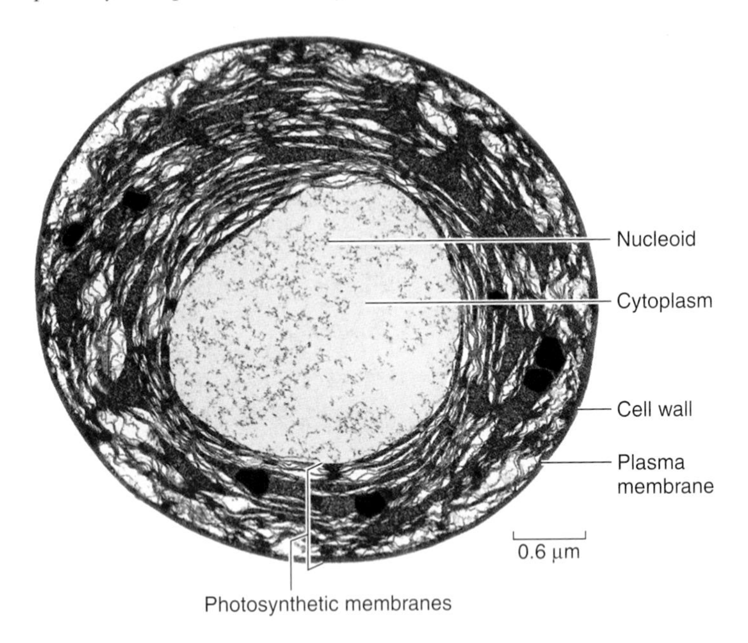

figure 4.4

ELECTRON MICROGRAPH OF A PHOTOSYNTHETIC BACTERIAL CELL. Extensive folded photosynthetic membranes are shown in green in this false color electron micrograph of a *Prochloron* cell.

inquiry

If you were designing a cell to be as large as possible, what modifications would facilitate large size?

Bacterial cell walls consist of peptidoglycan

Most bacterial cells are encased by a strong **cell wall.** This cell wall is composed of *peptidoglycan,* which consists of a carbohydrate matrix (polymers of sugars) that is cross-linked by short polypeptide units. Cell walls protect the cell, maintain its shape, and prevent excessive uptake or loss of water. Plants, fungi, and most protists also have cell walls of a different chemical structure that lack peptidoglycan, discussed in later chapters.

With the exception of the class Mollicutes, commonly called mycoplasma, which lack a cell wall, bacteria can be classified into two types based on differences in their cell walls that can be detected by the Gram staining procedure. **Gram-positive** bacteria have a thick, single-layered peptidoglycan cell wall that retains a violet dye from the Gram stain, causing the stained cells to appear purple under a microscope. **Gram-negative** bacteria have evolved a more complex, multilayered cell wall that does not retain the purple dye. These bacteria appear red after staining due to a second background dye.

The susceptibility of bacteria to antibiotics often depends on the structure of their cell walls. The drugs penicillin and vancomycin, for example, interfere with the ability of bacteria to cross-link the peptides in their peptidoglycan cell wall. Like removing all the nails from a wooden house, this destroys the integrity of the matrix, which can no longer prevent water from rushing in and swelling the cell to bursting.

Some bacteria also secrete a jellylike protective capsule of polysaccharide around the cell. Many disease-causing bacteria have such a capsule, which enables them to adhere to teeth, skin, food—or to practically any surface that will support their growth.

Archaea lack peptidoglycan

We are still learning about the physiology and structure of archaea. Many of these organisms are difficult to culture in the laboratory, and so this group has not yet been studied in detail. More is known about their genetic makeup than about any other feature.

The cell walls of archaea have various chemical compositions, including polysaccharides and proteins, and possibly even inorganic components. A common feature distinguishing archaea from bacteria is the nature of their membrane lipids. The chemical structure of archaeal lipids is distinctly different from that of lipids in bacteria and can include saturated hydrocarbons that are covalently attached to glycerol at both ends, such that their membrane is a monolayer. These features seem to confer greater thermal stability on archaeal membranes, although the tradeoff seems to be an inability to alter the degree of saturation of the hydrocarbons—meaning that archaea with this characteristic lose the ability to adapt to changing environmental temperatures.

The cellular machinery that replicates DNA and synthesized proteins in archaea is more closely related to eukaryotic systems than to bacterial systems. Even though they share a

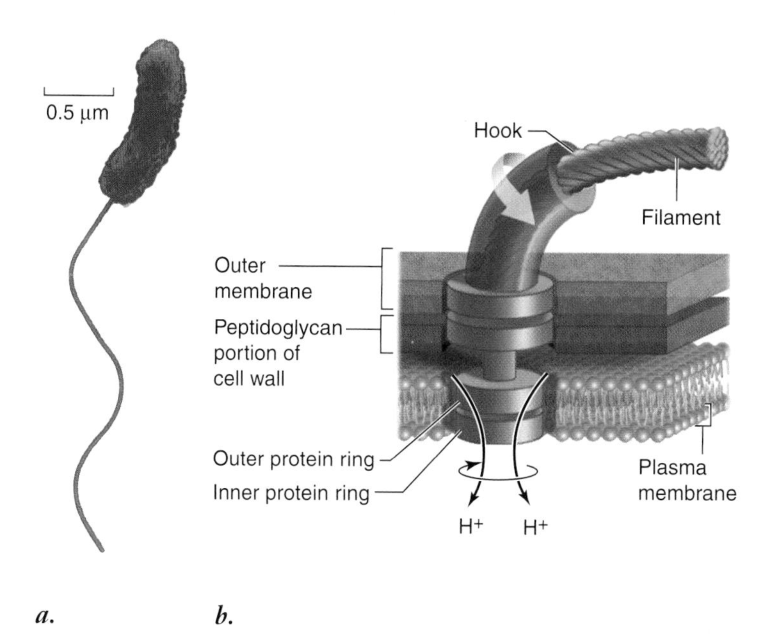

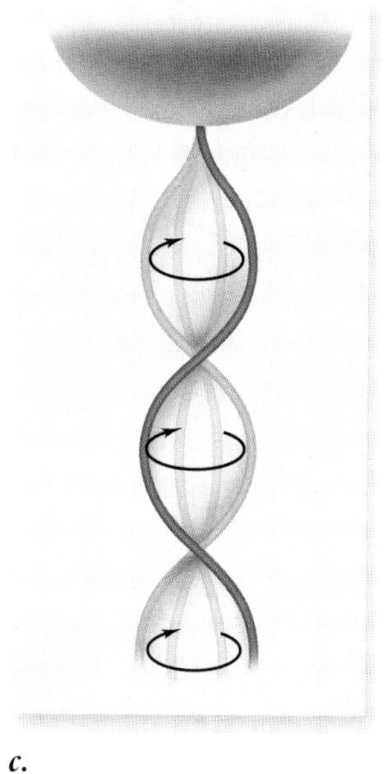

figure 4.5

SOME PROKARYOTES MOVE BY ROTATING THEIR FLAGELLA. ***a.*** The photograph shows *Vibrio cholerae*, the microbe that causes the serious disease cholera. ***b.*** The bacterial flagellum is a complex structure. The motor proteins, powered by a proton gradient, are anchored in the plasma membrane. Two rings are found in the cell wall. The motor proteins cause the entire structure to rotate. ***c.*** As the flagellum rotates it creates a spiral wave down the structure. This powers the cell forward.

similar overall cellular architecture with prokaryotes, archaea appear to be more closely related to eukaryotes.

Some prokaryotes move by means of rotating flagella

Flagella (singular, *flagellum*) are long, threadlike structures protruding from the surface of a cell that are used in locomotion. Prokaryotic flagella are protein fibers that extend out from the cell. There may be one or more per cell, or none, depending on the species. Bacteria can swim at speeds of up to 70 cell lengths per second by rotating their flagella like screws (figure 4.5). The rotary motor uses the energy stored in a gradient that transfers protons across the plasma membrane to power the movement of the flagellum. Interestingly, the same principle, in which a proton gradient powers the rotation of a molecule, is used in eukaryotic mitochondria and chloroplasts by an enzyme that synthesizes ATP (see chapter 7).

Prokaryotes are small cells that lack complex interior organization. The two domains of prokaryotes are archaea and bacteria. Bacteria are encased by a cell wall composed of peptidoglycan, and archaea have cell walls made from a variety of carbohydrate and peptides.

Bacteria can be separated into gram positive and gram negative based on staining that detects differences in cell wall architecture. Archaea have unusual membrane lipids.

Same prokaryotes are motile, propelled by external flagella that rotate.

4.3 Eukaryotic Cells

Eukaryotic cells (figures 4.6 and 4.7) are far more complex than prokaryotic cells. The hallmark of the eukaryotic cell is compartmentalization, which is achieved by an extensive **endomembrane system** that weaves through the cell interior and by numerous *organelles.* Organelles are the membrane-bounded structures that form compartments within which multiple biochemical processes can proceed simultaneously and independently.

Plant cells often have a large, membrane-bounded sac called a **central vacuole,** which stores proteins, pigments, and waste materials. Both plant and animal cells contain **vesicles,** smaller sacs that store and transport a variety of materials. Inside the nucleus, the DNA is wound tightly around proteins and packaged into compact units called **chromosomes.**

All eukaryotic cells are supported by an internal protein scaffold, the **cytoskeleton.** Although the cells of animals and some protists lack cell walls, the cells of fungi, plants, and many protists have strong cell walls composed of cellulose or chitin fibers embedded in a matrix of other polysaccharides and proteins. In the remainder of this chapter, we will examine the internal components of eukaryotic cells in more detail.

The nucleus acts as the information center

The largest and most easily seen organelle within a eukaryotic cell is the **nucleus** (Latin, "kernel" or "nut"), first described by the Scottish botanist Robert Brown in 1831. Nuclei are

figure 4.6

STRUCTURE OF AN ANIMAL CELL. In this generalized diagram of an animal cell, the plasma membrane encases the cell, which contains the cytoskeleton and various cell organelles and interior structures suspended in a semifluid matrix called the cytoplasm. Some kinds of animal cells possess fingerlike projections called microvilli. Other types of eukaryotic cells—for example, many protist cells—may possess flagella, which aid in movement, or cilia, which can have many different functions.

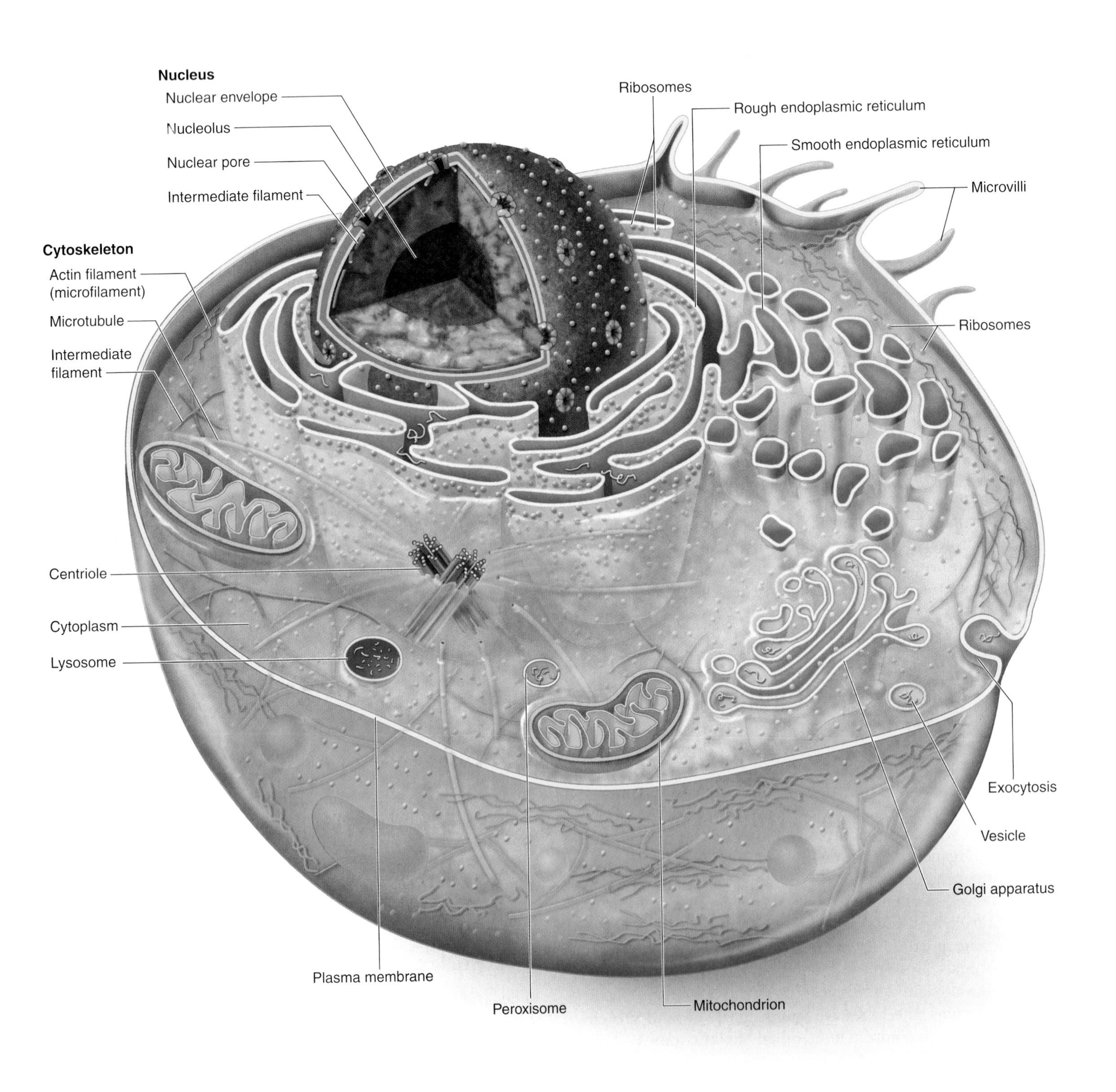

figure 4.7

STRUCTURE OF A PLANT CELL. Most mature plant cells contain a large central vacuole, which occupies a major portion of the internal volume of the cell, and organelles called chloroplasts, within which photosynthesis takes place. The cells of plants, fungi, and some protists have cell walls, although the composition of the walls varies among the groups. Plant cells have cytoplasmic connections to one another through openings in the cell wall called plasmodesmata. Flagella occur in sperm of a few plant species, but are otherwise absent from plant and fungal cells. Centrioles are also absent.

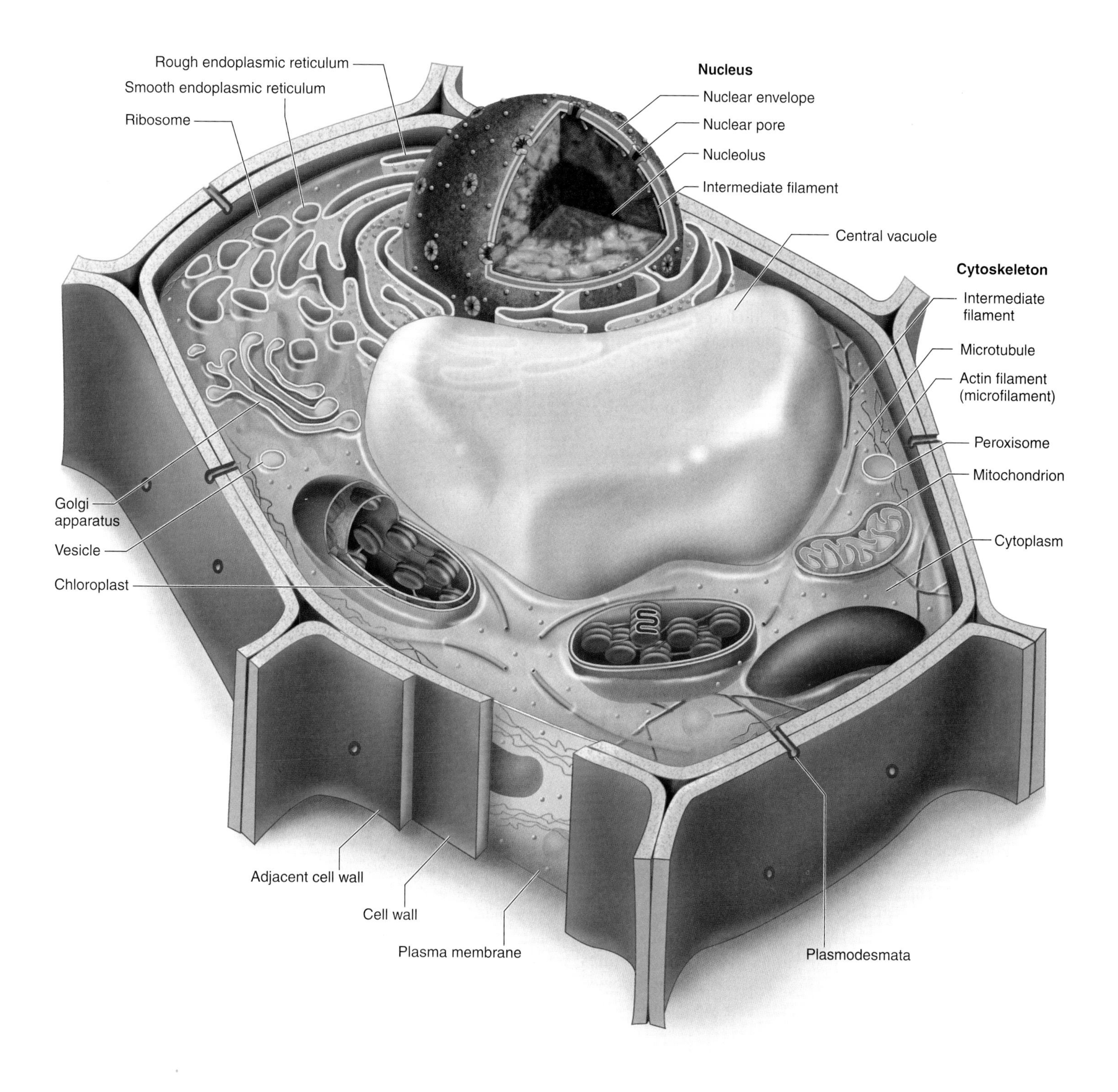

roughly spherical in shape, and in animal cells, they are typically located in the central region of the cell (figure 4.8*a*). In some cells, a network of fine cytoplasmic filaments seems to cradle the nucleus in this position.

The nucleus is the repository of the genetic information that leads to the synthesis of nearly all proteins of a living eukaryotic cell. Most eukaryotic cells possess a single nucleus, although the cells of fungi and some other groups may have several to many nuclei. Mammalian erythrocytes (red blood cells) lose their nuclei when they mature. Many nuclei exhibit a dark-staining zone called the **nucleolus,** which is a region where intensive synthesis of ribosomal RNA is taking place.

The nuclear envelope

The surface of the nucleus is bounded by *two* phospholipid bilayer membranes, which together make up the **nuclear envelope** (see figure 4.8). The outer membrane of the nuclear envelope is continuous with the cytoplasm's interior membrane system, called the *endoplasmic reticulum* (described later).

Scattered over the surface of the nuclear envelope are what appear as shallow depressions in the electron micrograph but are in fact structures called **nuclear pores** (see figure 4.8*b*, *c*). These pores form 50 to 80 nm apart at locations where the two membrane layers of the nuclear envelope pinch together. They have a complex structure with a cytoplasmic face, a nuclear face, and a central ring embedded in the membrane. The proteins that make up this nuclear pore complex are arranged radially with a large central hole. The complex allows small molecules to diffuse freely between nucleoplasm and cytoplasm while controlling the passage of proteins and RNA–protein complexes. Passage is restricted primarily to two kinds of molecules: (1) proteins moving into the nucleus to be incorporated into nuclear structures or to catalyze nuclear activities; and (2) RNA and RNA–protein complexes formed in the nucleus and exported to the cytoplasm.

The inner surface of the nuclear envelope is covered with a network of fibers that make up the nuclear lamina (see figure 4.8*d*). This is composed of intermediate filament fibers called *nuclear lamins.* This structure gives the nucleus its shape and is also involved in the deconstruction and reconstruction of the nuclear envelope that accompanies cell division.

Chromatin: DNA packaging

In both prokaryotes and eukaryotes, DNA contains the hereditary information specifying cell structure and function. In most prokaryotes, the DNA is organized into a single circular chromosome. In eukaryotes, the DNA is divided into multiple linear chromosomes. The DNA in these chromosomes is organized with proteins into a complex structure called **chromatin.** The chromosomes are packaged with proteins called **histones** into **nucleosomes** that are formed by DNA being wrapped about histones. These structures resemble beads on a string (see chapter 10).

Chromatin is usually in a more extended form that allows regulatory proteins to attach to specific nucleotide sequences along the DNA and regulate gene expression.

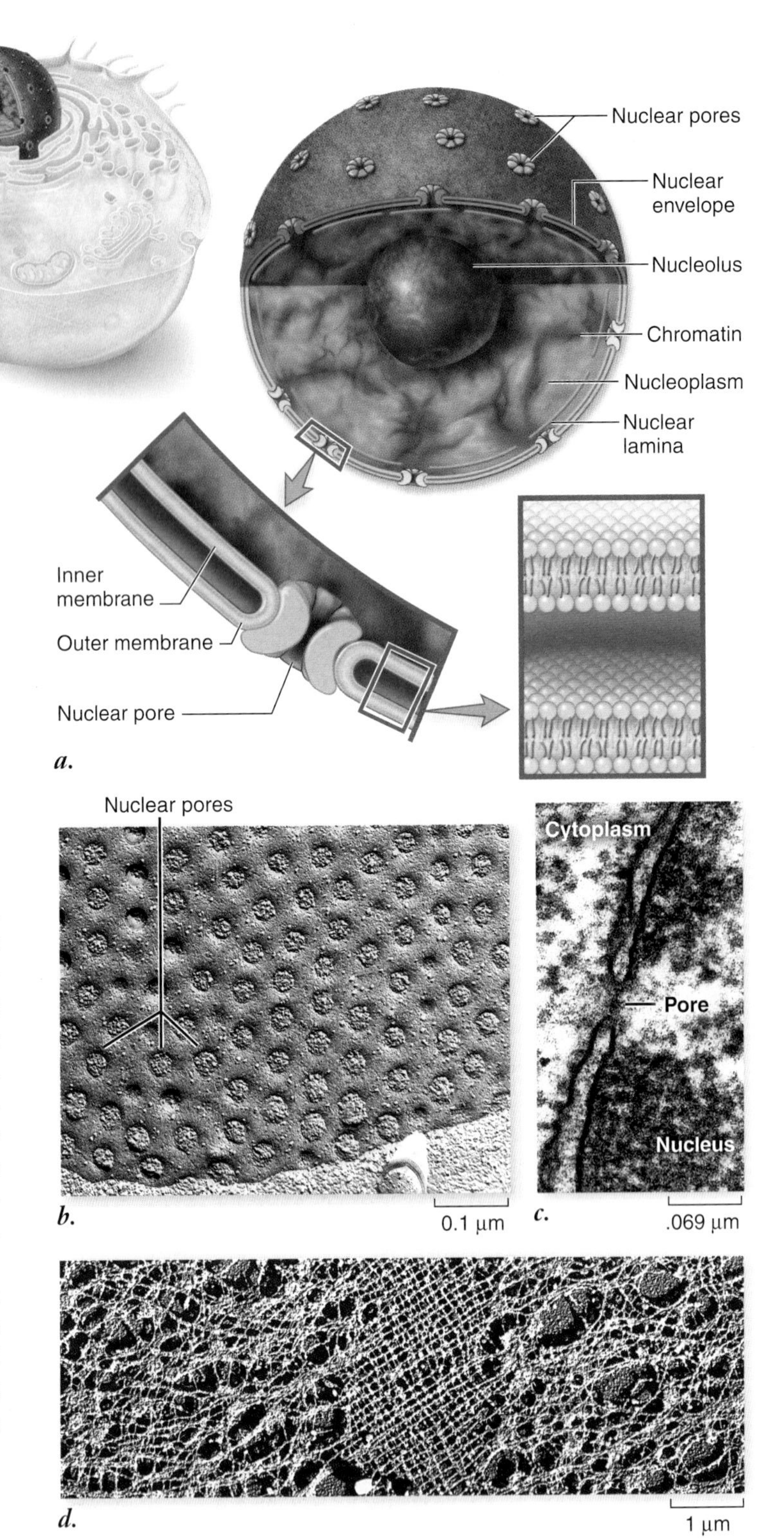

figure 4.8

THE NUCLEUS. ***a.*** The nucleus is composed of a double membrane called the nuclear envelope, enclosing a fluid-filled interior containing chromatin. The individual nuclear pores extend through the two membrane layers of the envelope. ***b.*** A freeze-fracture electron micrograph (see figure 6.6) of a cell nucleus, showing many nuclear pores. ***c.*** A transmission electron micrograph of the nuclear membrane showing a single nuclear pore; the dark material within the pore is protein, which acts to control access through the pore. ***d.*** The nuclear lamina is visible as a dense network of fibers made of intermediate filaments. The nucleus has been colored purple in the micrographs.

Without this access, DNA could not direct the day-to-day activities of the cell. When cells divide, it is necessary to further compact the chromatin into a more highly condensed form.

Under a light microscope, fully condensed chromosomes are readily seen as densely staining rods in dividing cells (figure 4.9). After cell division, eukaryotic chromosomes uncoil and can no longer be individually distinguished with a light microscope.

The nucleolus: Ribosomal subunit manufacturing

Before cells can synthesize proteins in large quantity, they must first construct a large number of ribosomes to carry out this synthesis. Hundreds of copies of the genes encoding the ribosomal RNAs are clustered together on the chromosome, facilitating ribosome construction. By transcribing RNA molecules from this cluster, the cell rapidly generates large numbers of the molecules needed to produce ribosomes.

The clusters of ribosomal RNA genes, the RNAs they produce, and the ribosomal proteins all come together within the nucleus during ribosome production. These ribosomal assembly areas are easily visible within the nucleus as one or more dark-staining regions called **nucleoli** (singular, *nucleolus*). Nucleoli can be seen under the light microscope even when the chromosomes are uncoiled.

Ribosomes are the cell's protein synthesis machinery

Although the DNA in a cell's nucleus encodes the amino acid sequence of each protein in the cell, the proteins are not assembled there. A simple experiment demonstrates this: If a brief pulse of radioactive amino acid is administered to a cell, the radioactivity shows up associated with newly made protein in the cytoplasm, not in the nucleus. When investigators first carried out these experiments, they found that protein synthesis is associated with large RNA–protein complexes (called ribosomes) outside the nucleus.

Ribosomes are among the most complex molecular assemblies found in cells. Each ribosome is composed of two subunits (figure 4.10), and each subunit is composed of a combination of RNA, called **ribosomal RNA (rRNA),** and proteins. The subunits join to form a functional ribosome only when they are actively synthesizing proteins. This complicated process requires the two other main forms of RNA: **messenger RNA (mRNA),** which carries coding information from DNA, and **transfer RNA (tRNA),** which carries amino acids. Ribosomes use the information in mRNA to direct the synthesis of a protein. This process will be described in more detail in chapter 15.

Ribosomes are found either free in the cytoplasm or associated with internal membranes, as described in the following section. Free ribosomes synthesize proteins that are found in the cytoplasm, nuclear proteins, mitochondrial proteins, and proteins found in other organelles not derived from the endomembrane system. Membrane-associated ribosomes synthesize membrane proteins, proteins found in the endomembrane system, and proteins destined for export from the cell.

Ribosomes can be thought of as "universal machines" because they are found in all cell types from all three domains of life. As we build a picture of the minimal essential functions for cellular life, ribosomes will be on the short list. Life is protein-based, and ribosomes are the factories that make proteins.

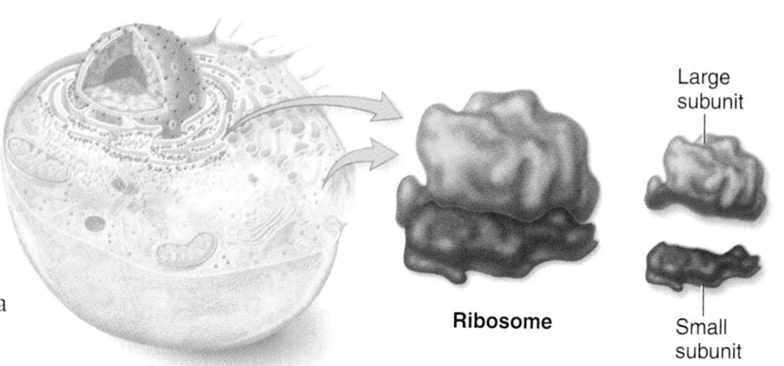

figure 4.10

A RIBOSOME. Ribosomes consist of a large and a small subunit composed of rRNA and protein. The individual subunits are synthesized in the nucleolus and then move through the nuclear pores to the cytoplasm, where they assemble to translate mRNA. Ribosomes serve as sites of protein synthesis.

figure 4.9

EUKARYOTIC CHROMOSOMES. These condensed chromosomes within an onion root tip cell in the process of dividing are visible under the light microscope.

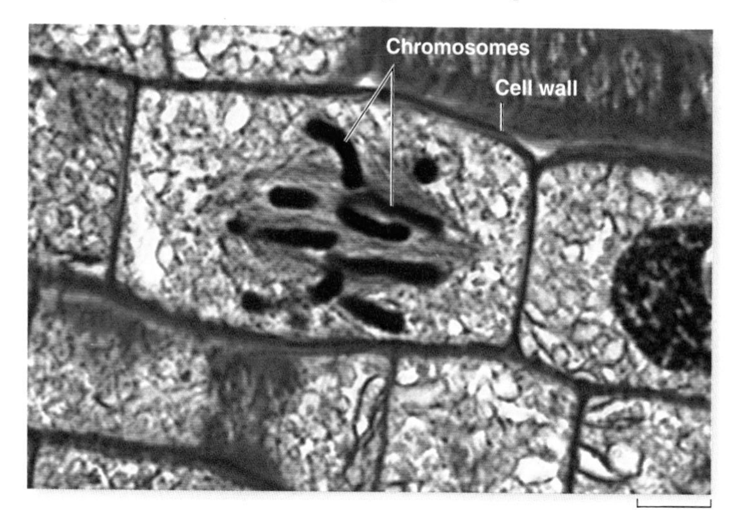

Eukaryotic cells contain membrane-bounded organelles that carry out specialized functions.

The nucleus of a eukaryotic cell contains the cell's genetic information. This nucleus is composed of a double membrane that is connected to the endomembrane system. Material moves between the nucleus and cytoplasm through nuclear pores. DNA found in the nucleus is complexed with proteins and organized into chromosomes.

Ribosomes are the cellular machines that synthesize proteins. Ribosomes are the only organelle that is truly universal, being found in all known cells.

4.4 The Endomembrane System

The interior of a eukaryotic cell is packed with membranes so thin that they are invisible under the low resolving power of light microscopes. This endomembrane system fills the cell, dividing it into compartments, channeling the passage of molecules through the interior of the cell, and providing surfaces for the synthesis of lipids and some proteins. The presence of these membranes in eukaryotic cells constitutes one of the most fundamental distinctions between eukaryotes and prokaryotes.

The largest of the internal membranes is called the **endoplasmic reticulum (ER).** *Endoplasmic* means "within the cytoplasm," and *reticulum* is Latin for "a little net." Like the plasma membrane, the ER is composed of a phospholipid bilayer embedded with proteins. It weaves in sheets through the interior of the cell, creating a series of channels between its folds (figure 4.11). Of the many compartments in eukaryotic cells, the two largest are the inner region of the ER, called the **cisternal space** or **lumen,** and the region exterior to it, the cytosol, which is the fluid component of the cytoplasm containing dissolved organic molecules such as proteins and ions.

The rough ER is a site of protein synthesis

The **rough ER (RER)** gets its name from its surface appearance, which is pebbly instead of smooth due to the presence of ribosomes. The RER is not easily visible with a light microscope, but it can be seen using the electron microscope. It appears to be composed of flattened sacs, the surfaces of which are bumpy with ribosomes (see figure 4.11).

The proteins synthesized on the surface of the RER are destined to be exported from the cell, sent to lysosomes or vacuoles (described in a later section), or embedded in the plasma membrane. These proteins enter the cisternal space as a first step in the pathway that will sort proteins to their eventual destinations. This pathway also involves vesicles and the Golgi apparatus, described later. The sequence of the protein being synthesized determines whether the ribosome will become associated with the ER or remain a cytoplasmic ribosome.

In the ER, newly synthesized proteins can be modified by the addition of short-chain carbohydrates to form **glycoproteins.** Those proteins destined for secretion are then kept separate from other products and are later packaged into vesicles. The ER also manufactures membranes by producing membrane proteins and phospholipid molecules. The membrane proteins are inserted into the ER's own membrane, which can then expand and pinch off in the form of vesicles to be transferred to other locations.

The smooth ER has multiple roles

Regions of the ER with relatively few bound ribosomes are referred to as **smooth ER (SER).** The SER appears more like a network of tubules than the flattened sacs of the RER. The membranes of the SER contain many embedded enzymes. Enzymes anchored within the ER, for example, catalyze the synthesis of a variety of carbohydrates and lipids. Steroid hormones are synthesized in the SER as well. The majority of membrane lipids are assembled in the SER and then sent to whatever parts of the cell need membrane components.

The SER is used to store Ca^{2+} in cells. This keeps the cytoplasmic level low, allowing Ca^{2+} to be used as a signaling molecule. In muscle cells, for example, Ca^{2+} is used to trigger muscle contraction. In other cells, Ca^{2+} release from SER stores is involved in diverse signalling pathways.

The ratio of SER to RER depends on a cell's function. In multicellular animals such as ourselves, great variation exists in this ratio. Cells that carry out extensive lipid synthesis, such as those in the testes, intestine, and brain have abundant SER. Cells that synthesize proteins that are secreted, such as antibodies, have much more extensive RER.

Another role of the SER is the modification of foreign substances to make them less toxic. In the liver, the enzymes of the SER carry out this detoxification. This action can include neu-

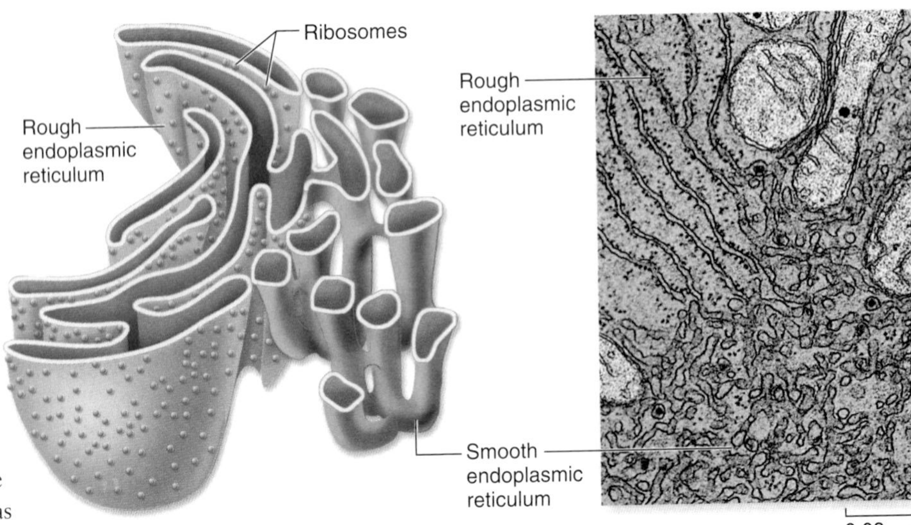

figure 4.11

THE ENDOPLASMIC RETICULUM. Rough ER (RER), blue in the drawing, is composed more of flattened sacs and forms a compartment throughout the cytoplasm. Ribosomes associated with the cytoplasmic face of the RER extrude newly made proteins into the interior, or lumen. The smooth ER (SER), green in the drawing, is a more tubelike structure connected to the RER. The micrograph has been colored to match the drawing.

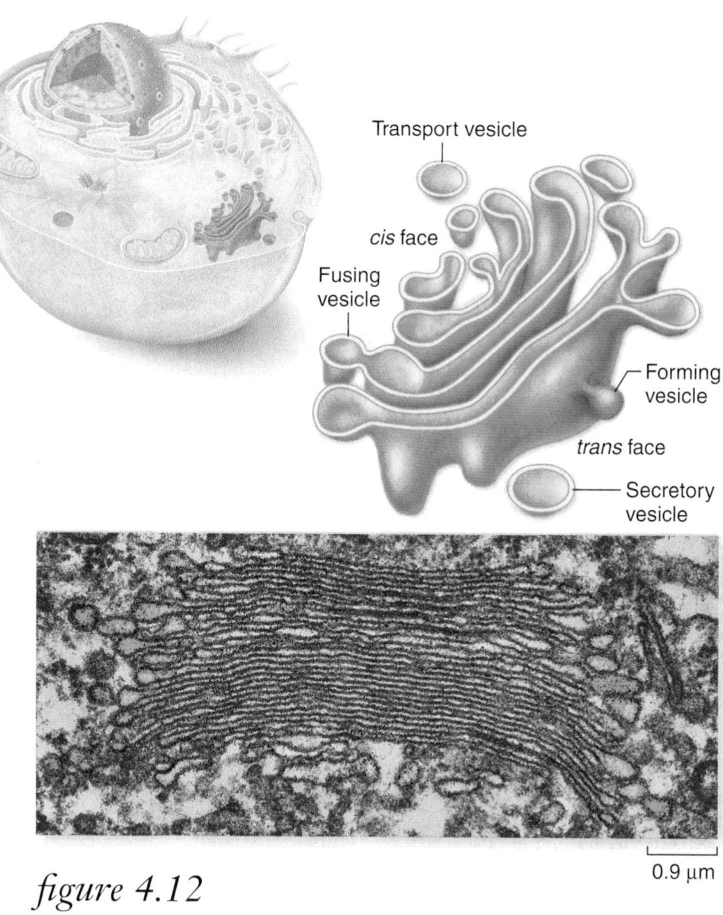

figure 4.12

THE GOLGI APPARATUS. The Golgi apparatus is a smooth, concave, membranous structure. It receives material for processing in transport vesicles on the *cis* face and sends the material packaged in transport or secretory vesicles off the *trans* face. The substance in a vesicle could be for export out of the cell or for distribution to another region within the same cell.

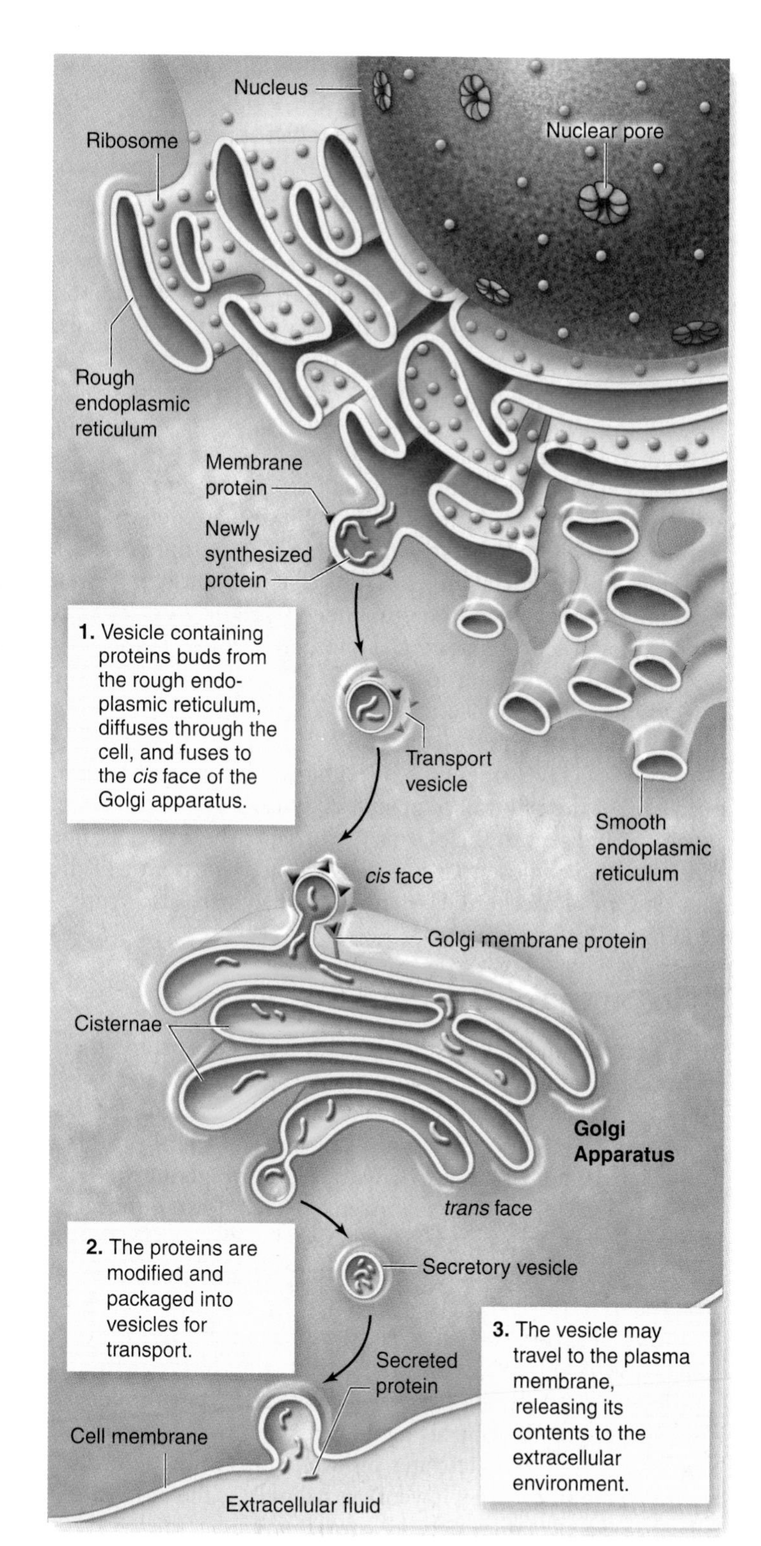

figure 4.13

PROTEIN TRANSPORT THROUGH THE ENDOMEMBRANE SYSTEM. Proteins synthesized by ribosomes on the RER are translocated into the internal compartment of the ER. These proteins may be used at a distant location in the cell or secreted. They are transported within vesicles that bud off the rough ER. These transport vesicles travel to the *cis* face of the Golgi apparatus. There they can be modified and packaged into vesicles that bud off the *trans* face of the Golgi apparatus. Vesicles leaving the *trans* face transport proteins to other locations in the cell, or fuse with the plasma membrane, releasing their contents to the extracellular environment.

tralizing substances that we have taken for a therapeutic reason, such as penicillin. Thus relatively high doses are prescribed for some drugs to offset our body's efforts to remove them. Liver cells have extensive SER as well as enzymes that can process a variety of substances by chemically modifying them.

The Golgi apparatus sorts and packages proteins

Flattened stacks of membranes called **Golgi bodies** can be found within the endomembrane system, often interconnected with one another. These structures are named for Camillo Golgi, the nineteenth-century Italian physician who first identified them. The number of Golgi bodies a cell contains ranges from 1 or a few in protists, to 20 or more in animal cells and several hundred in plant cells. They are especially abundant in glandular cells, which manufacture and secrete substances. Collectively, the Golgi bodies are referred to as the **Golgi apparatus** (figure 4.12).

The Golgi apparatus functions in the collection, packaging, and distribution of molecules synthesized at one location and utilized at another within the cell or even outside of it. A Golgi body has a front and a back, with distinctly different membrane compositions at these opposite ends. The front, or receiving end, is called the *cis* face, and it is usually located near ER. Materials move to the *cis* face in transport vesicles that bud off the ER. These vesicles fuse with the *cis* face, emptying their contents into the interior, or lumen, of the Golgi apparatus. The ER-synthesized molecules then pass through the channels of the Golgi apparatus until they reach the back, or discharging end, called the *trans* face, where they are discharged in secretory vesicles (figure 4.13).

Proteins and lipids manufactured on the rough and smooth ER membranes are transported into the Golgi apparatus and modified as they pass through it. The most common alteration is the addition or modification of short sugar chains, forming glycoproteins and glycolipids. In many instances, enzymes in the Golgi apparatus modify existing glycoproteins and glycolipids made in the ER by cleaving a sugar from a chain or by modifying one or more of the sugars.

The newly formed or altered glycoproteins and glycolipids collect at the ends of the Golgi bodies in flattened, stacked membrane folds called **cisternae** (Latin, "collecting vessels"). Periodically, the membranes of the cisternae push together, pinching off small, membrane-bounded secretory vesicles containing the glycoprotein and glycolipid molecules. These vesicles then diffuse to other locations in the cell, distributing the newly synthesized molecules to their appropriate destinations.

Another function of the Golgi apparatus is the synthesis of cell wall components. Noncellulose polysaccharides that form part of the cell wall of plants are synthesized in the Golgi apparatus and sent to the plasma membrane where they can be added to the cellulose that is assembled on the exterior of the cell. Other polysaccharides secreted by plants are also synthesized in the Golgi apparatus.

Lysosomes contain digestive enzymes

Membrane-bounded digestive vesicles, called **lysosomes,** are also components of the endomembrane system, and they arise from the Golgi apparatus. They contain high levels of degrading enzymes, which catalyze the rapid breakdown of proteins, nucleic acids, lipids, and carbohydrates. Throughout the lives of eukaryotic cells, lysosomal enzymes break down old organelles, recycling their component molecules and making room for newly formed organelles. For example, mitochondria are replaced in some tissues every 10 days.

The digestive enzymes in the lysosome are optimally active at acid pH. Lysosomes are activated by fusing with a food vesicle produced by *phagocytosis* (a specific type of endocytosis; see chapter 5) or by fusing with an old or worn-out organelle. The fusion event activates proton pumps in the lysosomal membrane, resulting in a lower internal pH. As the interior pH falls, the arsenal of digestive enzymes contained in the lysosome is activated. This leads to the degradation of macromolecules in the food vesicle or the destruction of the old organelle.

A number of human genetic disorders, collectively called lysosomal storage disorders, affect lysosomes. For example, Tay–Sachs disease is caused by the loss of function of a single lysosomal enzyme. This enzyme is necessary to break down a membrane glycolipid found in nerve cells. Accumulation of glycolipid in lysosomes affects nerve cell function, leading to a variety of clinical symptoms such as seizures and muscle rigidity.

In addition to breaking down organelles and other structures within cells, lysosomes eliminate other cells that the cell has engulfed by phagocytosis. When a white blood cell, for ex-

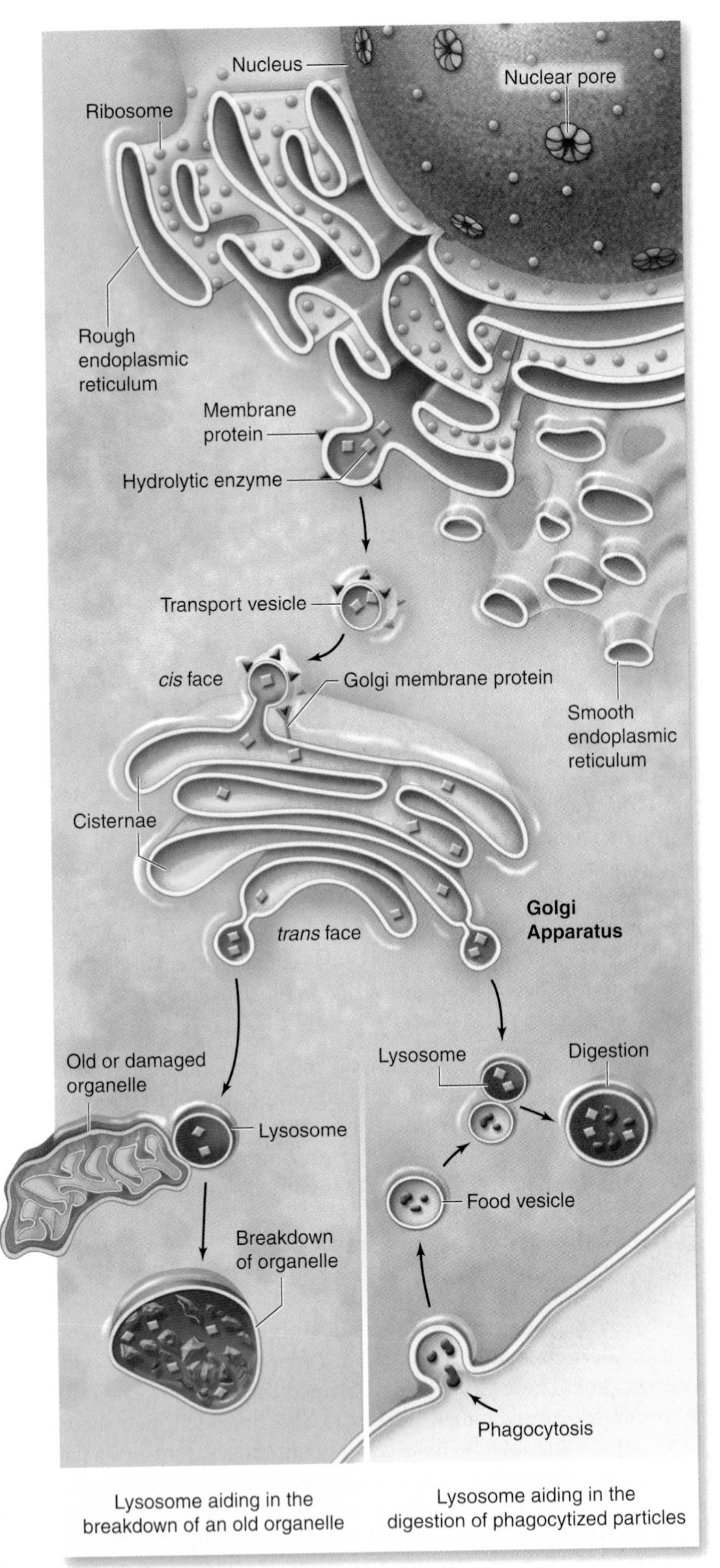

figure 4.14

LYSOSOMES. Lysosomes are formed from vesicles budding off the Golgi. They contain hydrolytic enzymes that digest particles or cells taken into the cell by phagocytosis, and break down old organelles.

figure 4.15

A PEROXISOME. Peroxisomes are spherical organelles that may contain a large crystal structure composed of protein. Peroxisomes contain digestive and detoxifying enzymes that produce hydrogen peroxide as a by-product. A peroxisome has been colored green in the electron micrograph.

ample, phagocytizes a passing pathogen, lysosomes fuse with the resulting "food vesicle," releasing their enzymes into the vesicle and degrading the material within (figure 4.14).

Microbodies are a diverse category of organelles

Eukaryotic cells contain a variety of enzyme-bearing, membrane-enclosed vesicles called **microbodies.** Although technically not considered part of the endomembrane system, we will consider them in this section. Microbodies are found in the cells of plants, animals, fungi, and protists. The distribution of enzymes into microbodies is one of the principal ways by which eukaryotic cells organize their metabolism.

Whereas lysosomes bud from the endomembrane system, microbodies grow by incorporating lipids and protein, and then dividing. Plant cells have a special type of microbody called a **glyoxysome,** which contains enzymes that convert fats into carbohydrates.

Peroxisomes: Peroxide utilization

Another type of microbody, a **peroxisome,** contains enzymes that catalyze the removal of electrons and associated hydrogen atoms (figure 4.15). If these oxidative enzymes were not isolated within microbodies, they would tend to short-circuit the metabolism of the cytoplasm, which often involves adding hydrogen atoms to oxygen. Peroxisomes were long thought to form by the addition of lipids and proteins, leading to growth, followed by budding to produce new peroxisomes. Although this still appears to be the case, recent evidence indicates that some peroxisomes may come from the fusion of ER-derived vesicles.

The name *peroxisome* refers to the hydrogen peroxide produced as a by-product of the activities of the oxidative enzymes in the microbody. Hydrogen peroxide is dangerous to cells because of its violent chemical reactivity. However, peroxisomes also contain the enzyme catalase, which breaks down hydrogen peroxide into harmless water and oxygen.

Plants use vacuoles for storage and water balance

Plant cells have specialized membrane-bounded structures called **vacuoles.** The most conspicuous example is the large central vacuole seen in most plant cells (figure 4.16). In fact, *vacuole* actually means blank space, referring to its appearance in the light microscope. The membrane surrounding this vacuole is called the **tonoplast** because it contains channels for water that are used to help the cell maintain its tonicity, or osmotic balance (see osmosis in chapter 5).

For many years biologists assumed that only one type of vacuole existed and that it served multiple functions. The functions

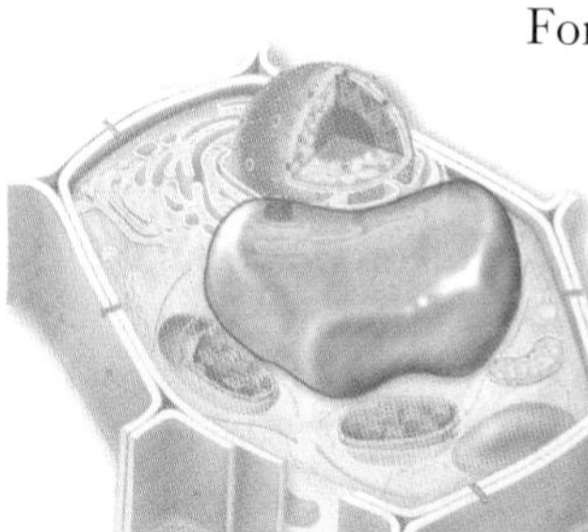

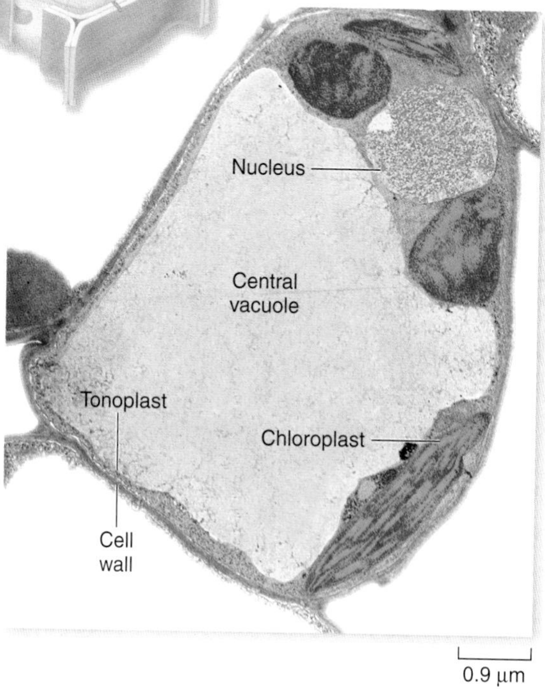

figure 4.16

THE CENTRAL VACUOLE. A plant's central vacuole stores dissolved substances and can expand in size to increase the tonicity of a plant cell. Micrograph shown with false color.

assigned to this vacuole included water balance and storage of both useful molecules, such as sugars, ions and pigments, and waste products. The vacuole was also thought to store enzymes involved in the breakdown of macromolecules and those used in detoxifying foreign substances. Old textbooks of plant physiology referred to vacuoles as the attic of the cell for the variety of substances thought to be stored there.

Studies of tonoplast transporters and the isolation of vacuoles from a variety of cell types have led to a more complex view of vacuoles. These studies have made it clear that different vacuolar types can be found in different cells. These vacuoles are specialized, depending on the function of the cell.

The central vacuole is clearly important for a number of roles in all plant cells. The central vacuole and the water channels of the tonoplast maintain the tonicity of the cell, allowing the cell to expand and contract depending on conditions. The central vacuole is also involved in cell growth by occupying most of the volume of the cell. Plant cells grow by expanding the vacuole, rather than by increasing cytoplasmic volume.

Vacuoles with a variety of functions are also found in some types of fungi and protists. One form is the contractile vacuole, found in some protists, which can pump water and is used to maintain water balance in the cell. Other vacuoles are used for storage or to segregate toxic materials from the rest of the cytoplasm. The number and kind of vacuoles found in a cell depends on the needs of the particular cell type.

The endoplasmic reticulum (ER) is an extensive system of folded membranes that spatially organize the cell's biosynthetic activities. The ER consists of smooth (SER) and rough (RER) components. RER is covered with ribosomes and is a site of protein synthesis.

Proteins synthesized on the RER are transported by vesicles to the Golgi apparatus where they are modified, packaged, and distributed to their final location.

Lysosomes are vesicles that contain digestive enzymes. These can fuse with vesicles containing extracelluar material or old organelles, causing the contents to be degraded.

Peroxisomes are a subclass of microbodies that carry out oxidative metabolism that generates peroxides. The isolation of these enzymes in vesicles protects the rest of the cell from the very reactive chemistry occurring inside.

Vacuoles are membrane-bounded structures that have a variety of roles ranging from storage to cell growth in plants. These are also found in some fungi and protists.

4.5 Mitochondria and Chloroplasts: Cellular Generators

Mitochondria and chloroplasts share structural and functional similarities. Structurally, they are both surrounded by a double membrane, and both contain their own DNA and protein-synthesis machinery. Functionally, they are both involved in energy metabolism, as we will explore in detail in later chapters on energy metabolism and photosynthesis.

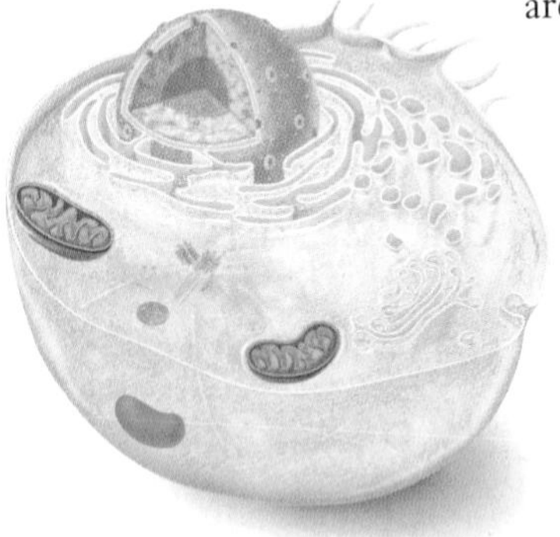

Mitochondria metabolize sugar to generate ATP

Mitochondria (singular, *mitochondrion*) are typically tubular or sausage-shaped organelles about the size of bacteria and found in all types of eukaryotic cells (figure 4.17). Mitochondria are bounded by two membranes: a smooth outer membrane, and an inner folded membrane with numerous contiguous layers called **cristae** (singular, *crista*).

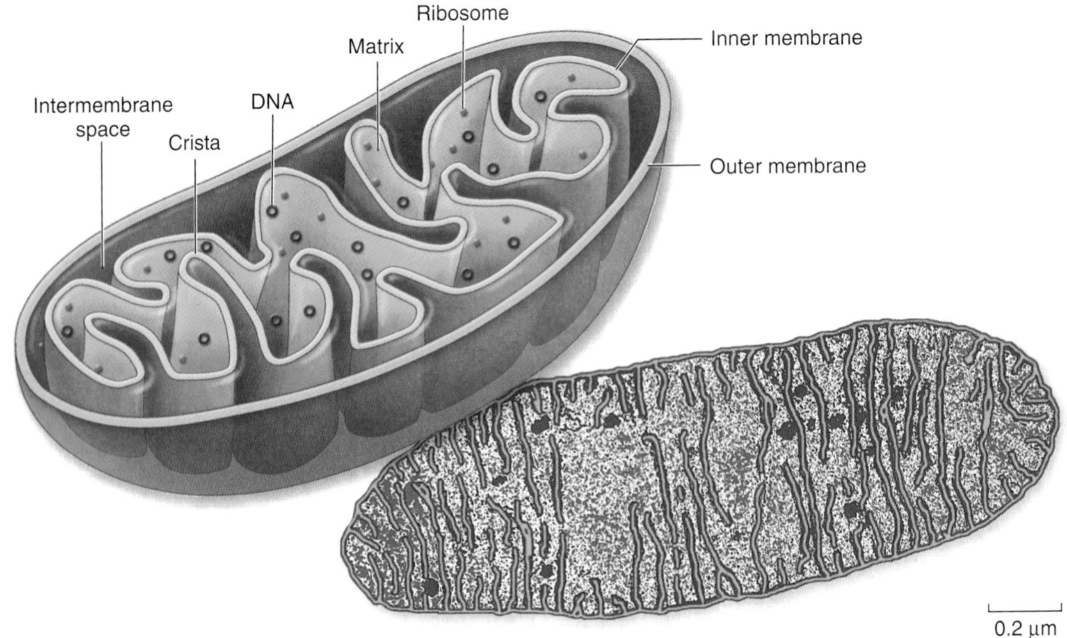

figure 4.17

MITOCHONDRIA. The inner membrane of a mitochondrion is shaped into folds called cristae that greatly increase the surface area for oxidative metabolism. A mitochondrion in cross section and cut lengthwise is shown colored red in the micrograph.

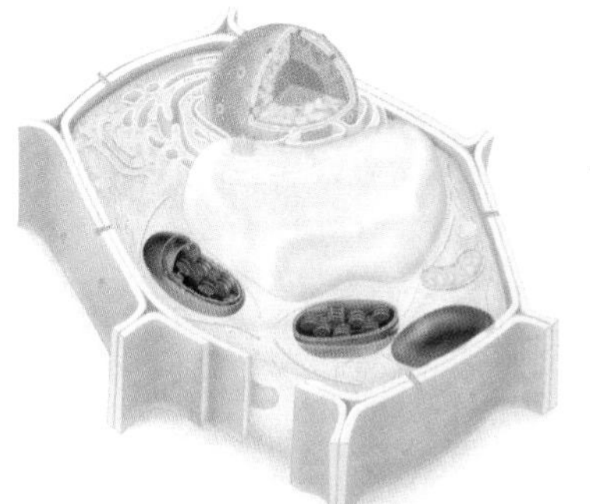

figure 4.18

CHLOROPLAST STRUCTURE. The inner membrane of a chloroplast surrounds a membrane system of stacks of closed chlorophyll-containing vesicles called thylakoids, within which photosynthesis takes place. Thylakoids are typically stacked one on top of the other in columns called grana. Chloroplast has been colored green in the micrograph on the right.

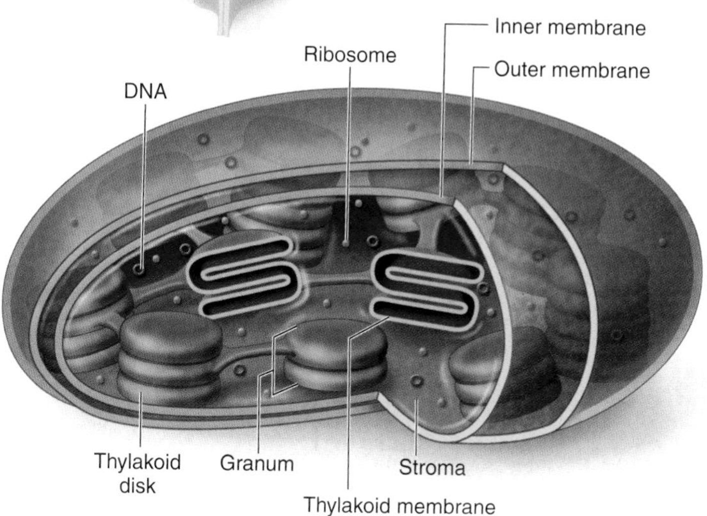

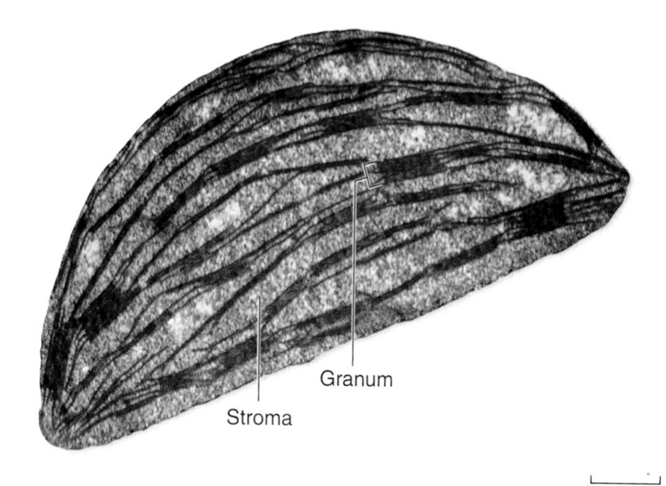

The cristae partition the mitochondrion into two compartments: a **matrix,** lying inside the inner membrane; and an outer compartment, or **intermembrane space,** lying between the two mitochondrial membranes. On the surface of the inner membrane, and also embedded within it, are proteins that carry out oxidative metabolism, the oxygen-requiring process by which energy in macromolecules is used to produce ATP (chapter 7).

Mitochondria have their own DNA; this DNA contains several genes that produce proteins essential to the mitochondrion's role in oxidative metabolism. Thus, the mitochondrion, in many respects, acts as a cell within a cell, containing its own genetic information specifying proteins for its unique functions. The mitochondria are not fully autonomous, however, because most of the genes that encode the enzymes used in oxidative metabolism are located in the cell nucleus.

A eukaryotic cell does not produce brand-new mitochondria each time the cell divides. Instead, the mitochondria themselves divide in two, doubling in number, and these are partitioned between the new cells. Most of the components required for mitochondrial division are encoded by genes in the nucleus and are translated into proteins by cytoplasmic ribosomes. Mitochondrial replication is, therefore, impossible without nuclear participation, and mitochondria thus cannot be grown in a cell-free culture.

Chloroplasts use light to generate ATP and sugars

Plant cells and cells of other eukaryotic organisms that carry out photosynthesis typically contain from one to several hundred **chloroplasts.** Chloroplasts bestow an obvious advantage on the organisms that possess them: They can manufacture their own food. Chloroplasts contain the photosynthetic pigment chlorophyll that gives most plants their green color.

The chloroplast, like the mitochondrion, is surrounded by two membranes (figure 4.18). However, chloroplasts are larger and more complex than mitochondria. In addition to the outer and inner membranes, which lie in close association with each other, chloroplasts have closed compartments of stacked membranes called **grana** (singular, *granum*), which lie inside the inner membrane.

A chloroplast may contain a hundred or more grana, and each granum may contain from a few to several dozen disk-shaped structures called **thylakoids.** On the surface of the thylakoids are the light-capturing photosynthetic pigments, to be discussed in depth in chapter 8. Surrounding the thylakoid is a fluid matrix called the *stroma.* The enzymes used to synthesize glucose during photosynthesis are found in the stroma.

Like mitochondria, chloroplasts contain DNA, but many of the genes that specify chloroplast components are also located in the nucleus. Some of the elements used in the photosynthetic process, including the specific protein components necessary to accomplish the reaction, are synthesized entirely within the chloroplast.

Other DNA-containing organelles in plants, called *leucoplasts,* lack pigment and a complex internal structure. In root cells and some other plant cells, leucoplasts may serve as starch storage sites. A leucoplast that stores starch (amylose) is sometimes termed an **amyloplast.** These organelles—chloroplasts, leucoplasts, and amyloplasts—are collectively called **plastids.** All plastids are produced by the division of existing plastids.

inquiry

Mitochondria and chloroplasts both generate ATP. What structural features do they share?

Mitochondria and chloroplasts arose by endosymbiosis

Symbiosis is a close relationship between organisms of different species that live together. As noted in chapter 29, the theory of **endosymbiosis** proposes that some of today's eukaryotic organelles evolved by a symbiosis arising between two cells that were each free-living. One cell, a prokaryote, was engulfed by and became part of another cell, which was the precursor to modern eukaryotes (figure 4.19).

According to the endosymbiont theory, the engulfed prokaryotes provided their hosts with certain advantages associated with their special metabolic abilities. Two key eukaryotic organelles are believed to be the descendants of these endosymbiotic prokaryotes: mitochondria, which are thought to have originated as bacteria capable of carrying out oxidative metabolism, and chloroplasts, which apparently arose from photosynthetic bacteria.

The endosymbiont theory is supported by a wealth of evidence, summarized as follows:

- Both mitochondria and chloroplasts are surrounded by two membranes; the inner membrane may have evolved from the plasma membrane of the engulfed prokaryote; the outer membrane is probably derived from the plasma membrane or endoplasmic reticulum of the host cell.
- Mitochondria are about the same size as most prokaryotes, and the cristae formed by their inner membranes resemble the folded membranes in various groups of bacteria.
- Mitochondrial ribosomes are also similar to prokaryotic ribosomes in size and structure.
- Both mitochondria and chloroplasts contain circular molecules of DNA similar to those in prokaryotes.
- The genomes of mitochondria and chloroplasts show similarities to the genomes of α-proteobacteria and cyanobacteria, respectively.
- Finally, mitochondria divide by simple fission, splitting in two just as prokaryotic cells do, and they apparently replicate and partition their DNA in much the same way as prokaryotes do.

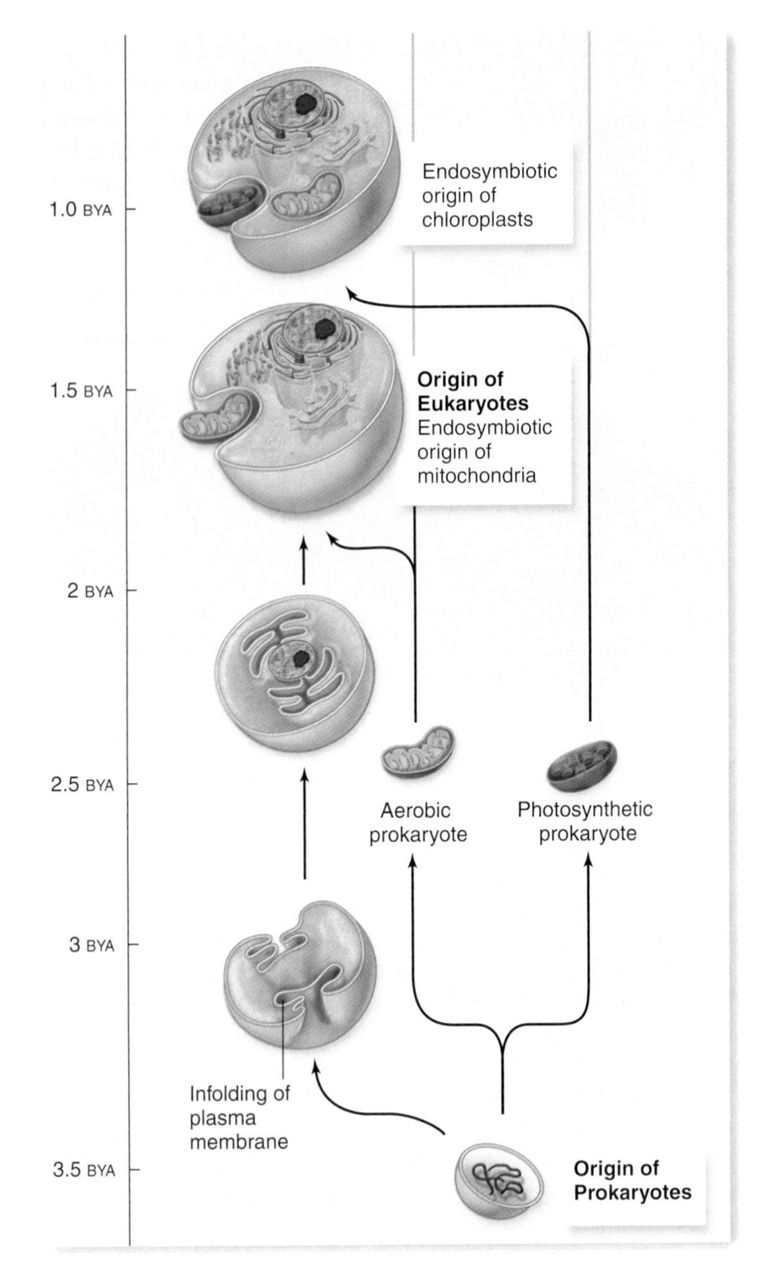

figure 4.19

PROPOSED ENDOSYMBIOTIC ORIGIN OF EUKARYOTIC CELLS. Both mitochondria and chloroplasts are thought to have arisen by endosymbiosis where a free-living cell is taken up but not digested. The cell that engulfed the future mitochondria and chloroplasts is thought to have first acquired a nuclear envelope and endomembrane system from infolding of the plasma membrane. The proposed series of events leading to modern eukaryotic cells is shown next to a rough time line of the history of Earth.

Mitochondria and chloroplasts are both involved in energy conversion. Both mitochondria and chloroplasts have their own DNA, which contains specific genes related to some of their functions, but both depend on nuclear genes for other functions.

Much evidence exists that both mitochondria and chloroplasts arose via endosymbiosis.

4.6 The Cytoskeleton

The cytoplasm of all eukaryotic cells is crisscrossed by a network of protein fibers that supports the shape of the cell and anchors organelles to fixed locations. This network, called the cytoskeleton, is a dynamic system, constantly forming and disassembling. Individual fibers consist of polymers of identical protein subunits that attract one another and spontaneously assemble into long chains. Fibers disassemble in the same way, as one subunit after another breaks away from one end of the chain.

Three types of fibers compose the cytoskeleton

Eukaryotic cells may contain the following three types of cytoskeletal fibers, each formed from a different kind of subunit: (1) actin filaments, sometimes called microfilaments, (2) microtubules, and (3) intermediate filaments.

Actin filaments (microfilaments)

Actin filaments are long fibers about 7 nm in diameter. Each filament is composed of two protein chains loosely twined together like two strands of pearls (figure 4.20). Each "pearl," or subunit, on the chains is the globular protein **actin.** Actin filaments exhibit polarity in that they have plus (+) and minus (–) ends. These designate the direction of growth of the filaments. Actin molecules spontaneously form these filaments, even in a test tube.

Cells regulate the rate of actin polymerization through other proteins that act as switches, turning on polymerization when appropriate. Actin filaments are responsible for cellular movements such as contraction, crawling, "pinching" during division, and formation of cellular extensions.

Microtubules

Microtubules, the largest of the cytoskeletal elements, are hollow tubes about 25 nm in diameter, each composed of a ring of 13 protein protofilaments (see figure 4.20). Globular proteins consisting of dimers of α- and β-*tubulin* subunits polymerize to form the 13 protofilaments. The protofilaments are arrayed side by side around a central core, giving the microtubule its characteristic tube shape.

In many cells, microtubules form from nucleation centers near the center of the cell and radiate toward the periphery. They are in a constant state of flux, continually polymerizing and depolymerizing. The average half-life of a microtubule ranges from as long as 10 minutes in a nondividing animal cell to as short as 20 seconds in a dividing animal cell. The ends of the microtubule are designated as plus (+) (away from the nucleation center) or minus (–) (toward the nucleation center).

Along with facilitating cellular movement, microtubules provide organization to the cytoplasm and are responsible for moving materials within the cell itself, as described shortly.

Intermediate filaments

The most durable element of the cytoskeleton in animal cells is a system of tough, fibrous protein molecules twined together in an overlapping arrangement (see figure 4.20). These **intermediate filaments** are characteristically 8 to 10 nm in diameter, intermediate in size between actin filaments and microtubules. Once formed, intermediate filaments are stable and usually do not break down.

Intermediate filaments constitute a heterogeneous group of cytoskeletal fibers. The most common type, composed of protein subunits called *vimentin,* provides structural stability for many kinds of cells. *Keratin,* another class of intermediate filament, is found in epithelial cells (cells that line organs and body cavities) and associated structures such as hair and fingernails. The intermediate filaments of nerve cells are called *neurofilaments.*

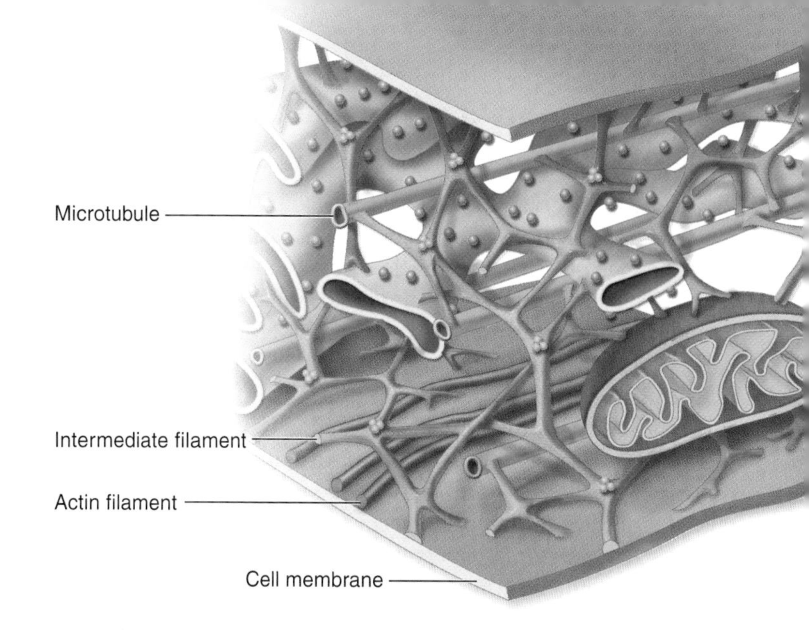

a. Actin filaments

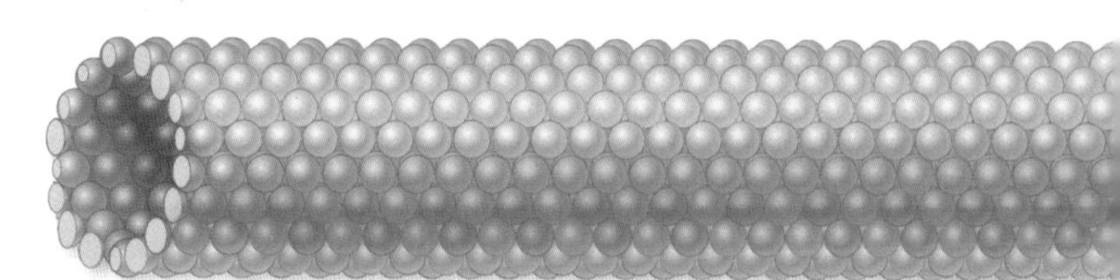

b. Microtubules

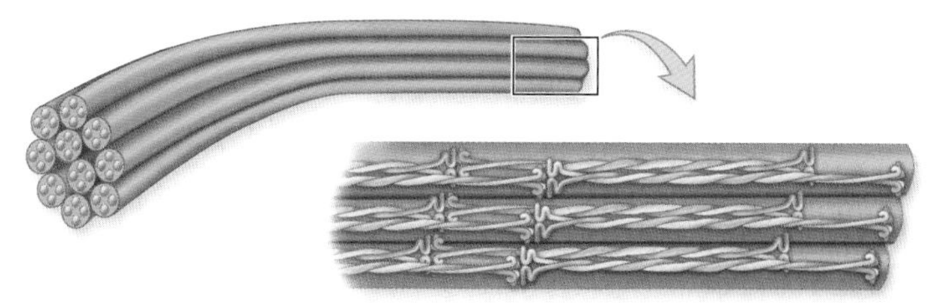

c. Intermediate filament

figure 4.20

MOLECULES THAT MAKE UP THE CYTOSKELETON. ***a.*** *Actin filaments:* Actin filaments, also called *microfilaments,* are made of two strands of the globular protein actin twisted together. They often are found in bundles, or a branching network. Actin filaments in many cells are concentrated below the plasma membrane in bundles known as stress fibers, which may have a contractile function. ***b.*** *Microtubules:* Microtubules are composed of α and β tubulin protein subunits arranged side by side to form a tube. Microtubules are comparatively stiff cytoskeletal elements with many functions in the cell including intracellular transport and the separation of chromosomes during mitosis. ***c.*** *Intermediate filaments:* Intermediate filaments are composed of overlapping staggered tetramers of protein. These tetramers are then bundled into cables. This molecular arrangement allows for a ropelike structure that imparts tremendous mechanical strength to the cell.

Centrosomes are microtubule-organizing centers

Centrioles are barrel-shaped organelles found in the cells of animals and most protists. They occur in pairs, usually located at right angles to each other near the nuclear membranes (figure 4.21); the region surrounding the pair in almost all animal cells is referred to as a **centrosome.** Surrounding the centrioles in the centrosome is the **pericentriolar material,** which contains ring-shaped structures composed of tubulin. The pericentriolar material can nucleate the assembly of microtubules in animal cells. Structures with this function are called **microtubule-organizing centers.** The centrosome is also responsible for the reorganization of microtubules that occurs during cell division. The centrosomes of plants and fungi lack centrioles, but still contain microtubule-organizing centers. You will learn more about the actions of the centrosomes when we describe the process of cell division in chapter 10.

The cytoskeleton helps move materials within cells

Actin filaments and microtubules often orchestrate their activities to affect cellular processes. For example, during cell reproduction (see chapter 10), newly replicated chromosomes move to opposite sides of a dividing cell because they are attached to shortening microtubules. Then, in animal cells, a belt of actin pinches the cell in two by contracting like a purse string.

Muscle cells also use actin filaments sliding relative to filaments of the motor protein myosin to contract. The fluttering of an eyelash, the flight of an eagle, and the awkward crawling of a baby all depend on these cytoskeletal movements within muscle cells.

Not only is the cytoskeleton responsible for the cell's shape and movement, but it also provides a scaffold that holds certain enzymes and other macromolecules in defined areas of the cytoplasm. For example, many of the enzymes involved in cell metabolism bind to actin filaments; so do ribosomes. By moving and anchoring particular enzymes near one another, the cytoskeleton, like the endoplasmic reticulum, helps organize the cell's activities.

Molecular motors

All eukaryotic cells must move materials from one place to another in the cytoplasm. One way that cells do this is by using the channels of the endoplasmic reticulum as an intracellular highway. Material can also be moved using vesicles loaded with cargo that can move along the cytoskeleton like a railroad track. For example, in a nerve cell with an axon that may extend far from the cell body, vesicles can be moved along tracks of microtubules from the cell body to the end of the axon.

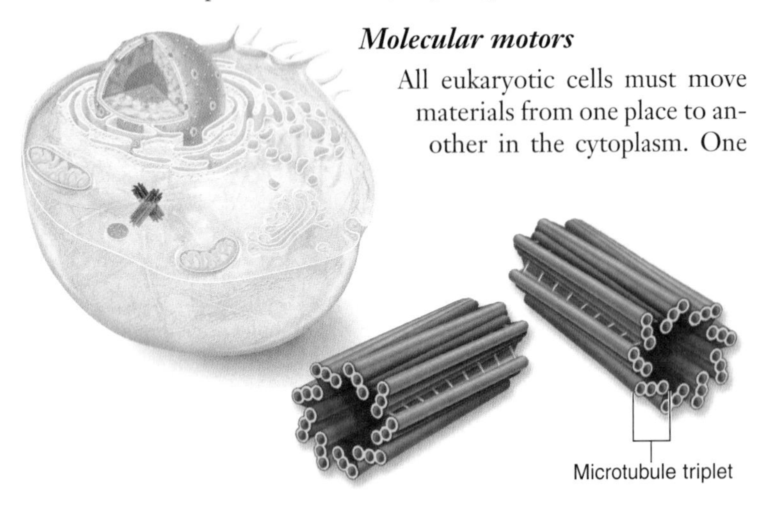

figure 4.21

CENTRIOLES. Each centriole is composed of nine triplets of microtubules. Centrioles are not found in plant cells. In animal cells they help to organize microtubules.

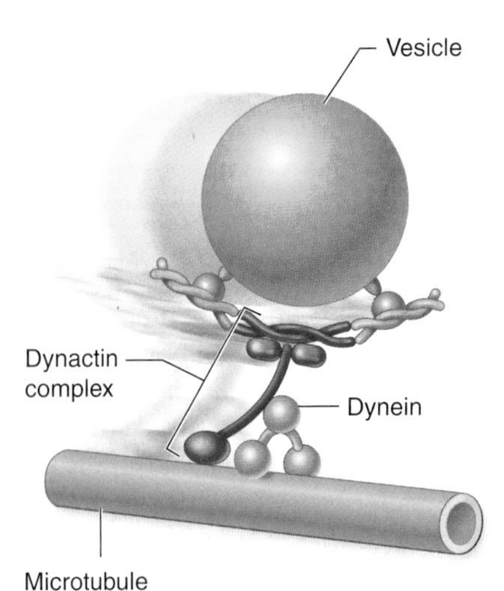

figure 4.22

MOLECULAR MOTORS. Vesicles can be transported along microtubules using motor proteins that use ATP to generate force. The vesicles are attached to motor proteins by connector molecules, such as the dynactin complex shown here. The motor protein dynein moves the connected vesicle along microtubules.

Four components are required: (1) a vesicle or organelle that is to be transported, (2) a motor protein that provides the energy-driven motion, (3) a connector molecule that connects the vesicle to the motor molecule, and (4) microtubules on which the vesicle will ride like a train on a rail (figure 4.22).

The direction that a vesicle is moved depends on the identity of the motor protein involved, and the fact that microtubules are organized with their plus ends toward the periphery of the cell. In one case, a protein called kinectin binds vesicles to the motor protein **kinesin.** Kinesin uses ATP to power its movement toward the cell periphery, dragging the vesicle with it as it travels along the microtubule toward the plus end. As nature's tiniest motors, these proteins literally pull the transport vesicles along the microtubular tracks. Another set of vesicle proteins, called the dynactin complex, binds vesicles to the motor protein **dynein** (see figure 4.22), which directs movement in the opposite direction along microtubules toward the minus end, inward toward the cell's center. (Dynein is also involved in the movement of eukaryotic flagella, as discussed later.) The destination of a particular transport vesicle and its content is thus determined by the nature of the linking protein embedded within the vesicle's membrane.

The major eukaryotic cell structures and their respective functions are summarized in table 4.2.

The three principal fibers of the cytoskeleton are actin filaments (microfilaments), microtubules, and intermediate filaments. These fibers interact to modulate cell shape and permit cell movement and act to move materials within the cytoplasm.

Material is also moved in large cells using vesicles and molecular motors. The motor proteins move the vesicles along tracks of microtubules.

TABLE 4.2 Eukaryotic Cell Structures and Their Functions

Structure		Description	Function
Plasma membrane		Phospholipid bilayer with embedded proteins	Regulates what passes into and out of cell; cell-to-cell recognition; connection and adhesion; cell communication
Nucleus		Structure (usually spherical) that contains chromosomes and is surrounded by double membrane	Instructions for protein synthesis and cell reproduction; contains genetic information
Chromosomes		Long threads of DNA that form a complex with protein	Contain hereditary information used to direct synthesis of proteins
Nucleolus		Site of genes for rRNA synthesis	Synthesis of rRNA and ribosome assembly
Ribosomes		Small, complex assemblies of protein and RNA, often bound to ER	Sites of protein synthesis
Endoplasmic reticulum (ER)		Network of internal membranes	Intracellular compartment forms transport vesicles; participates in lipid synthesis and synthesis of membrane or secreted proteins
Golgi apparatus		Stacks of flattened vesicles	Packages proteins for export from cell; forms secretory vesicles
Lysosomes		Vesicles derived from Golgi apparatus that contain hydrolytic digestive enzymes	Digest worn-out organelles and cell debris; digest material taken up by endocytosis
Microbodies		Vesicles that are formed from incorporation of lipids and proteins and that contain oxidative and other enzymes	Isolate particular chemical activities from rest of cell
Mitochondria		Bacteria-like elements with double membrane	"Power plants" of the cell; sites of oxidative metabolism
Chloroplasts		Bacteria-like elements with membranes containing chlorophyll, a photosynthetic pigment	Sites of photosynthesis
Cytoskeleton		Network of protein filaments	Structural support; cell movement; movement of vesicles within cells
Flagella (cilia)		Cellular extensions with 9 + 2 arrangement of pairs of microtubles	Motility or moving fluids over surfaces
Cell wall		Outer layer of cellulose or chitin; or absent	Protection; support

4.7 Extracellular Structures and Cell Movement

Essentially all cell motion is tied to the movement of actin filaments, microtubules, or both. Intermediate filaments act as intracellular tendons, preventing excessive stretching of cells, and actin filaments play a major role in determining the shape of cells. Because actin filaments can form and dissolve so readily, they enable some cells to change shape quickly.

Some cells crawl

The arrangement of actin filaments within the cell cytoplasm allows cells to crawl, literally! Crawling is a significant cellular phenomenon, essential to such diverse processes as inflammation, clotting, wound healing, and the spread of cancer. White blood cells in particular exhibit this ability. Produced in the bone marrow, these cells are released into the circulatory system and then eventually crawl out of venules and into the tissues to destroy potential pathogens.

At the leading edge of a crawling cell, actin filaments rapidly polymerize, and their extension forces the edge of the cell forward. This extended region is stabilized when microtubules polymerize into the newly formed region. Forward movement of the cell overall is then achieved through the action of the protein **myosin,** which is best known for its role in muscle contraction. Myosin motors along the actin filaments contract, pulling the contents of the cell toward the newly extended front edge.

Overall crawling of the cell takes place when these steps occur continuously, with a leading edge extending and stabilizing, and then motors contracting to pull the remaining cell contents along. Receptors on the cell surface can detect molecules outside the cell and stimulate extension in specific directions, allowing cells to move toward particular targets.

Flagella and cilia aid movement

Earlier in this chapter, we described the structure of prokaryotic flagella. Eukaryotic cells have a completely different kind of flagellum, consisting of a circle of nine microtubule pairs surrounding two central microtubules; this arrangement is referred to as the **9 + 2 structure** (figure 4.23).

As pairs of microtubules move past each other using arms composed of the motor protein dynein, the eukaryotic flagellum *undulates,* rather than rotates. When examined carefully, each flagellum proves to be an outward projection of the cell's interior, containing cytoplasm and enclosed by the plasma membrane. The microtubules of the flagellum are derived from a **basal body,** situated just below the point where the flagellum protrudes from the surface of the cell.

The flagellum's complex microtubular apparatus evolved early in the history of eukaryotes. Although today the cells of many multicellular and some unicellular eukaryotes no longer possess flagella and are nonmotile, an organization similar to the 9 + 2 arrangement of microtubules can still be found within them, in structures called **cilia** (singular, *cilium*). Cilia are short cellular projections that are often organized in rows. They are more numerous than flagella on the cell surface, but have the same internal structure.

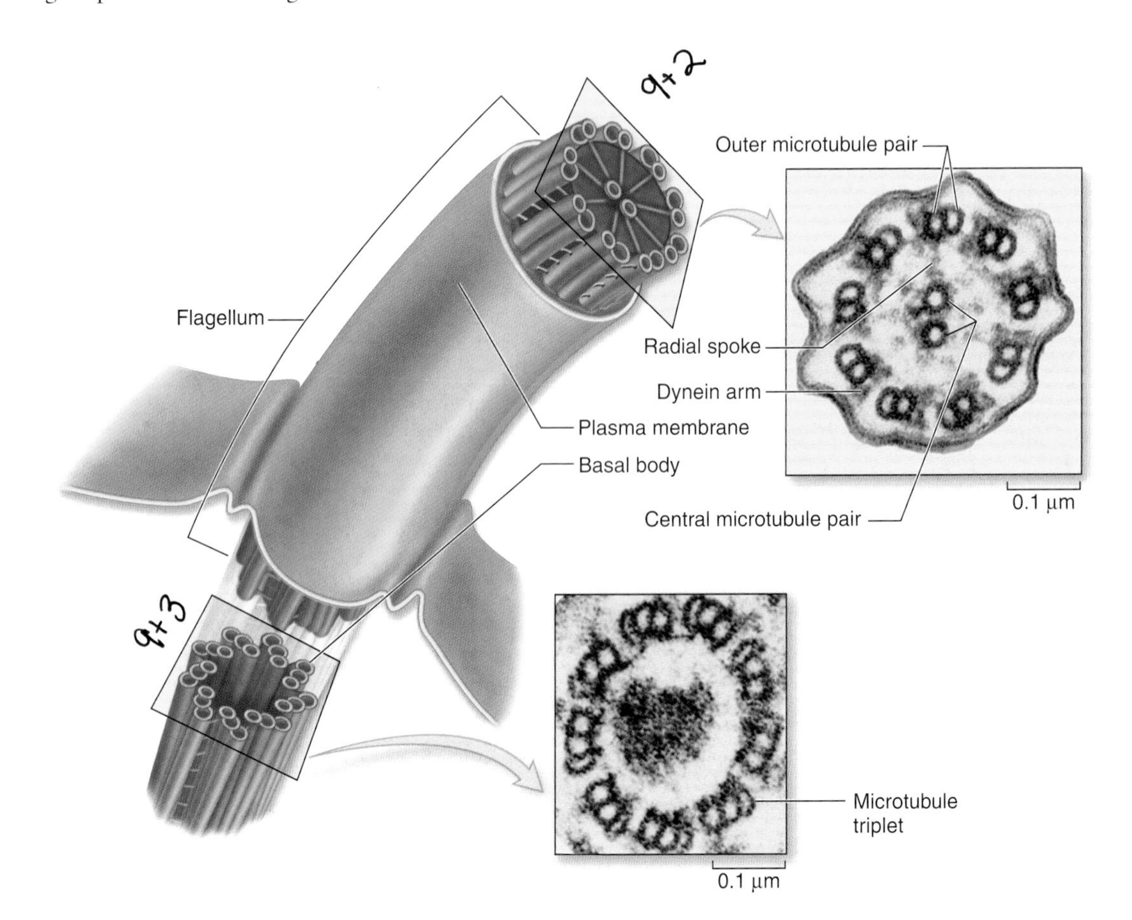

figure 4.23

FLAGELLA AND CILIA. A eukaryotic flagellum originates directly from a basal body. The flagellum has two microtubules in its core connected by radial spokes to an outer ring of nine paired microtubules with dynein arms (9 + 2 structure). The basal body consists of nine microtubule triplets connected by short protein segments. The structure of cilia is similar to that of flagella, but cilia are usually shorter.

In many multicellular organisms, cilia carry out tasks far removed from their original function of propelling cells through water. In several kinds of vertebrate tissues, for example, the beating of rows of cilia move water over the tissue surface. The sensory cells of the vertebrate ear also contain conventional cilia surrounded by actin-based stereocilia; sound waves bend these structures and provide the initial sensory input for hearing. Thus, the 9 + 2 structure of flagella and cilia appears to be a fundamental component of eukaryotic cells (figure 4.24).

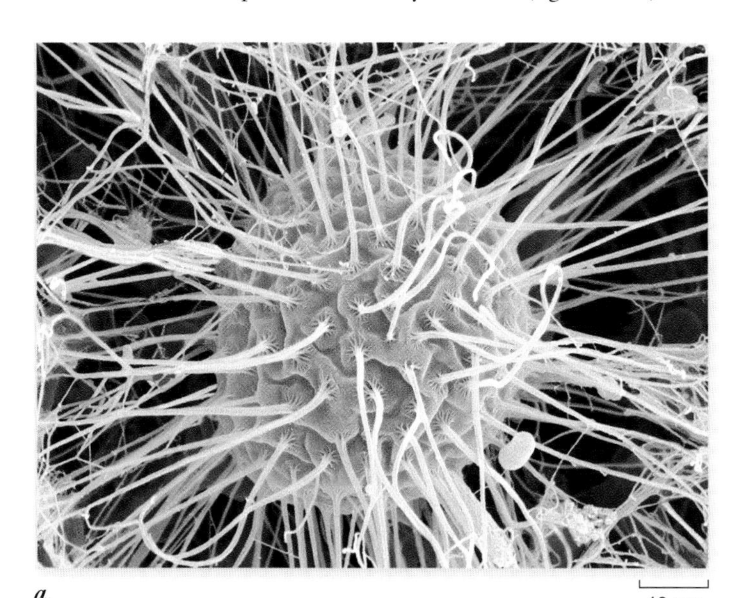

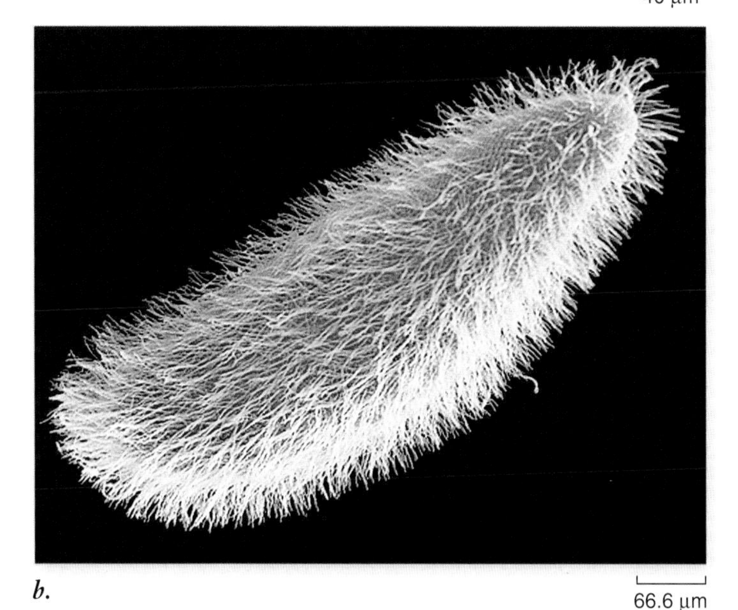

figure 4.24

FLAGELLA AND CILIA. ***a.*** A flagellated green alga with numerous flagella that allow it to move through the water. ***b.*** Paramecium are covered with many cilia, which beat in unison to move the cell. The cilia can also be used to move fluid into the mouth to ingest material.

inquiry

The passageways of human trachea (the path of airflow into and out of the lungs) are known to be lined with ciliated cells. What function could these cilia perform?

Plant Cell Walls Provide Protection and Support

The cells of plants, fungi, and many types of protists have cell walls, which protect and support the cells. The cell walls of these eukaryotes are chemically and structurally different from prokaryotic cell walls. In plants and protists, the cell walls are composed of fibers of the polysaccharide cellulose, whereas in fungi, the cell walls are composed of chitin.

In plants, **primary walls** are laid down when the cell is still growing, and between the walls of adjacent cells is a sticky substance called the **middle lamella,** which glues the cells together (figure 4.25). Some plant cells produce strong **secondary walls,** which are deposited inside the primary walls of fully expanded cells.

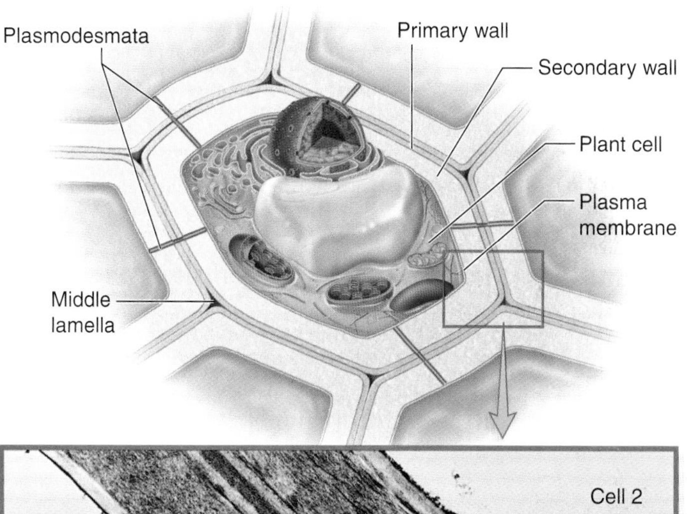

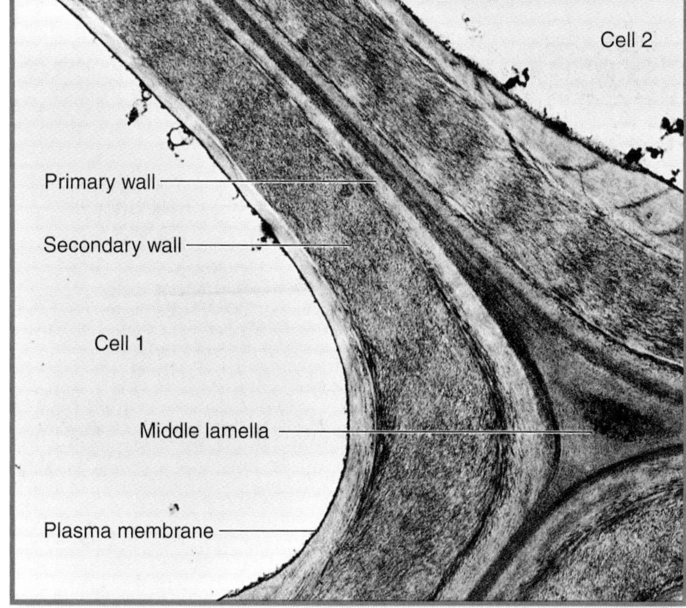

figure 4.25

CELL WALLS IN PLANTS. Plant cell walls are thick, strong, and rigid. Primary cell walls are laid down when the cell is young. Thicker secondary cell walls may be added later when the cell is fully grown.

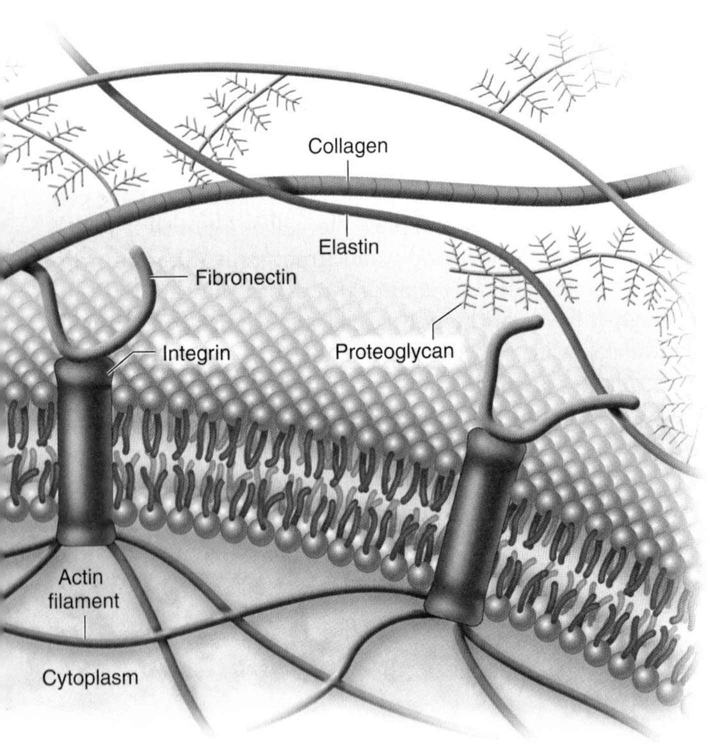

figure 4.26

THE EXTRACELLULAR MATRIX. Animal cells are surrounded by an extracellular matrix composed of various glycoproteins that give the cells support, strength, and resilience.

Animal cells secrete an extracellular matrix

Animal cells lack the cell walls that encase plants, fungi, and most protists. Instead, animal cells secrete an elaborate mixture of glycoproteins into the space around them, forming the **extracellular matrix (ECM)** (figure 4.26). The fibrous protein collagen, the same protein found in cartilage, tendons, and ligaments may be abundant in the ECM. Strong fibers of collagen and another fibrous protein, elastin, are embedded within a complex web of other glycoproteins, called proteoglycans, that form a protective layer over the cell surface.

The ECM of some cells is attached to the plasma membrane by a third kind of glycoprotein, **fibronectin.** Fibronectin molecules bind not only to ECM glycoproteins but also to proteins called **integrins,** which are an integral part of the plasma membrane. Integrins extend into the cytoplasm, where they are attached to the microfilaments and intermediate filaments of the cytoskeleton. Linking ECM and cytoskeleton, integrins allow the ECM to influence cell behavior in important ways, altering gene expression and cell migration patterns by a combination of mechanical and chemical signaling pathways. In this way, the ECM can help coordinate the behavior of all the cells in a particular tissue.

Table 4.3 compares and reviews the features of three types of cells.

Cell movement involves proteins. These can either be internal in the case of crawling cells that use actin and myosin or external in the case of cells powered by cilia or flagella.

Eukaryotic cilia and flagella are composed of bundles of microtubules in a 9 + 2 array. They undulate rather than rotate.

Plant cells have a cellulose-based cell wall.

In animal cells, which lack a cell wall, the cytoskeleton is linked by integrin proteins to a web of glycoproteins called the extracellular matrix.

TABLE 4.3 **A Comparison of Prokaryotic, Animal, and Plant Cells**

	Prokaryote	Animal	Plant
EXTERIOR STRUCTURES			
Cell wall	Present (protein-polysaccharide)	Absent	Present (cellulose)
Cell membrane	Present	Present	Present
Flagella/cilia	Flagella may be present	May be present (9 + 2 structure)	Absent except in sperm of a few species (9 + 2 structure)
INTERIOR STRUCTURES			
ER	Absent	Usually present	Usually present
Ribosomes	Present	Present	Present
Microtubules	Absent	Present	Present
Centrioles	Absent	Present	Absent
Golgi apparatus	Absent	Present	Present
Nucleus	Absent	Present	Present
Mitochondria	Absent	Present	Present
Chloroplasts	Absent	Absent	Present
Chromosomes	A single circle of DNA	Multiple; DNA–protein complex	Multiple; DNA–protein complex
Lysosomes	Absent	Usually present	Present
Vacuoles	Absent	Absent or small	Usually a large single vacuole

concept review

4.1 Cell Theory

Modern cell theory states that organisms are composed of one or more cells. Cells are the smallest unit of life and arise from preexisting cells.

- Cell size is constrained by the effective distance of diffusion within a cell, from the surface to the interior of the cell.
- As a cell increases in size the surface area increases as a square function and the volume increases as a cubic function.
- Large cells deal with the diffusion problem by having more than one nucleus or by becoming flattened or elongated.
- The visualization of cells and their components is facilitated by microscopes and staining cell structures.
- All cells have DNA, a cytoplasm, a plasma membrane, and ribosomes.

4.2 Prokaryotic Cells (figure 4.3)

Prokaryotic cells do not have a nucleus or an internal membrane system, and they lack membrane-bounded organelles.

- The plasma membrane is surrounded by a rigid cell wall that maintains shape and helps maintain osmotic balance.
- The plasma membrane in some prokaryotes is infolded and provides similar functions to eukaryotic internal membranes.
- Bacteria have a cell wall made up of peptidoglycan. Archaeal cell walls have different architecture.
- Archaeal plasma membranes differ from bacteria and eukaryotes.
- The plasma membrane in some archaea is a monolayer composed of saturated lipids attached to glycerol at each end.
- Structurally Archaea resemble prokaryotes, but functionally they more closely resemble eukaryotes.
- Prokaryotic flagella rotate because of proton transfer.

4.3 Eukaryotic Cells (figures 4.6 and 4.7)

Eukaryotic cells have a membrane-bounded nucleus, an endomembrane system, and many different organelles.

- The nucleus contains genetic information.
- The nuclear envelope consists of two phospholipid bilayers; the outer layer is contiguous with the ER.
- The inside of the nuclear envelope is covered with nuclear lamins, which maintain the shape of the nucleus.
- Nuclear pores allow exchange of small molecules between the nucleoplasm and the cytoplasm.
- DNA is organized with proteins into chromatin.
- The nucleolus is a region of the nucleoplasm where rRNA is transcribed and ribosomes are assembled.
- Ribosomes are composed of RNA and protein and use information in mRNA to direct the synthesis of proteins.

4.4 The Endomembrane System

The endomembrane system forms compartments and vesicles and provides channels to carry molecules and surfaces for synthesis of macromolecules.

- The endoplasmic reticulum (ER) creates channels and passages within the cytoplasm (figure 4.11).
- The interior compartment of the ER is called the cisternal space, or lumen.
- The rough endoplasmic reticulum (RER) has ribosomes on the surface and is composed mainly of flattened sacs. RER is involved in protein synthesis and modification.
- The smooth endoplasmic reticulum (SER) lacks ribosomes and is composed more of tubules. SER is involved in synthesis of carbohydrates and lipids and in detoxification.
- The Golgi apparatus receives vesicles from the ER on the *cis* face, modifies and packages macromolecules, and transports them in vesicles formed on the *trans* face (figure 4.13).
- Lysosomes are vesicles containing enzymes that break down macromolecules located in food vacuoles and recycle the components of old organelles (figure 4.14).
- Microbodies contain enzymes and grow by incorporating lipids and proteins before they divide.
- Peroxisomes contain enzymes that catalyze oxidation reactions, resulting in the formation of hydrogen peroxide.
- Plants have many specialized vacuoles. The conspicuous central vacuole, surrounded by the tonoplast membrane, is used for storage, maintaining water balance, and growth.

4.5 Mitochondria and Chloroplasts: Cellular Generators

Mitochondria and chloroplasts have a double-membrane structure, contain their own DNA, can synthesize proteins, can divide, and are involved in energy metabolism.

- Mitochondria produce ATP using energy-contained macromolecules (figure 4.17).
  - *The inner membrane is extensively folded into layers called cristae.*
  - *The intermembrane space is a compartment between the inner and outer membrane.*
  - *The mitochondrial matrix is a compartment consisting of the fluid within the inner membrane.*
- Chloroplasts use light to generate ATP and sugars (figure 4.18).
  - *In addition to a double membrane, chloroplasts also have stacked membranes called grana that contain vesicles called thylakoids.*
  - *The fluid stroma surrounds the thylakoids.*
- Evidence indicates that mitochondria and chloroplasts arose via endosymbiosis.

4.6 The Cytoskeleton

The cytoskeleton is composed of three different fibers that support cell shape and anchors organelles and enzymes (figure 4.20).

- Actin filaments, or microfilaments, are long, thin polymers responsible for cell movement, cytoplasmic division, and formation of cellular extensions.
- Microtubules are hollow structures that are used in cell movement and movement of materials within a cell.
- Intermediate filaments are stable structures that serve a wide variety of functions.
- Paired centrioles, located in the centrosome, help assemble the nuclear division apparatus of animal cells (figure 4.21).
- Molecular motors move vesicles along microtubules.

4.7 Extracellular Structures and Cell Movement

Extracellular structures provide protection, support, strength, and cell recognition.

- Plants have cell walls composed of cellulose fibers. Fungi have cell walls composed of chitin.
- Animals have a complex extracellular matrix.
- Cell crawling occurs as actin polymerization forces the cell membrane forward while myosin pulls the cell forward.
- Eukaryotic flagella have a 9 + 2 structure and arise from a basal body.
- Cilia are shorter and more numerous flagella.

review questions

SELF TEST

1. Which of the following statements is NOT part of the cell theory?
   a. All organisms are composed of one or more cells.
   b. Cells come from other cells by division.
   c. Cells are the smallest living things.
   d. Eukaryotic cells have evolved from prokaryotic cells.
2. The most important factor that limits the size of a cell is—
   a. the amount of proteins and organelles that can be made by a cell
   b. the rate of diffusion
   c. the surface-area-to-volume ratio of the cell
   d. the amount of DNA in the cell
3. What type of microscope would you use to examine the surface details of a cell?
   a. Compound light microscope
   b. Transmission electron microscope
   c. Scanning electron microscope
   d. Confocal microscope
4. All cells have all of the following except—
   a. Plasma membrane
   b. Genetic material
   c. Cytoplasm
   d. Cell wall
5. Eukaryotic cells are more complex than prokaryotic cells. Which of the following would you NOT find in a prokaryotic cell?
   a. Cell wall
   b. Plasma membrane
   c. Nucleus
   d. Ribosomes
6. The difference between a gram-positive and gram-negative bacteria is—
   a. the thickness of the peptidoglycan cell wall
   b. the type of polysaccharide present in the cell wall
   c. the type and amount of protein in the cell wall
   d. the layers of cellulose in the cell wall
7. Which of the following is not true of bacterial flagella?
   a. Bacterial flagella rotate producing a spiral wave.
   b. Bacterial flagella are anchored to a basal body.
   c. Bacterial flagella are powered by a proton gradient.
   d. Bacterial flagella are composed of microtubules.
8. All eukaryotic cells possess each of the following except—
   a. Mitochondria
   b. Cell wall
   c. Cytoskeleton
   d. Nucleus
9. Ribosomal RNAs are manufactured in which specific area of a eukaryotic cell?
   a. The nucleus
   b. The cytoplasm
   c. The nucleolus
   d. The chromatin
10. Which of these organelles is NOT associated with the production of proteins in a cell?
    a. Ribosomes
    b. Smooth endoplasmic reticulum (SER)
    c. Rough endoplasmic reticulum (RER)
    d. Golgi apparatus
11. Proteins can move from the Golgi apparatus to—
    a. the extracellular fluid
    b. transport vesicles
    c. lysosomes
    d. all of the above
12. Lysosomes function to—
    a. carry proteins to the surface of the cell
    b. add short-chain carbohydrates to make glycoproteins
    c. break down organelles, proteins, and nucleic acids
    d. remove electrons and hydrogen atoms from hydrogen peroxide
13. What do chloroplasts and mitochondria have in common?
    a. Both are present in animal cells.
    b. Both have an outer membrane and an elaborate inner membrane.
    c. Both are present in all eukaryotic cells.
    d. Both organelles function to produce glucose.
14. Eukaryotic cells are composed of three types of cytoskeletal filaments. How are these three filaments similar?
    a. They contribute to the shape of the cell.
    b. They are all made of the same type of protein.
    c. They are all the same size and shape.
    d. They are all equally dynamic and flexible.
15. Animal cells connect to the extracellular matrix through—
    a. glycoproteins
    b. fibronectins
    c. integrins
    d. collagen

CHALLENGE QUESTIONS

1. Eukaryotic cells are typically larger than prokaryotic cells (refer to figure 4.2). How might the difference in the cellular structure of eukaryotic versus a prokaryotic cell help to explain this observation?
2. The smooth endoplasmic reticulum is the site of synthesis of the phospholipids that make up all the membranes of a cell—especially the plasma membrane. Use the diagram of an animal cell (figure 4.6) to trace a pathway that would carry a phospholipid molecule from the SER to the plasma membrane. What endomembrane compartments would the phospholipids travel through? How can a phospholipid molecule move between membrane compartments?
3. Use the information provided in table 4.3 to develop a set of predictions about the properties of mitochondria and chloroplasts if these organelles were once free-living prokaryotic cells. How do your predictions match with the evidence for endosymbiosis?
4. In evolutionary theory, homologous traits are those with a similar structure and function derived from a common ancestor. Analogous traits represent adaptations to a similar environment, but from distantly related organisms. Consider the structure and function of the flagella found on eukaryotic and prokaryotic cells. Are the flagella an example of a homologous or analogous trait? Defend your answer.
5. The protist, *Giardia lamblia*, is the organism associated with water-borne diarrheal diseases. *Giardia* is an unusual eukaryote because it seems to lack mitochondria. Explain the existence of a mitochondria-less eukaryote in the context of the endosymbiotic theory.

Do you need additional review? *Visit* www.ravenbiology.com *for practice quizzes, animations, videos, and activities designed to help you master the material in this chapter.*

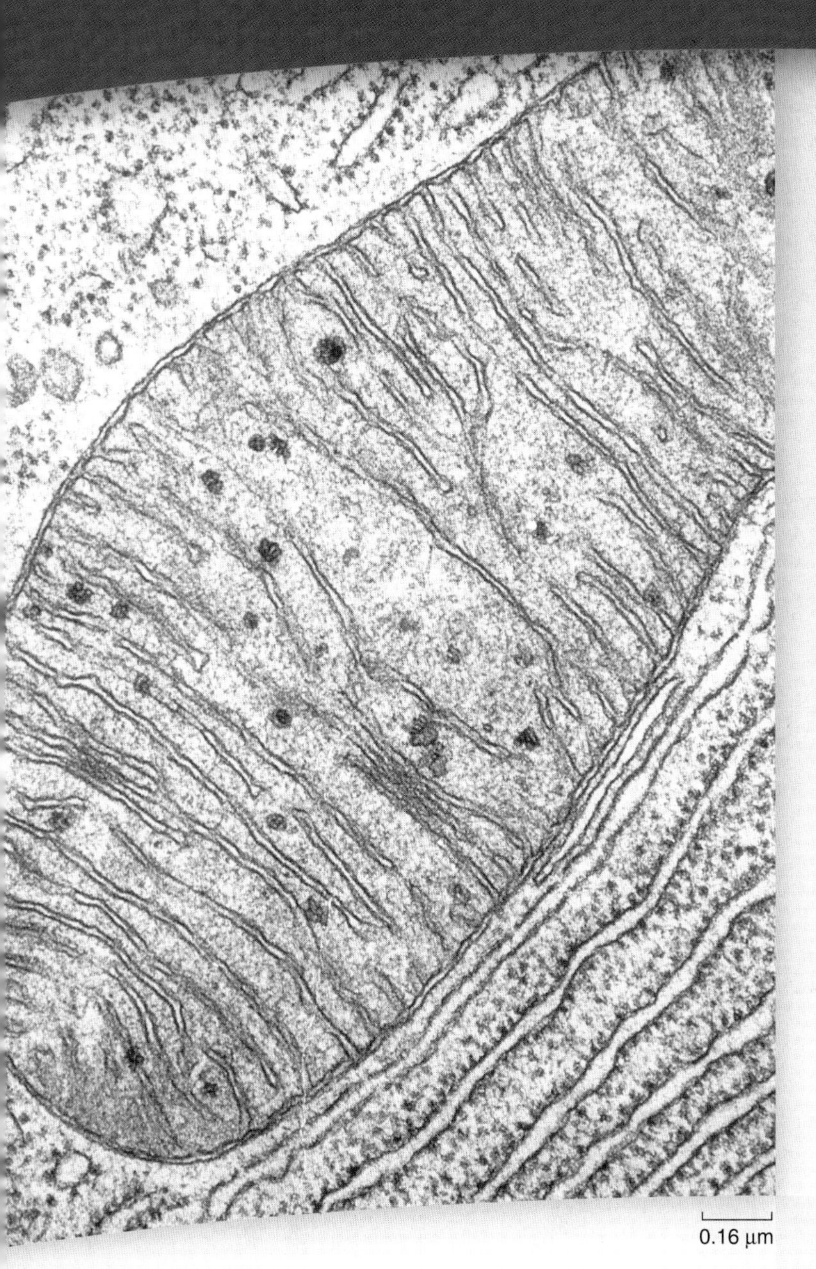

chapter 5

Membranes

introduction

AMONG A CELL'S MOST IMPORTANT ACTIVITIES are its interactions with the environment, a give-and-take that never ceases. Without it, life could not persist. Living cells are encased within a lipid membrane through which few water-soluble substances can pass; but at the same time, the membrane contains protein passageways that permit specific substances to move into and out of the cell and allow the cell to exchange information with its environment. Eukaryotic cells also contain internal membranes like those of the mitochondrion and endoplasmic reticulum pictured here. We call the delicate skin of lipids with embedded protein molecules that encase the cell a **plasma membrane.** This chapter examines the structure and function of this remarkable membrane.

concept outline

5.1 The Structure of Membranes

The membranes that encase all living cells are sheets of lipid only two molecules thick; more than 10,000 of these sheets piled on one another would just equal the thickness of this sheet of paper. Biologists established the components of membranes—not only lipids, but also proteins and other molecules—through biochemical assays, but the nature of the membrane structure remained elusive.

We begin by considering the theories that have been advanced about membrane structure. We then look at the individual components of membranes more closely.

The fluid mosaic model shows proteins embedded in a fluid lipid bilayer

The lipid layer that forms the foundation of a cell's membranes is a bilayer formed of **phospholipids** (figure 5.1). For many years, biologists thought that the protein components of the cell membrane covered the inner and outer surfaces of the phospholipid bilayer like a coat of paint. An early model portrayed the membrane as a sandwich; a phospholipid bilayer between two layers of globular protein. This model, however, was not consistent with what researchers were learning in the 1960s about the structure of membrane proteins.

Unlike most proteins found within cells, membrane proteins are not very soluble in water. These **globular proteins** possess long stretches of nonpolar hydrophobic amino acids. If such proteins indeed coated the surface of the lipid bilayer, then their nonpolar portions would separate the polar portions of the phospholipids from water, causing the bilayer to dissolve! This was clearly not the case, so the model required modification.

In 1972, S. Jonathan Singer and Garth J. Nicolson revised the model in a simple but profound way: They proposed that the globular proteins are *inserted* into the lipid bilayer, with their nonpolar segments in contact with the nonpolar interior of the bilayer and their polar portions protruding out from the membrane surface. In this model, called the **fluid mosaic model,** a mosaic of proteins floats in or on the fluid lipid bilayer like boats on a pond (figure 5.2).

figure 5.1

DIFFERENT VIEWS OF PHOSPHOLIPID STRUCTURE. Phospholipids are composed of glycerol (*pink*) linked to two fatty acids and a phosphate group. The phosphate group (*yellow*) can have additional molecules attached, such as the positively charged choline (*green*) shown. Phosphatidylcholine is a common component of membranes, it is shown in ***(a)*** with its chemical formula, ***(b)*** as a space-filling model and, ***(c)*** as the icon that is used in most of the figures in this chapter. The phosphate portion of the molecule is hydrophilic, and the fatty acid tails are hydrophobic. This allows them to associate into bilayers, with the hydrophobic tails in the middle, in water.

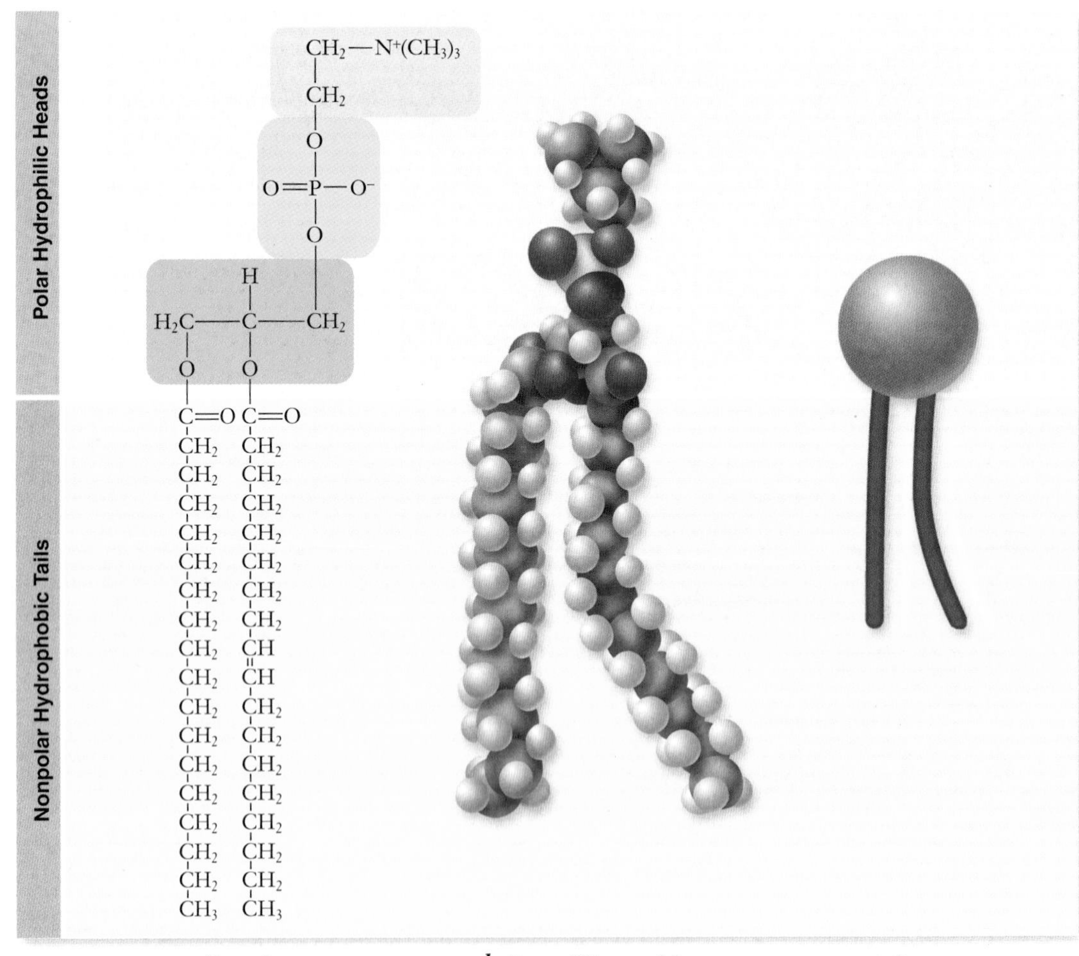

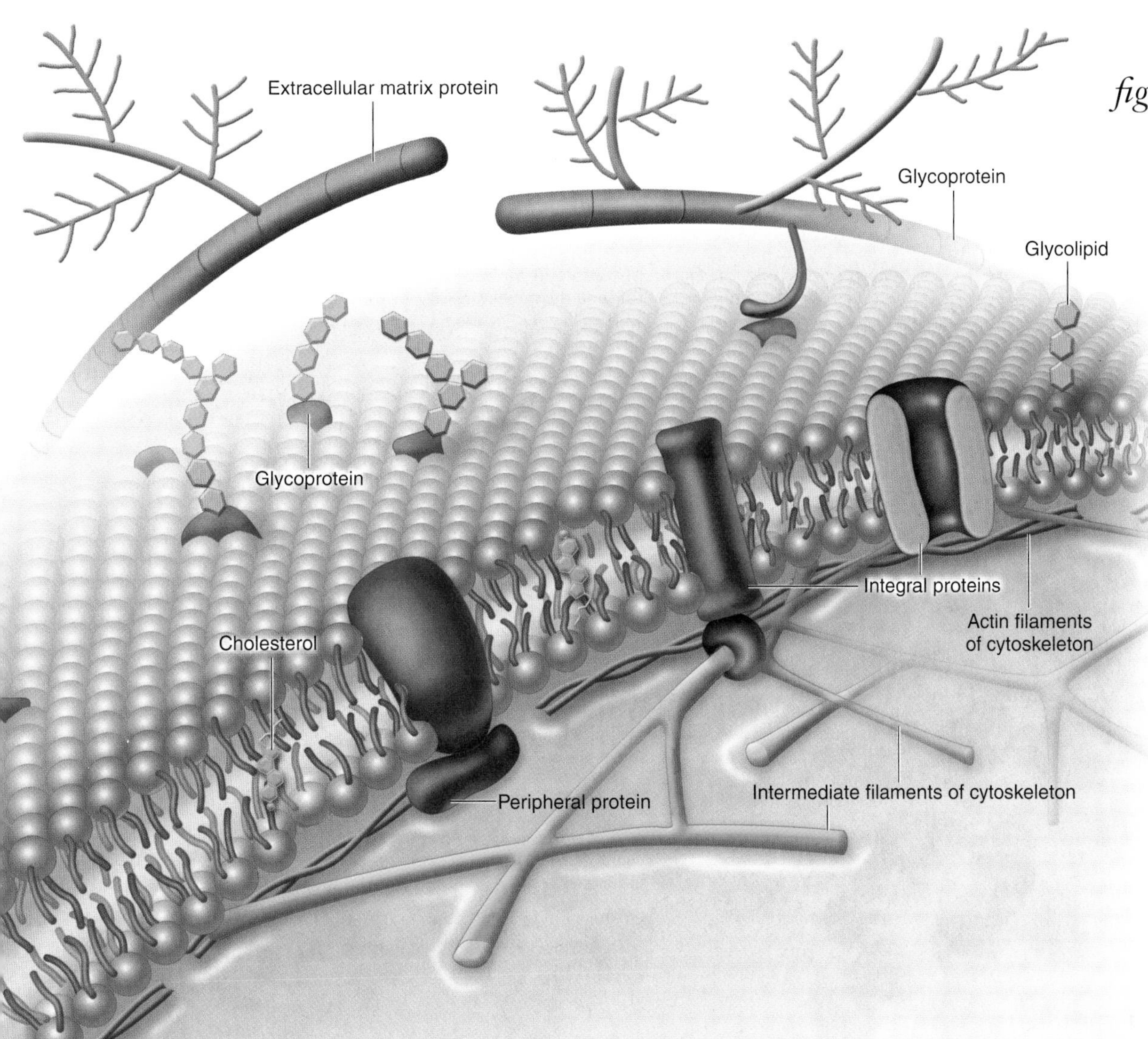

figure 5.2

THE FLUID MOSAIC MODEL OF CELL MEMBRANES. Integral proteins protrude through the plasma membrane, with nonpolar regions that tether them to the membrane's nonpolar interior. Carbohydrate chains are often bound to the extracellular portion of these proteins forming glycoproteins. Peripheral membrane proteins are associated with the surface of the membrane. Membrane phospholipids can be modified by the addition of carbohydrates to form glycolipids. In addition, a variety of associated proteins are found inside and outside the cell. Inside the cell, actin filaments and intermediate filaments interact with membrane proteins. Outside the cell, many animal cells have an elaborate extracellular matrix composed primarily of glycoproteins.

Cellular membranes donsist of four component groups

A eukaryotic cell contains many membranes. Although they are not all identical, they share the same fundamental architecture. Cell membranes are assembled from four components (table 5.1):

1. **Phospholipid bilayer.** Every cell membrane is composed of phospholipids in a bilayer. The other components of the membrane are embedded within the bilayer, which provides a flexible matrix and, at the same time, imposes a barrier to permeability. Animal cell membranes also contain cholesterol, a steroid with a polar hydroxyl group (—OH).
2. **Transmembrane proteins.** A major component of every membrane is a collection of proteins that float in the lipid bilayer. These proteins provide passageways that allow substances and information to cross the membrane. Many membrane proteins are not fixed in position; they can move about, just as the phospholipid molecules do. Some membranes are crowded with proteins, but in others, the proteins are more sparsely distributed. Because these proteins are embedded in the membrane structure they are also called **integral membrane proteins.**
3. **Interior protein network.** Membranes are structurally supported by intracellular proteins that reinforce the membrane's shape. For example, a red blood cell has a characteristic biconcave shape because a scaffold made of a protein called spectrin links proteins in the plasma membrane with actin filaments in the cell's cytoskeleton.

   Membranes use networks of other proteins to control the lateral movements of some key membrane proteins, anchoring them to specific sites. Proteins that are associated with the membrane but not part of its structure are usually called **peripheral membrane proteins.**
4. **Cell surface markers.** As you learned in the preceding chapter, membrane sections assemble in the endoplasmic reticulum, transfer to the Golgi apparatus, and then are transported to the plasma membrane. The ER adds chains of sugar molecules to membrane proteins and lipids, converting them into **glycoproteins** and **glycolipids.** Different cell types exhibit different varieties of these glycoproteins and glycolipids, which act as cell identity markers, on their surfaces.

TABLE 5.1 Components of the Cell Membrane

Component	Composition	Function	How It Works	Example
Phospholipid bilayer	Phospholipid molecules	Provides permeability barrier, matrix for proteins	Excludes water-soluble molecules from nonpolar interior of bilayer and cell	Bilayer of cell is impermeable to large water-soluble molecules, such as glucose
Transmembrane proteins	Carriers	Actively or passively transport molecules across membrane	Move specific molecules through the membrane in a series of conformational changes	Glycophorin carrier for sugar transport; sodium–potassium pump
	Channels	Passively transport molecules across membrane	Create a selective tunnel that acts as a passage through membrane	Sodium and potassium channels in nerve, heart, and muscle cells
	Receptors	Transmit information into cell	Signal molecules bind to cell surface portion of the receptor protein; this alters the portion of the receptor protein within the cell, inducing activity	Specific receptors bind peptide hormones and neurotransmitters
Interior protein network	Spectrins	Determine shape of cell	Form supporting scaffold beneath membrane, anchored to both membrane and cytoskeleton	Red blood cell
	Clathrins	Anchor certain proteins to specific sites, especially on the exterior plasma membrane in receptor-mediated endocytosis	Proteins line coated pits and facilitate binding to specific molecules	Localization of low-density lipoprotein receptor within coated pits
Cell surface markers	Glycoproteins	"Self" recognition	Create a protein/carbohydrate chain shape characteristic of individual	Major histocompatibility complex protein recognized by immune system
	Glycolipid	Tissue recognition	Create a lipid/carbohydrate chain shape characteristic of tissue	A, B, O blood group markers

Originally, it was believed that because of its fluidity, the plasma membrane was uniform, with lipids and proteins free to diffuse rapidly in the plane of the membrane. However, in the last decade evidence has accumulated suggesting the plasma membrane is not homogeneous and contains microdomains with distinct lipid and protein composition. One type of microdomain, the **lipid raft,** is heavily enriched in cholesterol, which fills space between the phospholipids, making them more tightly packed than the surrounding membrane.

Lipid rafts appear to be involved in many important biological processes, including signal reception and cell movement. The structural proteins of replicating HIV viruses are targeted to lipid raft regions of the plasma membrane during virus assembly within infected cells.

Electron microscopy has provided structural evidence

Electron microscopy allows biologists to examine the delicate, filmy structure of a cell membrane. We discussed two types of electron microscopes in chapter 4: the transmission electron microscope (TEM) and the scanning electron microscope (SEM). Both provide illuminating views of membrane structure.

When examining cell membranes with electron microscopy, specimens must be prepared for viewing. In one method of preparing a specimen, the tissue of choice is embedded in a hard epoxy matrix. The epoxy block is then cut with a microtome, a machine with a very sharp blade that makes incredibly thin, transparent "epoxy shavings" less than 1 μm thick that peel away from the block of tissue.

These shavings are placed on a grid, and a beam of electrons is directed through the grid with the TEM. At the high magnification an electron microscope provides, resolution is good enough to reveal the double layers of a membrane. False color can be added to the micrograph to enhance detail.

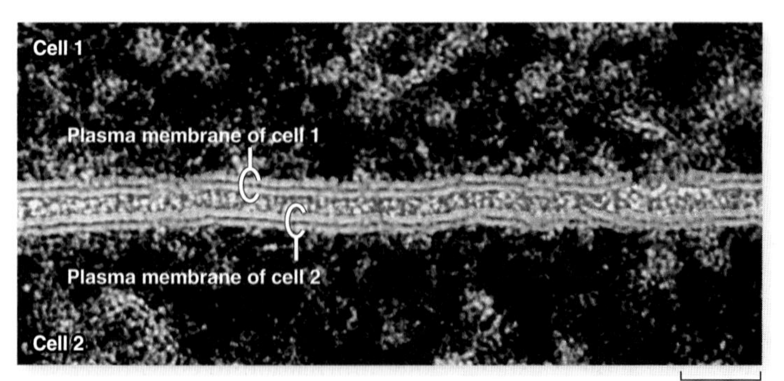

Freeze-fracturing a specimen is another way to visualize the inside of the membrane (figure 5.3). The tissue is embedded in a medium and quick frozen with liquid nitrogen. The frozen tissue is then "tapped" with a knife,

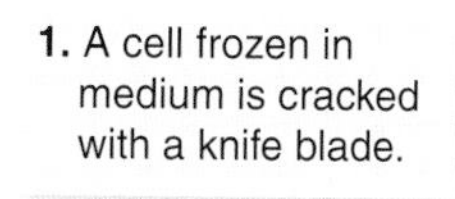

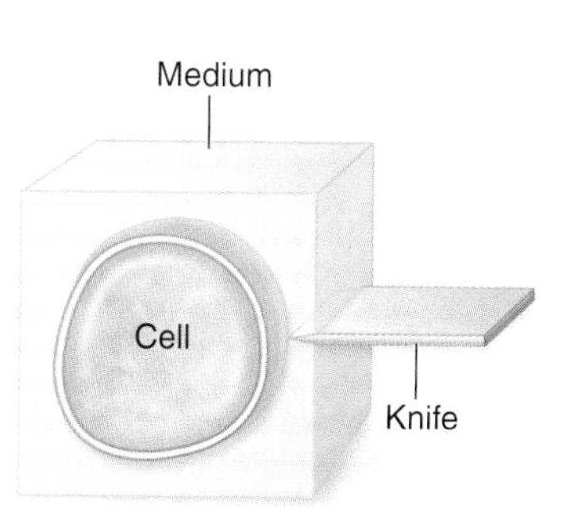

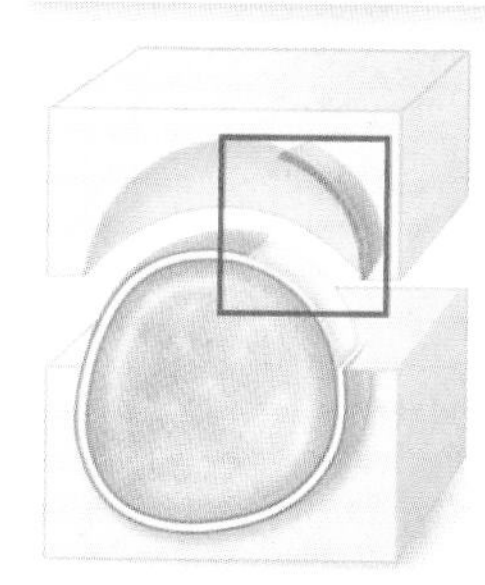

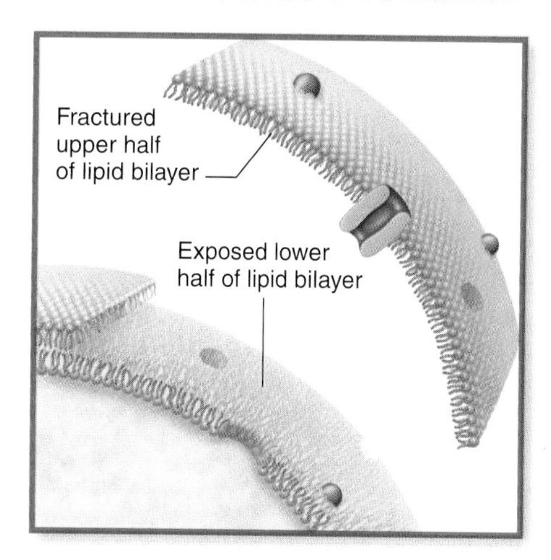

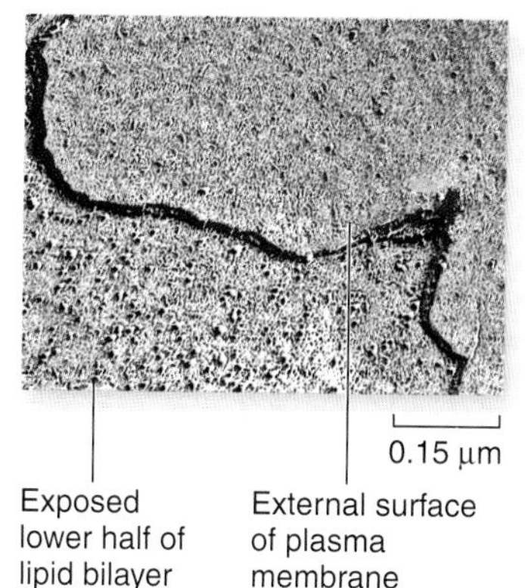

figure 5.3

VIEWING A PLASMA MEMBRANE WITH FREEZE-FRACTURE MICROSCOPY.

causing a crack between the phospholipid layers of membranes. Proteins, carbohydrates, pits, pores, channels, or any other structure affiliated with the membrane will pull apart (whole, usually) and stick with one or the other side of the split membrane.

Next, a very thin coating of platinum is evaporated onto the fractured surface, forming a replica or "cast" of the surface. After the topography of the membrane has been preserved in the cast, the actual tissue is dissolved away, and the cast is examined with electron microscopy, creating a strikingly different view of the membrane.

Cellular membranes contain four components: (1) a phospholipid bilayer, (2) transmembrane proteins, (3) an internal protein network providing structural support, and (4) cell surface markers composed of glycoproteins and glycolipids.

From the results of electron microscopy and molecular biology techniques, we have gained a much clearer view of the structure of membranes and the interactions of membrane components. The modern view of the membrane is called the fluid mosaic model. This refers to the fluid nature of the membrane and that it is a mosaic of phospholipids and proteins floating in the phospholipid bilayer.

5.2 Phospholipids: The Membrane's Foundation

Like the fat molecules (triglycerides) described in chapter 3, a phospholipid has a backbone derived from the three-carbon polyalcohol *glycerol.* Attached to this backbone are one to three fatty acids, long chains of carbon atoms ending in a carboxyl (—COOH) group. A triglyceride molecule has three such chains, one attached to each carbon in the backbone; because these chains are nonpolar, they do not form hydrogen bonds with water, and triglycerides are not water-soluble.

A phospholipid, by contrast, has only two fatty acid chains attached to its backbone. The third carbon of the glycerol carries a phosphate group, thus *phospho*lipid. An additional polar organic molecule is often added to the phosphate group as well.

From this simple molecular framework, a large variety of lipids can be constructed by varying the polar organic group attached to the phosphate and the fatty acid chains attached to the glycerol. Mammalian membranes, for example, contain hundreds of chemically distinct species of lipids.

Phospholipids spontaneously form bilayers

The phosphate groups are charged, and other molecules attached to them are polar or charged. This creates a huge change in the molecule's physical properties compared with a triglyceride. The strongly polar phosphate end is hydrophilic, or "water-loving," while the fatty acid end is strongly nonpolar and hydrophobic, or "water-fearing." The two nonpolar fatty acids extend in one direction, roughly parallel to each other, and the polar phosphate group points in the other direction. To represent this structure, phospholipids are often diagrammed as a polar head with two dangling nonpolar tails, as in figure 5.1*c*.

What happens when a collection of phospholipid molecules is placed in water? The polar water molecules repel the long, nonpolar tails of the phospholipids while seeking partners for hydrogen bonding. Because of the polar nature of the water molecules, the nonpolar tails of the phospholipids end up packed closely together, sequestered as far as possible from water. Every phospholipid molecule is oriented with its polar head toward water and its nonpolar tails away. When *two* layers form with the tails facing each other, no tails ever come in contact with water. The resulting structure is the phospholipid bilayer. Phospholipid bilayers form spontaneously, driven by the tendency of water molecules to form the maximum number of hydrogen bonds.

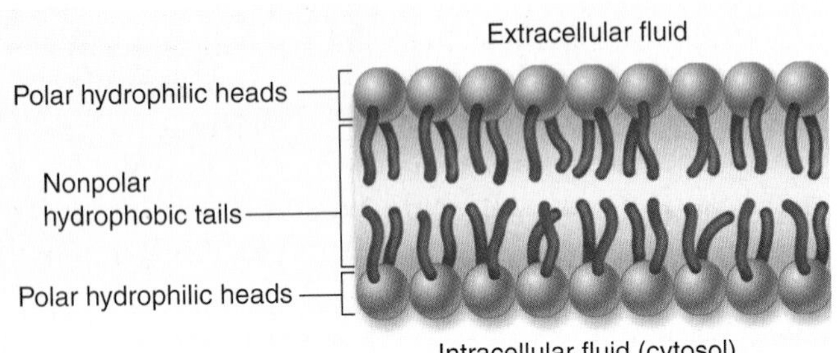

The nonpolar interior of a lipid bilayer impedes the passage of any water-soluble substances through the bilayer, just as a layer of oil impedes the passage of a drop of water. This barrier to the passage of water-soluble substances is the key biological property of the lipid bilayer.

The phospholipid bilayer is fluid

A lipid bilayer is stable because water's affinity for hydrogen bonding never stops. Just as surface tension holds a soap bubble together, even though it is made of a liquid, so the hydrogen bonding of water holds a membrane together. Although water continually drives phospholipid molecules into the bilayer configuration, it does not have any effect on the mobility of phospholipids relative to their lipid and nonlipid neighbors in the bilayer. Because the interactions of phospholipids with one another are relatively weak, individual phospholipids and unanchored proteins are relatively free to move about within the membrane. This can be demonstrated vividly by fusing cells and watching their proteins intermix with time (figure 5.4).

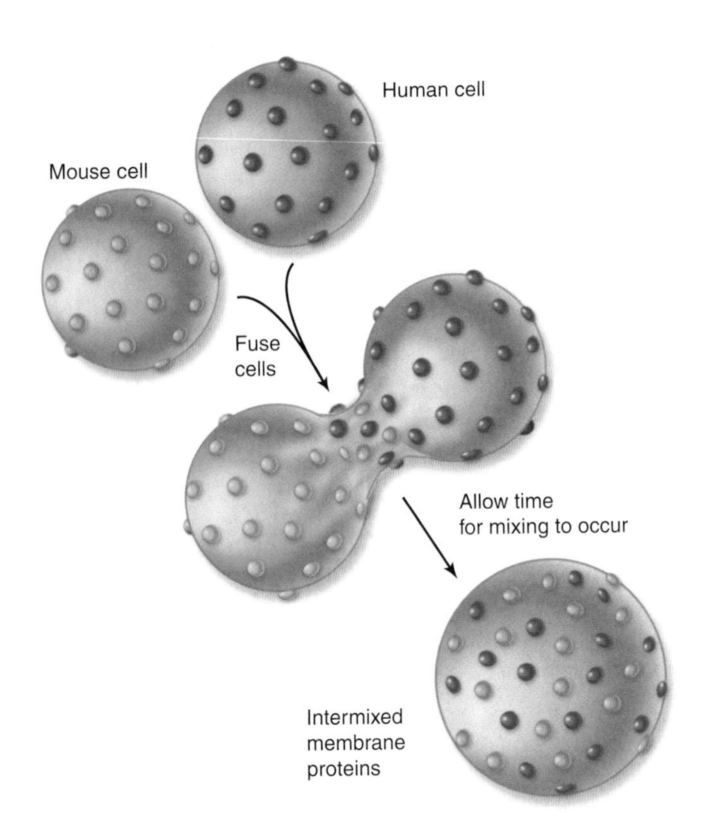

figure 5.4

PROTEIN MOVEMENT IN MEMBRANES. The way that proteins move about within membranes can be demonstrated by labeling the plasma membrane proteins of a mouse cell with fluorescent antibodies of one color and then fusing that cell with a human cell whose membrane proteins have been labeled with fluorescent antibodies of a different color. At first, all of the mouse proteins are located on the mouse side of the fused cell, and all of the human proteins are located on the human side. However, within an hour, the different-colored labeled proteins are intermixed throughout the hybrid cell's plasma membrane.

Membrane fluidity can change

The degree of membrane fluidity changes with the composition of the membrane itself. Much like triglycerides can be solid or liquid at room temperature, depending on their fatty acid composition, membrane fluidity can be altered by changing the membrane's fatty acid composition.

Saturated fats tend to make the membrane less fluid because they pack together well, while unsaturated fats make the membrane more fluid—the "kinks" introduced by the double bonds keep them from packing tightly. You saw this effect on fats and oils earlier in chapter 3. Most membranes also contain sterols such as cholesterol, which can either increase or decrease membrane fluidity, depending on the temperature.

For single-celled organisms such as bacteria, a fluctuating environment can have a large effect on their membranes. Increasing temperature makes a membrane more fluid, and decreasing temperature makes a membrane less fluid. Bacteria have evolved mechanisms to maintain a constant membrane fluidity despite fluctuating temperatures. Some bacteria contain enzymes called *fatty acid desaturases* that can introduce double bonds into fatty acids in membranes. Genetic studies, involving either the inactivation of these enzymes or the introduction of them into cells that normally lack them, indicate that the action of these enzymes confers cold tolerance. At colder temperatures, the double bonds introduced by fatty acid desaturase make the membrane more fluid, counteracting the environmental effect of reduced temperature.

Biological membranes are composed of phospholipid molecules organized into a bilayer, with phosphate groups facing out on each side and fatty acids occupying the interior. This organization is spontaneous and results from the polar phosphate groups interacting with water and the nonpolar fatty acids being sequestered from the water. This structure is also fluid, with the phospholipids able to diffuse laterally in the membrane.

5.3 Proteins: Multifunctional Components

Cell membranes contain a complex assembly of proteins enmeshed in the fluid array of phospholipid molecules. This enormously flexible design permits a broad range of interactions with the environment, some directly involving membrane proteins.

Proteins and protein complexes perform key functions

Although cells interact with their environment through their plasma membranes in many ways, we will focus on six key classes of membrane protein in this chapter and in chapter 9 (figure 5.5).

1. **Transporters.** Membranes are very selective, allowing only certain substances to enter or leave the cell, either through channels or carriers composed of proteins.
2. **Enzymes.** Cells carry out many chemical reactions on the interior surface of the plasma membrane, using enzymes attached to the membrane.
3. **Cell surface receptors.** Membranes are exquisitely sensitive to chemical messages, which are detected by receptor proteins on their surfaces.
4. **Cell surface identity markers.** Membranes carry cell surface markers that identify them to other cells. Most cell types carry their own ID tags, specific combinations of cell surface proteins and protein complexes such as glycoproteins, characteristic of that cell type.
5. **Cell-to-cell adhesion proteins.** Cells use specific proteins to glue themselves to one another. Some act by forming temporary interactions, and others form a more permanent bond. (See chapter 9.)
6. **Attachments to the cytoskeleton.** Surface proteins that interact with other cells are often anchored to the cytoskeleton by linking proteins.

Structural features of membrane proteins

As we've just detailed, membrane proteins can serve a variety of functions. Given these diverse functions, these proteins have diverse structures, yet they also have common structural features related to their role as membrane proteins.

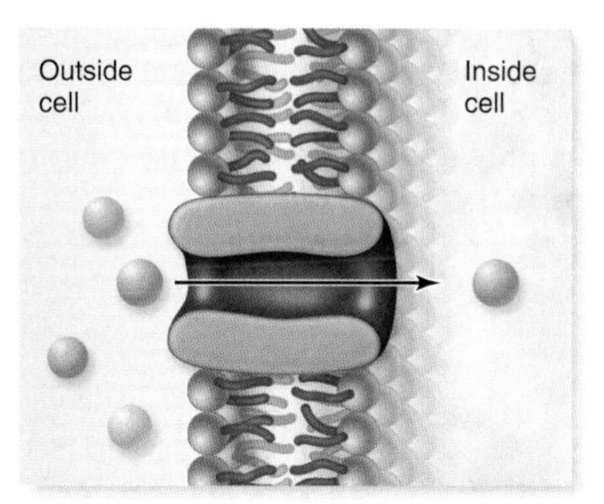

Transporter

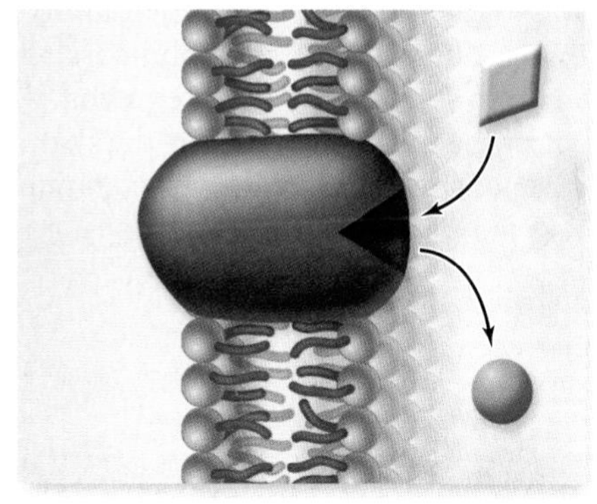
Enzyme

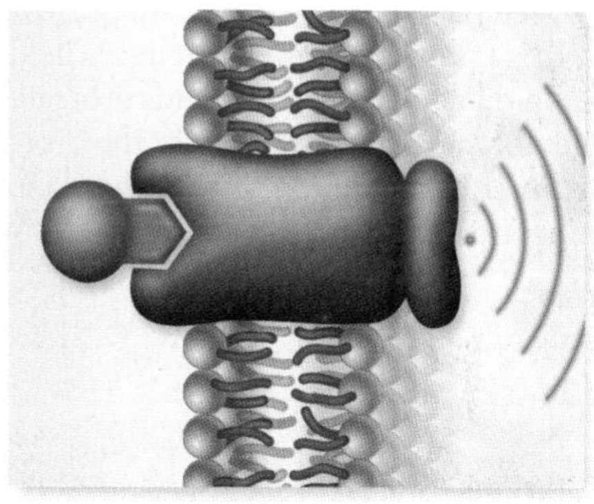
Cell surface receptor

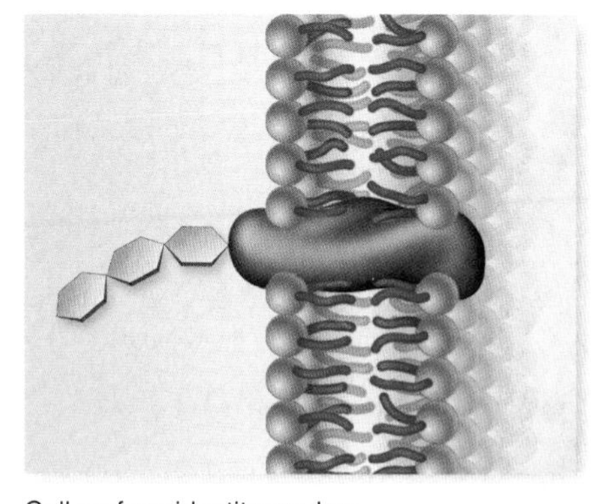
Cell surface identity marker

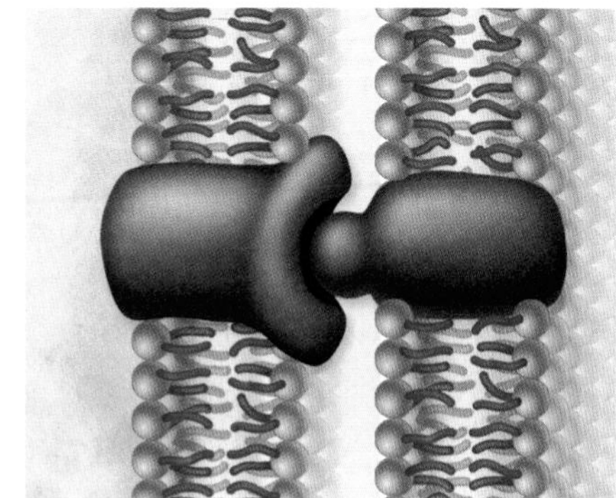
Cell-to-cell adhesion

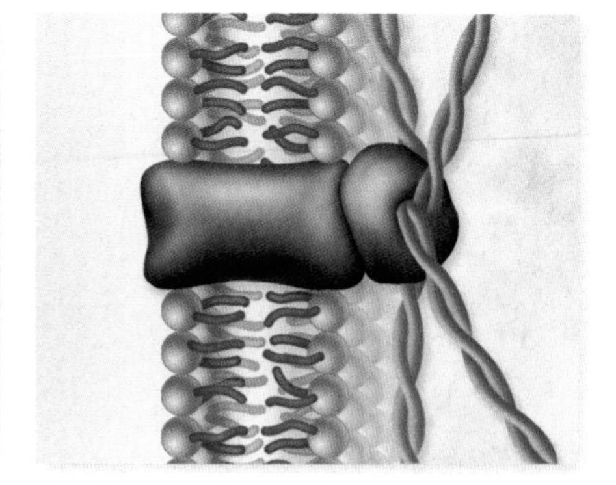
Attachment to the cytoskeleton

figure 5.5

FUNCTIONS OF PLASMA MEMBRANE PROTEINS. Membrane proteins act as transporters, enzymes, cell surface receptors, and cell surface markers, as well as aiding in cell-to-cell adhesion and securing the cytoskeleton.

inquiry

? ***According to the fluid mosaic model, membranes are held together by hydrophobic interactions. Considering the forces that some cells may experience, why do membranes not break apart every time an animal moves?***

The anchoring of proteins in the bilayer

Many membrane proteins are attached to the surface of the membrane by special molecules that associate strongly with phospholipids. Like a ship tied to a floating dock, these anchored proteins (peripheral proteins) are free to move about on the surface of the membrane tethered to a phospholipid. The anchoring molecules are modified lipids that have (1) nonpolar regions that insert into the internal portion of the lipid bilayer and (2) chemical bonding domains that link directly to proteins.

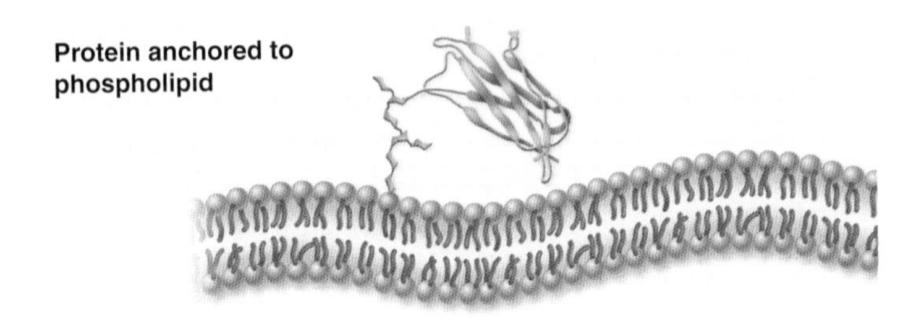

In contrast, other proteins actually span the lipid bilayer (integral membrane proteins). The part of the protein that extends through the lipid bilayer, in contact with the nonpolar interior, consists of nonpolar amino acid helices or β-pleated sheets (see chapter 3). Because water avoids nonpolar amino acids, these portions of the protein are held within the interior of the lipid bilayer. The polar ends protrude from both sides of the membrane. Any movement of the protein out of the membrane, in either direction, brings the nonpolar regions of the protein into contact with water, which "shoves" the protein back into the interior. These forces prevent the transmembrane proteins from simply popping out of the membrane and floating away.

Transmembrane domains

Cell membranes contain a variety of different transmembrane proteins, which differ in the way they traverse the lipid bilayer. The primary difference lies in the number of times that the protein crosses the membrane. Each membrane spanning region is called a **transmembrane domain.** These domains are composed of hydrophobic amino acids usually arranged into α-helices (figure 5.6).

Proteins need only a single transmembrane domain to be anchored in the membrane, but they often have more than one such domain. An example of a protein with a single transmembrane domain is the linking protein that attaches the spectrin network of the cytoskeleton to the interior of the plasma membrane.

Biologists classify some types of receptors based on the number of transmembrane domains they have, such as G protein–coupled signal receptors with seven membrane-spanning domains (chapter 9). These receptors respond to external molecules, such as epinephrine, and initiate a cascade of events inside the cell.

Another example is bacteriorhodopsin, one of the key transmembrane proteins that carries out photosynthesis in halophilic (salt-loving) archaea. It contains seven nonpolar helical segments that traverse the membrane, forming a structure within the membrane through which protons pass during the light-driven pumping of protons (figure 5.7).

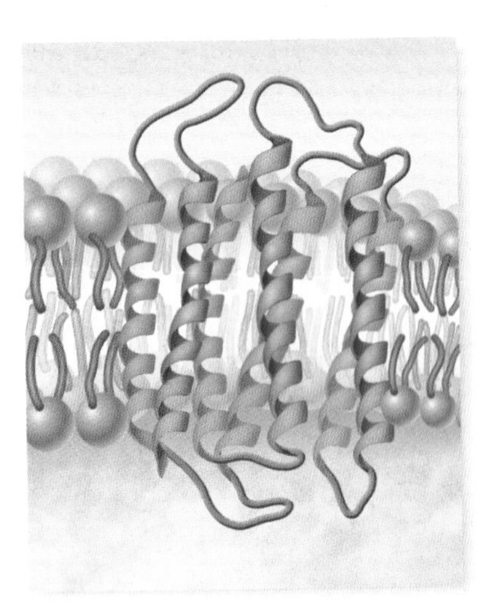

a.

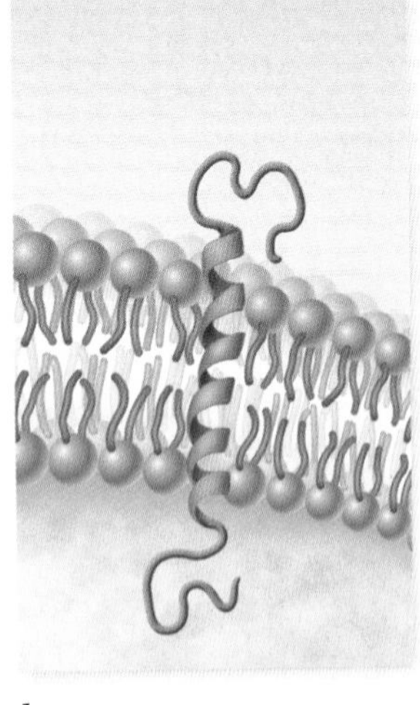

b.

figure 5.6

TRANSMEMBRANE DOMAINS. Integral membrane proteins have at least one hydrophobic transmembrane domain (shown in blue) to anchor them in the membrane. ***a.*** Receptor protein with seven transmembrane domains. ***b.*** Protein with single transmembrane domain.

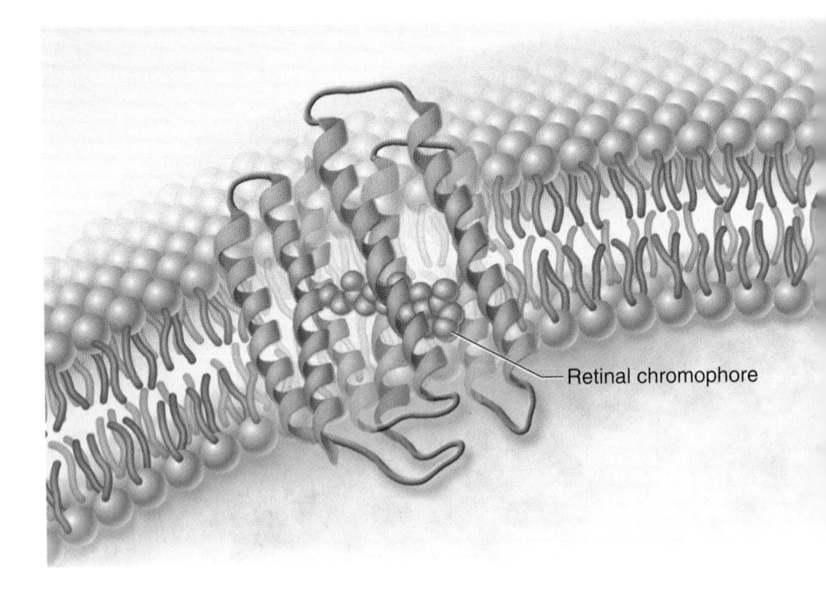

figure 5.7

BACTERIORHODOPSIN. This transmembrane protein mediates photosynthesis in the archaean *Halobacterium salinarium.* The protein traverses the membrane seven times with hydrophobic helical strands that are within the hydrophobic center of the lipid bilayer. The helical regions form a structure across the bilayer through which protons are pumped by the retinal chromophore (*green*) using energy from light.

Pores

Some transmembrane proteins have extensive nonpolar regions with secondary configurations of β-pleated sheets instead of α-helices (chapter 3). The β-sheets form a characteristic motif, folding back and forth in a cylinder so the sheets come to be arranged like a pipe through the membrane. This forms a polar environment in the interior of the β-sheets spanning the membrane. This so-called *β-barrel*, open on both ends, is a common feature of the porin class of proteins that are found within the outer membrane of some bacteria, where they allow molecules to pass through the membrane (figure 5.8).

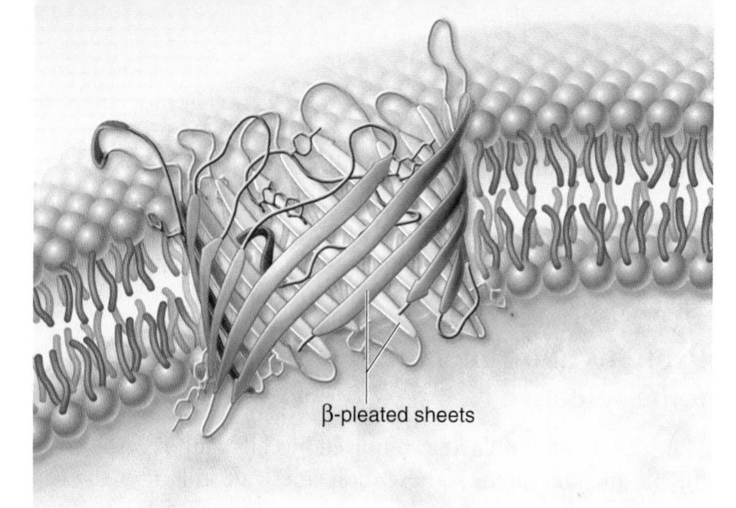

figure 5.8

A PORE PROTEIN. The bacterial transmembrane protein porin creates large open tunnels called pores in the outer membrane of a bacterium. Sixteen strands of β-pleated sheets run antiparallel to one another, creating a so-called β-barrel in the bacterial outer cell membrane. The tunnel allows water and other materials to pass through the membrane.

inquiry

Based only on amino acid sequence, how would you recognize an integral membrane protein?

Proteins in the membrane have a variety of functions and confer the main differences between membranes of different cells. These functions include transport, enzymatic functions, reception of extracellular signals, cell-to-cell interactions, and cell identity markers.

Proteins that are embedded in the membrane have one or more hydrophobic regions, called transmembrane domains, that anchor them in the membrane.

5.4 Passive Transport Across Membranes

Many substances can move in and out of the cell without the cell's having to expend energy. This type of movement is termed **passive transport.** Some ions and molecules can pass through the membrane fairly easily and do so because of a **concentration gradient**—a difference between the concentration on the inside of the membrane and that on the outside. Some substances also move in response to a gradient, but do so through specific channels formed by proteins in the membrane.

Transport can occur by simple diffusion

Molecules and ions dissolved in water are in constant random motion. This random motion causes a net movement of these substances from regions of high concentration to regions of lower concentration, a process called **diffusion** (figure 5.9).

Net movement driven by diffusion will continue until the concentration is the same in all regions. Consider adding a drop of colored ink to a bowl of water; what happens to the ink? Over time the ink becomes dispersed throughout the solution. This is due to diffusion of the ink molecules. In the context of cells, we are usually concerned with differences in concentration of molecules across the plasma membrane. We need to consider the relative concentrations both inside and outside the cell, as well as how readily a molecule can cross the membrane.

The major barrier to crossing a biological membrane is the hydrophobic interior that repels polar molecules but not nonpolar molecules. If a concentration difference exists for a nonpolar molecule, it will move across the membrane until the concentration is equal on both sides. At this point, movement

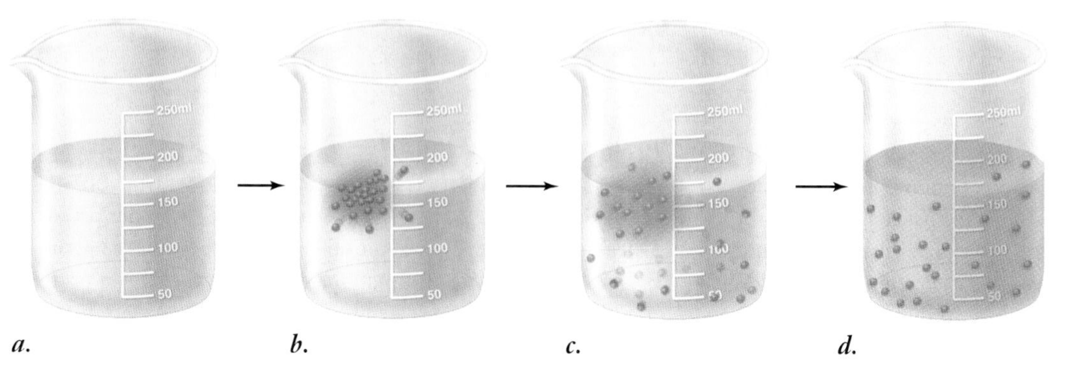

figure 5.9

DIFFUSION. If a drop of colored ink is dropped into a beaker of water *(a)* its molecules dissolve *(b)* and diffuse *(c)*. Eventually, diffusion results in an even distribution of ink molecules throughout the water *(d)*.

in both directions still occurs, but there is no net movement in either direction. This includes molecules like O_2 and nonpolar organic molecules such as steroid hormones.

The plasma membrane has limited permeability to small polar molecules and very limited permeability to larger polar molecules and ions. The movement of water, one of the most important polar molecules, is discussed in its own section later on.

Proteins allow membrane diffusion to be selective

Many important molecules required by cells cannot easily cross the plasma membrane. These molecules can still enter the cell by diffusion through specific channel proteins or carrier proteins embedded in the plasma membrane, provided there is a higher concentration of the molecule outside the cell than inside. **Channel proteins** have a hydrophilic interior that provides an aqueous channel through which polar molecules can pass when the channel is open. **Carrier proteins,** in contrast to channels, bind specifically to the molecule they assist, much like an enzyme binds to its substrate. These channels and carriers are usually selective for one type of molecule, and thus the cell membrane is said to be **selectively permeable.**

Diffusion of ions through channels

You saw in chapter 2 that atoms with an unequal number of protons and electrons have electric charge and are called ions. Those that carry positive charge are called *cations* and those that carry negative charge are called *anions.*

Because of their charge, ions interact well with polar molecules such as water, but are repelled by nonpolar molecules such as the interior of the plasma membrane. Therefore, ions cannot move between the cytoplasm of a cell and the extracellular fluid without the assistance of membrane transport proteins.

Ion channels possess a hydrated interior that spans the membrane. Ions can diffuse through the channel in either direction, depending on their relative concentration across the membrane (figure 5.10). Some channel proteins can be opened or closed in response to a stimulus. These channels are called **gated channels** and depending on the nature of the channel, the stimulus can be either chemical or electrical.

Three conditions determine the direction of net movement of the ions: (1) their relative concentrations on either side of the membrane, (2) the voltage difference across the membrane and for gated channels, (3) the state of the gate (open or closed). A voltage difference is an electrical potential difference across the membrane called a **membrane potential.** Changes in membrane potential form the basis for transmission of signals in the nervous system and some other tissues. (We discuss this topic in detail in chapter 44.) Each type of channel is specific for a particular ion, such as calcium (Ca^{2+}), sodium (Na^+), potassium (K^+), or chloride (Cl^-), or in some cases, for more than one cation or anion. Ion channels play an essential role in signaling by the nervous system.

Carrier proteins and facilitated diffusion

Carrier proteins can help transport both ions and other solutes, such as some sugars and amino acids, across the membrane. Transport through a carrier is still a form of diffusion and therefore requires a concentration difference across the membrane (figure 5.11). Because this process is facilitated by a carrier protein, it is often called **facilitated diffusion.**

Carriers must bind to the molecule they transport, so the relationship between concentration and rate of transport differs from that due to simple diffusion. As concentration increases, transport by simple diffusion shows a linear increase in rate of transport. But when a carrier protein is involved, a concentration increase means that more of the carriers are bound to the transported molecule. At high enough concentrations all carriers will be occupied, and the

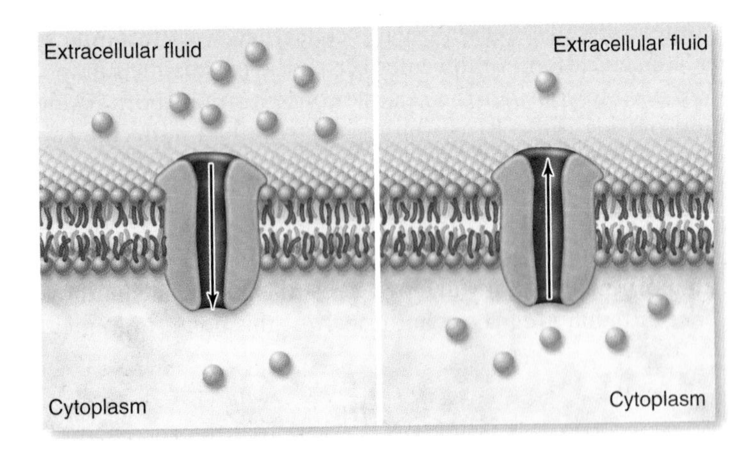

figure 5.10

ION CHANNELS. The movement of ions through a channel is shown. In the panel on the left the concentration is higher outside the cell, so the ions move into the cell. In the panel on the right the situation is reversed. In both cases, transport will continue until the concentration is equal on both sides of the membrane. At this point, ions continue to cross the membrane in both directions, but there is no net movement in either direction.

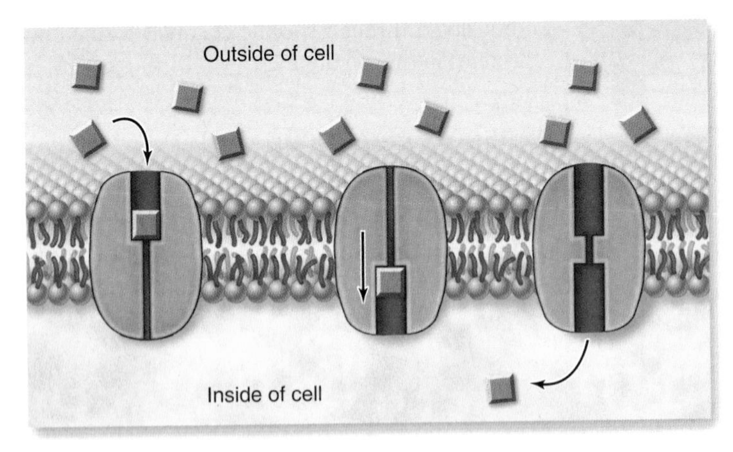

figure 5.11

FACILITATED DIFFUSION IS A CARRIER-MEDIATED TRANSPORT PROCESS. Molecules bind to a carrier protein on the extracellular side of the cell and pass through the plasma membrane via a conformational change in the carrier protein. This will only occur when there is a higher concentration outside the cell.

rate of transport will be constant. This means that the carrier exhibits **saturation.**

This situation is somewhat like that of a stadium (the cell) where a crowd must pass through turnstiles to enter. If there are unoccupied turnstiles, you can go right through, but when all are occupied, you must wait. When ticket holders are passing through the gates at maximum speed, the rate at which they enter cannot increase, no matter how many are waiting outside.

Facilitated diffusion provides the cell with a mechanism to prevent the buildup of unwanted molecules within the cell or to take up needed molecules that may be present outside the cell in high concentrations. Facilitated diffusion has three essential characteristics:

1. **It is specific.** A given carrier transports only certain molecules or ions.
2. **It is passive.** The direction of net movement is determined by the relative concentrations of the transported substance inside and outside the cell. The direction is always from high concentration to low concentration.
3. **It saturates.** If all relevant protein carriers are in use, increases in the concentration gradient do not increase the rate of movement.

Facilitated diffusion in red blood cells

Several examples of facilitated diffusion can be found in the plasma membrane of vertebrate red blood cells (RBCs). One RBC carrier protein, for example, transports a different molecule in each direction: chloride ion (Cl^-) in one direction and bicarbonate ion (HCO_3^-) in the opposite direction. As you will learn in chapter 49, this carrier is important in the uptake and release of carbon dioxide.

The glucose transporter is a second vital facilitated diffusion carrier in RBCs. Red blood cells keep their internal concentration of glucose low through a chemical trick: They immediately add a phosphate group to any entering glucose molecule, converting it to a highly charged glucose phosphate that can no longer bind to the glucose transporter, and therefore cannot pass back across the membrane. This maintains a steep concentration gradient for unphosphorylated glucose, favoring its entry into the cell.

The glucose transporter that assists the entry of glucose into the cell does not appear to form a channel in the membrane. Instead, this transmembrane protein appears to bind to a glucose molecule and then to flip its shape, dragging the glucose through the bilayer and releasing it on the inside of the plasma membrane. After it releases the glucose, the transporter reverts to its original shape and is then available to bind the next glucose molecule that comes along outside the cell.

Osmosis is the movement of water across membranes

The cytoplasm of a cell contains ions and molecules, such as sugars and amino acids, dissolved in water. The mixture of these substances and water is called an *aqueous solution.* Water is termed the **solvent,** and the substances dissolved in the water are **solutes.** Both water and solutes tend to diffuse from regions of high concentration to ones of low concentration; that is, they diffuse down their concentration gradients.

When two regions are separated by a membrane, what happens depends on whether the solutes can pass freely through that membrane. Most solutes, including ions and sugars, are not lipid-soluble and, therefore, are unable to cross the lipid bilayer. The concentration gradient of these solutes can lead to the movement of water.

Osmosis

Water molecules interact with dissolved solutes by forming hydration shells around the charged solute molecules. When a membrane separates two solutions having different concentrations of solutes, different concentrations of *free* water molecules exist on the two sides of the membrane. The side with higher solute concentration has tied up more water molecules in hydration shells and thus has fewer free water molecules.

As a consequence of this difference, free water molecules move down their concentration gradient, toward the higher solute concentration. This net diffusion of water across a membrane toward a higher solute concentration is called **osmosis** (figure 5.12).

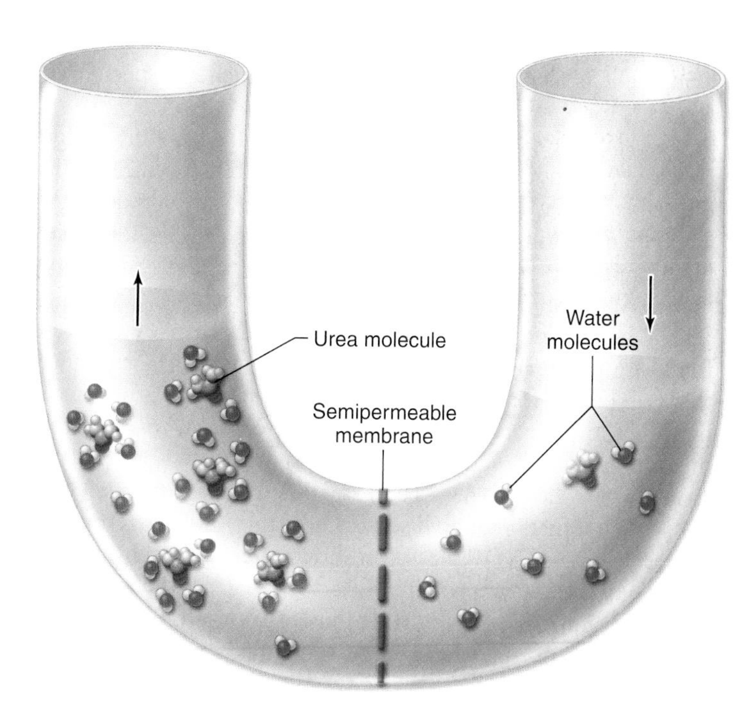

figure 5.12

OSMOSIS. Concentration differences in charged or polar molecules that cannot cross a semipermeable membrane will result in movement of water, which can cross the membrane. Water molecules form hydrogen bonds with charged or polar molecules forming a hydration shell around them in solution. A higher concentration of polar molecules (urea) shown on the left side of the membrane leads to water molecules gathering around each urea molecule. These water molecules are no longer free to diffuse across the membrane. The polar solute has reduced the concentration of free water molecules, creating a gradient. This causes a net movement of water by diffusion from right to left in the U-tube, raising the level on the left and lowering the level on the right.

The concentration of *all* solutes in a solution determines the **osmotic concentration** of the solution. If two solutions have unequal osmotic concentrations, the solution with the higher concentration is **hypertonic** (Greek *hyper,* "more than"), and the solution with the lower concentration is **hypotonic** (Greek *hypo,* "less than"). When two solutions have the same osmotic concentration, the solutions are **isotonic** (Greek *iso,* "the same"). The terms *hyperosmotic, hypoosmotic,* and *isosmotic* are also used to describe these conditions.

A cell in any environment can be thought of as a plasma membrane separating two solutions: the cytoplasm and the extracellular fluid. The direction and extent of any diffusion of water across the plasma membrane is determined by comparing the osmotic strength of these solutions. Put another way, water diffuses out of a cell in a hypertonic solution (that is, the cytoplasm of the cell is hypotonic, compared with the extracellular fluid). This loss of water causes the cell to shrink until the osmotic concentrations of the cytoplasm and the extracellular fluid become equal.

Aquaporins: water channels

The transport of water across the membrane is complex. Studies on artificial membranes show that water, despite its polarity, can cross the membrane, but this flow is limited. Water flow in living cells is facilitated by **aquaporins,** which are specialized channels for water.

A simple experiment demonstrates this. If an amphibian egg is placed in hypotonic spring water (the solute concentration in the cell is higher than that of the surrounding water), it does not swell. If aquaporin mRNA is then injected into the egg, the channel proteins are expressed and appear in the egg's plasma membrane. Water will now diffuse into the egg, causing it to swell.

More than 11 different kinds of aquaporins have been found in mammals. These fall into two general classes: those that are specific for only water, and those that allow other small hydrophilic molecules, such as glycerol or urea, to cross the membrane as well. This latter class explains how some membranes allow the easy passage of small hydrophilic substances.

The human genetic disease, hereditary (nephrogenic) diabetes insipidus (NDI), has been shown to be caused by a nonfunctional aquaporin protein. This disease causes the excretion of large volumes of dilute urine, illustrating the importance of aquaporins to our physiology.

Osmotic pressure

What happens to a cell in a hypotonic solution? (That is, the cell's cytoplasm is hypertonic relative to the extracellular fluid.) In this situation, water diffuses into the cell from the extracellular fluid, causing the cell to swell. The pressure of the cytoplasm pushing out against the cell membrane, or **hydrostatic pressure,** increases. The amount of water that enters the cell depends on the difference in solute concentration between the cell and the extracellular fluid. This is measured as **osmotic pressure,** defined as the force needed to stop osmotic flow.

If the membrane is strong enough, the cell reaches an equilibrium, at which the osmotic pressure, which tends to drive water into the cell, is exactly counterbalanced by the hydrostatic pressure, which tends to drive water back out of the cell. However, a plasma membrane by itself cannot withstand large internal pressures, and an isolated cell under such conditions would burst like an overinflated balloon (figure 5.13).

Accordingly, it is important for animal cells, which only have plasma membranes, to maintain osmotic balance. In contrast, the cells of prokaryotes, fungi, plants, and many protists are surrounded by strong cell walls, which can withstand high internal pressures without bursting.

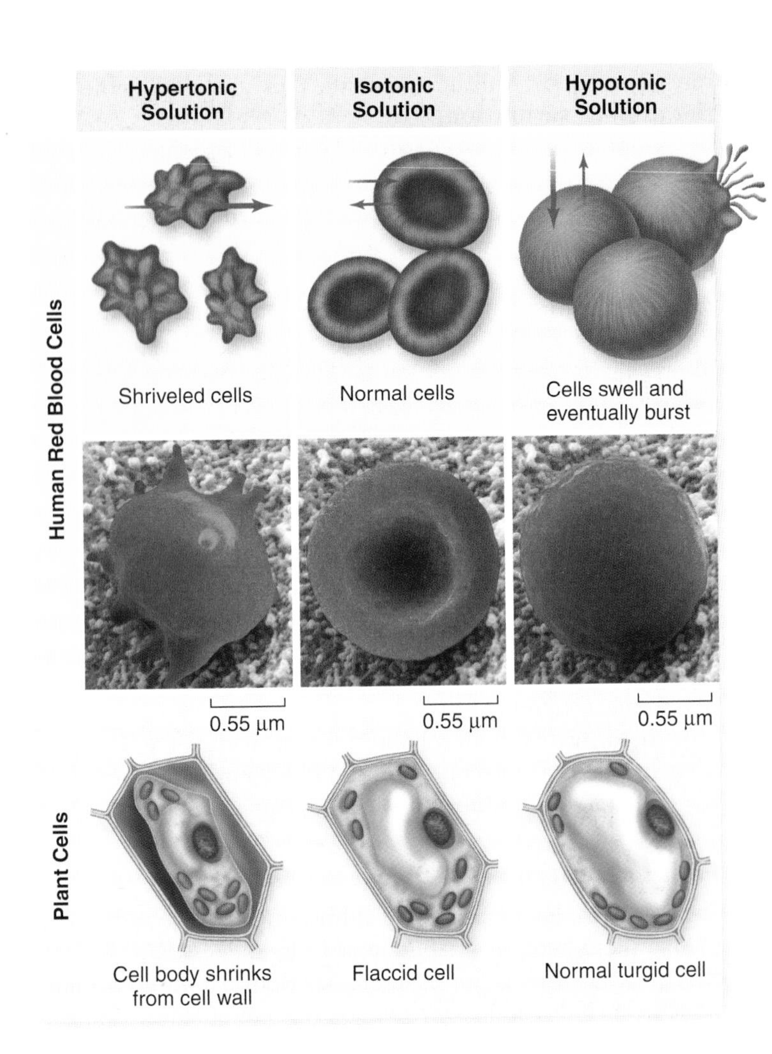

figure 5.13

HOW SOLUTES CREATE OSMOTIC PRESSURE. In a hypertonic solution, water moves out of the cell causing the cell to shrivel. In an isotonic solution, water diffuses into and out of the cell at the same rate, with no change in cell size. In a hypotonic solution, water moves into the cell. Direction and amount of water movement is shown with blue arrows (top). As water enters the cell from a hypotonic solution, pressure is applied to the plasma membrane until the cell ruptures. Water enters the cell due to osmotic pressure from the higher solute concentration in the cell. Osmotic pressure is measured as the force needed to stop osmosis. The strong cell wall of plant cells can withstand the hydrostatic pressure to keep the cell from rupturing. This is not the case with animal cells.

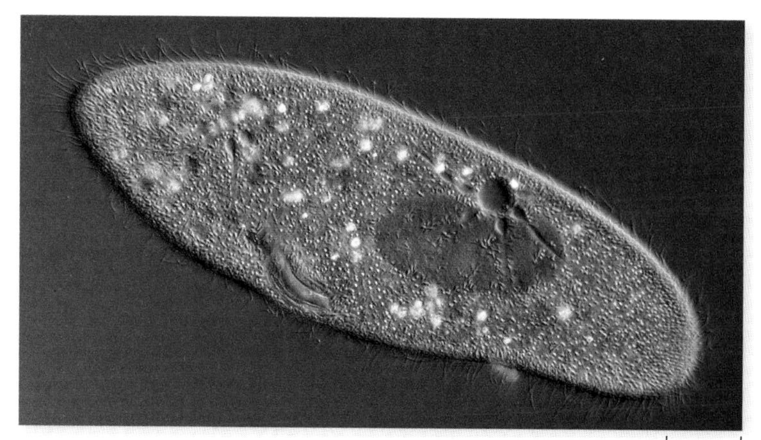

figure 5.14

CONTRACTILE VACUOLE. A micrograph of *Paramecium caudatum* with a large contractile vacuole visible near the center of the cell. The vacuole can contract to expel water. This helps the cell maintain osmotic balance in hypotonic solutions where water enters the cell by osmosis.

Maintaining osmotic balance

Organisms have developed many strategies for solving the dilemma posed by being hypertonic to their environment and therefore having a steady influx of water by osmosis.

Extrusion. Some single-celled eukaryotes, such as the protist *Paramecium*, use organelles called contractile vacuoles to remove water. Each vacuole collects water from various parts of the cytoplasm and transports it to the central part of the vacuole, near the cell surface. The vacuole possesses a small pore that opens to the outside of the cell. By contracting rhythmically, the vacuole pumps out (extrudes) through this pore the water that is continuously drawn into the cell by osmotic forces (figure 5.14).

Isosmotic Regulation. Some organisms that live in the ocean adjust their internal concentration of solutes to match that of the surrounding seawater. Because they are isosmotic with respect to their environment, no net flow of water occurs into or out of these cells.

Many terrestrial animals solve the problem in a similar way, by circulating a fluid through their bodies that bathes cells in an isotonic solution. The blood in your body, for example, contains a high concentration of the protein albumin, which elevates the solute concentration of the blood to match that of your cells' cytoplasm.

Turgor. Most plant cells are hypertonic to their immediate environment, containing a high concentration of solutes in their central vacuoles. The resulting internal hydrostatic pressure, known as **turgor pressure,** presses the plasma membrane firmly against the interior of the cell wall, making the cell rigid. Most green plants depend on turgor pressure to maintain their shape, and thus they wilt when they lack sufficient water.

Passive transport involves diffusion, which requires a concentration gradient. Diffusion will only cause net movement of molecules until equilibrium is reached, when concentrations on both sides are equal.

Diffusion can occur directly through the membrane for hydrophobic molecules, or through channels for ions. Facilitated diffusion occurs through carrier proteins that bind to the molecule being transported. The selective nature of both channel proteins and carrier proteins gives the membrane selective permeability.

Water is transported in response to concentration differences inside and outside the cell of solutes that cannot cross the membrane. This process is called osmosis. The transport of water occurs through the membrane but is aided by channel proteins called aquaporins.

5.5 Active Transport Across Membranes

While diffusion, facilitated diffusion, and osmosis are passive transport processes that move materials down their concentration gradients, cells can also move substances across a cell membrane *up* their concentration gradients. This process requires the expenditure of energy, typically from ATP, and is therefore called **active transport.**

Active transport uses energy to move materials against a concentration gradient

Like facilitated diffusion, active transport involves highly selective protein carriers within the membrane that bind to the transported substance, which could be an ion or a simple molecule, such as a sugar, an amino acid, or a nucleotide. These carrier proteins are called **uniporters** if they transport a single type of molecule and symporters or antiporters if they transport two different molecules together. **Symporters** transport two molecules in the same direction, and **antiporters** transport two molecules in opposite directions. These terms can also be used to describe facilitated diffusion carriers.

Active transport is one of the most important functions of any cell. It enables a cell to take up additional molecules of a substance that is already present in its cytoplasm in concentrations higher than in the extracellular fluid. Active transport also enables a cell to move substances out of its cytoplasm and into the extracellular fluid, despite higher external concentrations.

The use of energy from ATP in active transport may be direct or indirect. Let's first consider how ATP is used directly to move ions against their concentration gradients.

The sodium–potassium pump runs directly on ATP

More than one-third of all of the energy expended by an animal cell that is not actively dividing is used in the active transport of sodium (Na^+) and potassium (K^+) ions. Most animal cells have a low internal concentration of Na^+, relative to their surroundings, and a high internal concentration of K^+. They maintain these concentration differences by actively pumping Na^+ out of the cell and K^+ in.

The remarkable protein that transports these two ions across the cell membrane is known as the **sodium–potassium pump** (figure 5.15). This carrier protein uses the energy stored in ATP to move these two ions. In this case, the energy is used to change the conformation of the carrier protein, which changes its affinity for either Na^+ ions or K^+ ions. This is an excellent

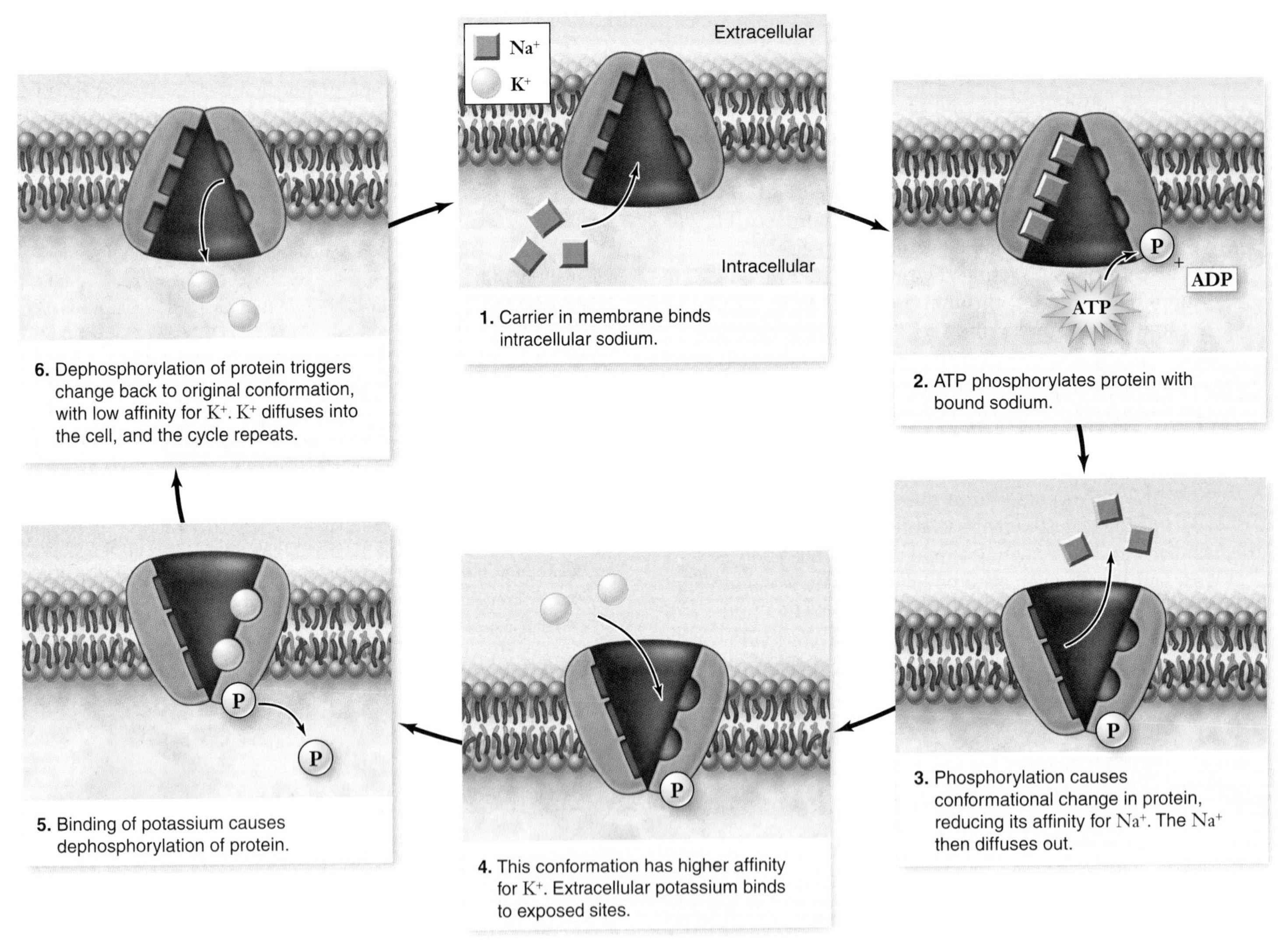

figure 5.15

THE SODIUM–POTASSIUM PUMP. The protein carrier known as the sodium–potassium pump transports sodium (Na^+) and potassium (K^+) ions across the plasma membrane. For every three Na^+ transported out of the cell, two K^+ are transported into it. The sodium–potassium pump is fueled by ATP hydrolysis. The affinity of the pump for Na^+ and K^+ is changed by adding or removing phosphate, which changes the conformation of the protein.

illustration of how subtle changes in the structure of a protein affect its function.

The important characteristic of the sodium–potassium pump is that it is an active transport process, transporting Na^+ and K^+ from areas of low concentration to areas of high concentration. This transport is the opposite of passive transport by diffusion; it is achieved only by the constant expenditure of metabolic energy. The sodium–potassium pump works through the following series of conformational changes in the transmembrane protein (summarized in figure 5.15):

Step 1. Three Na^+ bind to the cytoplasmic side of the protein, causing the protein to change its conformation.

Step 2. In its new conformation, the protein binds a molecule of ATP and cleaves it into adenosine diphosphate (ADP) and phosphate (P_i). ADP is released, but the phosphate group is covalently linked to the protein. The protein is now phosphorylated.

Step 3. The phosphorylation of the protein induces a second conformational change in the protein. This change translocates the three Na^+ across the membrane, so they now face the exterior. In this new conformation, the protein has a low affinity for Na^+, and the three bound Na^+ dissociate from the protein and diffuse into the extracellular fluid.

Step 4. The new conformation has a high affinity for K^+, two of which bind to the extracellular side of the protein as soon as it is free of the Na^+.

Step 5. The binding of the K^+ causes another conformational change in the protein, this time resulting in the hydrolysis of the bound phosphate group.

Step 6. Freed of the phosphate group, the protein reverts to its original conformation, exposing the two K^+ to the cytoplasm. This conformation has a low affinity for K^+, so the two bound K^+ dissociate from the protein and diffuse into the interior of the cell. The original conformation has a high affinity for Na^+; when these ions bind, they initiate another cycle.

In every cycle, three Na^+ leave the cell and two K^+ enter. The changes in protein conformation that occur during the cycle are rapid, enabling each carrier to transport as many as 300 Na^+ per second. The sodium–potassium pump appears to be ubiquitous in animal cells, although cells vary widely in the number of pump proteins they contain.

Coupled transport uses ATP indirectly

Some molecules are moved against their concentration gradient by using the energy stored in a gradient of a different molecule. In this process, called **coupled transport,** the energy released as one molecule moves down its concentration gradient is captured and used to move a different molecule against its gradient. As you just saw, the energy stored in ATP molecules can be used to create a gradient of Na^+ and K^+ across the membrane. These gradients can then be used to power the transport of other molecules across the membrane.

As one example, let's consider the active transport of glucose across the membrane in animal cells. Glucose is such an important molecule that there are a variety of transporters for it, one of which was discussed earlier under passive transport. In a multicellular organism, intestinal epithelial cells can have a higher concentration of glucose inside the cell than outside, so these cells need to be able to transport glucose against its concentration gradient. This requires energy and a different transporter than the one involved in facilitated diffusion of glucose.

The active glucose transporter uses the Na^+ gradient produced by the sodium–potassium pump as a source of energy to power the movement of glucose into the cell. In this system, both glucose and Na^+ bind to the transport protein, which allows Na^+ to pass into the cell down its concentration gradient, capturing the energy and using it to move glucose into the cell. In this kind of cotransport, both molecules are moving in the same direction across the membrane; therefore the transporter is a symporter (figure 5.16).

In a related process, called **countertransport,** the inward movement of Na^+ is coupled with the outward movement of another substance, such as Ca^{2+} or H^+. As in cotransport, both Na^+ and the other substance bind to the same transport protein, which in this case is an antiporter, as the substances bind on opposite sides of the membrane and are moved in opposite directions. In countertransport, the cell uses the energy

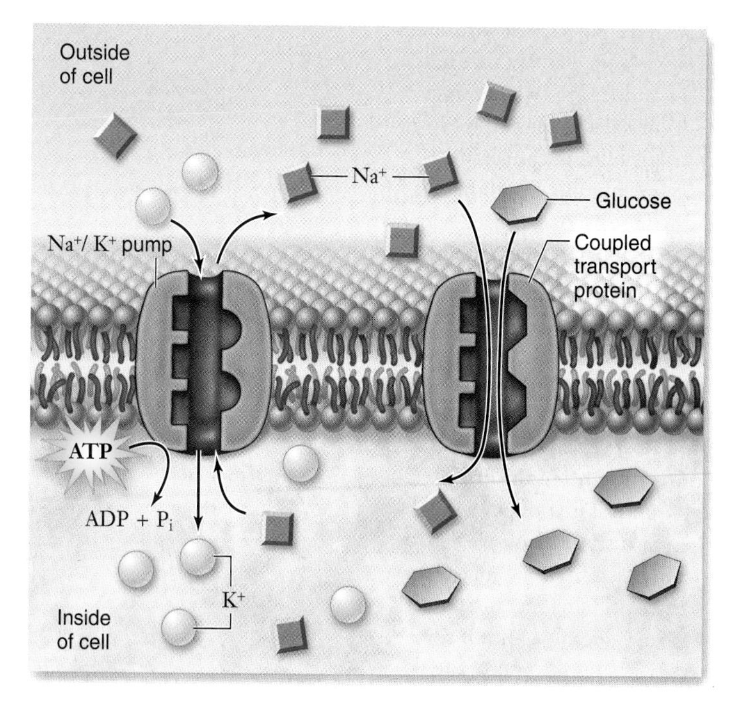

figure 5.16

COUPLED TRANSPORT. A membrane protein transports Na^+ into the cell, down its concentration gradient, at the same time it transports a glucose molecule into the cell. The gradient driving the Na^+ entry is so great that sugar molecules can be brought in against their concentration gradient. The Na^+ gradient is maintained by the Na^+/K^+ pump.

released as Na^+ moves down its concentration gradient into the cell to eject a substance against its concentration gradient. In both cotransport and countertransport, the potential energy in the concentration gradient of one molecule is used to transport another molecule against its concentration gradient. They differ only in the direction that the second molecule moves relative to the first.

Molecules that need to be concentrated in the cell or moved against a concentration gradient must be moved by active transport. This transport requires both a carrier protein and energy, usually in the form of ATP. Transport of a molecule against a concentration gradient can also be coupled to the transport of another molecule down its concentration gradient by the same transporter.

5.6 Bulk Transport by Endocytosis and Exocytosis

The lipid nature of cell plasma membranes raises a second problem. The substances cells require for growth are mostly large, polar molecules that cannot cross the hydrophobic barrier a lipid bilayer creates. How do these substances get into cells? Two processes are involved in this **bulk transport:** *endocytosis* and *exocytosis.*

Bulk material enters the cell in vesicles

In **endocytosis,** the plasma membrane envelops food particles and fluids. Cells use three major types of endocytosis: phagocytosis, pinocytosis, and receptor-mediated endocytosis (figure 5.17). Like active transport, these processes also require energy expenditure.

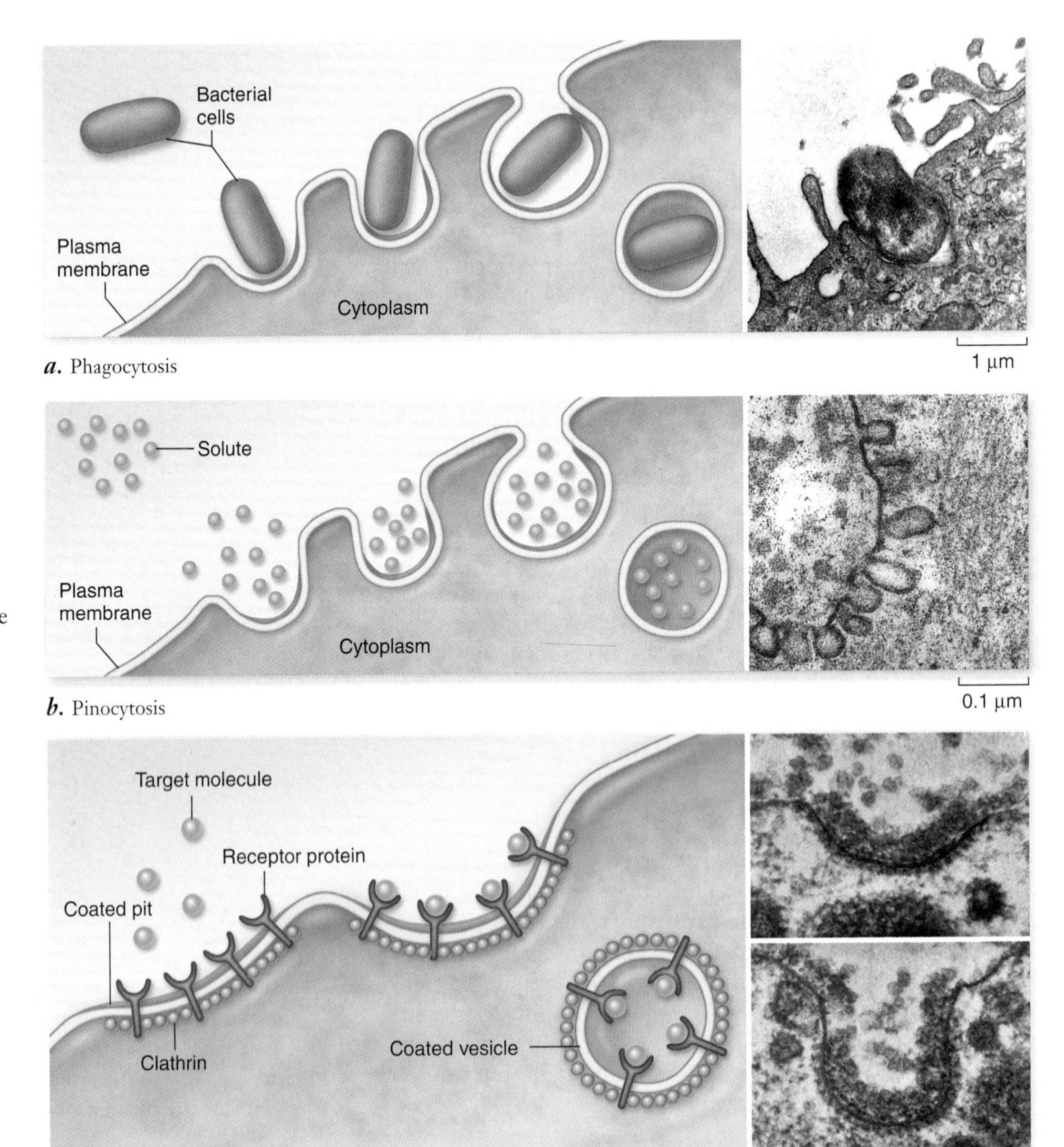

figure 5.17

ENDOCYTOSIS. Both **(*a*)** phagocytosis and **(*b*)** pinocytosis are forms of endocytosis. ***c.*** In receptor-mediated endocytosis, cells have pits coated with the protein clathrin that initiate endocytosis when target molecules bind to receptor proteins in the plasma membrane. Photo inserts (false color has been added to enhance distinction of structures): (*a*) A TEM of phagocytosis of a bacterium, *Rickettsia tsutsugamushi*, by a mouse peritoneal mesothelial cell. The bacterium enters the host cell by phagocytosis and replicates in the cytoplasm. (*b*) A TEM of pinocytosis in a smooth muscle cell. (*c*) A coated pit appears in the plasma membrane of a developing egg cell, covered with a layer of proteins. When an appropriate collection of molecules gathers in the coated pit, the pit deepens and will eventually seal off to form a vesicle.

Phagocytosis and pinocytosis

If the material the cell takes in is particulate (made up of discrete particles), such as an organism or some other fragment of organic matter (figure 5.17*a*), the process is called **phagocytosis** (Greek *phagein*, "to eat," + *cytos*, "cell"). If the material the cell takes in is liquid (figure 5.17*b*), the process is called **pinocytosis** (Greek *pinein*, "to drink"). Pinocytosis is common among animal cells. Mammalian egg cells, for example, "nurse" from surrounding cells; the nearby cells secrete nutrients that the maturing egg cell takes up by pinocytosis.

Virtually all eukaryotic cells constantly carry out these kinds of endocytotic processes, trapping particles and extracellular fluid in vesicles and ingesting them. Endocytosis rates vary from one cell type to another. They can be surprisingly high; some types of white blood cells ingest up to 25% of their cell volume each hour.

Receptor-mediated endocytosis

Molecules are often transported into eukaryotic cells through **receptor-mediated endocytosis.** These molecules first bind to specific receptors in the plasma membrane—they have a conformation that fits snugly into the receptor. Different cell types contain a characteristic battery of receptor types, each for a different kind of molecule in their membranes.

The portion of the receptor molecule that lies inside the membrane is trapped in an indented pit coated on the cytoplasmic side with the protein *clathrin.* Each pit acts like a molecular mousetrap, closing over to form an internal vesicle when the right molecule enters the pit (figure 5.17*c*). The trigger that releases the trap is the binding of the properly fitted target molecule to the embedded receptor. When binding occurs, the cell reacts by initiating endocytosis; the process is highly specific and very fast. The vesicle is now inside the cell carrying its cargo.

One type of molecule that is taken up by receptor-mediated endocytosis is low-density lipoprotein (LDL). LDL molecules bring cholesterol into the cell where it can be incorporated into membranes. Cholesterol plays a key role in determining the stiffness of the body's membranes. In the human genetic disease familial hypercholesterolemia, the LDL receptors lack tails, so they are never fastened in the clathrin-coated pits and as a result, do not trigger vesicle formation. The cholesterol stays in the bloodstream of affected individuals, accumulating as plaques inside arteries and leading to heart attacks.

It is important to understand that endocytosis in itself does not bring substances directly into the cytoplasm of a cell. The material taken in is still separated from the cytoplasm by the membrane of the vesicle.

Material can leave the cell by exocytosis

The reverse of endocytosis is **exocytosis,** the discharge of material from vesicles at the cell surface (figure 5.18). In plant cells, exocytosis is an important means of exporting the materials needed to construct the cell wall through the plasma membrane. Among protists, contractile vacuole discharge is considered a form of exocytosis. In animal cells, exocytosis provides a mechanism for secreting many hormones, neurotransmitters, digestive enzymes, and other substances.

The mechanisms for transport across cell membranes are summarized in table 5.2.

Large molecules and other bulky materials can enter and leave the cell by endocytosis and exocytosis, respectively. These processes require energy. Endocytosis may be mediated by receptor proteins in the membrane that trigger the formation of vesicles.

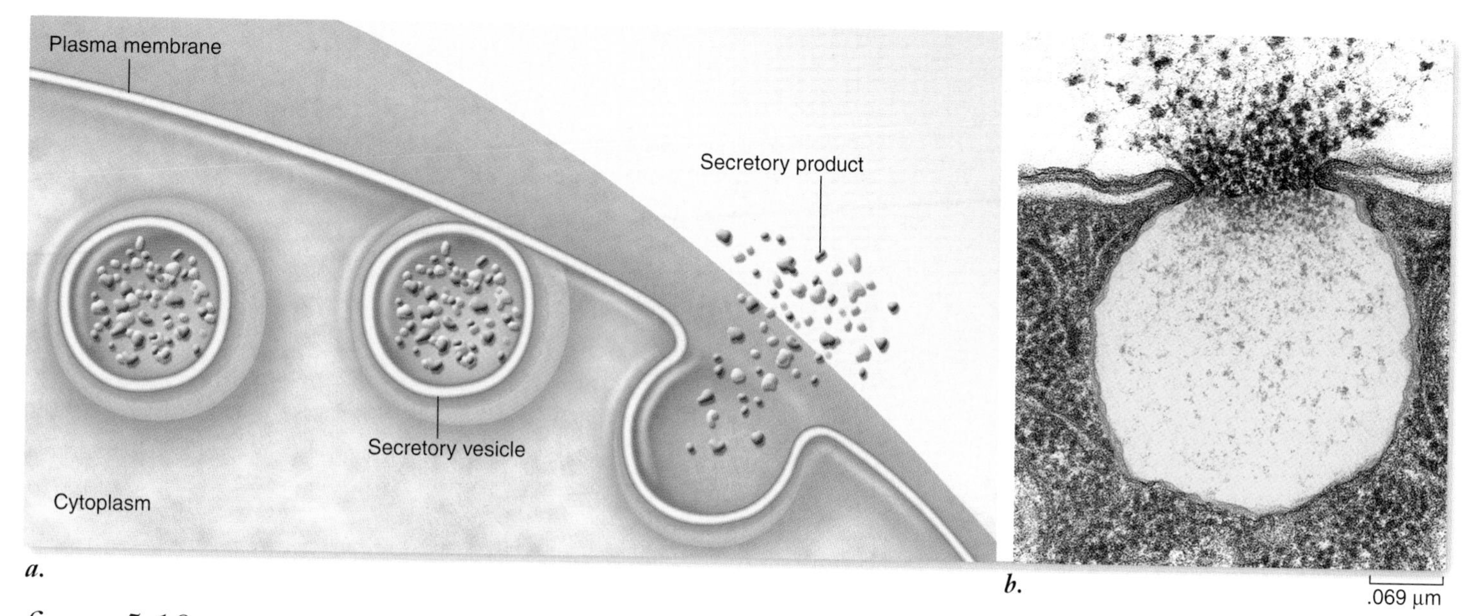

figure 5.18

EXOCYTOSIS. ***a.*** Proteins and other molecules are secreted from cells in small packets called vesicles, whose membranes fuse with the plasma membrane, releasing their contents outside the cell. ***b.*** A false-colored transmission electron micrograph showing exocytosis.

TABLE 5.2 Mechanisms for Transport Across Cell Membranes

Process	How It Works	Example
PASSIVE PROCESSES		
Diffusion		
Direct	Random molecular motion produces net migration of nonpolar molecules toward region of lower concentration	Movement of oxygen into cells
Protein channel	Polar molecules move through a protein channel; net movement is toward region of lower concentration	Movement of ions in or out of cell
Facilitated Diffusion		
Protein carrier	Molecule binds to carrier protein in membrane and is transported across; net movement is toward region of lower concentration	Movement of glucose into cells
Osmosis		
Aquaporins	Diffusion of water across the membrane via osmosis; requires osmotic gradient	Movement of water into cells placed in a hypotonic solution
ACTIVE PROCESSES		
Active Transport		
Protein carrier		
Na^+/K^+ pump	Carrier uses energy to move a substance across a membrane against its concentration gradient	Na^+ and K^+ against their concentration gradients
Coupled transport	Molecules are transported across a membrane against their concentration gradients by the cotransport of sodium ions or protons down their concentration gradients	Coupled uptake of glucose into cells against its concentration gradient using a Na^+ gradient
Endocytosis		
Membrane vesicle		
Phagocytosis	Particle is engulfed by membrane, which folds around it and forms a vesicle	Ingestion of bacteria by white blood cells
Pinocytosis	Fluid droplets are engulfed by membrane, which forms vesicles around them	"Nursing" of human egg cells
Receptor-mediated endocytosis	Endocytosis triggered by a specific receptor, forming clathrin-coated vesicles	Cholesterol uptake
Exocytosis		
Membrane vesicle	Vesicles fuse with plasma membrane and eject contents	Secretion of mucus; release of neurotransmitters

5.1 The Structure of Membranes

Membranes that encase all living cells are sheets of phospholipid bilayers with associated proteins (figure 5.2).

- Membranes are formed by two layers of phospholipids with the hydrophobic regions oriented inward and hydrophilic regions oriented outward.
- The fluid mosaic model consists of proteins floating on or in the lipid bilayer.
- Integral membrane proteins span the membrane and have an extracellular domain and an intracellular domain.
- Peripheral membrane proteins are associated with but are not an integral part of membranes.
- The interior protein network is composed of cytoskeletal filaments and peripheral membrane proteins.
- Membranes contain glycoproteins and glycolipids that act as cell identity markers.
- Membranes contain microdomains of distinct lipid and protein composition. One type is the lipid raft, which contains cholesterol.

5.2 Phospholipids: The Membrane's Foundation

Phospholipids are composed of two fatty acids and a phosphate group linked to a three-carbon glycerol molecule.

- The phosphate group is hydrophilic because it is polar and carries a negative charge.
- The fatty acids are nonpolar and hydrophobic and orient away from the polar head of the phospholipids.
- The nonpolar interior of the lipid bilayer impedes the passage of water and water-soluble substances.
- Phospholipids and unanchored proteins are loosely associated and can diffuse laterally within the membrane.
- Membrane fluidity can be altered by changing the fatty acid composition of the membrane.

5.3 Proteins: Multifunctional Components

Cell membranes contain many proteins serving different functions.

- Integral membrane proteins have one or more transmembrane domains that are hydrophobic.
- Transporters are integral membrane proteins that transport specific substances through the membrane.
- Enzymes for metabolic reactions are often found on the interior surface of the membrane.
- Cell surface receptors respond to external chemical messages and change conditions inside the cell.
- Cell surface identity markers identify cells to other cells.
- Cell-to-cell adhesion proteins glue cells together.
- Surface proteins that interact with other cells are anchored to the cytoskeleton.
- Surface proteins are attached to the surface by nonpolar regions imbedded into the middle of the bilayer and by interactions with polar regions of phospholipids.

5.4 Passive Transport Across Membranes

Transport across the membrane can occur by diffusion—either directly through the membrane for hydrophobic substances or through channels and carrier proteins for hydrophilic substances.

- Simple diffusion is the passive movement of a substance along a chemical or electrical gradient.
- Biological membranes pose a barrier because the hydrophobic interior of the membrane repels polar molecules.
- The proteins in the membrane allow diffusion to be selective.
- Channel proteins have a hydrophilic interior, allowing polar molecules to pass through.
- Ion channels have a hydrated interior, and ions can move in either direction based on their relative concentrations or a difference in voltage across the membrane.
- Carrier proteins bind to the specific molecule they assist and facilitate diffusion. The speed of facilitated diffusion is limited by the number of carrier proteins that are in use.
- Osmosis is the movement of water across a membrane, and the direction of movement depends on the solute concentration on either side of the membrane (figure 5.12).
- Solutions can be isotonic, hypotonic, or hypertonic. Cells in an isotonic solution do not take up water, cells in a hypotonic solution will take up water by osmosis, and cells in a hypertonic solution will lose water by osmosis.
- Aquaporins facilitate the diffusion of water. They are specific for water only or for small hydrated molecules.
- Cells maintain osmotic balance by extrusion of water, adjusting of solute concentration or, in the case of plant cells, turgor pressure due to the cell wall.

5.5 Active Transport Across Membranes

Energy is needed to transport compounds against a concentration gradient, that is, from an area of lower concentration to an area of higher concentration.

- Active transport uses highly specialized protein carriers.
- Uniporters transport a specific molecule in one direction.
- Symporters transport two molecules in the same direction.
- Antiporters transport two molecules in opposite directions.
- Coupled transport occurs when the energy released by a diffusing molecule is used to transport a different molecule against its concentration gradient.
- Countertransport is similar to coupled transport. They differ only in the direction that the second molecule moves relative to the first molecule.
- The sodium potassium pump is an example of an ATP-driven carrier protein that moves Na^+ out of the cell and K^+ into the cell against their concentration gradients.

5.6 Bulk Transport by Endocytosis and Exocytosis

Bulk transport is for moving substances that cannot pass through the cell membrane.

- Endocytosis brings materials into the cell by the membrane surrounding material and pinching off to form a vesicle.
- Phagocytosis is a form of endocytosis that brings in large particles.
- Pinocytosis is a form of endocytosis that brings in extracellular fluids.
- Receptor-mediated endocytosis brings in specific molecules that bind to receptors on the cell membrane.
- Exocytosis discharges material from the cell by a vesicle fusing with the membrane to release its contents.

review questions

SELF TEST

1. The description of a membrane as a "fluid mosaic" means—
   a. Water molecules make up part of the membrane.
   b. The membrane is a mosaic of phospholipids and proteins.
   c. The phospholipids that make up the membrane can move.
   d. Membranes are made of proteins and lipids that can freely move.
2. Membrane proteins are distinguished by what chemical property?
   a. They possess regions with hydrophobic amino acids.
   b. They possess regions with hydrophilic amino acids.
   c. They are polar.
   d. They form hydrogen bonds with the fatty acid tails of the lipid molecules.
3. What chemical property characterizes the interior of the phospholipid bilayer?
   a. It is hydrophobic.
   b. It is hydrophilic.
   c. It is polar.
   d. It is saturated.
4. How do some bacterial cells adapt to cold temperatures?
   a. By increasing the amount of cholesterol in their membranes
   b. By altering the amount of protein present in the membrane
   c. By increasing the number of carbon–carbon double bonds in the fatty acid portion of their phospholipids
   d. By decreasing the rate of diffusion of the phospholipids
5. The specific function of a membrane within a cell is determined by the—
   a. degree of saturation of the fatty acids within the phospholipids bilayer
   b. location of the membrane within the cell
   c. presence of lipid rafts and cholesterol
   d. type and number of membrane proteins
6. The transmembrane domain of an integral membrane protein—
   a. is composed of hydrophobic amino acids
   b. often forms an α-helical structure
   c. can cross the membrane multiple times
   d. all of the above
7. What variable(s) influences whether a nonpolar molecule can move across a membrane by passive diffusion?
   a. The structure of the phospholipids bilayer
   b. The difference in concentration of the molecule across the membrane
   c. The presence of transport proteins in the membrane
   d. All of the above
8. Which of the following does NOT contribute to the selective permeability of a biological membrane?
   a. Specificity of the carrier proteins in the membrane
   b. Selectivity of channel proteins in the membrane
   c. Hydrophobic barrier of the phospholipids bilayer
   d. Hydrogen bond formation between water and phosphate groups
9. Which of the following membrane proteins does NOT rely on diffusion to function?
   a. An ion channel protein
   b. A carrier protein
   c. Aquaporin
   d. Na^+/K^+ pump
10. The movement of water across a membrane is dependent on—
    a. the solvent concentration
    b. the solute concentration
    c. the presence of carrier proteins
    d. membrane potential
11. If a cell is in an isotonic environment, then which of the following is true?
    a. The cell will gain water and burst.
    b. No water will move across the membrane. There will be no effect.
    c. The cell will lose water and shrink.
    d. Osmosis still occurs, but there is no net gain or loss of cell volume.
12. How are *active* transport and *coupled* transport related?
    a. They both use ATP to move molecules.
    b. Active transport establishes a concentration gradient, but coupled transport doesn't.
    c. Coupled transport uses the concentration gradient established by active transport.
    d. Active transport moves one molecule, but coupled transport moves two.
13. In the process of receptor-mediated endocytosis, the receptor is a __________ protein, but the clathrin is an example of a __________ membrane protein.
    a. carrier; channel
    b. transmembrane; peripheral
    c. pump; uniporter
    d. phagocytosis; pinocytosis
14. Which of the following is NOT a mechanism for bringing material into a cell?
    a. exocytosis
    b. endocytosis
    c. pinocytosis
    d. phagocytosis

CHALLENGE QUESTIONS

1. Figure 5.4 describes a classic experiment demonstrating the ability of proteins to move within the plane of the cell's plasma membrane. The following table outlines three different experiments using the fusion of labeled mouse and human cells.

Experiment	Conditions	Temperature (°C)	Result
1	*Fuse human and mouse cells*	37	*Intermixed membrane proteins*
2	*Fuse human and mouse cells in presence of ATP inhibitors*	37	*Intermixed membrane proteins*
3	*Fuse human and mouse cells*	4	*No intermixing of membrane proteins*

   What conclusions can you reach about the movement of these proteins?

2. Each compartment of the endomembrane system of a cell is connected to the plasma membrane. Create a simple diagram of a cell including the RER, Golgi apparatus, vesicle, and the plasma membrane. Starting with the RER, use two different colors to represent the inner and outer halves of the bilayer for each of these membranes. What do you observe?

Do you need additional review? *Visit* www.ravenbiology.com *for practice quizzes, animations, videos, and activities designed to help you master the material in this chapter.*

chapter 6

Energy and Metabolism

introduction

LIFE CAN BE VIEWED AS A CONSTANT flow of energy, channeled by organisms to do the work of living. Each of the significant properties by which we define life—order, growth, reproduction, responsiveness, and internal regulation—requires a constant supply of energy. Both the lion and the giraffe need to eat to provide energy for a wide variety of cellular functions. Deprived of a source of energy, life stops. Therefore, a comprehensive study of life would be impossible without discussing **bioenergetics,** the analysis of how energy powers the activities of living systems. In this chapter, we focus on energy—what it is and how it changes during chemical reactions.

concept outline

6.1 The Flow of Energy in Living Systems

Thermodynamics is the branch of chemistry concerned with energy changes. Cells are governed by the laws of physics and chemistry, so we must understand these laws in order to understand how cells function.

Energy can take many forms

Energy is defined as the capacity to do work. We think of energy as existing in two states: kinetic energy and potential energy (figure 6.1). **Kinetic energy** is the energy of motion. Moving objects perform work by causing other matter to move. **Potential energy** is stored energy. Objects that are not actively moving but have the capacity to do so possess potential energy. A boulder perched on a hilltop has gravitational potential energy; as it begins to roll downhill, some of its potential energy is converted into kinetic energy. Much of the work that living organisms carry out involves transforming potential energy into kinetic energy.

Energy can take many forms: mechanical energy, heat, sound, electric current, light, or radioactive radiation. Because it can exist in so many forms, energy can be measured in many ways. The most convenient measure is in terms of heat, because all other forms of energy can be converted into heat. In fact, the term *thermodynamics* means "heat changes."

The unit of heat most commonly employed in biology is the kilocalorie (kcal). One kilocalorie is equal to 1000 calories (cal), and one calorie is the heat required to raise the temperature of one gram of water one degree Celsius (°C). (In dietetics and nutrition, the term *Calorie* with a capital C is used, and it is actually the same as kilocalorie.) Another energy unit, often used in physics, is the *joule;* one joule equals 0.239 cal.

The sun provides energy for living systems

Energy flows into the biological world from the Sun. It is estimated that the Sun provides the Earth with more than 13×10^{23} calories per year, or 40 million billion calories per second! Plants, algae, and certain kinds of bacteria capture a fraction of this energy through photosynthesis.

In photosynthesis, energy garnered from sunlight is used to combine small molecules (water and carbon dioxide) into more complex ones (sugars). This process converts carbon from an inorganic to an organic form, in the process storing energy from the Sun as potential energy in the covalent bonds between atoms in the sugar molecules.

Breaking the bonds between atoms requires energy; in fact, the strength of a covalent bond is measured by the amount of energy required to break it. For example, it takes

a. Potential energy

b. Kinetic energy

figure 6.1

POTENTIAL AND KINETIC ENERGY. ***a.*** Objects that have the capacity to move but are not moving have potential energy. The energy required for the girl to climb to the top of the slide is stored as potential energy. ***b.*** Objects that are in motion have kinetic energy. The stored potential energy is released as kinetic energy as the girl slides down.

98.8 kcal to break one mole (6.023×10^{23}) of the carbon–hydrogen (C—H) bonds found in organic molecules. In molecules like the fats used for energy storage, with many C—H bonds, this adds up. The oxidation of one mole of a 16-carbon fatty acid completely saturated with hydrogens yields 2340 kcal.

Oxidation–reduction reactions transfer electrons while bonds are made or broken

During a chemical reaction, the energy stored in chemical bonds may be used to make new bonds. In some of these reactions, electrons actually pass from one atom or molecule to another. An atom or molecule that loses an electron is said to be oxidized, and the process by which this occurs is called **oxidation** because in biological systems, oxygen is the most common electron acceptor. Conversely, an atom or molecule that gains an electron is said to be reduced, and the process is called **reduction.** The reduced form of a molecule has a higher level of energy than the oxidized form (figure 6.2).

Oxidation and reduction always take place together, because every electron that is lost by an atom through oxidation is gained by some other atom through reduction. Therefore, chemical reactions of this sort are called **oxidation–reduction,** or **redox, reactions.** Oxidation–reduction reactions play a key role in the flow of energy through biological systems.

In the next two chapters, you will learn the details of how organisms derive energy from the oxidation of organic compounds via respiration, and from the energy in sunlight via photosynthesis.

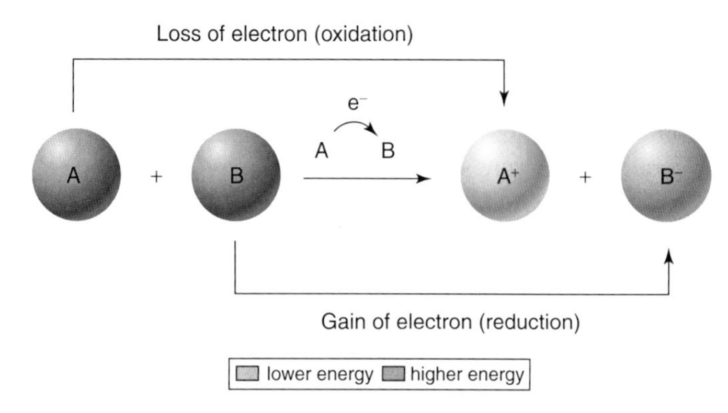

figure 6.2

REDOX REACTIONS. Oxidation is the loss of an electron; reduction is the gain of an electron. In this example, the charges of molecules A and B appear as superscripts in each molecule. Molecule A loses energy as it loses an electron, and molecule B gains energy as it gains an electron.

Energy is defined as the capacity to do work. This can either be energy of motion, kinetic energy, or stored energy, potential energy. The ultimate source of energy for living systems is the Sun. Organisms derive energy from oxidation–reduction reactions in which electrons are transferred from one molecule to another.

6.2 The Laws of Thermodynamics and Free Energy

All activities of living organisms—growing, running, thinking, singing, reading these words—involve changes in energy. A set of two universal laws we call the laws of thermodynamics govern all energy changes in the universe, from nuclear reactions to a bird flying through the air.

The First Law states that energy cannot be created or destroyed

The **First Law of Thermodynamics** concerns the amount of energy in the universe. Energy cannot be created or destroyed; it can only change from one form to another (from potential to kinetic, for example). The total amount of energy in the universe remains constant.

The lion eating a giraffe at the beginning of this chapter is in the process of acquiring energy. Rather than creating new energy or capturing the energy in sunlight, the lion is merely transferring some of the potential energy stored in the giraffe's tissues to its own body, just as the giraffe obtained the potential energy stored in the plants it ate while it was alive.

Within any living organism, chemical potential energy stored in some molecules can be shifted to other molecules and stored in different chemical bonds, or it can be converted into other forms, such as kinetic energy, light, or electricity. During each conversion, some of the energy dissipates into the environment as **heat,** which is a measure of the random motion of molecules (and therefore a measure of one form of kinetic energy). Energy continuously flows through the biological world in one direction, with new energy from the Sun constantly entering the system to replace the energy dissipated as heat.

Heat can be harnessed to do work only when there is a heat gradient—that is, a temperature difference between two areas. Cells are too small to maintain significant internal temperature differences, so heat energy is incapable of doing the work of cells. Instead, cells must rely on chemical reactions for energy.

Although the total amount of energy in the universe remains constant, the energy available to do work decreases as more of it progressively dissipates as heat.

The Second Law states that some energy is lost as disorder increases

The **Second Law of Thermodynamics** concerns the transformation of potential energy into heat, or random molecular motion. It states that the disorder in the universe, more formally called **entropy,** is continuously increasing. Put simply, disorder is more likely than order. For example, it is much more likely that a column of bricks will tumble over than that a pile of bricks will arrange themselves spontaneously to form a column.

In general, energy transformations proceed spontaneously to convert matter from a more ordered, less stable form to a less ordered, but more stable form (figure 6.3). For this reason, the second law is sometimes called "time's arrow." If you saw a photograph of the column of bricks, and a photograph of a random pile of the same bricks, you could put the pictures into correct sequence using the information that time had elapsed with only natural processes occurring.

The Second Law of Thermodynamics can also be stated simply as "entropy increases." When the universe formed, it held all the potential energy it will ever have. It has become progressively more disordered ever since, with every energy exchange increasing the amount of entropy.

Chemical reactions can be predicted based on changes in free energy

It takes energy to break the chemical bonds that hold the atoms in a molecule together. Heat energy, because it increases atomic motion, makes it easier for the atoms to pull apart. Both chemical bonding and heat have a significant influence on a molecule, the former reducing disorder, and the latter increasing it. The net effect, the amount of energy actually available to break and subsequently form other chemical bonds, is called the *free energy* of that molecule. In a more general sense, **free energy** is defined as the energy available to do work in any system.

In a molecule within a cell, where pressure and volume usually do not change, the free energy is denoted by the symbol G (for "Gibbs' free energy," which limits the system being considered to the cell). G is equal to the energy contained in a molecule's chemical bonds (called **enthalpy** and designated H) minus the energy unavailable because of disorder (called *entropy* and symbolized as S) times the absolute temperature, T, expressed in the Kelvin scale ($K = °C + 273$):

$$G = H - TS$$

Chemical reactions break some bonds in the reactants and form new ones in the products. Consequently, reactions can produce changes in free energy. When a chemical reaction occurs under conditions of constant temperature, pressure, and volume—as do most biological reactions—the change symbolized by the Greek capital letter delta, Δ, in free energy (ΔG) is simply:

$$\Delta G = \Delta H - T\Delta S$$

The change in free energy, or ΔG, is a fundamental property of chemical reactions. In some reactions, the ΔG is positive, which means that the products of the reaction contain *more* free energy than the reactants; the bond energy (H) is higher, or the disorder (S) in the system is lower. Such reactions do not proceed spontaneously because they require an input of energy. Any reaction that requires an input of energy is said to be **endergonic** ("inward energy").

In other reactions, the ΔG is negative. In this case, the products of the reaction contain less free energy than the reactants; either the bond energy is lower, or the disorder is higher, or both. Such reactions tend to proceed spontaneously. Any chemical reaction tends to proceed spontaneously if the difference in disorder ($T\Delta S$) is *greater* than the difference in bond energies between reactants and products (ΔH).

Note that spontaneous does not mean the same thing as instantaneous. A spontaneous reaction may proceed very slowly. These reactions release the excess free energy as heat and are thus said to be **exergonic** ("outward energy"). Figure 6.4 sums up endergonic and exergonic reactions.

Because chemical reactions are reversible, a reaction that is exergonic in the forward direction will be endergonic in the reverse direction. For each reaction, an equilibrium exists at

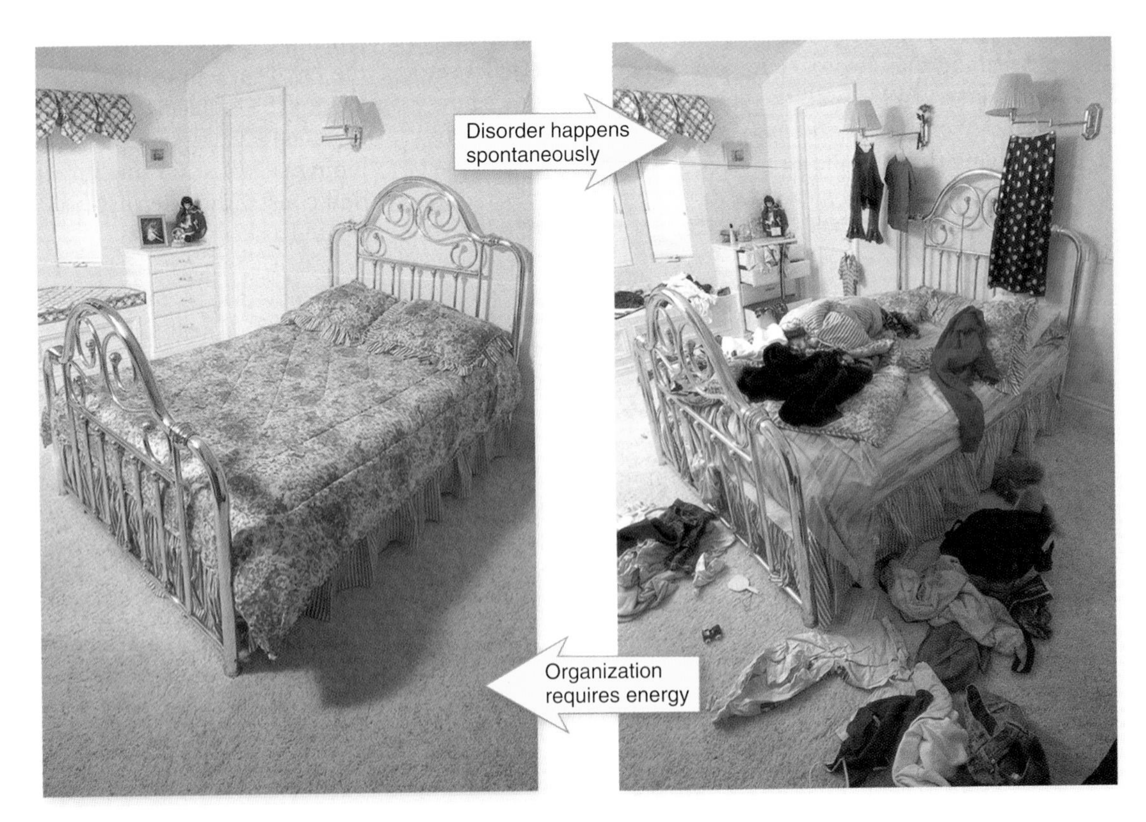

figure 6.3

ENTROPY IN ACTION. As time elapses, the room shown at right becomes more disorganized. Entropy has increased in the state of this room. It takes energy to restore it to the ordered state shown at left.

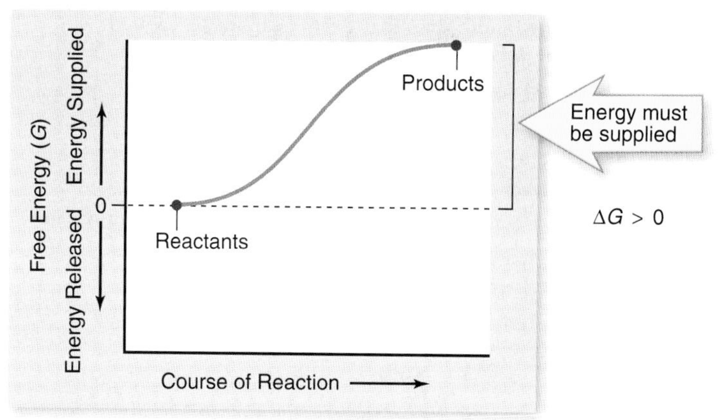

a.

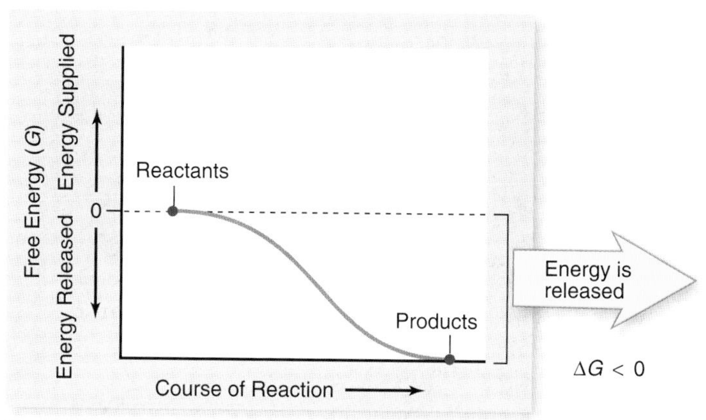

b.

figure 6.4

ENERGY IN CHEMICAL REACTIONS. ***a.*** In an endergonic reaction, the products of the reaction contain more energy than the reactants, and the extra energy must be supplied for the reaction to proceed. ***b.*** In an exergonic reaction, the products contain less energy than the reactants, and the excess energy is released.

some point between the relative amounts of reactants and products. This equilibrium has a numeric value and is called the *equilibrium constant.* This characteristic of reactions provides us with another way to think about free energy changes: an exergonic reaction has an equilibrium favoring the products, and an endergonic reaction has an equilibrium favoring the reactants.

Spontaneous chemical reactions require activation energy

If all chemical reactions that release free energy tend to occur spontaneously, why haven't all such reactions already occurred? Consider the gasoline tank of your car: The oxidation of the hydrocarbons in gasoline is an exergonic reaction, but your gas tank does not spontaneously explode. One reason is that most reactions require an input of energy to get started. In the case of your car, this input consists of the electrical sparks in the engine's cylinders, producing a controlled explosion.

Activation energy

Before new chemical bonds can form, even bonds that contain less energy, existing bonds must first be broken, and that requires energy input. The extra energy needed to destabilize existing chemical bonds and initiate a chemical reaction is called **activation energy** (figure 6.5).

The rate of an exergonic reaction depends on the activation energy required for the reaction to begin. Reactions with larger activation energies tend to proceed more slowly because fewer molecules succeed in overcoming the initial energy hurdle. The rate of reactions can be increased in two ways: by increasing the energy of reacting molecules, or by lowering activation energy. Chemists often drive important industrial reactions by increasing the energy of the reacting molecules, which is frequently accomplished simply by heating up the reactants. The other strategy is to use a catalyst to lower the activation energy.

How catalysts work

Activation energies are not constant. Stressing particular chemical bonds can make them easier to break. The process of influencing chemical bonds in a way that lowers the activation energy needed to initiate a reaction is called **catalysis,** and substances that accomplish this are known as **catalysts** (see figure 6.5).

Catalysts cannot violate the basic laws of thermodynamics; they cannot, for example, make an endergonic reaction proceed spontaneously. By reducing the activation energy, a catalyst accelerates both the forward and the reverse reactions by exactly the same amount. Therefore, a catalyst does not alter the proportion of reactant that is ultimately converted into product.

To grasp this, imagine a bowling ball resting in a shallow depression on the side of a hill. Only a narrow rim of dirt below the ball prevents it from rolling down the hill. Now imagine digging away that rim of dirt. If you remove enough dirt from below the ball, it will start to roll down the hill—but removing dirt from below the ball will *never* cause the ball to roll up the

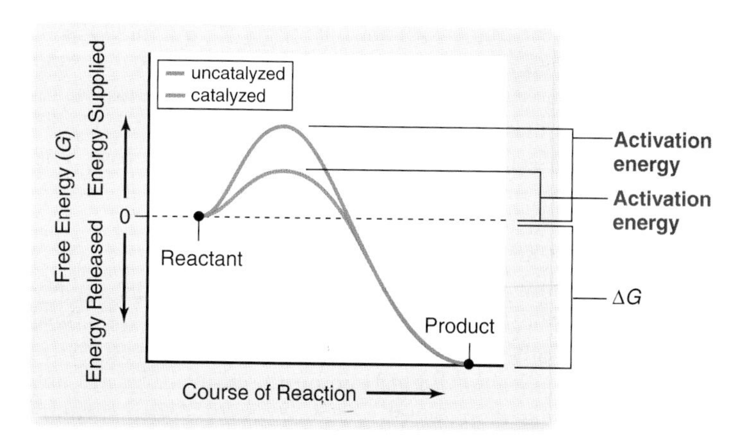

figure 6.5

ACTIVATION ENERGY AND CATALYSIS. Exergonic reactions do not necessarily proceed rapidly because energy must be supplied to destabilize existing chemical bonds. This extra energy is the activation energy for the reaction shown in the curve for the uncatalyzed reaction. Catalysts accelerate particular reactions by lowering the amount of activation energy required to initiate the reaction. Catalysts do not change the free energy of the reactants or products and therefore do not alter the free-energy change produced by the reaction.

hill. Removing the lip of dirt simply allows the ball to move freely; gravity determines the direction it then travels.

Similarly, the direction in which a chemical reaction proceeds is determined solely by the difference in free energy between reactants and products. Like digging away the soil below the bowling ball on the hill, catalysts reduce the energy barrier that is preventing the reaction from proceeding. Only exergonic reactions can proceed spontaneously, and catalysts cannot change that. What catalysts *can* do is make a reaction proceed much faster. In living systems, enzymes act as catalysts.

Thermodynamics states that energy cannot be created or destroyed, and that energy transactions are inherently inefficient. The loss of energy results in greater disorder, or entropy. Chemical reactions can be predicted based on free-energy changes (ΔG). Reactions with a negative ΔG occur spontaneously, and those with a positive ΔG do not. Even spontaneous reactions may occur slowly because they require an initial source of energy called activation energy. Catalysts, such as enzymes in living systems, lower this activation energy to speed up reactions.

6.3 ATP: The Energy Currency of Cells

The chief "currency" all cells use for their energy transactions is the nucleotide **adenosine triphosphate (ATP).** In cells ATP powers almost every energy-requiring process, from making sugars, to supplying activation energy for chemical reactions, to actively transporting substances across membranes, to moving through the environment and growing.

Cells store and release energy in the bonds of ATP

You saw in chapter 3 that nucleotides serve as the building blocks for nucleic acids, but they play other cellular roles as well. ATP is used as a building block for RNA molecules, and it also has a critical function as a portable source of energy on demand for endergonic cellular processes.

The structure of ATP

Like all nucleotides, ATP is composed of three smaller components (figure 6.6). The first component is a five-carbon sugar, ribose, which serves as the framework to which the other two subunits are attached. The second component is adenine, an organic molecule composed of two carbon–nitrogen rings. Each of the nitrogen atoms in the ring has an unshared pair of electrons and weakly attracts hydrogen ions, making adenine chemically a weak base. The third component of ATP is a chain of three phosphates.

How ATP stores energy

The key to how ATP stores energy lies in its triphosphate group. Phosphate groups are highly negatively charged, and thus they strongly repel one another. Due to this electrostatic repulsion, the covalent bonds joining the phosphates are unstable. The molecule is often referred to as a "coiled spring," with the phosphates straining away from one another.

The unstable bonds holding the phosphates together in the ATP molecule have a low activation energy and are easily broken by hydrolysis. When they break, they can transfer a considerable amount of energy. Another way of saying this is that the hydrolysis of ATP has a negative ΔG, and the energy released can be used to perform work.

In most reactions involving ATP, only the outermost high-energy phosphate bond is hydrolyzed, cleaving off the phosphate group on the end. When this happens, ATP becomes **adenosine diphosphate (ADP)** plus an **inorganic phosphate (P_i)**,

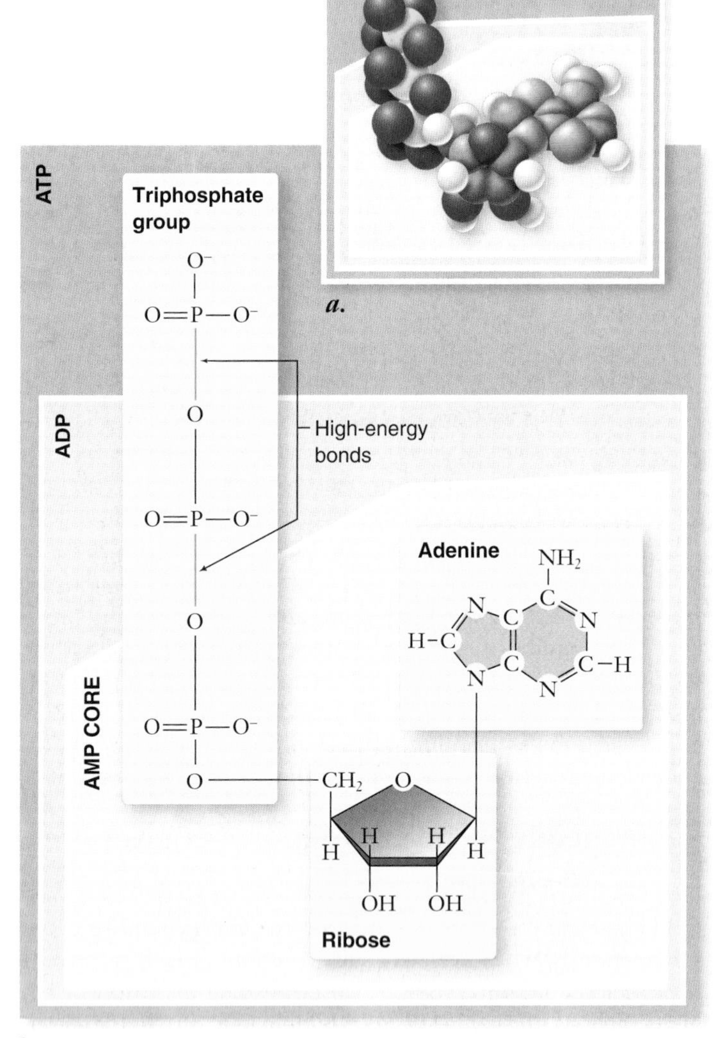

figure 6.6

THE ATP MOLECULE. ***a.*** The model and ***(b)*** the structural diagram both show that ATP has a core of AMP. Addition of one phosphate to AMP yields ADP, and addition of a second phosphate yields ATP. These two terminal phosphates are attached by high-energy bonds such that removal of either by hydrolysis is an exergonic reaction that releases energy.

and energy equal to 7.3 kcal/mol is released under standard conditions. The liberated phosphate group usually attaches temporarily to some intermediate molecule. When that molecule is dephosphorylated, the phosphate group is released as P_i.

Both of the two terminal phosphates can be hydrolyzed to release energy, leaving **adenosine monophosphate (AMP),** but the third phosphate is not attached by a high-energy bond. With only one phosphate group, AMP has no other phosphates to provide the electrostatic repulsion that makes the bonds holding the two terminal phosphate groups high-energy bonds.

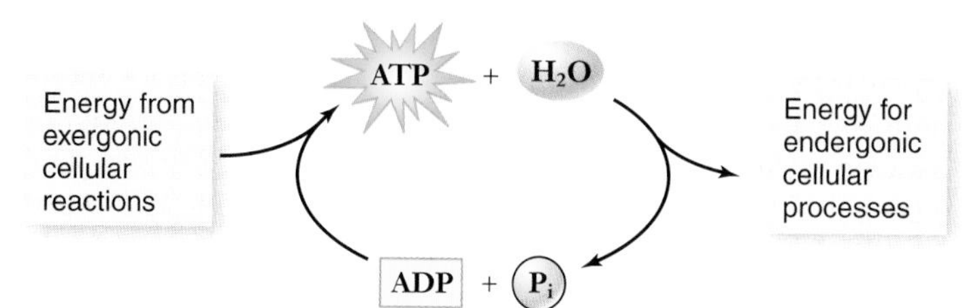

figure 6.7

THE ATP CYCLE. ATP is synthesized and used in a cyclic fashion. The synthesis of ATP from ADP + P_i is endergonic and is powered by exergonic cellular reactions. The hydrolysis of ATP to ADP + P_i is exergonic, and the energy released is used to power endergonic cellular functions such as muscle contraction.

ATP hydrolysis drives endergonic reactions

Cells use ATP to drive endergonic reactions. These reactions do not proceed spontaneously because their products possess more free energy than their reactants. However, if the cleavage of ATP's terminal high-energy bond releases more energy than the other reaction consumes, the two reactions can be coupled so that the energy released by the hydrolysis of ATP can be used to supply the endergonic reaction with the energy it needs. Coupled together, these reactions result in a net release of energy ($-\Delta G$) and are therefore exergonic and proceed spontaneously. Because almost all the endergonic reactions in cells require less energy than is released by the cleavage of ATP, ATP can provide most of the energy a cell needs.

inquiry

When ATP hydrolysis is coupled with an endergonic reaction, and supplies more than enough energy, is the overall process endergonic or exergonic? Would the ΔG for the overall process be a negative or positive?

ATP cycles continuously

The same feature that makes ATP an effective energy donor—the instability of its phosphate bonds—precludes it from being a good long-term energy-storage molecule. Fats and carbohydrates serve that function better.

The use of ATP can be thought of as a cycle: Cells use exergonic reactions to provide the energy needed to synthesize ATP from ADP + P_i; they then use the hydrolysis of ATP to provide energy to drive the endergonic reactions they need (figure 6.7).

Most cells do not maintain large stockpiles of ATP. Instead, they typically have only a few seconds' supply of ATP at any given time, and they continually produce more from ADP and P_i. It is estimated that even a sedentary individual turns over an amount of ATP in one day roughly equal to their body weight. This statistic makes clear the importance of ATP synthesis. In the next two chapters we will explore in detail the cellular mechanisms for synthesizing ATP.

Cells use the molecule ATP as a portable form of energy. ATP is a nucleotide with three phosphate groups. Removal of each of the two terminal phosphates by hydrolysis releases energy. The synthesis of ATP requires energy so the cell is constantly making and using ATP.

6.4 Enzymes: Biological Catalysts

The chemical reactions within living organisms are regulated by controlling the points at which catalysis takes place. Life itself, therefore, can be seen as regulated by catalysts. The agents that carry out most of the catalysis in living organisms are called enzymes. Most enzymes are proteins, although increasing evidence indicates that some enzymes are actually RNA molecules, as discussed later in this chapter.

An enzyme alters the activation energy of a reaction

The unique three-dimensional shape of an enzyme enables it to stabilize a temporary association between **substrates,** the molecules that will undergo the reaction. By bringing two substrates together in the correct orientation, or by stressing particular chemical bonds of a substrate, an enzyme lowers the activation energy required for new bonds to form. The reaction thus proceeds much more quickly than it would without the enzyme.

The enzyme itself is not changed or consumed in the reaction, so only a small amount of an enzyme is needed, and it can be used over and over.

As an example of how an enzyme works, let's consider the reaction of carbon dioxide and water to form carbonic acid. This important enzyme-catalyzed reaction occurs in vertebrate red blood cells:

$$\underset{\text{carbon dioxide}}{CO_2} + \underset{\text{water}}{H_2O} \rightleftharpoons \underset{\text{carbonic acid}}{H_2CO_3}$$

This reaction may proceed in either direction, but because it has a large activation energy, the reaction is very slow in the absence of an enzyme: Perhaps 200 molecules of carbonic acid form in an hour in a cell in the absence of any enzyme. Reactions that proceed this slowly are of little use to a cell. Vertebrate red blood cells overcome this problem by employing an enzyme within their cytoplasm called *carbonic anhydrase* (enzyme names usually end in "–ase"). Under the same conditions, but in the presence of carbonic anhydrase, an estimated 600,000 molecules of carbonic acid form every *second!* Thus, the enzyme increases the reaction rate by more than one million times.

Thousands of different kinds of enzymes are known, each catalyzing one or a few specific chemical reactions. By facilitating particular chemical reactions, the enzymes in a cell determine the course of metabolism—the collection of all chemical reactions—in that cell.

Different types of cells contain different sets of enzymes, and this difference contributes to structural and functional variations among cell types. For example, the chemical reactions taking place within a red blood cell differ from those that occur within a nerve cell, in part because different cell types contain different arrays of enzymes.

Active sites of enzymes conform to fit the shape of substrates

Most enzymes are globular proteins with one or more pockets or clefts, called **active sites,** on their surface (figure 6.8). Substrates bind to the enzyme at these active sites, forming an **enzyme–substrate complex** (figure 6.9). For catalysis to occur within the complex, a substrate molecule must fit precisely into an active site. When that happens, amino acid side groups of the enzyme end up in close proximity to certain bonds of the substrate. These side groups interact chemically with the substrate, usually stressing or distorting a particular bond and consequently lowering the activation energy needed to break the bond. After the bonds of the substrates are broken, or new bonds are formed, the substrates have been converted to products. These products then dissociate from the enzyme, leaving the enzyme ready to bind its next substrate and begin the cycle again.

Proteins are not rigid. The binding of a substrate induces the enzyme to adjust its shape slightly, leading to a better *induced fit* between enzyme and substrate (see figure 6.8). This interaction may also facilitate the binding of other substrates; in such cases, one substrate "activates" the enzyme to receive other substrates.

Enzymes occur in many forms

Although many enzymes are suspended in the cytoplasm of cells, not attached to any structure, other enzymes function as integral parts of cell membranes and organelles. Enzymes may also form associations termed *multienzyme complexes* to carry out reaction sequences. And, as mentioned earlier, evidence exists that some enzymes may consist of RNA rather than being only protein.

Multienzyme complexes

Often several enzymes catalyzing different steps of a sequence of reactions are associated with one another in noncovalently bonded assemblies called **multienzyme complexes.** The bacterial pyruvate dehydrogenase multienzyme complex, shown in figure 6.10, contains enzymes that carry out three sequential reactions in oxidative metabolism. Each complex has multiple copies of each of the three enzymes—60 protein subunits in all. The many subunits work in concert, forming a molecular machine that performs multiple functions.

Multienzyme complexes offer the following significant advantages in catalytic efficiency:

1. The rate of any enzyme reaction is limited by the frequency with which the enzyme collides with its substrate. If a series of sequential reactions occurs within a multienzyme complex, the product of one reaction can be delivered to the next enzyme without releasing it to diffuse away.
2. Because the reacting substrate never leaves the complex during its passage through the series of reactions, the possibility of unwanted side reactions is eliminated.
3. All of the reactions that take place within the multienzyme complex can be controlled as a unit.

In addition to pyruvate dehydrogenase, which controls entry to the Krebs cycle during aerobic respiration (chapter 7), several other key processes in the cell are catalyzed by multienzyme complexes. One well-studied system is the fatty acid synthetase complex that catalyzes the synthesis of fatty acids from two-carbon precursors. Seven different enzymes make up this multienzyme complex, and the reaction intermediates remain associated with the complex for the entire series of reactions.

Nonprotein enzymes

Until a few years ago, most biology textbooks contained statements such as "Proteins called enzymes are the catalysts of biological systems." We can no longer make that statement without qualification.

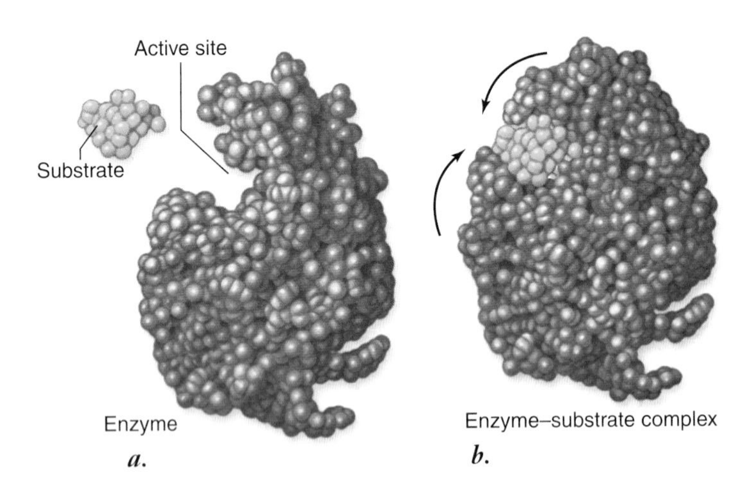

figure 6.8

ENZYME BINDING ITS SUBSTRATE. ***a.*** The active site of the enzyme lysozyme fits the shape of its substrate, a peptidoglycan that makes up bacterial cell walls. ***b.*** When the substrate, indicated in yellow, slides into the groove of the active site, its entry induces the protein to alter its shape slightly and bind the substrate more tightly. This alteration of the shape of the enzyme to better fit the substrate is called induced fit.

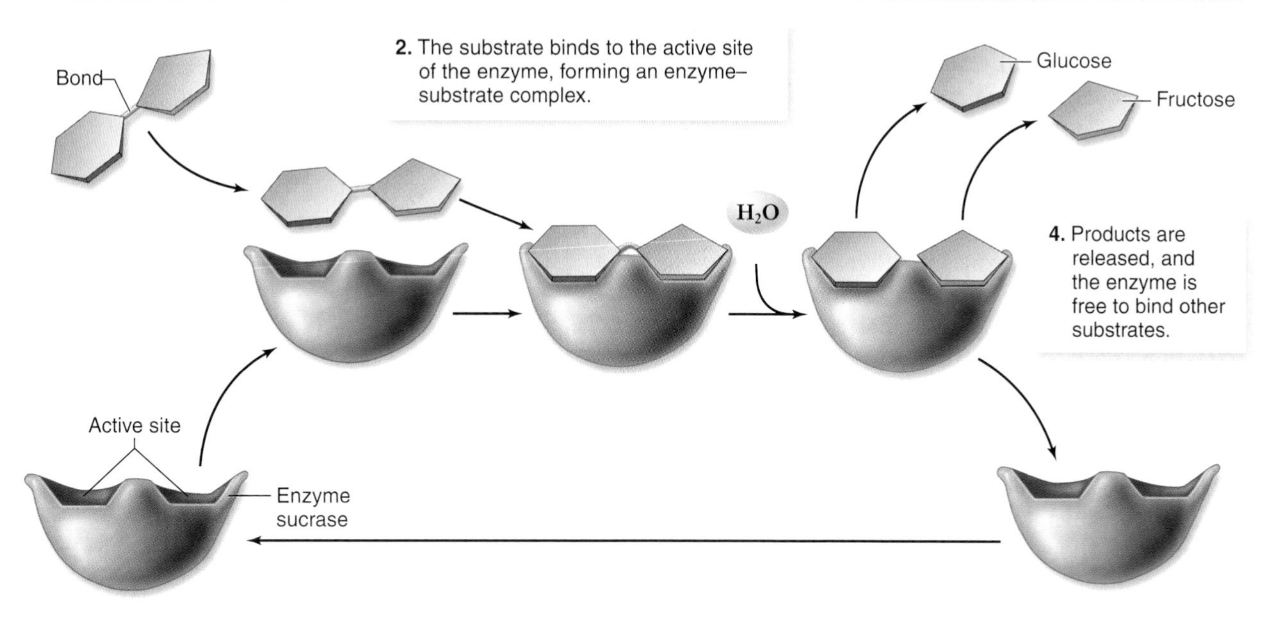

figure 6.9

THE CATALYTIC CYCLE OF AN ENZYME. Enzymes increase the speed at which chemical reactions occur, but they are not altered permanently themselves as they do so. In the reaction illustrated here, the enzyme sucrase is splitting the sugar sucrose into two simpler sugars: glucose and fructose.

Thomas J. Cech and colleagues at the University of Colorado reported in 1981 that certain reactions involving RNA molecules appear to be catalyzed in cells by RNA itself, rather than by enzymes. This initial observation has been corroborated by additional examples of RNA catalysis. Like enzymes, these RNA catalysts, which are loosely called "ribozymes," greatly accelerate the rate of particular biochemical reactions and show extraordinary substrate specificity.

Research has revealed at least two sorts of ribozymes. Some ribozymes have folded structures and catalyze reactions on themselves, a process called *intra*molecular catalysis. Other ribozymes act on other molecules without being changed themselves, a process called *inter*molecular catalysis.

The most striking example of the role of RNA as enzyme is emerging from recent work on the structure and function of the ribosome. For many years it was thought that RNA was a structural framework for this vital organelle, but it is now clear that ribosomal RNA plays a key role in ribosome function. The ribosome itself is a ribozyme.

The ability of RNA, an informational molecule, to act as a catalyst has stirred great excitement among biologists because it appears to provide a potential answer to the question. Which came first, the protein or the nucleic acid? It now seems at least possible that RNA may have evolved first and may have catalyzed the formation of the first proteins.

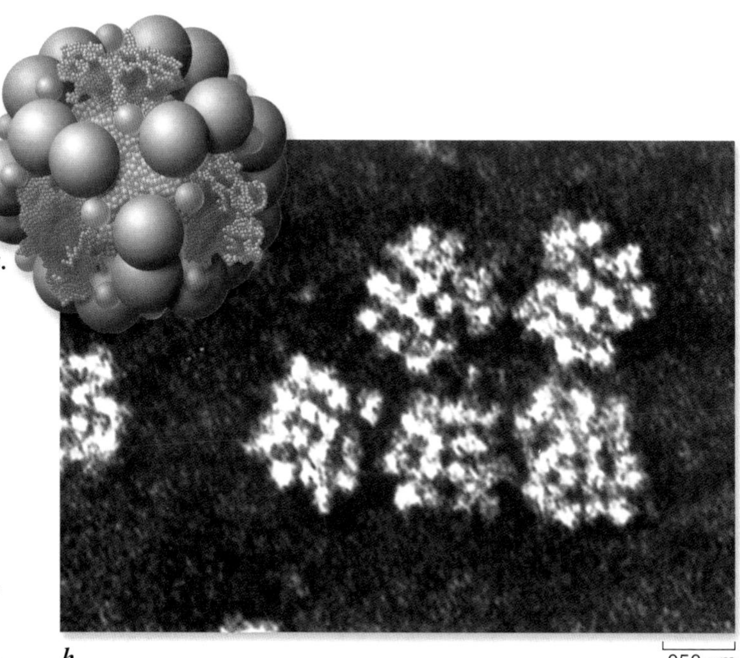

figure 6.10

A COMPLEX ENZYME: PYRUVATE DEHYDROGENASE. Pyruvate dehydrogenase, which catalyzes the oxidation of pyruvate, is one of the most complex enzymes known. ***a.*** A model of the enzyme showing the arrangement of the 60 protein subunits. ***b.*** Many of the protein subunits are clearly visible in the electron micrograph.

Environmental and other factors affect enzyme function

The rate of an enzyme-catalyzed reaction is affected by the concentrations of both the substrate and the enzyme that works on it. In addition, any chemical or physical factor that alters the enzyme's three-dimensional shape—such as temperature, pH, and the binding of regulatory molecules—can affect the enzyme's ability to catalyze the reaction.

Temperature

Increasing the temperature of an uncatalyzed reaction increases its rate because the additional heat increases random molecular movement. This motion can add stress to molecular bonds and affect the activation energy of a reaction.

The rate of an enzyme-catalyzed reaction also increases with temperature, but only up to a point called the *optimum temperature* (figure 6.11*a*). Below this temperature, the hydrogen bonds and hydrophobic interactions that determine the enzyme's shape are not flexible enough to permit the induced fit that is optimum for catalysis. Above the optimum temperature, these forces are too weak to maintain the enzyme's shape against the increased random movement of the atoms in the enzyme. At higher temperatures, the enzyme denatures, as described in chapter 3.

Most human enzymes have an optimum temperature between 35°C and 40°C, a range that includes normal body temperature. Prokaryotes that live in hotsprings have more stable enzymes (that is, enzymes held together more strongly), so the optimum temperature for those enzymes can be 70°C or higher. In each case the optimal temperature for the enzyme corresponds to the "normal" temperature usually encountered in the body or the environment, depending on the type of organism.

pH

Ionic interactions between oppositely charged amino acid residues, such as glutamic acid (–) and lysine (+), also hold enzymes together. These interactions are sensitive to the hydrogen ion concentration of the fluid in which the enzyme is dissolved, because changing that concentration shifts the balance between positively and negatively charged amino acid residues. For this reason, most enzymes have an *optimum pH* that usually ranges from pH 6 to 8.

Enzymes able to function in very acidic environments are proteins that maintain their three-dimensional shape even in the presence of high hydrogen ion concentrations. The enzyme pepsin, for example, digests proteins in the stomach at pH 2, a very acidic level (figure 6.11*b*).

Inhibitors and activators

Enzyme activity is sensitive to the presence of specific substances that can bind to the enzyme and cause changes in its shape. Through these substances, a cell is able to regulate which of its enzymes are active and which are inactive at a particular time. This ability allows the cell to increase its efficiency and to control changes in its characteristics during development. A substance that binds to an enzyme and *decreases* its activity is called an **inhibitor.** Very often, the end product of a biochemical pathway acts as an inhibitor of an early reaction in the pathway, a process called *feedback inhibition* (discussed later in this chapter).

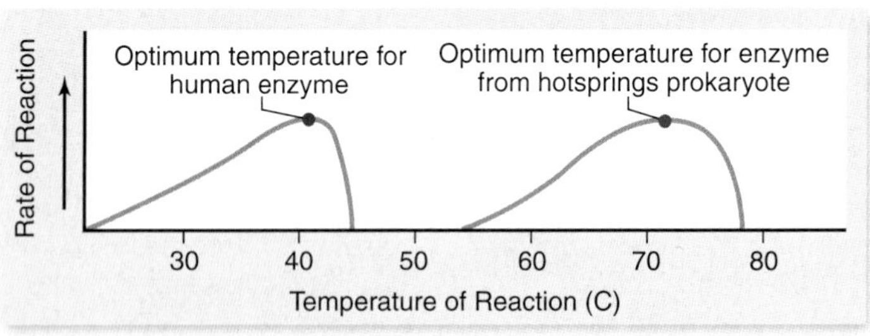

a.

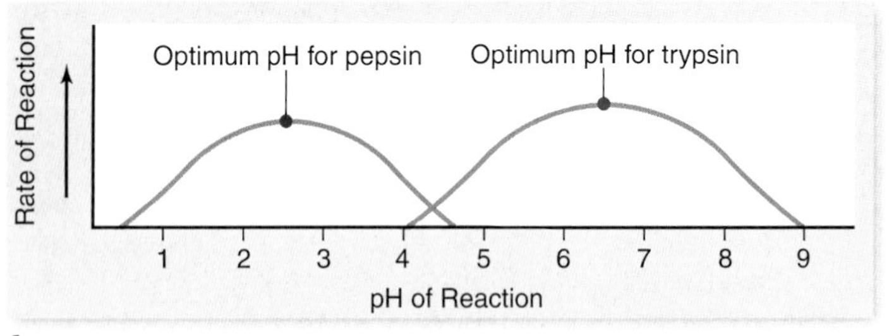

b.

figure 6.11

ENZYME SENSITIVITY TO THE ENVIRONMENT. The activity of an enzyme is influenced by both ***(a)*** temperature and ***(b)*** pH. Most human enzymes, such as the protein-degrading enzyme trypsin, work best at temperatures of about 40°C and within a pH range of 6 to 8. The hotsprings prokaryote has a higher environmental temperature and a correspondingly higher temperature optimum for enzymes. Pepsin works in the acidic environment of the stomach and has a lower optimum pH.

Enzyme inhibition occurs in two ways: **Competitive inhibitors** compete with the substrate for the same active site, occupying the active site and thus preventing substrates from binding; **noncompetitive inhibitors** bind to the enzyme in a location other than the active site, changing the shape of the enzyme and making it unable to bind to the substrate (figure 6.12).

Many enzymes can exist in either an active or inactive conformation; such enzymes are called **allosteric enzymes.** Most noncompetitive inhibitors bind to a specific portion of the enzyme called an **allosteric site.** These sites serve as chemical on/off switches; the binding of a substance to the site can switch the enzyme between its active and inactive configurations. A substance that binds to an allosteric site and reduces enzyme activity is called an **allosteric inhibitor** (figure 6.12*b*).

This kind of control is also used to activate enzymes. An **allosteric activator** binds to allosteric sites to keep an enzyme in its active configurations, thereby *increasing* enzyme activity.

Enzyme cofactors

Enzyme function is often assisted by additional chemical components known as **cofactors.** These can be metal ions that are often found in the active site participating directly in catalysis. For example, the metallic ion zinc is used by some enzymes, such as protein-digesting carboxypeptidase, to draw electrons away from their position in covalent bonds, making the bonds less stable and easier to break. Other metallic elements, such as molybdenum and manganese, are also used as cofactors. Like zinc, these substances are required in the diet in small amounts.

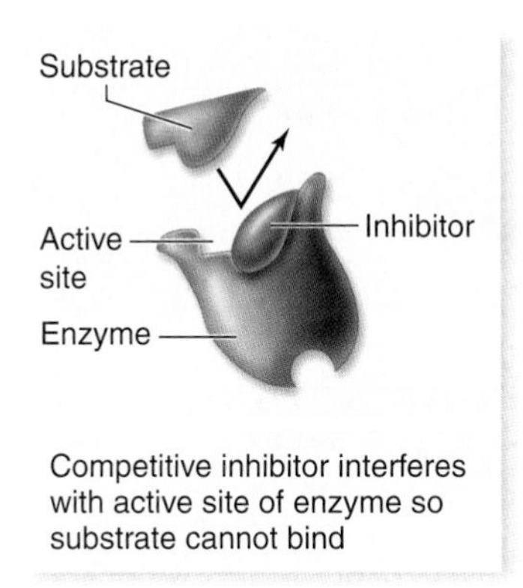

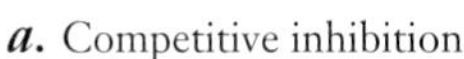

a. Competitive inhibition

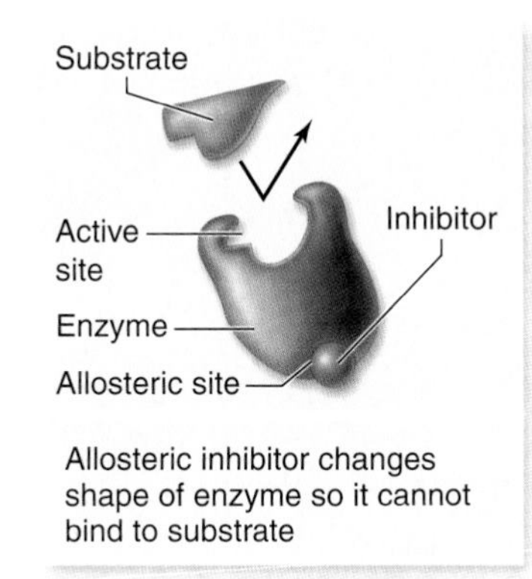

b. Noncompetitive inhibition

figure 6.12

HOW ENZYMES CAN BE INHIBITED. ***a.*** In competitive inhibition, the inhibitor has a shape similar to the substrate and competes for the active site of the enzyme. ***b.*** In noncompetitive inhibition, the inhibitor binds to the enzyme at the allosteric site, a place away from the active site, effecting a conformational change in the enzyme, making it unable to bind to its substrate.

When the cofactor is a nonprotein organic molecule, it is called a **coenzyme.** Many of the small organic molecules essential in our diets that we call vitamins function as coenzymes. For example the B vitamins B_6 and B_{12}, both function as coenzymes for a number of different enzymes. Modified nucleotides are also used as coenzymes.

In numerous oxidation–reduction reactions that are catalyzed by enzymes, the electrons pass in pairs from the active site of the enzyme to a coenzyme that serves as the electron acceptor. The coenzyme then transfers the electrons to a different enzyme, which releases them (and the energy they bear) to the substrates in another reaction. Often, the electrons combine with protons (H^+) to form hydrogen atoms. In this way, coenzymes shuttle energy in the form of hydrogen atoms from one enzyme to another in a cell. The role of coenzymes and the specifics of their action will be explored in detail in the following two chapters.

Biological catalysts are enzymes, which are usually proteins. These bind to their substrates based on shape, allowing great specificity. Enzyme activity is affected by environmental conditions such as temperature and pH. Some enzymes require an inorganic cofactor.

6.5 Metabolism: The Chemical Description of Cell Function

Living chemistry, the total of all chemical reactions carried out by an organism, is called **metabolism.** Those chemical reactions that expend energy to make or transform chemical bonds are called *anabolic* reactions, or **anabolism.** Reactions that harvest energy when chemical bonds are broken are called *catabolic* reactions, or **catabolism.** This section presents a general overview of metabolic processes that will be described in much greater detail in later chapters.

Biochemical pathways organize chemical reactions in cells

Organisms contain thousands of different kinds of enzymes that catalyze a bewildering variety of reactions. Many of these reactions in a cell occur in sequences called **biochemical pathways.** In such pathways, the product of one reaction becomes the substrate for the next (figure 6.13). Biochemical pathways are the organizational units of metabolism, the elements an organism controls to achieve coherent metabolic activity.

Many sequential enzyme steps in biochemical pathways take place in specific compartments of the cell; for example, the steps of the Krebs cycle (chapter 7), occur in the matrix inside mitochondria in eukaryotes. By determining where many of the enzymes that catalyze these steps are located, we can "map out" a model of metabolic processes in the cell.

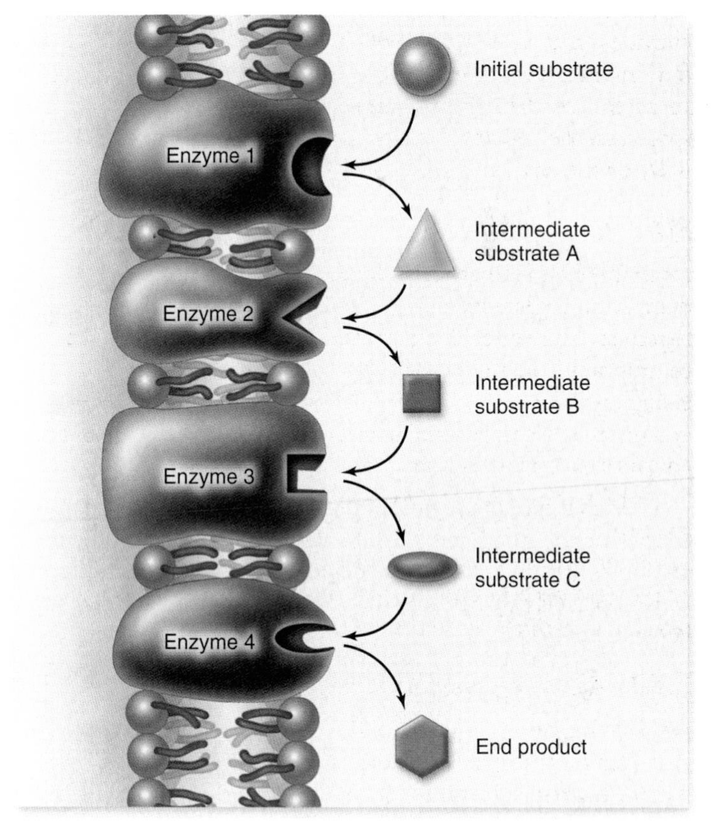

figure 6.13

A BIOCHEMICAL PATHWAY. The original substrate is acted on by enzyme ***1,*** changing the substrate to a new intermediate, substrate A, recognized as a substrate by enzyme ***2.*** Each enzyme in the pathway acts on the product of the previous stage. These enzymes may be either soluble or arranged in a membrane as shown.

Biochemical pathways may have evolved in stepwise fashion

In the earliest cells, the first biochemical processes probably involved energy-rich molecules scavenged from the environment. Most of the molecules necessary for these processes are

figure 6.14

FEEDBACK INHIBITION. ***a.*** A biochemical pathway with no feedback inhibition. ***b.*** A biochemical pathway in which the final end-product becomes the allosteric inhibitor for the first enzyme in the pathway. In other words, the formation of the pathway's final end-product stops the pathway. The pathway could be the synthesis of an amino acid, a nucleotide, or another important cellular molecule.

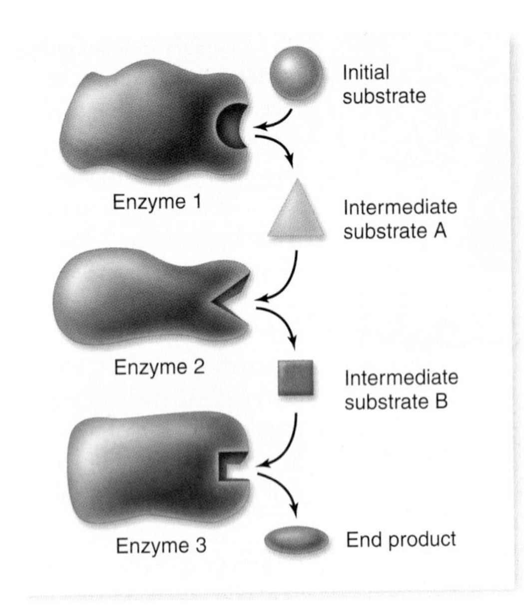

a.

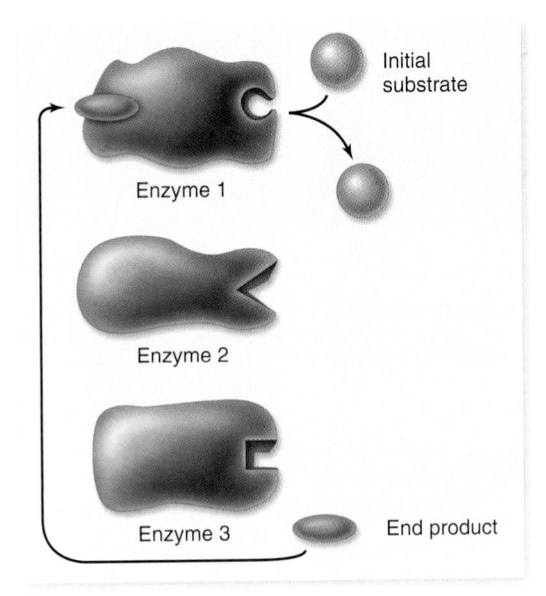

b.

thought to have existed independently in the "organic soup" of the early oceans.

The first catalyzed reactions were probably simple, one-step reactions that brought these molecules together in various combinations. Eventually, the energy-rich molecules became depleted in the external environment, and only organisms that had evolved some means of making those molecules from other substances could survive. Thus, a hypothetical reaction,

$$\begin{matrix} F \\ + \\ G \end{matrix} \longrightarrow H$$

where two energy-rich molecules (F and G) react to produce compound H and release energy, became more complex when the supply of F in the environment ran out.

A new reaction was added in which the depleted molecule, F, is made from another molecule, E, which was also present in the environment:

$$E \longrightarrow \begin{matrix} F \\ + \\ G \end{matrix} \longrightarrow H$$

When the supply of E in turn became depleted, organisms that were able to make E from some other available precursor, D, survived. When D became depleted, those organisms in turn were replaced by ones able to synthesize D from another molecule, C:

$$C \longrightarrow D \longrightarrow E \longrightarrow \begin{matrix} F \\ + \\ G \end{matrix} \longrightarrow H$$

This hypothetical biochemical pathway would have evolved slowly through time, with the final reactions in the pathway evolving first and earlier reactions evolving later.

Looking at the pathway now, we would say that the "advanced" organism, starting with compound C, is able to synthesize H by means of a series of steps. This is how the biochemical pathways within organisms are thought to have evolved—not all at once, but one step at a time, backwards.

Feedback inhibition regulates some biochemical pathways

For a biochemical pathway to operate efficiently, its activity must be coordinated and regulated by the cell. Not only is it unnecessary to synthesize a compound when plenty is already present, but doing so would waste energy and raw materials that could be put to use elsewhere. It is to the cell's advantage, therefore, to temporarily shut down biochemical pathways when their products are not needed.

The regulation of simple biochemical pathways often depends on an elegant feedback mechanism: The end-product of the pathway binds to an allosteric site on the enzyme that catalyzes the first reaction in the pathway. This mode of regulation is called **feedback inhibition** (figure 6.14).

In the hypothetical pathway we just described, the enzyme catalyzing the reaction $C \longrightarrow D$ would possess an allosteric site for H, the end-product of the pathway. As the pathway churned out its product and the amount of H in the cell increased, it would become more likely that an H molecule would encounter the allosteric site on the $C \longrightarrow D$ enzyme. Binding to the allosteric site would essentially shut down the reaction $C \longrightarrow D$ and in turn effectively shut down the whole pathway.

In this chapter we have reviewed the basics of energy and its transformations as carried out in living systems. Chemical bonds are the primary location of energy storage and release, and cells have developed elegant methods of making and breaking chemical bonds to create the molecules they need. Enzymes facilitate these reactions by serving as catalysts. In the following chapters you will learn the details of the mechanisms by which organisms harvest, store, and utilize energy.

The sum of all chemical reactions in a cell constitutes its metabolism. This is usually arranged in pathways where sequential reactions either build up increasingly complex molecules, or break down complex molecules in steps. These pathways are regulated, often by the end-product feeding back to stop its production when levels are high.

concept review

6.1 The Flow of Energy in Living Systems

Energy is defined as the capacity to do work, and it exists in two states: potential energy and kinetic energy.

- Kinetic energy is the energy of motion.
- Potential energy is stored energy.
- Energy can take many forms: mechanical, heat, sound, electrical current, light, or radioactive radiation.
- Energy can be measured in units of heat known as kilocalories.
- Photosynthesis stores light energy as potential energy in the covalent bonds of sugar molecules.
- Oxidation is a chemical reaction involving the loss of electrons, and reduction is the gain of electrons (figure 6.2).
- Oxidation–reduction reactions provide energy for living systems.

6.2 The Laws of Thermodynamics and Free Energy

All activities involve changes in energy, and these changes are governed by the laws of thermodynamics.

- The First Law of Thermodynamics states that energy cannot be created or destroyed, it only changes form.
- The Second Law of Thermodynamics states that the disorder, or entropy, in the universe is increasing. This also means that conversions of energy from one form to another are inefficient, with some energy being lost as heat.
- Free energy (G) is the energy available to do work.
- Changes in free energy (ΔG) predict the direction of reactions. Reactions with a negative ΔG are spontaneous (exergonic) reactions, whereas reactions with a positive ΔG are not spontaneous (endergonic).
- Endergonic chemical reactions absorb energy from the surroundings.
- Exergonic chemical reactions release energy to the surroundings.
- Activation energy is required to destabilize chemical bonds and initiate chemical reactions (figure 6.5).
- Catalysts speed up chemical reactions by decreasing the amount of activation energy needed.

6.3 ATP: The Energy Currency of Cells

Adenosine triphosphate (ATP) is the molecular currency used for cellular energy transactions.

- ATP is a nucleotide that stores energy in unstable bonds between terminal phosphate groups.
- ATP use is cyclical; it releases energy to drive endergonic reactions, and it is synthesized with the energy released in exergonic reactions (figure 6.7).

6.4 Enzymes: Biological Catalysts

Biological reactions are regulated by catalysts called enzymes.

- Enzymes lower the activation energy needed to initiate a chemical reaction.
- Enzymes are not changed or consumed in reactions, so only a small amount of enzyme is needed.
- Enzyme specificity contributes to the differences in structural and functional variations among cell types (figure 6.9).
- Substrates bind to the active site of an enzyme. Enzymes adjust their shape to the substrate so there is a better fit.
- Allosteric enzymes have a second site, located away from the active site, which binds effectors to activate or inhibit the enzyme.
- Enzymes can be free in the cytosol or exist as components bound to membranes and organelles.
- Enzymes that are involved in a series of chemical sequences can form multienzyme complexes. This facilitates sequential reactions, protects intermediate products, and allows control of the reactions within a unit.
- Not all enzymes are proteins. RNA molecules, called ribozymes, can catalyze some chemical reactions.
- An enzyme's functionality depends on maintenance of its three-dimensional shape. This can be affected by temperature and pH.
- Competitive inhibitors compete for the enzyme's active site, which leads to decreased enzyme activity (figure 6.12).
- Noncompetitive inhibitors and activators bind to the allosteric site of the enzyme, thereby changing the structure of the enzyme to inhibit or activate it.
- Cofactors are metals necessary for enzyme function.
- Coenzymes are nonprotein organic molecules necessary for enzyme function.

6.5 Metabolism: The Chemical Description of Cell Function

Metabolism is the sum total of all chemical reactions in an organism.

- Anabolic reactions require energy to create or transform chemical bonds.
- Catabolic reactions break chemical bonds and release energy.
- Chemical reactions in biochemical pathways use the product of one reaction as the substrate for the next.
- Biochemical pathways are often regulated by feedback inhibition where the end-product of the pathway is an allosteric inhibitor of the first enzyme in the sequence (figure 6.14).

review questions

SELF TEST

1. A covalent bond between two atoms represents what kind of energy?
   a. Kinetic energy
   b. Potential energy
   c. Mechanical energy
   d. Solar energy
2. During a redox reaction the molecule that gains an electron has been—
   a. reduced and now has a higher energy level
   b. oxidized and now has a lower energy level
   c. reduced and now has a lower energy level
   d. oxidized and now has a higher energy level
3. An endergonic reaction has the following properties—
   a. $+\Delta G$ and the reaction is spontaneous.
   b. $+\Delta G$ and the reaction is not spontaneous.
   c. $-\Delta G$ and the reaction is spontaneous.
   d. $-\Delta G$ and the reaction is not spontaneous.
4. A spontaneous reaction is one in which—
   a. the reactants have a higher free energy than the products
   b. the products have a higher free energy than the reactants
   c. energy is required
   d. entropy is decreased
5. What is *activation energy*?
   a. The thermal energy associated with random movements of molecules
   b. The energy released through the active breaking of chemical bonds
   c. The difference in free energy between reactants and products
   d. The energy required to initiate a chemical reaction
6. Which of the following is NOT a property of a catalyst?
   a. A catalyst reduces the activation energy of a reaction
   b. A catalyst lowers the free energy of the reactants
   c. A catalyst does not change as a result of the reaction
   d. A catalyst works in both the forward and reverse directions of a reaction
7. Where is the energy stored in a molecule of ATP?
   a. Within the bonds between nitrogen and carbon
   b. In the carbon bonds found in the ribose
   c. In the oxygen double bond
   d. In the bonds connecting the two terminal phosphate groups
8. Why is ATP capable of driving endergonic reactions?
   a. Because ATP is a catalyst
   b. Energy released by ATP makes the ΔG for coupled reactions more negative
   c. Energy released by ATP makes the ΔG for coupled reactions more positive
   d. Because the conversion of ATP to ADP is also endergonic
9. Which of the following statements is NOT true about enzymes?
   a. Enzymes use the three-dimensional shape of their active site to bind reactants.
   b. Enzymes lower the activation energy for a reaction.
   c. The process of the reactions alters the enzyme.
   d. Enzymes can catalyze the forward and reverse directions of a reaction.
10. What is the function of the *active site* of an enzyme?
   a. Bind the substrate, forming an enzyme–substrate complex.
   b. Side groups within the active site interact with the substrate.
   c. It binds to the product triggering induced fit of the protein.
   d. Both a and b.
11. A multienzyme complex is capable of—
   a. catalyzing a single reaction at a much greater rate
   b. catalyzing a series of reactions using multiple different enzymes
   c. lowering the activation energy for a reaction through the activity of multiple enzymes
   d. both a and c
12. What is the common factor that influences enzyme function at extreme temperatures or pH?
   a. The rate of movement of the substrate molecules
   b. The strength of the chemical bonds within the substrate
   c. The three-dimensional shape of the enzyme
   d. The rate of movement of the enzyme
13. The discovery of ribozymes meant that
   a. only proteins have catalytic function
   b. only nucleic acids have catalytic function
   c. some RNAs have enzymatic activity
   d. RNA could be destroyed by enzymes
14. Molecules that bind within the active site of an enzyme are ______________, whereas molecules that bind at a site distant from the active site are ________________.
   a. cofactors; products
   b. competitive inhibitors; allosteric inhibitors
   c. noncompetitive inhibitors; competitive inhibitors
   d. products; coenzymes

CHALLENGE QUESTIONS

1. Examine the graph showing the rate of reaction versus temperature for an enzyme-catalyzed reaction in a human.
   a. Describe what is happening to the enzyme at around 40°C.
   b. Explain why the line touches the x-axis at approximately 20°C and 45°C.
   c. Average body temperature for humans is 37°C. Suggest a reason why the temperature optimum of this enzyme is greater than 37°C.

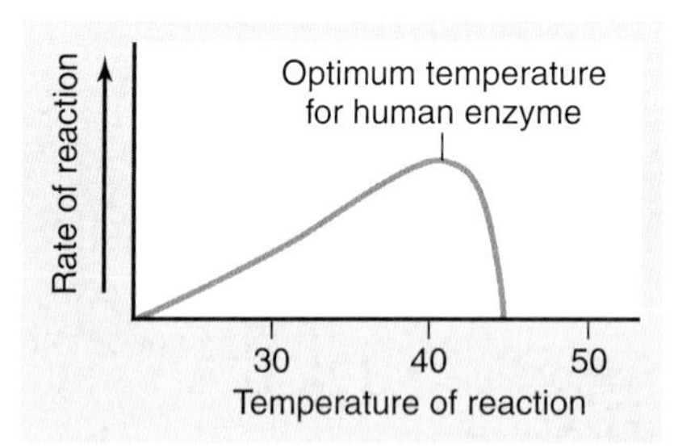

2. Phosphofructokinase functions to add a phosphate group to a molecule of fructose-6-phosphate. This enzyme functions early in glycolysis, an energy-yielding biochemical pathway discussed in chapter 7. The enzyme has an active site that binds fructose and ATP. An allosteric inhibitory site also binds ATP when cellular levels of ATP are very high.
   a. Predict the rate of the reaction if the levels of cellular ATP were low.
   b. Predict the rate of the reaction if levels of cellular ATP are very high.
   c. Describe what is happening to the enzyme when levels of ATP are very high.

Do you need additional review? *Visit* www.ravenbiology.com *for practice quizzes, animations, videos, and activities designed to help you master the material in this chapter.*

chapter 7

How Cells Harvest Energy

introduction

LIFE IS DRIVEN BY ENERGY. All the activities organisms carry out—the swimming of bacteria, the purring of a cat, your thinking about these words—use energy. In this chapter, we discuss the processes all cells use to derive chemical energy from organic molecules and to convert that energy to ATP. Then, in chapter 8, we will examine photosynthesis, which uses light energy to make chemical energy. We consider the conversion of chemical energy to ATP first because all organisms, both the plant, a photosynthesizer, and the caterpillar feeding on the plant, pictured in the photo are capable of harvesting energy from chemical bonds. Energy harvest via respiration is a universal process.

concept outline

7.1 Overview of Respiration

Plants, algae, and some bacteria harvest the energy of sunlight through photosynthesis, converting radiant energy into chemical energy. These organisms, along with a few others that use chemical energy in a similar way, are called **autotrophs** ("self-feeders"). All other organisms live on the organic compounds autotrophs produce, using them as food, and are called **heterotrophs** ("fed by others"). At least 95% of the kinds of organisms on Earth—all animals and fungi, and most protists and prokaryotes—are heterotrophs. Autotrophs also extract energy from organic compounds—they just have the additional capacity to use the energy from sunlight to synthesize these compounds. The process by which energy is harvested is **cellular respiration**—the oxidation of organic compounds to extract energy from chemical bonds.

Cells oxidize organic compounds to drive metabolism

Most foods contain a variety of carbohydrates, proteins, and fats, all rich in energy-laden chemical bonds. Carbohydrates and fats, as you recall from chapter 3, possess many carbon–hydrogen (C—H) bonds, as well as carbon–oxygen (C—O) bonds.

The job of extracting energy from the complex organic mixture in most foods is tackled in stages. First, enzymes break down the large molecules into smaller ones, a process called **digestion** (chapter 48). Then, other enzymes dismantle these fragments a little at a time, harvesting energy from C—H and other chemical bonds at each stage.

The reactions that break down these molecules share a common feature: They are oxidations. Energy metabolism is therefore concerned with redox reactions, and to understand the process we must follow the fate of the electrons lost from the food molecules.

These reactions are not the simple transfer of electrons, however; they are also **dehydrogenations.** That is, the electrons lost are accompanied by protons, so that what is really lost is a hydrogen atom, and not just an electron.

Cellular respiration is the complete oxidation of glucose

In chapter 6, you learned that an atom that loses electrons is said to be *oxidized*, and an atom accepting electrons is said to be *reduced*. Oxidation reactions are often coupled with reduction reactions in living systems, and these paired reactions are called *redox reactions*. Cells utilize enzyme-facilitated redox reactions to take energy from food sources and convert it to ATP.

Redox reactions

Oxidation–reduction reactions play a key role in the flow of energy through biological systems because the electrons that pass from one atom to another carry energy with them. The amount of energy an electron possesses depends on its orbital position, or energy level, around the atom's nucleus. When this electron departs from one atom and moves to another in a redox reaction, the electron's energy is transferred with it.

Figure 7.1 shows how an enzyme catalyzes a redox reaction involving an energy-rich substrate molecule, with the help of a cofactor, **nicotinamide adenine dinucleotide (NAD^+).** In this reaction, NAD^+ accepts a pair of electrons from the substrate, along with a proton, to form **NADH** (this process is described in more detail shortly). The oxidized product is now released from the enzyme's active site, as is NADH.

In the overall process of cellular energy harvest, dozens of redox reactions take place, and a number of molecules, including NAD^+, act as electron acceptors. During each transfer of electrons energy is released. This energy may be captured and used to make ATP or to form other chemical bonds; the rest is lost as heat.

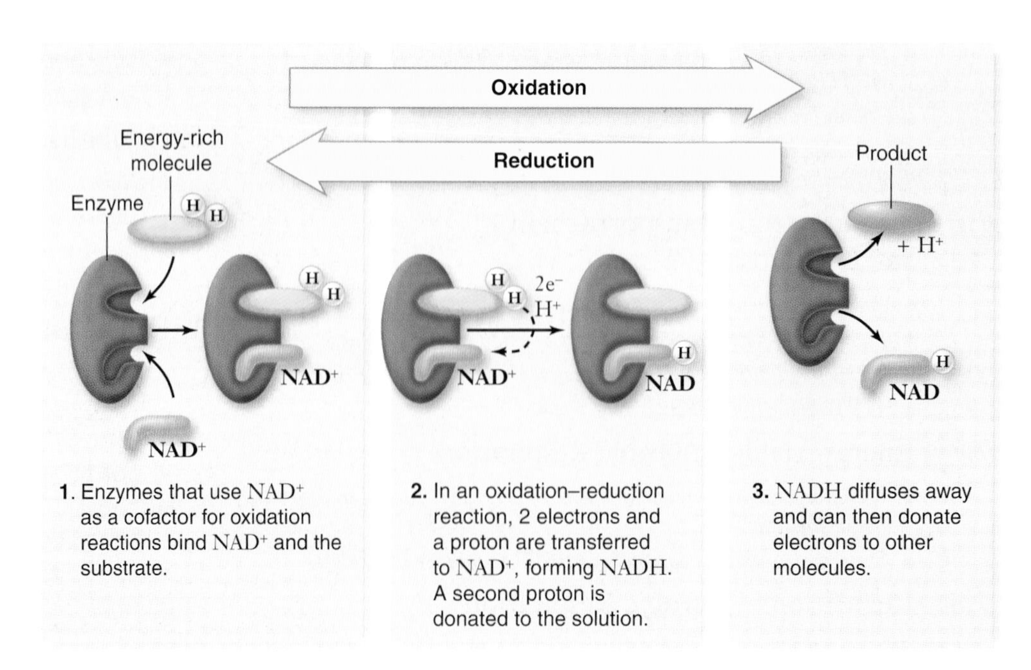

figure 7.1

OXIDATION–REDUCTION REACTIONS OFTEN EMPLOY COFACTORS. Cells use a chemical cofactor called NAD^+ to carry out many oxidation–reduction reactions. Two electrons and a proton are transferred to NAD^+ with another proton donated to the solution. Molecules that gain energetic electrons are said to be reduced, while ones that lose energetic electrons are said to be oxidized. NAD^+ oxidizes energy-rich molecules by acquiring their electrons (in the figure, this proceeds 1 ⟶ 2 ⟶ 3) and then reduces other molecules by giving the electrons to them (in the figure, this proceeds 3 ⟶ 2 ⟶ 1).

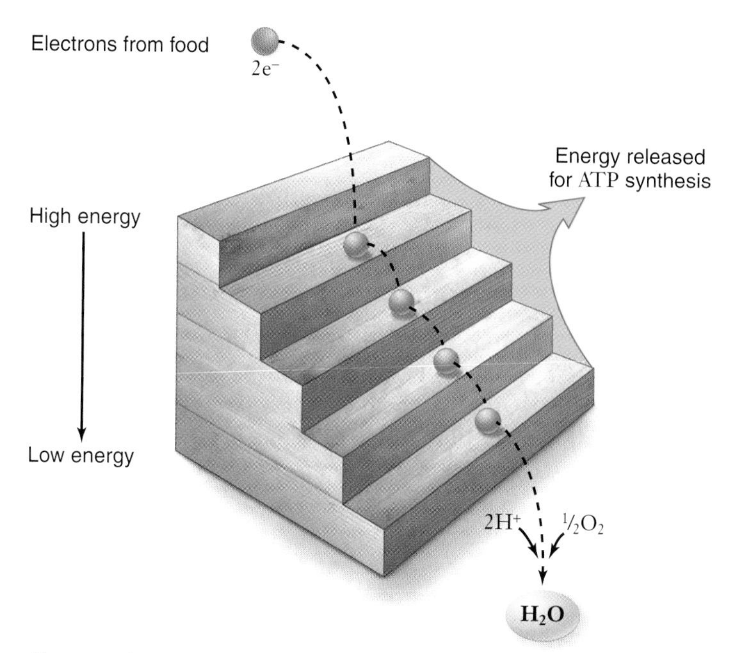

figure 7.2

HOW ELECTRON TRANSPORT WORKS. This diagram shows how ATP is generated when electrons transfer from one energy level to another. Rather than releasing a single explosive burst of energy, electrons "fall" to lower and lower energy levels in steps, releasing stored energy with each fall as they tumble to the lowest (most electronegative) electron acceptor, O_2.

At the end of this process, high-energy electrons from the initial chemical bonds have lost much of their energy, and these depleted electrons are transferred to a final electron acceptor (figure 7.2). When this acceptor is oxygen, the process is called **aerobic respiration.** When the final electron acceptor is an inorganic molecule other than oxygen, the process is called **anaerobic respiration,** and when it is an organic molecule, the process is called **fermentation.**

"Burning" carbohydrates

Chemically, there is little difference between the catabolism of carbohydrates in a cell and the burning of wood in a fireplace. In both instances, the reactants are carbohydrates and oxygen, and the products are carbon dioxide, water, and energy:

$$\underset{\text{glucose}}{C_6H_{12}O_6} + \underset{\text{oxygen}}{6\,O_2} \longrightarrow \underset{\text{carbon dioxide}}{6\,CO_2} + \underset{\text{water}}{6\,H_2O} + \text{energy (heat and ATP)}$$

The change in free energy in this reaction is –686 kcal/mol (or –2870 kJ/mol) under standard conditions (that is, at room temperature, 1 atm pressure, and so forth). In the conditions that exist inside a cell, the energy released can be as high as –720 kcal/mol (–3012 kJ/mol) of glucose. This means that under actual cellular conditions, more energy is released than under standard conditions.

The same amount of energy is released whether glucose is catabolized or burned, but when it is burned, most of the energy is released as heat. Cells harvest useful energy from the catabolism of glucose by using a portion of the energy to drive the production of ATP.

Electron carriers play a critical role in energy metabolism

During respiration, glucose is oxidized to CO_2, but if the electrons were given directly to O_2, the reaction would be combustion, and cells would burst into flames. Instead, as you have just seen, the cell transfers the electrons to intermediate electron carriers, then eventually to O_2.

Many forms of electron carriers are used in this process: soluble carriers that move electrons from one molecule to another, membrane-bound carriers that form a redox chain, and carriers that move within the membrane. The common feature of all of these carriers is that they can be reversibly oxidized and reduced. Some of these carriers, such as the iron-containing cytochromes, can carry just electrons, and some carry both electrons and protons.

NAD^+ is one of the most important electron (and proton) carriers. As shown on the left in figure 7.3, the NAD^+

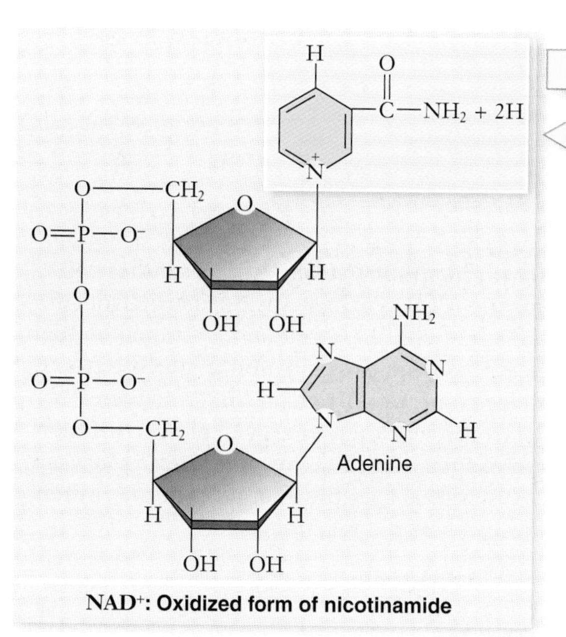

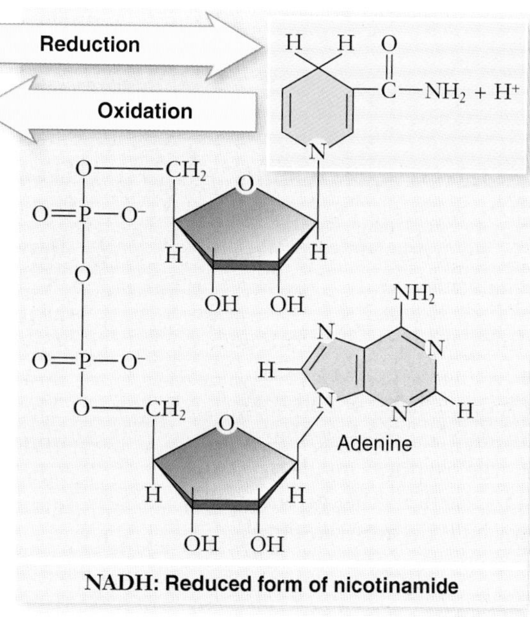

figure 7.3

NAD^+ AND NADH. This dinucleotide serves as an "electron shuttle" during cellular respiration. NAD^+ accepts a pair of electrons and a proton from catabolized macromolecules and is reduced to NADH.

molecule is composed of two nucleotides bound together. The two nucleotides that make up NAD^+, nicotinamide monophosphate (NMP) and adenine monophosphate (AMP), are joined head-to-head by their phosphate groups. The two nucleotides serve different functions in the NAD^+ molecule: AMP acts as the core, providing a shape recognized by many enzymes; NMP is the active part of the molecule, as it is readily reduced, that is, easily accepts electrons.

When NAD^+ acquires two electrons and a proton from the active site of an enzyme, it is reduced to NADH, shown on the right in figure 7.3. The NADH molecule now carries the two energetic electrons and can supply them to other molecules and reduce them.

This ability to supply high-energy electrons is critical to both energy metabolism and to the biosynthesis of many organic molecules, including fats and sugars. In animals, when ATP is plentiful, the reducing power of the accumulated NADH is diverted to supplying fatty acid precursors with high-energy electrons, reducing them to form fats and storing the energy of the electrons.

Metabolism harvests energy in stages

It is generally true that the larger the release of energy in any single step, the more of that energy is released as heat, and the less is available to be channeled into more useful paths. In the combustion of gasoline, the same amount of energy is released whether all of the gasoline in a car's gas tank explodes at once, or burns in a series of very small explosions inside the cylinders. By releasing the energy in gasoline a little at a time, the harvesting efficiency is greater, and more of the energy can be used to push the pistons and move the car.

The same principle applies to the oxidation of glucose inside a cell. If all of the electrons were transferred to oxygen in one explosive step, releasing all of the free energy at once, the cell would recover very little of that energy in a useful form. Instead, cells burn their fuel much as a car does, a little at a time.

The electrons in the C—H bonds of glucose are stripped off in stages in the series of enzyme-catalyzed reactions collectively referred to as glycolysis and the Krebs cycle. The electrons are removed by transferring them to NAD^+, as described earlier, or to other electron carriers.

The energy released by all of these oxidation reactions is also not all released at once (see figure 7.2). The electrons are passed to another set of electron carriers called the **electron transport chain,** which is located in the mitochondrial inner membrane. Movement of electrons through this chain produces potential energy in the form of an electrochemical gradient. We examine this process in more detail later in this chapter.

ATP plays a central role in metabolism

The previous chapter introduced ATP as the energy currency of the cell. Cells use ATP to power most of those activities that require work. One of the most obvious activities is movement. Tiny fibers within muscle cells pull against one another when muscles contract. Mitochondria can move a meter or more along the narrow nerve cells that extend from your spine to your feet. Chromosomes are pulled apart by microtubules during cell division. All of these movements require the expenditure of energy by ATP hydrolysis. Cells also use ATP to drive endergonic reactions that would otherwise not occur spontaneously (chapter 6).

How does ATP drive an endergonic reaction? The enzyme that catalyzes a particular reaction has two binding sites on its surface: one for the reactant and another for ATP. The ATP site splits the ATP molecule, liberating over 7 kcal ($\Delta G = -7.3$ kcal/mol) of chemical energy. This energy pushes the reactant at the second site "uphill," reaching the activation energy and driving the endergonic reaction. Thus endergonic reactions coupled to ATP hydrolysis become favorable.

The many steps of cellular respiration have as their ultimate goal the production of ATP. ATP synthesis is itself an endergonic reaction, which requires cells to perform exergonic reactions to drive this synthesis. The details of these reactions are presented in the following sections of this chapter.

Cells require energy to maintain their structure, for growth and metabolism. The process of cellular respiration achieves the complete oxidation of glucose. This process utilizes electron carriers that aid in the gradual release of the energy from the oxidation of glucose. The result of energy metabolism is the synthesis of ATP, which cells use as a portable source of energy.

7.2 The Oxidation of Glucose: A Summary

Cells are able to make ATP from the oxidation of glucose by two fundamentally different mechanisms.

1. In **substrate-level phosphorylation,** ATP is formed by transferring a phosphate group directly to ADP from a phosphate-bearing intermediate, or substrate (figure 7.4). During **glycolysis,** the initial breakdown of glucose (discussed later), the chemical bonds of glucose are shifted around in reactions that provide the energy required to form ATP by substrate-level phosphorylation.
2. In **oxidative phosphorylation,** ATP is synthesized by the enzyme **ATP synthase,** using energy from a proton (H^+) gradient. This gradient is formed by high-energy electrons from the oxidation of glucose passing down an electron transport chain (described later). These electrons, with their energy depleted, are then donated to oxygen, hence the term *oxidative phosphorylation*. ATP synthase uses the energy from the proton gradient to catalyze the reaction:

$$ADP + P_i \longrightarrow ATP$$

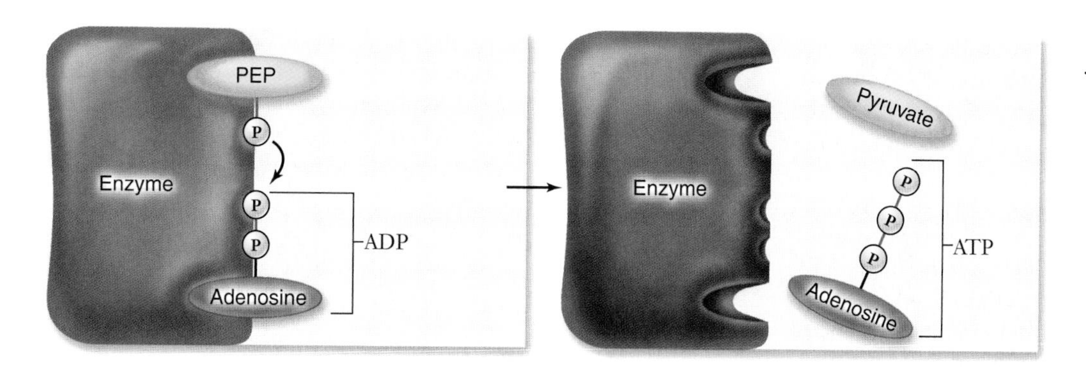

figure 7.4

SUBSTRATE-LEVEL PHOSPHORYLATION. Some molecules, such as phosphoenolpyruvate (PEP), possess a high-energy phosphate bond similar to the bonds in ATP. When PEP's phosphate group is transferred enzymatically to ADP, the energy in the bond is conserved, and ATP is created.

Eukaryotes and aerobic prokaryotes produce the vast majority of their ATP this way.

In most organisms, these two processes are combined. To harvest energy to make ATP from glucose in the presence of oxygen, the cell carries out a complex series of enzyme-catalyzed reactions that occur in four stages: The first stage captures energy by substrate-level phosphorylation through glycolysis; the next three stages carry out aerobic respiration by oxidizing the end-product of glycolysis and producing ATP. In this section, we provide an overview of these stages (figure 7.5); each topic is then discussed in depth in the sections that follow.

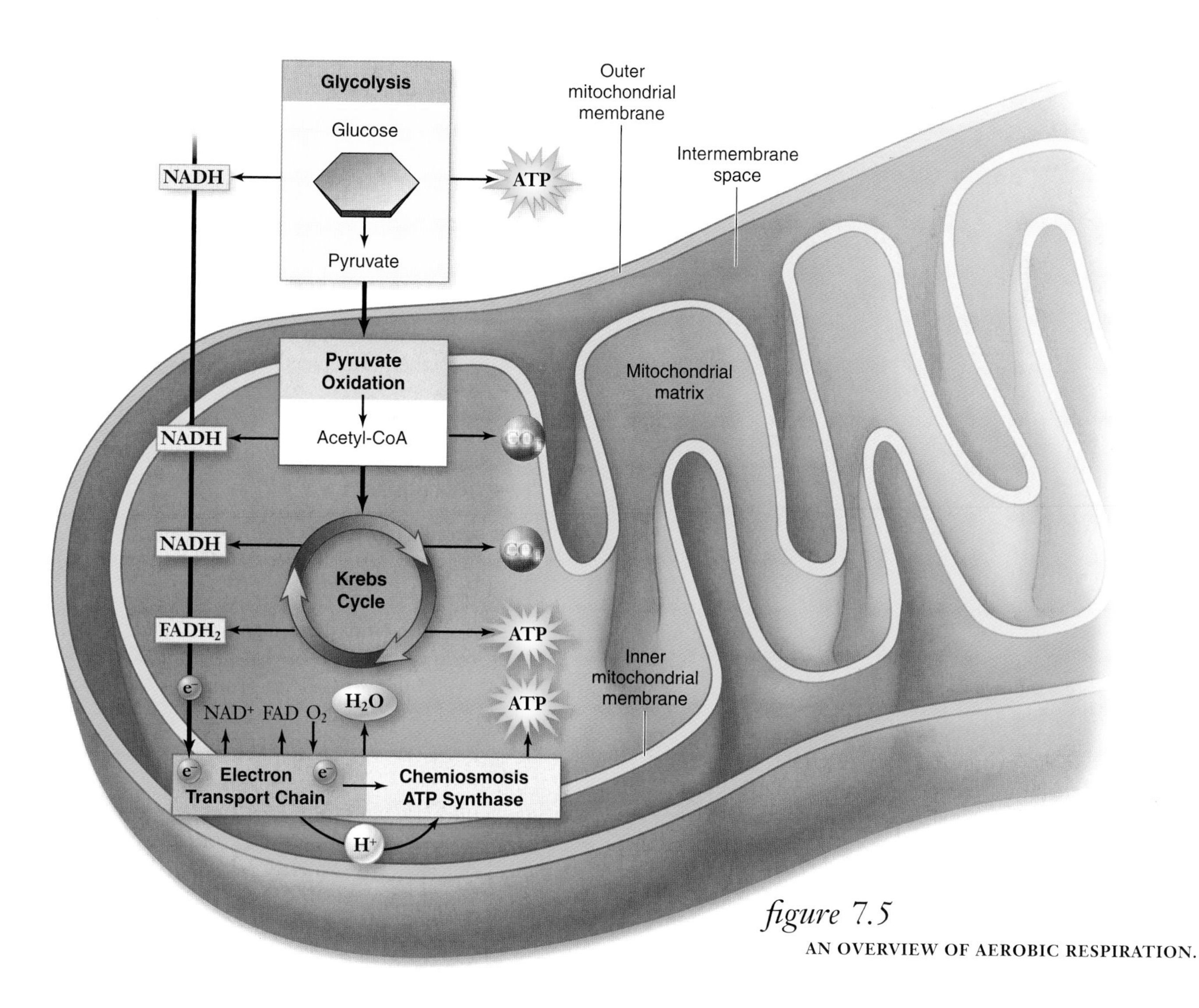

figure 7.5

AN OVERVIEW OF AEROBIC RESPIRATION.

Stage One: Glycolysis The first stage is a 10-reaction biochemical pathway called *glycolysis* that produces ATP by substrate-level phosphorylation. The enzymes that catalyze the glycolytic reactions are in the cytoplasm of the cell, not associated with any membrane or organelle.

For each glucose molecule, two ATP molecules are used up early in the pathway, and four ATP molecules are produced by substrate-level phosphorylation. The net yield is two ATP molecules for each molecule of glucose catabolized. In addition, four electrons are harvested from the chemical bonds of glucose and carried by NADH for oxidative phosphorylation. Glycolysis yields two energy-rich **pyruvate** molecules for each glucose entering the pathway. This remaining energy can be harvested in later stages.

Stage Two: Pyruvate Oxidation In the second stage, pyruvate is converted into carbon dioxide and a two-carbon molecule called **acetyl-CoA.** For each molecule of pyruvate converted, one molecule of NAD^+ is reduced to NADH, again to carry electrons that can be used to make ATP. Remember that two pyruvate molecules result from each glucose.

Stage Three: The Krebs Cycle The third stage introduces acetyl-CoA into a cycle of nine reactions called the **Krebs cycle,** named after the German biochemist Hans Krebs, who discovered it. The Krebs cycle is also called the *citric acid cycle,* for the citric acid, or citrate, formed in its first step, and less commonly, the *tricarboxylic acid cycle,* because citrate has three carboxyl groups.

For each turn of the Krebs cycle, one ATP molecule is produced by substrate-level phosphorylation, and a large number of electrons are removed by the reduction of NAD^+ to NADH and FAD to $FADH_2$. Each glucose provides two acetyl-CoA to the Krebs cycle allowing two turns.

Stage Four: Electron Transport Chain and Chemiosmosis In the fourth stage, energetic electrons carried by NADH are transferred to a series of electron carriers that progressively extract the electrons' energy and use it to pump protons across a membrane.

The proton gradient created by electron transport is used by ATP synthase to produce ATP. This utilization of a proton gradient to drive the synthesis of ATP is called **chemiosmosis** and is the basis for oxidative phosphorylation.

Pyruvate oxidation, the reactions of the Krebs cycle, and ATP production by electron transport chains occur within many forms of prokaryotes and inside the mitochondria of all eukaryotes. Figure 7.5 provides an overview of the complete process of aerobic respiration beginning with glycolysis.

The oxidation of glucose can be broken down into stages. These include glycolysis, which produces pyruvate, the oxidation of pyruvate, and the Krebs cycle. The electrons derived from oxidation reactions are used in the electron transport chain to produce a proton gradient that can be used by the enzyme ATP synthase to make ATP in the process called chemiosmosis.

7.3 Glycolysis: Splitting Glucose

Glucose molecules can be dismantled in many ways, but primitive organisms evolved a glucose-catabolizing process that releases enough free energy to drive the synthesis of ATP in enzyme-coupled reactions. Glycolysis occurs in the cytoplasm and converts glucose into two 3-carbon molecules of pyruvate (figure 7.6). For each molecule of glucose that passes through this transformation, the cell nets two ATP molecules.

Priming changes glucose into an easily cleaved form

The first half of glycolysis consists of five sequential reactions that convert one molecule of glucose into two molecules of the 3-carbon compound **glyceraldehyde 3-phosphate (G3P).** These reactions require the expenditure of ATP, so they are an endergonic process.

Step A: Glucose priming Three reactions "prime" glucose by changing it into a compound that can be cleaved readily into two 3-carbon phosphorylated molecules. Two of these reactions transfer a phosphate from ATP, so this step requires the cell to use two ATP molecules.

Step B: Cleavage and rearrangement In the first of the remaining pair of reactions, the 6-carbon product of step A is split into two 3-carbon molecules. One is G3P, and the other is then converted to G3P by the second reaction (figure 7.7).

ATP is synthesized by substrate-level phosphorylation

In the second half of glycolysis, five more reactions convert G3P into pyruvate in an energy-yielding process that generates ATP.

Step C: Oxidation Two electrons (and one proton) are transferred from G3P to NAD^+, forming NADH. A molecule of P_i is also added to G3P to produce 1,3-bisphosphoglycerate. The phosphate incorporated will later be transferred to ADP by substrate-level phosphorylation to allow a net yield of ATP.

Step D: ATP generation Four reactions convert 1,3-bisphosphoglycerate into pyruvate. This process generates two ATP molecules per G3P (see figures 7.4 and 7.7) produced in Step B.

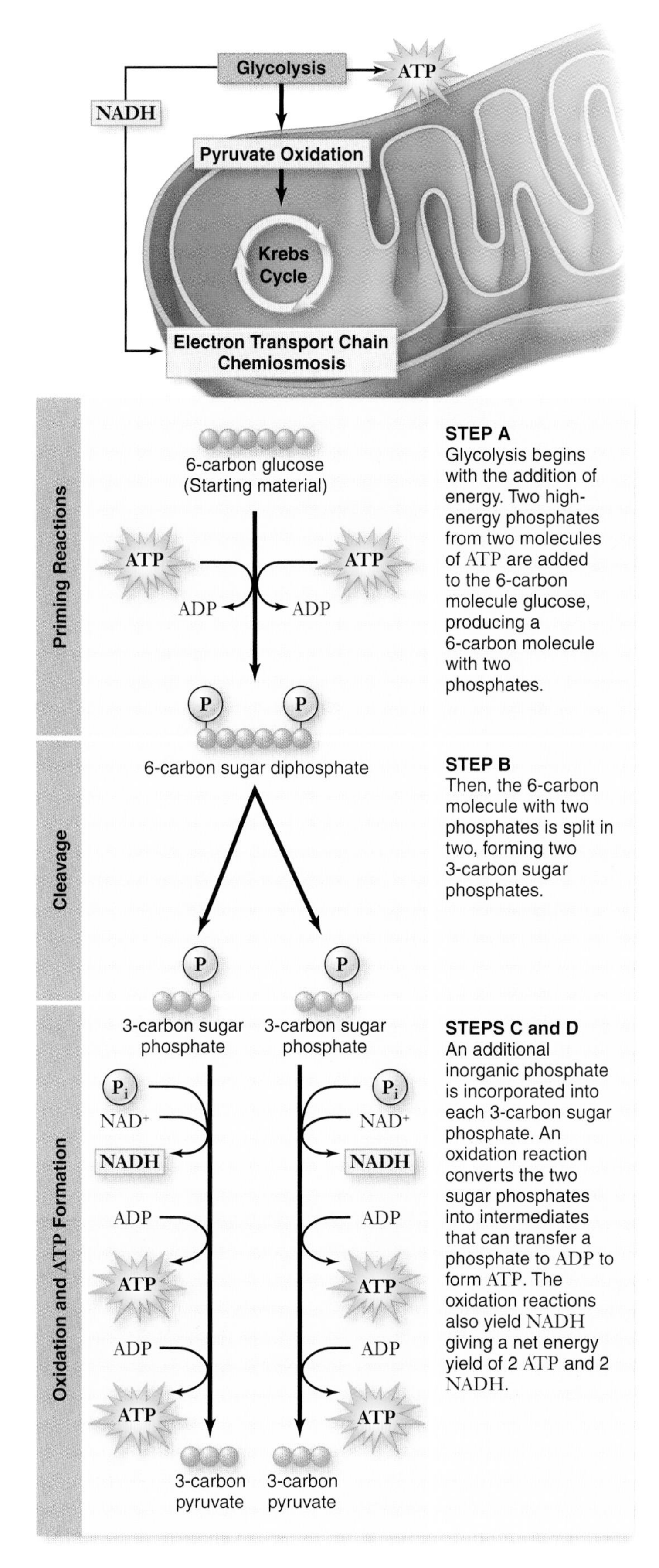

figure 7.6

HOW GLYCOLYSIS WORKS.

Because each glucose molecule is split into two G3P molecules, the overall reaction sequence has a net yield of two molecules of ATP, as well as two molecules of NADH and two of pyruvate:

4 ATP (2 ATP for each of the 2 G3P molecules in step D)
– 2 ATP (used in the two reactions in step A)
2 ATP (net yield for entire process)

The hydrolysis of one molecule of ATP yields a ΔG of –7.3 kcal/mol under standard conditions. Thus cells harvest a maximum of 14.6 kcal of energy per mole of glucose from glycolysis.

A brief history of glycolysis

Although far from ideal in terms of the amount of energy it releases, glycolysis does generate ATP. For more than a billion years during the anaerobic first stages of life on Earth, glycolysis was the primary way heterotrophic organisms generated ATP from organic molecules.

Like many biochemical pathways, glycolysis is believed to have evolved backward, with the last steps in the process being the most ancient. Thus, the second half of glycolysis, the ATP-yielding breakdown of G3P, may have been the original process. The synthesis of G3P from glucose would have appeared later, perhaps when alternative sources of G3P were depleted.

Why does glycolysis take place in modern organisms, since its energy yield in the absence of oxygen is comparatively little? The answer is that evolution is an incremental process: Change occurs by improving on past successes. In catabolic metabolism, glycolysis satisfied the one essential evolutionary criterion—it was an improvement. Cells that could not carry out glycolysis were at a competitive disadvantage, and only cells capable of glycolysis survived. Later improvements in catabolic metabolism built on this success. Metabolism evolved as one layer of reactions added to another. Nearly every present-day organism carries out glycolysis, as a metabolic memory of its evolutionary past.

The last section of this chapter discusses the evolution of metabolism in more detail.

NADH must be recycled to continue respiration

Inspect for a moment the net reaction of the glycolytic sequence:

$$\text{glucose} + 2\ \text{ADP} + 2\ \text{P}_i + 2\ \text{NAD}^+ \longrightarrow 2\ \text{pyruvate} + 2\ \text{ATP} + 2\ \text{NADH} + 2\ \text{H}^+ + 2\ \text{H}_2\text{O}$$

You can see that three changes occur in glycolysis: (1) Glucose is converted into two molecules of pyruvate; (2) two molecules of ADP are converted into ATP via substrate-level phosphorylation; and (3) two molecules of NAD^+ are reduced to NADH. This leaves the cell with two problems: extracting the energy that remains in the two pyruvate molecules, and regenerating NAD^+ to be able to continue glycolysis.

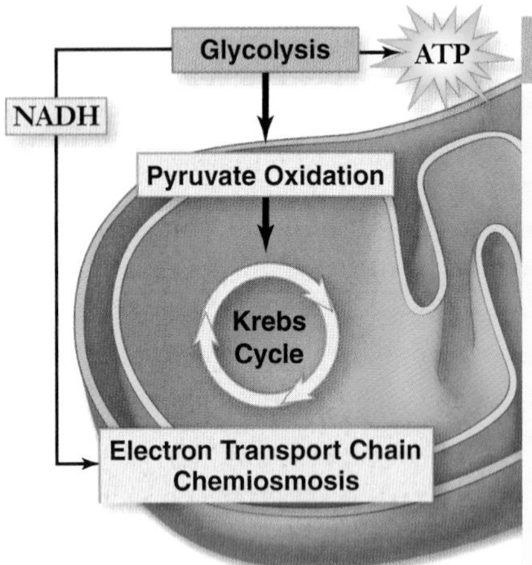

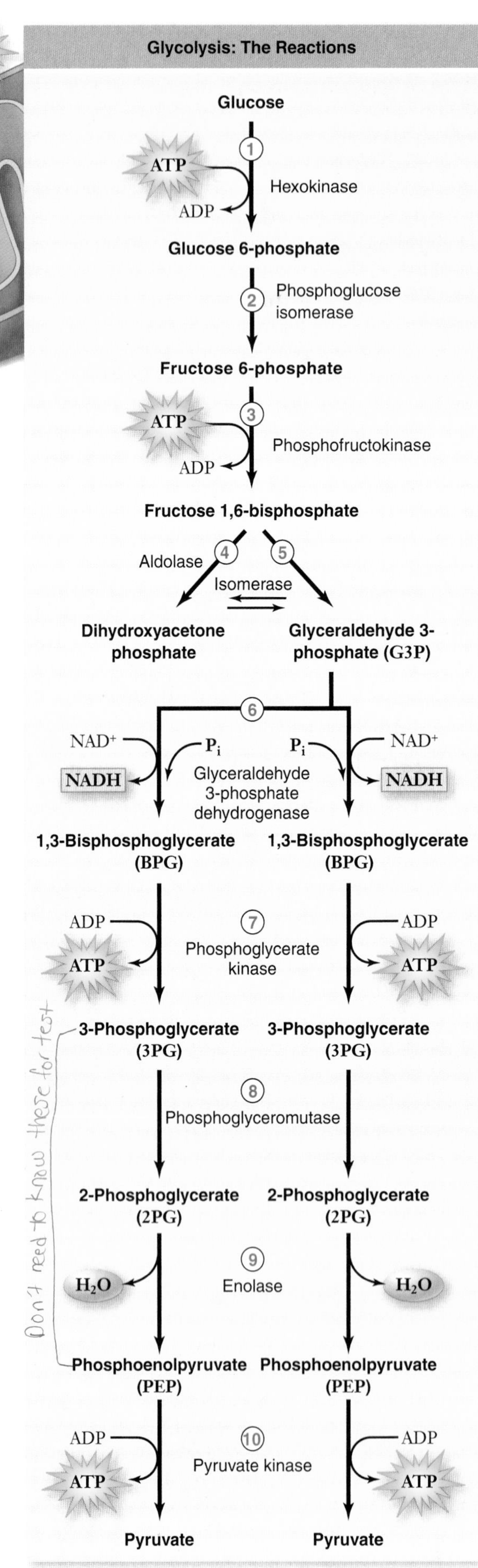

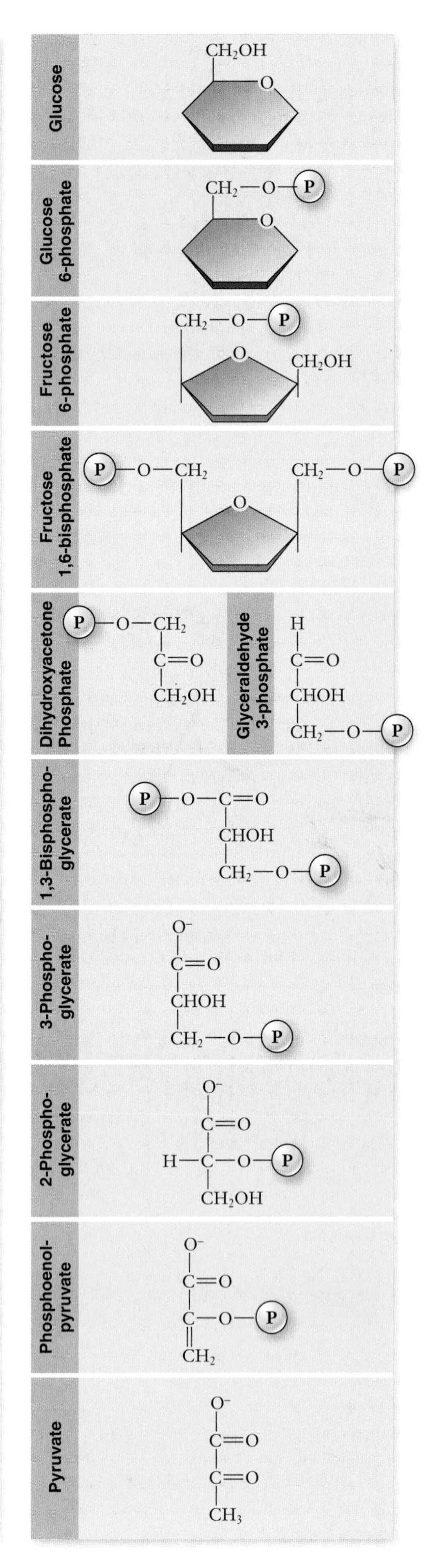

1. Phosphorylation of glucose by ATP.

2–3. Rearrangement, followed by a second ATP phosphorylation.

4–5. The 6-carbon molecule is split into two 3-carbon molecules—one G3P, another that is converted into G3P in another reaction.

6. Oxidation followed by phosphorylation produces two NADH molecules and two molecules of BPG, each with one high-energy phosphate bond.

7. Removal of high-energy phosphate by two ADP molecules produces two ATP molecules and leaves two 3PG molecules.

8–9. Removal of water yields two PEP molecules, each with a high-energy phosphate bond.

10. Removal of high-energy phosphate by two ADP molecules produces two ATP molecules and two pyruvate molecules.

figure 7.7

THE GLYCOLYTIC PATHWAY. The first five reactions convert a molecule of glucose into two molecules of G3P. The second five reactions convert G3P into pyruvate.

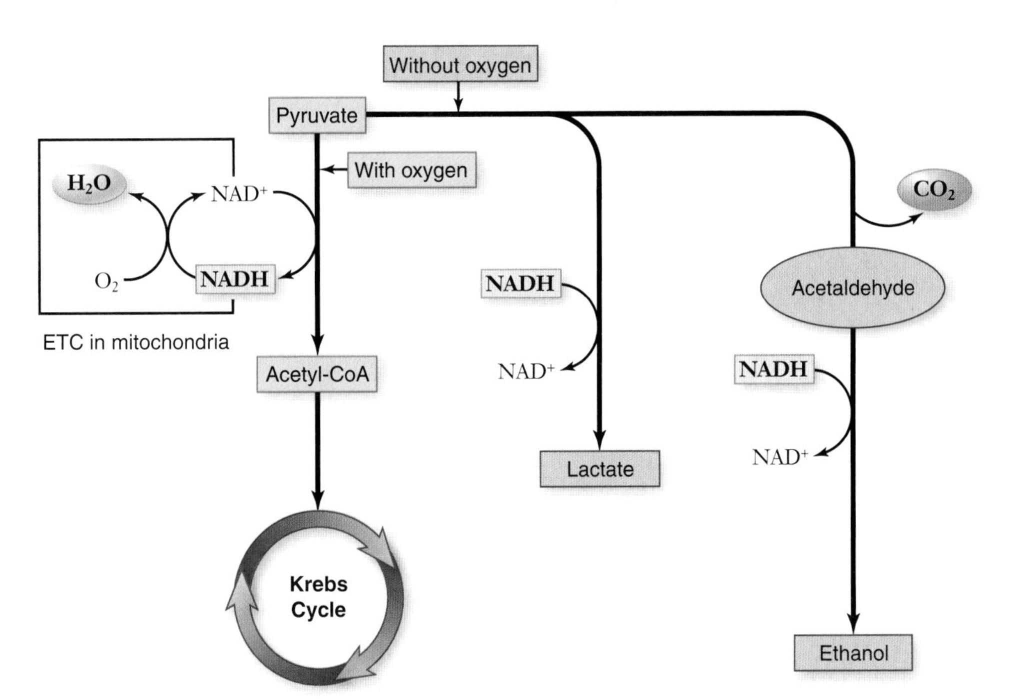

figure 7.8

THE FATE OF PYRUVATE AND NADH PRODUCED BY GLYCOLYSIS. In the presence of oxygen, NADH is oxidized by the electron transport chain (ETC) in mitochondria using oxygen as the final electron acceptor. This regenerates NAD^+ allowing glycolysis to continue. The pyruvate produced by glycolysis is oxidized to acetyl-CoA, which enters the Krebs cycle. In the absence of oxygen, pyruvate is instead reduced, oxidizing NADH and regenerating NAD^+ to allow glycolysis to continue. Direct reduction of pyruvate, as in muscle cells, produces lactate. In yeast, carbon dioxide is first removed from pyruvate producing acetaldehyde, which is then reduced to ethanol.

Recycling NADH

As long as food molecules that can be converted into glucose are available, a cell can continually churn out ATP to drive its activities. In doing so, however, it accumulates NADH and depletes the pool of NAD^+ molecules. A cell does not contain a large amount of NAD^+, and for glycolysis to continue, NADH must be recycled into NAD^+. Some molecule other than NAD^+ must ultimately accept the electrons taken from G3P and be reduced. Two processes can carry out this key task (figure 7.8):

1. **Aerobic respiration.** Oxygen is an excellent electron acceptor. Through a series of electron transfers, electrons taken from G3P can be donated to oxygen, forming water. This process occurs in the mitochondria of eukaryotic cells in the presence of oxygen. Because air is rich in oxygen, this process is also referred to as *aerobic metabolism.* A significant amount of ATP is also produced.
2. **Fermentation.** When oxygen is unavailable, an organic molecule, such as acetaldehyde in wine fermentation, can accept electrons instead (figure 7.9). This reaction plays an important role in the metabolism of most organisms, even those capable of aerobic respiration.

figure 7.9

HOW WINE IS MADE. The conversion of pyruvate to ethanol takes place naturally in grapes left to ferment on vines, as well as in fermentation vats of crushed grapes. Yeasts carry out the process to continue glycolysis under anaerobic conditions. When their conversion increases the ethanol concentration to about 12%, the toxic effects of the alcohol kill the yeast cells. What is left is wine.

The fate of pyruvate

The fate of the pyruvate that is produced by glycolysis depends on which of these two processes takes place. The aerobic respiration path starts with the oxidation of pyruvate to produce acetyl-CoA, which is then further oxidized in a series of reactions called the Krebs cycle. The fermentation path, by contrast, uses the reduction of all or part of pyruvate to oxidize NADH back to NAD^+. We examine aerobic respiration next; fermentation is described in detail in a later section.

Glycolysis splits the 6-carbon molecule glucose into two 3-carbon molecules of pyruvate. This process requires first using two ATP molecules in "priming" reactions eventually producing four molecules of ATP per glucose for a net yield of two ATP. The oxidation reactions of glycolysis require NAD^+ and produce NADH. This NAD^+ must be regenerated either by oxidation in the electron transport chain using O_2, or by using an organic molecule in a fermentation reaction.

7.4 The Oxidation of Pyruvate to Produce Acetyl-CoA

In the presence of oxygen, the oxidation of glucose that begins in glycolysis continues where glycolysis leaves off—with pyruvate. In eukaryotic organisms, the extraction of additional energy from pyruvate takes place exclusively inside mitochondria. In prokaryotes similar reactions take place in the cytoplasm and at the plasma membrane.

The cell harvests pyruvate's considerable energy in two steps. First, pyruvate is oxidized to produce a two-carbon compound and CO_2, with the electrons transferred to NAD^+ to produce NADH. Next, the two-carbon compound is oxidized to CO_2 by the reactions of the Krebs cycle.

Pyruvate is oxidized in a "decarboxylation" reaction that cleaves off one of pyruvate's three carbons. This carbon departs as CO_2 (figure 7.10). The remaining two-carbon compound, called an acetyl group, is then attached to coenzyme A; this entire molecule is called *acetyl-CoA.* A pair of electrons and one associated proton is transferred to the electron carrier NAD^+, reducing it to NADH, with a second proton donated to the solution.

The reaction involves three intermediate stages, and it is catalyzed within mitochondria by a *multienzyme complex.* As chapter 6 noted, a multienzyme complex organizes a series of enzymatic steps so that the chemical intermediates do not diffuse away or undergo other reactions. Within the complex, component polypeptides pass the substrates from one enzyme to the next, without releasing them. *Pyruvate dehydrogenase*, the complex of enzymes that removes CO_2 from pyruvate, is one of the largest enzymes known; it contains 60 subunits! The reaction can be summarized as:

$$\text{pyruvate} + NAD^+ + \text{CoA} \longrightarrow \text{acetyl-CoA} + NADH + CO_2 + H^+$$

The molecule of NADH produced is used later to produce ATP. The acetyl group is fed into the Krebs cycle, with the CoA being recycled for another oxidation of pyruvate. The Krebs cycle then completes the oxidation of the original carbons from glucose.

Pyruvate is oxidized in the mitochondria to produce acetyl-CoA and CO_2. This reaction is a link between glycolysis and the reactions of the Krebs cycle as acetyl-CoA is used by the Krebs cycle.

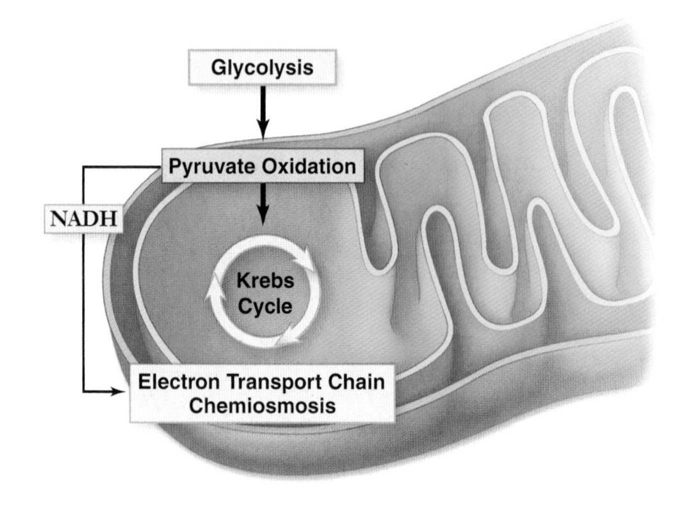

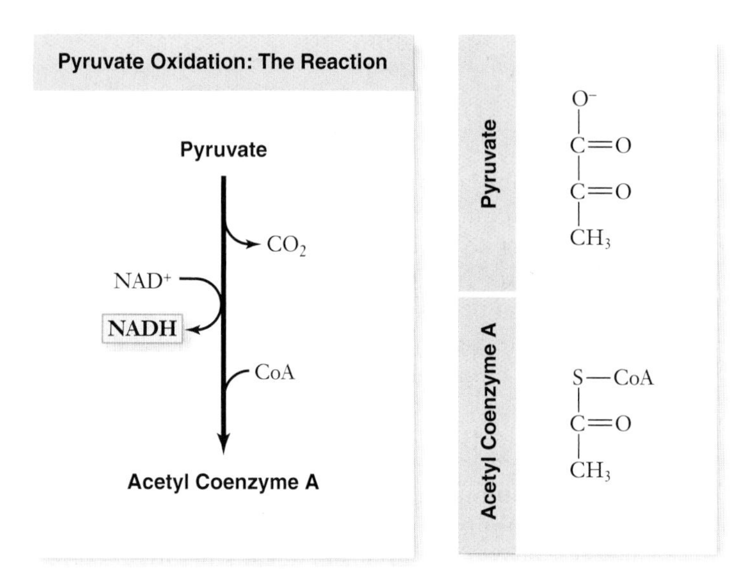

figure 7.10

THE OXIDATION OF PYRUVATE. This complex reaction uses NAD^+ to accept electrons reducing it to NADH. The product, acetyl-CoA, feeds the acetyl unit into the Krebs cycle, and the CoA is recycled for another oxidation of pyruvate. NADH provides energetic electrons for the electron transport chain.

7.5 The Krebs Cycle

In this third stage, the acetyl group from pyruvate is oxidized in a series of nine reactions called the *Krebs cycle.* These reactions occur in the matrix of mitochondria.

In this cycle, the two-carbon acetyl group of acetyl-CoA combines with a four-carbon molecule called oxaloacetate. The resulting six-carbon molecule, citrate, then goes through a several-step sequence of electron-yielding oxidation reactions, during which two CO_2 molecules split off, restoring oxaloacetate. The regenerated oxaloacetate is used to bind to another acetyl group for the next round of the cycle.

In each turn of the cycle, a new acetyl group is added and two carbons are lost as two CO_2 molecules, and more electrons are transferred to electron carriers. These electrons are then used by the electron transport chain to drive *proton pumps* that generate ATP.

The Krebs cycle has three segments: An overview

The nine reactions of the Krebs cycle can be grouped into three overall segments. These are described in the following sections and summarized in figure 7.11.

Segment A: Acetyl-CoA plus oxaloacetate This reaction produces the 6-carbon citrate molecule.

Segment B: Citrate rearrangement and decarboxylation Five more steps, which have been simplified in figure 7.11, reduce citrate to a 5-carbon intermediate and then to 4-carbon succinate. During these reactions, two NADH and one ATP are produced.

Segment C: Regeneration of oxaloacetate Succinate undergoes three additional reactions, also simplified in the figure, to become oxaloacetate. During these reactions, one NADH is produced; in addition, a molecule of **flavin adenine dinucleotide (FAD),** another cofactor, becomes reduced to $FADH_2$.

The specifics of each reaction are described next.

The Krebs cycle is geared to extract electrons and synthesize one ATP

Figure 7.12 summarizes the sequence of the Krebs cycle reactions. A 2-carbon group from acetyl-CoA enters the cycle at the beginning, and two CO_2 molecules, one ATP, and four pairs of electrons are produced.

Reaction 1: Condensation Citrate is formed from acetyl-CoA and oxaloacetate. This condensation reaction is irreversible, committing the 2-carbon acetyl group to the Krebs cycle. The reaction is inhibited when the cell's ATP concentration is high and stimulated when it is low. The result is that when the cell possesses ample amounts of ATP, the Krebs cycle shuts down, and acetyl-CoA is channeled into fat synthesis.

Reactions 2 and 3: Isomerization Before the oxidation reactions can begin, the hydroxyl (—OH) group of citrate must be repositioned. This rearrangement is done in two steps: First, a water molecule is removed from one carbon; then water is added to a different carbon. As a result, an —H group and an —OH group change positions. The product is an isomer

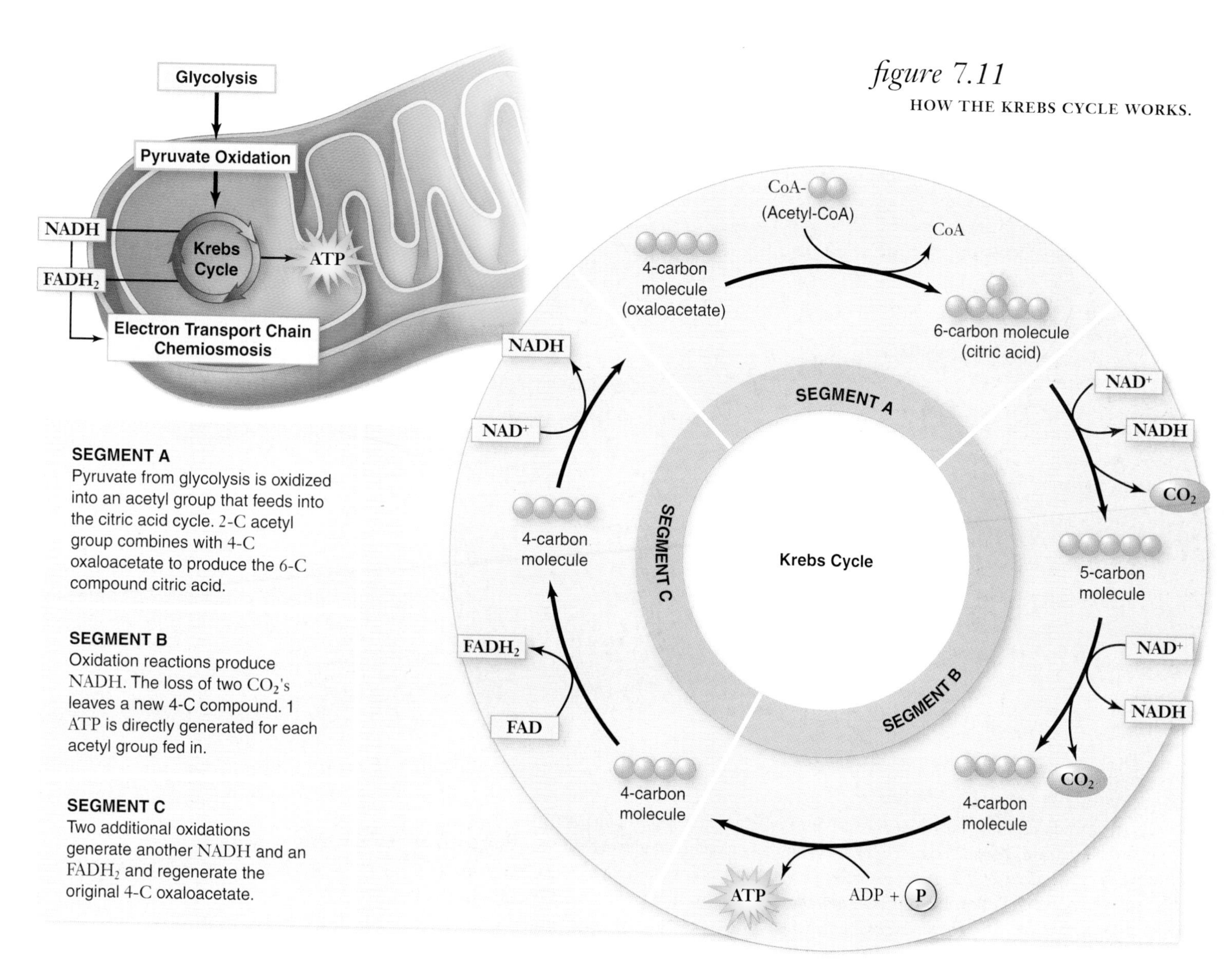

figure 7.11

HOW THE KREBS CYCLE WORKS.

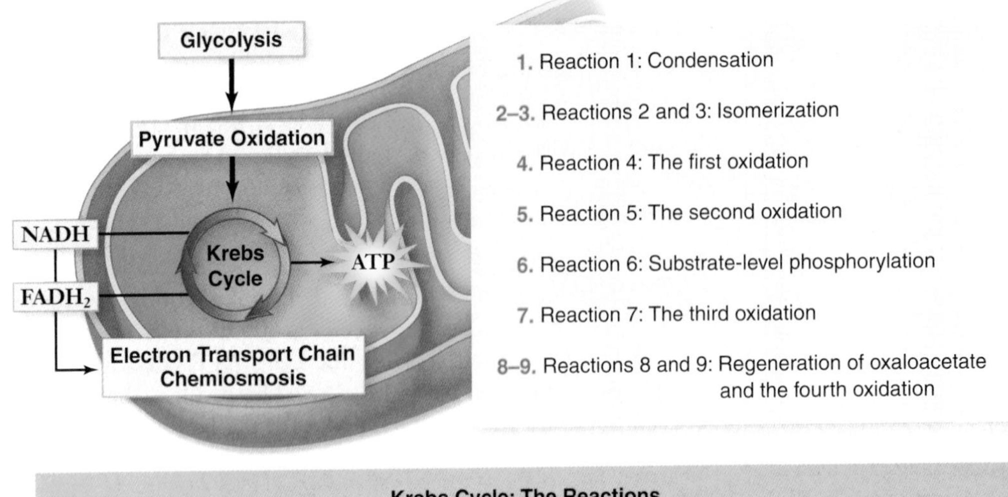

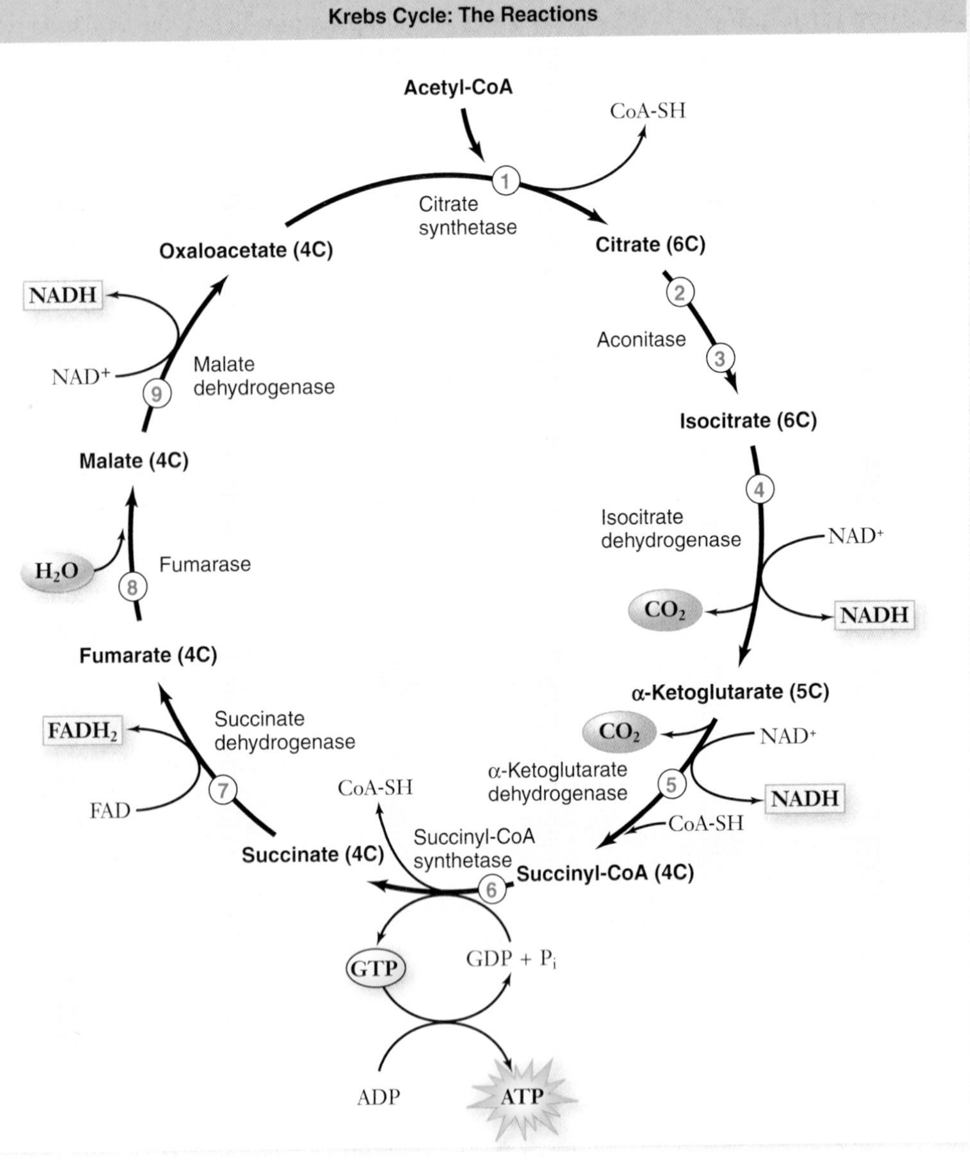

Acetyl-CoA	S—CoA, C=O, CH$_3$
Citrate	COO$^-$, CH$_2$, HO—C—COO$^-$, CH$_2$, COO$^-$
Isocitrate	COO$^-$, CH$_2$, HC—COO$^-$, HO—CH, COO$^-$
α-Ketoglutarate	COO$^-$, CH$_2$, CH$_2$, C=O, COO$^-$
Succinyl-CoA	COO$^-$, CH$_2$, CH$_2$, C=O, S—CoA
Succinate	COO$^-$, CH$_2$, CH$_2$, COO$^-$
Fumarate	COO$^-$, CH, HC, COO$^-$
Malate	COO$^-$, HO—CH, CH$_2$, COO$^-$
Oxaloacetate	COO$^-$, O=C, CH$_2$, COO$^-$

figure 7.12

THE KREBS CYCLE. This series of reactions takes place within the matrix of the mitochondrion. For the complete breakdown of a molecule of glucose, the two molecules of acetyl-CoA produced by glycolysis and pyruvate oxidation each have to make a trip around the Krebs cycle. Follow the different carbons through the cycle, and notice the changes that occur in the carbon skeletons of the molecules and where oxidation reactions take place as they proceed through the cycle.

of citrate called *isocitrate.* This rearrangement facilitates the subsequent reactions.

Reaction 4: The First Oxidation In the first energy-yielding step of the cycle, isocitrate undergoes an oxidative decarboxylation reaction. First, isocitrate is oxidized, yielding a pair of electrons that reduce a molecule of NAD^+ to NADH. Then the oxidized intermediate is decarboxylated; the central carboxyl group splits off to form CO_2, yielding a 5-carbon molecule called *α-ketoglutarate.*

Reaction 5: The Second Oxidation Next, α-ketoglutarate is decarboxylated by a multienzyme complex similar to pyruvate dehydrogenase. The succinyl group left after the removal of CO_2 joins to coenzyme A, forming *succinyl-CoA.* In the process, two electrons are extracted, and they reduce another molecule of NAD^+ to NADH.

Reaction 6: Substrate-Level Phosphorylation The linkage between the 4-carbon succinyl group and CoA is a high-energy bond. In a coupled reaction similar to those that take place in glycolysis, this bond is cleaved, and the energy released drives the phosphorylation of guanosine diphosphate (GDP), forming guanosine triphosphate (GTP). GTP can transfer a phosphate to ADP converting it into ATP. The 4-carbon molecule that remains is called *succinate.*

Reaction 7: The Third Oxidation Next, succinate is oxidized to *fumarate* by an enzyme located in the inner mitochondrial membrane. The free-energy change in this reaction is not large enough to reduce NAD^+. Instead, FAD is the electron acceptor. Unlike NAD^+, FAD is not free to diffuse within the mitochondrion; it is tightly associated with its enzyme in the inner mitochondrial membrane. Its reduced form, $FADH_2$, can only contribute electrons to the electron transport chain in the membrane.

Reactions 8 and 9: Regeneration of Oxaloacetate In the final two reactions of the cycle, a water molecule is added to fumarate, forming *malate.* Malate is then oxidized, yielding a 4-carbon molecule of *oxaloacetate* and two electrons that reduce a molecule of NAD^+ to NADH. Oxaloacetate, the molecule that began the cycle, is now free to combine with another 2-carbon acetyl group from acetyl-CoA and reinitiate the cycle.

Glucose becomes CO_2 and potential energy

In the process of aerobic respiration, glucose is entirely consumed. The six-carbon glucose molecule is cleaved into a pair of 3-carbon pyruvate molecules during glycolysis. One of the carbons of each pyruvate is then lost as CO_2 in the conversion of pyruvate to acetyl-CoA. The two other carbons from acetyl-CoA are lost as CO_2 during the oxidations of the Krebs cycle.

All that is left to mark the passing of a glucose molecule into six CO_2 molecules is its energy, some of which is preserved in four ATP molecules and in the reduced state of 12 electron carriers. Ten of these carriers are NADH molecules; the other two are $FADH_2$.

Following the electrons in the reactions reveals the direction of transfer

As you examine the changes in electrical charge in the reactions that oxidize glucose, a good strategy for keeping the transfers clear is always to *follow the electrons.* For example, in glycolysis, an enzyme extracts two hydrogens—that is, two electrons and two protons—from glucose and transfers both electrons and one of the protons to NAD^+. The other proton is released as a hydrogen ion, H^+, into the surrounding solution. This transfer converts NAD^+ into NADH; that is, two negative electrons ($2e^-$) and one positive proton (H^+) are added to one positively charged NAD^+ to form NADH, which is electrically neutral.

As mentioned earlier, energy captured by NADH is not harvested all at once. The two electrons carried by NADH are passed along the electron transport chain, which consists of a series of electron carriers, mostly proteins, embedded within the inner membranes of mitochondria.

NADH delivers electrons to the beginning of the electron transport chain, and oxygen captures them at the end. The oxygen then joins with hydrogen ions to form water. At each step in the chain, the electrons move to a slightly more electronegative carrier, and their positions shift slightly. Thus, the electrons move *down* an energy gradient.

The entire process of electron transfer releases a total of 53 kcal/mol (222 kJ/mol) under standard conditions. The transfer of electrons along this chain allows the energy to be extracted gradually. Next, we will discuss how this energy is put to work to drive the production of ATP.

The Krebs cycle completes the oxidation of glucose begun with glycolysis. Units of the 2-carbon molecule acetyl-CoA are added to the 2-carbon molecule oxaloacetate to produce citric acid (another name for the cycle is the citric acid cycle). The cycle then uses a series of oxidation, decarboxylation, and rearrangement reactions to return to oxaloacetate. This process produces NADH and $FADH_2$, which provide electrons and protons for the electron transport chain. It also produces one ATP per turn of the cycle.

7.6 The Electron Transport Chain and Chemiosmosis

The NADH and $FADH_2$ molecules formed during aerobic respiration each contain a pair of electrons that were gained when NAD^+ and FAD were reduced. The NADH molecules carry their electrons to the inner mitochondrial membrane, where they transfer the electrons to a series of membrane-associated proteins collectively called the *electron transport chain (ETC).*

The electron transport chain produces a proton gradient

The first of the proteins to receive the electrons is a complex, membrane-embedded enzyme called **NADH dehydrogenase.** A carrier called *ubiquinone* then passes the electrons to a protein–cytochrome complex called the *bc_1 complex*. Each complex in the chain operates as a proton pump, driving a proton out across the membrane into the intermembrane space (figure 7.13*a*).

The electrons are then carried by another carrier, *cytochrome c*, to the cytochrome oxidase complex. This complex uses four electrons to reduce a molecule of oxygen. Each oxygen then combines with two protons to form water:

$$O_2 + 4\ H^+ + 4\ e^- \longrightarrow 2\ H_2O$$

In contrast to NADH, which contributes its electrons to NADH dehydrogenase, $FADH_2$, which is located in the inner mitochondrial membrane, feeds its electrons to ubiquinone, which is also in the membrane. Electrons from $FADH_2$ thus "skip" the first step in the electron transport chain.

The plentiful availability of a strong electron acceptor, oxygen, is what makes oxidative respiration possible. As you'll see in chapter 8, the electron transport chain used in aerobic respiration is similar to, and may well have evolved from, the chain employed in photosynthesis.

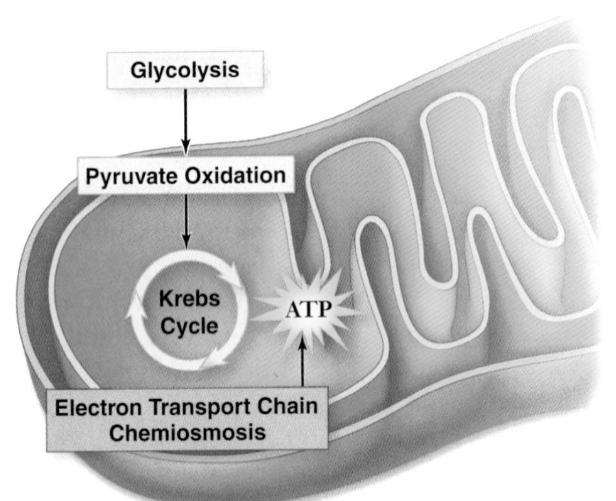

The gradient forms as electrons move through electron carriers

Respiration takes place within the mitochondria present in virtually all eukaryotic cells. The internal compartment, or matrix, of a mitochondrion contains the enzymes that carry out the reactions of the Krebs cycle. As mentioned earlier, protons (H^+) are produced when electrons are transferred to NAD^+. As the electrons harvested by oxidative respiration are passed along the electron transport chain, the energy they release transports protons out of the matrix and into the outer compartment called the intermembrane space.

Three transmembrane complexes of the electron transport chain in the inner mitochondrial membrane actually accomplish the proton transport (see figure 7.13*a*). The flow of highly energetic electrons induces a change in the shape of pump proteins, which causes them to transport protons across the membrane. The electrons contributed by NADH activate all three of these proton pumps, whereas those contributed by $FADH_2$ activate only two because of where they enter the chain. In this way a proton gradient is formed between the intermembrane space and the matrix.

Chemiosmosis utilizes the electrochemical gradient to produce ATP

The internal negativity of the matrix with respect to the intermembrane space attracts the positively charged protons and induces them to reenter the matrix. The higher outer concentration of protons also tends to drive protons back in by diffusion, but because membranes are relatively impermeable to ions, this process occurs only very slowly. Most of the protons that reenter the matrix instead pass through ATP synthase, an

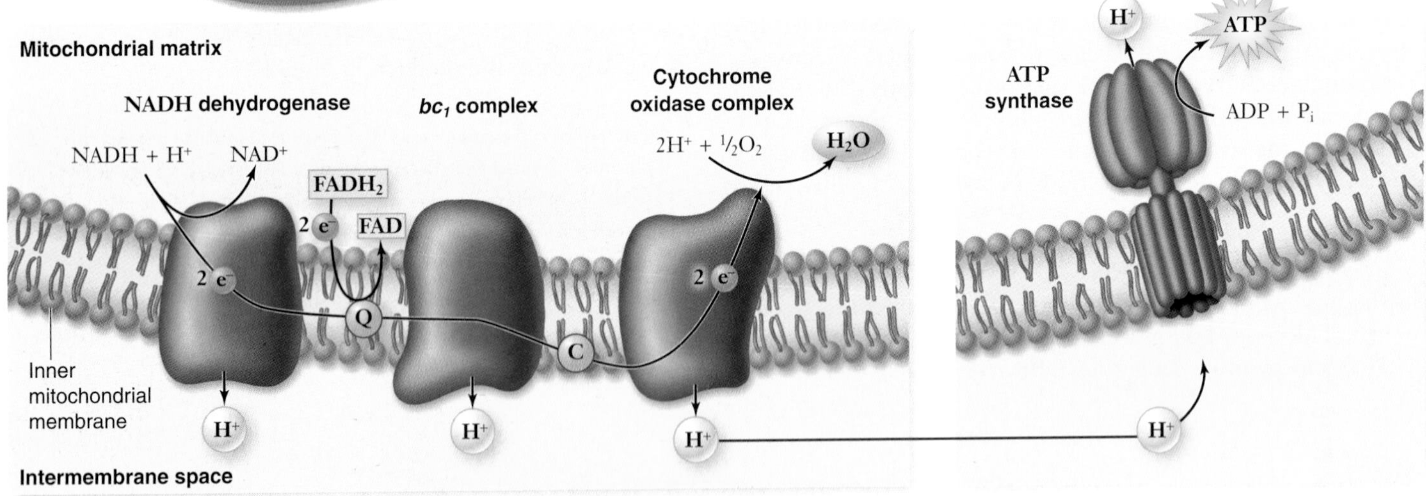

a. The electron transport chain

b. Chemiosmosis

figure 7.13

THE ELECTRON TRANSPORT CHAIN AND CHEMIOSMOSIS. ***a.*** High-energy electrons harvested from catabolized molecules are transported by mobile electron carriers (ubiquinone, marked Q, and cytochrome *c*, marked C) between three complexes of membrane proteins. These three complexes use portions of the electrons' energy to pump protons out of the matrix and into the intermembrane space. The electrons are finally used to reduce oxygen forming water. ***b.*** This creates a concentration gradient of protons across the inner membrane. This electrochemical gradient is a form of potential energy that can be used by ATP synthase. This enzyme couples the reentry of protons to the phosphorylation of ADP to form ATP.

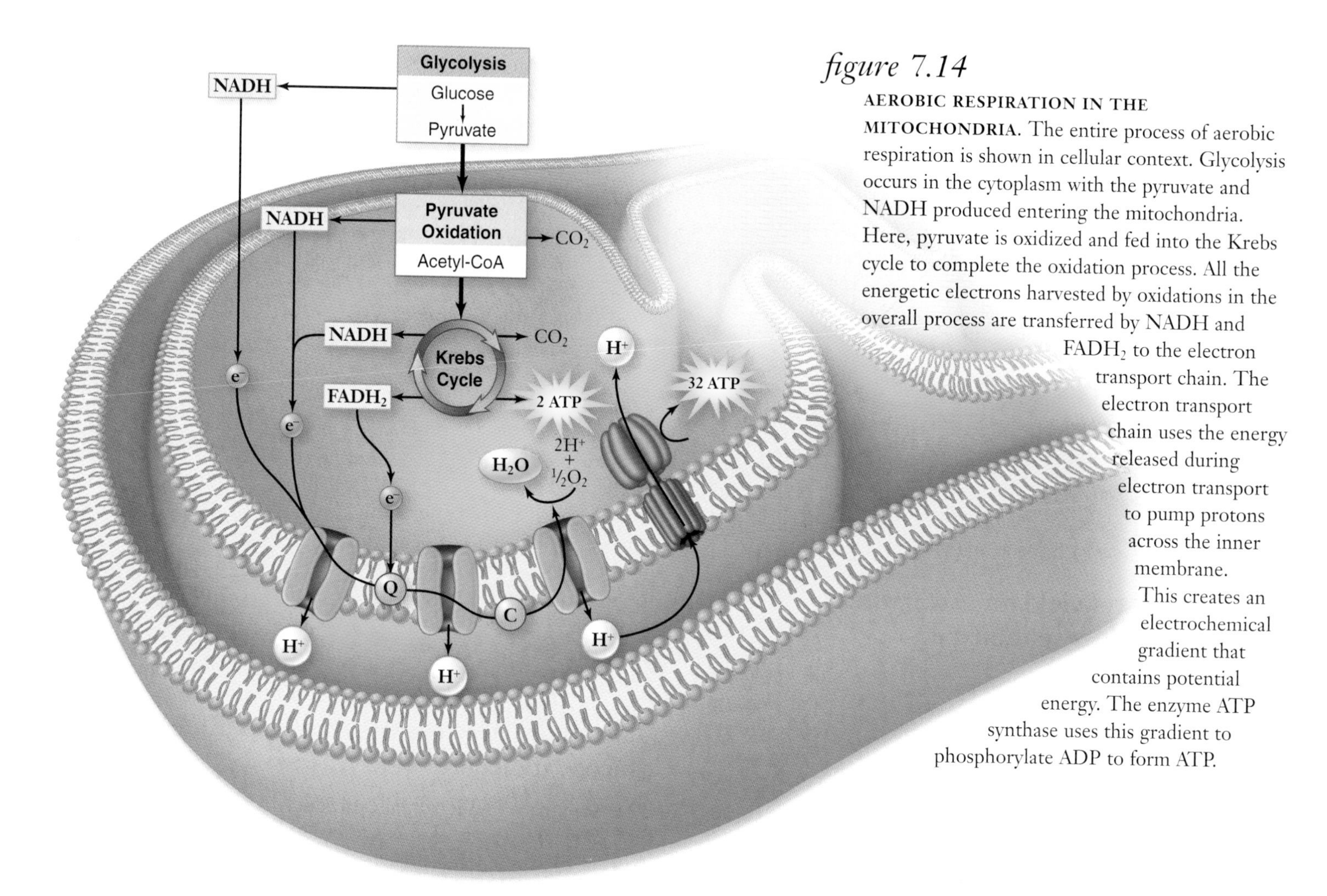

figure 7.14

AEROBIC RESPIRATION IN THE MITOCHONDRIA. The entire process of aerobic respiration is shown in cellular context. Glycolysis occurs in the cytoplasm with the pyruvate and NADH produced entering the mitochondria. Here, pyruvate is oxidized and fed into the Krebs cycle to complete the oxidation process. All the energetic electrons harvested by oxidations in the overall process are transferred by NADH and $FADH_2$ to the electron transport chain. The electron transport chain uses the energy released during electron transport to pump protons across the inner membrane. This creates an electrochemical gradient that contains potential energy. The enzyme ATP synthase uses this gradient to phosphorylate ADP to form ATP.

enzyme that uses the energy of the gradient to catalyze the synthesis of ATP from ADP and P_i. Because the chemical formation of ATP is driven by a diffusion force similar to osmosis, this process is referred to as *chemiosmosis* (figure 7.13*b*). The newly formed ATP is transported by facilitated diffusion to the many places in the cell where enzymes require energy to drive endergonic reactions.

The energy released by the reactions of cellular respiration ultimately drives the proton pumps that produce the proton gradient. The proton gradient provides the energy required for the synthesis of ATP. Figure 7.14 summarizes the overall process.

ATP synthase is a molecular rotary motor

ATP synthase uses a fascinating molecular mechanism to perform ATP synthesis (figure 7.15). Structurally, the enzyme has a membrane-bound portion and a narrow stalk that connects the membrane portion to a knoblike catalytic portion. This complex can be dissociated into two subportions: the F_0 membrane-bound complex, and the F_1 complex composed of the stalk and a knob, or head domain.

The F_1 complex has enzymatic activity. The F_0 complex contains a channel through which protons move across the membrane down their concentration gradient. As they do so, their movement causes part of the F_0 complex and the stalk to rotate relative to the knob. The mechanical energy of this

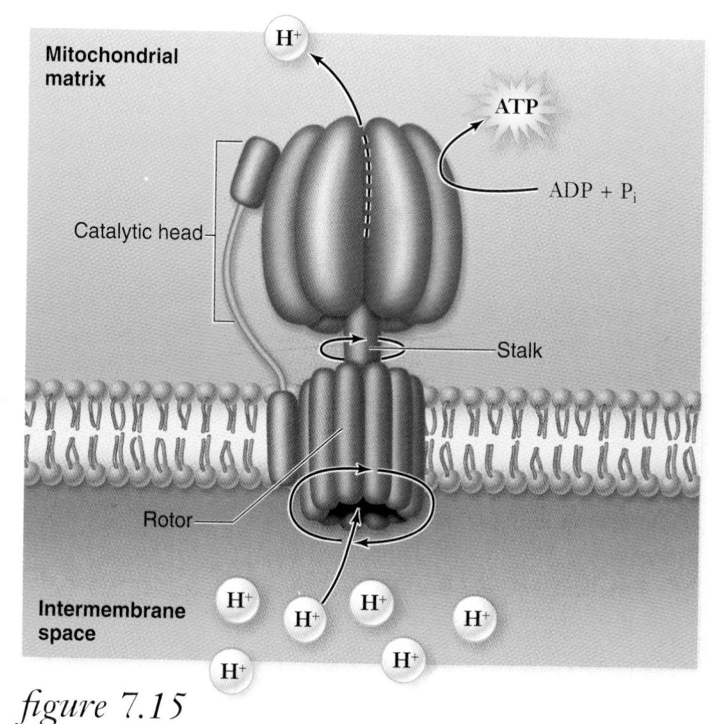

figure 7.15

THE ATP ROTARY ENGINE. Protons move across the membrane down their concentration gradient. The energy released causes the rotor and stalk structures to rotate. This mechanical energy alters the conformation of the ATP synthase enzyme to catalyze the formation of ATP.

rotation is used to change the conformation of the catalytic domain in the F_1 complex.

Thus, the synthesis of ATP is achieved by a tiny rotary motor, the rotation of which is driven directly by a gradient of protons. The process is like that of a water mill, in which the flow of water due to gravity causes a millwheel to turn and accomplish work or produce energy. The flow of protons is similar to the flow of water that causes the wheel to turn. Of course, ATP synthase is a much more complex motor, and it produces a chemical end-product.

The electron transport chain uses electrons from oxidation reactions carried by NADH and $FADH_2$ to create a proton gradient across the inner membrane of the mitochondria. The protein complexes of the electron transport chain are located in the inner membrane and use the energy from electron transfer to pump protons across the membrane, creating an electrochemical gradient. The enzyme ATP synthase can then use this gradient of protons to drive the endergonic reaction of phosphorylating ADP to ATP.

7.7 Energy Yield of Aerobic Respiration

How much metabolic energy in the form of ATP does a cell gain from aerobic breakdown of glucose? Knowing the steps involved in the process, we can calculate the theoretical yield of ATP and compare it with the actual yield.

The theoretical yield for eukaryotes is 36 ATP per glucose molecule

The chemiosmotic model suggests that one ATP molecule is generated for each proton pump activated by the electron transport chain. Because the electrons from NADH activate three pumps and those from $FADH_2$ activate two, we would expect each molecule of NADH and $FADH_2$ to generate three and two ATP molecules, respectively.

In doing this accounting, remember that everything downstream of glycolysis must be multiplied by 2 because two pyruvates are produced per molecule of glucose. A total of 10 NADH molecules is generated by respiration: 2 from glycolysis, 2 from the oxidation of pyruvate (1×2), and another 6 from the Krebs cycle (3×2). Also, two $FADH_2$ are produced (1×2). Finally, two ATP are generated directly by glycolysis and another two ATP from the Krebs cycle (1×2). This gives a total of $10 \times 3 = 30$ ATP from NADH, plus $2 \times 2 = 4$ ATP from $FADH_2$, plus four ATP, for a total of 38 ATP (figure 7.16).

This number is accurate for bacteria, but it does not hold for eukaryotes because the NADH produced in the cytoplasm by glycolysis needs to be transported into the mitochondria by active transport, which costs 1 ATP per NADH transported. This reduces the predicted yield for eukaryotes to 36 ATP.

The actual yield for eukaryotes is 30 ATP per glucose molecule

The amount of ATP actually produced in a eukaryotic cell during aerobic respiration is somewhat lower than 36, for two reasons. First, the inner mitochondrial membrane is somewhat "leaky" to protons, allowing some of them to reenter the matrix without passing through ATP synthase. Second, mitochondria often use the proton gradient generated by chemiosmosis for purposes other than ATP synthesis (such as transporting pyruvate into the matrix).

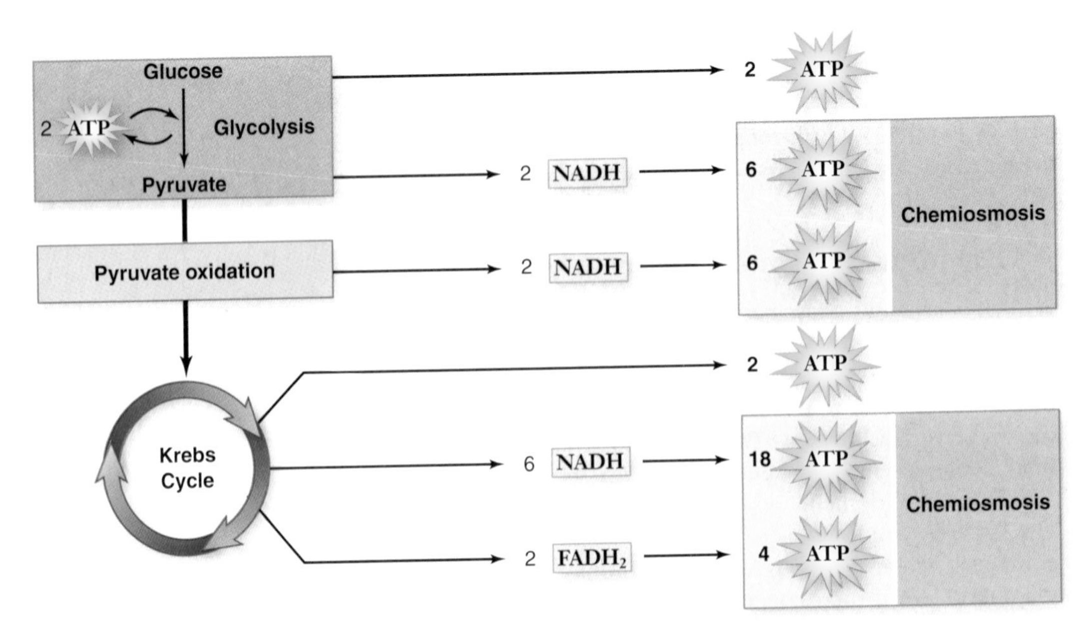

figure 7.16

THEORETICAL ATP YIELD. The theoretical yield of ATP harvested from glucose by aerobic respiration totals 38 molecules. In eukaryotes this is reduced to 36 as the NADH generated by glycolysis in the cytoplasm has to be actively transported into the mitochondria costing the cell 1 ATP per NADH transported.

Consequently, the actual measured values of ATP generated by NADH and $FADH_2$ are closer to 2.5 for each NADH, and 1.5 for each $FADH_2$. With these corrections, the overall harvest of ATP from a molecule of glucose in a eukaryotic cell is calculated as: 4 ATP from substrate-level phosphorylation + 25 ATP from NADH (2.5×10) + 3 ATP from $FADH_2$ (1.5×2) − 2 ATP for transport of glycolytic NADH = 30 molecules of ATP.

We mentioned earlier that the catabolism of glucose by aerobic respiration, in contrast to that by glycolysis alone, has a large energy yield. Aerobic respiration in a eukaryotic cell harvests about $(7.3 \times 30)/686 = 32\%$ of the energy available in glucose. (By comparison, a typical car converts only about 25% of the energy in gasoline into useful energy.)

The higher yield of aerobic respiration was one of the key factors that fostered the evolution of heterotrophs. As this mechanism for producing ATP evolved, nonphotosynthetic organisms could more successfully base their metabolism on the exclusive use of molecules derived from other organisms. As long as some organisms captured energy by photosynthesis, others could exist solely by feeding on them.

Passage of electrons down the electron transport chain produces roughly 3 ATP per NADH. This can result in a maximum of 38 ATP for all of the NADH generated by the complete oxidation of glucose, plus the ATP generated by substrate-level phosphorylation. NADH generated in the cytoplasm only lead to 2 ATP/NADH due to the cost of transporting the NADH into the mitochondria, leading to a total of 36 ATP for the mitochondria per glucose

7.8 Regulation of Aerobic Respiration

When cells possess plentiful amounts of ATP, the key reactions of glycolysis, the Krebs cycle, and fatty acid breakdown are inhibited, slowing ATP production. The regulation of these biochemical pathways by the level of ATP is an example of feedback inhibition. Conversely, when ATP levels in the cell are low, ADP levels are high, and ADP activates enzymes in the pathways of carbohydrate catabolism to stimulate the production of more ATP.

Control of glucose catabolism occurs at two key points in the catabolic pathway, namely at a point in glycolysis and at the beginning of the Krebs cycle (figure 7.17). The control point in glycolysis is the enzyme phosphofructokinase, which catalyzes the conversion of fructose phosphate to fructose bisphosphate. This is the first reaction of glycolysis that is not readily reversible, committing the substrate to the glycolytic sequence. ATP itself is an allosteric inhibitor (chapter 6) of phosphofructokinase, as is the Krebs cycle intermediate citrate. High levels of both ATP and citrate inhibit phosphofructokinase. Thus, under conditions when ATP is in excess, or when the Krebs cycle is producing citrate faster than it is being consumed, glycolysis is slowed.

The main control point in the oxidation of pyruvate occurs at the committing step in the Krebs cycle with the enzyme pyruvate dehydrogenase, which converts pyruvate to acetyl-CoA. This enzyme is inhibited by high levels of NADH, a key product of the Krebs cycle.

Another control point in the Krebs cycle is the enzyme citrate synthetase, which catalyzes the first reaction, the conversion of oxaloacetate and acetyl-CoA into citrate. High levels of ATP inhibit citrate synthetase (as well as phosphofructokinase, pyruvate dehydrogenase, and two other Krebs cycle enzymes), slowing down the entire catabolic pathway.

Respiration is controlled by levels of ATP in the cell and levels of key intermediates in the process. The control point for glycolysis is the enzyme phosphofructokinase.

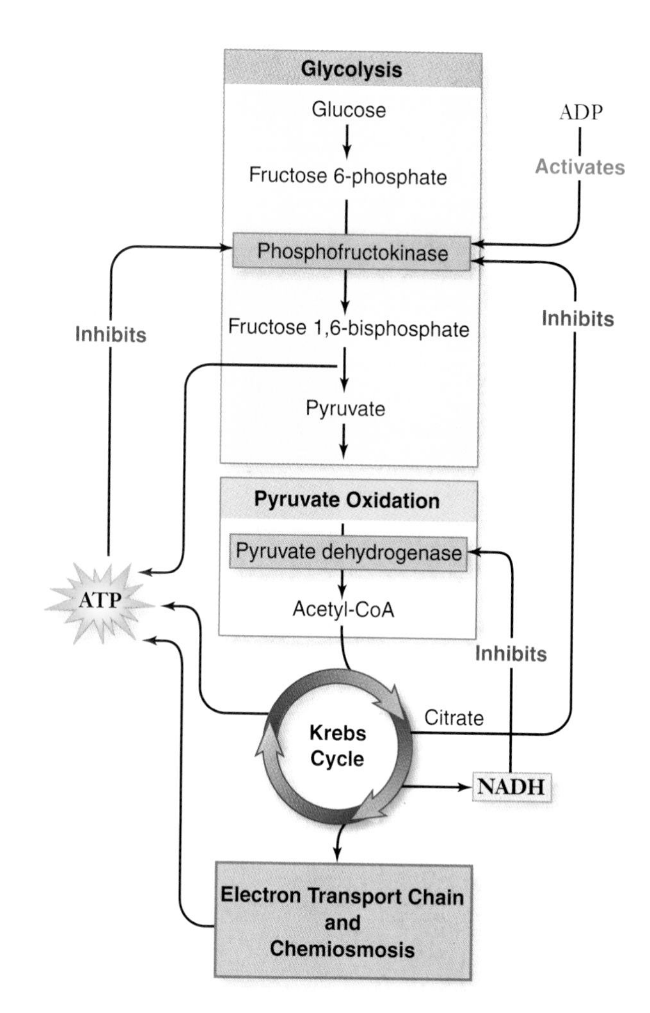

figure 7.17

CONTROL OF GLUCOSE CATABOLISM. The relative levels of ADP and ATP and key intermediates NADH and citrate control the catabolic pathway at two key points: the committing reactions of glycolysis and the Krebs cycle.

7.9 Oxidation Without O_2

In the presence of oxygen, cells can use oxygen to produce a large amount of ATP. But even when no oxygen is present to accept electrons, some organisms can still respire *anaerobically*, using inorganic molecules as final electron acceptors for an electron transport chain.

For example, many prokaryotes use sulfur, nitrate, carbon dioxide, or even inorganic metals as the final electron acceptor in place of oxygen (figure 7.18). The free energy released by using these other molecules as final electron acceptors is not as great as that using oxygen because they have a lower affinity for electrons. Less total ATP is produced, but the process is still respiration and not fermentation.

Methanogens use carbon dioxide

Among the heterotrophs that practice anaerobic respiration are primitive Archaea such as thermophiles and methanogens. Methanogens use carbon dioxide (CO_2) as the electron acceptor, reducing CO_2 to CH_4 (methane). The hydrogens are derived from organic molecules produced by other organisms. Methanogens are found in diverse environments, including soil and the digestive systems of ruminants like cows.

Sulfur bacteria use sulfate

Evidence of a second anaerobic respiratory process among primitive bacteria is seen in a group of rocks about 2.7 BYA, known as the Woman River iron formation. Organic material in these rocks is enriched for the light isotope of sulfur, ^{32}S, relative to the heavier isotope, ^{34}S. No known geochemical process produces such enrichment, but biological sulfur reduction does, in a process still carried out today by certain primitive prokaryotes.

In this sulfate respiration, the prokaryotes derive energy from the reduction of inorganic sulfates (SO_4) to hydrogen sulfide (H_2S). The hydrogen atoms are obtained from organic molecules other organisms produce. These prokaryotes thus are similar to methanogens, but they use SO_4 as the oxidizing (that is, electron-accepting) agent in place of CO_2.

The early sulfate reducers set the stage for the evolution of photosynthesis, creating an environment rich in H_2S. As dis-

figure 7.18

SULFUR-RESPIRING PROKARYOTE. ***a.*** The micrograph shows the archaeal species *Thermoproteus tenax*. This organism can use elemental sulfur as a final electron acceptor for anaerobic respiration. ***b.*** *Thermoproteus* is often found in sulfur containing hot springs such as the Norris Geyser Basin in Yellowstone National Park shown here.

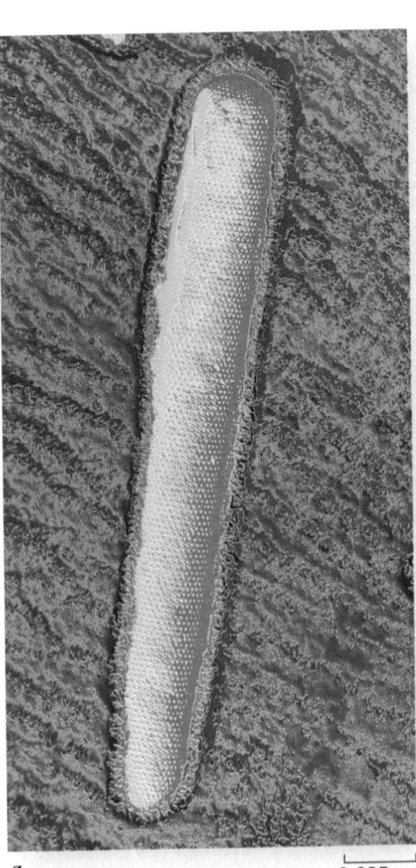

a.

b.

cussed in chapter 8, the first form of photosynthesis obtained hydrogens from H_2S using the energy of sunlight.

Fermentation uses organic compounds as electron acceptors

In the absence of oxygen, cells that cannot utilize an alternative electron acceptor for respiration must rely exclusively on glycolysis to produce ATP. Under these conditions, the electrons generated by glycolysis are donated to organic molecules in a process called *fermentation.* This process recycles NAD^+, the electron acceptor that allows glycolysis to proceed.

Bacteria carry out more than a dozen kinds of fermentation reactions, often using pyruvate or a derivative of pyruvate to accept the electrons from NADH. Organic molecules other than pyruvate and its derivatives can be used as well; the important point is that the process regenerates NAD^+:

$$\text{organic molecule} + \text{NADH} \longrightarrow \text{reduced organic molecule} + \text{NAD}^+$$

Often the reduced organic compound is an organic acid—such as acetic acid, butyric acid, propionic acid, or lactic acid—or an alcohol.

Ethanol fermentation

Eukaryotic cells are capable of only a few types of fermentation. In one type, which occurs in yeast, the molecule that accepts electrons from NADH is derived from pyruvate, the end-product of glycolysis.

Yeast enzymes remove a terminal CO_2 group from pyruvate through decarboxylation, producing a 2-carbon molecule called acetaldehyde. The CO_2 released causes bread made with yeast to rise; bread made without yeast (unleavened bread) does not rise. The acetaldehyde accepts a pair of electrons from NADH, producing NAD^+ and ethanol (ethyl alcohol) (figure 7.19).

This particular type of fermentation is of great interest to humans, because it is the source of the ethanol in wine and beer. Ethanol is a by-product of fermentation that is actually toxic to yeast; as it approaches a concentration of about 12%, it begins to kill the yeast. That explains why naturally fermented wine contains only about 12% ethanol.

Lactic acid fermentation

Most animal cells regenerate NAD^+ without decarboxylation. Muscle cells, for example, use the enzyme lactate dehydrogenase to transfer electrons from NADH back to the pyruvate that is produced by glycolysis. This reaction converts pyruvate into lactic acid and regenerates NAD^+ from NADH (see figure 7.19). It therefore closes the metabolic circle, allowing glycolysis to continue as long as glucose is available.

Circulating blood removes excess lactate, the ionized form of lactic acid, from muscles, but when removal cannot keep pace with production, the accumulating lactic acid interferes with muscle function and contributes to muscle fatigue.

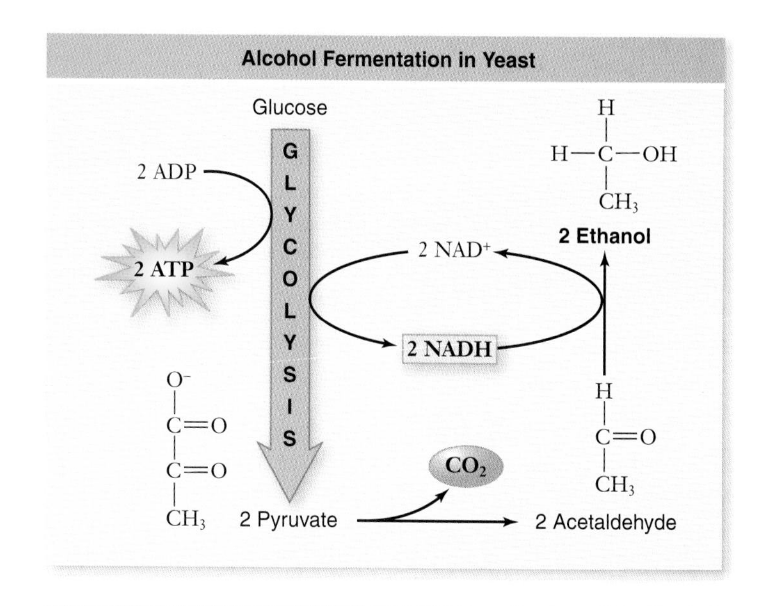

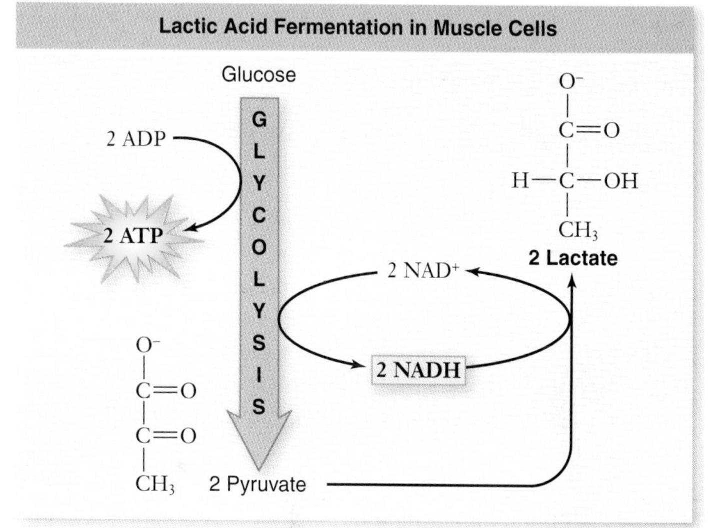

figure 7.19

FERMENTATION. Yeasts carry out the conversion of pyruvate to ethanol. Muscle cells convert pyruvate into lactate, which is less toxic than ethanol. In each case, the reduction of a metabolite of glucose has oxidized NADH back to NAD^+ to allow glycolysis to continue under anaerobic conditions.

O_2 is the electron acceptor for aerobic respiration. Due to oxygen's high affinity for electrons this results in the highest yield of ATP, but this is not the only form of respiration found in living systems. Nitrate, sulfur, and CO_2 are among other terminal electron acceptors used in anaerobic respiration. Organic molecules can also be used in fermentation reactions but allow only a partial oxidation of glucose via glycolysis. Fermentation reactions produce a variety of compounds, including ethanol in yeast and lactic acid in humans.

7.10 Catabolism of Proteins and Fats

Thus far we have focused on the aerobic respiration of glucose, which organisms obtain from the digestion of carbohydrates or from photosynthesis. Organic molecules other than glucose, particularly proteins and fats, are also important sources of energy (figure 7.20).

Catabolism of proteins removes amino groups

Proteins are first broken down into their individual amino acids. The nitrogen-containing side group (the amino group) is then removed from each amino acid in a process called **deamination.** A series of reactions convert the carbon chain that remains into a molecule that enters glycolysis or the Krebs cycle. For example, alanine is converted into pyruvate, glutamate into α-ketoglutarate (figure 7.21), and aspartate into oxaloacetate. The reactions of glycolysis and the Krebs cycle then extract the high-energy electrons from these molecules and put them to work making ATP.

Catabolism of fatty acids produces acetyl groups

Fats are broken down into fatty acids plus glycerol. Long-chain fatty acids typically have an even number of carbons, and the many C—H bonds provide a rich harvest of energy. Fatty acids are oxidized in the matrix of the mitochondrion. Enzymes remove the 2-carbon acetyl groups from the end of each fatty acid until the entire fatty acid is converted into acetyl groups (figure 7.22). Each acetyl group is combined with coenzyme A to form acetyl-CoA. This process is known as **β-oxidation.** This process is oxygen-dependent, which explains why aerobic exercise burns fat, but anaerobic exercise does not.

How much ATP does the catabolism of fatty acids produce? Let's compare a hypothetical 6-carbon fatty acid with the six-carbon glucose molecule, which we've said yields about 30 molecules of ATP in a eukaryotic cell. Two rounds of β-oxidation would convert the fatty acid into three molecules of acetyl-CoA. Each round requires one molecule of ATP to prime the process, but it also produces one molecule of NADH and one of $FADH_2$. These molecules together yield four molecules of ATP (assuming 2.5 ATPs per NADH, and 1.5 ATPs per $FADH_2$).

The oxidation of each acetyl-CoA in the Krebs cycle ultimately produces an additional 10 molecules of ATP. Overall, then, the ATP yield of a 6-carbon fatty acid would be approximately: 8 (from two rounds of β-oxidation) − 2 (for priming those two rounds) + 30 (from oxidizing the three acetyl-CoAs) = 36 molecules of ATP. Therefore, the respiration of a 6-carbon fatty acid yields 20% more ATP than the respiration of glucose.

Moreover, a fatty acid of that size would weigh less than two-thirds as much as glucose, so a gram of fatty acid contains more than twice as many kilocalories as a gram of glucose. You can see from this fact why fat is a storage molecule for excess energy in many types of animals. If excess energy were stored instead as carbohydrate, as it is in plants, animal bodies would have to be much bulkier.

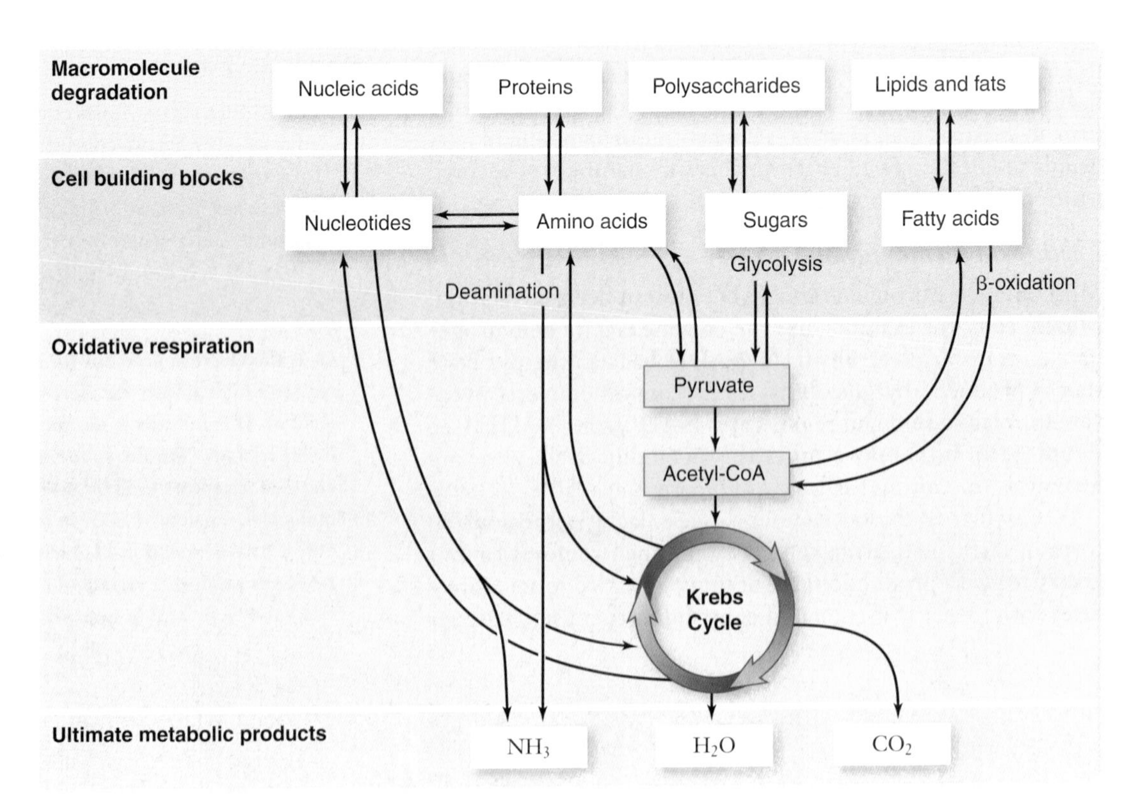

figure 7.20

HOW CELLS EXTRACT CHEMICAL ENERGY. All eukaryotes and many prokaryotes extract energy from organic molecules by oxidizing them. The first stage of this process, breaking down macromolecules into their constituent parts, yields little energy. The second stage, oxidative or aerobic respiration, extracts energy, primarily in the form of high-energy electrons, and produces water and carbon dioxide. Key intermediates in these energy pathways are also used for biosynthetic pathways, shown by reverse arrows.

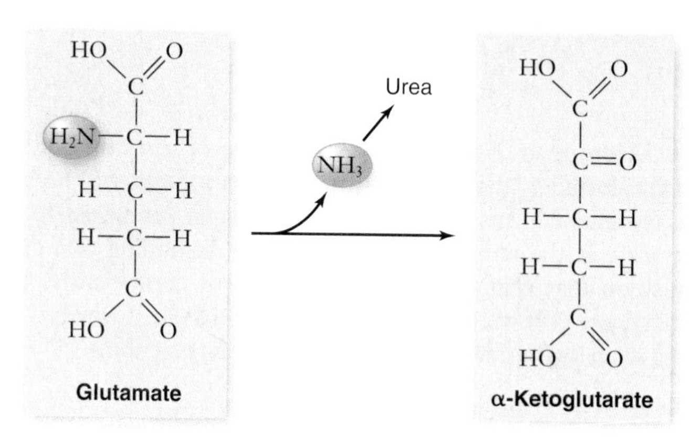

figure 7.21

DEAMINATION. After proteins are broken down into their amino acid constituents, the amino groups are removed from the amino acids to form molecules that participate in glycolysis and the Krebs cycle. For example, the amino acid glutamate becomes α-ketoglutarate, a Krebs cycle intermediate, when it loses its amino group.

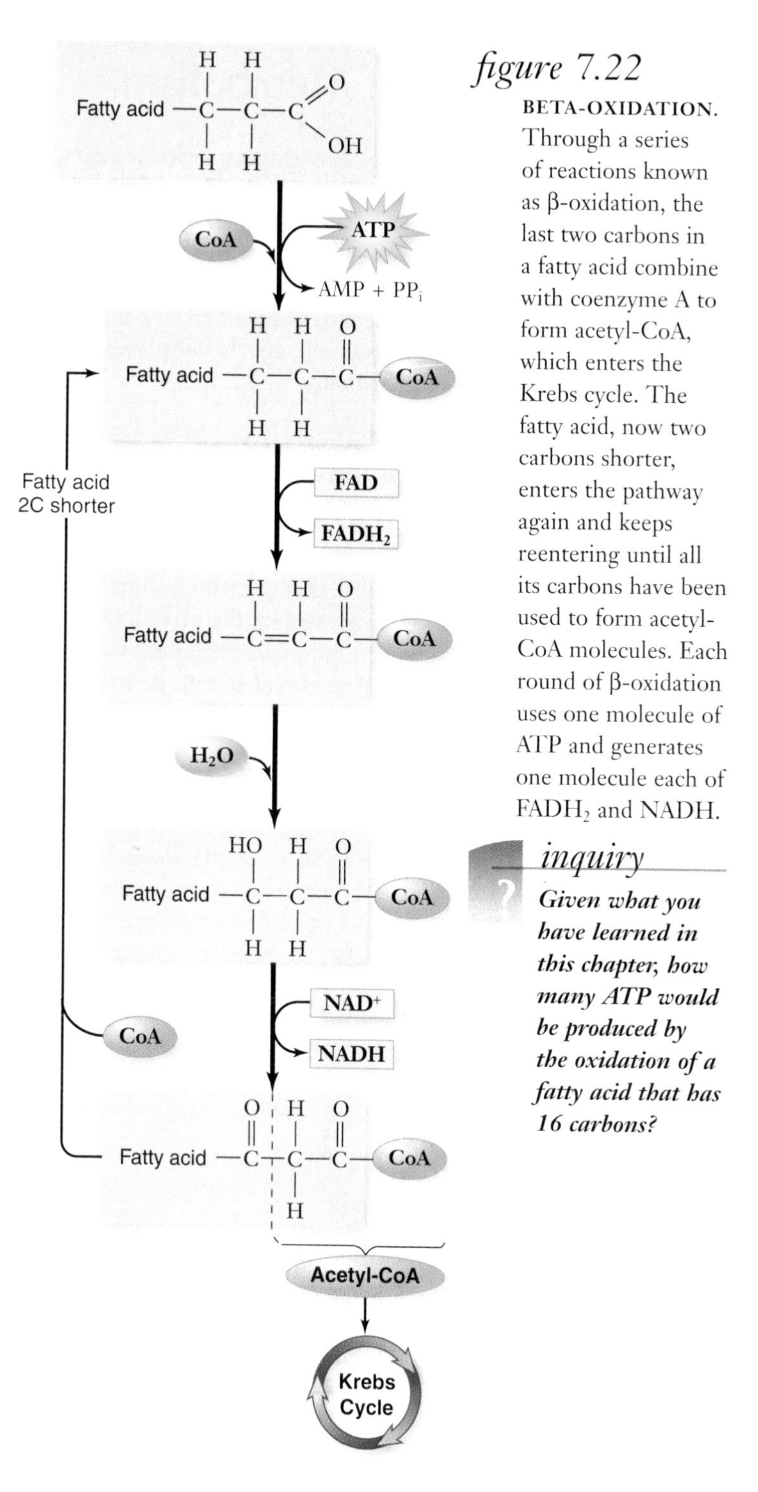

figure 7.22

BETA-OXIDATION. Through a series of reactions known as β-oxidation, the last two carbons in a fatty acid combine with coenzyme A to form acetyl-CoA, which enters the Krebs cycle. The fatty acid, now two carbons shorter, enters the pathway again and keeps reentering until all its carbons have been used to form acetyl-CoA molecules. Each round of β-oxidation uses one molecule of ATP and generates one molecule each of $FADH_2$ and NADH.

inquiry

Given what you have learned in this chapter, how many ATP would be produced by the oxidation of a fatty acid that has 16 carbons?

A small number of key intermediates connect metabolic pathways

Oxidation pathways of food molecules are interrelated in that a small number of key intermediates, such as pyruvate and acetyl-CoA, link the breakdown from different starting points. These key intermediates allow the interconversion of different types of molecules, such as sugars and amino acids (see figure 7.20).

Cells can make glucose, amino acids, and fats, as well as getting them from external sources, and they use reactions similar to those that break down these substances. In many cases, the reverse pathways even share enzymes if the free-energy changes are small. For example, gluconeogenesis, the process of making new glucose, uses all but three enzymes of the glycolytic pathway. Thus, much of glycolysis runs forward or backward, depending on the concentrations of the intermediates—with only three key steps having different enzymes for forward and reverse directions.

Acetyl-CoA has many roles

Many different metabolic processes generate acetyl-CoA. Not only does the oxidation of pyruvate produce it, but the metabolic breakdown of proteins, fats, and other lipids also generates acetyl-CoA. Indeed, almost all molecules catabolized for energy are converted into acetyl-CoA.

Acetyl-CoA has a role in anabolic metabolism as well. Units of two carbons derived from acetyl-CoA are used to build up the hydrocarbon chains in fatty acids. Acetyl-CoA produced from a variety of sources can therefore be channeled into fatty acid synthesis or into ATP production, depending on the organism's energy requirements. Which of these two options is taken depends on the level of ATP in the cell.

When ATP levels are high, the oxidative pathway is inhibited, and acetyl-CoA is channeled into fatty acid synthesis. This explains why many animals (humans included) develop fat reserves when they consume more food than their activities require. Alternatively, when ATP levels are low, the oxidative pathway is stimulated, and acetyl-CoA flows into energy-producing oxidative metabolism.

Fats are a major energy storage molecule that can be broken down into units of acetyl-CoA by β-oxidation and fed into the Krebs cycle. The major metabolic pathways are connected by a number of key intermediates. This allows many processes to be used to either build up (anabolism) or break down (catabolism) the major biological macromolecules and allows interconversion of different types of molecules.

7.11 Evolution of Metabolism

We talk about cellular respiration as a continuous series of stages, but it is important to note that these stages evolved over time, and metabolism has changed a great deal in that time. Both anabolic processes and catabolic processes evolved in concert with each other. We do not know the details of this biochemical evolution, or the order of appearance of these processes. Therefore the following timeline is based on the available geochemical evidence and represents a hypothesis rather than a strict timeline.

The earliest life forms degraded carbon-based molecules present in the environment

The most primitive forms of life are thought to have obtained chemical energy by degrading, or breaking down, organic molecules that were abiotically produced, that is, carbon-containing molecules formed by inorganic processes on the early Earth.

The first major event in the evolution of metabolism was the origin of the ability to harness chemical bond energy. At an early stage, organisms began to store this energy in the bonds of ATP.

The evolution of glycolysis also occurred early

The second major event in the evolution of metabolism was glycolysis, the initial breakdown of glucose. As proteins evolved diverse catalytic functions, it became possible to capture a larger fraction of the chemical bond energy in organic molecules by breaking chemical bonds in a series of steps.

Glycolysis undoubtedly evolved early in the history of life on Earth, because this biochemical pathway has been retained by all living organisms. It is a chemical process that does not appear to have changed for well over 2 billion years.

Anaerobic photosynthesis allowed the capture of light energy

The third major event in the evolution of metabolism was anaerobic photosynthesis. Early in the history of life, a different way of generating ATP evolved in some organisms. Instead of obtaining energy for ATP synthesis by reshuffling chemical bonds, as in glycolysis, these organisms developed the ability to use light to pump protons out of their cells, and to use the resulting proton gradient to power the production of ATP through chemiosmosis.

Photosynthesis evolved in the absence of oxygen and works well without it. Dissolved H_2S, present in the oceans of the early Earth beneath an atmosphere free of oxygen gas, served as a ready source of hydrogen atoms for building organic molecules. Free sulfur was produced as a by-product of this reaction.

Oxygen-forming photosynthesis used a different source of hydrogen

The substitution of H_2O for H_2S in photosynthesis was the fourth major event in the history of metabolism. Oxygen-forming photosynthesis employs H_2O rather than H_2S as a source of hydrogen atoms and their associated electrons. Because it garners its electrons from reduced oxygen rather than from reduced sulfur, it generates oxygen gas rather than free sulfur.

More than 2 BYA, small cells capable of carrying out this oxygen-forming photosynthesis, such as cyanobacteria, became the dominant forms of life on Earth. Oxygen gas began to accumulate in the atmosphere. This was the beginning of a great transition that changed conditions on Earth permanently. Our atmosphere is now 20.9% oxygen, every molecule of which is derived from an oxygen-forming photosynthetic reaction.

Nitrogen fixation provided new organic nitrogen

Nitrogen is available from dead organic matter, and from chemical reactions that generated the original organic molecules. For life to expand, a new source of nitrogen was needed. Nitrogen fixation was the fifth major step in the evolution of metabolism. Proteins and nucleic acids cannot be synthesized from the products of photosynthesis because both of these biologically critical molecules contain nitrogen. Obtaining nitrogen atoms from N_2 gas, a process called *nitrogen fixation*, requires breaking an $N{\equiv}N$ triple bond.

This important reaction evolved in the hydrogen-rich atmosphere of the early Earth, where no oxygen was present. Oxygen acts as a poison to nitrogen fixation, which today occurs only in oxygen-free environments or in oxygen-free compartments within certain prokaryotes.

Aerobic respiration utilized oxygen

Aerobic respiration is the sixth and final event in the history of metabolism. Aerobic respiration employs the same kind of proton pumps as photosynthesis and is thought to have evolved as a modification of the basic photosynthetic machinery.

Biologists think that the ability to carry out photosynthesis without H_2S first evolved among purple nonsulfur bacteria, which obtain their hydrogens from organic compounds instead. It was perhaps inevitable that among the descendants of these respiring photosynthetic bacteria, some would eventually do without photosynthesis entirely, subsisting only on the energy and electrons derived from the breakdown of organic molecules. The mitochondria within all eukaryotic cells are thought to be descendants of these bacteria.

The complex process of aerobic metabolism developed over geological time, as natural selection favored organisms with more efficient methods of obtaining energy from organic molecules. The process of photosynthesis, as you have seen in this concluding section, has also developed over time, and the rise of photosynthesis changed life on Earth forever. The next chapter explores photosynthesis in detail.

Although the evolution of metabolism is not known in detail, major milestones can be recognized. These include the evolution of metabolic pathways that allow extraction of energy from organic compounds, the evolution of photosynthesis, and the evolution of nitrogen fixation. Photosynthesis began as an anoxygenic process that later evolved to produce oxygen, thus allowing the evolution of aerobic metabolism.

7.1 Overview of Respiration

Respiration occurs when carbohydrates and oxygen are converted to carbon dioxide, water, and energy.

- Autotrophs convert energy from sunlight to organic molecules.
- Heterotrophs use organic compounds made by autotrophs.
- Energy-rich molecules are degraded by oxidation reactions.
- Electron carriers can be reversibly oxidized and reduced.
- Energy released from redox reactions is used to make ATP.
- NAD^+ is an important electron carrier that can act as a coenzyme.
- Aerobic respiration uses oxygen as the final electron acceptor.
- Oxidizing food molecules in stages is more efficient than one step.

7.2 The Oxidation of Glucose: A Summary

Cells make ATP from the oxidation of glucose by two fundamentally different mechanisms.

- Substrate-level phosphorylation transfers a phosphate from a phosphate-bearing intermediate directly to ADP (figure 7.4).
- In oxidative phosphorylation ATP is generated by the enzyme ATP synthase, which is powered by a proton gradient.

7.3 Glycolysis: Splitting Glucose (figure 7.6)

Glycolysis is a series of chemical reactions that occur in the cell cytoplasm. Glucose yields 2 pyruvate, 2 NADH and 2 ATP.

- Priming reactions add two phosphates to glucose.
- This 6-carbon diphosphate is cleaved into two 3-carbon molecules of glyceraldehyde-3-phosphate (G3P).
- Oxidation of G3P transfers electrons to NAD^+ yielding NADH.
- The final product is two molecules of pyruvate.
- Glycolysis produces a net of 2 ATP, 2 NADH, and 2 pyruvate.
- NADH must be recycled into NAD^+ to continue glycolysis.
- In the presence of oxygen NADH is oxidized during respiration.

7.4 The Oxidation of Pyruvate to Produce Acetyl-CoA

Pyruvate from glycolysis is transported into the mitochondria where it is oxidized, and the product is fed into the Krebs cycle.

- Pyruvate oxidation results in 1 CO_2, 1 NADH, and 1-acetyl-CoA per pyruvate.
- Acetyl-CoA enters the Krebs cycle as two-carbon acetyl units.

7.5 The Krebs Cycle

Each acetyl compound that enters the Krebs cycle yields 2 CO_2, 1 ATP, 3 NADH and 1 $FADH_2$.

- An acetyl group combines with oxaloacetate producing citrate.
- Citrate is oxidized, removing CO_2 and generating NADH.

7.6 The Electron Transport Chain and Chemiosmosis (figure 7.13)

The electron transport chain is located on the inner membrane of mitochondria. It produces a proton gradient used in ATP synthesis.

- NADH is oxidized to NADH by NADH dehydrogenase.
- The electrons are transferred sequentially through 3 complexes to cytochrome oxidase, where electrons join with H^+ and oxygen.
- As the electrons move down the electron transport chain, three protons are pumped into the intermembrane space.
- This provides sufficient energy to produce 3 ATP.
- Protons diffuse back into the mitochondrial matrix through the ATP synthase channel, which phosphorylates ADP to ATP.
- As each proton passes through ATP synthase the energy causes the rotor and rod to rotate, altering the conformation of ATP synthase and catalyzing the formation of one ATP (figure 7.15).
- $FADH_2$ transfers electrons to ubiquinone. Only two protons are transported into the intermembrane space and 2 ATP are produced.

7.7 Energy Yield of Aerobic Respiration (figure 7.16)

Aerobic respiration theoretically yields 38 ATP per molecule of glucose.

- Eukaryotes yield 36 ATP per glucose molecule because it costs ATP to transport NADH formed during glycolysis into mitochondria.

7.8 Regulation of Aerobic Respiration (figure 7.17)

Glucose catabolism is controlled by the concentration of ATP molecules and products of the Krebs cycle.

- High ATP concentrations inhibit phosphofructokinase, the third enzyme in glycolysis; low ATP levels activate this enzyme.
- High concentrations of NADH inhibit pyruvate dehydrogenase.

7.9 Oxidation Without O_2 (figure 7.8)

In the absence of oxygen another final electron acceptor is necessary for respiration. For normally aerobic organisms, in the absence of oxygen, ATP can only be produced by glycolysis.

- In many prokaryotes inorganic molecules are used as final electron acceptors for an electron transport chain.
- The regeneration of NAD^+ by the oxidation of NADH and reduction of an organic molecule is called fermentation.
- In yeast, pyruvate is decarboxylated, then reduced to ethanol as NADH is oxidized to NAD^+.
- In animals, pyruvate is reduced to lactate as NADH is oxidized.

7.10 Catabolism of Proteins and Fats (figure 7.20)

Proteins, fats, and nucleic acids are built up and broken down through key intermediates.

- Nucleic acids are metabolized through the Krebs cycle.
- Amino acids are deaminated before they are metabolized.
- The fatty acids are converted to acetyl-CoA by β-oxidation.
- With high ATP, acetyl-CoA is converted into fatty acids.

7.11 Evolution of Metabolism

Major milestones are recognized in the evolution of metabolism, the order of events is hypothetical.

- Five major metabolic processes evolved before atmospheric oxygen was present.
  - *Early life-forms metabolized organic molecules that were abiotically produced and began to store energy as ATP.*
  - *Glycolysis evolved incrementally.*
  - *Early photosynthesis used H_2S to make organic molecules from CO_2.*
  - *The substitution of H_2O for H_2S resulted in the formation of oxygen.*
  - *Nitrogen fixation made N available.*

review questions

SELF TEST

1. An *autotroph* is an organism that—
   a. Extracts energy from organic sources
   b. Converts energy from sunlight into chemical energy
   c. Relies on the energy produced by other organisms as an energy source
   d. Both a and b
2. Which of the following processes is (are) required for the complete oxidation of glucose?
   a. The Krebs cycle
   b. Glycolysis
   c. Pyruvate oxidation
   d. All of the above
3. The energy associated with a molecule of glucose is stored in its—
   a. carbon atoms
   b. chemical bonds
   c. electrons
   d. protons
4. How is ATP produced by glycolysis?
   a. Through the priming reactions
   b. Through the production of glyceraldehyde-3-phosphate
   c. By substrate level phosphorylation
   d. As a result of the reduction of NAD^+ to NADH
5. Which of the following is NOT a true statement regarding cellular respiration?
   a. Enzymes catalyze reactions that transfer electrons.
   b. Electrons have a higher potential energy at the end of the process.
   c. Carbon dioxide gas is a by-product.
   d. The process involves multiple redox reactions.
6. The majority of the ATP produced during aerobic respiration is made by—
   a. the electrons carried by NADH
   b. the movement of hydrogen ions through an ATP synthase enzyme
   c. substrate-level phosphorylation
   d. autophosphorylation
7. What is the role of NAD^+ in the process of cellular respiration?
   a. It functions as an electron carrier.
   b. It functions as an enzyme.
   c. It is the final electron acceptor for anaerobic respiration.
   d. It is a nucleotide source for the synthesis of ATP.
8. Which of the following is NOT a product of glycolysis?
   a. ATP
   b. Pyruvate
   c. CO_2
   d. NADH
9. Why is fermentation an important metabolic function in cells?
   a. It generates glucose for the cell in the absence of O_2.
   b. It oxidizes NADH to NAD^+.
   c. It oxidizes pyruvate.
   d. It produces ATP.
10. Which of the following statements is NOT true about the oxidation of pyruvate?
    a. Pyruvate oxidation occurs in the cytoplasm.
    b. Pyruvate oxidation only occurs if oxygen is present.
    c. Pyruvate is converted into acetyl-CoA.
    d. Pyruvate oxidation results in the production of NADH.
11. The Krebs cycle occurs in which region of a mitochondrion?
    a. The inner membrane
    b. The intermembrane space
    c. The outer membrane
    d. The matrix
12. What happens to the electrons carried by NADH and $FADH_2$?
    a. They are pumped into the intermembrane space.
    b. They are transferred to the ATP synthase.
    c. They are moved between proteins in the inner membrane of the mitochondrion.
    d. They are transported into the matrix of the mitochondrion.
13. Can cellular respiration occur in the absence of O_2?
    a. No, O_2 is required as the final electron acceptor.
    b. No, anaerobic organisms only need glycolysis and fermentation.
    c. Yes, because oxygen can be generated by splitting H_2O.
    d. Yes, but only when another final electron acceptor is available.

CHALLENGE QUESTIONS

1. Use the following table to outline the relationship between the molecules and the metabolic reactions.

Molecules	Glycolysis	Cellular Respiration
Glucose		
Pyruvate		
Oxygen		
ATP		
CO_2		

2. Human babies and hibernating or cold-adapted animals are able to maintain body temperature (a process called *thermogenesis*) due to the presence of brown fat. Brown fat is characterized by a high concentration of mitochondria. These brown fat mitochondria have a special protein located within their inner membranes. *Thermogenin* is a protein that functions as a passive proton transporter. Propose a likely explanation for the role of brown fat in thermogenesis based on your knowledge of metabolism, transport, and the structure and function of mitochondria.

ARIS™ **Do you need additional review?** *Visit* www.ravenbiology.com *for practice quizzes, animations, videos, and activities designed to help you master the material in this chapter.*

chapter 8

Photosynthesis

introduction

THE RICH DIVERSITY OF LIFE that covers our Earth would be impossible without photosynthesis. Almost every oxygen atom in the air we breathe was once part of a water molecule, liberated by photosynthesis. All the energy released by the burning of coal, firewood, gasoline, and natural gas, and by our bodies' burning of all the food we eat—directly or indirectly—has been captured from sunlight by photosynthesis. It is vitally important, then, that we understand photosynthesis. Research may enable us to improve crop yields and land use, important goals in an increasingly crowded world. In chapter 7, we described how cells extract chemical energy from food molecules and use that energy to power their activities. In this chapter, we examine photosynthesis, the process by which organisms such as the aptly named sunflowers in the picture capture energy from sunlight and use it to build food molecules that are rich in chemical energy.

concept outline

8.1 Overview of Photosynthesis

Life is powered by sunshine. The energy used by most living cells comes ultimately from the Sun, captured by plants, algae, and bacteria through the process of photosynthesis.

The diversity of life is only possible because our planet is awash in energy streaming Earthward from the Sun. Each day, the radiant energy that reaches Earth equals about 1 million Hiroshima-sized atomic bombs. Photosynthesis captures about 1% of this huge supply of energy (an amount equal to 10,000 Hiroshima bombs) and uses it to provide the energy that drives all life.

Photosynthesis combines CO_2 and H_2O, producing glucose and O_2

Photosynthesis occurs in a wide variety of organisms, and it comes in different forms. These include a form of photosynthesis that does not produce oxygen (anoxygenic) and a form that does (oxygenic). Anoxygenic photosynthesis is found in four different bacterial groups: purple bacteria, green sulfur bacteria, green nonsulfur bacteria, and heliobacteria. Oxygenic photosynthesis is found in cyanobacteria, seven groups of algae, and essentially all land plants. These two types of photosynthesis share similarities in the types of pigments used to trap light energy, but they differ in the arrangement and action of these pigments.

In the case of plants, photosynthesis takes place primarily in the leaves. Figure 8.1 illustrates the levels of organization in a plant leaf. As you learned in chapter 4, the cells of plant leaves contain organelles called chloroplasts,

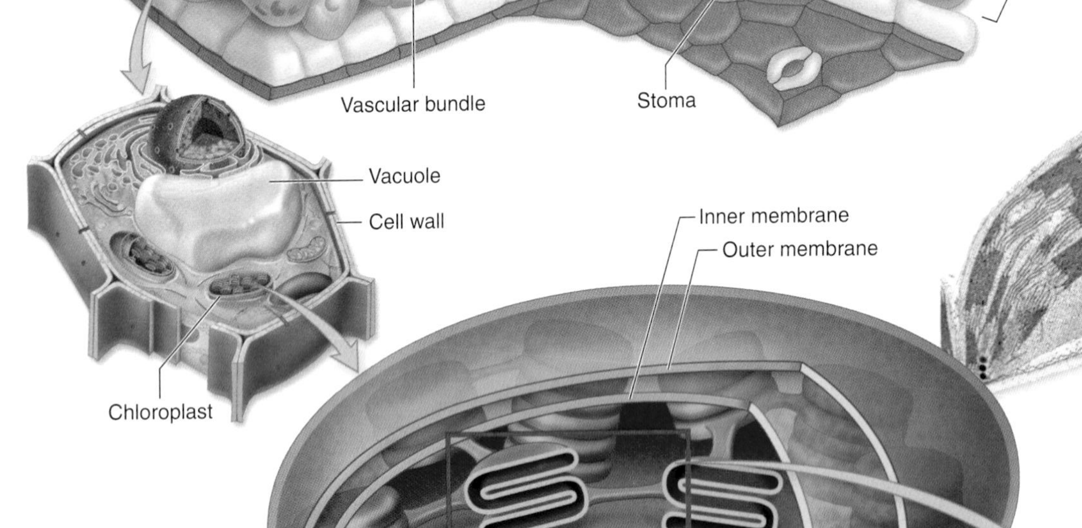

figure 8.1

JOURNEY INTO A LEAF. A plant leaf possesses a thick layer of cells (the mesophyll) rich in chloroplasts. The chloroplast consists of two membranes: an outer membrane enclosing the entire organelle and an inner one organized into flattened structures called thylakoid disks. The flattened thylakoids in the chloroplast are stacked into columns called grana. The rest of the interior is filled with a semifluid substance called stroma.

which carry out the photosynthetic process. No other structure in a plant cell is able to carry out photosynthesis (figure 8.2). Photosynthesis takes place in three stages:

1. capturing energy from sunlight;
2. using the energy to make ATP and to reduce the compound $NADP^+$, an electron carrier, to NADPH; and
3. using the ATP and NADPH to power the synthesis of organic molecules from CO_2 in the air.

The first two stages require light and are commonly called the **light-dependent reactions.**

The third stage, the formation of organic molecules from CO_2, is called **carbon fixation,** and this process takes place via a cyclic series of reactions. As long as ATP and NADPH are available, the carbon fixation reactions can occur either in the presence or in the absence of light, and so these reactions are also called the **light-independent reactions.**

The following simple equation summarizes the overall process of photosynthesis:

$$\underset{\text{carbon dioxide}}{6\ CO_2} + \underset{\text{water}}{12\ H_2O} + \text{light} \longrightarrow \underset{\text{glucose}}{C_6H_{12}O_6} + \underset{\text{water}}{6\ H_2O} + \underset{\text{oxygen}}{6\ O_2}$$

You may notice that this equation is the reverse of the reaction for respiration. In respiration, glucose is oxidized to CO_2 using O_2 as an electron acceptor. In photosynthesis, CO_2 is reduced to glucose using electrons gained from the oxidation of water. The oxidation of H_2O and the reduction of CO_2 requires energy that is provided by light. Although this statement is an oversimplification, it provides a useful "global perspective."

In plants, photosynthesis takes place in chloroplasts

In the preceding chapter, you saw that a mitochondrion's complex structure of internal and external membranes contribute to its function. The same is true for the structure of the chloroplast.

The internal membrane of chloroplasts, called the **thylakoid membrane,** is a continuous phospholipid bilayer organized into flattened sacs that are found stacked on one another in columns called **grana** (singular, *granum*). The thylakoid membrane contains **chlorophyll** and other photosynthetic pigments for capturing light energy along with the machinery to make ATP. Connections between grana are termed *stroma lamella.*

Surrounding the thylakoid membrane system is a semiliquid substance called **stroma.** The stroma houses the enzymes needed to assemble organic molecules from CO_2 using energy from ATP coupled with reduction via NADPH. In the thylakoid membrane, photosynthetic pigments are clustered together to form **photosystems,** which show distinct organization within the thylakoid.

Each pigment molecule within the photosystem is capable of capturing photons, which are packets of energy. When light of a proper wavelength strikes a pigment molecule in the photosystem, the resulting excitation passes from one pigment molecule to another.

The excited electron is not transferred physically—rather, its *energy* passes from one molecule to another. The passage is similar to the transfer of kinetic energy along a row of upright dominoes. If you push the first one over, it falls against the next, and that one against the next, and so on, until all of the dominoes have fallen down.

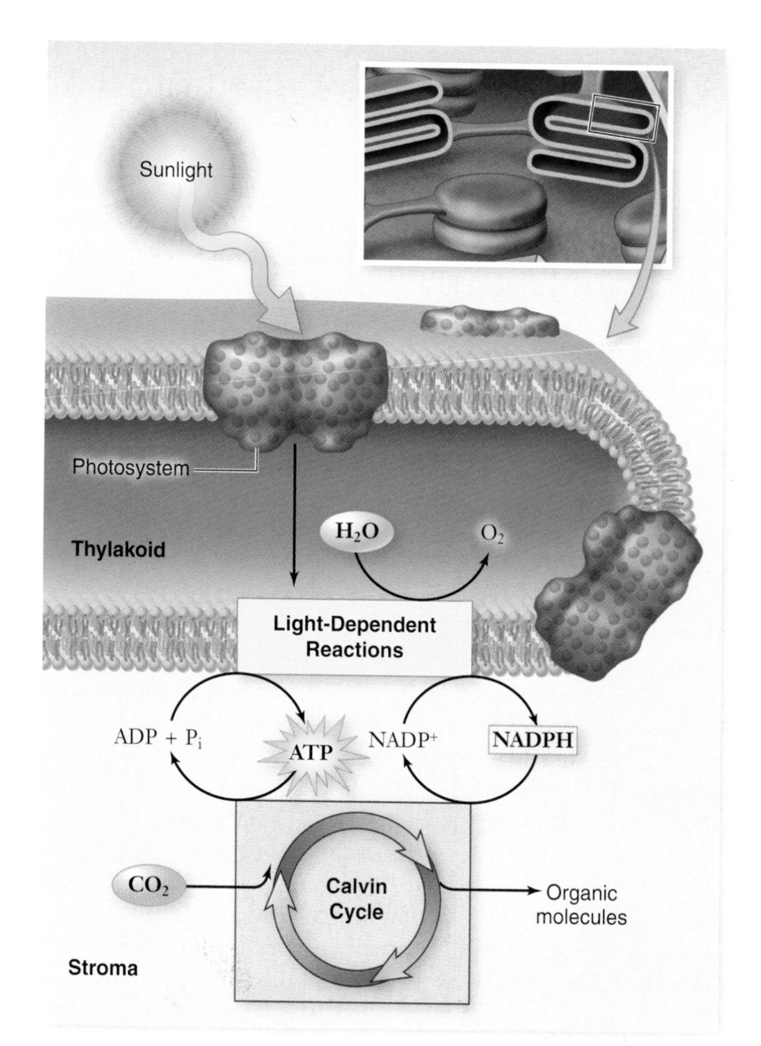

figure 8.2

The light-dependent reactions take place on the thylakoid membrane where photosystems absorb photons of light and use this energy to generate ATP and NADPH. Electrons lost from the photosystems are replaced by the oxidation of water, which produces O_2 as a by-product. The ATP and NADPH produced by the light reactions is used to fuel carbon fixation via the Calvin cycle. The stroma contains enzymes that carry out the Calvin cycle.

Eventually, the energy arrives at a key chlorophyll molecule that is in contact with a membrane-bound protein. The energy is transferred as an excited electron to that protein, which passes it on to a series of other membrane proteins that put the energy to work making ATP and NADPH. These compounds are then used to build organic molecules. The photosystem thus acts as a large antenna, gathering the light energy harvested by many individual pigment molecules.

Photosynthesis converts light energy into organic molecules. This process uses reactions, called light-dependent reactions, that require sunlight and others, that convert CO_2 into organic molecules in a process called carbon fixation. The overall reaction is essentially the reverse of respiration. The energy of sunlight is used to oxidize water with the electrons and protons used to reduce CO_2 to glucose. This also produces O_2 as a by-product.

8.2 The Discovery of Photosynthetic Processes

The story of how we learned about photosynthesis begins over 300 years ago, and it continues to this day. It starts with curiosity about how plants manage to grow, often increasing their organic mass considerably.

Plants do not increase mass from soil and water alone

From the time of the Greeks, plants were thought to obtain their food from the soil, literally sucking it up with their roots. A Belgian doctor, Jan Baptista van Helmont (1580–1644) thought of a simple way to test this idea.

He planted a small willow tree in a pot of soil, after first weighing the tree and the soil. The tree grew in the pot for several years, during which time van Helmont added only water. At the end of five years, the tree was much larger, its weight having increased by 74.4 kg. However, the soil in the pot weighed only 57 g less than it had five years earlier. With this experiment, van Helmont demonstrated that the substance of the plant was not produced only from the soil. He incorrectly concluded, however, that the water he had been adding mainly accounted for the plant's increased mass.

A hundred years passed before the story became clearer. The key clue was provided by the English scientist Joseph Priestly (1733–1804). On the 17th of August, 1771, Priestly put a living sprig of mint into air in which a wax candle had burnt out. On the 27th of the same month, Priestly found that another candle could be burned in this same air. Somehow, the vegetation seemed to have restored the air. Priestly found that while a mouse could not breathe candle-exhausted air, air "restored" by vegetation was not "at all inconvenient to a mouse." The key clue was that *living vegetation adds something to the air.*

How does vegetation "restore" air? Twenty-five years later, the Dutch physician Jan Ingen-Housz (1730–1799) solved the puzzle. He demonstrated that air was restored only in the presence of sunlight and only by a plant's green leaves, not by its roots. He proposed that the green parts of the plant carry out a process that uses sunlight to split carbon dioxide into carbon and oxygen. He suggested that the oxygen was released as O_2 gas into the air, while the carbon atom combined with water to form carbohydrates. Other research refined his conclusions, and by the end of the nineteenth century, the overall reaction for photosynthesis could be written as:

$$CO_2 + H_2O + \text{light energy} \longrightarrow (CH_2O) + O_2$$

It turns out, however, that there's more to it than that. When researchers began to examine the process in more detail in the twentieth century, the role of light proved to be unexpectedly complex.

Photosynthesis includes both light-dependent and light-independent reactions

At the beginning of the twentieth century, the English plant physiologist F. F. Blackman (1866–1947) came to the startling conclusion that photosynthesis is in fact a multistage process, only one portion of which uses light directly.

Blackman measured the effects of different light intensities, CO_2 concentrations, and temperatures on photosynthesis. As long as light intensity was relatively low, he found photosynthesis could be accelerated by increasing the amount of light, but not by increasing the temperature or CO_2 concentration (figure 8.3). At high light intensities, however, an increase in temperature or CO_2 concentration greatly accelerated photosynthesis.

Blackman concluded that photosynthesis consists of an initial set of what he called "light" reactions, that are largely independent of temperature but depend on light, and a second set of "dark" reactions (more properly called light-independent reactions), that seemed to be independent of light but limited by CO_2.

Do not be confused by Blackman's labels—the so-called "dark" reactions occur in the light (in fact, they require the products of the light-dependent reactions); his use of the word *dark* simply indicates that light is not *directly* involved in those reactions.

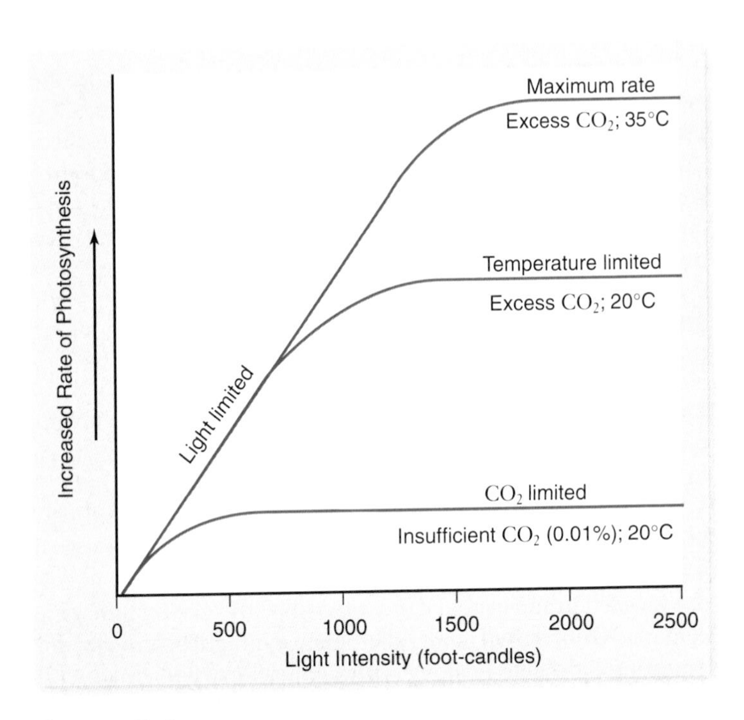

figure 8.3

DISCOVERY OF THE LIGHT-INDEPENDENT REACTIONS. Blackman measured photosynthesis rates under differing light intensities, CO_2 concentrations, and temperatures. As this graph shows, light is the limiting factor at low light intensities, but temperature and CO_2 concentration are the limiting factors at higher light intensities. This implies the existence of reactions using CO_2 that involve enzymes.

inquiry

Blackman found that increasing light intensity above 2000 foot-candles did not lead to any further increase in the rate of photosynthesis. Can you suggest a hypothesis that would explain this?

Blackman found that increased temperature increased the rate of the light-independent reactions, but only up to about 35°C. Higher temperatures caused the rate to fall off rapidly. Because many plant enzymes begin to be denatured at 35°C, Blackman concluded that enzymes must carry out the light-independent reactions.

O_2 comes from water, not from CO_2

In the 1930s, C. B. van Niel (1897–1985), then a graduate student at Stanford, discovered that purple sulfur bacteria do not release oxygen during photosynthesis; instead, they convert hydrogen sulfide (H_2S) into globules of pure elemental sulfur that accumulate inside them. The process van Niel observed was:

$$CO_2 + 2\ H_2S + \text{light energy} \longrightarrow (CH_2O) + H_2O + 2\ S$$

The striking parallel between this equation and Ingenhousz's equation led van Niel to propose that the generalized process of photosynthesis can be shown as:

$$CO_2 + 2\ H_2A + \text{light energy} \longrightarrow (CH_2O) + H_2O + 2\ A$$

In this equation, the substance H_2A serves as an electron donor. In photosynthesis performed by green plants, H_2A is water, whereas in purple sulfur bacteria, H_2A is hydrogen sulfide. The product, A, comes from the splitting of H_2A. Therefore, the O_2 produced during green plant photosynthesis results from splitting water, not carbon dioxide.

When isotopes came into common use in the early 1950s, van Niel's revolutionary proposal was tested. Investigators examined photosynthesis in green plants supplied with water containing radioactive oxygen (^{18}O); they found that the ^{18}O label ended up in oxygen gas rather than in carbohydrate, just as van Niel had predicted:

$$CO_2 + 2\ H_2{}^{18}O + \text{light energy} \longrightarrow (CH_2O) + H_2O + {}^{18}O_2$$

In algae and green plants, the carbohydrate typically produced by photosynthesis is glucose, which has six carbons. The complete balanced equation for photosynthesis in these organisms thus becomes:

$$6\ CO_2 + 12\ H_2O + \text{light energy} \longrightarrow C_6H_{12}O_6 + 6\ H_2O + 6\ O_2$$

ATP and NADPH from light-dependent reactions reduce CO_2 to make glucose

In his pioneering work on the light-dependent reactions, van Niel had further proposed that the H^+ ions and electrons generated by the splitting of water were used to convert CO_2 into organic matter in a process he called *carbon fixation*. In the 1950s, researcher Robin Hill (1899–1991) demonstrated that van Niel was indeed right, and that light energy could be harvested and used in a reduction reaction. Chloroplasts isolated from leaf cells were able to reduce a dye and release oxygen in response to light. Later experiments showed that the electrons released from water were transferred to $NADP^+$, and that illuminated chloroplasts deprived of CO_2 accumulate ATP. If CO_2 is then introduced, neither ATP nor NADPH accumulate, and the CO_2 is assimilated into organic molecules.

These experiments are important for three reasons: First, they firmly demonstrate that photosynthesis in plants occurs within chloroplasts. Second, they show that the light-dependent reactions use light energy to reduce $NADP^+$ and to manufacture ATP. Third, they confirm that the ATP and NADPH from this early stage of photosynthesis are then used in the subsequent reactions to reduce carbon dioxide, forming simple sugars.

Knowledge of photosynthesis has accumulated over many years of study. Early experiments indicated that plants do not grow entirely based on nutrients from soil.

These experiments led to the recognition of two types of reactions that produce organic material. The light-dependent reactions produce O_2 from H_2O, and generate ATP and NADPH that can be used in a second set of reactions to synthesize organic compounds.

Pigments

For plants to utilize the energy of sunlight, some biochemical structure must be present in chloroplasts and the thylakoids that can absorb this energy. Molecules that absorb light energy in the visible range are termed **pigments,** and we are most familiar with them as dyes that impart a certain color to clothing or other materials. The color that we see is the color that is not absorbed—that is, it is reflected. To understand how plants use pigments to capture light energy, we must first review current knowledge about the nature of light.

Light is a form of energy

The wave nature of light produces an electromagnetic spectrum that differentiates light based on its wavelength (figure 8.4). We are most familiar with the visible range of this spectrum because we can actually see it, but visible light is only a small part of the entire spectrum. Visible light can be divided into its separate colors by the use of a prism, which separates light based on wavelength.

A particle of light, termed a **photon,** acts like a discrete bundle of energy. We use the wave concept of light to understand different colors of light and the particle nature of light to understand the energy transfers that occur during photosynthesis. Thus, we will refer both to wavelengths of light and to photons of light throughout the chapter.

The energy in photons

The energy content of a photon is inversely proportional to the wavelength of the light: Short-wavelength light contains photons of higher energy than long-wavelength light (see figure 8.4). X-rays, which contain a great deal of energy,

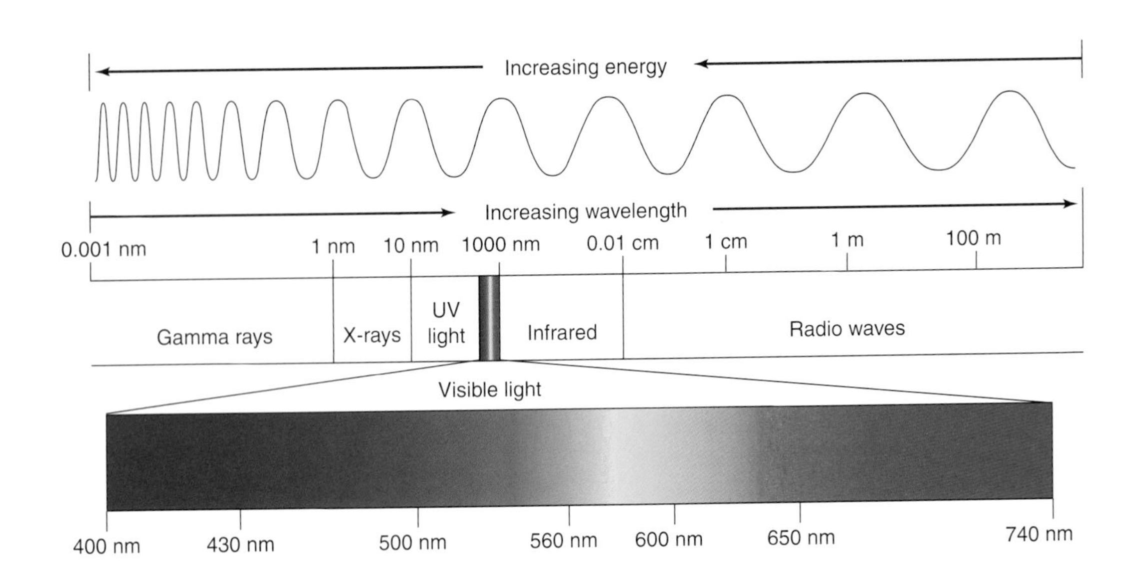

figure 8.4

THE ELECTROMAGNETIC SPECTRUM. Light is a form of electromagnetic energy conveniently thought of as a wave. The shorter the wavelength of light, the greater its energy. Visible light represents only a small part of the electromagnetic spectrum between 400 and 740 nm.

have very short wavelengths—much shorter than those of visible light.

A beam of light is able to remove electrons from certain molecules, creating an electrical current. This phenomenon is called the **photoelectric effect,** and it occurs when photons transfer energy to electrons. The strength of the photoelectric effect depends on the wavelength of light; that is, short wavelengths are much more effective than long ones in producing the photoelectric effect because they have more energy.

In photosynthesis, chloroplasts are acting as photoelectric devices: They absorb sunlight and transfer the excited electrons to a carrier. As we unravel the details of this process, it will become clear how this process traps energy and uses it to synthesize organic compounds.

Each pigment has a characteristic absorption spectrum

When a photon strikes a molecule, its energy is either lost as heat or absorbed by the electrons of the molecule, boosting those electrons into higher energy levels. Whether the photon's energy is absorbed depends on how much energy it carries (defined by its wavelength), and also on the chemical nature of the molecule it hits.

As described in chapter 2, electrons occupy discrete energy levels in their orbits around atomic nuclei. To boost an electron into a different energy level requires just the right amount of energy, just as reaching the next rung on a ladder requires you to raise your foot just the right distance. A specific atom, therefore, can absorb only certain photons of light—namely, those that correspond to the atom's available energy levels. As a result, each molecule has a characteristic

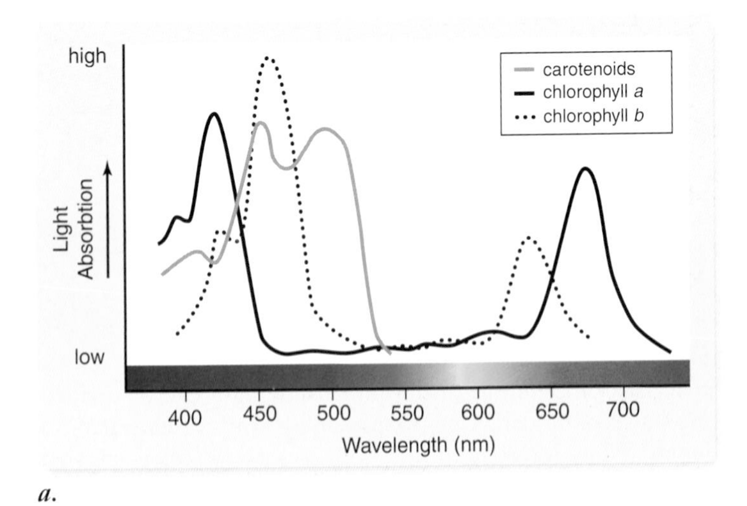

a.

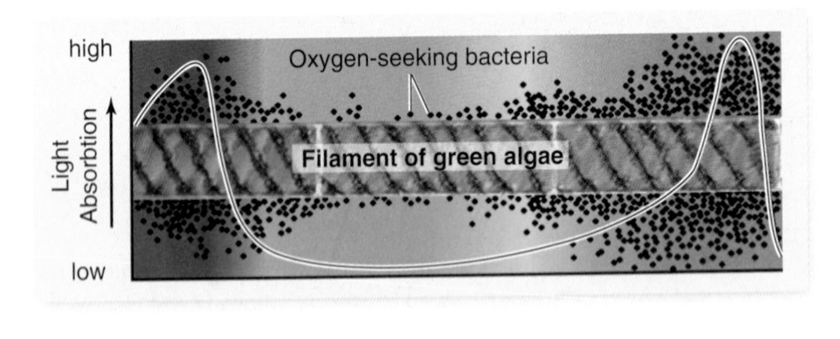

b.

figure 8.5

RELATIONSHIP BETWEEN CHLOROPHYLL ABSORPTION SPECTRUM AND ACTION SPECTRUM FOR PHOTOSYNTHESIS. ***a.*** The peaks represent wavelengths of light of sunlight absorbed by the two common forms of photosynthetic pigment, chlorophylls *a* and *b*, and the carotenoids. Chlorophylls absorb predominantly violet-blue and red light in two narrow bands of the spectrum and reflect green light in the middle of the spectrum. Carotenoids absorb mostly blue and green light and reflect orange and yellow light. ***b.*** A simple experiment allows the construction of an action spectrum for photosynthesis showing what wavelengths of light are maximally absorbed by the process. A filament of green algal cells on a slide is illuminated with light split by a prism. Wavelengths that are used for photosynthesis lead to the production of oxygen by the algal cells. Oxygen production is indicated by the presence on the slide of oxygen-seeking bacteria. The bacteria move to regions of the slide with the most oxygen, corresponding to the most photosynthetically active areas. The bacteria accumulate in the areas where chlorophyll absorbs maximally.

absorption spectrum, the range and efficiency of photons it is capable of absorbing.

As mentioned earlier, pigments are molecules that are good absorbers of light in the visible range. Organisms have evolved a variety of different pigments, but only two general types are used in green plant photosynthesis: chlorophylls and carotenoids. In some organisms, other molecules also absorb light energy.

Chlorophyll absorption spectra

Chlorophylls absorb photons within narrow energy ranges. Two kinds of chlorophyll in plants, chlorophyll *a* and chlorophyll *b*, preferentially absorb violet-blue and red light (figure 8.5*a*). Neither of these pigments absorbs photons with wavelengths between about 500 and 600 nm; light of these wavelengths is reflected. When these reflected photons are subsequently absorbed by the retinal pigment in our eyes, we perceive them as green.

Chlorophyll *a* is the main photosynthetic pigment in plants and cyanobacteria and the only pigment that can act directly to convert light energy to chemical energy. **Chlorophyll *b*,** acting as an **accessory pigment,** or secondary light-absorbing pigment, complements and adds to the light absorption of chlorophyll *a*.

Chlorophyll *b* has an absorption spectrum shifted toward the green wavelengths. Therefore, chlorophyll *b* can absorb photons that chlorophyll *a* cannot, greatly increasing the proportion of the photons in sunlight that plants can harvest. In addition, a variety of different accessory pigments are found in plants, bacteria, and algae.

Structure of chlorophylls

Chlorophylls absorb photons by means of an excitation process analogous to the photoelectric effect. These pigments contain a complex ring structure, called a *porphyrin ring*, with alternating single and double bonds. At the center of the ring is a magnesium atom (figure 8.6).

Photons excite electrons in the porphyrin ring, which are then channeled away through the alternating carbon single- and double-bond system. Different small side groups attached to the outside of the ring alter the absorption properties of the molecule in the different kinds of chlorophyll (see figure 8.6). The precise absorption spectrum is also influenced by the local microenvironment created by the association of chlorophyll with different proteins.

The **action spectrum** of photosynthesis—that is, the relative effectiveness of different wavelengths of light in promoting photosynthesis—corresponds to the absorption spectrum for chlorophylls. This is demonstrated in the experiment in figure 8.5*b*. All plants, algae, and cyanobacteria use chlorophyll *a* as their primary pigments.

It is reasonable to ask why these photosynthetic organisms do not use a pigment like retinal (the pigment in our eyes), which has a broad absorption spectrum that covers the range of 500 to 600 nm. The most likely hypothesis involves *photoefficiency*. Although retinal absorbs a broad range of wavelengths, it does so with relatively low efficiency. Chlorophyll, in contrast, absorbs in only two narrow bands, but does so with high efficiency. Therefore, plants and most other photosynthetic organisms achieve far higher overall energy capture rates with chlorophyll than with other pigments.

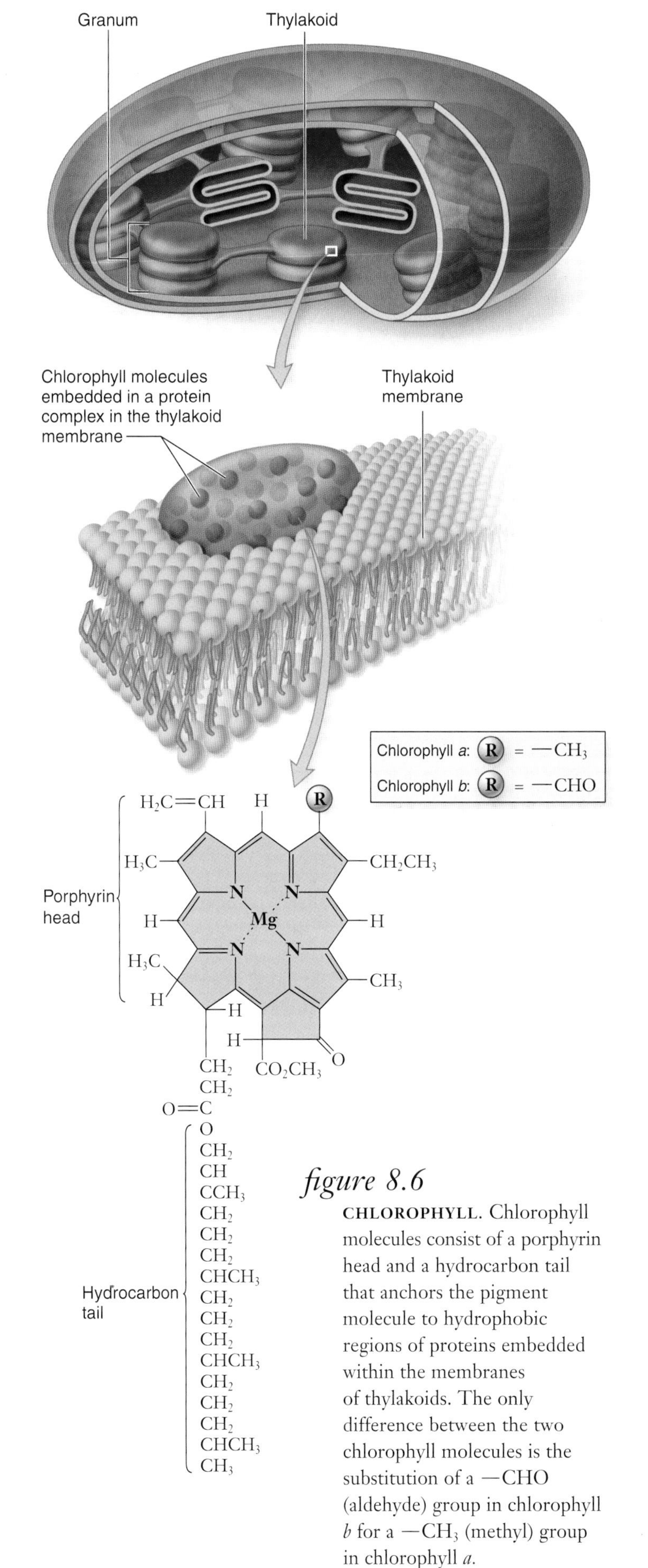

figure 8.6

CHLOROPHYLL. Chlorophyll molecules consist of a porphyrin head and a hydrocarbon tail that anchors the pigment molecule to hydrophobic regions of proteins embedded within the membranes of thylakoids. The only difference between the two chlorophyll molecules is the substitution of a —CHO (aldehyde) group in chlorophyll *b* for a $-CH_3$ (methyl) group in chlorophyll *a*.

figure 8.7

FALL COLORS ARE PRODUCED BY CAROTENOIDS AND OTHER ACCESSORY PIGMENTS. During the spring and summer, chlorophyll in leaves masks the presence of carotenoids and other accessory pigments. When cool fall temperatures cause leaves to cease manufacturing chlorophyll, the chlorophyll is no longer present to reflect green light, and the leaves reflect the orange and yellow light that carotenoids and other pigments do not absorb.

Carotenoids and other accessory pigments

Carotenoids consist of carbon rings linked to chains with alternating single and double bonds. They can absorb photons with a wide range of energies, although they are not always highly efficient in transferring this energy. Carotenoids assist in photosynthesis by capturing energy from light composed of wavelengths that are not efficiently absorbed by chlorophylls (figure 8.7; see also figure 8.5).

Carotenoids also perform a valuable role in scavenging free radicals. The oxidation–reduction reactions that occur in the chloroplast can generate destructive free radicals. Carotenoids can act as general-purpose antioxidants to lessen damage. Thus carotenoids have a protective role in addition to their role as light-absorbing molecules. This protective role is not surprising, because unlike the chlorophylls, carotenoids are found in many different kinds of organisms, including members of all three domains of life.

A typical carotenoid is β-carotene, which contains two carbon rings connected by a chain of 18 carbon atoms with alternating single and double bonds. Splitting a molecule of β-carotene into equal halves produces two molecules of vitamin A. Oxidation of vitamin A produces retinal, the pigment used in vertebrate vision. This connection explains why eating carrots, which are rich in β-carotene, may enhance vision.

Phycobiloproteins are accessory pigments found in cyanobacteria and some algae. These pigments are composed of proteins attached to a tetrapyrrole group. These pyrrole rings contain a system of alternating double bonds similar to those found in other pigments and molecules that transfer electrons. Phycobiloproteins can be organized into complexes called phycobilisomes to form another light-harvesting complex that can absorb green light, which is typically reflected by chlorophyll. These complexes are probably ecologically important to cyanobacteria, helping them to exist in low-light situations in oceans. In this habitat, green light remains because red and blue light has been absorbed by green algae closer to the surface.

Light is a form of energy that can behave as both a wave and a particle. Photosynthesis depends on molecules that can absorb light energy: pigments. The main photosynthetic pigment is chlorophyll, which exists in several forms. Pigment molecules are characterized by their absorption spectrum, the wavelengths of light absorbed most efficiently. Efficiency of light capture can also be increased by accessory pigments that have absorption spectra different from chlorophyll.

8.4 Photosystem Organization

One way to study the role that pigments play in photosynthesis is to measure the correlation between the output of photosynthesis and the intensity of illumination—that is, how much photosynthesis is produced by how much light. Experiments on plants show that the output of photosynthesis increases linearly at low light intensities, but finally becomes saturated (no further increase) at high-intensity light. Saturation occurs because all of the light-absorbing capacity of the plant is in use.

Production of one O_2 molecule requires many chlorophyll molecules

Given the saturation observed with increasing light intensity, the next question is how many chlorophyll molecules have actually absorbed a photon. The question can be phrased this way: "Does saturation occur when all chlorophyll molecules have absorbed photons?" Finding an answer required being able to measure both photosynthetic output (on the basis of O_2 production) and the number of chlorophyll molecules present. Using the unicellular algae *Chlorella*, investigators could obtain these values. Illuminating a *Chlorella* culture with pulses of light with increasing intensity should increase the yield of O_2 per pulse until the system becomes saturated. Then O_2 production can be compared with the number of chlorophyll molecules present in the culture.

The observed level of O_2 per chlorophyll molecule at saturation, however, turned out to be only one molecule of O_2 per 2500 chlorophyll molecules (figure 8.8). This result was very different from what was expected, and it led to the idea that light is absorbed not by independent pigment molecules, but rather by clusters of chlorophyll and accessory pigment molecules (photosystems). Light is absorbed by any one of hundreds of pigment molecules in a photosystem, and each pigment molecule transfers its excitation energy to a single molecule with a lower energy level than the others.

A generalized photosystem contains an antenna complex and a reaction center

In chloroplasts and all but one class of photosynthetic prokaryotes, light is captured by photosystems. Each photosystem is a network of chlorophyll *a* molecules, accessory pigments, and associated proteins held within a protein matrix on the surface of the photosynthetic membrane. Like a magnifying glass focusing light on a precise point, a photosystem channels the excitation energy gathered by any one of its pigment molecules to a specific molecule, the reaction center chlorophyll. This molecule then passes the energy out of the photosystem as excited electrons that are put to work driving the synthesis of ATP and organic molecules.

A photosystem thus consists of two closely linked components: (1) an *antenna complex* of hundreds of pigment molecules that gather photons and feed the captured light energy to the reaction center; and (2) a *reaction center* consisting of one or more chlorophyll *a* molecules in a matrix of protein, that passes excited electrons out of the photosystem.

The antenna complex

The **antenna complex** is also called a light-harvesting complex, which accurately describes its role. This light-harvesting complex captures photons from sunlight (figure 8.9) and channels them to the reaction center chlorophylls.

In chloroplasts, light-harvesting complexes consist of a web of chlorophyll molecules linked together and held tightly in the thylakoid membrane by a matrix of proteins. Varying amounts of carotenoid accessory pigments may also be present. The protein matrix holds individual pigment molecules in orientations that are optimal for energy transfer.

The excitation energy resulting from the absorption of a photon passes from one pigment molecule to an adjacent molecule on its way to the reaction center. After the transfer, the excited electron in each molecule returns to the low-energy level it had before the photon was absorbed. Consequently, it is energy, not the excited electrons themselves, that passes from

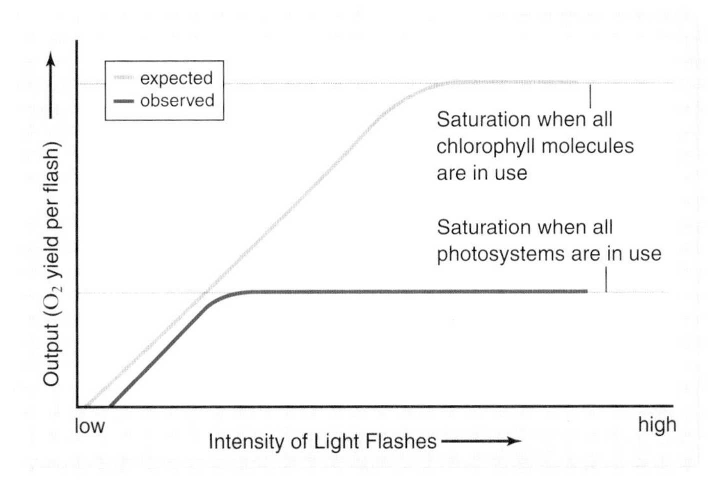

figure 8.8

SATURATION OF PHOTOSYNTHESIS. When photosynthetic saturation is achieved, further increases in intensity cause no increase in output. This saturation occurs far below the level expected for the number of individual chlorophyll molecules present. This led to the idea of organized photosystems each containing many chlorophyll molecules. These photosystems saturate at a lower O_2 yield than that expected for the number of individual chlorophyll molecules.

inquiry

Under what experimental conditions would you expect the saturation levels for a given number of chlorophyll molecules to be higher?

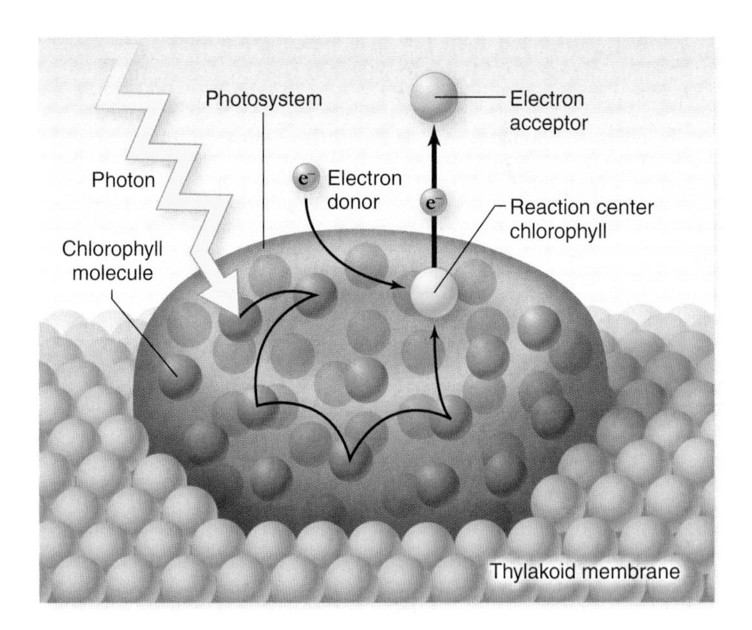

figure 8.9

HOW THE ANTENNA COMPLEX WORKS. When light of the proper wavelength strikes any pigment molecule within a photosystem, the light is absorbed by that pigment molecule. The excitation energy is then transferred from one molecule to another within the cluster of pigment molecules until it encounters the reaction center chlorophyll *a*. When excitation energy reaches the reaction center chlorophyll, electron transfer is initiated.

one pigment molecule to the next. The antenna complex funnels the energy from many electrons to the reaction center.

The reaction center

The **reaction center** is a transmembrane protein–pigment complex. In the reaction center of purple photosynthetic bacteria, which is simpler than in chloroplasts but is better understood, a pair of bacteriochlorophyll *a* molecules acts as a trap for photon energy, passing an excited electron to an acceptor precisely positioned as its neighbor. Note that here in the reaction center, the excited electron itself is transferred, and not just the energy, as was the case in the pigment–pigment transfers of the antenna complex. This difference allows the energy absorbed from photons to move away from the chlorophylls, and it is the key conversion of light into chemical energy.

Figure 8.10 shows the transfer of excited electrons from the reaction center to the primary electron acceptor. By energizing an electron of the reaction center chlorophyll, light creates a strong electron donor where none existed before. The chlorophyll transfers the energized electron to the primary acceptor, a molecule of quinone, reducing the quinone and converting it to a strong electron donor. A nearby weak electron donor then passes a low-energy electron to the chlorophyll, restoring it to its original condition. The quinone transfers its electrons to another acceptor and the process is repeated.

In plant chloroplasts, water serves as this weak electron donor. When water is oxidized in this way, oxygen is released along with two protons (H^+).

Chlorophylls and accessory pigments are organized into photosystems found in the thylakoid membrane of the chloroplast. The photosystem can be subdivided into an antenna complex, which is involved in light harvesting, and a reaction center, where the photochemical reactions occur. The reaction center uses the light energy to excite an electron that can then be transferred to an electron acceptor.

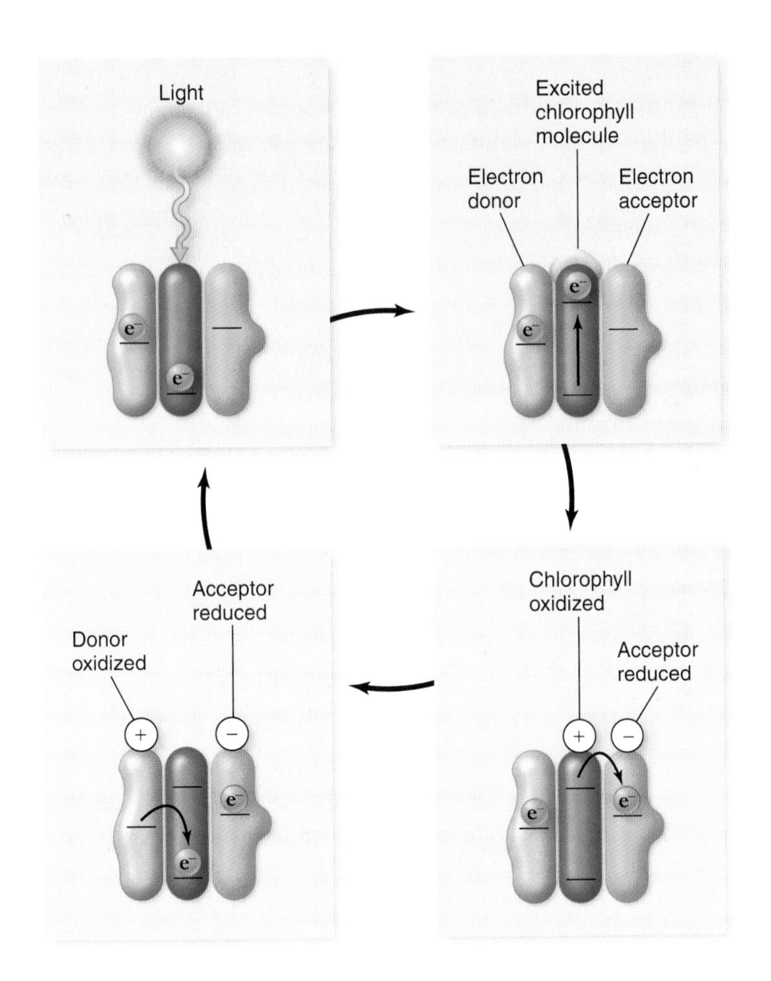

figure 8.10

CONVERTING LIGHT TO CHEMICAL ENERGY. When a chlorophyll in the reaction center absorbs a photon of light, an electron is excited to a higher energy level. This light-energized electron can be transferred to the primary electron acceptor, reducing it. The oxidized chlorophyll then fills its electron "hole" by oxidizing a donor molecule. The source of this donor varies with the photosystem as discussed in the text.

8.5 The Light-Dependent Reactions

As you have seen, the light-dependent reactions of photosynthesis occur in membranes. In photosynthetic bacteria, the plasma membrane itself is the photosynthetic membrane. In many bacteria, the plasma membrane is highly infolded to produce an increased surface area. In plants and algae, photosynthesis is carried out by chloroplasts, which are thought to be the evolutionary descendants of photosynthetic bacteria.

The internal thylakoid membrane is highly organized and contains the structures involved in the light-dependent reactions. For this reason, the reactions are also referred to as the thylakoid reactions. The thylakoid reactions take place in four stages:

1. **Primary photoevent.** A photon of light is captured by a pigment. The result of this primary photoevent is the excitation of an electron within the pigment.
2. **Charge separation.** This excitation energy is transferred to the reaction center, which transfers an energetic electron to an acceptor molecule, initiating electron transport.
3. **Electron transport.** The excited electrons are shuttled along a series of electron carrier molecules embedded within the photosynthetic membrane. Several of them react by transporting protons across the membrane, generating a proton gradient. Eventually the electrons are used to reduce a final acceptor, NADPH.
4. **Chemiosmosis.** The protons that accumulate on one side of the membrane now flow back across the membrane through ATP synthase where chemiosmotic synthesis of ATP takes place, just as it does in aerobic respiration (chapter 7).

These four processes make up the two stages of the light-dependent reactions mentioned at the beginning of this chapter. Steps 1 through 3 represent the stage of capturing energy from light; step 4 is the stage of producing ATP (and, as you'll see, NADPH). In the rest of this section we discuss the evolution of photosystems and the details of photosystem function in the light-dependent reactions.

Some bacteria use a single photosystem

Photosynthetic pigment arrays are thought to have evolved more than 2 BYA in bacteria similar to the sulfur bacteria alive today. A two-stage process takes place within these bacterial photosystems.

1. **An electron is joined with a proton to make hydrogen.** In these bacteria, peak absorption of light having a wavelength of 840 nm (near infrared, not visible to the human eye), and thus the reaction center pigment is called P_{840}. Absorption of a photon results in the transfer of an energetic electron along an electron transport chain. Eventually the electron combines with a proton to form a hydrogen atom. In the sulfur bacteria, the electron is extracted from hydrogen sulfide, leaving elemental sulfur and protons as by-products. In bacteria that evolved later, as well as in plants and algae, the electron comes from water, producing oxygen and protons as by-products.
2. **An electron is recycled to chlorophyll.** The ejection of an electron from the bacterial reaction center leaves it short one electron. Before the photosystem of the sulfur bacteria can function again, an electron must be returned. These bacteria channel the electron back to the pigment through an electron transport system similar to the one described in chapter 7; the electron's passage drives a proton pump that promotes the chemiosmotic synthesis of ATP. One molecule of ATP is produced for every three electrons that follow this path.

Viewed overall (figure 8.11), the path of the electron is a circle. Chemists therefore call the electron transfer process leading to ATP formation **cyclic photophosphorylation.**

Note that the electron that left the P_{840} reaction center was a high-energy electron, boosted by the absorption of a photon of light. In contrast, the electron that returns has only as much energy as it had before the photon was absorbed. The difference in the energy of that electron is the photosynthetic payoff, the energy that drives the proton pump.

For more than a billion years, cyclic photophosphorylation was the only photosynthetic light-dependent reaction. Its major limitation, however, is that it is geared only toward energy production and not toward biosynthesis of carbohydrates.

Because carbohydrate molecules are more reduced (that is, have more hydrogen atoms) than carbon dioxide, a source of hydrogens for reduction must be provided. Cyclic photophosphorylation does not do this. The hydrogen atoms extracted from H_2S are used as a source of protons for driving ATP pumps and are not available to join to carbon. Bacteria that are restricted to this process therefore must scavenge hydrogens from other sources, an inefficient undertaking.

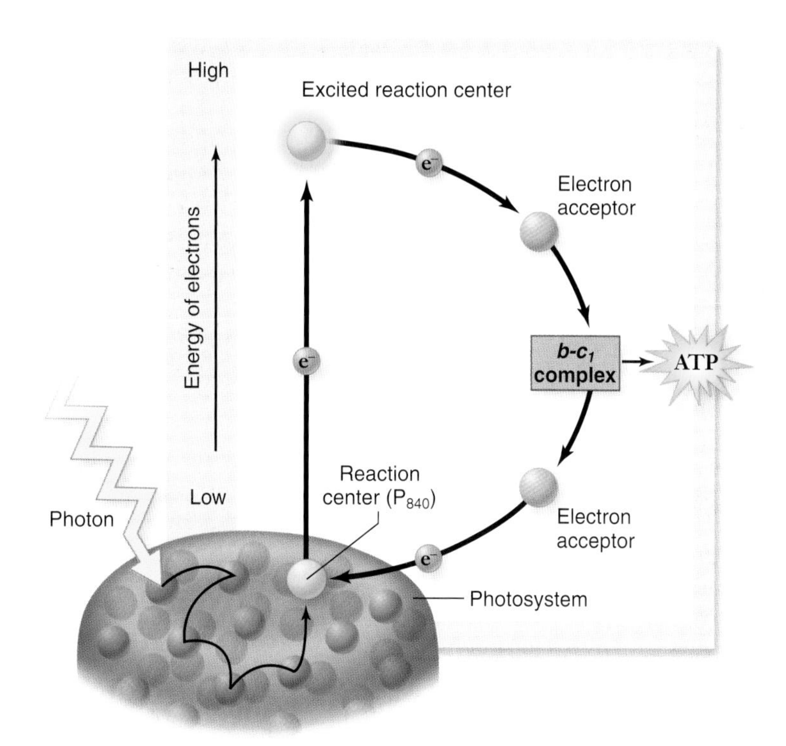

figure 8.11

THE PATH OF AN ELECTRON IN SULFUR BACTERIA. When a light-energized electron is ejected from the photosystem reaction center (P_{840}) it returns to the photosystem via a cyclic path that produces ATP but not NADPH.

Chloroplasts have two connected photosystems

In contrast to the sulfur bacteria, plants have two linked photosystems. This overcomes the limitations of cyclic photophosphorylation by providing an alternative source of electrons from the oxidation of water. The oxidation of water also generates O_2, thus oxygenic photosynthesis. The noncyclic transfer of electrons also produces NADPH, which can be used in the biosynthesis of carbohydrates.

One photosystem, called **photosystem I,** has an absorption peak of 700 nm, so its reaction center pigment is called P_{700}. This photosystem functions in a way analogous to the photosystem found in the sulfur bacteria discussed earlier. The other photosystem, called **photosystem II,** has an absorption peak of 680 nm, so its reaction center pigment is called P_{680}. This photosystem can generate an oxidation potential high enough to oxidize water. Working together, the two photosystems carry out a noncyclic transfer of electrons that is used to generate both ATP and NADPH.

The photosystems were named I and II in the order of their discovery, and not in the order in which they operate in the light-dependent reactions. In plants and algae, the two photosystems are specialized for different roles in the overall process of oxygenic photosynthesis. Photosystem I transfers electrons ultimately to $NADP^+$, producing NADPH. The electrons lost from photosystem I are replaced by electrons from photosystem II. Photosystem II with its high oxidation potential can oxidize water to replace the electrons transferred to photosystem I. Thus there is an overall flow of electrons from water to NADPH.

These two photosystems are connected by a complex of electron carriers called the **cytochrome/b_6-f complex** (explained shortly). This complex can use the energy from the passage of electrons to move protons across the thylakoid membrane to generate the proton gradient used by an ATP synthase enzyme.

The two photosystems work together in noncyclic photophosphorylation

Evidence for the action of two photosystems came from experiments that measured the rate of photosynthesis using two light beams of different wavelengths: one red and the other far-red. Using both beams produced a rate greater than the sum of the rates using individual beams of these wavelengths (figure 8.12). This surprising result, called the **enhancement effect,** can be explained by a mechanism involving two photosystems acting in series (that is, one after the other), one photosystem absorbs preferentially in the red, the other in the far-red.

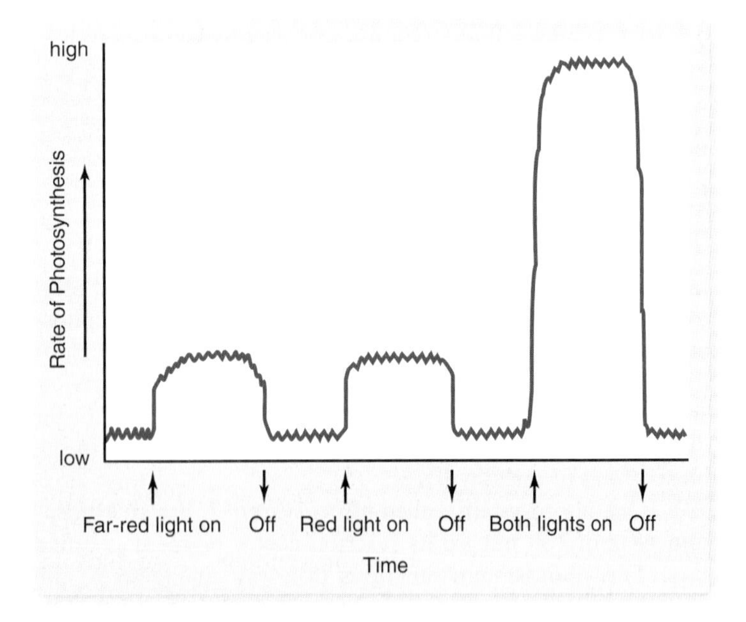

figure 8.12

THE ENHANCEMENT EFFECT. The rate of photosynthesis when red and far-red light are provided together is greater than the sum of the rates when each wavelength is provided individually. This result baffled researchers in the 1950s. Today, it provides key evidence that photosynthesis is carried out by two photochemical systems that act in series. One absorbs maximally in the far red, the other in the red portion of the spectrum.

inquiry

What would you conclude if "both lights on" did not change the relative rate of photosynthesis?

Plants use photosystems II and I in series, first one and then the other, to produce both ATP and NADPH. This two-stage process is called **noncyclic photophosphorylation** because the path of the electrons is not a circle—the electrons ejected from the photosystems do not return to them, but rather end up in NADPH. The photosystems are replenished with electrons obtained by splitting water.

The scheme shown in figure 8.13, called a *Z diagram*, illustrates the two electron-energizing steps, one catalyzed by each photosystem. The horizontal axis shows the progress of the light reactions and the relative positions of the complexes, and the vertical axis shows relative energy levels of electrons. The electrons originate from water, which holds onto its electrons very tightly (redox potential = +820 mV), and end up in NADPH, which holds its electrons much more loosely (redox potential = −320 mV).

Photosystem II acts first. High-energy electrons generated by photosystem II are used to synthesize ATP and are then passed to photosystem I to drive the production of NADPH. For every pair of electrons obtained from a molecule of water, one molecule of NADPH and slightly more than one molecule of ATP are produced.

Photosystem II

The reaction center of photosystem II closely resembles the reaction center of purple bacteria. It consists of a core of 10 transmembrane protein subunits with electron transfer components and two P_{680} chlorophyll molecules arranged around this core. The light-harvesting antenna complex consists of molecules of chlorophyll *a* and accessory pigments bound to several protein chains. The reaction center of photosystem II differs from the reaction center of the purple bacteria in that it also contains four manganese atoms. These manganese atoms are essential for the oxidation of water.

Although the chemical details of the oxidation of water are not entirely clear, the outline is emerging. Four manganese atoms are bound in a cluster to reaction center proteins. Two water molecules are also bound to this cluster of manganese atoms. When the reaction center of photosystem II absorbs a photon, this excites an electron in a P_{680} chlorophyll molecule, which transfers this electron to an acceptor. The oxidized P_{680} then removes an electron from a manganese atom. The oxidized manganese atoms, with the aid of reaction center proteins, remove electrons from oxygen atoms in the two water molecules. This process requires the reaction center to absorb four photons to complete the oxidation of two water molecules, producing one O_2 in the process.

The role of the b_6-f complex

The primary electron acceptor for the light-energized electrons leaving photosystem II is a quinone molecule. The reduced quinone that results from accepting a pair of electrons (*plastoquinone*, symbolized PQ) is a strong electron donor; it passes the excited electron pair to a proton pump called the **b_6-f complex** embedded within the thylakoid membrane

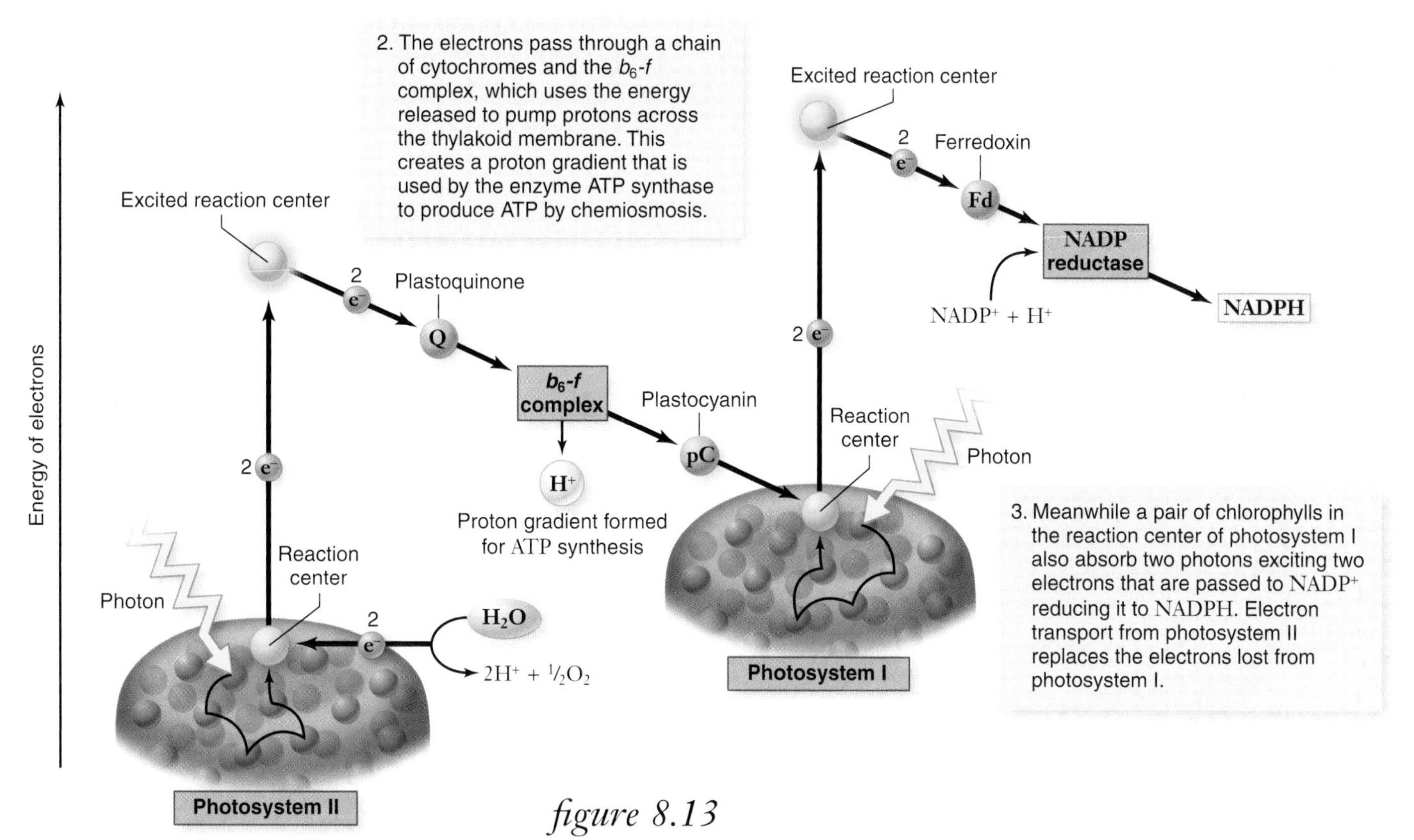

figure 8.13

A Z DIAGRAM OF PHOTOSYSTEMS I AND II. Two photosystems work sequentially and have different roles. Photosystem II passes energetic electrons to photosystem I via an electron transport chain. The electrons lost are replaced by oxidizing water. Photosystem I uses energetic electrons to reduce $NADP^+$ to NADPH.

(figure 8.14). This complex closely resembles the bc_1 complex in the respiratory electron transport chain of mitochondria, discussed in chapter 7.

Arrival of the energetic electron pair causes the b_6-f complex to pump a proton into the thylakoid space. A small, copper-containing protein called *plastocyanin* (symbolized PC) then carries the electron pair to photosystem I.

Photosystem I

The reaction center of photosystem I consists of a core transmembrane complex consisting of 12 to 14 protein subunits with two bound P_{700} chlorophyll molecules. Energy is fed to it by an antenna complex consisting of chlorophyll *a* and accessory pigment molecules.

Photosystem I accepts an electron from plastocyanin into the "hole" created by the exit of a light-energized electron. The absorption of a photon by photosystem I boosts the electron leaving the reaction center to a very high energy level. Unlike photosystem II and the bacterial photosystem, the plant photosystem I does not rely on quinones as electron acceptors. Instead, it passes electrons to an iron–sulfur protein called *ferredoxin* (Fd).

Making NADPH

Photosystem I passes electrons to ferredoxin on the stromal side of the membrane (outside the thylakoid). The reduced ferredoxin carries an electron with very high potential. Two of them, from two molecules of reduced ferredoxin, are then donated to a molecule of $NADP^+$ to form NADPH. The reaction is catalyzed by the membrane-bound enzyme *NADP reductase*.

Because the reaction occurs on the stromal side of the membrane and involves the uptake of a proton in forming NADPH, it contributes further to the proton gradient established during photosynthetic electron transport. The function of the two photosystems is summarized in figure 8.14.

ATP is generated by chemiosmosis

Protons are pumped from the stroma into the thylakoid compartment by the b_6-f complex. The splitting of water also produces added protons that contribute to the gradient. The thylakoid membrane is impermeable to protons, so this creates an electrochemical gradient that can be used to synthesize ATP.

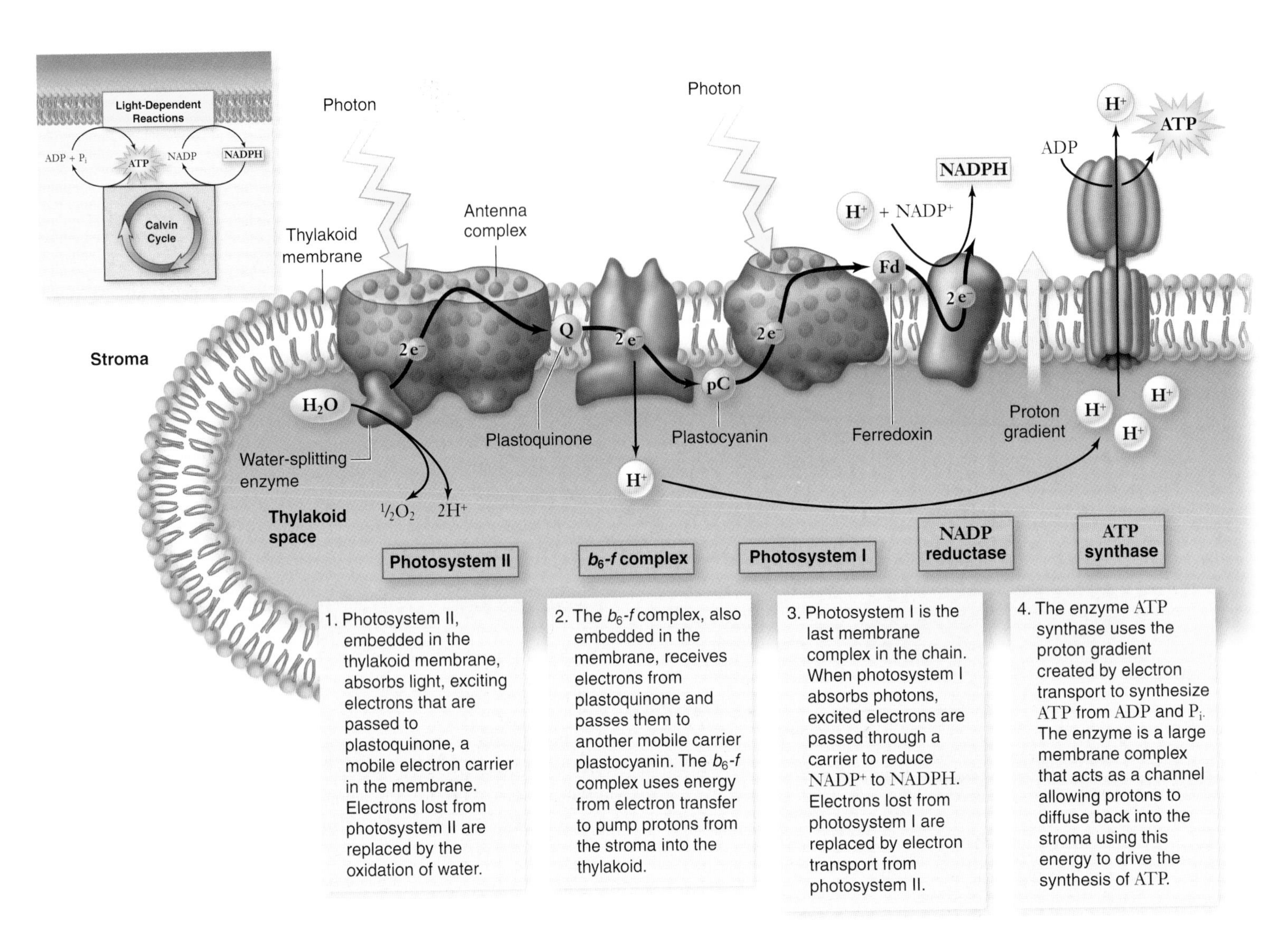

figure 8.14

THE PHOTOSYNTHETIC ELECTRON TRANSPORT SYSTEM AND ATP SYNTHASE. The two photosystems are arranged in the thylakoid membrane joined by an electron transport system that includes the b_6-f complex. These function together to create a proton gradient that is used by ATP synthase to synthesize ATP.

ATP synthase

The chloroplast has ATP synthase enzymes in the thylakoid membrane that form a channel, allowing protons to cross back out into the stroma. These channels protrude like knobs on the external surface of the thylakoid membrane. As protons pass out of the thylakoid through the ATP synthase channel, ADP is phosphorylated to ATP and released into the stroma (figure 8.14). The stroma contains the enzymes that catalyze the reactions of carbon fixation, the Calvin cycle reactions.

This mechanism is the same as that seen in the mitochondrial ATP synthase, and in fact, the two enzymes are evolutionarily related. This similarity in generating a proton gradient by electron transport and ATP by chemiosmosis illustrates the similarities in structure and function in mitochondria and chloroplasts.

The production of additional ATP

The passage of an electron pair from water to NADPH in noncyclic photophosphorylation generates one molecule of NADPH and slightly more than one molecule of ATP. But as you will learn later in this chapter, building organic molecules takes more energy than that—it takes 1.5 ATP molecules per NADPH molecule to fix carbon.

To produce the extra ATP, many plant species are capable of short-circuiting photosystem I, switching photosynthesis into a *cyclic photophosphorylation* mode, so that the light-excited electron leaving photosystem I is used to make ATP instead of NADPH. The energetic electrons are simply passed back to the b_6-f complex, rather than passing on to $NADP^+$. The b_6-f complex pumps protons into the thylakoid space, adding to the proton gradient driving the chemiosmotic synthesis of ATP. The relative proportions of cyclic and noncyclic photo-

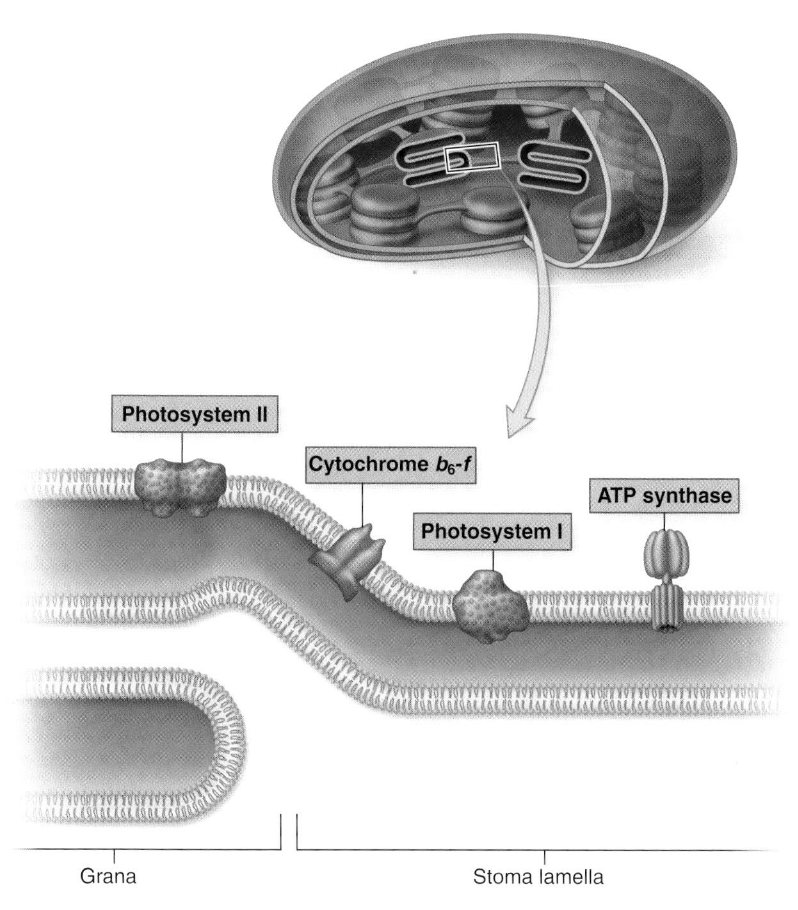

figure 8.15

MODEL FOR THE ARRANGEMENT OF COMPLEXES WITHIN THE THYLAKOID. The arrangement of the two kinds of photosystems and the other complexes involved in photosynthesis is not random. Photosystem II is concentrated within grana, especially in stacked areas. Photosystem I and ATP synthase are concentrated in stroma lamella and the edges of grana. The cytochrome b_6-f complex is in the margins between grana and stroma lamella. This is one possible model for this arrangement.

phosphorylation in these plants determine the relative amounts of ATP and NADPH available for building organic molecules.

Thylakoid structure reveals components' locations

The four complexes responsible for the light-dependent reactions—namely photosystems I and II, cytochrome b_6-f, and ATP synthase—are not randomly arranged in the thylakoid. Researchers are beginning to image these complexes with the atomic force microscope, and a picture is emerging in which photosystem II is found primarily in the grana, whereas photosystem I and ATP synthase are found primarily in the stroma lamella. Photosystem I and ATP synthase may also be found in the edges of the grana that are not stacked. The cytochrome b_6-f complex is found in the borders between grana and stroma lamella. One possible model for the arrangement of the complexes is shown in figure 8.15.

The thylakoid itself is no longer thought of only as stacked disks. Some models of the thylakoid, based on electron microscopy and other imaging, depict the grana as folds of the interconnecting stroma lamella. This kind of arrangement would be more similar to the folds seen in bacterial photosynthesis, and it would allow for more flexibility in how the various complexes are arranged relative to one another.

The light-dependent reactions produce ATP and NADPH. The sulfur bacteria use a simplified form of photosynthesis involving the cyclic transfer of electrons to generate a proton gradient to synthesize ATP, but that does not directly produce NADPH. The chloroplast has two photosystems located in the thylakoid membrane that are connected by an electron transport chain. Each photosystem can absorb a photon of light, exciting an electron that is then passed to an electron carrier. Photosystem I passes an electron to NADPH. This electron is replaced by one from photosystem II via an electron transport chain. Photosystem II then oxidizes water to replace the electron it has lost. ATP is generated by chemiosmosis using a proton gradient created by electron transport between the two photosystems.

8.6 Carbon Fixation: The Calvin Cycle

Carbohydrates contain many C—H bonds and are highly reduced compared with CO_2. To build carbohydrates, cells use energy and a source of electrons produced by the light-dependent reactions of the thylakoids:

1. **Energy.** ATP (provided by cyclic and noncyclic photophosphorylation) drives the endergonic reactions.
2. **Reduction potential.** NADPH (provided by photosystem I) provides a source of protons and the energetic electrons needed to bind them to carbon atoms. Much of the light energy captured in photosynthesis ends up invested in the energy-rich C—H bonds of sugars.

Calvin cycle reactions convert inorganic carbon into organic molecules

Because early research showed temperature dependence, photosynthesis was predicted to involve enzyme-catalyzed reactions. These reactions form a cycle of enzyme-catalyzed steps much like the Krebs cycle of respiration. Unlike the Krebs cycle, however, carbon fixation is geared toward producing new compounds, so the nature of the cycles is quite different.

The cycle of reactions that allow carbon fixation is called the **Calvin cycle,** after its discoverer, Melvin Calvin (1911–1997). Because the first intermediate of the cycle, phosphoglycerate, contains three carbon atoms, this process is also called **C_3 photosynthesis.**

The key step in this process—the event that makes the reduction of CO_2 possible—is the attachment of CO_2 to a highly specialized organic molecule. Photosynthetic cells produce this molecule by reassembling the bonds of two intermediates in glycolysis—fructose 6-phosphate and glyceraldehyde 3-phosphate (G3P)—to form the energy-rich five-carbon sugar **ribulose 1,5-bisphosphate (RuBP).**

CO_2 reacts with RuBP to form a transient six-carbon intermediate that immediately splits into two molecules of the three-carbon *3-phosphoglycerate* (*PGA*). This overall reaction is called the *carbon fixation reaction* because inorganic carbon (CO_2) has been incorporated into an organic form: the acid PGA. The enzyme that carries out this reaction, **ribulose bisphosphate carboxylase/oxygenase** (usually abbreviated **rubisco**) is a large, 16-subunit enzyme found in the chloroplast stroma.

Carbon is transferred through cycle intermediates, eventually producing glucose

We will consider how the Calvin cycle can produce one molecule of glucose, although this glucose is not produced directly by the cycle (figure 8.16). In a series of reactions, six molecules of CO_2 are bound to six RuBP by rubisco to produce 12 molecules of PGA (containing $12 \times 3 = 36$ carbon atoms in all, six from CO_2 and 30 from RuBP). The 36 carbon atoms then undergo a cycle of reactions that regenerates the six molecules of RuBP used in the initial step (containing $6 \times 5 = 30$ carbon atoms). This leaves two molecules of *glyceraldehyde 3-phosphate*

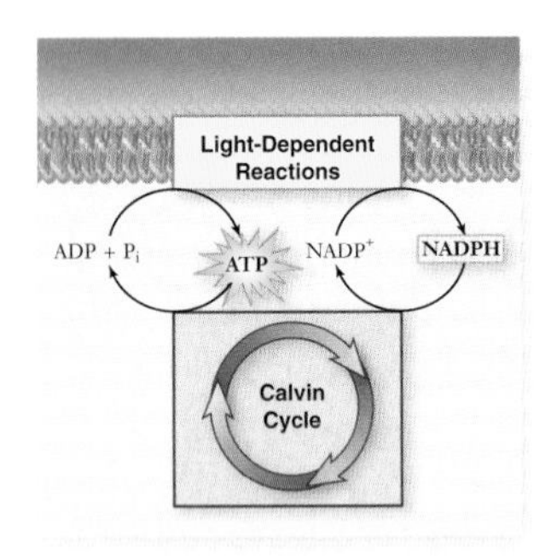

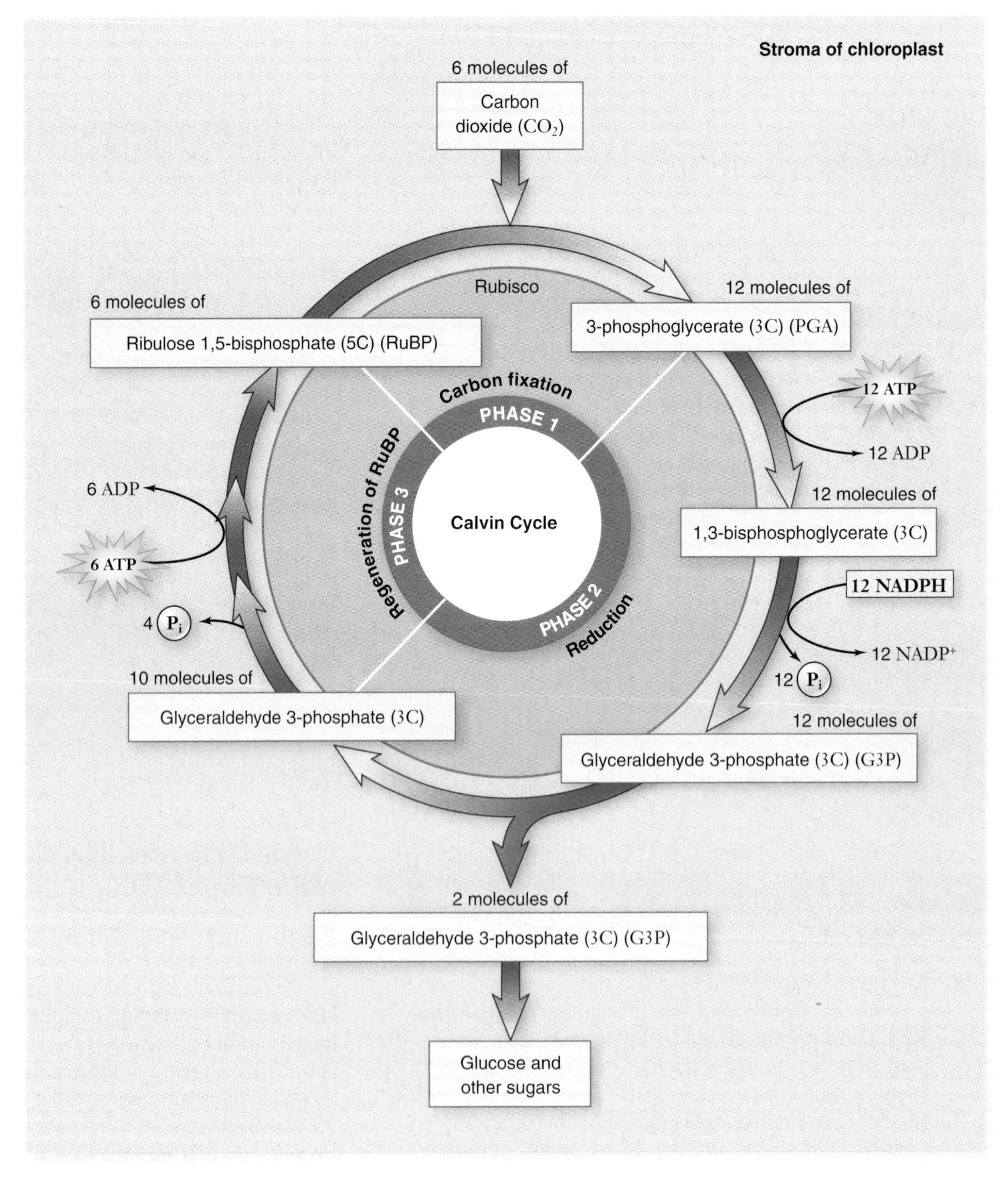

figure 8.16

THE CALVIN CYCLE. The Calvin cycle accomplishes carbon fixation: converting inorganic carbon in the form of CO_2 into organic carbon in the form of carbohydrates. The cycle can be broken down into three phases: carbon fixation, reduction, and regeneration of RuBP. For every six CO_2 molecules fixed by the cycle, a molecule of glucose can be synthesized from the products of the reduction reactions, G3P. The cycle uses the ATP and NADPH produced by the light reactions.

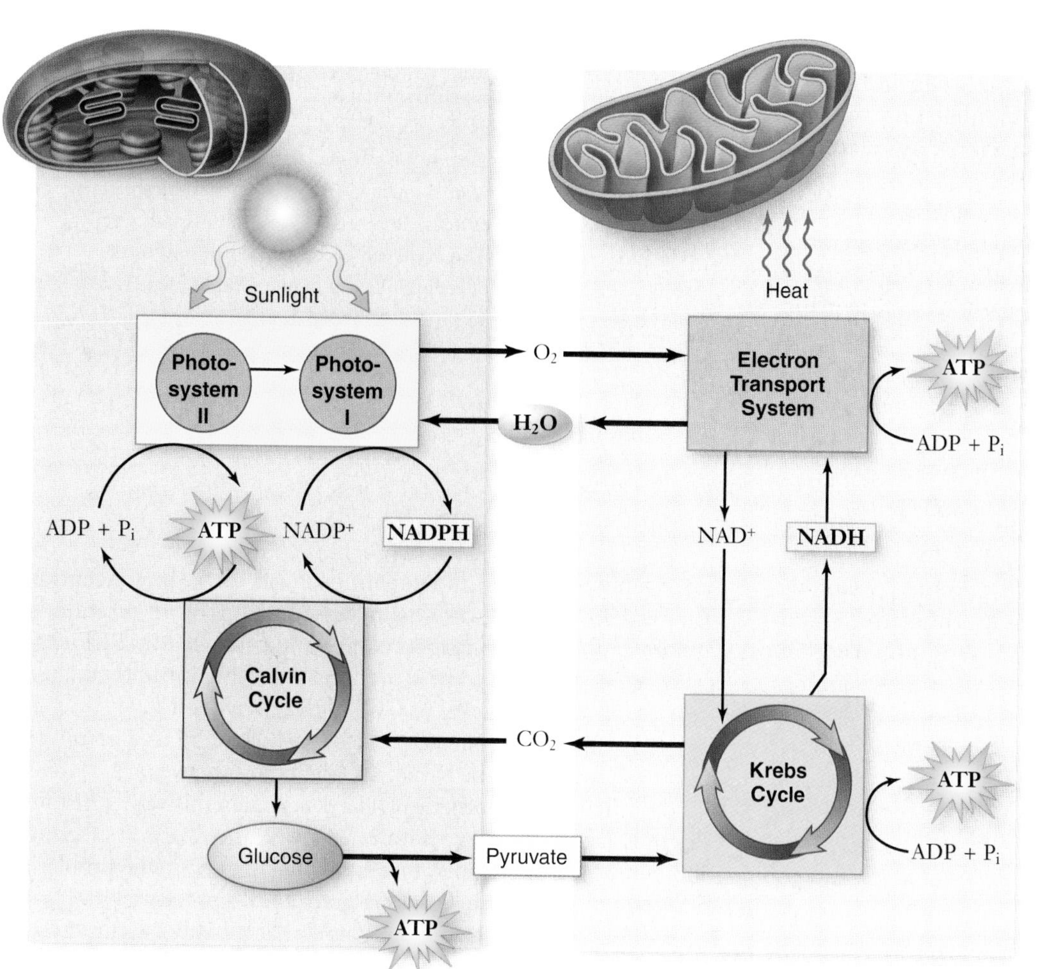

figure 8.17

CHLOROPLASTS AND MITOCHONDRIA: COMPLETING AN ENERGY CYCLE. Water and O_2 cycle between chloroplasts and mitochondria within a plant cell, as do glucose and CO_2. Cells with chloroplasts require an outside source of CO_2 and H_2O and generate glucose and O_2. Cells without chloroplasts, such as animal cells, require an outside source of glucose and O_2 and generate CO_2 and H_2O.

(*G3P*) (each with three carbon atoms) as the net gain. (You may recall G3P as also being the product of the first half of glycolysis, described in chapter 7.) These two molecules of G3P can then be used to make one molecule of glucose.

The net equation of the Calvin cycle is:

$$6\ CO_2 + 18\ ATP + 12\ NADPH + \text{water} \longrightarrow 2\ \text{glyceraldehyde 3-phosphate} + 16\ P_i + 18\ ADP + 12\ NADP^+$$

With six full turns of the cycle, six molecules of carbon dioxide enter, two molecules of G3P are produced, and six molecules of RuBP are regenerated. Thus six turns of the cycle produce two G3P that can be used to make a single glucose molecule. The six turns of the cycle also incorporated six CO_2 molecules, providing enough carbon to synthesize glucose, although the six carbon atoms do not all end up in this molecule of glucose.

Phases of the cycle

The Calvin cycle can be thought of as divided into three phases: (1) carbon fixation, (2) reduction, and (3) regeneration of RuBP. The carbon fixation reaction generates two molecules of the three-carbon acid PGA; PGA is then reduced to G3P by reactions that are essentially a reverse of part of glycolysis; finally, the PGA is used to regenerate RuBP. Three turns around the cycle incorporate enough carbon to produce a new molecule of G3P, and six turns incorporate enough carbon to synthesize one glucose molecule.

We now know that light is required *indirectly* for different segments of the CO_2 reduction reactions. Five of the Calvin cycle enzymes—including rubisco—are light-activated; that is, they become functional or operate more efficiently in the presence of light. Light also promotes transport of required three-carbon intermediates across chloroplast membranes. And finally, light promotes the influx of Mg^{2+} into the chloroplast stroma, which further activates the enzyme rubisco.

Output of the Calvin cycle

Glyceraldehyde 3-phosphate is a three-carbon sugar, a key intermediate in glycolysis. Much of it is transported out of the chloroplast to the cytoplasm of the cell, where the reversal of several reactions in glycolysis allows it to be converted to fructose 6-phosphate and glucose 1-phosphate. These products can then be used to form sucrose, a major transport sugar in plants. (Sucrose, table sugar, is a disaccharide made of fructose and glucose.)

In times of intensive photosynthesis, G3P levels rise in the stroma of the chloroplast. As a consequence, some G3P in the chloroplast is converted to glucose 1-phosphate, in a set of reactions analogous to those occurring in the cytoplasm, by reversing several reactions similar to those of glycolysis. The glucose 1-phosphate is then combined into an insoluble polymer, forming long chains of starch stored as bulky starch grains in the cytoplasm. These starch grains represent stored glucose for later use.

The energy cycle

The energy-capturing metabolisms of the chloroplasts studied in this chapter and the mitochondria studied in chapter 7 are intimately related (figure 8.17). Photosynthesis uses the products of respiration as starting substrates, and respiration uses the products of photosynthesis as starting substrates. The production of glucose from G3P even uses part of the ancient glycolytic pathway, run in reverse. Also, the principal proteins involved in electron transport and ATP production in plants are evolutionarily related to those in mitochondria.

Photosynthesis is but one aspect of plant biology, although it is an important one. In chapters 36 through 42, we examine plants in more detail. We have discussed photosynthesis as a part of cell biology because photosynthesis arose long before plants did, and because most organisms depend directly or indirectly on photosynthesis for the energy that powers their lives.

Carbon fixation is the incorporation of inorganic CO_2 into an organic molecule. This is accomplished by the Calvin cycle reactions that takes place in the stroma of the chloroplast. The key intermediate is the five-carbon sugar RuBP that combines with CO_2 in the carbon fixation reaction. This reaction is catalyzed by the enzyme rubisco. The cycle can be broken down into three stages: carbon fixation, reduction, and regeneration of RuBP. The products of the light reactions, ATP and NADPH, provide energy and electrons for the reduction reactions.

8.7 Photorespiration

Evolution does not necessarily result in optimum solutions. Rather, it favors workable solutions that can be derived from features that already exist. Photosynthesis is no exception. Rubisco, the enzyme that catalyzes the key carbon-fixing reaction of photosynthesis, provides a decidedly suboptimal solution. This enzyme has a second enzymatic activity that interferes with carbon fixation, namely that of *oxidizing* RuBP. In this process, called **photorespiration,** O_2 is incorporated into RuBP, which undergoes additional reactions that actually release CO_2. Hence, photorespiration releases CO_2, essentially undoing carbon fixation.

Photorespiration reduces the yield of photosynthesis

The carboxylation and oxidation of RuBP are catalyzed at the same active site on rubisco, and CO_2 and O_2 compete with each other at this site. Under normal conditions at 25°C, the rate of the carboxylation reaction is four times that of the oxidation reaction, meaning that 20% of photosynthetically fixed carbon is lost to photorespiration.

This loss rises substantially as temperature increases, because under hot, arid conditions, specialized openings in the leaf called *stomata* (singular, *stoma*) (figure 8.18) close to conserve water. This closing also cuts off the supply of CO_2 entering the leaf and does not allow O_2 to exit (figure 8.19). As a result, the low-CO_2 and high-O_2 conditions within the leaf favor photorespiration.

Plants that fix carbon using only C_3 photosynthesis (the Calvin cycle) are called **C_3 plants** (figure 8.20*a*). Other plants add CO_2 to phosphoenolpyruvate (PEP) to form a four-carbon molecule. This reaction is catalyzed by the enzyme PEP *carboxylase.* This enzyme has two advantages over rubisco: it has a much greater affinity for CO_2 than rubisco, and it does not have oxidase activity.

The four-carbon compound produced by PEP carboxylase undergoes further modification, only to be eventually decarboxylated. The CO_2 released by this decarboxylation is then used by rubisco in the Calvin cycle. Because the source of CO_2 was an organic compound and not atmospheric CO_2, the concentration of CO_2 relative to O_2 is increased and photorespiration is minimized. The four-carbon compound produced by PEP carboxylase allows CO_2 to be stored in an organic form, then later released to keep the level of CO_2 high relative to O_2.

The reduction in the yield of carbohydrate as a result of photorespiration is not trivial. C_3 plants lose between 25% and 50% of their photosynthetically fixed carbon in this way. The rate depends largely on temperature. In tropical climates, especially those in which the temperature is often above 28°C, the problem is severe, and it has a major effect on tropical agriculture.

The two main groups of plants that initially capture CO_2 using PEP carboxylase differ in how they maintain high levels of CO_2 relative to O_2. In **C_4 plants** (figure 8.20*b*), the capture of CO_2 occurs in one cell and the decarboxylation occurs in an

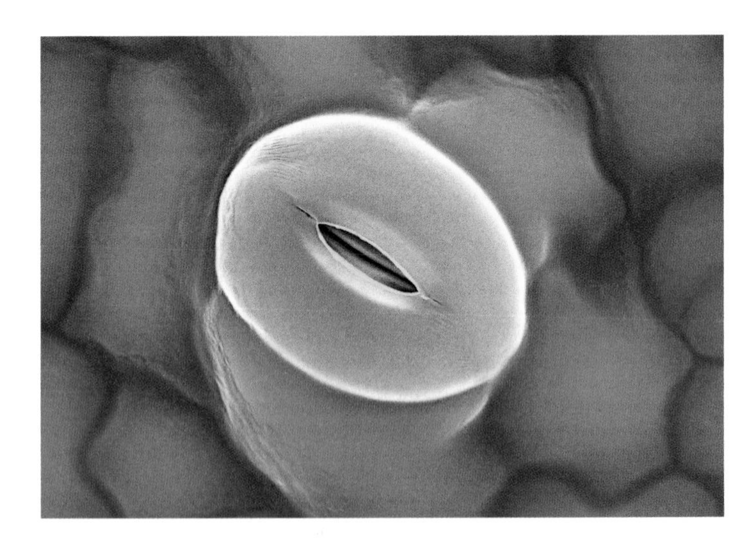

figure 8.18

STOMA. A closed stoma in the leaf of a tobacco plant. Each stoma is formed from two guard cells whose shape changes with turgor pressure to open and close. Under dry conditions plants close their stomata to conserve water.

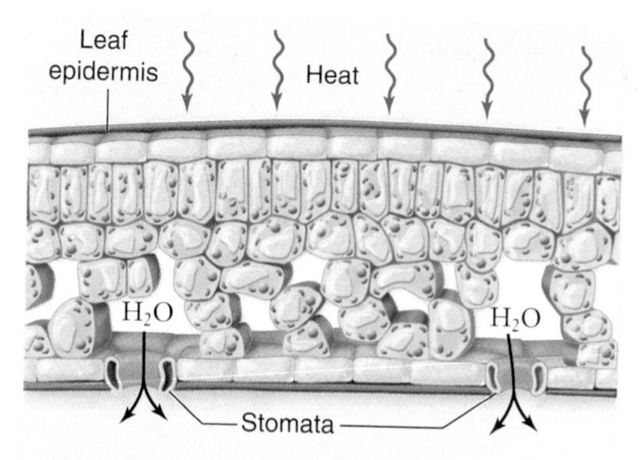

figure 8.19

CONDITIONS FAVORING PHOTORESPIRATION. In hot, arid environments, stomata close to conserve water, which also prevents CO_2 from entering and O_2 from exiting the leaf. The high-O_2/low-CO_2 conditions favor photorespiration.

adjacent cell. This represents a spatial solution to the problem of photorespiration. The second group, **CAM plants,** perform both reactions in the same cell, but capture CO_2 using PEP carboxylase at night, then decarboxylate during the day. CAM stands for **crassulacean acid metabolism,** after the plant family Crassulaceae (the stonecrops, or hens-and-chicks), in which it was first discovered. This mechanism represents a temporal solution to the photorespiration problem.

C_4 plants have evolved to minimize photorespiration

The C_4 plants include corn, sugarcane, sorghum and a number of other grasses. These plants initially fix carbon using PEP carboxylase in mesophyll cells. This reaction produces the organic acid oxaloacetate, which is converted to malate and transported to bundle-sheath cells that surround the leaf veins. Within the bundle-

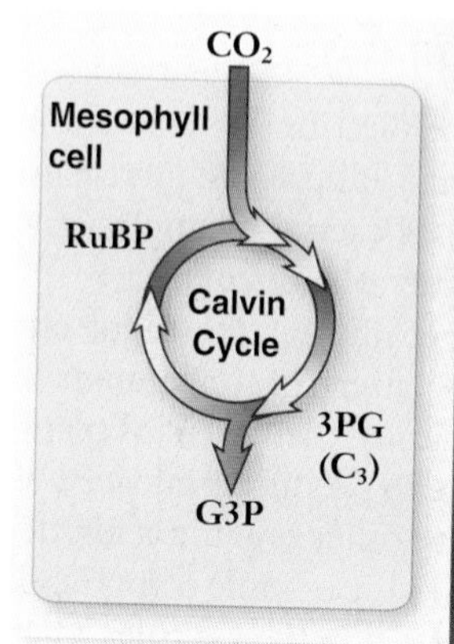

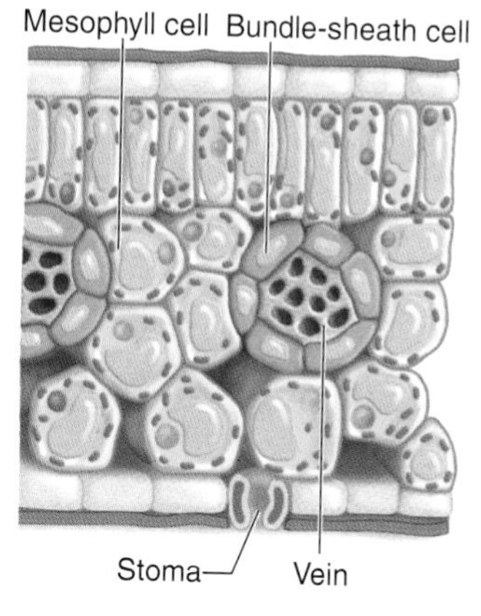

a. C_3 pathway

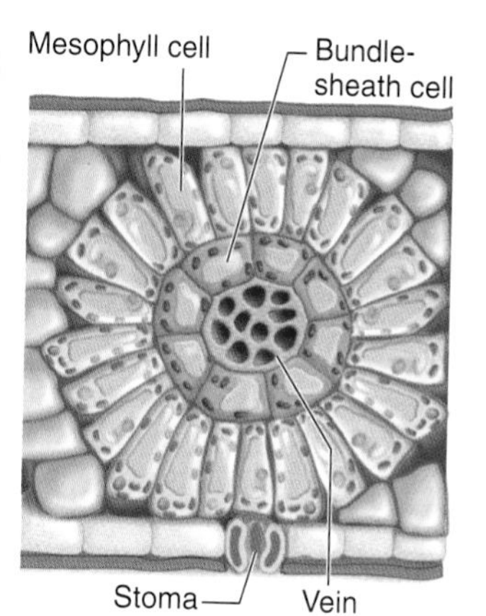

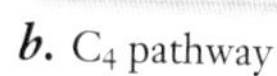

b. C_4 pathway

figure 8.20

COMPARISON OF C_3 AND C_4 PATHWAYS OF CARBON FIXATION. ***a.*** Carbon fixation by the C_3 pathway using the Calvin cycle as described in the text. All reactions occur in mesophyll cells using CO_2 that diffuses in through stomata. ***b.*** The C_4 pathway of carbon fixation in which one cell fixes CO_2 to produce a four-carbon molecule of malate in mesophyll cells. This is transported to the bundle sheath cells where it is converted back into CO_2 and pyruvate, creating a high local level of CO_2. This allows efficient carbon fixation by the Calvin cycle and avoids photorespiration.

sheath cells, malate is decarboxylated to produce pyruvate and CO_2 (figure 8.21). Because the bundle-sheath cells are impermeable to CO_2, the local level of CO_2 is high and carbon fixation by rubisco and the Calvin cycle is efficient. The pyruvate produced by decarboxylation is transported back to the mesophyll cells, where it is converted back to PEP, thereby completing the cycle.

The C_4 pathway, although it overcomes the problems of photorespiration, does have a cost. The conversion of pyruvate back to PEP requires breaking two high-energy bonds in ATP. Thus each CO_2 transported into the bundle-sheath cells cost the equivalent of two ATP. To produce a single glucose, this requires 12 additional ATP compared with the Calvin cycle alone. Despite this additional cost, C_4 photosynthesis is advantageous in hot dry climates where photorespiration would remove more than half of the carbon fixed by the usual C_3 pathway alone.

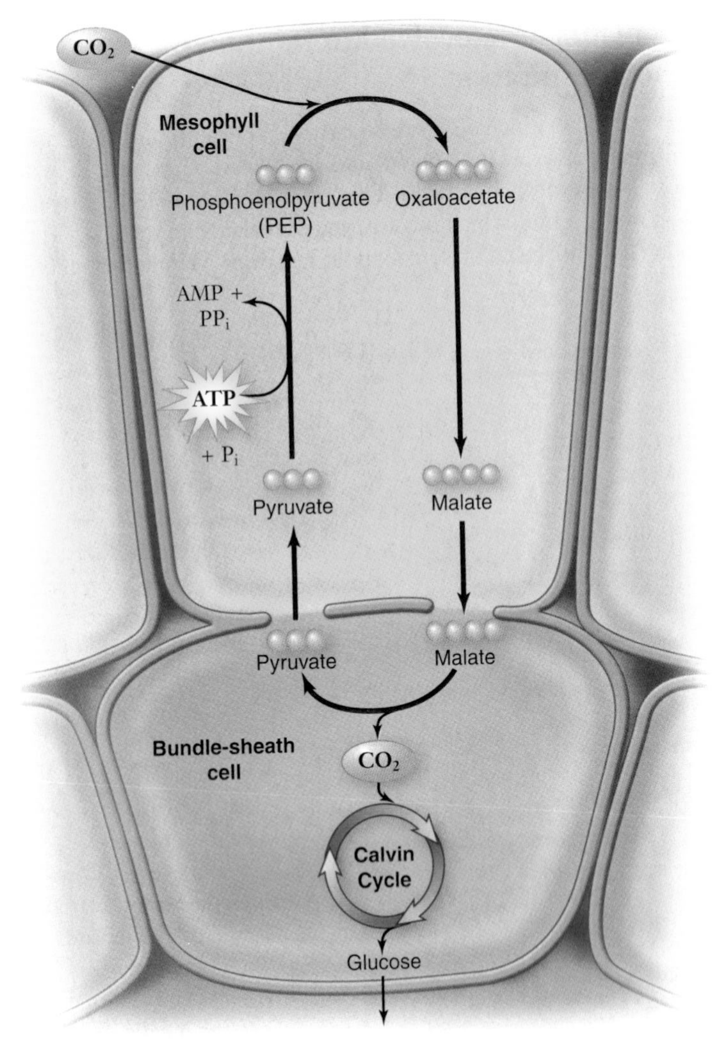

figure 8.21

CARBON FIXATION IN C_4 PLANTS. This process is called the C_4 pathway because the first molecule formed, oxaloacetate, contains four carbons. The oxaloacetate is converted to malate, which moves into bundle-sheath cells where it is decarboxylated back to CO_2 and pyruvate. This produces a high level of CO_2 in the bundle-sheath cells that can be fixed by the usual C_3 Calvin cycle with little photorespiration. The pyruvate diffuses back into the mesophyll cells, where it is converted back to PEP to be used in another C_4 fixation reaction.

figure 8.22

CARBON FIXATION IN CAM PLANTS. CAM plants also use both C_4 and C_3 pathways to fix carbon and minimize photorespiration. In CAM plants, the two pathways occur in the same cell but are separated in time: The C_4 pathway is utilized to fix carbon at night, then CO_2 is released from these accumulated stores during the day to drive the C_3 pathway. This achieves the same effect of minimizing photorespiration while also minimizing loss of water by opening stomata at night when temperatures are lower.

The Crassulacean acid pathway splits photosynthesis into night and day

A second strategy to decrease photorespiration in hot regions has been adopted by the CAM plants. These include many succulent (water-storing) plants, such as cacti, pineapples, and some members of about two dozen other plant groups.

In these plants, the stomata open during the night and close during the day (figure 8.22). This pattern of stomatal opening and closing is the reverse of that in most plants. CAM plants initially fix CO_2 using PEP carboxylase to produce oxaloacetate. The oxaloacetate is often converted into other organic acids, depending on the particular CAM plant. These organic compounds accumulate during the night and are stored in the vacuole. Then during the day, when the stomata are closed, the organic acids are decarboxylated to yield high levels of CO_2. These high levels of CO_2 drive the Calvin cycle and minimize photorespiration.

Like C_4 plants, CAM plants use both C_3 and C_4 pathways. They differ in that they use both of these pathways in the same cell: the C_4 pathway at night and the C_3 pathway during the day. In C_4 plants the two pathways occur in different cells.

In addition to catalyzing the carbon fixation reaction, the enzyme rubisco can oxidize RuBP. This photorespiration short-circuits photosynthesis under conditions of high O_2 and low CO_2. Plants in hot, dry environments have adapted to this by altering the usual dark reactions to store CO_2 in an organic molecule. This process is called C_4 metabolism to distinguish it from the usual C_3 metabolism. This involves CO_2 being stored in a four-carbon molecule in one cell, then transferred to another cell to release the CO_2 for use in the Calvin cycle. A further adaptation that splits photosynthesis into night and day reactions that occur in a single cell is found in CAM plants.

concept review

8.1 Overview of Photosynthesis

Photosynthesis is the conversion of light energy into chemical energy (figure 8.10).

- Photosynthesis has three stages: absorbing light energy, using the absorbed energy to synthesize ATP and NADPH, and using the ATP and NADPH to convert CO_2 to organic molecules.
- Photosynthesis has both light-dependent and light-independent reactions.
- In plant and algal cells, photosynthesis takes place in chloroplasts.
- Chloroplasts contain internal thylakoid membranes and a fluid matrix called stroma.

8.2 The Discovery of the Photosynthetic Processes

Knowledge about the photosynthetic process has accumulated over the last 300 years.

- Plants do not increase mass from soil and water alone as was originally thought.
- The rate of photosynthesis depends on the relative amounts of light, CO_2 concentration, and temperature.
- The O_2 produced by photosynthesis comes from H_2O.
- Light-dependent reactions produce O_2 from H_2O and generate ATP and NADPH that are used in carbon fixation reactions.

8.3 Pigments

For plants to utilize sunlight for photosynthesis pigments must be available that absorb light.

- Light is a form of energy that exists both as a wave and a particle called a photon.
- Light can remove electrons from some metals by the photoelectric effect.
- Photosynthetic pigments include chlorophyll *a*, chlorophyll *b*, and carotenoids; each has a characteristic absorption spectrum.
- Chlorophyll *a* is the primary pigment and the only pigment that can convert light energy into chemical energy.
- Chlorophyll *b* is an accessory pigment that increases the proportion of photons that can be harvested for photosynthesis.
- Carotenoids and other accessory pigments further increase a plants ability to harvest photons.

8.4 Photosystem Organization (figure 8.9)

Photosynthetic pigments are organized into photosystems that absorb photons of light and transfer electrons.

- A photosystem is a network of chlorophyll *a*, accessory pigments, and proteins embedded in the thylakoid membrane.
- Photosystems contain an antenna complex and a reaction center.
  - *The antenna complex is composed of pigment molecules that harvest photons and feed light energy to the reaction center.*
  - *The reaction center is composed of two chlorophyll* a *molecules in a protein matrix that pass an excited electron to an electron acceptor.*

8.5 The Light-dependent Reactions

Plants, algae, and photosynthetic bacteria use photosystems to trap and transfer energy.

- Anoxygenic photosynthetic bacteria use a single photosystem to generate ATP by cyclic photophosphorylation (figure 8.11).
- Cyclic photophosphorylation cycles excited electrons back to the pigment molecule to generate a proton gradient.
- Plants utilize two linked photosystems, photosystem I and photosystem II, acting to generate ATP and NADPH (figure 8.14).
- Photosystem I transfers electrons to $NADP^+$, reducing it to NADPH.
- Electrons lost by photosystem I are replaced by electrons from photosystem II.
- Electrons lost from photosystem II are replaced by electrons from the oxidation of water, which also produces O_2.
- Photosystem II and photosystem I are linked by an electron transport chain that pumps protons into the thylakoid space.
- The proton gradient is used by ATP synthase to phosphorylate ADP to ATP by a chemiosmotic mechanism similar to mitochondria.
- Plants can also make additional ATP by cyclic photophosphorylation.

8.6 Carbon Fixation: The Calvin Cycle (figure 8.16)

The Calvin cycle synthesizes organic molecules from inorganic carbon (CO_2).

- The Calvin cycle requires CO_2, ATP, and NADPH.
- The Calvin cycle occurs in three stages: carbon fixation, reduction, and regeneration.
  - *Carbon fixation involves the enzyme rubisco, combining CO_2 and the five-carbon ribulose 1,5-bisphosphate (RuBP). The resulting compound splits into two 3-carbon 3-phosphoglycerates (PGA).*
  - *Reduction converts the 3-carbon PGA to glyceraldehyde 3-phosphate (G3P) in a series of reactions that use ATP and NADPH.*
  - *Regeneration involves using G3P to synthesize more RuBP.*
- Three turns of the cycle fix enough carbon for one new G3P.
- It takes six turns of the cycle to fix enough carbon to have two excess G3Ps that can be used to make one molecule of glucose.

8.7 Photorespiration

The enzyme rubisco, which catalyzes the carbon fixation reaction, can also catalyze the oxidation of RuBP, reversing carbon fixation.

- Dry, hot conditions cause the leaf stomata to close, resulting in lower CO_2 and higher O_2 concentrations within the leaf.
- As O_2 concentration increases, rubisco tends to bind oxygen and catalyze the oxidation of RuBP, eventually producing CO_2.
- Photorespiration can result in up to 50% reductions in glucose production.
- C_4 plants use an alternative mechanism of carbon fixation.
- C_4 plants fix carbon by adding CO_2 to a three-carbon phosphoenolpyruvate to form a four-carbon oxaloacetate. The CO_2 is released later for carbon fixation by the Calvin cycle.
- C_4 plants fix carbon in one cell by the C_4 pathway, then release CO_2 in another cell for the Calvin cycle (figure 8.21).
- CAM plants use the C_4 pathway during the day and the Calvin cycle at night in the same cell.

review questions

SELF TEST

1. The *light-dependent* reactions of photosynthesis are responsible for the production of—
   a. glucose
   b. CO_2
   c. ATP and NADPH
   d. H_2O
2. Which region of a chloroplast is associated with the capture of light energy?
   a. Thylakoid membrane
   b. Outer membrane
   c. Stroma
   d. Both a and c
3. The colors of light that are most effective for photosynthesis are—
   a. Red, blue, and violet
   b. Green, yellow, and orange
   c. Infrared and ultraviolet
   d. All colors of light are equally effective
4. The colors associated with pigments such as chlorophyll or carotenoids are a product of—
   a. The wavelengths of light absorbed by the pigment
   b. The wavelengths of light reflected by the pigment
   c. The energy transferred between pigments
   d. The wavelengths of light emitted by the pigment
5. Which of the following best describes a photosystem?
   a. A collection of pigment molecules
   b. A collection of pigments that transfer energy captured from light to a reaction center pigment
   c. A collection of thylakoid membranes assembled into a structure called a granum
   d. A collection of chlorophyll molecules that capture light energy and use it to make ATP
6. How is a reaction center pigment different from a pigment in the antenna complex?
   a. The reaction center pigment is a chlorophyll molecule.
   b. The antenna complex pigment can only reflect light.
   c. The reaction center pigment loses an electron when it absorbs light energy.
   d. The antenna complex pigments are not attached to proteins.
7. What happens to the energy from an excited reaction center electron in the cyclic photophosphorylation of sulfur bacteria?
   a. It is used to make ATP.
   b. It is used to phosphorylate proteins in an electron transport chain.
   c. It is used to generate a new pigment molecule.
   d. It is used to excite another pigment molecule within the photosystem.
8. During noncyclic photosynthesis, photosystem I functions to __________, and photosystem II functions to ____________.
   a. synthesize ATP; produce O_2
   b. reduce $NADP^+$; oxidize H_2O
   c. reduce CO_2; oxidize NADPH
   d. restore an electron to its reaction center; gain an electron from water
9. Where in a chloroplast would you find the highest concentration of protons?
   a. In the stroma
   b. In the lumen of the thylakoid
   c. In the intermembrane space
   d. In the antenna complex
10. How does the reaction center of photosystem I regain an electron during noncyclic photosynthesis?
    a. The electron is recycled directly back to the reaction center pigment.
    b. The electron is donated from H_2O.
    c. The electron is donated from photosystem II.
    d. The electron is donated from NADPH.
11. Which of the following is NOT associated with the thylakoid membrane?
    a. Photosystem II
    b. ATP synthase
    c. Rubisco
    d. b_6-f complex
12. Carbon fixation occurs when a molecule of CO_2 reacts with a molecule of—
    a. ribulose 1,5-bisphosphate (RuBP)
    b. glyceraldehyde 3-phosphate (G3P)
    c. 3-phosphoglycerate (PGA)
    d. pyruvate
13. The function of the Calvin cycle is to—
    a. absorb light energy
    b. synthesize RuBP
    c. fix carbon
    d. convert glucose to CO_2, yielding energy
14. What is photorespiration?
    a. The production of chemical energy (ATP) using light energy (photons)
    b. Carbon fixation using the energy gained from the light reactions of photosynthesis
    c. The use of O_2 by plants as a final electron acceptor for photosynthesis
    d. The addition of O_2 to RuBP, leading to the loss of CO_2 and ATP
15. The adaptation of fixing CO_2 from the atmosphere at night is characteristic of—
    a. C_3 plants
    b. C_4 plants
    c. CAM plants
    d. all of the above

CHALLENGE QUESTIONS

1. Study the process of the Calvin cycle diagrammed in figure 8.16. Where do the ATP and NADPH used in this reaction come from? How can a chloroplast generate enough ATP to support the needs of the Calvin cycle?
2. Compare the process of photosynthesis in green plants and anoxygenic bacteria.
3. Do plant cells need mitochondria? Explain your answer.

Do you need additional review? *Visit* www.ravenbiology.com *for practice quizzes, animations, videos, and activities designed to help you master the material in this chapter.*

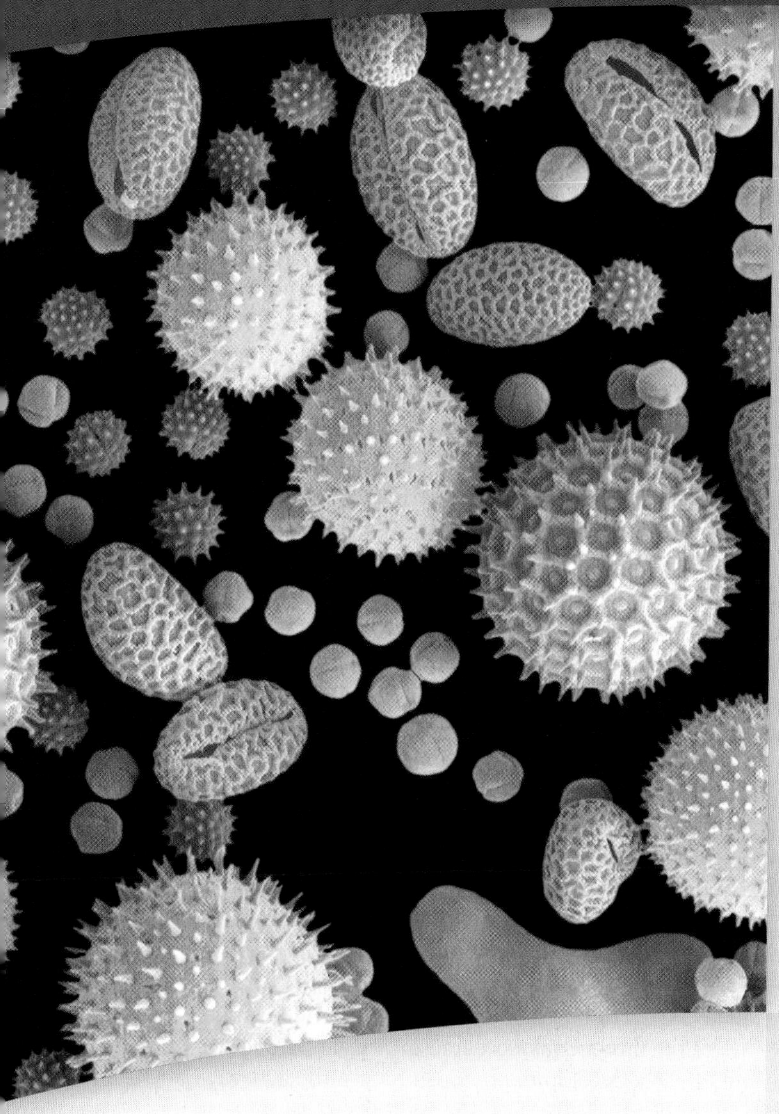

chapter 9

Cell Communication

introduction

SPRINGTIME IS A TIME OF REBIRTH and renewal. Trees that have appeared dead produce new leaves and buds, and flowers sprout from the ground. For sufferers of seasonal allergy, this is not quite such a pleasant time. The pollen in the micrograph and other allergens produced stimulate the immune system to produce the molecule histamine and other molecules that form cellular signals. These signals cause inflammation, mucus secretion, vasodilation, and other responses that together cause the runny nose, itching watery eyes, and other symptoms that make up the allergic reaction. We treat allergy symptoms by using drugs called antihistamines that interfere with this cellular signaling. The popular drug loratadine (better known as Claritin), for example, acts by blocking the receptor for histamine, thus preventing its action.

We will begin this chapter with a general overview of signaling, and the kinds of receptors cells use to respond to signals. Then we will look in more detail at how these different types of receptors can elicit a response from cells and finally how cells make connections with one another.

concept outline

9.1 Overview of Cell Communication

Communication between cells is common in nature. Cell signaling occurs in all multicellular organisms, providing an indispensable mechanism for cells to influence one another. Effective signaling requires a signaling molecule, called a **ligand,** and a molecule to which the signal binds, called a **receptor protein.** The interaction of these two components initiates the process of *signal transduction*, which converts the information in the signal into a cellular response (figure 9.1).

The cells of multicellular organisms use a variety of molecules as signals, including but not limited to, peptides, large proteins, individual amino acids, nucleotides, and steroids and other lipids. Even dissolved gases such as NO (nitric oxide) are used as signals.

Any cell of a multicellular organism is exposed to a constant stream of signals. At any time, hundreds of different chemical signals may be present in the environment surrounding the cell. Each cell responds only to certain signals, however, and ignores the rest, like a person following the conversation of one or two individuals in a noisy, crowded room.

How does a cell "choose" which signals to respond to? The number and kind of receptor molecules determine this. When a ligand approaches a receptor protein that has a complementary shape, the two can bind, forming a complex. This binding induces a change in the receptor protein's shape, ultimately producing a response in the cell via a signal transduction pathway. In this way, a given cell responds to the signaling molecules that fit the particular set of receptor proteins it possesses and ignores those for which it lacks receptors.

Signaling is defined by the distance from source to receptor

Cells can communicate through any of four basic mechanisms, depending primarily on the distance between the signaling and responding cells (figure 9.2). These mechanisms are (1) direct contact, (2) paracrine signaling, (3) endocrine signaling, and (4) synaptic signaling.

In addition to using these four basic mechanisms, some cells actually send signals to themselves, secreting signals that bind to specific receptors on their own plasma membranes. This process, called *autocrine signaling*, is thought to play an important role in reinforcing developmental changes, and it is an important component of signaling in the immune system (chapter 51).

Direct contact

As you saw in chapter 5, the surface of a eukaryotic cell is richly populated with proteins, carbohydrates, and lipids attached to and extending outward from the plasma membrane. When cells are very close to one another, some of the molecules on the plasma membrane of one cell can be recognized by receptors on the plasma membrane of an adjacent cell. Many of the important interactions between cells in early development occur by means of direct contact between cell surfaces. Cells also signal through gap junctions (figure 9.2*a*). We'll examine contact-dependent interactions more closely later in this chapter.

Paracrine signaling

Signal molecules released by cells can diffuse through the extracellular fluid to other cells. If those molecules are taken up by neighboring cells, destroyed by extracellular enzymes, or quickly removed from the extracellular fluid in some other way, their influence is restricted to cells in the immediate vicinity of the releasing cell. Signals with such short-lived, local effects are called **paracrine** signals (figure 9.2*b*).

Like direct contact, paracrine signaling plays an important role in early development, coordinating the activities of clusters of neighboring cells. The immune response in vertebrates also involves paracrine signaling between immune cells (chapter 51).

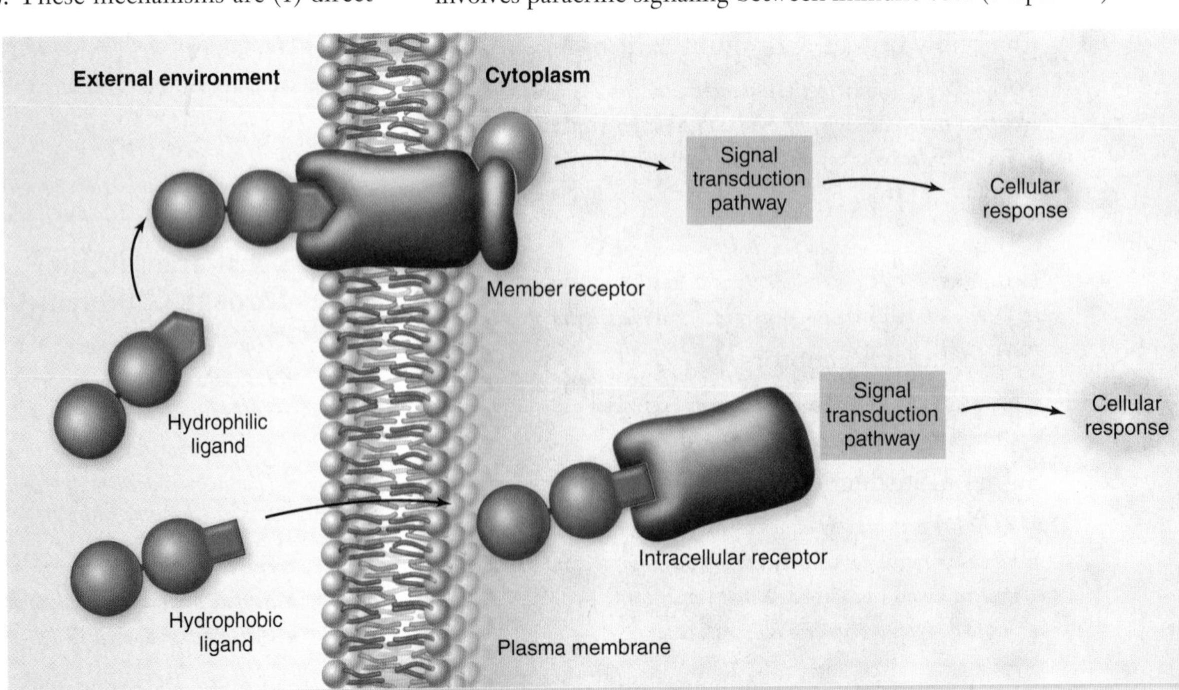

figure 9.1

OVERVIEW OF CELL SIGNALING. Cell signaling involves a signal molecule called a ligand, a receptor, and a signal transduction pathway that produces a cellular response. The location of the receptor can either be intracellular, for hydrophobic ligands that can cross the membrane, or in the plasma membrane, for hydrophilic ligands that cannot cross the membrane.

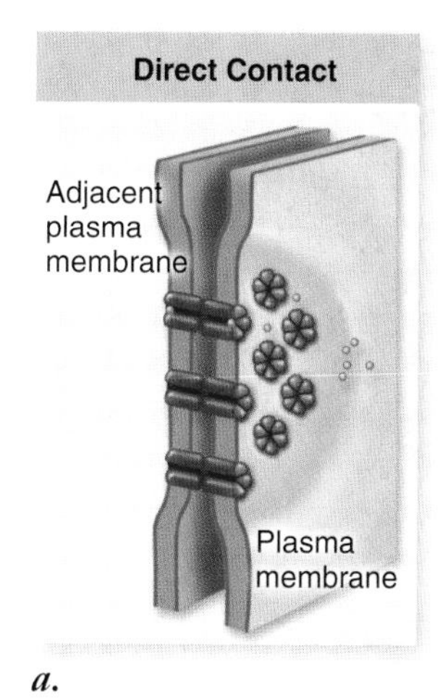

a.

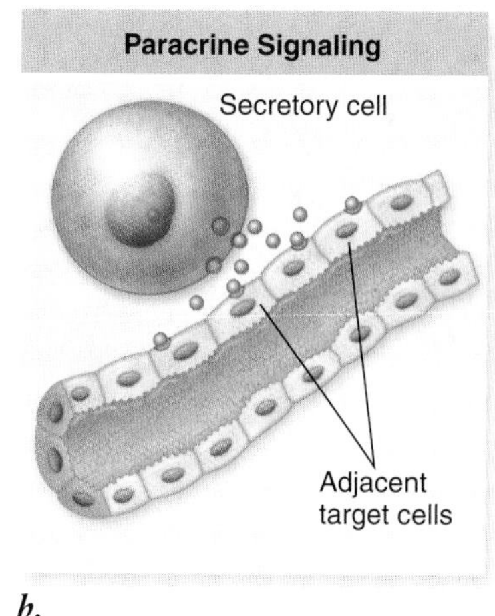

b.

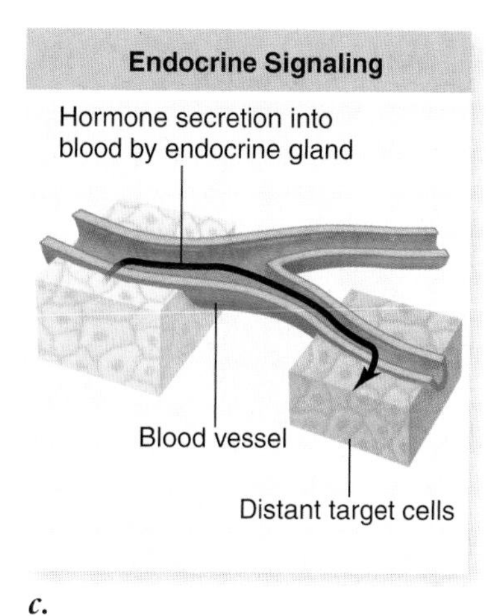

c.

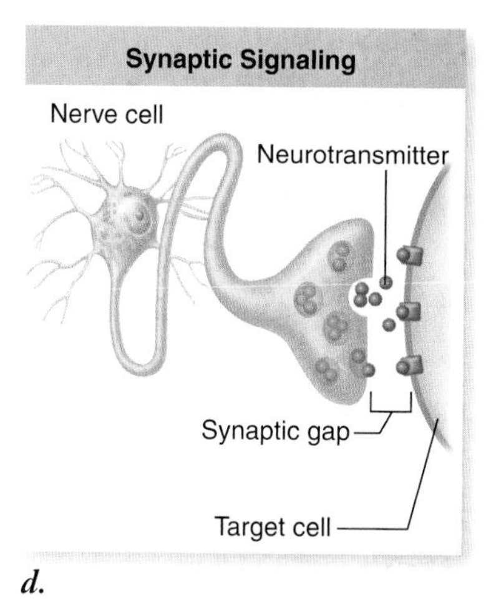

d.

figure 9.2

FOUR KINDS OF CELL SIGNALING. Cells communicate in several ways. ***a.*** Two cells in direct contact with each other may send signals across gap junctions. ***b.*** In paracrine signaling, secretions from one cell have an effect only on cells in the immediate area. ***c.*** In endocrine signaling, hormones are released into the organism's circulatory system, which carries them to the target cells. ***d.*** Chemical synapse signaling involves transmission of signal molecules, called neurotransmitters, from a neuron over a small synaptic gap to the target cell.

Endocrine signaling

A released signal molecule that remains in the extracellular fluid may enter the organism's circulatory system and travel widely throughout the body. These longer lived signal molecules, which may affect cells very distant from the releasing cell, are called **hormones,** and this type of intercellular communication is known as **endocrine signaling** (figure 9.2*c*). Chapter 46 discusses endocrine signaling in detail. Both animals and plants use this signaling mechanism extensively.

Synaptic signaling

In animals, the cells of the nervous system provide rapid communication with distant cells. Their signal molecules, **neurotransmitters,** do not travel to the distant cells through the circulatory system as hormones do. Rather, the long, fiberlike extensions of nerve cells release neurotransmitters from their tips very close to the target cells (figure 9.2*d*). The association of a neuron and its target cell is called a **chemical synapse,** and this type of intercellular communication is called **synaptic signaling.** Whereas paracrine signals move through the fluid between cells, neurotransmitters cross the synaptic gap and persist only briefly. We will examine synaptic signaling more fully in chapter 44.

Signal transduction pathways lead to cellular responses

The types of signaling outlined earlier are descriptive and say nothing about how cells respond to signals. The events that occur within the cell on receipt of a signal are called **signal transduction.** These events form discrete pathways that lead to a cellular response to the signal received by receptors. Knowledge of these signal transduction pathways has exploded in recent years and indicates a high degree of complexity that explains how in some cases different cell types can have the same response to different signals, and in other cases different cell types can have a different response to the same signal.

For example, a variety of cell types respond to the hormone glucagon by mobilizing glucose as part of the body's mechanism to control blood glucose (chapter 46). This involves breaking down stored glycogen into glucose and turning on the genes that encode the enzymes necessary to synthesize glucose. In contrast, the hormone epinephrine has diverse effects on different cell types. We have all been startled or frightened by a sudden event. Your heart beats faster, you feel more alert, and you can even feel the hairs on your skin stand up. All of this is due in part to your body releasing the hormone epinephrine (also called adrenaline) into the bloodstream. This leads to the heightened state of alertness and increased heart rate and energy that prepare us to respond to extreme situations.

These differing effects of epinephrine depend on the different cell types with receptors for this hormone. In the liver, cells are stimulated to mobilize glucose while in the heart muscle cells contract more forcefully to increase blood flow. In addition, blood vessels respond by expanding in some areas and contracting in others to redirect blood flow to the liver, heart, and skeletal muscles. These different reactions depend on the fact that each cell type has a receptor for epinephrine, but different sets of proteins that respond to this signal.

Phosphorylation is key in control of protein function

The function of a signal transduction pathway is to change the behavior or nature of a cell. This action may require changing the composition of proteins that make up a cell or altering the activity of cellular proteins. Many proteins are inactive or nonfunctional as they are initially synthesized and require modification after synthesis for activation. In other cases, a protein may require modification for deactivation. A major source of control for protein function is the addition or removal of phosphate groups, called **phosphorylation** or **dephosphorylation,** respectively.

figure 9.3

PHOSPHORYLATION OF PROTEINS. Many proteins are controlled by their phosphorylation state: that is, they are activated by phosphorylation and deactivated by dephosphorylation or the reverse. The enzymes that add phosphate groups are called kinases. These form two classes depending on the amino acid the phosphate is added to, either serine/threonine kinases or tyrosine kinases. The action of kinases is reversed by another enzyme, a protein phosphatase.

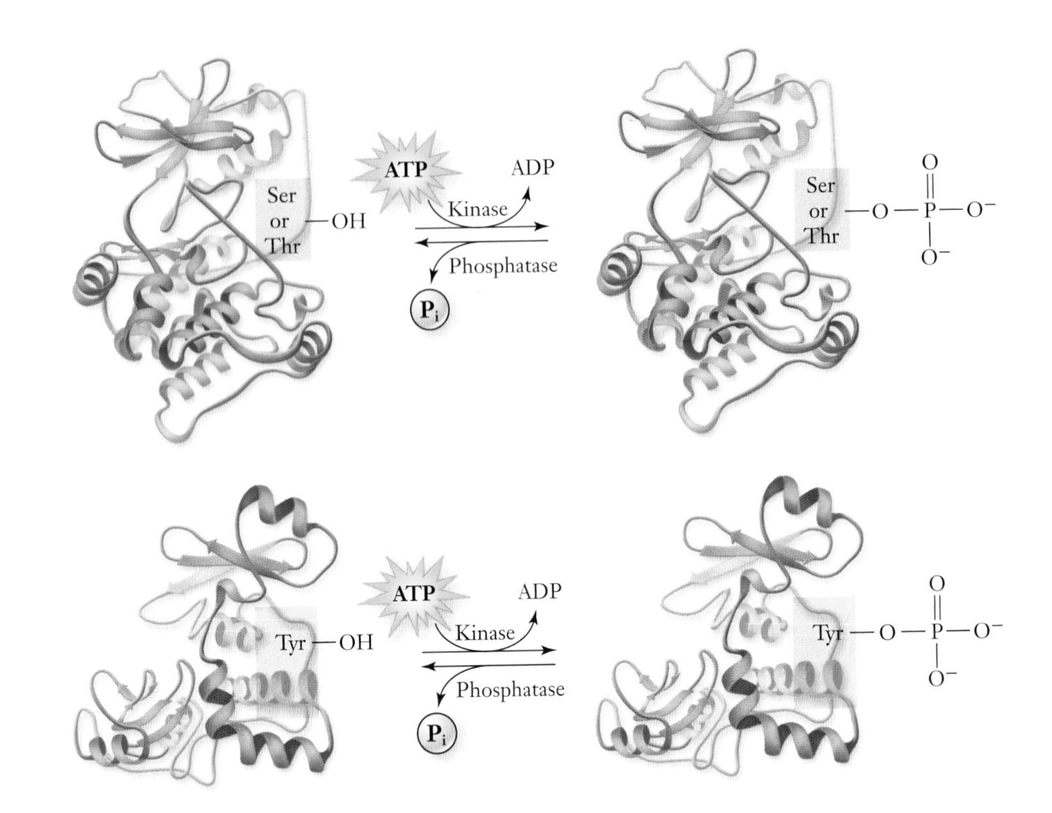

As you learned in preceding chapters, the end result of the metabolic pathways of cellular respiration and photosynthesis was the phosphorylation of ADP to ATP. The ATP synthesized by these processes can donate phosphate groups to proteins. The phosphorylation of proteins alters their function, which allows them to transmit information from an extracellular signal through a signal transduction pathway.

Protein kinases

The class of enzyme that adds phosphate groups from ATP to proteins is called a *protein kinase.* These phosphate groups can be added to the three amino acids that have an OH as part of their R group, namely serine, threonine, and tyrosine. We categorize protein kinases based on which of these three substrates they alter (figure 9.3). Most cytoplasmic protein kinases fall into the serine/threonine kinase class.

Phosphatases

Part of the reason for the versatility of phosphorylation as a form of protein modification is that it is reversible. Another class of enzymes called **phosphatases** removes phosphate groups, reversing the action of kinases (see figure 9.3). Thus, a protein activated by a kinase can be deactivated by a phosphatase, or the reverse.

Cell communication involves chemical signals, or ligands, that bind to cellular receptors. Binding of ligand to receptor initiates signal transduction pathways that lead to a cellular response. Different cells may have the same response to one signal and the same signal can also elicit different responses in different cells. The phosphorylation–dephosphorylation of proteins is a common mechanism of controlling protein function found in signaling pathways.

9.2 Receptor Types

The first step in understanding cell signaling is to consider the receptors themselves. Cells must have a specific receptor to be able to respond to a particular signaling molecule. The interaction of a receptor and its ligand is an example of molecular recognition, a process in which one molecule fits specifically based on its complementary shape with another molecule. This interaction causes subtle changes in the structure of the receptor, thereby activating it. This is the beginning of any signal transduction pathway.

Receptors are defined by location

The nature of these receptor molecules depends on their location and on the kind of ligands they bind. The broadest classes of receptors are those that bind a ligand inside the cell **(intracellular receptors)**, and those that bind a ligand outside the cell **(cell surface receptors** or **membrane receptors)** (figure 9.1). Membrane receptors consist of transmembrane proteins that are in contact with both the cytoplasm and the extracellular environment. Table 9.1 summarizes the types of receptors and other communication mechanisms we will discuss in this chapter.

Membrane receptors include three subclasses

When a receptor is a transmembrane protein, the ligand can bind to the receptor outside of the cell and never actually cross the plasma membrane. In this case, the receptor itself, and not the signaling molecule is responsible for information crossing the membrane. Membrane receptors can be subdivided based on their structure and function.

TABLE 9.1	Receptors Involved in Cell Signaling		
Receptor Type	**Structure**	**Function**	**Example**
Intracellular Receptors	No extracellular signal-binding site	Receives signals from lipid-soluble or noncharged, nonpolar small molecules	Receptors for NO, steroid hormone, vitamin D, and thyroid hormone
Cell Surface Receptors			
Chemically gated ion channels	Multipass transmembrane protein forming a central pore	Molecular "gates" triggered chemically to open or close	Neurons
Enzymatic receptors	Single-pass transmembrane protein	Binds signal extracellularly; catalyzes response intracellularly	Phosphorylation of protein kinases
G protein-coupled receptors	Seven-pass transmembrane protein with cytoplasmic binding site for G protein	Binding of signal to receptor causes GTP to bind a G protein; G protein, with attached GTP, detaches to, deliver the signal inside the cell	Peptide hormones, rod cells in the eyes

Channel-linked receptors

Chemically gated ion channels are receptor proteins that allow the passage of ions (figure 9.4*a*). The receptor proteins that bind many neurotransmitters have the same basic structure. Each is a membrane protein with multiple transmembrane domains, meaning that the chain of amino acids threads back and forth across the plasma membrane several times. In the center of the protein is a pore that connects the extracellular fluid with the cytoplasm. The pore is big enough for ions to pass through, so the protein functions as an **ion channel.**

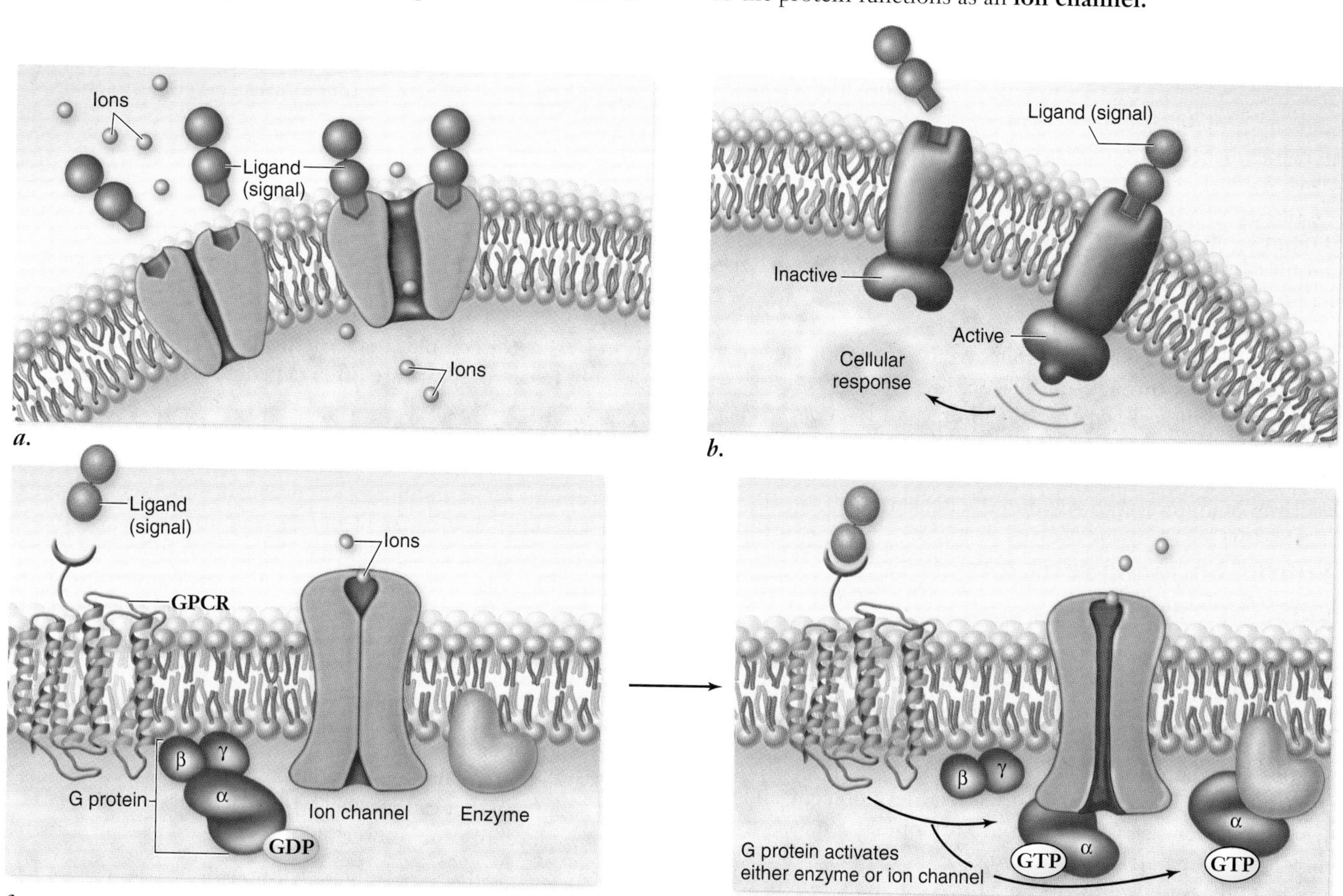

figure 9.4

CELL SURFACE RECEPTORS. ***a.*** Chemically gated ion channels are proteins that form a pore in the plasma membrane. This pore is opened or closed by chemical signals. They are usually selective, allowing the passage of only one type of ion. ***b.*** Enzymatic receptors are integral membrane proteins that bind the signal on the extracellular surface. A catalytic region on their cytoplasmic portion then transmits the signal by acting as an enzyme in the cytoplasm. ***c.*** G protein-coupled receptors (GPCR) bind to the signal outside the cell and to G proteins inside the cell. The G protein then activates an enzyme or ion channel, mediating the passage of a signal from the cell's surface to its interior.

The channel is said to be chemically gated because it opens only when a chemical (the neurotransmitter) binds to it. The type of ion that flows across the membrane when a chemically gated ion channel opens depends on the shape and charge structure of the channel. Sodium, potassium, calcium, and chloride ions all have specific ion channels.

The acetylcholine receptor found in muscle cell membranes functions as an Na^+ channel. When the receptor binds to its ligand, the neurotransmitter acetylcholine, the channel opens allowing Na^+ to flow into the muscle cell. This is a critical step linking the signal from a motor neuron to muscle cell contraction (chapter 44).

Enzymatic receptors

Many cell surface receptors either act as enzymes or are directly linked to enzymes (figure 9.4*b*). When a signal molecule binds to the receptor, it activates the enzyme. In almost all cases, these enzymes are **protein kinases,** enzymes that add phosphate groups to proteins. We discuss these receptors in detail in a later section of this chapter.

G Protein-coupled receptors

A third class of cell surface receptors acts indirectly on enzymes or ion channels in the plasma membrane with the aid of an assisting protein, called a **G protein.** The G protein, which is so named because it binds the nucleotide *guanosine triphosphate* (GTP), can be thought of as being inserted between the receptors and the enzyme (effector). That is, the ligand binds to the receptor, activating it, which activates the G protein, which in turn activates the effector protein (figure 9.4*c*). These receptors are also discussed in detail later on.

Membrane receptors can generate second messengers

Some enzymatic receptors and most G protein-coupled receptors utilize other substances to relay the message within the cytoplasm. These other substances, small molecules or ions called **second messengers,** alter the behavior of cellular proteins by binding to them and changing their shape. (The original signal molecule is considered the "first messenger.") Two common second messengers are **cyclic adenosine monophosphate (cyclic-AMP, or cAMP)** and calcium ions. The role of these second messengers will be explored in more detail in a later section.

Receptors may be internal (intracellular receptors) or external (membrane receptors). Membrane receptors include channel-linked receptors, enzymatic receptors, and G protein-coupled receptors. Signal transduction through membrane receptors often involves the production of a second signaling molecule, or second messenger, inside the cell.

9.3 Intracellular Receptors

Many cell signals are lipid-soluble or very small molecules that can readily pass through the plasma membrane of the target cell and into the cell, where they interact with an *intracellular receptor.* Some of these ligands bind to protein receptors located in the cytoplasm; others pass across the nuclear membrane as well and bind to receptors within the nucleus.

Steroid hormone receptors affect gene expression

Of all of the receptor types discussed in this chapter, the action of the steroid hormone receptors is the simplest and most direct.

Steroid hormones form a large class of compounds, including cortisol, estrogen, progesterone, and testosterone, that share a common nonpolar structure. Estrogen, progesterone, and testosterone are involved in sexual development and behavior (chapter 52). Other steroid hormones, such as cortisol, also have varied effects depending on the target tissue, ranging from the mobilization of glucose to the inhibition of white blood cells to control inflammation. Their antiinflammatory action is the basis of their use in medicine.

The nonpolar structure allows these hormones to cross the membrane and bind to intracellular receptors. The location of steroid hormone receptors prior to hormone binding is cytoplasmic, but their primary site of action is in the nucleus. Binding of the hormone to the receptor causes the complex to shift from the cytoplasm to the nucleus (figure 9.5). As the ligand–receptor complex makes it all the way to the nucleus of the cell, these receptors are often called **nuclear receptors.**

Steroid receptor action

The primary function of steroid hormone receptors, as well as receptors for a number of other small, lipid-soluble signal molecules such as vitamin D and thyroid hormone, is to act as regulators of gene expression (see chapter 16).

All of these receptors have similar structures; the genes that code for them appear to be the evolutionary descendants of a single ancestral gene. Because of their structural similarities, they are all part of the *nuclear receptor superfamily.*

Each of these receptors has three functional domains:

(1) a hormone-binding domain,
(2) a DNA-binding domain, and
(3) a domain that can interact with coactivators to affect the level of gene transcription.

In its inactive state, the receptor typically cannot bind to DNA because an inhibitor protein occupies the DNA-binding site. When the signal molecule binds to the hormone-binding site, the conformation of the receptor changes, releasing the inhibitor and exposing the DNA-binding site, allowing the receptor to attach to specific nucleotide sequences on the DNA (see figure 9.5). This binding activates (or, in a few instances, suppresses) particular genes, usually located adjacent to the hormone-binding sequences. In the case of cortisol, which is

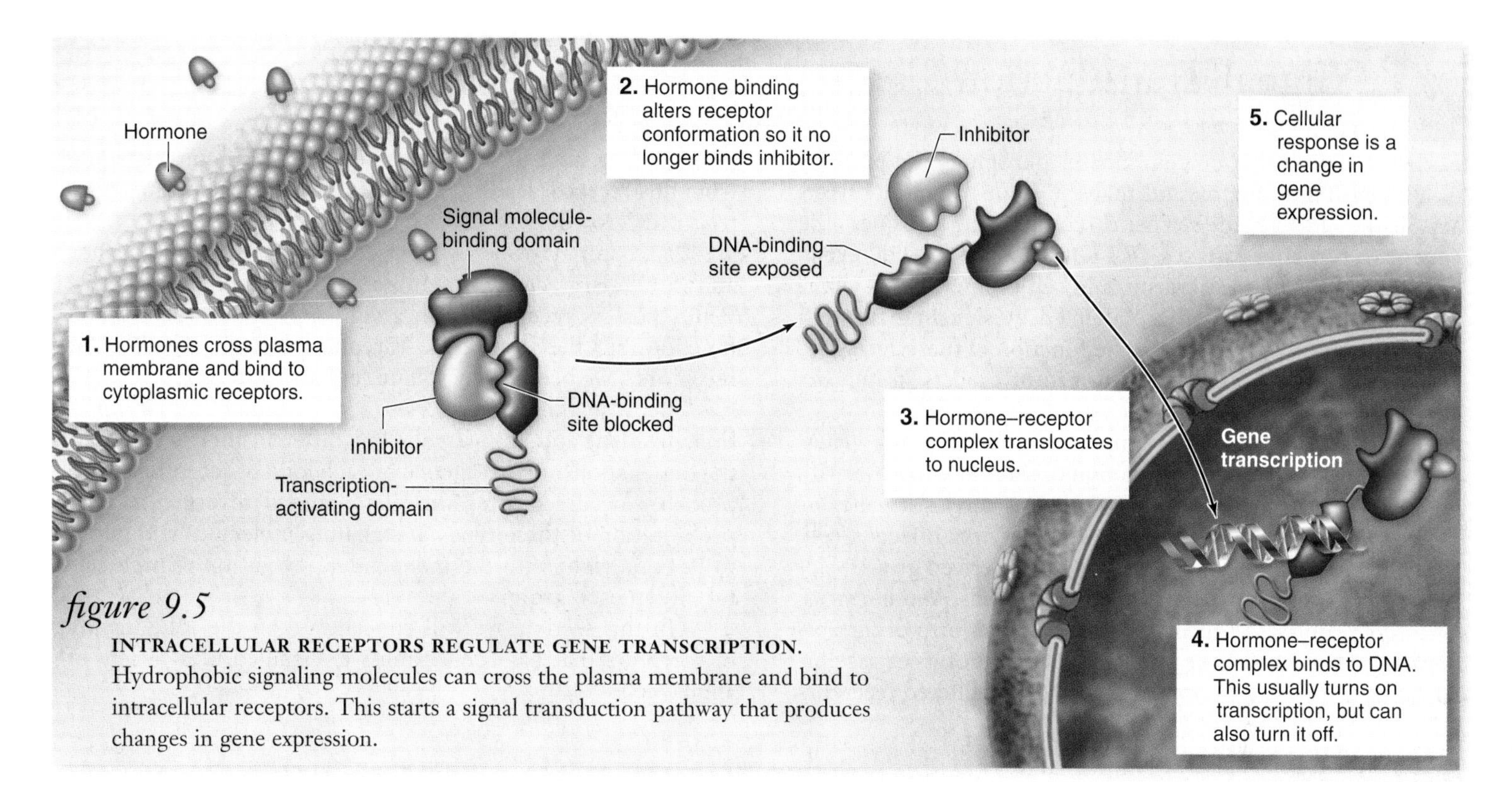

figure 9.5

INTRACELLULAR RECEPTORS REGULATE GENE TRANSCRIPTION. Hydrophobic signaling molecules can cross the plasma membrane and bind to intracellular receptors. This starts a signal transduction pathway that produces changes in gene expression.

a glucocorticoid hormone that can increase levels of glucose in cells, a number of different genes involved in the synthesis of glucose have binding sites for the hormone receptor complex.

The lipid-soluble ligands that intracellular receptors recognize tend to persist in the blood far longer than water-soluble signals. Most water-soluble hormones break down within minutes, and neurotransmitters break down within seconds or even milliseconds. In contrast, a steroid hormone such as cortisol or estrogen persists for hours.

Specificity and the role of coactivators

The target cell's response to a lipid-soluble cell signal can vary enormously, depending on the nature of the cell. This characteristic is true even when different target cells have the same intracellular receptor. Given that the receptor proteins bind to specific DNA sequences, which are the same in all cells, this may seem puzzling. It is explained in part by the fact that the receptors act in concert with **coactivators,** and the number and nature of these molecules can differ from cell to cell. Thus, a cell's response depends on not only the receptors but also the coactivators present.

The hormone estrogen has different effects in uterine tissue than in mammary tissue. This differential response is mediated by coactivators and not by the presence or absence of a receptor in the two tissues. In mammary tissue, a critical coactivator is lacking and the hormone–receptor complex instead interacts with another protein that acts to reduce gene expression. In uterine tissue, the coactivator is present, and the expression of genes that encode proteins involved in preparing the uterus for pregnancy are turned on.

Other intracellular receptors act as enzymes

A very interesting example of a receptor acting as an enzyme is found in the receptor for nitric oxide (NO). This small gas molecule diffuses readily out of the cells where it is produced and passes directly into neighboring cells, where it binds to the enzyme guanylyl cyclase. Binding of NO activates this enzyme, enabling it to catalyze the synthesis of **cyclic guanosine monophosphate (cGMP),** an intracellular messenger molecule that produces cell-specific responses such as the relaxation of smooth muscle cells.

When the brain sends a nerve signal to relax the smooth muscle cells lining the walls of vertebrate blood vessels, acetylcholine released by the nerve cell binds to receptors on epithelial cells. This causes an increase in intracellular Ca^{2+} in the epithelial cell that stimulates nitric oxide synthase to produce NO. The NO diffuses into the smooth muscle, where it increases the level of cGMP, leading to relaxation. This relaxation allows the vessel to expand and thereby increases blood flow. This explains the use of nitroglycerin to treat the pain of angina caused by constricted blood vessels to the heart. The nitroglycerin is converted by cells to NO, which then acts to relax the blood vessels.

The drug sildenafil (better known as Viagra) also functions via this signal transduction pathway by binding to and inhibiting the enzyme cGMP phosphodiesterase, which breaks down cGMP. This keeps levels of cGMP high, thereby stimulating production of NO. The reason for Viagra's selective effect is that it binds to a form of cGMP phosphodiesterase found in cells in the penis. This allows relaxation of smooth muscle in erectile tissue, thereby increasing blood flow.

Hydrophobic signaling molecules can cross the membrane and bind to intracellular receptors. The steroid hormone receptors act by directly affecting gene expression. On binding hormone, the hormone–receptor moves into the nucleus to turn on (or sometimes turn off) gene expression. This also requires another protein called a coactivator that functions with the hormone–receptor. Thus, the cell's response to a hormone depends on the presence of a receptor and coactivators as well.

9.4 Signal Transduction Through Receptor Kinases

Earlier you read that protein kinases phosphorylate proteins to alter protein function and that the most common kinases act on the amino acids serine, threonine, and tyrosine. The **receptor tyrosine kinases (RTK)** influence the cell cycle, cell migration, cell metabolism, and cell proliferation—virtually all aspects of the cell are affected by signaling through these receptors. Alterations to the function of these receptors and their signaling pathways can lead to cancers in humans and other animals.

Some of the earliest examples of cancer-causing genes, or oncogenes, involve RTK function (discussed in chapter 10). The cancer-causing simian sarcoma virus carries a gene for platelet-derived growth factor. When the virus infects a cell, the cell overproduces and secretes platelet-derived growth factor, causing overgrowth of the surrounding cells. Another virus, avian erythroblastosis virus, carries an altered form of the epidermal growth factor receptor that lacks most of its extracellular domain. When this virus infects a cell the altered receptors produced are stuck in the "on" state. The continuous signaling from this receptor leads to cells that have lost the normal controls over growth.

Receptor tyrosine kinases recognize hydrophilic ligands and form a large class of membrane receptors in animal cells. Plants possess receptors with a similar overall structure and function, but they are serine–threonine kinases. These plant receptors have been named **plant receptor kinases.**

Because these receptors are performing similar functions in plant and animal cells but differ in their substrates, the duplication and divergence of each kind of receptor kinase probably occurred after the plant–animal divergence. The proliferation of these types of signaling molecules is thought to be coincident with the independent evolution of multicellularity in each group.

In this section, we will concentrate on the RTK family of receptors that has been extensively studied in a variety of animal cells.

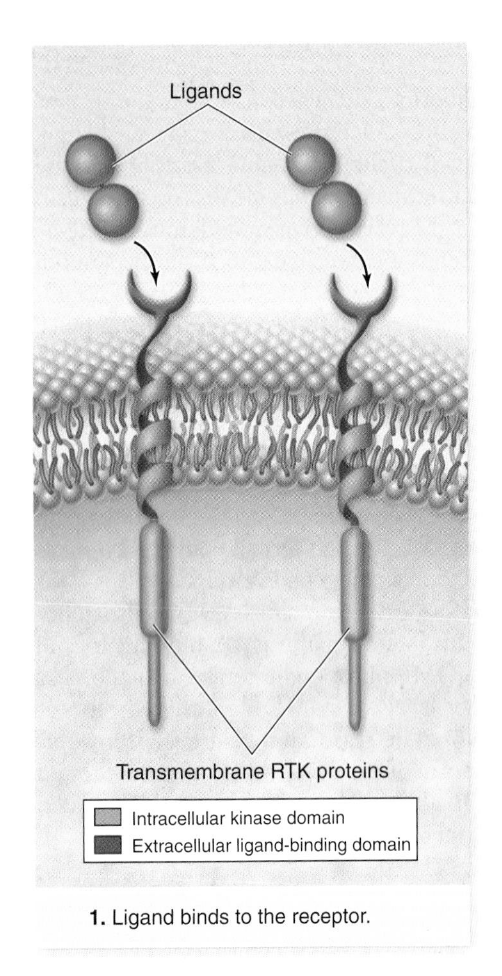

1. Ligand binds to the receptor.

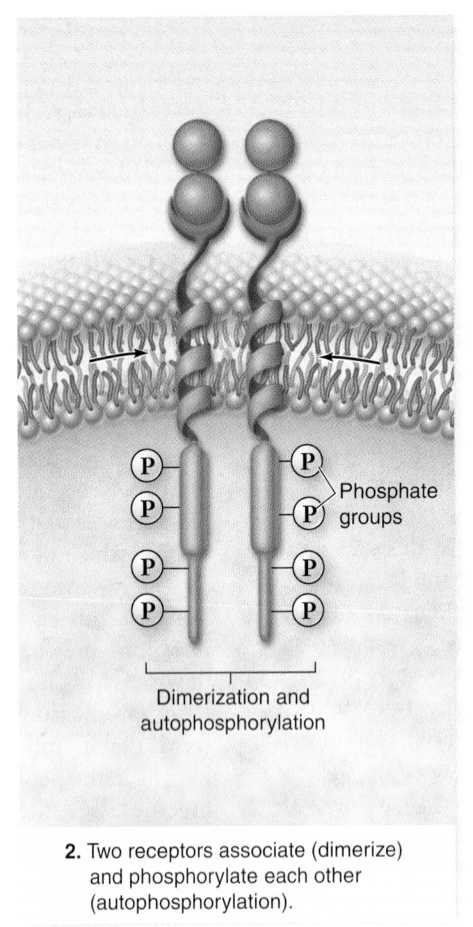

2. Two receptors associate (dimerize) and phosphorylate each other (autophosphorylation).

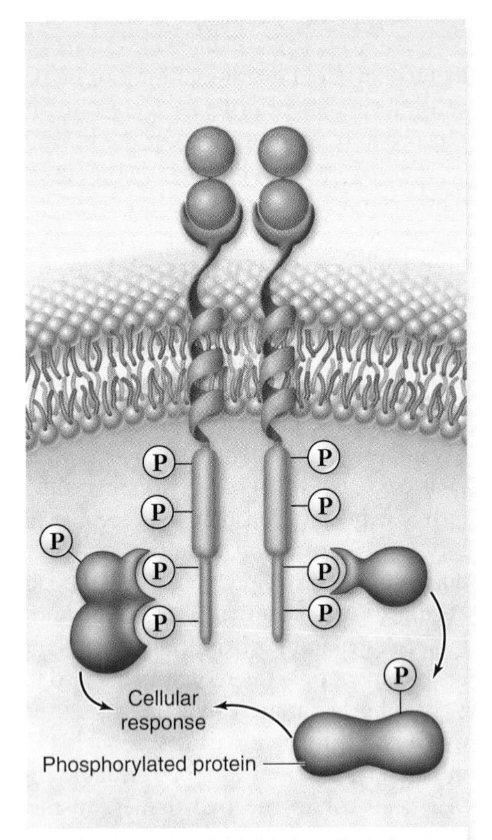

3. Response proteins bind to phosphotyrosine on receptor. Receptor can phosphorylate other response proteins.

figure 9.6

ACTIVATION OF A RECEPTOR TYROSINE KINASE (RTK). These membrane receptors bind hormones or growth factors that are hydrophilic and cannot cross the membrane. The receptor is a transmembrane protein with an extracellular ligand binding domain and an intracellular kinase domain. Signal transduction pathways begin with response proteins binding to phosphotyrosine on receptor, and by receptor phosphorylation of response proteins.

RTKs are activated by autophosphorylation

Receptor tyrosine kinases have a relatively simple structure consisting of a single transmembrane domain that anchors them in the membrane, an extracellular ligand-binding domain, and an intracellular kinase domain. This kinase domain contains the catalytic site of the receptor, which acts as a protein kinase that adds phosphate groups to tyrosines. On ligand binding to a specific receptor, two of these receptor–ligand complexes associate together (often referred to as dimerization) and phosphorylate each other, a process called *autophosphorylation* (figure 9.6).

The autophosphorylation event transmits across the membrane the signal that began with the binding of the ligand to the receptor. The next step, propagation of the signal in the cytoplasm, can take a variety of different forms. These forms include activation of the tyrosine kinase domain to phosphorylate other intracellular targets or interaction of other proteins with the phosphorylated receptor.

The cellular response after activation depends on the possible response proteins in the cell. Two different cells can have the same receptor yet a different response, depending on what response proteins are present in the cytoplasm. For example fibroblast growth factor stimulates cell division in fibroblasts but stimulates nerve cells to differentiate rather than to divide.

Phosphotyrosine domains mediate protein–protein interactions

One way that the signal from the receptor can be propagated in the cytoplasm is via proteins that bind specifically to phosphorylated tyrosines in the receptor. When the receptor is activated, regions of the protein outside of the catalytic site are phosphorylated. This creates "docking" sites for proteins that bind specifically to phosphotyrosine. The proteins that bind to these phosphorylated tyrosines can initiate intracellular events to convert the signal from the ligand into a response (see figure 9.6).

The insulin receptor

The use of docking proteins is illustrated by the insulin receptor. The hormone insulin is part of the body's control system to maintain a constant level of blood glucose. The role of insulin is to lower blood glucose, acting by binding to an RTK. Another protein called the *insulin response protein* binds to the phosphorylated receptor and is itself phosphorylated. The insulin response protein passes the signal on by binding to additional proteins that lead to the activation of the enzyme glycogen synthase, which converts glucose to glycogen (figure 9.7), thereby lowering blood glucose. Other proteins act to inhibit the synthesis of enzymes involved in making glucose.

Adapter proteins

Another class of proteins, **adapter proteins,** can also bind to phosphotyrosines. These proteins themselves do not participate in signal transduction but act as a link between the receptor and proteins that initiate downstream signaling events. For example, the Ras protein discussed later, is activated by adapter proteins binding to a receptor.

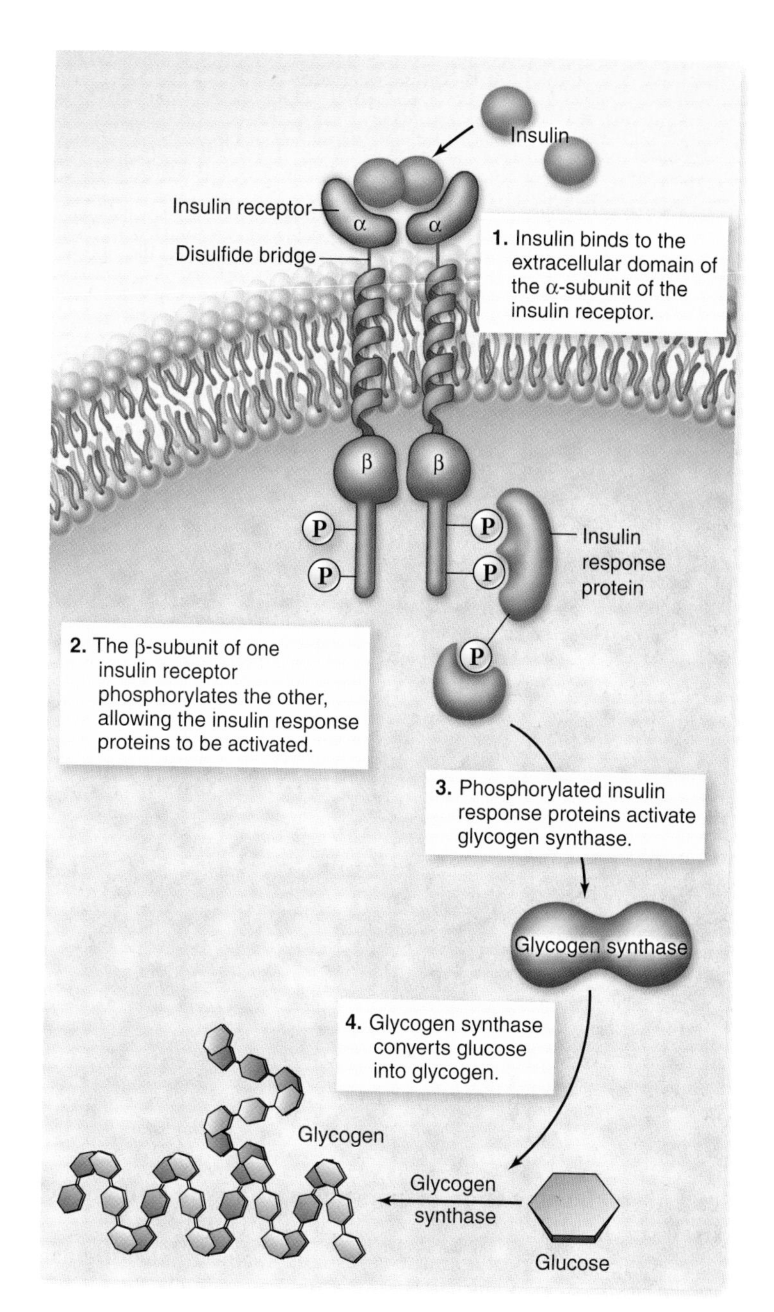

figure 9.7

THE INSULIN RECEPTOR. The insulin receptor is a receptor tyrosine kinase that initiates a variety of cellular responses related to glucose metabolism. One signal transduction pathway that this receptor mediates leads to the activation of the enzyme glycogen synthase. This enzyme converts glucose to glycogen.

Protein kinase cascades can amplify a signal

One important class of cytoplasmic kinases are **mitogen-activated protein (MAP) kinases.** A *mitogen* is a chemical that stimulates cell division by activating the normal pathways that control division. The MAP kinases are activated by a signaling module called a *phosphorylation cascade* or a **kinase cascade.** This module is a series of protein kinases that phosphorylate each other in succession. The final step in the

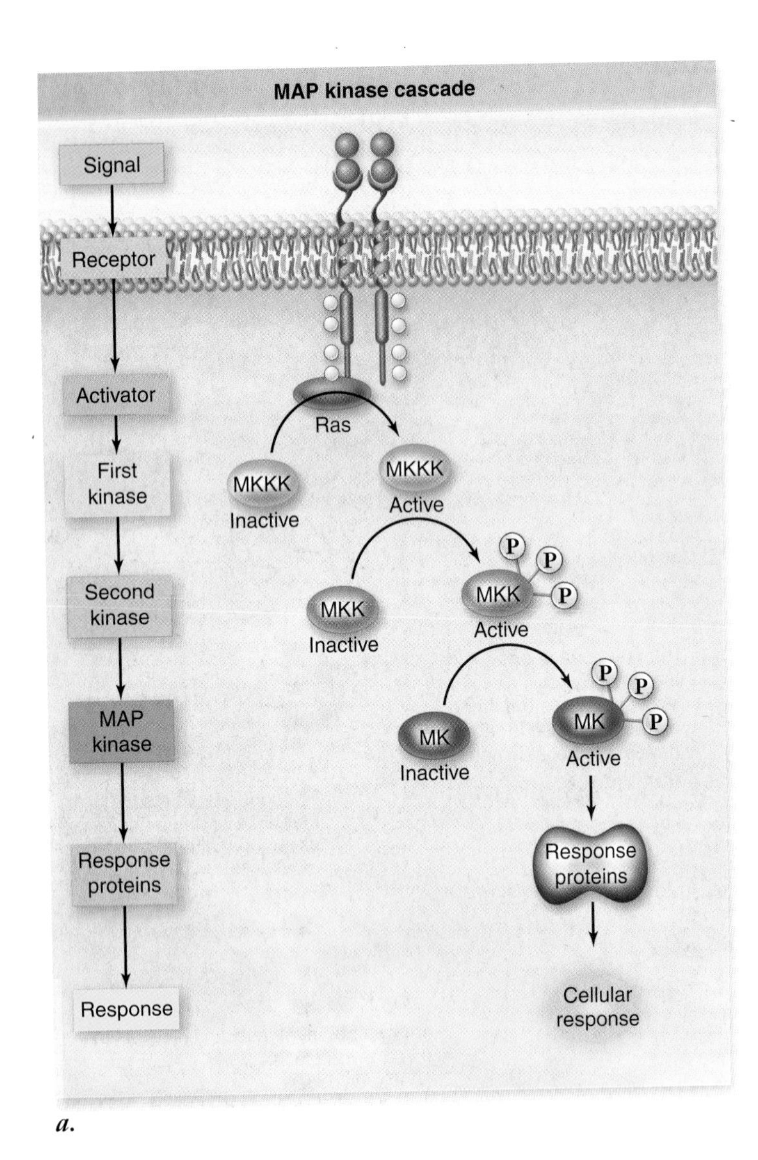

a.

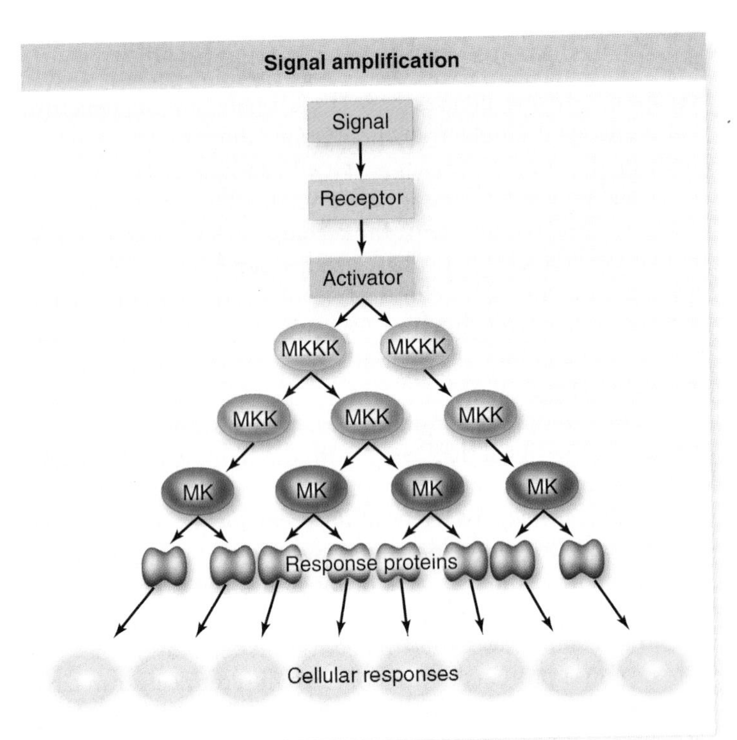

b.

figure 9.8

MAP KINASE CASCADE LEADS TO SIGNAL AMPLIFICATION. ***a.*** Phosphorylation cascade is shown as a flowchart on the left. Next to the flowchart the corresponding cellular events are shown, beginning with the receptor in the plasma membrane. Each kinase is named starting with the last, the MAP kinase (MK), which is phosphorylated by a MAP kinase kinase (MKK), which is in turn phosphorylated by a MAP kinase kinase kinase (MKKK). The cascade is linked to the receptor protein by an activator protein. The proteins are color-coded to link the action in the cell with the flowchart. ***b.*** At each step the enzymatic action of the kinase on multiple substrates leads to amplification of the signal.

cascade is the activation by phosphorylation of MAP kinase itself (figure 9.8).

One function of a kinase cascade is to amplify the original signal. Because each step in the cascade is an enzyme, it can act on a number of substrate molecules. With each enzyme in the cascade acting on many substrates this produces a large amount of the final product (see figure 9.8). This allows a small number of initial signaling molecules to produce a large response.

The cellular response to this cascade in any particular cell depends on the targets of the MAP kinase, but usually involves phosphorylating transcription factors that then activate gene expression (chapter 16). An example of this kind of signaling through growth factor receptors is provided in chapter 10 and illustrates how signal transduction initiated by a growth factor can control the process of cell division through a kinase cascade.

Scaffold proteins organize kinase cascades

The proteins in a kinase cascade need to act sequentially to be effective. One way the efficiency of this process can be increased is to organize them in the cytoplasm. Proteins called **scaffold proteins** are thought to organize the components of a kinase cascade into a single protein complex, the ultimate in a signaling module. The scaffold protein binds to each individual kinase such that they are spatially organized for optimal function (figure 9.9).

The advantages of this kind of organization are many. A physically arranged sequence is clearly more efficient than one that depends on diffusion to produce the appropriate order of events. This organization also allows the segregation of signaling modules in different cytoplasmic locations.

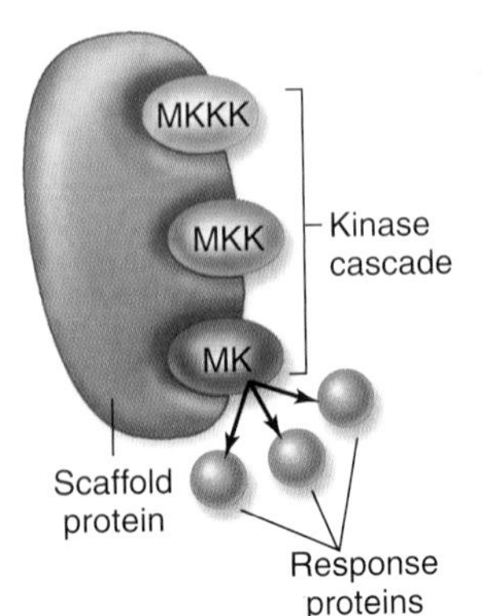

figure 9.9

KINASE CASCADE CAN BE ORGANIZED BY SCAFFOLD PROTEINS. The scaffold protein binds to each kinase in the cascade, organizing them so each substrate is next to its enzyme. This organization also sequesters the kinases from other signaling pathways in the cytoplasm.

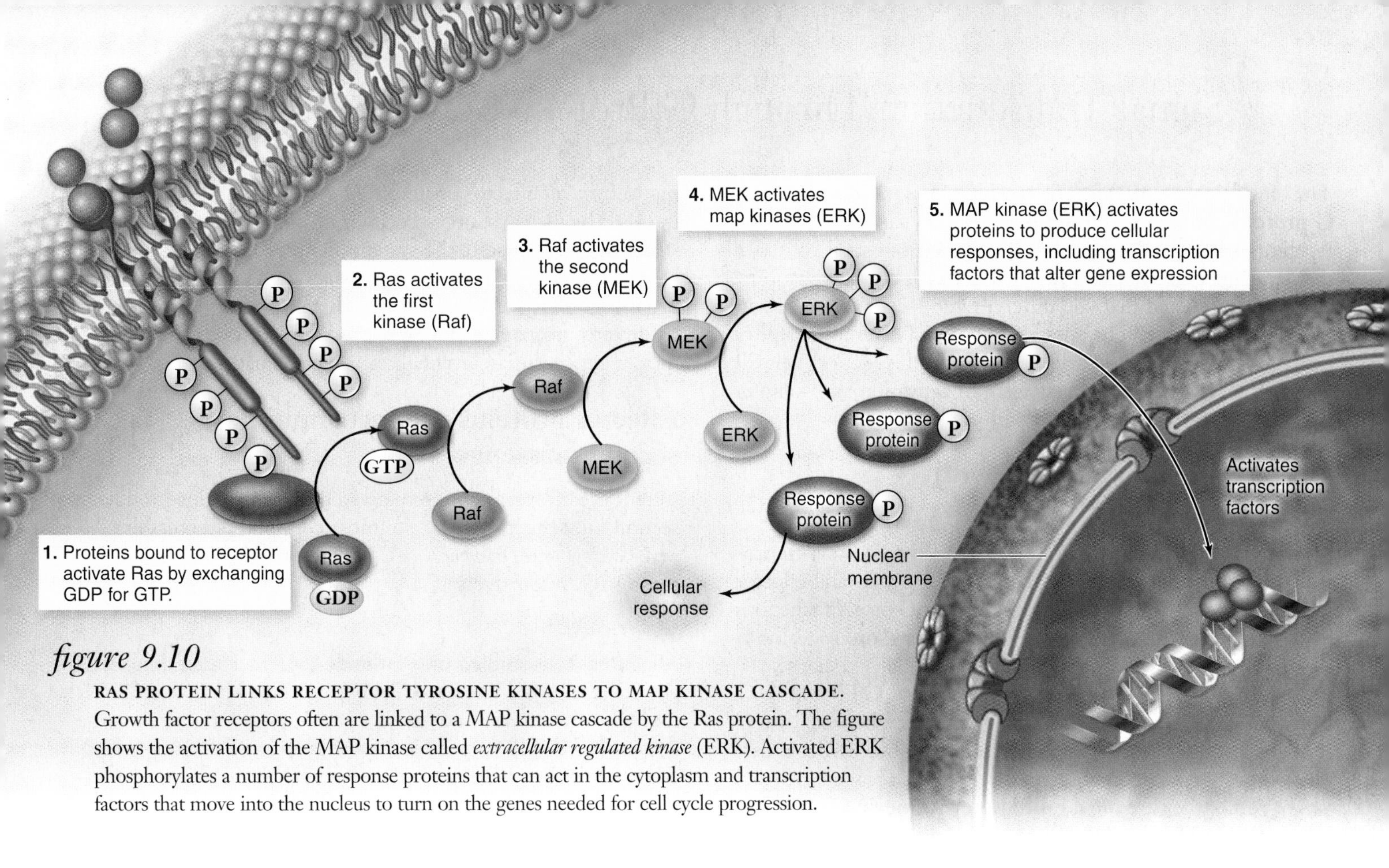

figure 9.10

RAS PROTEIN LINKS RECEPTOR TYROSINE KINASES TO MAP KINASE CASCADE. Growth factor receptors often are linked to a MAP kinase cascade by the Ras protein. The figure shows the activation of the MAP kinase called *extracellular regulated kinase* (ERK). Activated ERK phosphorylates a number of response proteins that can act in the cytoplasm and transcription factors that move into the nucleus to turn on the genes needed for cell cycle progression.

The disadvantage of this kind of organization is that it reduces the amplification effect of the kinase cascade. Enzymes held in one place are not free to find new substrate molecules, but must rely on substrates being nearby.

The best studied example of a scaffold protein comes from mating behavior in budding yeast. Yeast cells respond to mating pheromones with changes in cell morphology and gene expression, mediated by a protein kinase cascade. A protein called Ste5 was originally identified as a protein required for mating behavior, but no enzymatic activity could be detected for this protein. It has now been shown that this protein interacts with all of the members of the kinase cascade and acts as a scaffold protein that organizes the cascade and insulates it from other signaling pathways.

Ras proteins link receptors with kinase cascades

The link between the RTK and the MAP kinase cascade is a small GTP-binding protein (G protein) called **Ras.** The Ras protein is mutated in many human tumors, indicative of its central role in linking growth factor receptors to their cellular response.

The Ras protein is active when bound to GTP and inactive when bound to GDP. When an RTK, such as a growth factor receptor, is activated, it binds to adapter proteins that then act on Ras to stimulate the exchange of GDP for GTP, activating Ras. The Ras protein then activates the first kinase in the MAP kinase cascade (figure 9.10).

One key to signaling through this pathway is that Ras can regulate itself. The Ras protein has intrinsic GTPase activity, hydrolyzing GTP to GDP and P_i, leaving the GDP bound to Ras, which is now inactivated. The action of the RTK turns on Ras, which can be thought of as a switch that can turn itself off. This is one reason that stimulation of cell division by growth factors is short-lived.

RTKs are inactivated by internalization

It is important to cells that signaling pathways are only activated transiently. Continued activation could render the cell unable to respond to other signals or to respond inappropriately to a signal that is no longer relevant. Consequently, inactivation is as important for the control of signaling as activation. Receptor tyrosine kinases can be inactivated by two basic mechanisms: dephosphorylation and internalization. Internalization is by endocytosis, in which the receptor is taken up into the cytoplasm in a vesicle where it can be degraded or recycled.

The enzymes in the kinase cascade are all controlled by dephosphorylation by phosphatase enzymes. This leads to termination of the response at both the level of the receptor and the response proteins.

Receptor tyrosine kinases (RTK) are membrane receptors that can phosphorylate tyrosine. When activated, they autophosphorylate, creating binding domains for other proteins. These proteins transmit the signal inside the cell. One form of signaling pathway involves the MAP kinase cascade, a series of kinases that each activate the next in the series. This ends with a MAP kinase that activates transcription factors to alter gene expression.

9.5 Signal Transduction Through G Protein-Coupled Receptors

The single largest category of receptor type in animal cells is **G protein-coupled receptors (GPCRs),** so named because the receptors act by coupling with a G protein. G proteins are proteins that bind guanosine nucleotides, such as Ras discussed previously.

It is estimated that 750 different genes encode GPCRs in the human genome. The sheer number and diversity of function of these receptors is overwhelming. In this section, we will concentrate on the basic mechanism of activation and some of the possible signal transduction pathways.

G proteins link receptors with effector proteins

The function of the G protein in signaling by GPCRs is to provide a link between a receptor that receives signals and effector proteins that produce cellular responses. The G protein functions as a switch that is turned on by the receptor. In its "on" state, the G protein activates effector proteins to cause a cellular response.

All G proteins are active when bound to GTP and inactive when bound to GDP. The main difference between the G proteins in GPCRs and the Ras protein described earlier is that these G proteins are composed of three subunits, called α, β, and γ. As a result, they are often called **heterotrimeric G proteins.** When a ligand binds to a GPCR and activates its associated G protein, the G protein exchanges GDP for GTP and dissociates into two parts consisting of the G_α subunit bound to GTP, and the G_β and G_γ subunits together ($G_{\beta\gamma}$). The signal can then be propagated by either the G_α or the $G_{\beta\gamma}$ components, thereby acting to turn on effector proteins. The hydrolysis of bound GTP to GDP by G_α causes reassociation of the heterotrimer and restores the "off" state of the system (figure 9.11).

The effector proteins are usually enzymes. An effector protein might be a protein kinase that phosphorylates proteins to directly propagate the signal, or it may produce a second messenger to initiate a signal transduction pathway.

Effector proteins produce multiple second messengers

Often, the effector proteins activated by G proteins produce a second messenger. Two of the most common effectors are *adenylyl cyclase* and *phospholipase C,* which produce cAMP and IP_3 plus DAG, respectively.

Cyclic-AMP

All animal cells studied thus far use cAMP as a second messenger (chapter 46). When a signaling molecule binds to a GPCR that uses the enzyme **adenylyl cyclase** as an effector, a large amount of cAMP is produced within the cell (figure 9.12*a*). The cAMP then binds to and activates the enzyme protein kinase A (PKA), which adds phosphates to specific proteins in the cell (figure 9.13).

The effect of this phosphorylation on cell function depends on the identity of the cell and the proteins that are phosphorylated. In muscle cells, for example, PKA activates an enzyme necessary to break down glycogen and inhibits another enzyme

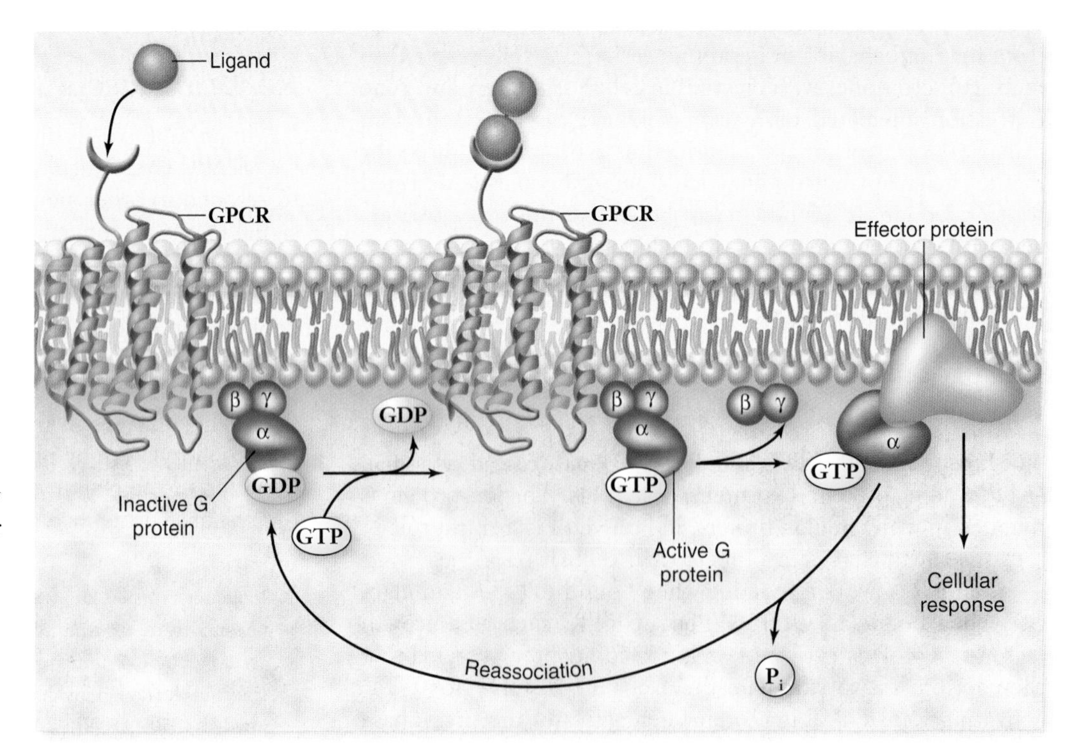

figure 9.11

THE ACTION OF G PROTEIN-COUPLED RECEPTORS. G protein-coupled receptors act through a heterotrimeric G protein that links the receptor to an effector protein. When ligand binds to the receptor, it activates an associated G protein, exchanging GDP for GTP. The active G protein complex dissociates into G_α and $G_{\beta\gamma}$. The G_α subunit (bound to GTP) is shown activating an effector protein. The effector protein may act directly on cellular proteins or produce a second messenger to cause a cellular response. G_α can hydrolyze GTP inactivating the system, then reassociate with $G_{\beta\gamma}$.

ATP

Adenylyl cyclase

cAMP + PP_i

cAMP

PP_i

a.

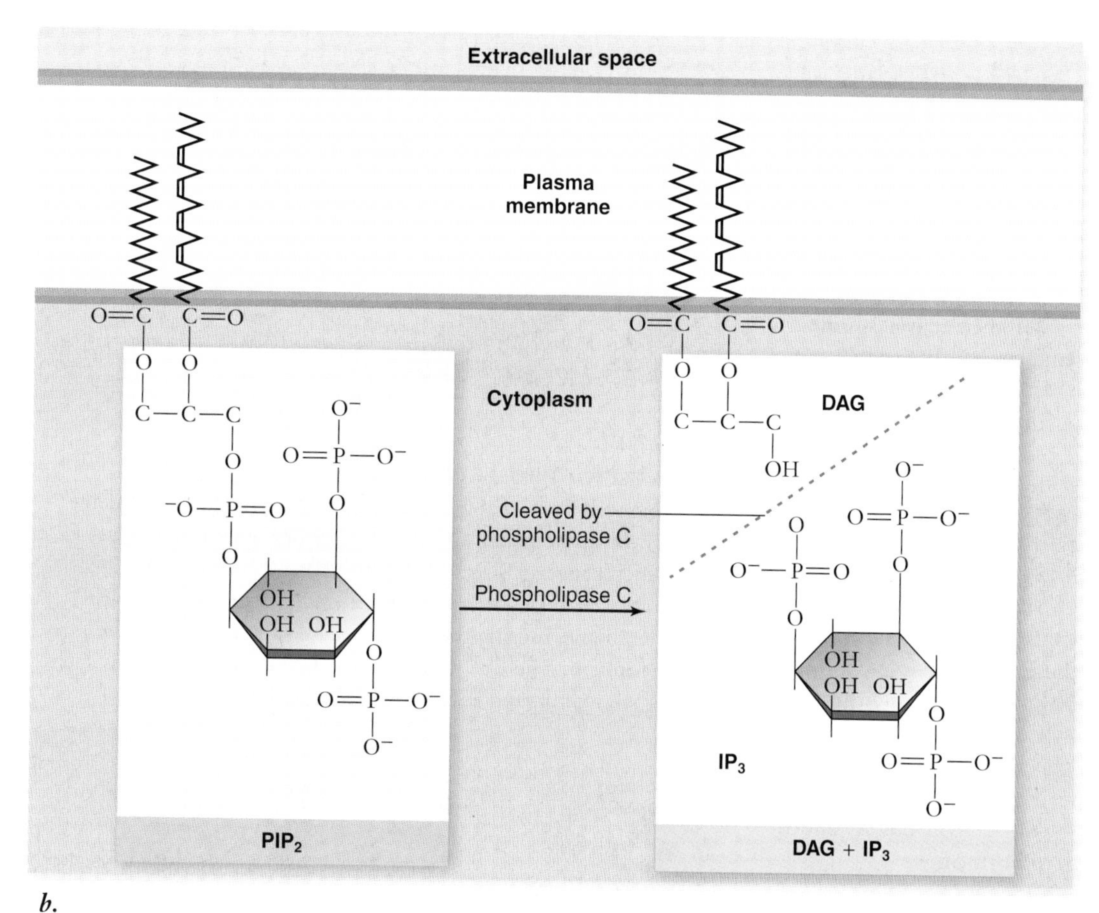

b.

figure 9.12

PRODUCTION OF SECOND MESSENGERS. Second messengers are signaling molecules produced within the cell. ***a.*** The nucleotide ATP is converted by the enzyme adenylyl cyclase into cyclic AMP, or cAMP. ***b.*** The inositol phospholipid is composed of two lipids and a phosphate attached to glycerol. The phosphate is also attached to the sugar inositol. This molecule can be cleaved by the enzyme phospholipase C to produce two different second messengers: DAG, made up of the glycerol with the two lipids, and IP_3, inositol triphosphate.

necessary to synthesize glycogen. This leads to an increase in glucose available to the muscle. By contrast, in the kidney the action of PKA leads to the production of water channels that can increase the permeability of tubule cells to water.

Disruption of cAMP signaling can have a variety of effects. The symptoms of the disease cholera are due to altered cAMP levels in cells in the gut. The bacterium *Vibrio cholerae* produces a toxin that binds to a GPCR in the epithelium of the gut, causing it to be locked into an "on" state. This causes a large increase in intracellular cAMP that, in these cells, causes Cl^- ions to be transported out of the cell. Water follows the Cl^-, leading to the diarrhea and dehydration characteristic of the disease.

Inositol phosphates

A common second messenger is produced from the molecules called inositol phospholipids. These are inserted into the plasma membrane by their lipid ends and have the **inositol phosphate** portion protruding into the cytoplasm. The most common inositol phospholipid is phosphatidylinositol-4,5-bisphosphate (PIP_2). This molecule is a substrate of the effector protein phospholipase C, which cleaves PIP_2 to yield **diacylglycerol (DAG)** and **inositol-1,4,5-trisphosphate (IP_3)** (see figure 9.12*b*).

Both of these compounds then act as second messengers with a variety of cellular effects. DAG, like cAMP, can activate a protein kinase, in this case protein kinase C (PKC).

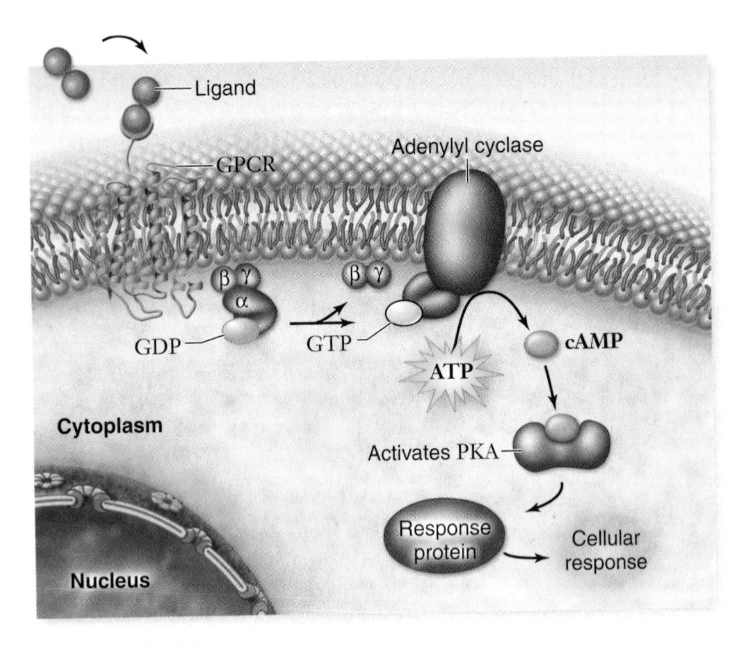

figure 9.13

cAMP SIGNALING PATHWAY. Extracellular signal binds to a GPCR, activating a G protein. The G protein then activates the effector protein adenylyl cyclase, which catalyzes the conversion of ATP to cAMP. The cAMP then activates protein kinase A (PKA), which phosphorylates target proteins to cause a cellular response.

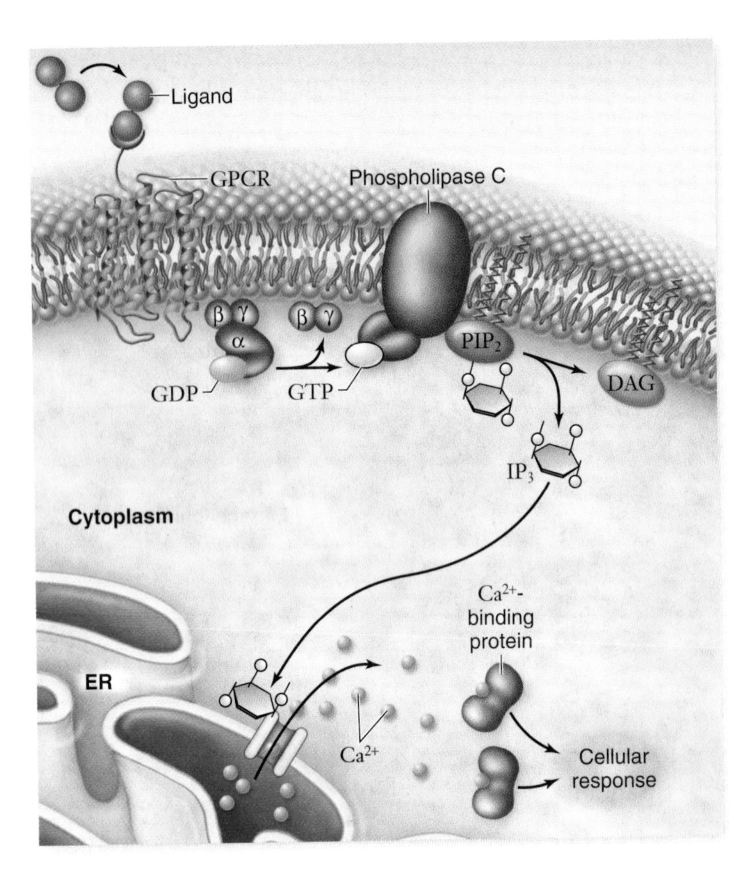

figure 9.14

INOSITOL PHOSPHOLIPID AND Ca^{2+} SIGNALING. Extracellular signal binds to a GPCR activating a G protein. The G protein activates the effector protein phospholipase C, which converts PIP_2 to DAG and IP_3. IP_3 is then bound to a channel-linked receptor on the endoplasmic reticular (ER) membrane, causing the ER to release stored Ca^{2+} into the cytoplasm. The Ca^{2+} then binds to Ca^{2+}-binding proteins such as calmodulin and PKC to cause a cellular response.

Calcium

Calcium ions (Ca^{2+}) serve widely as second messengers. Ca^{2+} levels inside the cytoplasm are normally very low (less than 10^{-7} M), whereas outside the cell and in the endoplasmic reticulum, Ca^{2+} levels are quite high (about 10^{-3} M). The endoplasmic reticulum has receptor proteins that act as ion channels to release Ca^{2+}. One of the most common of these receptors can bind the second messenger IP_3 to release Ca^{2+}, linking signaling through inositol phosphates with signaling by Ca^{2+} (figure 9.14).

The result of the outflow of Ca^{2+} from the endoplasmic reticulum depends on the cell type. For example, in skeletal muscle cells Ca^{2+} stimulates muscle contraction but in endocrine cells it stimulates the secretion of hormones.

Ca^{2+} initiates some cellular responses by binding to *calmodulin*, a 148-amino-acid cytoplasmic protein that contains four binding sites for Ca^{2+} (figure 9.15). When four Ca^{2+} ions are bound to calmodulin, the calmodulin/Ca^{2+} complex is able to bind to other proteins to activate them. These proteins include protein kinases, ion channels, receptor proteins, and cyclic nucleotide phosphodiesterases. These many uses of Ca^{2+} make it one of the most versatile second messengers in cells.

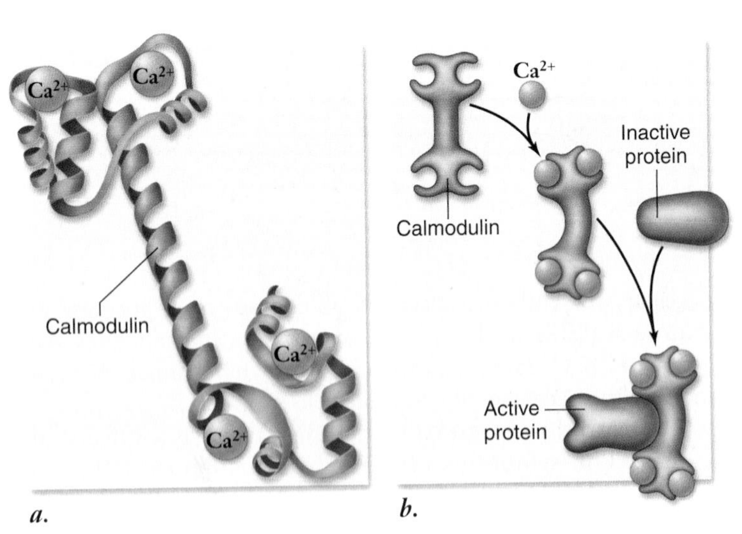

figure 9.15

CALMODULIN. ***a.*** Calmodulin is a protein containing 148 amino acid residues that mediates Ca^{2+} function. ***b.*** When four Ca^{2+} are bound to the calmodulin molecule, it undergoes a conformational change that allows it to bind to other cytoplasmic proteins and effect cellular responses.

Different receptors can produce the same second messengers

As mentioned previously, the two hormones glucagon and epinephrine can both stimulate liver cells to mobilize glucose. The reason that these different signals have the same effect is that they both act by the same signal transduction pathway to stimulate the breakdown and inhibit the synthesis of glycogen.

The binding of either hormone to its receptor activates a G protein that simulates adenylyl cyclase. The production of cAMP leads to the activation of PKA, which in turn activates another protein kinase called phosphorylase kinase. Activated phosphorylase kinase then activates glycogen phosphorylase, which cleaves off units of glucose-6-phosphate from the glycogen polymer (figure 9.16). The action of multiple kinases again leads to amplification such that a few signaling molecules result in a large number of glucose molecules being released.

At the same time, PKA also phosphorylates the enzyme glycogen synthase, but in this case it inhibits the enzyme, thus preventing the synthesis of glycogen. In addition, PKA phosphorylates other proteins that activate the expression of genes encoding the enzymes needed to synthesize glucose. This convergence of signal transduction pathways from different receptors leads to the same result: Glucose is mobilized.

Receptor subtypes can lead to different effects in different cells

We also saw earlier how a single signaling molecule, epinephrine, can have different effects in different cells. One way this happens is through the existence of multiple forms of the same receptor. The receptor for epinephrine actually has nine different subtypes, or isoforms. These are encoded by different genes and are actually different receptor molecules. The sequences of these proteins are very similar, especially in the ligand-binding domain, which allows them to bind epinephrine. They differ mainly in their cytoplasmic domains, which interact with G proteins. This leads to different isoforms activating different G proteins, thereby leading to different signal transduction pathways.

Thus, in the heart, muscle cells have one isoform of the receptor that, when bound to epinephrine, activates a G protein that activates adenylyl cyclase, leading to increased cAMP. This increases the rate and force of contraction. In the intestine, smooth muscle cells have a different isoform of the receptor that, when bound to epinephrine, activates a different G protein that inhibits adenylyl cyclase, which decreases cAMP. This has the result of relaxing the muscle.

G protein-coupled receptors and receptor tyrosine kinases can activate the same pathways

Different receptor types can affect the same signaling module. For example, RTKs were shown to activate the MAP kinase cascade, but GPCRs can also activate this same cascade. Similarly, the activation of phospholipase C was mentioned previously in the context of GPCR signaling, but it can also be activated by RTKs.

This cross-reactivity may appear to introduce complications into cell function, but in fact it provides the cell with an incredible amount of flexibility. Cells have a large, but limited number of intracellular signaling modules, which can be turned on and off by different kinds of membrane receptors. This leads to signaling networks that interconnect possible cellular effectors with multiple incoming signals.

The Internet represents an example of a network in which many different kinds of computers are connected globally. This network can be broken down into subnetworks that are connected to the overall network. Because of the nature of the connections, when you send an e-mail message across the Internet, it can reach its destination through many different pathways. Likewise, the cell has interconnected networks of signaling pathways in which many different signals, receptors, and response proteins are interconnected. Specific pathways like the MAP kinase cascade, or signaling through second messengers like cAMP and Ca^{2+}, represent subnetworks within the global signaling network. A specific signal can activate different pathways in different cells, or different signals can activate the same pathway. We do not yet understand the cell at this level, but the field of systems biology is moving toward such global understanding of cell function.

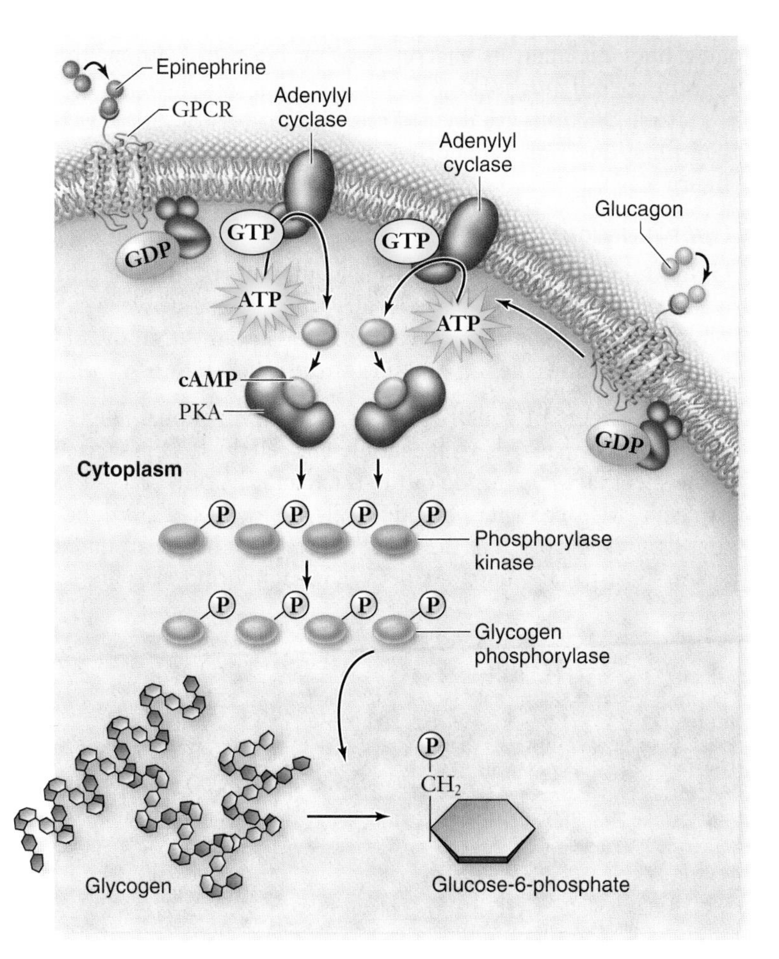

figure 9.16

DIFFERENT RECEPTORS CAN ACTIVATE THE SAME SIGNALING PATHWAY. The hormones glucagon and epinephrine both act through GPCRs. Each of these receptors acts via a G protein that activates adenylyl cyclase, producing cAMP. The activation of PKA begins a kinase cascade that leads to the breakdown of glycogen.

Signaling through GPCRs uses a three-part system: a receptor, a G protein, and an effector protein. G proteins are active when bound to GTP and inactive when bound to GDP. A ligand binding to the receptor activates the G protein, which then activates the effector protein. Effector proteins include adenylyl cyclase, which produces the second messenger cAMP. Another effector protein, phospholipase C, cleaves the inositol phosphates and results in the release of Ca^{2+} from the ER.

9.6 Cell-to-Cell Interactions

In multicellular organisms, not only must cells be able to communicate with one another, but they must also be organized in specific ways. With the exception of a few primitive types of organisms, the hallmark of multicellular life is the organization of highly specialized groups of cells into *tissues*, such as blood and muscle. Remarkably, each cell within a tissue performs the functions of that tissue and no other, even though all cells of the body are derived from a single fertilized cell and contain the same genetic information—all of the genes found in the genome.

This kind of tissue organization requires that cells have both identity and specific kinds of cell-to-cell connections. As an organism develops, the cells acquire their identities by carefully controlling the *expression* of those genes, turning on the specific set of genes that encode the functions of each cell type. How do cells sense where they are, and how do they "know" which type of tissue they belong to? Table 9.2 provides a summary of the kinds of connections seen between cells that are explored in the following sections.

Surface proteins give cells identity

One key set of genes functions to mark the surfaces of cells, identifying them as being of a particular type. When cells make contact, they "read" each other's cell surface markers and react accordingly. Cells that are part of the same tissue type recognize each other, and they frequently respond by forming connections between their surfaces to better coordinate their functions.

Glycolipids

Most tissue-specific cell surface markers are glycolipids, that is, lipids with carbohydrate heads. The glycolipids on the surface of red blood cells are also responsible for the A, B, and O blood types.

MHC proteins

One example of the function of cell surface markers is the recognition of "self" and "nonself" cells by the immune system. This function is vital for multicellular organisms, which need to defend themselves against invading or malignant cells. The immune system of vertebrates uses a particular set of markers to distinguish self from nonself cells, encoded by genes of the *major histocompatibility complex* (*MHC*). Cell recognition and the immune system is covered in chapter 51.

Cell connections mediate cell-to-cell adhesion

Most cells in a multicellular organism are in physical contact with other cells at all times, usually as members of organized tissues such as those in a leaf or those in your lungs, heart, or gut. These cells and the mass of other cells clustered around them form long-lasting or permanent connections with one another called **cell junctions.**

The nature of the physical connections between the cells of a tissue in large measure determines what the tissue is like. Indeed, a tissue's proper functioning often depends critically on how the individual cells are arranged within it. Just as a house cannot maintain its structure without nails and cement, so a tissue cannot maintain its characteristic architecture without the appropriate cell junctions.

Cell junctions are divided into three categories, based on their functions: tight, anchoring, and communicating junctions (figure 9.17).

Tight junctions

Tight junctions connect the plasma membranes of adjacent cells in a sheet, preventing small molecules from leaking between the cells. This allows the sheet of cells to act as a wall within the organ, keeping molecules on one side or the other (figure 9.17*a*).

Creating sheets of cells. The cells that line an animal's digestive tract are organized in a sheet only one cell thick. One surface of the sheet faces the inside of the tract, and the other faces the extracellular space, where blood vessels are located. Tight junctions encircle each cell in the sheet, like a belt cinched around a person's waist. The junctions between neighboring cells are so

TABLE 9.2 Cell-to-Cell Connections and Cell Identity

Type of Connection	Structure	Function	Example
Surface markers	Variable, integral proteins or glycolipids in plasma membrane	Identify the cell	MHC complexes, blood groups, antibodies
Tight junctions	Tightly bound, leakproof, fibrous protein seal that surrounds cell	Organizing junction; holds cells together such that materials pass *through* but not *between* the cells	Junctions between epithelial cells in the gut
Anchoring junction (Desmosome)	Intermediate filaments of cytoskeleton linked to adjoining cells through cadherins	Anchoring junction: binds cells together	Epithelium
Anchoring junction (Adherens junction)	Transmembrane fibrous proteins	Anchoring junction: connects extracellular matrix to cytoskeleton	Tissues with high mechanical stress, such as the skin
Communicating junction (Gap junction)	Six transmembrane connexon proteins creating a pore that connects cells	Communicating junction: allows passage of small molecules from cell to cell in a tissue	Excitable tissue such as heart muscle
Communicating junction (Plasmodesmata)	Cytoplasmic connections between gaps in adjoining plant cell walls	Communicating junction between plant cells	Plant tissues

securely attached that there is no space between them for leakage. Hence, nutrients absorbed from the food in the digestive tract must pass directly through the cells in the sheet to enter the blood because they cannot pass through spaces between cells.

Partitioning the sheet. The tight junctions between the cells lining the digestive tract also partition the plasma membranes of these cells into separate compartments. Transport proteins in the membrane facing the inside of the tract carry nutrients from that side to the cytoplasm of the cells. Other proteins, located in the membrane on the opposite side of the cells, transport those nutrients from the cytoplasm to the extracellular fluid, where they can enter the blood.

For the sheet to absorb nutrients properly, these proteins must remain in the correct locations within the fluid membrane. Tight junctions effectively segregate the proteins on opposite sides of the sheet, preventing them from drifting within the membrane from one side of the sheet to the other. When tight junctions are experimentally disrupted, just this sort of migration occurs.

Anchoring junctions

Anchoring junctions mechanically attach the cytoskeleton of a cell to the cytoskeletons of other cells or to the extracellular matrix. These junctions are most common in tissues subject to mechanical stress, such as muscle and skin epithelium.

Cadherin and intermediate filaments *Desmosomes* connect the cytoskeletons of adjacent cells (figure 9.17*b*), and *hemidesmosomes* anchor epithelial cells to a basement membrane. Proteins called **cadherins,** most of which are single-pass transmembrane

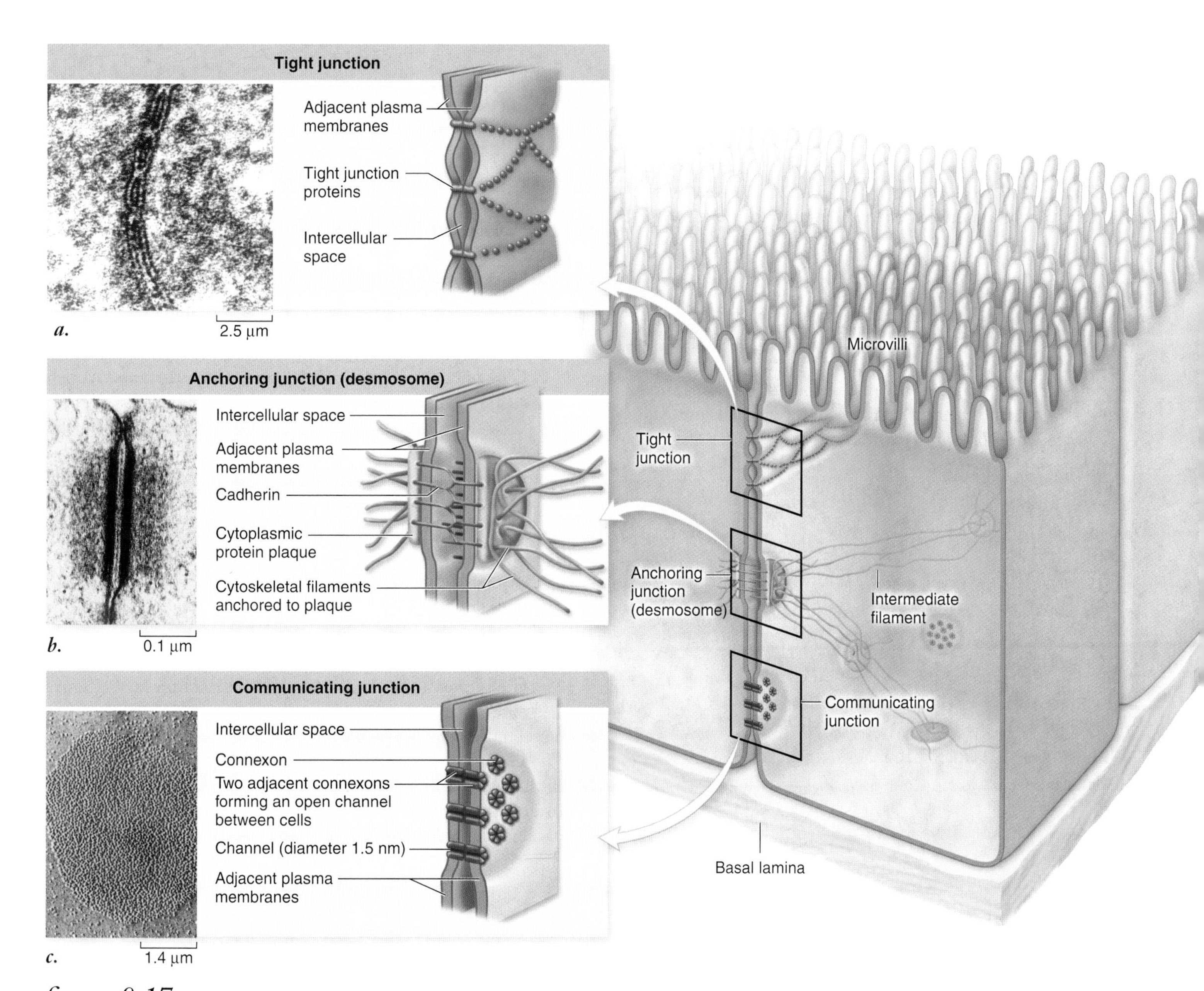

figure 9.17

AN OVERVIEW OF CELL JUNCTION TYPES. Here, the diagram of gut epithelial cells on the right illustrates the comparative structures and locations of common cell junctions. The detailed models on the left show the structures of the three major types of cell junctions: *(a)* tight junction; *(b)* anchoring junction, the example shown is a desmosome; *(c)* communicating junction, the example shown is a gap junction.

glycoproteins, create the critical link. Proteins link the short cytoplasmic end of a cadherin to the intermediate filaments in the cytoskeleton. The other end of the cadherin molecule projects outward from the plasma membrane, joining directly with a cadherin protruding from an adjacent cell similar to a firm handshake, binding the cells together. Connections between proteins tethered to the intermediate filaments are much more secure than connections between free-floating membrane proteins.

Cadherin and actin filaments. Cadherins can also connect the actin frameworks of cells in cadherin-mediated junctions (figure 9.18). When they do, they form less stable links between cells than when they connect intermediate filaments. Many kinds of actin-linking cadherins occur in different tissues. For example, during vertebrate development, the migration of neurons in the embryo is associated with changes in the type of cadherin expressed on their plasma membranes.

Integrin-mediated links. Anchoring junctions called **adherens junctions** connect the actin filaments of one cell with those of neighboring cells or with the extracellular matrix. The linking proteins in these junctions are members of a large superfamily of cell surface receptors called **integrins** that bind to a protein component of the extracellular matrix. At least 20 different integrins exist, having differently shaped binding domains.

Communicating junctions

Many cells communicate with adjacent cells through direct connections called **communicating junctions.** In these junctions, a chemical or electrical signal passes directly from one cell to an adjacent one. Communicating junctions permit small molecules or ions to pass from one cell to the other. In animals, these direct communication channels between cells are called *gap junctions*, and in plants, *plasmodesmata.*

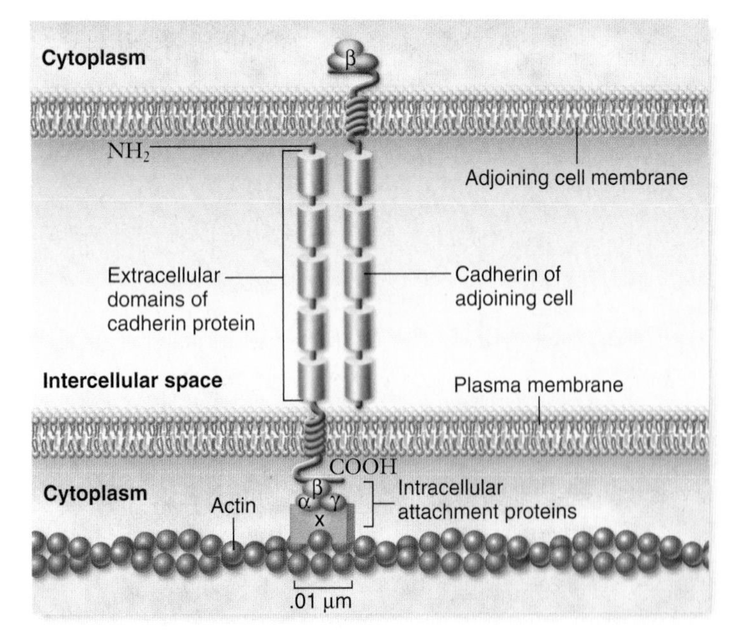

figure 9.18

A CADHERIN-MEDIATED JUNCTION. The cadherin molecule is anchored to actin in the cytoskeleton and passes through the membrane to interact with the cadherin of an adjoining cell.

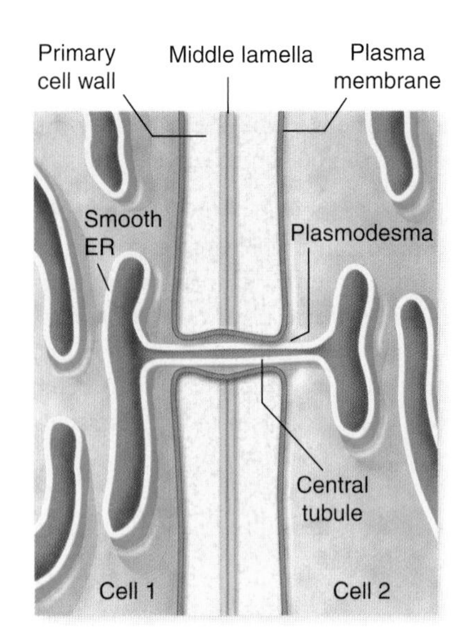

figure 9.19

PLASMODESMATA. Plant cells can communicate through specialized openings in their cell walls, called plasmodesmata, where the cytoplasm of adjoining cells are connected.

Gap junctions in animals. Gap junctions are composed of structures called connexons, complexes of six identical transmembrane proteins (see figure 9.17*c*). The proteins in a connexon are arranged in a circle to create a channel through the plasma membrane that protrudes several nanometers from the cell surface. A gap junction forms when the connexons of two cells align perfectly, creating an open channel that spans the plasma membranes of both cells.

Gap junctions provide passageways large enough to permit small substances, such as simple sugars and amino acids, to pass from one cell to the next, yet small enough to prevent the passage of larger molecules, such as proteins.

Gap junction channels are dynamic structures that can open or close in response to a variety of factors, including Ca^{2+} and H^+ ions. This gating serves at least one important function. When a cell is damaged, its plasma membrane often becomes leaky. Ions in high concentrations outside the cell, such as Ca^{2+}, flow into the damaged cell and close its gap junction channels. This isolates the cell and so prevents the damage from spreading to other cells.

Plasmodesmata in plants. In plants, cell walls separate every cell from all others. Cell–cell junctions occur only at holes or gaps in the walls, where the plasma membranes of adjacent cells can come into contact with one another. Cytoplasmic connections that form across the touching plasma membranes are called **plasmodesmata** (singular, plasmodesma) (figure 9.19). The majority of living cells within a higher plant are connected with their neighbors by these junctions.

Plasmodesmata function much like gap junctions in animal cells, although their structure is more complex. Unlike gap junctions, plasmodesmata are lined with plasma membrane and contain a central tubule that connects the endoplasmic reticulum of the two cells.

Cells in multicellular organisms are usually organized into tissues. This requires that cells have distinct identity and connections. Cell identity is conferred by glycoproteins, including MHC proteins, which are important in the immune system. Cell connections fall into three basic categories: tight, anchoring, and communicating junctions. Tight junctions help to make sheets of cells that form watertight seals, and anchoring junctions provide strength and flexibility. Communicating junctions include gap junctions in animals and plasmodesmata in plants.

concept review

9.1 Overview of Cell Communication (figure 9.1)

Cell communication requires signal molecules, called ligands, binding to specific receptor proteins producing a cellular response.

- Depending on the distance between signaling and responding cells there are four basic signaling mechanisms (figure 9.2)
  - *Direct contact—molecules on the plasma membrane of one cell contact the receptor molecules on an adjacent cell.*
  - *Paracrine signaling—short-lived signal molecules are released into the extracellular fluid and influence neighboring cells.*
  - *Endocrine signaling—long-lived hormones enter the circulatory system and are carried to target cells some distance away.*
  - *Synaptic signaling—short-lived neurotransmitters are released by neurons into the gap, called a synapse, between nerves and target cells.*
- Signal transduction refers to the intracellular events that result from a receptor binding a signal molecule.
- Proteins can be controlled by phosphate added by kinase and removed by phosphatase enzymes.

9.2 Receptor Types (figure 9.4)

Receptor proteins can be classified by both function and location.

- Receptors are broadly defined as intracellular or cell-surface receptors (membrane receptors).
- Membrane receptors are transmembrane proteins that transfer information across the membrane, but not the signal molecule.
- Membrane receptors are divided into three types:
  - *Channel-linked receptors are chemically gated ion channels that allow specific ions to pass through a central pore.*
  - *Enzymatic receptors are enzymes activated by binding a ligand; these enzymes are usually protein kinases.*
  - *G protein-coupled receptors interact with G proteins that control the function of effector proteins: enzymes or ion channels.*
- Some enzymatic and most G protein-coupled receptors produce second messengers, to relay messages in the cytoplasm.

9.3 Intracellular Receptors (figure 9.5)

Many cell signals are lipid-soluble and readily pass through the plasma membrane and bind to receptors in the cytoplasm or nucleus.

- Steroid hormones bind cytoplasmic receptors then are transported to the nucleus. Thus, they are called nuclear receptors.
- Nuclear receptors can directly affect gene expression, usually activating transcription of the genes they control.
- Nuclear receptors have three functional domains: hormone-binding, DNA-binding, and transcription-activating domains.
- Ligand binding changes receptor shape, releasing an inhibitor occupying the DNA-binding site.
- A cell's response to a lipid-soluble signal depends on the hormone–receptor complex and the other protein coactivators present.
- Some intracellular receptors activate cellular enzymes and do not affect gene expression.

9.4 Signal Transduction Through Receptor Kinases

Receptor kinases in plants and animals recognize hydrophilic ligands and influence the cell cycle, cell migration, cell metabolism, and cell proliferation.

- Because they are involved in growth control, alterations of receptor kinases and their signaling pathways can lead to cancer.
- Ligand binding causes the receptor to phosphorylate itself on specific tyrosines: this is called autophosphorylation.
- The activated receptor can also phosphorylate other intracellular proteins.
- Adapter proteins can bind to phosphotryrosine and act as links between the receptors and downstream signaling events.
- Protein kinase can amplify a signal because at each step of the cascade an enzyme acts on a number of substrate molecules.
- Scaffold proteins organize cascade proteins into a single complex so they can act sequentially and be optimally functional.
- Receptor tyrosine kinases are inactivated by dephosphorylation or internalization where they are degraded or recycled.

9.5 Signal Transduction Through G Protein-Coupled Receptors (figure 9.11)

G protein-coupled receptors function through activation of G proteins that link receptors to effector proteins.

- G proteins are active bound to GTP and inactive bound to GDP. Receptors promote exchange of GDP for GTP.
- The activated G protein dissociates into two parts, G_α and $G_{\beta\gamma}$, each of which can act on effector proteins.
- G_α also hydrolyzes GTP to GDP to inactivate the G protein.
- Effector proteins can produce second messengers.
- Two common effector proteins are adenylyl cyclase and phospholipase C, which produce second messengers known as cAMP, and DAG and IP_3, respectively.
- Ca^{2+} is also a second messenger. Ca^{2+} release is triggered by IP_3 binding to channel-linked receptors in the ER.
- Ca^{2+} can bind to a cytoplasmic protein calmodulin, which in turn activates other proteins, producing a variety of responses.
- Different receptors can activate the same effector, which will produce the same second messenger leading to the same response.
- Different receptor subtypes or isoforms lead to different effects in different cells.
- Different receptor types such as G protein-coupled receptors and receptor tyrosine kinases can activate the same signaling pathways.

9.6 Cell-to-Cell Interactions (figure 9.17)

In multicellular organisms, cells are organized in specific ways so they can function as tissues.

- Cells form long-lasting or permanent connections with one another called cell junctions.
- There are three categories of cell junctions:
  - *Tight junctions connect the plasma membranes of adjacent cells into sheets and prevent small molecules from leaking out of the cells.*
  - *Anchoring junctions connect adjacent cells.*
    - Desmosomes link the intermediate and actin filaments of neighboring cells by cadherin-mediated links.
    - Adherens junctions connect actin filaments of adjacent cells or attach to the extracellular matrix by integrin-mediated links.
  - *Communicating junctions allow cytoplasm to move between cells.*
    - In animals gap junctions composed of connexons allow the passage of small molecules between cells.
    - In plants junctions lined with plasma membrane called plasmodesmata penetrate the cell wall and connect cells.

SELF TEST

1. What is a ligand?
   a. An integral membrane protein associated with G proteins
   b. A DNA-binding protein that alters gene expression
   c. A cytoplasmic second messenger molecule
   d. A molecule or protein that can bind to a receptor
2. In the case of paracrine signaling the ligand is—
   a. produced by the cell itself
   b. secreted by neighboring cells
   c. present on the plasma membrane of neighboring cells
   d. secreted by distant cells
3. A neurotransmitter functions as a ligand in which type of signaling?
   a. Direct contact
   b. Endocrine
   c. Synaptic signaling
   d. Autocrine
4. The function of a ________ is to add phosphates to proteins, whereas a _________ functions to remove the phosphates.
   a. tyrosine; serine
   b. protein phosphatase; protein dephosphatase
   c. protein kinase; protein phosphatase
   d. receptor; ligand
5. Which of the following type(s) of membrane receptors functions by changing the phosphorylation state of proteins in the cell?
   a. Channel-linked receptor
   b. Enzymatic receptor
   c. G protein-coupled receptor
   d. Both b and c
6. How does the function of an intracellular receptor differ from that of a membrane receptor?
   a. The intracellular receptor binds a ligand.
   b. The intracellular receptor binds DNA.
   c. The intracellular receptor activates a kinase.
   d. The intracellular receptor functions as a second messenger.
7. During a protein kinase signal cascade—
   a. Sequential phosphorylation of different kinases leads to a change in gene expression.
   b. Multiple G proteins become activated.
   c. Phosphorylation of adapter proteins leads to the formation of second messengers.
   d. The number of MAP kinase proteins present in the cytoplasm is amplified.
8. What is the function of Ras during tyrosine kinase cell signaling?
   a. It activates the opening of channel-linked receptors.
   b. It synthesizes the formation of second messengers.
   c. It phosphorylates other enzymes as part of a pathway.
   d. It links the receptor protein to the MAP kinase pathway.
9. Which of the following best describes the immediate effect of ligand binding to a G protein-coupled receptor?
   a. The G protein trimer releases a GDP and binds a GTP.
   b. The G protein trimer dissociates from the receptor.
   c. The G protein trimer interacts with an effector protein.
   d. The α subunit of the G protein becomes phosphorylated.
10. The amplification of a cellular signal requires all of the following except—
    a. A ligand
    b. DNA
    c. A second messenger
    d. A protein kinase
11. *Adenylyl cyclase* is responsible for the production of which second-messenger molecule?
    a. Cyclic-AMP
    b. Calcium
    c. IP_3
    d. Calmodulin
12. The response to signaling through G protein-coupled receptors can vary in different cells because
    a. All receptors act through the same G protein.
    b. Different isoforms of a receptor bind the same ligand but activate different effectors.
    c. The amount of receptor in the membrane differs in different cell types.
    d. Different receptors can activate the same effector.
13. What is the function of the tight junctions in the formation of a tissue?
    a. Tight junctions connect one cell to the next, creating a barrier between the cells.
    b. Tight junctions form a strong anchor between two cells.
    c. Tight junctions allow for the movement of small molecules between cells.
    d. Tight junctions connect the cell to the extracellular matrix.
14. Cadherins and intermediate filament proteins are associated with _________, whereas connexons are associated with ___________
    a. tight junctions; anchoring junctions
    b. cell surface markers; tight junctions
    c. desmosomes; gap junctions
    d. adherens junctions; plasmodesmata
15. Cells are able to anchor themselves to the extracellular matrix through the activity of—
    a. connexon proteins
    b. MHC proteins
    c. cadherin proteins
    d. integrin proteins

CHALLENGE QUESTIONS

1. Describe the common features found in all examples of cellular signaling discussed in this chapter. Provide examples to illustrate your answer.
2. The sheet of cells that form the gut epithelium folds into peaks called villi and valleys called crypts. The cells within the crypt region secrete a protein, Netrin-1, that becomes concentrated within the crypts. Netrin-1 is the ligand for a receptor protein that is found on the surface of all gut epithelial cells. Netrin-1 binding triggers a signal pathway that promotes cell growth. Gut epithelial cells undergo apoptosis (cell death) in the absence of Netrin-1 ligand binding.
   a. How would you characterize the type of signaling (autocrine, paracrine, endocrine) found in this system?
   b. Predict where the greatest amount of cell growth and cell death would occur in the epithelium.
   c. The loss of the Netrin-1 receptor is associated with some types of colon cancer. Suggest an explanation for the link between this signaling pathway and tumor formation.

Do you need additional review? *Visit* www.ravenbiology.com *for practice quizzes, animations, videos, and activities designed to help you master the material in this chapter.*

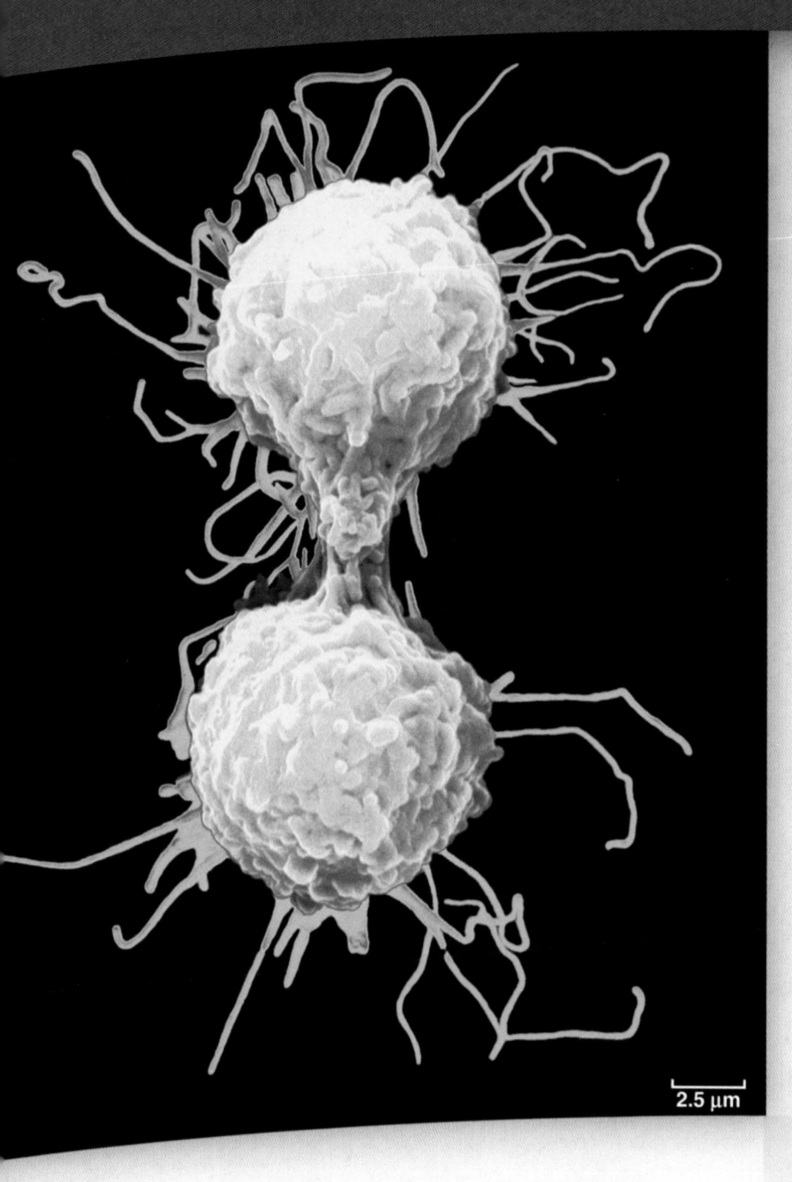

chapter 10

How Cells Divide

introduction

ALL SPECIES OF ORGANISMS—bacteria, alligators, the weeds in a lawn—grow and reproduce. From the smallest creature to the largest, all species produce offspring like themselves and pass on the hereditary information that makes them what they are. In this chapter, we examine how cells, like the white blood cell shown in the figure, divide and reproduce. Cell division is necessary for the growth of organisms, for wound healing, and to replace cells lost regularly, like the cells in your skin and in the lining of your gut. The mechanism of cell reproduction and its biological consequences have changed significantly during the evolution of life on Earth. The process is complex in eukaryotes, involving both the replication of chromosomes and their separation into daughter cells. Much of what we are learning about the causes of cancer relates to how cells control this process, and in particular their propensity to divide, a mechanism that in broad outline remains the same in all eukaryotes.

concept outline

10.1 Bacterial Cell Division

Bacteria divide as a means to reproduce themselves. The reproduction of bacteria is clonal—that is, each cell produced by cell division is an identical copy of the original cell. Although mechanisms of DNA exchange can be found in bacteria, they do not have a sexual cycle like eukaryotes. Thus all growth in a bacterial population is due to division to produce new cells.

Binary fission is a simple form of cell division

Both bacterial and eukaryotic cells reproduce by mechanisms that produce two new cells with the same genetic information as the original cell. The differences between these two basic cell types lead to large differences in how this process occurs. Despite these differences, the essentials of the process are the same: duplication and segregation of genetic information into daughter cells and division of cellular contents. We begin by looking at the simpler process, **binary fission,** which occurs in bacteria.

Most bacteria have a genome made up of a single, circular DNA molecule. In spite of its apparent simplicity, the DNA molecule of the bacterium *Escherichia coli* is actually on the order of 500 times longer than the cell itself! Thus, this "simple" structure is actually packaged very tightly to fit into the cell. Although not found in a nucleus, the DNA is in a compacted form called a *nucleoid* that is distinct from the cytoplasm around it.

During binary fission, the chromosome must be replicated, and then the products partitioned to each end of the cell prior to the actual division of the cell into daughter cells. One key feature of bacterial cell division is that replication and partitioning of the chromosome occur as a concerted process. In contrast, eukaryotic cells' DNA replication occurs early in division, and chromosome separation occurs much later.

Proteins control chromosome separation and septum formation

For many years, geneticists believed that newly replicated *E. coli* DNA molecules were passively segregated by attachment to the membrane and growth of the membrane as the cell elongated. More recently, a more complex picture is emerging that involves both active partitioning of the DNA and enzyme-controlled selection of a site at which the elongated cell will be divided in two. Species as different as *E. coli* and *Bacillus subtilis* both exhibit active partitioning of the newly replicated DNA molecules during division.

Binary fission begins with the replication of the bacterial DNA at a specific site—the origin of replication (see chapter 14)—and proceeds bidirectionally around the circular DNA to a specific site of termination (figure 10.1). Growth of the cell results in elongation, and the newly replicated DNA molecules are actively partitioned to one-quarter and three-quarters of the cell length. This process appears to require specific DNA sequences near the origin of replication and proteins that bind to these sequences.

The process of replication itself may contribute to chromosome partitioning. One model involves a fixed replication complex in the midcell region such that the action of synthesis of the DNA pushes the two daughter chromosomes apart. The

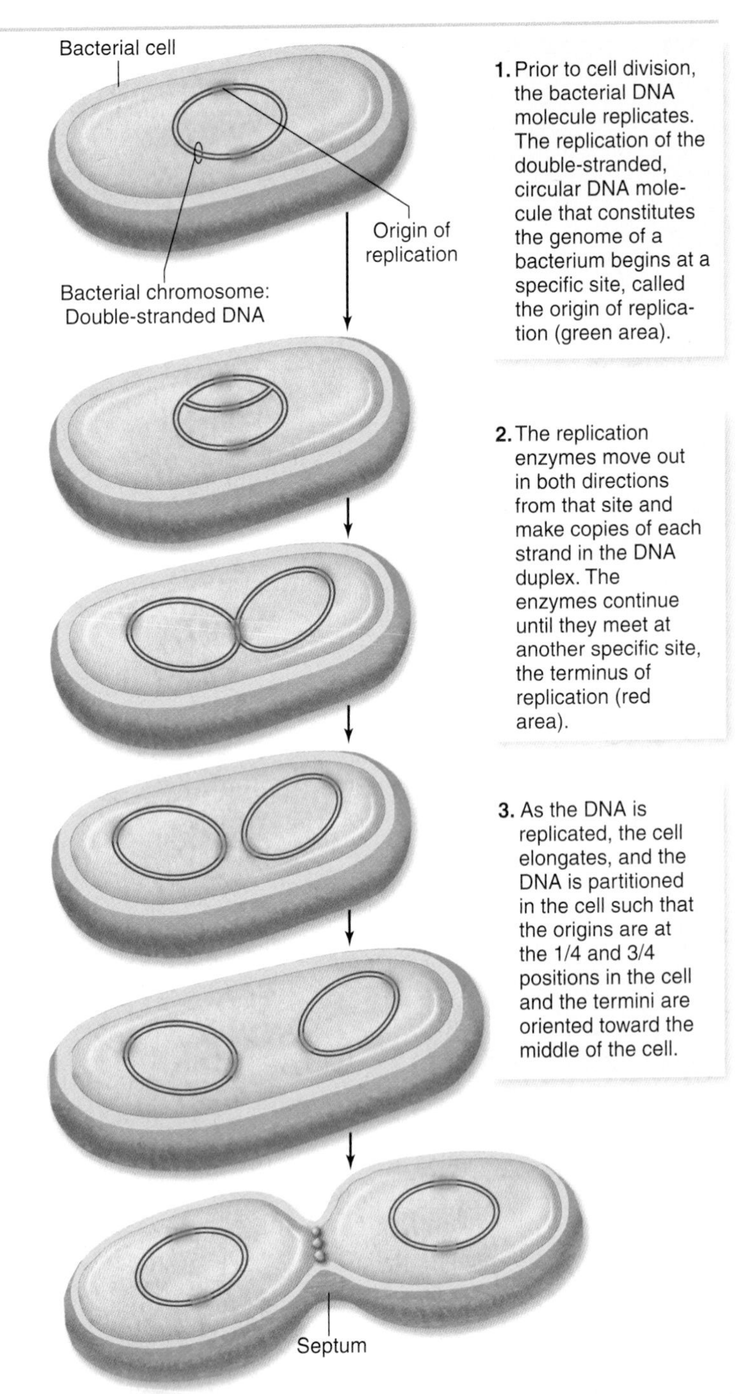

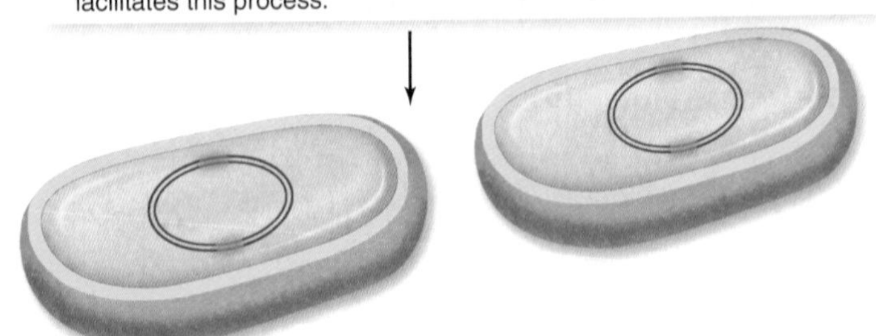

figure 10.1

BINARY FISSION.

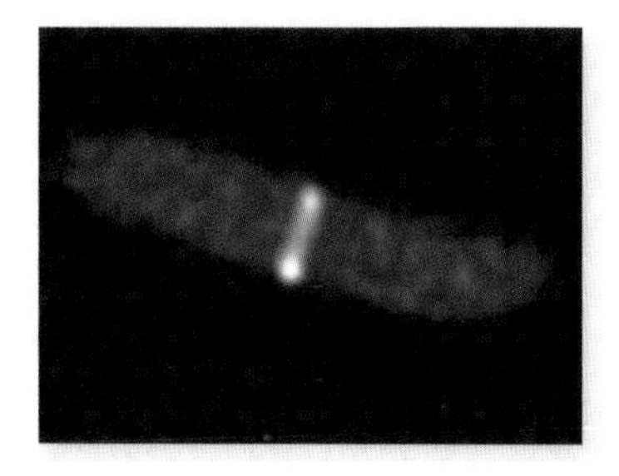
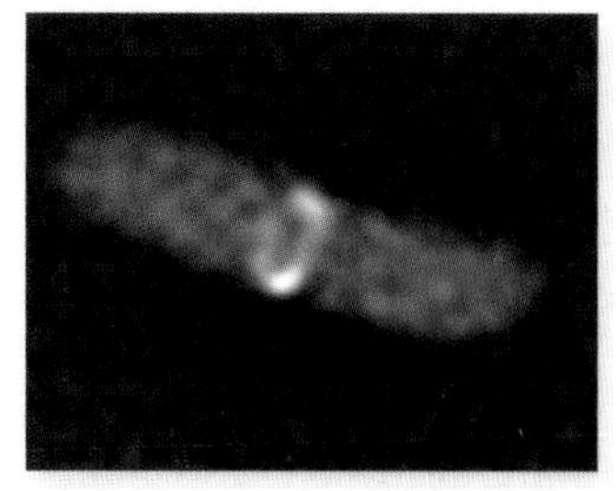

figure 10.2

THE FtsZ PROTEIN. In these dividing *E. coli* bacteria, the FtsZ protein is labeled with fluorescent dye to show its location during binary fission. The protein assembles into a ring at approximately the midpoint of the cell, where it facilitates septation and cell division. Bacteria carrying mutations in the *FtsZ* gene are unable to divide.

final positioning would then involve proteins that bind to the origin sequences.

The cell's other components are partitioned by the growth of new membrane and production of the **septum** (see figure 10.1). This process, termed **septation,** is complex and is under the control of cellular proteins. The site of septation is usually the midpoint of the cell, and it begins with the formation of a ring composed of many copies of the protein FtsZ (figure 10.2). Next, accumulation of a number of other proteins occurs, including ones embedded in the membrane. The exact mechanism of septation is not known, but this structure contracts inward radially until the cell pinches off into two new cells.

The FtsZ protein is interesting for a number of reasons. It has a long evolutionary history, having been identified in most prokaryotes, including archaea. It can form filaments and rings, and recent three-dimensional crystals show a high degree of similarity to eukaryotic tubulin as well. In addition, molecules similar to actin, another component of the eukaryote cytoskeleton, have been discovered. Tubulin and actin play a role in eukaryotic cell division, and these similar molecules in prokaryotes may provide evidence of an evolutionary link between the two processes.

The evolution of eukaryotic cells included much more complex genomes composed of multiple linear chromosomes housed in a membrane-bounded nucleus. These complex genomes may have been made possible by the evolution of mechanisms that delay chromosome separation after replication. Although it is unclear how this ability to keep chromosomes together evolved, it does seem more closely related to binary fission than we once thought (figure 10.3).

Most bacteria divide by binary fission, a simple form of cell division in which DNA replication and segregation occur simultaneously. This process involves active partitioning of the chromosome and positioning of the site of septation.

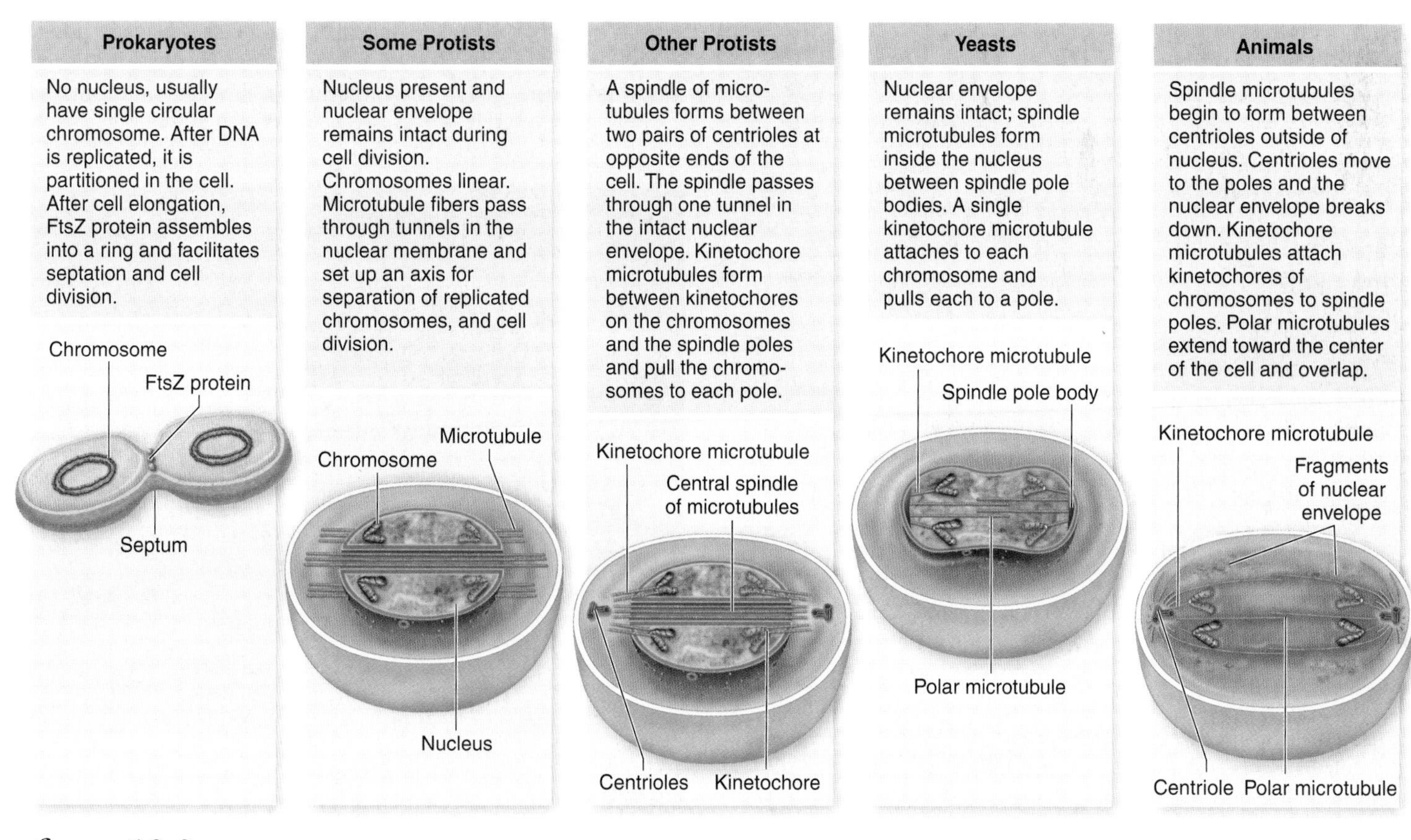

figure 10.3

A COMPARISON OF PROTEIN ASSEMBLIES DURING CELL DIVISION AMONG DIFFERENT ORGANISMS. The prokaryotic protein FtsZ has a structure that is similar to that of the eukaryotic protein tubulin. Tubulin is the protein component of microtubules, which are fibers that are used to separate chromosomes in eukaryotic cell division.

10.2 Eukaryotic Chromosomes

Chromosomes were first observed by the German embryologist Walther Flemming (1843–1905) in 1879, while he was examining the rapidly dividing cells of salamander larvae. When Flemming looked at the cells through what would now be a rather primitive light microscope, he saw minute threads within their nuclei that appeared to be dividing lengthwise. Flemming called their division **mitosis,** based on the Greek word *mitos*, meaning "thread."

Chromosome number varies among species

Since their initial discovery, chromosomes have been found in the cells of all eukaryotes examined. Their number may vary enormously from one species to another. A few kinds of organisms have only a single pair of chromosomes, whereas some ferns have more than 500 pairs (table 10.1). Most eukaryotes have between 10 and 50 chromosomes in their body cells.

Human cells each have 46 chromosomes, consisting of 23 nearly identical pairs (figure 10.4). Each of these 46 chromosomes contains hundreds or thousands of genes that play important roles in determining how a person's body develops and functions. For this reason, possession of all the chromosomes is essential to survival. Human embryos missing even one chromosome, a condition called *monosomy*, do not survive in most cases. Having an extra copy of any one chromosome, a condition called *trisomy*, is usually fatal except where the smallest chromosomes are involved. (You'll learn more about human chromosome abnormalities in chapter 13.)

TABLE 10.1 Chromosome Number in Selected Eukaryotes

Group	Total Number of Chromosomes
FUNGI	
Neurospora (haploid)	7
Saccharomyces (a yeast)	16
INSECTS	
Mosquito	6
Drosophila	8
Honeybee	diploid females 32, haploid males 16
Silkworm	56
PLANTS	
Haplopappus gracilis	2
Garden pea	14
Corn	20
Bread wheat	42
Sugarcane	80
Horsetail	216
Adder's tongue fern	1262
VERTEBRATES	
Opossum	22
Frog	26
Mouse	40
Human	46
Chimpanzee	48
Horse	64
Chicken	78
Dog	78

Eukaryotic chromosomes exhibit complex structure

Researchers have learned a great deal about chromosome structure and composition in the more than 125 years since their discovery. But despite intense research, the exact structure of eukaryotic chromosomes during the cell cycle remains unclear. The structures described in this chapter represent the currently accepted model.

Composition of chromatin

Chromosomes are composed of **chromatin,** a complex of DNA and protein; most chromosomes are about 40% DNA and 60% protein. A significant amount of RNA is also associated with chromosomes because chromosomes are the sites of RNA synthesis.

The DNA of a single chromosome is one very long, double-stranded fiber that extends unbroken through the chromosome's entire length. A typical human chromosome contains

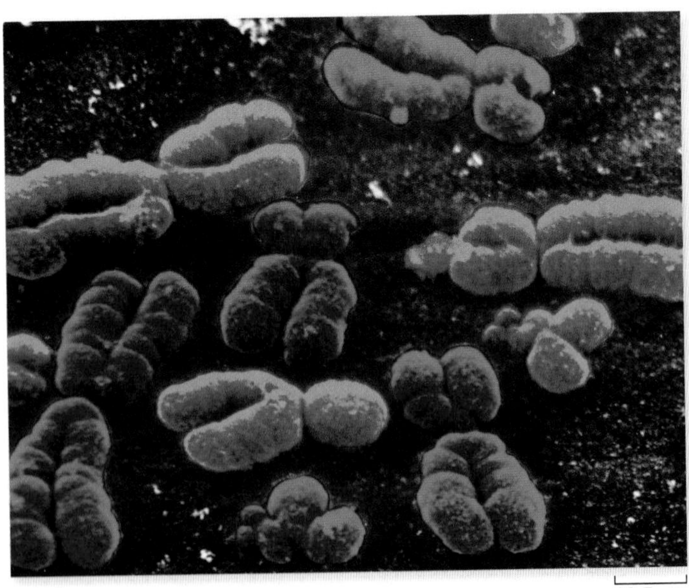

figure 10.4

HUMAN CHROMOSOMES. This SEM micrograph shows human chromosomes as they appear immediately before nuclear division. Each DNA molecule has already replicated, forming identical copies held together at a visible constriction called the centromere. False color has been added to the chromosomes.

about 140 million (1.4×10^8) nucleotides in its DNA. If we think of each nucleotide as a "word," then the amount of information an average chromosome contains would fill about 280 printed books of 1000 pages each, with 500 "words" per page. If we could lay out the strand of DNA from a single chromosome in a straight line, it would be about 5 cm (2 in.) long. Fitting such a strand into a cell nucleus is like cramming a string the length of a football field into a baseball—and that's only 1 of 46 chromosomes! In the cell, however, the DNA is coiled, allowing it to fit into a much smaller space than would otherwise be possible.

The organization of chromatin in the nondividing nucleus is not well understood, but geneticists have recognized for years that some domains of chromatin, called **heterochromatin,** are not expressed, and other domains of chromatin, called **euchromatin,** are expressed. This genetically measurable state is also related to the physical state of chromatin, although researchers are just beginning to see the details.

Chromosome structure

If we gently disrupt a eukaryotic nucleus and examine the DNA with an electron microscope, we find that it resembles a string of beads (figure 10.5). Every 200 nucleotides (nt), the DNA duplex (double strand) is coiled around a core of eight **histone proteins.** Unlike most proteins, which have an overall negative charge, histones are positively charged because of an abundance of the basic amino acids arginine and lysine. Thus, they are strongly attracted to the negatively charged phosphate groups of the DNA, and the histone cores act as "magnetic forms" that promote and guide the coiling of the DNA. The complex of DNA and histone proteins is termed a **nucleosome.**

Further coiling occurs when the string of nucleosomes is wrapped into higher order coils called *solenoids.* The precise path of this higher order folding of chromatin is still a subject of some debate, but it leads to a fiber with a diameter of 30 nm and thus is often called the 30-nm fiber. This 30-nm fiber, or solenoid, is the usual state of interphase (nondividing) chromatin.

During mitosis the chromatin in the solenoid is arranged around a scaffold of protein assembled at this time to achieve maximum compaction of the chromosomes. This process prepares the chromosomes for the events of mitosis described later on. The exact nature of this compaction is unknown, but one long-standing model involves radial looping of the solenoid about the protein scaffold, aided by a complex of proteins called **condensin.** The protein scaffold itself is actually what gives mitotic chromosomes their distinctive shape.

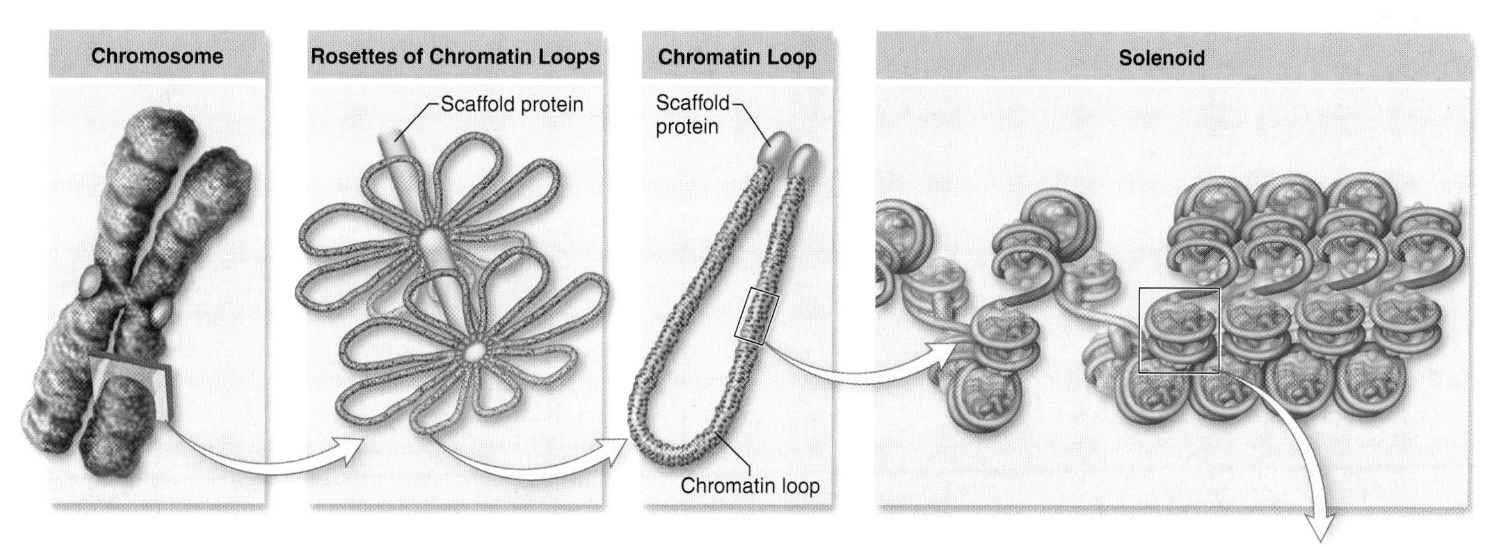

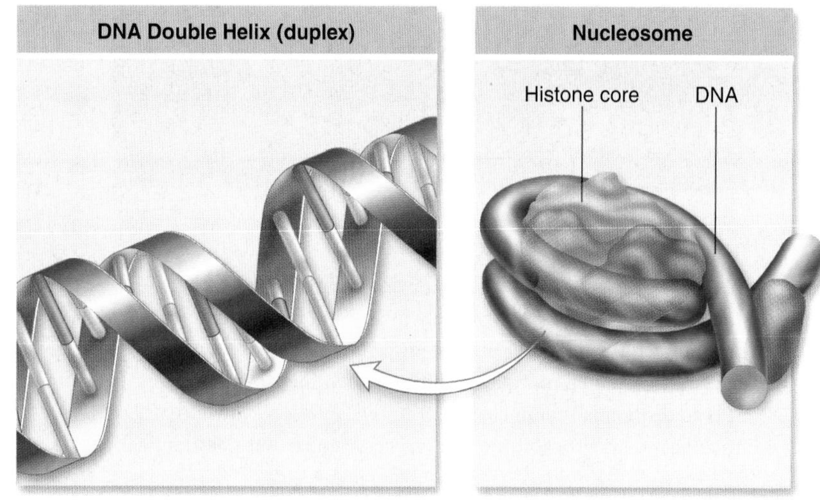

figure 10.5

LEVELS OF EUKARYOTIC CHROMOSOMAL ORGANIZATION. Each chromosome consists of a long double-stranded DNA molecule. These strands require further packaging to fit into the cell nucleus. The DNA duplex is tightly bound to and wound around proteins called histones. The DNA-wrapped histones are called nucleosomes. The nucleosomes are further coiled into a solenoid. This solenoid is then organized into looped domains. The final organization of the chromosome is unknown, but it appears to involve further radial looping into rosettes around a preexisting scaffolding of protein. The arrangement illustrated here is one of many possibilities.

Chromosome karyotypes

Chromosomes vary in size, staining properties, the location of the centromere (a constriction found on all chromosomes, described shortly), the relative length of the two arms on either side of the centromere, and the positions of constricted regions along the arms. The particular array of chromosomes an individual organism possesses is called its **karyotype.** The karyotype in figure 10.6 shows the set of chromosomes from a single human individual, exhibiting variations in size and structure.

When defining the number of chromosomes in a species, geneticists count the **haploid (*n*)** number of chromosomes. This refers to one complete set of chromosomes necessary to define an organism. For humans and many other species, the normal number of chromosomes in a cell is called the **diploid (2*n*)** number, which is twice the haploid number. For humans, the haploid number is 23 and the diploid number is 46. Diploid chromosomes reflect the equal genetic contribution that parents make to offspring. We refer to the maternal and paternal chromosomes as being **homologous,** and each one of the pair is termed a **homologue.**

Chromosome replication

Chromosomes as seen in a karyotype are only present for a brief period during cell division. Prior to replicating, each chromosome is composed of a single DNA molecule that is arranged into the 30-nm fiber described earlier. After replication, each chromosome is composed of two identical DNA molecules held together by a complex of proteins called **cohesins.** As the chromosomes become more condensed and arranged about the protein scaffold, they become visible as two strands that are held together. At this point, we still call this one chromosome, but it is composed of two sister **chromatids** (figure 10.7).

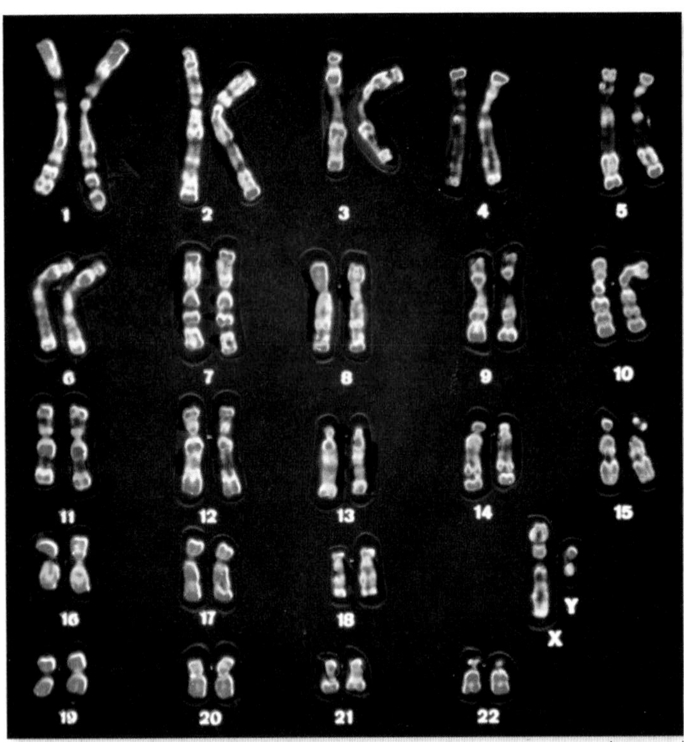

figure 10.6

A HUMAN KARYOTYPE. The individual chromosomes that make up the 23 pairs differ widely in size and in centromere position. In this preparation, the chromosomes have been specifically stained to indicate differences in their composition and to distinguish them clearly from one another. Notice that members of a chromosome pair are very similar but not identical.

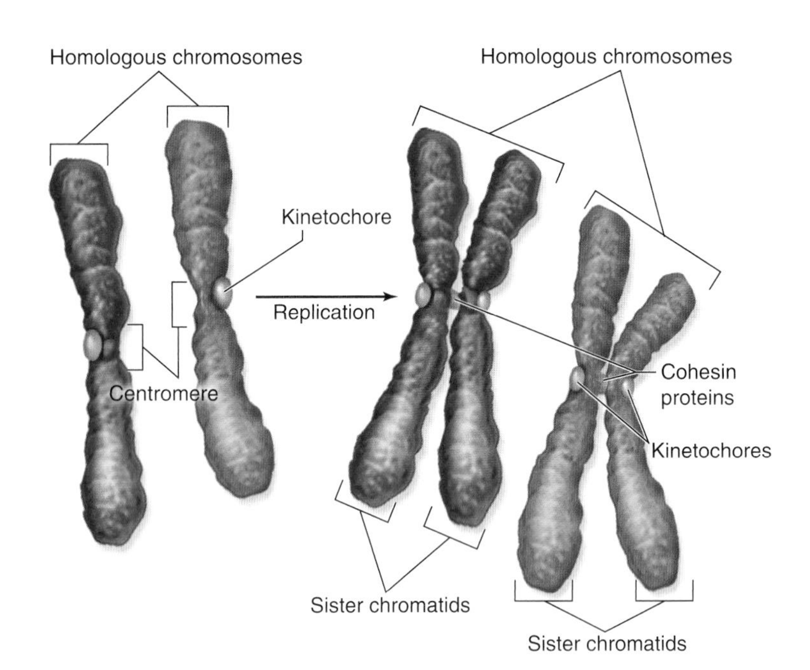

figure 10.7

THE DIFFERENCE BETWEEN HOMOLOGOUS CHROMOSOMES AND SISTER CHROMATIDS. Homologous chromosomes are the maternal and paternal copies of the same chromosome—say, chromosome number 16. Sister chromatids are the two replicas of a single chromosome held together at their centromeres by cohesin proteins after DNA replication. The kinetochore (described later in the chapter) is composed of proteins found at the centromere that attach to microtubules during mitosis.

The fact that the products of replication are held together is critical to the division process. One problem that a cell must solve is how to ensure that each new cell receives a complete set of chromosomes. If we were designing a system, we might use some kind of label to identify each chromosome, much like most of us use when we duplicate files on a computer. The cell has no mechanism to label chromosomes; instead, it keeps the products of replication together until the moment of chromosome segregation, ensuring that one copy of each chromosome goes to each daughter cell. This separation of sister chromatids is the key event in the mitotic process described in detail shortly.

Eukaryotic chromosomes are complex structures that can be compacted for cell division. During interphase, DNA is coiled around proteins into a structure called a nucleosome. The string of nucleosomes is further coiled into a solenoid (30-nm fiber). After chromosome replication, the resulting chromatids are held together for the division process by proteins called cohesins.

10.3 Overview of the Eukaryotic Cell Cycle

Compared with prokaryotes, the increased size and more complex organization of eukaryotic genomes required radical changes in the partitioning of the two replicas of the genome into daughter cells. The overall process involves the duplication of the genome, its accurate segregation, and the division of cellular contents. These events make up the **cell cycle.**

The cell cycle is divided into five phases

The cell cycle is divided into phases based on the key events of genome duplication and segregation. The cell cycle is usually diagrammed using the metaphor of a clock face (figure 10.8).

Biologists separate the cell cycle into five main phases:

- **G_1 (gap phase 1)** is the primary growth phase of the cell. The term *gap phase* refers to its filling the gap between cytokinesis and DNA synthesis. For most cells, this phase encompasses the major portion of the cell cycle.
- **S (synthesis)** is the phase in which the cell synthesizes a replica of the genome.
- **G_2 (gap phase 2)** is the second growth phase, in which preparations are made for separation of the newly replicated genome. This phase fills the gap between DNA synthesis and the beginning of mitosis. During this phase, mitochondria and other organelles replicate, chromosomes prepare to condense, and microtubules begin to assemble at a spindle.

  G_1, S, and G_2 together constitute **interphase,** the portion of the cell cycle between cell divisions.
- **M (mitosis)** is the phase of the cell cycle in which the spindle apparatus assembles, binds to the chromosomes, and moves the sister chromatids apart. Mitosis is the essential step in the separation of the two daughter genomes. Although mitosis is a continuous process, it is traditionally subdivided into five stages: prophase, prometaphase, metaphase, anaphase, and telophase.
- **C (cytokinesis)** is the phase of the cell cycle when the cytoplasm divides, creating two daughter cells. In animal cells, the microtubule spindle helps position a contracting ring of actin that constricts like a drawstring to pinch the cell in two. In cells with a cell wall, such as plant cells, a plate forms between the dividing cells.

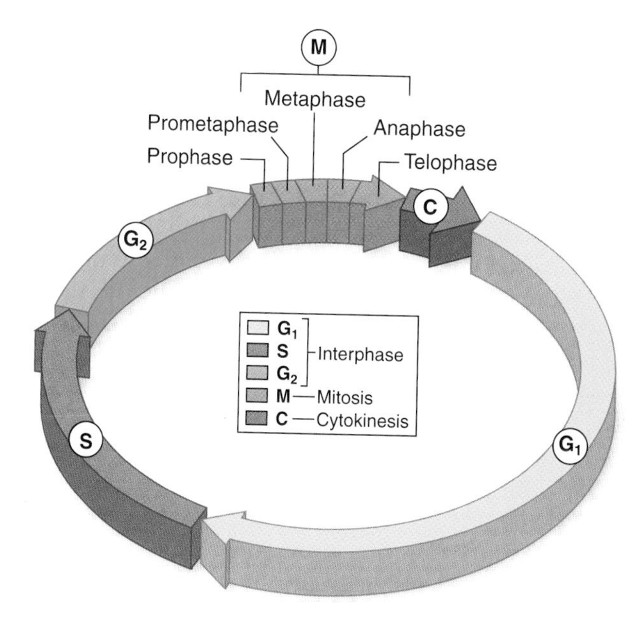

figure 10.8

THE CELL CYCLE. The cell cycle is depicted as a circle. The first gap phase, G_1, involves growth and preparation for DNA synthesis. During S phase, a copy of the genome is synthesized. The second gap phase, G_2, prepares the cell for mitosis. During mitosis, replicated chromosomes are partitioned. Cytokinesis divides the cell into two cells with identical genomes.

The duration of the cell cycle varies depending on cell type

The time it takes to complete a cell cycle varies greatly. Cells in animal embryos can complete their cell cycle in under 20 min; the shortest known animal nuclear division cycles occur in fruit fly embryos (8 min). Cells such as these simply divide their nuclei as quickly as they can replicate their DNA, without cell growth. Half of their cycle is taken up by S, half by M, and essentially none by G_1 or G_2.

Because mature cells require time to grow, most of their cycles are much longer than those of embryonic tissue. Typically, a dividing mammalian cell completes its cell cycle in about 24 h, but some cells, such as certain cells in the human liver, have cell cycles lasting more than a year. During the cycle, growth occurs throughout the G_1 and G_2 phases (referred to as gap phases because they separate S from M), as well as during the S phase. The M phase takes only about an hour, a small fraction of the entire cycle.

Most of the variation in the length of the cell cycle between one organism or cell type and another occurs in the G_1 phase. Cells often pause in G_1 before DNA replication and enter a resting state called the **G_0 phase;** cells may remain in this phase for days to years before resuming cell division. At any given time, most of the cells in an animal's body are in G_0 phase. Some, such as muscle and nerve cells, remain there permanently; others, such as liver cells, can resume G_1 phase in response to factors released during injury.

Cell division in eukaryotes is a complex process that involves five phases: a first gap phase (G_1); DNA synthesis phase (S); a second gap phase (G_2); mitosis (M), in which chromosomes are separated; and cytokinesis (C) in which a cell becomes two separate cells.

10.4 Interphase: Preparation for Mitosis

The events that occur during interphase—the G_1, S, and G_2 phases—are very important for the successful completion of mitosis. During G_1, cells undergo the major portion of their growth. During the S phase, each chromosome replicates to produce two sister chromatids, which remain attached to each other at the centromere. In the G_2 phase, the chromosomes coil even more tightly.

The **centromere** is a point of constriction on the chromosome containing certain repeated DNA sequences that bind specific proteins. These proteins make up a disklike structure called the **kinetochore.** This disk functions as an attachment site for microtubules necessary to separate the chromosomes during cell division (figure 10.9). As seen in figure 10.6, each chromosome's centromere is located at a characteristic site along the length of the chromosome.

After the S phase, the sister chromatids appear to share a common centromere, but at the molecular level the DNA of the centromere has actually already replicated, so there are two complete DNA molecules. Functionally, however, the two chromatids have a single centromere due to their being attached by cohesin proteins at the centromere site (figure 10.10). In metazoan animals, the cohesins that hold sister chromatids together after replication appear to be replaced by condensin during the process of chromosome compaction. This leaves the chromosomes still attached tightly at the centromere, but loosely attached elsewhere.

figure 10.10

PROTEINS FOUND AT THE CENTROMERE. In this image DNA, a cohesin protein and a kinetochore protein have all been labeled with a different colored fluorescent dye. Cohesin (*red*), which holds centromeres together, lies between the sister chromatids (*blue*). Each sister chromatid has its own separate kinetochore (*green*).

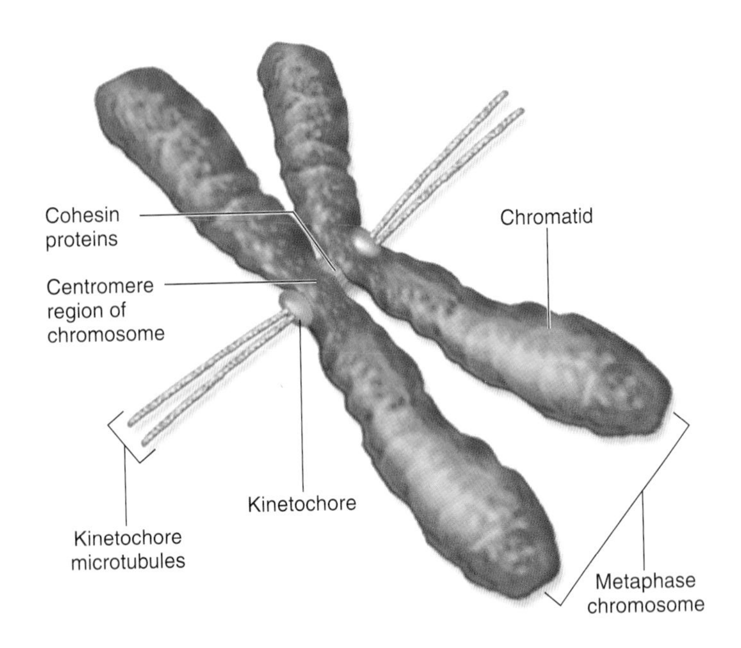

figure 10.9

KINETOCHORES. Separation of sister chromatids during mitosis depends on microtubules attaching to proteins found in the kinetochore. These kinetochore proteins are assembled on the centromere of chromosomes. The centromeres of the two sister chromatids are held together by cohesin proteins.

The cell grows throughout interphase. The G_1 and G_2 segments of interphase are periods of active growth, during which proteins are synthesized and cell organelles are produced. The cell's DNA replicates only during the S phase of the cell cycle.

After the chromosomes have replicated in S phase, they remain fully extended and uncoiled, which makes them invisible when viewed with the light microscope. In G_2 phase, they begin the process of **condensation,** coiling ever more tightly. Special *motor proteins* are involved in the rapid final condensation of the chromosomes that occurs early in mitosis. Also during G_2 phase, the cells begin to assemble the machinery they will later use to move the chromosomes to opposite poles of the cell. In animal cells, a pair of microtubule-organizing centers called **centrioles** replicate, producing one for each pole. All eukaryotic cells undertake an extensive synthesis of **tubulin,** the protein that forms microtubules.

Interphase includes the G_1, S, and G_2 phases of the cell cycle. During interphase, the cell grows; replicates chromosomes, organelles, and centrioles; and synthesizes components needed for mitosis, including tubulin.

10.5 Mitosis: Chromosome Segregation

The process of mitosis is one of the most dramatic and beautiful biological processes that can be readily observed. In our attempts to understand this process, we have divided it into discrete phases but it should always be remembered that this is a dynamic, continuous process, not a set of discrete steps. This process is shown both schematically and in micrographs in figure 10.11.

During prophase, the mitotic apparatus forms

When the chromosome condensation initiated in G_2 phase reaches the point at which individual condensed chromosomes first become visible with the light microscope, the first stage of mitosis, **prophase,** has begun. The condensation process continues throughout prophase; consequently, chromosomes that start prophase as minute threads appear quite bulky before its conclusion. Ribosomal RNA synthesis ceases when the portion of the chromosome bearing the rRNA genes is condensed.

The spindle and centrioles

The assembly of the **spindle** apparatus that will later separate the sister chromatids occurs during prophase. The normal microtubule structure in the cell disassembled in the G_2 phase is replaced by the spindle. In animal cells, the two centriole pairs formed during G_2 phase begin to move apart early in prophase, forming between them an axis of microtubules referred to as spindle fibers. By the time the centrioles reach the opposite poles of the cell, they have established a bridge of microtubules, called the spindle apparatus, between them. In plant cells, a similar bridge of microtubular fibers forms between opposite poles of the cell, although centrioles are absent in plant cells.

In animal cell mitosis, the centrioles extend a radial array of microtubules toward the nearby plasma membrane when they reach the poles of the cell. This arrangement of microtubules is called an **aster.** Although the aster's function is not fully understood, it probably braces the centrioles against the membrane and stiffens the point of microtubular attachment during the retraction of the spindle. Plant cells, which have rigid cell walls, do not form asters.

Breakdown of the nuclear envelope

During the formation of the spindle apparatus, the nuclear envelope breaks down, and the endoplasmic reticulum reabsorbs its components. At this point, the microtubular spindle fibers extend completely across the cell, from one pole to the other. Their orientation determines the plane in which the cell will subsequently divide, through the center of the cell at right angles to the spindle apparatus.

During prometaphase, chromosomes attach to the spindle

The transition from prophase to **prometaphase** occurs following the disassembly of the nuclear envelope. During prometaphase the condensed chromosomes become attached to the spindle by their kinetochores. Each chromosome possesses two kinetochores, one attached to the centromere region of each sister chromatid (see figure 10.9).

Microtubule attachment

As prometaphase continues, a second group of microtubules grow from the poles of the cell toward the centromeres. These microtubules are captured by the kinetochores on each pair of sister chromatids. This results in the kinetochores of each sister chromatid being connected to opposite poles of the spindle.

This bipolar attachment is critical to the process of mitosis; any mistakes in microtubule positioning can be disastrous. For example, the attachment of the kinetochores of both sister chromatids to the same pole leads to a failure of sister chromatid separation, and they will be pulled to the same pole ending up in the same daughter cell, with the other daughter cell missing that chromosome.

Movement of chromosomes to the cell center

With each chromosome attached to the spindle by microtubules from opposite poles to the kinetochores of sister chromatids, the chromosomes begin to move to the center of the cell. This movement is jerky, as if a chromosome is being pulled toward both poles at the same time. This process is called *congression,* and it eventually leads to all of the chromosomes being arranged at the equator of the cell with the sister chromatids of each chromosome oriented to opposite poles by their kinetochore microtubules.

The force that moves chromosomes has been of great interest since the process of mitosis was first observed. Two basic mechanisms have been proposed to explain this: (1) assembly and disassembly of microtubules provides the force to move chromosomes, and (2) motor proteins located at the kinetochore and poles of the cell pull on microtubules to provide force. Data have been obtained that support both mechanisms.

In support of the microtubule-shortening proposal isolated chromosomes can be pulled by microtubule disassembly. The spindle is a very dynamic structure, with microtubules being added to at the kinetochore and shortened at the poles, even during metaphase. In support of the motor protein proposal, multiple motor proteins have been identified as kinetochore proteins, and inhibition of the motor protein dynein slows chromosome separation at anaphase. Like many phenomena that we analyze in living systems, the answer is not a simple either–or choice; both mechanisms are probably at work.

In metaphase, the centromeres align

The alignment of the chromosomes in the center of the cell signals the third stage of mitosis, **metaphase.** When viewed with a light microscope, the chromosomes appear to array themselves in a circle along the inner circumference of the cell, just as the equator girdles the Earth (figure 10.12). An imaginary plane perpendicular to the axis of the spindle that passes through this circle is called the *metaphase plate.* The metaphase plate is not an actual structure, but rather an indication of the future axis of cell division.

Positioned by the microtubules attached to the kinetochores of their centromeres, all of the chromosomes line up on the metaphase plate. At this point their centromeres are neatly arrayed in

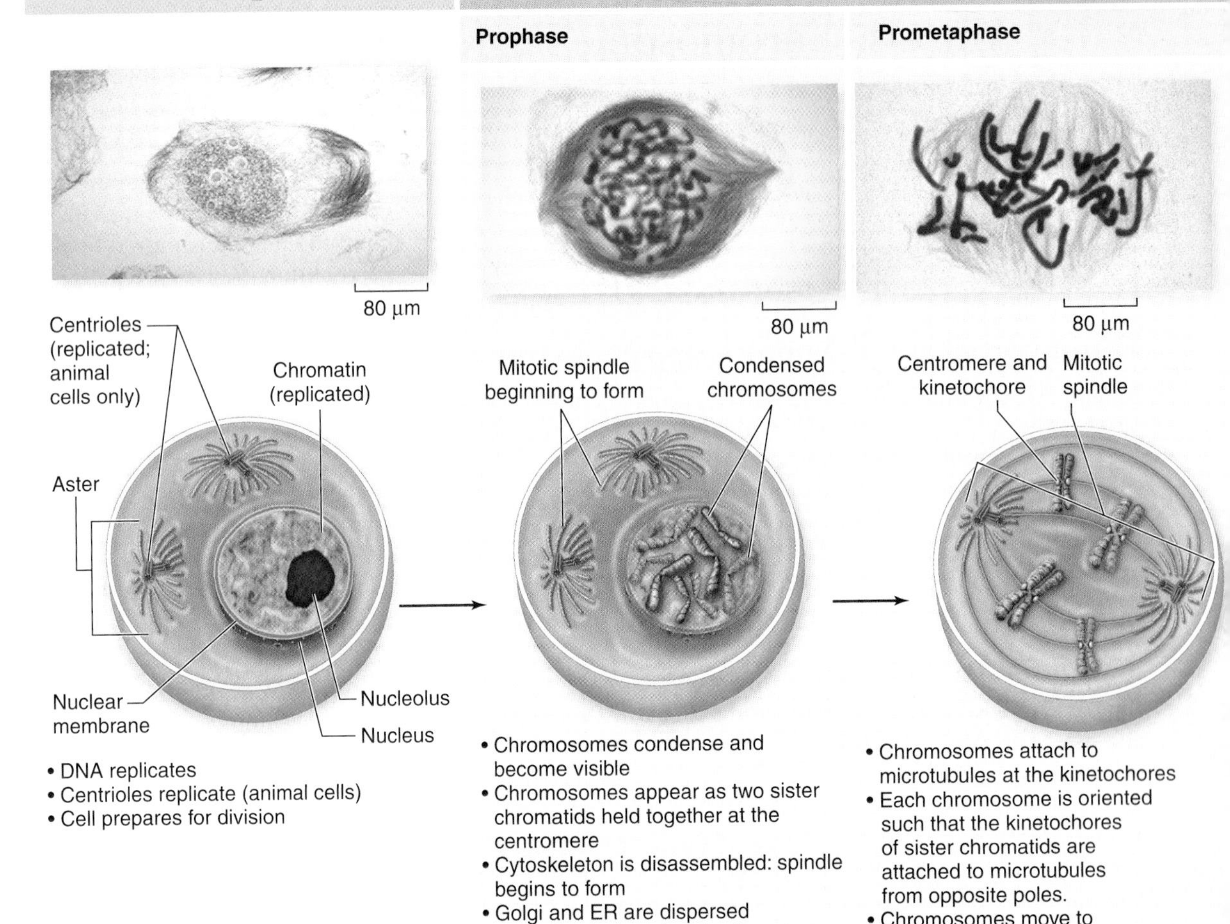

figure 10.11

MITOSIS AND CYTOKINESIS. Mitosis is conventionally divided into five stages—prophase, prometaphase, metaphase, anaphase, and telophase—which together act to separate duplicated chromosomes. This is followed by cytokinesis, which divides the cell into two separate cells. Photos depict mitosis and cytokinesis in a plant, the African blood lily (*Haemanthus katharinae*), with chromosomes stained blue and microtubules stained red. Drawings depict mitosis and cytokinesis in animal cells.

a circle, equidistant from the two poles of the cell, with microtubules extending back toward the opposite poles of the cell. At this point the cell is prepared to properly separate sister chromatids, such that each daughter cell will receive a complete set of chromosomes. Thus metaphase is really a transitional phase in which all the preparations are checked before the action continues.

At anaphase, the chromatids separate

Of all the stages of mitosis, shown in figure 10.11, **anaphase** is the shortest and the most amazing to watch. It begins when the centromeres split, freeing the two sister chromatids from each other. Up to this point in mitosis, sister chromatids have been held together by cohesin proteins concentrated at the centromere, as mentioned earlier. The key event in anaphase, then, is the simultaneous removal of these proteins from all of the chromosomes. The control and details of this process are discussed later on in the context of control of the entire cell cycle.

Freed from each other, the sister chromatids are pulled rapidly toward the poles to which their kinetochores are attached. In the process, two forms of movement take place simultaneously, each driven by microtubules. These movements are often called anaphase A and anaphase B to distinguish them.

First, during anaphase A, the *kinetochores are pulled toward the poles* as the microtubules that connect them to the poles shorten. This shortening process is not a contraction; the microtubules do not get any thicker. Instead, tubulin subunits are removed from the kinetochore ends of the microtubules. As more subunits are removed, the chromatid-bearing microtubules are progressively disassembled, and the chromatids are pulled ever closer to the poles of the cell.

Second, during anaphase B, the *poles move apart* as microtubular spindle fibers physically anchored to opposite poles slide past each other, away from the center of the cell (figure 10.13).

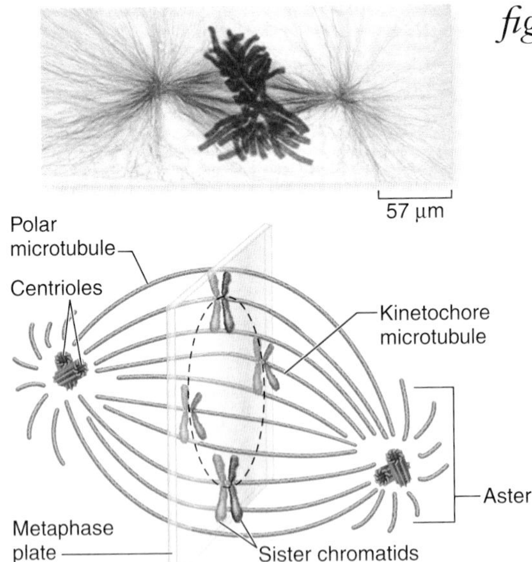

figure 10.12

METAPHASE. In metaphase, the chromosomes are arrayed at the midpoint of the cell. The imaginary plane through the equator of the cell is called the metaphase plate. As the spindle itself is a three dimensional structure, the chromosomes are arrayed in a rough circle on the metaphase plate.

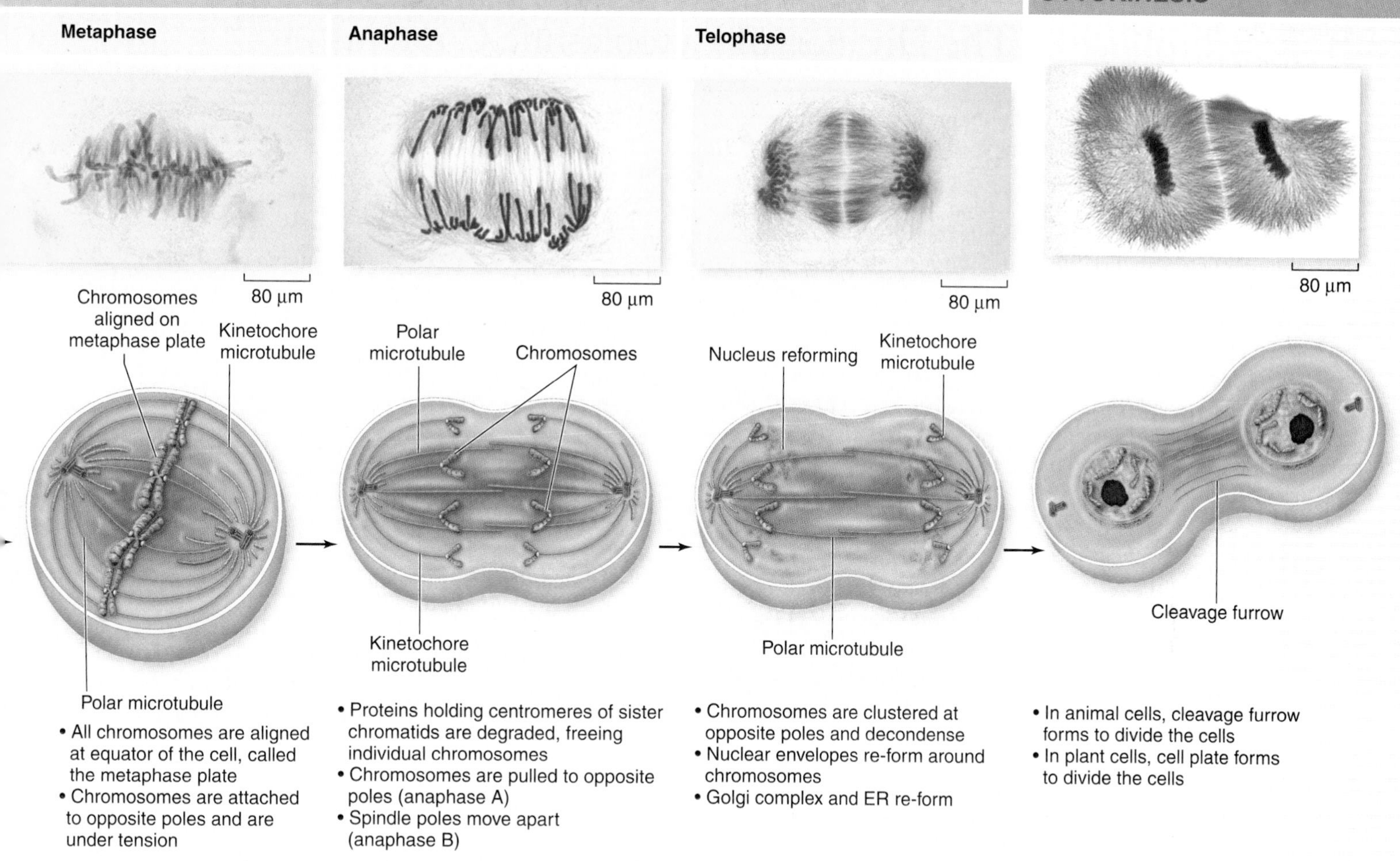

Because another group of microtubules attach the chromosomes to the poles, the chromosomes move apart, too. If a flexible membrane surrounds the cell, it becomes visibly elongated.

When the sister chromatids separate in anaphase, the accurate partitioning of the replicated genome—the essential element of mitosis—is complete.

During telophase, the nucleus re-forms

In **telophase,** the spindle apparatus disassembles as the microtubules are broken down into tubulin monomers that can be used to construct the cytoskeletons of the daughter cells. A nuclear envelope forms around each set of sister chromatids, which can now be called chromosomes because they are no longer attached at the centromere. The chromosomes soon begin to uncoil into the more extended form that permits gene expression. One of the early group of genes expressed after mitosis is complete are the rRNA genes, resulting in the reappearance of the nucleolus.

Telophase can be viewed as a reversal of the process of prophase, bringing the cell back to the state of interphase.

Mitosis is divided into phases: prophase, prometaphase, metaphase, anaphase, and telophase. The early phases involve restructuring the cell to create a spindle made of microtubules that is used to pull chromosomes to opposite poles during anaphase. The nucleus is re-formed in telophase.

Metaphase

Pole | Overlapping microtubules | Pole

Late Anaphase

Pole | Overlapping microtubules | Pole | 2 μm

figure 10.13

MICROTUBULES SLIDE PAST EACH OTHER AS THE CHROMOSOMES SEPARATE. In these electron micrographs of dividing diatoms, the overlap of the microtubules lessens markedly during spindle elongation as the cell passes from metaphase to anaphase. During anaphase B the poles move farther apart as the chromosomes move toward the poles.

10.6 Cytokinesis: The Division of Cytoplasmic Contents

Mitosis is complete at the end of telophase. The eukaryotic cell has partitioned its replicated genome into two new nuclei positioned at opposite ends of the cell.

The replication of organelles takes place before cytokinesis, often in the S or G_2 phase. While mitosis was going on, the cytoplasmic organelles, including mitochondria and chloroplasts (if present), were reassorted to areas that will separate and become the daughter cells.

Cell division is still not complete at the end of mitosis, however, because the division of the cell body proper has not yet begun. The phase of the cell cycle when the cell actually divides is called **cytokinesis.** It generally involves the cleavage of the cell into roughly equal halves.

In animal cells, a belt of actin pinches off the daughter cells

In animal cells and the cells of all other eukaryotes that lack cell walls, cytokinesis is achieved by means of a constricting belt of actin filaments. As these filaments slide past one another, the diameter of the belt decreases, pinching the cell and creating a **cleavage furrow** around the cell's circumference (figure 10.14*a*).

As constriction proceeds, the furrow deepens until it eventually slices all the way into the center of the cell. At this point, the cell is divided in two (figure 10.14*b*).

In plant cells, a cell plate divides the daughter cells

Plant cell walls are far too rigid to be squeezed in two by actin filaments. Instead, these cells assemble membrane components in their interior, at right angles to the spindle apparatus. This expanding membrane partition, called a **cell plate,** continues to grow outward until it reaches the interior surface of the plasma membrane and fuses with it, effectively dividing the cell in two (figure 10.15). Cellulose is then laid down on the new membranes, creating two new cell walls. The space between the daughter cells becomes impregnated with pectins and is called a **middle lamella.**

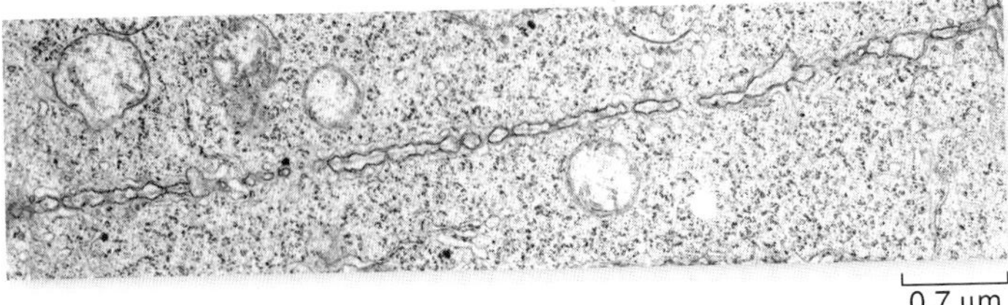

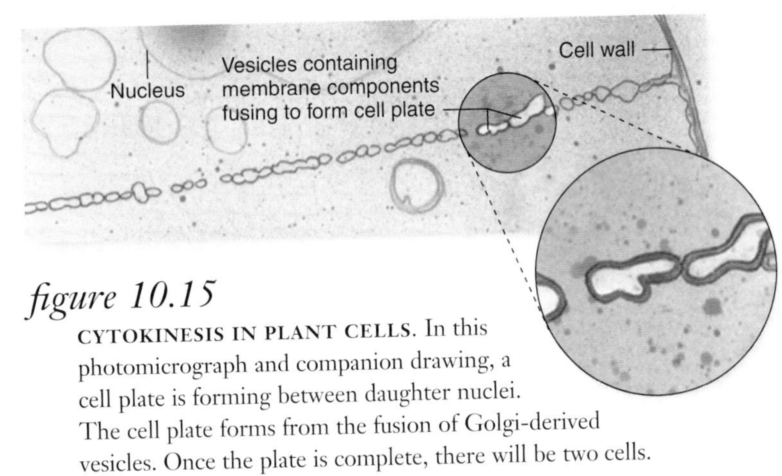

figure 10.15

CYTOKINESIS IN PLANT CELLS. In this photomicrograph and companion drawing, a cell plate is forming between daughter nuclei. The cell plate forms from the fusion of Golgi-derived vesicles. Once the plate is complete, there will be two cells.

In fungi and some protists, daughter nuclei are separated during cytokinesis

In most fungi and some groups of protists, the nuclear membrane does not dissolve, and as a result, all the events of mitosis occur entirely *within* the nucleus. Only after mitosis is complete in these organisms does the nucleus divide into two daughter nuclei; then, during cytokinesis, one nucleus goes to each daughter cell. This separate nuclear division phase of the cell cycle does not occur in plants, animals, or most protists.

After cytokinesis in any eukaryotic cell, the two daughter cells contain all the components of a complete cell. Whereas mitosis ensures that both daughter cells contain a full complement of chromosomes, no similar mechanism ensures that organelles such as mitochondria and chloroplasts are distributed equally between the daughter cells. But as long as at least one of each organelle is present in each cell, the organelles can replicate to reach the number appropriate for that cell.

Cytokinesis divides the cell cytoplasm and organelles into separate daughter cells. In animal cells, actin pinches the cell in two; in plant cells, a cell plate forms in the middle of the dividing cell. In fungi and some protists, the nucleus divides after mitosis has been completed, and the resulting nuclei separate during cytokinesis.

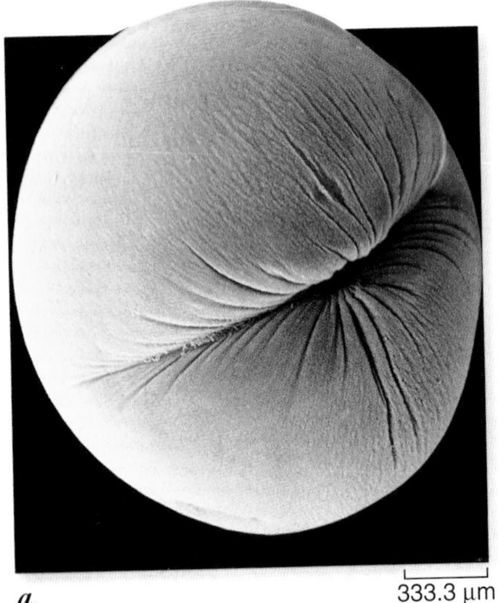

a. *b.*

figure 10.14

CYTOKINESIS IN ANIMAL CELLS. ***a.*** A cleavage furrow forms around a dividing frog egg. ***b.*** The completion of cytokinesis in an animal cell. The two daughter cells are still joined by a thin band of cytoplasm occupied largely by microtubules.

10.7 Control of the Cell Cycle

Our knowledge of how the cell cycle is controlled, although still incomplete, has grown enormously in the past 30 years. Our current view integrates two basic concepts. First, the cell cycle has two irreversible points: the replication of genetic material and the separation of the sister chromatids. Second, the cell cycle can be put on hold at specific points called *checkpoints.* At any of these checkpoints, the process is checked for accuracy and can be halted if there are errors. This leads to extremely high fidelity overall for the entire process. The checkpoint organization also allows the cell cycle to respond to both the internal state of the cell, including nutritional state and integrity of genetic material, and to signals from the environment, which are integrated at major checkpoints.

A brief history of cell cycle control

The history of investigation into control of the cell cycle is instructive in two ways. First, it allows us to place modern observations into context; second we can see how biologists using very different approaches often end up at the same place. The following brief history introduces three observations and then shows how they can be integrated into a single mechanism.

Discovery of MPF

Research on the activation of frog oocytes led to the discovery of a substance that was first called *maturation-promoting factor* (*MPF*). Frog oocytes, which go on to become egg cells, become arrested near the end of their development at the G_2 stage of meiosis I, which is the division leading to the production of gametes (chapter 11). They remain in this arrested state and await hormonal signaling to complete this division process.

Cytoplasm taken from a variety of actively dividing cells could prematurely induce cell division when injected into oocytes (figure 10.16*a*). These experiments indicated the presence of a positive regulator of cell cycle progression in the cytoplasm of dividing cells: MPF. These experiments also fit well with cell fusion experiments done with mitotic and interphase cells that also indicated a cytoplasmic positive regulator that could induce mitosis (figure 10.16*b*).

Further studies highlighted two key aspects of MPF. First, MPF activity varied during the cell cycle: low in early G_2, rising throughout this phase, and then peaking in mitosis (figure 10.17). Second, the enzymatic activity of MPF involved the phosphorylation of proteins. This second point is not surprising given the importance of phosphorylation as a reversible

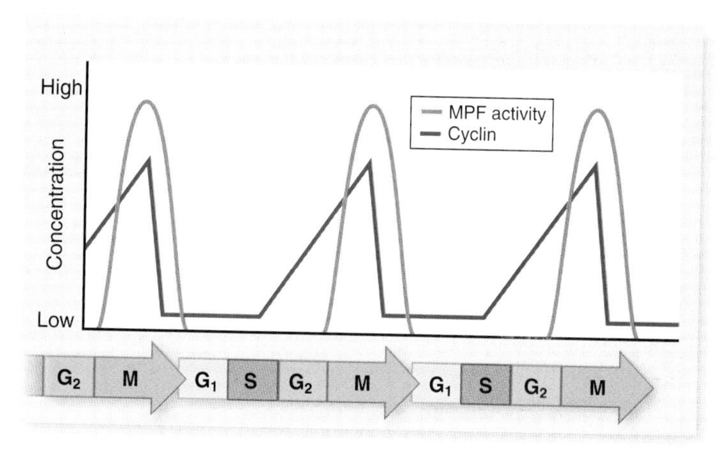

figure 10.17

CORRELATION OF MPF ACTIVITY, AMOUNT OF CYCLIN PROTEIN AND STAGES OF THE CELL CYCLE. Cyclin concentration and MPF activity are shown plotted vs. stage of the cell cycle. MPF activity (stimulation of oocyte maturation) changes in a repeating pattern through the cell cycle. This also correlates with the level of mitotic cyclin in the cell, which shows a similar pattern. The reason for this correlation is that cyclin is actually one component of MPF, the other being a cyclin-dependent kinase (Cdk). Together these act as a positive regulator of cell division.

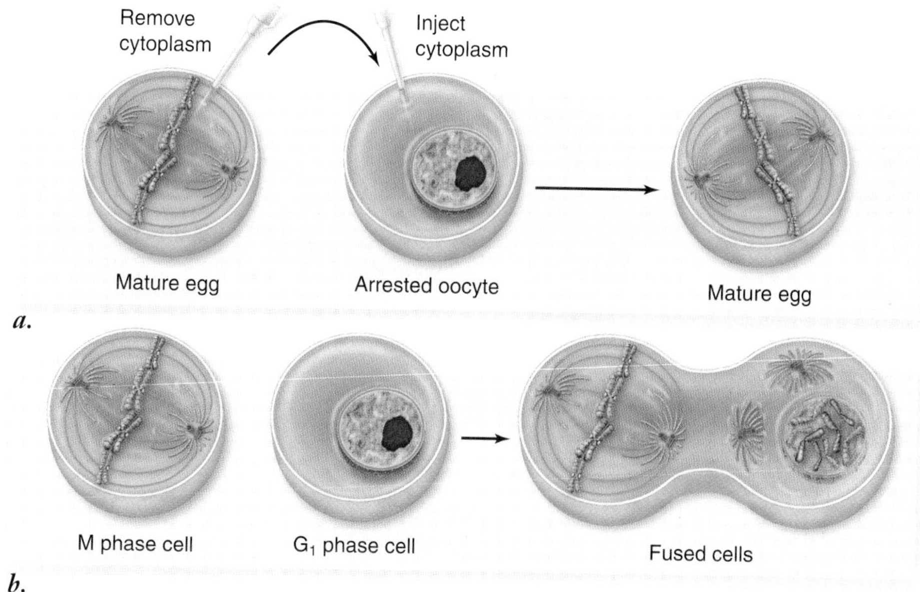

figure 10.16

DISCOVERY OF POSITIVE REGULATOR OF CELL DIVISION. ***a.*** Frog oocytes are arrested at an early stage of meiosis. A hormonal signal is required to mature into an egg. If cytoplasm removed from a mature egg is injected into an oocyte, it will continue meiosis to become a mature egg. This indicates the cytoplasm of mature eggs contains a positive regulator of maturation: maturation promoting factor (MPF). ***b.*** When M-phase cells are fused with interphase cells, the interphase nucleus breaks down and chromosomes condense as though the cells were entering mitosis. This indicates dividing cells contain a positive regulator of mitosis.

switch on the activity of proteins (see chapter 9). The first observation indicated that MPF itself was not always active, but rather was being regulated with the cell cycle, and the second showed the possible enzymatic activity of MPF.

Discovery of cyclins

Other researchers examined proteins produced during the early divisions in sea urchin embryos. They identified proteins that were produced in synchrony with the cell cycle, and named them **cyclins** (see figure 10.17). These observations were extended in another marine invertebrate, the surf clam. Two forms of cyclin were found that cycled at slightly different times, reaching peaks at the G_1/S and G_2/M boundaries. Despite much effort, no identified enzymatic activity was associated with these proteins. Their hallmark was the timing of their production and not any intrinsic activity.

Genetic analysis of the cell cycle

Geneticists using two different yeasts, budding yeast and fission yeast, as model systems set out to determine the genes necessary for control of the cell cycle. By isolating mutants that were halted during division, they identified genes that were necessary for cell cycle progression. These studies indicated that in yeast, there were two critical control points: the commitment to DNA synthesis, called START, as it meant committing to divide, and the commitment to mitosis. One particular gene, named ***cdc2***, from fission yeast, was shown to be critical for passing both of these boundaries.

MPF is cyclin plus cdc2

All of these findings came together in an elegant fashion with the following observations. First, the protein encoded by the *cdc2* gene was shown to be a protein kinase. Second, the purification and identification of MPF showed that it was composed of both a cyclin component and a kinase component. Last, the kinase itself was the cdc2 protein!

The cdc2 protein was the first identified **cyclin-dependent kinase (Cdk),** that is, a protein kinase enzyme that is only active when complexed with cyclin. This finding led to the renaming of MPF as *mitosis*-promoting factor, as its role was clearly more general than simply promoting the maturation of frog oocytes.

These Cdk enzymes are the key positive drivers of the cell division cycle. They are often called the engine that drives cell division. The control of the cell cycle in higher eukaryotes is much more complex than the simple single-engine cycle of yeast, but the yeast model remains a useful framework for understanding more complex regulation. The discovery of Cdks and their role in the cell cycle is an excellent example of the progressive nature of science.

The cell cycle can be halted at three checkpoints

Although we have divided the cell cycle into phases and subdivided mitosis into stages, the cell recognizes three points at which the cycle can be delayed or halted. The cell uses these three checkpoints to both assess its internal state and integrate external signals (figure 10.18): G_1/S, G_2/M, and late metaphase (the spindle checkpoint). Passage through these checkpoints is controlled by the Cdk enzymes described earlier and also in the following section.

G_1/S checkpoint

The **G_1/S checkpoint** is the primary point at which the cell "decides" whether or not to divide. This checkpoint is therefore the primary point at which external signals can influence events of the cycle. It is the phase during which growth factors (discussed later on) affect the cycle and also the phase that links cell division to cell growth and nutrition.

In yeast systems, where the majority of the genetic analysis of the cell cycle has been performed, this checkpoint is called START. In animals, it is called the restriction point (R point). In all systems, once a cell has made this irreversible commitment to replicate its genome, it has committed to divide. Damage to DNA can halt the cycle at this point, as can starvation conditions or lack of growth factors.

G_2/M checkpoint

The **G_2/M checkpoint** has received a large amount of attention because of its complexity and its importance as the stimulus for the events of mitosis. Historically, Cdks active at this checkpoint were first identified as MPFs, a term that has now evolved into **M phase-promoting factor (MPF).**

Passage through this checkpoint represents the commitment to mitosis. This checkpoint assesses the success of DNA replication and can stall the cycle if DNA has not been accurately replicated. DNA-damaging agents result in arrest at this checkpoint as well as at the G_1/S checkpoint.

Spindle checkpoint

The **spindle checkpoint** ensures that all of the chromosomes are attached to the spindle in preparation for anaphase. The second irreversible step in the cycle is the separation of chromosomes during anaphase, and therefore it is critical that they are properly arrayed at the metaphase plate.

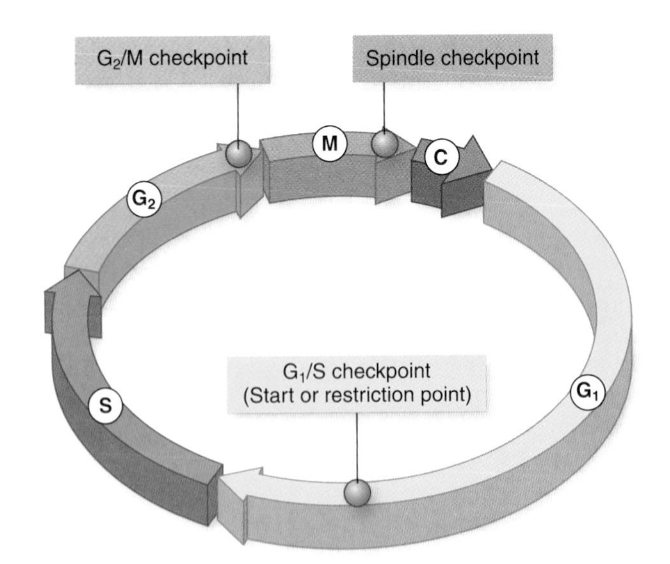

figure 10.18

CONTROL OF THE CELL CYCLE. Cells use a centralized control system to check whether proper conditions have been achieved before passing three key checkpoints in the cell cycle.

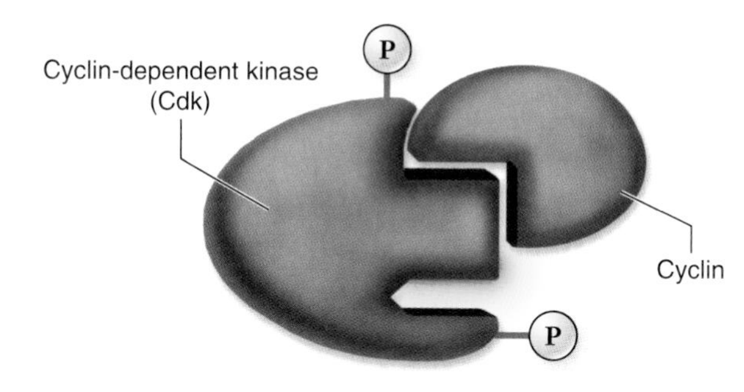

figure 10.19

Cdk ENZYME FORMS A COMPLEX WITH CYCLIN. Cdk is a protein kinase that activates numerous cell proteins by phosphorylating them. Cyclin is a regulatory protein required to activate Cdk. This complex is also called mitosis-promoting factor (MPF). The activity of Cdk is also controlled by the pattern of phosphorylation: phosphorylation at one site (represented by the red site) inactivates the Cdk, and phosphorylation at another site (represented by the green site) activates the Cdk.

Cyclin-dependent kinases drive the cycle

The primary molecular mechanism of cell cycle control is phosphorylation, which you may recall is the addition of a phosphate group to the amino acids serine, threonine, and tyrosine in proteins (chapter 9). The enzymes that accomplish this phosphorylation are the Cdks (figure 10.19).

The action of Cdks

The first important cell cycle kinase was identified in fission yeast and named Cdc2 (now also called Cdk1). In yeast this Cdk can partner with different cyclins at different points in the cell cycle (figure 10.20).

Even in the simplified cycle of the yeasts we are left with the important question of what controls the activity of the Cdks during the cycle. For many years, a common view was that cyclins drove the cell cycle—that is, the periodic synthesis and destruction of cyclins acted as a clock. More recently, it has become clear that the Cdc2 kinase is also itself controlled by phosphorylation: Phosphorylation at one site activates Cdc2, and phosphorylation at another site inactivates it (see figure 10.19). Full activation of the Cdc2 kinase requires complexing with a cyclin and the appropriate pattern of phosphorylation.

As the G_1/S checkpoint is approached, the triggering signal in yeast appears to be the accumulation of G_1 cyclins. These form a complex with Cdc2 to create the active G_1/S Cdk, which phosphorylates a number of targets that bring about the increased enzyme activity for DNA replication.

The action of MPF

MPF and its role at the G_2/M checkpoint has been extensively analyzed in a number of different experimental systems. The control of MPF is sensitive to agents that disrupt or delay replication and to agents that damage DNA. It was once thought that MPF was controlled solely by the level of the M phase-specific cyclins, but it has now become clear that this is not the case.

Although M phase cyclin is necessary for MPF function, activity is controlled by inhibitory phosphorylation of the kinase component, Cdc2. The critical signal in this process is the removal of the inhibitory phosphates by a protein, phosphatase. This action forms a molecular switch based on positive feedback because the active MPF further activates its own activating phosphatase.

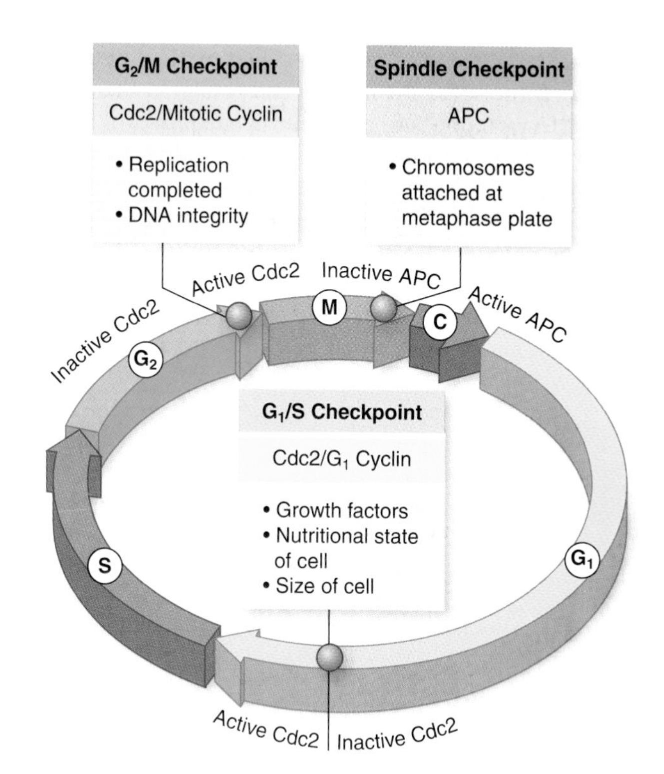

figure 10.20

CHECKPOINTS OF THE YEAST CELL CYCLE. The simplest cell cycle that has been studied in detail is the fission yeast. This is controlled by three main checkpoints and a single Cdk enzyme, called Cdc2. The Cdc2 enzyme partners with different cyclins to control the G_1/S and G_2/M checkpoints. The spindle checkpoint is controlled by the anaphase-promoting complex (APC).

The checkpoint assesses the balance of the kinase that adds inhibitory phosphates with the phosphatase that removes them. Damage to DNA acts through a complex pathway that includes damage sensing and a response to tip the balance toward the inhibitory phosphorylation of MPF. Later on, we describe how some cancers overcome this inhibition.

The anaphase-promoting complex

The molecular details of the sensing system at the spindle checkpoint are not clear. The presence of all chromosomes at the metaphase plate and the tension on the microtubules between opposite poles are both important. The signal is transmitted through the **anaphase-promoting complex (APC).**

The function of the APC is to trigger anaphase itself. As described earlier, the sister chromatids at metaphase are still held together by the protein complex cohesin. The APC does not act directly on cohesin, but rather acts by marking a protein called *securin* for destruction. The securin protein acts as an inhibitor of another protease called *separase* that appears to be specific for the cohesin complex. Once inhibition is lifted, separase destroys cohesin.

This process has been analyzed in detail in budding yeast, where it has been shown that the separase enzyme specifically degrades a component of cohesin called Scc1. This leads to the release of the sister chromatids and results in their sudden movement toward opposite poles during anaphase.

In vertebrates, most cohesin is removed from the sister chromatids during chromosome condensation, possibly with cohesin being replaced by condensin. At metaphase, the majority of the cohesin that remains on vertebrate chromatids is concentrated at the centromere (figure 10.10). The removal of this cohesin by the mechanism described earlier would then explain the anaphase movement of chromosomes and the apparent "division" of the centromeres.

The APC has a number of roles in mitosis: it activates the proteins that remove the cohesins holding sister chromatids together, and it is necessary for the destruction of mitotic cyclins to drive the cell out of mitosis. The APC complex marks proteins for destruction by the proteosome, the organelle responsible for the controlled degradation of proteins (chapter 16). The signal to degrade a protein is the addition of a molecule called *ubiquitin.*

In multicellular eukaryotes, many Cdks and external signals act on the cell cycle

The major difference between more complex animals and single-celled eukaryotes such as fungi and protists is twofold: First, multiple Cdks control the cycle as opposed to the single Cdk in yeasts; and second, animal cells respond to a greater variety of external signals than do yeasts, which primarily respond to signals necessary for mating.

In higher eukaryotes there are more Cdk enzymes and more cyclins that can partner with these multiple Cdks, but their basic role is the same as in the yeast cycle. A more complex cell cycle is shown in figure 10.21. These more complex controls allow the integration of more input into control of the cycle. With the evolution of more complex forms of organization (tissues, organs, and organ systems), more complex forms of cell cycle control evolved as well.

A multicellular body's organization cannot be maintained without severely limiting cell proliferation—so that only certain cells divide, and only at appropriate times. The way cells inhibit individual growth of other cells is apparent in mammalian cells growing in tissue culture: A single layer of cells expands over a culture plate until the growing border of cells comes into contact with neighboring cells, and then the cells stop dividing. If a sector of cells is cleared away, neighboring cells rapidly refill that sector and then stop dividing again on cell contact.

How are cells able to sense the density of the cell culture around them? When cells come in contact with one another, receptor proteins in the plasma membrane activate a signal transduction pathway that acts to inhibit Cdk action. This prevents entry into the cell cycle.

Growth factors and the cell cycle

Growth factors act by triggering intracellular signaling systems. Fibroblasts, for example, possess numerous receptors on their plasma membranes for one of the first growth factors to be identified, **platelet-derived growth factor (PDGF).** The PDGF receptor is a receptor tyrosine kinase (RTK) that initiates a MAP kinase cascade to stimulate cell division (discussed in chapter 9).

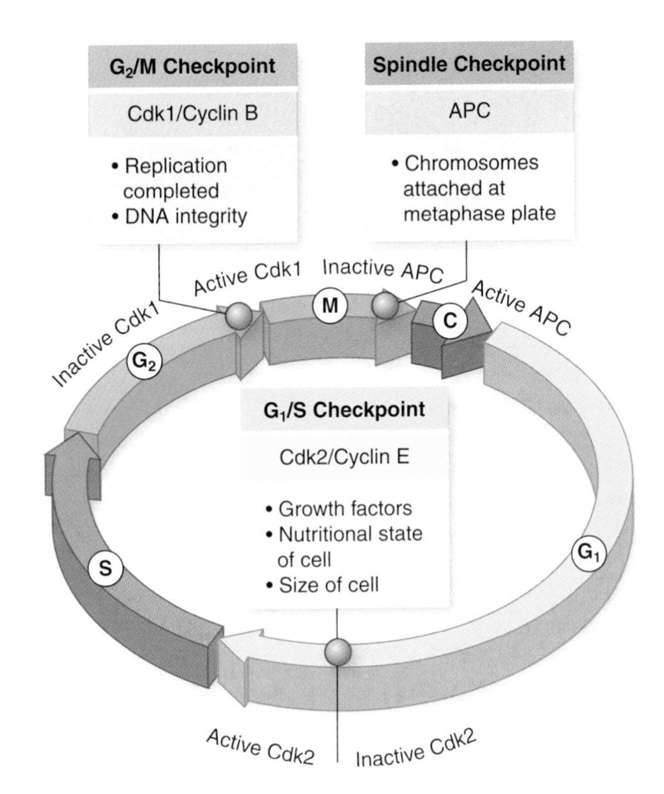

figure 10.21

CHECKPOINTS OF THE MAMMALIAN CELL CYCLE. The more complex mammalian cell cycle is shown. This cycle is still controlled through three main checkpoints. These integrate internal and external signals to control progress through the cycle. These inputs control the state of two different Cdk–cyclin complexes and the anaphase-promoting complex (APC). The arrows represent inputs, which can be complex networks such as the signal transduction cascade seen in growth factor signaling.

PDGF was discovered when investigators found that fibroblasts would grow and divide in tissue culture only if the growth medium contained blood serum. Serum is the liquid that remains in blood after clotting; blood plasma, the liquid from which cells have been removed without clotting, would not work. The researchers hypothesized that platelets in the blood clots were releasing into the serum one or more factors required for fibroblast growth. Eventually, they isolated such a factor and named it PDGF.

Growth factors such as PDGF can override cellular controls that otherwise inhibit cell division. When a tissue is injured, a blood clot forms, and the release of PDGF triggers neighboring cells to divide, helping to heal the wound. Only a tiny amount of PDGF (approximately 10^{-10} M) is required to stimulate cell division in cells with PDGF receptors.

Characteristics of growth factors

Over 50 different proteins that function as growth factors have been isolated, and more undoubtedly exist. A specific cell surface receptor recognizes each growth factor, its binding site fitting that growth factor precisely. These growth factor receptors often initiate MAP kinase cascades in which the final kinase enters the nucleus and activates transcription factors by phosphorylation. These transcription factors stimulate the

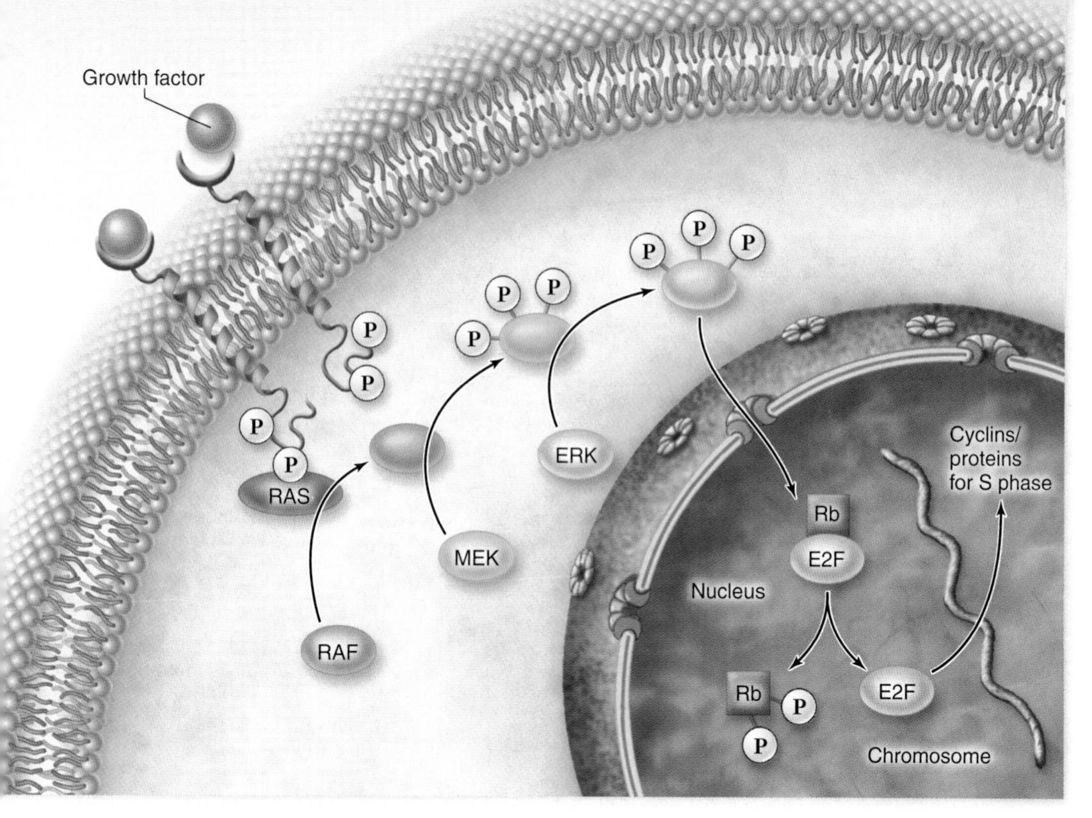

figure 10.22

THE CELL PROLIFERATION-SIGNALING PATHWAY. Binding of a growth factor sets in motion a MAP kinase intracellular signaling pathway (described in chapter 9), which activates nuclear regulatory proteins that trigger cell division. In this example, when the nuclear retinoblastoma protein (Rb) is phosphorylated, another nuclear protein (the transcription factor E2F) is released and is then able to stimulate the production of cyclin and other proteins necessary for S phase.

production of G_1 cyclins and the proteins that are necessary for cell cycle progression (figure 10.22).

The cellular selectivity of a particular growth factor depends on which target cells bear its unique receptor. Some growth factors, such as PDGF and epidermal growth factor (EGF), affect a broad range of cell types, but others affect only specific types. For example, nerve growth factor (NGF) promotes the growth of certain classes of neurons, and erythropoietin triggers cell division in red blood cell precursors. Most animal cells need a combination of several different growth factors to overcome the various controls that inhibit cell division.

The G_0 phase

If cells are deprived of appropriate growth factors, they stop at the G_1 checkpoint of the cell cycle. With their growth and division arrested, they remain in this dormant G_0 phase.

The ability to enter G_0 accounts for the incredible diversity seen in the length of the cell cycle in different tissues. Epithelial cells lining the human gut divide more than twice a day, constantly renewing this lining. By contrast, liver cells divide only once every year or two, spending most of their time in the G_0 phase. Mature neurons and muscle cells usually never leave G_0.

Cancer is a failure of cell cycle control

The unrestrained, uncontrolled growth of cells in humans leads to the disease called **cancer.** Cancer is essentially a disease of cell division—a failure of cell division control.

The p53 *gene*

Recent work has identified one of the culprits in cancer. Officially dubbed ***p53,*** this gene plays a key role in the G_1 checkpoint of cell division.

The gene's product, the p53 protein, monitors the integrity of DNA, checking that it is undamaged. If the p53 protein detects damaged DNA, it halts cell division and stimulates the activity of special enzymes to repair the damage. Once the DNA has been repaired, p53 allows cell division to continue. In cases where the DNA damage is irreparable, p53 then directs the cell to kill itself.

By halting division in damaged cells, the *p53* gene prevents the development of many mutated cells, and it is therefore considered a **tumor-suppressor gene** although its activities are not limited to cancer prevention. Scientists have found that *p53* is entirely absent or damaged beyond use in the majority of cancerous cells they have examined. It is precisely because *p53* is nonfunctional that cancer cells are able to repeatedly undergo cell division without being halted at the G_1 checkpoint (figure 10.23).

Proto-oncogenes

The disease we call cancer is actually many different diseases, depending on the tissue affected. The common theme in all cases is the loss of control over the cell cycle. Research has identified numerous so-called **oncogenes,** genes that can, when introduced into a cell, cause it to become a cancer cell. This identification then led to the discovery of **proto-oncogenes,** which are normal cellular genes that become oncogenes when mutated.

The action of proto-oncogenes is often related to signalling by growth factors, and their mutation can lead to loss of growth control in multiple ways. Some proto-oncogenes encode receptors for growth factors, and others encode proteins involved in signal transduction that act after growth factor receptors. If a receptor for a growth factor becomes mutated such that it is permanently "on," the cell is no longer dependent on the presence of the growth factor for cell division. This is analogous to a light switch that is stuck on: The light will always be on. PDGF and EGF receptors both fall into the category of proto-oncogenes. Only one copy of a proto-oncogene needs to undergo this mutation for uncontrolled division to take place; thus, this change acts like a dominant mutation (chapter 13).

figure 10.23

CELL DIVISION, CANCER AND p53 PROTEIN. Normal p53 protein monitors DNA, destroying cells that have irreparable damage to their DNA. Abnormal p53 protein fails to stop cell division and repair DNA. As damaged cells proliferate, cancer develops.

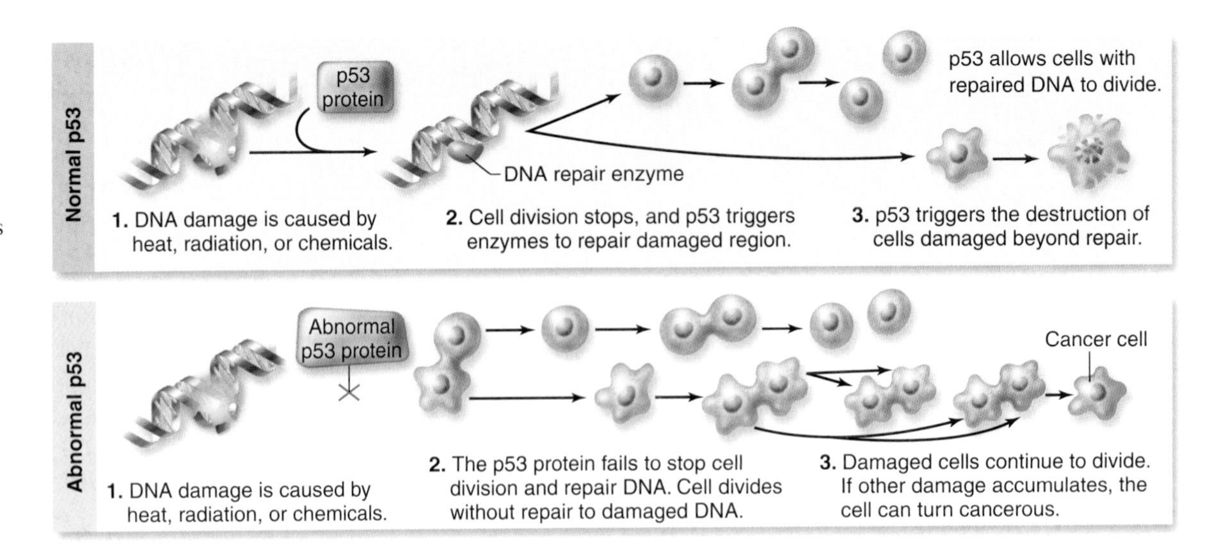

The number of proto-oncogenes identified has grown to more than 50 over the years. This line of research connects our understanding of cancer with our understanding of the molecular mechanisms governing cell cycle control.

Tumor-suppressor genes

After the discovery of proto-oncogenes, a second category of genes related to cancer was identified: the tumor-suppressor genes. We mentioned earlier that the *p53* gene acts as a tumor-suppressor gene, and a number of other such genes exist.

Both copies of a tumor-suppressor gene must lose function for the cancerous phenotype to develop, in contrast to the mutations in proto-oncogenes. Put another way, the proto-oncogenes act in a dominant fashion, and tumor suppressors act in a recessive fashion.

The first tumor-suppressor identified was the **retinoblastoma susceptibility gene (*Rb*),** which predisposes individuals for a rare form of cancer that affects the retina of the eye. Despite the fact that a cell heterozygous for a mutant *Rb* allele is normal, it is inherited as a dominant in families. The reason is that inheriting a single mutant copy of *Rb* means the individual has only one "good" copy left, and during the hundreds of thousands of divisions that occur to produce the retina, any error that damages the remaining good copy leads to a cancerous cell. A single cancerous cell in the retina then leads to the formation of a retinoblastoma tumor.

The role of the Rb protein in the cell cycle is to integrate signals from growth factors. The Rb protein is called a "pocket protein" because it has binding pockets for other proteins. Its role is therefore to bind important regulatory proteins and prevent them from stimulating the production of the necessary cell cycle proteins, such as cyclins or Cdks (see figure 10.21) discussed previously.

The binding of Rb to other proteins is controlled by phosphorylation: When it is dephosphorylated, it can bind a variety of regulatory proteins, but loses this capacity when phosphorylated. The action of growth factors results in the phosphorylation of Rb protein by a Cdk. This then brings us full circle, because the phosphorylation of Rb releases previously bound regulatory proteins, resulting in the production of S phase cyclins that are necessary for the cell to pass the G_1/S boundary and begin chromosome replication.

Figure 10.24 summarizes the types of genes that can cause cancer when mutated.

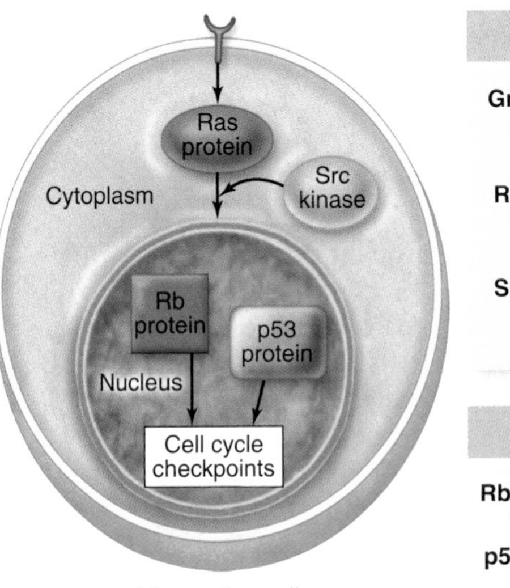

Proto-oncogenes

Growth factor receptor: more per cell in many breast cancers.

Ras protein: activated by mutations in 20–30% of all cancers.

Src kinase: activated by mutations in 2–5% of all cancers.

Tumor-suppressor Genes

Rb protein: mutated in 40% of all cancers.

p53 protein: mutated in 50% of all cancers.

figure 10.24

KEY PROTEINS ASSOCIATED WITH HUMAN CANCERS. Mutations in genes encoding key components of the cell division-signaling pathway are responsible for many cancers. Among them are protooncogenes encoding growth factor receptors, protein relay switches such as Ras protein, and kinase enzymes such as Src, which act after Ras and growth factor receptors. Mutations that disrupt tumor-suppressor proteins, such as Rb and p53, also foster cancer development.

The cell cycle is driven by positive regulators in the cytoplasm called cyclin-dependent kinases (Cdks) composed of a protein kinase and a cyclin protein. Inhibition of Cdks will halt the cell cycle if errors are encountered or a process is not complete. Three checkpoints exist: the G_1/S checkpoint, the G_2/M checkpoint, and the spindle checkpoint. In multicellular organisms, inputs such as growth factors affect control of the cycle at these checkpoints. The loss of cell cycle control leads to cancer, which can occur by a combination of two basic mechanisms: proto-oncogenes that gain function to become oncogenes, and tumor-suppressor genes that lose function and allow cell proliferation.

concept review

10.1 Prokaryotic Cell Division

Prokaryotic cell division is clonal and results in two cells identical to the original cell.

- During binary fission, the circular DNA replicates and is actively segregated.
- DNA replication begins at a specific point, the origin, and proceeds bidirectionally to a specific termination site.
- Septation involves insertion of new cell membrane and other cellular materials at the midpoint of the cell.
- A ring of FtsZ and proteins embedded in the cell membrane expands radially inward, pinching the cell into two new cells.

10.2 Eukaryotic Chromosomes

All eukaryotic cells have linear chromosomes and divide by mitosis.

- Chromosomes are composed of chromatin (a complex of DNA, protein, and RNA).
- Chromatin contains two domains: heterochromatin that is not expressed and euchromatin that is expressed.
- Newly replicated chromosomes remain attached at a constricted area called a centromere, consisting of repeated DNA sequences.
- The DNA of a single chromosome is a very long, double-stranded fiber.
- A nucleosome is a complex consisting of DNA wrapped around a core of eight positively charged histones.
- Nucleosomes are further coiled into a 30-nm fiber. This state is common in interphase when the cells are not dividing.
- During mitosis chromosomes are further condensed by arranging coiled 30-nm fibers radially around a protein scaffold.
- Haploid cells have one complete set of chromosomes, whereas diploid cells have two complete chromosome sets.
- Replicated DNA strands are held together at their centromeres by a complex of proteins called cohesins.
- After replication, a chromosome consists of two sister chromatids held together at the centromere (figure 10.7).

10.3 Overview of the Eukaryotic Cell Cycle (figure 10.8)

The cell cycle requires the duplication of the genome, its accurate segregation and the division of cellular contents.

- The cell cycle is divided into five phases: gap 1 (G_1), synthesis (S), gap 2 (G_2), mitosis (M), and cytokinesis (C).
- Cells can exit G_1 and enter a nondividing phase called G_0; the G_0 phase can be temporary or permanent.
- The length of a cell cycle varies with age, cell type, and species of organism.

10.4 Interphase: Preparation for Mitosis

Interphase is a preparatory stage that includes the G_1, S, and G_2 phases.

- G_1 is the primary growth phase of the cell and occurs between cytokinesis and synthesis.
- DNA synthesis occurs during S phase.
- G_2 phase occurs between the synthesis and mitosis phases.
- The centromere binds proteins assembled into a disklike structure called a kinetochore where microtubules attach during mitosis.

10.5 Mitosis: Chromosome Segregation (figure 10.11)

Mitosis, or M phase, is divided into five stages: prophase, prometaphase, metaphase, anaphase, and telophase.

- During prophase chromosomes condense, the spindle is formed, and the nuclear envelope disintegrates.
- In animals centriole pairs separate and migrate to opposite ends of the cell, establishing the axis of nuclear division.
- Chromosomes become attached to polar spindle fibers during prometaphase.
- The chromosomes migrate to the middle of the cell; at metaphase all chromosomes align at the equator of the cell due to tension from opposite poles.
- During anaphase the centromeres of sister chromatids separate and are then pulled to opposite poles; the poles also move apart.
- Telophase reverses the events of prophase and prepares the cell for cytokinesis.

10.6 Cytokinesis: The Division of Cytoplasmic Contents

During cytokinesis the cytoplasm divides roughly in half, forming two identical daughter cells.

- In animals cytokinesis occurs by constriction of actin filaments pinching off two daughter cells.
- In plants an expanding membrane partition called the cell plate fuses with the outer cell membrane to form two cells.

10.7 Control of the Cell Cycle (figure 10.18)

The cell cycle, a highly organized sequence of events, can be delayed or stopped at any one of three checkpoints.

- The primary molecular mechanism of cell cycle control is phosphorylation by cyclin-dependent kinases (Cdk's).
- Cdk's are complexes of a kinase and a regulatory molecule called cyclin.
- The cycle is driven forward by the action of Cdks. In yeast, this is only one enzyme, in vertebrates it is more than four enzymes.
- Entry into the cycle requires passing the G_1/S checkpoint. The commitment to DNA replication irreversibly commits the cell to division.
- During the G_1 phase, G_1 cyclin combines with Cdc2 kinase to trigger entry into S phase.
- The G_2/M checkpoint ensures DNA integrity; the cell assesses the accuracy of DNA replication.
- The spindle checkpoint ensures that all chromosomes are attached to spindle fibers, with bipolar orientation.
- Separation of chromosomes at anaphase is also irreversible.
- The anaphase-promoting complex (APC) activates a protease that removes cohesins holding the centromeres of sister chromatids together to trigger anaphase.
- The APC triggers destruction of mitotic cyclins to exit mitosis.
- In higher eukaryotes there are multiple Cdks and external signals such as growth factors that affect mitosis.
- Mutation in cell cycle control genes can lead to cancer.
- Mutations in proto-oncogenes have dominant, gain-of-function effects leading to cancer.
- Mutations in tumor suppressor genes are recessive; loss of function of both copies leads to cancer.

review questions

SELF TEST

1. Which of the following is NOT involved in binary fission in prokaryotes?
   a. Replication of DNA
   b. Elongation of the cell
   c. Separation of daughter cells by septum formation
   d. Assembly of the nuclear envelope
2. Chromatin is composed of—
   a. RNA and protein
   b. DNA and protein
   c. sister chromatids
   d. chromosomes
3. What is a nucleosome?
   a. It is a region in the cell's nucleus that contains euchromatin.
   b. It is a region of DNA wound around a collection of histone proteins.
   c. It is a region of a chromosome made up of multiple loops of chromatin.
   d. It is a 30-nm fiber found in chromatin.
4. How do *sister chromatids* differ from *homologous chromosomes*?
   a. Sister chromatids only represent the maternal genetic contribution.
   b. Homologous chromosomes are exact copies, but sister chromatids are just similar.
   c. Homologous chromosomes are similar, but sister chromatids are exact copies.
   d. Sister chromatids represent only half the genetic information stored in a chromosome.
5. What is the role of cohesin proteins in cell division?
   a. They organize the DNA of the chromosomes into highly condensed structures.
   b. They hold the DNA of the sister chromatids together.
   c. They help the cell divide into two daughter cells.
   d. They hold the microtubules onto the chromosome.
6. Replication of the organelles of the eukaryotic cell occurs during—
   a. interphase
   b. G_1
   c. S
   d. G_2
7. Replicated copies of each chromosome are called _______ and are connected at the _________.
   a. homologues; centromere
   b. sister chromatids; kinetochores
   c. sister chromatids; centromere
   d. homologues; kinetochore
8. Kinetochores are associated with which of the following on a sister chromatid?
   a. Centromere
   b. Centriole
   c. Condensation
   d. Cohesion
9. Separation of the sister chromatids and elongation of the cell occurs during—
   a. prophase
   b. prometaphase
   c. anaphase
   d. telophase
10. Why is cytokinesis an important part of cell division?
    a. It is responsible for the proper separation of genetic information.
    b. It is responsible for the proper separation of the cytoplasmic contents.
    c. It triggers the movement of a cell through the cell cycle.
    d. It is responsible for the elongation of the cell.
11. At which stage in the cell cycle does a cell make a commitment to undergo cell division?
    a. At the G_1/S checkpoint
    b. At the G_2/M checkpoint
    c. At the spindle checkpoint
    d. At cytokinesis
12. How is the activity of cyclin-dependent kinases (Cdks) regulated?
    a. Presence or absence of cyclins
    b. Phosphorylation
    c. Growth factors
    d. All of the above
13. Why is the function of the protein p53 important for preventing cancer?
    a. It ensures that the chromosomes are properly aligned for anaphase.
    b. It integrates signals in the cell leading to the production of cyclins.
    c. It checks for DNA damage before allowing replication during S phase.
    d. It functions as a phosphatase, removing phosphates to regulate Cdk.
14. What is a proto-oncogene?
    a. A mutant gene associated with the cell cycle that leads to cancer
    b. A normal gene that can result in cancer if it becomes mutated
    c. A gene introduced into a cell that can cause cancer
    d. A gene that suppresses unregulated cell divisions

CHALLENGE QUESTIONS

1. Regulation of the cell cycle is very complex and involves multiple proteins, all interacting to control the process of cell division. MPF is the cyclin–Cdk complex that is responsible for moving the cell past the G_2/M checkpoint. The activity of the cyclin-dependent kinase of MPF is regulated by phosphorylation. Cdk is inhibited when it is phosphorylated by the kinase, Wee-1. Predict what would happen to a cell if Wee-1 were absent.
2. Review your knowledge of signaling pathways (chapter 9). Create an outline illustrating how a growth factor (ligand) can lead to the production of a cyclin protein that would trigger S phase.

Do you need additional review? *Visit* www.ravenbiology.com *for practice quizzes, animations, videos, and activities designed to help you master the material in this chapter.*

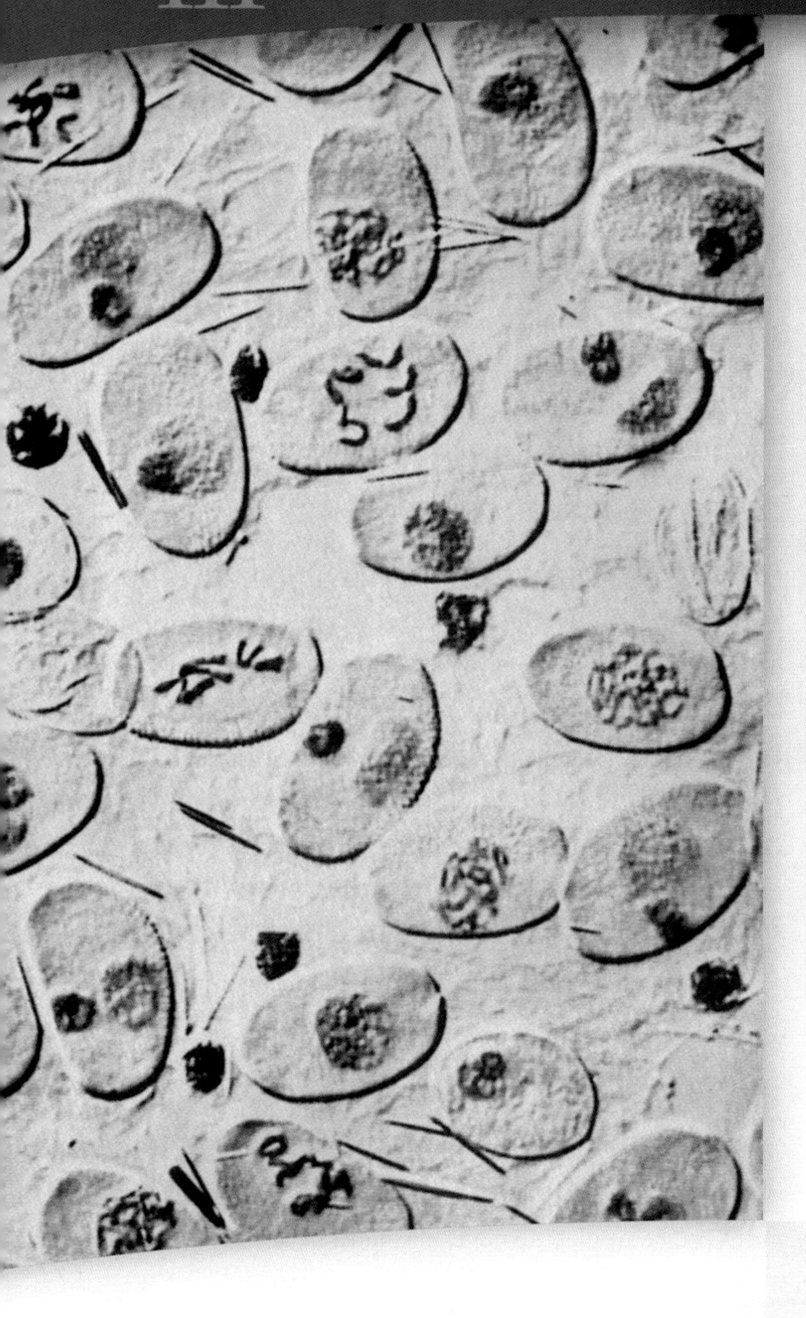

chapter 11

Sexual Reproduction and Meiosis

introduction

MOST ANIMALS AND PLANTS reproduce sexually. Gametes of opposite sex unite to form a cell that, dividing repeatedly by mitosis, eventually gives rise to an adult body with some 100 trillion cells. The gametes that form the initial cell are the products of a special form of cell division called *meiosis,* visible in the photo to the left, and the subject of this chapter. Meiosis is far more intricate than mitosis, and the details behind it are not as well understood. The basic process, however, is clear. Also clear are the profound consequences of sexual reproduction: It plays a key role in generating the tremendous genetic diversity that is the raw material of evolution.

concept outline

11.1 Sexual Reproduction Requires Meiosis

The essence of sexual reproduction is the genetic contribution of two cells. This mode of reproduction imposes difficulties for sexually reproducing organisms that biologists recognized early on. We are only recently making progress on the underlying mechanism for the elaborate behavior of chromosomes during meiosis. To begin, we briefly consider the history of meiosis and its relationship to sexual reproduction.

Meiosis reduces the number of chromosomes

Only a few years after Walther Flemming's discovery of chromosomes in 1879, Belgian cytologist Edouard van Beneden was surprised to find different numbers of chromosomes in different types of cells in the roundworm *Ascaris.* Specifically, he observed that the **gametes** (eggs and sperm) each contained two chromosomes, while all of the nonreproductive cells, or **somatic cells,** of embryos and mature individuals each contained four.

From his observations, van Beneden proposed in 1883 that an egg and a sperm, each containing half the complement of chromosomes found in other cells, fuse to produce a single cell called a **zygote.** The zygote, like all of the cells ultimately derived from it, contains two copies of each chromosome. The fusion of gametes to form a new cell is called **fertilization,** or **syngamy.**

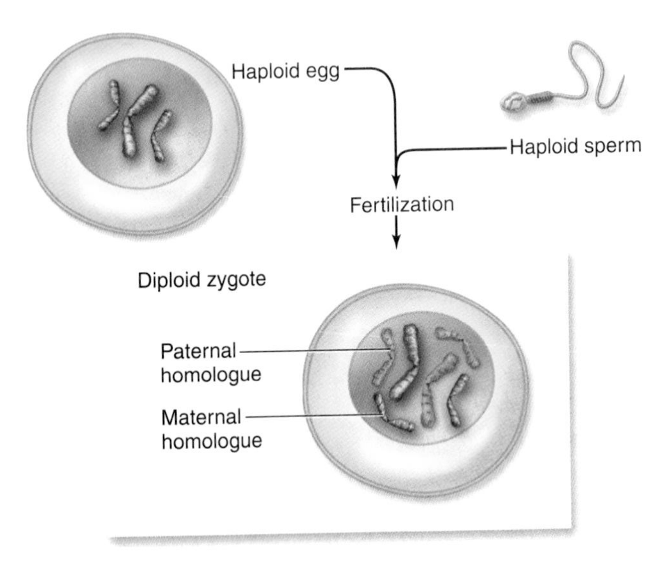

figure 11.1

DIPLOID CELLS CARRY CHROMOSOMES FROM TWO PARENTS. A diploid cell contains two versions of each chromosome, a maternal homologue contributed by the haploid egg of the mother, and a paternal homologue contributed by the haploid sperm of the father.

It was clear even to early investigators that gamete formation must involve some mechanism that reduces the number of chromosomes to half the number found in other cells. If it did not, the chromosome number would double with each fertilization, and after only a few generations, the number of chromosomes in each cell would become impossibly large. For example, in just 10 generations, the 46 chromosomes present in human cells would increase to over 47,000 (46×2^{10}).

The number of chromosomes does not explode in this way because of a special reduction division, **meiosis,** that occurs during gamete formation, producing cells with half the normal number of chromosomes. The subsequent fusion of two of these cells ensures a consistent chromosome number from one generation to the next.

Sexual life cycles have both haploid and diploid stages

Meiosis and fertilization together constitute a cycle of reproduction. Two sets of chromosomes are present in the somatic cells of adult individuals, making them **diploid** cells, but only one set is present in the gametes, which are thus **haploid.** Reproduction that involves this alternation of meiosis and fertilization is called **sexual reproduction.** Its outstanding characteristic is that offspring inherit chromosomes from *two* parents (figure 11.1). You, for example, inherited 23 chromosomes from your mother (maternal homologue), and 23 from your father (paternal homologue).

The life cycles of all sexually reproducing organisms follow a pattern of alternation between diploid and haploid chromosome numbers, but there is some variation in the life cycles. Many types of algae, for example, spend the majority of their life cycle in a haploid state, the zygote undergoing meiosis to produce haploid cells that then undergo mitosis (figure 11.2*a*). In most animals, the diploid state dominates; the zygote first undergoes mitosis to produce diploid cells, and later in the life cycle, some of these diploid cells undergo meiosis to produce haploid gametes (figure 11.2*b*). Some plants and some algae alternate between a multicellular haploid phase and a multicellular diploid phase (figure 11.2*c*).

Germ-line cells are set aside early in animal development

In animals, the single diploid zygote undergoes mitosis to give rise to all of the cells in the adult body. The cells that will eventually undergo meiosis to produce gametes are set aside from somatic cells early in the course of development. These cells are referred to as **germ-line cells.**

Both the somatic cells and the gamete-producing germ-line cells are diploid, but whereas somatic cells undergo mitosis

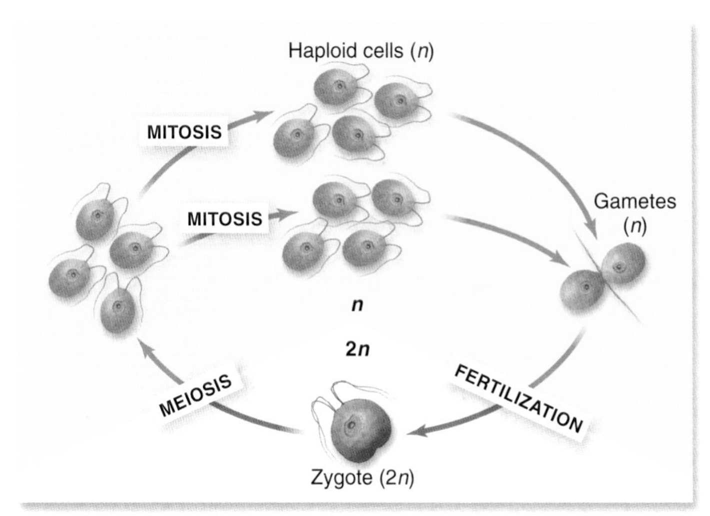

a. Algae and fungi

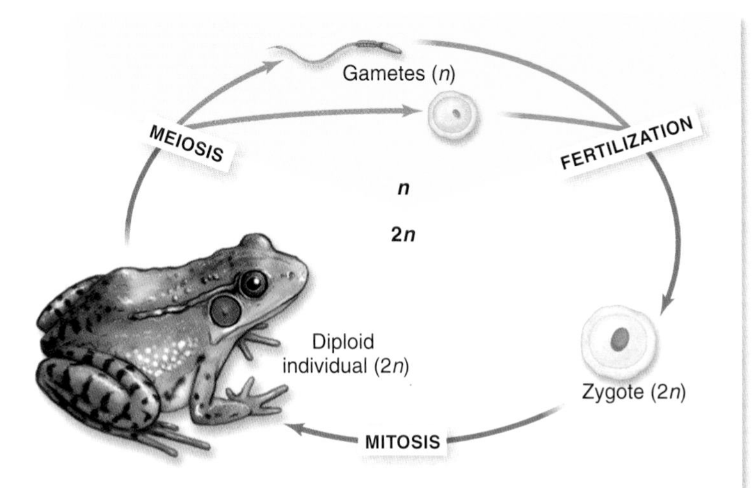

b. Most animals

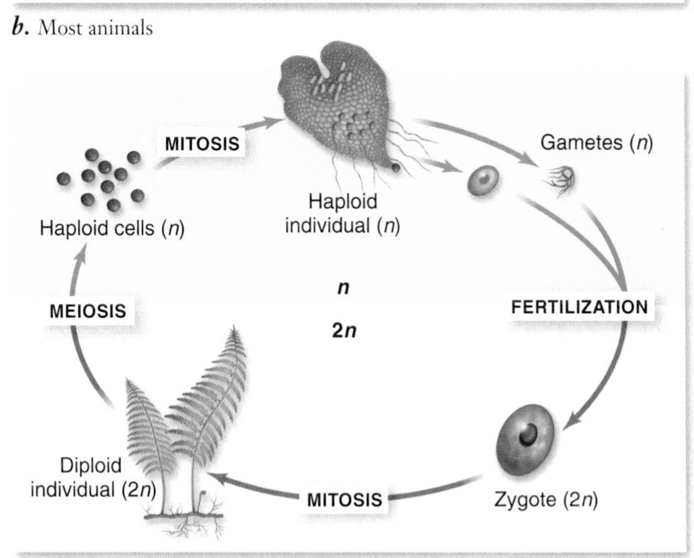

c. Some plants and some algae

figure 11.2

THREE TYPES OF SEXUAL LIFE CYCLES. In sexual reproduction, haploid cells or organisms alternate with diploid cells or organisms.

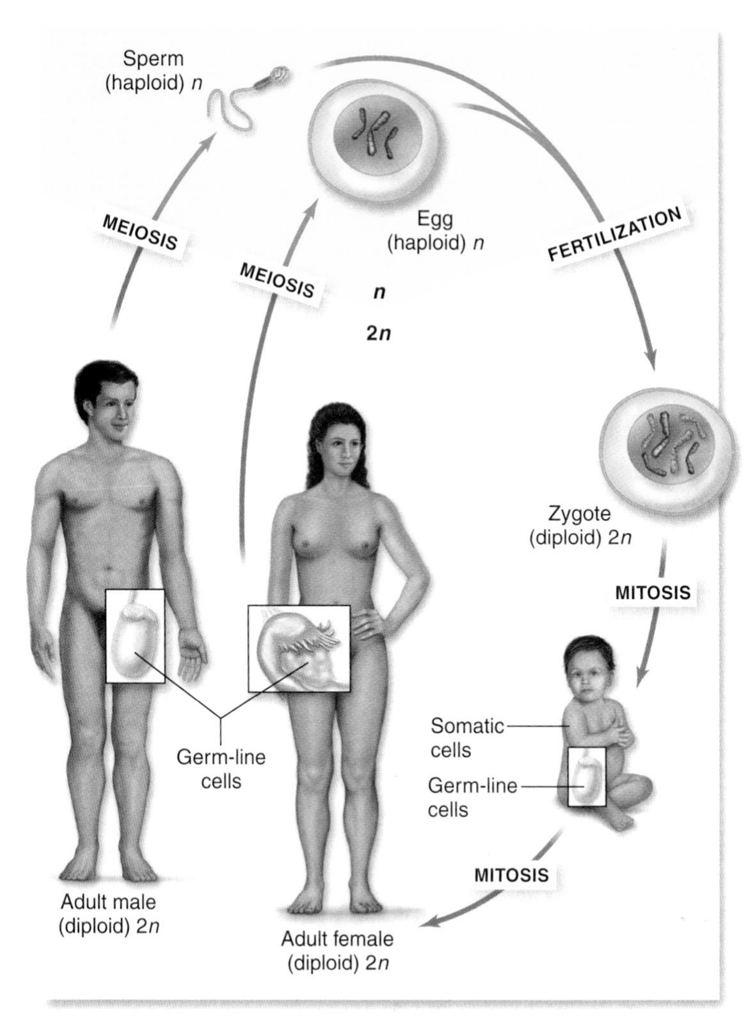

figure 11.3

THE SEXUAL LIFE CYCLE IN ANIMALS. In animals, the zygote undergoes mitotic divisions and gives rise to all the cells of the adult body. Germ-line cells are set aside early in development and undergo meiosis to form the haploid gametes (eggs or sperm). The rest of the body cells are called somatic cells.

to form genetically identical, diploid daughter cells, gamete-producing germ-line cells undergo meiosis to produce haploid gametes (figure 11.3).

Sexual reproduction involves the genetic contribution of two cells, each from a different individual. Meiosis produces haploid cells with half the number of chromosomes, making sexual reproduction possible. Fertilization then unites these haploid cells to restore the diploid state of the next generation. Meiosis occurs only in specialized cells called germ-line cells. All other cells in the body are called somatic cells and can undergo only mitotic division.

11.2 Features of Meiosis

The mechanism of meiotic cell division varies in important details in different organisms. These variations are particularly evident in the chromosomal separation mechanisms: Those found in protists and fungi are very different from those in plants and animals, which we describe here.

Meiosis in a diploid organism consists of two rounds of division, called **meiosis I** and **meiosis II,** with each round containing prophase, metaphase, anaphase, and telophase stages. Before describing the details of this process, we first examine the features of meiosis that distinguish it from mitosis.

Homologous chromosomes pair during meiosis

During early prophase I of meiosis, homologous chromosomes find each other and become closely associated, a process called pairing, or **synapsis** (figure 11.4*a*). Despite a long history of investigation, molecular details remain unclear. Biologists have used electron microscopy, data from genetic crosses, and biochemical analysis to shed light on synapsis. Thus far this knowledge has not been integrated into a complete picture.

The synaptonemal complex

It is clear that homologous chromosomes find their proper partners and become intimately associated during prophase I. This process includes the formation in many species of an elaborate structure called the **synaptonemal complex,** consisting of the homologues paired closely along a lattice of proteins between them (see figure 11.4*b*). The components of the synaptonemal complex include a meiosis-specific form of cohesin, a type of protein that joins sister chromatids during mitosis (described in the preceding chapter). This form of cohesin helps to join homologues as well as sister chromatids. The result is that all four chromatids of the two homologues are closely associated during this phase of meiosis. This structure is also sometimes called a *tetrad* or *bivalent.*

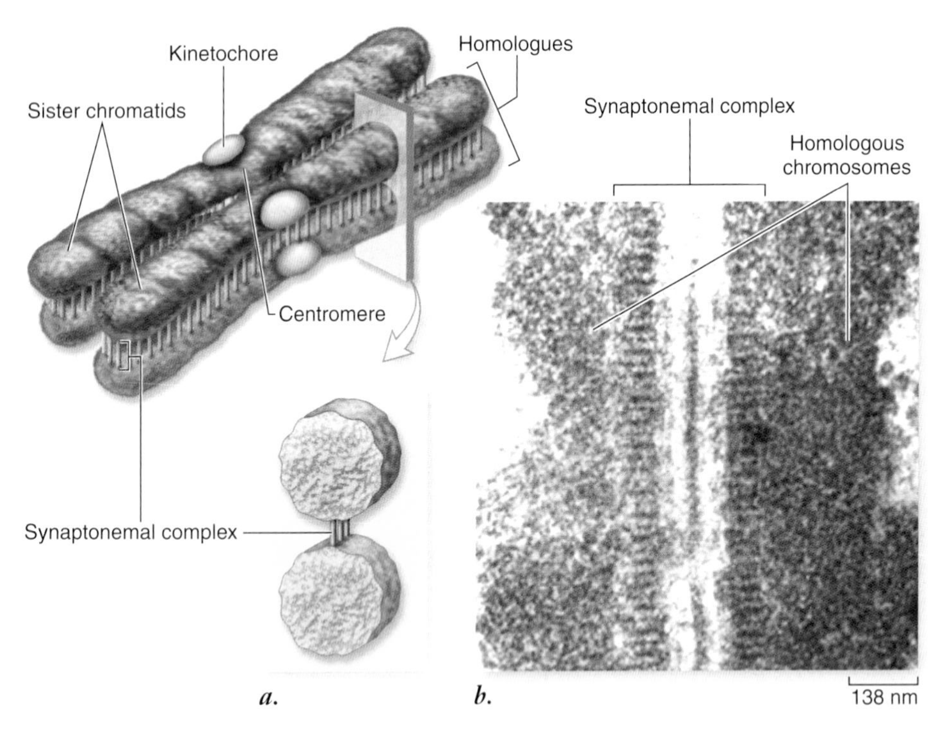

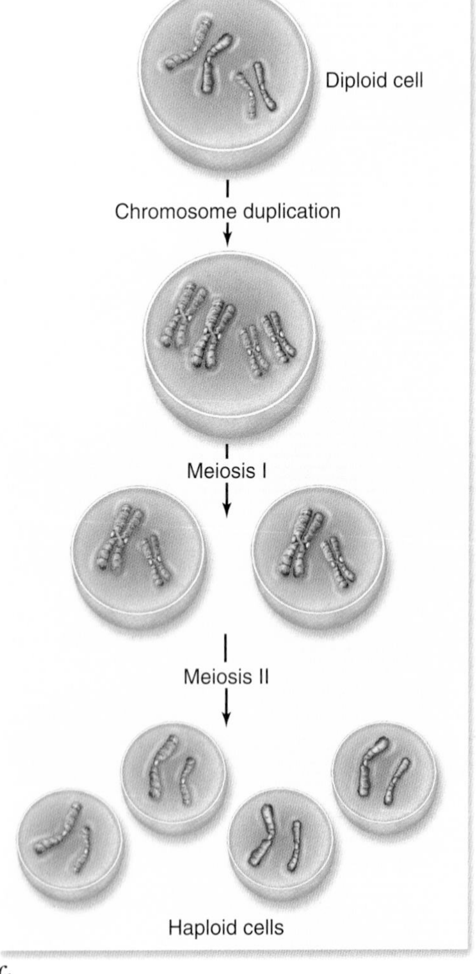

figure 11.4

UNIQUE FEATURES OF MEIOSIS. ***a.*** Homologous chromosomes pair during prophase I of meiosis. This process, called synapsis, produces homologues connected by a structure called the synaptonemal complex. The paired homologues can physically exchange parts, a process called crossing over ***b.*** A portion of the synaptonemal complex of the ascomycete *Neotiella rutilans*, a cup fungus. ***c.*** This pairing allows the disjunction of homologues, and not sister chromatids during meiosis I. The suppression of chromosome duplication before meiosis II leads to disjunction of sister chromatids producing the final haploid products.

The exchange of genetic material between homologues

While homologues are paired during prophase I, another process unique to meiosis occurs: genetic **recombination,** or **crossing over.** This process literally allows the homologues to exchange chromosomal material. The cytological observation of this phenomenon is called crossing over, and its detection genetically is called recombination—because alleles of genes that were formerly on separate homologues can now be found on the same homologue. (Genetic recombination is covered in detail in the next chapter.)

The sites of crossing over are called **chiasmata** (singular, chiasma), and these sites of contact are maintained until anaphase I. The physical connection of homologues due to crossing over and the continued connection of the sister chromatids lock homologues together.

Homologue association and separation

The association between the homologues persists throughout meiosis I and dictates the behavior of the chromosomes. During metaphase I, the paired homologues move to the metaphase plate and become oriented with homologues of each pair attached to opposite poles of the spindle. By contrast, in mitosis homologues behave independently of one another.

Then, during anaphase I, homologues are pulled to opposite poles for each pair of chromosomes. This again is in contrast to mitosis, in which sister chromatids, not homologues, are pulled to opposite poles.

You can now see why the first division is termed the "reduction division"—it results in daughter cells that contain one homologue from each chromosome pair. The second meiotic division will not further reduce the number of chromosomes; it will merely separate the sister chromatids for each homologue.

Meiosis features two divisions with one round of DNA replication

The most obvious distinction between meiosis and mitosis is the simple observation that meiosis involves two successive divisions with no replication of genetic material between them. One way to view this is that DNA replication must be suppressed between the two meiotic divisions. The behavior of chromosomes during meiosis I puts the resulting cells into a position where a division that acts like mitosis without DNA replication produces cells with half the original number of chromosomes (figure 11.4*c*). This is the last key to understanding meiosis: The second meiotic division is like mitosis with no chromosome duplication.

Meiosis is characterized by the pairing of homologous chromosomes during prophase I, usually accompanied by the formation of an elaborate structure between homologues called the synaptonemal complex. During this pairing, homologues may exchange chromosomal material at sites called chiasmata. Meiosis allows the segregation of homologues at the first division, followed by a second division without replication to segregate sister chromatids and produce haploid cells.

11.3 The Process of Meiosis

To understand meiosis, it is necessary to carefully follow the behavior of chromosomes during each division. The events of meiosis depend on homologues exchanging chromosomal material by crossing over. This allows sister chromatid cohesion around the sites of exchange to hold homologues together. The loss of sister chromatid cohesion is then different on the chromosome arms and at the centromeres: it is lost at anaphase I on the chromosome arms but is retained at the centromeres until anaphase II.

Prophase I sets the stage for the reductive division

Meiotic cells have an interphase period that is similar to mitosis with G_1, S, and G_2 phases. After interphase, germ-line cells enter meiosis I. In prophase I, the DNA coils tighter, and individual chromosomes first become visible under the light microscope as a matrix of fine threads. Because the DNA has already replicated before the onset of meiosis, each of these threads actually consists of two sister chromatids joined at their centromeres. In prophase I, homologous chromosomes become closely associated in synapsis, exchange segments by crossing over, and then separate.

Synapsis

During interphase in germ-line cells, the ends of the chromatids seem to be attached to the nuclear envelope at specific sites. The sites the homologues attach to are adjacent, so that during prophase I the members of each homologous pair of chromosomes are brought close together. Homologous pairs then align side by side, apparently guided by heterochromatin sequences, in the process of synapsis.

This association joins homologues along their entire length. The sister chromatids of each homologue are also joined by the cohesin complex in a process called *sister chromatid cohesion* (similar to what happens during mitosis). This brings all four chromatids for each set of paired homologues into close association.

Crossing over

Along with the synaptonemal complex that forms during prophase I (see figure 11.4), another kind of structure appears that correlates in timing with the recombination process. These are called *recombination nodules,* and they are thought to contain the enzymatic machinery necessary to break and rejoin chromatids of homologous chromosomes.

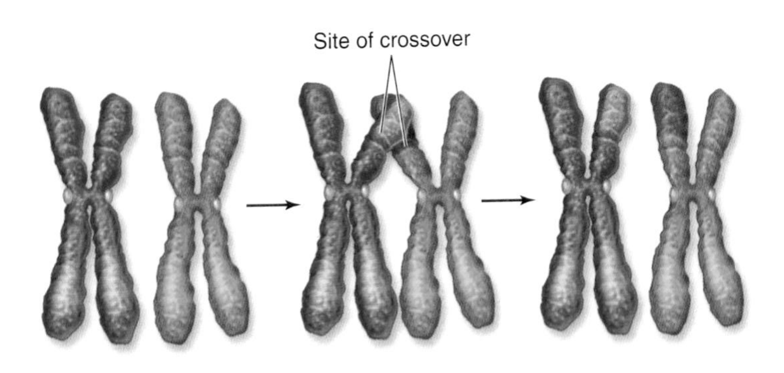

figure 11.5

THE RESULTS OF CROSSING OVER. During crossing over, homologous chromosomes may exchange segments.

Crossing over involves a complex series of events in which DNA segments are exchanged between nonsister chromatids (figure 11.5). Crossing over between sister chromatids is suppressed during meiosis. Reciprocal crossovers between nonsister chromatids are controlled such that each chromosome arm usually has one or a few crossovers per meiosis, no matter what the size of the chromosome. Human chromosomes typically have two or three.

When crossing over is complete, the synaptonemal complex breaks down, and the homologous chromosomes become less tightly associated but remain attached by chiasmata. At this point, there are four chromatids for each type of chromosome (two homologous chromosomes, each of which consists of two sister chromatids).

The four chromatids do not separate completely because they are held together in two ways: (1) The two sister chromatids of each homologue, recently created by DNA replication, are held together by their common centromeres; and (2) the paired homologues are held together at the points where crossing over occurred by sister chromatid cohesion around the site of exchange. These points are the chiasmata that can be observed microscopically. Like small rings moving down two strands of rope, the chiasmata move to the end of the chromosome arm before metaphase I.

While the elaborate behavior of chromosome pairing is taking place, other events must take place during prophase I. The nuclear envelope must be dispersed, along with the interphase structure of microtubules. The microtubules are formed into a spindle, just as in mitosis.

During metaphase I, paired homologues align

By metaphase I, the second stage of meiosis I, the chiasmata have moved down the paired chromosomes to the ends. At this point, they are called *terminal chiasmata*. Terminal chiasmata hold the homologous chromosomes together in metaphase I so that homologues can be aligned at the equator of the cell.

The capture of microtubules by kinetochores occurs such that the kinetochores of sister chromatids act as a single unit. This results in microtubules from opposite poles becoming attached to the kinetochores of *homologues*, and not to those of sister chromatids (figure 11.6).

The ability of sister centromeres to behave as a unit during meiosis I is not understood. It has been suggested, based on electron microscope data, that the centromere/kinetochore complex of sister chromatids is compacted during meiosis I, allowing them to function as a single unit.

The monopolar attachment of centromeres of sister chromatids would be disastrous in mitosis, but it is critical to meiosis I. It produces tension on the homologues, which are joined by chiasmata and sister chromatid cohesion, pulling paired homologues to the equator of the cell. In this way, each joined pair of homologues lines up on the metaphase plate (see figure 11.6).

The orientation of each pair on the spindle axis is random; either the maternal or the paternal homologue may be oriented toward a given pole (figure 11.7; see also figure 11.8).

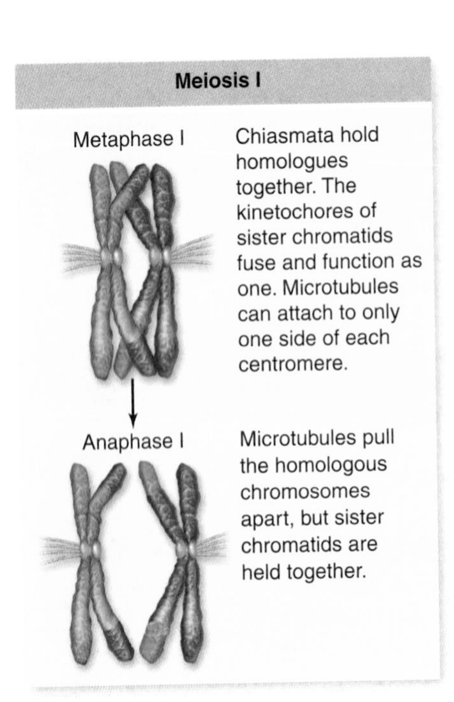

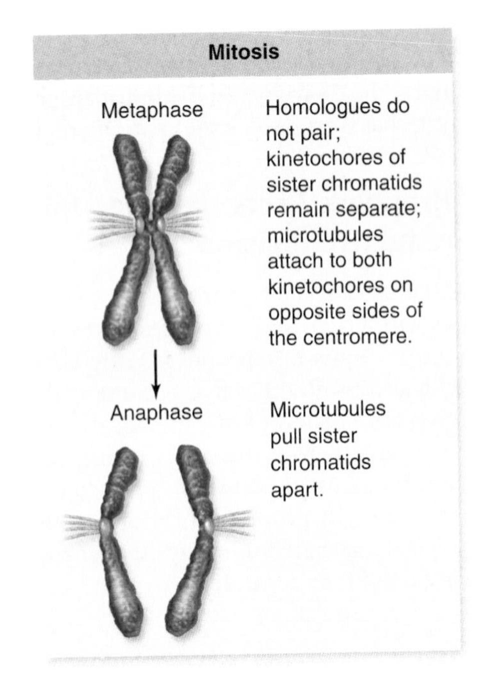

figure 11.6

ALIGNMENT OF CHROMOSOMES DIFFERS BETWEEN MEIOSIS I AND MITOSIS. In metaphase I, the chiasmata and connections between sister chromatids hold homologous chromosomes together; paired kinetochores for sister chromatids of each homologue become attached to microtubules from one pole. By the end of meiosis I, connections between sister chromatid arms are broken as microtubules shorten, pulling the homologous chromosomes apart. The sister chromatids remain joined by their centromeres. In mitosis, microtubules from opposite poles attach to the kinetochore of each sister centromere; when the connections between sister centromeres are broken microtubules shorten, pulling the sister chromatids to opposite poles.

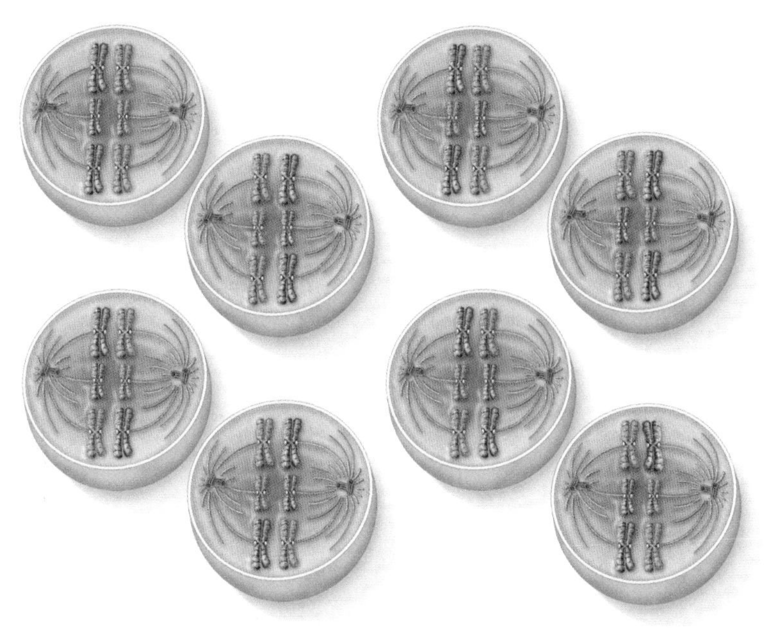

figure 11.7

RANDOM ORIENTATION OF CHROMOSOMES ON THE METAPHASE PLATE. The number of possible chromosome orientations equals 2 raised to the power of the number of chromosome pairs. In this hypothetical cell with three chromosome pairs, eight (2^3) possible orientations exist. Each orientation produces gametes with different combinations of parental chromosomes.

Anaphase I results from the differential loss of sister chromatid cohesion along the arms

In anaphase I, the microtubules of the spindle fibers begin to shorten. As they shorten, they break the chiasmata and pull the centromeres toward the poles, dragging the chromosomes along with them.

Anaphase I comes about by the release of sister chromatid cohesion along the chromosome arms, but not at the centromeres. This release is thought to be the result of the destruction of meiosis-specific cohesin in a process analogous to anaphase in mitosis. The difference is that the destruction is inhibited at the centromeres by an as yet unknown mechanism.

As a result of this release, the homologues are pulled apart, but not the sister chromatids. Each homologue moves to one pole, taking both sister chromatids with it. When the spindle fibers have fully contracted, each pole has a complete haploid set of chromosomes consisting of one member of each homologous pair.

Because of the random orientation of homologous chromosomes on the metaphase plate, a pole may receive either the maternal or the paternal homologue from each chromosome pair. As a result, the genes on different chromosomes assort independently; that is, meiosis I results in the **independent assortment** of maternal and paternal chromosomes into the gametes (see chapter 12).

Telophase I completes meiosis I

By the beginning of telophase I, the chromosomes have segregated into two clusters, one at each pole of the cell. Now the nuclear membrane re-forms around each daughter nucleus.

Because each chromosome within a daughter nucleus had replicated before meiosis I began, each now contains two sister chromatids attached by a common centromere. Note that *the sister chromatids are no longer identical* because of the crossing over that occurred in prophase I (figure 11.8); as you will see, this change has important implications for genetic variability.

Cytokinesis, the division of the cytoplasm and its contents, may or may not occur after telophase I. The second meiotic division, meiosis II, occurs after an interval of variable length.

Achiasmate segregation of homologues is possible

The preceeding description of meiosis I relies on the observation that homologues are held together by chiasmata and by sister chromatid cohesion. This connection produces the critical behavior of chromosomes during metaphase I and anaphase I, when homologues move to the metaphase plate and then move to opposite poles.

Although this connection of homologues is the rule, there are exceptions. In *Drosophila* males for example, there is no recombination, and yet meiosis proceeds accurately, a process called **achiasmate segregation** ("without chiasmata"). This seems to involve an alternative mechanism for joining homologues and then allowing their segregation during anaphase I. Telomeres and other heterochromatic sequences have been implicated, but the details remain unclear.

Despite these exceptions, the vast majority of species that have been examined use the formation of chiasmata and sister chromatid cohesion to hold homologues together for segregation during anaphase I.

Meiosis II is like a mitotic division without DNA replication

Typically, interphase between meiosis I and meiosis II is brief and does not include an S phase: Meiosis II resembles a normal mitotic division. Prophase II, metaphase II, anaphase II, and telophase II follow in quick succession (see figure 11.8).

Prophase II. At the two poles of the cell, the clusters of chromosomes enter a brief prophase II, each nuclear envelope breaking down as a new spindle forms.

Metaphase II. In metaphase II, spindle fibers from opposite poles bind to kinetochores of each sister chromatid, allowing each chromosome to migrate to the metaphase plate as a result of tension on the chromosomes from polar microtubules pulling on sister centromeres. This process is the same as metaphase during a mitotic division.

Anaphase II. The spindle fibers contract, and the cohesin complex joining the centromeres of sister chromatids is destroyed, splitting the centromeres and pulling the sister chromatids to opposite poles. This process is also the same as anaphase during a mitotic division.

Telophase II. Finally, the nuclear envelope re-forms around the four sets of daughter chromosomes. Cytokinesis then follows.

The final result of this division is four cells containing haploid sets of chromosomes. The cells that contain these haploid nuclei may develop directly into gametes, as they do in animals. Alternatively, they may themselves divide mitotically, as they do in plants, fungi, and many protists, eventually producing greater numbers of gametes or, as in some plants and insects, adult individuals with varying numbers of chromosome sets.

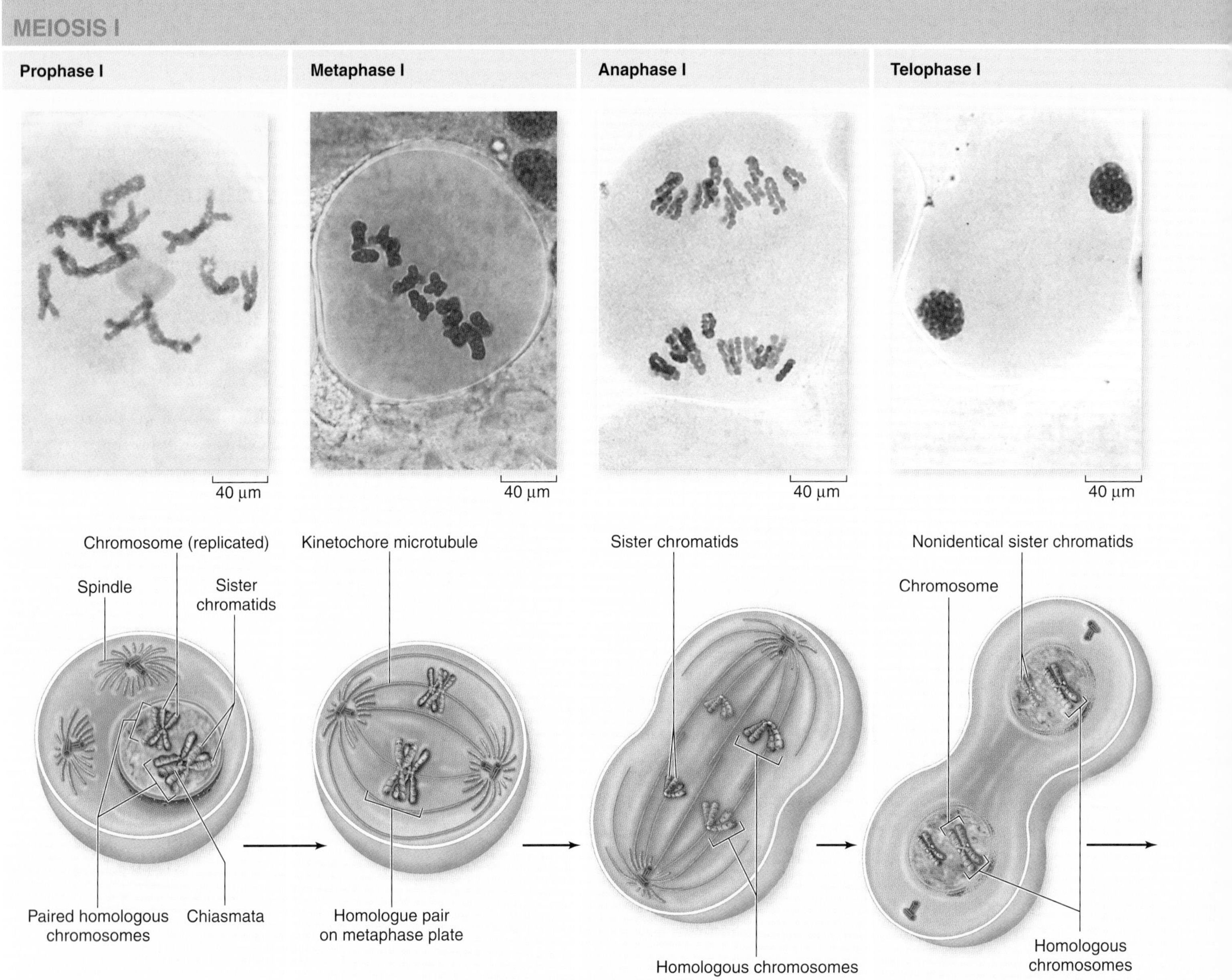

MEIOSIS I

Prophase I

In prophase I of meiosis I, the chromosomes begin to condense, and the spindle of microtubules begins to form. The DNA has been replicated, and each chromosome consists of two sister chromatids attached at the centromere. In the cell illustrated here, there are four chromosomes, or two pairs of homologues. Homologous chromosomes pair up and become closely associated during synapsis. Crossing over occurs, forming chiasmata, which hold homologous chromosomes together.

Metaphase I

In metaphase I, the pairs of homologous chromosomes align along the metaphase plate. Chiasmata help keep the pairs together and produce tension when microtubules from opposite poles attach to sister kinetochores of each homologue. A kinetochore microtubule from one pole of the cell attaches to one homologue of a chromosome, while a kinetochore microtubule from the other cell pole attaches to the other homologue of a pair.

Anaphase I

In anaphase I, kinetochore microtubules shorten, and homologous pairs are pulled apart. One duplicated homologue goes to one pole of the cell, while the other duplicated homologue goes to the other pole. Sister chromatids do not separate.This is in contrast to mitosis, where duplicated homologues line up individually on the metaphase plate, kinetochore microtubules from opposite poles of the cell attach to opposite sides of one homologue's centromere, and sister chromatids are pulled apart in anaphase.

Telophase I

In telophase I, the separated homologues form a cluster at each pole of the cell, and the nuclear envelope re-forms around each daughter cell nucleus. Cytokinesis may occur. The resulting two cells have half the number of chromosomes as the original cell: In this example, each nucleus contains two chromosomes (versus four in the original cell). Each chromosome is still in the duplicated state and consists of two sister chromatids, but sister chromatids are not identical because crossing over has occurred.

figure 11.8

THE STAGES OF MEIOSIS. Meiosis in plant cells (photos) and animal cells (drawings) is shown.

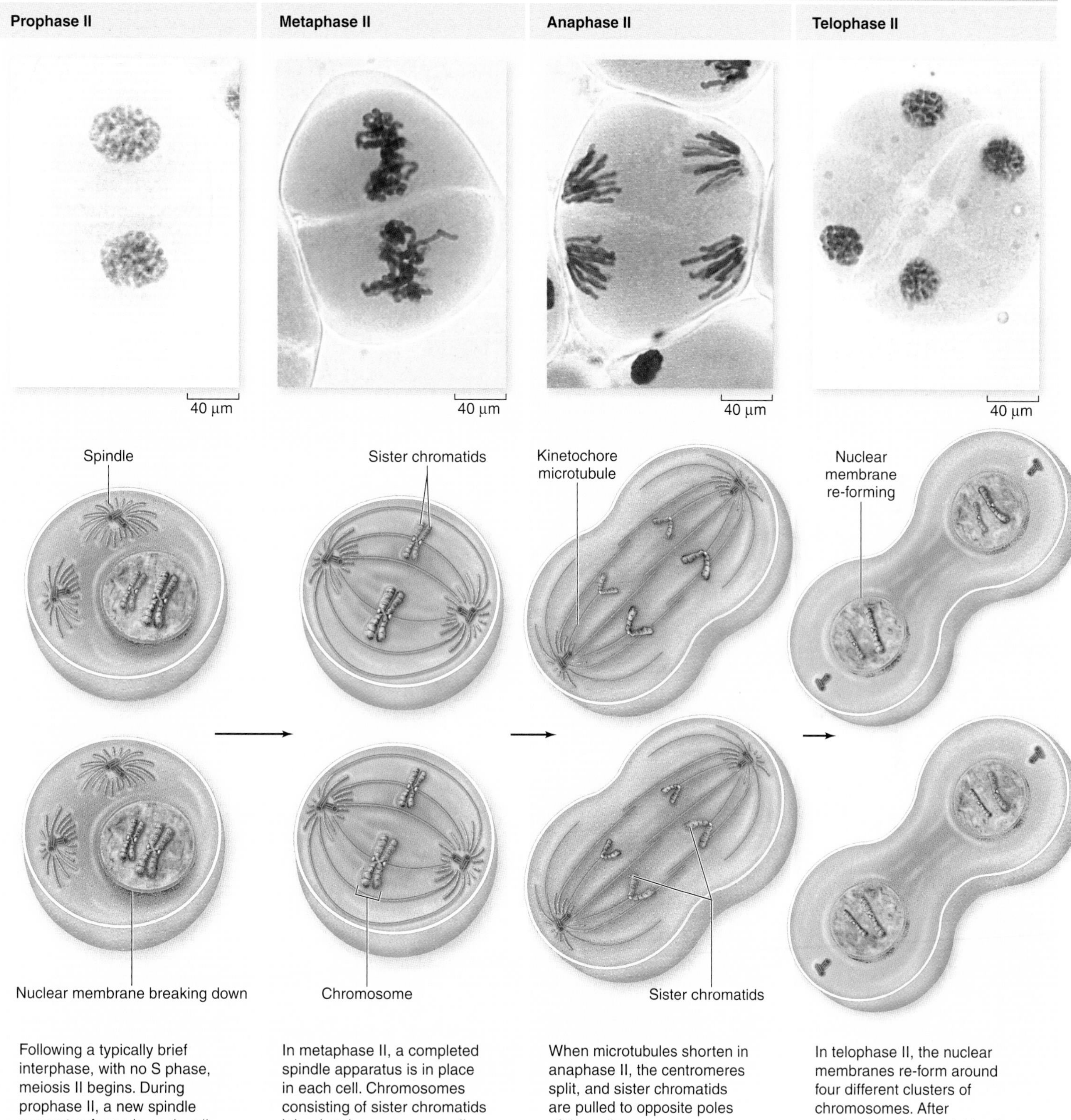

Following a typically brief interphase, with no S phase, meiosis II begins. During prophase II, a new spindle apparatus forms in each cell, and the nuclear envelope breaks down. In some species the nuclear envelope does not re-form in telophase I removing the need for nuclear envelope breakdown.

In metaphase II, a completed spindle apparatus is in place in each cell. Chromosomes consisting of sister chromatids joined at the centromere align along the metaphase plate in each cell. Now, kinetochore microtubules from opposite poles attach to kinetochores of sister chromatids.

When microtubules shorten in anaphase II, the centromeres split, and sister chromatids are pulled to opposite poles of the cells.

In telophase II, the nuclear membranes re-form around four different clusters of chromosomes. After cytokinesis, four haploid cells result. No two cells are alike due to the random alignment of homologous pairs at metaphase I and crossing over during prophase I.

Errors in meiosis produce aneuploid gametes

It is critical that the process of meiosis is accurate because any failure will produce gametes that do not have the correct number of chromosomes. Failure of chromosomes to move to opposite poles during either meiotic division is called *nondisjunction*, and it produces one gamete that lacks a chromosome and one that has two copies. Gametes with an improper number of chromosomes are called **aneuploid gametes.** In humans, this condition is the most common cause of spontaneous abortion. The implications of aneuploid gametes are explored in more detail in chapter 13.

During prophase I, homologues for each chromosome find each other and pair along their entire length. The exchange of material between homologues combined with sister chromatid cohesion joins homologues such that they can be segregated during anaphase I. The loss of sister chromatid cohesion on the arms but not the centromere results in homologues being pulled to opposite poles, effectively halving the number of chromosomes. Meiosis I is followed by a second division, Meiosis II, that begins without replication but is otherwise like mitosis with sister chromatids being pulled to opposite poles.

11.4 Summing Up: Meiosis Versus Mitosis

The key to meiosis is understanding the differences between meiosis and mitosis. The basic machinery in both processes is the same, but the behavior of chromosomes is distinctly different during the first meiotic division (figure 11.9).

Meiosis is characterized by four distinct features:

1. Pairing of homologues and crossing over that joins maternal and paternal homologues.
2. Cosegregation of sister centromeres during anaphase I due to maintenance of sister chromatid cohesion at sister centromeres.
3. The attachment of sister kinetochores to the same pole in meiosis I and to opposite poles in mitosis.
4. Suppression of replication between the two meiotic divisions.

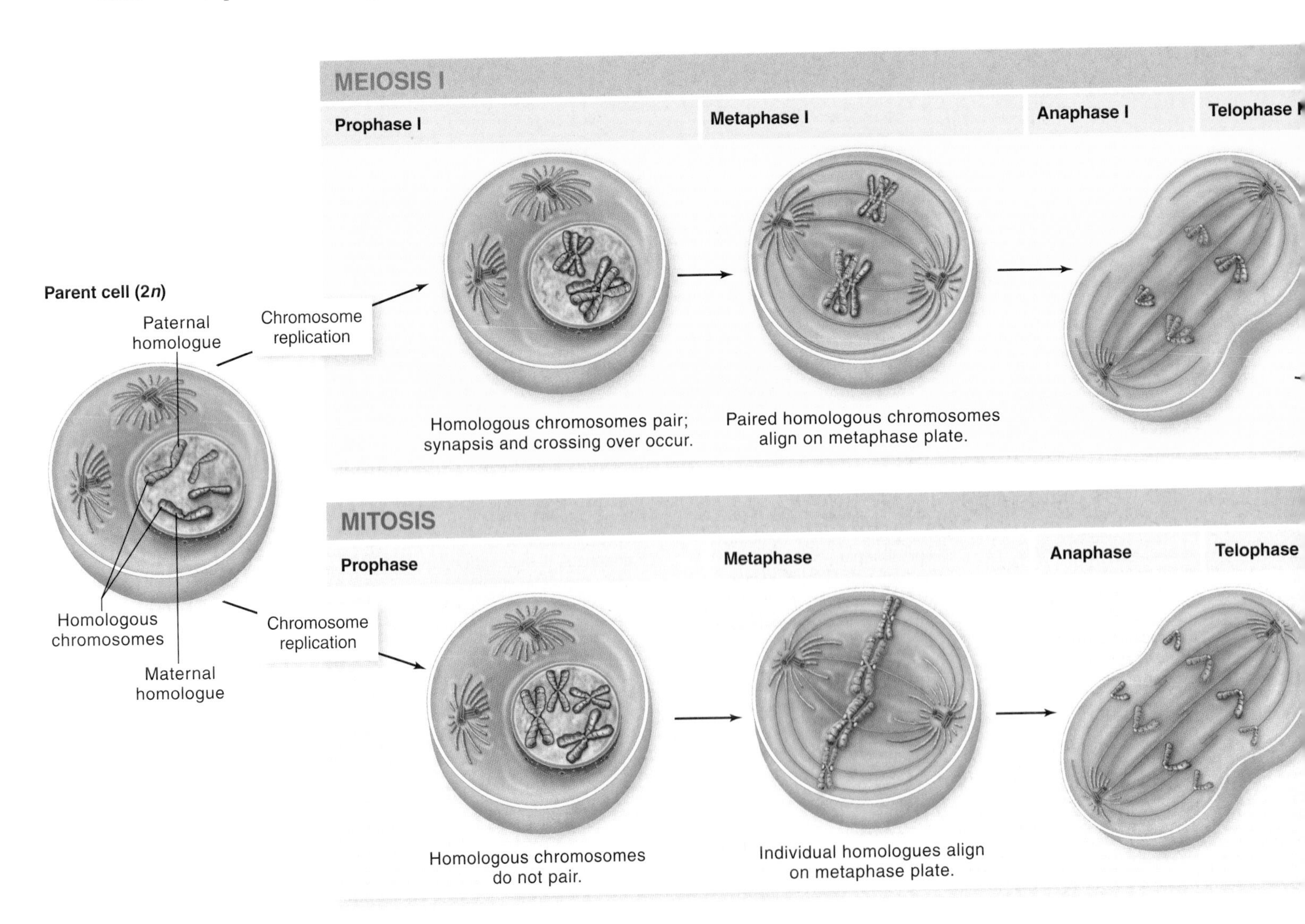

Although the underlying molecular mechanisms are unclear, we will consider what we know of each of these features in the following sections.

Homologous pairing and crossing over may involve meiosis-specific cohesins

The pairing of homologues during prophase I of meiosis is the first deviation from mitosis and sets the stage for all of the subsequent differences (figure 11.9). How homologues find each other and become aligned is one of the great mysteries of meiosis. Some cytological evidence implicates telomeres and other specific sites as being necessary for pairing, but this finding does little to clarify the essential process.

Some light has been shed on the mechanisms with the discovery of meiosis-specific cohesin proteins. In yeast, the protein Rec8 replaces the mitotic Scc1 protein as part of the cohesin complex. You saw in chapter 10 that Scc1 is destroyed during anaphase of mitosis to allow sister chromatids to be pulled to opposite poles; the role of Rec8 is similar but more complex, as we will see in the next section.

In other species, including mice, other components of the cohesin complex are also replaced during meiosis. Proteins that form part of the cohesin complex have also been found associated with the synaptonemal complex. Although these data do not explain how homologues find each other, it does put pairing on the familiar ground of sister chromatid cohesion via cohesin proteins.

The molecular details of the recombination process that produces crossing over are complex, but many of the enzymes involved have been identified. Of interest is the observation that a clear overlap exists between the machinery necessary for meiotic recombination and the machinery involved in the repair of double-strand breaks in DNA. Recombination probably first evolved as a repair mechanism and was later co-opted for use in disjoining chromosomes. The importance of recombination for proper disjunction is clear from the observation in many organisms that loss of function for recombination enzymes also results in higher levels of nondisjunction.

Sister chromatid cohesion is maintained through meiosis I but released in meiosis II

Meiosis I is characterized by the segregation of homologues, not sister chromatids, during anaphase. For this separation to occur, the centromeres of sister chromatids must move to the same pole, or cosegregate, during anaphase I. This means that meiosis-specific cohesin proteins must first be removed from the chromosome arms, then later from sister centromeres.

Homologues are joined by chiasmata, and sister chromatid cohesion around the site of exchange then holds homologues

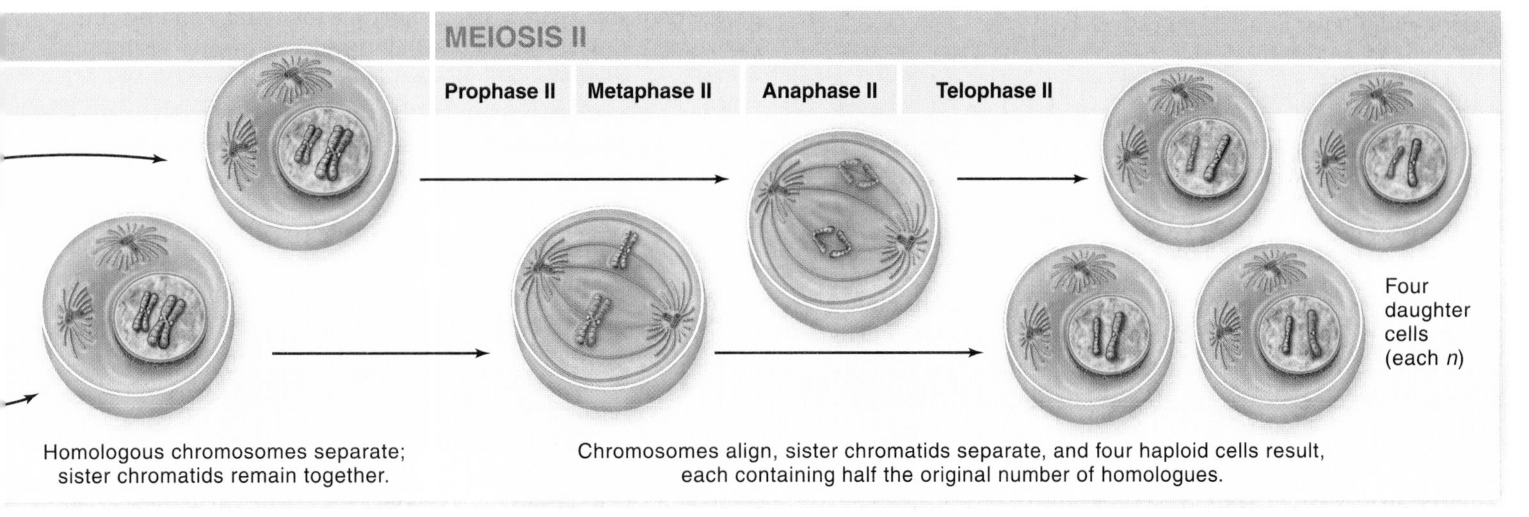

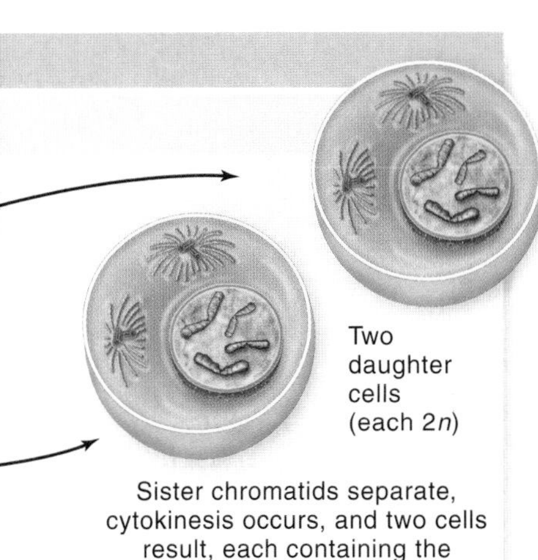

figure 11.9

A COMPARISON OF MEIOSIS AND MITOSIS. Meiosis involves two nuclear divisions with no DNA replication between them. It thus produces four daughter cells, each with half the original number of chromosomes. Crossing over occurs in prophase I of meiosis. Mitosis involves a single nuclear division after DNA replication. It thus produces two daughter cells, each containing the original number of chromosomes.

inquiry

If the chromosomes of a mitotic cell behaved the same as chromosomes in meiosis I, would the resulting cells have the proper chromosomal constitution?

together. The destruction of Rec8 protein on the chromosome arms appears to be what allows homologues to be pulled apart at anaphase I.

This leaves the key distinction between meiosis and mitosis being the maintenance of sister chromatid cohesion at the centromere during all of meiosis I, while cohesion is lost from the chromosome arms during anaphase I (see figure 11.9). It is unclear how the cohesin complex at the centromere is protected from destruction, but it apparently depends on the replacement of Scc1 by the Rec8 protein. This meiosis-specific cohesin is critical to maintaining sister chromatid cohesion at the centromere until anaphase II.

Sister kinetochores are attached to the same pole during meiosis I

The cosegregation of sister centromeres requires that the kinetochores of sister chromatids are attached to the same pole during meiosis I. This attachment is in contrast to both mitosis (see figure 11.9) and meiosis II, in which sister kinetochores must become attached to opposite poles.

The underlying basis of this monopolar attachment of sister kinetochores is unclear, but it seems to be based on structural differences between centromere/kinetochore complexes in meiosis I and in mitosis. Mitotic kinetochores visualized with the electron microscope appear to be recessed, making bipolar attachment more likely. Meiosis I kinetochores protrude more, making monopolar attachment easier.

It is clear that both the maintenance of sister chromatid cohesion at the centromere and monopolar attachment are required for the segregation of homologues that distinguishes meiosis I from mitosis.

Replication is suppressed between meiotic divisions

After a mitotic division, a new round of DNA replication must occur before the next division. For meiosis to succeed in halving the number of chromosomes, this replication must be suppressed between the two divisions. The detailed mechanism of suppression of replication between meiotic division is unknown. One clue is the observation that the level of one of the cyclins, cyclin B, is reduced between meiotic divisions, but is not lost completely, as it is between mitotic divisions.

During mitosis, the destruction of mitotic cyclin is necessary for a cell to enter another division cycle. The result of this maintenance of cyclin B between meiotic divisions in germ-line cells is the failure to form initiation complexes necessary for DNA replication to proceed. This failure to form initiation complexes appears to be critical to suppressing DNA replication.

Meiosis produces cells that are not identical

The daughter cells produced by mitosis are identical to the parental cell, at least in terms of their chromosomal constitution. This exact copying is critical to producing new cells for growth, for development, and for wound healing. Meiosis, because of the random orientation of different chromosomes at the first meiotic division and because of crossing over, rarely produces cells that are identical (figure 11.10). The gametes from meiosis all carry an entire haploid set of chromosomes, but these chromosomes are a mixture of maternal and paternal homologues; furthermore, the homologues themselves have exchanged material by crossing over. The resulting variation is essential for evolution and is the reason that sexually reproducing populations have much greater variation than asexually reproducing ones.

Meiosis is not only critical for the process of sexual reproduction, but is also the foundation for understanding the basis of heredity. The different cells produced by meiosis form the basis for understanding the behavior of observable traits in genetic crosses. In the next two chapters we will follow the behavior of traits in genetic crosses and see how this correlates with the behavior of chromosomes in meiosis.

Meiosis is characterized by homologue pairing and exchange; by loss of sister chromatid cohesion in the arms during the first division, but not at the centromere until the second division; and by the suppression of DNA replication between the two meiotic divisions. The haploid cells that result from meiosis are not identical and therefore allow variation in offspring.

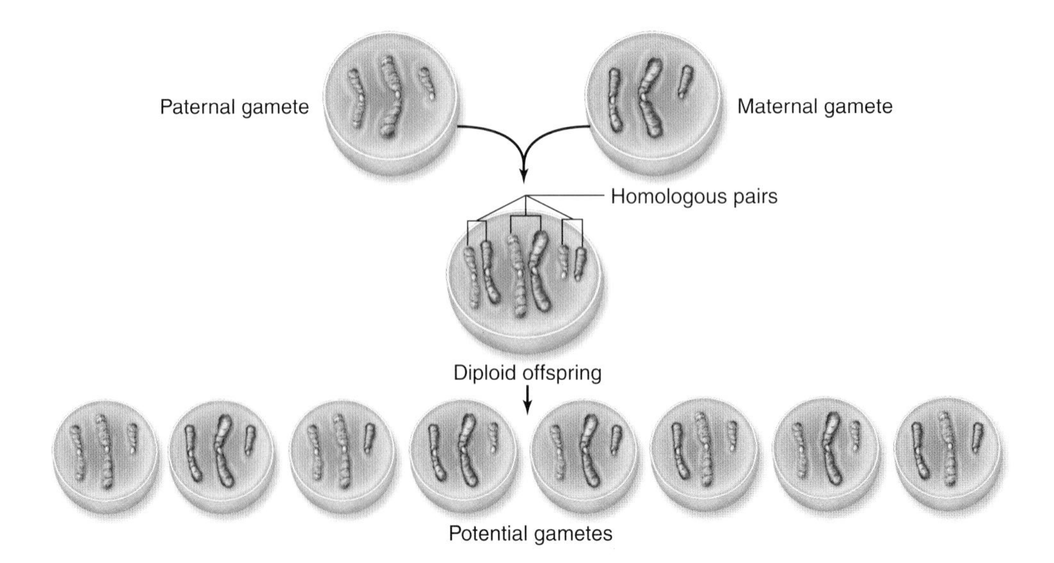

figure 11.10

INDEPENDENT ASSORTMENT INCREASES GENETIC VARIABILITY. Independent assortment contributes new gene combinations to the next generation because the orientation of chromosomes on the metaphase plate is random. For example, in cells with three chromosome pairs, eight different gametes can result, each with different combinations of parental chromosomes. This is also increased by crossing over, or genetic recombination as this further shuffles the arrangements of genes on chromosomes.

concept review

11.1 Sexual Reproduction Requires Meiosis (figure 11.2*b*)

Meiosis is a reductive division that converts a diploid cell into four haploid cells, each with one complete set of chromosomes.

- Egg and sperm are haploid and contain one set of all chromosomes.
- During fertilization, or syngamy, the fusion of two haploid gametes results in a diploid zygote, which contains two sets of chromosomes.
- Meiosis and fertilization constitute a cycle of reproduction in sexual organisms as they alternate between diploid and haploid chromosome numbers.
- Somatic cells divide by mitosis and form the body of an organism.
- Cells that form haploid gametes by meiosis are called germ-line cells.

11.2 Features of Meiosis

In diploid organisms meiosis consists of two rounds of division called meiosis I and meiosis II with no replication between divisions.

- The pairing of homologous chromosomes, called synapsis, occurs during early prophase I; synapsis does not occur in mitosis.
- During synapsis homologous chromosomes pair along their entire length, often joined by a structure called the synatonemal complex (figure 11.4).
- During synapsis, crossing over occurs between homologous chromosomes exchanging chromosomal material (figure 11.5).
- Because the homologues are paired, they move as a unit to the metaphase plate during metaphase I.
- During anaphase I, homologues of each pair are pulled to opposite poles, producing two cells each with one complete set of chromosomes.
- Meiosis II is like mitosis with no replication of DNA.

11.3 The Process of Meiosis (figures 11.8 and 11.9)

Meiosis depends on homologues exchanging chromosomal material by crossing over, which assists in holding homologues together during nuclear division.

- Meiotic cells have an interphase period similar to mitosis with G_1, S, and G_2 phases.
- During prophase I homologous chromosomes pair along their entire length in a process called synapsis.
  - *The sister chromatids of each homologue are held together by cohesin proteins in a process called sister chromatid cohesion.*
  - *While joined by the synaptonemal complex, recombinant nodules form, allowing homologues to exchange genetic material; crossing over between sister chromatids (of the same homologue) is suppressed.*
  - *Sites of crossing over are called chiasmata.*
  - *When crossing over is complete, the synaptonemal complex breaks down, leaving paired homologues joined together only at chiasmata.*
  - *Sister chromatids of each homologue remain joined at their centromeres.*
  - *Prior to metaphase the chiasmata between homologous pairs move to the ends of the chromosome arms where they become terminal chiasmata, holding the homologues together.*
- The nuclear envelope disperses and the spindle apparatus forms.
- During metaphase I homologous pairs align at the cell equator, or metaphase plate.
  - *Spindle fibers attach to the kinetochores of the homologues and not those of the sister chromatids.*
  - *Homologues of each pair become attached by kinetochore microtubules to opposite poles.*
  - *The orientation of each homologous pair on the equator is random; either the maternal or paternal homologue may be oriented toward a given pole.*
- During anaphase I the homologues of each pair are pulled to opposite poles.
  - *During anaphase I cohesin proteins joining sister chromatids are lost on the arms, but not between the centromeres that connect sister chromatids.*
  - *Loss of sister chromatid cohesion on the arms but not the centromeres allows homologues to separate.*
  - *Kinetochore microtubules shorten, pulling each homologue with its two sister chromatids to opposite poles.*
  - *At the end of anaphase I each pole has a complete set of haploid chromosomes, consisting of one member of each homologous pair.*
  - *As a result of the random orientation of homologous pairs at metaphase I, meiosis I results in the independent assortment of maternal and paternal chromosomes in gametes.*
- Telophase I is characterized by the reforming of the nuclear envelope around each daughter nucleus. This does not occur in all species.
- Cytokinesis may or may not occur after telophase I.
- A brief interphase II then occurs during which no DNA replication takes place.
- Meiosis II is similar to mitosis.
- The cohesin proteins at the centromeres holding sister chromatids together are destroyed, allowing them to migrate to opposite poles of the cell.
- The result of meiosis I and II is four cells, each containing haploid sets of chromosomes that are not identical.
- Once completed, the haploid cells may produce gametes or divide mitotically to produce even more gametes or haploid adults.
- Errors occur during meiosis because of nondisjunction, the failure of chromosomes to move to opposite poles.
- Nondisjunction results in one gamete with no chromosome and another gamete with two copies of a chromosome.
- Gametes with an improper number of chromosomes are called aneuploid gametes.

11.4 Summing Up: Meiosis Versus Mitosis

The basic machinery of meiosis and mitosis is the same, but the behavior of chromosomes is distinctly different during the first meiotic division.

- Four distinct features of meiosis I are not found in mitosis:
  - *Maternal and paternal homologues pair, and exchange genetic information by crossing over.*
  - *The kinetochores of sister chromatids function as a unit during meiosis I, allowing sister chromatids to cosegregate during anaphase I.*
  - *Kinetochores of sister chromatids are connected to a single pole in meiosis I and to opposite poles in mitosis.*
  - *DNA replication is suppressed between meiosis I and meiosis II.*
- Daughter cells produced by meiosis are not genetically identical because of independent assortment of homologues and crossing over.
- Meiosis is the foundation for understanding the basis of heredity.

SELF TEST

1. Gametes contain ________ the number of chromosomes found in somatic cells.
   a. the same
   b. twice
   c. half
   d. one fourth
2. Somatic cells are _________, whereas gametes are _________.
   a. haploid; diploid
   b. diploid; polyploid
   c. polyploidy; haploid
   d. diploid; haploid
3. An organism is said to be diploid if—
   a. It contains genetic information from two parents
   b. It is multicellular
   c. It reproduces
   d. Undergoes mitotic cell division
4. What are *homologous* chromosomes?
   a. The two halves of a replicated chromosome
   b. Two identical chromosomes from one parent
   c. Two genetically identical chromosomes, one from each parent
   d. Two genetically similar chromosomes, one from each parent
5. When homologous chromosomes form chiasmata, they are—
   a. exchanging genetic information
   b. reproducing their DNA
   c. separating their sister chromatids
   d. replicating their chromatids
6. Crossing over involves each of the following with the exception of—
   a. the transfer of DNA between two nonsister chromatids
   b. the transfer of DNA between two sister chromatids
   c. the formation of a synaptonemal complex
   d. the alignment of homologous chromosomes
7. Terminal chiasmata are seen during which phase of meiosis?
   a. Anaphase I
   b. Prophase I
   c. Metaphase I
   d. Metaphase II
8. Which of the following occurs during anaphase I?
   a. Sister chromatids are separated and move to the poles.
   b. Homologous chromosomes move to opposite poles.
   c. Homologous chromosomes align at the middle of the cell.
   d. All the chromosomes align independently at the middle of the cell.
9. Telophase I results in the production of—
   a. four cells containing one homologue of each homologous pair
   b. two cells containing both homologues of each homologous pair
   c. four cells containing both homologues of each homologous pair
   d. two cells containing one homologue of each homologous pair
10. Which of the following does *not* contribute to genetic diversity?
    a. Independent assortment
    b. Recombination
    c. Metaphase of meiosis II
    d. Metaphase of meiosis I
11. How does S phase following meiosis I differ from S phase in mitosis?
    a. DNA replication takes less time because the cell is haploid.
    b. DNA does not replicate during S phase following meiosis I.
    c. DNA replication takes more time due to cohesin proteins.
    d. There is no difference.
12. What occurs during anaphase of meiosis II?
    a. The homologous chromosomes align.
    b. Sister chromatids are pulled to opposite poles.
    c. Homologous chromosomes are pulled to opposite poles.
    d. The haploid chromosomes line up.
13. What is an aneuploid gamete?
    a. A diploid gamete cell
    b. A haploid gamete cell
    c. A gamete cell with the wrong number of chromosomes
    d. A haploid somatic cell
14. Which of the following is *not* a distinct feature of meiosis?
    a. Pairing and exchange of genetic material between homologous chromosomes
    b. Attachment of sister kinetochores to spindle microtubules
    c. Movement of sister chromatids to the same pole
    d. Suppression of DNA replication
15. Which phase of meiosis I is most similar to the comparable phase in mitosis?
    a. Prophase I
    b. Metaphase I
    c. Anaphase I
    d. Telophase I

CHALLENGE QUESTIONS

1. Diagram the process of meiosis for an imaginary cell with six chromosomes in a diploid cell.
   a. How many homologous pairs are present in this cell? Create a drawing that distinguishes between homologous pairs.
   b. Label each homologue to indicate whether it is maternal (M) or paternal (P).
   c. Draw a new cell showing how these chromosomes would arrange themselves during metaphase of meiosis I. Do all the maternal homologues have to line up on the same side of the cell?
   d. How would this picture differ if you were diagramming anaphase of meiosis II?
2. Mules are the offspring of the mating of a horse and a donkey. Mules are unable to reproduce. A horse has a total of 64 chromosomes, whereas donkeys have 62 chromosomes. Use your knowledge of meiosis to predict the diploid chromosome number of a mule. Propose a possible explanation for the inability of mules to reproduce.
3. Compare the processes of *independent assortment* and *crossing over*. Which process has the greatest influence on genetic diversity?
4. Aneuploid gametes are cells that contain the wrong number of chromosomes. Aneuploidy occurs as a result of *nondisjunction*, or lack of separation of the chromosomes during either phase of meiosis.
   a. At what point in meiotic cell division would nondisjunction occur?
   b. Imagine a cell had a diploid chromosome number of 4. Create a diagram to illustrate the effects of nondisjunction of one pair of homologous chromosomes in meiosis I versus meiosis II.

Do you need additional review? *Visit* www.ravenbiology.com *for practice quizzes, animations, videos, and activities designed to help you master the material in this chapter.*

chapter 12

Patterns of Inheritance

introduction

EVERY LIVING CREATURE IS A PRODUCT of the long evolutionary history of life on Earth. All organisms share this history, but as far as we know, only humans wonder about the processes that led to their origin and investigate the possibilities. We are far from understanding everything about our origins, but we have learned a great deal. Like a partially completed jigsaw puzzle, the boundaries of this elaborate question have fallen into place, and much of the internal structure is becoming apparent. In this chapter, we discuss one piece of the puzzle—the enigma of heredity. Why do individuals, like the children in this picture, differ so much in appearance despite the fact that we are all members of the same species? And, why do members of a single family tend to resemble one another more than they resemble members of other families?

concept outline

12.1 The Mystery of Heredity

As far back as written records go, patterns of resemblance among the members of particular families have been noted and commented on (figure 12.1), but there was no coherent model to explain these patterns. Before the 20th century, two concepts provided the basis for most thinking about heredity. The first was that heredity occurs within species. The second was that traits are transmitted directly from parents to offspring. Taken together, these ideas led to a view of inheritance as resulting from a blending of traits within fixed, unchanging species.

Inheritance itself was viewed as traits being borne through fluid, usually identified as blood, that led to their blending in offspring. This older idea persists today in the use of the term "bloodlines" when referring to the breeding of domestic animals such as horses.

Taken together, however, these two classical assumptions led to a paradox. If no variation enters a species from outside, and if the variation within each species blends in every generation, then all members of a species should soon have the same appearance. It is clear that this does not happen—individuals within most species differ from one another, and they differ in characteristics that are transmitted from generation to generation.

Early plant biologists produced hybrids and saw puzzling results

The first investigator to achieve and document successful experimental **hybridizations** was Josef Kölreuter, who in 1760 cross-fertilized (or crossed, for short) different strains of tobacco and obtained fertile offspring. The hybrids differed in appearance from both parent strains. When individuals within the hybrid generation were crossed, their offspring were highly variable. Some of these offspring resembled plants of the hybrid generation (their parents), but a few resembled the original strains (their grandparents).

Kölreuter's work represents the beginning of modern genetics. The patterns of inheritance observed in his hybrids contradicted the theory of direct transmission because of the variation observed in second-generation offspring.

Over the next hundred years, other investigators elaborated on Kölreuter's work. In one such series of experiments, carried out in 1823, T. A. Knight, an English landholder, crossed two varieties of the garden pea, *Pisum sativum* (figure 12.2). One of these varieties had green seeds, and the other had yellow seeds. Both varieties were **true-breeding,** meaning that the offspring produced from self-fertilization would remain uniform from one generation to the next.

All of the progeny (offspring) of the cross between the two varieties had yellow seeds. Among the offspring of these hybrids, however, some plants produced yellow seeds and others, less common, produced green seeds.

Other investigators made observations similar to Knight's, namely that alternative forms of observed characters were being distributed among the offspring. Referring to a heritable feature as a *character*, a modern geneticist would say the alternative forms of each character were **segregating** among the progeny of a mating, meaning that some offspring exhibited one form of a character (yellow seeds), and other offspring from the same mating exhibited a different form (green seeds). This segregation of alternative forms of a character, or **trait,** provided the clue that led Gregor Mendel to his understanding of the nature of heredity.

Within these deceptively simple results were the makings of a scientific revolution. Nevertheless, another century passed before the process of segregation was fully appreciated.

Mendel used mathematics to analyze his crosses

Born in 1822 to peasant parents, Gregor Mendel (figure 12.3) was educated in a monastery and went on to study science and mathematics at the University of Vienna, where he failed his examina-

figure 12.1

HEREDITY AND FAMILY RESEMBLANCE. Family resemblances are often strong—a visual manifestation of the mechanism of heredity.

figure 12.2

THE GARDEN PEA, *Pisum sativum*. Easy to cultivate and able to produce many distinctive varieties, the garden pea was a popular experimental subject in investigations of heredity as long as a century before Gregor Mendel's experiments.

tions for a teaching certificate. He returned to the monastery and spent the rest of his life there, eventually becoming abbot. In the garden of the monastery, Mendel initiated his own series of experiments on plant hybridization. The results of these experiments would ultimately change our views of heredity irrevocably.

Practical considerations for use of the garden pea

For his experiments, Mendel chose the garden pea, the same plant Knight and others had studied. The choice was a good one for several reasons. First, many earlier investigators had produced hybrid peas by crossing different varieties, so Mendel knew that he could expect to observe segregation of traits among the offspring.

Second, a large number of pure varieties of peas were available. Mendel initially examined 34 varieties. Then, for further study, he selected lines that differed with respect to seven easily distinguishable traits, such as round versus wrinkled seeds and yellow versus green seeds, the latter a trait that Knight had studied.

Third, pea plants are small and easy to grow, and they have a relatively short generation time. A researcher can therefore conduct experiments involving numerous plants, grow several generations in a single year, and obtain results relatively quickly.

A fourth advantage of studying peas is that both the male and female sexual organs are enclosed within each pea flower (figure 12.3), and gametes produced by the male and female parts of the same flower can fuse to form viable offspring, a process termed **self-fertilization.** This self-fertilization takes place automatically within an individual flower if it is not disturbed, resulting in offspring that are the progeny from a single individual. It is also possible to prevent self-fertilization by removing a flower's male parts before fertilization occurs, then introduce pollen from a different strain, thus performing *cross-pollination* that results in **cross-fertilization** (see figure 12.3).

Mendel's experimental design

Mendel was careful to focus on only a few specific differences between the plants he was using and to ignore the countless other differences he must have seen. He also had the insight to realize that the differences he selected must be comparable. For example, he appreciated that trying to study the inheritance of round seeds versus tall height would be useless.

Mendel usually conducted his experiments in three stages:

1. By allowing plants of a given variety to self-cross for multiple generations, Mendel was able to assure himself that the traits he was studying were indeed true-breeding, that is, transmitted unchanged from generation to generation.

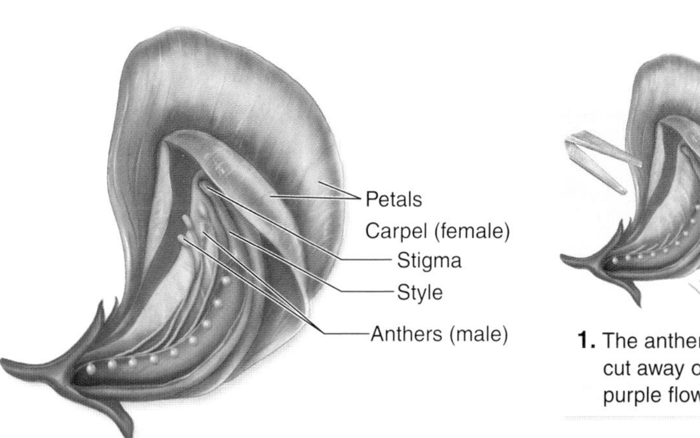

1. The anthers are cut away on the purple flower.

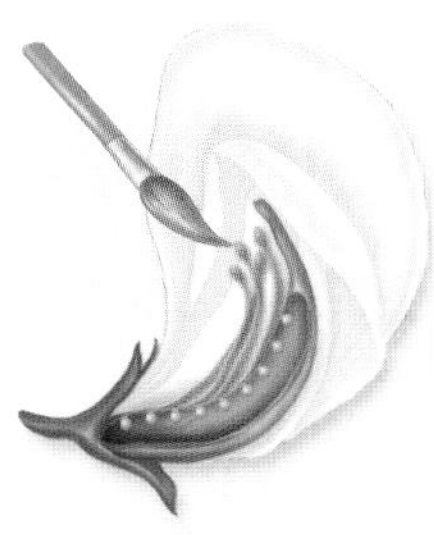

2. Pollen is obtained from the white flower.

3. Pollen is transferred to the purple flower.

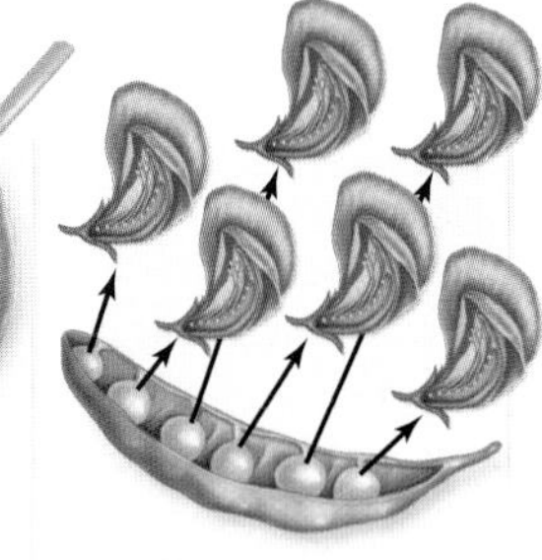

4. All progeny result in purple flowers.

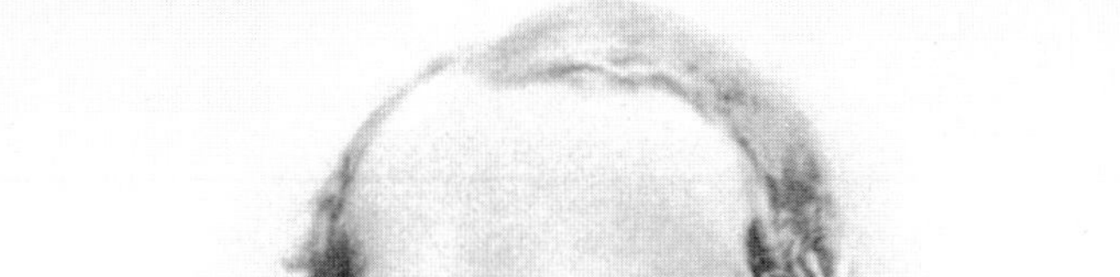

figure 12.3

HOW MENDEL CONDUCTED HIS EXPERIMENTS. In a pea plant flower, petals enclose the male anther (containing pollen grains, which give rise to haploid sperm) and the female carpel (containing ovules, which give rise to haploid eggs). This ensures self-fertilization will take place unless the flower is disturbed. Mendel collected pollen from the anthers of a white flower, then placed that pollen onto the stigma of a purple flower with anthers removed. This cross fertilization yields all hybrid seeds that give rise to purple flowers. Using pollen from a white flower to fertilize a purple flower gives the same result.

inquiry

What confounding problems could have been seen if Mendel had chosen another plant with exposed male and female structures?

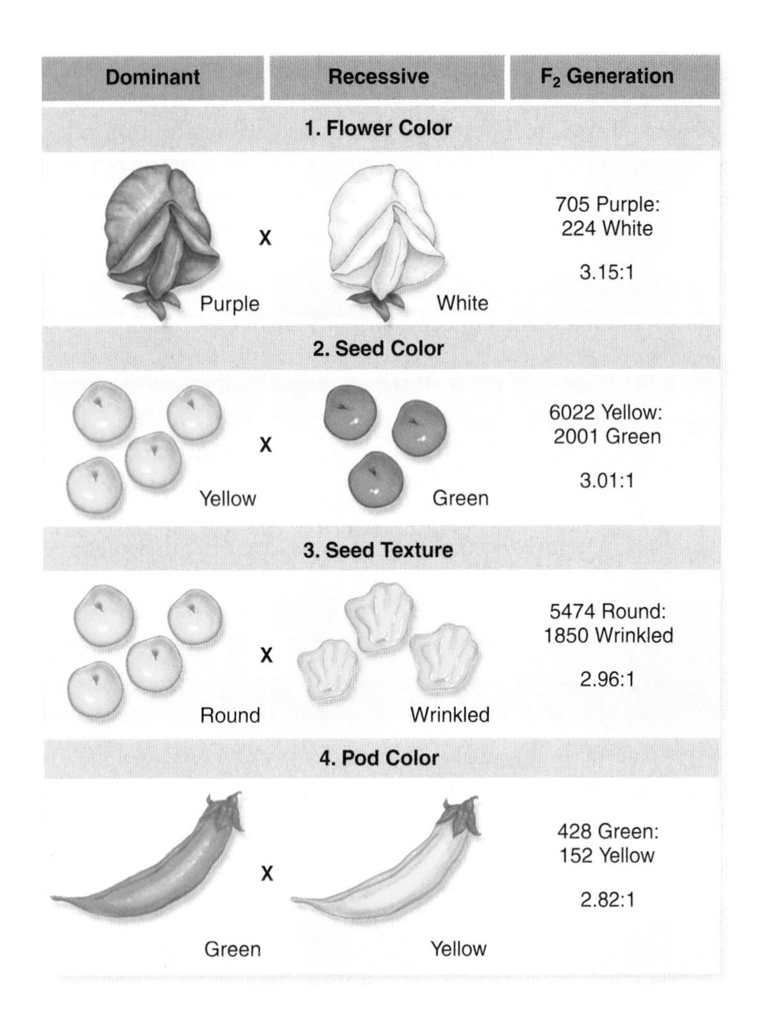

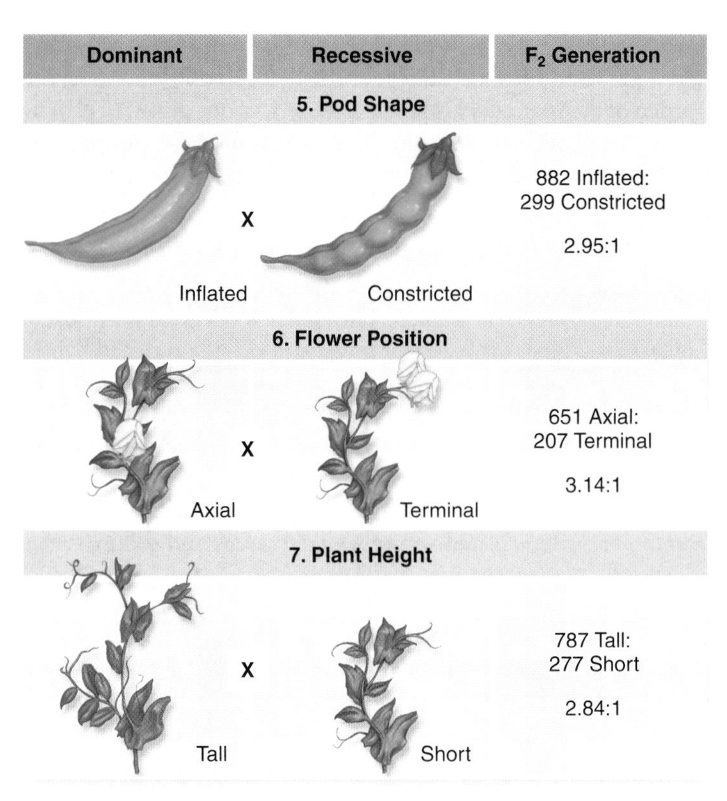

figure 12.4

MENDEL'S SEVEN TRAITS. Mendel studied how differences among varieties of peas were inherited when the varieties were crossed. Similar experiments had been done before, but Mendel was the first to quantify the results and appreciate their significance. Results are shown for seven different monohybrid crosses. The F_1 generation is not shown in the table.

2. Mendel then performed crosses between true-breeding varieties exhibiting alternative forms of traits. He also performed **reciprocal crosses:** using pollen from a white-flowered plant to fertilize a purple-flowered plant, then using pollen from a purple-flowered plant to fertilize a white-flowered plant.
3. Finally, Mendel permitted the hybrid offspring produced by these crosses to self-fertilize for several generations, allowing him to observe the inheritance of alternative forms of a trait. Most important, he counted the numbers of offspring exhibiting each trait in each succeeding generation.

This quantification of results is what distinguished Mendel's research from that of earlier investigators, who only noted differences in a qualitative way. Mendel's mathematical analysis of experimental results led to the inheritance model that we still use today.

Ideas about inheritance before Mendel did not form a consistent model. The dominant view was of blending inheritance, but plant hybridizers before Mendel had already cast doubt on this model. Mendel followed up on the work of early plant hybridizers by systematizing and quantifying his observations.

12.2 Monohybrid Crosses: The Principle of Segregation

A **monohybrid cross** is a cross that follows only two variations on a single trait, such as white- and purple-colored flowers. This deceptively simple kind of cross can lead to important conclusions about the nature of inheritance.

The seven characters Mendel studied in his experiments possessed two variants that differed from one another in ways that were easy to recognize and score (see figure 12.4). We examine in detail Mendel's crosses with flower color. His experiments with other characters were similar, and they produced similar results.

The F_1 generation exhibits only one of two traits, without blending

When Mendel crossed white-flowered and purple-flowered plants, the hybrid offspring he obtained did not have flowers of intermediate color, as the hypothesis of blending inheritance would predict. Instead, in every case the flower color of the offspring resembled that of one of their parents. These offspring are customarily referred to as the **first filial generation,**

figure 12.5

SEED SHAPE: A MENDELIAN CHARACTER.
One of the differences Mendel studied involved the shape of pea plant seeds. In some varieties, the seeds were round, but in others, they were wrinkled.

or $\mathbf{F_1}$. In a cross of white-flowered and purple-flowered plants, the F_1 offspring all had purple flowers, as others had reported earlier.

Mendel referred to the form of each trait expressed in the F_1 plants as **dominant,** and to the alternative form that was not expressed in the F_1 plants as **recessive.** For each of the seven pairs of contrasting traits that Mendel examined, one of the pair proved to be dominant and the other recessive.

The F_2 generation exhibits both traits in a 3:1 ratio

After allowing individual F_1 plants to mature and self-fertilize, Mendel collected and planted the seeds from each plant to see what the offspring in the **second filial generation,** or $\mathbf{F_2}$, would look like. He found that although most F_2 plants had purple flowers, some exhibited white flowers, the recessive trait. Although hidden in the F_1 generation, the recessive trait had reappeared among some F_2 individuals.

Believing the proportions of the F_2 types would provide some clue about the mechanism of heredity, Mendel counted the numbers of each type among the F_2 progeny. In the cross between the purple-flowered F_1 plants, he obtained a total of 929 F_2 individuals. Of these, 705 (75.9%) had purple flowers, and 224 (24.1%) had white flowers (see figure 12.4). Approximately 1/4 of the F_2 individuals, therefore, exhibited the recessive form of the character.

Mendel obtained the same numerical result with the other six characters he examined: Of the F_2 individuals, 3/4 exhibited the dominant trait, and 1/4 displayed the recessive trait. In other words, the dominant-to-recessive ratio among the F_2 plants was always close to 3:1. Mendel carried out similar experiments with other traits, such as wrinkled versus round seeds (figure 12.5), and obtained the same result.

The 3:1 ratio is actually 1:2:1

Mendel went on to examine how the F_2 plants passed traits to subsequent generations. He found that plants exhibiting the recessive trait were always true-breeding. For example, the white-flowered F_2 individuals reliably produced white-flowered offspring when they were allowed to self-fertilize. By contrast, only 1/3 of the dominant, purple-flowered F_2 individuals (1/4 of all F_2 offspring) proved pure-breeding, but 2/3 were not. This last class of plants produced dominant and recessive individuals in the third filial generation (F_3) in a 3:1 ratio.

This result suggested that, for the entire sample, the 3:1 ratio that Mendel observed in the F_2 generation was really a disguised 1:2:1 ratio: 1/4 true-breeding dominant individuals, 1/2 not-true-breeding dominant individuals, and 1/4 true-breeding recessive individuals (figure 12.6).

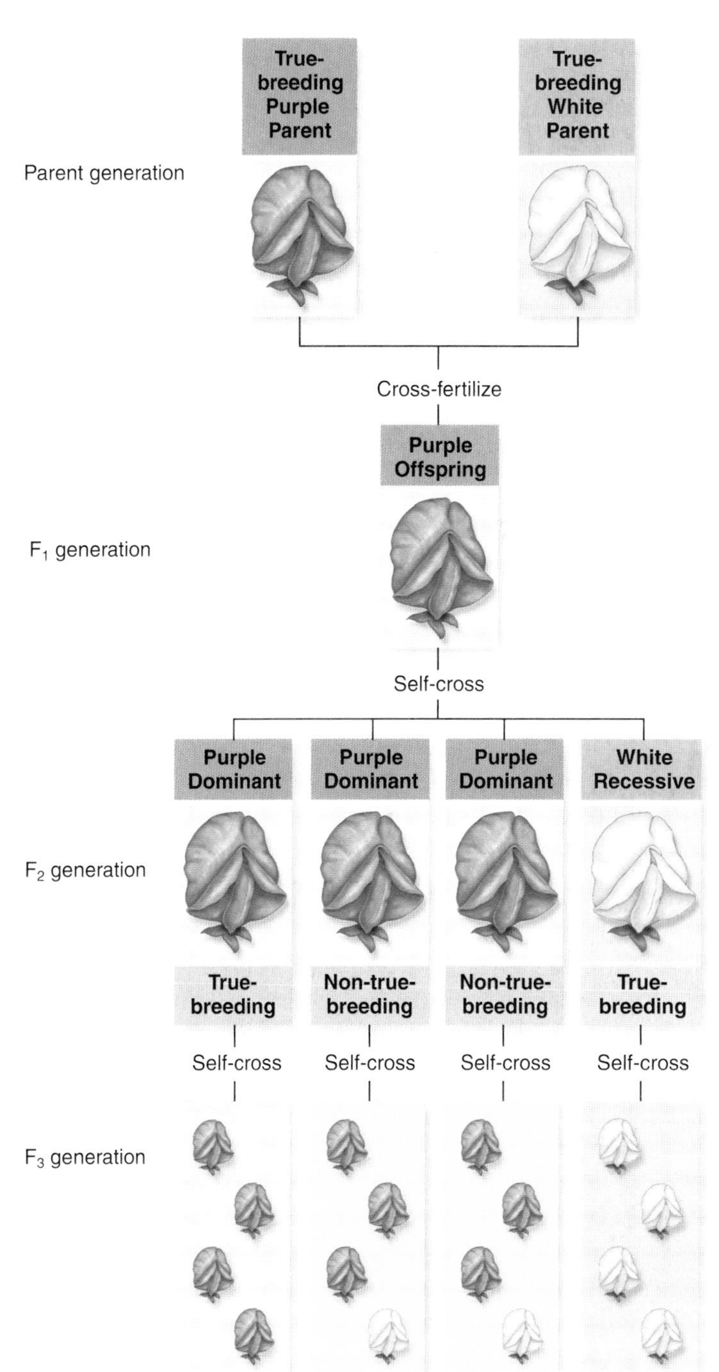

figure 12.6

THE F_2 GENERATION IS A DISGUISED 1:2:1 RATIO.
By allowing the F_2 generation to self-fertilize, Mendel found from the offspring (F_3) that the ratio of F_2 plants was 1 pure-breeding dominant: 2 not-pure-breeding dominant: and 1 pure-breeding recessive.

Mendel's Principle of Segregation explains monohybrid observations

From his experiments, Mendel was able to understand four things about the nature of heredity:

- The plants he crossed did not produce progeny of intermediate appearance, as a hypothesis of blending inheritance would have predicted. Instead, different plants inherited each trait intact, as a discrete characteristic.
- For each pair of alternative forms of a trait, one alternative was not expressed in the F_1 hybrids, although it reappeared in some F_2 individuals. *The trait that "disappeared" must therefore be latent (present but not expressed) in the F_1 individuals.*
- The pairs of alternative traits examined were segregated among the progeny of a particular cross, some individuals exhibiting one trait and some the other.
- These alternative traits were expressed in the F_2 generation in the ratio of 3/4 dominant to 1/4 recessive. This characteristic 3:1 segregation is referred to as the **Mendelian ratio** for a monohybrid cross.

Mendel's five-element model

To explain these results, Mendel proposed a simple model that has become one of the most famous in the history of science, containing simple assumptions and making clear predictions. The model has five elements:

1. Parents do not transmit physiological traits directly to their offspring. Rather, they transmit discrete information for the traits, what Mendel called "factors." We now call these factors **genes.**
2. Each individual receives two genes that encode each trait. We now know that the two factors are carried on chromosomes, and each adult individual is diploid. Gametes, produced by meiosis, are haploid.
3. Not all copies of a gene are identical. The alternative forms of a gene are called **alleles.** When two haploid gametes containing the same allele fuse during fertilization, the resulting offspring is said to be **homozygous.** When the two haploid gametes contain different alleles, the resulting offspring is said to be **heterozygous.**
4. The two alleles remain discrete—they neither blend with nor alter each other. Therefore, when the individual matures and produces its own gametes, the alleles segregate randomly into these gametes.
5. The presence of a particular allele does not ensure that the trait it encodes will be expressed. In heterozygous individuals, only one allele is expressed (the dominant one), and the other allele is present but unexpressed (the recessive one).

Geneticists now refer to the total set of alleles that an individual contains as the individual's **genotype.** The physical appearance or other observable characteristics of that individual, which result from an allele's expression, is termed the individual's **phenotype.** In other words, the genotype is the blueprint, and the phenotype is the visible outcome.

This also allows us to present Mendel's ratios in more modern terms. The 3:1 ratio of dominant to recessive is the monohybrid phenotypic ratio. The 1:2:1 ratio of homozygous dominant to heterozygous to homozygous recessive is the monohybrid genotypic ratio. The genotypic ratio "collapses" into the phenotypic ratio due to the action of the dominant allele making the heterozygote appear the same as homozygous dominant.

The principle of segregation

Mendel's model accounts for the ratios he observed in a neat and satisfying way. His main conclusion—that alternative alleles for a character segregate from each other during gamete formation and remain distinct—has since been verified in many other organisms. It is commonly referred to as Mendel's first law of heredity, or the **Principle of Segregation.** It can be simply stated as: *The two alleles for a gene segregate during gamete formation and are rejoined at random, one from each parent, during fertilization.*

The physical basis for allele segregation is the behavior of chromosomes during meiosis. As you saw in chapter 11, homologues for each chromosome disjoin during anaphase I of meiosis. The second meiotic division then produces gametes that contain only one homologue for each chromosome.

It is a tribute to Mendel's intellect that his analysis arrived at the correct scheme, even though he had no knowledge of the cellular mechanisms of inheritance; neither chromosomes nor meiosis had yet been described.

The Punnett square allows symbolic analysis

To test his model, Mendel first expressed it in terms of a simple set of symbols. He then used the symbols to interpret his results.

Consider again Mendel's cross of purple-flowered with white-flowered plants. By convention, we assign the symbol P (uppercase) to the dominant allele, associated with the production of purple flowers, and the symbol p (lowercase) to the recessive allele, associated with the production of white flowers.

In this system, the genotype of an individual that is true-breeding for the recessive white-flowered trait would be designated pp. Similarly, the genotype of a true-breeding purple-flowered individual would be designated PP. In contrast, a heterozygote would be designated Pp (dominant allele first). Using these conventions and denoting a cross between two strains with ×, we can symbolize Mendel's original purple × white cross as $PP \times pp$.

Because a white-flowered parent (pp) can produce only p gametes, and a true-breeding purple-flowered parent (PP, *homozygous dominant*) can produce only P gametes, the union of these gametes can produce only heterozygous Pp offspring in the F_1 generation. Because the P allele is dominant, all of these F_1 individuals are expected to have purple flowers.

When F_1 individuals are allowed to self-fertilize, the P and p alleles segregate during gamete formation to produce both P gametes and p gametes. Their subsequent union at fertilization to form F_2 individuals is random.

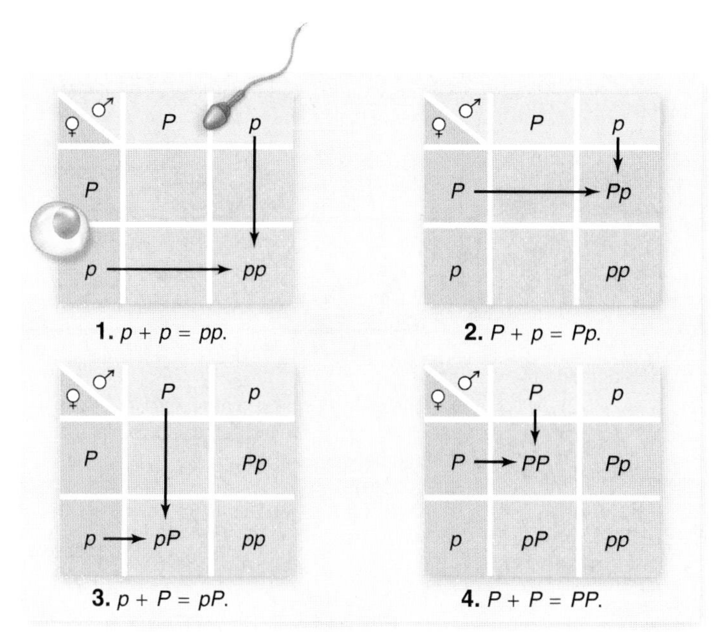

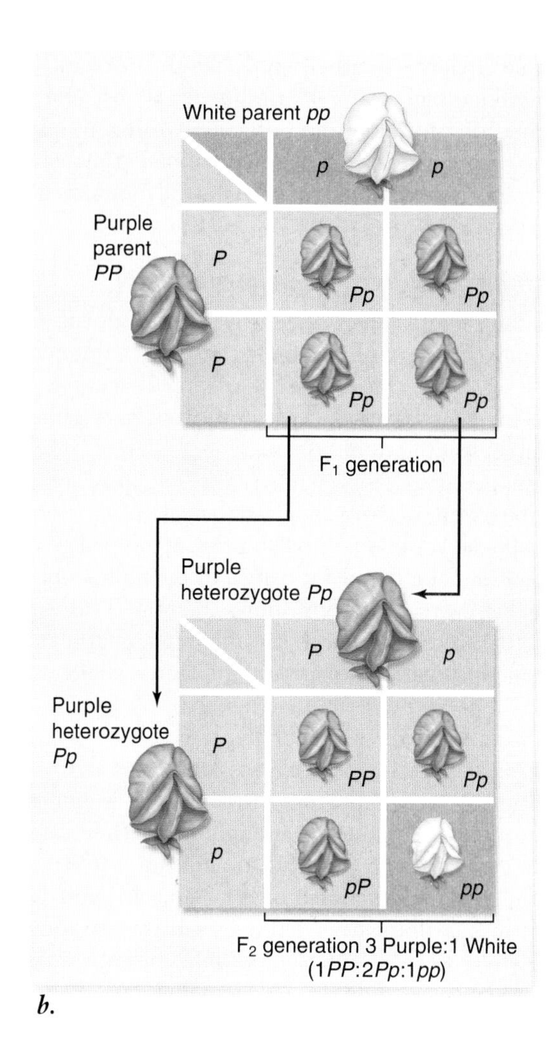

figure 12.7

USING A PUNNETT SQUARE TO ANALYZE MENDEL'S CROSS. ***a.*** To make a Punnett square, place the different possible types of female gametes along the side of a square and the different possible types of male gametes along the top. Each potential zygote is represented as the intersection of a vertical line and a horizontal line. ***b.*** In Mendel's cross of purple by white flowers, the original parents each only make one type of gamete. The resulting F_1 generation are all *Pp* heterozygotes with purple flowers. These F_1 then each make two types of gametes that can be combined to produce three kinds of F_2 offspring: *PP* homozygotes (purple flowers); *Pp* heterozygotes (also purple flowers); and *pp* homozygotes (white flowers). The ratio of dominant to recessive phenotypes is 3:1. The ratio of genotypes is 1:2:1 (1 *PP*: 2 *Pp*: 1 *pp*).

The F_2 possibilities may be visualized in a simple diagram called a **Punnett square,** named after its originator, the English geneticist R. C. Punnett (figure 12.7*a*). Mendel's model, analyzed in terms of a Punnett square, clearly predicts that the F_2 generation should consist of 3/4 purple-flowered plants and 1/4 white-flowered plants, a phenotypic ratio of 3:1 (figure 12.7*b*).

Some human traits exhibit dominant/recessive inheritance

A number of human traits have been shown to display both dominant and recessive inheritance (table 12.1 provides a sample of these). Researchers cannot perform controlled crosses in humans the way Mendel did with pea plants, so to analyze human inheritance, geneticists study crosses that have been

TABLE 12.1 Some Dominant and Recessive Traits in Humans

Recessive Traits	Phenotypes	Dominant Traits	Phenotypes
Albinism	Lack of melanin pigmentation	Middigital hair	Presence of hair on middle segment of fingers
Alkaptonuria	Inability to metabolize homogentisic acid	Brachydactyly	Short fingers
Red-green color blindness	Inability to distinguish red or green wavelengths of light	Huntington disease	Degeneration of nervous system, starting in middle age
Cystic fibrosis	Abnormal gland secretion, leading to liver degeneration and lung failure	Phenylthiocarbamide (PTC) sensitivity	Ability to taste PTC as bitter
Duchenne muscular dystrophy	Wasting away of muscles during childhood	Camptodactyly	Inability to straighten the little finger
Hemophilia	Inability of blood to clot properly, some clots form but the process is delayed	Hypercholesterolemia (the most common human Mendelian disorder)	Elevated levels of blood cholesterol and risk of heart attack
Sickle cell anemia	Defective hemoglobin that causes red blood cells to curve and stick together	Polydactyly	Extra fingers and toes

performed already—in other words, family histories. The organized methodology we use is a **pedigree,** a consistent graphical representation of matings and offspring over multiple generations for a particular trait. The information in the pedigree may allow geneticists to deduce a model for the mode of inheritance of the trait.

A dominant pedigree: Juvenile glaucoma

One of the most extensive pedigrees yet produced traced the inheritance of a form of blindness caused by a dominant allele. The disease allele causes a form of hereditary juvenile glaucoma. The disease causes degeneration of nerve fibers in the optic nerve, leading to blindness.

This pedigree followed inheritance over three centuries following the origin back to a couple in a small town in northwestern France who died in 1495. A small portion of this pedigree is shown in figure 12.8. The dominant nature of the trait is obvious from the fact that every generation shows the trait. This is extremely unlikely for a recessive trait as it would require large numbers of unrelated individuals to be carrying the disease allele.

A recessive pedigree: Albinism

An example of inheritance of a recessive human trait is albinism, a condition in which the pigment melanin is not produced. Long thought to be due to a single gene, there are actually multiple genes that can all lead to albinism, the common feature is the loss of pigment from hair, skin, and eyes. The loss of pigment makes albinistic individuals sensitive to the sun. The tanning effect we are all familiar with from exposure to the sun is due to increased numbers of pigment-producing cells, and

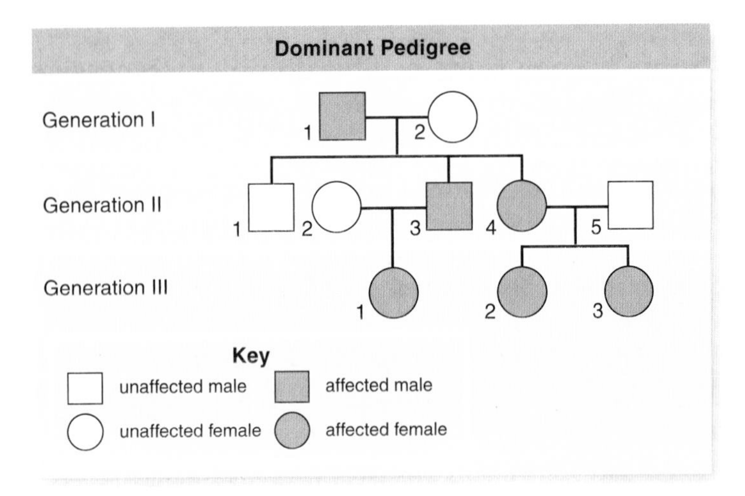

figure 12.8

DOMINANT PEDIGREE FOR HEREDITARY JUVENILE GLAUCOMA. Males are shown as squares and females are shown as circles. Affected individuals are shown shaded. The dominant nature of this trait can be seen in the trait appearing in every generation, a feature of dominant traits.

inquiry

If one of the affected females in the third generation married an unaffected male, could she produce unaffected offspring? If so, what are the chances of having unaffected offspring?

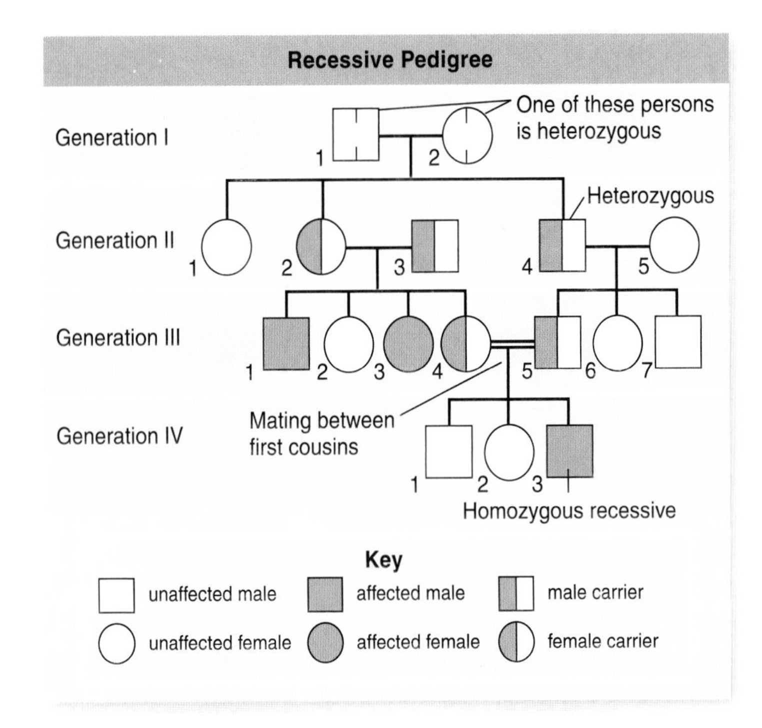

figure 12.9

RECESSIVE PEDIGREE FOR ALBINISM. One of the two individuals in the first generation must be heterozygous and individuals II-2 and II-4 must be heterozygous. Notice that for each affected individual, neither parent is affected, but both must be heterozygous (carriers). The double line indicates a consanguineous mating (between relatives) that, in this case, produced affected offspring.

inquiry

From a genetic disease standpoint, why is it never advisable for close relatives to mate and have children?

increased production of pigment. This is lacking in albinistic individuals due to the lack of any pigment to begin with.

The pedigree in figure 12.9 is for a form of albinism due to a nonfunctional allele of the enzyme tyrosinase, which is required for the formation of melanin pigment. The genetic characteristics of this form of albinism are: females and males are affected equally, most affected individuals have unaffected parents, a single affected parent usually does not have affected offspring, and affected offspring are more frequent when parents are related. Each of these features can be see in figure 12.9, and all of this fits a recessive mode of inheritance quite well.

Monohybrid crosses show that traits are due to factors inherited intact with no blending. Traits that appear in the F_1 generation are called dominant; traits that are not observed are called recessive. In the F_2 generation, both traits are observed in a predictable ratio of 3 dominant to 1 recessive. The Principle of Segregation states that during gamete formation, alleles segregate into different gametes and are then randomly combined during fertilization. Dominant/recessive inheritance is analyzed in humans using pedigrees.

12.3 Dihybrid Crosses: The Principle of Independent Assortment

The Principle of Segregation explains the behavior of alternative forms of a single trait in a monohybrid cross. The next step is to extend this to follow the behavior of two different traits in a single cross: a **dihybrid cross.**

With an understanding of the behavior of single traits, Mendel went on to ask if different traits behaved independently in hybrids. He first established a series of true-breeding lines of peas that differed in two of the seven characters he had studied. He then crossed contrasting pairs of the true-breeding lines to create heterozygotes. These heterozygotes are now doubly heterozygous, or dihybrid. Finally, he self-crossed the dihybrid F_1 plants to produce an F_2 generation, and counted all progeny types.

The F_1 generation displays two of four traits, without blending

Consider a cross involving different seed shape alleles (round, *R*, and wrinkled, *r*) and different seed color alleles (yellow, *Y*, and green, *y*). Crossing round yellow (*RR YY*) with wrinkled green (*rr yy*), produces heterozygous F_1 individuals having the same phenotype (namely round and yellow) and the same genotype (*Rr Yy*). Allowing these dihybrid F_1 individuals to self-fertilize produces an F_2 generation.

The F_2 generation exhibits four types of progeny in a 9:3:3:1 ratio

In analyzing these results, we first consider the number of possible phenotypes. We expect to see the two parental phenotypes: round yellow and wrinkled green. If the traits behave independently, then we can also expect one trait from each parent to produce plants with round green seeds and others with wrinkled yellow seeds.

Next consider what types of gametes the F_1 individuals can produce. Again, we expect the two types of gametes found in the parents: *R Y* and *r y*. If the traits behave independently, then we can also expect the gametes *R y* and *r Y*. Using modern language, two genes each with two alleles can be combined four ways to produce these gametes: *R Y*, *r y*, *R y*, and *r Y*.

A dihybrid Punnett square

We can then construct a Punnett square with these gametes to generate all possible progeny. This is a 4 × 4 square with 16 possible outcomes. Filling in the Punnett square produces all possible offspring (figure 12.10). From this we can see that there are 9 round yellow, 3 wrinkled yellow, 3 round green, and 1 wrinkled green. This predicts a phenotypic ratio of 9:3:3:1 for traits that behave independently.

Mendel's data

What did Mendel actually observe? From a total of 556 seeds from self-fertilized dihybrid plants, he observed the following results:

- 315 round yellow (signified *R__ Y__*, where the underscore indicates the presence of either allele),
- 108 round green *(R__ yy)*,
- 101 wrinkled yellow *(rr Y__)*, and
- 32 wrinkled green *(rr yy)*.

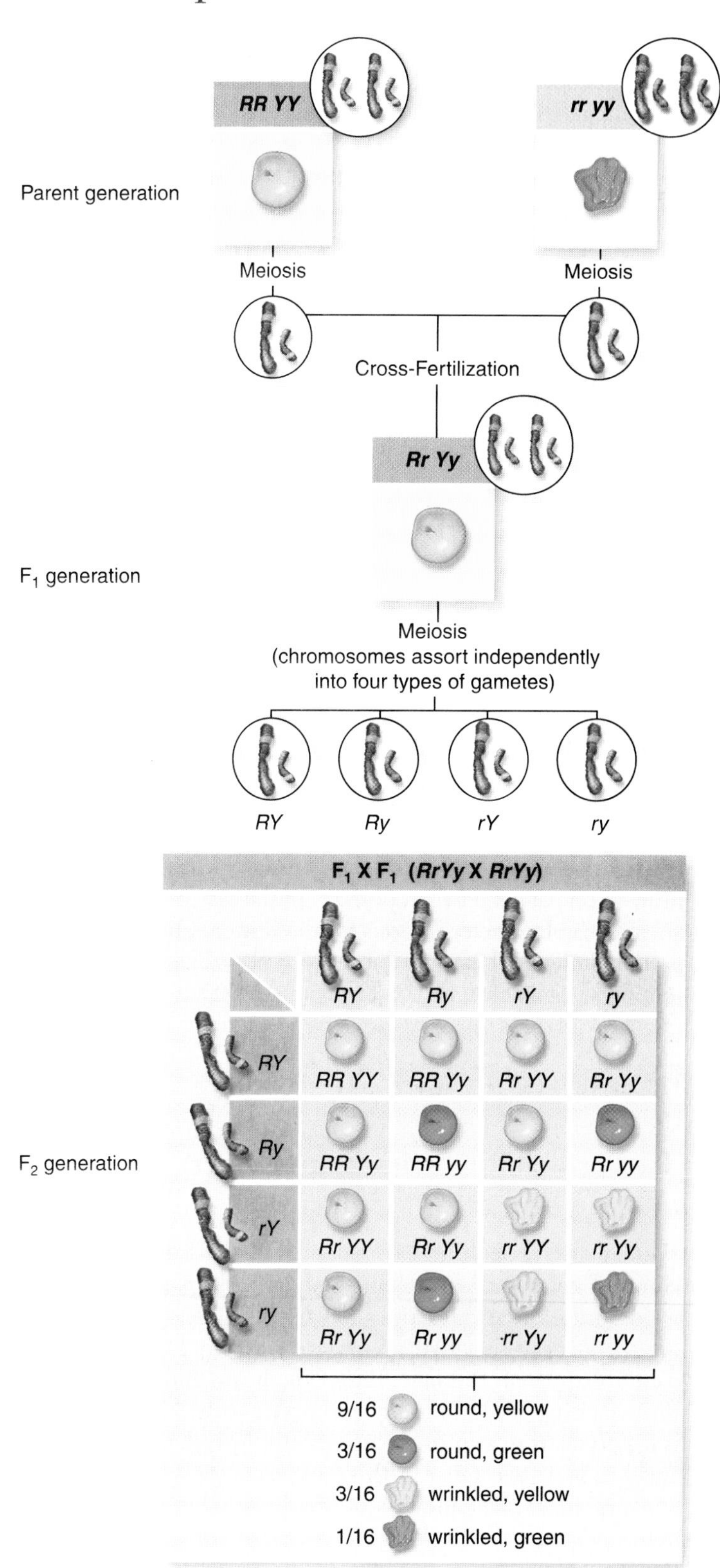

figure 12.10

ANALYZING A DIHYBRID CROSS. This Punnett square shows the results of Mendel's dihybrid cross between plants with round yellow seeds and plants with wrinkled green seeds. The ratio of the four possible combinations of phenotypes is predicted to be 9:3:3:1, the ratio that Mendel found.

These results are very close to a 9:3:3:1 ratio. (The expected 9:3:3:1 ratio from this many offspring would be 313:104:104:35.)

The alleles of two genes appeared to behave independently of each other. Mendel referred to this phenomenon as the traits assorting independently. Note that this **independent assortment** of different genes in no way alters the segregation of individual pairs of alleles for each gene. Round versus wrinkled seeds occur in a ratio of approximately 3:1 (423:133); so do yellow versus green seeds (416:140). Mendel obtained similar results for other pairs of traits.

Mendel's Principle of Independent Assortment explains dihybrid results

Mendel's discovery is often referred to as Mendel's second law of heredity, or the **Principle of Independent Assortment.** This can also be stated simply: *In a dihybrid cross, the alleles of each gene assort independently.* Like segregation, independent assortment arises from the behavior of chromosomes during meiosis to produce haploid gametes (chapter 11)—in this case, the independent alignment of different homologous pairs during metaphase I.

Mendel's analysis of dihybrid crosses revealed that the segregation of allele pairs for different genes is independent, known as Mendel's Principle of Independent Assortment. When individuals that differ in two traits are crossed, and their progeny are intercrossed, the result is four different types that occur in a ratio of 9:3:3:1, or Mendel's dihybrid ratio.

12.4 Probability: Predicting the Results of Crosses

Probability allows us to predict the likelihood of the outcome of random events. Because the behavior of different chromosomes during meiosis is independent, we can use probability to predict the outcome of crosses. The probability of an event that is certain to happen is equal to 1. In contrast, an event that can never happen has a probability of 0. Therefore, probabilities for all other events have fractional values, between 0 and 1. For instance, when you flip a coin, two outcomes are possible; there is only one way to get the event "heads" so the probability of heads is one divided by two, or 1/2. In the case of genetics, consider a pea plant heterozygous for the flower color alleles *P* and *p*. This individual can produce two types of gametes in equal numbers, again due to the behavior of chromosomes during meiosis. There is one way to get a *P* gamete, so the probability of any particular gamete carrying a *P* allele is 1 divided by 2 or 1/2, just like the coin toss.

Two probability rules help predict monohybrid cross results

We can use probability to make predictions about the outcome of genetic crosses using only two simple rules. Before we describe these rules and their uses, we need another definition. We say that two events are *mutually exclusive* if both cannot happen at the same time. The heads and tails of a coin flip are examples of mutually exclusive events. Notice that this is different from two consecutive coin flips where you can get two heads or two tails. In this case, each coin flip represents an *independent event* and it is the distinction between independent and mutually exclusive events that forms the basis for our two rules.

The rule of addition

If we consider a six-sided die instead of a coin, for any roll of the die, only one outcome is possible; each of the possible outcomes are mutually exclusive. The probability of any particular number coming up is 1/6. The probability of either of two different numbers is the sum of the individual probabilities, or to restate as the **rule of addition:**

For two mutually exclusive events, the probability of either event occurring is the sum of the individual probabilities.

$$\text{Probability of rolling either a 2 or a 6 is} = 1/6 + 1/6 = 2/6 = 1/3$$

To apply this to our cross of heterozygous purple F_1, four mutually exclusive outcomes are possible: *PP*, *Pp*, *pP*, and *pp*. The probability of being heterozygous is the same as the probability of being either *Pp* or *pP*, or 1/4 plus 1/4, or 1/2.

$$\text{Probability of } F_2 \text{ heterozygote} = 1/4Pp + 1/4pP = 1/2$$

In the previous example, of 379 total offspring, we would expect about 190 to be heterozygotes. (The actual number is 189.5.)

The rule of multiplication

The second rule, and by far the most useful for genetics, deals with the outcome of independent events. This is called the **product rule,** or **rule of multiplication,** and it states that the probability of two independent events both occurring is the **product** of their individual probabilities.

We can apply this to a monohybrid cross where offspring are formed by gametes from each of two parents. For any particular outcome then, this is due to two independent events: the formation of two different gametes. Consider the purple F_1 parents from earlier. They are all *Pp* (heterozygotes), so the probability that a particular F_2 individual will be *pp* (homozygous recessive) is the probability of receiving a *p* gamete from the male (1/2) times the probability of receiving a *p* gamete from the female (1/2), or 1/4:

$$\text{Probability of } pp \text{ homozygote} = 1/2p \text{ (male parent)} \times 1/2p \text{ (female parent)} = 1/4pp$$

This is actually the basis for the Punnett square that we used before. Each cell in the square was the product of the probabilities of the gametes that contribute to the cell. We then use the addition rule to sum the probabilities of the mutually exclusive events that make up each cell.

We can use the result of a probability calculation to predict the number of homozygous recessive offspring in a cross between heterozygotes. For example, out of 379 total offspring, we would expect about 95 to exhibit the homozygous recessive phenotype. (The actual calculated number is 94.75.)

Dihybrid cross probabilities are based on monohybrid cross probabilities

Probability analysis can be extended to the dihybrid case. For our purple F_1 by F_1 cross, there are four possible outcomes, three of which show the dominant phenotype. Thus the probability of any offspring showing the dominant phenotype is 3/4, and the probability of any offspring showing the recessive phenotype is 1/4. Now we can use this and the product rule to predict the outcome of a dihybrid cross. We will use our example of seed shape and color from earlier, but now examine it using probability.

If the alleles affecting seed shape and seed color segregate independently, then the probability that a particular pair of seed shape alleles would occur together with a particular pair of seed color alleles is the product of the individual probabilities for each pair. For example, the probability that an individual with wrinkled green seeds (*rryy*) would appear in the F_2 generation would be equal to the probability of obtaining wrinkled seeds (1/4) times the probability of obtaining green seeds (1/4), or 1/16.

$$\text{Probability of } rryy = 1/4\ rr \times 1/4\ yy = 1/16\ rryy$$

Because of independent assortment, we can think of the dihybrid cross of consisting of two independent monohybrid crosses; since these are independent events, the product rule applies. So, we can calculate the probabilities for each dihybrid phenotype:

$$\text{Probability of round yellow } (R\_\_\ Y\_\_) = 3/4\ R\_\_ \times 3/4\ Y\_\_ = 9/16$$

$$\text{Probability of round green } (R\_\_\ yy) = 3/4\ R\_\_ \times 1/4\ yy = 3/16$$

$$\text{Probability of wrinkled yellow } (rr\ Y\_\_) = 1/4\ rr \times 3/4\ Y\_ = 3/16$$

$$\text{Probability of wrinkled green } (rryy) = 1/4\ rr \times 1/4\ yy = 1/16$$

The hypothesis that color and shape genes are independently assorted thus predicts that the F_2 generation will display a 9:3:3:1 phenotypic ratio. These ratios can be applied to an observed total offspring to predict the expected number in each phenotypic group. The underlying logic and the results are the same as obtained using the Punnett square.

The probability of either of two events occurring is the sum of the individual probabilities. The probability of two independent events both occurring is the product of the individual probabilities. These can be applied to genetic crosses to determine the probability of particular genotypes and phenotypes.

The Testcross: Revealing Unknown Genotypes

To test his model further, Mendel devised a simple and powerful procedure called the **testcross.** In a testcross, an individual with unknown genotype is crossed with the homozygous recessive genotype—that is, the recessive parental variety. The contribution of the homozygous recessive parent can be ignored, because this parent can contribute only recessive alleles.

Consider a purple-flowered pea plant. It is impossible to tell whether such a plant is homozygous or heterozygous simply by looking at it. To learn its genotype, you can perform a testcross to a white-flowered plant. In this cross, the two possible test plant genotypes will give different results (figure 12.11):

Alternative 1: Unknown individual is homozygous dominant (*PP*)
PP × *pp:* All offspring have purple flowers (*Pp*).

Alternative 2: Unknown individual is heterozygous (*Pp*)
Pp × *pp:* 1/2 of offspring have white flowers (*pp*), and 1/2 have purple flowers (*Pp*).

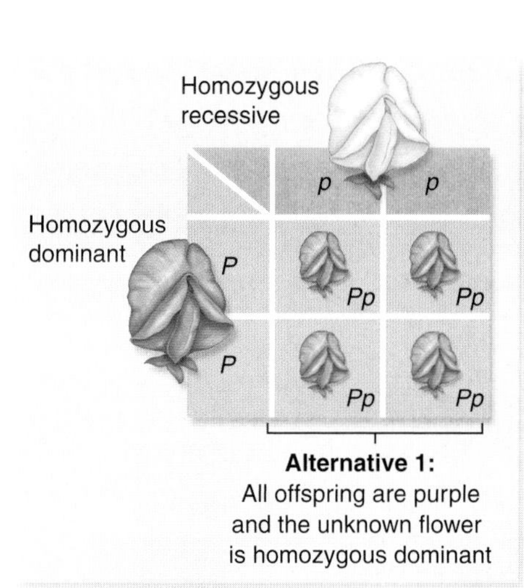

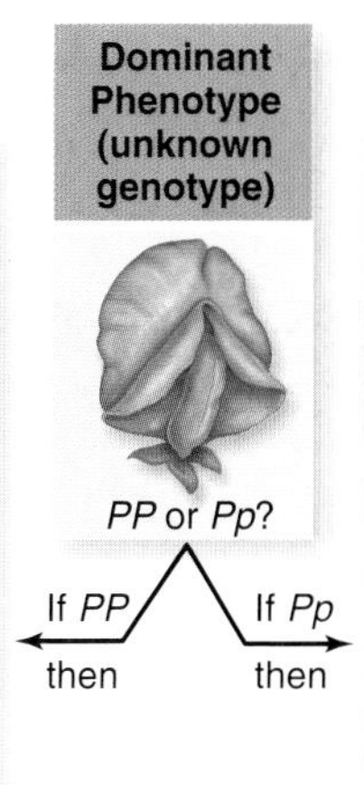

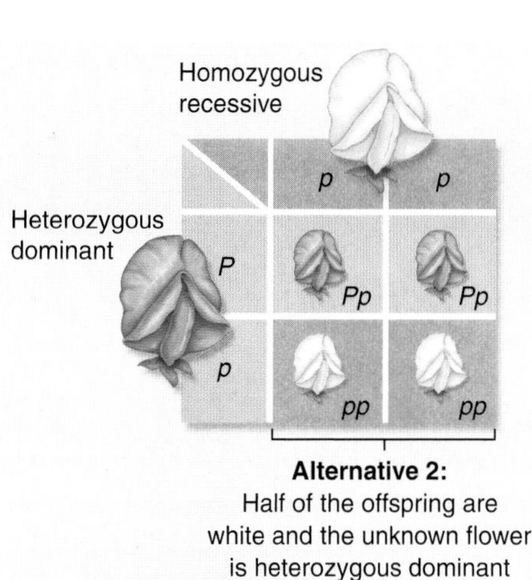

figure 12.11

A TESTCROSS. To determine whether an individual exhibiting a dominant phenotype, such as purple flowers, is homozygous or heterozygous for the dominant allele, Mendel crossed the individual in question with a plant that he knew to be homozygous recessive—in this case, a plant with white flowers.

Put simply, the appearance of the recessive phenotype in the offspring of a testcross indicates that the test individual's genotype is heterozygous.

For each pair of alleles Mendel investigated, he observed phenotypic F_2 ratios of 3:1 (see figure 12.4) and testcross ratios of 1:1, just as his model had predicted. Testcrosses can also be used to determine the genotype of an individual when two genes are involved. Mendel often performed testcrosses to verify the genotypes of dominant-appearing F_2 individuals.

An F_2 individual exhibiting both dominant traits (*A__ B__*) might have any of the following genotypes: *AABB*, *AaBB*, *AABb*, or *AaBb*. By crossing dominant-appearing F_2 individuals with homozygous recessive individuals (that is, *A__ B__* × *aabb*), Mendel was able to determine whether either or both of the traits bred true among the progeny, and so to determine the genotype of the F_2 parent (table 12.2).

Testcrossing is a powerful tool that simplifies genetic analysis. We will use this method of analysis in the next chapter, when we explore genetic mapping.

TABLE 12.2 Dihybrid Testcross

Actual Genotype	Results of Testcross	
	Trait A	**Trait B**
AABB	Trait A breeds true	Trait B breeds true
AaBB	———	Trait B breeds true
AABb	Trait A breeds true	———
AaBb	———	———

Individuals showing the dominant phenotype can be either homozygous dominant or heterozygous. The genotype can be determined using a testcross, which involves crossing the individual of unknown genotype to a homozygous recessive individual. Heterozygous individuals produce both dominant and recessive phenotypes in equal numbers as a result of the testcross.

12.6 Extensions to Mendel

Although Mendel's results did not receive much notice during his lifetime, three different investigators independently rediscovered his pioneering paper in 1900, 16 years after his death. They came across it while searching the literature in preparation for publishing their own findings, which closely resembled those Mendel had presented more than 30 years earlier.

In the decades following the rediscovery of Mendel's ideas, many investigators set out to test them. However, scientists attempting to confirm Mendel's theory often had trouble obtaining the same simple ratios he had reported.

The reason that Mendel's simple ratios were not obtained had to do with the traits that others examined. A number of assumptions are built into Mendel's model that are oversimplifications. These assumptions include that each trait is specified by a single gene with two alternative alleles; that there are no environmental effects; and that gene products act independently. The idea of dominance also hides a wealth of biochemical complexity. In the following sections, you'll see how Mendel's simple ideas can be extended to provide a more complete view of genetics (table 12.3).

In polygenic inheritance, more than one gene can affect a single trait

Often, the relationship between genotype and phenotype is more complicated than a single allele producing a single trait. Most phenotypes also do not reflect simple two-state cases like purple or white flowers.

Consider Mendel's crosses between tall and short pea plants. In reality, the "tall" plants actually have normal height, and the "short" plants are dwarfed by an allele at a single gene. But in most species, including humans, height varies over a continuous range, rather than having discrete values. This continuous distribution of a phenotype has a simple genetic explanation: the action of more than one gene. The mode of inheritance that takes place in this case is often called **polygenic inheritance.**

TABLE 12.3 When Mendel's Laws/Results May Not Be Observed

Genetic Occurrence	Definition	Examples
Polygenic inheritance	More than one gene can affect a single trait.	• Four genes are involved in determining eye color. • Human height
Pleiotropy	A single gene can affect more than one trait.	• A pleiotropic allele dominant for yellow fur in mice is recessive for a lethal developmental defect. • Cystic fibrosis • Sickle cell anemia
Multiple alleles for one gene	Genes may have more than two alleles.	ABO blood types in humans
Dominance is not always complete	• In incomplete dominance the heterozygote is intermediate. • In codominance no single allele is dominant, and the heterozygote shows some aspect of both homozygotes.	• Japanese four o'clocks • Human blood groups
Environmental factors	Genes may be affected by the environment.	Siamese cats
Gene interaction	Products of genes can interact to alter genetic ratios.	• The production of a purple pigment in corn • Coat color in mammals

In reality, few phenotypes result from the action of only one gene. Instead, most characters reflect multiple additive contributions to the phenotype by several genes. When multiple genes act jointly to influence a character, such as height or weight, the character often shows a range of small differences. When these genes segregate independently, a gradation in the degree of difference can be observed when a group consisting of many individuals is examined (figure 12.12). We call this gradation **continuous variation,** and we call such traits **quantitative traits.** The greater the number of genes that influence a character, the more continuous the expected distribution of the versions of that character.

This continuous variation in traits is similar to blending different colors of paint: Combining one part red with seven parts white, for example, produces a much lighter pink shade than does combining five parts red with three parts white. Different ratios of red to white result in a continuum of shades, ranging from pure red to pure white.

Often, variations can be grouped into categories, such as different height ranges. Plotting the numbers in each height category produces a curve called a *histogram,* such as that shown in figure 12.12. The bell-shaped histogram approximates an idealized *normal distribution,* in which the central tendency is characterized by the mean, and the spread of the curve indicates the amount of variation.

Even simple-appearing traits can have this kind of polygenic basis. For example, human eye colors are often described in simple terms with brown dominant to blue, but this is actually incorrect. Extensive analysis indicates that at least four genes are involved in determining eye color. This leads to more complex inheritance patterns than initially reported. For example, blue-eyed parents can have brown-eyed offspring, although it is rare.

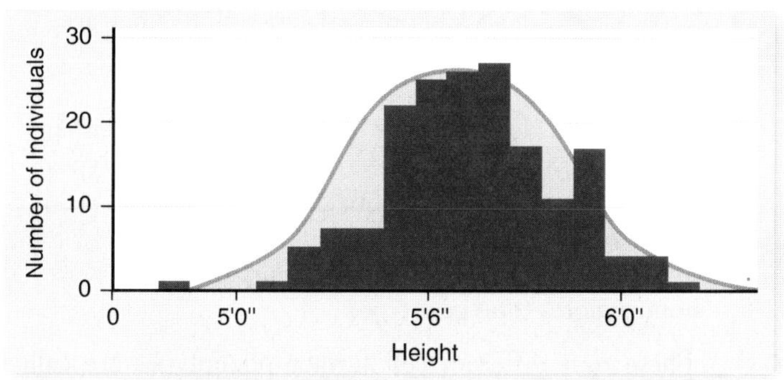

figure 12.12

HEIGHT IS A CONTINUOUSLY VARYING TRAIT. The photo and accompanying graph show variation in height among students of the 1914 class at the Connecticut Agricultural College. Because many genes contribute to height and tend to segregate independently of one another, the cumulative contribution of different combinations of alleles to height forms a *continuous* distribution of possible heights, in which the extremes are much rarer than the intermediate values. Variation can also arise due to environmental factors such as nutrition.

In pleiotropy, a single gene can affect more than one trait

Not only can more than one gene affect a single trait, but a single gene can affect more than one trait. Considering the complexity of biochemical pathways and the interdependent nature of organ systems in multicellular organisms, this should be no surprise.

An allele that has more than one effect on phenotype is said to be **pleiotropic.** The pioneering French geneticist Lucien Cuenot studied yellow fur in mice, a dominant trait, and found he was unable to obtain a pure-breeding yellow strain by crossing individual yellow mice with each other. Individuals homozygous for the yellow allele died, because the yellow allele was pleiotropic: One effect was yellow coat color, but another was a lethal developmental defect.

A pleiotropic allele may be dominant with respect to one phenotypic consequence (yellow fur) and recessive with respect to another (lethal developmental defect). Pleiotropic effects are difficult to predict, because a gene that affects one trait often performs other, unknown functions.

Pleiotropic effects are characteristic of many inherited disorders in humans, including cystic fibrosis and sickle cell anemia (discussed in the following chapter). In these disorders, multiple symptoms (phenotypes) can be traced back to a single gene defect. Cystic fibrosis patients exhibit clogged blood vessels, overly sticky mucus, salty sweat, liver and pancreas failure, and a battery of other symptoms. It is often difficult to deduce the nature of the primary defect from the range of a gene's pleiotropic effects. As it turns out, all these symptoms of cystic fibrosis are pleiotropic effects of a single defect, a mutation in a gene that encodes a chloride ion transmembrane channel.

Genes may have more than two alleles

Mendel always looked at genes with two alternative alleles. Although any diploid individual can carry only two alleles for a gene, there may be more than two alleles in a population. The example of ABO blood types in humans, described later on, involves an allelic series with three alleles.

If you think of a gene as a sequence of nucleotides in a DNA molecule, then the number of possible alleles is huge because even a single nucleotide change could produce a new allele. In reality, the number of alleles possible for any gene is constrained, but usually more than two alleles exist for any gene in an outbreeding population. The dominance relationships of these alleles cannot be predicted, but can be determined by observing the phenotypes for the various heterozygous combinations.

Dominance is not always complete

Mendel's idea of dominant and recessive traits can seem hard to explain in terms of modern biochemistry. For example, if a recessive trait is caused by the loss of function of an enzyme encoded by the recessive allele, then why should a heterozygote, with only half the activity of this enzyme, have the same appearance as a homozygous dominant individual?

The answer is that enzymes usually act in pathways and not alone. These pathways, as you have seen in earlier chapters, can be highly complex in terms of inputs and outputs, and they can sometimes tolerate large reductions in activity of single enzymes in the pathway without reductions in the level of the end-product. When this is the case, complete dominance will be observed; however, not all genes act in this way.

Incomplete dominance

In **incomplete dominance,** the heterozygote is intermediate in appearance between the two homozygotes. For example, in a cross between red- and white-flowering Japanese four o'clocks, described in figure 12.13, all the F_1 offspring have pink flowers—indicating that neither red nor white flower color was dominant. Looking only at the F_1, we might conclude that this is a case of blending inheritance. But when two of the F_1 pink flowers are crossed, they produce red-, pink-, and white-flowered plants in a 1:2:1 ratio. In this case the phenotypic ratio is the same as the genotypic ratio because all three genotypes can be distinguished.

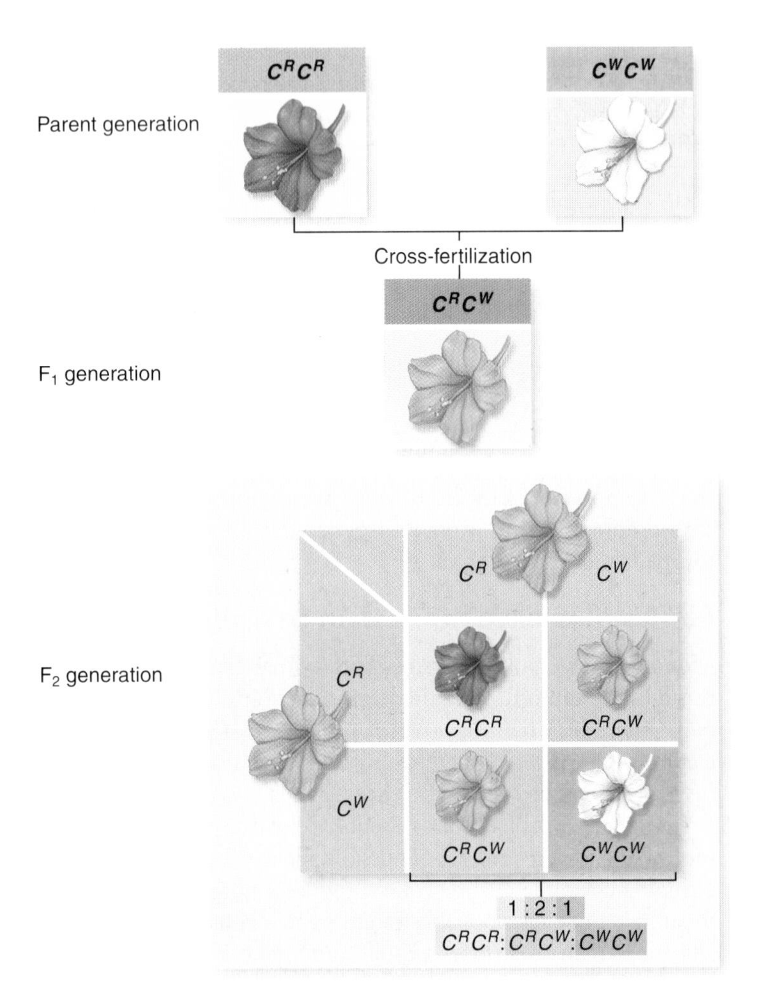

figure 12.13

INCOMPLETE DOMINANCE. In a cross between a red-flowered (genotype C^RC^R) Japanese four o'clock and a white-flowered one (C^WC^W), neither allele is dominant. The heterozygous progeny have pink flowers and the genotype C^RC^W. If two of these heterozygotes are crossed, the phenotypes of their progeny occur in a ratio of 1:2:1 (red:pink:white).

Codominance

Most genes in a population possess several different alleles, and often no single allele is dominant; instead, each allele has its own effect, and the heterozygote shows some aspect of the phenotype of both homozygotes. The alleles are said to be **codominant.**

Codominance can be distinguished from incomplete dominance by the appearance of the heterozygote. In incomplete dominance, the heterozygote is intermediate between the two homozygotes, whereas in codominance, some aspect of both alleles is seen in the heterozygote. One of the clearest human examples is found in the human blood groups.

The different phenotypes of human blood groups are based on the response of the immune system to proteins on the surface of red blood cells. In homozygotes a single type of protein is found on the surface of cells, and in heterozygotes, two kinds of protein are found, leading to codominance.

The human ABO blood group system

The gene that determines ABO blood types encodes an enzyme that adds sugar molecules to proteins on the surface of red blood cells. These sugars act as recognition markers for the immune system (chapter 51). The gene that encodes the enzyme, designated *I*, has three common alleles: I^A, whose product adds galactosamine; I^B, whose product adds galactose; and *i*, which codes for a protein that does not add a sugar.

The three alleles of the *I* gene can be combined to produce six different genotypes. An individual heterozygous for the I^A and I^B alleles produces both forms of the enzyme and exhibits both galactose and galactosamine on red blood cells. Because both alleles are expressed simultaneously in heterozygotes, the I^A and I^B alleles are codominant. Both I^A and I^B are dominant over the *i* allele, because both I^A and I^B alleles lead to sugar addition, whereas the *i* allele does not. The different combinations of the three alleles produce four different phenotypes (figure 12.14):

1. Type A individuals add only galactosamine. They are either I^AI^A homozygotes or I^Ai heterozygotes (two genotypes).
2. Type B individuals add only galactose. They are either I^BI^B homozygotes or I^Bi heterozygotes (two genotypes).
3. Type AB individuals add both sugars and are I^AI^B heterozygotes (one genotype).
4. Type O individuals add neither sugar and are *ii* homozygotes (one genotype).

These four different cell surface phenotypes are called the **ABO blood groups.**

A person's immune system can distinguish among these four phenotypes. If a type A individual receives a transfusion of type B blood, the recipient's immune system recognizes the "foreign" antigen (galactose) and attacks the donated blood cells, causing them to clump, or agglutinate. The same thing would happen if the donated blood is type AB. However, if the donated blood is type O, no immune attack occurs, because there are no galactose antigens.

In general, any individual's immune system will tolerate a transfusion of type O blood, and so type O is termed the "universal donor." Because neither galactose nor galactosamine

	Alleles	Blood Type	Sugars Exhibited	Donates and Receives
	I^AI^A, I^Ai (I^A dominant to i)	A	Galactosamine	Receives A and O Donates to A and AB
	I^BI^B, I^Bi (I^B dominant to i)	B	Galactose	Receives B and O Donates to B and AB
	I^AI^B (codominant)	AB	Both galactose and galactosamine	Universal receiver Donates to AB
	ii (i is recessive)	O	None	Receives O Universal donor

figure 12.14

ABO BLOOD GROUPS ILLUSTRATE BOTH CODOMINANCE AND MULTIPLE ALLELES. There are three alleles of the I gene: I^A, I^B and i. I^A and I^B are both dominant to i (see types A and B), but codominant to each other (see type AB). The genotypes that give rise to each blood type are shown with the associated phenotypes in terms of sugars added to surface proteins and the behavior in blood transfusions.

is foreign to type AB individuals (whose red blood cells have both sugars), those individuals may receive any type of blood, and type AB is termed the "universal recipient." Nevertheless, matching blood is preferable for any transfusion.

Genes may be affected by the environment

Another assumption, implicit in Mendel's work, is that the environment does not affect the relationship between genotype and phenotype. For example, the soil in the abbey yard where Mendel performed his experiments was probably not uniform, and yet its possible effect on the expression of traits was ignored. But in reality, although the expression of genotype produces phenotype, the environment can affect this relationship.

Environmental effects are not limited to the external environment. For example, the alleles of some genes encode heat-sensitive products, that are affected by differences in internal body temperature. The *ch* allele in Himalayan rabbits and Siamese cats encodes a heat-sensitive version of the enzyme tyrosinase, which as you may recall is involved in albinism (figure 12.15). The Ch version of the enzyme is inactivated at temperatures above about 33°C. At the surface of the torso and head of these animals, the temperature is above 33°C and tyrosinase is inactive, producing a whitish coat. At the extremities, such as the tips of the ears and tail, the temperature is usually below 33°C and the enzyme is active allowing production of melanin that turns the coat in these areas a dark color.

inquiry

Many studies of identical twins separated at birth have revealed phenotypic differences in their development (height, weight, etc.). If these are identical twins, can you propose an explanation for these differences?

In epistasis, interactions of genes alter genetic ratios

The last simplifying assumption in Mendel's model is that the products of genes do not interact. But the products of genes may not act independently of one another, and the interconnected behavior of gene products can change the ratio expected by independent assortment, even if the genes are on different chromosomes that do exhibit independent assortment.

Given the interconnected nature of metabolism, it should not come as a surprise that many gene products are not independent. Genes that act in the same metabolic pathway, for example, should show some form of dependence at the level of function. In such cases, the ratio Mendel would predict is not readily observed, but it is still there in an altered form.

Epistasis in corn

In the tests of Mendel's ideas that followed the rediscovery of his work, scientists had trouble obtaining Mendel's simple ratios, particularly with dihybrid crosses. Sometimes, it was not possible to identify successfully each of the four phenotypic classes expected, because two or more of the classes looked alike.

An example of this comes from the analysis of particular varieties of corn, *Zea mays*. Some commercial varieties exhibit a purple pigment called anthocyanin in their seed coats, whereas others do not. In 1918, geneticist R. A. Emerson crossed two true-breeding corn varieties, each lacking anthocyanin pigment. Surprisingly, all of the F_1 plants produced purple seeds.

When two of these pigment-producing F_1 plants were crossed to produce an F_2 generation, 56% were pigment producers and 44% were not. This is clearly not the Mendelian expectation. Emerson correctly deduced that two genes were involved in producing pigment, and that the second cross had thus been a dihybrid cross. According to Mendel's theory, gametes in a dihybrid cross could combine in 16 equally possible ways—so the puzzle was to figure out how these 16 combinations could occur in the two phenotypic groups of progeny. Emerson multiplied the fraction that were pigment producers (0.56) by 16 to obtain 9, and multiplied the fraction that lacked pigment (0.44) by 16 to obtain 7. Emerson therefore had a **modified ratio** of 9:7 instead of the usual 9:3:3:1 ratio (figure 12.16).

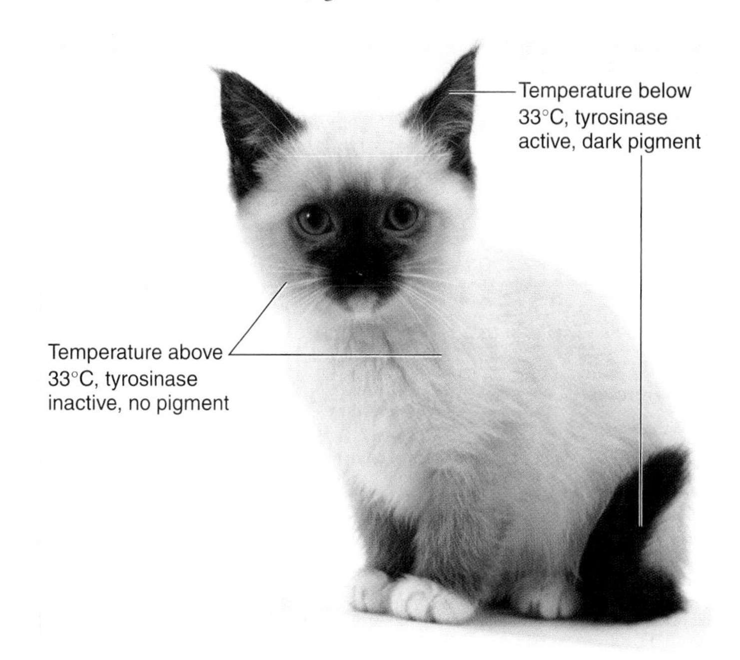

figure 12.15

SIAMESE CAT. The pattern of coat color is due to an allele that encodes a temperature-sensitive form of the enzyme tyrosinase.

This modified ratio is easily rationalized by considering the function of the products encoded by these genes. When gene products act sequentially, as in a biochemical pathway, an allele expressed as a defective enzyme early in the pathway blocks the flow of material through the rest of the pathway. In this case, it is impossible to judge whether the later steps of the pathway are functioning properly. This type of gene interaction, in which one gene can interfere with the expression of another, is the basis of the phenomenon called **epistasis.**

The pigment anthocyanin is the product of a two-step biochemical pathway:

$$\underset{\text{(colorless)}}{\textbf{starting molecule}} \xrightarrow{\text{enzyme 1}} \underset{\text{(colorless)}}{\textbf{intermediate}} \xrightarrow{\text{enzyme 2}} \underset{\text{(purple)}}{\textbf{anthocyanin}}$$

To produce pigment, a plant must possess at least one functional copy of each enzyme's gene. The dominant alleles encode functional enzymes, and the recessive alleles encode nonfunctional enzymes. Of the 16 genotypes predicted by random assortment, 9 contain at least one dominant allele of both genes; they therefore produce purple progeny. The remaining 7 genotypes lack dominant alleles at *either or both* loci (3 + 3 + 1 = 7) and so produce colorless progeny, giving the phenotypic ratio of 9:7 that Emerson observed (see figure 12.16).

You can see that although this ratio is not the expected dihybrid ratio, it is a modification of the expected ratio.

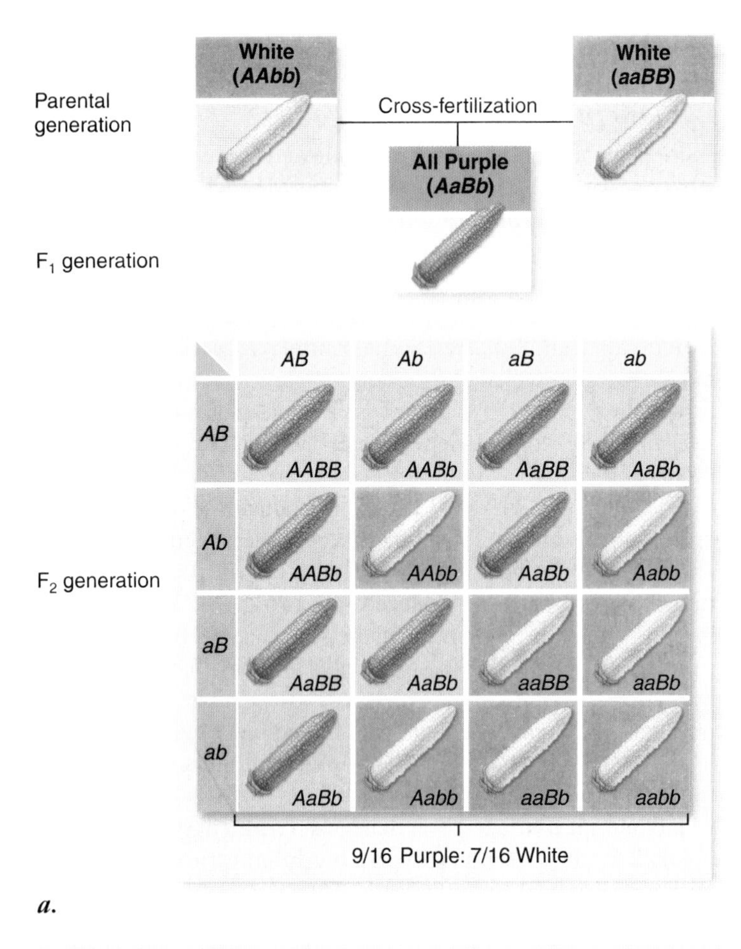

a.

b.

figure 12.16

HOW EPISTASIS AFFECTS GRAIN COLOR. ***a.*** Crossing some white varieties of corn yields an all purple F_1. If the white kernels were due to a recessive allele for a single gene we would expect white offspring. Self-crossing the F_1 yields 9 purple:7 white. This can be explained by the presence of two genes, each encoding an enzyme necessary for pigment production. Unless both enzymes are active (the plant has a dominant allele for each of the two genes, *A_B_*), no pigment is expressed. ***b.*** The biochemical pathway for pigment production with enzymes encoded by A and B genes.

Epistasis in Labrador retrievers

In many animals, coat color is the result of epistatic interactions among genes. Coat color in Labrador retrievers, a breed of dog, is due primarily to the interaction of two genes. The *E* gene determines whether a dark pigment, eumelanin, will be deposited in the fur. A dog having the genotype *ee* has no dark pigment deposited, and its fur is yellow. A dog having the genotype *EE* or *Ee* (*E*__) does have dark pigment deposited in the fur.

A second gene, the *B* gene, determines how dark the pigment will be. This gene controls the distribution of melanosomes in a hair. Dogs with the genotype *E__bb* have brown fur and are called chocolate labs. Dogs with the genotype *E__B__* have black fur.

Even in yellow dogs, however, the *B* gene does have some effect. Yellow dogs with the genotype *eebb* exhibit brown pigment on their nose, lips, and eye rims, but yellow dogs with the genotype *eeB__* have black pigment in these areas.

Mendel's model is correct, but incomplete. Some traits are produced by the action of multiple genes (polygenic inheritance), and one gene can affect more than one trait (pleiotropy). Genes may have more than two alleles that may not show simple dominance. In incomplete dominance, the heterozygote is intermediate between the two homozygotes, and in codominance the heterozygote shows aspects of both homozygotes. The action of genes is also not always independent. This can lead to modified dihybrid ratios although the alleles of each gene are segregating independently. In epistasis, the action of one gene obscures the action of other genes.

concept review

12.1 The Mystery of Heredity

Our understanding of inheritance is the result of the scientific observations and Mendel's pea hybridization research.

- Traits, or characters, are transmitted directly to offspring, but they do not necessarily blend.
- Inherited characters can disappear in one generation only to reappear later, that is, the traits segregate among the offspring of a cross.
- Some traits are observed more often in the offspring of crosses.
- Mendel's experiments involved reciprocal crosses between pure-breeding pea varieties followed by one or more generations of self-fertilization.
- Mendel's mathematic analysis of experimental results lead to the present model of inheritance.

12.2 Monohybrid Crosses: The Principle of Segregation (figure 12.6)

A monohybrid cross follows only two forms of a single trait.

- Traits are determined by discrete factors we now call genes.
- Alleles are alternative forms of a gene that produce alternative forms of a trait.
- A genotype refers to the total set of alleles possessed by an individual.
- A phenotype refers to the physical appearance or other observable characteristic of the individual that is the result of the genotype's expression.
- The offspring of a parental cross (P) are the first filial generation (F_1).
- In crosses between pure-breeding parents the dominant trait is expressed and the alternative or recessive trait is not expressed until the F_2 generation.
- In the F_2 generation, the Mendelian ratio is expressed as 75% dominant to 25% recessive; also expressed as a 3:1 ratio.
- The F_2 generation disguises a 1:2:1 ratio in which 1/4 are true-breeding dominants, 2/4 (1/2) are not true-breeding and 1/4 are true-breeding recessives.
- The Principle of Segregation states that alternative alleles for a gene segregate during gamete formation and are randomly rejoined during fertilization.
- A homozygous individual carries two alleles of a gene that are the same.
- A heterozygous individual carries two alleles of a gene that are different.
- A trait determined by a dominant allele will be seen in both the homozygous dominant and the heterozygote.
- A trait determined by a recessive allele will only be seen in the homozygous recessive.
- The results of Mendelian crosses can be predicted with a Punnett square or by probability theory (figure 12.7).
- Human inheritance is studied using family pedigrees.

12.3 Dihybrid Crosses: The Principle of Independent Assortment (figure 12.10)

During meiosis, the segregation of different pairs of alleles is independent of each other.

- A dihybrid cross follows the behavior of two different traits during a single cross.
- During a dihybrid cross the F_1 generation displays only two of four possible combination of traits with no blending.
- The F_2 generation of a dihybrid cross exhibits all four possible combinations of traits in a 9:3:3:1 ratio.
- The Principle of Independent Assortment states that the alleles of each gene assort independently.

12.4 Probability: Predicting the Behavior of Crosses

Because the behavior of different chromosomes during meiosis is independent we can use probability to predict the outcome of crosses.

- Two rules of probability help predict genotypes and phenotypes from monohybrid cross results.
- The rule of addition states that the probability of two mutually exclusive events occuring is the **sum** of the individual probabilities.
- The rule of multiplication states that the probability of two independent events both occurring is the **product** of individual probabilities.
- Dihybrid cross probabilities are based on monohybrid cross probabilities using the product rule.

12.5 The Testcross: Revealing Unknown Genotypes (figure 12.11)

In a testcross, an unknown genotype is crossed with a homozygous recessive genotype.

- If the unknown genotype is homozygous dominant, the F_1 offspring will be the same.
- If the unknown genotype is heterozygous, the F_1 offspring will exhibit a 1:1 ratio.
- The results of a testcross support the Principle of Segregation.

12.6 Extensions to Mendel

In subsequent research scientists concluded that Mendel's basic model is correct, but it is incomplete and makes assumptions that are not valid.

- In polygenic inheritance more than one gene contributes to a phenotype.
- Many complex traits are due to multiple additive contributions by several genes, resulting in a continuous variation of quantitative traits.
- A pleiotropic effect occurs when an allele affects more than one trait and their effects are difficult to predict.
- Genes may have more than two (multiple), alleles.
- Incomplete dominance occurs when the heterozygous condition exhibits an intermediate phenotype resulting in a 1:2:1 ratio (figure 12.13).
- Codominant alleles each exhibit their own effect on the phenotype because one allele is not dominant over the other.
- The environment may affect the expression of a genotype, resulting in different phenotypes.
- In epistasis genes interact, and one gene interferes with the expression of a second.

review questions

SELF TEST

1. A true-breeding plant is one that—
   a. produces offspring that are different from the parent
   b. forms hybrid offspring through cross pollination
   c. produces offspring that are always the same as the parent
   d. can only reproduce with itself
2. What property distinguished Mendel's investigation from previous studies?
   a. Mendel used true-breeding pea plants.
   b. Mendel quantified his results.
   c. Mendel examined many different traits.
   d. Mendel examined the segregation of traits.
3. A monohybrid cross—
   a. is the same as self-fertilization
   b. examines a single variant of a trait
   c. produces a single offspring
   d. examines two variants of a single trait
4. What was the appearance of the F_1 generation of a monohybrid cross of purple (*PP*) and white (*pp*) flower pea plants?
   a. All the F_1 plants had white flowers.
   b. The F_1 plants had a light purple or blended appearance.
   c. All the F_1 plants had purple flowers.
   d. The most of the F_1 (3/4) had purple flowers, but 1/4 of the plants had white.
5. The F_1 plants from the previous question are allowed to self-fertilize. What will the phenotypic ratio be for this F_2?
   a. All purple
   b. 1 purple:1 white
   c. 3 purple:1 white
   d. 3 white:1 purple
6. Which of the following is *not* a part of Mendel's five-element model?
   a. Traits have alternative forms (what we now call alleles).
   b. Parents transmit discrete traits to their offspring.
   c. If an allele is present it will be expressed.
   d. Traits do not blend.
7. A *heterozygous* individual is one that carries—
   a. two completely different sets of genes
   b. two identical alleles for a particular gene
   c. only one functional allele
   d. two different alleles for a given gene
8. An organism's __________ is determined by its ___________
   a. genotype; phenotype
   b. phenotype; genotype
   c. alleles; phenotype
   d. F_1 generation; alleles
9. Which of the following represent the phenotype for the recessive human trait, *albinism*?
   a. Absence of the pigment melanin
   b. Presence of a nonfunctional allele for the enzyme tyrosinase
   c. Absence of the enzyme tyrosinase from the individual's cells
   d. Both a and c
10. A dihybrid cross between a plant with long smooth leaves and a plant with short hairy leaves produces a long smooth F_1. If this F_1 is allowed to self-cross to produce an F_2, what would you predict for the ratio of F_2 phenotypes?
    a. 9 long smooth:3 long hairy:3 short hairy:1 short smooth
    b. 9 long smooth:3 long hairy:3 short smooth:1 short hairy
    c. 9 short hairy:3 long hairy:3 short smooth:1 long smooth
    d. 1 long smooth:1 long hairy:1 short smooth:1 short hairy
11. A testcross is used to determine if an individual is—
    a. homozygous dominant or heterozygous
    b. homozygous recessive or homozygous dominant
    c. heterozygous or homozygous recessive
    d. true-breeding
12. What is a polygenic trait?
    a. A set of multiple phenotypes determined by a single gene
    b. A single phenotypic trait determined by two alleles
    c. A single phenotypic trait determined by more than one gene
    d. The collection of traits possessed by an individual
13. When a single gene influences multiple phenotypic traits the effect is called—
    a. Codominance
    b. Epistasis
    c. Incomplete dominance
    d. Pleiotropy
14. What is the probability of obtaining an individual with the genotype *bb* from a cross between two individuals with the genotype *Bb*?
    a. 1/2
    b. 1/4
    c. 1/8
    d. 0
15. What is the probability of obtaining an individual with the genotype *CC* from a cross between two individuals with the genotypes *CC* and *Cc*?
    a. 1/2
    b. 1/4
    c. 1/8
    d. 1/16

CHALLENGE QUESTIONS

1. Create a Punnett square for the following crosses and use this to predict phenotypic ratio for dominant and recessive traits. Dominant alleles are indicated by uppercase letters and recessive are indicated by lowercase letters.
   a. A monohybrid cross between individuals with the genotype *Aa* and *Aa*
   b. A dihybrid cross between two individuals with the genotype *AaBb*
   c. A dihybrid cross between individuals with the genotype *AaBb* and *aabb*
2. Use probability to predict the following.
   a. What is the probability of obtaining an individual with the genotype *rr* from the self-fertilization of a plant with the genotype *Rr*?
   b. What is the probability that a testcross with a heterozygous individual will produce homozygous recessive offspring?
   c. A plant with the genotype *Gg* is self-fertilized. Use probability to determine the proportion of the offspring that will have the dominant phenotype.
   d. Use probability to determine the proportion of offspring from a dihybrid cross (*GgRr* × *GgRr*) that will have the phenotype *ggR_*.

Do you need additional review? *Visit* www.ravenbiology.com *for practice quizzes, animations, videos, and activities designed to help you master the material in this chapter.*

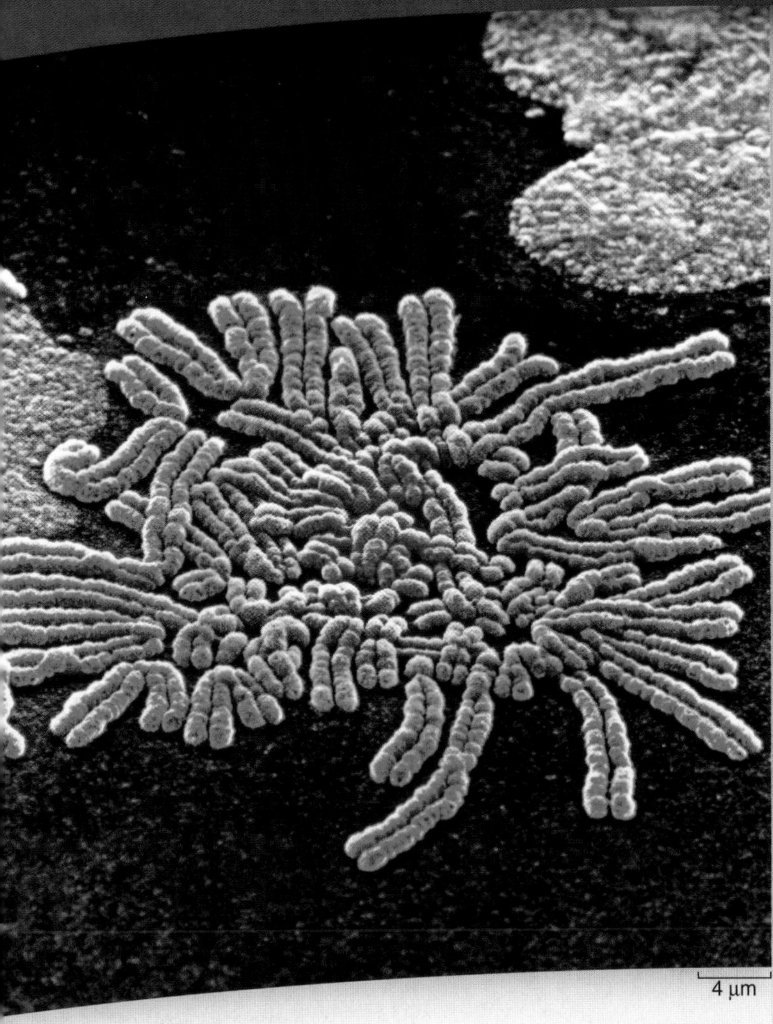

chapter 13

Chromosomes, Mapping, and the Meiosis–Inheritance Connection

introduction

MENDEL'S EXPERIMENTS OPENED the door to understanding inheritance, but many questions remained. In the early part of the 20th century, we did not know the nature of the factors whose behavior Mendel had described. The next step, which involved many researchers in the early part of the century, was uniting information about the behavior of chromosomes, seen in the picture, and the inheritance of traits. The basis for Mendel's principles of segregation and independent assortment lie in events that occur during meiosis.

The behavior of chromosomes during meiosis not only explains Mendel's principles, but leads to new and different approaches to the study of heredity. The ability to construct genetic maps is one of the most powerful tools of classical genetic analysis. The tools of genetic mapping developed in flies and other organisms in combination with information from the human genome project now allow us to determine the location and isolate genes that are involved in genetic diseases.

concept outline

13.1 Sex Linkage and the Chromosomal Theory of Inheritance

A central role for chromosomes in heredity was first suggested in 1900 by the German geneticist Carl Correns, in one of the papers announcing the rediscovery of Mendel's work. Soon after, observations that similar chromosomes paired with one another during meiosis led directly to the **chromosomal theory of inheritance,** first formulated by the American Walter Sutton in 1902.

Morgan correlated the inheritance of a trait with sex chromosomes

In 1910, Thomas Hunt Morgan, studying the fruit fly *Drosophila melanogaster*, discovered a mutant male fly with white eyes instead of red (figure 13.1).

Morgan immediately set out to determine whether this new trait would be inherited in a Mendelian fashion. He first crossed the mutant male to a normal red-eyed female to see whether the red-eyed or white-eyed trait was dominant. All of the F_1 progeny had red eyes, so Morgan concluded that red eye color was dominant over white.

The F_1 cross

Following the experimental procedure that Mendel had established long ago, Morgan then crossed the red-eyed flies from the F_1 generation with each other. Of the 4252 F_2 progeny Morgan examined, 782 (18%) had white eyes. Although the ratio of red eyes to white eyes in the F_2 progeny was greater than 3:1, the results of the cross nevertheless provided clear evidence that eye color segregates. However, something about the outcome was strange and totally unpredicted by Mendel's theory—*all of the white-eyed F_2 flies were males!* (figure 13.2)

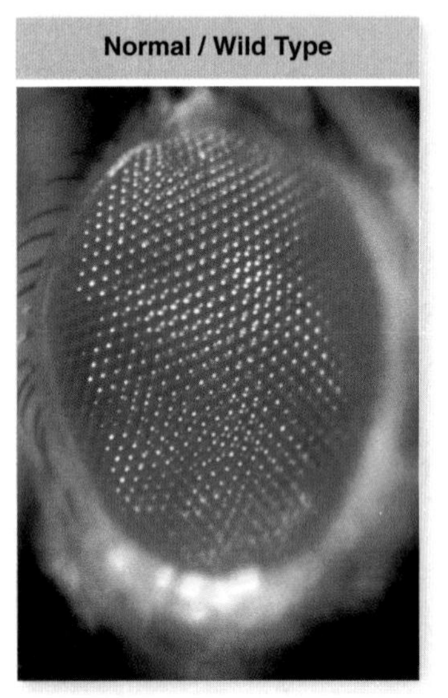

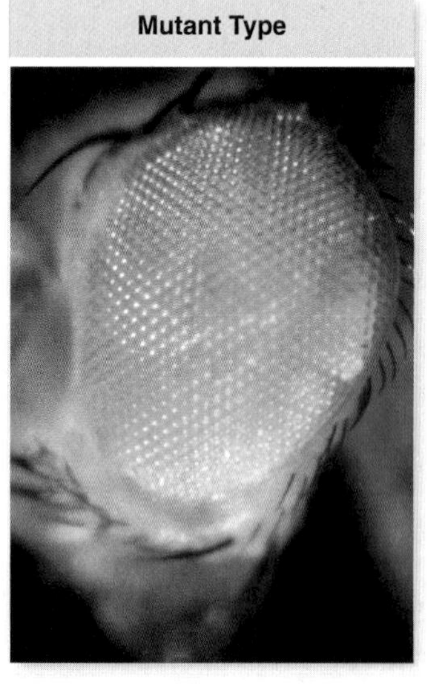

figure 13.1

RED-EYED (NORMAL) AND WHITE-EYED (MUTANT) *Drosophila*. Mutations are heritable alterations in genetic material. By studying the inheritance pattern of white and red alleles (located on the X chromosome), Morgan first demonstrated that genes are on chromosomes.

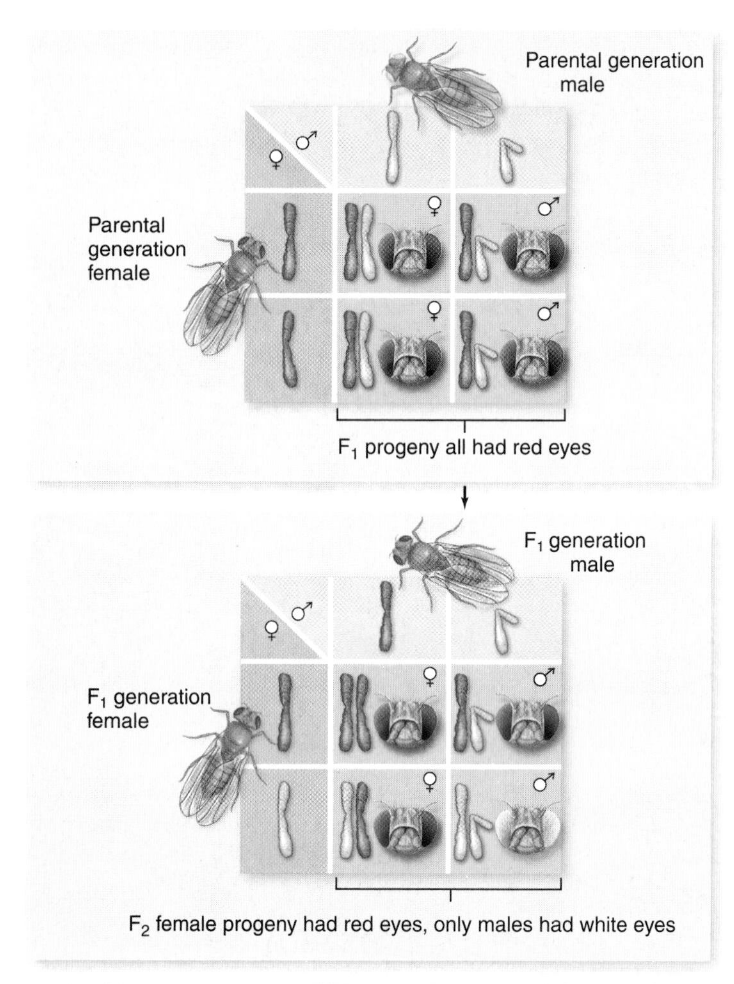

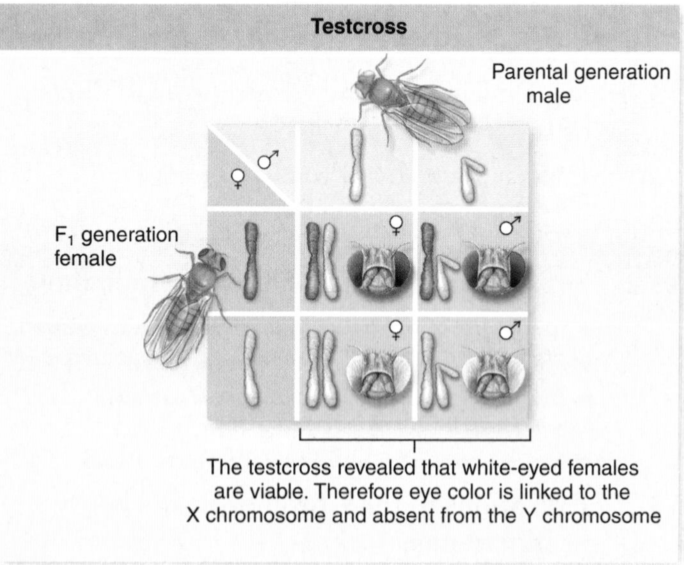

figure 13.2

THE CHROMOSOMAL BASIS OF SEX LINKAGE. White-eyed male flies are crossed to red-eyed females. The F_1 flies all have red eyes, as expected for a recessive white-eye allele. In the F_2, all of the white-eyed flies are male because the Y chromosome lacks the white gene. Inheritance of the sex chromosomes correlates with eye color, showing the *white* gene is on the X chromosome.

The testcross

Morgan sought an explanation for this result. One possibility was simply that white-eyed female flies don't exist; such individuals might not be viable for some unknown reason. To test this idea, Morgan testcrossed the female F_1 progeny with the original white-eyed male. He obtained white-eyed and red-eyed flies of both sexes in a 1:1:1:1 ratio, just as Mendel's theory had predicted. Therefore, white-eyed female flies are viable. Given that white-eyed females can exist, Morgan turned to the nature of the chromosomes in males and females for an explanation.

The gene for eye color lies on the X chromosome

In *Drosophila*, the sex of an individual is determined by the number of copies it has of a particular chromosome, the **X chromosome.** Observations of *Drosophila* chromosomes revealed that female flies have two X chromosomes, but male flies have only one. In males, the single X chromosome pairs in meiosis with a dissimilar partner called the **Y chromosome.** These two chromosomes are termed **sex chromosomes** because of their association with sex.

During meiosis, a female produces only X-bearing gametes, but a male produces both X-bearing and Y-bearing gametes. When fertilization involves an X sperm, the result is an XX zygote, which develops into a female; when fertilization involves a Y sperm, the result is an XY zygote, which develops into a male.

The solution to Morgan's puzzle is that the gene causing the white-eye trait in *Drosophila* resides only on the X chromosome—it is absent from the Y chromosome. (We now know that the Y chromosome in flies carries almost no functional genes.) A trait determined by a gene on the X chromosome is said to be **sex-linked,** or X-linked, because it is associated with the sex of the individual. Knowing the white-eye trait is recessive to the red-eye trait, we can now see that Morgan's result was a natural consequence of the Mendelian segregation of chromosomes (figure 13.2).

Morgan's experiment was one of the most important in the history of genetics because it presented the first clear evidence that the genes determining Mendelian traits do indeed reside on the chromosomes, as Sutton had proposed. Mendelian traits segregate in genetic crosses because homologues separate during gamete formation.

Walter Sutton proposed the chromosomal theory of inheritance, which states that hereditary traits are carried on chromosomes. Thomas Hunt Morgan's discovery of the trait for white eyes in *Drosophila* allowed the association of traits with chromosomes because the white-eye allele was clearly shown to reside on the X chromosome.

13.2 Sex Chromosomes and Sex Determination

The structure and number of sex chromosomes vary in different species (table 13.1). In the fruit fly, *Drosophila*, females are XX and males XY, as in humans and other mammals. However, in birds, the male has two Z chromosomes, and the female has a Z and a W chromosome. Some insects, such as grasshoppers, have no Y chromosome—females are XX and males are characterized as XO (the O indicating the absence of a chromosome).

TABLE 13.1 Sex Determination in Some Organisms

	Female	Male
Humans, *Drosophila*	XX	XY
Birds	ZW	ZZ
Grasshoppers	XX	XO
Honeybees	Diploid	Haploid

In humans, the Y chromosome generally determines maleness

In chapter 10, you learned that humans have 46 chromosomes, or 23 pairs. Twenty-two of these pairs are perfectly matched in both males and females and are called **autosomes.** The remaining pair are the sex chromosomes: XX in females, and XY in males.

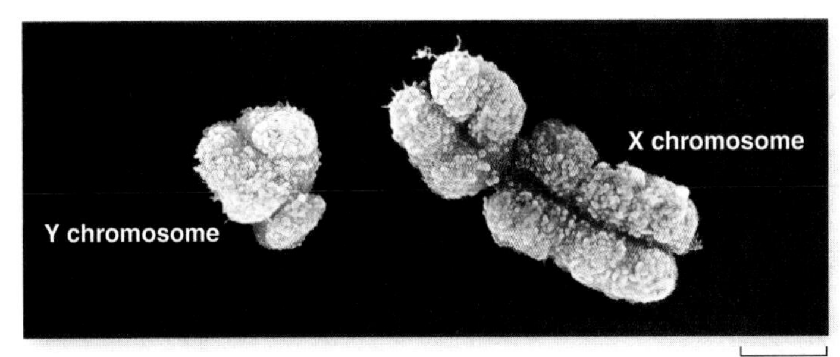

The Y chromosome in males is highly condensed. Because few genes on the Y chromosome are expressed, recessive alleles on a male's single X chromosome have no *active* counterpart on the Y chromosome.

The "default" setting in human embryonic development is for production of a female. Some of the active genes on the Y chromosome, notably the *SRY* gene, are responsible for the masculinization of genitalia and secondary sex organs, producing features associated with "maleness" in humans. Consequently, any individual with *at least one Y chromosome* is normally a male.

The exceptions to this rule actually provide support for this mechanism of sex determination. For example, movement of part of the Y chromosome to the X chromosome can cause otherwise XX individuals to develop as male. There is also a genetic disorder that causes a failure to respond to the androgen hormones (androgen insensitivity syndrome) that causes XY individuals to develop as female. Lastly, mutations in *SRY* itself can cause XY individuals to develop as females.

This form of sex determination seen in humans is shared among mammals, but is not universal in vertebrates. Among fishes and some species of reptiles, environmental factors can cause changes in the expression of this sex-determining gene, and thus in the sex of the adult individual.

Some human genetic disorders display sex linkage

From ancient times, people have noted conditions that seem to affect males to a greater degree than females. Red-green color blindness is one well-known condition that is more common in males because the gene affected is carried on the X chromosome.

Another example is **hemophilia,** a disease that affects a single protein in a cascade of proteins involved in the formation of blood clots. Thus, in an untreated hemophiliac, even minor cuts will not stop bleeding. This form of hemophilia is caused by an X-linked recessive allele; women who are heterozygous for the allele are asymptomatic carriers, and men who receive an X chromosome with the recessive allele exhibit the disease.

The allele for hemophilia was introduced into a number of different European royal families by Queen Victoria of England. Because these families kept careful genealogical records, we have an extensive pedigree for this condition. In the five generations after Victoria, ten of her male descendants have had hemophilia as shown in the pedigree in figure 13.3.

The Russian house of Romanov inherited this condition through Alexandra Feodorovna, a granddaughter of Queen Victoria. She married Czar Nicholas II, and their only son, Alexis,

figure 13.3

THE ROYAL HEMOPHILIA PEDIGREE. Queen Victoria, shown at the bottom center of the photo, was a carrier for hemophilia. Two of Victoria's four daughters, Alice and Beatrice, inherited the hemophilia allele from Victoria. Two of Alice's daughters are standing behind Victoria (wearing feathered boas): Princess Irene of Prussia (*right*) and Alexandra (*left*), who would soon become czarina of Russia. Both Irene and Alexandra were also carriers of hemophilia. From the pedigree, it is clear that Alice introduced hemophilia into the Russian and Prussian royal houses, and Victoria's daughter Beatrice introduced it into the Spanish royal house. Victoria's son Leopold, himself a victim, also transmitted the disorder in a third line of descent. Half-shaded symbols represent carriers with one normal allele and one defective allele; fully shaded symbols represent affected individuals.

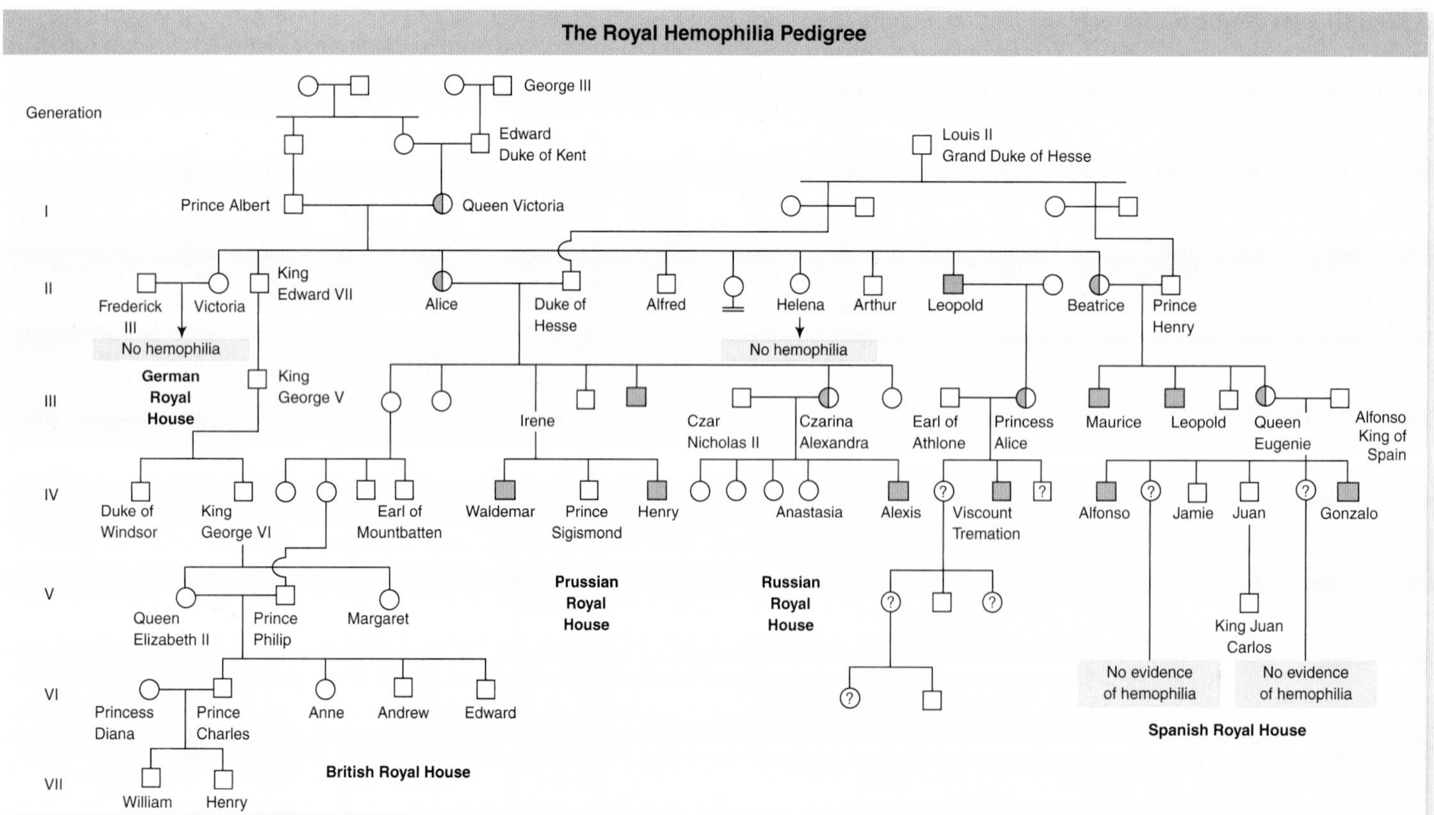

was afflicted with the disease. The entire family was executed during the Russian revolution. (Recently, a woman who had long claimed to be Anastasia, a surviving daughter, was shown not to be a Romanov using modern genetic techniques to test her remains.)

Ironically, this condition has not affected the current British royal family, because Victoria's son Edward, who became King Edward VII, did not receive the hemophilia allele. All of the subsequent rulers of England are his descendants.

Dosage compensation prevents doubling of sex-linked gene products

Although males have only one copy of the X chromosome and females have two, female cells do not produce twice as much of the proteins encoded by genes on the X chromosome. Instead, one of the X chromosomes in females is inactivated early in embryonic development, shortly after the embryo's sex is determined. This inactivation is an example of **dosage compensation,** which ensures an equal level of expression from the sex chromosomes despite a differing number of sex chromosomes in males and females. (In *Drosophila*, by contrast, dosage compensation is achieved by increasing the level of expression on the male X chromosome.)

Which X chromosome is inactivated in females varies randomly from cell to cell. If a woman is heterozygous for a sex-linked trait, some of her cells will express one allele and some the other. The inactivated and highly condensed X chromosome is visible as a darkly staining **Barr body** attached to the nuclear membrane.

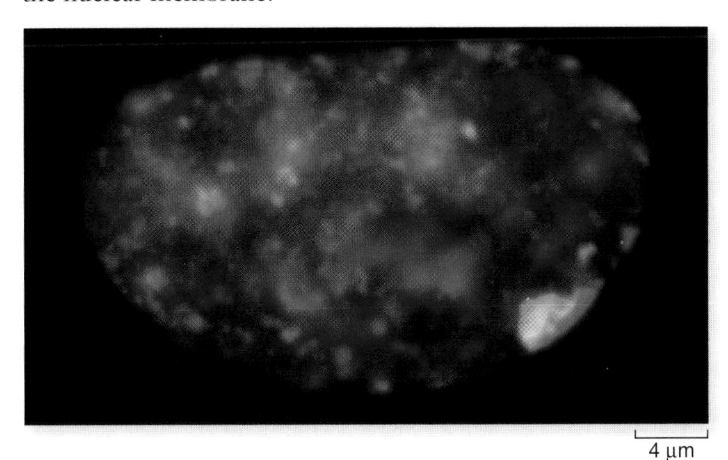

X chromosome inactivation can lead to genetic mosaics

X chromosome inactivation to produce dosage compensation is not unique to humans but is true of all mammals. Females that are heterozygous for X chromosome alleles are **genetic mosaics:** Their individual cells may express different alleles, depending on which chromosome is inactivated.

One example is the calico cat, a female that has a patchy distribution of dark fur, orange fur, and white fur (figure 13.4). The dark fur and orange fur are due to heterozygosity for a gene on the X chromosome that determines pigment type. One allele results in dark fur, and another allele results in orange fur. Which of these colors is observed in any particular patch is due to inactivation of one X chromosome: If the chromosome containing the orange allele is inactivated, then the fur will be dark, and vice versa.

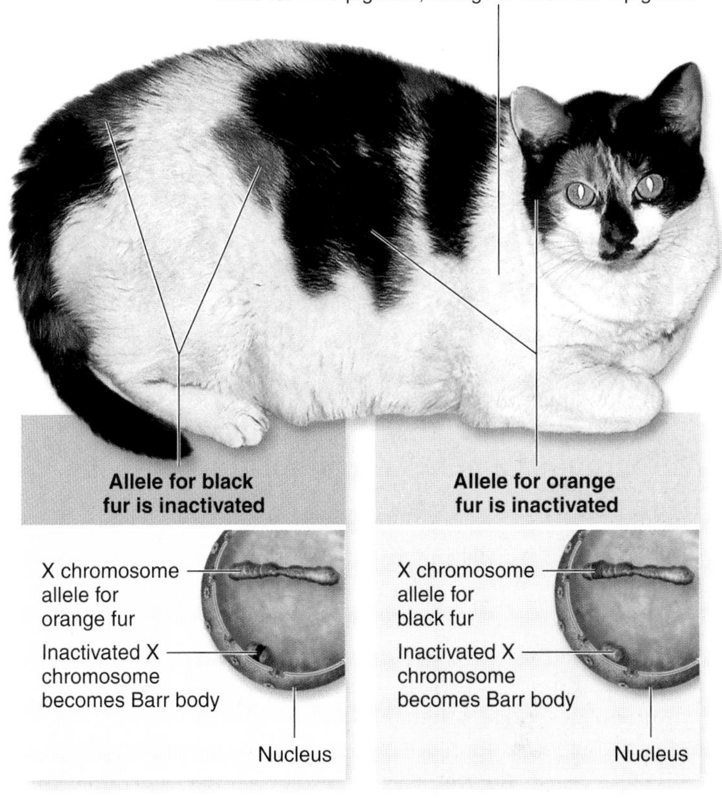

figure 13.4

A CALICO CAT. The cat is heterozygous for alleles of a coat color gene that produce either black fur or orange fur. This gene is on the X chromosome, so the different-colored fur is due to inactivation of one X chromosome. The patchy distribution and white color is due to a second gene that is epistatic to the coat color gene and thus masks its effects.

The patchy distribution of color, and the presence of white fur, is due to a second gene that is epistatic to the fur color gene (chapter 12). That is, the presence of this second gene produces a patchy distribution of pigment, with some areas totally lacking pigment. In the areas that lack pigment, the effect of either fur color allele is masked. Thus, in this one animal we can see an excellent example of both epistasis and X inactivation.

Not all organisms have the same sex chromosomes, but all have some difference in chromosomes between the sexes. In humans, male development depends on the presence of the Y chromosome. XY males will show recessive traits for alleles on the X chromosome, leading to sex-linked inheritance. Mammalian females inactivate one X chromosome to balance the levels of gene expression in males and females. This random inactivation can lead to genetic mosaics in females heterozygous for X chromosome genes.

13.3 Exceptions to the Chromosomal Theory of Inheritance

Although the chromosomal theory explains most inheritance, there are exceptions. Primarily, these are due to the presence of DNA in organelle genomes, specifically in mitochondria and chloroplasts. Non-Mendelian inheritance via organelles was studied in depth by Ruth Sager, who in the face of universal skepticism constructed the first map of chloroplast genes in *Chlamydomonas*, a unicellular green alga, in the 1960s and 1970s.

Mitochondria and chloroplasts are not partitioned with the nuclear genome by the process of meiosis. Thus any trait that is due to the action of genes in these organelles will not show Mendelian inheritance.

Mitochondrial genes are inherited from the female parent

Organelles are usually inherited from only one parent, generally the mother. When a zygote is formed, it receives an equal contribution of the nuclear genome from each parent, but it gets all of its mitochondria from the egg cell, which contains a great deal more cytoplasm (and thus the organelles). As the zygote divides, these original mitochondria divide as well, and are partitioned randomly.

As a result, the mitochondria in every cell of an adult organism can be traced back to the original maternal mitochondria present in the egg. This mode of uniparental (one-parent) inheritance from the mother is called **maternal inheritance.**

In humans, the disease Leber's hereditary optic neuropathy (LHON) shows maternal inheritance. The genetic basis of this disease is a mutant allele for a subunit of NADH dehydrogenase. The mutant allele reduces the efficiency of electron flow in the electron transport chain in mitochondria (see chapter 7), in turn reducing overall ATP production. Some nerve cells in the optic system are particularly sensitive to reduction in ATP production, resulting in neural degeneration.

A mother with this disease will pass it on to all of her progeny, whereas a father with the disease will not pass it on to any of his progeny. Note that this condition differs from sex-linked inheritance because males and females are equally affected.

Chloroplast genes may also be passed on uniparentally

The inheritance pattern of chloroplasts is also usually maternal, although both paternal and biparental inheritance of chloroplasts is also observed depending on the species. Carl Correns first hypothesized in 1909 that chloroplasts were responsible for inheritance of variegation (mixed green and white leaves) in the plant commonly known as the four o'clock (*Mirabilis jalapa*). The offspring exhibited the phenotype of the female parent, regardless of the male's phenotype.

In Sager's work on *Chlamydomonas*, resistance to the antibiotic streptomycin was shown to be transmitted via the DNA of chloroplasts from only one of the organism's two mating types (termed plus and minus).

Organelles such as mitochondria and chloroplasts contain their own genomes. These organelles divide independently of the nucleus, and they are carried in the cytoplasm of the egg cell. Inheritance of traits in these genomes are therefore said to be maternally inherited. In some species, however, chloroplasts may also be passed on paternally or biparentally.

13.4 Genetic Mapping

We have seen that Mendelian characters are determined by genes located on chromosomes and that the independent assortment of Mendelian traits reflects the independent assortment of chromosomes in meiosis. This is fine as far as it goes, but it is still incomplete. Of Mendel's seven traits in figure 12.4, six are on different chromosomes and two are on the same chromosome, yet all show independent assortment with one another. The two on the same chromosome should not behave the same as those that are on different chromosomes. In fact, organisms will generally have many more genes that assort independently than the number of chromosomes. This means that independent assortment cannot be due only to the random alignment of chromosomes during meiosis.

inquiry

Mendel did not examine plant height and pod shape in his dihybrid crosses. The genes for these traits are very close together on the same chromosome. How would this have changed Mendel's results?

The solution to this problem is found in an observation that was introduced in chapter 11: the crossing over of homologues during meiosis. In prophase I of meiosis, homologues appear to physically exchange material by crossing over (figure 13.5). In chapter 11, you saw how this was part of the mechanism that allows homologues, and not sister chromatids, to disjoin at anaphase I.

Genetic recombination exchanges alleles on homologues

Consider a dihybrid cross performed using the Mendelian framework. Two true-breeding parents that each differ with respect to two traits are crossed, producing doubly heterozygous F_1 progeny. If the genes for the two traits are on a single chromosome, then during meiosis we would expect alleles for both loci to segregate together and produce only gametes that resemble the two parental types. But if a crossover occurs between the two loci, then each homologue would carry one allele from each parent and produce gametes that combine these parental traits (see figure 13.5). We call gametes with this new combination of alleles *recombinant* gametes as they are formed by recombining the parental alleles.

The first investigator to provide evidence for this was Morgan, who studied three genes on the X chromosome of *Drosophila*. He found an excess of parental types, which he explained as due to the genes all being on the X chromosome and therefore coinherited (inherited together). He went further suggesting that the recombinant genotypes were due to crossing over between homologues during meiosis.

Experiments performed independently by Barbara McClintock and Harriet Creighton in maize and by Curt Stern in *Drosophila*, provided evidence for this physical exchange of genetic material. The experiment done by Creighton and McClintock is detailed in figure 13.6. In this experiment, they used a chromosome with two alterations visible under a microscope: a knob on one end of the chromosome and a part of a different chromosome attached to the other end. In addition to these cytological markers, this chromosome also carried two genetic markers: a gene that determines kernel color and one that determines kernel texture.

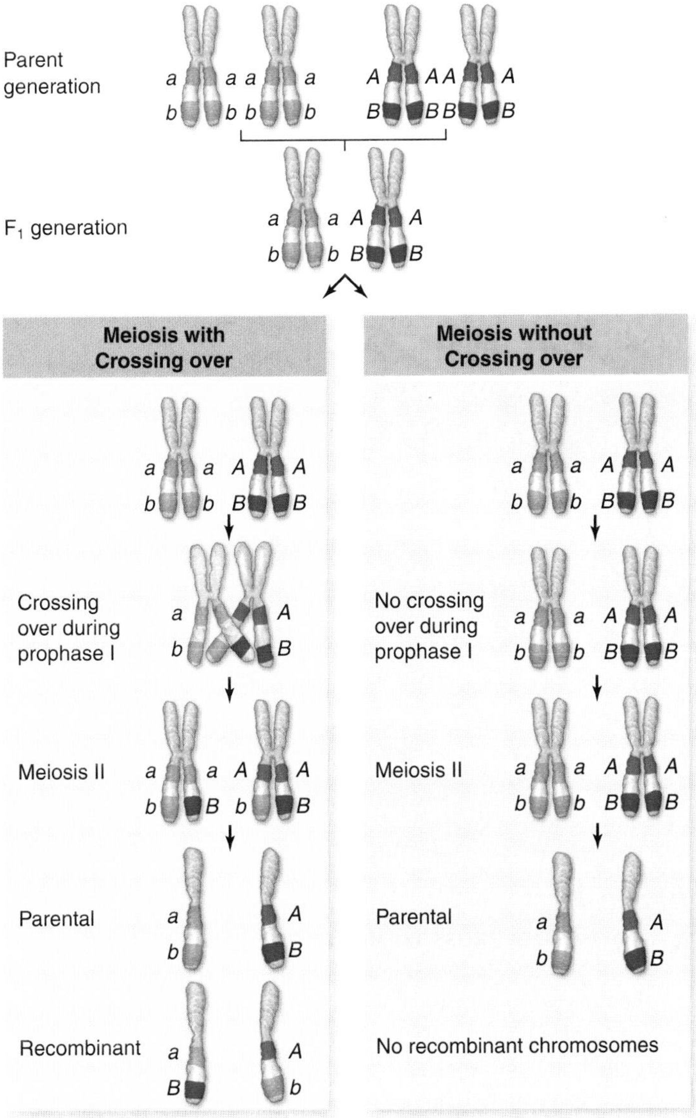

figure 13.5

CROSSING OVER EXCHANGES ALLELES ON HOMOLOGUES. When a crossover occurs between two loci, it leads to the production of recombinant chromosomes. If no crossover occurs, then the chromosomes will carry the parental combination of alleles.

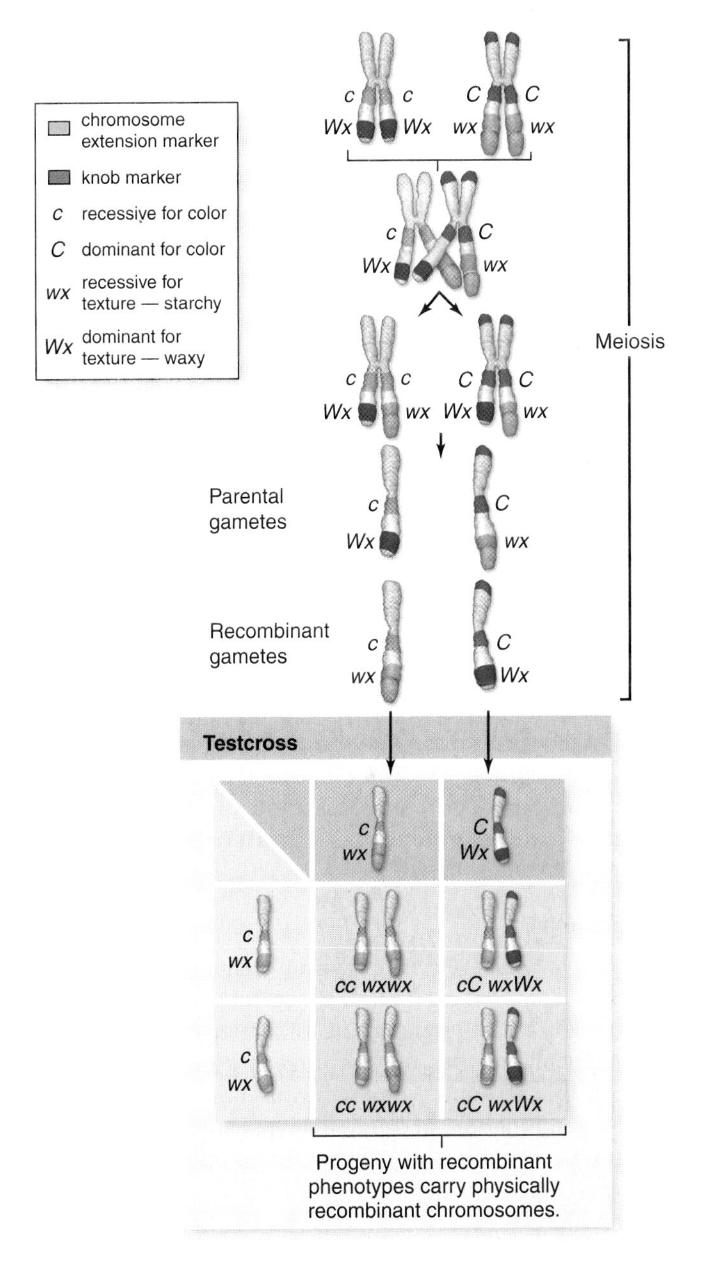

figure 13.6

THE CREIGHTON AND MCCLINTOCK EXPERIMENT. This experiment first demonstrated that chromosomes physically exchange genetic material during recombination. The experimental design was to use chromosomal differences visible in the microscope and two unrelated genes on the same chromosome. When plants heterozygous for visible and genetic markers were testcrossed, progeny that are genetically recombinant have also exchanged visible markers. This shows that the chromosomes have physically exchanged genetic material.

The longer chromosome, which had the knob, carried the dominant colored allele for kernel color (*C*) and the recessive waxy allele for kernel texture (*wx*). Heterozygotes were produced with the altered chromosome paired with a normal chromosome carrying the recessive colorless allele for kernel color (*c*) and the dominant starchy allele for kernel texture (*Wx*) (see figure 13.6). These plants appeared colored and starchy because they were heterozygous for both loci, and they were also heterozygous for the two visibly distinct chromosomes.

A testcross was performed with these F_1 plants and colorless waxy plants. The progeny were analyzed for both physical recombination (using a microscope to observe chromosomes) and genetic recombination (by examining the phenotype of progeny). The results were striking: All of the progeny that exhibited the recombinant phenotype also now had only one of the chromosomal markers. That is, physical exchange was accompanied by the recombinant phenotype.

Recombination is the basis for genetic maps

The ability to map the location of genes on chromosomes using data from genetic crosses is one of the most powerful tools of genetics. The insight that allowed this technique, like many great insights, is so simple as to seem obvious in retrospect.

Morgan had already suggested that the frequency with which a particular group of recombinant progeny appeared was a reflection of the relative location of genes on the chromosome. An undergraduate in Morgan's laboratory, Alfred Sturtevant put this observation on a quantitative basis. Sturtevant reasoned that the frequency of recombination observed in crosses could be used as a measure of genetic distance. That is, as physical distance on a chromosome increases, so does the probability of recombination (crossover) occurring between the gene loci. Using this logic, the frequency of recombinant gametes produced is a measure of their distance apart on a chromosome.

Linkage data

To be able to measure recombination frequency easily, investigators used a testcross instead of intercrossing the F_1 progeny to produce an F_2 generation. In a testcross, as described earlier, the phenotypes of the progeny reflect the gametes produced by the doubly heterozygous F_1 individual. In the case of recombination, progeny that appear parental have not undergone crossover, and progeny that appear recombinant have experienced a crossover between the two loci in question (see figure 13.5).

When genes are close together, the number of recombinant progeny is much lower than the number of parental progeny, and the genes are defined on this basis as being **linked.** The number of recombinant progeny divided by total progeny gives a value defined as the **recombination frequency.** This value is converted to a percentage, and each 1% of recombination is termed a **map unit.** This unit has been named the centimorgan (cM) for T. H. Morgan, although it is also called simply a map unit (m.u.) as well.

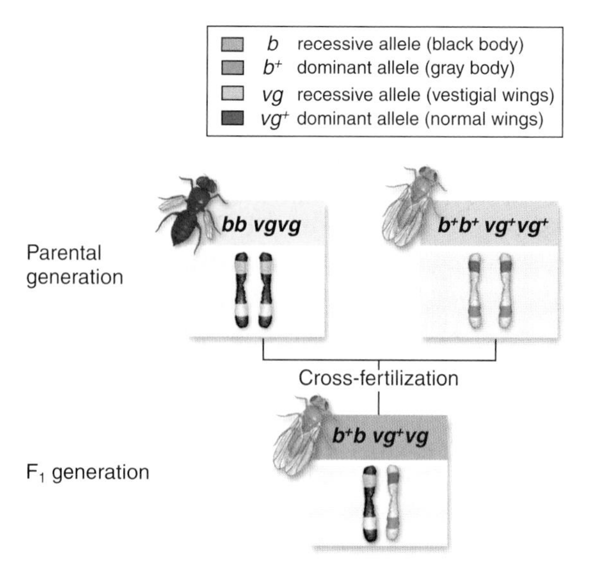

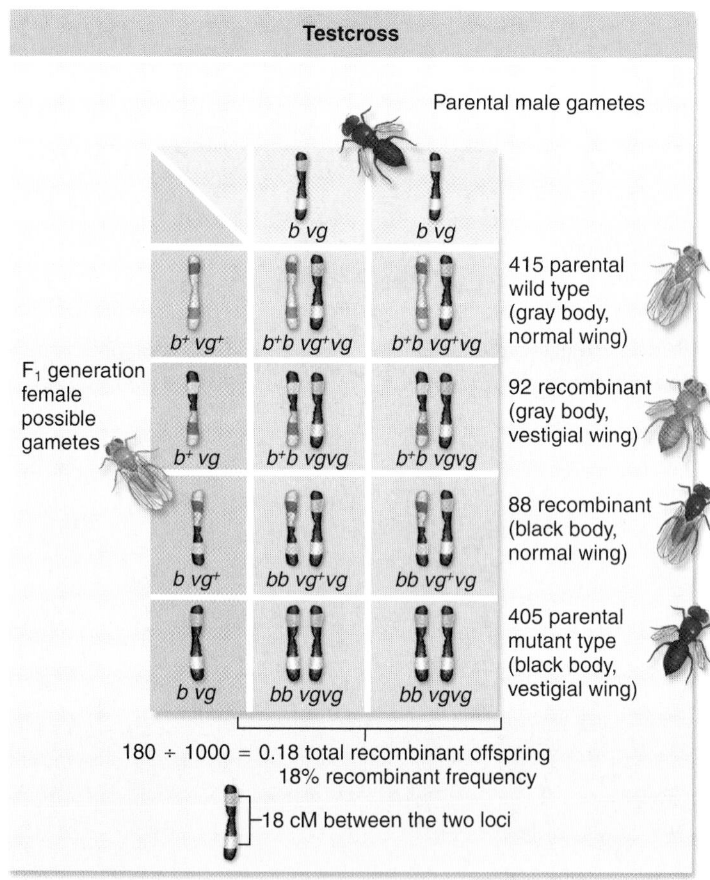

figure 13.7

TWO-POINT CROSS TO MAP GENES. Flies homozygous for long wings (vg^+) and gray bodies (b^+) are crossed to flies homozygous for vestigial wings (*vg*) and black bodies (*b*). Both vestigial wing and black body are recessive to the normal (wild type) long wing and grey body. The F_1 progeny are then testcrossed to homozygous vestigial black to produce the progeny for mapping. Data are analyzed in the text.

Constructing maps

Constructing genetic maps then becomes a simple process of performing testcrosses with doubly heterozygous individuals and counting progeny to determine percent recombination. This is best shown with an example using a two-point cross.

Drosophila homozygous for two mutations, vestigial wings (*vg*) and black body (*b*), are crossed to flies homozygous for the wild type, or normal alleles, of these genes ($vg^+ b^+$). The doubly heterozygous F_1 progeny are then testcrossed to homozygous recessive individuals (*vg b*/*vg b*), and progeny are counted (figure 13.7). The data are shown below:

vestigial wings, black body (*vg b*)	405 (parental)
long wings, gray body ($vg^+ b^+$)	415 (parental)
vestigial wings, gray body ($vg\ b^+$)	92 (recombinant)
long wings, black body ($vg^+ b$)	88 (recombinant)
Total Progeny	1000

The recombination frequency is 92 + 88 divided by 1000, or 0.18. Converting this number to a percentage yields 18 cM as the map distance between these two loci.

Multiple crossovers can yield independent assortment results

As the distance separating loci increases, the probability of recombination occurring between them during meiosis also increases. What happens when more than one recombination event occurs?

If homologues undergo two crossovers between loci, then the parental combination is restored. This leads to an underestimate of the true genetic distance because not all events can be noted. As a result, the relationship between true distance on a chromosome and the recombination frequency is not linear. It begins as a straight line, but the slope decreases; the curve levels off at a recombination frequency of 0.5 (figure 13.8).

At long distances, multiple events between loci become frequent. In this case, odd numbers of crossovers (1, 3, 5) produce recombinant gametes, and no crossover or even numbers of crossovers (0, 2, 4) produce parental gametes. At large enough distances, these frequencies are about equal, leading to the number of recombinant gametes being equal to the number of parental gametes, and the loci exhibit independent assortment! This is how Mendel could use two loci on the same chromosome and have them assort independently.

inquiry

What would Mendel have observed in a dihybrid cross if the two loci were 10 cM apart on the same chromosome? Is this likely to have led him to the idea of independent assortment?

Three-point crosses can be used to put genes in order

Because multiple crossovers reduce the number of observed recombinant progeny, longer map distances are not accurate. As a result, when geneticists try to construct maps from a series of two-point crosses, determining the order of genes is problematic. Using three loci instead of two, or a three-point cross, can help solve the problem.

In a three-point cross, the gene in the middle allows us to see recombination events on either side. For example, a double crossover for the two outside loci is actually a single crossover between each outside locus and the middle locus (figure 13.9).

The probability of two crossovers is equal to the product of the probability of each individual crossover, each of which is relatively low. Therefore, in any three-point cross, the class of offspring with two crossovers is the least frequent class. Analyzing these individuals to see which locus is recombinant identifies the locus that lies in the middle of the three loci in the cross (see figure 13.9).

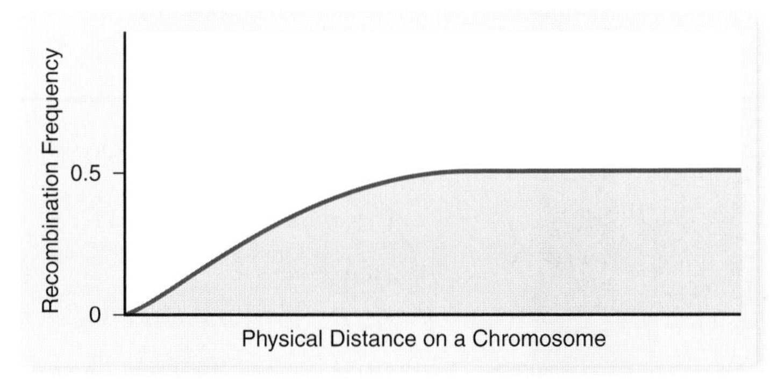

figure 13.8

RELATIONSHIP BETWEEN TRUE DISTANCE AND RECOMBINATION FREQUENCY. As distance on a chromosome increases, the recombinants are not all detected due to double crossovers. This leads to a curve that levels off at 0.5.

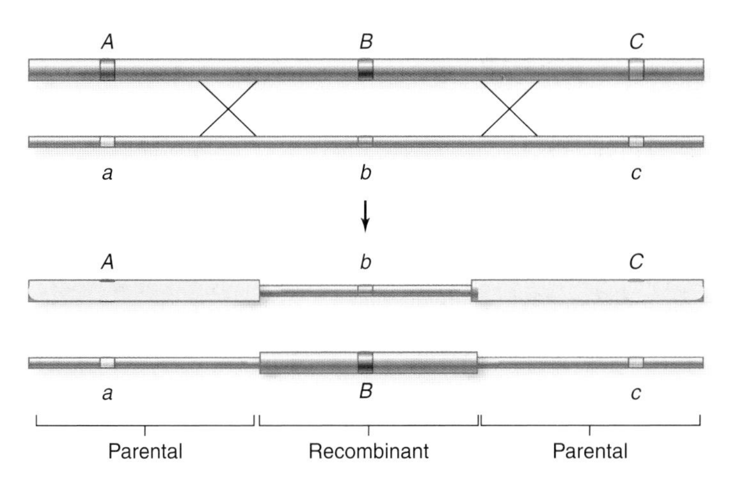

figure 13.9

USE OF A THREE-POINT CROSS TO ORDER GENES. In a two-point cross, the outside loci will appear parental for double crossovers. With the addition of a third locus, the two crossovers can still be detected because the middle locus will be recombinant. This double crossover class should be the least frequent, so whatever locus has recombinant alleles in this class must be in the middle.

In practice, geneticists use three-point crosses to determine the order of genes, then use data from the closest two-point crosses to determine distances. Longer distances are generated by simple addition of shorter distances. This avoids using inaccurate measures from two-point crosses between distant loci.

Genetic maps can be constructed for the human genome

Human genes can be mapped, but the data must be derived from historical pedigrees, such as those of the royal families of Europe mentioned earlier. The principle is the same—genetic distance is still proportional to recombination frequency—but the analysis requires the use of complex statistics and summing data from many families.

The difficulty of mapping in humans

Looking at nonhuman animals with extensive genetic maps, the majority of genetic markers have been found at loci where alleles cause morphological changes, such as variant eye color, body color, or wing morphology in flies. In humans, such alleles generally, but not always, correspond to what we consider disease states. As recently as the early 1980s, the number of markers for the human genome numbered in the hundreds. Because the human genome is so large, however, this low number of markers would never allow very dense coverage to use for mapping.

Another consideration is that the disease-causing alleles are those that we wish to map, but they occur at low frequencies in the population. Any one family would be highly unlikely to carry multiple disease alleles, the segregation of which would allow for mapping.

Anonymous markers

This situation changed with the development of **anonymous markers,** genetic markers that can be detected using molecular techniques, but that do not cause a detectable phenotype. The nature of these markers has evolved with technology, leading to a standardized set of markers scattered throughout the genome. These markers can be detected using techniques that are easy to automate, and they have a relatively high density. As a result of analysis, geneticists now have several thousand markers to work with, instead of hundreds, and have produced a human genetic map that would have been unthinkable 25 years ago (figure 13.10). (In the following chapters of this unit, you'll learn about some of the molecular techniques that have been developed for use with genomes.)

Single nucleotide polymorphisms (SNPs)

The information developed from sequencing the human genome can then be used to identify and map single bases that differ between individuals. Any differences between individuals in populations are termed *polymorphisms*; polymorphisms affecting a single base of a gene locus are called **single-nucleotide polymorphisms (SNPs).** Over 2 million such differences have been identified and are being placed on both the genetic map and the human genome sequence. This confluence of techniques will allow the ultimate resolution of genetic analysis.

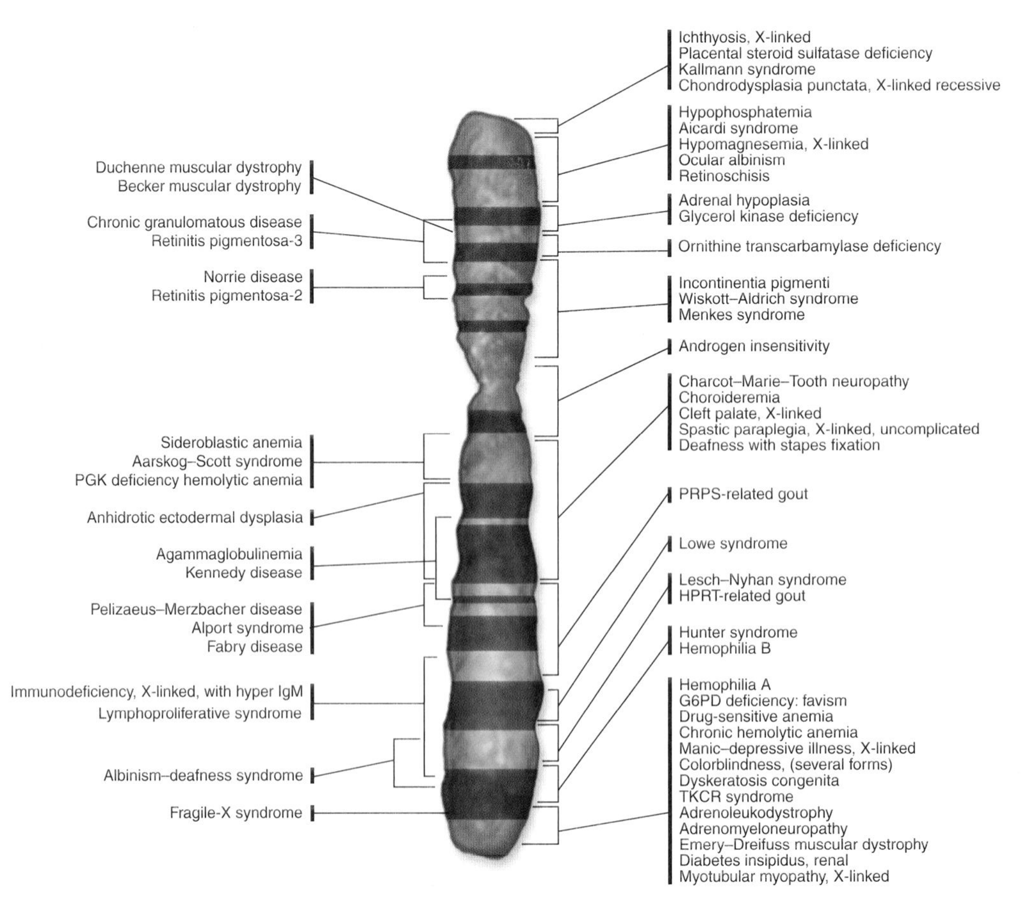

figure 13.10

THE HUMAN X CHROMOSOME GENE MAP. A partial map for the human X chromosome, a more detailed map would require a much larger figure. The black bands represent staining patterns that can be seen in the microscope, and the constriction represents the centromere. Analysis of the sequence of the X chromosome indicates 1098 genes on the X chromosome. Many of these may have mutant alleles that can affect disease states. The 59 diseases shown have been traced to specific segments of the X chromosome, indicated by brackets, by analyzing inheritance patterns of affected and unaffected individuals.

The recent progress in gene mapping applies to more than just the relatively small number of genes that show simple Mendelian inheritance. The development of a high-resolution genetic map, and the characterization of millions of SNPs, opens up the possibility of being able to characterize complex quantitative traits in humans as well.

On a more immediate practical level, the kinds of molecular markers described earlier are used in forensic analysis. Although not quite as rapid as some television programs would have you believe, this does allow rapid DNA testing of crime scene samples to help eliminate or confirm crime suspects and for paternity testing.

Genetic mapping takes advantage of the phenomenon of crossing over during meiosis, which exchanges alleles on homologues. Genes that are close together are said to be linked, and they show a greater frequency of parental types upon testcrossing. The frequency of recombination due to crossing over is a measure of genetic distance. Loci separated by large distances will have multiple crossovers between them. This can lead to independent assortment for loci on the same chromosome.

13.5 Selected Human Genetic Disorders

Diseases that run in families have been known for many years. These can be nonlife-threatening like albinism, or may result in premature death like Huntington's, which were used as examples of recessive and dominant traits in humans previously. We now want to consider the kind of genetic change that results in these disorders. This can range from the alteration of a single base to the loss of genetic material (deletion) to the loss of an entire chromosome. In this section we discuss some of the genetic disorders that have been found in human populations.

Genetic disorders can be due to altered proteins

The change of a single amino acid in a protein can lead to a debilitating clinical phenotype. As you will see in chapter 14, this situation can sometimes result from a change in a single base of the DNA chain that encodes the protein. A small sample of diseases due to alterations of alleles of a single gene is provided in table 13.2.

The first human disease in which such a change has been shown to occur is **sickle cell anemia.** It is caused by a defect in the oxygen carrier molecule, hemoglobin, that leads to impaired oxygen delivery to tissues. The defective hemoglobin molecules stick to one another, leading to stiff, rodlike structures that alter the shape of the red blood cells that carry them. These red blood cells take on a characteristic shape that led to the name "sickle cell" (figure 13.11).

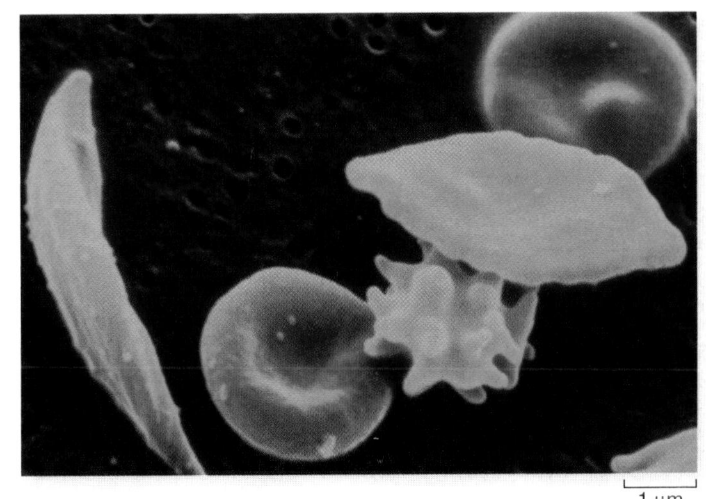

figure 13.11

SICKLE CELL ANEMIA. In individuals homozygous for the sickle cell trait, many of the red blood cells have sickled or irregular shapes, such as the cell on the far right.

TABLE 13.2 Some Important Genetic Disorders

Disorder	Symptom	Defect	Dominant/ Recessive	Frequency Among Human Births
Cystic fibrosis	Mucus clogs lungs, liver, and pancreas	Failure of chloride ion transport mechanism	Recessive	1/2500 (Caucasians)
Sickle cell anemia	Blood circulation is poor	Abnormal hemoglobin molecules	Recessive	1/600 (African Americans)
Tay–Sachs disease	Central nervous system deteriorates in infancy	Defective enzyme (hexosaminidase A)	Recessive	1/3500 (Ashkenazi Jews)
Phenylketonuria	Brain fails to develop in infancy	Defective enzyme (phenylalanine hydroxylase)	Recessive	1/12,000
Hemophilia	Blood fails to clot	Defective blood-clotting factor VIII	X-linked recessive	1/10,000 (Caucasian males)
Huntington disease	Brain tissue gradually deteriorates in middle age	Production of an inhibitor of brain cell metabolism	Dominant	1/24,000
Muscular dystrophy (Duchenne)	Muscles waste away	Degradation of myelin coating of nerves stimulating muscles	X-linked recessive	1/3700 (males)
Hypercholesterolemia	Excessive cholesterol levels in blood lead to heart disease	Abnormal form of cholesterol cell surface receptor	Dominant	1/500

Individuals with sickle-shaped red blood cells exhibit intermittent illness and reduced life span. At the molecular level, this condition is caused by a single amino acid, glutamic acid, in the 146-amino-acid β-globin chain being changed to another amino acid, valine. The affected amino acid site is not in the oxygen-binding region of the protein, but the change still has a catastrophic effect on hemoglobin's function. The replacement of glutamic acid, a charged amino acid, with nonpolar valine on the surface of the protein makes the protein sticky. This stickiness is due to the tendency of nonpolar amino acids to aggregate in water-based solutions such as blood plasma, leading to the stiff, rodlike structures seen in sickled red blood cells (figure 13.12).

Individuals heterozygous for the sickle cell allele are indistinguishable from normal individuals in a normal oxygen environment, although their red cells do exhibit reduced ability to carry oxygen.

The sickle cell allele is particularly prevalent in people of African descent. In some regions of Africa, up to 45% of the population is heterozygous for the trait, and 6% are homozygous. This proportion of heterozygotes is higher than would be expected on the basis of chance alone. It turns out that heterozygosity confers a greater resistance to the blood-borne parasite that causes the disease malaria. In regions of central Africa where malaria is endemic, the sickle cell allele also occurs at a high frequency.

The sickle cell allele is not the end of the story for the β-globin gene; a large number of other alterations of this gene have been observed that lead to anemias. In fact, for hemoglobin, which is composed of two α-globins and two β-globins, over 700 structural variants have been cataloged. It is estimated that 7% of the human population worldwide are carriers for different inherited hemoglobin disorders.

The Human Gene Mutation Database has cataloged the nature of many disease alleles, including the sickle cell allele. The majority of alleles seem to be simple changes. Almost 60% of the close to 28,000 alleles in the Human Gene

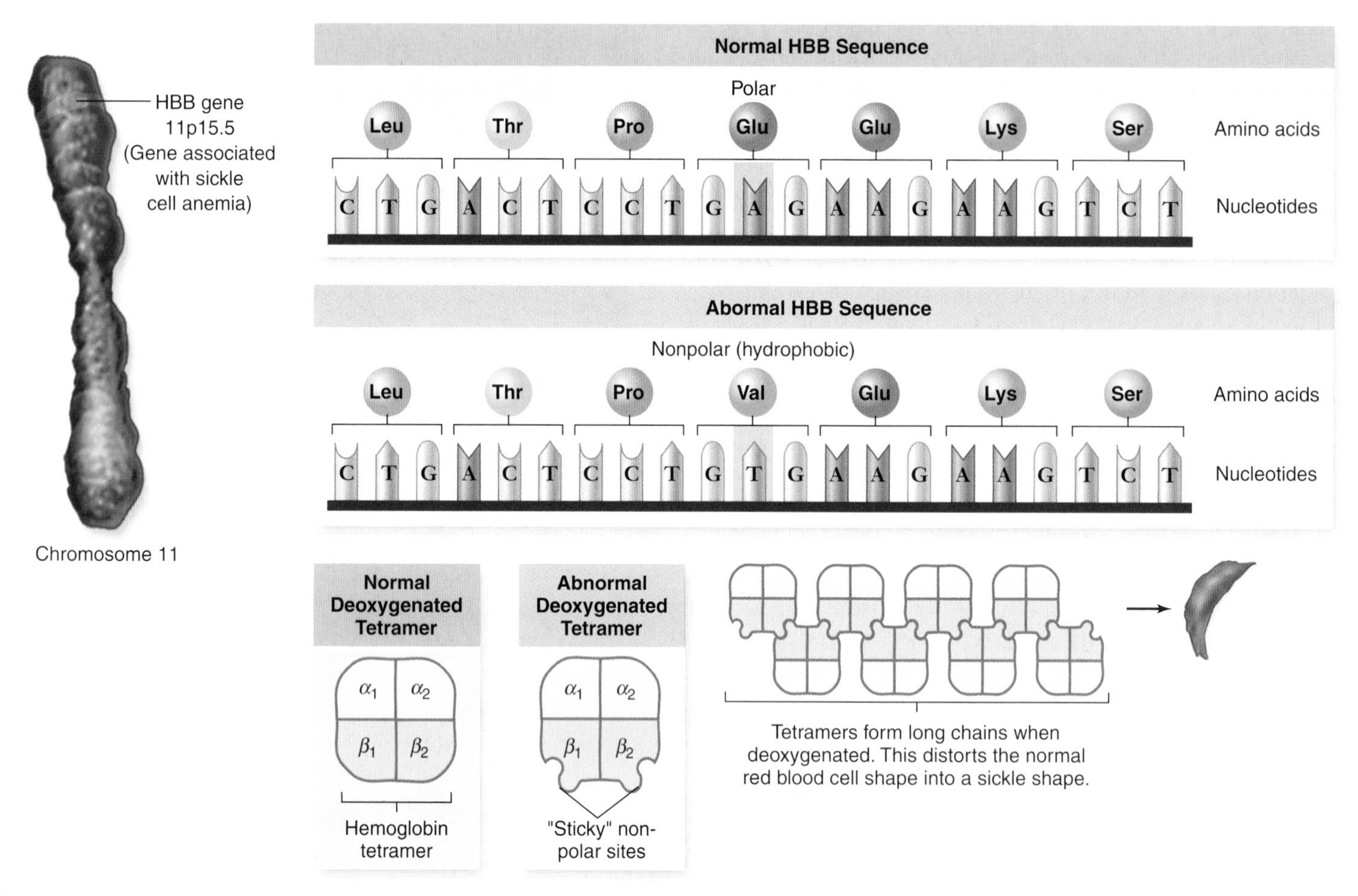

figure 13.12

SICKLE CELL ANEMIA IS CAUSED BY AN ALTERED PROTEIN. Hemoglobin is composed of a tetramer of two α-globin chains and two β-globin chains. Protein sequences are encoded in DNA in groups of three nucleotides (chapter 15 details the genetic code). The sickle cell allele of the β-globin gene has a single change in the DNA sequence that results in an amino acid substitution of valine for glutamic acid. The valine is hydrophobic creating regions on the surface of the protein that are "sticky." In an individual with sickle cell, mutant β-globin chains associate to form long chains that distort the red blood cell.

Mutation Database are single-base substitutions. Another 23% are due to small insertions or deletions of less than 20 bases. The rest of the alleles are made of more complex alterations. It is clear that simple changes in genes can have profound effects.

Nondisjunction of chromosomes changes chromosome number

The failure of homologues or sister chromatids to separate properly during meiosis is called **nondisjunction.** This failure leads to the gain or loss of a chromosome, a condition called **aneuploidy.** The frequency of aneuploidy in humans is surprisingly high, being estimated to occur in 5% of conceptions.

Nondisjunction of autosomes

Humans who have lost even one copy of an autosome are called **monosomics,** and generally they do not survive embryonic development. In all but a few cases, humans who have gained an extra autosome (called **trisomics**) also do not survive. Data from clinically recognized spontaneous abortions indicate levels of aneuploidy as high as 35%.

Five of the smallest human autosomes, however—those numbered 13, 15, 18, 21, and 22—can be present as three copies and still allow the individual to survive, at least for a time. The presence of an extra chromosome 13, 15, or 18 causes severe developmental defects, and infants with such a genetic makeup die within a few months. In contrast, individuals who have an extra copy of chromosome 21 or, more rarely, chromosome 22, usually survive to adulthood. In these people, the maturation of the skeletal system is delayed, so they generally are short and have poor muscle tone. Their mental development is also affected, and children with trisomy 21 are always mentally retarded to some degree.

The developmental defect produced by trisomy 21 (figure 13.13) was first described in 1866 by J. Langdon Down; for this reason, it is called **Down syndrome.** About 1 in every 750 children exhibits Down syndrome, and the frequency is comparable in all racial groups. Similar conditions also occur in chimpanzees and other related primates.

In humans, the defect occurs when a particular small portion of chromosome 21 is present in three copies instead of two. In 97% of the cases examined, all of chromosome 21 is present in three copies. In the other 3%, a small portion of chromosome 21 containing the critical segment has been added to another chromosome by a process called *translocation* (see chapter 15); it exists along with the normal two copies of chromosome 21. This latter condition is known as *translocation Down syndrome.*

In mothers younger than 20 years of age, the risk of giving birth to a child with Down syndrome is about 1 in 1700; in mothers 20 to 30 years old, the risk is only about 1 in 1400. However, in mothers 30 to 35 years old, the risk rises to 1 in 750, and by age 45, the risk is as high as 1 in 16 (figure 13.14).

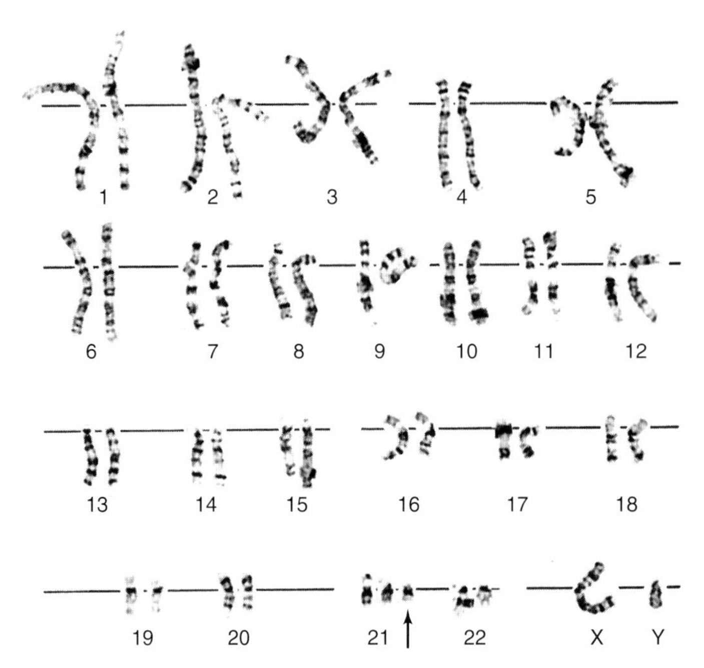

figure 13.13

DOWN SYNDROME. As shown in this male karyotype, Down syndrome is associated with trisomy of chromosome 21 (arrow shows third copy of chromosome 21).

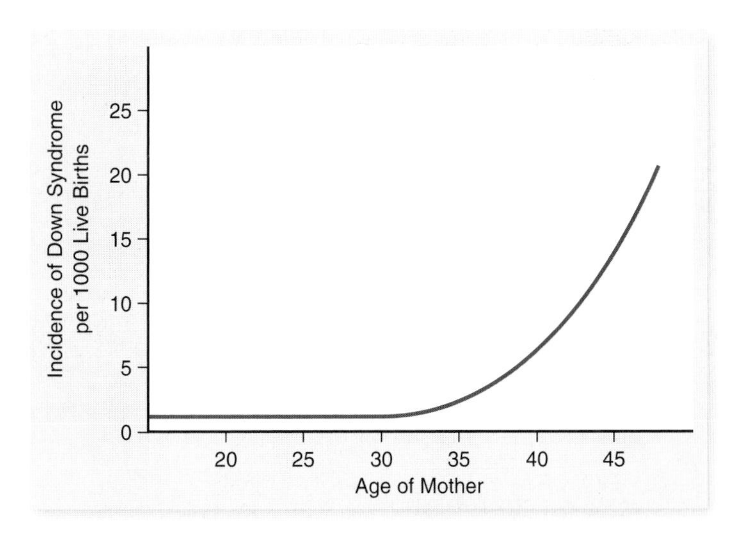

figure 13.14

CORRELATION BETWEEN MATERNAL AGE AND THE INCIDENCE OF DOWN SYNDROME. As women age, the chances they will bear a child with Down syndrome increase. After a woman reaches 35, the frequency of Down syndrome rises rapidly.

inquiry

Over a five-year period between ages 20 and 25, the incidence of Down syndrome increases 0.1 per thousand; over a five-year period between ages 35 and 40, the incidence increases to 8.0 per thousand, 80 times as great. The period of time is the same in both instances. What has changed?

Primary nondisjunctions are far more common in women than in men because all of the eggs a woman will ever produce have developed to the point of prophase in meiosis I by the time she is born. By the time a woman has children, her eggs are as old as she is. Therefore, there is a much greater chance for cell-division problems of various kinds, including those that cause primary nondisjunction, to accumulate over time in female gametes. In contrast, men produce new sperm daily. For this reason, the age of the mother is more critical than that of the father for couples contemplating childbearing.

Nondisjunction of sex chromosomes

Individuals who gain or lose a sex chromosome do not generally experience the severe developmental abnormalities caused by similar changes in autosomes. Although such individuals have somewhat abnormal features, they often reach maturity and in some cases may be fertile.

X chromosome nondisjunction. When X chromosomes fail to separate during meiosis, some of the gametes produced possess both X chromosomes, and so are XX gametes; the other gametes have no sex chromosome and are designated "O" (figure 13.15).

If an XX gamete combines with an X gamete, the resulting XXX zygote develops into a female with one functional X chromosome and two Barr bodies. She may be taller in stature but is otherwise normal in appearance.

If an XX gamete instead combines with a Y gamete, the effects are more serious. The resulting XXY zygote develops into a male who has many female body characteristics and, in some cases but not all, diminished mental capacity. This condition, called *Klinefelter syndrome*, occurs in about 1 out of every 500 male births.

If an O gamete fuses with a Y gamete, the resulting OY zygote is nonviable and fails to develop further; humans cannot survive when they lack the genes on the X chromosome. But if an O gamete fuses with an X gamete, the XO zygote develops into a sterile female of short stature, with a webbed neck and sex organs that never fully mature during puberty. The mental abilities of an XO individual are in the low-normal range. This condition, called *Turner syndrome*, occurs roughly once in every 5000 female births.

Y chromosome nondisjunction. The Y chromosome can also fail to separate in meiosis, leading to the formation of YY gametes. When these gametes combine with X gametes, the XYY zygotes develop into fertile males of normal appearance. The frequency of the XYY genotype (*Jacob syndrome*) is about 1 per 1000 newborn males.

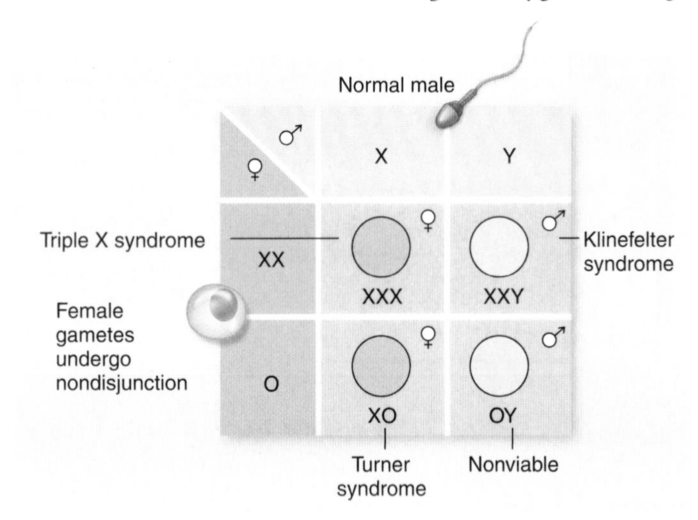

figure 13.15

HOW NONDISJUNCTION CAN PRODUCE ABNORMALITIES IN THE NUMBER OF SEX CHROMOSOMES. When nondisjunction occurs in the production of female gametes, the gamete with two X chromosomes (XX) produces Klinefelter males (XXY) and triple-X females (XXX). The gamete with no X chromosome (O) produces Turner females (XO) and nonviable OY males lacking any X chromosome.

inquiry

Can you think of two nondisjunction scenarios that would produce an XXY male?

Genomic imprinting depends on the parental origin of alleles

By the late 20th century, geneticists were confident that they understood the basic mechanisms governing inheritance. It came as quite a surprise when mouse geneticists found an important exception to classical Mendelian genetics that appears to be unique to mammals. In **genomic imprinting,** the phenotype caused by a specific allele is exhibited when the allele comes from one parent, but not from the other.

The basis for genomic imprinting is the expression of a gene depending on passage through maternal or paternal germ lines. Some genes are inactivated in the paternal germ line and therefore are not expressed in the zygote. Other genes are inactivated in the maternal germ line, with the same result. This condition makes the zygote effectively haploid for an imprinted gene. The expression of variant alleles of imprinted genes depends on the parent of origin. Furthermore, imprinted genes seem to be concentrated in particular regions of the genome. These regions include genes that are both maternally and paternally imprinted.

Prader–Willi and Angelman syndromes

An example of genomic imprinting in humans involves the two diseases Prader–Willi syndrome (PWS) and Angelman syndrome (AS). The effects of PWS include respiratory distress, obesity, short stature, mild mental retardation, and obsessive–compulsive behavior. The effects of AS include developmental delay, severe mental retardation, hyperactivity, aggressive behavior, and inappropriate laughter.

Genetic studies have implicated genes on chromosome 15 for both disorders, but the pattern of inheritance is complementary. The most common cause of both syndromes is a deletion of material on chromosome 15 and, in fact, the same deletion can cause either syndrome. The determining factor is the parental origin of the normal and deleted chromosomes. If the chromosome with the deletion is paternally inherited it causes PWS, if the chromosome with the deletion is maternally inherited it causes AS.

The region of chromosome 15 that is lost is subject to imprinting, with some genes being inactivated in the maternal germ line, and others in the paternal germ line. In PWS, genes are inactivated in the maternal germ line, such that deletion or other functional loss of paternally derived alleles produces the syndrome. The opposite is true for AS syndrome: Genes are inactivated in the paternal germ line, such that loss of maternally derived alleles leads to the syndrome.

Molecular basis of genomic imprinting

Although genomic imprinting is not well understood, at least one aspect seems clear: The basis for inactivating genes appears to be linked to modifications of the DNA itself. DNA can be modified by the addition of methyl groups, termed *methylation.* This modification is correlated with inactivity of genes. The proteins that are associated with chromosomes can also be modified, leading to effects on gene expression. The control of gene expression is discussed in more detail in the following chapters.

Some genetic defects can be detected early in pregnancy

Although most genetic disorders cannot yet be cured, we are learning a great deal about them, and progress toward successful therapy is being made in many cases. In the absence of a cure, however, the only recourse is to try to avoid producing children with these conditions. The process of identifying parents at risk for having children with genetic defects and of assessing the genetic state of early embryos is called **genetic counseling.**

Pedigree analysis

One way of assessing risks is through pedigree analysis, often employed as an aid in genetic counseling. By analyzing a person's pedigree, it is sometimes possible to estimate the likelihood that the person is a carrier for certain disorders. For example, if a counseling client's family history reveals that a relative has been afflicted with a recessive genetic disorder, such as cystic fibrosis, it is possible that the client is a heterozygous carrier of the recessive allele for that disorder.

When a couple is expecting a child, and pedigree analysis indicates that both of them have a significant chance of being heterozygous carriers of a deleterious recessive allele, the pregnancy is said to be high-risk. In such cases, a significant probability exists that their child will exhibit the clinical disorder.

Another class of high-risk pregnancy is that in which the mothers are more than 35 years old. As discussed earlier, the frequency of infants with Down syndrome increases dramatically in the pregnancies of older women (see figure 13.14).

Amniocentesis

When a pregnancy is diagnosed as high-risk, many women elect to undergo **amniocentesis,** a procedure that permits the prenatal diagnosis of many genetic disorders. In the fourth month of pregnancy, a sterile hypodermic needle is inserted into the expanded uterus of the mother, removing a small sample of the amniotic fluid that bathes the fetus (figure 13.16). Within the fluid are free-floating cells derived from the fetus; once removed, these cells can be grown in cultures in the laboratory.

During amniocentesis, the position of the needle and that of the fetus are usually observed by means of *ultrasound.*

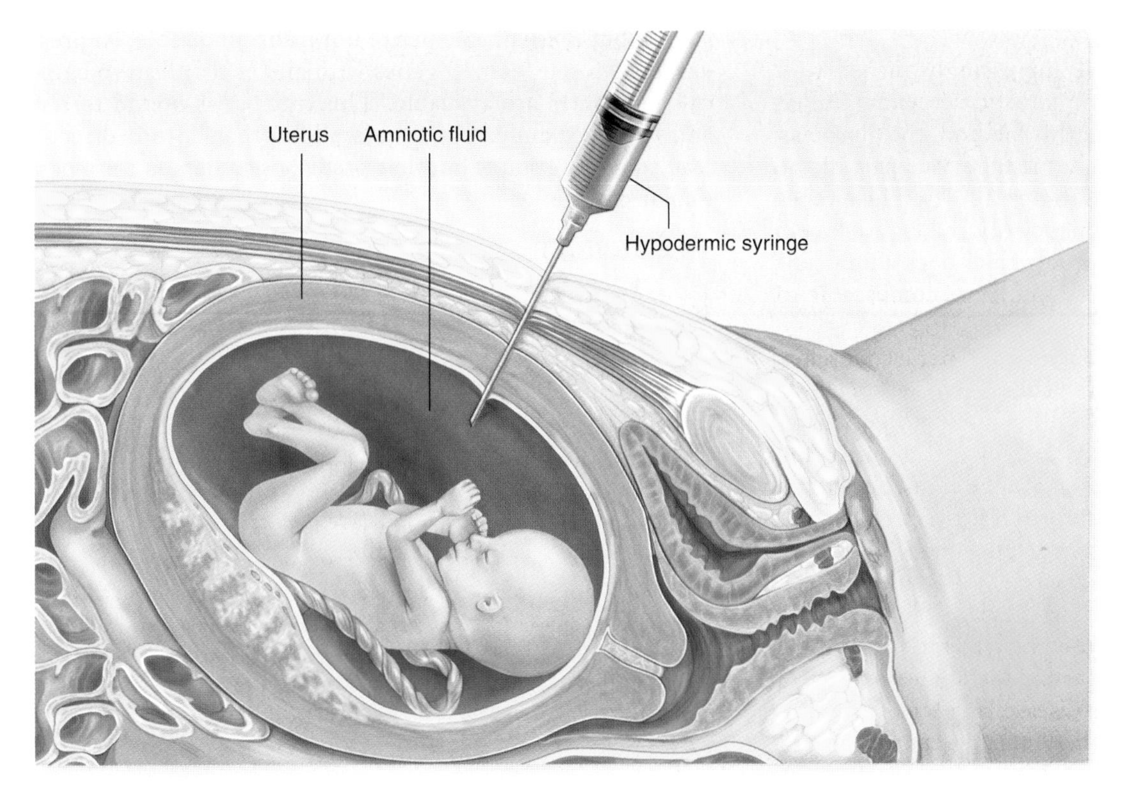

figure 13.16

AMNIOCENTESIS. A needle is inserted into the amniotic cavity, and a sample of amniotic fluid, containing some free cells derived from the fetus, is withdrawn into a syringe. The fetal cells are then grown in culture, and their karyotype and many of their metabolic functions are examined.

figure 13.17

CHORIONIC VILLI SAMPLING. Cells can be taken from the chorionic villi as early as the eighth to tenth week of pregnancy. Cells are removed by suction with a tube inserted through the cervix. These cells can then be grown in culture and examined for karyotypes and tested biochemically for defects.

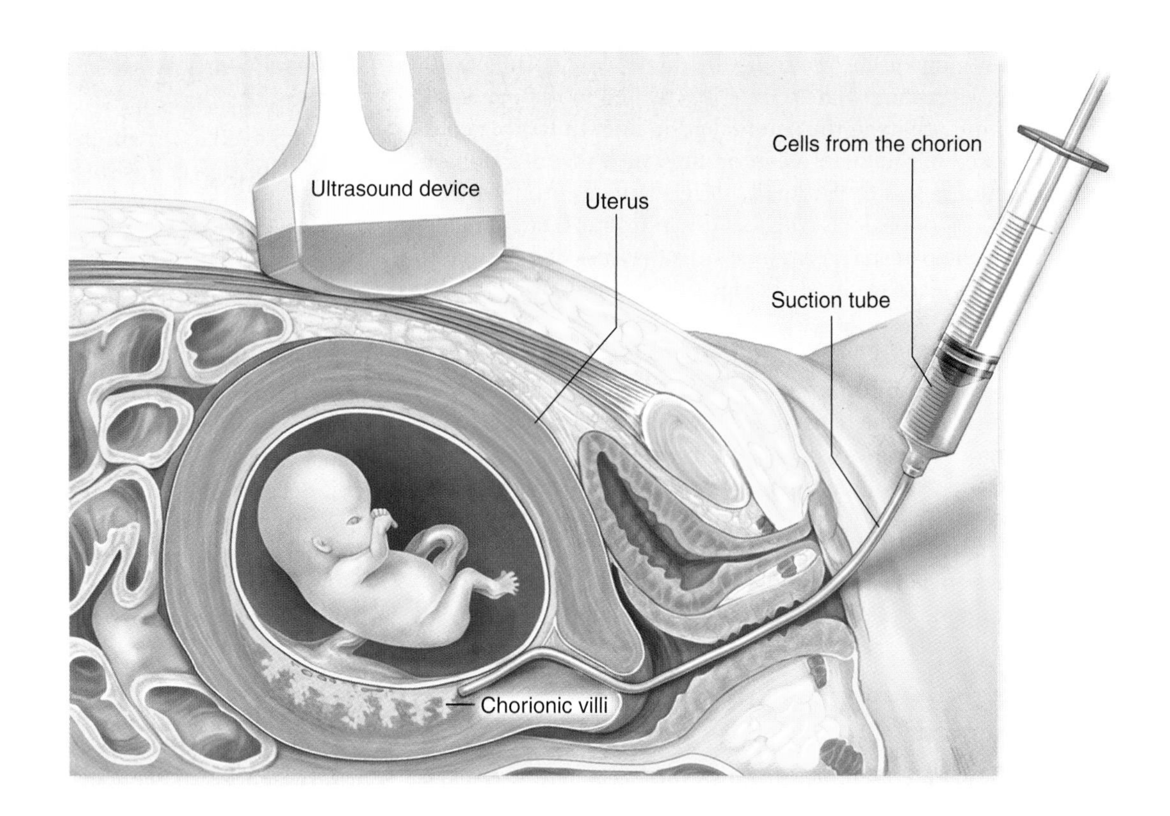

The sound waves used in ultrasound are not harmful to mother or fetus, and they permit the person withdrawing the amniotic fluid to do so without damaging the fetus. In addition, ultrasound can be used to examine the fetus for signs of major abnormalities. However, about 1 out of 200 amniocentesis procedures may result in fetal death and miscarriage.

Chorionic villi sampling

In recent years, physicians have increasingly turned to a new, less invasive procedure for genetic screening called **chorionic villi sampling.** Using this method, the physician removes cells from the chorion, a membranous part of the placenta that nourishes the fetus (figure 13.17). This procedure can be used earlier in pregnancy (by the eighth week) and yields results much more rapidly than does amniocentesis. Risks from chorionic villi sampling are comparable to that for amniocentesis.

To test for certain genetic disorders, genetic counselors look for three characteristics in the cultures of cells obtained from amniocentesis or chorionic villi sampling. First, analysis of the karyotype can reveal aneuploidy (extra or missing chromosomes) and gross chromosomal alterations. Second, in many cases it is possible to test directly for the proper functioning of enzymes involved in genetic disorders. The lack of normal enzymatic activity signals the presence of the disorder. As examples, the lack of the enzyme responsible for breaking down phenylalanine indicates phenylketonuria (PKU); the absence of the enzyme responsible for the breakdown of gangliosides indicates Tay–Sachs disease; and so forth. Additionally, with information from the Human Genome Project, more disease alleles for genetic disorders are known. If there are a small number of alleles for a specific disease in the population, these can be identified as well.

With the changes in human genetics brought about by the Human Genome Project (chapter 18), it is possible to design tests for many more diseases. Difficulties still exist in discerning the number and frequency of disease-causing alleles, but these problems are not insurmountable. At present, tests for at least 13 genes with alleles that lead to clinical syndromes are available. This number is bound to rise and to be expanded to include alleles that do not directly lead to disease states but that predispose a person for a particular disease.

inquiry

Based on what you read in this chapter, what reasons could a mother have to undergo CVS, considering its small but potential risks?

Human genetic disorders may be caused by single-base mutations or by multiple changes, additions, or deletions in a gene's DNA. On the chromosome level, nondisjunction during meiosis can result in gametes with too few or too many chromosomes, most of which produce inviable offspring. Imprinting refers to inactivation of alleles depending on which parent the alleles come from. In parents with a high risk of bearing children with a genetic defect, testing can help provide information about the genetic health of a fetus.

concept review

13.1 Sex Linkage and the Chromosomal Theory of Inheritance

The chromosomal theory of inheritance proposed by Sutton states that hereditary traits are carried on chromosomes.

- Morgan showed the trait for white eyes in *Drosophila* cosegregated with sex chromosomes, indicating that traits are associated with chromosomes (figure 13.2).
- Traits associated with sex chromosomes are referred to as sex-linked.

13.2 Sex Chromosomes and Sex Determination

The structure and number of sex chromosomes vary in different organisms.

- Sex determination in animals is usually associated with a chromosomal difference.
- In some animals, for example mammals and flies, females have two similar sex chromosomes and males have sex chromosomes that differ.
- In other species, for example birds and some reptiles, males have two similar sex chromosomes and females have sex chromosomes that differ (table 13.1).
- The "default setting" in human embryonic development is for production of a female.
- In humans the Y chromosome determines maleness.
- The Y chromosome is highly condensed and does not have an active counterpart to most of the genes on the X chromosome.
- The *SRY* gene on the Y chromosome is responsible for the masculinization of genitalia and secondary sex organs.
- If an XX individual has part of the Y chromosome translocated to one of the X chromosomes, the embryo will develop as a male.
- Mutations in the *SRY* gene or failure of the embryo to respond to androgen hormones can cause an XY individual to develop into sterile females.
- Genetic disorders such as color blindness and hemophilia are sex linked (figure 13.3).
- In mammalian females one of the X chromosomes is randomly inactivated during development.
- This inactivated condensed chromosome, or Barr body, is an example of dosage compensation, which balances levels of gene expression in males and females.
- X-chromosome inactivation can lead to genetic mosaics if the female is heterozygous for X-chromosome alleles. An example is calico cats (figure 13.4).

13.3 Exceptions to the Chromosomal Theory of Inheritance

Not all inheritance is explained by chromosomes.

- Mitochondrial genes are usually inherited from the female parent.
- Chloroplast genes are usually maternally inherited although paternal and biparental inheritance has also been observed.

13.4 Genetic Mapping

If two genes are linked, they both occur on a chromosome, and their inheritance behavior differ if the genes were on separate chromosomes.

- Homologous chromosomes may exchange alleles during crossing over (figure 13.5).
- The recombination of alleles during crossing over is the basis for constructing genetic maps.
- The further apart two linked genes are, the greater the frequency of recombination due to crossing over between gene loci.
- A map unit is expressed as the percentage of recombinant progeny out of total progeny.
- The probability of multiple crossovers increases with distance between two linked genes and results in an underestimate of recombination frequency.
- Maps constructed using crosses between three linked genes can be used to determine the order of genes (figure 13.9).
- Longer map distances are calculated by adding shorter, more accurate, distances.
- Human genetic mapping was difficult and usually involved disease-causing alleles until anonymous markers were developed.
- Single-nucleotide polymorphisms (SNPs) can be used to detect differences among individuals.

13.5 Selected Human Genetic Disorders

Causes of human diseases range from an altered single base to the deletion of genetic material to the loss of an entire chromosome.

- A change in a single amino acid can result in a debilitating clinical phenotype.
- Nondisjunction, the failure of homologues or sister chromatids to separate during meiosis, results in a condition called aneuploidy.
- Monosomics lose at least one copy of an autosome and generally do not survive embryonic development.
- Trisomics gain an extra autosome and most often do not survive development.
- Translocation occurs when part of one chromosome is joined to another chromosome, resulting in three copies of a chromosomal segment.
- X-chromosome nondisjunction occurs when X chromosomes fail to separate during meiosis. The resulting gamete has either XX or O (zero sex chromosomes) (figure 13.15).
- Y-chromosome nondisjunction results in YY gametes.
- In genomic imprinting the expression of a gene depends on whether it passes through the maternal or paternal germ-line.
- Imprinted genes are inactivated by methylation.
- Genetic defects in populations can be determined by pedigree analysis, amniocentesis, or chorionic villi sampling.

SELF TEST

1. Why is the white-eye phenotype always observed in males carrying the white-eye allele?
   a. Because the trait is dominant
   b. Because the trait is recessive
   c. Because the allele is located on the X chromosome and males only have one X
   d. Because the allele is located on the Y chromosome and only males have Y chromosomes
2. An *autosome* is a chromosome that—
   a. contains genetic information to determine the sex of an organism
   b. determines all other traits of an organism other than sex
   c. is only inherited from the mother (maternal inheritance)
   d. has no matching chromosome within an organism's genome
3. Sex linkage in humans occurs when—
   a. an allele is located on both the X and Y chromosomes
   b. all allele is located on the X chromosome
   c. an allele is located on an autosome
   d. a phenotype is only observed in females
4. What are Barr bodies?
   a. X chromosomes inactivated to prevent overexpression of the alleles found on the X chromosome in females
   b. Highly condensed Y chromosomes in males
   c. X chromosomes inactivated to allow for expression of the males-specific phenotype
   d. Inactive autosomal chromosomes specific to females
5. How does maternal inheritance of mitochondrial genes differ from sex linkage?
   a. Mitochondrial genes do not contribute to the phenotype of an individual.
   b. Because mitochondria are inherited from the mother, only females are affected.
   c. Since mitochondria are inherited from the mother, females and males are equally affected.
   d. Mitochondrial genes must be dominant. Sex-linked traits are typically recessive.
6. What cellular process in responsible for genetic recombination?
   a. Independent assortment
   b. Separation of the homologues in meiosis 1
   c. Separation of the chromatids during meiosis II
   d. Crossing over
7. The number of map units between two genes is determined by—
   a. the recombination frequency
   b. the frequency of parental types
   c. the total number of genes within a given piece of DNA
   d. the number of linked genes within a chromosome
8. How many map units separate two alleles if the recombination frequency is 0.07?
   a. 700 cM
   b. 70 cM
   c. 7 cM
   d. 0.7 cM
9. Multiple crossovers lead to—
   a. restoration of the paternal combination of genes
   b. increased genetic diversity
   c. increased numbers of recombinant progeny
   d. anueploidy
10. The disease sickle cell anemia is caused by—
    a. altered expression of the HBB gene
    b. a change of a single amino acid in the protein hemoglobin
    c. a change in the HBB gene
    d. Both (b) and (c)
11. What determines whether an individual is a genetic mosaic?
    a. The presence of different alleles on the autosomal chromosomes
    b. The inactivation of an allele on an autosomal chromosome
    c. The inactivation of an allele on the X chromosome of a heterozygous female
    d. The inactivation of an allele on the X chromosome of a homozygous male
12. Down syndrome is the result of—
    a. a single-base substitution on human chromosome 21
    b. nondisjunction of chromosome 21 during meiosis
    c. inactivation of chromosome 21
    d. nondisjunction of chromosome 21 during mitosis in early development
13. Which of the following examples of nondisjunction of sex chromosomes is lethal?
    a. XXX
    b. XXY
    c. OY
    d. XO
14. What is genomic imprinting?
    a. The blending of a phenotype due to the genetic contribution of both parents
    b. The expression of a dominant allele
    c. The development of a phenotype in response to interactions between distinct alleles
    d. The expression of different phenotypes dependent upon the parental origin of alleles
15. Which of the following is *not* a method used in genetic counseling?
    a. ultrasound
    b. chorionic villi sampling
    c. amniocentesis
    d. pedigree analysis

CHALLENGE QUESTIONS

1. Color blindness is caused by a sex-linked, recessive gene. If a woman, heterozygous for the color blind allele, marries a man with normal color vision, what percentage of their children will be color blind? What sex will the color blind children be?
2. What conditions would have to exist to produce a color blind female?
3. Imagine that the genes for seed color and seed shape are located on the same chromosome. A cross was made between two, true-breeding plants. One plant produces green wrinkled seed *(rryy)* and the second parent produced round yellow seeds *(RRYY)*. A testcross is made between the F_1 generation with the following results:

green, wrinkled	645
green, round	36
yellow, wrinkled	29
yellow, round	590

   Calculate the distance between the two loci.
4. Is it possible to have a calico cat that is male? Why or why not?

Do you need additional review? *Visit* www.ravenbiology.com *for practice quizzes, animations, videos, and activities designed to help you master the material in this chapter.*

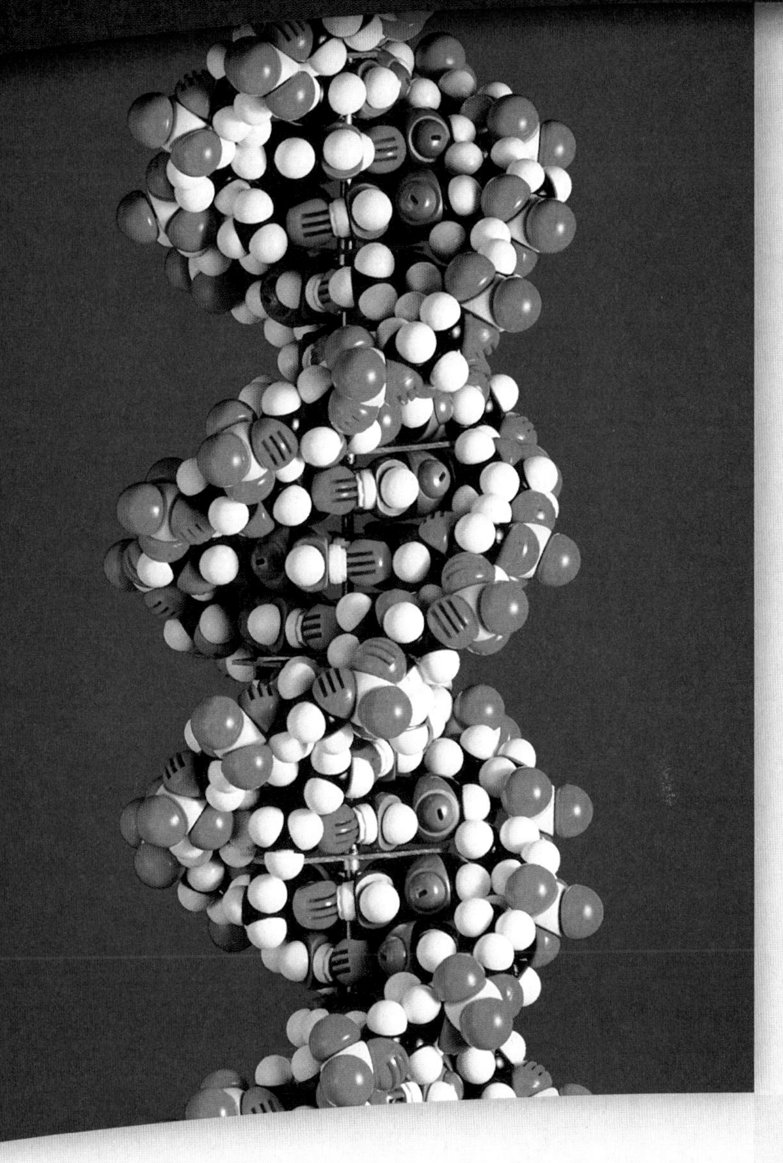

chapter 14

DNA: The Genetic Material

introduction

THE REALIZATION THAT PATTERNS OF heredity can be explained by the segregation of chromosomes in meiosis raised a question that occupied biologists for over 50 years: What is the exact nature of the connection between hereditary traits and chromosomes? This chapter describes the chain of experiments that led to our current understanding of DNA, modeled in the picture, and of the molecular mechanisms of heredity. These experiments are among the most elegant in science. And, just as in a good detective story, each discovery has led to new questions. But however erratic and lurching the course of the experimental journey may appear, our picture of heredity has become progressively clearer, the image more sharply defined.

concept outline

14.1 The Nature of the Genetic Material

In the previous two chapters, you learned about the nature of inheritance and how genes, which contain the information to specify traits, are located on chromosomes. This finding led to the question of what part of the chromosome actually contains the genetic information. Specifically, biologists wondered about the chemical identity of the genetic information. They knew that chromosomes are composed primarily of both protein and DNA. Which of these organic molecules actually comprises the genes?

Starting in the late 1920s and continuing for about 30 years, a series of investigations addressed this question. DNA consists of four chemically similar nucleotides. In contrast, protein contains 20 different amino acids that are much more chemically diverse than nucleotides. These characteristics seemed initially to indicate greater informational capacity in protein than in DNA.

Experiments began to reveal evidence in favor of DNA; we describe three of these major findings in this section.

Griffith finds that bacterial cells can be transformed

The first clue came in 1928 with the work of the British microbiologist Frederick Griffith. Griffith studied a pathogenic bacteria, *Streptococcus pneumoniae*, that causes pneumonia in mice. There are two forms of this bacteria: The normal virulent form that causes pneumonia, and a mutant, nonvirulent form that does not. The normal virulent form of this bacterium is referred to as the S form because it forms smooth colonies on a culture dish. The mutant, nonvirulent form, which lacks an enzyme needed to manufacture the polysaccharide coat, is called the R form because it forms rough colonies.

Griffith performed a series of simple experiments in which mice were infected with these bacteria, then monitored for disease symptoms (figure 14.1). Mice infected with the virulent S form died from pneumonia, whereas infection with the nonvirulent R form had no effect. This result shows that the polysaccharide coat is necessary for virulence. If the virulent S form is first heat-killed, infection does not harm the mice, showing that the coat itself is not sufficient to cause disease. Lastly, infecting mice with a mixture of heat-killed S form with live R form caused pneumonia and death in the mice. This was unexpected as neither treatment alone caused disease. Furthermore, high levels of live S form bacteria were found in the lungs of the dead mice.

Somehow, the information specifying the polysaccharide coat had passed from the dead, virulent S bacteria to the live, coatless R bacteria in the mixture, permanently altering the coatless R bacteria into the virulent S variety. Griffith called this transfer of virulence from one cell to another, **transformation.** Our modern interpretation is that genetic material was actually transferred between the cells.

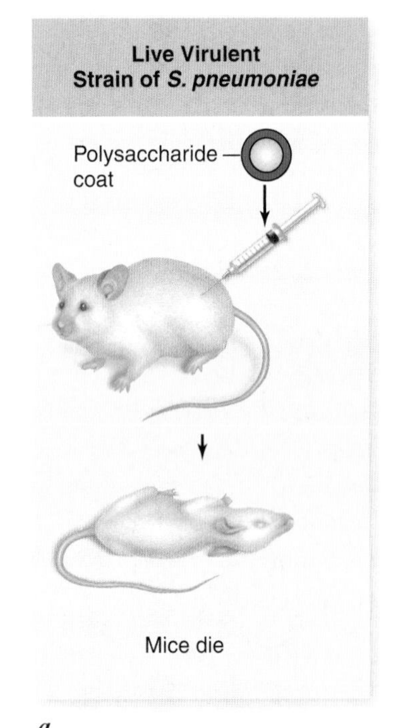

a.

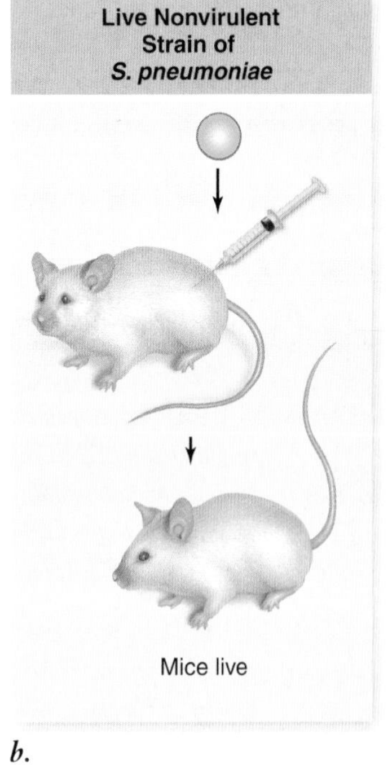

b.

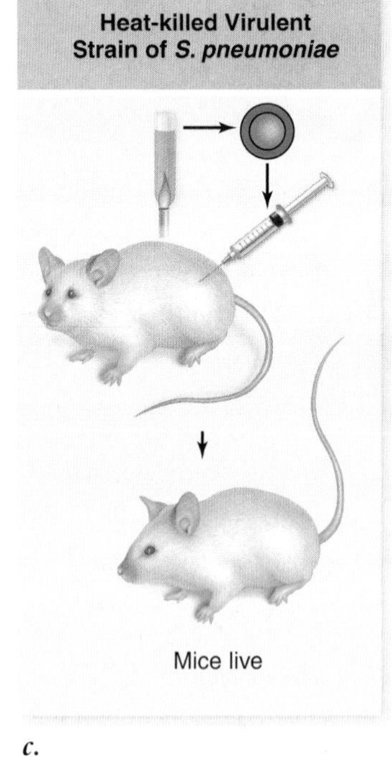

c.

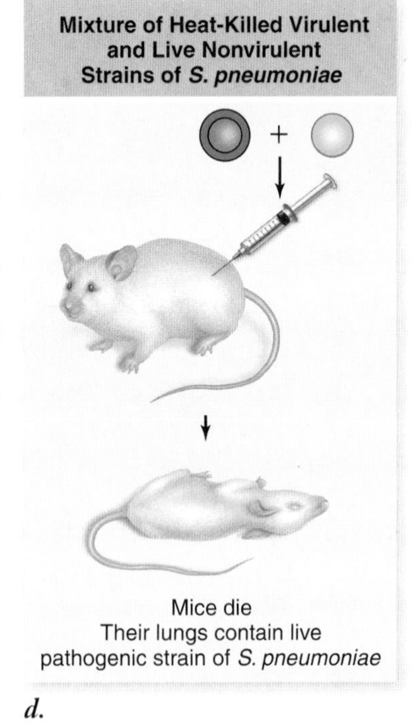

d.

figure 14.1

GRIFFITH'S EXPERIMENT. Griffith was trying to make a vaccine against pneumonia and instead discovered transformation. ***a.*** Injecting live virulent bacteria into mice produces pneumonia. Injection of nonvirulent bacteria ***(b)*** or heat killed virulent bacteria ***(c)*** had no effect. ***d.*** However, a mixture of heat killed virulent and live nonvirulent bacteria produced pneumonia in the mice. This indicates the genetic information for virulence was transferred from dead, virulent cells to live, nonvirulent cells transforming them from nonvirulent to virulent.

Avery, MacLeod, and McCarty identify the transforming principle

The agent responsible for transforming *Streptococcus* went undiscovered until 1944. In a classic series of experiments, Oswald Avery and his coworkers Colin MacLeod and Maclyn McCarty identified the substance responsible for transformation in Griffith's experiment.

They first prepared the mixture of dead S *Streptococcus* and live R *Streptococcus* that Griffith had used. Then they removed as much of the protein as they could from their preparation, eventually achieving 99.98% purity. They found that despite the removal of nearly all protein, the transforming activity was not reduced.

Moreover, the properties of this substance resembled those of DNA in several ways:

1. The elemental composition agreed closely with that of DNA.
2. When spun at high speeds in an ultracentrifuge, it migrated to the same level (density) as DNA.
3. Extracting lipids and proteins did not reduce transforming activity.
4. Protein-digesting enzymes did not affect transforming activity, nor did RNA-digesting enzymes.
5. DNA-digesting enzymes destroyed all transforming activity.

These experiments supported the identity of DNA as the substance transferred between cells by transformation and indicated that the genetic material, at least in this bacterial species, is DNA.

Hershey and Chase demonstrate that phage genetic material is DNA

Avery's results were not widely accepted at first because many biologists continued to believe that proteins were the repository of hereditary information. But additional evidence supporting Avery's conclusion was provided in 1952 by Alfred Hershey and Martha Chase, who experimented with viruses that infect bacteria. These viruses are called **bacteriophages,** or more simply, **phages.**

Viruses, described in more detail in chapter 27, are much simpler than cells; they generally consist of genetic material (DNA or RNA) surrounded by a protein coat. The phage used in these experiments is called a *lytic* phage because infection causes the cell to burst, or lyse. When such a phage infects a bacterial cell, it first binds to the cell's outer surface and then injects its genetic information into the cell. There, the viral genetic information is expressed by the bacterial cell's machinery, leading to production of thousands of new viruses. The buildup of viruses eventually causes the cell to lyse, releasing progeny phage.

The phage used by Hershey and Chase, contains only DNA and protein, and therefore it provides the simplest possible system to differentiate the roles of DNA and protein. Hershey and Chase set out to identify the molecule that the phage injects into the bacterial cells. To do this, they needed a method to label both DNA and protein in unique ways that would allow them to be distinguished. Nucleotides contain phosphorus, but proteins do not, and some amino acids contain sulfur, but DNA does not. Thus, the radioactive ^{32}P isotope can be used to label DNA specifically, and the isotope ^{35}S can be used to label proteins specifically. The two isotopes are easily distinguished based on the particles they emit when they decay.

Two experiments were performed (figure 14.2). In one, viruses were grown on a medium containing ^{32}P, which was incorporated into DNA; in the other, viruses were grown on medium containing ^{35}S, which was incorporated into coat proteins. Each group of labeled viruses was then allowed to infect separate bacterial cultures.

After infection, the bacterial cell suspension was agitated in a blender to remove the infecting viral particles from the surfaces of the bacteria. This step ensured that only the part of the virus that had been injected into the bacterial cells—that is, the genetic material—would be detected.

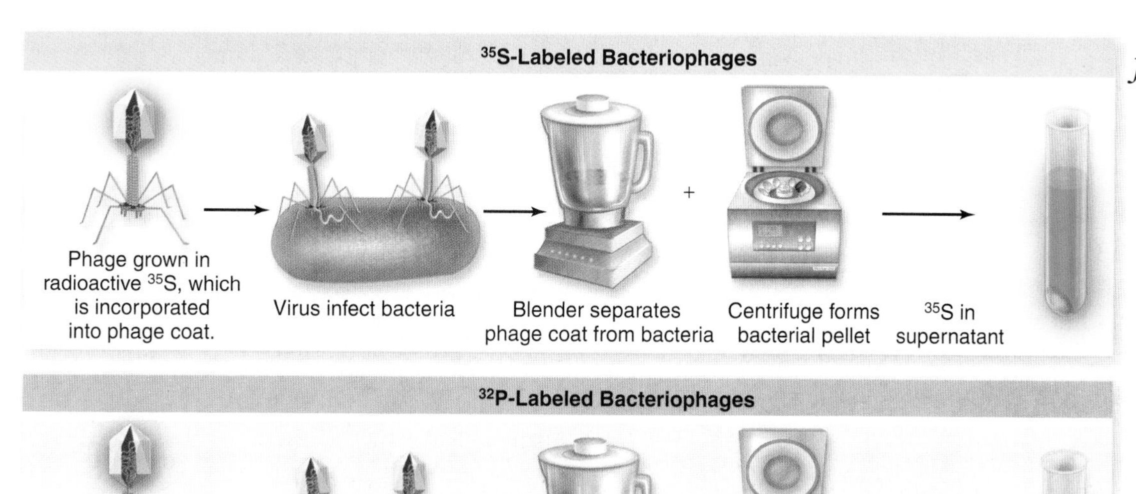

figure 14.2

HERSHEY–CHASE EXPERIMENT. ^{35}S and ^{32}P are used as specific labels for protein and DNA respectively. The virus injects genetic material into the bacterial cell reprogramming it to produce progeny virus. The blender is used to separate phage coats from cells with injected genetic material. The presence of ^{32}P and lack of ^{35}S in cell pellet indicates that injected genetic material used to reprogram cell is DNA and not protein.

Each bacterial suspension was then centrifuged to produce a pellet of cells for analysis. In the ^{32}P experiment, a large amount of radioactive phosphorus was found in the cell pellet, but in the ^{25}S experiment, very little radioactive sulfur was found in the pellet (see figure 14.2). Hershey and Chase deduced that DNA, and not protein, constituted the genetic information that viruses inject into bacteria.

DNA and protein were both considered as possible candidates for genetic material. Experiments with pneumonia-causing bacteria showed that virulence could be passed from one cell to another, termed transformation. When the factor responsible for transformation was purified, it was shown to be DNA. Labeling experiments with phage also showed the genetic material to be DNA.

14.2 DNA Structure

A Swiss chemist, Friedrich Miescher, discovered DNA in 1869, only four years after Mendel's work was published—although it is unlikely that Miescher knew of Mendel's experiments.

Miescher extracted a white substance from the nuclei of human cells and fish sperm. The proportion of nitrogen and phosphorus in the substance was different from that found in any other known constituent of cells, which convinced Miescher that he had discovered a new biological substance. He called this substance "nuclein" because it seemed to be specifically associated with the nucleus. Because Miescher's nuclein was slightly acidic, it came to be called **nucleic acid.**

DNA's components were known, but its three-dimensional structure was a mystery

Although the three-dimensional structure of the DNA molecule was not elucidated until Watson and Crick, it was known that it contained three main components (figure 14.3):

1. a five-carbon sugar
2. a phosphate (PO_4) group
3. a nitrogen-containing (nitrogenous) base. The base may be a **purine** (adenine, A, or guanine, G), a two-ringed structure, or a **pyrimidine** (thymine, T, or cytosine, C), a single-ringed structure. RNA contains the pyrimidine uracil (U) in place of thymine.

The convention in organic chemistry is to number the carbon atoms of a molecule and then to use these numbers to refer to any functional group attached to a carbon atom (chapter 3). In the ribose sugars found in nucleic acids, four of the carbon atoms together with an oxygen atom form a five-membered ring. As illustrated in figure 14.4, the carbon atoms are numbered 1′ to 5′,

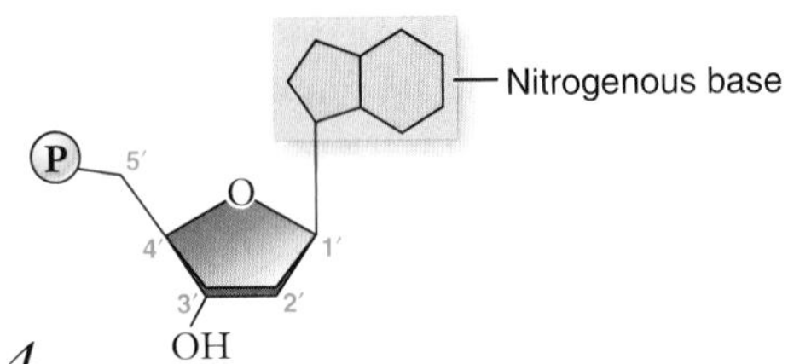

figure 14.4

NUMBERING THE CARBON ATOMS IN A NUCLEOTIDE. The carbon atoms in the sugar of the nucleotide are numbered 1′ to 5′, clockwise from the oxygen atom. The "prime" symbol (′) indicates that the carbon belongs to the sugar rather than to the base.

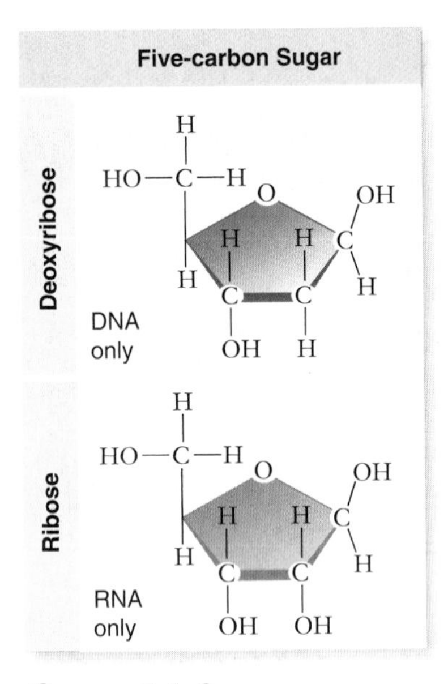

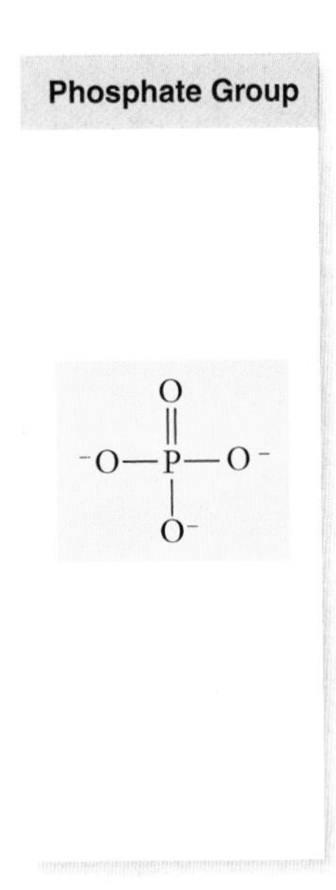

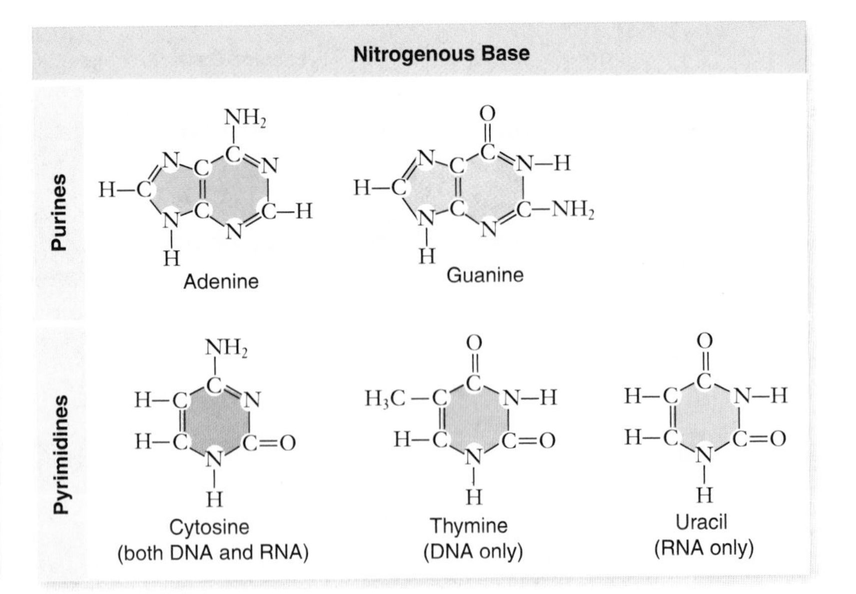

figure 14.3

NUCLEOTIDE SUBUNITS OF DNA AND RNA. The nucleotide subunits of DNA and RNA are composed of three components: (*left*) a five-carbon sugar (deoxyribose in DNA and ribose in RNA); (*middle*) a phosphate group; and (*right*) a nitrogenous base (either a purine or a pyrimidine).

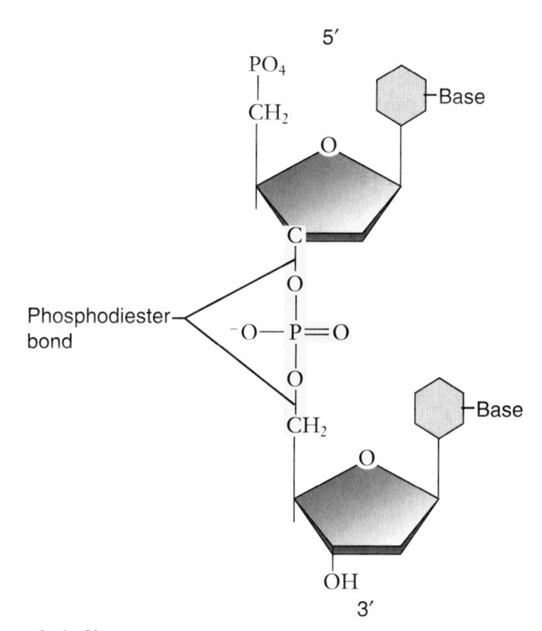

figure 14.5

A PHOSPHODIESTER BOND.

proceeding clockwise from the oxygen atom; the prime symbol (′) indicates that the number refers to a carbon in a sugar rather than to the atoms in the bases attached to the sugars.

Under this numbering scheme, the phosphate group is attached to the 5′ carbon atom of the sugar, and the base is attached to the 1′ carbon atom. In addition, a free hydroxyl (—OH) group is attached to the 3′ carbon atom.

The 5′ phosphate and 3′ hydroxyl groups allow DNA and RNA to form long chains of nucleotides, by dehydration synthesis (chapter 3). The linkage is called a **phosphodiester bond** because the phosphate group is now linked to the two sugars by means of a pair of ester bonds (figure 14.5). Many thousands of nucleotides can join together via these linkages to form long nucleic acid polymers.

Linear strands of DNA or RNA, no matter how long, will almost always have a free 5′ phosphate group at one end and a free 3′ hydroxyl group at the other. Therefore, every DNA and RNA molecule has an intrinsic polarity, and we can refer unambiguously to each end of the molecule. By convention, the sequence of bases is usually written in the 5′ to 3′ direction.

Chargaff, Franklin, and Wilkins obtained some structural evidence

To understand the model that Watson and Crick proposed, we need to review the evidence that they had available to construct their model.

Chargaff's rules

A careful study carried out by Erwin Chargaff showed that the nucleotide composition of DNA molecules varied in complex ways, depending on the source of the DNA. This strongly suggested that DNA was not a simple repeating polymer and that it might have the information-encoding properties genetic material requires. Despite DNA's complexity, however, Chargaff observed an important underlying regularity in the ratios of the bases found in native DNA: *The amount of adenine present in DNA always equals the amount of thymine, and the amount of guanine always equals the amount of cytosine.* These findings are commonly referred to as **Chargaff's rules:**

1. The proportion of A always equals that of T, and the proportion of G always equals that of C, or: A = T, and G = C.
2. It follows that there is always an equal proportion of purines (A and G) and pyrimidines (C and T).

As mounting evidence indicated that DNA stored the hereditary information, investigators began to puzzle over how such a seemingly simple molecule could carry out such a complex coding function.

Franklin: X-ray diffraction patterns of DNA

Another line of evidence provided more direct information about the possible structure of DNA. The British chemist Rosalind Franklin (figure 14.6*a*) used the technique of X-ray diffraction to analyze DNA. In X-ray diffraction, a molecule is bombarded with a beam of X-rays. The rays are bent, or diffracted, by the molecules they encounter, and the diffraction pattern is recorded on photographic film. The patterns resemble the ripples created by tossing a rock into a smooth lake (figure 14.6*b*). When analyzed mathematically, the diffraction pattern can yield information about the three-dimensional structure of a molecule.

X-ray diffraction works best on substances that can be prepared as perfectly regular crystalline arrays. At the time Franklin conducted her analysis, it was impossible to obtain true crystals of natural DNA, so she had to use DNA in the form of fibers. Maurice Wilkins, another researcher working in the same laboratory, had been able to prepare more uniformly oriented DNA fibers than anyone else at the time. Using these fibers, Franklin succeeded in obtaining crude diffraction information on natural DNA. The diffraction patterns she obtained

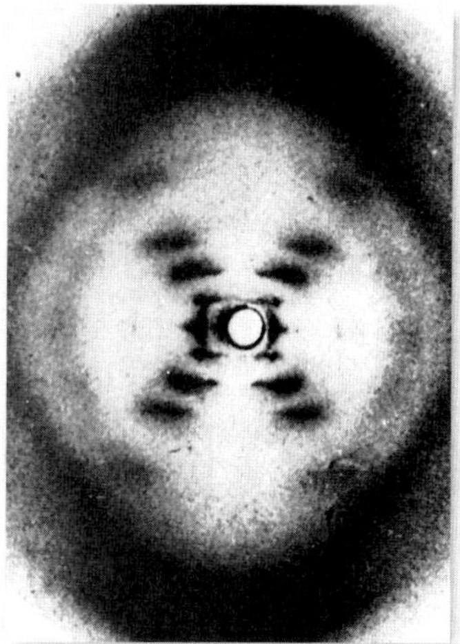

a. ***b.***

figure 14.6

ROSALIND FRANKLIN'S X-RAY DIFFRACTION PATTERNS. ***a.*** Rosalind Franklin. ***b.*** This X-ray diffraction photograph of DNA fibers, made in 1953 by Rosalind Franklin was interpreted to show helical structure of DNA.

suggested that the DNA molecule had the shape of a helix, or corkscrew, with a consistent diameter of about 2 nm and a complete helical turn every 3.4 nm.

Tautomeric forms of bases

One piece of evidence important to Watson and Crick was the form of the bases themselves. Because of the alternating double and single bonds in the bases, they actually exist in equilibrium between two different forms when in solution. The different forms have to do with keto (C=O) versus enol (C—OH) groups and amino ($—NH_2$) versus imino (=NH) groups that are attached to the bases. These structural forms are called *tautomers.*

The importance of this distinction is that the two forms exhibit very different hydrogen-bonding possibilities. The predominant forms of the bases contain the keto and amino groups (see figure 14.3), but a prominent biochemistry text of the time actually contained the opposite, and incorrect, information. Legend has it that Watson learned the correct forms while having lunch with a biochemist friend.

The Watson–Crick model fit the evidence available

Learning informally of Franklin's results before they were published in 1953, James Watson and Francis Crick, two young investigators at Cambridge University, quickly worked out a likely structure for the DNA molecule (figure 14.7), which we now know was substantially correct. Watson and Crick did not perform a single experiment themselves related to DNA structure; rather, they built detailed molecular models based on the information available.

The key to their model was Watson and Crick's understanding that each DNA molecule is actually made up of *two* chains of nucleotides that are intertwined—the double helix.

The phosphodiester backbone

The two strands of the double helix are made up of long polymers of nucleotides, and as described earlier, each strand is made up of repeating sugar and phosphate units joined by phosphodiester bonds (figure 14.8). We call this the *phosphodiester backbone* of the molecule. The two strands of the backbone are then wrapped about a common axis forming a double helix (figure 14.9). The helix is often compared to a spiral staircase, in which the two strands of the double helix are the handrails on the staircase.

Complementarity of bases

Watson and Crick proposed that the two strands were held together by formation of hydrogen bonds between bases on opposite strands. These bonds would result in specific **base-pairs:** Adenine (A) can form two hydrogen bonds with thymine (T) to form an A–T base-pair, and guanine (G) can form three hydrogen bonds with cytosine (C) to form a G–C base-pair (figure 14.10).

Note that this configuration also pairs a two-ringed purine with a single-ringed pyrimidine in each case, so that the di-

figure 14.7

THE DNA DOUBLE HELIX. James Watson (*left*) and Francis Crick (*right*) deduced the structure of DNA in 1953 from Chargaff's rules, knowing the proper tautomeric forms of the bases and Franklin's diffraction studies.

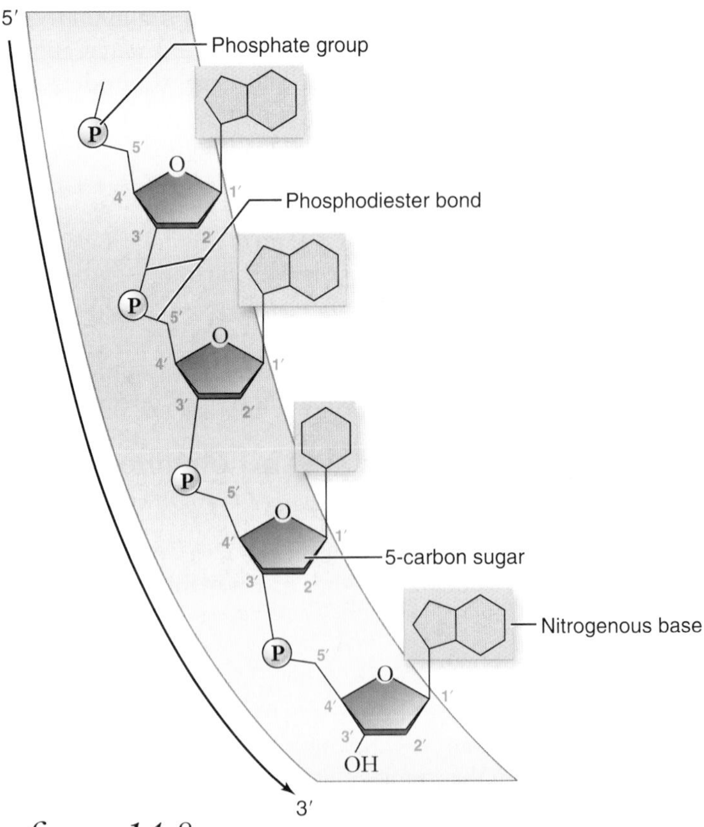

figure 14.8

STRUCTURE OF A SINGLE STRAND OF DNA. The phosphodiester backbone is composed of alternating sugar and phosphate groups. The bases are attached to each sugar.

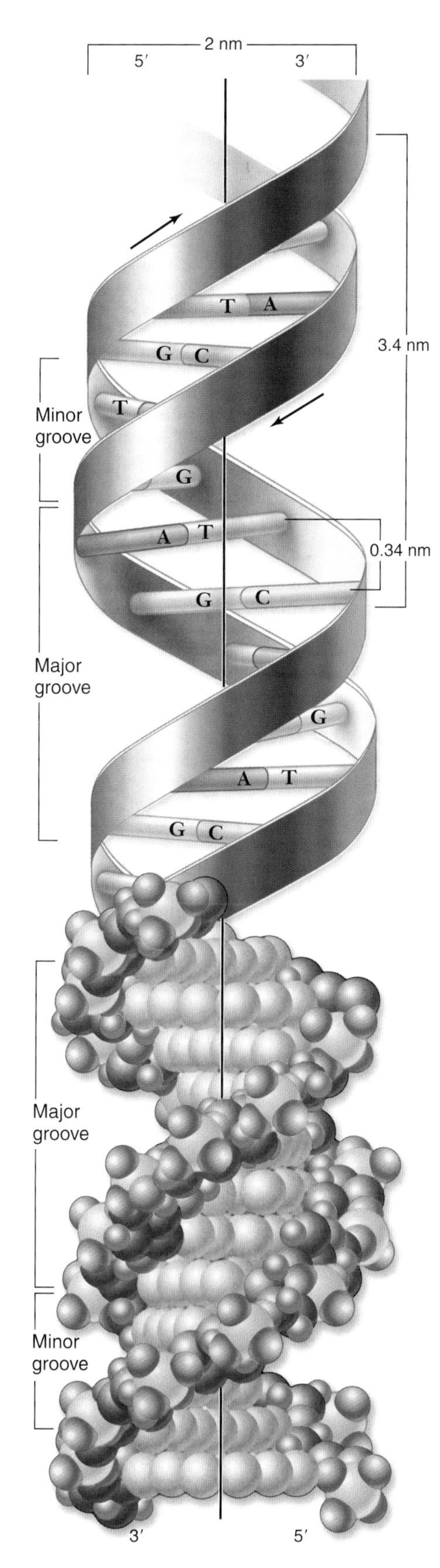

figure 14.9

THE DOUBLE HELIX. Shown with the phosphodiester backbone as a ribbon on top and a space-filling model on the bottom. The bases protrude into the interior of the helix where they hold it together by base-pairing. The backbone forms two grooves, the larger major groove and the smaller minor groove.

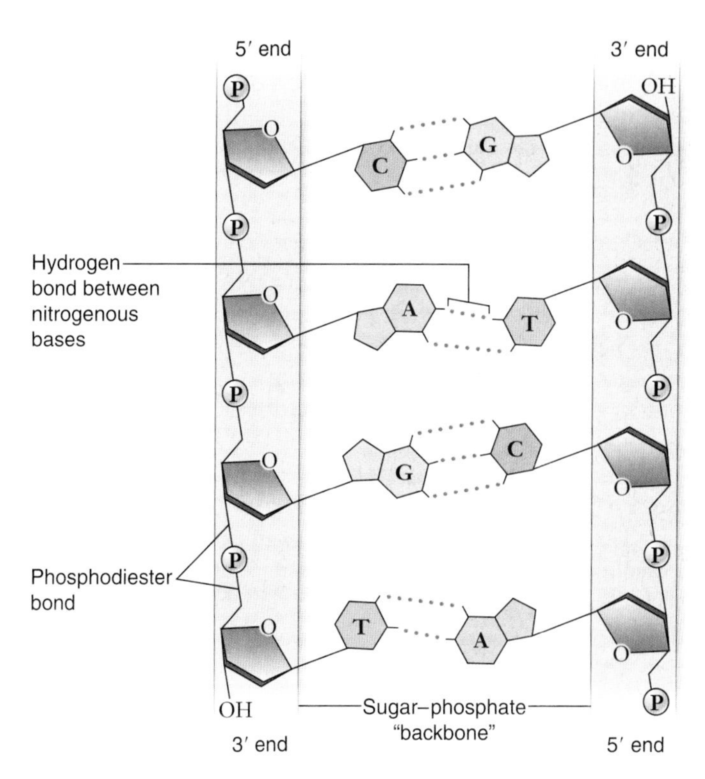

figure 14.10

BASE-PAIRING HOLDS STRANDS TOGETHER. The H-bonds that form between A and T and between G and C are shown with dashed lines. These produce AT and GC base-pairs that hold the two strands together. This always pairs a purine with a pyrimidine, keeping the diameter of the double helix constant.

inquiry

Does the Watson–Crick model account for all of the data discussed in the text?

ameter of each base-pair is the same. This consistent diameter is indicated by the X-ray diffraction data.

We refer to this pattern of base-pairing as **complementary,** which means that although the strands are not identical, they each can be used to specify the other by base-pairing. If the sequence of one strand is ATGC, then the complementary strand sequence must be TACG. This characteristic becomes critical for DNA replication and expression, as you will see later in this chapter.

The Watson–Crick model also explained Chargaff's results: In a double helix, adenine forms two hydrogen bonds with thymine, but it will not form hydrogen bonds properly with cytosine. Similarly, guanine forms three hydrogen bonds with cytosine, but it will not form hydrogen bonds properly with thymine. Because of this base-pairing, adenine and thymine will always occur in the same proportions in any DNA molecule, as will guanine and cytosine.

Antiparallel configuration

As stated earlier, a single phosphodiester strand has an inherent polarity, meaning that one end terminates in a 3′ OH and the other end terminates in a 5′ PO_4. Strands are thus referred to as having either a 5′ to 3′ or a 3′ to 5′ polarity. Two strands could be put together in two ways: with the polarity the same in each (parallel) or with the polarity opposite (antiparallel).

Native double-stranded DNA always has the antiparallel configuration, with one strand running 5′ to 3′ and the other running 3′ to 5′ (see figure 14.10). In addition to its complementarity, this antiparallel nature also has important implications for DNA replication.

The Watson–Crick DNA molecule

In the Watson and Crick model, each DNA molecule is composed of two complementary phosphodiester strands that each form a helix with a common axis. These strands are antiparallel, with the bases extending into the interior of the helix. The bases from opposite strands form base-pairs with each other to join the two complementary strands (see figures 14.9 and 14.10).

Although the hydrogen bonds between each individual base-pair are low-energy bonds, the sum of bonds between the many base-pairs of the polymer has enough energy that the entire molecule is stable. To return to our spiral staircase analogy—the backbone is the handrails, the base-pairs are the stairs.

Although the Watson–Crick model provided a rational structural for DNA, researchers had to answer further questions about how DNA could be replicated, a crucial step in cell division, and also about how cells could repair damaged or otherwise altered DNA. We explore these questions in the rest of this chapter. (In the following chapter, we continue with the genetic code and the connection between the code and protein synthesis.)

Chargaff's experiments on DNA structure revealed that the amount of adenine was equivalent to the amount of thymine, and similarly, the amount of guanosine was equivalent to the amount of cytosine. X-ray diffraction studies by Franklin and Wilkins indicated that DNA appeared to be helical. Watson and Crick built a model to elucidate the structure, consisting of two helical strands wrapped about a common axis, held together by hydrogen bonds. Adenosine pairs with thymine and guanosine pairs with cytosine, such that the strands are complementary.

14.3 Basic Characteristics of DNA Replication

The accurate replication of DNA prior to cell division is a basic and crucial function. Research has revealed that this complex process requires the participation of a large number of cellular proteins. Before geneticists could look for these details, however, they needed to perform some groundwork on the general mechanisms.

Meselson and Stahl demonstrate the semiconservative mechanism

The Watson–Crick model immediately suggested that the basis for copying the genetic information is complementarity. One chain of the DNA molecule may have any conceivable base sequence, but this sequence completely determines the sequence of its partner in the duplex.

In replication, the sequence of parental strands must be duplicated in daughter strands. That is, one parental helix with two strands must yield two daughter helices with four strands. The two daughter molecules are then separated during the course of cell division.

Three models of DNA replication are possible (figure 14.11):

1. In a **conservative model,** both strands of the parental duplex would remain intact (conserved), and new DNA copies would consist of all-new molecules. Both daughter strands would contain all-new molecules.
2. In a **semiconservative model,** one strand of the parental duplex remains intact in daughter strands (semiconserved); a new complementary strand is built for each parental strand consisting of new molecules. Daughter strands would consist of one parental strand and one newly synthesized strand.
3. In a **dispersive model,** copies of DNA would consist of mixtures of parental and newly synthesized strands; that is, the new DNA would be dispersed throughout each strand of both daughter molecules after replication.

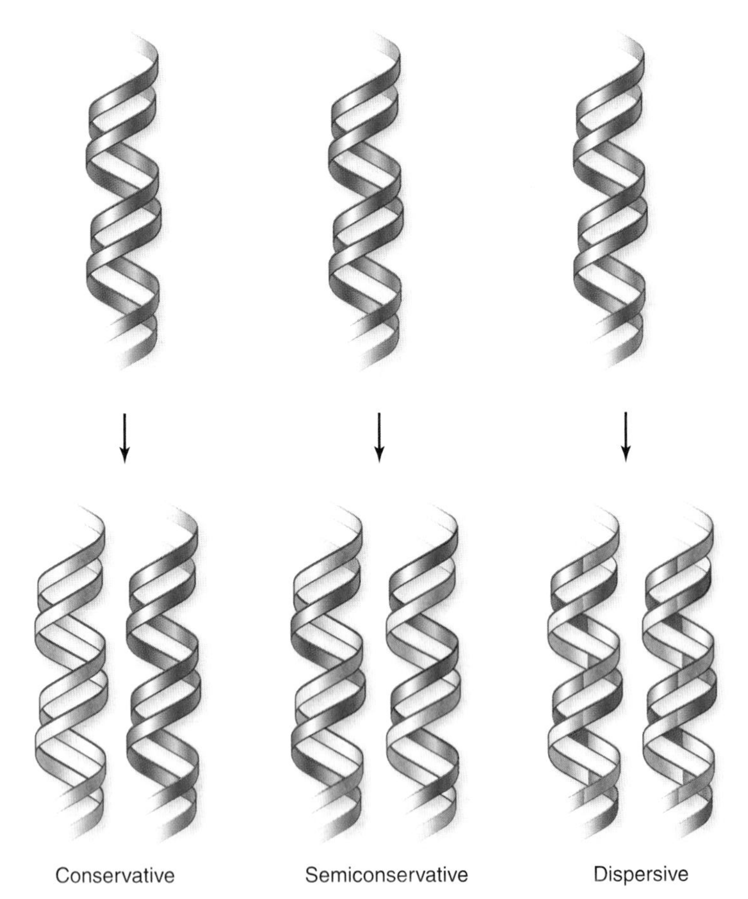

figure 14.11

THREE POSSIBLE MODELS FOR DNA REPLICATION. The conservative model produces one entirely new molecule and conserves the old. The semiconservative model produces two hybrid molecules of old and new strands. The dispersive model produces hybrid molecules with each strand a mixture of old and new.

Notice that these three models suggest general mechanisms of replication, without specifying any molecular details of the process.

The Meselson–Stahl experiment

The three models for DNA replication were evaluated in 1958 by Matthew Meselson and Franklin Stahl. To distinguish between these models, they labeled DNA and then followed the labeled DNA through two rounds of replication (figure 14.12).

The label Meselson and Stahl used though was a heavy isotope of nitrogen (^{15}N), not a radioactive label. Molecules containing ^{15}N have a greater density than those containing the common ^{14}N isotope. The technique of ultracentrifugation can be used to separate molecules that have different densities.

Bacteria were grown in a medium containing ^{15}N, which became incorporated into the bases of the bacterial DNA. After several generations, the DNA of these bacteria was denser than that of bacteria grown in a medium containing the normally available ^{14}N. Meselson and Stahl then transferred the bacteria from the ^{15}N medium to ^{14}N medium and collected the DNA at various time intervals.

The DNA for each interval was dissolved in a solution containing a heavy salt, cesium chloride. This solution was spun at very high speeds in an ultracentrifuge. The enormous centrifugal forces caused cesium ions to migrate toward the bottom of the centrifuge tube, creating a gradient of cesium concentration, and thus of density. Each DNA strand floated or sank in the gradient until it reached the point at which its density exactly matched the density of the cesium at that location. Because ^{15}N strands are denser than ^{14}N strands, they migrated farther down the tube.

The DNA collected immediately after the transfer of bacteria to new ^{14}N medium was all of one density equal to that of ^{15}N DNA alone. However, after the bacteria completed a first round of DNA replication, the density of their DNA had decreased to a value intermediate between ^{14}N-DNA alone and ^{15}N-DNA. After the second round of replication, two density classes of DNA were observed: one intermediate and one equal to that of ^{14}N-DNA (see figure 14.12).

Interpretation of the Meselson–Stahl findings

Meselson and Stahl compared their experimental data with the results that would be predicted on the basis of the three models.

1. The conservative model was not consistent with the data because after one round of replication, two densities should have been observed: DNA strands would either be all-heavy (parental) or all-light (daughter). This model is rejected.
2. The semiconservative model is consistent with all observations: After one round of replication, a single density would be predicted because all DNA molecules would have a light strand and a heavy strand. After two rounds of replication, half of the molecules would have two light strands, and half would have a light strand and a heavy strand—and so two densities would be observed. Therefore, the results support the semiconservative model.
3. The dispersive model was consistent with the data from the first round of replication, because in this model, every DNA

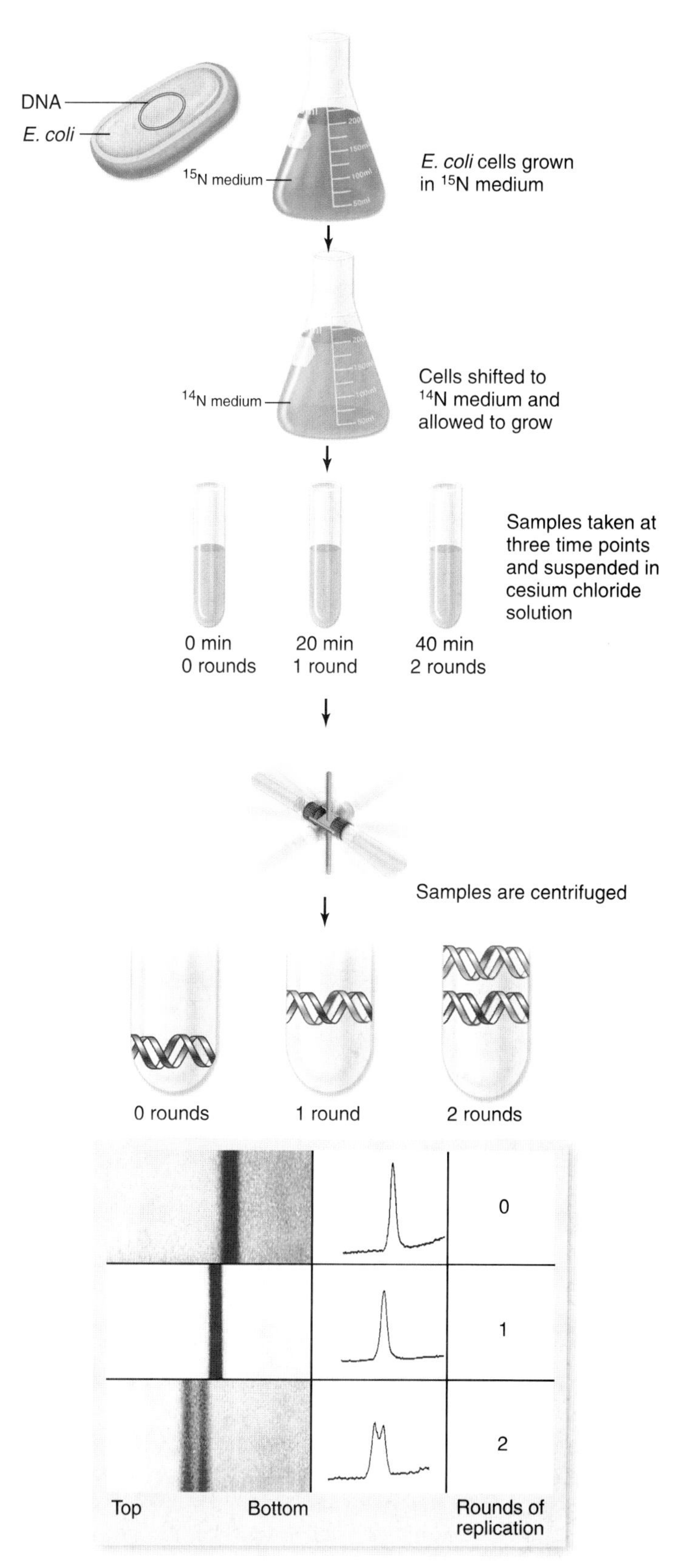

figure 14.12

THE MESELSON-STAHL EXPERIMENT. Bacteria grown in heavy ^{15}N medium are shifted to light ^{14}N medium and grown for two rounds of replication. Samples are taken at time points corresponding to zero, one, and two rounds of replication and centrifuged in cesium chloride to form a gradient. The actual data are shown at the bottom with the interpretation of semiconservative replication shown schematically.

helix would consist of strands that are mixtures of 1/2 light (new) and 1/2 heavy (old) molecules. But after two rounds of replication, the dispersive model would still yield only a single density; DNA strands would be composed of 3/4 light and 1/4 heavy molecules. Instead, two densities were observed. Therefore, this model is also rejected.

The basic mechanism of DNA replication is semiconservative. At the simplest level, then, DNA is replicated by opening up a DNA helix and making copies of both strands to produce two daughter helices, each consisting of one old strand and one new strand.

The replication process: An overview

Replication requires three things: something to copy, something to do the copying, and the building blocks to make the copy. The parental DNA molecules serve as a template, enzymes perform the actions of copying the template, and the building blocks are nucleotide triphosphates.

The process of replication can be thought of as having a beginning where the process starts; a middle where the majority of building blocks are added; and an end where the process is terminated. We use the terms *initiation*, *elongation*, and *termination* to describe a biochemical process. Although this may seem overly simplistic, in fact, discrete functions are usually required for initiation and termination that are not necessary for elongation.

Initiation: The origin

Replication does not start at a random point on a DNA double strand, but rather has been found to begin at one or more sites called origins of replication. Initiator proteins recognize and bind to the origin, forming a complex that opens the helix to expose single-stranded templates used for the process of building a new strand.

Elongation: DNA polymerase

A number of enzymes work together to accomplish the task of assembling a new strand, but the enzyme that actually matches the existing DNA bases with complementary nucleotides and

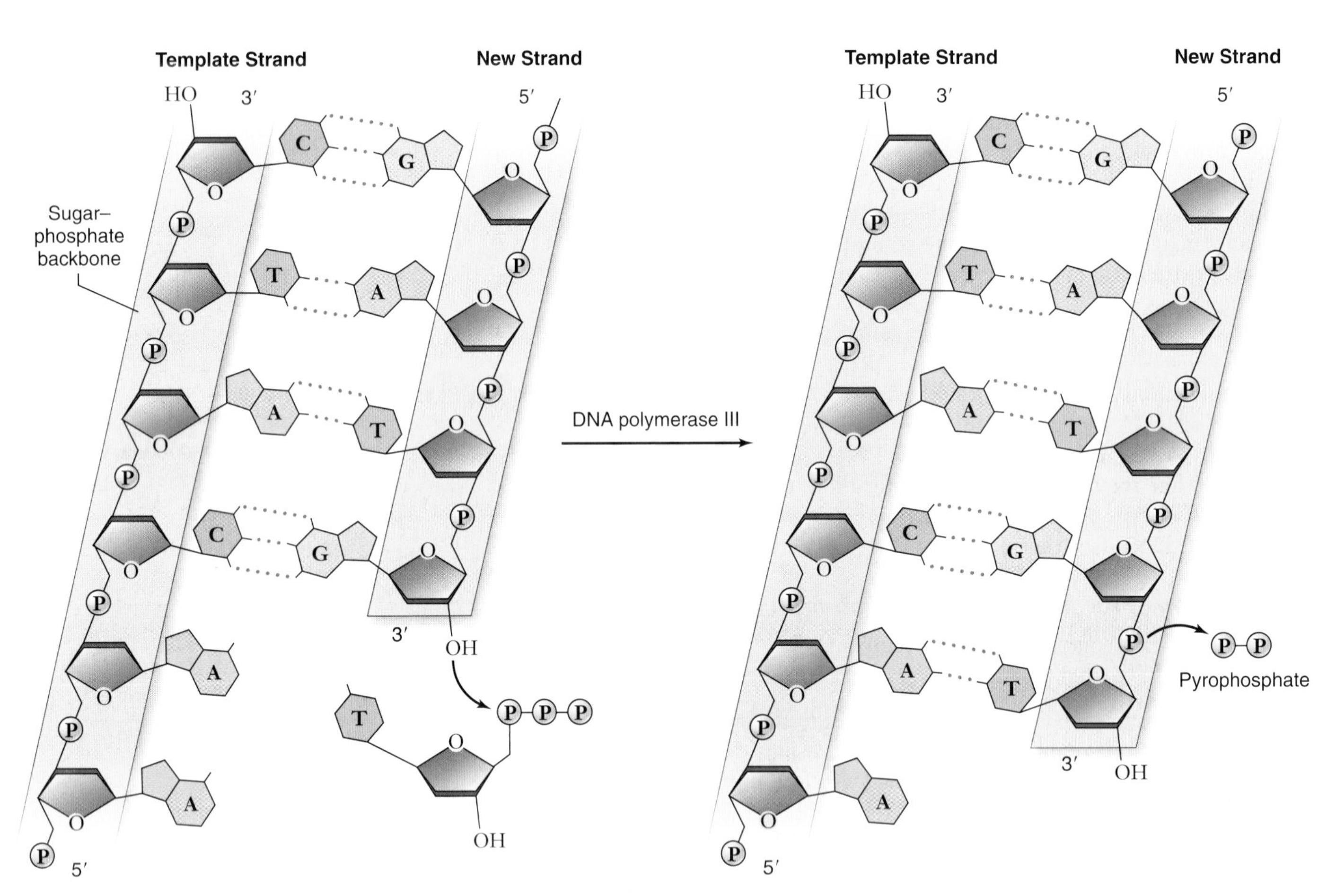

figure 14.13

ACTION OF DNA POLYMERASE. DNA polymerases add nucleotides to the 3′ end of a growing chain. The nucleotide added depends on the base that is in the template strand. Each new base must be complementary to the base in the template strand. With the addition of each new nucleotide triphosphate, two of its phosphates are cleaved off as pyrophosphate.

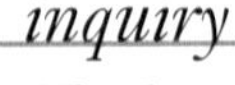

inquiry

Why do you think it is important that the sugar–phosphate backbone of DNA is held together by covalent bonds, and the cross-bridges between the two strands are held together by hydrogen bonds?

then links the nucleotides together to make the new strand is **DNA polymerase** (figure 14.13). As described shortly, many different types of DNA polymerases have been discovered.

All DNA polymerases that have been examined have several common features. They all add new bases to the 3′ end of existing strands. That is, they synthesize in a 5′ to 3′ direction by extending a strand base-paired to the template. All DNA polymerases also require a **primer** to begin synthesis; they cannot begin without a strand of RNA or DNA base-paired to the template. RNA polymerases do not have this requirement, so they usually synthesize the primers.

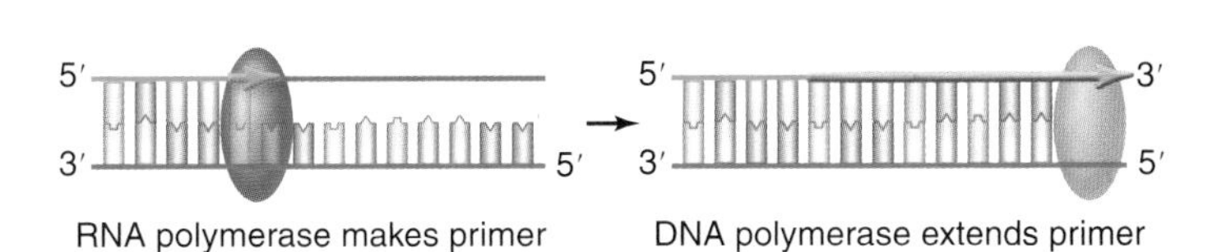

Termination

An end point for replication is as important as a starting point. In prokaryotes, which have circular DNA, the replication ends when the process comes around to the origin again. In eukaryotes, end points for each chromosome are indicated by *telomeres*, specific regions of repeated bases.

The details of this process are described in the sections that follow—first for prokaryotes, used for much of the initial research on DNA replication, and then for eukaryotes.

Meselson and Stahl showed that the basic mechanism of replication is semiconservative: Each new DNA helix is composed of one old strand and one new strand. The process of replication involves three phases: initiation, elongation, and termination. Initiation and termination both occur at specific sites and require functions not found in elongation. Elongation is accomplished by DNA polymerase enzymes that synthesize in a 5′ to 3′ direction from a primer, usually RNA.

14.4 Prokaryotic Replication

To build up a more detailed picture of replication, we first concentrate on prokaryotic replication using *E. coli* as a model. We can then look at eukaryotic replication primarily in how it differs from the prokaryotic system.

Prokaryotic replication starts at a single origin

Replication in *E. coli* initiates at a specific site, the origin (called *oriC*), and ends at a specific site, the terminus. The sequence of *oriC* consists of repeated nucleotides that bind an initiator protein and an AT-rich sequence that can be opened easily during initiation of replication. (A–T base-pairs have only two hydrogen bonds, compared with the three hydrogen bonds in G–C base-pairs.)

After initiation, replication proceeds bidirectionally from this unique origin to the unique terminus (figure 14.14). The complete chromosome plus the origin can be thought of as a single functional unit called a **replicon.**

E. coli has at least three different DNA polymerases

DNA polymerase, as mentioned earlier, refers to a group of enzymes responsible for the building of a new DNA strand from the template. The first DNA polymerase isolated in *E. coli* was given the name **DNA polymerase I (pol I).** At first, investigators assumed this polymerase was responsible for the bulk synthesis of DNA during replication. A mutant was isolated, however, that had no pol I activity, but that still showed replication activity. Two additional polymerases were isolated from this strain of *E. coli* and were named **DNA polymerase II (pol II)** and **DNA polymerase III (pol III).** As with all other known polymerases, all three of these enzymes synthesize polynucleotide strands only in the 5′ to 3′ direction, and require a primer.

Many DNA polymerases have additional enzymatic activity that aids their function. This activity is a nuclease activity, or

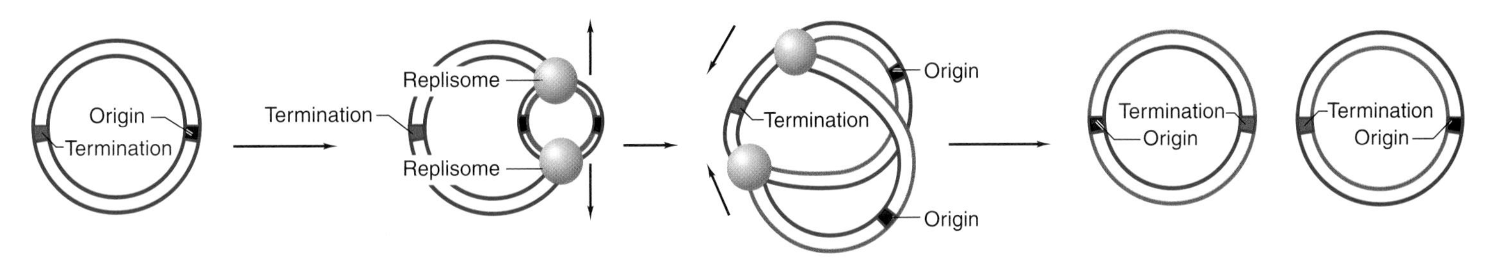

figure 14.14

REPLICATION IS BIDIRECTIONAL FROM A UNIQUE ORIGIN. Replication initiates from a unique origin. Two separate replisomes are loaded onto the origin and initiate synthesis in the opposite directions on the chromosome. These two replisomes continue in opposite directions until they come to a unique termination site.

the ability to break phosphodiester bonds between nucleotides. Nucleases are classified as either **endonucleases** (cut DNA internally) or **exonucleases** (chew away at an end of DNA). DNA pol I, pol II, and pol III have 3′ to 5′ exonuclease activity, which serves as a proofreading function because it allows the enzyme to remove a mispaired base. In addition, the DNA pol I enzyme also has a 5′ to 3′ exonuclease activity, the importance of which will become clear shortly.

The three different polymerases have different roles in the replication process. DNA pol III is the main replication enzyme, it is responsible for the bulk of DNA synthesis. DNA pol I acts on the lagging strand to remove primers and replace them with DNA. The pol II enzyme does not appear to play a role in replication but is involved in DNA repair processes.

For many years, these three polymerases were thought to be the only DNA polymerases in *E. coli*, but recently several new polymerases have been identified. There are now five known polymerases, although all are not active in DNA replication.

Unwinding DNA requires energy and causes torsional strain

Although some DNA polymerases can unwind DNA as they synthesize new DNA, another class of enzymes has the single function of unwinding DNA strands to make this process more efficient. Enzymes that use energy from ATP to unwind the DNA template are called **helicases.**

The single strands of DNA produced by helicase action are unstable because the process exposes the hydrophobic bases to water. Cells solve this problem by using a protein to coat exposed single strands called single-strand binding protein (SSB).

The unwinding of the two strands introduces torsional strain in the DNA molecule. Imagine two rubber bands twisted together. If you now unwind the rubber bands, what happens? The rubber bands, already twisted about each other, will further coil in space. When this happens with a DNA molecule it is called **supercoiling** (figure 14.15). The branch of mathematics that studies how forms twist and coil in space is called *topology*, and therefore we describe this coiling of the double helix as the *topological state* of DNA. This state describes how the double helix itself coils in space. You have already seen an example of this coiling with DNA wrapped about histone proteins in the nucleosomes of eukaryotic chromosomes (chapter 10).

Enzymes that can alter the topological state of DNA are called **topoisomerases.** Topoisomerase enzymes act to relieve the torsional strain caused by unwinding and to prevent this supercoiling from happening. **DNA gyrase** is the topoisomerase involved in DNA replication (see figure 14.15).

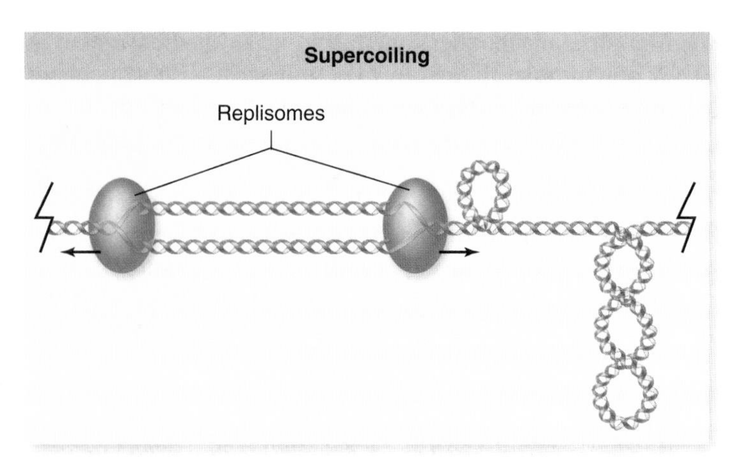

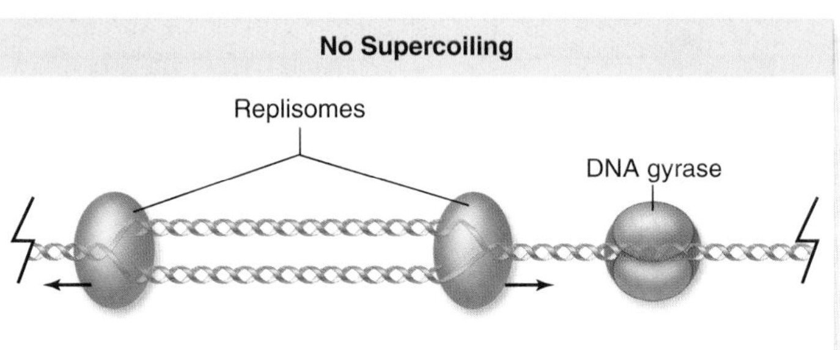

figure 14.15

UNWINDING THE HELIX CAUSES TORSIONAL STRAIN. If the ends of a linear DNA molecule are constrained, as they are in the cell, unwinding the helix produces torsional strain. This can cause the double helix to further coil in space (supercoiling). The enzyme DNA gyrase can relieve supercoiling.

Replication is semidiscontinuous

Earlier, DNA was described as being antiparallel—meaning that one strand runs in the 3′ to 5′ direction, and its complementary strand runs in the 5′ to 3′ direction. The antiparallel nature of DNA combined with the nature of the polymerase enzymes puts constraints on the replication process. Because polymerases can synthesize DNA in only one direction, and the two DNA strands run in opposite directions, polymerases on the two strands must be synthesizing DNA in opposite directions (figure 14.16).

The requirement of DNA polymerases for a primer means that on one strand primers will need to be added as the helix is opened up (see figure 14.16). This means that one strand can be synthesized in a continuous fashion from an initial primer, but the other strand must be synthesized in a discontinuous fashion with multiple priming events and short sections of DNA being assembled. The strand that is continuous is called the **leading strand,** and the strand that is discontinuous is the **lagging strand.** DNA fragments synthesized on the lagging strand are named **Okazaki fragments** in honor of the man who first experimentally demonstrated discontinuous synthesis. They introduce a need for even more enzymatic activity on the lagging strand, as is described next.

Synthesis occurs at the replication fork

The partial opening of a DNA helix to form two single strands has a forked appearance, and is thus called the **replication fork.** All of the enzymatic activities that we have discussed plus a few more are found at the replication fork (table 14.1). Synthesis on the leading strand and on the lagging strand proceed in different ways, however.

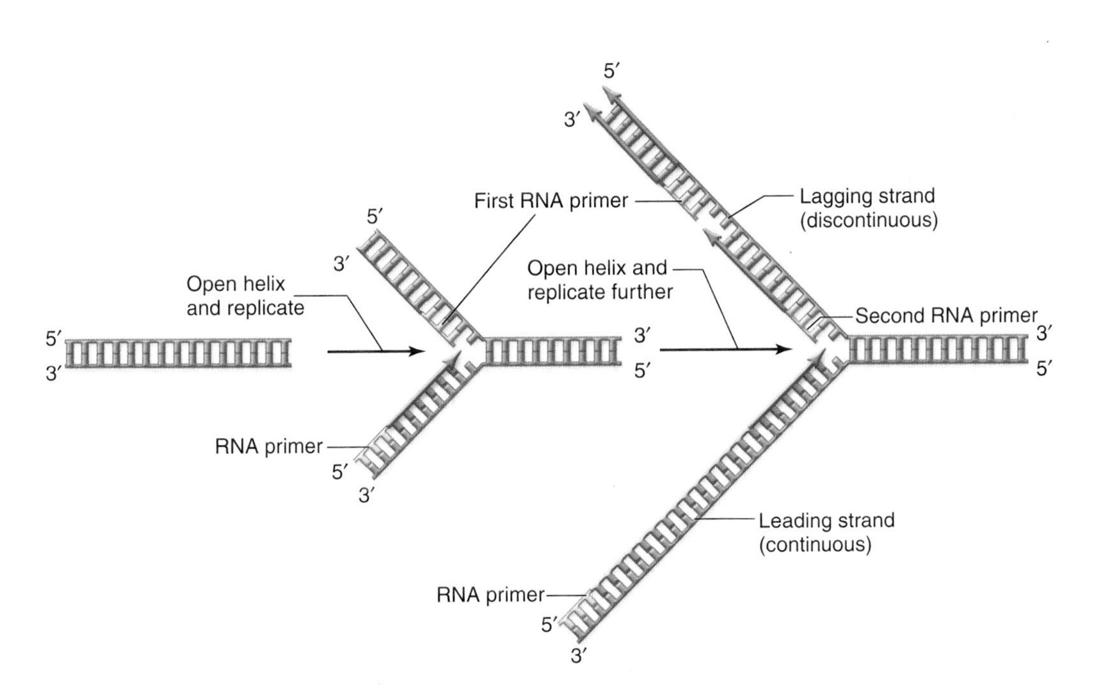

figure 14.16

REPLICATION IS SEMIDISCONTINUOUS. The 5′ to 3′ synthesis of the polymerase and the antiparallel nature of DNA mean that only one strand, the leading strand, can be synthesized continuously. The other strand, the lagging strand, must be made in pieces, each with its own primer.

Priming

The primers required by DNA polymerases during replication are synthesized by the enzyme DNA **primase.** This enzyme is an RNA polymerase that synthesizes short stretches of RNA 10–20 bp (base-pairs) long that function as primers for DNA polymerase. Later on, the RNA primer is removed and replaced with DNA.

Leading strand synthesis

Synthesis on the leading strand is relatively simple. A single priming event is required, and then the strand can be extended indefinitely by the action of DNA pol III. If the enzyme remains attached to the template, it can synthesize around the entire circular *E. coli* chromosome.

The ability of a polymerase to remain attached to the template is called **processivity.** The pol III enzyme is a large multisubunit enzyme that has high processivity due to the action of one subunit of the enzyme, called the β *subunit* (figure 14.17*a*).

TABLE 14.1 DNA Replication Enzymes of *E. coli*

Protein	Role	Size (kDa)	Molecules per Cell
Helicase	Unwinds the double helix	300	20
Primase	Synthesizes RNA primers	60	50
Single-strand binding protein	Stabilizes single-stranded regions	74	300
DNA gyrase	Relieves torque	400	250
DNA polymerase III	Synthesizes DNA	≈900	20
DNA polymerase I	Erases primer and fills gaps	103	300
DNA ligase	Joins the ends of DNA segments; DNA repair	74	300

The β subunit is made up of two identical protein chains that come together to form a circle. This circle can be loaded onto the template like a clamp to hold the pol III enzyme to the DNA (figure 14.17*b*). This structure is therefore referred to as the "sliding clamp," and a similar structure is found in eukaryotic polymerases as well.

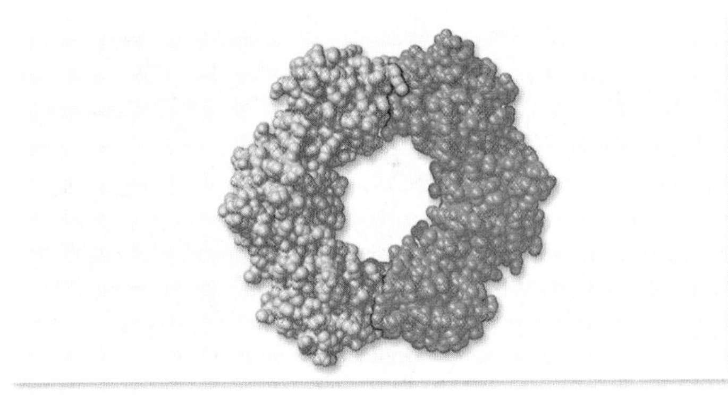

a.

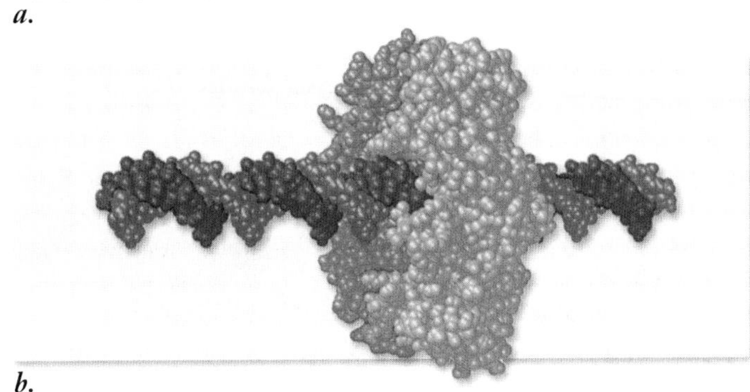

b.

figure 14.17

THE DNA POLYMERASE SLIDING CLAMP. ***a.*** The β subunit forms a ring that can encircle DNA. ***b.*** The β subunit is shown attached to the DNA. This forms the "sliding clamp" that keeps the polymerase attached to the template.

Lagging strand synthesis

The discontinuous nature of synthesis on the lagging strand requires the cell to do much more work than on the leading strand (see figure 14.16). Primase is needed to synthesize primers for each Okazaki fragment, and then all these RNA primers need to be removed and replaced with DNA. Finally, the fragments need to be stitched together.

DNA pol III accomplishes the synthesis of Okazaki fragments. The removal and replacement of primer segments, however, is accomplished by DNA pol I. Using its 5′ to 3′ exonuclease activity, it can remove primers in front and then replace them by using its usual 5′ to 3′ polymerase activity. The synthesis is primed by the previous Okazaki fragment, which is composed of DNA and has a free 3′ OH that can be extended.

This leaves only the last phosphodiester bond to be formed where synthesis by pol I ends. This is done by **DNA ligase,** which seals this "nick," eventually joining the Okazaki fragments into complete strands. All of this activity on the lagging strand is summarized in figure 14.18.

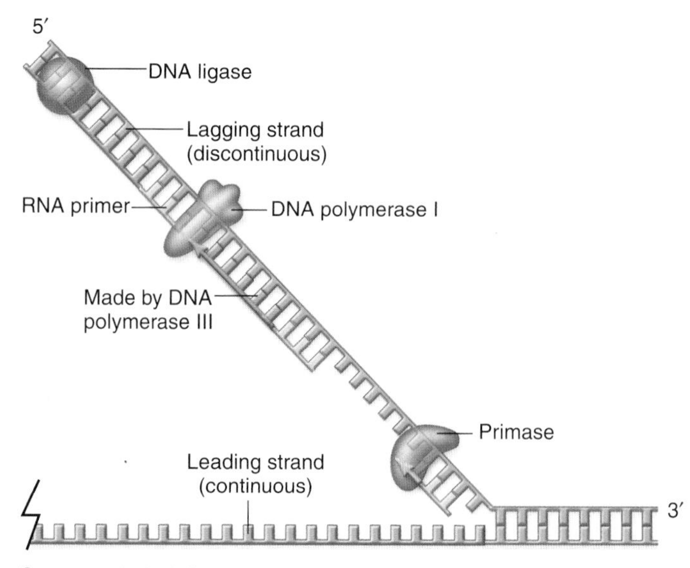

figure 14.18

LAGGING STRAND SYNTHESIS. The action of primase synthesizes the primers needed by DNA polymerase III (not shown). These primers are removed by DNA polymerase I using its 5′ to 3′ exonuclease activity and extending the previous Okazaki fragment to replace the RNA. The nick between Okazaki fragments after primer removal is sealed by DNA ligase.

inquiry

What is the role of DNA ligase? What would happen to DNA replication in a cell where this enzyme is not functional?

Termination

Termination occurs at a specific site located roughly opposite *oriC* on the circular chromosome. The last stages of replication produce two daughter molecules that are intertwined like two rings in a chain. These intertwined molecules are unlinked by the same enzyme that relieves torsional strain at the replication fork: DNA gyrase.

The replisome contains all the necessary enzymes for replication

The enzymes involved in DNA replication form a macromolecular assembly called the **replisome.** This assembly can be thought of as the "replication organelle," just as the ribosome is the protein synthesis organelle. The replisome is a protein machine capable of fast and accurate replication of DNA during cell division.

The replisome has two main subcomponents: the *primosome*, and a complex of two DNA pol III enzymes, one for each strand. The primosome is composed of primase and helicase, along with a number of accessory proteins. The need for constant priming on the lagging strand explains the need for the primosome complex as part of the entire replisome at the replication fork.

The two pol III complexes include two synthetic core subunits, each with its own β subunit, and a number of other proteins that hold the entire complex together.

Even given the difficulties with lagging-strand synthesis, the two pol III enzymes in the replisome are active on both leading and lagging

figure 14.19

THE REPLICATION FORK. A model for the structure of the replication fork with two polymerase III enzymes held together by a large complex of accessory proteins. These include the "clamp loader," which loads the β-subunit sliding clamp periodically on the lagging strand. The polymerase III on the lagging strand periodically releases its template and reassociate along with the β-clamp. The loop in the lagging strand template allows both polymerases to move in the same direction despite DNA being antiparallel. Primase, which makes primers for the lagging strand fragments, and helicase are also associated with the central complex. Polymerase I removes primers and ligase joins the fragments together.

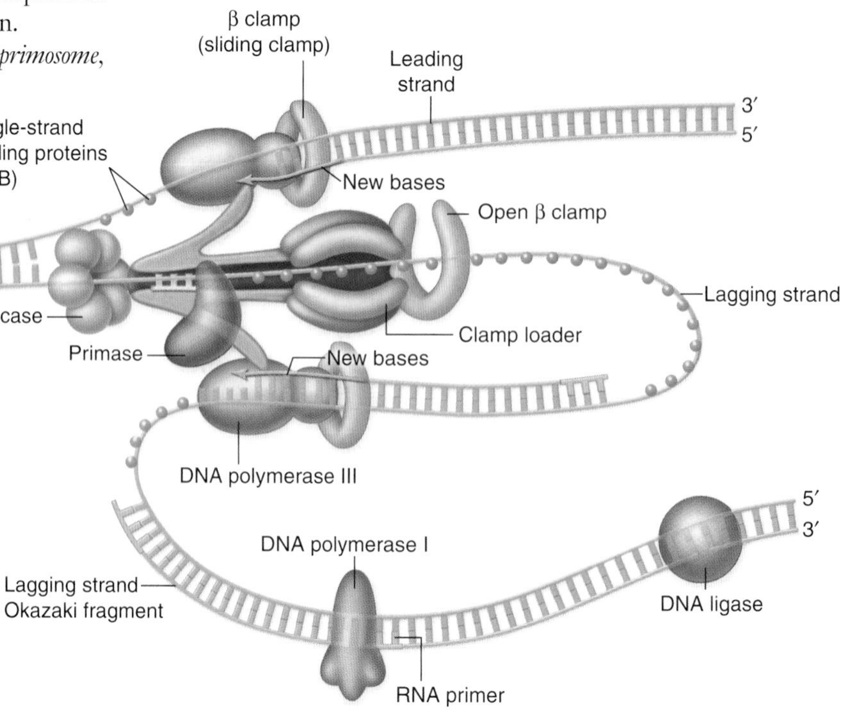

strands simultaneously. How can the two strands be synthesized in the same direction when the strands are antiparallel? The model first proposed, still with us in some form, involves a loop formed in the lagging strand, so that the polymerases can move in the same direction (figure 14.19). Current evidence also indicates that this replication complex is probably stationary, with the DNA strand moving through it like thread in a sewing machine, rather than the complex moving along the DNA strands. This stationary complex also pushes the newly synthesized DNA outward, which may aid in chromosome segregation. This process is summarized in figure 14.20.

DNA replication occurs at the replication fork, where the two strands are separated. Assembled here is a massive complex, the replisome, containing DNA polymerase III, primase, helicase, and other proteins. The lagging strand requires DNA polymerase I to remove the primers and replace them with DNA and ligase to join Okazaki fragments. The polymerases for each strand are actually part of a single complex made possible by a loop in the lagging strand. Replication starts at a unique site, the origin *(oriC)* and continues bidirectionally to a unique termination site.

1. A DNA polymerase III enzyme is active on each strand. Primase synthesizes new primers for the lagging strand.

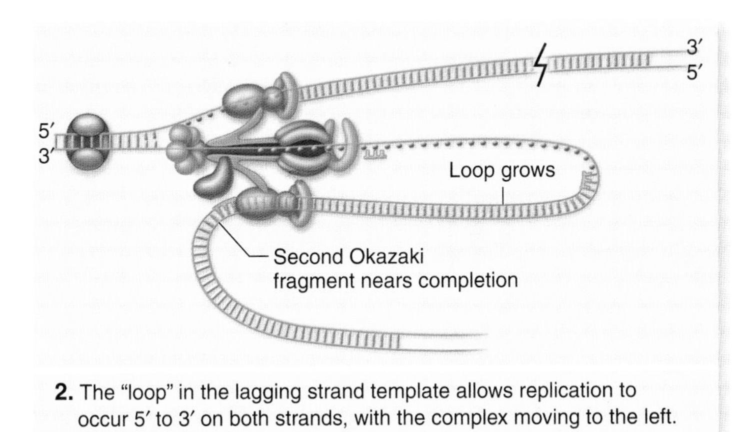

2. The "loop" in the lagging strand template allows replication to occur 5′ to 3′ on both strands, with the complex moving to the left.

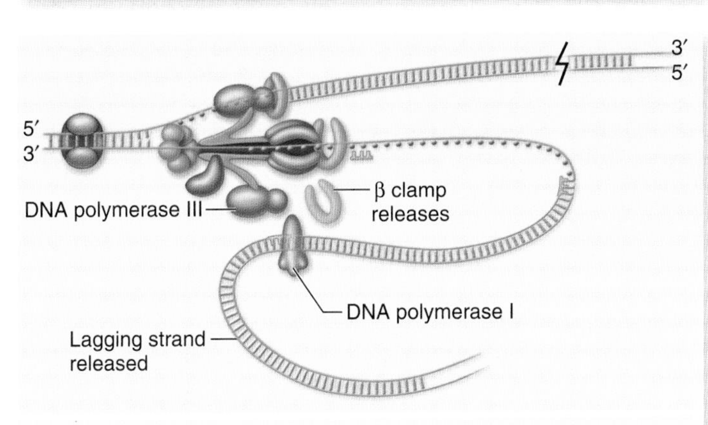

3. When the polymerase III on the lagging strand hits the previously synthesized fragment, it releases the β clamp and the template strand. DNA polymerase I attaches to remove the primer.

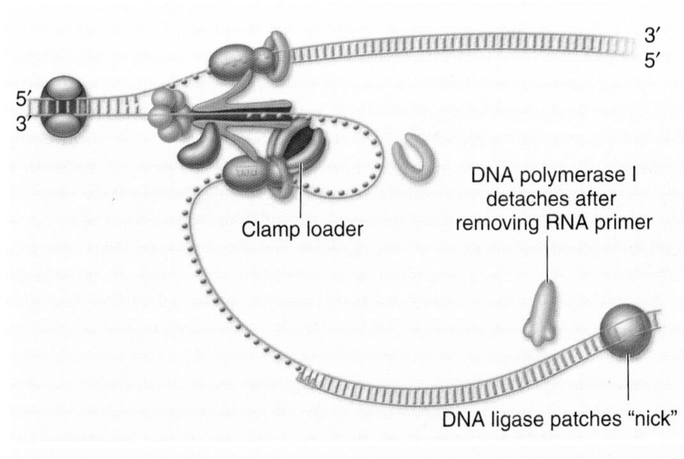

4. The clamp loader attaches the β clamp and transfers this to polymerase III, creating a new loop in the lagging strand template. DNA ligase joins the fragments after DNA polymerase I removes the primers.

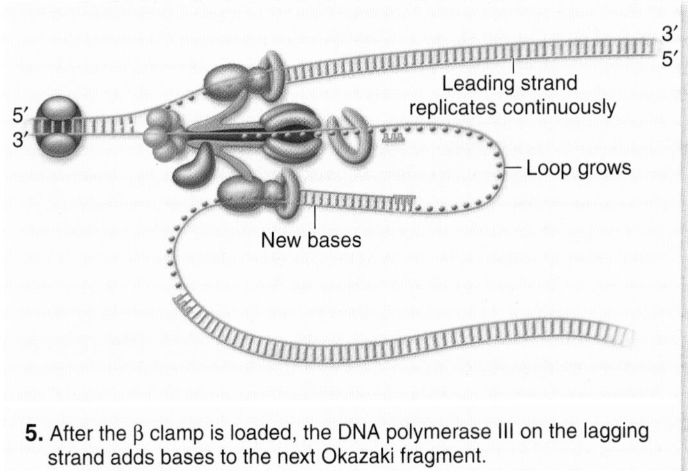

5. After the β clamp is loaded, the DNA polymerase III on the lagging strand adds bases to the next Okazaki fragment.

figure 14.20

DNA SYNTHESIS BY THE REPLISOME. The semidiscontinuous synthesis of DNA is illustrated in stages using the model from figure 14.19.

14.5 Eukaryotic Replication

Eukaryotic replication is complicated by two main factors: the larger amount of DNA organized into multiple chromosomes, and the linear structure of the chromosomes. This process requires new enzymatic activities only for dealing with the ends of chromosomes; otherwise the basic enzymology is the same.

figure 14.21

DNA OF A SINGLE HUMAN CHROMOSOME. This chromosome has been relieved of most of its packaging proteins leaving the DNA in its native form. The residual protein scaffolding appears as the dark material in the lower part of the micrograph.

Eukaryotic replication requires multiple origins

The sheer amount of DNA and how it is packaged constitute a problem for eukaryotes (figure 14.21). Eukaryotes usually have multiple chromosomes that are each larger than the *E. coli* chromosome. The machinery could in principle be the same; however, if only a single unique origin existed for each chromosome, the length of the time necessary for replication would be prohibitive. This problem is solved by the use of multiple origins of replication for each chromosome, resulting in multiple *replicons*, sections of DNA replicated from individual origins (figure 14.22).

The origins are not as sequence-specific as *oriC*, and their recognition seems to depend on chromatin structure as well as on sequence. The number of origins that "fire" can also be adjusted during the course of development, so that early on, when cell divisions need to be rapid, more origins are activated.

The enzymology of eukaryotic replication is more complex

The replication machinery of eukaryotes is similar to that found in *E. coli*, but it is larger and more complex. The initiation phase of replication requires more factors to assemble both

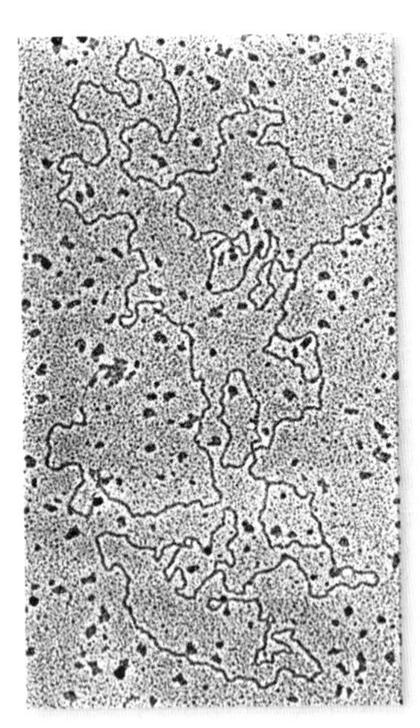

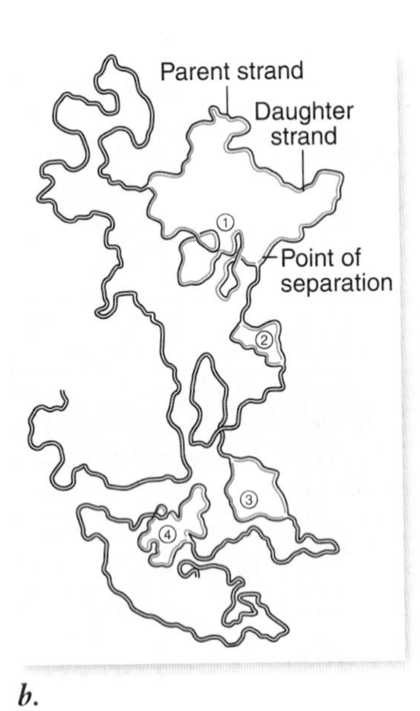

a. *b.*

figure 14.22

EUKARYOTIC CHROMOSOMES POSSESS NUMEROUS REPLICATION UNITS. ***a.*** The electron micrograph shows eukaryotic DNA with four replication units, each with two replication forks, being replicated. ***b.*** The drawing shows the four replication units with newly synthesized strands in red and parental strands in black.

helicase and primase complexes onto the template, then load the polymerase with its sliding clamp unit.

The eukaryotic primase is interesting in that it is a complex of both an RNA polymerase and a DNA polymerase. It first makes short RNA primers, then extends these with DNA to produce the final primer. The reason for this added complexity is unclear.

The main replication polymerase itself is also a complex of two different enzymes that work together. One is called *DNA polymerase epsilon* (pol ε) and the other *DNA polymerase delta* (pol δ). The sliding clamp subunit that allows the enzyme complex to stay attached to the template also exists in eukaryotes, but it is called PCNA (for proliferating cell nuclear antigen). This unusual name reflects the fact that PCNA was first identified as an antibody-inducing protein in proliferating (dividing) cells. Despite the additional complexity, the action of the replisome seems to be similar to that described earlier in *E. coli*, and the replication fork has essentially the same components as well.

Linear chromosomes require different termination

The specialized structures found on the ends of eukaryotic chromosomes are called **telomeres.** These structures protect the ends of chromosomes from nucleases and maintain the integrity of linear chromosomes. These telomeres are composed of specific DNA sequences, but they are not made by the replication complex.

Replicating ends

The very structure of a linear chromosome causes a cell problems in replicating the ends. The directionality of polymerases, combined with their requirement for a primer, create this problem.

Consider a simple linear molecule like the one in figure 14.23. Replication of one end of each template strand is simple, namely the 5′ end of the leading-strand template. When the polymerase reaches this end, synthesizing in the 5′ to 3′ direction, it eventually runs out of template and is finished.

But on the other strand's end, the 3′ end of the lagging strand, removal of the last primer on this end leaves a gap. This gap cannot be primed, meaning that the polymerase complex cannot finish this end properly. The result would be a gradual shortening of chromosomes with each round of cell division (see figure 14.23).

The action of telomerase

When the sequence of telomeres was determined, they were found to be composed of short repeated sequences of DNA. This repeating nature is easily explained by their synthesis. They are made by an enzyme called **telomerase,**

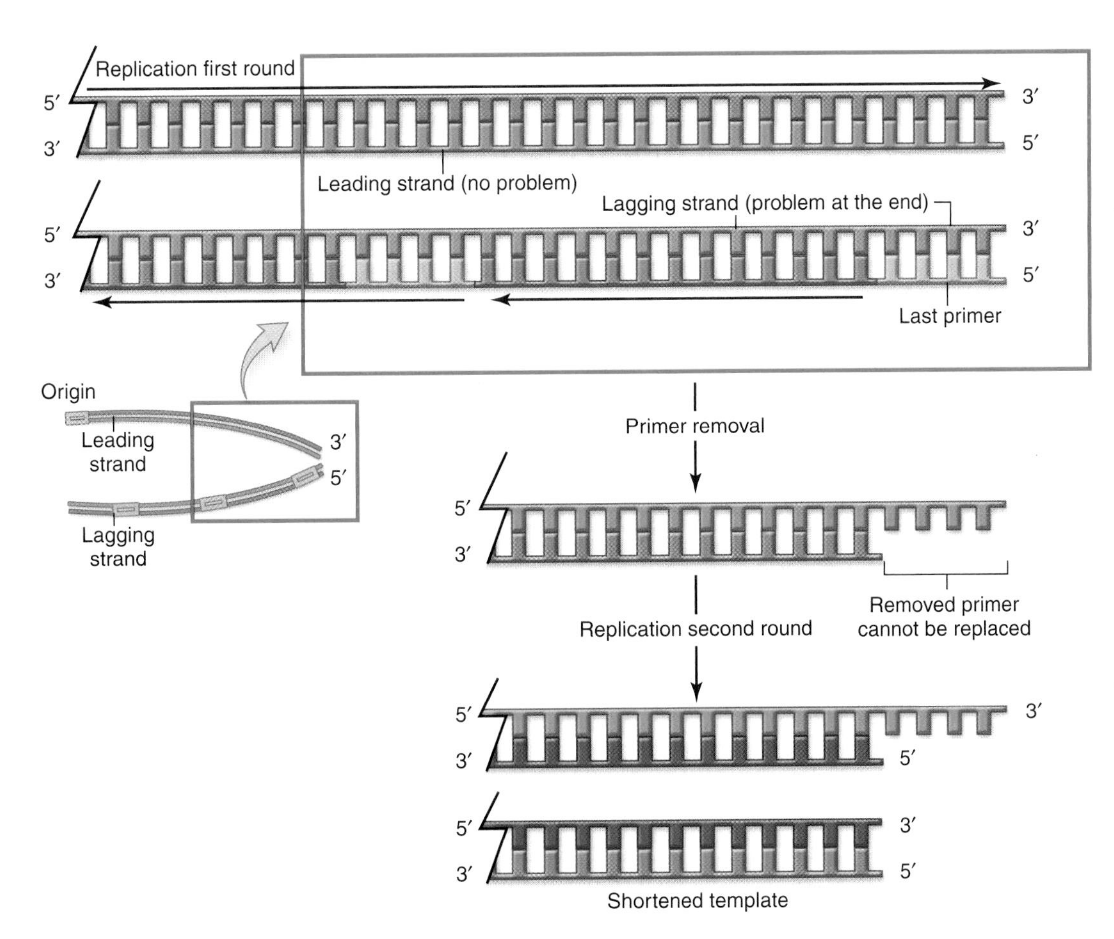

figure 14.23

REPLICATION OF THE END OF LINEAR DNA. Only one end is shown for simplicity, the problem exists at both ends. The leading strand can be completely replicated, but the lagging strand cannot be finished. When the last primer is removed, it cannot be replaced. During the next round of replication, when this shortened template is replicated, it will produce a shorter chromosome.

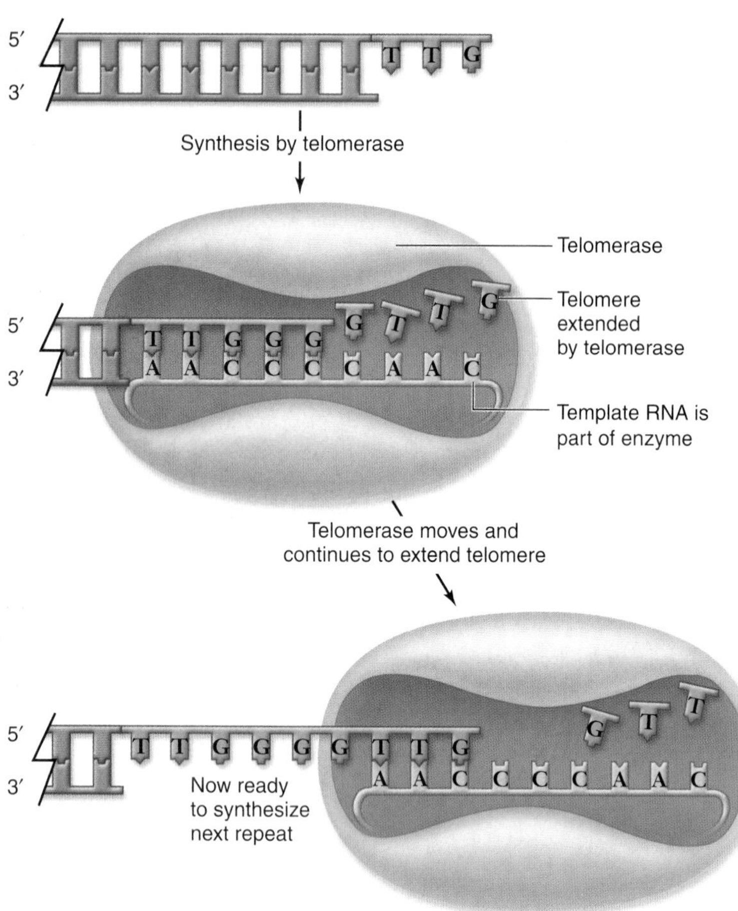

figure 14.24

ACTION OF TELOMERASE. Telomerase contains an internal RNA the enzyme uses as a template to extend the DNA of the chromosome end. Multiple rounds of synthesis by telomerase produce repeated sequences. This single strand is completed by normal synthesis using it as a template (not shown).

which uses an internal RNA as a template and not the DNA itself (figure 14.24).

The use of the internal RNA template allows short stretches of DNA to be synthesized, composed of repeated nucleotide sequences complementary to the RNA of the enzyme. The other strand of these repeated units is synthesized by the usual action of the replication machinery copying the strand made by telomerase.

Telomerase, aging, and cancer

A gradual shortening of the ends of chromosomes occurs in the absence of telomerase activity. During embryonic and childhood development in humans, telomerase activity is high, but it is low in most somatic cells of the adult. The exceptions are cells that must divide as part of their function, such as lymphocytes. The activity of telomerase in somatic cells is kept low by preventing the expression of the gene encoding this enzyme.

Evidence for the shortening of chromosomes in the absence of telomerase was obtained by producing mice with no telomerase activity. These mice appear to be normal for up to six generations, but they show steadily decreasing telomere length that eventually leads to inviable offspring.

This finding indicates a relationship between cell senescence (aging) and telomere length. Normal cells undergo only a specified number of divisions when grown in culture. This limit is at least partially based on telomere length.

Support for the relationship between senescence and telomere length comes from experiments in which telomerase was introduced into fibroblasts in culture. These cells have their lifespan increased relative to controls with no added telomerase. Interestingly, these cells do not show the hallmarks of malignant cells, indicating that activation of telomerase alone does not make cells malignant.

A relationship has been found, however, between telomerase and cancer. Cancer cells do continue to divide indefinitely, and this would not be possible if their chromosomes were being continually shortened. Cancer cells generally show activation of telomerase, which allows them to maintain telomere length; but this is clearly only one aspect of conditions that allow them to escape normal growth controls.

inquiry

How does the structure of eukaryotic genomes affect replication? Does this introduce problems that are not faced by prokaryotes?

Eukaryotic replication uses the same basic enzymology as that in prokaryotes. Eukaryotes can replicate a large amount of DNA in a short time by using multiple origins of replication. Linear chromosomes end in telomeres, which cannot be constructed by the replication machinery. Another enzyme, telomerase, synthesizes the ends of chromosomes. Cancer cells show activation of telomerase.

14.6 DNA Repair

As you learned earlier, many DNA polymerases have 3′ to 5′ exonuclease activity that allows "proofreading" of added bases. This action increases the accuracy of replication, but errors still occur. Without error correction mechanisms, cells would accumulate errors at an unacceptable rate, leading to high levels of deleterious or lethal mutations. A balance must exist between the introduction of new variation by mutation, and the effects of deleterious mutations on the individual.

Cells are constantly exposed to DNA-damaging agents

In addition to errors in DNA replication, cells are constantly exposed to agents that can damage DNA. These agents include radiation, such as UV light and X-rays, and chemicals in the environment. Agents that damage DNA can lead to mutations, and any agent that increases the number of mutations above background levels is called a **mutagen.**

The number of potentially mutagenic agents that organisms encounter is huge. Sunlight itself includes radiation in the UV range and is thus mutagenic. Ozone normally screens out much of the harmful UV radiation in sunlight, but some remains. The relationship between sunlight and mutations is shown clearly by the increase in skin cancer in regions of the southern hemisphere that are underneath a seasonal "ozone hole."

Organisms also may encounter mutagens in their diet in the form of either contaminants in food or natural plant products that can damage DNA. When a simple test was designed to detect mutagens, screening of possible sources indicated an amazing diversity of mutagens in the environment and in natural sources. As a result, consumer products are now screened to reduce the load of mutagens we are exposed to, but we cannot escape natural sources.

DNA repair restores damaged DNA

Cells cannot escape exposure to mutagens, but systems have evolved that enable cells to repair some damage. These DNA repair systems are vital to continued existence, whether a cell is a free-living, single-celled organism or part of a complex multicellular organism.

The importance of DNA repair is indicated by the multiplicity of repair systems that have been discovered and characterized. All cells that have been examined show multiple pathways for repairing damaged DNA and for reversing errors that occur during replication. These systems are not perfect, but they do reduce the mutational load on organisms to an acceptable level. In the rest of this section, we illustrate the action of DNA repair by concentrating on two examples drawn from these multiple repair pathways.

Repair can be either specific or nonspecific

DNA repair falls into two general categories: specific and nonspecific. Specific repair systems target a single kind of lesion in DNA and repair only that damage. Nonspecific forms of repair use a single mechanism to repair multiple kinds of lesions in DNA.

Photorepair: A specific repair mechanism

Photorepair is specific for one particular form of damage caused by UV light, namely the **thymine dimer.** Thymine dimers are formed by a photochemical reaction of UV light and adjacent thymine bases in DNA. The UV radiation causes the thymines to react, covalently linking them together: a thymine dimer (figure 14.25).

Repair of these thymine dimers can be accomplished by multiple pathways, including photorepair. In photorepair, an enzyme called a *photolyase* absorbs light in the visible range and uses this energy to cleave the thymine dimer. This action restores the two thymines to their original state (see figure 14.25). It is interesting that sunlight in the UV range can

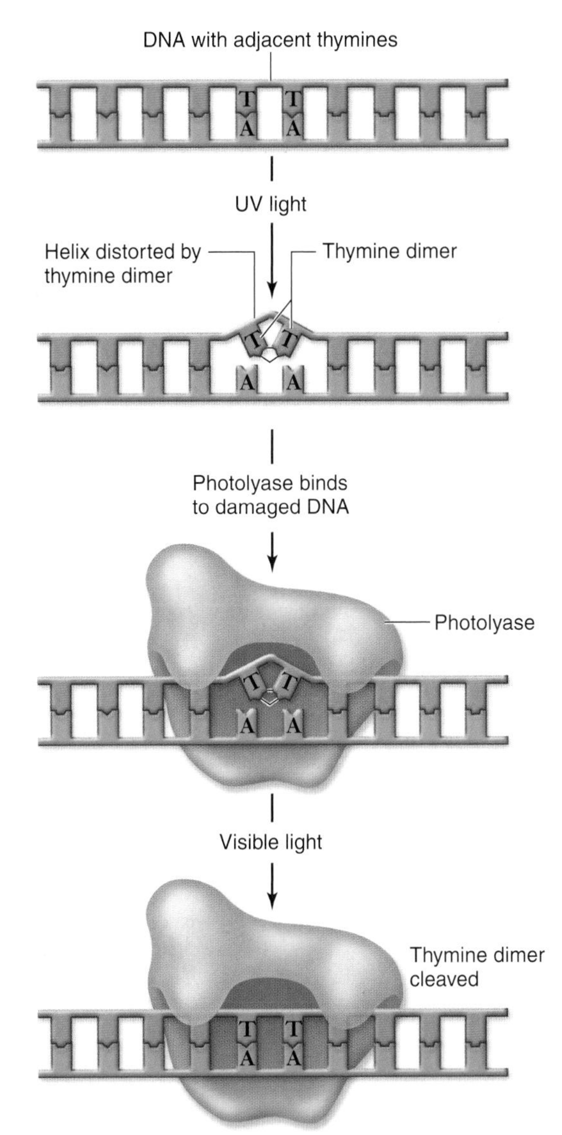

figure 14.25

REPAIR OF THYMINE DIMER BY PHOTOREPAIR. UV light can catalyze a photochemical reaction to form a covalent bond between two adjacent thymines, thereby creating a thymine dimer. A photolyase enzyme recognizes the damage and binds to the thymine dimer. The enzyme absorbs visible light and uses the energy to cleave the thymine dimer.

cause this damage, and sunlight in the visible range can be used to repair the damage. Photorepair does not occur in cells deprived of visible light.

The photolyase enzyme has been found in many different species, ranging from bacteria, to single-celled eukaryotes, to humans. The ubiquitous nature of this enzyme illustrates the importance of this form of repair. For as long as cells have existed on Earth, they have been exposed to UV light and its potential to damage DNA.

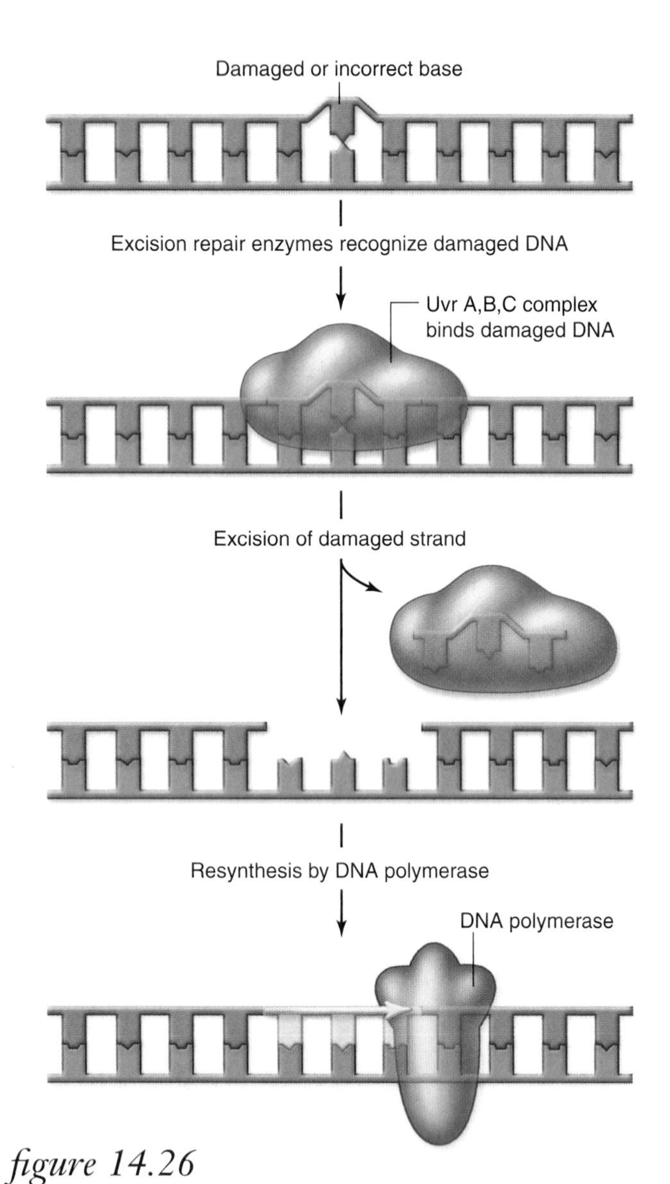

figure 14.26

REPAIR OF DAMAGED DNA BY EXCISION REPAIR. Damaged DNA is recognized by the uvr complex, which binds to the damaged region and removes it. Synthesis by DNA polymerase replaces the damaged region. DNA ligase finishes the process (not shown).

Excision repair: A nonspecific repair mechanism

A common form of nonspecific repair is **excision repair.** In this pathway, a damaged region is removed, or excised, and is then replaced by DNA synthesis (figure 14.26). In *E. coli*, this action is accomplished by proteins encoded by the *uvr A, B*, and *C* genes. Although these genes were identified based on mutations that increased sensitivity of the cell to UV light (hence the "uvr" in their names), their proteins can act on damage due to other mutagens.

Excision repair follows three steps: (1) recognition of damage, (2) removal of the damaged region, and (3) resynthesis using the information on the undamaged strand as a template (see figure 14.26). Recognition and excision are accomplished by the UvrABC complex. The UvrABC complex binds to damaged DNA and then cleaves a single strand on either side of the damage, removing it. In the synthesis stage, DNA pol I or pol II replaces the damaged DNA. This restores the original information in the damaged strand by using the information in the complementary strand.

Other repair pathways

Cells have other forms of nonspecific repair, and these fall into two categories: error-free and error-prone. It may seem strange to have an error-prone pathway, but it can be thought of as a last-ditch effort to save a cell that has been exposed to such massive damage that it has overwhelmed the error-free systems. In fact, this system in *E. coli* is part of what is called the "SOS response."

Cells can also repair damage that produces breaks in DNA. These systems use enzymes related to those that are involved in recombination during meiosis (chapter 11). It is thought that recombination uses enzymes that originally evolved for DNA repair.

The number of different systems and the wide spectrum of damage that can be repaired illustrate the importance of maintaining the integrity of the genome. Accurate replication of the genome is useless if a cell cannot reverse errors that can occur during this process or repair damage due to environmental causes.

inquiry

Cells are constantly exposed to DNA-damaging agents, ranging from UV light to by-products of oxidative metabolism. How does the cell deal with this, and what would happen if the cell had no way of dealing with this?

Cells have multiple repair pathways to reverse damage to DNA. Some of these systems are specific for one type of damage, such as photorepair that reverses thymine dimers caused by UV light. Other systems are nonspecific, such as excision repair that removes and replaces damaged regions.

concept review

14.1 The Nature of Genetic Material

Our knowledge of the molecular basis of the genetic material was built over a long history of experimentation.

- Griffith's experiments showed genetic material can be transferred between cells, a process call transformation.
- Avery, MacLeod, and McCarty demonstrated that DNA is the substance transferred between bacteria.
- Hershey and Chase's research showed that DNA was the genetic material of bacteriophages.

14.2 DNA Structure

Miescher discovered nucleic acids, which contain three components: a five-carbon sugar, a phosphate group, and a nitrogenous base.

- The sugar found in DNA is deoxyribose.
- The nitrogenous bases in DNA are two-ringed purines—adenine (A) and guanine (G), and single-ringed pyrimidines—cytosine (C) and thymine (T).
- Phosphodiester bonds are formed by linking the 5′ phosphate of one nucleotide to the 3′ hydroxyl of another nucleotide (figure 14.5).
- Chargaff found that the proportion of adenine equals that of thymine and the proportion of cytosine equals that of guanine.
- The bases can exist in two different tautomeric forms. The keto and enol forms predominate and this affects hydrogen bonding.
- X-ray diffraction studies by Franklin and Wilkins indicated that DNA molecules had a helical structure.
- Watson and Crick provided a rational structural model for DNA using available data and building models.
- The Watson–Crick model consists of the following features (see figures 14.9 and 14.10).
  - *DNA consists of two polynucleotide strands that form a double helix.*
  - *The two strands are held together by hydrogen bonds, forming specific base pairs between adenine and thymine, and guanosine and cytosine.*
  - *We say that the two strands are complementary because each strand determines the other by base pairing.*
  - *Complementary phosphodiester strands are antiparallel to each other.*

14.3 Basic Characteristics of DNA Replication

Meselson and Stahl showed that DNA replication is semiconservative, resulting in two identical DNA molecules, each of which is composed of one original strand and one new strand (figure 14.11).

- DNA replication can be divided into three phases:
  - *Initiation of replication begins at a specific site called the origin.*
  - *Elongation uses DNA polymerases that synthesize a new strand complementary to the template. These enzymes require a short base-paired primer and only synthesize in the 5′-to-3′ direction.*
  - *Termination ends replication at a specific site, the terminus.*

14.4 Prokaryotic Replication

Prokaryotic replication involves a circular DNA template.

- Prokaryotic replication begins at a unique site, the origin, and proceeds bidirectionally with two replication forks.
- The complete prokaryotic chromosome with a single origin forms a single functional unit called a replicon.
- There are three prokaryotic DNA polymerases, polymerase I, II and III, all of which synthesize DNA in a 5′-to-3′ direction.
- DNA polymerases may also have the ability to degrade DNA from one end, a process called exonuclease activity.
- Unwinding DNA uses the enzyme DNA helicase and energy.
- Unwinding DNA introduces torsional strain that is removed by the enzyme DNA gyrase.
- The antiparallel nature of DNA and the fact that the DNA polymerases only synthesize DNA in a 5′-to-3′ direction mean that replication is discontinuous on one strand (figure 14.16).
  - *Only one strand, called the leading strand, is synthesized continuously.*
  - *The other strand, called the lagging strand, is synthesized discontinuously.*
- Synthesis occurs at the replication fork, where the two strands are being opened.
- DNA polymerases require primers that are synthesized by DNA primase.
- DNA polymerase III can remain attached to the template because of a subunit that acts as a sliding clamp.
- Synthesis on the lagging strand is much more complex.
  - *Polymerase III is still the main polymerase.*
  - *DNA primase periodically synthesizes short primers.*
  - *Each primer is extended by DNA polymerase III until it hits the previous fragment.*
  - *The RNA primers are removed by DNA polymerase I and replaced with DNA.*
  - *The fragments are joined by DNA ligase.*
- All of the activities are coordinated in a complex called the replisome containing two copies of polymerase III, DNA primase, DNA helicase, and a number of accessory proteins.
- The replisome moves in one direction by creating a loop in the lagging strand, allowing the antiparallel template strands to be copied in the same direction (figure 14.19).

14.5 Eukaryotic Replication

Eukaryotic replication is complicated by the large amount of DNA organized into multiple linear chromosomes.

- Eukaryote chromosomes have multiple origins of replication.
- The enzymology of eukaryotic replication is more complex, involving more enzymes.
- The main replication polymerase is a complex of two enzymes.
- The ends of linear chromosomes are called telomeres, and they protect the ends of chromosomes.
- Linear chromosomes introduce problems finishing replication.
- Telomeres are specialized structures that are made by the enzyme telomerase and not by the replication complex.
- Telomerase contains an internal RNA that acts as a template to extend the DNA of the chromosome end.
- Adult cells lack telomerase activity, and telomere shortening correlates with senescence.

14.6 DNA Repair

Detection and correction of DNA errors to reduce rates of mutations.

- The error rate during replication is reduced by the proofreading ability of DNA polymerases.
- Environmental mutagens damage DNA and increase the rate of mutations above background levels.
- Cells have multiple specific and nonspecific pathways for repairing DNA damage.
- In photorepair the enzyme photolyase absorbs visible light and uses this energy to specifically cleave thymine dimers caused by UV light.
- Excision repair is nonspecific, and in prokaryotes a damaged region of DNA is removed by enzymes of the *uvr* system.

review questions

SELF TEST

1. What was the key finding from Griffith's experiments using live and heat-killed pathogenic bacteria?
   a. Bacteria with a smooth coat could kill mice
   b. Bacteria with a rough coat are not lethal
   c. Heat-killed smooth-coat bacteria would not cause death in mice.
   d. Heat-killed smooth-coat bacteria could transform the nonlethal live bacteria.
2. When Hershey and Chase differentially tagged the DNA and proteins of bacteriophages and allowed them to infect bacteria, what did the viruses transfer to the bacteria?
   a. Radioactive phosphorous and sulfur
   b. Radioactive sulfur
   c. DNA
   d. Both (b) and (c) are correct.
3. Which of the following is *not* a component of DNA?
   a. The pyrimidine uracil
   b. Five-carbon sugars
   c. The purine adenine
   d. Phosphate groups
4. What type of chemical bond allows DNA or RNA to form a long polymer?
   a. Hydrogen bonds
   b. Peptide bonds
   c. Ionic bonds
   d. Phosphodiester bonds
5. What is *Chargaff's rule*?
   a. The number of phosphate groups always equals the number of five-carbon sugars.
   b. The proportions of A equal that of C and G equals T.
   c. The proportions of A equal that of T and G equals C.
   d. Purines bind to pyrimidines.
6. The bonds that hold two complementary strands of DNA together are—
   a. Hydrogen bonds
   b. Peptide bonds
   c. Ionic bonds
   d. Phosphodiester bonds
7. If one strand of a DNA molecule has the sequence ATTGCAT, then the complementary strand will have the sequence—
   a. ATTGCAT
   b. TACGTTA
   c. TAACGTA
   d. GCCTAGC
8. Which of the following is *not* part of the Watson–Crick model of the structure of DNA?
   a. DNA is composed of two strands.
   b. The two DNA strands are oriented in parallel (5′ to 3′).
   c. Purines bind to pyrimidines.
   d. DNA forms a double helix.
9. Meselson and Stahl demonstrated that—
   a. DNA replication occurs in bacteria.
   b. DNA replication is dispersive.
   c. DNA replication is conservative.
   d. DNA replication is semiconservative.
10. Which of the following steps in DNA replication involves the formation of new phosphodiester bonds?
    a. Initiation at an origin of replication
    b. Elongation by a DNA polymerase
    c. Unwinding of the double helix
    d. Termination
11. The difference in leading- versus lagging-strand synthesis is a consequence of—
    a. the antiparallel configuration of DNA
    b. DNA polymerase III synthesizing only in the 5′-to-3′ direction
    c. the activity of DNA gyrase
    d. both (a) and (b)
12. Okazaki fragments are—
    a. synthesized in the 3′-to-5′ direction
    b. found on the lagging strand
    c. found on the leading strand
    d. made of RNA
13. Successful DNA synthesis requires all of the following *except*—
    a. helicase
    b. endonuclease
    c. DNA primase
    d. DNA ligase
14. What is a telomere?
    a. An A–T-rich region of DNA
    b. The point of DNA termination on a bacterial chromosome
    c. Regions of repeated sequences of DNA on the ends of eukaryotic chromosomes
    d. The sequence of RNA found on a replicating molecule of DNA
15. Which type of enzyme is involved in excision repair?
    a. photolyase
    b. DNA polymerase III
    c. endonuclease
    d. telomerase

CHALLENGE QUESTIONS

1. The work by Griffith provided the first indication that DNA was the genetic material. Review the four experiments outlined in figure 14.1. Predict the likely outcome for the following variations on this classic research.
   a. Heat-killed pathogenic and heat-killed nonpathogenic
   b. Heat-killed pathogenic and live nonpathogenic in the presence of an enzyme that digests proteins (proteases)
   c. Heat-killed pathogenic and live nonpathogenic in the presence of an enzyme that digests DNA (endonuclease)
2. Imagine that you have identified the sequence 5′-TTATAAAGCAATAGT-3′ in a eukaryotic chromosome. Could this region of the chromosome function as an origin of replication? Predict the sequence of an RNA primer that would be formed in association with this sequence.
3. Enzyme function is critically important for the proper replication of DNA. Predict the consequence of a loss of function for each of the following enzymes.
   a. DNA gyrase
   b. DNA polymerase III
   c. DNA ligase
   d. DNA polymerase I

Do you need additional review? *Visit* www.ravenbiology.com *for practice quizzes, animations, videos, and activities designed to help you master the material in this chapter.*

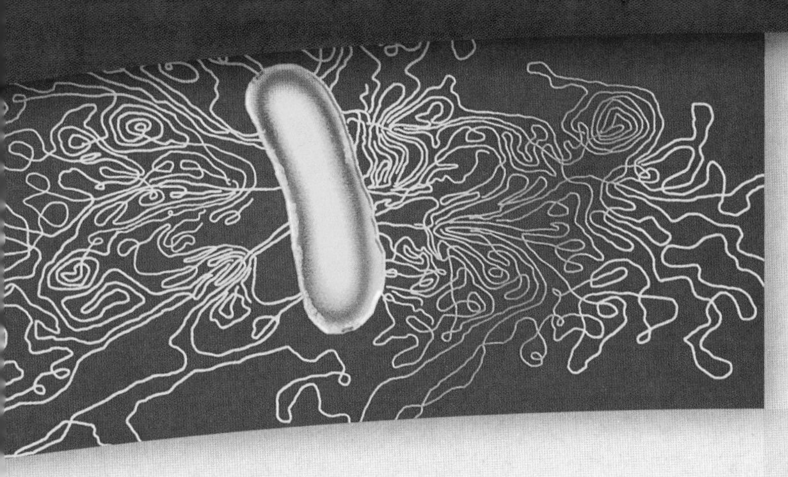

chapter 15

Genes and How They Work

concept outline

15.1 The Nature of Genes
- *Garrod concluded that inherited disorders can involve specific enzymes*
- *Beadle and Tatum showed that genes specify enzymes*
- *The central dogma describes information flow in cells as DNA to RNA to protein*

15.2 The Genetic Code
- *The code is read in groups of three*
- *Nirenberg and others deciphered the code*
- *The code is degenerate but specific*
- *The code is practically universal, but not quite*

15.3 Overview of Gene Expression
- *Transcription makes an RNA copy of DNA*
- *Translation uses information in RNA to synthesize proteins*
- *RNA has multiple roles in gene expression*

15.4 Prokaryotic Transcription
- *Prokaryotes have a single RNA polymerase*
- *Initiation occurs at promoters*
- *Elongation adds successive nucleotides*
- *Termination occurs at specific sites*
- *Prokaryotic transcription is coupled to translation*

15.5 Eukaryotic Transcription
- *Eukaryotes have three RNA polymerases*
- *Each polymerase has its own promoter*
- *Initiation and termination differ from that in prokaryotes*
- *Eukaryotic transcripts are modified*

15.6 Eukaryotic pre-mRNA Splicing
- *Eukaryotic genes may contain interruptions*
- *The spliceosome is the splicing organelle*
- *Splicing can produce multiple transcripts from the same gene*

15.7 The Structure of tRNA and Ribosomes
- *Aminoacyl-tRNA synthetases attach amino acids to tRNA*
- *The ribosome has multiple tRNA-binding sites*
- *The ribosome has both decoding and enzymatic functions*

15.8 The Process of Translation
- *Initiation requires accessory factors*
- *Elongation adds successive amino acids*
- *Termination requires accessory factors*
- *Proteins may be targeted to the ER*

15.9 Summarizing Gene Expression

15.10 Mutation: Altered Genes
- *Point mutations affect a single site in the DNA*
- *Chromosomal mutations change the structure of chromosomes*
- *Mutations are the starting point of evolution*
- *Our view of the nature of genes has changed with new information*

introduction

YOU'VE SEEN HOW GENES SPECIFY TRAITS and how these traits can be followed in genetic crosses. You've also seen that the information in genes resides in the DNA molecule; the picture on the left shows the DNA that comprises the entire *E. coli* chromosome. Information in DNA is replicated by the cell and then partitioned equally during the process of cell division. The information in DNA is much like a blueprint for a building. The construction of the building uses the information in the blueprint, but requires building materials and, carpenters and other skilled laborers using a variety of tools and working together to actually construct the building. Similarly, the information in DNA requires nucleotide and amino acid building blocks, multiple forms of RNA, and many proteins acting in a coordinated fashion to make up the structure of a cell.

We now turn to the nature of the genes themselves and how cells extract the information in DNA in the process of **gene expression.** Gene expression can be thought of as the conversion of genotype into the phenotype.

15.1 The Nature of Genes

We know that DNA encodes proteins, but this knowledge alone tells us little about how the information in DNA can control cellular functions. Researchers had evidence that genetic mutations affected proteins, and in particular enzymes, long before the structure and code of DNA was known. In this section we review the evidence of the link between genes and enzymes.

Garrod concluded that inherited disorders can involve specific enzymes

In 1902, the British physician Archibald Garrod noted that certain diseases among his patients seemed to be more prevalent in particular families. By examining several generations of these families, he found that some of the diseases behaved as though they were the product of simple recessive alleles. Garrod concluded that these disorders were Mendelian traits, and that they had resulted from changes in the hereditary information in an ancestor of the affected families.

Garrod investigated several of these disorders in detail. In alkaptonuria, patients produced urine that contained homogentisic acid (alkapton). This substance oxidized rapidly when exposed to air, turning the urine black. In normal individuals, homogentisic acid is broken down into simpler substances. With considerable insight, Garrod concluded that patients suffering from alkaptonuria lack the enzyme necessary to catalyze this breakdown. He speculated that many other inherited diseases might also reflect enzyme deficiencies.

Beadle and Tatum showed that genes specify enzymes

From Garrod's finding, it took but a short leap of intuition to surmise that the information encoded within the DNA of chromosomes acts to specify particular enzymes. This point was not actually established, however, until 1941, when a series of experiments by George Beadle and Edward Tatum at Stanford University provided definitive evidence. Beadle and Tatum deliberately set out to create mutations in chromosomes and verified that they behaved in a Mendelian fashion in crosses. These alterations to single genes were analyzed for their effects on the organism (figure 15.1).

Neurospora crassa, *the bread mold*

One of the reasons Beadle and Tatum's experiments produced clear-cut results was their choice of experimental organism, the bread mold *Neurospora crassa*. This fungus can be grown readily in the laboratory on a defined medium consisting of only a carbon source (glucose), a vitamin (biotin), and inorganic salts. This type of medium is called "minimal" because it represent the minimal requirements to support growth. Any cells that can grow on minimal medium must be able to synthesize all necessary biological molecules.

Beadle and Tatum exposed *Neurospora* spores to X-rays, expecting that the DNA in some of the spores would experience damage in regions encoding the ability to make compounds needed for normal growth (see figure 15.1). Such a mutation would cause cells to be unable to grow on minimal medium.

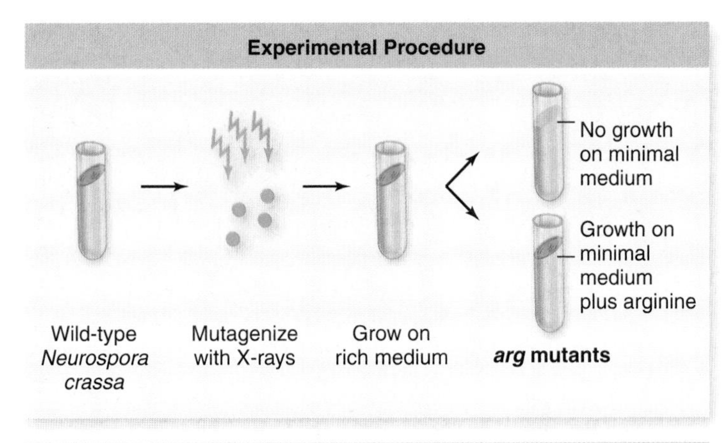

Results

Mutation in Enzyme	Plus Ornithine	Plus Citruline	Plus Arginosuccinate	Plus Arginine
E				
F				
G				
H				

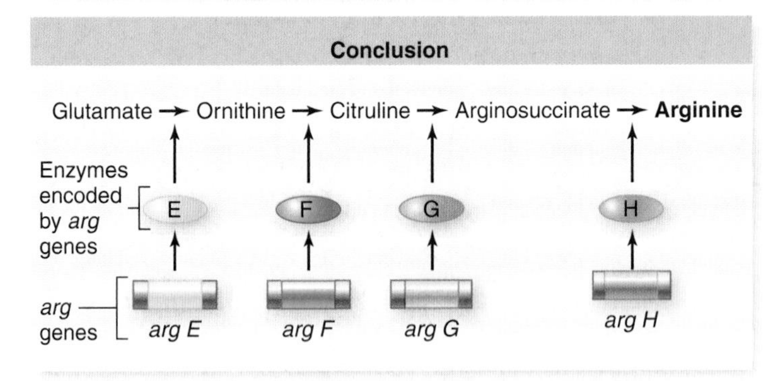

figure 15.1

THE BEADLE AND TATUM EXPERIMENT. Wild-type *Neurospora* were mutagenized with X-rays to produce mutants deficient in the synthesis of arginine (top panel). The specific defect in each mutant was identified by growing on medium supplemented with intermediates in the biosynthetic pathway for arginine (middle panel). A mutant will grow only on media supplemented with an intermediate produced after the defective enzyme in the pathway for each mutant. The enzymes in the pathway can then be correlated with genes on chromosomes (bottom panel).

Such mutations are called **nutritional mutations** because cells carrying them grow only if the medium is supplemented with additional nutrients.

Nutritional mutants

To identify mutations causing metabolic deficiencies, Beadle and Tatum placed subcultures of individual fungal cells grown on a rich medium onto minimal medium. Any cells that had lost the ability to make compounds necessary for growth would not grow on minimal medium. Using this approach, Beadle and Tatum succeeded in isolating and identifying many nutritional mutants.

Next, the researchers supplemented the minimal medium with different compounds to identify the deficiency in each mutant. This step allowed them to pinpoint the nature of the strain's biochemical deficiency. They concentrated in particular on mutants that would grow only in the presence of the amino acid arginine, dubbed *arg* mutants. When their chromosomal positions were located, the *arg* mutations were found to cluster in three areas.

One gene/one polypeptide

The next step was to determine where each mutation was blocked in the biochemical pathway for arginine biosynthesis. To do this, they supplemented the medium with each intermediate in the pathway to see which intermediate would support a mutant's growth. If the mutation affects an enzyme in the pathway that acts prior to the intermediate used as a supplement, then growth should be supported—but not if the mutation affects a step after the intermediate used (see figure 15.1). For each enzyme in the arginine biosynthetic pathway, Beadle and Tatum were able to isolate a mutant strain with a defective form of that enzyme. The mutation was always located at one of a few specific chromosomal sites, and each mutation had a unique location. Thus, each of the mutants they examined had a defect in a single enzyme, caused by a mutation at a single site on a chromosome.

Beadle and Tatum concluded that genes specify the structure of enzymes, and that each gene encodes the structure of one enzyme (see figure 15.1). They called this relationship the *one-gene/one-enzyme hypothesis.* Today, because many enzymes contain multiple polypeptide subunits, each encoded by a separate gene, the relationship is more commonly referred to as the **one-gene/one-polypeptide hypothesis.** This hypothesis clearly states the molecular relationship between genotype and phenotype.

As you learn more about genomes and gene expression, you'll find that this clear relationship is overly simple. As described later in this chapter, eukaryotic genes are more complex. In addition, some enzymes are composed at least in part of RNA, itself an intermediate in the production of proteins. Nevertheless, this one-gene/one-polypeptide concept is a useful starting point to think about gene expression.

The central dogma describes information flow in cells as DNA to RNA to protein

The conversion of genotype to phenotype requires information stored in DNA to be converted to protein. The nature of information flow in cells was first described by Francis Crick as the **central dogma of molecular biology.** Information passes in one direction from the gene (DNA) to an RNA copy of the gene, and the RNA copy directs the sequential assembly of a chain of amino acids into a protein (figure 15.2). Stated briefly,

DNA ⟶ RNA ⟶ protein

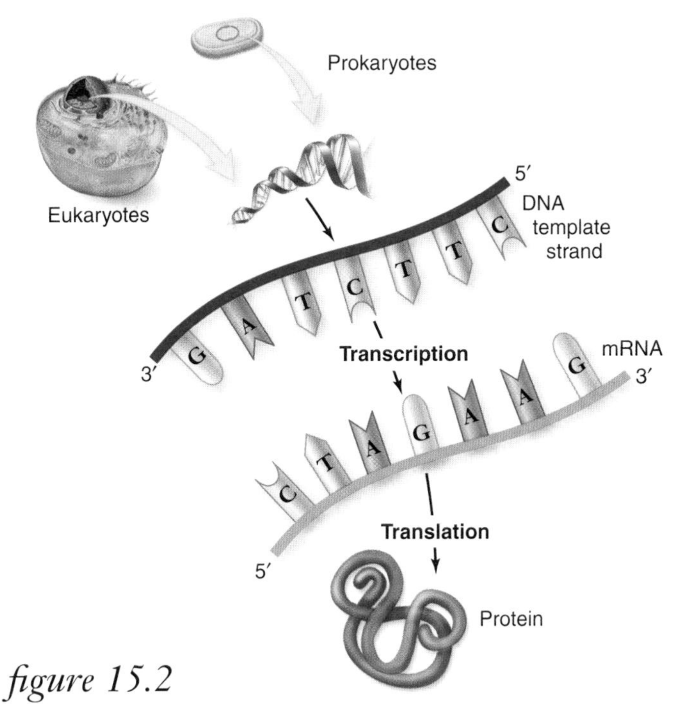

figure 15.2

THE CENTRAL DOGMA OF MOLECULAR BIOLOGY. DNA is transcribed to make mRNA, which is translated to make a protein.

We can view this as a concise description of the process of gene expression, or the conversion of genotype to phenotype. We call the DNA to RNA step **transcription,** and the RNA to protein step **translation** (see figure 15.2). These processes will be explored in detail throughout the chapter.

This, again, is an oversimplification of how information flows in eukaryotic cells. A class of viruses called **retroviruses** has been discovered that can convert their RNA genome into a DNA copy, using the viral enzyme **reverse transcriptase.** This conversion violates the direction of information flow of the central dogma, and the discovery forced an updating of the dogma to include this "reverse" flow of information.

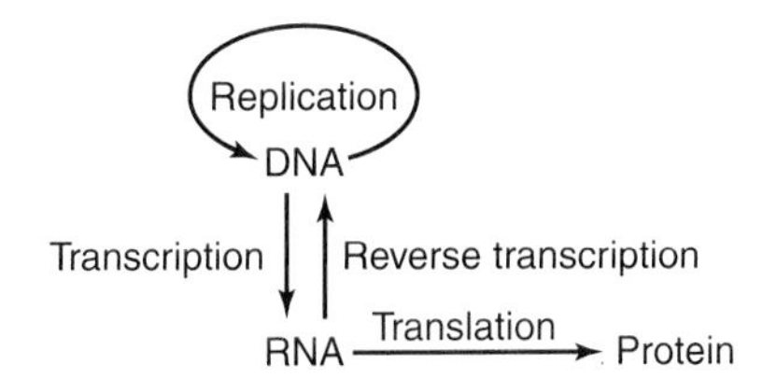

Metabolic disorders can be due to the presence of altered enzymes. Each gene encodes the information to make one polypeptide. The flow of information in cells, according to the central dogma, begins with information in a gene in DNA. DNA is transcribed into RNA, and this RNA copy is used to direct the synthesis of proteins.

15.2 The Genetic Code

How does the order of nucleotides in a DNA molecule encode the information that specifies the order of amino acids in a polypeptide? The answer to this essential question came in 1961, through an experiment led by Francis Crick and Sydney Brenner. That experiment was so elegant and the result so critical to understanding the genetic code that we describe it in detail.

The code is read in groups of three

Crick and Brenner reasoned that the genetic code most likely consisted of a series of blocks of information called **codons,** each corresponding to an amino acid in the encoded protein. They further hypothesized that the information within one codon was probably a sequence of three nucleotides. With four DNA nucleotides (G, C, T, and A), using two in each codon will produce only 4^2, or 16, different codons—not enough to code for 20 amino acids. However, three nucleotides results in 4^3, or 64, different combinations of three, more than enough.

Spaced or unspaced codons?

In theory, the sequence of codons in a gene could be punctuated with nucleotides between the codons that are not used, like the spaces that separate the words in this sentence. Alternatively, the codons could lie immediately adjacent to each other, forming a continuous sequence of nucleotides.

If the information in the genetic message is separated by spaces, then altering any single word would not affect the entire sentence. In contrast, if all of the words are run together but read in groups of three, then any alteration that is not in groups of three would alter the entire sentence. These two ways of using information in DNA imply different methods of translating the information into protein.

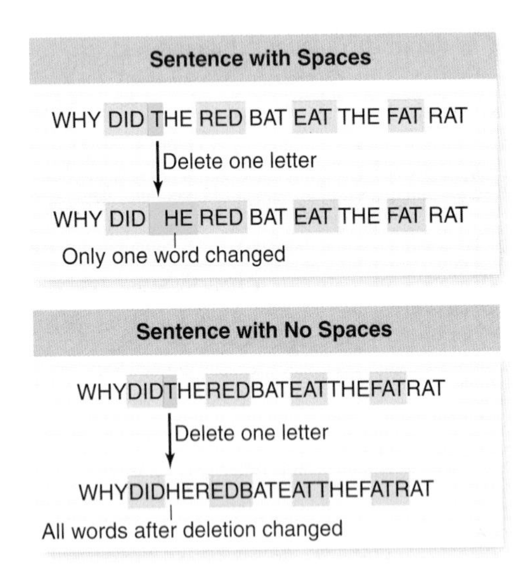

Determining that codons are unspaced

To choose between these alternative mechanisms, Crick and his colleagues used a chemical to create mutations that would delete one, two, or three nucleotides from a viral DNA molecule, which was then transcribed and translated into a polypeptide. They then asked whether this action altered only a single amino acid or all amino acids after the deletions.

When they made a single deletion or two deletions near each other, the genetic message shifted, altering all of the amino acids after the deletion. When they made three deletions, however, the protein after the deletions was normal. They obtained the same results when they made additions to the DNA consisting of 1, 2, or 3 nt (nucleotides).

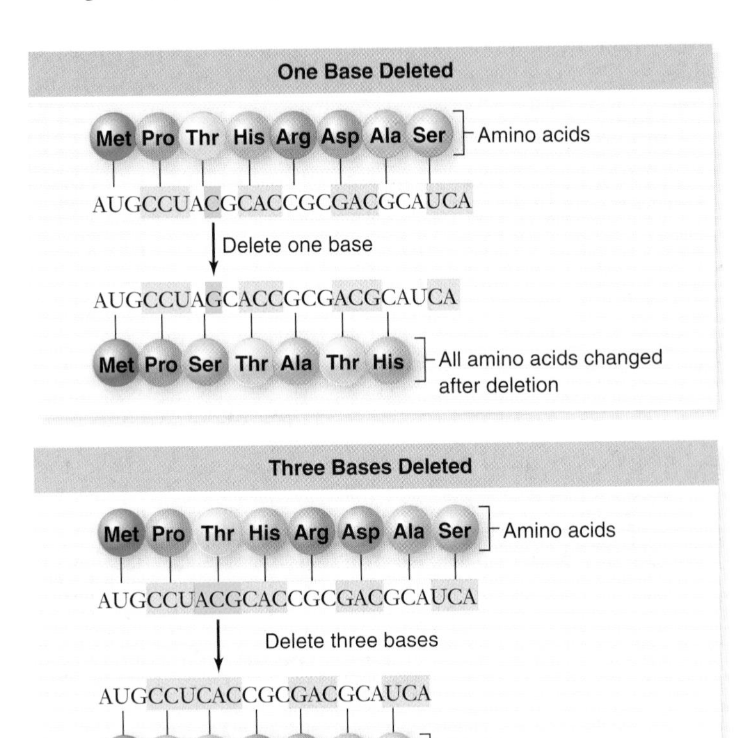

Thus, Crick and Brenner concluded that the genetic code is read in increments of three nucleotides (in other words, it is a triplet code), and that reading occurs continuously without punctuation between the 3-nt units.

These experiments indicate the importance of the **reading frame** for the genetic message. Because there is no punctuation, the reading frame established by the first codon in the sequence determines how all subsequent codons are read. We now call the kinds of mutations that Crick and Brenner used **frameshift mutations** because they alter the reading frame of the genetic message.

Nirenberg and others deciphered the code

The determination of which of the 64 possible codons encoded each particular amino acids was one of the greatest triumphs of 20th-century biochemistry. Accomplishing this decryption required two main developments to succeed. First, cell-free biochemical systems that would support protein synthesis from a defined RNA were needed. Second, the ability to produce synthetic, defined RNAs that could be used in the cell-free system was necessary.

During a five-year period from 1961 to 1966, work performed primarily in Marshall Nirenberg's laboratory led to the elucidation of the genetic code. Nirenberg's group first showed that adding the synthetic RNA molecule polyU (an RNA molecule consisting of a string of uracil nucleotides) to their cell-free systems produced the polypeptide polyphenylalanine (a string of phenylalanine amino acids). Therefore, UUU encodes phenylalanine.

Next they used enzymes to produce RNA polymers with more than one nucleotide. These polymers allowed them to infer the composition of many of the possible codons, but not the order of bases in each codon.

The researchers then were able to use enzymes to synthesize defined 3-base sequences that could be tested for binding to the protein synthetic machinery. This so-called *triplet-binding assay* allowed them to identify 54 of the 64 possible triplets.

The organic chemist H. Gobind Khorana provided the final piece of the puzzle by using organic synthesis to produce artificial RNA molecules of defined sequence, and then examining what polypeptides they directed in cell-free systems. The combination of all of these methods allowed the determination of all 64 possible three-nucleotide sequences, and the full genetic code was determined (table 15.1).

The code is degenerate but specific

Some obvious features of the code jump out of table 15.1. First, 61 of the 64 possible codons are used to specify amino acids. Three codons, UAA, UGA, and UAG, are reserved for another function: they signal "stop" and are known as **stop codons.** The only other form of "punctuation" in the code is that AUG is used to signal "start" and is therefore the **start codon.** In this case the codon has a dual function as it also encodes the amino acid methionine (Met).

You can see that with 61 codons to encode only 20 amino acids, there are many more codons than amino acids. One way to deal with this abundance would be to use only 20 of the 61 codons, but this is not what cells do. In reality, all 61 codons are used, making the code **degenerate,** which means that some amino acids are specified by more than one codon. The reverse, however, in which a single codon would specify more than one amino acid, is never found.

This degeneracy is not uniform. Some amino acids have only one codon, and some have up to six codons. In addition, the degenerate base usually occurs in position 3 of a codon, such that the first two positions are the same, and two or four of the possible nucleic acids at position 3 encode the same amino acid. (The nature of protein synthesis on ribosomes explains how this codon usage works, and it is discussed later.)

The code is practically universal, but not quite

The genetic code is the same in almost all organisms. The universality of the genetic code is among the strongest evidence that all living things share a common evolutionary heritage. Because the code is universal, genes can be transferred from one organism to another and can be successfully expressed in

TABLE 15.1 The Genetic Code

First Letter	SECOND LETTER: U		C		A		G		Third Letter
U	UUU	Phe Phenylalanine	UCU	Ser Serine	UAU	Tyr Tyrosine	UGU	Cys Cysteine	U
	UUC		UCC		UAC		UGC		G
	UUA	Leu Leucine	UCA		UAA	"Stop"	UGA	"Stop"	A
	UUG		UCG		UAG	"Stop"	UGG	Trp Tryptophan	G
C	CUU	Leu Leucine	CCU	Pro Proline	CAU	His Histidine	CGU	Arg Arginine	U
	CUC		CCC		CAC		CGC		C
	CUA		CCA		CAA	Gln Glutamine	CGA		A
	CUG		CCG		CAG		CGG		G
A	AUU	Ile Isoleucine	ACU	Thr Threonine	AAU	Asn Asparagine	AGU	Ser Serine	U
	AUC		ACC		AAC		AGC		C
	AUA		ACA		AAA	Lys Lysine	AGA	Arg Arginine	A
	AUG	Met Methionine; "Start"	ACG		AAG		AGG		G
G	GUU	Val Valine	GCU	Ala Alanine	GAU	Asp Aspartate	GGU	Gly Glycine	U
	GUC		GCC		GAC		GGC		C
	GUA		GCA		GAA	Glu Glutamate	GGA		A
	GUG		GCG		GAG		GGG		G

A codon consists of three nucleotides read in the sequence shown. For example, ACU codes for threonine. The first letter, A, is in the First Letter column; the second letter, C, is in the Second Letter column; and the third letter, U, is in the Third Letter column. Each of the mRNA codons is recognized by a corresponding anticodon sequence on a tRNA molecule. Many amino acids are specified by more than one codon. For example, threonine is specified by four codons, which differ only in the third nucleotide (ACU, ACC, ACA, and ACG).

figure 15.3

TRANSGENIC PIG. The piglet on the right is a conventional piglet. The piglet on the left was engineered to express a gene from jellyfish that encodes green fluorescent protein. The color of this piglet's nose is due to expression of this introduced gene. Such transgenic animals indicate the universal nature of the genetic code.

their new host (figure 15.3). This universality of gene expression is central to many of the advances of genetic engineering discussed in chapter 17.

In 1979, investigators began to determine the complete nucleotide sequences of the mitochondrial genomes in humans, cattle, and mice. It came as something of a shock when these investigators learned that the genetic code used by these mammalian mitochondria was not quite the same as the "universal code" that has become so familiar to biologists.

In the mitochondrial genomes, what should have been a stop codon, UGA, was instead read as the amino acid tryptophan; AUA was read as methionine rather than as isoleucine; and AGA and AGG were read as stop codons rather than as arginine. Furthermore, minor differences from the universal code have also been found in the genomes of chloroplasts and in ciliates (certain types of protists).

Thus, it appears that the genetic code is not quite universal. Some time ago, presumably after they began their endosymbiotic existence, mitochondria and chloroplasts began to read the code differently, particularly the portion of the code associated with "stop" signals.

inquiry

The genetic code is almost universal. Why do you think it is nearly universal?

The genetic codes was shown to be triplets with no punctuation: three bases determine one amino acid, and these groups of three are read in order with no "spaces." With 61 codons that specify amino acids (plus 3 that mean "stop," for 64 total), the code is degenerate: Some amino acids have more than one codon, but all codons encode only one amino acid. The code is practically universal, with some exceptions.

15.3 Overview of Gene Expression

The central dogma provides an intellectual framework that describes information flow in biological systems. We call the DNA-to-RNA step *transcription* because it produces an exact copy of the DNA, much as a legal transcription contains the exact words of a court proceeding. The RNA-to-protein step is termed *translation* because it requires translating from the nucleic acid to the protein "languages."

Transcription makes an RNA copy of DNA

The process of transcription produces an RNA copy of the information in DNA. That is, transcription is the DNA-directed synthesis of RNA. This process uses the principle of complementarity, described in the previous chapter, to use DNA as a template to make RNA (figure 15.4).

Because DNA is double-stranded and RNA is single-stranded, only one of the two DNA strands needs to be copied. We call the strand that is copied the **template strand.** The RNA transcript's sequence is complementary to the template strand. The strand of DNA not used as a template is called the **coding strand.** It has the same sequence as the RNA transcript, except that U in the RNA is T in the DNA-coding strand.

Coding 5′–TCAGCCGTCAGCT–3′ ⎤ DNA
Template 3′–AGTCGGCAGTCGA–5′ ⎦

Transcription ↓

Coding 5′–UCAGCCGUCAGCU–3′ mRNA

The RNA transcript used to direct the synthesis of polypeptides is termed **messenger RNA (mRNA).** Its name reflects the recognition that some molecule must carry the DNA message to the ribosome for processing.

As with replication, DNA transcription can be thought of as having three parts: *initiation*, *elongation*, and *termination.*

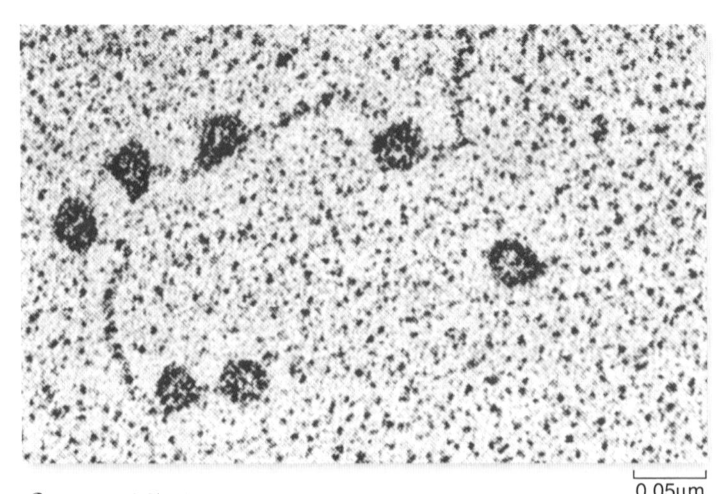

figure 15.4

RNA POLYMERASE. In this electron micrograph, the dark circles are RNA polymerase molecules synthesizing RNA from a DNA template.

Initiation of transcription

Initiation involves a number of components, which differ in prokaryotes and eukaryotes:

- DNA sequences called *promoters* provide attachment sites for the enzyme, *RNA polymerase*, that makes the RNA transcript.
- The *start site* on the DNA is the first base transcribed.
- In eukaryotes, one or more *transcription factors* are also involved in initiation.

Once RNA polymerase has become bound to the DNA at the promoter, transcription can begin at the start site.

Elongation of the transcript

During elongation, the RNA transcript is synthesized:

- New nucleotide triphosphates complementary to the template strand are joined by phosphodiester bonds in the 5′-to-3′ direction by the enzyme *RNA polymerase*, producing the new RNA chain.
- As transcription proceeds, DNA is unwound by RNA polymerase to allow transcription and rewound behind the enzyme. We call the region that is unwound by the enzyme a *transcription bubble.*

Termination of transcription

Elongation proceeds until a stop sequence is reached:

- DNA sequences called *terminators*, described later, cause the RNA polymerase to stop and release the DNA.
- The newly synthesized RNA transcript dissociates from the DNA, and the DNA rewinds.

Translation uses information in RNA to synthesize proteins

The process of translation is by necessity much more complex than transcription. In this case, RNA cannot be used as a direct template for a protein because there is no complementarity—that is, a sequence of amino acids cannot be aligned to an RNA template based on any kind of "chemical fit." Molecular geneticists suggested that some kind of adapter molecule must exist that can interact with both RNA and amino acids, and **transfer RNA (tRNA)** was found to fill this role. This need for an intermediary adds a level of complexity to the process that is not seen in either DNA replication or transcription of RNA.

Translation takes place on the ribosome, the cellular protein-synthetic machinery, and it requires the participation of multiple kinds of RNA and many proteins. Here we provide an outline of the processes; all are described in detail in the sections that follow.

Initiation of translation

Initiation depends on the presence of a start codon and the formation of an initiation complex:

- An *initiation complex* is formed containing the ribosome, mRNA, and the *initiator tRNA* bound to the amino acid methionine.
- The assembly of this complex requires the participation of a number of initiation factors.

Elongation of the polypeptide

The polypeptide grows as tRNA intermediaries bring individual amino acids to the ribosome complex. A tRNA molecule that is carrying an amino acid is called a *charged tRNA*. The ribosome must move along the mRNA strand and bind to charged tRNAs so their anticodon can hydrogen-bond to codons in the mRNA. The ribosome can bind to two tRNAs and form a peptide bond between the amino acids they are carrying.

- Charged tRNAs are brought to the ribosome. The charged tRNA's anticodon must be complementary to each new codon in the mRNA.
- The enzyme *peptidyl transferase* catalyzes formation of a peptide bond between each new amino acid and the growing chain.
- The ribosome complex moves along the mRNA, ejecting the empty tRNA and positioning the binding site for the new tRNA over the next codon in the mRNA.

Termination of translation

- Elongation proceeds until a *stop codon* is encountered.
- *Release factors* recognize the stop codon and cause dissociation of the peptide chain, releasing the last tRNA from the ribosome complex.

The sites at which transcription and translation occur are different in prokaryotes and eukaryotes because eukaryotes have a membrane-bounded nucleus—mRNA must exit the nucleus before translation can begin. In prokaryotes, by contrast, transcription and translation often occur in tandem. We will examine each of these processes in detail in the succeeding sections.

inquiry

It is widely accepted that RNA polymerase has no proofreading capacity. Would you expect high or low levels of error in transcription as compared with DNA replication? Why do you think it is more important for DNA polymerase than for RNA polymerase to proofread?

RNA has multiple roles in gene expression

All RNAs are synthesized from a DNA template by transcription. Gene expression requires the participation of multiple kinds of RNA, each with different roles in the overall process. Here is a brief summary of these roles, which are described in detail later on.

Messenger RNA. Even before the details of gene expression were unraveled, geneticists recognized that there must be an intermediate form of the information in DNA that can be transported out of the eukaryotic nucleus to the cytoplasm for ribosomal processing. This hypothesis was called the "messenger hypothesis," and we retain this language in the name messenger RNA (mRNA).

Ribosomal RNA. The class of RNA found in ribosomes is called **ribosomal RNA (rRNA).** There are multiple forms of rRNA, and rRNA is found in both ribosomal subunits. This rRNA is critical to the function of the ribosome.

Transfer RNA. The intermediary adapter molecule between mRNA and amino acids, as mentioned earlier, is transfer RNA **(tRNA).** Transfer RNA molecules have amino acids covalently attached to one end, and an anticodon that can base-pair with an mRNA codon at the other. The tRNAs act to interpret information in mRNA and to help position the amino acids on the ribosome.

Small nuclear RNA. Small nuclear RNAs **(snRNAs)** are part of the machinery that is involved in nuclear processing of eukaryotic "pre-mRNA." We discuss this splicing reaction later in the chapter.

SRP RNA. In eukaryotes where some proteins are synthesized by ribosomes on the RER, this process is mediated by the **signal recognition particle,** or **SRP,** described later in the chapter. The SRP is composed of both RNA and proteins.

Micro-RNA. A new class of RNA recently discovered is **micro-RNA (miRNA).** These very short RNAs escaped detection for many years because they were lost during the techniques typically used in nucleic acid purification. Their role is unclear, but one class, **small interfering RNAs (siRNAs)** appear to be involved in controlling gene expression and are part of a system to protect cells from viral attack.

In transcription, the enzyme RNA polymerase synthesizes an RNA strand using DNA as a template. Only one strand of the DNA, the template strand, is copied; the other strand, which has the same sequence as the transcribed RNA, is called the coding strand. Translation takes place on ribosomes and utilizes tRNA as an adapter between mRNA codons and amino acids.

15.4 Prokaryotic Transcription

We begin a detailed examination of gene expression by describing the process of transcription in prokaryotes. The later description of eukaryotic transcription will concentrate on their differences from prokaryotes.

Prokaryotes have a single RNA polymerase

The single **RNA polymerase** of prokaryotes exists in two forms called *core polymerase* and *holoenzyme.* The core polymerase can synthesize RNA using a DNA template, but it cannot initiate synthesis accurately. The holoenzyme can initiate accurately.

The core polymerase is composed of four subunits: two identical α subunits, a β subunit, and a β′ subunit (figure 15.5*a*). The two α subunits help to hold the complex together and can bind to regulatory molecules. The active site of the enzyme is formed by the β and β′ subunits, which bind to the DNA template and the ribonucleotide triphosphate precursors.

The *holoenzyme* that can properly initiate synthesis is formed by the addition of a σ (sigma) subunit to the core polymerase (see figure 15.5*a*). Its ability to recognize specific signals in DNA allows RNA polymerase to locate the beginning of genes, which is critical to its function. Note that initiation of mRNA synthesis does not require a primer, in contrast to DNA replication.

Initiation occurs at promoters

Accurate initiation of transcription requires two sites in DNA: one called a **promoter** that forms a recognition and binding site for the RNA polymerase, and the actual **start site.** The polymerase also needs a signal to end transcription, which we call a **terminator.** We then refer to the region from promoter to terminator as a **transcription unit.**

The action of the polymerase moving along the DNA can be thought of as analogous to water flowing in a stream. We can speak of sites on the DNA as being "upstream" or "downstream" of the start site. We can also use this comparison to form a simple system for numbering bases in DNA to refer to positions in the transcription unit. The first base transcribed is called **+1,** and this numbering continues downstream until the last base transcribed. Any bases upstream of the start site receive negative numbers, starting at **–1.**

The promoter is a short sequence found upstream of the start site and is therefore not transcribed by the polymerase. Two 6-base sequences are common to bacterial promoters: One is located 35 nt upstream of the start site (–35), and the other is located 10 nucleotides upstream of the start site (–10) (figure 15.5*b*). These two sites provide the promoter with asymmetry; they indicate not only the site of initiation, but also the direction of transcription.

The binding of RNA polymerase to the promoter is the first step in transcription. Promoter binding is controlled by the σ subunit of the RNA polymerase holoenzyme, which recognizes the –35 sequence in the promoter and positions the RNA polymerase at the correct start site, oriented to transcribe in the correct direction.

figure 15.5

BACTERIAL RNA POLYMERASE AND TRANSCRIPTION INITIATION. ***a.*** RNA polymerase has two forms: core polymerase and holoenzyme. ***b.*** The σ subunit of holoenzyme recognizes promoter elements at –35 and –10 and binds to the DNA. The helix is opened at the –10 region, and transcription will begin at the start site at +1.

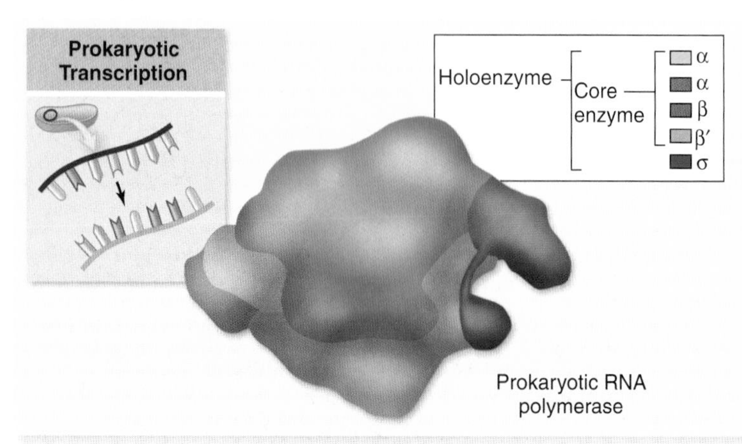

a.

Once bound to the promoter, the RNA polymerase begins to unwind the DNA helix at the –10 site (see figure 15.5*b*). The polymerase covers a region of about 75 bp but only unwinds about 12–14 bp.

inquiry

The prokaryotic promoter has two distinct elements that are not identical. How is this important to the initiation of transcription?

Elongation adds successive nucleotides

In prokaryotes, the transcription of the RNA chain usually starts with ATP or GTP. One of these forms the 5′ end of the chain, which grows in the 5′ to 3′ direction as ribonucleotides are added. As the RNA polymerase molecule leaves the promoter region, the σ factor is no longer required, although it may remain in association with the enzyme.

This process of leaving the promoter, called *clearance*, or *escape*, involves more than just synthesizing the first few nucleotides of the transcript and moving on, because the enzyme has made strong contacts to the DNA during initiation. It is necessary to break these contacts with the promoter region to be able to move progressively down the template. The enzyme goes through conformational changes during this clearance stage, and subsequently contacts less of the DNA than it does during the initial promoter binding.

The region containing the RNA polymerase, the DNA template, and the growing RNA transcript is called the **transcription bubble** because it contains a locally unwound "bubble" of DNA (figure 15.6). Within the bubble, the first 9 bases of the newly synthesized RNA strand temporarily form a helix with the template DNA strand. This stabilizes the positioning of the 3′ end of the RNA so it can interact with an incoming ribonucleotide triphosphate. The enzyme itself covers about 50 bp of DNA around this transcription bubble.

The transcription bubble created by RNA polymerase moves down the bacterial DNA at a constant rate, about 50 nt/sec, with the growing RNA strand protruding from the bubble. After the transcription bubble passes, the now-transcribed DNA is rewound as it leaves the bubble.

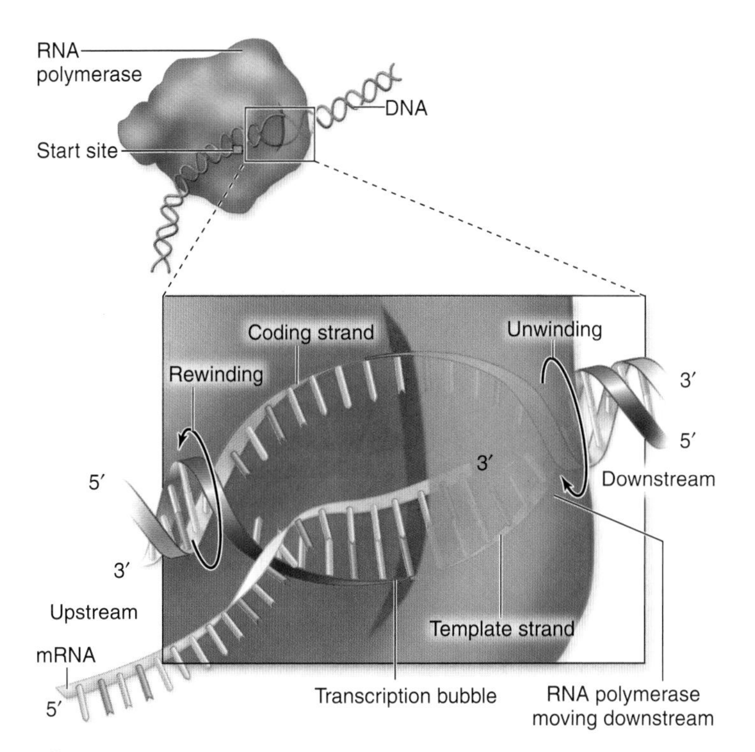

figure 15.6

MODEL OF A TRANSCRIPTION BUBBLE. The DNA duplex is unwound by the RNA polymerase complex, rewinding at the end of the bubble. One of the strands of DNA functions as a template, and nucleotide building blocks are assembled into RNA from this template. There is a short region of RNA–DNA hybrid within the bubble.

Termination occurs at specific sites

The end of a bacterial transcription unit is marked by terminator sequences that signal "stop" to the polymerase. Reaching these sequences causes the formation of phosphodiester bonds to cease, the RNA–DNA hybrid within the transcription bubble to dissociate, the RNA polymerase to release the DNA, and the DNA within the transcription bubble to rewind.

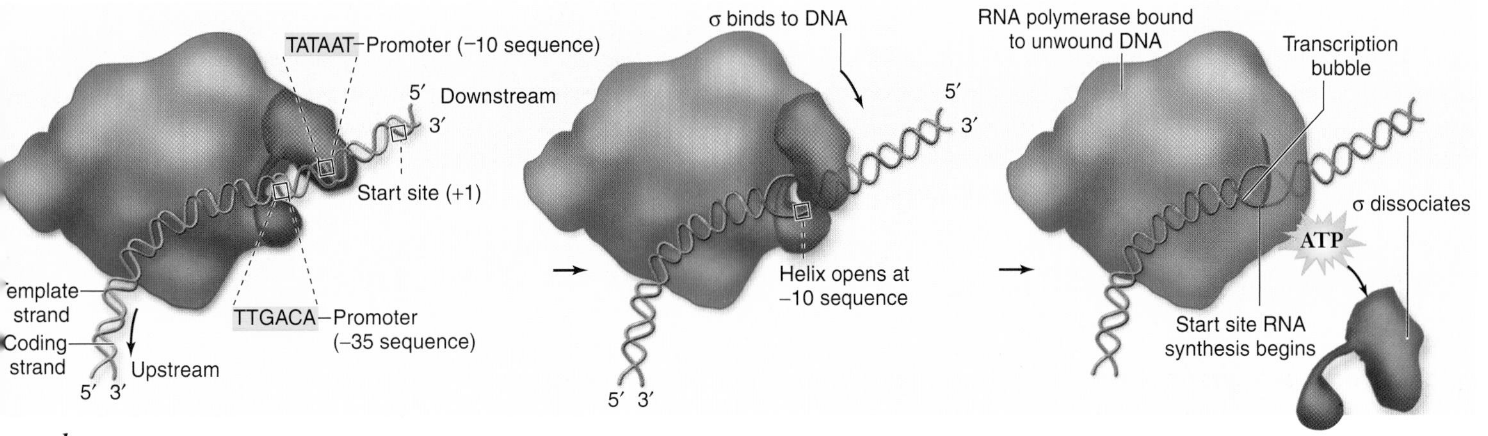

b.

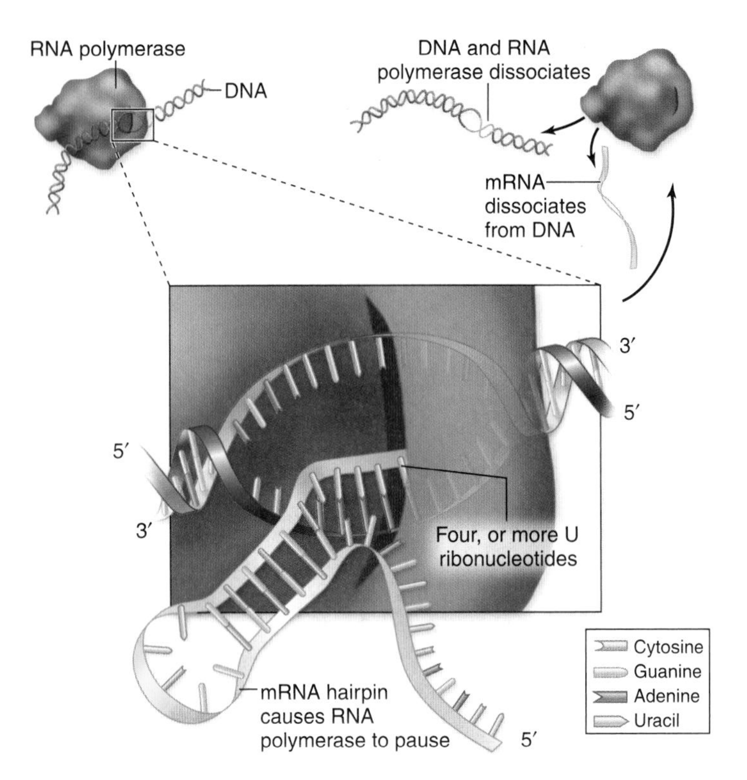

figure 15.7

BACTERIAL TRANSCRIPTION TERMINATOR. The self-complementary G–C region forms a double-stranded stem with a single stranded loop called a hairpin. The stretch of U's form a less stable RNA–DNA hybrid that falls off the enzyme.

The simplest terminators consist of a series of G–C base pairs followed by a series of A–T base pairs. The RNA transcript of this stop region can form a double-stranded structure in the GC region called a *hairpin,* which is followed by four or more uracil (U) ribonucleotides (figure 15.7). Formation of the hairpin causes the RNA polymerase to pause, placing it directly over the run of four uracils. The pairing of U with the DNA's A is the weakest of the four hybrid base-pairs, and it is not strong enough to hold the hybrid strands when the polymerase pauses. Instead, the RNA strand dissociates from the DNA within the transcription bubble, and transcription stops. A variety of protein factors also act at these terminators to aid in terminating transcription.

Prokaryotic transcription is coupled to translation

In prokaryotes, the mRNA produced by transcription begins to be translated before transcription is finished—that is, they are *coupled* (figure 15.8). As soon as a 5′ end of the mRNA becomes available, ribosomes are loaded onto this to begin translation.

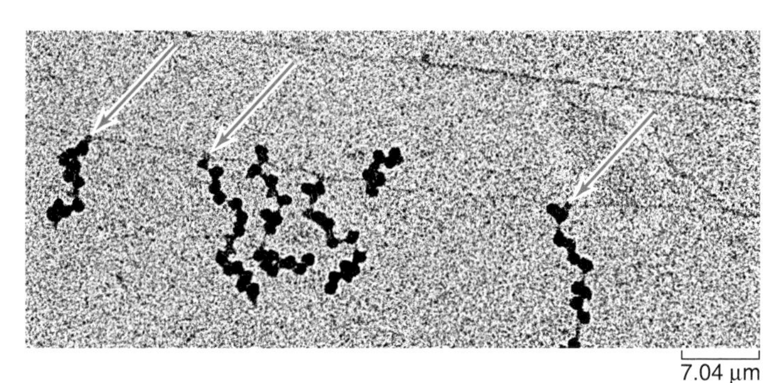

figure 15.8

TRANSCRIPTION AND TRANSLATION ARE COUPLED IN BACTERIA. In this micrograph of gene expression in *E. coli*, translation is occurring during transcription. The arrows point to RNA polymerase enzymes and ribosomes are attached to the mRNAs extending from the polymerase. Polypeptides being synthesized by ribosomes, which are not visible in the micrograph, have been added to last mRNA in the drawing.

(This coupling cannot occur in eukaryotes because transcription occurs in the nucleus, and translation occurs in the cytoplasm.)

Another difference between prokaryotic and eukaryotic gene expression is that the mRNA produced in prokaryotes may contain multiple genes. Prokaryotic genes are often organized such that genes encoding related functions are clustered together. This grouping of functionally related genes is referred to as an **operon.** An operon is a single transcripting unit that encodes multiple enzymes necessary for a biochemical pathway. By clustering genes by function, they can be regulated together, a topic that we return to in the next chapter.

The bacterial RNA polymerase has two forms: a core polymerase with all synthetic activity, and a holoenzyme that can initiate accurately. Prokaryotic transcription starts at sites called promoters that are recognized by the holoenzyme. Elongation consists of synthesis by the core enzyme until it reaches a terminator where synthesis stops, and the transcript dissociates from the enzyme.

15.5 Eukaryotic Transcription

The basic mechanism of transcription by RNA polymerase is the same in eukaryotes as in prokaryotes; however, the details of the two processes differ enough that it is necessary to consider them separately. Here we concentrate only on how eukaryotic systems differ from prokaryotic systems, such as the bacterial system just discussed. All other features may be assumed to be the same.

Eukaryotes have three RNA polymerases

Unlike prokaryotes, which have a single RNA polymerase enzyme, eukaryotes have three different RNA polymerases that are distinguished by both structure and function. The enzyme **RNA polymerase I** transcribes rRNA, **RNA polymerase II** transcribes mRNA and some small nuclear RNAs, and **RNA polymerase III** transcribes tRNA and some other small RNAs. Together, these three enzymes accomplish all transcription in the nucleus of eukaryotic cells.

Each polymerase has its own promoter

The existence of three different RNA polymerases requires different signals in the DNA to allow each polymerase to recognize where to begin transcription; each polymerase recognizes a different promoter structure.

RNA polymerase I promoters

RNA polymerase I promoters at first puzzled biologists, because comparisons of rRNA genes between species showed no similarities outside the coding region. The current view is that these promoters are also specific for each species, and for this reason, cross-species comparisons do not yield similarities.

RNA polymerase II promoters

The RNA polymerase II promoters are the most complex of the three types, probably a reflection of the huge diversity of genes that are transcribed by this polymerase. When the first eukaryotic genes were isolated, many had a sequence called the **TATA box** upstream of the start site. This sequence was similar to the prokaryotic –10 sequence, and it was assumed that the TATA box was the primary promoter element. With the sequencing of entire genomes, many more genes have been analyzed, and this assumption has proved overly simple. It has been replaced by the idea of a "core promoter" that can be composed of a number of different elements, including the TATA box. Additional control elements allow for tissue-specific and developmental time–specific expression (chapter 16).

RNA polymerase III promoters

Promoters for RNA polymerase III also were a source of surprise for biologists in the early days of molecular biology who were examining the control of eukaryotic gene expression. A common technique for analyzing regulatory regions was to make successive deletions from the 5′ end of genes until enough was deleted to abolish specific transcription. The logic followed experiences with prokaryotes, in which the regulatory regions had been found at the 5′ end of genes. But in the case of tRNA genes, the 5′ deletions had no effect on expression! The promoters were found to actually be internal to the gene itself.

Initiation and termination differ from that in prokaryotes

The initiation of transcription at RNA polymerase II promoters is analogous to prokaryotic initiation but is more complex. Instead of a single factor allowing promoter recognition, eukaryotes use a host of **transcription factors.** These proteins are necessary to get the RNA polymerase II enzyme to a promoter and to initiate gene expression. A number of these transcription factors interact with RNA polymerase II to form an *initiation complex* at the promoter (figure 15.9). We explore this complex in detail in chapter 16 when we describe the control of gene expression.

The termination of transcription for RNA polymerase II also differs from that in prokaryotes. Although termination sites exist, they are not as well defined as are prokaryotic terminators. The end of the mRNA is also not formed by RNA polymerase II because the primary transcript is modified after transcription.

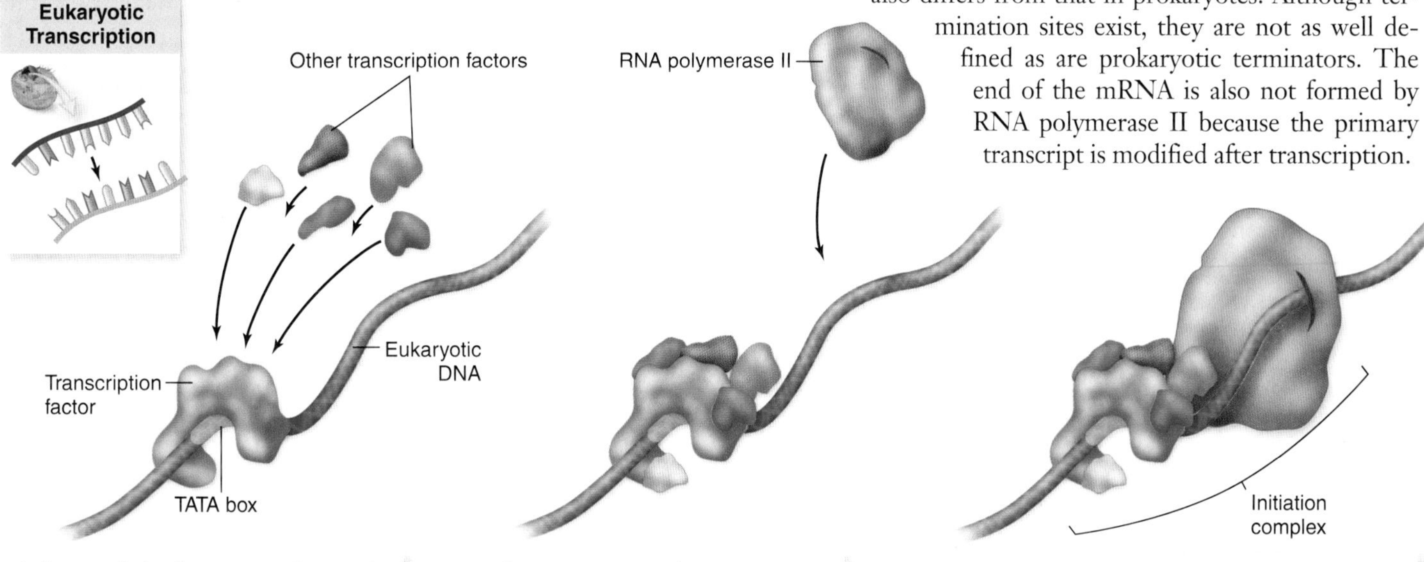

1. A transcription factor recognizes and binds to the TATA box sequence, which is part of the core promoter.

2. Other transcription factors are recruited, and the initiation complex begins to build.

3. Ultimately, RNA polymerase II associates with the transcription factors and the DNA, forming the initiation complex, and transcription begins.

figure 15.9

EUKARYOTIC INITIATION COMPLEX. Unlike transcription in prokaryotic cells, where the RNA polymerase recognizes and binds to the promoter, eukaryotic transcription requires the binding of transcription factors to the promoter before RNA polymerase II binds to the DNA. The association of transcription factors and RNA polymerase II at the promoter is called the initiation complex.

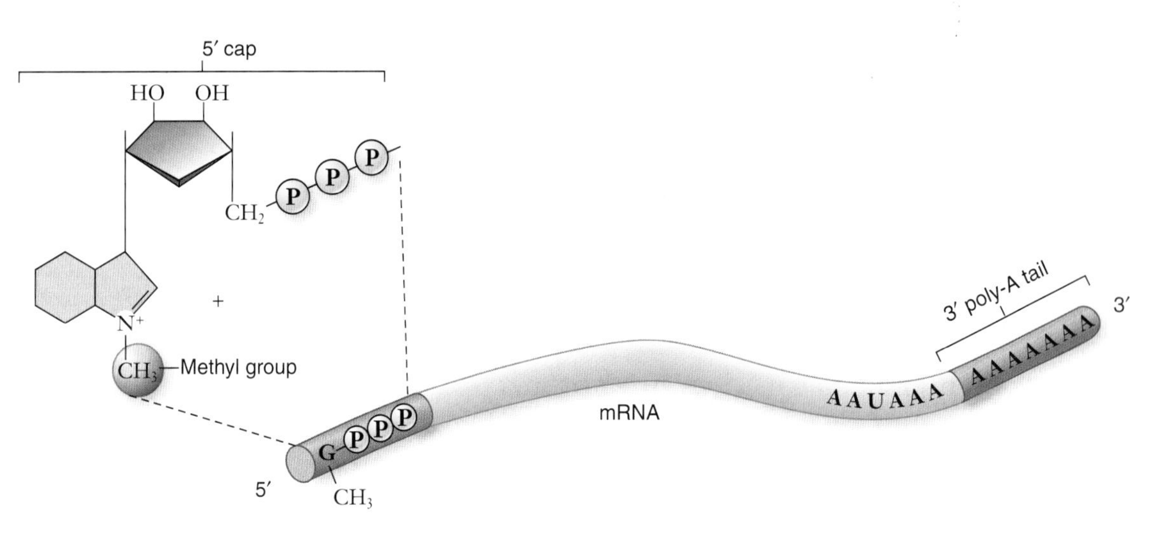

figure 15.10

POSTTRANSCRIPTIONAL MODIFICATIONS TO 5′ AND 3′ ENDS. Eukaryotic mRNA molecules are modified in the nucleus with the addition of a methylated GTP to the 5′ end of the transcript, called the 5′ cap, and a long chain of adenine residues to the 3′ end of the transcript, called the poly-A tail.

Eukaryotic transcripts are modified

A primary difference between prokaryotes and eukaryotes is the fate of the transcript itself. Between transcription in the nucleus and export of a mature mRNA to the cytoplasm, a number of modifications occur to the initial transcripts made by RNA polymerase II. We call the RNA synthesized by RNA polymerase II the **primary transcript** and the final processed form the **mature mRNA.**

The 5′ cap

Eukaryotic transcripts have an unusual structure that is added to the 5′ end of mRNAs. The first base in the transcript is usually an adenine (A) or a guanine (G), and this is further modified by the addition of GTP to the 5′ PO_4 group, forming what is known as a **5′ cap** (figure 15.10). The G nucleotide in the cap is joined to the transcript by its 5′ end; the only such 5′-to-5′ bond found in nucleic acids. The G in the GTP is also modified by the addition of a methyl group, so it is often called a *methyl-G cap.* The cap is added while transcription is still in progress. This cap protects the 5′ end of the mRNA from degradation and is also involved in translation initiation.

The 3′ poly-A tail

A major difference between transcription in prokaryotes and eukaryotes is that in eukaryotes, the end of the transcript is not the end of the mRNA. The eukaryotic transcript is cleaved downstream of a specific site (AAUAAA) prior to the termination site for transcription. A series of adenine (A) residues, called the **3′ poly-A tail,** are added after this cleavage by the enzyme poly-A polymerase. Thus the end of the mRNA is not created by RNA polymerase II and is not the end of the transcript (see figure 15.10).

The enzyme poly-A polymerase is part of a complex that recognizes the poly-A site, cleaves the transcript, then adds 1–200 A's to the end. The poly-A tail appears to play a role in the stability of mRNAs by protecting them from degradation (chapter 16).

Splicing of primary transcripts

Eukaryotic genes may contain noncoding sequences that have to be removed to produce the final mRNA. This process, called pre-mRNA splicing, is accomplished by an organelle called the *spliceosome.* This complex topic is discussed in the next section.

Eukaryotes have three RNA polymerases, called polymerase I, II, and III. Each is responsible for the synthesis of different cellular RNAs and recognizes its own promoter. Polymerase II is responsible for mRNA synthesis. The primary mRNA transcript is modified by addition of a 5′ cap and a 3′ poly-A tail consisting of 1–200 adenines. Noncoding regions are removed by splicing.

15.6 Eukaryotic pre-mRNA Splicing

The first genes isolated were prokaryotic genes found in *E. coli* and its viruses. A clear picture of the nature and some of the control of gene expression emerged from these systems before any eukaryotic genes were isolated. It was assumed that although details would differ, the outline of gene expression in eukaryotes would be similar. The world of biology was in for a shock with the isolation of the first genes from eukaryotic organisms.

Eukaryotic genes may contain interruptions

Many eukaryotic genes appeared to contain sequences that were not represented in the mRNA. It is hard to exaggerate how unexpected this finding was. A basic tenet of molecular biology based on *E. coli* was that a gene was *colinear* with its protein product, that is, the sequence of bases in the gene corresponds to the sequence of bases in the mRNA, which in turn corresponds to the sequence of amino acids in the protein.

In the case of eukaryotes, genes can be interrupted by sequences that are not represented in the mRNA and the protein. The term "split genes" was used at the time, but the nomenclature that has stuck describes the unexpected nature of these sequences. We call the noncoding DNA that interrupts the sequence of the gene "intervening sequences," or **introns,** and we call the coding sequences **exons** because they are *ex*pressed (figure 15.11).

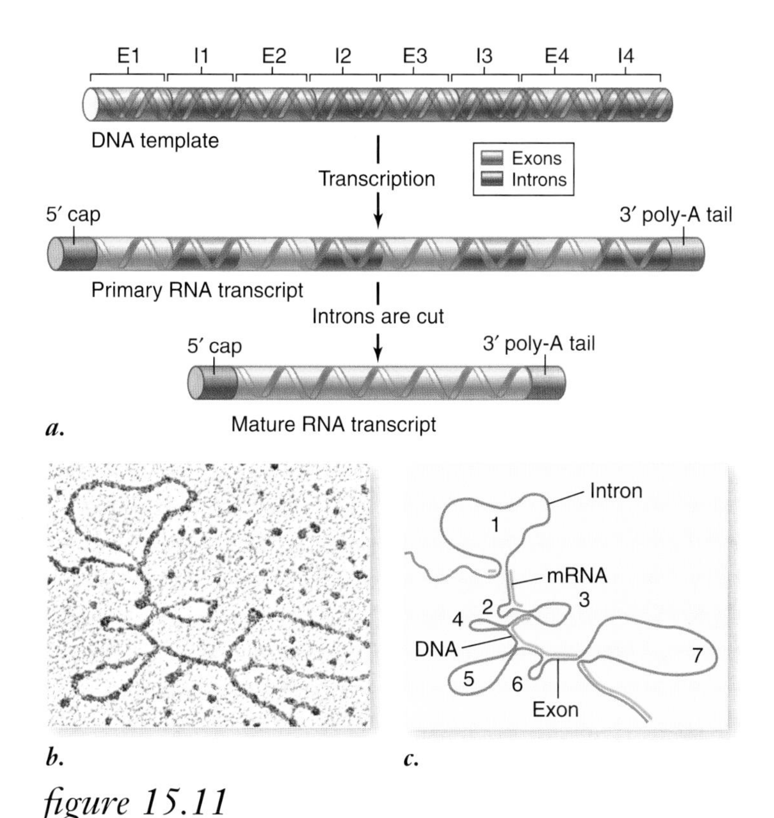

figure 15.11

EUKARYOTIC GENES CONTAIN INTRONS AND EXONS. ***a.*** Eukaryotic genes contain sequences that will form the coding sequence called exons and intervening sequences called introns. ***b.*** An electron micrograph showing hybrids formed with the mRNA and the DNA of the ovalbumin gene, which has seven introns. Introns within the DNA sequence have no corresponding sequence in the mRNA and thus appear as seven loops. ***c.*** A schematic drawing of the micrograph.

inquiry

How can the same gene encode different transcripts?

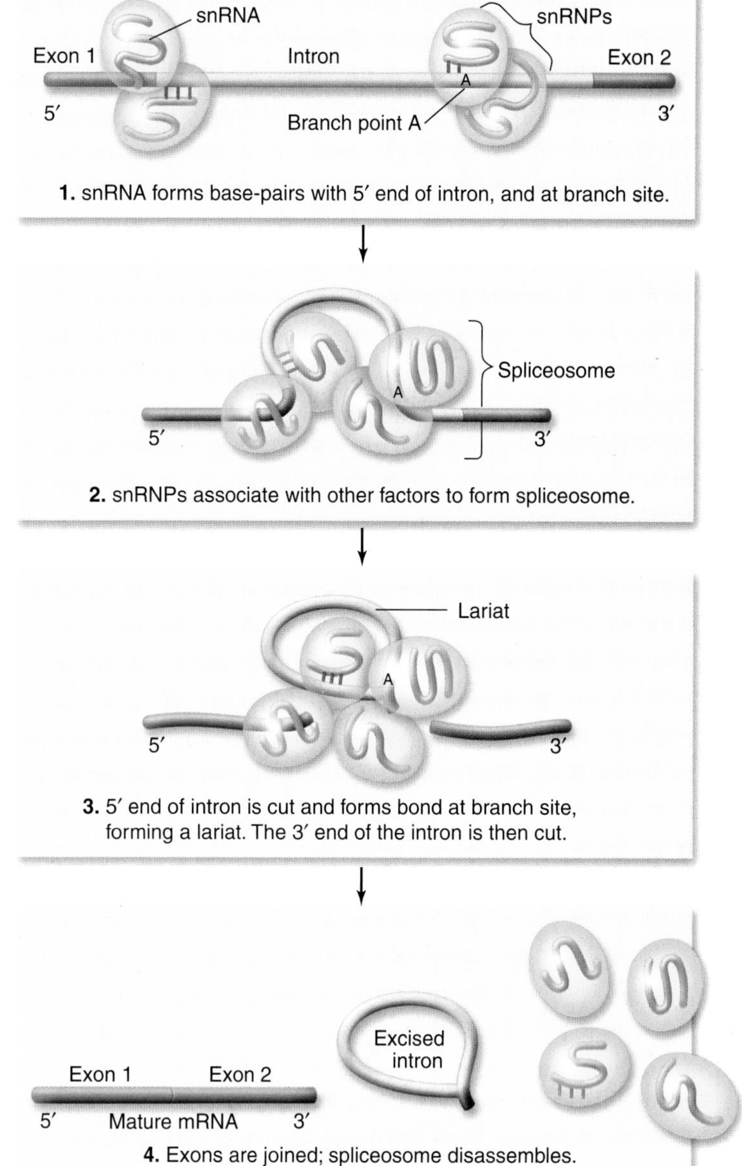

figure 15.12

PRE-mRNA SPLICING BY THE SPLICEOSOME. Particles called snRNPs contain snRNA that interacts with the 5′ end of an intron and with a branch site internal to the intron. Several snRNPs come together with other proteins to form the spliceosome. As the intron forms a loop, the 5′ end is cut and linked to a site near the 3′ end of the intron. The intron forms a lariat that is excised, and the exons are spliced together. The spliceosome then disassembles and releases the spliced mRNA.

The spliceosome is the splicing organelle

It is still true that the mature eukaryotic mRNA is colinear with its protein product, but a gene that contains introns is not. Imagine looking at an interstate highway from a satellite. Scattered randomly along the thread of concrete would be cars, some moving in clusters, others individually; most of the road would be bare. That is what a eukaryotic gene is like—scattered exons embedded within much longer sequences of introns.

In humans, only 1 to 1.5% of the genome is devoted to the exons that encode proteins; 24% is devoted to the noncoding introns within which these exons are embedded.

The splicing reaction

The obvious question is, how do eukaryotic cells deal with the noncoding introns? The answer is that the primary transcript is cut and put back together to produce the mature mRNA. The latter process is referred to as **pre-mRNA splicing,** and it occurs in the nucleus prior to the export of the mRNA to the cytoplasm.

The intron–exon junctions are recognized by **small nuclear ribonucleoprotein particles,** called **snRNPs** (pronounced "snurps"). The snRNPs are complexes composed of snRNA and protein. These snRNPs then cluster together with other associated proteins to form a larger complex called the **spliceosome,** which is responsible for the splicing, or removal, of the introns.

For splicing to occur accurately, the spliceosome must be able to recognize intron–exon junctions. Introns all begin with the same 2-base sequence and end with another 2-base sequence that tags them for removal. In addition, within the intron there is a conserved A nucleotide, called the *branch point,* that is important for the splicing reaction (figure 15.12).

The splicing process begins with cleavage of the 5′ end of the intron. This 5′ end becomes attached to the 2′ OH of the branch point A, forming a branched structure called a *lariat* for its resemblance to a cowboy's lariat in a rope (see figure 15.12). The 3′ end of the first exon is then used to displace the 3′ end of the intron, joining the two exons together and releasing the intron as a lariat.

The process of transcription and RNA processing do not occur in a linear sequence, but are rather all part of a concerted process that produces the mature mRNA. The capping reaction occurs during transcription, as does the splicing process. The RNA polymerase II enzyme itself helps to recruit the other factors necessary for modification of the primary transcript, and in this way the process of transcription and pre-mRNA processing forms an integrated system.

Distribution of introns

No rules govern the number of introns per gene or the sizes of introns and exons. Some genes have no introns; others may have 50 introns. The sizes of exons range from a few nucleotides to 7500 nt, and the sizes of introns are equally variable. The presence of introns partly explains why so little of a eukaryotic genome is actually composed of "coding sequences" (see chapter 18 for results from the Human Genome Project).

One explanation for the existence of introns suggests that exons represent functional domains of proteins, and that the intron–exon arrangements found in genes represent the shuffling of these functional units over long periods of evolutionary time. This hypothesis, called *exon shuffling,* was proposed soon after the discovery of introns and has been the subject of much debate over the years.

The recent flood of genomic data has shed light on this issue by allowing statistical analysis of the placement of introns and on intron–exon structure. This analysis has provided support for the exon shuffling hypothesis for many genes; however, it is also clearly not universal, because all proteins do not show this kind of pattern. It is possible that introns do not have a single origin, and therefore cannot be explained by a single hypothesis.

Splicing can produce multiple transcripts from the same gene

One consequence of the splicing process is greater complexity in gene expression in eukaryotes. A single primary transcript can be spliced into different mRNAs by the inclusion of different sets of exons, a process called **alternative splicing.**

Evidence indicates that the normal pattern of splicing is important to an organism's function. It has been estimated that 15% of known human genetic disorders are due to altered splicing. Mutations in the signals for splicing can introduce new splice sites or can abolish normal patterns of splicing. (In chapter 16 we consider how alternative splicing can be used to regulate gene expression.)

Although many cases of alternative splicing have been documented, the recent completion of the draft sequence of the human genome, along with other large data sets of expressed sequences, now allow large-scale comparisons between sequences found in mRNAs and in the genome. Three different computer-based analyses have been performed, producing results that are in rough agreement. These initial genomic assessments indicate a range of 35 to 59% for human genes that exhibit some form of alternative splicing. If we pick the middle ground of around 40%, this result still vastly increases the number of potential proteins encoded by the 25,000 genes in the human genome.

It is important to note that these analyses are primarily computer-based, and the functions of the possible spliced products have been investigated for only a small part of the potentially spliced genes. These analyses, however, do explain how the 25,000 genes of the human genome can encode the more than 80,000 different mRNAs reported to exist in human cells. The emerging field of proteomics addresses the number and functioning of proteins encoded by the human genome.

Eukaryotic genes contain exon regions that are expressed and intervening sequences, introns, that interrupt the exons. The introns are removed by the spliceosome, leaving the exons joined together. Alternative splicing can generate different mRNAs, and thus different proteins, from the same gene. Recent estimates are that as many as half of human genes may be alternatively spliced.

15.7 The Structure of tRNA and Ribosomes

The ribosome is the key organelle in translation, but it also requires the participation of mRNA, tRNA and a host of other factors. Critical to this process is the interaction of the ribosomes with tRNA and mRNA. To understand this, we first examine the structure of the tRNA adapter molecule and the ribosome itself.

Aminoacyl-tRNA synthetases attach amino acids to tRNA

Each amino acid must be attached to a tRNA with the correct anticodon for protein synthesis to proceed. This connection is accomplished by the action of activating enzymes called **aminoacyl-tRNA synthetases.** One of these enzymes is present for each of the 20 common amino acids.

tRNA structure

Transfer RNA is a bifunctional molecule that must be able to interact with mRNA and with amino acids. The structure of tRNAs is highly conserved in all living systems, and it can be formed into a cloverleaf type of structure based on intramolecular base-pairing that produces double-stranded regions. This primary structure is then folded in space to form an L-shaped molecule that has two functional ends: the **acceptor stem** and the **anticodon loop** (figure 15.13).

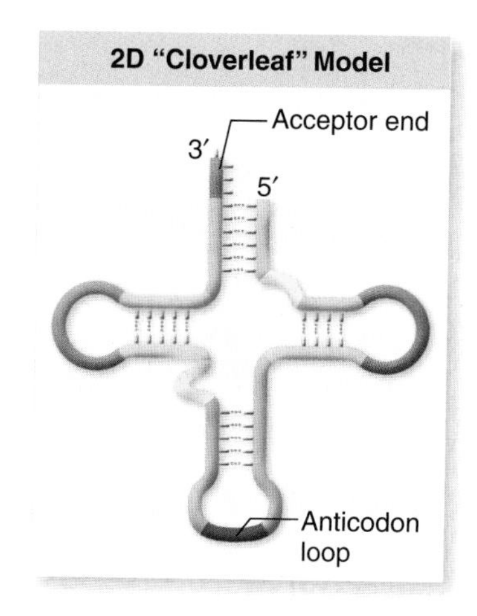

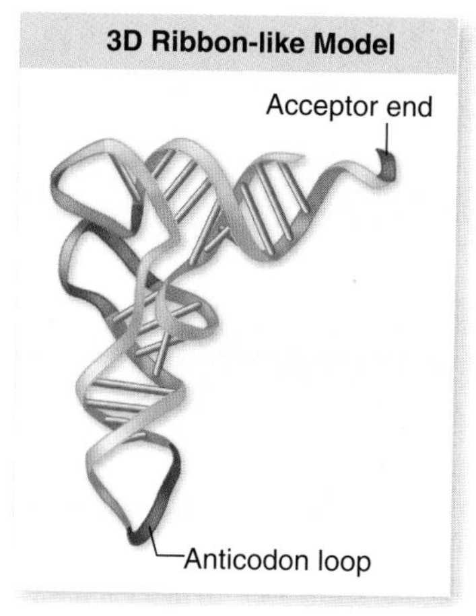

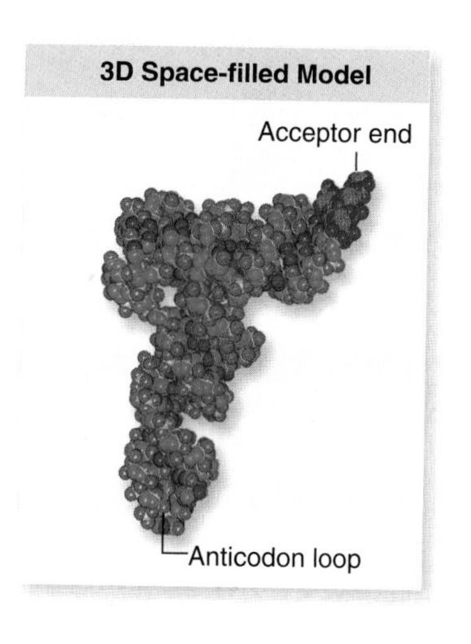

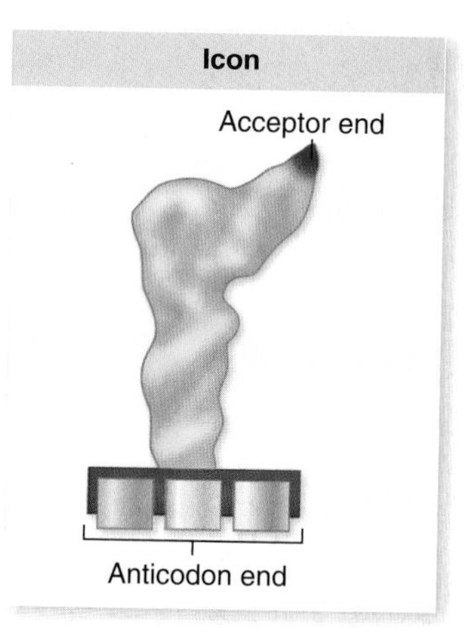

figure 15.13

THE STRUCTURE OF tRNA. Base-pairing within the molecule creates three stem and loop structures in a characteristic cloverleaf shape. The loop at the bottom of the cloverleaf contains the anticodon sequence, which can base-pair with codons in the mRNA. Amino acids are attached to the free, single-stranded —OH end of the acceptor stem. In its final three-dimensional structure, the loops of tRNA are folded into the final L-shaped structure.

The acceptor stem is the 3′ end of the molecule, which always ends in 5′ CCA 3′. The amino acid can be attached to this end of the molecule. The anticodon loop is the bottom loop of the cloverleaf, and it can base-pair with codons in mRNA.

The charging reaction

The aminoacyl-tRNA synthetases must be able to recognize specific tRNA molecules as well as their corresponding amino acids. Although 61 codons code for amino acids, there are actually not 61 tRNAs in cells, although the number varies from species to species. Therefore some aminoacyl-tRNA synthetases must be able to recognize more than one tRNA—but each recognizes only a single amino acid.

The reaction catalyzed by the enzymes is called the tRNA **charging reaction,** and the product is an amino acid joined to a tRNA, now called a *charged tRNA*. An ATP molecule provides energy for this endergonic reaction. The charged tRNA produced by the reaction is an activated intermediate that can undergo the peptide bond-forming reaction without an additional input of energy.

The charging reaction joins the acceptor stem to the carboxyl terminus of an amino acid (figure 15.14). Keeping this

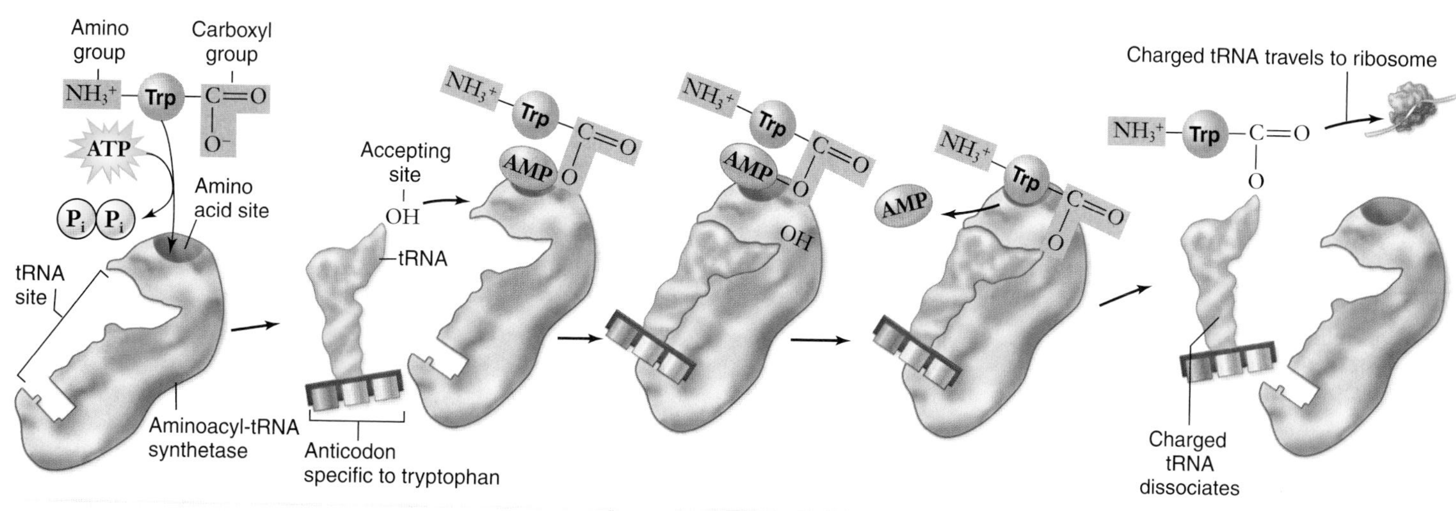

figure 15.14

tRNA CHARGING REACTION. There are 20 different aminoacyl-tRNA synthetase enzymes each specific for one amino acid, such as tryptophan (Trp). The enzyme must also recognize and bind to the tRNA molecules with anticodons specifying that amino acid, ACC for tryptophan. The reaction uses ATP and produces an activated intermediate that will not require further energy for peptide bond formation.

figure 15.15

RIBOSOMES HAVE TWO SUBUNITS. Ribosome subunits come together and apart as part of a ribosome cycle. The smaller subunit fits into a depression on the surface of the larger one. Ribosomes have three tRNA-binding sites: aminoacyl site (A), peptidyl site (P), and empty site (E).

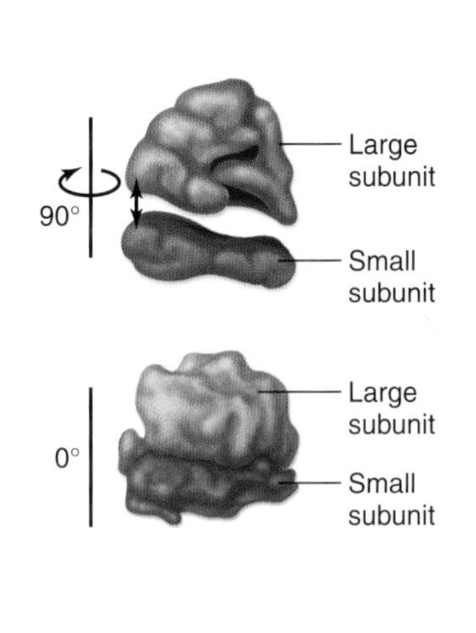

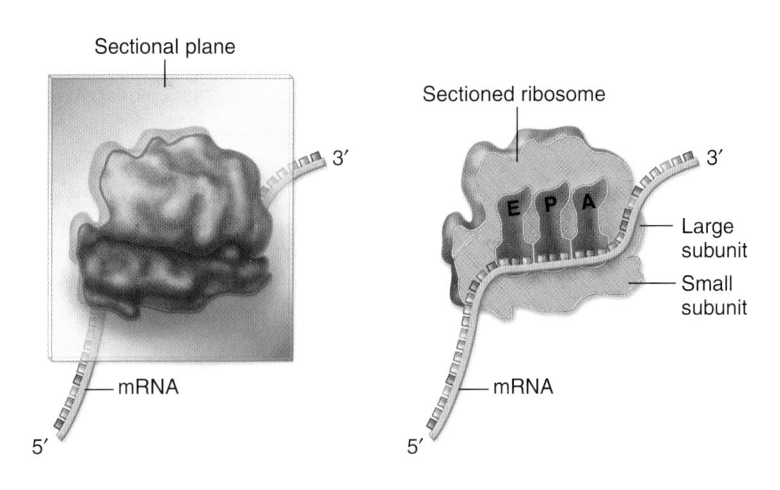

directionality in mind is critical to understanding the function of the ribosome, because each peptide bond will be formed between the amino group of one amino acid and the carboxyl group of another amino acid.

The correct attachment of amino acids to tRNAs is important because the ribosome does not verify this attachment. Ribosomes can only ensure that the codon–anticodon pairing is correct. In an elegant experiment, cysteine was converted chemically to alanine after the charging reaction, when the amino acid was already attached to tRNA. When this charged tRNA was used in an in vitro protein synthesis system, alanine was incorporated in the place of cysteine, showing that the ribosome cannot "proofread" the amino acids attached to tRNA.

In a very real sense, therefore, the charging reaction is the real translation step; amino acids are incorporated into a peptide based solely on the tRNA anticodon and its interaction with the mRNA.

The ribosome has multiple tRNA-binding sites

The synthesis of any biopolymer can be broken down into initiation, elongation, and termination—you have seen this division for DNA replication as well as for transcription. In the case of translation, or protein synthesis, all three of these steps take place on the ribosome, a large macromolecular assembly consisting of rRNA and proteins. Details of the process by which the two ribosome subunits are assembled during initiation are described shortly.

For the ribosome to function it must be able to bind to at least two charged tRNAs at once so that a peptide bond can be formed between their amino acids, as described in the previous overview. In reality, the bacterial ribosome contains three binding sites, summarized in figure 15.15:

- The **P site** (peptidyl) binds to the tRNA attached to the growing peptide chain.
- The **A site** (aminoacyl) binds to the tRNA carrying the next amino acid to be added.
- The **E site** (exit) binds the tRNA that carried the previous amino acid added (see figure 15.15).

Transfer RNAs move through these sites successively during the process of elongation. Relative to the mRNA, the sites are arranged 5′ to 3′ in the order E, P, and A. The incoming charged tRNAs enter the ribosome at the A site, transit through the P site, and then leave via the E site.

The ribosome has both decoding and enzymatic functions

The two functions of the ribosome involve decoding the transcribed message and forming peptide bonds. The decoding function resides primarily in the small subunit of the ribosome. The formation of peptide bonds requires the

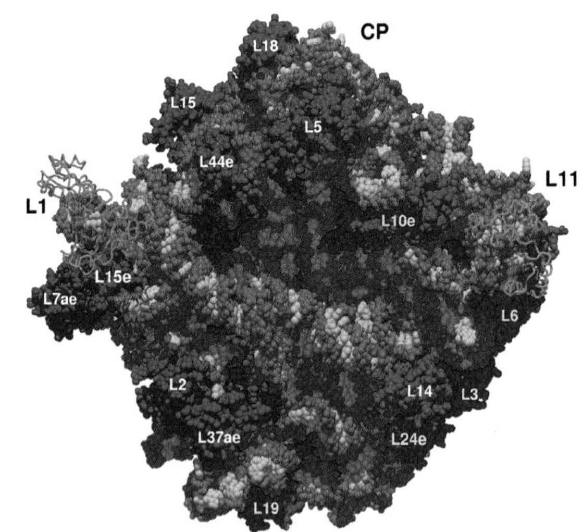

figure 15.16

3-D STRUCTURE OF PROKARYOTIC RIBOSOME. The complete atomic structure of a prokaryotic large ribosomal subunit has been determined at 2.4-Å resolution. Bases of RNA are white, the polynucleotide backbone is red and proteins are blue. The faces of each ribosomal subunit are lined with rRNA such that their interaction with tRNAs, amino acids, and mRNA all involve rRNA. Proteins are absent from the active site but abundant everywhere on the surface. The proteins stabilize the structure by interacting with adjacent RNA strands.

enzyme **peptidyl transferase,** which resides in the large subunit.

Our view of the ribosome has changed dramatically over time. Initially, molecular biologists assumed that the proteins in the ribosome carried out its function, and that the rRNA was a structural scaffold necessary to hold the proteins in the correct position. Now this view has mostly been reversed; the ribosome is seen instead as rRNAs that are held in place by proteins. The faces of the two subunits that interact with each other are lined with rRNA, and the parts of both subunits that interact with mRNA, tRNA, and amino acids are also primarily rRNA (figure 15.16). It is now thought that the peptidyl transferase activity resides in an rRNA in the large subunit.

The tRNA is a bifunctional molecule with one end that can form a bond to an amino acid and another end that can base-pair with mRNA. The tRNA charging reaction joins the carboxyl end of an amino acid to the 3′ acceptor stem of its tRNA. This reaction is catalyzed by 20 different aminoacyl-tRNA synthetases, one for each amino acid. The ribosome has three different binding sites for tRNA, one for the growing chain (P site), one for the next charged tRNA (A site), and one for the last tRNA used (E site). The ribosome can be thought of as having a decoding function and an enzymatic function.

15.8 The Process of Translation

The process of translation is one of the most complex and energy-expensive tasks that cells perform. An overview of the process, as you saw earlier, is perhaps deceptively simple: The mRNA is threaded through the ribosome, while tRNAs carrying amino acids bind to the ribosome, where they interact with mRNA by base-pairing with the mRNA's codons. The ribosome and tRNAs position the amino acids such that peptide bonds can be formed between each new amino acid and the growing polypeptide.

Initiation requires accessory factors

As mentioned earlier, the start codon is AUG, which also encodes the amino acid methionine. The ribosome usually uses the first AUG it encounters in an mRNA strand to signal the start of translation.

Prokaryotic initiation

In prokaryotes, the **initiation complex** includes a special **initiator tRNA** molecule charged with a chemically modified methionine, *N-formylmethionine.* The initiator tRNA is shown as $tRNA^{fMet}$. The initiation complex also includes the small ribosomal subunit and the mRNA strand (figure 15.17). The small subunit is positioned correctly on the mRNA due to a conserved sequence in the 5′ end of the mRNA called the **ribosome-binding sequence (RBS)** that is complementary to the 3′ end of a small subunit rRNA.

A number of initiation factors mediate this interaction of the ribosome, mRNA and $tRNA^{fMet}$ to form the initiation complex. These factors are involved in initiation only and are not part of the ribosome.

Once the complex of mRNA, initiator tRNA, and small ribosomal subunit is formed, the large subunit is added, and translation can begin. With the formation of the complete ribosome, the initiator tRNA is bound to the P site with the A site empty.

Eukaryotic initiation

Initiation in eukaryotes is similar, although it differs in two important ways. First, in eukaryotes, the initiating amino acid is methionine rather than *N*-formylmethionine. Second, the initiation complex is far more complicated than in prokaryotes, containing nine or more protein factors, many consisting of several subunits.

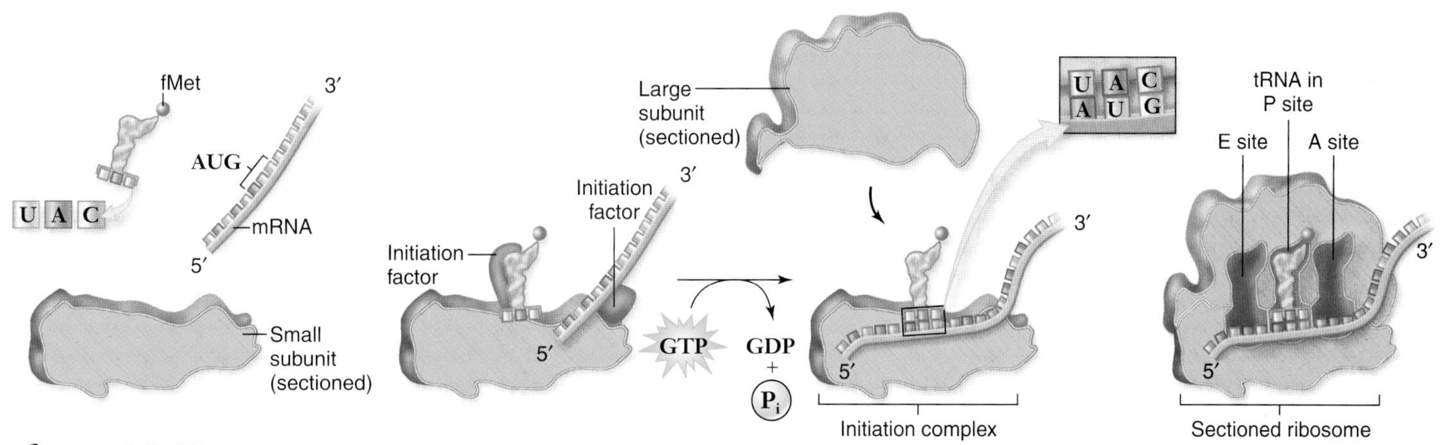

figure 15.17

INITIATION OF TRANSLATION. In prokaryotes, proteins called initiation factors play key roles in positioning the small ribosomal subunit and the initiator $tRNA^{fMet}$, molecule at the beginning of the mRNA. When the $tRNA^{fMet}$ is positioned over the first AUG codon of the mRNA, the large ribosomal subunit binds, forming the E, P, and A sites where successive tRNA molecules bind to the ribosomes, and polypeptide synthesis begins.

Eukaryotic mRNAs also lack an RBS. The small subunit binds to the mRNA initially by binding to the 5′ cap of the mRNA.

Elongation adds successive amino acids

When the entire ribosome is assembled around the initiator tRNA and mRNA, the second charged tRNA can be brought to the ribosome and bind to the empty A site. This requires an **elongation factor** called **EF-Tu,** which binds to the charged tRNA and to GTP.

A peptide bond can then form between the amino acid of the initiator tRNA and the newly arrived charged tRNA in the A site. The geometry of this bond relative to the two charged tRNAs is critical to understanding the process. Remember that an amino acid is attached to a tRNA by its carboxyl terminus. The peptide bond is formed between the amino end of the incoming amino acid (in the A site) and the carboxyl end of the growing chain (in the P site) (figure 15.18).

The addition of successive amino acids is a series of events that occur in a cyclic fashion. Figure 15.19 shows the details of the elongation cycle.

1. **Matching tRNA anticodon with mRNA codon**
   Each new charged tRNA comes to the ribosome bound to EF-Tu and GTP. The charged tRNA binds to the A site if its anticodon is complementary to the mRNA codon in the A site.
   After binding, GTP is hydrolyzed, and EF-Tu–GDP dissociates from the ribosome where it is recycled by another factor. This two-step binding and hydrolysis of GTP is thought to increase the accuracy of translation since the codon–anticodon pairing can be checked twice.
2. **Peptide bond formation**
   Peptidyl transferase, located in the large subunit, forms a peptide bond between the amino group of the amino acid in the A site and the carboxyl group of the growing chain. This reaction has the effect of transferring the growing chain to the tRNA in the A site, leaving the tRNA in the P site empty (no longer charged).
3. **Translocation of the ribosome**
   After the peptide bond has been formed, the ribosome moves relative to the mRNA and the tRNAs. The next codon in the mRNA shifts into the A site, and the tRNA with the growing chain moves to the P site. The uncharged tRNA formerly in the P site is now in the E site, and it will be ejected in the next cycle. This translocation step requires the accessory factor EF-G and the hydrolysis of another GTP.

This elongation cycle continues with each new amino acid added. The ribosome moves down the mRNA in a 5′ to 3′ direction, reading successive codons. The tRNAs move through the ribosome in the opposite direction, from the A site to the P site and finally the E site, before they are ejected as empty tRNAs, which can be charged with another amino acid and then used again.

Wobble pairing

As mentioned, there are fewer tRNAs than codons. This situation is workable because the pairing between the 3′ base of the codon and the 5′ base of the anticodon is less stringent than normal. In some tRNAs, the presence of modified bases with less accurate pairing in the 5′ position of the anticodon enhances this flexibility. This effect is referred to as **wobble pairing** because these tRNAs can "wobble" a bit in the ribosome, so that a single tRNA can "read" more than one codon in the mRNA.

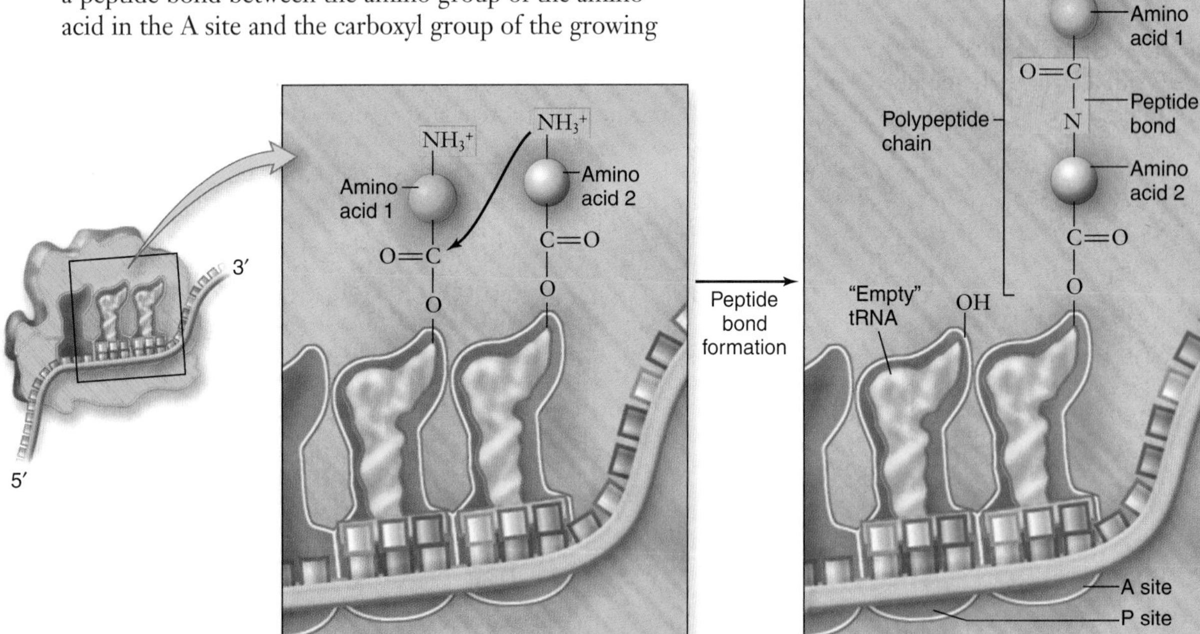

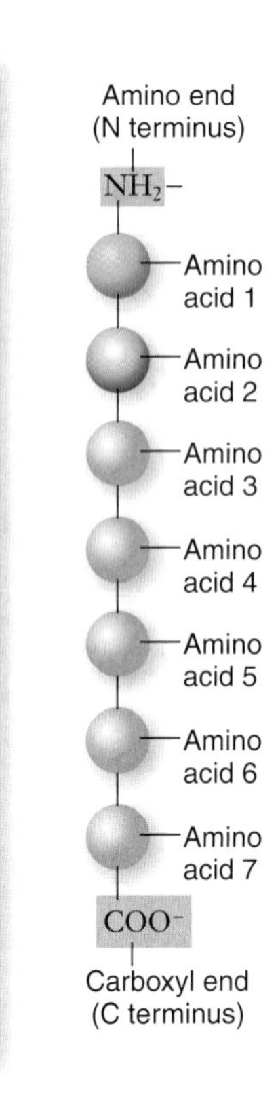

figure 15.18

PEPTIDE BOND FORMATION. Peptide bonds are formed on the ribosome between a "new" charged tRNA in the A site, and the growing chain attached to the tRNA in the P site. The bond forms between the amino group of the new amino acid and the carboxyl group of the growing chain. This transfers the growing chain to the A site as the new amino acid remains attached to its tRNA by its carboxyl terminus.

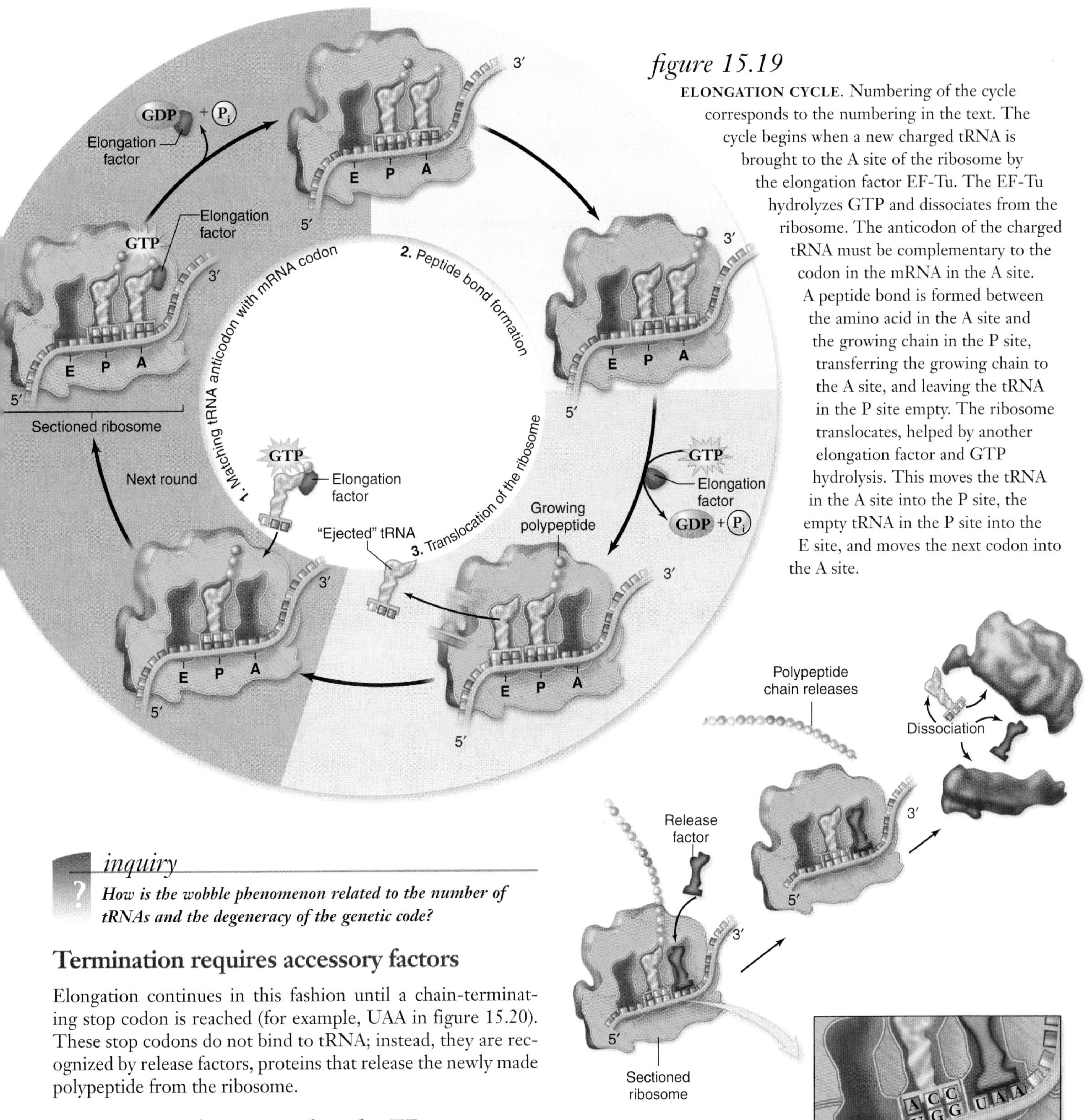

figure 15.19

ELONGATION CYCLE. Numbering of the cycle corresponds to the numbering in the text. The cycle begins when a new charged tRNA is brought to the A site of the ribosome by the elongation factor EF-Tu. The EF-Tu hydrolyzes GTP and dissociates from the ribosome. The anticodon of the charged tRNA must be complementary to the codon in the mRNA in the A site. A peptide bond is formed between the amino acid in the A site and the growing chain in the P site, transferring the growing chain to the A site, and leaving the tRNA in the P site empty. The ribosome translocates, helped by another elongation factor and GTP hydrolysis. This moves the tRNA in the A site into the P site, the empty tRNA in the P site into the E site, and moves the next codon into the A site.

inquiry

How is the wobble phenomenon related to the number of tRNAs and the degeneracy of the genetic code?

Termination requires accessory factors

Elongation continues in this fashion until a chain-terminating stop codon is reached (for example, UAA in figure 15.20). These stop codons do not bind to tRNA; instead, they are recognized by release factors, proteins that release the newly made polypeptide from the ribosome.

Proteins may be targeted to the ER

In eukaryotes, translation can occur either in the cytoplasm or on the rough endoplasmic reticulum (RER). The proteins being translated are targeted to the ER, based on their own initial amino acid sequence. The ribosomes found on the RER are not permanently bound to it.

A polypeptide that starts with a short series of amino acids called a **signal sequence** is specifically recognized and bound by a cytoplasmic complex of proteins called the **signal recognition particle (SRP).** The complex of signal sequence and SRP is in turn recognized by a receptor protein in the ER membrane. The

figure 15.20

TERMINATION OF PROTEIN SYNTHESIS. There is no tRNA with an anticodon complementary to any of the three termination signal codons, such as the UAA codon illustrated here. When a ribosome encounters a termination codon, it therefore stops translocating. A specific protein release factor facilitates the release of the polypeptide chain by breaking the covalent bond that links the polypeptide to the P site tRNA.

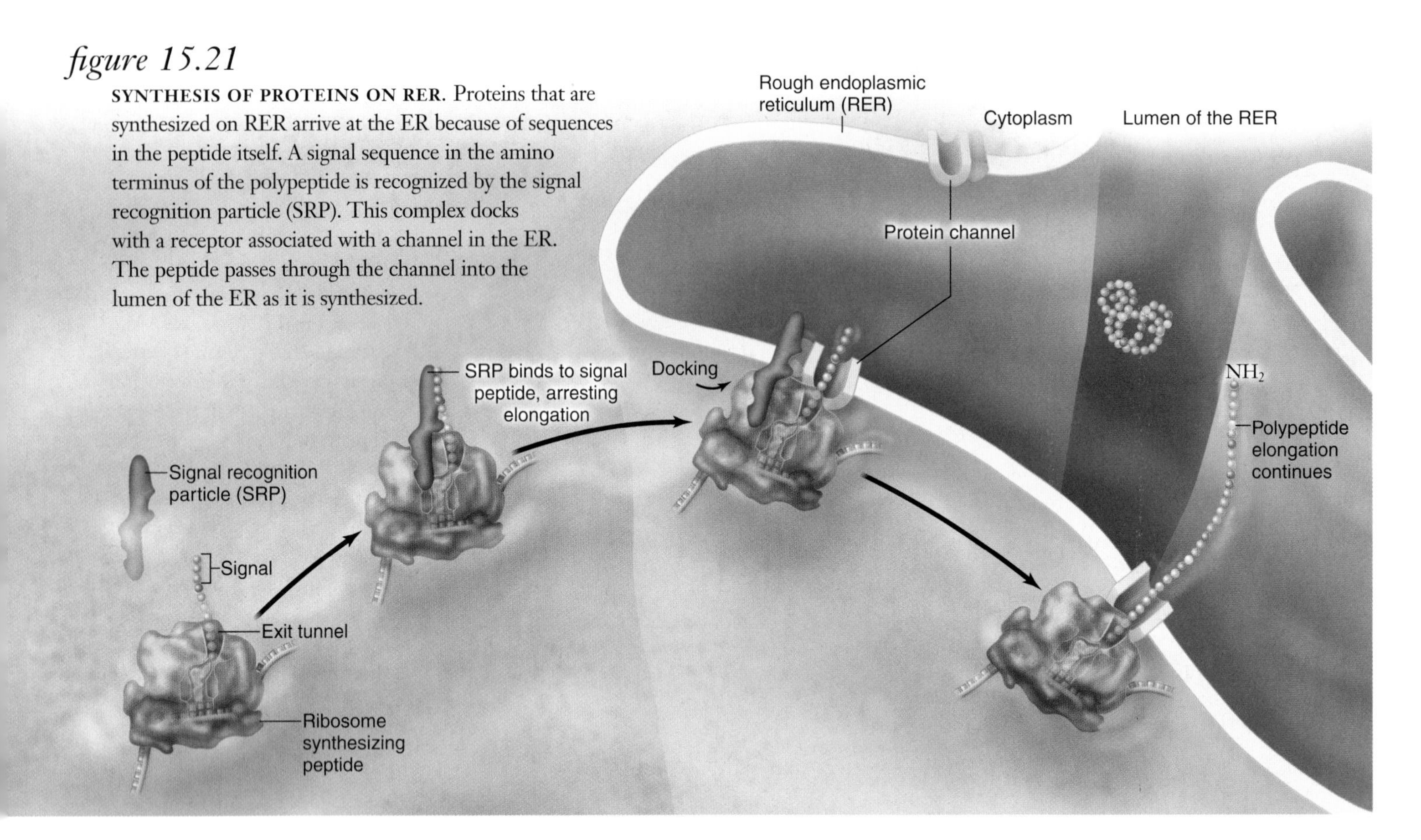

figure 15.21

SYNTHESIS OF PROTEINS ON RER. Proteins that are synthesized on RER arrive at the ER because of sequences in the peptide itself. A signal sequence in the amino terminus of the polypeptide is recognized by the signal recognition particle (SRP). This complex docks with a receptor associated with a channel in the ER. The peptide passes through the channel into the lumen of the ER as it is synthesized.

binding of the ER receptor to the signal sequence/SRP complex holds the ribosome engaged in translation of the protein on the ER membrane, a process called *docking* (figure 15.21).

As the protein is assembled, it passes through a channel formed by the docking complex and into the interior ER compartment, the cisternal space. This is the basis for the docking metaphor—the ribosome is not actually bound to the ER itself, but with the newly synthesized protein entering the ER, the ribosome is like a boat tied to a dock with a rope.

The basic mechanism of protein translocation across membranes by the SRP and its receptor and channel complex has been conserved across all three cell types: eukaryotes, bacteria, and archaea. Given that only eukaryotic cells have an endomembrane system, this universality may seem curious; however, bacteria and archaea both export proteins through their plasma membrane, and the mechanism used is similar to the way in which eukaryotes move proteins into the cisternal space of the ER.

Once within the ER cisternal space, or lumen, the newly synthesized protein can be modified by the addition of sugars (glycosylation) and transported by vesicles to the Golgi apparatus (see chapter 5). This is the beginning of the protein-trafficking pathway that can lead to other intracellular targets, to incorporation into the plasma membrane, or to release outside of the cell itself.

Translation initiation involves the interaction of the small ribosomal subunit with mRNA, and an initiator tRNA. The elongation cycle involves bringing in new charged tRNAs to the ribosomes A site, forming peptide bonds, and translocating the ribosome along the mRNA. The tRNAs transit through the ribosome from A to P to E sites during the process. In eukaryotes, signal sequences of newly formed polypeptides may target them to be moved to the endoplasmic reticulum, where they enter the cisternal space during synthesis.

15.9 Summarizing Gene Expression

Because of the complexity of the process of gene expression, it is worth stepping back to summarize some key points:

- The process of gene expression converts information in the genotype into the phenotype.
- A copy of the gene in the form of mRNA is produced by transcription, and the mRNA is used to direct the synthesis of a protein by translation.
- Both transcription and translation can be broken down into initiation, elongation, and termination cycles that produce their respective polymers. (The same is true for DNA replication.)
- Eukaryotic gene expression is much more complex than that of prokaryotes.

The nature of eukaryotic genes with their intron and exon components greatly complicates the process of gene expression by requiring additional steps between transcription and translation. The production and processing of eukaryotic mRNAs also takes place in the nucleus, whereas translation takes place in the cytoplasm. This necessitates the transport of the mRNA through the nuclear pores to the cytoplasm before translation can take place. The entire eukaryotic process is summarized in figure 15.22.

A number of differences can be highlighted between gene expression in prokaryotes and in eukaryotes. Table 15.2 (on p. 298) summarizes these main points.

The greater complexity of eukaryotic gene expression is related to the functional organization of the cell, with DNA in the nucleus and ribosomes in the cytoplasm. The differences in gene expression between prokaryotes and eukaryotes is mainly in detail, but some differences have functional significance.

1. The primary transcript is produced by RNA polymerase II in the nucleus. The transcription reaction proceeds in the 5′ to 3′ direction by copying one strand of a DNA template.

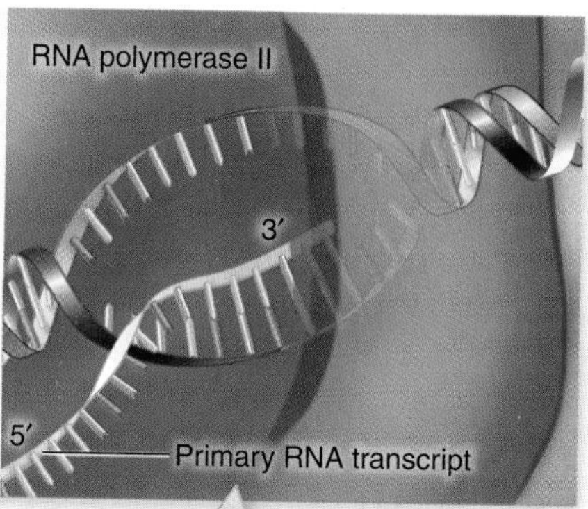

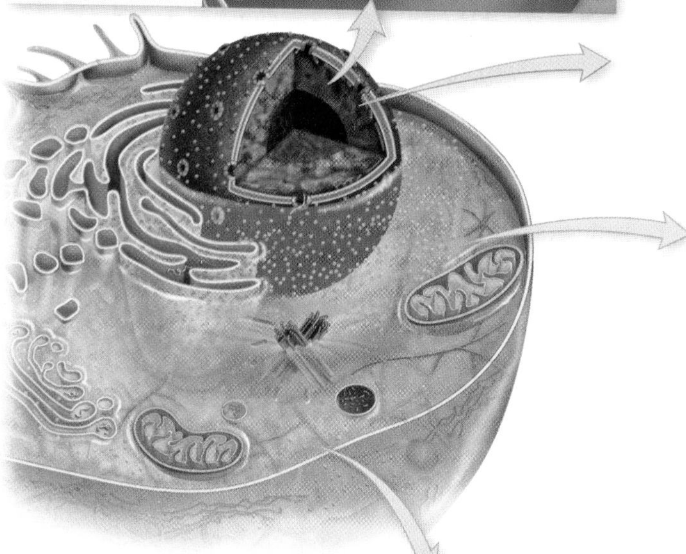

2. The primary transcript is processed to produce the mature mRNA. This involves addition of a 5′ methyl-G cap, cleavage and polyadenylation of the 3′ end and removal of introns by the spliceosome. All of these events occur in the nucleus. The mature mRNA is then exported through nuclear pores to the cytoplasm for translation.

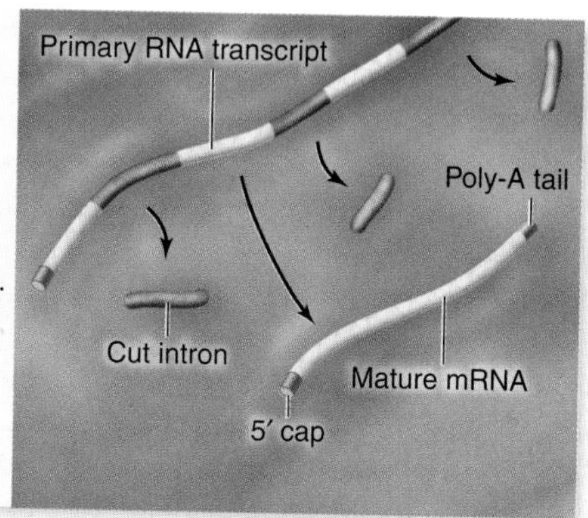

3. The mRNA associates with the ribosome in the cytoplasm. The 5′ cap binds to the small ribosomal subunit to begin the process of initiation. The initiator tRNA and the large subunit are added to complete initiation.

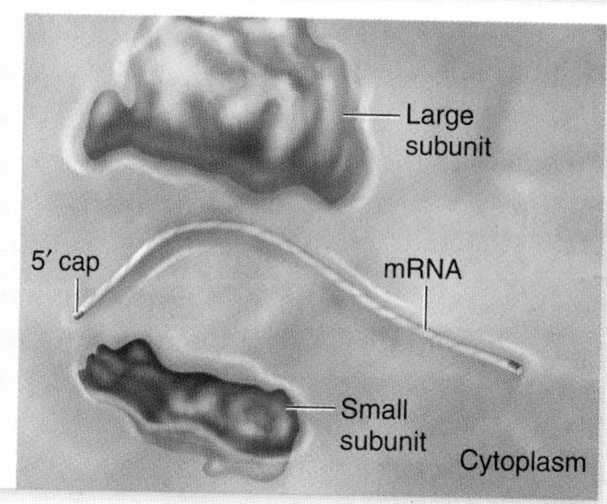

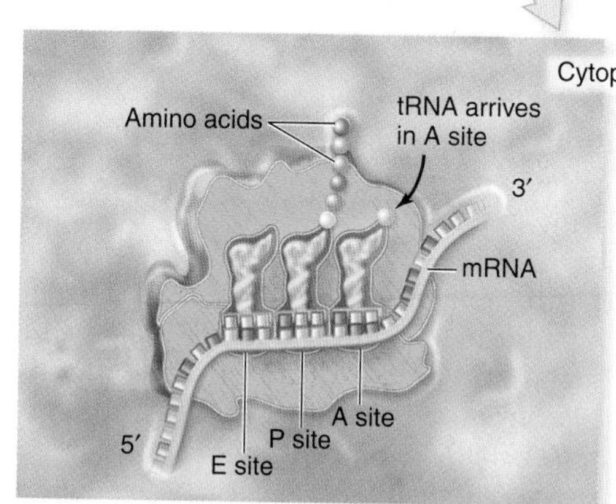

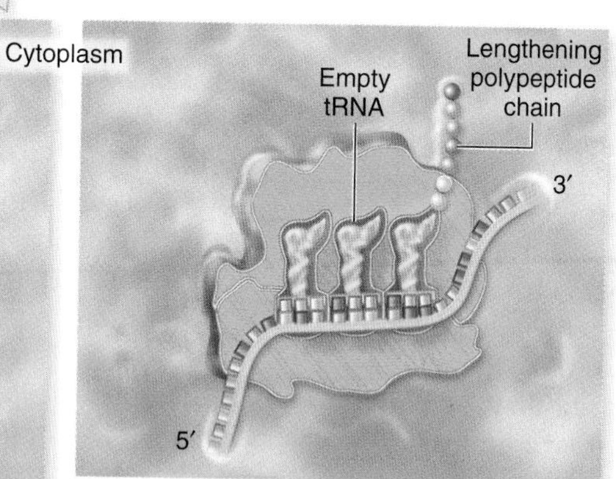

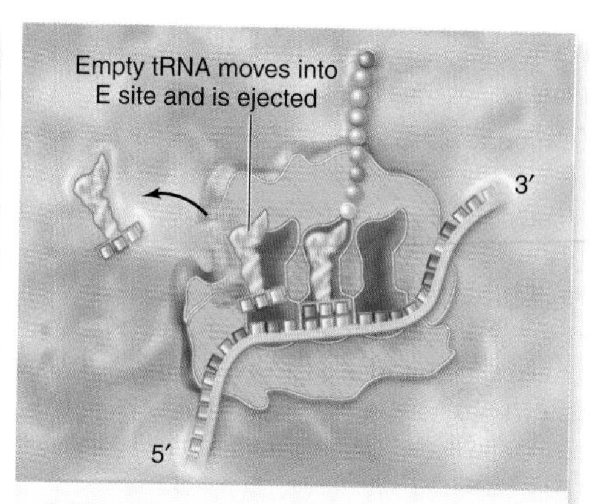

4. Protein synthesis involves the ribosome cycle. The cycle begins with the growing peptide attached to the tRNA in the P site. The next tRNA binds to the A site with its anticodon complementary to the codon in the mRNA in the A site.

5. Peptide bonds are formed between the amino terminus of the incoming tRNA and the carboxyl terminus of the growing peptide. This breaks the bond between the growing peptide and the tRNA in the P site leaving this tRNA "empty" and shifts the growing chain to the tRNA in the A site.

6. Ribosome translocation moves the ribosome relative to the mRNA and its bound tRNAs. This moves the empty tRNA into the E site, the tRNA with the growing peptide into the P site and leaves the A site unoccupied ready to accept the next charged tRNA.

figure 15.22

AN OVERVIEW OF GENE EXPRESSION IN EUKARYOTES.

TABLE 15.2 Differences Between Prokaryotic and Eukaryotic Gene Expression

Characteristic	Prokaryotes	Eukaryotes
Introns	No introns, although some archaeal genes possess them.	Most genes contain introns.
Number of genes in mRNA	Several genes may be transcribed into a single mRNA molecule. Often these have related functions and form an operon. This coordinates regulation of biochemical pathways.	Only one gene per mRNA molecule; regulation of pathways accomplished in other ways.
Site of transcription and translation	No membrane-bounded nucleus, transcription and translation are coupled.	Transcription in nucleus; mRNA moves out of nucleus for translation.
Initiation of translation	Begins at AUG codon preceded by special sequence that binds the ribosome.	Begins at AUG codon preceded by the 5′ cap (methylated GTP) that binds the ribosome.
Modification of mRNA after transcription	None; translation begins before transcription is completed.	A number of modifications while the mRNA is in the nucleus: Introns are removed and exons are spliced together; a 5′ cap is added; a poly-A tail is added.

15.10 Mutation: Altered Genes

One way to analyze the function of genes is to find or to induce mutations in a gene to see how this affects its function. In terms of the organism, however, inducing mutations is usually negative; most mutations have deleterious effects on the phenotype of the organism. In chapter 13, you saw how a number of genetic diseases, such as sickle cell anemia, are due to single base changes. We now consider mutations from the perspective of how the DNA itself is altered.

Point mutations affect a single site in the DNA

A mutation that alters a single base is termed a **point mutation.** The mutation can be either the substitution of one base for another, or the deletion or addition of a single base (or a small number of bases).

Base substitution

The substitution of one base pair for another in DNA is called a **base substitution mutation.** These are sometimes called **missense mutations** as the "sense" of the codon produced after transcription of the mutant gene will be altered (figure 15.23*c*). These fall into two classes, *transitions* and *transversions.* A transition does not change the type of bases in the base pair, that is, a pyrimidine is substituted for a pyrimidine, or purine for purine. In contrast, a transversion does change the type of bases in a base pair, that is, pyrimidine to purine or the reverse. Because of the degenerate nature of the genetic code, base substitution may or may not alter the amino acid encoded. If the new codon from the base substitution still encodes the same amino acid, we say the mutation is *silent* (figure 15.23*b*). A variety of human genetic diseases, including sickle cell anemia, are caused by base substitution.

Nonsense mutations

A special category of base substitution arises when a base is changed such that the transcribed codon is converted to a stop codon (figure 15.23*d*). We call these **nonsense mutations** because the mutation does not make "sense" to the translation apparatus. The stop codon results in premature termination of translation and leads to a truncated protein. How short the resulting protein is depends on where in the gene a stop codon has been introduced.

Frameshift mutations

The addition or deletion of a single base has much more profound consequences than does the substitution of one base for another. These mutations are called **frameshift mutations** because they alter the reading frame in the mRNA downstream of the mutation. This class of mutations was used by Crick and Brenner, as described earlier in the chapter, to infer the nature of the genetic code.

Changing the reading frame early in a gene, and thus in its mRNA transcript, means that the majority of the protein will be altered. Frameshifts also can cause premature termination of translation because 3 in 64 codons are stop codons, which represents a high probability in the sequence that has been randomized by the frameshift.

Triplet repeat expansion mutations

Given the long history of molecular genetics, and the relatively short time that molecular analysis has been possible on humans, it is surprising that a new kind of mutation was discovered in humans. However, one of the first genes isolated that was associated with human disease, the gene for *Huntington disease*, provided a new kind of mutation. The gene for Huntington contains a triplet sequence of DNA that is repeated, and this repeat unit is expanded in the disease allele relative to the normal allele. Since this initial discovery, at least 20 other human genetic diseases appear to be due to this mechanism. The prevalence of this kind of mutation is unknown, but at present humans and mice are the only organisms in which they have been observed, implying that they may be limited to vertebrates, or even mammals. No such mutation has ever been found in *Drosophila* for example.

The expansion of the triplet can occur in the coding region or in noncoding transcribed DNA. In the case of Huntington disease, the repeat unit is actually in the coding region of the gene where the triplet encodes glutamine and expansion results in a polyglutamine region in the protein. A num-

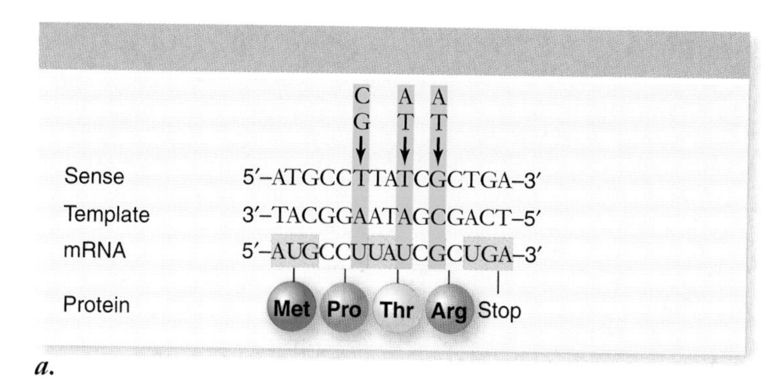

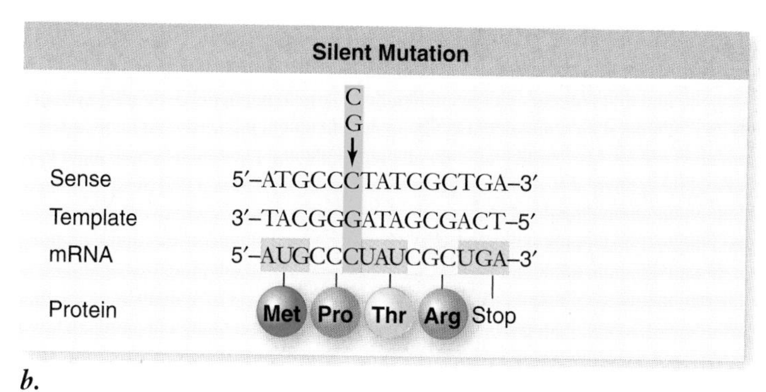

Missense Mutation

A T

Sense 5′–ATGCCCTATCACTGA–3′

Template 3′–TACGGGATAGTGACT–5′

mRNA 5′–AUGCCCUAUCACUGA–3′

Protein

Met Pro Thr His Stop

c.

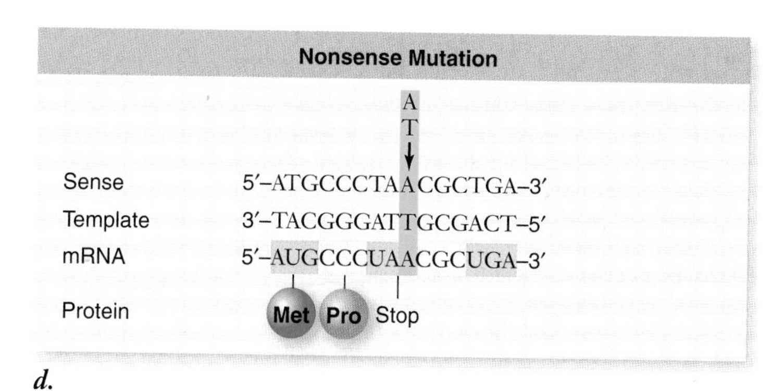

figure 15.23

TYPES OF MUTATIONS. ***a.*** A hypothetical gene is shown with encoded mRNA and protein. Arrows above the gene indicate sites of mutations described in the rest of the figure. ***b.*** Silent mutation. A change in the third position of a codon is often silent due to degeneracy in the genetic code. In this case T/A to C/G mutation does not change the amino acid encoded (proline). ***c.*** Missesense mutation. The G/C to A/T mutation changes the amino acid encoded from arginine to histidine. ***d.*** Nonsense mutation. The T/A to A/T mutation produces a UAA stop codon in the mRNA.

ber of other neurodegenerative disorders also show this kind of mutation. In the case of fragile-X syndrome, an inherited form of mental retardation, the repeat is in noncoding DNA.

Chromosomal mutations change the structure of chromosomes

Point mutations affect a single site in a chromosome, but more extensive changes can alter the structure of the chromosome itself, resulting in **chromosomal mutations.** Many human cancers are associated with chromosomal abnormalities, so these are of great clinical relevance. We briefly consider possible alterations to chromosomal structure, all of which are summarized in figure 15.24.

Deletions

A **deletion** is the loss of a portion of a chromosome. Frameshifts can be caused by one or more small deletions, but much larger regions of a chromosome may also be lost. If too much information is lost, the deletion is usually fatal to the organism.

One human syndrome that is due to deletion is *cri-du-chat*, which is French for "cry of the cat" after the noise made by children with this syndrome. Cri-du-chat syndrome is caused by a large deletion from the short arm of chromosome 5. It usually results in early death, although many affected individuals show a normal lifespan. It has a variety of effects, including respiratory problems.

Duplications

The **duplication** of a region of a chromosome may or may not lead to phenotypic consequences. Effects depend upon the location of the "breakpoints" where the duplication occurred. If the duplicated region does not lie within a gene, there may be no effect. If the duplication occurs next to the original region, it is termed a *tandem duplication.* These tandem duplications are important in the evolution of families of related genes, such as the globin family that encode the protein hemoglobin.

Inversions

An **inversion** results when a segment of a chromosome is broken in two places, reversed, and put back together. An inversion may not have an effect on phenotype if the sites where the inversion occurs do not break within a gene. In fact, although humans all have the "same" genome, the order of genes in all individuals in a population is not precisely the same due to inversions that occur in different lineages.

Translocations

If a piece of one chromosome is broken off and joined to another chromosome, we call this a **translocation.** Translocations are complex because they can cause problems during meiosis, particularly when two different chromosomes try to pair with each other during meiosis I.

Translocations can also move genes from one chromosomal region to another in a way that changes the expression of genes in the region involved. Two forms of leukemia have been shown to be associated with translocations that move oncogenes into regions of a chromosome where they are expressed inappropriately in blood cells (see chapter 10).

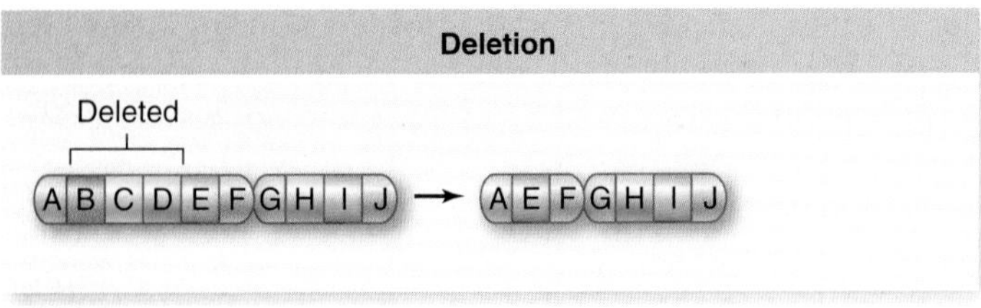

a.

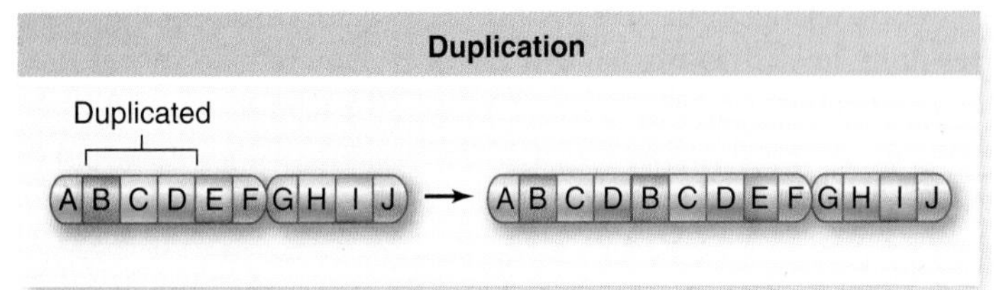

b.

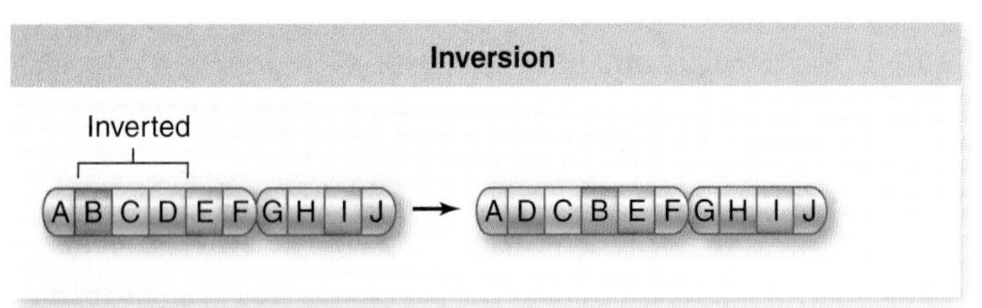

c.

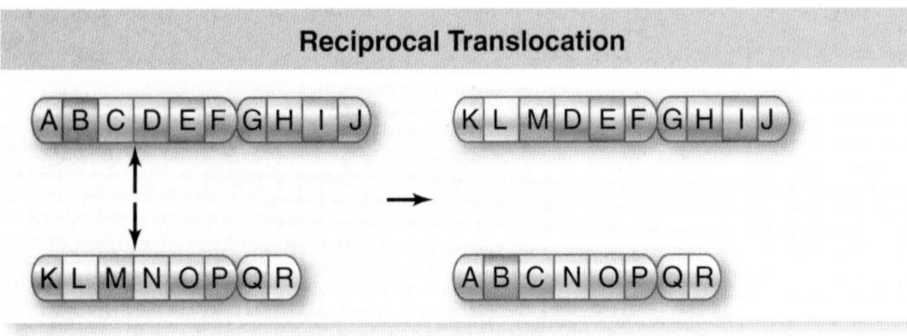

d.

figure 15.24

CHROMOSOMAL MUTATIONS. Larger-scale changes in chromosomes are also possible. Material can be deleted ***(a)***, duplicated ***(b)***, and inverted ***(c)***. Translocations occur when one chromosome is broken and becomes part of another chromosome. This often occurs where both chromosomes are broken and exchange material, an event called a reciprocal translocation ***(d)***.

Mutations are the starting point of evolution

If no changes occurred in genomes over time, then there could be no evolution. Too much change, however, is harmful to the individual with a greatly altered genome. Thus a balance must exist between the amount of new variation that arises in a species and the health of individuals in the species. This topic is explored in more detail later in the book when we consider evolution and population genetics (chapter 20).

The larger scale alteration of chromosomes has also been important in evolution, although its role is poorly understood. It is clear that gene families arise by the duplication of an ancestral gene, followed by the functional divergence of the duplicated copies. It is also clear that even among closely related species, the number and arrangements of genes on chromosomes can differ. Large-scale rearrangements may have occurred.

Our view of the nature of genes has changed with new information

In this and the preceding chapters, we have seen multiple views of genes. Mendel followed traits determined by what we now call genes in his crosses. The behavior of these genes can be predicted based on the behavior of chromosomes during meiosis. Morgan and others learned to map the location of genes on chromosomes. These findings led to the view of genes as abstract entities that could be followed through generations and mapped to chromosomal locations like "beads on a string," with the beads being genes and the string the chromosome.

The original molecular analysis of genes led to the simple one-gene/one-polypeptide paradigm. This oversimplification was changed when geneticists observed the alternative splicing of eukaryotic genes, which can lead to multiple protein products from the same genetic information. Furthermore, some genes do not encode proteins at all, but only RNA, which can either be a part of the gene expression machinery (rRNA, tRNA, and other forms) or can itself act as an enzyme. Other stretches of DNA are important for regulating genes but are not expressed. All of these findings make a simple definition of genes difficult.

We are left with the rich complexity of the nature of genes, which defies simple definition. To truly understand the nature of genes we must consider both their molecular nature as well as their phenotypic expression. This brings us full circle, back to the relationship between genotype and phenotype, with a much greater appreciation for the complexity of this relationship.

15.1 The Nature of Genes (figure 15.2)

Evidence shows that genetic mutations affect proteins.

- Garrod showed alkaptonuria is due to an altered enzyme.
- Beadle and Tatum showed that genes specify enzymes.
- Information in cells flows from DNA through RNA to proteins.

15.2 The Genetic Code

The order of nucleotides in DNA encodes information to specify the order of amino acids in polypeptides.

- A codon consists of 3 nucleotides. There are $4^3 = 64$ possible codons.
- The code uses adjacent codons with no spaces.
- Three codons signal "stop," one codon signals "start" and also encodes methionine. Thus 61 codons encode the 20 amino acids.
- The code is degenerate (usually in the third position) but specific.
- The code is essentially universal.

15.3 Overview of Gene Expression

Transcription produces an RNA copy from a DNA template, and translation uses the RNA template to direct the synthesis of a protein.

- The strand copied during transcription is called the template strand.
- Transcription of RNA involves initiation at a promoter, elongation of the transcript, and termination at a terminator site.
- Translation involves formation of initiation complex, elongation by addition of amino acids, and termination at a stop codon.
- RNA plays multiple roles in gene expression (see pps. 283, 284).

15.4 Prokaryotic Transcription

Prokaryote gene expression is similar to that in eukaryotes, with some important differences.

- Prokaryotes have a single RNA polymerase that exists in two forms: core polymerase and holoenzyme.
- Core polymerase can synthesize RNA. Holoenzyme, core plus σ factor, can initiate RNA at a promoter (figure 15.5) .
- A transcription unit begins with a promoter, contains one or more genes, and ends with a terminator.
- RNA polymerase unwinds a short region of DNA at promoters.
- Transcription of the mRNA chain grows in the 5′ to 3′ direction.
- A transcription bubble contains RNA polymerase, DNA template, and the growing mRNA transcript (figure 15.6).
- Terminators consist of complementary sequences that form a double-stranded hairpin loop where the polymerase pauses (figure 15.7).
- In prokaryotes the mRNA is translated into a polypeptide while it is being synthesized (transcription–translation coupling).

15.5 Eukaryotic Transcription

The transcription reaction in eukaryotes is the same as prokaryotes, but there are some distinct differences.

- Eukaryotes have three RNA polymerases: I transcribes rRNA; II transcribes mRNA and some snRNAs; III transcribes tRNA.
- Transcription by RNA polymerase II requires a host of transcription factors to form an initiation complex at the promoter.
- RNA polymerase III trascribes tRNA, and its promoters are found internal to the gene, not at the 5′ end.
- In eukaryotes, the primary RNA transcript is modified (figure 15.10).
  - *A methyl-GTP cap is added to the 5′ end.*
  - *A 3′ poly-A tail is added by polyA polymerase at a specific site.*
  - *Noncoding internal regions are also removed by splicing.*

15.6 Eukaryotic pre-mRNA Splicing

In eukaryotes introns are removed by splicing (figure 15.12).

- Coding DNA (exons) is interrupted by noncoding introns.
- Intron–exon junctions are recognized by snRNPs.
- The snRNPs recruit a larger complex called the spliceosome.
- During splicing the 5′ end of the intron is cut and becomes bound to the branch site, forming a structure called a lariat.
- The 3′ end of the first exon is joined to the 5′ end of the next exon.
- One transcript can produce different mRNAs by alternative splicing.

15.7 The Structure of tRNA and Ribosomes

Although the ribosome is a key organelle in translation, it requires the participation of mRNA, tRNA, and other factors.

- The charging reaction attaches the carboxyl terminus of an amino acid to the 3′ end of the correct tRNA (figure 15.14).
- This is catalyzed by enzymes called aminoacyl-tRNA synthetases.
- The anticodon loop of tRNAs can base-pair to codons in mRNA.
- The ribosome consists of two subunits: large and small.
- The small subunit binds to mRNA and is involved in decoding, while the large subunit contains the enzyme peptidyl transferase.
- The ribosome has three tRNA-binding sites (figure 15.15).
  - *The P site binds to tRNA attached to the growing peptide chain.*
  - *The A site binds to the tRNA carrying the next amino acid to be added.*
  - *The E site binds to the tRNA that carried the previous amino acid.*

15.8 The Process of Translation

Protein synthesis is complex and energetically expensive.

- In prokaryotes the initiation complex forms with the small ribosomal subunit, mRNA, and a special initiator tRNA.
- The RBS in prokaryotic mRNA is complementary to rRNA in the small subunit. Eukaryotes use the 5′ cap for the same function.
- Peptide bonds form between the amino end of the new amino acid and the carboxyl end of the growing chain (figure 15.18).
- Protein synthesis involves a cycle of events (figure 15.19):
  - *New charged tRNAs are brought to the ribosomes by EF-Tu.*
  - *A peptide bond forms between new amino acid and growing chain.*
  - *The ribosome moves relative to the mRNA and bound tRNAs.*
- One tRNA can bind multiple codons by wobble pairing.
- Stop codons are recognized by termination factors.
- Proteins targeted to the ER have a signal sequence in their amino terminus that binds the SRP, and this complex docks the ribosome.
- The signal recognition particle (SRP) binds the signal sequence, and this complex is recognized by a receptor protein in the ER.

15.9 Summarizing Gene Expression

Gene expression converts information in the genome into proteins. This process differs between prokaryotes and eukaryotes (figure 15.22).

15.10 Mutation: Altered Genes

Mutations can be used to understand the function of genes.

- Point mutations involve the alteration of a single base.
- Nonsense mutations convert codons into stop codons.
- Frameshift mutations involve the addition or deletion of a base.
- Triplet-repeat expansion mutations can cause genetic diseases.
- Chromosomal mutations alter the structure of chromosomes.
- Mutations are the starting point of evolution.

SELF TEST

1. The experiments with nutritional mutants in *Neurospora* by Beadle and Tatum provided evidence that—
   a. Bread mold can be grown in a lab on minimal media.
   b. X-rays can damage DNA.
   c. Cells need enzymes.
   d. Genes specify enzymes.
2. What is the *central dogma* of molecular biology?
   a. DNA is the genetic material.
   b. Information passes from DNA to protein.
   c. Information passes from DNA to RNA to protein.
   d. One gene encodes only one polypeptide.
3. The manufacture of new proteins is termed __________, and the production of a messenger RNA corresponding to a specific gene is called __________.
   a. translation; transcription
   b. termination; translation
   c. transcription; translation
   d. transfer; translation
4. Each amino acid in a protein is specified by—
   a. multiple genes
   b. a promoter
   c. a codon
   d. a molecule of mRNA
5. The TATA box in eukaryotes is part of a—
   a. core promoter
   b. –35 sequence
   c. –10 sequence
   d. 5′ cap
6. What is the *coding strand*?
   a. The single DNA strand copied to produce a molecule of RNA
   b. The single-stranded RNA molecule that is transcribed from the DNA
   c. The DNA strand that is not copied to synthesize a molecule of RNA
   d. The region of a chromosome that contains a gene
7. An anticodon would be found on which of the following types of RNA?
   a. snRNA (small nuclear RNA)
   b. mRNA (messenger RNA)
   c. tRNA (transfer RNA)
   d. rRNA (ribosomal RNA)
8. RNA polymerase binds to a ________ to initiate ________.
   a. mRNA; translation
   b. promoter; transcription
   c. primer; transcription
   d. transcription factor; translation
9. Which of the following functions as a "stop" signal for a prokaryotic RNA polymerase?
   a. Formation of a transcription bubble
   b. Addition of a long chain of adenine nucleotides to the 3′ end
   c. Addition of a 5′ cap
   d. Formation of a GC hairpin
10. An *exon* is a sequence of RNA that—
    a. codes for protein
    b. is removed through the action of a spliceosome
    c. is part of a noncoding DNA sequence
    d. Both (b) and (c) are correct.
11. The job of a ribosome during translation can best be described as—
    a. targeting proteins to the rough endoplasmic reticulum
    b. determining the sequence of amino acids
    c. carrying amino acids to the mRNA
    d. catalyzing peptide bond formation between amino acids
12. What is the function of the *signal sequence*?
    a. It initiates transcription by triggering RNA polymerase binding.
    b. It initiates translation.
    c. It is the binding site of signal recognition particle.
    d. It signals the end of translation, resulting in the disassembly of the ribosome.
13. How can a point mutation lead to a nonsense mutation?
    a. Changing a single base has no effect on the protein.
    b. Changing a single base leads to a premature termination of translation.
    c. Changing a single base within a codon from an A to a C.
    d. The addition or deletion of a base alters the reading frame for the gene.
14. Which of the following is a consequence of a translocation?
    a. Genes move from one chromosome to another.
    b. RNA polymerase produces a molecule of mRNA.
    c. A molecule of mRNA interacts with a ribosome to produce a protein.
    d. A segment of a chromosome is broken, reversed, and reinserted.
15. What is the relationship between mutations and evolution?
    a. Mutations make genes better.
    b. Mutations can create new alleles.
    c. Mutations happened early in evolution, but not now.
    d. There is no relationship between evolution and genetic mutations.

CHALLENGE QUESTIONS

1. A template strand of DNA has the following sequence:
   3′ – CGTTACCCGAGCCGTACGATTAGG – 5′
   Use the sequence information to determine—
   a. the predicted sequence of the mRNA for this gene
   b. the predicted amino acid sequence of the protein
2. Describe how each of the following mutations will affect the final protein product. Name the type of mutation.
   Original template strand:
   3′ – CGTTACCCGAGCCGTACGATTAGG – 5′
   a. 3′ – CGTTACCCGAGCCGTAACGATTAGG – 5′
   b. 3′ – CGTTACCCGATCCGTACGATTAGG – 5′
   c. 3′ – CGTTACCCGAGCCGTTCGATTAGG – 5′
3. Predict whether gene expression (from initiation of transcription to final protein product) would be faster in a prokaryotic or eukaryotic cell. Explain your answer.

Do you need additional review? *Visit* www.ravenbiology.com *for practice quizzes, animations, videos, and activities designed to help you master the material in this chapter.*

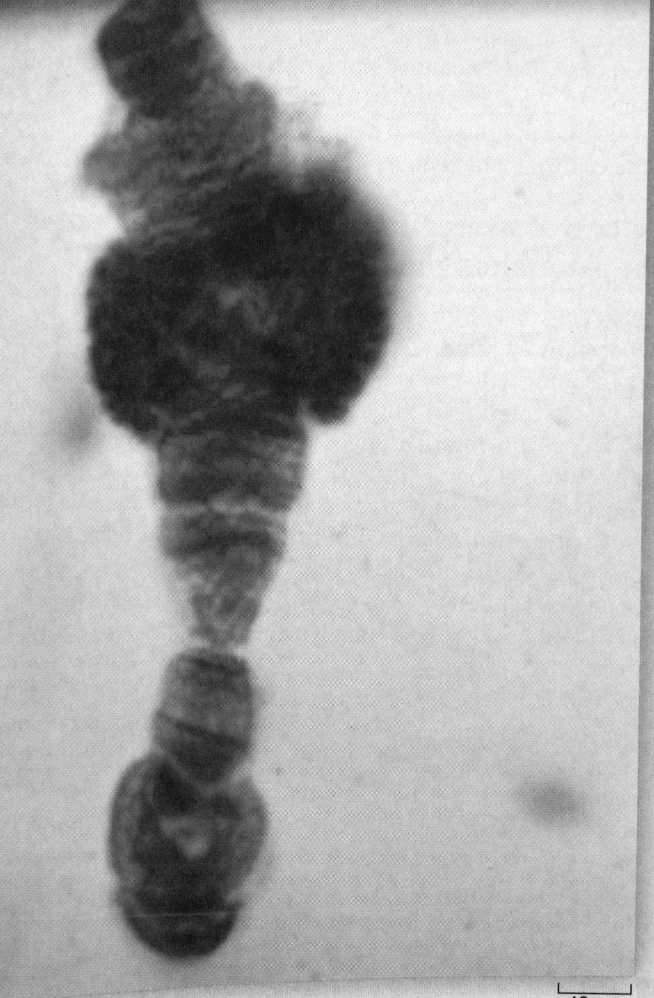

chapter 16

Control of Gene Expression

introduction

IN MUSIC, DIFFERENT INSTRUMENTS PLAY their own parts at different times during a piece; a musical score determines which instruments play when. Similarly, in an organism different genes are expressed at different times, with a "genetic score," written in regulatory regions of the DNA, determining which genes are active when. The picture shows the expanded "puff" of this *Drosophila* chromosome, which represents genes that are being actively expressed. Gene expression and how it is controlled is our topic in this chapter.

concept outline

16.1 Control of Gene Expression

- *Control usually occurs at the level of transcription initiation*
- *Control strategies in prokaryotes are geared to adjust to environmental changes*
- *Control strategies in eukaryotes are aimed at maintaining homeostasis*

16.2 Regulatory Proteins

- *Proteins can interact with DNA through the major groove*
- *DNA-binding domains interact with specific DNA sequences*
- *Several common DNA-binding motifs are shared by many proteins*

16.3 Prokaryotic Regulation

- *Control of transcription can be either positive or negative*
- *Prokaryotes adjust gene expression in response to environmental conditions*
- *The* lac *operon is negatively regulated by the* lac *repressor*
- *The presence of glucose prevents induction of the* lac *operon*
- *The* trp *operon is controlled by the* trp *repressor*

16.4 Eukaryotic Regulation

- *Transcription factors can be either general or specific*
- *Promoters and enhancers are binding sites for transcription factors*
- *Coactivators and mediators link transcription factors to RNA polymerase II*
- *The transcription complex brings things together*

16.5 Eukaryotic Chromatin Structure

- *Both DNA and histone proteins can be modified*
- *Some transcription activators alter chromatin structure*
- *Chromatin-remodeling complexes also change chromatin structure*

16.6 Eukaryotic Posttranscriptional Regulation

- *Small RNAs can affect gene expression*
- *Alternative splicing can produce multiple proteins from one gene*
- *RNA editing alters mRNA after transcription*
- *mRNA must be transported out of the nucleus for translation*
- *Initiation of translation can be controlled*
- *The degradation of mRNA is controlled*

16.7 Protein Degradation

- *Addition of ubiquitin marks proteins for destruction*
- *The proteasome degrades polyubiquitinated proteins*

16.1 Control of Gene Expression

Control of gene expression is essential to all organisms. In prokaryotes, it allows the cell to take advantage of changing environmental conditions. In multicellular eukaryotes, it is critical for directing development and maintaining homeostasis.

Control usually occurs at the level of transcription initiation

You learned in the previous chapter that gene expression is the conversion of genotype to phenotype—the flow of information from DNA to produce functional proteins that control cellular activities. We could envision controlling this process at any step along the way, and in fact, examples of control occur at most steps. The most logical place to control this process, however, is at the first step: production of mRNA from DNA by transcription.

Transcription itself could be controlled at any step, but again, the beginning is the most logical place. Although cells do not always behave in ways that conform to human logic, control of the initiation of transcription is common.

RNA polymerase is key to transcription, and it must have access to the DNA helix and must be capable of binding to the gene's promoter for transcription to begin. **Regulatory proteins** act by modulating the ability of RNA polymerase to bind to the promoter. This idea of controlling the access of RNA polymerase to a promoter is common to both prokaryotes and eukaryotes, but the details differ greatly, as you will see.

These regulatory proteins bind to specific nucleotide sequences on the DNA that are usually only 10–15 nt in length. (Even a large regulatory protein has a "footprint," or binding area, of only about 20 nt.) Hundreds of these regulatory sequences have been characterized, and each provides a binding site for a specific protein that is able to recognize the sequence. Binding of the protein either *blocks* transcription by getting in the way of RNA polymerase or *stimulates* transcription by facilitating the binding of RNA polymerase to the promoter.

Control strategies in prokaryotes are geared to adjust to environmental changes

Control of gene expression is accomplished very differently in prokaryotes than it is in eukaryotes. Prokaryotic cells have been shaped by evolution to grow and divide as rapidly as possible, enabling them to exploit transient resources. Proteins in prokaryotes turn over rapidly, allowing these organisms to respond quickly to changes in their external environment by changing patterns of gene expression.

In prokaryotes, the primary function of gene control is to adjust the cell's activities to its immediate environment. Changes in gene expression alter which enzymes are present in response to the quantity and type of available nutrients and the amount of oxygen. Almost all of these changes are fully reversible, allowing the cell to adjust its enzyme levels up or down as the environment changes.

Control strategies in eukaryotes are aimed at maintaining homeostasis

The cells of multicellular organisms, in contrast, have been shaped by evolution to be protected from transient changes in their immediate environment. Most of them experience fairly constant conditions. Indeed, **homeostasis**—the maintenance of a constant internal environment—is considered by many to be the hallmark of multicellular organisms. Cells in such organisms respond to signals in their immediate environment (such as growth factors and hormones) by altering gene expression, and in doing so they participate in regulating the body as a whole.

Some of these changes in gene expression compensate for changes in the physiological condition of the body. Others mediate the decisions that actually produce the body, ensuring that the correct genes are expressed in the right cells at the right time during development. Later chapters deal with the details, but for now we can simplify by saying that the growth and development of multicellular organisms entail a long series of biochemical reactions, each catalyzed by a specific enzyme. Once a particular developmental change has occurred, these enzymes cease to be active, lest they disrupt the events that must follow.

To produce this sequence of enzymes, genes are transcribed in a carefully prescribed order, each for a specified period of time, following a fixed genetic program that may even lead to programmed cell death **(apoptosis).** The one-time expression of the genes that guide a developmental program is fundamentally different from the reversible metabolic adjustments prokaryotic cells make to the environment. In all multicellular organisms, changes in gene expression within particular cells serve the needs of the whole organism, rather than the survival of individual cells.

Unicellular eukaryotes also use different control mechanisms than prokaryotes. All eukaryotes have a membrane-bounded nucleus, use similar mechanisms to condense DNA into chromosomes, and have the same gene expression machinery, all of which differ from those of prokaryotes.

Gene expression is usually controlled at the level of transcription initiation. Regulatory proteins that can bind to specific sites on DNA affect the binding of RNA polymerase to promoters. Prokaryotes and eukaryotes differ in the details of this process.

16.2 Regulatory Proteins

The ability of certain proteins to bind to *specific* DNA regulatory sequences provides the basic tool of gene regulation, the key ability that makes transcriptional control possible. To understand how cells control gene expression, it is first necessary to gain a clear picture of this molecular recognition process.

Proteins can interact with DNA through the major groove

Molecular biologists formerly thought that the DNA helix had to unwind before proteins could distinguish one DNA sequence from another; only in this way, they reasoned, could regulatory proteins gain access to the hydrogen bonds between base-pairs. We now know it is unnecessary for the helix to unwind because proteins can bind to its outside surface, where the edges of the base-pairs are exposed.

Careful inspection of a DNA molecule reveals two helical grooves winding around the molecule, one deeper than the other. Within the deeper groove, called the **major groove,** the nucleotides' hydrogen bond donors and acceptors are accessible. The pattern created by these chemical groups is unique for each of the four possible base-pair arrangements, providing a ready way for a protein nestled in the groove to read the sequence of bases (figure 16.1).

DNA-binding domains interact with specific DNA sequences

Protein–DNA recognition is an area of active research; so far, the structures of over 30 regulatory proteins have been analyzed. Although each protein is unique in its fine details, the part of the protein that actually binds to the DNA is much less variable. Almost all of these proteins employ one of a small set of **DNA-binding motifs.** A motif, as described in chapter 3, is a form of three-dimensional substructure that is found in many proteins. These DNA-binding motifs share the property of interacting with specific sequences of bases, usually through the major groove of the DNA helix.

DNA-binding motifs are the key structure within the DNA-binding domain of these proteins. This domain is a functionally distinct part of the protein necessary to bind to DNA in a sequence-specific manner. Regulatory proteins also need to be able to interact with the transcription apparatus, which is accomplished by a different regulatory domain.

Note that two proteins that share the same DNA-binding domain do not necessarily bind to the same DNA sequence. The similarities in the DNA-binding motifs appear in their 3-D structure, and not in the specific contacts that they make with DNA.

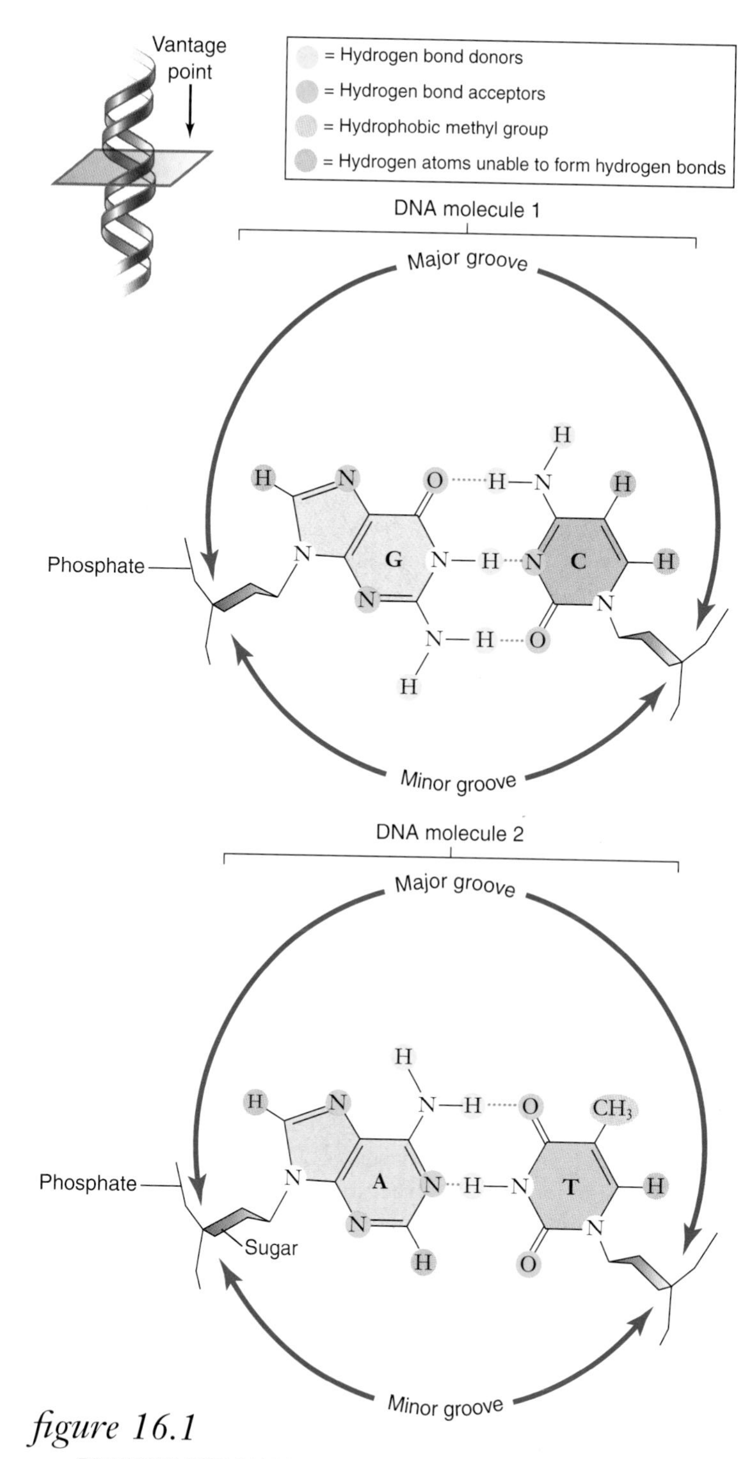

figure 16.1

READING THE MAJOR GROOVE OF DNA. Looking down into the major groove of a DNA helix, we can see the edges of the bases protruding into the groove. Each of the four possible base-pair arrangements (two are shown here) extends a unique set of chemical groups into the groove, indicated in this diagram by differently colored circles. A regulatory protein can identify the base-pair arrangement by this characteristic signature.

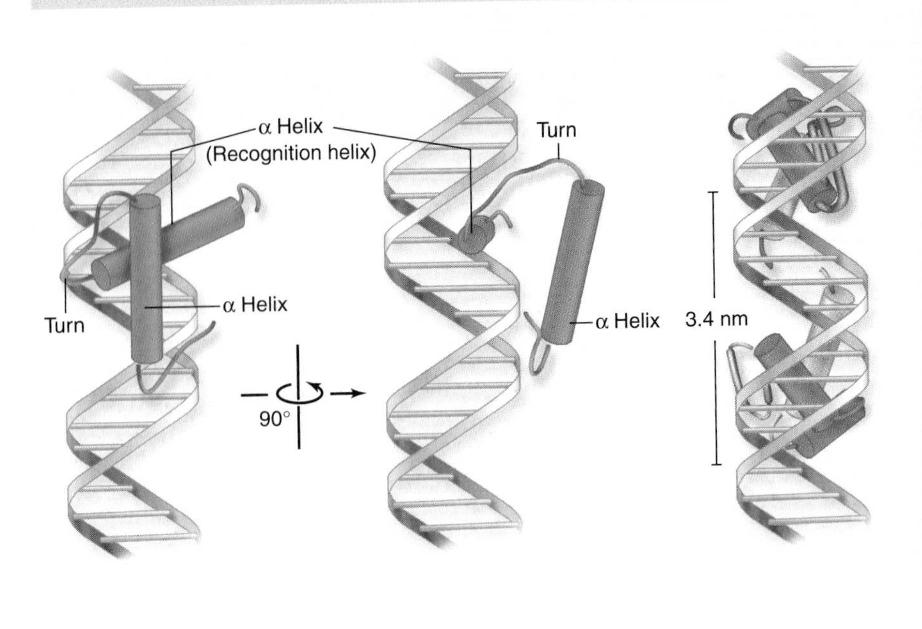

figure 16.2

MAJOR DNA-BINDING MOTIFS. A number of common DNA-binding motifs are pictured interacting with DNA. ***a.*** The helix-turn-helix motif illustrated here binds to DNA using one α helix, the recognition helix, to interact with the major groove and the other to position the recognition helix. Proteins with this motif are usually dimers, with two identical subunits, each containing the DNA-binding motif. The two copies of the motif (*red*) are separated by 3.4 nm, precisely the spacing of one turn of the DNA helix. This allows the regulatory proteins to slip into two adjacent portions of the major groove in DNA, providing a strong attachment. ***b.*** The homeodomain motif is common in proteins that regulate development and shares some structural similarity to the helix-turn-helix in (***a***). ***c.*** The zinc finger motif has two α helices that interact with the major groove. These DNA-binding motifs act like the fingers of a hand holding the DNA. ***d.*** The leucine zipper acts to hold two subunits in a multisubunit protein together allowing α-helical regions to interact with DNA.

Several common DNA-binding motifs are shared by many proteins

A limited number of common DNA-binding motifs have been described that are found in a wide variety of different proteins. Four of the best known are detailed in the following sections to give the flavor of how DNA-binding proteins interact with DNA.

The helix-turn-helix motif

The most common DNA-binding motif is the **helix-turn-helix,** constructed from two α-helical segments of the protein linked by a short, nonhelical segment, the "turn" (figure 16.2*a*). As the first motif recognized, the helix-turn-helix motif has since been identified in hundreds of DNA-binding proteins.

A close look at the structure of a helix-turn-helix motif reveals how proteins containing such motifs interact with the major groove of DNA. The helical segments of the motif interact with one another, so that they are held at roughly right angles. When this motif is pressed against DNA, one of the helical segments (called the *recognition helix*) fits snugly in the major groove of the DNA molecule, and the other butts up against the outside of the DNA molecule, helping to ensure the proper positioning of the recognition helix.

Most DNA regulatory sequences recognized by helix-turn-helix motifs occur in symmetrical pairs. Such sequences are bound by proteins containing two helix-turn-helix motifs separated by 3.4 nanometers (nm), the distance required for one turn of the DNA helix (figure 16.2*a*). Having *two* protein–DNA-binding sites doubles the zone of contact between protein and DNA and greatly strengthens the bond between them.

The homeodomain motif

A special class of helix-turn-helix motifs, the **homeodomain,** plays a critical role in development in a wide variety of eukaryotic organisms, including humans. These motifs were discovered when researchers began to characterize a set of homeotic mutations in *Drosophila* (mutations that cause one body part to be replaced by another). They found that the mutant genes encoded regulatory proteins. Normally these proteins would initiate key stages of development by binding to developmental switch-point genes. More than 50 of these regulatory proteins have been analyzed, and they all contain a nearly identical sequence of 60 amino acids, which was termed the *homeodomain* (figure 16.2*b*). The most conserved part of the homeodomain contains a recognition helix of a helix-turn-helix motif. The rest of the homeodomain forms the other two helices of this motif.

The zinc finger motif

A different kind of DNA-binding motif uses one or more zinc atoms to coordinate its binding to DNA. Called **zinc fingers** (figure 16.2*c*), these motifs exist in several forms. In one form, a zinc atom links an α-helical segment to a β-sheet segment (chapter 3) so that the helical segment fits into the major groove of DNA.

This sort of motif often occurs in clusters, the β sheets spacing the helical segments so that each helix contacts the major groove. The more zinc fingers in the cluster, the stronger the protein binds to the DNA. In other forms of the zinc finger motif, the β sheet's place is taken by another helical segment.

The leucine zipper motif

In yet another DNA-binding motif, two different protein subunits cooperate to create a single DNA-binding site. This motif is created where a region on one subunit containing several hydrophobic amino acids (usually leucines) interacts with a similar region on the other subunit. This interaction holds the two subunits together at those regions, while the rest of

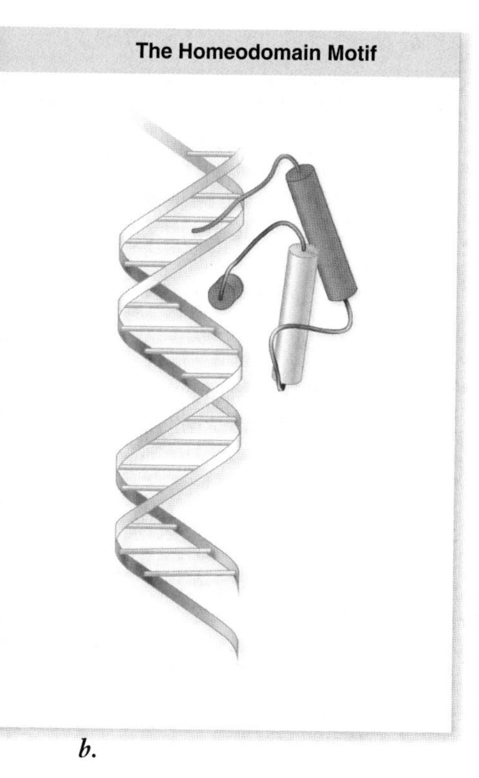

b.

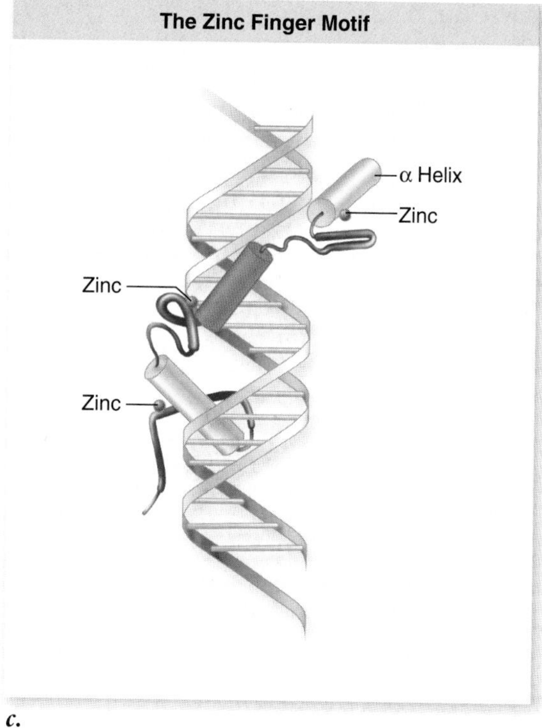

c.

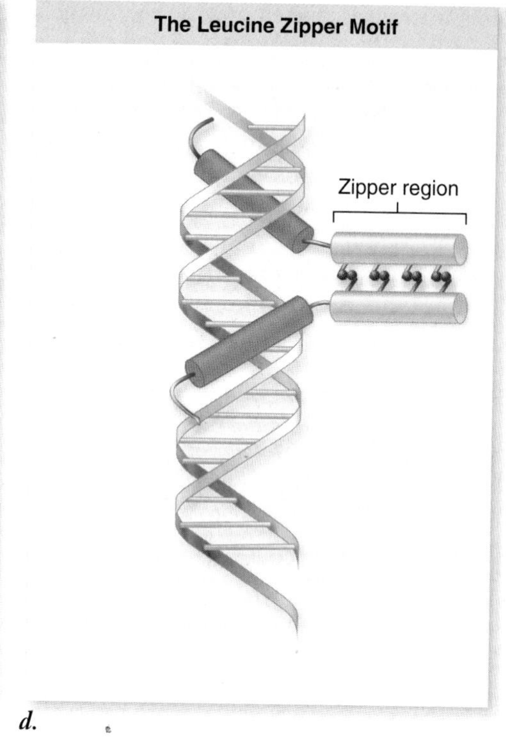

d.

the subunits remain separated. Called a **leucine zipper,** this structure has the shape of a Y, with the two arms of the Y being helical regions that fit into the major groove of DNA (figure 16.2*d*). Because the two subunits can contribute quite different helical regions to the motif, leucine zippers allow for great flexibility in controlling gene expression.

Regulatory proteins must be able to bind to DNA to affect transcription. These proteins all contain one of a relatively small set of common DNA-binding motifs. These form the active part of the DNA-binding domain, and another domain of the protein interacts with the transcription apparatus.

16.3 Prokaryotic Regulation

The details of regulation can be revealed by examining mechanisms used by prokaryotes to control the initiation of transcription. Prokaryotes and eukaryotes share some common themes, but they have some profound differences as well. Later on we discuss eukaryotic systems and concentrate on how they differ from the simpler prokaryotic systems.

Control of transcription can be either positive or negative

Control at the level of transcription initiation can be either positive or negative. **Positive control** increases the frequency of initiation, and **negative control** decreases the frequency of initiation. Each of these forms of control are mediated by regulatory proteins, but the proteins have opposite effects.

Negative control by repressors

Negative control is mediated by proteins called **repressors.** Repressors are proteins that bind to regulatory sites on DNA called **operators** to prevent or decrease the initiation of transcription. They act as a kind of roadblock to prevent the polymerase from initiating effectively.

Repressors do not act alone; each responds to specific effector molecules. Effector binding can alter the conformation of the repressor to either enhance or abolish its binding to DNA. These repressor proteins are allosteric proteins with an active site that binds DNA and a regulatory site that binds effectors. Binding changes the conformation of allosteric proteins, as described in chapter 6.

Positive control by activators

Positive control is mediated by another class of regulatory, allosteric proteins called **activators** that can bind to DNA and stimulate the initiation of transcription. These activators enhance the binding of RNA polymerase to the promoter to increase the level of transcription initiation.

Activators are the logical and physical opposites to repressors. Effector molecules can either enhance or decrease activator binding.

Prokaryotes adjust gene expression in response to environmental conditions

Changes in the environments that bacteria and archaea encounter often result in changes in gene expression. In general, genes encoding proteins involved in catabolic pathways (breaking down molecules) respond oppositely from genes encoding proteins involved in anabolic pathways (building up molecules). In the discussion that follows, we describe enzymes in the catabolic pathway that transports and utilizes the sugar lactose. Later we describe the anabolic pathway that synthesizes the amino acid tryptophan.

As mentioned in the preceding chapter, prokaryotic genes are often organized into operons, multiple genes that are part of a single transcription unit having a single promoter. Genes that are involved in the same metabolic pathway are often organized in this fashion. The proteins necessary for the utilization of lactose are encoded by the ***lac* operon,** and the proteins necessary for the synthesis of tryptophan are encoded by the ***trp* operon.**

inquiry

What advantage might it be to a bacterium to link several genes, all of whose products contribute to a single biochemical pathway, into a single operon?

Induction and repression

If a bacterium encounters lactose, it begins to make the enzymes necessary to utilize lactose. When lactose is not present, however, there is no need to make these proteins. Thus, we say that the synthesis of the proteins is *induced* by the presence of lactose. **Induction** therefore occurs when enzymes for a certain pathway are produced in response to a substrate.

When tryptophan is available in the environment, a bacterium will not synthesize the enzymes necessary to make tryptophan. If tryptophan ceases to be available, then the bacterium begins to make these enzymes. **Repression** occurs when bacteria capable of making biosynthetic enzymes do not produce them. In the case of both induction and repression, the bacterium is adjusting to produce the enzymes that are optimal for its immediate environment.

Negative control

Knowing that control of gene expression is probably at the level of initiation of transcription does not tell us the nature of the control—it might be either positive or negative. On the surface, repression may appear to be negative and induction positive; but in the case of both the *lac* and *trp* operons, control is negative by a repressor protein. The key is that the effector proteins have opposite effects on the repressor in induction with those seen in repression.

For either mechanism to work, the molecule in the environment, such as lactose or tryptophan, must produce the proper effect on the gene being regulated. In the case of *lac* induction, the presence of lactose must *prevent* a repressor protein from binding to its regulatory sequence. In the case of *trp* repression, by contrast, the presence of tryptophan must *cause* a repressor protein to bind to its regulatory sequence.

These responses are opposite because the needs of the cell are opposite for anabolic versus catabolic pathways. Each pathway is examined in detail in the following sections to show how protein–DNA interactions allow the cell to respond to environmental conditions.

The *lac* operon is negatively regulated by the *lac* repressor

The control of gene expression in the *lac* operon was elucidated by the pioneering work of Jaques Monod and François Jacob.

The *lac* operon consists of the genes that encode functions necessary to utilize lactose: β-galactosidase (*lacZ*), lactose permease (*lacY*), and lactose transacetylase (*lacA*), plus the regulatory regions necessary to control the expression of these genes (figure 16.3). In addition, the gene for the *lac* repressor (*lacI*) is linked to the rest of the *lac* operon and is thus considered part of the operon although it has its own promoter. The arrangement of the control regions upstream of the coding region is typical of most prokaryotic operons, although the linked repressor is not.

Action of the repressor

Initiation of transcription of the *lac* operon is controlled by the *lac* repressor. The repressor binds to the operator, which is adjacent to the promoter (figure 16.4*a*). This binding prevents RNA polymerase from binding to the promoter. This DNA binding is sensitive to the presence of lactose: The repressor binds DNA in the absence of lactose, but not in the presence of lactose.

Interaction of repressor and effector

In the absence of lactose, the *lac* repressor binds to the operator, and the operon is repressed (figure 16.4*a*). The effector that controls the DNA binding of the repressor is a metabolite of lactose, allolactose, which is produced when lactose is available. Allolactose binds to the repressor, altering its conformation so that it no longer can bind to the operator (figure 16.4*b*). Induction of the operon begins.

As the level of lactose falls, allolactose will no longer be available to bind to the repressor, allowing the repressor to bind to DNA again. Thus this system of negative control by the *lac* repressor and its effector, allolactose, allow the cell to respond to changing levels of lactose in the environment.

Even in the absence of lactose, the *lac* operon is expressed at a very low level. When lactose becomes available, it is transported into the cell and enough allolactose is produced that induction of the operon can occur.

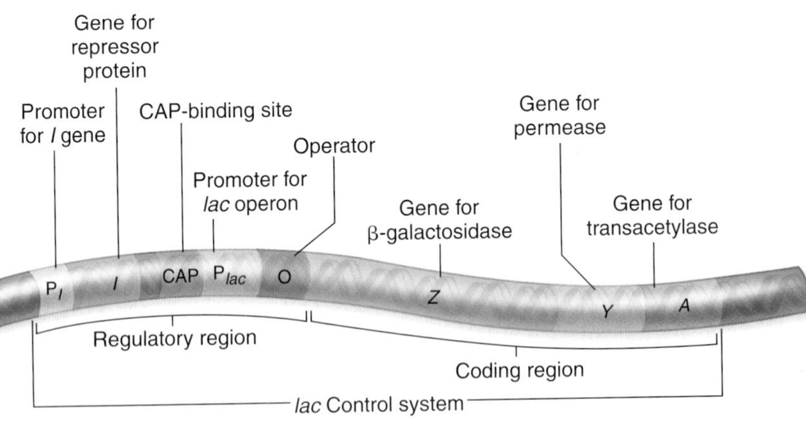

figure 16.3

THE *LAC* REGION OF THE *ESCHERICHIA COLI* CHROMOSOME. The *lac* operon consists of a promoter, an operator, and three genes (*lac Z, Y,* and *A*) that code for proteins required for the metabolism of lactose. In addition, there is a binding site for the catabolite activator protein (CAP), which affects RNA polymerase binding to the promoter. The *I* gene encodes the repressor protein, which will bind to the operator and block transcription of the *lac* genes.

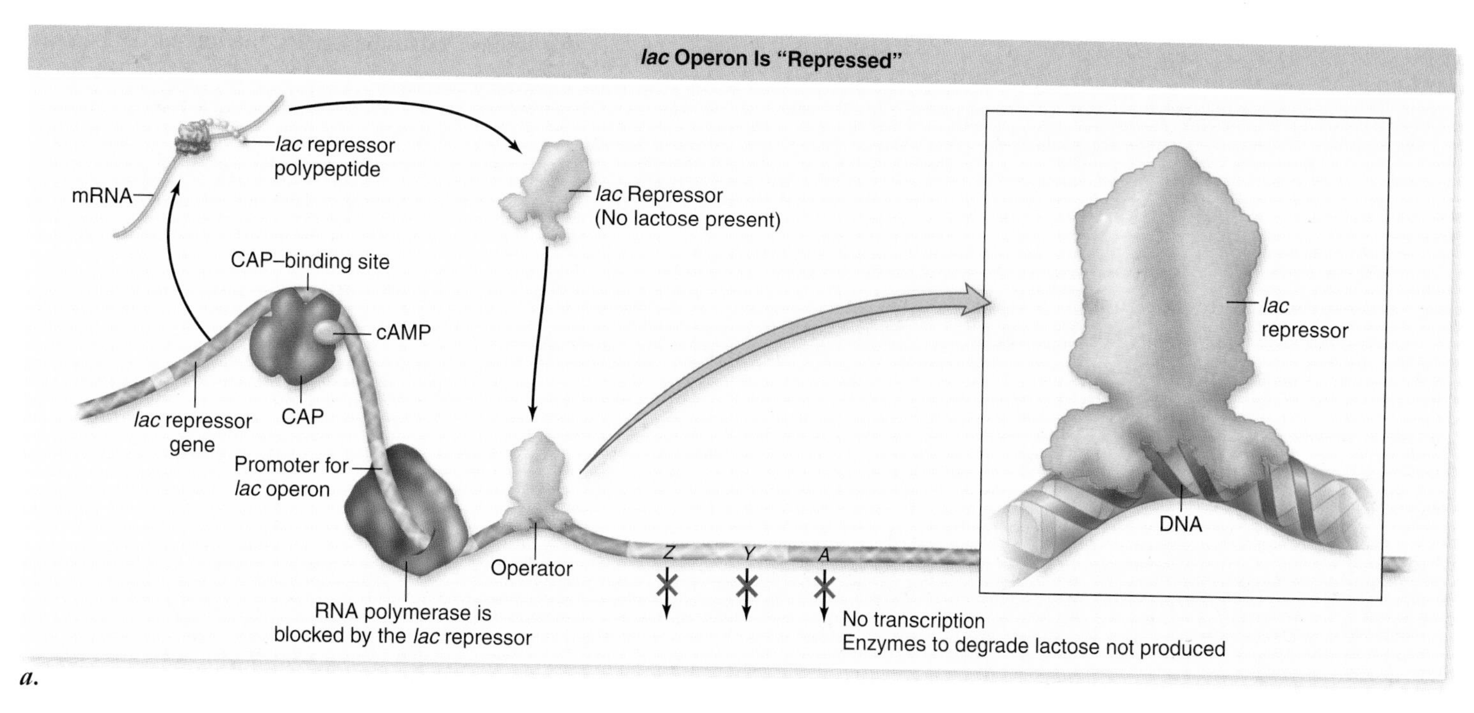

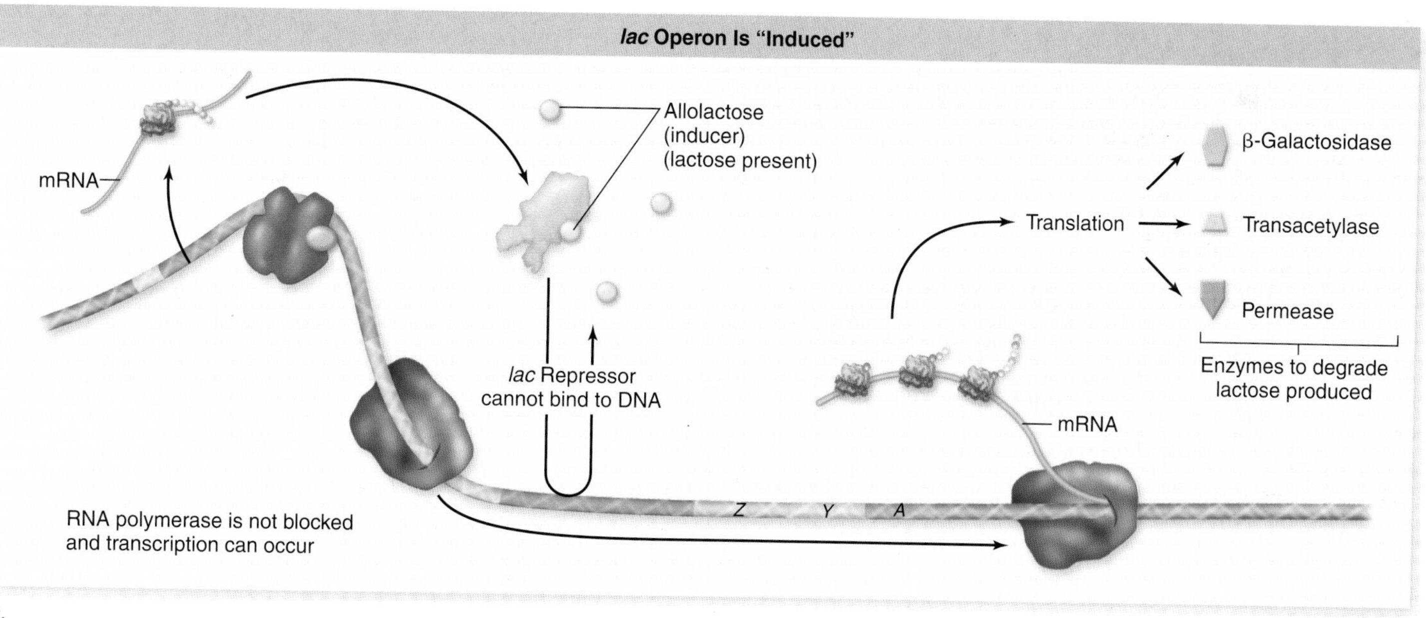

figure 16.4

INDUCTION OF THE *LAC* OPERON. ***a.*** The *lac* repressor. Because the repressor fills the major groove of the DNA helix, RNA polymerase cannot fully attach to the promoter, and transcription is blocked. When the repressor protein is bound to the operator site the *lac* operon is shut down (repressed). Because promoter and operator sites overlap, RNA polymerase and the repressor cannot functionally bind at the same time, any more than two cars can occupy the same parking space. ***b.*** The *lac* operon is transcribed (induced) when CAP is bound and when the repressor is not bound. Allolactose binding to the repressor alters the repressor's shape such that it cannot bind to the operator site and block RNA polymerase activity.

The presence of glucose prevents induction of the *lac* operon

Glucose repression is the preferential use of glucose in the presence of other sugars such as lactose. If bacteria are grown in the presence of both glucose and lactose, the *lac* operon is not induced. When the glucose is used up, the *lac* operon is induced, allowing lactose to be used as an energy source.

Despite the name glucose repression, this mechanism involves an activator protein that can stimulate transcription from multiple catabolic operons, including the *lac* operon. This activator, **catabolite activator protein (CAP),** is an allosteric protein that has cAMP as an effector. This protein is also called **cAMP response protein (CRP)** because it binds cAMP, but we will use the name CAP to emphasize its role as a positive regulator. CAP alone does not bind to DNA, but binding of the effector cAMP to CAP changes its conformation such that it can bind to DNA (figure 16.5). The level of cAMP in cells is reduced in the presence of glucose so that no stimulation of transcription from CAP-responsive operons takes place.

The CAP–cAMP system was long thought to be the sole mechanism of glucose repression. But more recent research has indicated that the presence of glucose inhibits the transport of lactose into the cell. This deprives the cell of the *lac* operon inducer, allolactose, allowing the repressor to bind to the operator. This mechanism, called **inducer exclusion,** is now thought to be the main form of glucose repression of the *lac* operon.

Given that inducer exclusion occurs, the role of CAP in the absence of glucose seems superfluous. But in fact, the action of CAP–cAMP allows maximal expression of the operon in the absence of glucose. The positive control of CAP–cAMP is necessary because the promoter of the *lac* operon alone is not efficient in binding RNA polymerase. This inefficiency is overcome by the action of the positive control of the CAP–cAMP activator (see figure 16.5).

The *trp* operon is controlled by the *trp* repressor

The organization of the *trp* operon is similar to the *lac* operon in that a series of genes arranged in a sequence encode enzymes necessary to synthesize tryptophan. The regulatory region that controls transcription of these genes is located upstream of the

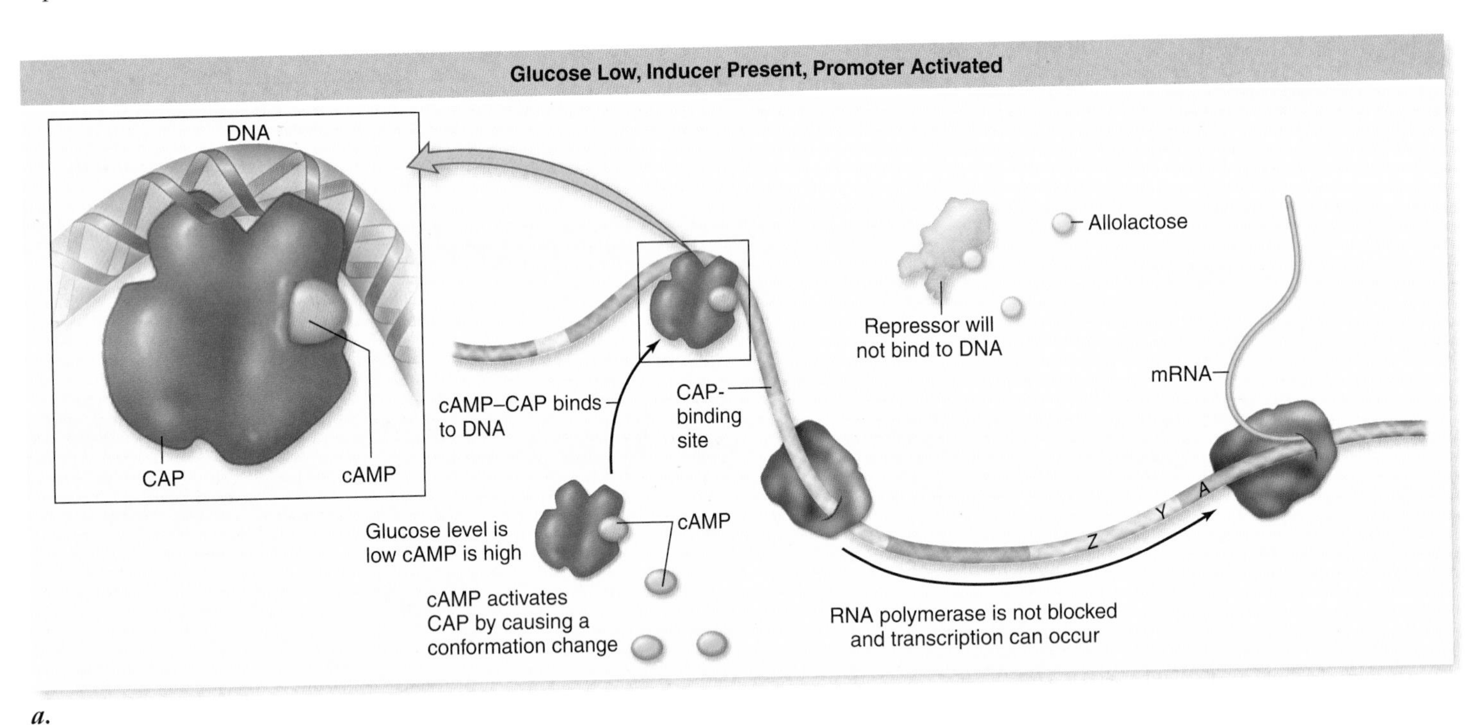

a.

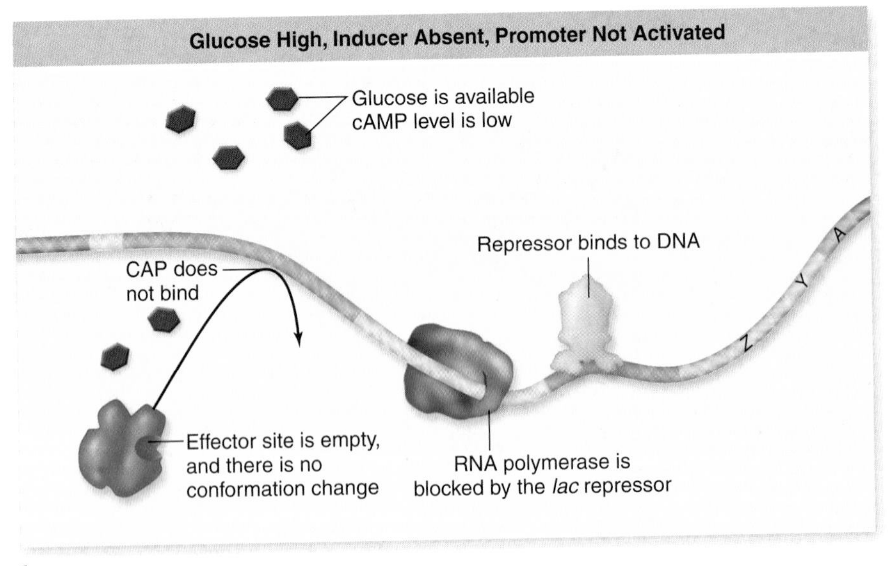

b.

figure 16.5

EFFECT OF GLUCOSE ON THE *LAC* OPERON. Expression of the *lac* operon is controlled by a negative regulator (repressor) and a positive regulator (CAP). The action of CAP is sensitive to glucose levels. ***a.*** For CAP to bind to DNA, it must bind to cAMP. When glucose levels are low, cAMP is abundant and binds to CAP. The CAP–cAMP complex causes the DNA to bend around it. This brings CAP into contact with RNA polymerase (not shown) making polymerase binding to the promoter more efficient. ***b.*** When glucose levels are high, there are two effects: cAMP is scarce so CAP is unable to activate the promoter, and the transport of lactose is blocked (inducer exclusion).

genes. The *trp* repressor is encoded by a gene located outside the *trp* operon. The *trp* operon is expressed in the absence of tryptophan and is not expressed in the presence of tryptophan.

The *trp* repressor is a helix-turn-helix regulatory protein that binds to the operator site located adjacent to the *trp* promoter (figure 16.6). This repressor behaves in a manner opposite to the *lac* repressor. The *trp* repressor alone does not bind to its operator, but when it is bound to tryptophan (the *corepressor*) its conformation alters, allowing it to bind to its operator and prevent RNA polymerase from binding to the promoter. The binding of tryptophan to the repressor alters the orientation of a pair of helix-turn-helix motifs,

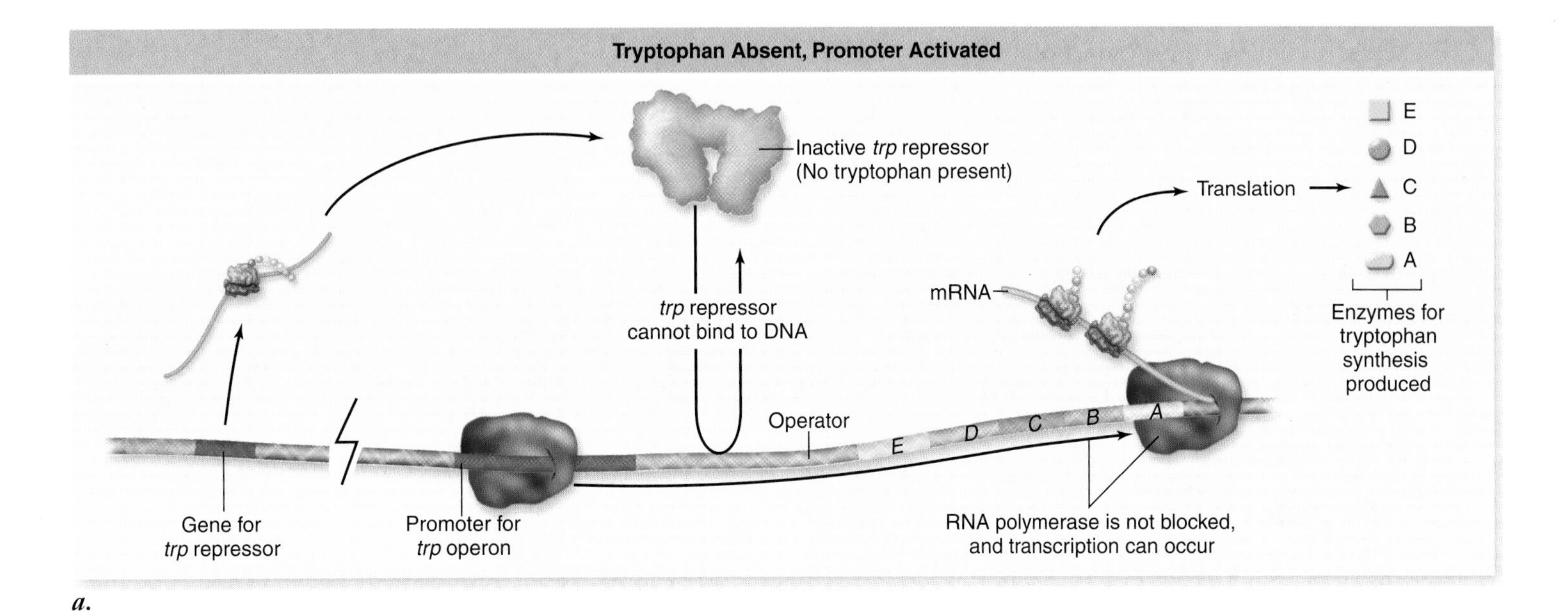

figure 16.6

HOW THE *TRP* OPERON IS CONTROLLED. The tryptophan operon encodes the enzymes necessary to synthesize tryptophan. ***a.*** The tryptophan repressor alone cannot bind to DNA. The promoter is free to function, and RNA polymerase transcribes the operon. ***b.*** When tryptophan is present, it binds to the repressor altering the repressor's conformation such that it now binds DNA. The tryptophan-repressor complex binds tightly to the operator, preventing RNA polymerase from initiating transcription.

causing their recognition helices to fit into adjacent major grooves of the DNA (figure 16.7).

When tryptophan is present and bound to the repressor, the repressor in turn is bound to the operator, the operon is said to be **repressed.** Transcription of the operon is shut off. As tryptophan levels fall, tryptophan is no longer bound to the repressor, so that the repressor can no longer bind to the operator. In this state, the operon is said to be **derepressed,** distinguishing this state from induction (see figure 16.6).

The key to understanding how both induction and repression can be due to negative regulation is knowledge of the behavior of repressor proteins and their effectors. In induction, the repressor alone can bind to DNA, and the inducer prevents DNA binding. In the case of repression, the repressor only binds DNA when bound to the corerepressor. Induction and repression are excellent examples of how interactions of molecules can affect their structures, and how molecular structure is critical to function.

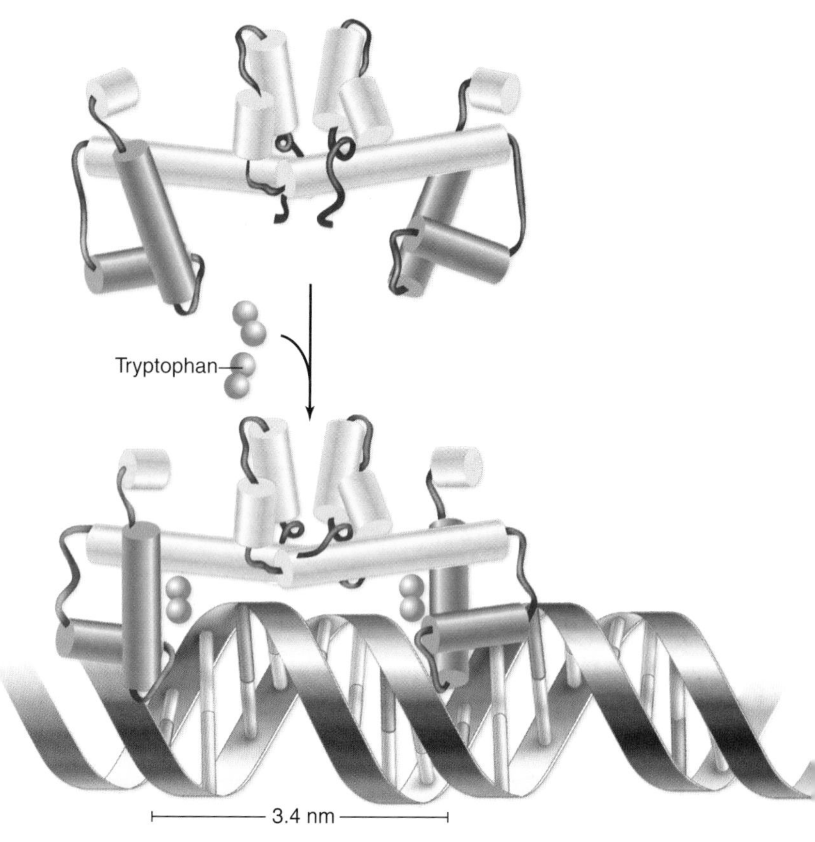

figure 16.7

HOW THE TRYPTOPHAN REPRESSOR WORKS. The binding of tryptophan to the repressor increases the distance between the two recognition helices in the repressor, allowing the repressor to fit snugly into two adjacent portions of the major groove in DNA.

Prokaryotes control gene expression to conform to their environment. The *lac* operon is controlled by a repressor protein that can bind to DNA to prevent transcription. When lactose is available, allolactose binds to the repressor, which then no longer binds to DNA, inducing the synthesis of the *lac* operon proteins. This operon is also positively regulated by an activator protein. The *trp* operon is turned off by a repressor protein that must be bound to tryptophan to bind with DNA. In the absence of tryptophan, the repressor cannot bind with DNA, derepressing the operon.

16.4 Eukaryotic Regulation

The control of transcription in eukaryotes is much more complex than in prokaryotes. The basic concepts of protein–DNA interactions are still valid, but the nature and number of interacting proteins is much greater due to some obvious differences. First, eukaryotes have their DNA organized into chromatin, complicating protein–DNA interactions considerably.

Second, eukaryotic transcription occurs in the nucleus, and translation occurs in the cytoplasm; in prokaryotes, these processes are spatially and temporally coupled. As a consequence, recruiting RNA polymerase II to a promoter is considerably more complex in eukaryotes than is the case for prokaryotic RNA polymerase.

Because of these differences, the amount of DNA involved in regulating eukaryotic genes is much greater. The need for a fine degree of flexible control is especially important for multicellular eukaryotes, with their complex developmental programs and multiple tissue types. General themes, however, emerge from this complexity.

Transcription factors can be either general or specific

In the preceding chapter we introduced the concept of transcription factors. Eukaryotic transcription requires a variety of these protein factors, which fall into two categories: *general transcription factors* and *specific transcription factors.* General factors are necessary for the assembly of a transcription apparatus and recruitment of RNA polymerase II to a promoter. Specific factors increase the level of transcription in certain cell types or in response to signals.

General transcription factors

Transcription of RNA polymerase II templates (that is, genes that encode protein products) requires more than just RNA polymerase II to initiate transcription. A host of **general transcription factors** are also necessary to establish productive initiation. These factors are required for transcription to occur, but they do not increase the rate above this basal rate.

General transcription factors are named with letter designations that follow the abbreviation TFII, for "transcription factor RNA polymerase II." The most important of these factors, TFIID, contains the TATA-binding protein that recognizes the TATA box sequence found in many eukaryotic promoters (figure 16.8).

Binding of TFIID is followed by binding of TFIIE, TFIIF, TFIIA, TFIIB, and TFIIH and a host of accessory factors called *transcription-associated factors*, TAFs. The *initiation complex* that results (figure 16.9) is clearly much more complex than the bacterial RNA polymerase holoenzyme binding to a promoter. And there is yet another level of complexity: The initiation complex, although capable of initiating synthesis at a basal level, does not achieve transcription at a high level without the participation of other, specific factors.

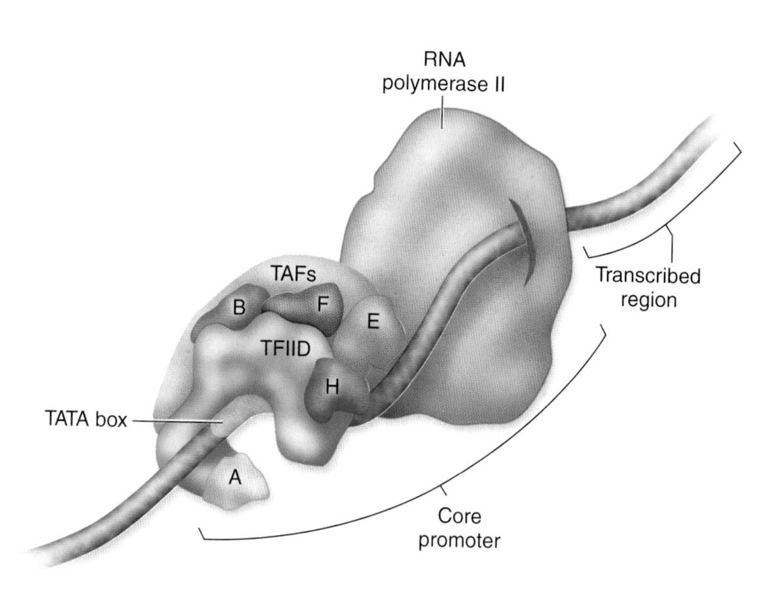

figure 16.9

FORMATION OF A EUKARYOTIC INITIATION COMPLEX. The general transcription factor, TFIID, binds to the TATA box and is joined by the other general factors, TFIIE, TFIIF, TFIIA, TFIIB, and TFIIH. This complex is added to by a number of transcription-associated factors (TAFs) that together recruit the RNA pol II molecule to the core promoter.

Specific transcription factors

Specific transcription factors act in a tissue- or time-dependent manner to stimulate higher levels of transcription than the basal level. The number and diversity of these factors are overwhelming. Some sense can be made of this proliferation of factors by concentrating on the DNA-binding motif, as opposed to the specific factors.

A key common theme that emerges from the study of these factors is that specific transcription factors, called ***activators,*** have a domain organization, that is, each factor consists of a DNA-binding domain and a separate activating domain that interacts with the transcription apparatus. These domains are essentially independent in the protein, such that they can be "swapped" between different factors and still retain their function.

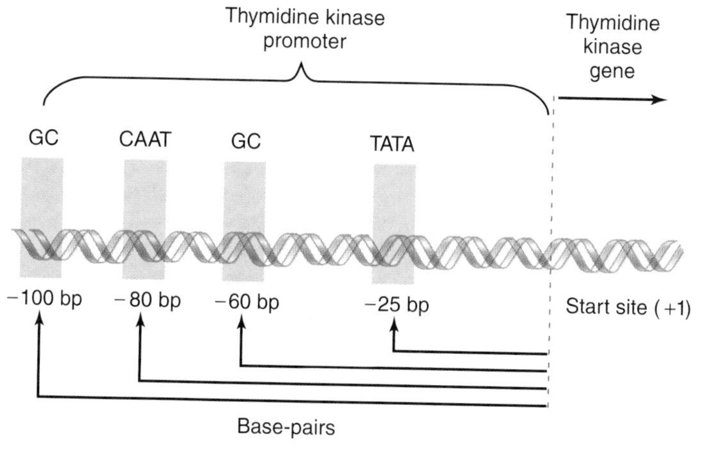

figure 16.8

A EUKARYOTIC PROMOTER. This promoter for the gene encoding the enzyme thymidine kinase. Formation of the transcription initiation complex begins with a general transcription factor binding to the TATA box. There are three other DNA sequences that direct the binding of other specific transcription factors.

Promoters and enhancers are binding sites for transcription factors

Promoters, as mentioned in the preceding chapter, form the binding sites for general transcription factors. These factors then mediate the binding of RNA polymerase II to the promoter (and also the binding of RNA polymerases I and III to their specific promoters). In contrast, the holoenzyme portion of the RNA polymerase of prokaryotes can directly recognize a promoter and bind to it.

Enhancers were originally defined as DNA sequences necessary for high levels of transcription that can act independently of position or orientation. At first, this concept seemed counterintuitive, especially since molecular biologists had been conditioned by prokaryotic systems to expect control regions to be immediately upstream of the coding region. It turns out that enhancers are the binding site of the specific transcription factors. The ability of enhancers to act over large distances was at first puzzling, but investigators now think this action is accomplished by DNA bending to form a loop, positioning the enhancer closer to the promoter.

Although more important in eukaryotic systems, this looping was first demonstrated using prokaryotic DNA-binding proteins (figure 16.10). The important point is that the linear distance separating two sites on the chromosome does not have to translate to great physical distance, because the flexibility of DNA allows bending and looping. An activator bound to an enhancer can thus be brought into contact with the transcription factors bound to a distant promoter (figure 16.11).

Coactivators and mediators link transcription factors to RNA polymerase II

Other factors specifically mediate the action of transcription factors. These **coactivators** and **mediators** are also necessary for activation of transcription by the transcription factor. They act by binding the transcription factor and then binding to another part of the transcription apparatus. Mediators are essential to the function of some transcription factors, but not all transcription factors require them. The number of coactivators is much smaller than the number of transcription factors because the same coactivator can be used with multiple transcription factors.

The transcription complex brings things together

Although a few general principles apply to a broad range of situations, nearly every eukaryotic gene—or group of genes with coordinated regulation—represents a unique case. Virtu-

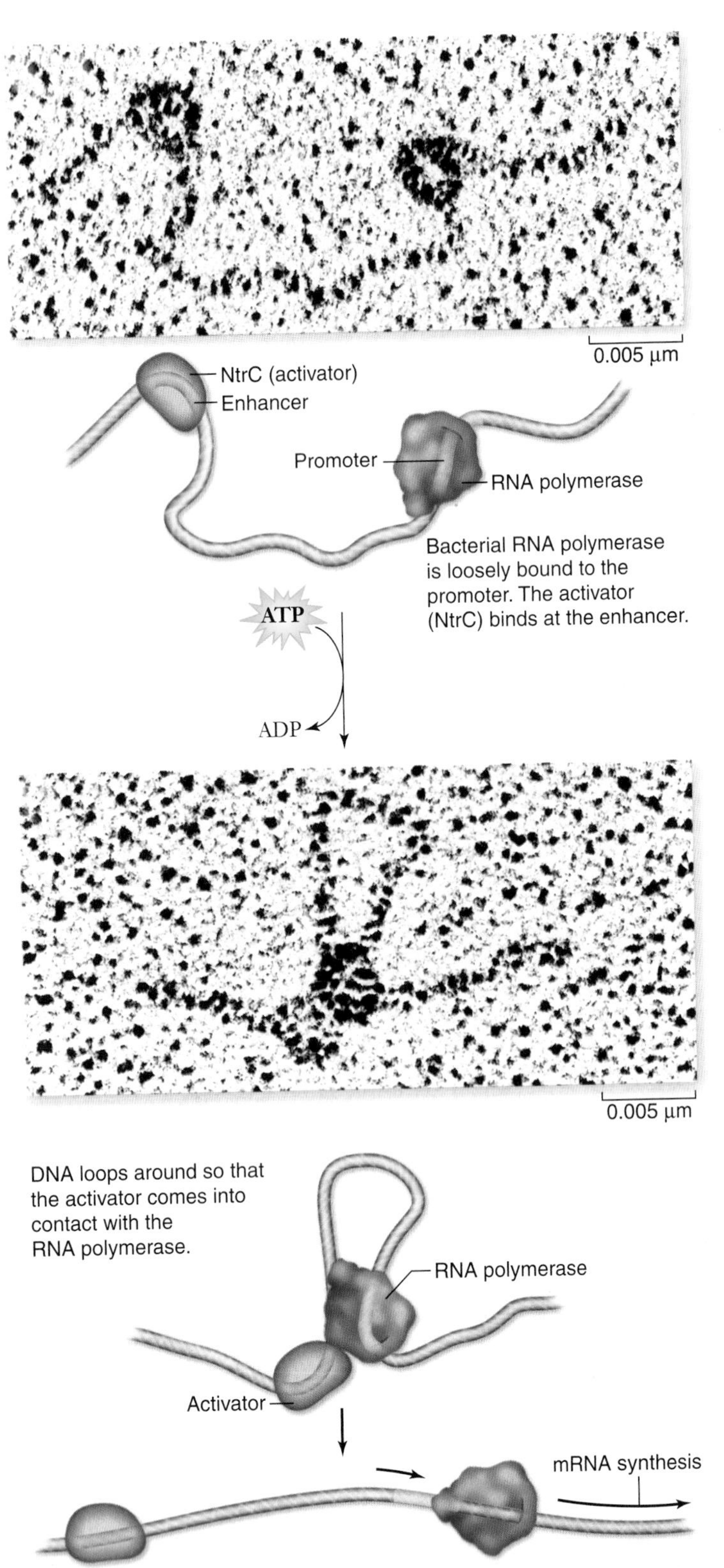

figure 16.10

DNA LOOPING CAUSED BY PROTEINS. When the bacterial activator NtrC binds to an enhancer, it causes the DNA to loop over to a distant site where RNA polymerase is bound, thereby activating transcription. Although such enhancers are rare in prokaryotes, they are common in eukaryotes.

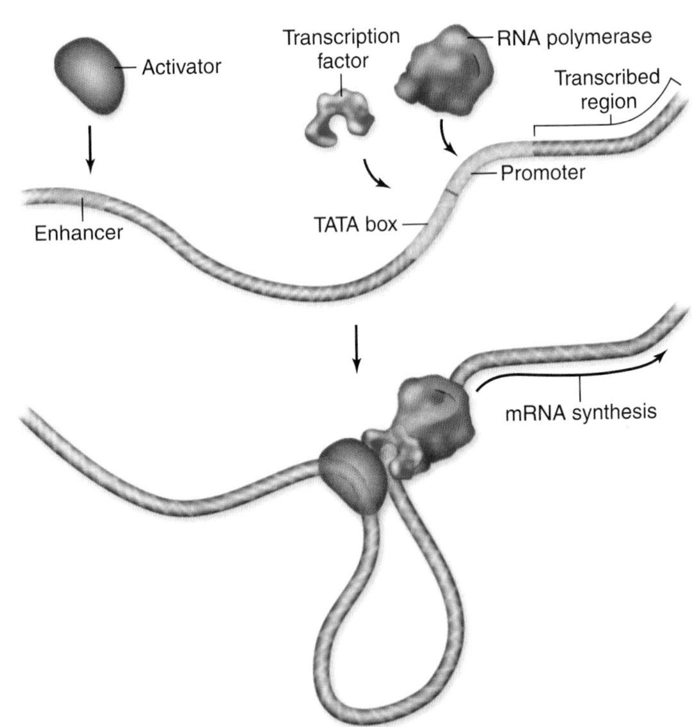

figure 16.11

HOW ENHANCERS WORK. The enhancer site is located far away from the gene being regulated. Binding of an activator (*gray*) to the enhancer allows the activator to interact with the transcription factors (*blue*) associated with RNA polymerase, stimulating transcription.

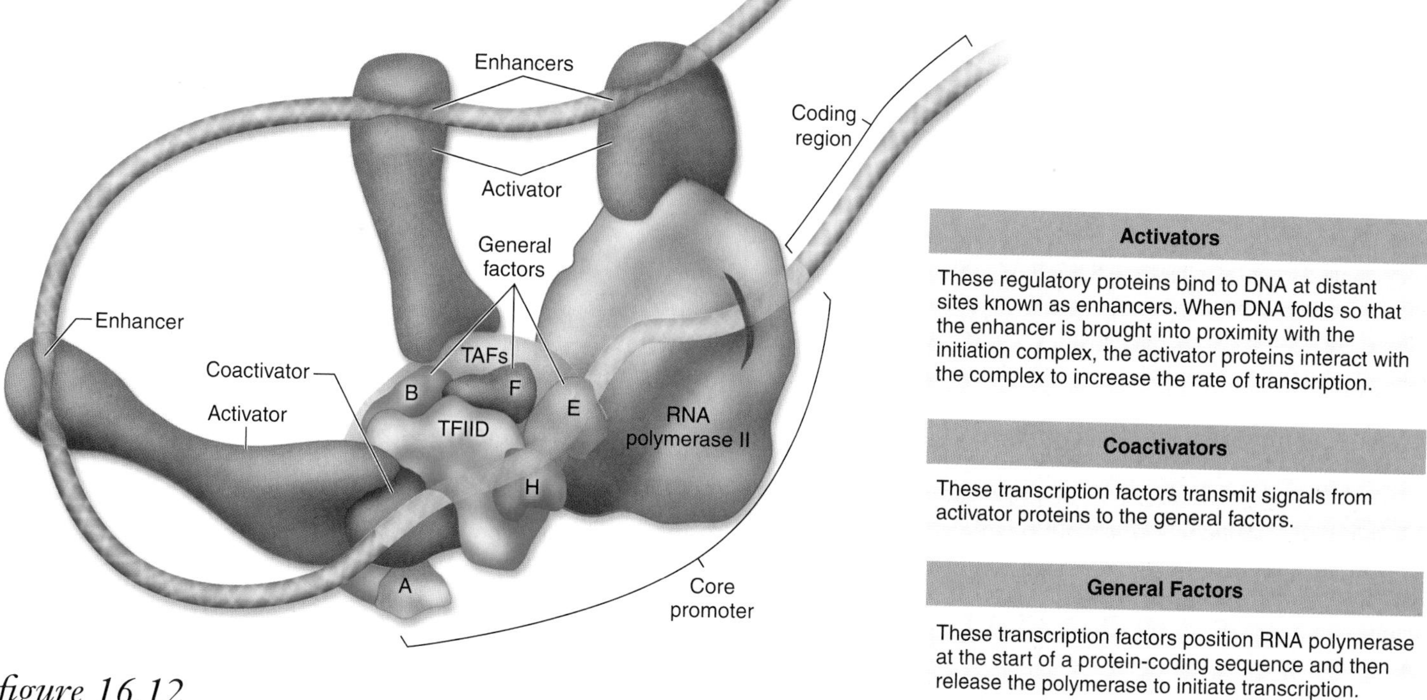

figure 16.12

INTERACTIONS OF VARIOUS FACTORS WITHIN THE TRANSCRIPTION COMPLEX. All specific transcription factors bind to enhancer sequences that may be distant from the promoter. These proteins can then interact with the initiation complex by DNA looping to bring the factors into proximity with the initiation complex. As detailed in the text, some transcription factors, called activators, can directly interact with the RNA polymerase II or the initiation complex, whereas others require additional coactivators.

ally all genes that are transcribed by RNA polymerase II need the same suite of general factors to assemble an initiation complex, but the assembly of this complex and its ultimate level of transcription depend on specific transcription factors that in combination make up the **transcription complex** (figure 16.12).

The makeup of eukaryotic promoters, therefore, is either very simple, if we consider only what is needed for the initiation complex, or very complicated, if we consider all factors that may bind in a complex and affect transcription. This kind of combinatorial gene regulation leads to great flexibility because it can respond to the many signals a cell may receive affecting transcription, allowing integration of these signals.

inquiry

How do eukaryotes coordinate the activation of many genes whose transcription must occur at the same time?

In eukaryotes, initiation of transcription requires general transcription factors that bind to the promoter and recruit RNA polymerase II to form an initiation complex. General factors result in the general level of transcription that can then be increased by the action of specific transcription factors that bind to enhancer sequences. Additional coactivators and mediators interact with both specific transcription factors and the rest of the transcription apparatus.

16.5 Eukaryotic Chromatin Structure

Eukaryotes have the additional gene expression hurdle of possessing DNA that is packaged into chromatin. The packaging of DNA first into nucleosomes and then into higher order chromatin structures is now thought to be directly related to the control of gene expression.

Chromatin structure at its lowest level is the organization of DNA and histone proteins into *nucleosomes* (see chapter 10). These nucleosomes may block binding of transcription factors and RNA polymerase II at the promoter.

The higher order organization of chromatin, which is not completely understood, appears to depend on the state of the histones in nucleosomes. Histones can be modified to result in a greater condensation of chromatin, making promoters even less accessible for protein–DNA interactions. A chromatin remodeling complex exists that can make DNA more accessible.

Both DNA and histone proteins can be modified

Chemical **methylation** of the DNA was once thought to play a major role in gene regulation in vertebrate cells. The addition of a methyl group to cytosine creates 5-methylcytosine, but this change has no effect on its base-pairing with guanine (figure 16.13). Similarly, the addition of a methyl group to uracil produces thymine, which clearly does not affect base-pairing with adenine.

Many inactive mammalian genes are methylated, and it was tempting to conclude that methylation caused the inactivation. But methylation is now viewed as having a less direct role, blocking the accidental transcription of "turned-off" genes. Vertebrate cells apparently possess a protein that binds to clusters of 5-methylcytosine, preventing transcriptional activators from gaining access to the DNA. DNA methylation in vertebrates thus ensures that once a gene is turned off, it stays off.

The histone proteins that form the core of the nucleosome (chapter 10) can also be modified. This modification is correlated with active versus inactive regions of chromatin, similar to the methylation of DNA just described. Histones can also be methylated, and this alteration is generally found in inactive regions of chromatin. Finally, histones can be modified by the addition of an acetyl group, and this addition is correlated with active regions of chromatin.

Some transcription activators alter chromatin structure

The control of eukaryotic transcription requires the presence of many different factors to activate transcription. Some activators seem to interact directly with the initiation complex or with coactivators that themselves interact with the initiation complex, as described earlier. Other cases are not so clear. The emerging consensus is that some coactivators act by modifying the structure of chromatin by adding acetyl groups to amino acids, making DNA accessible to transcription factors.

Recently, some coactivators have been shown to be histone acetylases. In these cases, it appears that transcription is increased by removing higher order chromatin structure that would prevent transcription (figure 16.14). Some corepressors have been shown to be histone deacetylases as well.

These observations have led to the suggestion that a "histone code" might exist, analogous to the genetic code. This histone code is postulated to underlie the control of chromatin structure and, thus, of access of the transcription machinery to DNA.

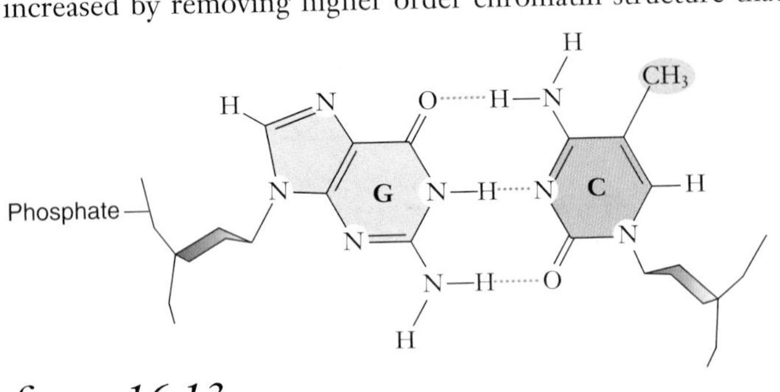

figure 16.13

DNA METHYLATION. Cytosine is methylated, creating 5-methylcytosine. Because the methyl group (*green*) is positioned to the side, it does not interfere with the hydrogen bonds of a G–C base-pair, but it can be recognized by proteins.

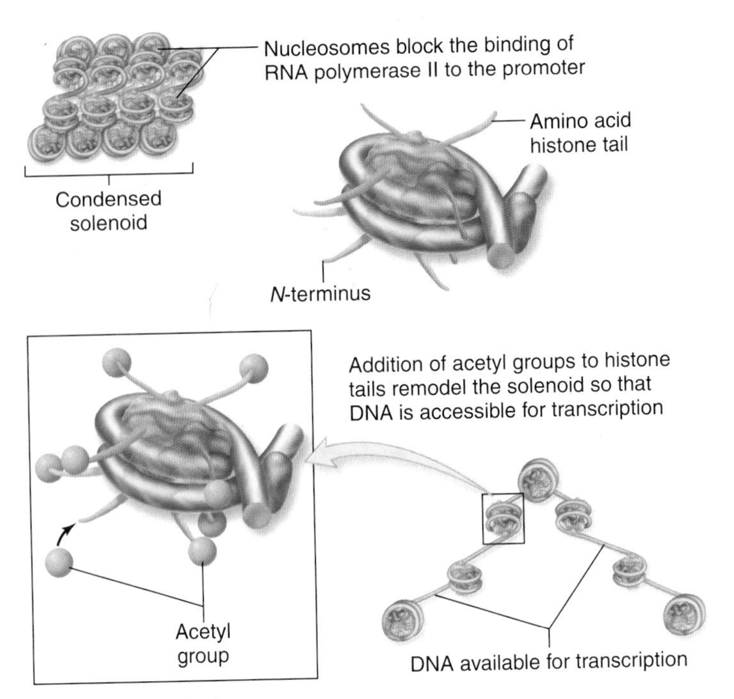

figure 16.14

HISTONE MODIFICATION AFFECTS CHROMATIN STRUCTURE. DNA in eukaryotes is organized first into nucleosomes and then into higher order chromatin structures. The histones that make up the nucleosome core have amino tails that protrude. These amino tails can be modified by the addition of acetyl groups. The acetylation alters the structure of chromatin, making it accessible to the transcription apparatus.

Chromatin-remodeling complexes also change chromatin structure

The outline of how alterations to chromatin structure can regulate gene expression are beginning to emerge. A key discovery is the existence of so-called **chromatin-remodeling complexes.** These large complexes of proteins include enzymes that modify histones and DNA and that also change chromatin structure itself. Chromatin remodeling complexes can move nucleosomes on DNA, repositioning them, and can even transfer nucleosomes from one part of DNA to another.

Eukaryotic DNA is packaged into chromatin, adding another structural challenge to transcription. Changes in chromatin structure correlate with modification of DNA and histones, and access to DNA by transcriptional regulators requires changes in chromatin structure. Some transcriptional activators modify histones by acetylation. Large chromatin-remodeling complexes may alter the structure of chromatin and thus affect gene expression.

16.6 Eukaryotic Posttranscriptional Regulation

Up to this point we have discussed gene regulation entirely in terms of transcription initiation—that is, when and how often RNA polymerase starts "reading" a particular gene. Most gene regulation appears to occur at this point. However, in principle many other points after transcription exist where gene expression could be regulated, and all of them serve as control points for at least some eukaryotic genes. In general, these posttranscriptional control processes involve the recognition of specific sequences on the primary RNA transcript by regulatory proteins and small RNA molecules.

Small RNAs can affect gene expression

Recent experiments indicate that a class of RNA molecules loosely called *small RNAs* may play a major role in regulating gene expression by interacting directly with primary gene transcripts. Small RNAs are short segments of RNA, ranging in length from 21 to 28 nt—small interfering RNA and micro-RNA are two types that were described in the preceding chapter. Researchers focusing on far larger mRNA, tRNA, and rRNA had not noticed these far smaller bits, tossing them out during experiments.

The first hints of the existence of small RNAs emerged in 1993, when researchers reported in the nematode *Caenorhabditis elegans* the presence of tiny RNA molecules that don't encode any protein. These small RNAs appeared to regulate the activity of specific *C. elegans* genes.

Soon researchers found evidence of similar small RNAs in a wide range of other organisms. In the plant *Arabidopsis thaliana*, small RNAs seemed to be involved in the regulation of genes critical to early development, and in yeasts they were identified as the agents that silence genes in tightly packed regions of the genome. In the ciliated protozoan *Tetrahymena thermophila*, the loss of major blocks of DNA during development seems guided by small RNA molecules.

RNA interference

How can small fragments of RNA act to regulate gene expression? The first clue emerged in 1998, when researchers injected small stretches of double-stranded RNA into *C. elegans*. Double-stranded RNA forms when a single strand with ends having complementary nucleotide sequences folds back in a hairpin loop; base-pairing holds the strands together much as it does the strands of a DNA duplex (figure 16.15). The injected double-stranded RNA strongly inhibited the expression of the genes from which the double-stranded RNA had been generated. This kind of gene silencing, since seen in *Drosophila* and other organisms, is called **RNA interference.**

Mechanism of RNA interference

In 2001, researchers identified an enzyme, dubbed **Dicer,** that appears to generate the small RNAs in the cell. Dicer chops double-stranded RNA molecules into little pieces. Two types of small RNA result: **micro-RNAs (miRNAs)** and **small interfering RNAs (siRNAs).**

Micro-RNAs appear to act by binding directly to mRNAs and preventing their translation. Researchers have identified over 100 different miRNAs and are still trying to sort out how each functions and which miRNAs occur in which species.

siRNAs appear to be the main agents of RNA interference, acting to degrade particular messenger RNAs after they have been transcribed, but before they can be translated by the ribosomes. The exact way they achieve this degradation of selected gene transcripts is not yet known.

Current data suggest that dicer delivers siRNAs to an enzyme complex called the *RNA-induced silencing complex (RISC)*, which searches out and degrades any mRNA molecules with a complementary sequence (figure 16.16).

figure 16.15

SMALL RNAs FORM DOUBLE-STRANDED LOOPS. These three RNA molecules all contain self-complementary regions. The molecules fold back to form hairpin loops due to base-pairing of complementary regions.

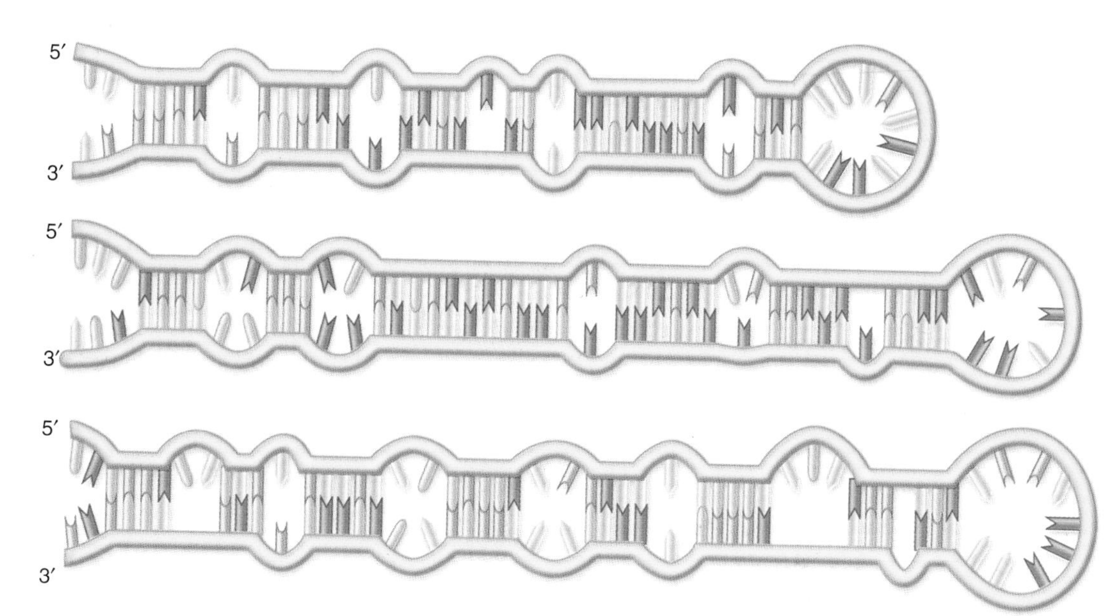

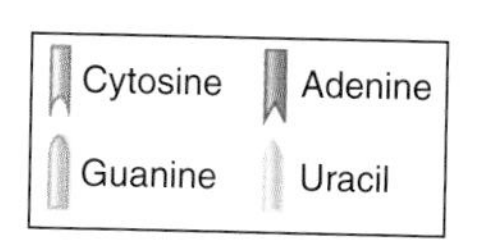

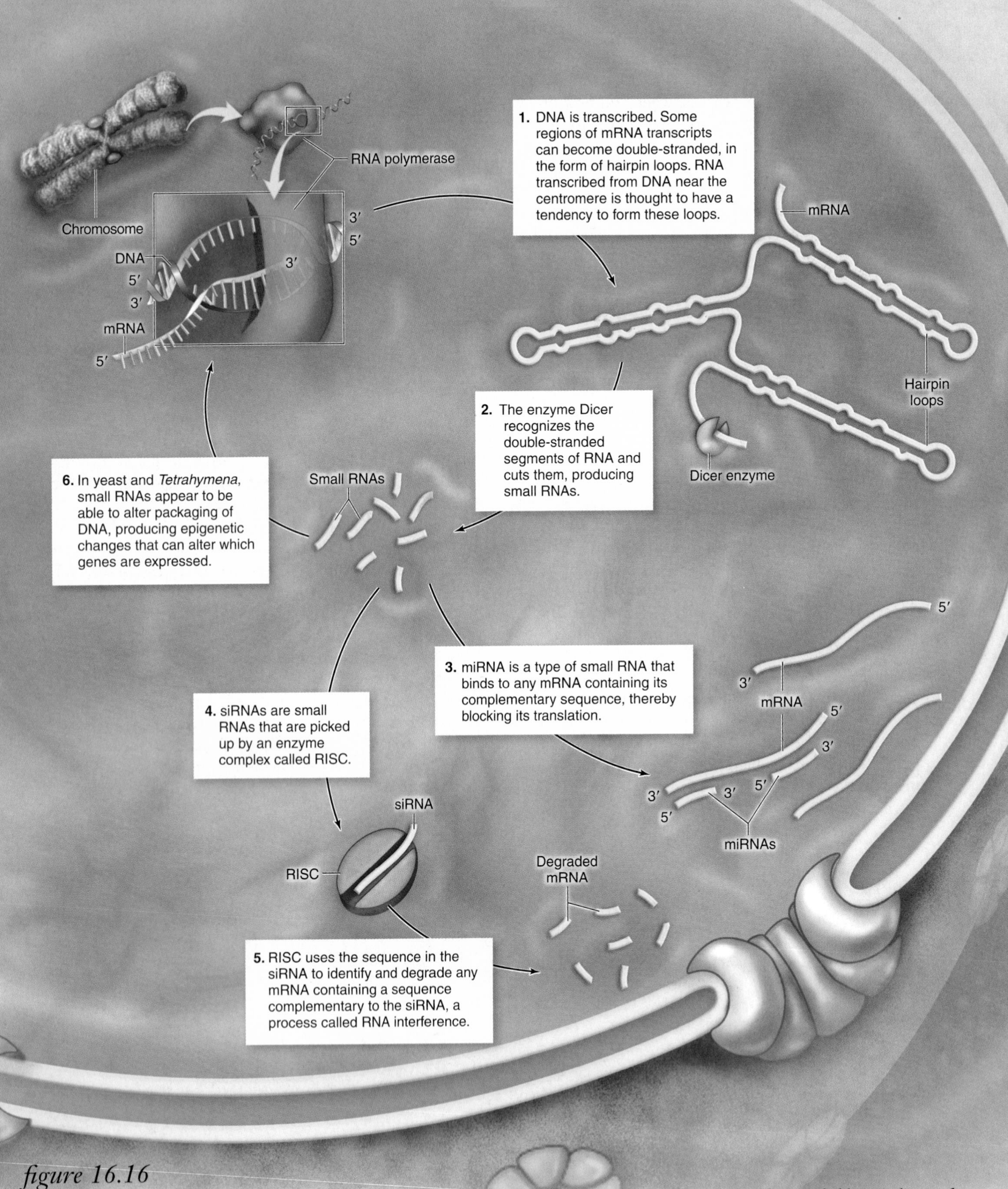

figure 16.16

HOW SMALL RNAs MAY ACT TO REGULATE GENE EXPRESSION. Small RNAs are produced when double-stranded hairpin loops of an RNA transcript are cut. Although the details are not well understood, two types of small RNA, miRNA and siRNA, are thought to block gene expression within the nucleus at the level of the mRNA gene transcript, a process called RNA interference. As this diagram suggests, small RNAs are also thought to influence chromatin packaging in some organisms.

Alternative splicing can produce multiple proteins from one gene

As noted in the preceding chapter, splicing of pre-mRNA is one of the processes leading to mature mRNA. Many of these splicing events may produce different mRNAs from a single primary transcript by alternative splicing. This mechanism allows another level of control of gene expression.

Alternative splicing can change the splicing events that occur during different stages of development or in different tissues. An example of developmental differences is found in *Drosophila*, in which sex determination is the result of a complex series of alternative splicing events that differ in males and females.

An excellent example of tissue-specific alternative splicing in action is found in two different human organs: the thyroid gland and the hypothalamus. The thyroid gland is responsible for producing hormones that control processes such as metabolic rate. The hypothalamus, located in the brain, collects information from the body (for example, salt balance) and releases hormones that in turn regulate the release of hormones from other glands, such as the pituitary gland. (You'll learn more about these glands in chapter 46.)

These two organs produce two distinct hormones: *calcitonin* and *CGRP* (calcitonin gene-related peptide) as part of their function. Calcitonin controls calcium uptake and the balance of calcium in tissues such as bones and teeth. CGRP is involved in a number of neural and endocrine functions. Although these two hormones are used for very different physiological purposes, they are produced from the same transcript (figure 16.17).

The synthesis of one product versus another is determined by tissue-specific factors that regulate the processing of the primary transcript. In the case of calcitonin and CGRP, pre-mRNA splicing is controlled by different factors that are present in the thyroid and in the hypothalamus.

RNA editing alters mRNA after transcription

In some cases, the editing of mature mRNA transcripts can produce an altered mRNA that is not truly encoded in the genome—an unexpected possibility. RNA editing was first discovered as the insertion of uracil residues into some RNA transcripts in protozoa, and it was thought to be an anomaly.

RNA editing of a different sort has since been found in mammalian species, including humans. In this case, the editing involves chemical modification of a base to change its base-pairing properties, usually by deamination. For example, both deamination of cytosine to uracil and deamination of adenine to inosine have been observed (inosine pairs as G would during translation).

Apolipoprotein B

The human protein apolipoprotein B is involved in the transport of cholesterol and triglycerides. The gene that encodes this protein, *apoB*, is large and complex, consisting of 29 exons scattered across almost 50 kilobases (kb) of DNA.

The protein exists in two isoforms: a full-length APOB100 form and a truncated APOB48 form. The truncated form is due to an alteration of the mRNA that changes a codon for glutamine to one that is a stop codon. Furthermore, this editing occurs in a tissue-specific manner; the edited form appears only in the intestine, whereas the liver makes only the full-length form. The full-length APOB100 form is part of the low-density lipoprotein (LDL) particle that carries cholesterol. High levels of serum LDL are thought to be a major predictor of atherosclerosis in humans. It does not appear that editing has any effect on the levels of the intestine-specific transcript.

The 5-HT serotonin receptor

RNA editing has also been observed in some brain receptors for opiates in humans. One of these receptors, the serotonin (5-HT) receptor, is edited at multiple sites to produce a total of 12 different isoforms of the protein.

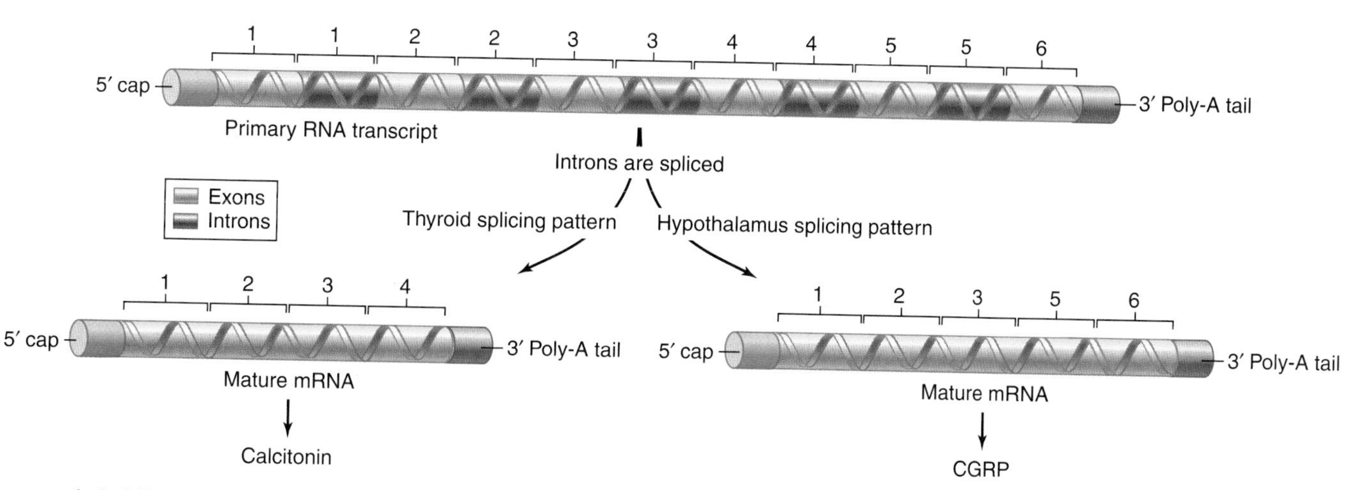

figure 16.17

ALTERNATIVE SPLICING. Many primary transcripts can be spliced in different ways to give rise to multiple mRNAs. In this example, in the thyroid the primary transcript is spliced to contain four exons encoding the protein calcitonin. In the hypothalamus the fourth exon, which contains the polyA site used in the thyroid, is skipped and two additional exons are added to encode the protein calcitonin gene related product (CGRP).

It is unclear how widespread these forms of RNA editing are, but they are further evidence that the information encoded within genes is not the end of the story for protein production.

mRNA must be transported out of the nucleus for translation

Processed mRNA transcripts exit the nucleus through the nuclear pores (described in chapter 5). The passage of a transcript across the nuclear membrane is an active process that requires the transcript to be recognized by receptors lining the interior of the pores. Specific portions of the transcript, such as the poly-A tail, appear to play a role in this recognition.

The transcript cannot move through a pore as long as any of the splicing enzymes remain associated with the transcript, ensuring that partially processed transcripts are not exported into the cytoplasm.

There is little hard evidence that gene expression is regulated at this point, although it could be. On average, about 10% of primary transcripts consists of exons that will make up

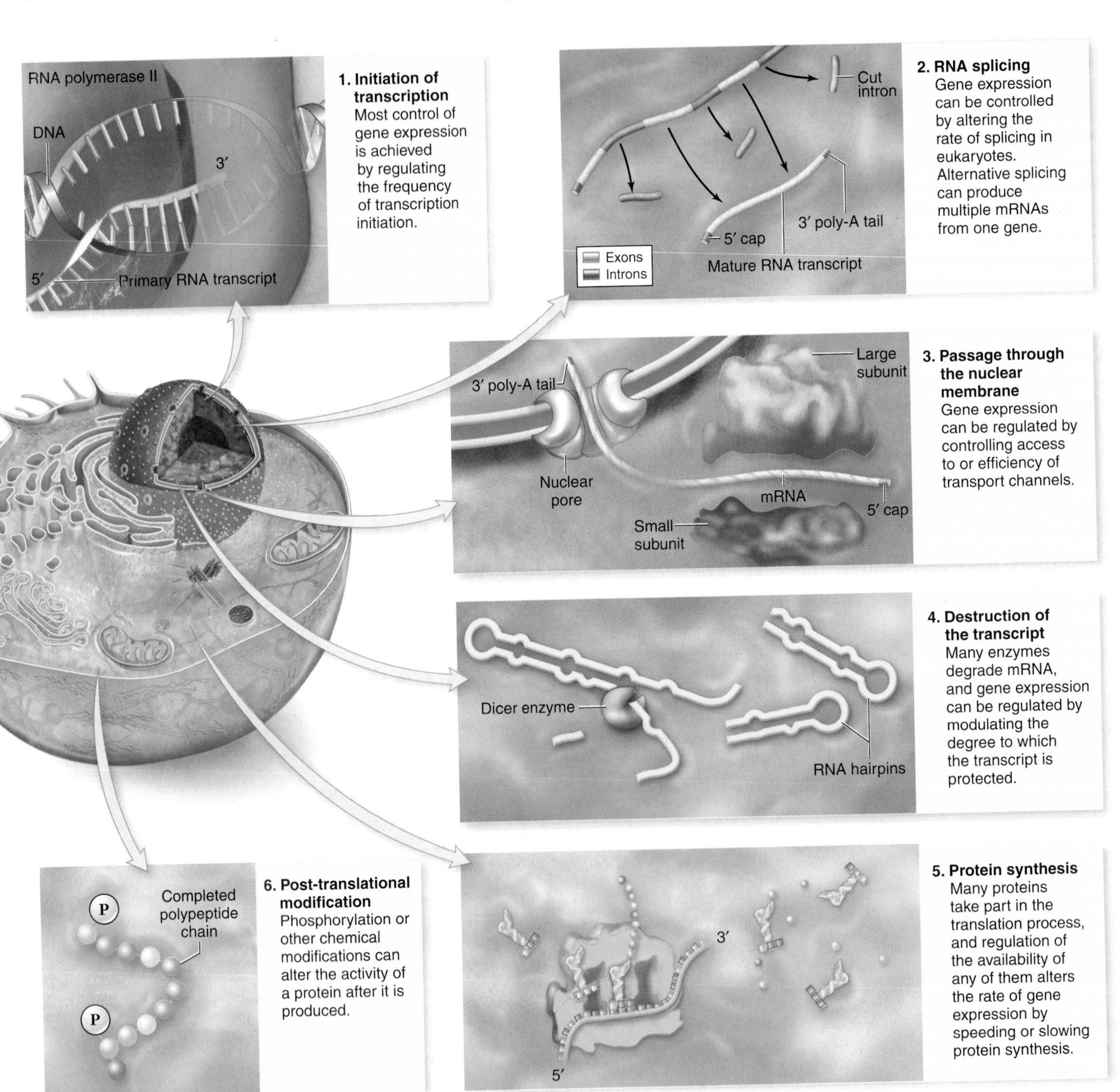

figure 16.18

MECHANISMS FOR CONTROL OF GENE EXPRESSION IN EUKARYOTES.

mRNA sequences, but only about 5% of the total mRNA produced as primary transcript ever reaches the cytoplasm. This observation suggests that about half of the exons in primary transcripts never leave the nucleus, but it is unclear whether the disappearance of this mRNA is selective.

Initiation of translation can be controlled

The translation of a processed mRNA transcript by ribosomes in the cytoplasm involves a complex of proteins called *translation factors.* In at least some cases, gene expression is regulated by modification of one or more of these factors. In other instances, **translation repressor proteins** shut down translation by binding to the beginning of the transcript, so that it cannot attach to the ribosome.

In humans, the production of ferritin (an iron-storing protein) is normally shut off by a translation repressor protein called aconitase. Aconitase binds to a 30-nt sequence at the beginning of the ferritin mRNA, forming a stable loop to which ribosomes cannot bind. When iron enters the cell, the binding of iron to aconitase causes the aconitase to dissociate from the ferritin mRNA, freeing the mRNA to be translated and increasing ferritin production 100-fold.

The degradation of mRNA is controlled

Another aspect that affects gene expression is the stability of mRNA transcripts in the cell cytoplasm. Unlike prokaryotic mRNA transcripts, which typically have a half-life of about 3 min, eukaryotic mRNA transcripts are very stable. For example, β-globin gene transcripts have a half-life of over 10 hr, an eternity in the fast-moving metabolic life of a cell.

The transcripts encoding regulatory proteins and growth factors, however, are usually much less stable, with half-lives of less than 1 hr. What makes these particular transcripts so unstable? In many cases, they contain specific sequences near their 3′ ends that make them targets for enzymes that degrade mRNA. A sequence of A and U nucleotides near the 3′ poly-A tail of a transcript promotes removal of the tail, which destabilizes the mRNA.

Histone transcripts, for example, have a half-life of about 1 hr in cells that are actively synthesizing DNA; at other times during the cell cycle, the poly-A tail is lost, and the transcripts are degraded within minutes.

Other mRNA transcripts contain sequences near their 3′ ends that are recognition sites for endonucleases, which causes these transcripts to be digested quickly. The short half-lives of the mRNA transcripts of many regulatory genes are critical to the function of those genes because they enable the levels of regulatory proteins in the cell to be altered rapidly.

A review of various methods of posttranscriptional control of gene expression is provided in figure 16.18.

Small RNAs may help to control gene expression by either selective degradation of mRNA, inhibition of translation, or alteration of chromatin structure. Multiple mRNAs can be formed from a single gene via alternative splicing, which can be tissue- and developmentally specific. The sequence of an mRNA can also be altered by RNA editing. All of these processes allow control of gene expression after transcription.

16.7 Protein Degradation

If all of the proteins produced by a cell during its lifetime remained in the cell, serious problems would arise. Protein labeling studies in the 1970s indicated that eukaryotic cells turn over proteins in a controlled manner. That is, proteins are continually being synthesized and degraded. Although this protein turnover is not as rapid as in prokaryotes, it indicates that a system regulating protein turnover is important.

Proteins can become altered chemically, rendering them nonfunctional; in addition, the need for any particular protein may be transient. Proteins also do not always fold correctly, or they may become improperly folded over time. These changes can lead to loss of function or other chemical behaviors, such as aggregating into insoluble complexes. In fact, a number of neurodegenerative diseases, such as Alzheimer dementia, Parkinson disease, and mad cow disease, are related to proteins that aggregate, forming characteristic plaques in brain cells. Thus, in addition to normal turnover of proteins, cells need a mechanism to get rid of old, unused, and incorrectly folded proteins.

Enzymes called **proteases** can degrade proteins by breaking peptide bonds, converting a protein into its constituent amino acids. Although there is an obvious need for these enzymes, they clearly cannot be floating around in the cytoplasm active at all times.

One way that eukaryotic cells handle such problems is to confine destructive enzymes to a specific cellular compartment. You may recall from chapter 4 that lysosomes are vesicles that contain digestive enzymes, including proteases. Lysosomes are used to remove proteins and old or nonfunctional organelles, but this system is not specific for particular proteins. Cells need another regulated pathway to remove proteins that are old or unused, but leave the rest of cellular proteins intact.

Addition of ubiquitin marks proteins for destruction

Eukaryotic cells solve this problem by marking proteins for destruction, then selectively degrading them. The mark that cells use is the attachment of a **ubiquitin** molecule. Ubiquitin, so named because it is found in essentially all eukaryotic cells (that is, it is ubiquitous), is a 76–amino-acid protein that can exist as an isolated molecule or in longer chains that are attached to other proteins.

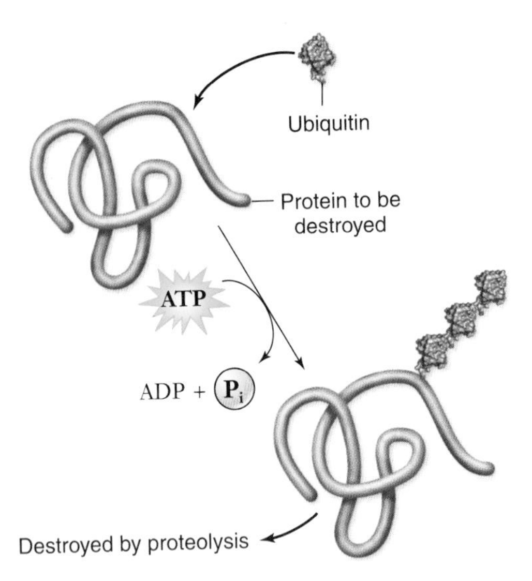

figure 16.19

UBIQUITINATION OF PROTEINS. Proteins that are to be degraded are marked with ubiquitin. The enzyme ubiquitin ligase, uses ATP to add ubiquitin to a protein. When a series of these have been added, the polyubiquitinated protein is destroyed.

The longer chains are added to proteins in a stepwise fashion by an enzyme called *ubiquitin ligase* (figure 16.19). This reaction requires ATP and other proteins, and it takes place in a multistep, regulated process. Proteins that have a ubiquitin chain attached are called *polyubiquitinated*, and this state is a signal to the cell to destroy this protein.

Two basic categories of proteins become ubiquitinated: those that need to be removed because they are improperly folded or nonfunctional, and those that are produced and degraded in a controlled fashion by the cell. An example of the latter are the cyclin proteins that help to drive the cell cycle (chapter 10). When these proteins have fulfilled their role in active division of the cell, they become polyubiquitinated and are removed. In this way, a cell can control entry into cell division or maintain a nondividing state.

The proteasome degrades polyubiquitinated proteins

The cellular organelle that degrades proteins marked with ubiquitin is the **proteasome,** a large cylindrical complex that proteins enter at one end and exit the other as amino acids or peptide fragments (figure 16.20).

The proteasome complex contains a central region that has protease activity and regulatory components at each end. Although not membrane-bounded, this organelle can be thought of as a form of compartmentalization on a very small scale. By using a two-step process, first to mark proteins for destruction, then to process them through a large complex, proteins to be degraded are isolated from the rest of the cytoplasm.

The process of ubiquitination followed by degradation by the proteasome is called the *ubiquitin–proteasome pathway.* It can be thought of as a cycle in that the ubiquitin added to proteins is not itself destroyed in the proteasome. As the proteins are degraded, the ubiquitin chain itself is simply cleaved back into ubiquitin units that can then be reused (figure 16.21).

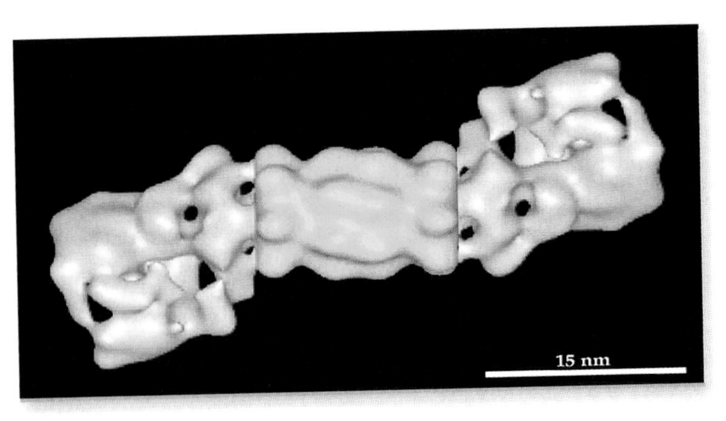

figure 16.20

THE *DROSOPHILA* PROTEASOME. The central complex contains the proteolytic activity, and the flanking regions act as regulators. Proteins enter one end of the cylinder and are cleaved to peptide fragments that exit the other end.

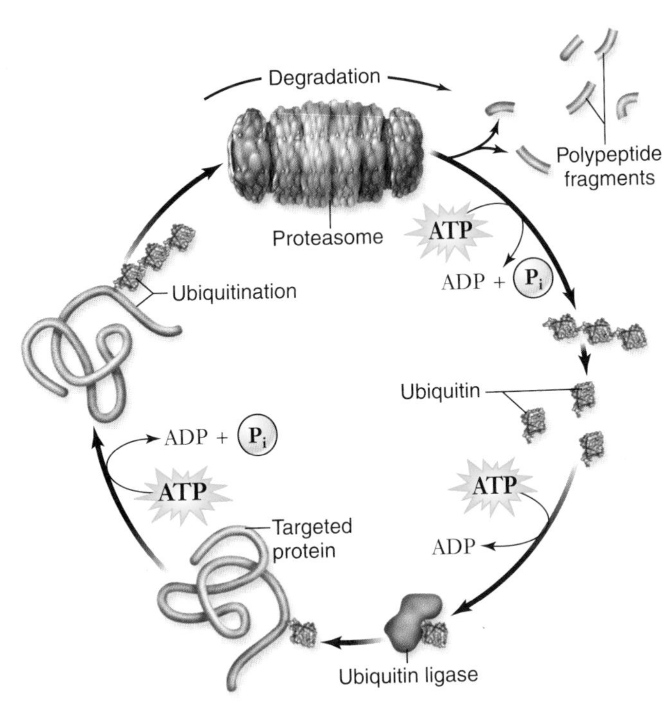

figure 16.21

DEGRADATION BY THE UBIQUITIN–PROTEASOME PATHWAY. Proteins are first ubiquitinated, then enter the proteasome to be degraded. In the proteasome, the polyubiquitin is removed and then is later "deubiquitinated" to produce single ubiquitin molecules that can be reused.

inquiry

What are two reasons a cell would polyubiquitinate a polypeptide?

Proteins are synthesized and degraded in a regulated fashion. Control of protein degradation in eukaryotes involves addition of the protein ubiquitin followed by proteolysis in the proteasome. The ubiquitin–proteasome pathway is a the major pathway to recycle old and improperly folded proteins.

16.1 Control of Gene Expression

Control of gene expression is essential to living organisms, allowing cells to respond to changing environmental conditions and enabling the development of complex multicellular organisms.

- Transcription is initiated by regulatory proteins that modulate the ability of RNA polymerase to bind to the promoter.
- Prokaryotes and eukaryotes differ in how they control gene expression.
- Control strategies in prokaryotes respond quickly to changing environmental conditions.
- In eukaryotes control strategies act to maintain homeostasis.

16.2 Regulatory Proteins

Regulatory proteins bind to specific DNA sequences and control whether transcription will occur.

- Regulatory proteins bind to the surface of the double helix and interact with base-pairs in the major groove.
- A DNA-binding motif refers to the three-dimensional structure of the region of the regulatory protein that binds to the DNA.
- Several different DNA-binding motifs are found in regulatory proteins (figure 16.2).

16.3 Prokaryotic Regulation

Although there are differences, prokaryotes and eukaryotes share many similarities in control of transcription.

- Negative control is mediated by allosteric proteins called repressors that prevent or decrease transcription.
- Positive control is mediated by another class of regulatory allosteric proteins called activators that stimulate transcription.
- Prokaryotes adjust gene expression in response to environmental conditions.
- The *lac* operon is induced in the presence of lactose; that is, the enzymes to utilize lactose are only produced in the presence of lactose.
- The *trp* operon is repressed; that is, the enzymes needed to produce trp are turned off in the presence of trp.
- In induction, the effector (allolactose) binds to the repressor, altering its conformation such that it no longer binds DNA (figure 16.4).
- In repression, the effector (called a corepressor) binds to the repressor, altering its conformation such that it can bind to DNA
- The presence of glucose prevents induction of the *lac* operon, a process callled glucose repression.
- Inducer exclusion occurs when an inducer molecule is prevented from entering the cell so operon activity remains suppressed. In the *lac* operon inducer exclusion is one mediator of glucose repression.
- The maximum expression of the *lac* operon requires positive control by the activitor: catabolite activator protein (CAP).
- CAP is active bound to cAMP, levels of which are high when glucose levels are low.
- The *trp* operon is also negatively controlled. The *trp* repressor does not bind to DNA, allowing expression in the absence of *trp*.
- The repressor binds to trp (the corepressor) and can then bind to DNA and shut off the operon when trp levels are high.

16.4 Eukaryotic Regulation

The control of transcription in eukaryotes is much more complex than in prokaryotes. Eukaryotic DNA is organized into chromatin and the nuclear membrane separates transcription and translation.

- Transcription factors can be either general or specific.
- General factors are necessary to assemble the transcription apparatus and recruitment of RNA polymerase II to the promoter.
- Specific factors act in a tissue- or time-dependent manner to stimulate higher rates of transcription.
- Promoters are binding sites for general transcription factors; enhancers are binding sites for specific transcription factors.
- Coactivators and mediators interact with specific transcription factors and the rest of the transcription apparatus (figure 16.12).
- Some transcription factors require a mediator but not all of them do.
- The number of coactivators is small because the same coactivator can be used with multiple transcription factors.
- Transcription by RNA polymerase II requires an initiation complex and specific transcription factors.

16.5 Eukaryotic Chromatin Structure

Expression of genes in eukaryotes is further complicated because their DNA is packaged into chromatin.

- In eukaryotes DNA is wrapped around proteins, called histones, forming nucleosomes and is not accessible for transcription.
- Methylation of DNA base-pairs, primarily cytosine, correlates with genes that have been "turned off."
- Methylation of histones is associated with inactive regions of chromatin.
- Acetylated histones are associated with active regions of chromatin.
- Transcription activators such as histone acetylases and deacetylases alter chromatin structure and accessibility for transcription.
- Chromatin-remodeling complexes contain enzymes that move, reposition, and transfer nucleosomes.

16.6 Eukaryotic Posttranscriptional Regulation

Control of eukaryotic gene expression can occur after the initiation of transcription (figure 16.16).

- RNA interference is mediated by small RNAs that folds back on themselves to form a double-stranded RNA with a hairpin loop.
- An enzyme, called Dicer, chops double-stranded RNA into micro-RNA (miRNAs) and small interfering RNA (siRNAs).
- Micro-RNAs bind directly to mRNA and prevent translation.
- Small interfering RNAs degrade particular mRNAs after they have been formed by transcription.
- Currently it is thought that siRNA works with an enzyme called RISC, which degrades mRNA complementary to siRNA.
- In response to tissue-specific factors alternative splicing of pre-mRNA from one gene can result in many different proteins.
- RNA editing involves modification of the mRNA, changing the base-pairing properties.
- An mRNA must be transported out of the nucleus for translation.
- The initiation of translation is controlled by translation factors and translation repressor proteins.
- Translation can be regulated by degredation of mRNA.

16.7 Protein Degradation

Proteins are continuously synthesized and degraded.

- In eukaryotes, proteins targeted for destruction have ubiquitin added to them.
- Proteins are ubiquitinated when old, nonfunctional, or produced in a controlled fashion such as cyclins.
- A cell organelle—the proteasome—degrades ubiquitinated proteins.

review questions

SELF TEST

1. Control of gene expression can occur at which of the following steps?
   a. Splicing of pre-mRNA into mature mRNA
   b. Initiation of translation
   c. Initiation of transcription
   d. All of the above
2. Regulatory proteins interact with DNA by—
   a. unwinding the helix and changing the pattern of base-pairing
   b. the sugar–phosphate backbone of the double helix
   c. unwinding the helix and disrupting base-pairing
   d. binding to the major groove of the double helix and interacting with base-pairs
3. The two proteins subunits of a leucine zipper are held together—
   a. by β-sheet domains
   b. by the interactions of hydrophobic amino acids
   c. by two α-helical domains separated by a turn
   d. the interaction of atoms of zinc
4. Which domain of a helix-turn-helix protein is directly involved with binding a specific DNA sequence?
   a. The recognition helix
   b. The homeodomain
   c. The zinc finger
   d. The leucine zipper
5. Negative control of transcription in a prokaryotic cell involves ________ molecules that alter the conformation of ________ proteins that bind to DNA and prevent transcription.
   a. operator; repressor
   b. activator; RNA polymerase
   c. activator; operator
   d. effector; repressor
6. What is an operon?
   a. A region of DNA involved in regulation of transcription
   b. A cluster of genes that are expressed as a single unit
   c. A DNA-binding motif
   d. A regulator protein that enhances transcription
7. What effect would the presence of lactose have on a *lac* operon?
   a. The repressor would bind to the operator site of the operon.
   b. Lactose will bind to the operator site of the operon.
   c. The *lac* operon would be transcribed.
   d. It would have no effect.
8. How does the presence of glucose influence the regulation of the *lac* operon?
   a. Glucose reduces the amount of cAMP, which CAP requires for action.
   b. Glucose prevents the transport of lactose into the cell.
   c. Glucose binds to the repressor protein.
   d. Both (a) and (b)
9. What effect does the amino acid tryptophan have on the *trp* operon?
   a. Tryptophan binds to the repressor, resulting in transcription of the *trp* operon.
   b. Tryptophan binds to the repressor and prevents transcription.
   c. Tryptophan increases cAMP, activating catabolite activator protein (CAP).
   d. Tryptophan interacts with repressor, leading to derepression of the *trp* operon.
10. How do specific transcription factors differ from general transcription factors?
    a. Specific transcription factors increase the rate of transcription.
    b. Specific transcription factors bind to the TATA box sequence.
    c. Specific transcription factors form an initiation complex.
    d. Specific transcription factors bind to RNA polymerase.
11. DNA methylation—
    a. inhibits transcription by blocking base-pairing of cytosine and guanine
    b. inhibits transcription by blocking base-pairing of uracil and adenine
    c. prevents transcription by blocking the TATA box sequence
    d. is correlated with genes that are turned off
12. What is the function of small interfering RNAs?
    a. They bind to mRNA and block translation.
    b. They block transcription of complementary mRNAs.
    c. They trigger the destruction of complementary mRNAs.
    d. They compete with transfer RNAs during translation.
13. RNA editing is a consequence of—
    a. base-pair substitutions caused by a mutation in the DNA
    b. splicing of a pre-mRNA
    c. methylation of the mRNA
    d. modifications of a base within the mRNA
14. What is *ubiquitin*?
    a. A type of protease
    b. A posttranslational modification that targets proteins for destruction
    c. A protein involved in the transport of mRNAs out of the nucleus
    d. A posttranslational modification that degrades mRNAs
15. Which of the following is *not* a true statement about proteasomes?
    a. They are membrane-bounded organelles.
    b. They break proteins into amino acids.
    c. They do not degrade ubiquitin.
    d. They remove ubiquitin from proteins.

CHALLENGE QUESTIONS

1. Examples of positive and negative control of transcription can be found in the regulation of expression of the bacterial operons *lac* and *trp*. Use these two operon systems to describe the difference between positive and negative regulation.
2. How is an operator different from a promoter in the regulation of prokaryotic gene expression?
3. What forms of eukaryotic control of gene expression are unique to eukaryotes? Could prokaryotes use the mechanisms, or are they due to differences in these cell types?
4. The number and type of proteins found in a cell can be influenced by genetic mutation and regulation of gene expression. Discuss how these two processes differ.

Do you need additional review? *Visit* www.ravenbiology.com *for practice quizzes, animations, videos, and activities designed to help you master the material in this chapter.*

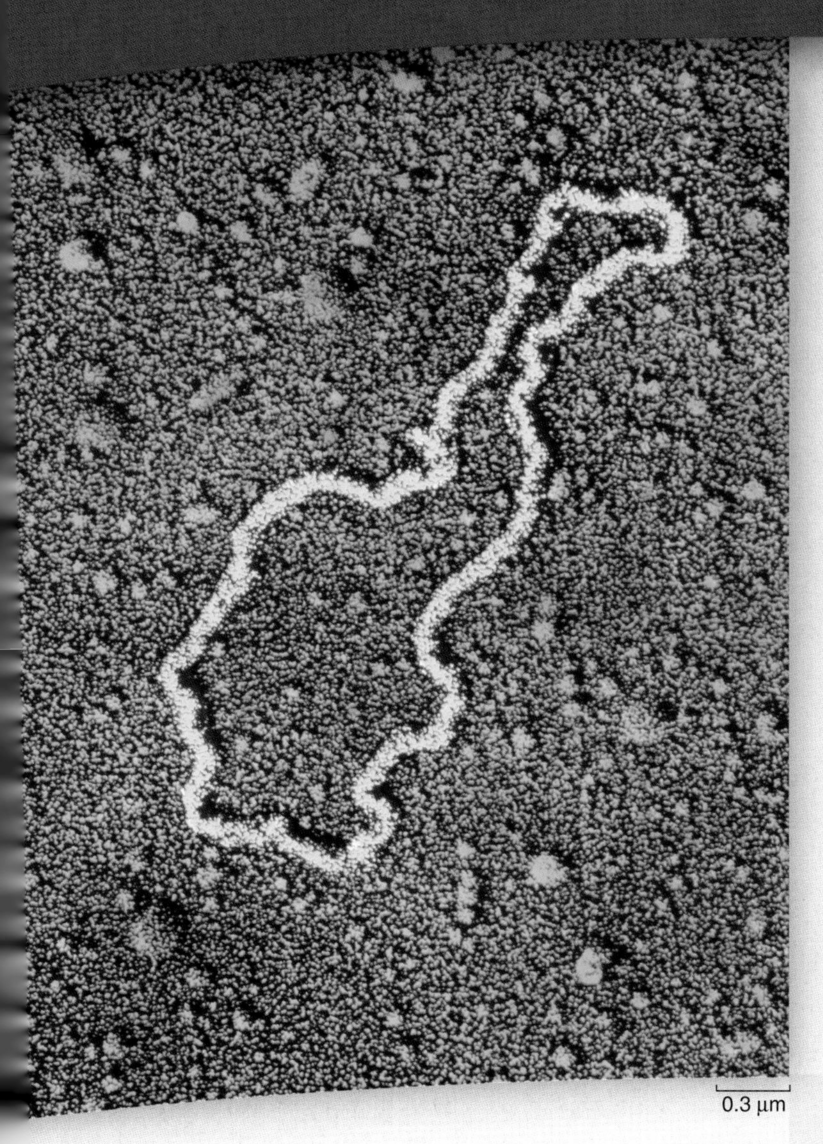

chapter 17

Biotechnology

introduction

OVER THE PAST DECADES, the development of new and powerful techniques for studying and manipulating DNA has revolutionized biology. The knowledge gained in the last 25 years is greater than the rest of the history of biology. Biotechnology also affects more aspects of everyday life than any other area of biology. From the food on your table to the future of medicine, biotechnology touches your life.

The ability to isolate specific DNA sequences arose from the study and use of small DNA molecules found in bacteria, like the plasmid pictured here. In this chapter, we explore these technologies and consider how they apply to specific problems of practical importance.

concept outline

17.1 DNA Manipulation

The ability to directly isolate and manipulate genetic material was one of the most profound changes in the field of biology in the late 20th century. The construction of **recombinant DNA** molecules, that is, a single DNA molecule made from two different sources, began in the mid-1970s. The development of this technology, which has led to the entire field of biotechnology, is based on enzymes that can be used to manipulate DNA.

Restriction enzymes cleave DNA at specific sites

The enzymes that catalyzed the molecular biology revolution were those able to cleave DNA at specific sites: these are called **restriction endonucleases.** As described in chapter 14, nucleases are enzymes that degrade DNA, and many were known prior to the isolation of the first restriction enzyme. But restriction endonucleases are different because they are able to fragment DNA at specific sites. If a DNA sequence were a rope, then restriction enzymes would be a knife that always cut that rope into specific lengths.

Discovery and significance of restriction endonucleases

This site-specific cleavage activity, long sought by molecular biologists, came out of basic research into why bacterial viruses can infect some cells but not others. This phenomenon was termed *host restriction.* The bacteria produce enzymes that can cleave the invading viral DNA at specific sequences. The host cells protect their own DNA from cleavage by modifying their DNA at the cleavage sites; the restriction enzymes do not cleave the modified DNA. Since the initial discovery of these restriction endonucleases, hundreds more have been isolated that recognize and cleave different **restriction sites.**

The ability to cut DNA at specific sites is significant in two ways: First, it allows a form of physical mapping that was previously impossible. Physical maps can be constructed based on the positioning of cleavage sites for restriction enzymes. These restriction maps provide crucial data for identifying and working with DNA molecules.

Second, restriction endonuclease cleavage allows the creation of recombinant molecules. The ability to construct recombinant molecules is critical to research, because many steps in the process of cloning and manipulating DNA require the ability to combine molecules from different sources.

How restriction enzymes work

There are two types of restriction enzymes: type I and type II. Type I enzymes make simple cuts across both DNA strands near, but not at, the recognition site. Because these do not cleave at precise locations, they are not often used in cloning and manipulating DNA.

Type II enzymes allow creation of recombinant molecules; these enzymes recognize a specific DNA sequence, ranging from 4 bases to 12 bases, and cleave the DNA at a specific base within this sequence (figure 17.1).

The recognition sites for type II enzymes are palindromes. A *palindrome* in language reads the same forward and in reverse, such as the sentence: "Madam I'm Adam." The palindromic DNA sequence reads the same from 5′ to 3′ on one strand as it does on the complementary strand (see figure 17.1).

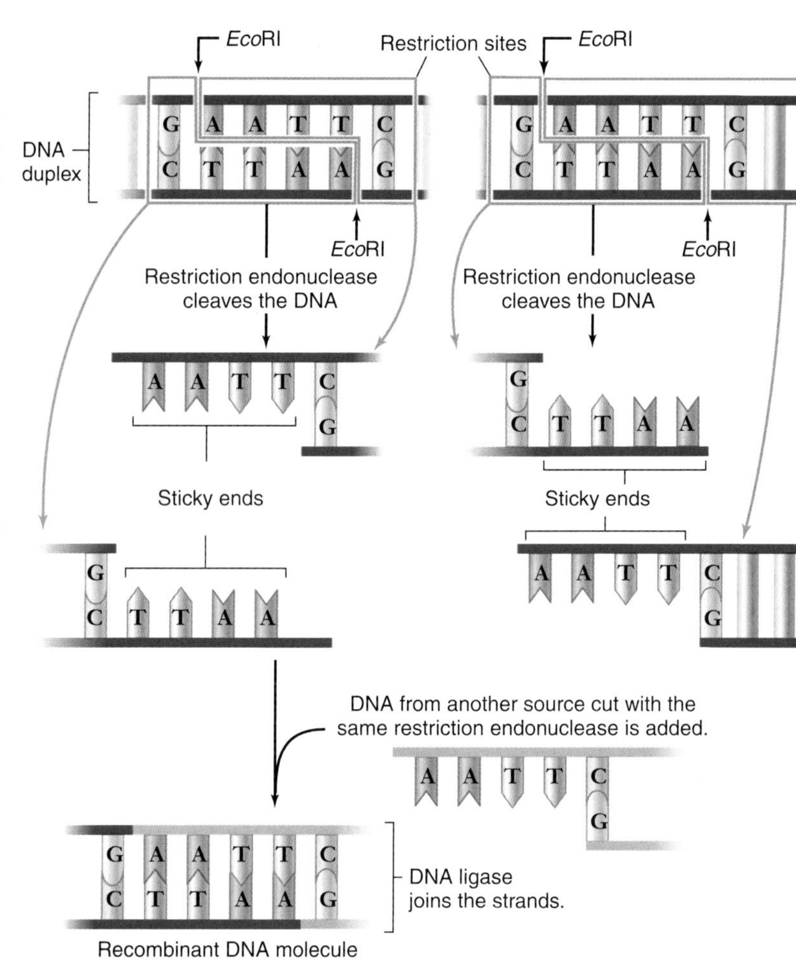

figure 17.1

MANY RESTRICTION ENDONUCLEASES PRODUCE DNA FRAGMENTS WITH "STICKY ENDS." The restriction endonuclease *Eco*RI always cleaves the sequence GAATTC between G and A. Because the same sequence occurs on both strands, both are cut. However, the two sequences run in opposite directions on the two strands. As a result, single-stranded tails called "sticky ends" are produced that are complementary to each other. These complementary ends can then be joined to a fragment from another DNA that is cut with the same enzyme. These two molecules can then be joined by DNA ligase to produce a recombinant molecule.

Given this kind of sequence, cutting the DNA at the same base on either strand can lead to staggered cuts that produce "sticky ends." These short, unpaired sequences will be the same for any DNA that is cut by this enzyme. Thus, these sticky ends allow DNAs from different sources to be easily joined together (see figure 17.1).

DNA ligase allows construction of recombinant molecules

As just described, the two ends of a DNA molecule cut by a type II restriction enzyme have complementary sequences, and so can pair to form a duplex. But to form a stable DNA molecule from the two fragments, an enzyme is needed to join the molecules. The enzyme **DNA ligase** catalyzes the formation of a phosphodiester bond between adjacent phosphate and hydroxyl groups of DNA nucleotides. The action of ligase is to seal nicks in one or both strands (see figure 17.1). This is the same enzyme that joins Okazaki fragments on the lagging strand during DNA replication (see chapter 14).

In the toolbox of the molecular biologist, the action of ligase is necessary to create stable recombinant molecules from the fragments that restriction enzymes make possible.

Gel electrophoresis separates DNA fragments

The fragments produced by restriction enzymes would not be much use if we could not also easily separate them for analysis. The most common separation technique used is gel electrophoresis. This technique takes advantage of the negative charge on DNA molecules by using an electrical field to provide the force necessary to separate DNA molecules based on size.

The gel, which is made of either agarose or polyacrylamide and spread thinly on supporting material, provides a three-dimensional matrix that separates molecules based on size. The gel is submerged in a buffer solution containing ions that can carry current and is subjected to an electrical field.

The strong negative charges from the phosphate groups in the backbone on DNA cause it to migrate toward the positive pole (figure 17.2*b*). The gel acts as a sieve to separate DNA molecules based on size: The larger the molecule, the slower it will move through the gel matrix. Over a given time period for electrophoresis, smaller molecules will migrate farther than larger molecules. The DNA in gels can be visualized using a fluorescent dye that binds to DNA (figure 17.2*c*, *d*).

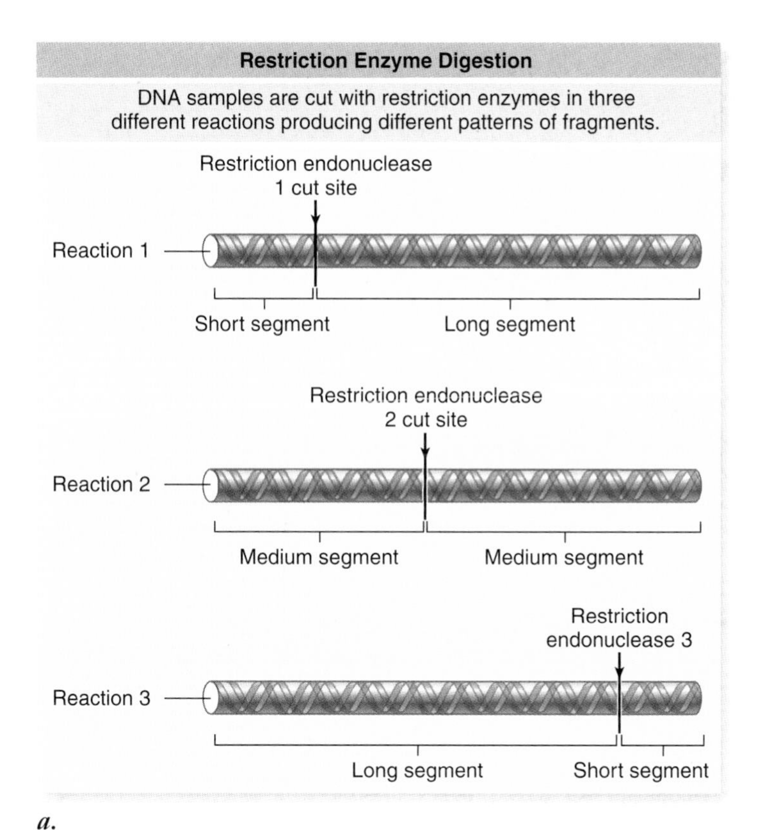

a.

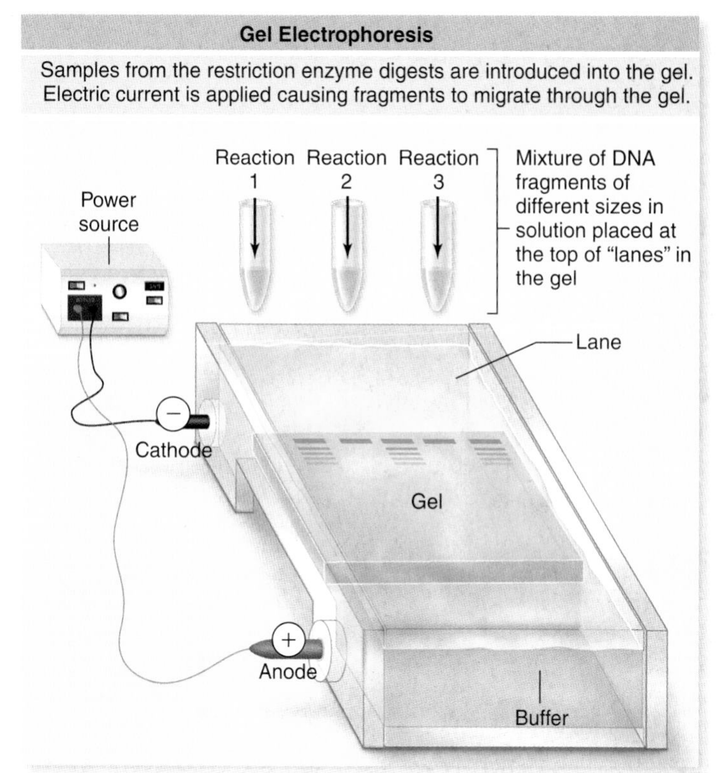

b.

Visualizing Stained Gel

Gel is stained with a dye to allow the fragments to be visualized.

Longer fragments

Shorter fragments

c.

Electrophoresis in the Laboratory

d.

figure 17.2

GEL ELECTROPHORESIS. ***a.*** Three restriction enzymes are used to fragment DNA into specific pieces depending on each enzyme's recognition sequence. ***b.*** The fragments are loaded into a gel (agarose or polyacrylamide), and an electrical current is applied. The DNA fragments migrate through the gel based on size, with larger ones moving more slowly. ***c.*** This results in a pattern of fragments separated based on size, with the smaller fragments migrating farther than larger ones. ***d.*** The fragments can be visualized by staining with the dye ethidium bromide. When the gel is exposed to UV light, the DNA with bound dye will fluoresce, appearing as pink bands in the gel. In the photograph, one band of DNA has been excised from the gel for further analysis and can be seen glowing in the tube the technician holds.

Electrophoresis is one of the most important methods in molecular biology, with uses ranging from DNA fingerprinting to DNA sequencing, both of which are described later on.

Transformation allows introduction of foreign DNA into *E. coli*

The construction of recombinant molecules is the first step toward genetic engineering. It is also necessary to be able to reintroduce these molecules into cells. In chapter 14 you learned that genetic material could be transferred between bacterial cells, as demonstrated by Frederick Griffith. This process, called *transformation*, is a natural process in the cells that Griffith was studying.

The bacterium *E. coli*, used routinely in molecular biology laboratories, does not undergo natural transformation; but artificial transformation techniques have been developed to allow introduction of foreign DNA into *E. coli*. In this way, recombinant molecules can be propagated in a cell that will make many copies of the constructed molecules.

In general, the introduction of DNA from an outside source into a cell is referred to as transformation, and it is important in *E. coli* for molecular cloning and the propagation of cloned DNA. Researchers also want to be able to reintroduce DNA into the original cells from which it was isolated. In this case, if the transformed cell can also be used to form all or part of an organism, we call this a **transgenic** organism. Later in this chapter we explore the construction and uses of transgenic plants and animals.

Techniques to manipulate DNA include use of restriction endonucleases to cleave DNA and DNA ligase to construct recombinant molecules. Gel electrophoresis is used to separate DNA fragments. Foreign DNA can be introduced into *E. coli* through the process of artificial transformation.

17.2 Molecular Cloning

The term **clone** refers to a genetically identical copy. The technique of propagating plants by growing a new plant from a cutting of a donor plant is an early method of cloning widely used in agriculture and horticulture. The topic of cloning entire organisms is discussed in chapter 19. For now, we explore the idea of molecular cloning.

Molecular cloning involves the isolation of a specific sequence of DNA, usually one that encodes a particular protein product. This is sometimes called *gene cloning*, but the term *molecular cloning* is more accurate.

Host–vector systems allow propagation of foreign DNA in bacteria

Although short sequences of DNA can be synthesized in vitro, the cloning of large unknown sequences requires propagation of recombinant DNA molecules in vivo (in a cell). The enzymes and methods described earlier allow biologists to produce, separate, and then introduce foreign DNA into cells.

The ability to propagate DNA in a host cell requires a **vector** (something to carry the recombinant DNA molecule) that can replicate in the host when it has been introduced. Such host–vector systems are crucial to molecular biology.

The most flexible and common host used for molecular cloning is the bacterium *E. coli*, but many other hosts are now possible. Investigators routinely reintroduce cloned eukaryotic DNA, using mammalian tissue culture cells, yeast cells, and insect cells as host systems. Each kind of host–vector system allows particular uses of the cloned DNA.

The two most commonly used vectors are plasmids and phages. *Plasmids* are small, circular extrachromosomal DNAs that are dispensable to the bacterial cell. *Phages* are viruses that infect bacterial cells.

Plasmid vectors

Plasmid vectors (small, circular chromosomes) are typically used to clone relatively small pieces of DNA, up to a maximum of about 10 kilobases (kb). A plasmid vector must have two components:

1. an *origin of replication* to allow it to be replicated in *E. coli* independently of the chromosome, and
2. a *selectable marker*, usually antibiotic resistance.

The selectable marker allows the presence of the plasmid to be easily identified through genetic selection. For example, cells that contain a plasmid with an antibiotic resistance gene continue to live when plated on antibiotic-containing growth media, whereas cells that lack the plasmid will die (they are killed by the antibiotic).

A fragment of DNA is inserted by the techniques described into a region of the plasmid called the **multiple cloning site (MCS).** This region contains a number of unique restriction sites such that when the plasmid is cut with restriction enzymes for these sites, the result is a linear plasmid. When DNA of interest is cut with the same restriction enzyme, it can then be ligated into this site. The plasmid is then introduced into cells by transformation (see figure 17.3*a*).

This region of the vector often has been engineered to contain another gene that becomes inactivated because it is now interrupted by the inserted DNA, so-called *insertional inactivation*. One of the first cloning vectors, pBR322, used another antibiotic resistance gene for insertional activation; resistance to one antibiotic and sensitivity to the other indicated the presence of inserted DNA.

More recent vectors use the gene for β-galactosidase, an enzyme that cleaves galactoside sugars such as lactose. When the enzyme cleaves the artificial substrate X-gal, a blue color is produced. In these plasmids, insertion of foreign DNA interrupts the β-galactosidase gene and a functional enzyme cannot be produced. When transformant cells are plated on medium containing both antibiotic (to select for plasmid-containing cells) and X-gal, they remain white, whereas transformants with no inserted DNA are blue (see figure 17.3*a*).

Phage vectors

Phage vectors are larger than plasmid vectors, and can take inserts up to 40 kb. Most phage vectors are based on the well-studied **phage lambda (λ).** Lambda-based vectors are used today primarily for *cDNA libraries*—collections of DNA fragments produced from mRNA (see below for details).

Although useful for cloning large fragments, the lambda vector has two requirements not shared with plasmid vectors:

- Both lambda and plasmids requires live cells for replication, but the virus will kill the cells such that you end up with a collection of viruses, not bacterial cells.
- The lambda genome is linear, so instead of "opening up" a circle, the middle part of the lambda genome is removed and replaced with the inserted DNA.

After the inserted DNA is ligated to the two lambda "arms," the lambda genome must be packaged into a phage head in vitro, and then used to infect *E. coli.* The two "arms" lacking any inserted DNA are not packaged efficiently into phage heads. This difference provides a kind of selection for recombinant phages, because the arms alone are not propagated (figure 17.3*b*).

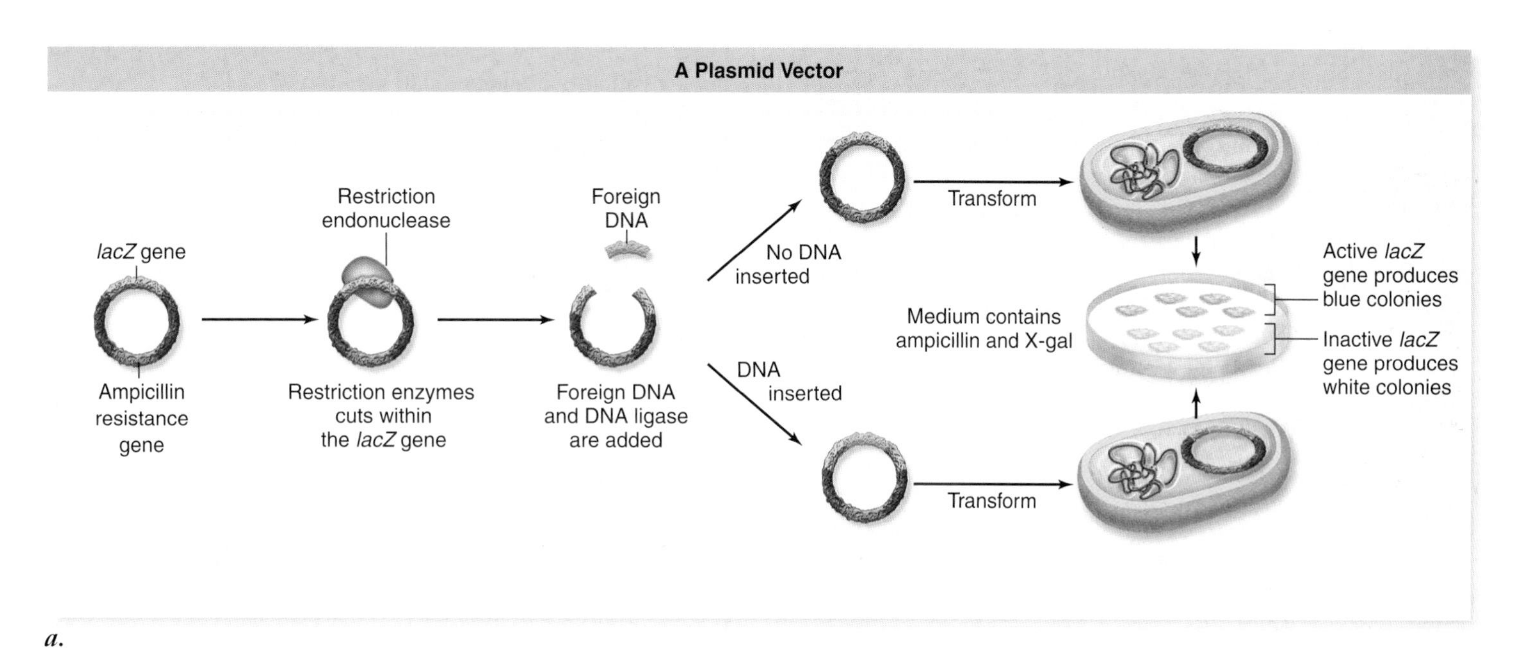

a.

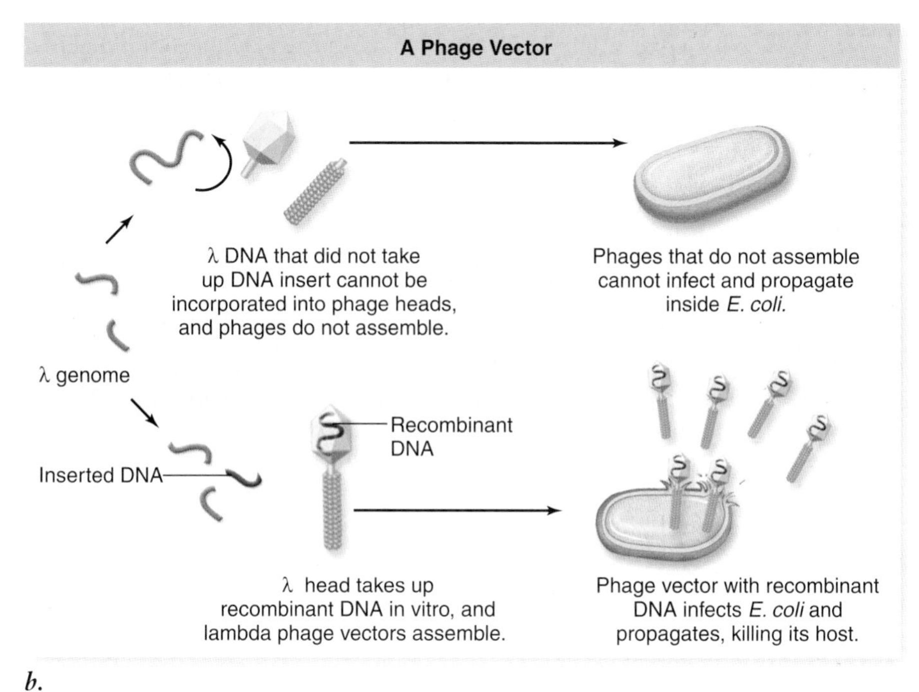

b.

figure 17.3

USING PLASMID AND PHAGE VECTORS. ***a.*** Plasmids are cut within the β-galactosidase gene (*lacZ*), and foreign DNA and DNA ligase are added. Foreign DNA inserted into *lacZ* interrupts the coding sequence inactivating the gene. Plating cells on medium containing the antibiotic ampicillin selects for plasmid-containing cells. The medium also contains X-gal and when *lacZ* is intact (*top*), the expressed enzyme cleaves the X-gal producing blue colonies. When *lacZ* is inactivated (*bottom*), X-gal is not cleaved and colonies remain white. ***b.*** Phage vectors are selected for the presence of recombinant DNA by the ability of the phage to assemble in vitro, infect a host, and propagate inside its host. The phage has been engineered such that only phage genomes with inserted DNA are long enough for the packaging machinery to produce mature phage that can infect cells.

Artificial Chromosomes

The size of DNAs that can be cloned in either plasmid or phage vectors has limited the large-scale analysis of genomes. To deal with this, geneticists decided to follow the strategy of cells and construct chromosomes, leading to the development of yeast artificial chromosomes (YACs) and bacterial artificial chromosomes (BACs). Progress has also been made on mammalian artificial chromosomes. Use of artificial chromosomes is described in the next chapter.

inquiry

An investigator wishes to clone a 32-kb recombinant molecule. What do you think is the best vector to use?

DNA libraries contain the entire genome of an organism

The idea of molecular cloning depends on the ability to construct a representation of very complex mixtures in DNA, such as an entire genome, in a form that is easier to work with than the enormous chromosomes within a cell. If the huge DNA molecules in chromosomes can be converted into random fragments, and inserted into a vector such as plasmids or phages, then when they are propagated in a host they will together represent the whole genome. This aggregate is termed a **DNA library,** a collection of DNAs in a vector that taken together represent the complex mixture of DNA (figure 17.4).

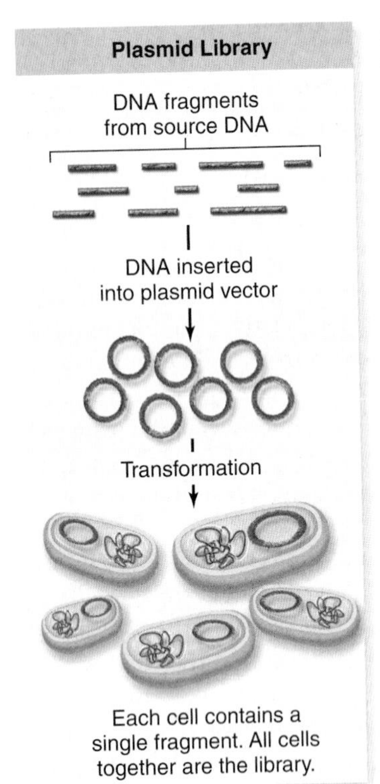

figure 17.4

CREATING DNA LIBRARIES. DNA libraries can be produced using *(a)* plasmid vectors or *(b)* phage vectors.

Conceptually the simplest kind of DNA library that can be made is a **genomic library,** a representation of the entire genome in a vector. This genome is randomly fragmented by partially digesting it with a restriction enzyme that cuts frequently. By not cutting the DNA to completion, not all sites are cleaved, and which sites are cleaved is random. The random fragments are then inserted into a vector and introduced into host cells.

Genomic libraries were originally made in phage λ because of the larger insertion sizes. Now with the advent of genomics and the analysis of entire genomes, these libraries are usually constructed in bacterial artificial chromosomes (BACs).

A variety of different kinds of libraries can be made depending on the source DNA used. Any particular clone in the library contains only a single DNA, and all of them together make up the library. Keep in mind that unlike a library full of books, which is organized and catalogued, a DNA library is a random collection of overlapping DNA fragments. We explore how to find a sequence of interest in this random collection later in the chapter.

Reverse transcriptase can make a DNA copy of RNA

In addition to genomic libraries, investigators often wish to isolate only the *expressed* part of genes. The structure of eukaryotic genes is such that the mRNA may be much smaller than the gene itself due to the presence of introns in the gene. After transcription by RNA polymerase II, the primary transcript is spliced to produce the mRNA (chapter 15). Because of this, genomic libraries are crucial to understanding the structure of the gene, but are not of much use if we want to express the gene in a bacterial species, whose genes do not contain introns and has no mechanism for splicing.

A library of only expressed sequences represents a much smaller amount of DNA than the entire genome, but it requires using mRNA as a starting point. Such a library of expressed sequences is made possible by the use of another enzyme: **reverse transcriptase.**

Reverse transcriptase was isolated from a class of viruses called retroviruses. The life cycle of a retrovirus requires making a DNA copy from its RNA genome. We can take advantage of the activity of the retrovirus enzyme to make DNA copies from isolated mRNA. DNA copies of mRNA are called **complementary DNA (cDNA)** (figure 17.5). A cDNA library is made by first isolating mRNA from genes being expressed and then using the reverse transcriptase enzyme to make cDNA from the mRNA. The cDNA is then used to make a library, as mentioned earlier. These cDNA libraries are extremely useful and are commonly made to represent the genes expressed in many different tissues or cells.

inquiry

Suppose you wanted a copy of a section of a eukaryotic genome that included the introns and exons. Would the creation of cDNA be a good way to go about this?

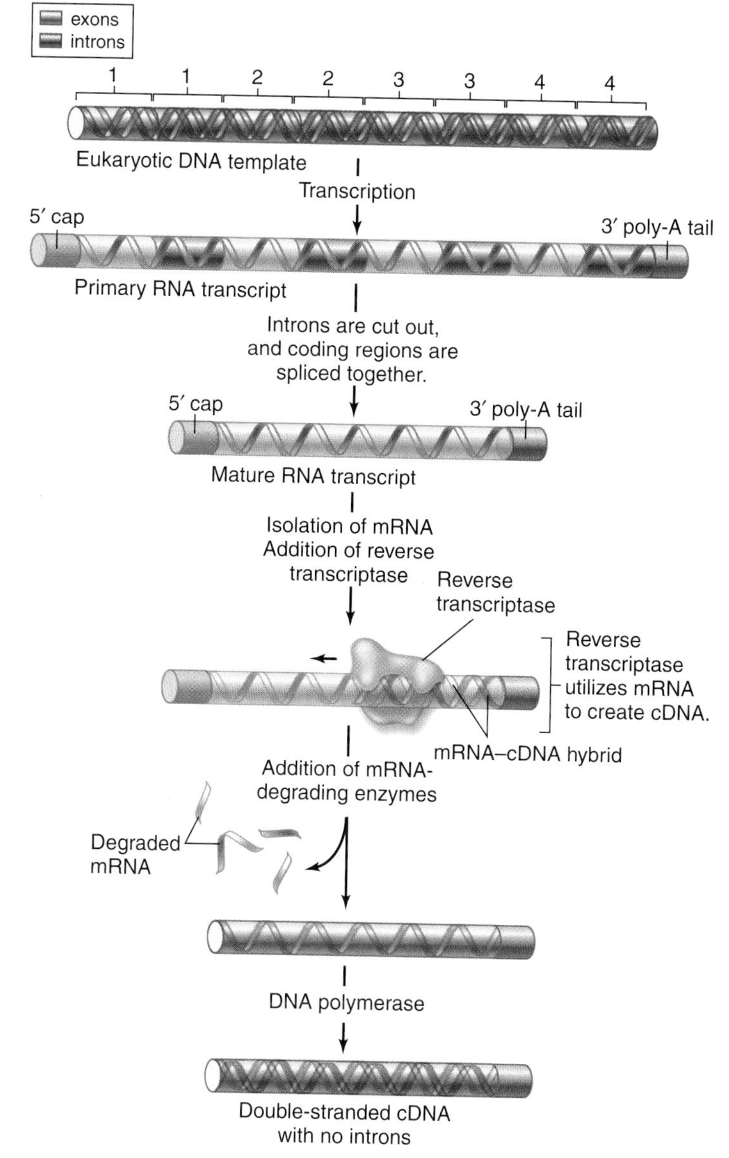

figure 17.5

THE FORMATION OF cDNA. A mature mRNA transcript is usually much smaller than the gene due to the loss of intron sequences by splicing. When mRNA is isolated from the cytoplasm of a cell the enzyme reverse transcriptase can use this as a template to make a DNA strand complementary to the mRNA. That newly made strand of DNA is the template for the enzyme DNA polymerase, which assembles a complementary DNA strand along it, producing cDNA, a double-stranded DNA version of the intron-free mRNA.

Hybridization allows identification of specific DNAs in complex mixtures

The technique of **molecular hybridization** is commonly used to identify specific DNAs in complex mixtures such as libraries. Hybridization, also called annealing, takes advantage of the specificity of base-pairing between the two strands of DNA. If a DNA molecule is denatured, that is, the two strands are separated, the strands can only reassociate with partners having the correct complementary sequence. Molecular biologists can take advantage of this feature experimentally to use a known, specific DNA molecule to find its partner in a complex mixture.

Any single-stranded nucleic acid (DNA or RNA) can be labeled with radioactivity or another detectable label, such as a fluorescent dye. This can then be used as a probe to identify its complement in a complex mixture of DNA or RNA. This renaturing is termed *hybridization* because the combination of labeled probe and unlabeled DNA form a hybrid molecule through base-pairing.

Probes have been made historically by a variety of techniques. One technique involved isolating a protein of interest and then chemically sequencing the protein. With the protein sequence in hand, the DNA sequence could be predicted using the genetic code. This information could then be used to make a synthetic DNA for use as a probe.

Specific clones can be isolated from a library

The isolation of a specific clone from the random collection that is a DNA library is akin to finding the proverbial needle in a haystack. It requires some information about the gene of interest. For example, many of the first genes isolated were those that are highly expressed in a specific cell type, such as the globin genes that encode the proteins found in the oxygen carrier hemoglobin.

Hybridization is the most common way of identifying a clone within a DNA library. This procedure is outlined for a DNA library in a plasmid vector in figure 17.6.

In the early days of molecular biology, individual investigators made their own DNA libraries, as is shown earlier in figure 17.4. Now, genomic and cDNA libraries are commercially available for a large number of organisms. Screening such a library involves growing the library on agar plates, making a replica of the library, and screening for the cloned sequence of interest.

Stage 1: Plating the library

Physically, the library is either a collection of bacterial viruses that each contain an inserted DNA, or bacterial cells that each harbor a plasmid or artificial chromosome with inserted DNA. To find a specific clone, the library needs to be represented in an organized fashion. Figure 17.6 shows this representation for a plasmid vector. The library of bacteria containing plasmids is grown on agar plates at a high density, but not so high that individual colonies cannot be distinguished.

Stage 2: Replicating the library

Once the library has been grown on plates, a replica can be made by laying a piece of filter paper on the plate; some of the viruses or cells in each colony will stick to the filter, and some will be left on the plate. The result is a copy of the library on a piece of filter paper. The DNA can be affixed to the filter paper by baking or by cross-linking it to the filter using UV light.

Stage 3: Screening the library

Once a replica of the library has been formed on a filter, a specific clone can be identified by hybridization. The probe, which represents the specific sequence of interest, is labeled with a

figure 17.6

SCREENING A LIBRARY USING HYBRIDIZATION. This takes advantage of the ability of DNA to be denatured and renatured, with complementary strands finding each other. Cells containing the library are plated on agar plates. A replica of the plates is made using special filter paper, nitrocellulose or nylon, that binds to single-stranded DNA. The filter paper with replica colonies is treated to lyse the cells and denature the DNA such that there will now be a pattern of DNA bound to the filter corresponding to the pattern of colonies. When a radioactive probe is added, it will find complementary DNA and form hybrids at the site of colonies that contained the gene of interest.

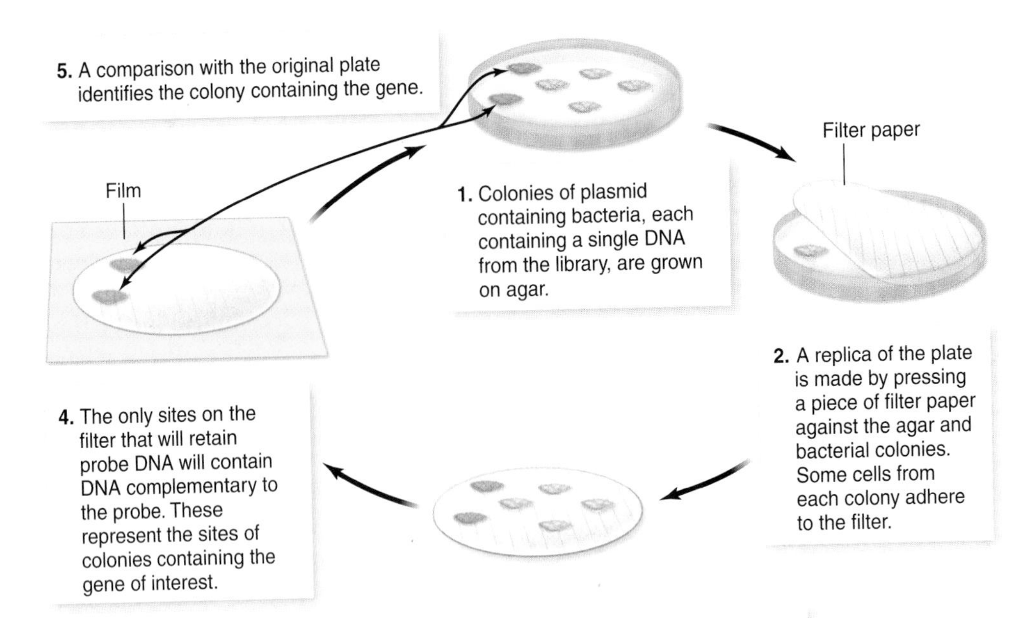

radioactive nucleotide. The probe is then added to the filters with the library replicated on them. Film sensitive to radioactive emissions is then placed in contact with the filters; where radioactivity is present, a dark spot appears on the film. When the film is aligned with the original plate, the clone of interest can be identified (see figure 17.6).

Molecular cloning is the isolation of a specific DNA sequence. Host–vector systems allow us to propagate DNA in *E. coli* and other organisms. DNA libraries are representations of complex mixtures of DNA, such as an entire genome, in a host–vector system. DNA libraries are often screened for specific clones using molecular hybridization, which uses a labeled probe to find DNA complementary to the probe.

17.3 DNA Analysis

Molecular cloning provides specific DNA for further manipulation and analysis. The number of ways that DNA can be manipulated could fill the rest of this book, so for our purposes, we highlight a few important methods of analysis and uses of molecular clones.

Restriction maps provide molecular "landmarks"

If you are new to a city, the easiest way to find your way around is to obtain a map and compare that map with your surroundings. In a similar fashion, molecular biologists need maps to analyze and compare cloned DNAs.

The first kind of physical maps were restriction maps, composed of the location and order of sites cut by the battery of restriction enzymes available. Initially, these maps were created by cutting the DNA with different enzymes, separating the fragments by gel electrophoresis, and analyzing the resulting patterns. Although this method is still in use, many restriction maps are now generated by computer searching of known DNA sequences for the sites cut by restriction enzymes.

Southern blotting reveals DNA differences

Once a gene has been cloned, it may be used as a probe to identify the same or a similar gene in DNA isolated from a cell or tissue (figure 17.7). In this procedure, called a **Southern blot,** DNA from the sample is cleaved into fragments with a restriction endonuclease, and the fragments are separated by gel electrophoresis. The double-stranded helix of each DNA fragment is then denatured into single strands by making the pH of the gel basic, and the gel is "blotted" with a sheet of filter paper, transferring some of the DNA strands to the sheet.

Next, the filter is incubated with a labeled probe consisting of purified, single-stranded DNA corresponding to a specific gene (or mRNA transcribed from that gene). Any fragment that has a nucleotide sequence complementary to the probe's sequence hybridizes with the probe (see figure 17.7).

This kind of blotting technique has also been adapted for use with RNA and proteins. When mRNA is separated by electrophoresis, the technique is called a **Northern blot,** and the methodology is the same except for the starting material (mRNA instead of DNA) and that no denaturation step

1. Electrophoresis is performed, using radioactively labeled markers as a size guide in the first lane.

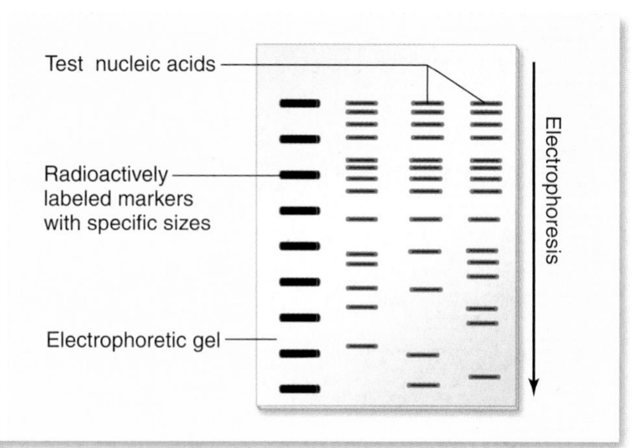

2. The gel is covered with a sheet of nitrocellulose and placed in a tray of buffer on top of a sponge. Alkaline chemicals in the buffer denature the DNA into single strands. The buffer wicks its way up through the gel and nitrocellulose into a stack of paper towels placed on top of the nitrocellulose.

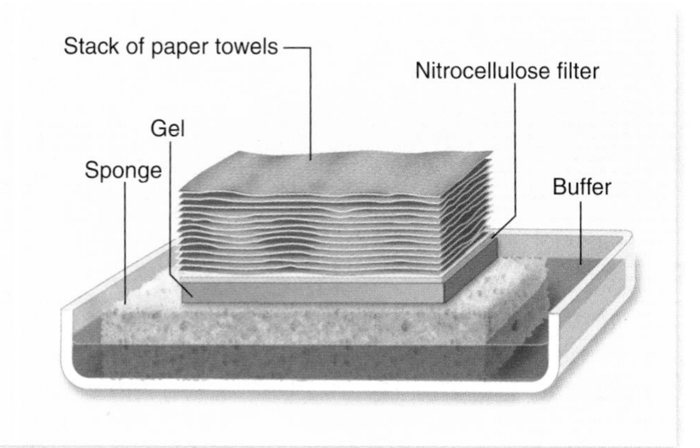

3. DNA in the gel is transferred, or "blotted," onto the nitrocellulose.

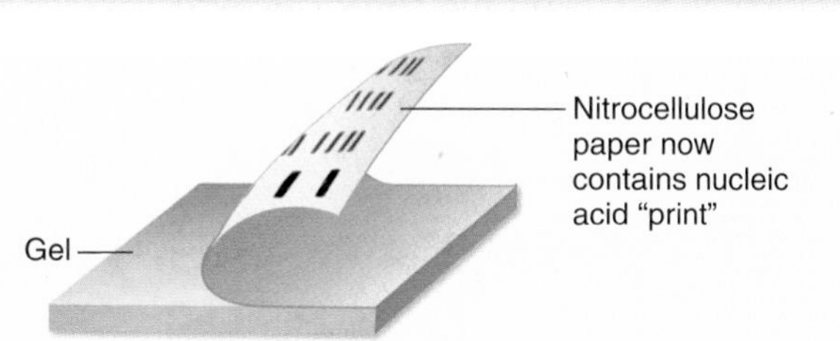

4. Nitrocellulose with bound DNA is incubated with radioactively labeled nucleic acids and is then rinsed.

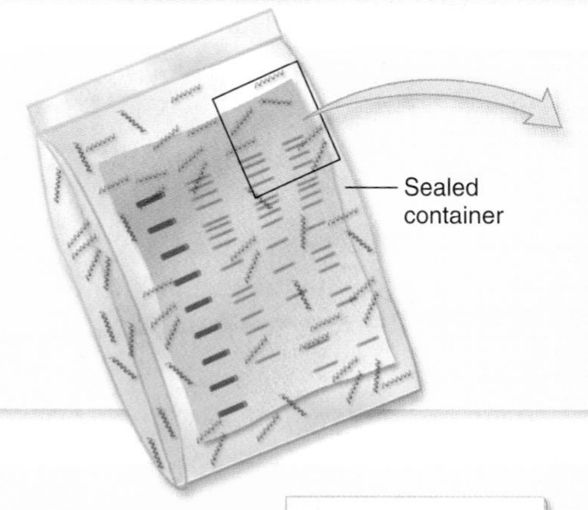

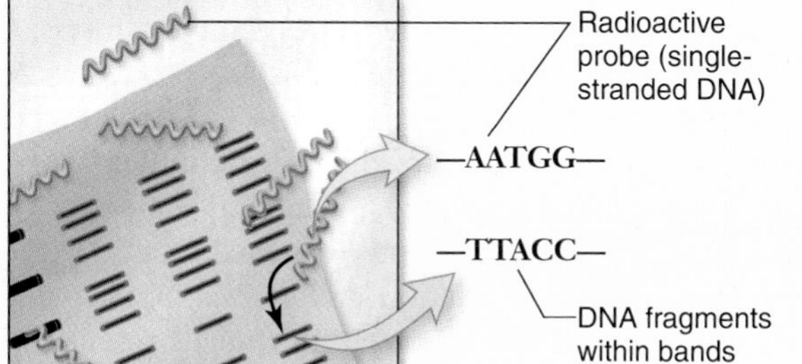

5. Photographic film is laid over the filter and is exposed only in areas that contain radioactivity (autoradiography). Bands on the film represent DNA in the gel that is complementary to the probe sequence.

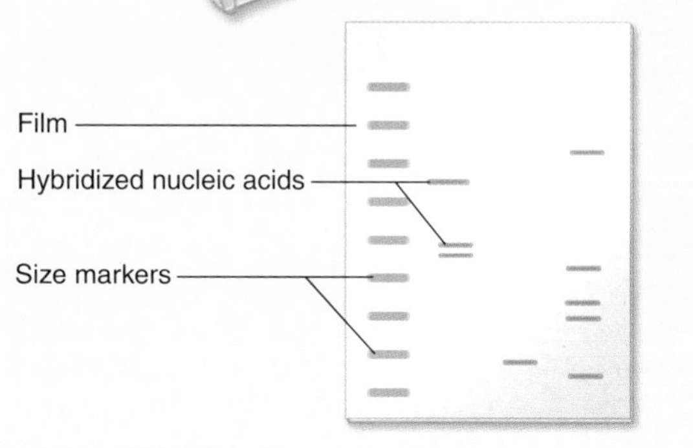

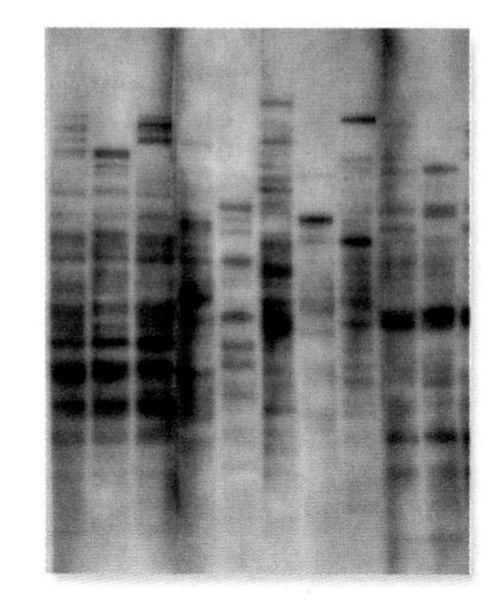

figure 17.7

THE SOUTHERN BLOT PROCEDURE. E. M. Southern developed this procedure in 1975 to enable DNA fragments of interest to be visualized in a complex sample containing many other fragments of similar size. In steps 1–3, the DNA is separated on a gel, and then transferred ("blotted") onto a solid support medium such as nitrocellulose paper or a nylon membrane. Sequences of interest can be detected by using a radioactively labeled probe. This probe (usually several hundred nucleotides in length) of single-stranded DNA (or an mRNA complementary to the gene of interest) is incubated with the filter containing the DNA fragments. All DNA fragments that contain nucleotide sequences complementary to the probe will form hybrids with the probe. Only a short segment of the probe and the complementary sequence are shown in panel 4. The fragments differ in size, with the smallest running the farthest in the gel. The fragments of interest are then detected using photographic film. A representative image is shown in panel 5. The use of film for detection is being replaced by phosphor imagers, computer-controlled devices that have electronic sensors for light or radioactive emissions.

figure 17.8

RESTRICTION FRAGMENT LENGTH POLYMORPHISM (RFLP) ANALYSIS. ***a.*** Three samples of DNA differ in their restriction sites due to a single base-pair substitution in one case and a sequence duplication in another case. ***b.*** When the samples are cut with a restriction endonuclease, different numbers and sizes of fragments are produced. ***c.*** Gel electrophoresis separates the fragments, and different banding patterns result.

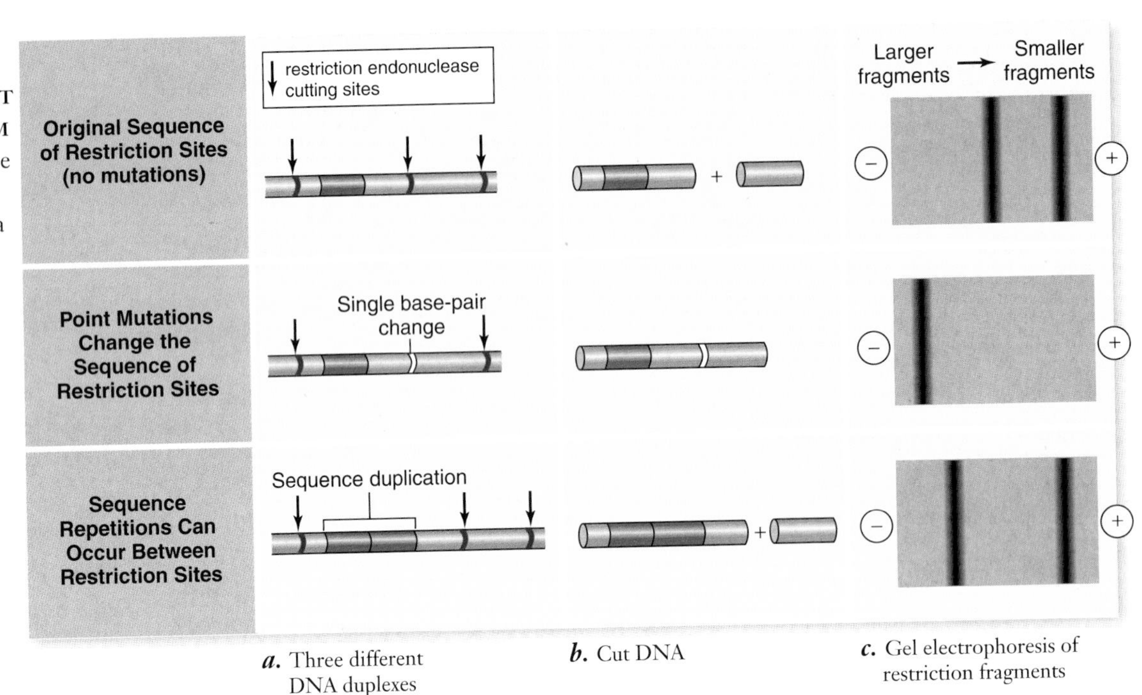

is required. Proteins can also be separated by electrophoresis and blotted by a procedure called a **Western blot.** In this case both the electrophoresis and the detection step are different from Southern blotting. The detection, in this case, requires an antibody that can bind to one protein.

The names of these techniques all go back to the original investigator, whose last name was Southern; the Northern and Western blotting names were word play on Southern's name using the cardinal points of the compass.

RFLP analysis

In some cases, an investigator wants to do more than find a specific gene, but instead is looking for variation in the genes of different individuals. One powerful way to do this is by analyzing **restriction fragment length polymorphisms,** or **RFLPs,** using Southern blotting (figure 17.8).

Point mutations that change the sequence of DNA can eliminate sequences recognized by restriction enzymes or create new recognition sequences, changing the pattern of fragments seen in a Southern blot. Sequence repetitions may also occur between the restriction endonuclease sites, and differences in repeat number between individuals can also alter the length of the DNA fragments. These differences can all be detected with Southern blotting.

When a genetic disease has an associated RFLP, the RFLP can be used to diagnose the disease. Huntington disease, cystic fibrosis, and sickle cell anemia all have associated RFLPs that have been used as molecular markers for diagnosis.

DNA fingerprinting

This technique has been used in **DNA fingerprinting.** When a probe is made for DNA that is repetitive, it often detects a large number of fragments. These fragments are often not identical in different individuals. We say that the population is **polymorphic** for these molecular markers. These markers can be used as DNA "fingerprints" in criminal investigations and other identification applications.

Figure 17.9 shows the DNA fingerprints a prosecuting attorney presented in a rape trial in 1987. They consist of autoradiographs, parallel bars on X-ray film. These bars can be thought of as being similar to the product price codes on consumer goods in that they may provide unique identifica-

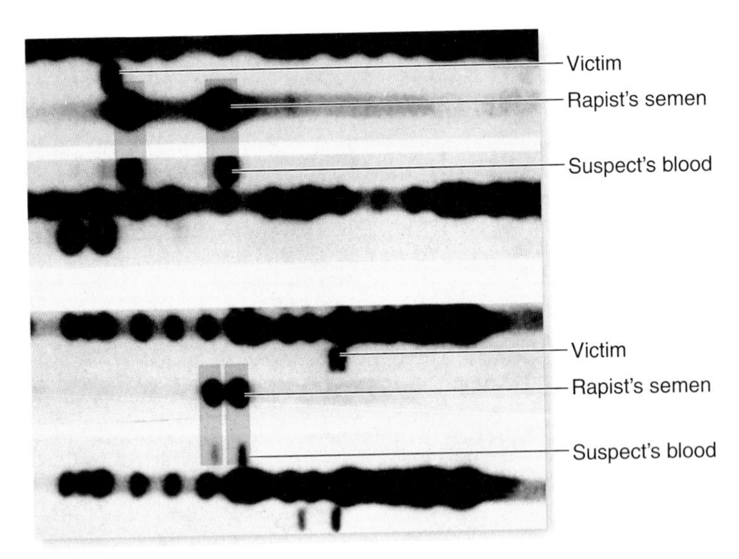

figure 17.9

TWO OF THE DNA PROFILES THAT LED TO THE CONVICTION OF TOMMIE LEE ANDREWS FOR RAPE IN 1987. The two DNA probes seen here were used to characterize DNA isolated from the victim, the semen left by the rapist, and the suspect. The dark channels are multiband controls. There is a clear match between the suspect's DNA and the DNA of the rapist's semen in these two profiles.

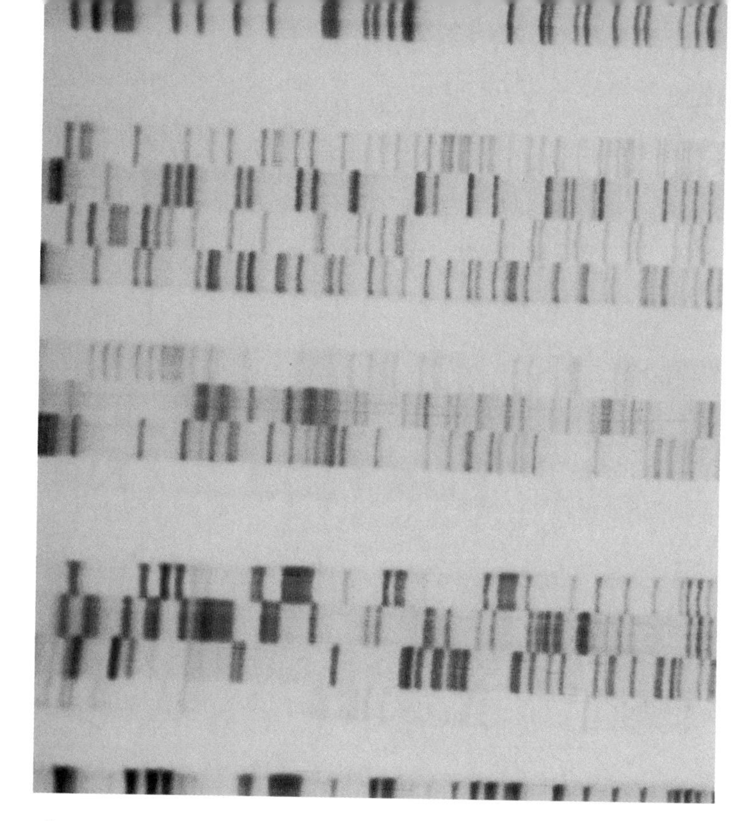

figure 17.10

LADDER OF FRAGMENTS USED IN DNA SEQUENCING. The photo shows the autoradiograph of the fragments generated by DNA-sequencing reactions. These fragments are generated by either organic reactions that cleave at specific bases or enzymatic reactions that terminate in specific bases. The gel can separate fragments that differ by a single base.

tion. Each bar represents the position of a DNA restriction endonuclease fragment produced by techniques similar to those described in figures 17.7 and 17.8. The lane with many bars represents a standardized control.

Two different probes were used to identify the restriction fragments. A vaginal swab had been taken from the victim within hours of her attack; from it, semen was collected and its DNA analyzed for restriction endonuclease patterns.

Compare the restriction endonuclease patterns of the semen to that of blood from the suspect. You can see that the suspect's two patterns match that of the rapist (and are not at all like those of the victim). The suspect was Tommie Lee Andrews, and on November 6, 1987, the jury returned a verdict of guilty. Andrews became the first person in the United States to be convicted of a crime based on DNA evidence.

Since the Andrews verdict, DNA fingerprinting has been admitted as evidence in more than 2000 court cases. Although some probes highlight profiles shared by many people, others are quite rare. Using several probes, the probability of identity can be calculated or identity can be ruled out.

Laboratory analyses of DNA samples, however, must be carried out properly—sloppy procedures could lead to a wrongful conviction. After widely publicized instances of questionable lab procedures, national standards are being developed.

DNA sequencing provides information about genes and genomes

The ultimate level of analysis is determination of the actual sequence of bases in a DNA molecule. The development of sequencing technology has paralleled the advancement of molecular biology. The field of genomics was born out of the ability to determine the sequence of an entire genome relatively rapidly.

The basic idea used in DNA sequencing is to generate a set of nested fragments that each begin with the same sequence and end in a specific base. When this set of fragments is separated by high-resolution gel electrophoresis, the result is a "ladder" of fragments (figure 17.10) in which each band consists of fragments that end in a specific base. By starting with the shortest fragment, one can then read the sequence by moving up the ladder.

The problem then became how to generate the sets of fragments that end in specific bases. In the early days of sequencing, both a chemical method and an enzymatic method were utilized. The chemical method involved organic reactions specific for the different bases that made breaks in the DNA chains at specific bases. The enzymatic method used DNA polymerase to synthesize chains, but it also included in the reaction modified nucleotides that could be incorporated but not extended: so-called *chain terminators.* The enzymatic method has proved more versatile, and it is easier to adapt to different uses.

Enzymatic sequencing

The enzymatic method of sequencing was developed by Fredrick Sanger, who also was the first to determine the complete sequence of a protein. This method uses dideoxynucleotides as chain terminators in DNA synthesis reactions. A **dideoxynucleotide** has H in place of OH at both the 2′ position and at the 3′ position.

All DNA nucleotides lack —OH at the 2′ carbon of the sugar, but dideoxynucleotides have no 3′ —OH at which the enzyme can add new nucleotides. Thus the chain is terminated.

The experimenter must perform four separate reactions, each with a single dideoxynucleotide, to generate a set of fragments that terminate in specific bases. Thus all of the fragments produced in the A reaction incorporate dideoxyadenosine and must end in A, and the same for the other three reactions with different terminators. When these fragments are separated by high-resolution gel electrophoresis, each reaction is run in a different track, or lane, to generate a pattern of nested fragments that can be read from the smallest fragment to fragments that are each longer by one base (figure 17.11*a*).

Notice that since this is a DNA polymerase reaction, it requires a primer to begin synthesis. The vectors used for DNA sequencing have known regions next to the site where DNA is inserted. Short DNAs are then synthesized that are complementary to these regions that can be used as primers. This serves the dual purposes of providing a primer and ensuring that the first few bases sequenced are known because they are known in the vector itself. This allows the investigator to determine where the sequence of interest begins. As the se-

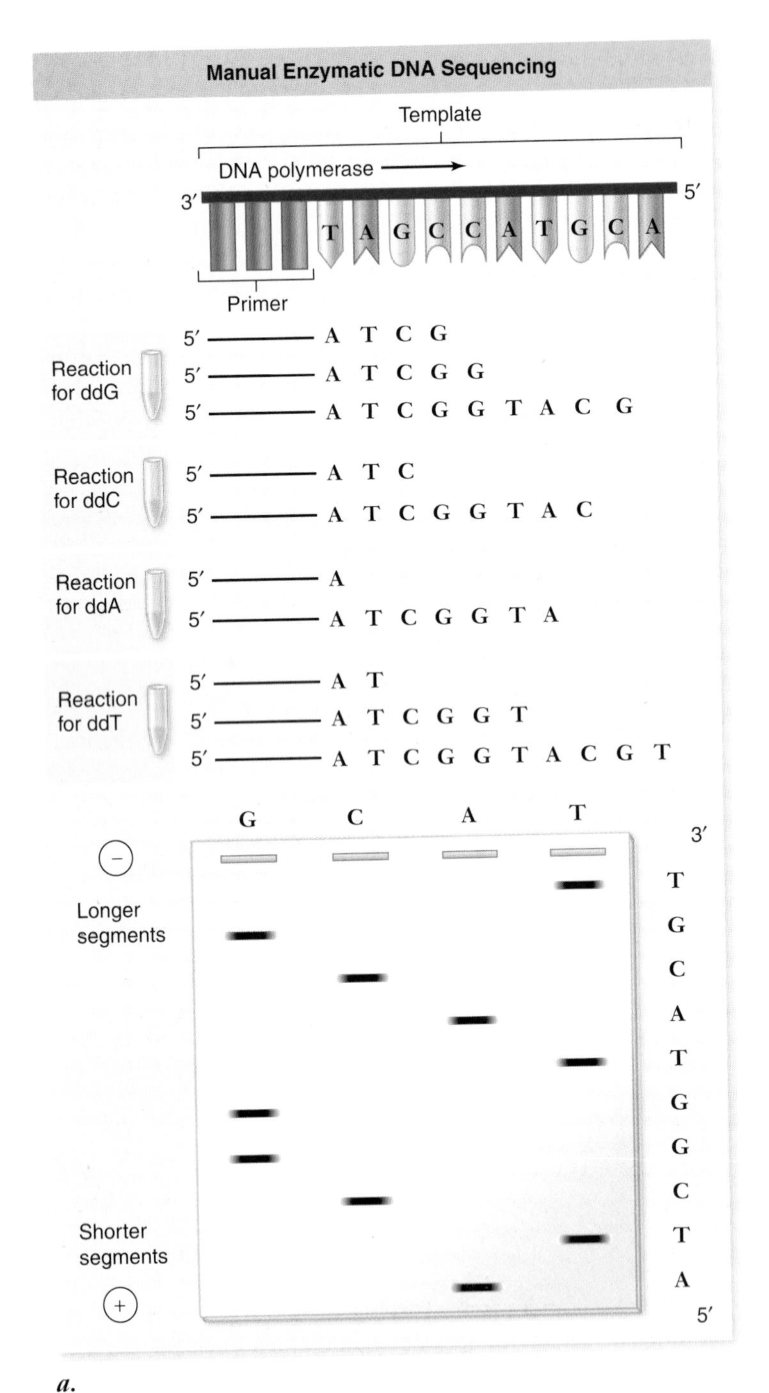

a.

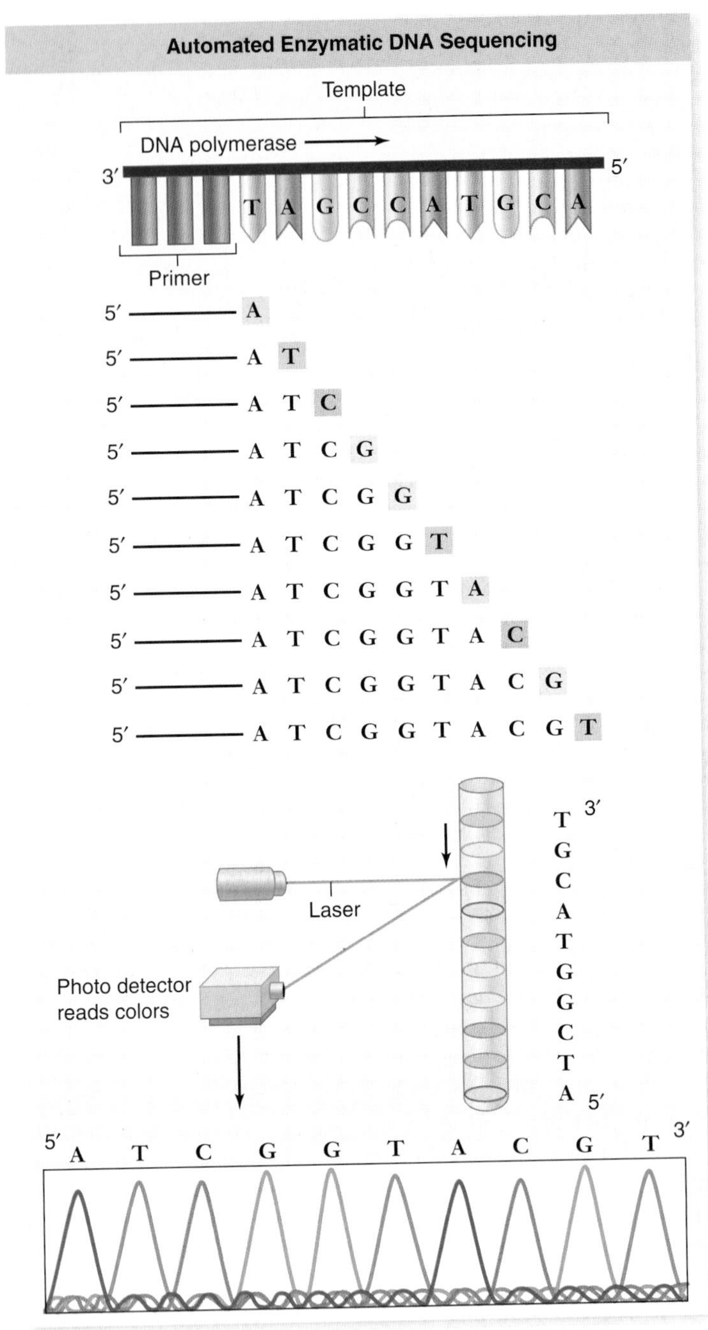

b.

figure 17.11

MANUAL AND AUTOMATED ENZYMATIC DNA SEQUENCING. The sequence to be determined is shown at the top as a template strand for DNA polymerase with a primer attached. ***a.*** In the manual method, four reactions were done, one for each nucleotide. For example, the A tube would contain dATP, dGTP, dCTP, dTTP, and ddATP. This leads to fragments that end in A due to the dideoxy terminator. The fragments generated in each reaction are shown along with the results of gel electrophoresis. ***b.*** In automated sequencing, each ddNTP is labeled with a different color fluorescent dye, which allows the reaction to be done in a single tube. The fragments generated by the reactions are shown. When these are electrophoresed in a capillary tube, a laser at the bottom of the tube excites the dyes, and each will emit a different color that is detected by a photodetector.

quence is generated, new primers can be designed near the end of the known sequence and DNA synthesized to use as a primer to extend the region sequenced in the next set of reactions.

Automated sequencing

The technique of enzymatic sequencing is very powerful, but it is also labor-intensive and takes a significant amount of time. It requires a series of enzymatic manipulations, time for electrophoresis, then time to expose the gel to film. At the end of this, a skilled researcher can read around 300 bases of sequence reliably. The development of automated techniques made sequencing a much more practical and less human-intensive procedure.

Automated sequencing machines use fluorescent dyes instead of a radioactive label and separate the products of the sequencing reactions using gels in thin capillary tubes instead of the large slab gels. The tubes run in front of a laser that excites the dyes, causing them to fluoresce. With a different colored dye for each base, a photodetecter can determine the identity of each base by its color.

The data are assembled by a computer that generates a visual image consisting of different colored peaks; these are converted into the raw sequence data (figure 17.11*b*). The sequence data come directly from the electrophoresis, eliminating the time needed for exposing gels to film and for manual reading of the sequences. The use of different colored dyes also reduces handling and allows more sequence to be produced at one time.

With increases in the number of samples per run and the length of sequences able to be read, along with decreases in handling time, the amount of sequence information that can be generated is limited mainly by the number of machines that can be run at once.

The polymerase chain reaction accelerated the process of analysis

The next revolution in molecular biology was the development of the **polymerase chain reaction (PCR).** Kary Mullis developed PCR in 1983 while he was a staff chemist at the Cetus Corporation; in 1993, he was awarded the Nobel Prize in chemistry for his discovery.

The idea of the polymerase chain reaction is simple: Two primers are used that are complementary to the opposite strands of a DNA sequence, oriented toward each other. When DNA polymerase acts on these primers and the sequence of interest, the primers produce complementary strands, each containing the other primer. If this procedure is done cyclically, the result is a large quantity of a sequence corresponding to the DNA that lies between the two primers (figure 17.12).

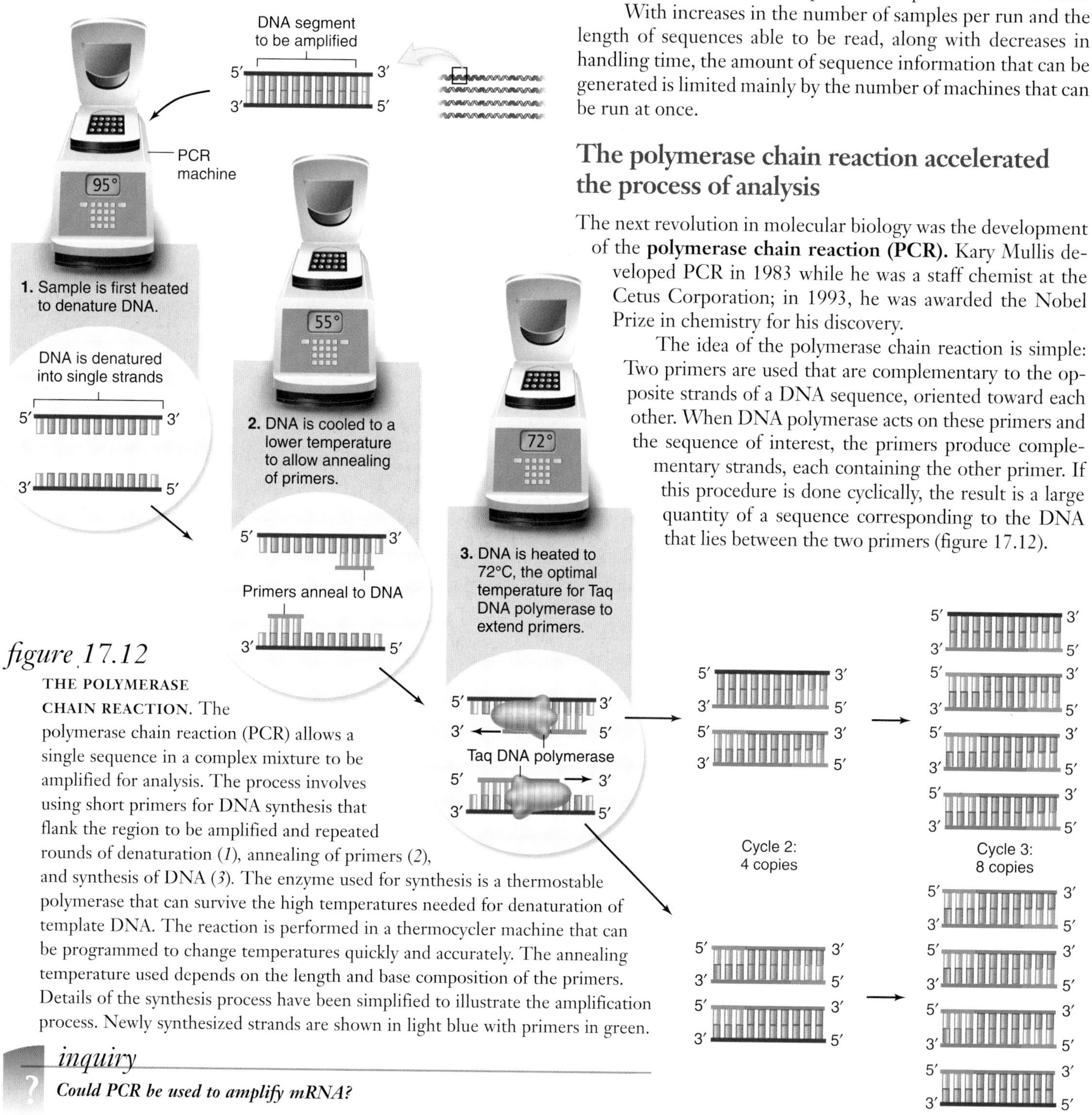

figure 17.12

THE POLYMERASE CHAIN REACTION. The polymerase chain reaction (PCR) allows a single sequence in a complex mixture to be amplified for analysis. The process involves using short primers for DNA synthesis that flank the region to be amplified and repeated rounds of denaturation (*1*), annealing of primers (*2*), and synthesis of DNA (*3*). The enzyme used for synthesis is a thermostable polymerase that can survive the high temperatures needed for denaturation of template DNA. The reaction is performed in a thermocycler machine that can be programmed to change temperatures quickly and accurately. The annealing temperature used depends on the length and base composition of the primers. Details of the synthesis process have been simplified to illustrate the amplification process. Newly synthesized strands are shown in light blue with primers in green.

inquiry

Could PCR be used to amplify mRNA?

The PCR procedure

Two developments turned this simple concept into a powerful technique. First, each cycle requires denaturing the DNA after each round of synthesis, which is easily done by raising the temperature; however, this destroys most polymerase enzymes. The solution was to isolate a DNA polymerase from a thermophilic, or heat-loving bacteria, *Thermus aquaticus*. This enzyme, called **Taq polymerase,** allows the reaction mixture to be repeatedly heated without destroying enzyme activity.

The second innovation was the development of machines with heating blocks that can be rapidly cycled over large temperature ranges with very accurate temperature control.

Thus each cycle of PCR involves three steps:

1. Denaturation (high temperature)
2. Annealing of primers (low temperature)
3. Synthesis (intermediate temperature)

Steps 1 to 3 are now repeated, and the two copies become four. It is not necessary to add any more polymerase, because the heating step does not harm Taq polymerase. Each complete cycle, which takes only 1–2 min, doubles the number of DNA molecules. After 20 cycles, a single fragment produces more than one million (2^{20}) copies!

In this way, the process of PCR allows the **amplification** of a single DNA fragment from a small amount of a complex mixture of DNA. This result is similar to what is isolated using molecular cloning, but in the case of PCR, the DNA cannot be reintroduced directly into a cell. The PCR product can be analyzed using electrophoresis, cloned into a vector for other manipulations, or directly sequenced. There are limitations on the size of the fragment that can be synthesized in this way, but it has been adapted for an amazing number of uses.

Applications of PCR

PCR, now fully automated, has revolutionized many aspects of science and medicine because it allows the investigation of minute samples of DNA. In criminal investigations, DNA fingerprints can now be prepared from the cells in a tiny speck of dried blood or at the base of a single human hair. In medicine, physicians can detect genetic defects in very early embryos by collecting a single cell and amplifying its DNA. Due to its sensitivity, speed, and ease of use, technicians now routinely use PCR methods for these applications.

PCR has even been used to analyze mitochondrial DNA from the early human species *Homo neanderthalensis*. This application provides the first glimpse of data from extinct related species. The amplification of ancient DNA has been a controversial field because contamination with modern DNA is difficult to avoid. But it remains an active area of genetic research.

Protein interactions can be detected with the two-hybrid system

Protein–protein interactions form the basis of many biological structures. Just as human society is ultimately dependent on interactions between people, cells are dependent on interactions between proteins. This observation has led to the large-scale goal of determining all interactions among proteins in different cells. This goal once would have been a dream, but it is now becoming a reality. The yeast two-hybrid system is one of the workhorses of this kind of analysis (figure 17.13).

The yeast two-hybrid system integrates much of the technology discussed in this chapter. It takes advantage of one feature of eukaryotic gene regulation, namely that the structure of proteins that turn on eukaryotic gene expression, transcription factors, have a modular structure.

The Gal4 gene of yeast encodes a transcriptional activator with modular structure consisting of a DNA-binding domain that binds sequences in Gal4-responsive promoters, and an activation domain that interacts with the transcription apparatus to turn on transcription. The system uses two vectors: one containing a fragment of the Gal4 gene that

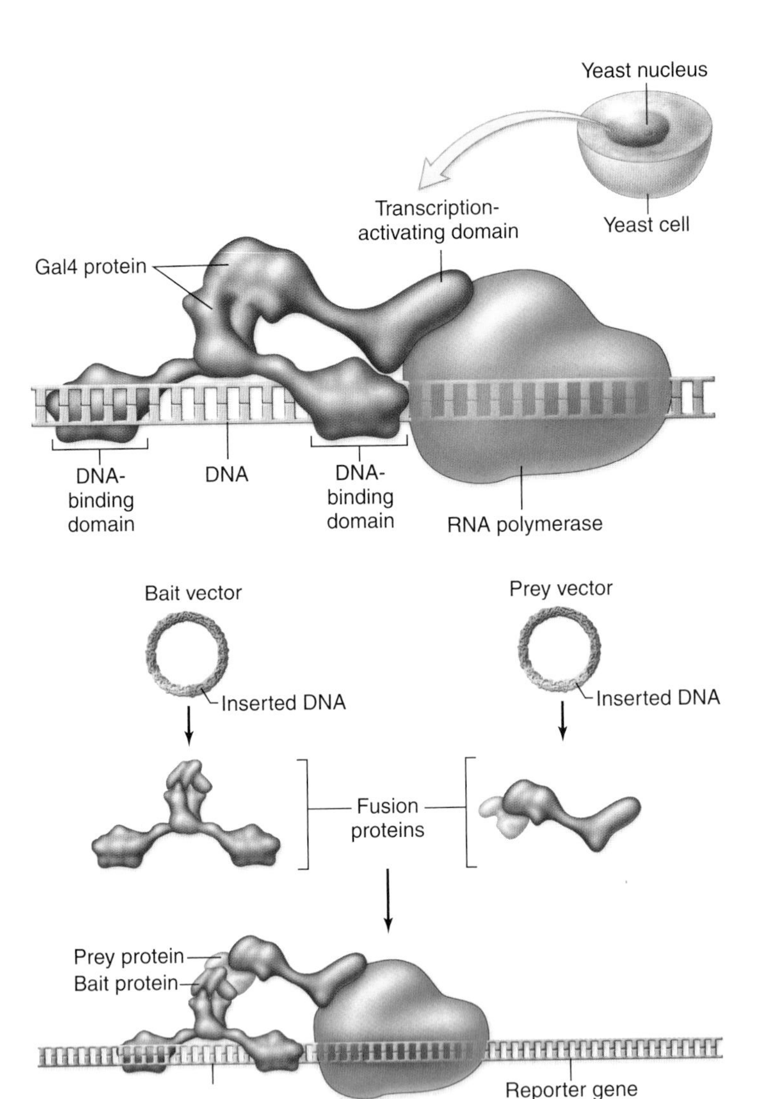

figure 17.13

THE YEAST TWO-HYBRID SYSTEM DETECTS INTERACTING PROTEINS. The Gal4 protein is a transcriptional activator (*top*). The Gal4 gene has been split and engineered into two different vectors such that one will encode only the DNA-binding domain (bait vector) and the other the transcription-activating domain (prey vector). When other genes are spliced into these vectors, they produce fusion proteins containing part of Gal4 and the proteins to be tested. If the proteins being tested interact, this will restore Gal4 function and activate expression of a reporter gene.

encodes the DNA binding domain, and another containing a fragment of the Gal4 gene that encodes the transcription activation domain. Neither of these alone can activate transcription.

When cDNAs are inserted into each of these two vectors in the proper reading frame, they are expressed as a single protein consisting of the protein of interest and part of the Gal4 activator protein (figure 17.13). These hybrid proteins are called *fusion proteins* since they are literally fused in the same polypeptide chain. The DNA-binding hybrid is called the *bait*, and the activating domain hybrid is called the *prey*.

These vectors are inserted into cells of different mating types that can be crossed. One of these vectors also contains a so-called *reporter gene* encoding a protein that can be assayed for enzymatic activity. The reporter gene is under control of a Gal4-responsive regulatory region, so that when active Gal4 is present, the reporter gene is expressed and can be detected by an enzymatic assay.

The DNA-binding hybrid binds to DNA adjacent to the reporter gene. When the two proteins in bait and prey interact, the prey hybrid brings the activating domain into position to turn on gene expression from the reporter gene (see figure 17.13).

The beauty of this system is that it is both simple and flexible. It can be used with two known proteins or with a known protein in the bait vector and entire cDNA libraries in the prey vector. In the latter case, all of the possible interactions in a cell type can be mapped.

It is already clear that there are even more protein interactions in cells than anticipated. In the future these data will form the basis for understanding the networks of protein interactions that make up the normal activities of a cell.

Restriction enzymes can be used to construct physical maps of DNA. The technique of Southern blotting allows the detection of DNA in complex mixtures, such as DNA isolated form cells or tissues, and can be used to anaylze differences between individuals. The ultimate level of analysis is to determine the actual DNA sequence. DNA sequencing uses a modified DNA polymerase reaction that contains chain terminators. The polymerase chain reaction (PCR) has changed molecular analysis allowing the production of a large amount of a specific DNA from a small amount of starting material. The yeast two-hybrid system is used to detect protein–protein interactions.

17.4 Genetic Engineering

The ability to clone individual genes for analysis ushered in an era of unprecedented research advancement. At the time, these advancements were not accompanied by grand announcements of potential medical breakthroughs and other applications. The ability to truly genetically engineer any kind of cell or organism was a long way off. But we are now approaching this ability, and it has generated much excitement and also controversy.

Expression vectors allow production of specific gene products

A variety of specialized vectors have been constructed since the development of cloning technology. One very important type of vector are the **expression vectors.** These vectors contain the sequences necessary to drive expression of inserted DNA in a specific cell type, namely the correct sequences to permit transcription and translation of the sequences. The production of recombinant proteins in bacteria, for example, uses expression vectors with bacterial promoters and other control regions. The bacteria transformed by such vectors synthesize large amounts of the protein encoded by the inserted DNA. A number of pharmaceuticals have been produced in this way, including the first, insulin, used to treat diabetes. (This type of application is discussed in more detail in the next section.)

Genes can be introduced across species barriers

The ability to reintroduce genes into an original host cell, or to introduce genes into another host, is true genetic engineering. An animal containing a gene that has been introduced without the use of conventional breeding is called a **transgenic animal.** We will explore a number of uses of transgenic animals in medicine and agriculture, but it is important to realize that their original use was for basic research.

The ability to engineer genes in context or out of context allows an experimenter to ask questions that could never be asked otherwise. A dramatic example was the use of the *eyeless* gene from mice in *Drosophila.* When this mouse gene was introduced into *Drosophila*, it was shown to be able to substitute for a *Drosophila* gene in organizing the formation of eyes. It could even cause the formation of eyes in incorrect locations when expressed in tissue that did not normally form eyes. This amazing result shows that the formation of the compound eye in an insect is not so different from the formation of the complex vertebrate eye.

Cloned genes can be used to construct "knockout" mice

One of the most important technologies for research purposes is **in vitro mutagenesis,** the ability to create mutations at any site in a cloned gene to examine the effect on function. Rather than depending on mutations induced by chemical agents or radiation in intact organisms, which is time- and labor-intensive, the DNA itself is directly manipulated. The ultimate use of this approach is to be able to replace the wild-type gene with a mutant copy to test the function of the mutated gene. Developed first in yeast, this technique has now been extended to the mouse.

In mice, this technique has produced **knockout mice** in which a known gene is inactivated ("knocked out"). The effect of loss of this function is then assessed in the adult mouse, or if it is lethal, the stage of development at which function fails can be determined. The idea is simple, but the technology is quite

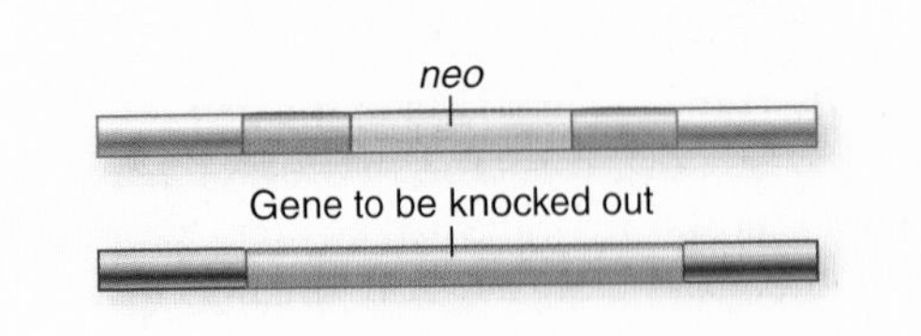

1. Using recombinant DNA techniques, the gene encoding resistance to *neomycin (neo)* is inserted into the gene of interest, disrupting it. The *neo* gene also confers resistance to the drug G418, which kills mouse cells. This construct is then introduced into ES cells.

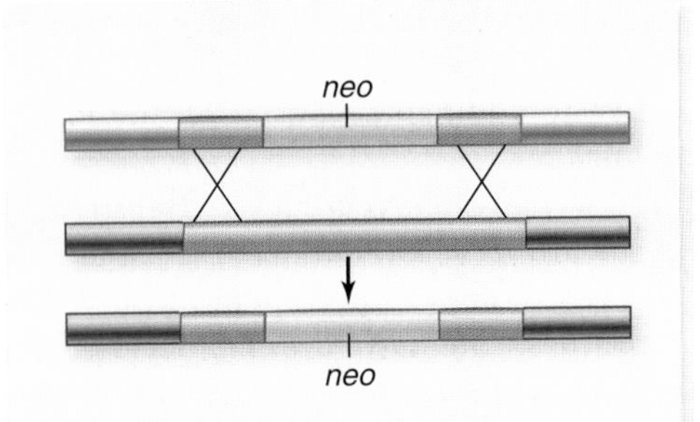

2. In some ES cells, the construct will recombine with the chromosomal copy of the gene to be knocked out. This replaces the chromosomal copy with the *neo* disrupted construct. This is the equivalent to a double crossover event in a genetic cross.

figure 17.14

CONSTRUCTION OF A KNOCKOUT MOUSE. Steps in the construction of a knockout mouse. Some technical details have been omitted, but the basic concept is shown.

complex. A streamlined description of the steps in this type of experiment are outlined below and illustrated in figure 17.14:

1. The cloned gene is disrupted by replacing it with a marker gene using recombinant DNA techniques described earlier. The marker gene codes for resistance to the antibiotic neomycin in bacteria, and allows mouse cells to survive when grown in a medium containing the related drug G418. The construction is done such that the marker gene is flanked by the DNA normally flanking the gene of interest in the chromosome.
2. The interrupted gene is introduced into **embryonic stem cells (ES cells).** These cells are derived from early embryos and can develop into different adult tissues. In these cells, the gene can recombine with the chromosomal copy of the gene based on the flanking DNA. This is the same kind of recombination used to map genes (chapter 13). The knockout gene with the drug resistance gene does not have an origin of replication, and thus it will be lost if no recombination occurs. Cells are grown in medium containing G418 to select for recombination events. (Only those containing the marker gene can grow in the presence of G418.)
3. The embryonic stem cells containing the knocked-out gene is injected into a blastocyst stage embryo, which is then implanted into a pseudopregnant female (one that has been mated with a vasectomized male and as a result has a receptive uterus). The pups from this female have one copy of the gene of interest knocked out. Transgenic animals can then be crossed to generate homozygous lines. These homozygous lines can be analyzed for phenotypes.

In conventional genetics, genes are identified based on mutants that show a particular phenotype. Molecular genetic techniques are then used to find the gene and isolate a molecular clone for analysis. The use of knockout mice is an example of **reverse genetics:** One takes a cloned gene of unknown function, then uses it to make a mutant deficient in that gene. A geneticist can then assess the effect on the entire organism of eliminating a single gene.

Sometimes this approach leads to surprises, such as when the gene for the p53 tumor suppressor was knocked out. Because this protein is found mutated in many human cancers, and plays a key role in the regulation of the cell cycle (chapter 10), it was thought to be essential—the knockout was expected to be lethal. Instead, the mice were born normal; that is, development had proceeded normally. These mice do have a phenotype however, they exhibit an increased incidence of tumors in a variety of tissues as they age.

Expression vectors that contain cloned genes allow the production of known proteins in different cells. This can be done for research purposes or to produce pharmaceuticals. Genes can also be introduced across species barriers. In mouse, mutations can be engineered in cloned genes, then reintroduced into the animal to create "knockout" mice deficient in specific genes.

17.5 Medical Applications

The early days of genetic engineering led to a rash of startup companies, many of which are no longer in business. At the same time, all of the major pharmaceutical companies either began research in this area or actively sought smaller companies with promising technology. The number of applications of this technology are far too numerous to mention here, so a few are highlighted; the section following discusses agricultural applications.

Human proteins can be produced in bacteria

The first and perhaps most obvious commercial application of genetic engineering was the introduction of genes that encode clinically important proteins into bacteria. Because bacterial cells can be grown cheaply in bulk, bacteria that incorporate recombinant genes can synthesize large amounts of the proteins those genes specify. This method has been used to produce several forms of

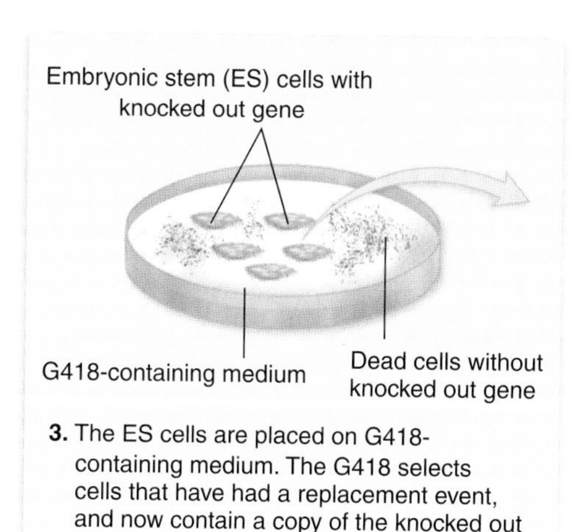

3. The ES cells are placed on G418-containing medium. The G418 selects cells that have had a replacement event, and now contain a copy of the knocked out gene.

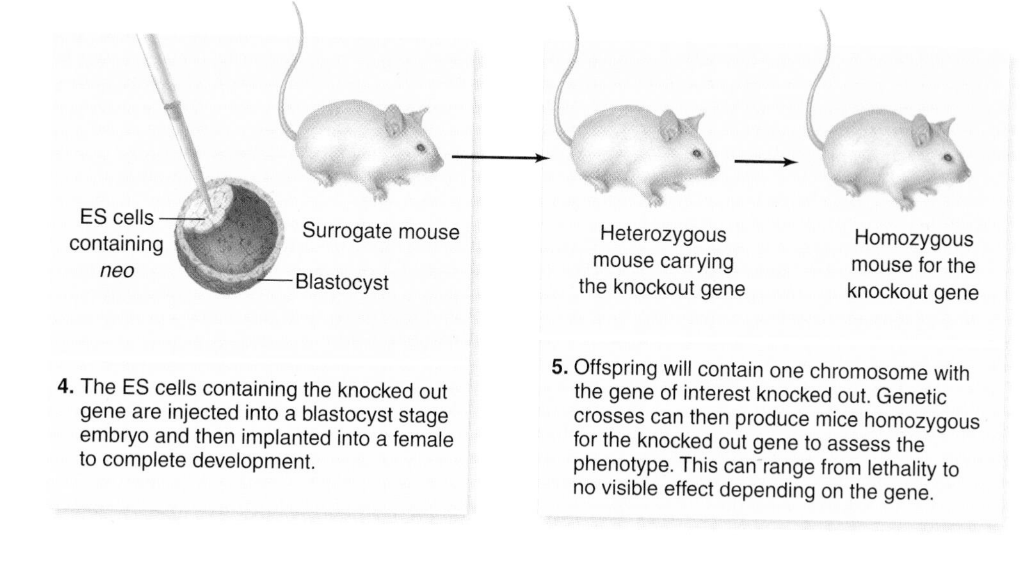

4. The ES cells containing the knocked out gene are injected into a blastocyst stage embryo and then implanted into a female to complete development.

5. Offspring will contain one chromosome with the gene of interest knocked out. Genetic crosses can then produce mice homozygous for the knocked out gene to assess the phenotype. This can range from lethality to no visible effect depending on the gene.

human insulin and interferon, as well as other commercially valuable proteins, such as human growth hormone (figure 17.15) and erythropoietin, which stimulates red blood cell production.

Among the medically important proteins now manufactured by these approaches are **atrial peptides,** small proteins that may provide a new way to treat high blood pressure and kidney failure. Another is **tissue plasminogen activator (TPA),** a human protein synthesized in minute amounts that causes blood clots to dissolve and that if used within the first 3 hr after an ischemic stroke (i.e., one that blocks blood to the brain) can prevent catastrophic disability.

A problem with this approach has been the difficulty of separating the desired protein from the others the bacteria make. The purification of proteins from such complex mixtures is both time-consuming and expensive, but it is still easier than isolating the proteins from bulk processing of the tissues of animals, which is how such proteins were formerly obtained. For example, insulin previously was extracted from hog pancreases because hog insulin was similar to human insulin.

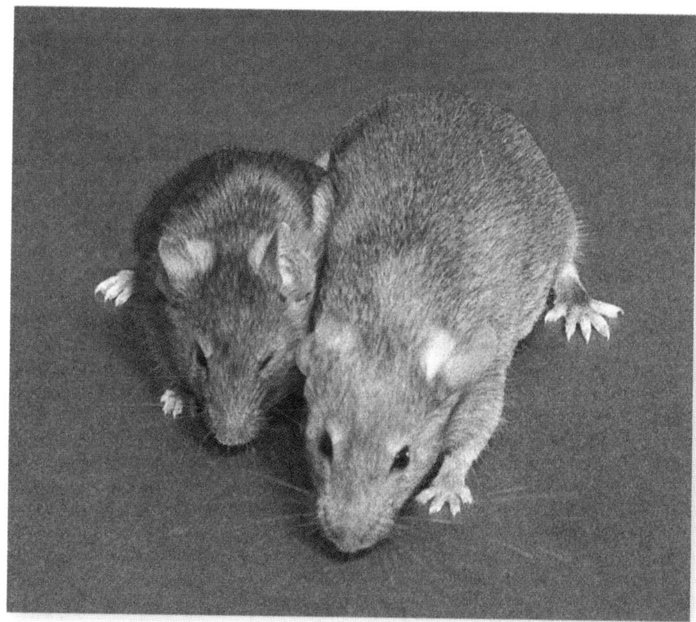

figure 17.15

GENETICALLY ENGINEERED MOUSE WITH HUMAN GROWTH HORMONE. These two mice are from an inbred line and differ only in that the large one has one extra gene: the gene encoding human growth hormone. The gene was added to the mouse's genome and is now a stable part of the mouse's genetic endowment.

Recombinant DNA may simplify vaccine production

Another area of potential significance involves the use of genetic engineering to produce vaccines against communicable diseases. Two types of vaccines are under investigation: *subunit vaccines* and *DNA vaccines.*

Subunit vaccines

Subunit vaccines may be developed against viruses such as those that cause herpes and hepatitis. Genes encoding a part, or subunit, of the protein polysaccharide coat of the herpes simplex virus or hepatitis B virus are spliced into a fragment of the vaccinia (cowpox) virus genome (figure 17.16).

The vaccinia virus, which British physician Edward Jenner used more than 200 years ago in his pioneering vaccinations against smallpox, is now used as a vector to carry the herpes or hepatitis viral coat gene into cultured mammalian cells. These cells produce many copies of the recombinant vaccinia virus, which has the outside coat of a herpes or hepatitis virus. When this recombinant virus is injected into a mouse or rabbit, the immune system of the infected animal produces antibodies directed against the coat of the recombinant virus. It therefore develops an immunity to herpes or hepatitis virus.

Vaccines produced in this way are harmless because the vaccinia virus is benign, and only a small fragment of the DNA from the disease-causing virus is introduced via the recombinant virus.

The great attraction of this approach is that it does not depend on the nature of the viral disease. In the future, similar recombinant viruses may be used in humans to confer resistance to a wide variety of viral diseases.

DNA vaccines

In 1995, the first clinical trials began to test a novel new kind of **DNA vaccine,** one that depends not on antibodies but rather on the second arm of the body's immune defense, the so-called

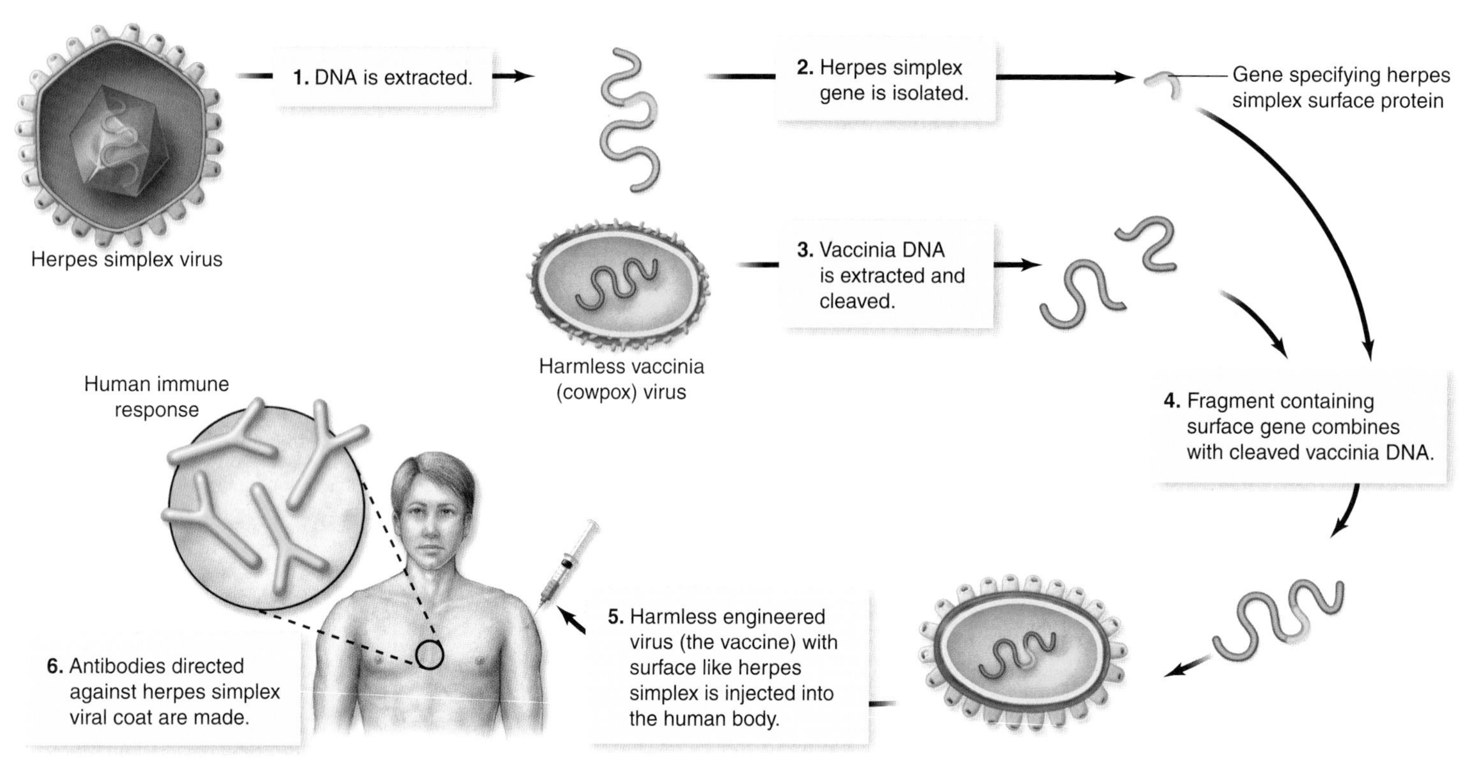

figure 17.16

STRATEGY FOR CONSTRUCTING A SUBUNIT VACCINE FOR HERPES SIMPLEX. Recombinant DNA techniques can be used to construct vaccines for a single protein from a virus or bacterium. In this example, the protein is a surface protein from the herpes simplex virus.

cellular immune response, in which blood cells known as killer T cells attack infected cells (chapter 51). The first DNA vaccines spliced an influenza virus gene encoding an internal nucleoprotein into a plasmid, which was then injected into mice. The mice developed a strong cellular immune response to influenza. Although new and controversial, the approach offers great promise.

Gene therapy can treat genetic diseases directly

In 1990, researchers first attempted to combat genetic defects by the transfer of human genes. When a hereditary disorder is the result of a single defective gene, an obvious way to cure the disorder would be to add a working copy of the gene. This approach is being used in an attempt to combat cystic fibrosis, and it offers potential for treating muscular dystrophy and a variety of other disorders (table 17.1).

One disease that illustrates both the potential and the problems with gene therapy is **severe combined immunodeficiency disease (SCID).** There are multiple forms of this disease, including an X-linked form (X-SCID) and a form that lacks the enzyme adenosine deaminase (ADA-SCID).

Recent trials for both of these forms showed great initial promise, with patients exhibiting restoration of immune function. But then problems arose in the case of the X-SCID trial when a patient developed a rare leukemia. Since that time, two other patients have developed the same leukemia, and it appears to be due to the gene therapy itself. The vector used to introduce the X-SCID gene integrated into the genome next to a proto-oncogene called *LMO2* in all three cases. Activation of this gene can cause childhood leukemias.

The insertion of a gene during gene therapy has always been a random event, and it has been a concern that the insertion could inactivate an essential gene, or turn on a gene

TABLE 17.1 Diseases Being Treated in Clinical Trials of Gene Therapy

Disease
Cancer (melanoma, renal cell, ovarian, neuroblastoma, brain, head and neck, lung, liver, breast, colon, prostate, mesothelioma, leukemia, lymphoma, multiple myeloma)
SCID (severe combined immunodeficiency)
Cystic fibrosis
Gaucher disease
Familial hypercholesterolemia
Hemophilia
Purine nucleoside phosphorylase deficiency
Alpha$_1$-antitrypsin deficiency
Fanconi anemia
Hunter syndrome
Chronic granulomatous disease
Rheumatoid arthritis
Peripheral vascular disease
Acquired immunodeficiency syndrome (AIDS)

inappropriately. That effect had not been observed prior to the X-SCID trial, despite a large number of genes introduced into blood cells in particular. For leukemia to occur in 15% of the patients treated implies that some influence of the genetic background associated with X-SCID potentiates this development. This possibility is supported by the observation that the ADA-SCID patients treated have not been affected thus far.

On the positive side, 15 children treated successfully are still alive, 14 of them after more than four years, with functioning immune systems. On the negative side, three other children treated have developed leukemia.

When we understand the basis of the preferential integration in the case of X-SCID, it should be possible to overcome this unfortunate result. In the meantime, the investigators have halted the trial and are working on new vectors to reduce the possibility of this preferential integration.

Medical applications of biotechnology include the production of proteins for pharmaceuticals and new ways to make vaccines. Genetic engineering can also be used to replace genes that cause genetic disease, a process called gene therapy. The technology for gene therapy has been controversial with some successes but some recent failures as well.

17.6 Agricultural Applications

Perhaps no area of genetic engineering touches all of us so directly as the applications that are being used in agriculture today. Crops are being modified to resist disease, to be tolerant to herbicides, and for changes in nutritional and other content in a variety of ways. Plant systems are also being used to produce pharmaceuticals by "biopharming," and domesticated animals are being genetically modified to produce biologically active compounds.

The Ti plasmid can transform broadleaf plants

In plants, the primary experimental difficulty has been identifying a suitable vector for introducing recombinant DNA. Plant cells do not possess the many plasmids that bacteria have, so the choice of potential vectors is limited.

The Ti plasmid

The most successful results thus far have been obtained with the **Ti (tumor-inducing) plasmid** of the plant bacterium *Agrobacterium tumefaciens*, which normally infects broadleaf plants such as tomato, tobacco, and soybean. Part of the Ti plasmid integrates into the plant DNA, and researchers have succeeded in attaching other genes to this portion of the plasmid (figure 17.17). The characteristics of a number of plants have been altered using this technique, which should be valuable in improving crops and forests.

Among the features scientists would like to affect are resistance to disease, frost, and other forms of stress; nutritional balance and protein content; and herbicide resistance. All of these traits have either been modified or are being modified.

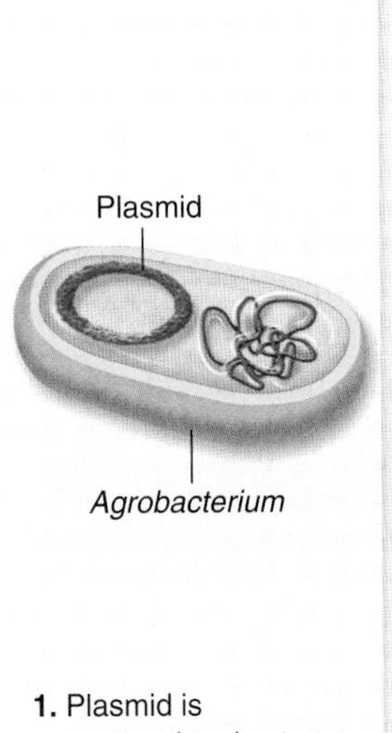

1. Plasmid is removed and cut open with restriction endonuclease.

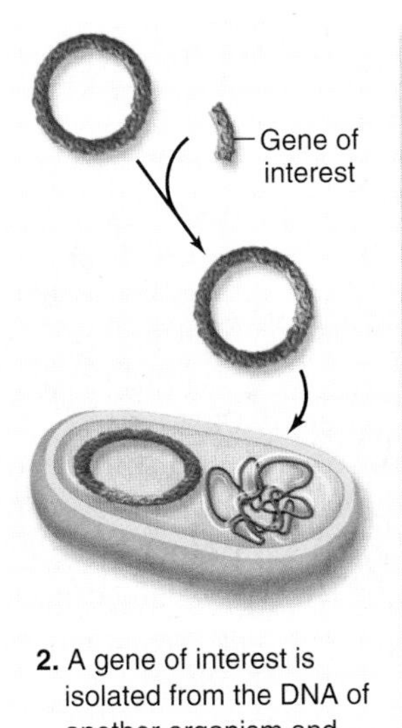

2. A gene of interest is isolated from the DNA of another organism and inserted into the plasmid. The plasmid is put back into the *Agrobacterium*.

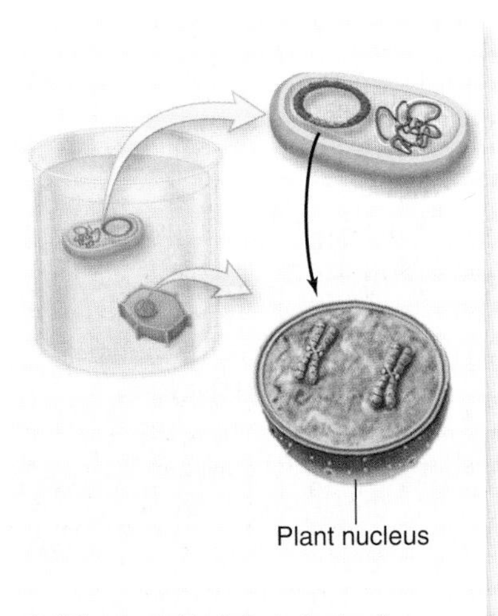

3. When used to infect plant cells, *Agrobacterium* duplicates part of the plasmid and transfers the new gene into a chromosome of the plant cell.

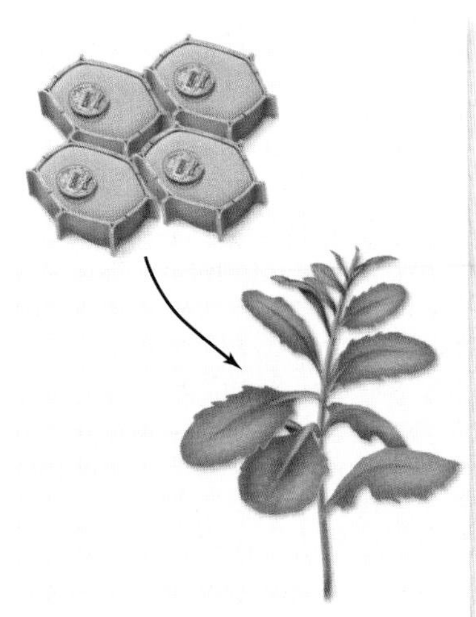

4. The plant cell divides, and each daughter cell receives the new gene. These cultured cells can be used to grow a new plant with the introduced gene.

figure 17.17

THE TI PLASMID. This *Agrobacterium tumefaciens* plasmid is used in plant genetic engineering.

Unfortunately, *Agrobacterium* normally does not infect cereals such as corn, rice, and wheat, but alternative methods can be used to introduce new genes into them.

Other methods of gene insertion

For cereal plants that are not normally infected by *Agrobacterium*, other methods have been used. One popular method, "the gene gun" uses bombardment with tiny gold or tungsten particles coated with DNA. This technique has the advantage of being possible for any species, but it does not allow as precise an engineering because the copy number of introduced genes is much harder to control.

Recently, modifications of the *Agrobacterium* system have allowed it to be used with cereal plants, so the gene gun technology may not be used as much in the future. A new bacterium has also been manipulated to function like *Agrobacterium*, offering another potential alternative method of engineering cereal crops.

It is clear that genetic modification of crop plants of all sorts has become a mature technology, which should accelerate the production of a variety of transgenic crops.

Case study: A better tomato?

One example of genetically manipulated fruit is Calgene's "Flavr Savr" tomato, which was genetically engineered to inhibit genes that cause cells to produce ethylene. In tomatoes and other plants, ethylene acts as a hormone to speed fruit ripening (chapter 41). In Flavr Savr tomatoes, inhibition of ethylene production delays ripening. The result is a tomato that can stay on the vine longer and resists overripening and rotting during transport to market.

The Flavr Savr tomato was a genetic engineering success, but was not a success in the marketplace. Its taste was not as good as other varieties, and it grew only in a limited area of the country and was not widely planted. The tomato was pulled from the market in 1997. Clearly there is more to producing a viable product than the genetic engineering of specific traits.

Herbicide-resistant crops allow no-till planting

Recently, broadleaf plants have been genetically engineered to be resistant to **glyphosate,** a powerful, biodegradable herbicide that kills most actively growing plants (figure 17.18). Glyphosate works by inhibiting an enzyme called EPSP synthetase, which plants require to produce aromatic amino acids.

Humans do not make aromatic amino acids; we get them from our diet, so we are unaffected by glyphosate. To make glyphosate-resistant plants, scientists used a Ti plasmid to insert extra copies of the EPSP synthetase gene into plants. These engineered plants produce 20 times the normal level of EPSP synthetase, enabling them to synthesize proteins and grow despite glyphosate's suppression of the enzyme. In later experiments, a bacterial form of the EPSP synthetase gene that differs from the plant form by a single nucleotide was introduced into plants via Ti plasmids; the bacterial enzyme is not inhibited by glyphosate.

These advances are of great interest to farmers because a crop resistant to glyphosate would not have to be weeded—the field could simply be treated with the herbicide. Because glyphosate is a broad-spectrum herbicide, farmers would no longer need to employ a variety of different herbicides, most of which kill only a few kinds of weeds. Furthermore, glyphosate breaks down readily in the environment, unlike many other herbicides commonly used in agriculture. A plasmid is actively being sought for the introduction of the EPSP synthetase gene into cereal plants, making them also glyphosate-resistant.

figure 17.18

GENETICALLY ENGINEERED HERBICIDE RESISTANCE. All four of these petunia plants were exposed to equal doses of the herbicide glyphosate. The two on the right were genetically engineered to be resistant to glyphosate, the active ingredient in glyphosate, but the two on the left were not.

At this point four important crop plants have been modified to be glyphosate-resistant: maize (corn), cotton, soybeans, and canola. The use of glyphosate-resistant soy has been especially popular, accounting for 60% of the global area of GM (genetically modified) crops grown in nine countries worldwide. In the United States, 90% of soy currently grown is GM soy. Global variation in the use of GM crops has occurred, with the Americas, led by the United States, the largest adopter. The area currently with the largest growth in the use of GM crops is Asia, while Europe has been the slowest to move to their use.

Bt crops are resistant to some insect pests

Many commercially important plants are attacked by insects, and the usual defense against such attacks has been to apply insecticides. Over 40% of the chemical insecticides used today are targeted against boll weevils, bollworms, and other insects that eat cotton plants. Scientists have produced plants that are resistant to insect pests, removing the need to use many externally applied insecticides.

The approach is to insert into crop plants genes encoding proteins that are harmful to the insects that feed on the plants, but harmless to other organisms. The most commonly used protein is a toxin produced by the soil bacterium *Bacillus thuringiensis* (*Bt toxin*). When insects ingest Bt toxin, endogenous enzymes convert it into an insect-specific toxin, causing paralysis and death. Because these enzymes are not found in other animals, the protein is harmless to them.

The same four crops that have been modified for herbicide resistance have also been modified for insect resistance using the Bt toxin. The use of Bt maize is the second most common GM crop globally, representing 14% of global area of GM crops in nine countries. The global distribution of these crops is also similar to the herbicide resistant relatives.

Given the popularity of both of these types of crop modifications, it is not surprising that they have also been combined, so-called *stacked GM crops*, in both maize and cotton. Stacked crops now represent 9% of global area of GM crops.

Golden Rice shows potential of GM crops

One of the successes of GM crops is the development of Golden Rice. This rice has been genetically modified to produce β-carotene (provitamin A). The World Health Organization (WHO) estimates that vitamin A deficiency affects between 140 and 250 million preschool children worldwide. The deficiency is especially severe in developing countries where the major staple food is rice. Provitamin A in the diet can be converted by enzymes in the body to vitamin A, alleviating the deficiency.

The science that led to Golden Rice, so named for its distinctive color imparted by the presence of β-carotene in the endosperm (the outer layer of rice that has been milled). Rice does not normally make β-carotene in endosperm tissue, but does produce a precursor, geranyl geranyl diphosphate, that can be converted by three enzymes, phytoene synthase, phytoene desaturase, and lycopene β-cyclase, to β-carotene. These three genes were engineered to be expressed in endosperm and introduced into rice to complete the biosynthetic pathway producing β-carotene in endosperm (figure 17.19).

This is an interesting case of genetic engineering for two reasons. First, it introduces a new biochemical pathway in tissue of the transgenic plants. Second, it could not have been done by conventional breeding as no rice cultivar known produces these enzymes in endosperm. The original constructs used two genes from daffodil and one from a bacterium (see figure 17.19). There are many reasons to expect failure in the introduction of a biochemical pathway without disrupting normal metabolism. That the original form of Golden Rice makes significant amounts of β-carotene in an otherwise healthy plant is impressive. A second-generation version that makes much higher levels of β-carotene has also been produced by using the gene for phytoene synthase from maize in place of the original daffodil gene.

Golden Rice was originally constructed in a public facility in Switzerland and made available for free with no commercial entanglements. Since its inception, Golden Rice has been improved both by public groups and by industry scientists, and these improved versions are also being made available without commercial strings attached.

GM crops raise a number of social issues

The adoption of GM crops has been resisted in some areas for a variety of reasons. Questions are asked about the safety of these crops for human consumption, the movement of the introduced genes into wild relatives, and the possible loss of biodiversity associated with these crops.

Powerful forces form opposing sides in this debate. On the side in favor of the use of GM crops are the multinational companies that are utilizing this technology to produce seeds for the various GM crops. On the side questioning the use of GM crops are a variety of political organizations that are opposed to genetically modified foods. Scientists can be found on both sides of the controversy.

Issues originally centered on the safety of introduced genes for human consumption. In the United States, this issue has been "settled" for the crops already mentioned, and a large amount of GM soy and maize is consumed in this country. Although some still raise the issue of long-term use and allergic reactions, no negative effects have been documented. Existing crops will be monitored for adverse effects, and each new modification will require regulatory approval for human consumption.

Another issue has been fear about the spread of genes outside of the GM crops, but at this point there is no evidence for the introgression of introduced genes into wild relatives. A recent study indicated no evidence for the movement of genes from GM crops into native species in Mexico, despite earlier studies indicating significant movement of introduced genes.

This finding does not mean that such movement is impossible, but it does indicate that it seems not to have occurred at present. It is clear that this area requires more study. This issue will likely have to be considered on a case-by-case basis because the number of wild relatives, and the ease of hybridization, varies greatly among crop plants.

Pharmaceuticals can be produced by "biopharming"

The medicinal use of plants goes back as far as recorded history. In modern times, the pharmaceutical industry began by isolating biologically active compounds from plants. This approach began to change when in 1897, the Bayer company introduced acetyl salicylic acid, otherwise known as aspirin. This compound was

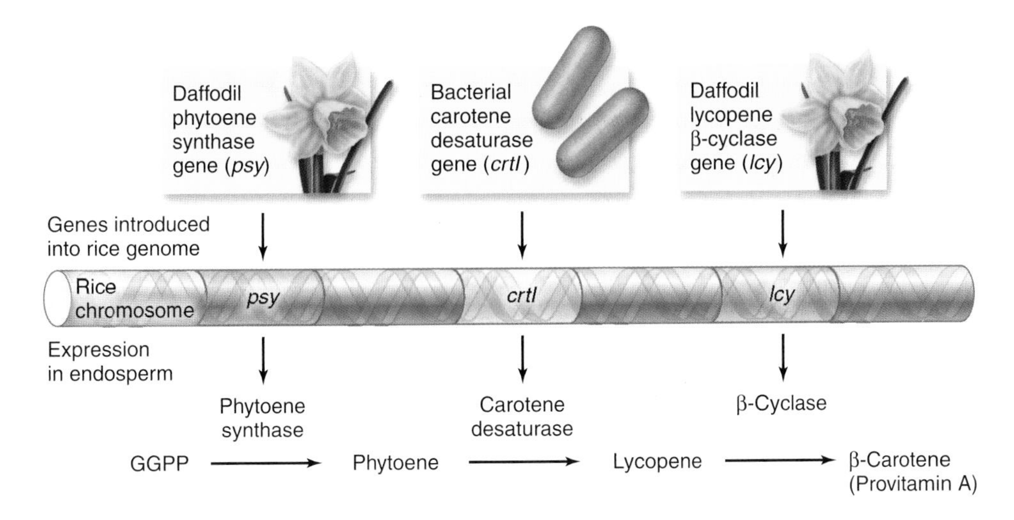

figure 17.19

CONSTRUCTION OF GOLDEN RICE. Rice does not normally express the enzymes needed to synthesize β-carotene in endosperm. Three genes were added to the rice genome to allow expression of the pathway for β-carotene in endosperm. The source of the genes and the pathway for synthesis of β-carotene is shown. The result is Golden Rice, which contains enriched levels of β-carotene in endosperm.

Calvin and Hobbes

by Bill Watterson

© 1995 Watterson/Dist. by Universal Press Syndicate

CALVIN AND HOBBES © 1995 Watterson. Dist. by Universal Press Syndicate. Reprinted with permission. All rights reserved.

a synthetic version of the compound salicylic acid, which was isolated from white willow bark. The production of pharmaceuticals has since been dominated more by organic synthesis and less by the isolation of plant products.

One exception to this trend is cancer chemotherapeutic agents such as taxol, vinblastine, and vincristine, all of which are isolated from plant sources. In an interesting closing of the historical loop, the industry is now looking at using transgenic plants for the production of useful compounds.

The first human protein to be produced in plants was human serum albumin, which was produced in 1990 by both genetically engineered tobacco and potato plants. Since that time more than 20 proteins have been produced in transgenic plants. This first crop of transgenic pharmaceuticals are now in the regulatory pipeline.

Recombinant subunit vaccines

One promising aspect of plant genetic engineering is the production of recombinant subunit vaccines discussed earlier. One of these, being produced in genetically modified potatoes, is a vaccine for Norwalk virus. Norwalk virus is not commonly known, but it reached the public consciousness when cruise ships were forced to cancel cruises due to outbreaks of the virus. The vaccine is now in clinical trials. A vaccine for rabies produced in transgenic spinach is also in clinical trials.

One obvious advantage of using plants for vaccine production is scalability. It has been estimated that 250 acres of greenhouse space could produce enough transgenic potato plants to supply Southeast Asia's need for hepatitis B vaccine.

Recombinant antibodies

Combining molecular cloning with immunology can be used to produce in transgenic plants the antibodies normally made by blood cells in vertebrates. The synthesis of monoclonal antibodies in plant systems is a promising use of transgenic plants.

A number of potentially therapeutic antibodies are being produced in plants, and some of these have reached clinical trial stage. One interesting example is an antibody against the bacterium responsible for dental caries, commonly known as tooth decay. It would make a visit to the dentist more pleasant to have a topical antibody applied instead of a drill.

Domesticated animals can also be genetically modified

Humans have been breeding and selecting domestic animals for hundreds of years. With the advent of genetic engineering, this process can be accelerated, and genes can be introduced from other species.

The production of transgenic livestock is in an early stage, and it is hard to predict where it will go. At this point, one of the uses of biotechnology is not to construct transgenics, but to use DNA markers to identify animals, and to map genes that are involved in such traits as palatability in food animals, texture of hair or fur, and other features of animal products. Molecular techniques combined with the ability to clone domestic animals (chapter 19) could produce improved animals for economically desirable traits.

Transgenic animal technology has not been as successful as initially predicted. Early on, pigs were engineered to overproduce growth hormone in the hope that this would lead to increased and faster growth. These animals proved to have only slightly increased growth, and they had lower fat levels, which reduces flavor, as well as showing other deleterious effects. The main use thus far has been engineering animals to produce pharmaceuticals in milk—another example of the biopharming concept.

One interesting idea for transgenics is the EnviroPig. This animal has been engineered with the introduction of the gene for phytase under the control of a salivary gland-specific promoter. The enzyme phytase breaks down phosphorus in the feed and can reduce phosphate excretion by up to 70%. Because phosphate is a major problem in pig waste, reducing its excretion could be a large environmental benefit.

As with GM crops, fears exist about the consumption of meat from transgenic animals. At this point, these fears do not seem to be based on sound science; nevertheless, every transgenic animal produced that is intended for consumption will need to be considered on a case-by-case basis.

Genes can be introduced into plants with the Ti plasmid. Crops have been modified to resist herbicide, and to produce a bacterial toxin to kill insects. Golden Rice was modified to produce provitamin A in endosperm. Biopharming is the use of plants and animals to produce useful proteins. These technologies have raised ethical issues.

concept review

17.1 DNA Manipulation

The construction of recombinant DNA from molecules of two different sources led to the field of molecular biotechnology.

- Restriction endonucleases are used to fragment DNA molecules at specific restriction sites.
- Restriction endonucleases allowed for a form of physical mapping of DNA and the creation of recombinant molecules.
- Type II restriction enzymes recognize DNA sequences that are 4–12 bases long with a central axis of symmetry, and read the same way (5′ to 3′) on one strand that they do in the opposite direction (also 5′ to 3′).
- Cleaving such sequences at the same base on each strand results in fragments with "sticky ends" or overhanging complementary ends (figure 17.1).
- DNA ligase joins two fragments to form a stable DNA molecule.
- Gel electrophoresis separates fragments based on size, using an electric field to cause DNA to migrate through a gel matrix. Smaller fragments migrate farther than large fragments (figure 17.2).
- Foreign DNA is introduced into cells by a process called transformation.

17.2 Molecular Cloning

A clone is an identical copy. Molecular cloning involves the isolation of a specific sequence of DNA and the making of many identical copies.

- A vector is used to propagate recombinant DNA molecules in host cells.
- Plasmid vectors are small, extrachromosomal DNAs used to clone relatively small pieces of DNA.
- Phage λ vectors have a linear genome that can accept larger DNA molecules.
- Artificial chromosomes are used to clone very large DNA molecules.
- A DNA, or genomic, library is a collection of fragments from an entire genome that have been inserted into host cells.
- To obtain only the expressed parts of a genome, complementary DNA (cDNA) can be made from mRNA by using the enzyme reverse transcriptase (figure 17.5).
- DNA can be reversibly denatured and renatured. Renaturation of complementary strands from different sources is called hybridization.
- Hybridization is a powerful tool for finding DNAs in complex mixtures. Known DNA can be labeled, then used to find complementary strands by hybridization.
- Hybridization is the most common way of identifying a clone within a DNA library.

17.3 DNA Analysis

Molecular cloning provides specific DNA for further manipulation.

- The first physical maps of DNA molecules were the sites of restriction enzyme cleavage.
- These can be generated by enzyme digestion or by computers searching known DNA sequences for cleavage sites.
- Blotting is a procedure in which a complex mixture separated by electrophoresis is transferred to a piece of filter paper.
- The Southern blot uses DNA isolated from a cell or tissue with a filter that is hybridized with labeled, cloned DNA as a probe.
- Northern blots use mRNA instead of DNA, and Western blots use protein.
- Restriction fragment length polymorphisms (RFLPs) were the first method used to detect individual differences in DNA.
- DNA fingerprinting is a technique that uses probes to locate polymorphic repetitive DNA fragments.
- Determination of the actual sequence of bases in a DNA molecule is the ultimate level of analysis. This uses chain-terminating reagents and high-resolution electrophoresis (figure 17.11).
- The polymerase chain reaction (PCR) is used to amplify a single small DNA fragment using two short primers that flank the region to be amplified.
- The yeast two-hybrid system is used to study protein–protein interactions (figure 17.13).

17.4 Genetic Engineering

We can now genetically modify most plant and animal systems by introducing new DNA or modifying existing DNA in cells.

- Expression vectors contain the promoters and enhancers necessary to drive expression of inserted DNA.
- Transgenic organisms have DNA introduced across species "barriers."
- In vitro mutagenesis allows directed alteration of cloned genes, which can then be used to study gene function.
- Knockout mice are engineered to lack function for a specific gene. This allows a researcher to remove function for a gene and assess the phenotype (figure 17.14).

17.5 Medical Applications

There are many medical applications of genetic engineering.

- Human proteins, such as insulin, are produced by bacteria.
- Recombinant DNA may simplify vaccine production by producing harmless subunit vaccines and DNA vaccines that depend on the body's cellular immune response.
- Gene therapy, the adding of a copy of a functional gene, can be used to treat human genetic diseases.
- One problem with recent gene therapy trials was that the therapy caused leukemias in some patients.

17.6 Agricultural Applications

Crops are being modified to resist disease, tolerate herbicides, change nutritional value, and produce pharmaceuticals and biologically active compounds.

- The tumor-inducing plasmid from a plant bacterium is used to transfer genes into broad-leaf plants.
- Crops resistant to the herbicide glyphosate are a common genetic alteration. This allows for no-till planting.
- Bacterially derived insecticidal proteins have been transferred into crop plants to create pest-resistant crops.
- Golden Rice has been modified to have a higher concentration of provitamin A, important for diets in less-developed countries.
- The adoption of genetically modified (GM) crops has raised societal issues.
- "Biopharming" uses transgenic plants to produce pharmaceuticals such as serum albumin, subunit vaccines, and antibodies.
- Transgenic plant technologies are more successful than transgenic animal technologies.
- One recent successful GM pig produces an enzyme that reduces the excretion of harmful phosphates into the environment.

review questions

SELF TEST

1. A recombinant DNA molecule is one that is—
   a. produced through the process of crossing over that occurs in meiosis.
   b. constructed from DNA from different sources.
   c. constructed from novel combinations of DNA from the same source.
   d. produced through mitotic cell division.
2. Type II restriction endonucleases are useful because—
   a. they degrade DNA from the 5′ end.
   b. they cleave the DNA at random locations.
   c. they cleave the DNA at specific sequences.
   d. they only cleave modified DNA.
3. What is the basis of separation of DNA fragments by gel electrophoresis?
   a. The negative charge on DNA
   b. The size of the DNA fragments
   c. The sequence of the fragments
   d. The presence of a dye
4. How is the gene for β-galactosidase used in the construction of a plasmid?
   a. The gene is a promoter that is sensitive to the presence of the sugar, galactose.
   b. It is an origin of replication.
   c. It is a cloning site.
   d. It is a marker for insertion of DNA.
5. The basic logic of enzymatic DNA sequencing is to produce—
   a. a nested set of DNA fragments produced by restriction enzymes.
   b. a nested set of DNA fragments that each begin with different bases.
   c. primers to allow PCR amplification of the region between the primers.
   d. a nested set of DNA fragments that end with known bases.
6. A DNA library is—
   a. an orderly array of all the genes within an organism.
   b. a collection of vectors.
   c. the collection of plasmids found within a single *E. coli.*
   d. a collection of DNA fragments representing the entire genome of an organism.
7. Molecular hybridization is used to—
   a. generate cDNA from mRNA.
   b. introduce a vector into a bacterial cell.
   c. screen a DNA library.
   d. introduce mutations into genes.
8. The enzyme used in the polymerase chain reaction is—
   a. a restriction endonuclease.
   b. heat-resistant RNA polymerase.
   c. reverse transcriptase.
   d. a heat-resistant DNA polymerase.
9. How does the yeast two-hybrid system detect protein–protein interactions?
   a. Binding of fusion partners triggers a signal cascade that alters gene expression.
   b. Fusion partners are detected using radioactive probes of Western blots.
   c. Protein–protein binding of fusion partners triggers expression of a reporter gene.
   d. Protein–protein binding of fusion partners triggers expression of the *Gal4* gene.
10. In vitro mutagenesis is used to—
    a. produce large quantities of mutant proteins.
    b. create mutations at specific sites within a gene.
    c. create random mutations within multiple genes.
    d. create organisms that carry foreign genes.
11. Insertion of a gene for a surface protein from a medically important virus such as herpes into a harmless virus is an example of—
    a. a DNA vaccine.
    b. reverse genetics.
    c. gene therapy.
    d. a subunit vaccine.
12. What is a Ti plasmid?
    a. A vector that can transfer recombinant genes into plant genomes.
    b. A vector that can be used to produce recombinant proteins in yeast.
    c. A vector that is specific to cereal plants like rice and corn.
    d. A vector that is specific to embryonic stem cells.
13. Which of the following is *not* a possible benefit of genetically modified crops?
    a. Increased nutritional value for people.
    b. Enhanced resistance to insect pests.
    c. Enhanced resistance to broad-spectrum herbicides.
    d. Enhanced resistance to insecticides.

CHALLENGE QUESTIONS

1. Many human proteins, such as hemoglobin, are only functional as an assembly of multiple subunits. Assembly of these functional units occurs within the endoplasmic reticulum and Golgi apparatus of a eukaryotic cell. Discuss what limitations, if any exist to the large-scale production of genetically engineered hemoglobin.
2. Enzymatic sequencing of a short strand of DNA was completed using dideoxynucleotides. Use the gel shown to determine the sequence of that DNA?

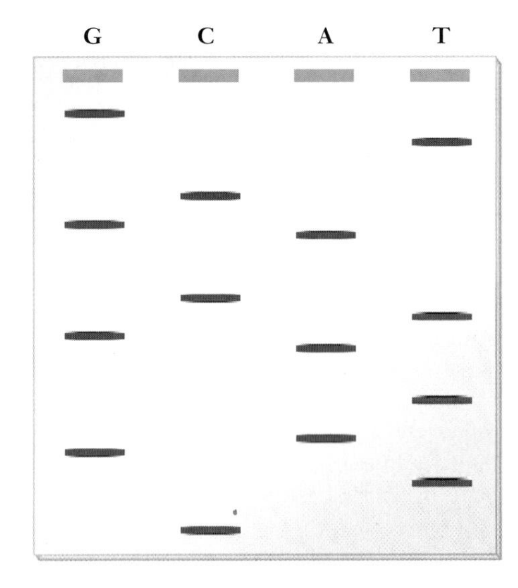

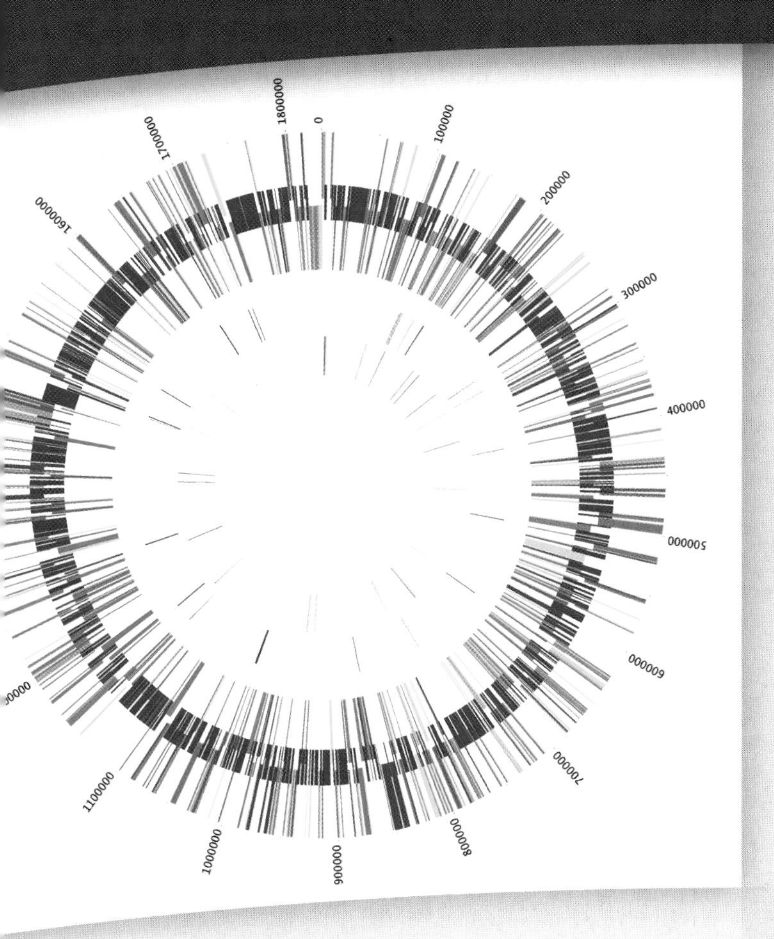

chapter 18

Genomics

introduction

THE PACE OF DISCOVERY IN BIOLOGY in the last 30 years has been like the exponential growth of a population. Starting with the isolation of the first genes in the mid-1970s, researchers had accomplished the first complete genome sequence by the mid-1990s—that of the bacteria species *Haemophilus influenzae*, shown in the picture (genes with similar functions are shown in the same color). By the turn of the twenty-first century, the molecular biology community had completed a draft sequence of the human genome.

Put another way, scientific accomplishments moved from cloning a single gene, to determining the sequence of a million base pairs in 20 years, then determining the sequence of a billion base pairs in another 5 years. The sequence of events is not linear since all are overlapping to some extent—but it is as though the first automobile had been invented on a Monday, then manufactured on an assembly line the following Wednesday, and by Friday, a formula 1 race car had been produced.

In the previous chapter you learned about the basic techniques of molecular biology. In this chapter you will see how those techniques have been applied to the analysis of whole genomes. This analysis integrates ideas from classical and molecular genetics with biotechnology.

concept outline

18.1 Mapping Genomes
- *Different kinds of physical maps can be generated*
- *Sequence-tagged sites provide a common language for physical maps*
- *Genetic maps provide a link to phenotypes*
- *Physical maps can be correlated with genetic maps*

18.2 Whole-Genome Sequencing
- *Genome sequencing requires larger molecular clones*
- *Whole-genome sequencing is approached in two ways: clone-by-clone and shotgun*
- *The Human Genome Project used both sequencing methods*

18.3 Characterizing Genomes
- *The Human Genome Project found fewer genes than expected*
- *Finding genes in sequence data requires computer searches*
- *Genomes contain both coding and noncoding DNA*
- *Expressed sequence tags identify genes that are transcribed*
- *SNPs are single-base differences between individuals*

18.4 Genomics and Proteomics
- *Comparative genomics reveals conserved regions in genomes*
- *Synteny allows comparison of unsequenced genomes*
- *Organelle genomes have exchanged genes with the nuclear genome*
- *Functional genomics reveals gene function at the genome level*
- *Proteomics moves from genes to proteins*
- *Large-scale screens reveal protein–protein interactions*

18.5 Applications of Genomics
- *Genomics can help to identify infectious diseases*
- *Genomics can help improve agricultural crops*
- *Genomics raises ethical issues over ownership of genomic information*

18.1 Mapping Genomes

We use maps to find out location, and depending on how accurately we wish to do this, we may use multiple maps with different resolutions. In genomics, we can locate a gene on a chromosome, in a subregion of a chromosome, and finally its precise location in the chromosome's DNA sequence. The DNA sequence level requires knowing the entire sequence of the genome, once out of our reach technologically. Knowing the entire sequence is useless, however, without other kinds of maps: Finding a single gene within the sequence of the human genome is like trying to find your house on a map of the world.

To overcome this difficulty, maps of genomes are constructed at different levels of resolution and using different kinds of information. We can make a distinction between *genetic maps* and *physical maps*. **Genetic maps** are abstract maps that place the relative location of genes on chromosomes based on recombination frequency (see chapter 13). **Physical maps** use landmarks within DNA sequences, ranging from restriction sites (described in the preceding chapter) to the ultimate level of detail: the actual DNA sequence.

Different kinds of physical maps can be generated

To make sense of genome mapping, it is important to have physical landmarks on the genome that are at a lower level of resolution than the entire sequence. In fact, long before the Human Genome Project was even conceived, physical maps of DNA were needed as landmarks on cloned DNA. Three types of physical maps are restriction maps, constructed using restriction enzymes; chromosome-banding patterns, generated by cytological dye methods; and radiation hybrid maps, created by using radiation to fragment chromosomes.

Restriction maps

Distances between "landmarks" on a physical map are measured in base-pairs (1000 base-pairs [bp] equal 1 kilobase, kb). It is not necessary to know the DNA sequence of a segment of DNA in order to create a physical map, or to know whether the DNA encompasses information for a specific gene.

The first physical maps were created by cutting genomic DNA with different restriction enzymes, both singly and with combinations of different enzymes (figure 18.1). The analysis of the patterns of fragments generated were used to generate a map.

In terms of larger pieces of DNA, this process is repeated and then used to put the pieces back together, based on size and overlap, into a contiguous segment of the genome called a **contig.** In an example of biological coincidence, the very first restriction enzymes to be isolated came from *Haemophilus*, which was also the first free-living genome to be completely sequenced.

Chromosome banding patterns

Cytologists studying chromosomes with light microscopes found that by using different stains, they could produce reproducible patterns of bands on the chromosomes. In this way, they could identify all of the chromosomes and divide them into subregions based on banding pattern.

1. Multiple copies of a segment of DNA are cut with restriction enzymes.

DNA
enzyme A
enzyme B
enzyme A + B

Molecular weight marker

2. The fragments produced by enzyme A only, by enzyme B only, and by enzymes A and B together are run side-by-side on a gel, which separates them according to size.

14 kb
10 kb
6 kb
2 kb
9 kb
8 kb
2 kb
14 kb
5 kb
9 kb
5 kb
3 kb
2 kb

3. The fragments are arranged so that the smaller ones produced by the simultaneous cut can be grouped to generate the larger ones produced by the individual enzymes.

2 kb 8 kb 9 kb
A A
5 kb 14 kb
B
2 kb 3 kb 5 kb 9 kb
A B A

4. A physical map is constructed.

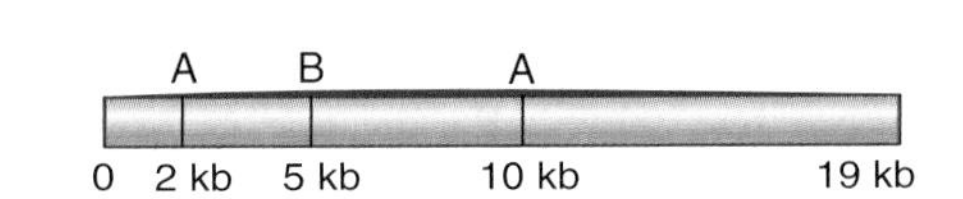

figure 18.1

RESTRICTION ENZYMES CAN BE USED TO CREATE A PHYSICAL MAP. DNA is digested with two different restriction enzymes singly and in combination, then electrophoresed to separate the fragments. The location of sites can be deduced by comparing the sizes of fragments from the individual reactions with the combined reaction.

The use of different stains allow the construction of a cytological map of the entire genome. These large-scale physical maps are like a map of an entire country, in that they encompass the whole genome, but at low resolution.

Cytological maps have been used to characterize chromosomal abnormalities associated with human diseases, such as chronic myelogenous leukemia. In this disease, a reciprocal translocation has occurred between chromosome 9 and chromosome 22, resulting in an altered form of tyrosine kinase that is always turned on, causing white-cell proliferation.

The use of hybridization with cloned DNA has added to the utility of chromosome-banding analysis. In this case, because the hybridization involves whole chromosomes, it is called *in situ hybridization.* It is done using fluorescently labeled probes, and so its complete name is **fluorescent in situ hybridization (FISH)** (figure 18.2).

Radiation hybrid maps

Radiation hybrid maps use radiation to fragment chromosomes randomly, then recover the fragments by fusing the irradiated cell to another cell. To construct a radiation hybrid map of the human genome, a human cell in culture is lethally irradiated, then fused to a rodent cell. The chromosome fragments produced by the radiation become integrated into the rodent-cell chromosomes. These fragments can be identified based on their banding patterns and by using known genes for FISH.

For the purposes of mapping, a series of these hybrid cells have been constructed that have overlapping fragments of human chromosomes representing the entire genome. The use of radiation hybrids in mapping is discussed in more detail later on.

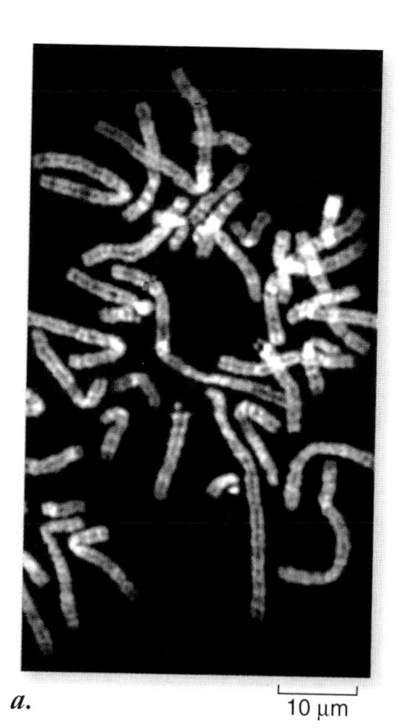

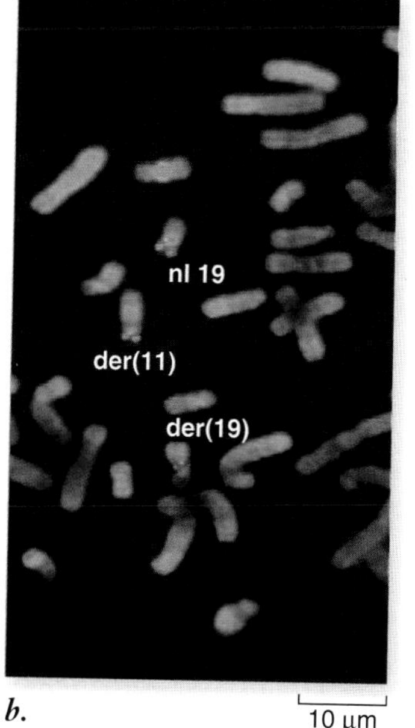

figure 18.2

USE OF FLUORESCENT IN SITU HYBRIDIZATION TO CORRELATE CLONED DNA WITH CYTOLOGICAL MAPS.
a. Part of a karyotype of human chromosomes using G banding. The red bands indicate hybridization with cloned DNA. ***b.*** The probe used in panel ***(a)*** shows a translocation in this patient that led to multiple congenital malformations and mental retardation.

Sequence-tagged sites provide a common language for physical maps

The construction of a physical map for a large genome requires the efforts of many laboratories in different locations. A variety of difficulties arose in comparing data from different labs, as well as integrating different types of landmarks used on physical and genetic maps.

In the early days of the Human Genome Project, this problem was addressed by the creation of a common molecular language that could be used to describe the different types of landmarks.

Defining common markers

Since all genetic information is ultimately based on DNA sequence, it was important for this common language to be sequence-based, but not to require generating a large amount of sequence for any landmark. The solution was the **sequence-tagged site,** or **STS.** This site is a small stretch of DNA that is unique in the genome, that is, it only occurs once.

The boundary of the STS is defined by PCR primers, so the presence of the STS can be identified by PCR using any DNA as a template (see chapter 17). These sites need to be only 200–500 bp long, an amount of sequence that can be determined easily. The STS can contain any other kind of landmark—for example, part of a cloned gene that has been genetically mapped, or a restriction site that is polymorphic. Any marker that has been mapped can be converted to an STS by sequencing only 200–500 bp.

The use of STSs

As maps are generated, new STSs are identified, and added to a database. For each STS, the database indicates the sequence of the STS, its location in the genome, and the PCR primers needed to identify it. Any researcher is then able to identify the presence or absence of any STS in the DNA that he or she is analyzing.

Fragments of DNA can be pieced together using STSs by identifying overlapping regions in fragments. Because of the high density of STSs in the human genome and the relative ease of identifying an STS in a DNA clone, investigators were able to develop physical maps on the huge scale of the 3.2-gigabase genome in the mid-1990s (figure 18.3). STSs essentially provide a scaffold for assembling genome sequences.

Genetic maps provide a link to phenotypes

The first genetic (linkage) map was made in 1911 when Alfred Sturtevant mapped five genes in *Drosophila.* Distances on a genetic map are measured in centimorgans (cM) in honor of the geneticist T. H. Morgan. One centimorgan corresponds to 1% recombination between two loci. Today, 14,065 genes have been mapped on the *Drosophila* genome.

Linkage mapping can be done without knowing the DNA sequence of a gene. Computer programs make it possible to create a linkage map for a thousand genes at a time. But a few limitations to genetic maps exist. One is that distances between genes determined by recombination frequencies do not directly correspond with physical distance on a chromosome. The conformation of DNA between genes varies, and this conformation can affect the frequency

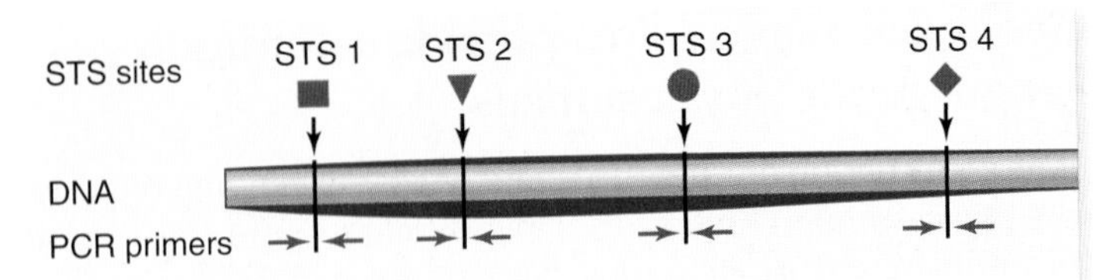

1. The location of 4 STSs in the genome is shown. PCR is used to amplify each STS from different clones in a library. Amplifying each STS by PCR generates a unique fragment that can be identified.

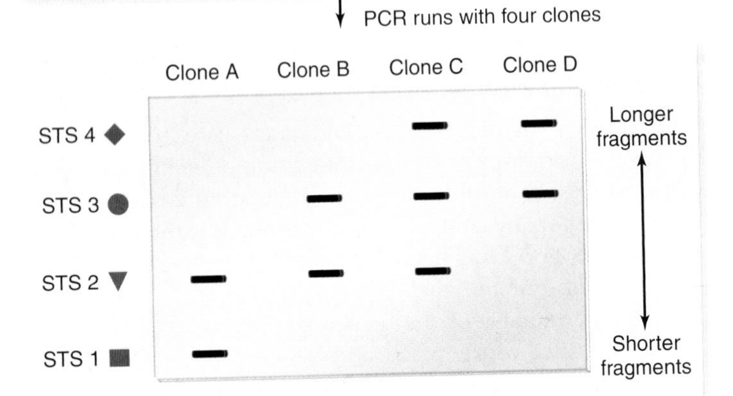

2. The products of the PCR reactions are separated by gel electrophoresis producing a different size fragment for each STS.

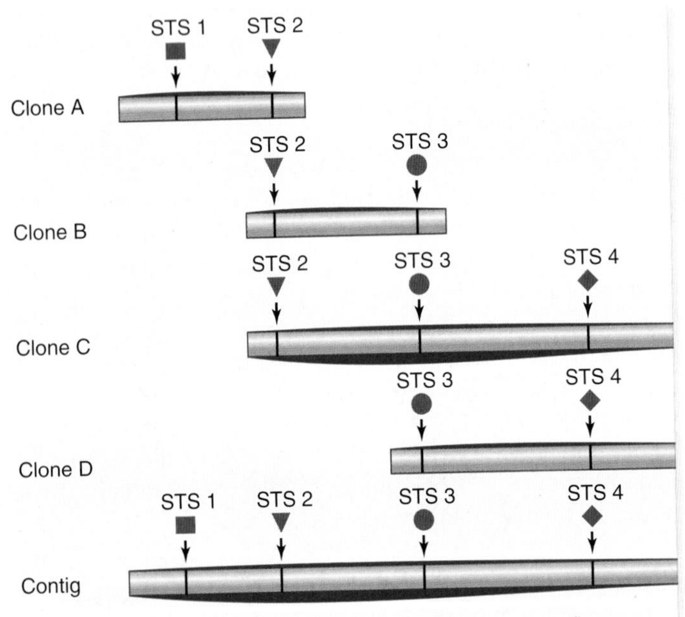

3. The presence or absence of each STS in the clones identifies regions of overlap. The final result is a contiguous sequence (contig) of overlapping clones.

figure 18.3

CREATING A PHYSICAL MAP WITH SEQUENCE-TAGGED SITES. The presence of landmarks called sequence-tagged sites, or STSs, in the human genome made it possible to begin creating a physical map large enough in scale to provide a foundation for sequencing the entire genome. (*1*) Primers (*green arrows*) that recognize unique STSs are added to a cloned segment of DNA, followed by DNA replication via PCR. (*2*) The PCR products from each reaction are separated based on size on a DNA gel, and the STSs contained in each clone are identified. (*3*) The cloned DNA segments are then aligned based on overlapping STSs to create a contig.

of recombination. Another limitation is that not all genes have obvious phenotypes that can be followed in segregating crosses.

As described in chapter 13, the human genetic map is quite dense, with a marker roughly every 1 cM. This level of detail would have been unheard of 20 years ago, and it was made possible by development of molecular markers that do not cause a phenotype change.

The most common type of markers are short repeated sequences, called short tandem repeats, or STR loci, that differ in repeat length between individuals. These repeats are identified by using PCR to amplify the region containing the repeat, then analyzing the products using electrophoresis. Once a map is constructed using these markers, genes with alleles that cause a disease state can be mapped relative to the molecular landmarks. Thirteen of these STR loci form the basis for modern DNA fingerprinting developed by the FBI. The alleles for these 13 loci are what is cataloged in the CODIS database used to identify criminal offenders.

Physical maps can be correlated with genetic maps

We need to be able to correlate genetic maps with physical maps, particularly genome sequences, to aid in finding physical sequences for genes that have been mapped genetically.

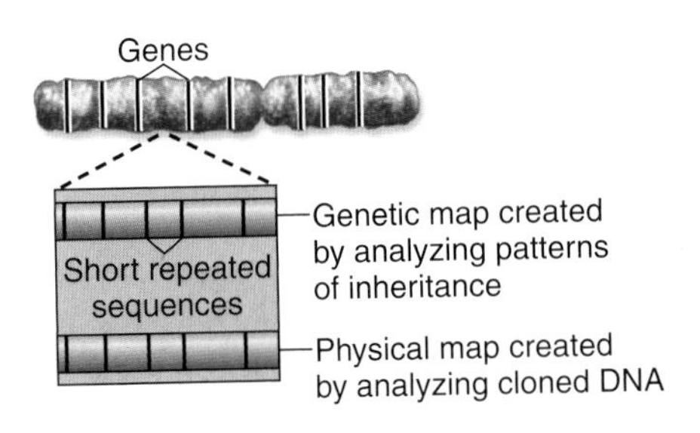

The problem in finding genes is that the resolution of genetic maps at present is not nearly as fine-grained as the genome sequence. Markers that are 1 cM apart may be as much as a million base pairs apart.

Since the markers used to construct genetic maps are now primarily molecular markers, they can be easily located within a genome sequence. Similarly, any gene that has been cloned can be placed within the genome sequence and can also be mapped genetically. This provides an automatic correlation between the two maps. The problem in terms of finding genes that have been mapped genetically but not isolated as molecular clones lies in the nature of genetic maps. Distances measured on genetic maps are not uniform due to variation in recombination frequency along the chromosome. So 1 cM of genetic distance will translate to different numbers of basepairs in different regions.

Radiation hybrid maps provide an alternative to genetic maps and are easily correlated with physical maps. These radiation hybrid maps consist of simple binary data: the presence or absence of a particular molecular marker in each cell in a radiation hybrid panel (described earlier). The more similar the score for any two markers, the closer they are on a chromosome. This is due to the random nature of the fragmentation by radiation. If two markers are close together, they will often be found on the same fragment, the farther apart they are, the less likely they will be on the same fragment.

This technique allows any kind of molecular marker to be ordered in the genome, including markers that are not polymorphic and therefore not suitable for genetic mapping. This also allows the integration of genetic and physical maps as both types of markers can be placed on the same radiation hybrid map. This is most useful in the early stages of a large-scale sequencing project. Such maps have been constructed for most animal species of interest to basic researchers, as well as domestic animals that are of economic importance and even companion animals such as dogs and cats. Most animal genome-sequencing projects include this kind of analysis. Currently in humans, this technique is being used to identify the location in the genome of all known transcripts.

All of these different kinds of maps are then stored in databases so they can be aligned and viewed. The National Center for Biotechnology Information (NCBI) is a branch of the National Library of Medicine, and it serves as the U.S. repository for these data and more. Similar databases exist in Europe and Japan, and all are kept current. An enormous storehouse of information is available for use by biological researchers worldwide.

Maps of genomes can by either physical maps or genetic maps. Physical maps include cytogenic maps of chromosome banding, restriction maps, or radiation hybrid maps. Genetic maps are correlated with physical maps by using DNA markers. Radiation hybrids can also be used to construct maps based on the probability of breakage by radiation occurring between two sites.

18.2 Whole-Genome Sequencing

The ultimate physical map is the base-pair sequence of the entire genome. In the early days of molecular biology, all sequencing was done manually, and was both time- and labor-intensive. As mentioned in chapter 17, the development of machines to automate this process increased the rate of sequence generation.

Large-scale genome sequencing requires the use of high-throughput automated sequencing and computer analysis (figure 18.4). Genome sequencing is one case in which technology drove the science, rather than the other way around. In a few hours, an automated sequencer can sequence the same number of base-pairs that a technician could manually sequence in a year—up to 50,000 bp. Without the automation of sequencing, it would have been impossible to sequence large, eukaryotic genomes like that of humans.

Genome sequencing requires larger molecular clones

Although it would be ideal to isolate DNA from an organism, add it to a sequencer, and then come back in a week or two to pick up a computer-generated printout of the genome sequence for that organism, scientific life is currently not quite that simple. Sequencers provide accurate sequences for DNA segments up to 800 bp long. Even then, errors are possible. So, 5–10 copies of a genome are sequenced to reduce errors.

Even with reliable sequence data in hand, each individual sequencing run produces a relatively small amount of sequence. Thus, the genome must be fragmented, and then individual molecular clones isolated for sequencing (see chapter 17).

Artificial chromosomes

As described in chapter 17, the development of artificial chromosomes has allowed scientists to clone larger pieces of DNA. The first generation of these new vectors were yeast artificial chromosomes (YACs). These are constructed by using a yeast origin of replication and centromere sequence, then adding foreign DNA to this construct. The origin of replication allows the artificial chromosome to replicate independently of the rest of the genome, and the centromere sequences make the chromosome mitotically stable.

YACs were useful for cloning larger pieces of DNA but they had many drawbacks, including a tendency to rearrange, or to lose portions of DNA by deletion. Despite the difficulties, the

figure 18.4

AUTOMATED SEQUENCING. This sequence facility simultaneously runs multiple automated sequencers, each processing 96 samples at a time.

YACs were used early on to construct physical maps by restriction enzyme digestion of the YAC DNA.

The artificial chromosomes most commonly used now, particularly for large-scale sequencing, are made in *E. coli.* These bacterial artificial chromosomes (BACs) are a logical extension of the use of bacterial plasmids. BAC vectors accept DNA inserts between 100 and 200 kb long. The downside of BAC vectors is that, like the bacterial chromosome, they are maintained as a single copy while plasmid vectors exist at high copy numbers.

Human artificial chromosomes

Human artificial chromosomes have been constructed to introduce large segments of human DNA into cultured cells. These artificial chromosomes are usually constructed by fragmentation of chromosomes and centromeric sequences. At present, they are circular, but some can still segregate correctly during mitosis up to 98% of the time. The construction of linear human artificial chromosomes is not yet possible.

Whole-genome sequencing is approached in two ways: clone-by-clone and shotgun

Sequencing an entire genome is an enormous task. Two ways of approaching this challenge have been developed, one a step-by-step approach, and one that attempts to take on the whole thing at once and depends on computers to sort out the data. The two techniques grew out of competing projects to sequence the human genome, as described in the next section.

Clone-by-clone sequencing

The cloning of large inserts in BACs facilitates the analysis of entire genomes. The strategy most commonly pursued is to construct a physical map first, and then use it to place the site of BAC clones for later sequencing.

Aligning large portions of a chromosome requires identifying regions that overlap between clones. This can be accomplished either by constructing restriction maps of each BAC clone, or by identifying STSs found in clones. If two BAC clones have the same STS, then they must overlap.

The alignment of a number of BAC clones results in a contiguous stretch of DNA called a *contig*. The individual BAC clones can then be sequenced 500 bp at a time to produce the sequence of the entire contig (figure 18.5*a*). This strategy of physical mapping followed by sequencing is called **clone-by-clone sequencing.**

Shotgun sequencing

The idea of **shotgun sequencing** is simply to randomly cut the DNA into small fragments, sequence all cloned fragments, and then use a computer to put together the overlaps (figure 18.5*b*). This terminology actually goes back to the early days of molecular cloning when the construction of a library of random cloned fragments was referred to as *shotgun cloning*. This approach is much less labor-intensive than the clone-by-clone method, but it requires much greater computer power to assemble the final sequence and very efficient algorithms to find overlaps.

Unlike the clone-by-clone approach, shotgun sequencing does not tie the sequence to any other information about the genome. Many investigators have used both clone-by-clone and shotgun-sequencing techniques, and such hybrid approaches are becoming the norm. This combination has the strength of tying the sequence to a physical map while greatly reducing the labor involved. The two methods are shown graphically in figure 18.5.

Assembler programs compare multiple copies of sequenced regions in order to assemble a **consensus sequence,** that is, a sequence that is consistent across all copies. Although computer assemblers are incredibly powerful, final human analysis is required after both clone-by-clone and shotgun sequencing to determine when a genome sequence is sufficiently accurate to be useful to researchers.

The Human Genome Project used both sequencing methods

The vast scale of genomics has ushered in a new way of doing biological research involving large teams. Although a single individual can isolate and manually sequence a molecular clone for a single gene, a huge genome like the human genome requires the collaborative efforts of hundreds of researchers.

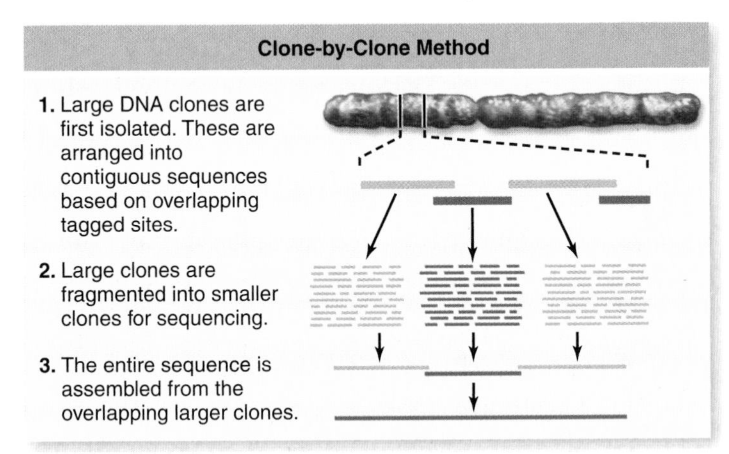

a.

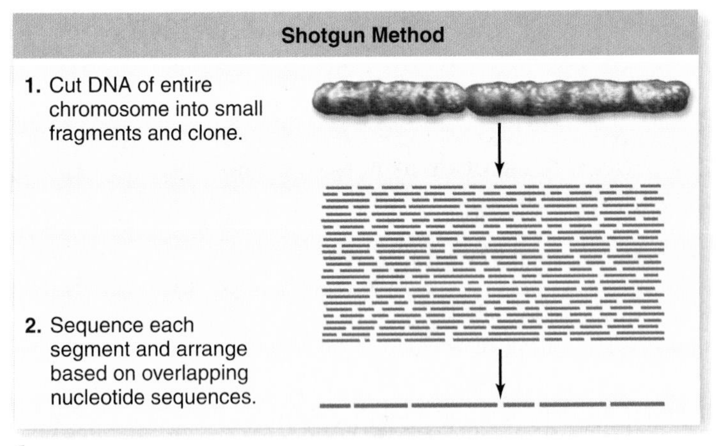

b.

figure 18.5

COMPARISON OF SEQUENCING METHODS. ***a.*** The clone-by-clone method uses large clones assembled into overlapping regions by STSs. Once assembled, these can be fragmented into smaller clones for sequencing. ***b.*** In the shotgun method the entire genome is fragmented into small clones and sequenced. Computer algorithms assemble the final DNA sequence based on overlapping nucleotide sequences.

The Human Genome Project originated in 1990 when a group of American scientists formed the International Human Genome Sequencing Consortium. The goal of this publicly funded effort was to use a clone-by-clone approach to sequence the human genome. Both genetic and physical maps were enhanced and published in the 1990s and used as scaffolding to sequence each chromosome.

Then, in May, 1998, Craig Venter, who had sequenced *Haemophilus influenzae*, announced that he had formed a private company to sequence the human genome. He proposed to shotgun-sequence the 3.2-gigabase genome in only two years. The consortium rose to the challenge, and the race to sequence the human genome began. The upshot was a tie of sorts. On June 26, 2000, the two groups jointly announced success, and each published its findings simultaneously in 2001. The consortium's draft alone included 248 names on its partial list of authors.

The draft sequence of the human genome was just the beginning. Gaps in the sequence are still being filled, and the map is constantly being refined. In 2004, the "finished" sequence was published and announced as the reference sequence (REFSEQ) in the databases. This sequence is down to only 341 gaps, a 400-fold reduction in gaps, and it now includes 99% of the euchromatic sequence, up from 95%. The reference sequence has an error rate of 1 per 100,000 bases.

More significantly, research on the whole genome can move ahead. Now that the ultimate physical map is in place and is being integrated with the genetic map, diseases that result from defects in more than one gene, such as diabetes, can be addressed. Comparisons with other genomes are already changing our understanding of genome evolution (see chapter 24).

Sequencing of entire genomes requires the use of automated sequencers running many samples in parallel. Larger fragments of DNA are needed for sequencing, and artificial chromosomes have provided a way of working with these large fragments. Two approaches were developed for whole genome sequencing, one that uses clones already aligned by physical mapping (clone-by-clone sequencing), and one that involves sequencing random clones and using a computer to assemble the final sequence (shotgun sequencing). In either case, significant computing power is necessary to assemble a final sequence.

18.3 Characterizing Genomes

Automated sequencing technology has produced huge amounts of sequence data, eventually allowing the sequencing of entire genomes. This has allowed researchers studying complex problems to move beyond approaches restricted to the analysis of individual genes. Sequencing projects in themselves are descriptive analysis that tells us nothing about the organization of genomes, let alone the function of gene products and how they may be interrelated. Additional research and evaluation has given us both answers and new puzzles.

The Human Genome Project found fewer genes than expected

For many years, geneticists had estimated the number of human genes to be around 100,000. This estimate, although based on some data, was really just a guess. Imagine researchers' surprise to find that the number now appears to be only around 25,000! This represents about twice as many genes as *Drosophila* and fewer genes than rice (figure 18.6). Clearly the complexity of an organism is not a simple function of the number of genes in its genome.

Finding genes in sequence data requires computer searches

Once a genome has been sequenced, the next step is to determine which regions of the genome contain which genes, and what those genes do. A lot of information can be mined from the sequence data. Using markers from physical maps and information from genetic maps, it is possible to find the sequence of the small percentage of genes that are identified by mutations with an observable (phenotypic) effect. Genes can

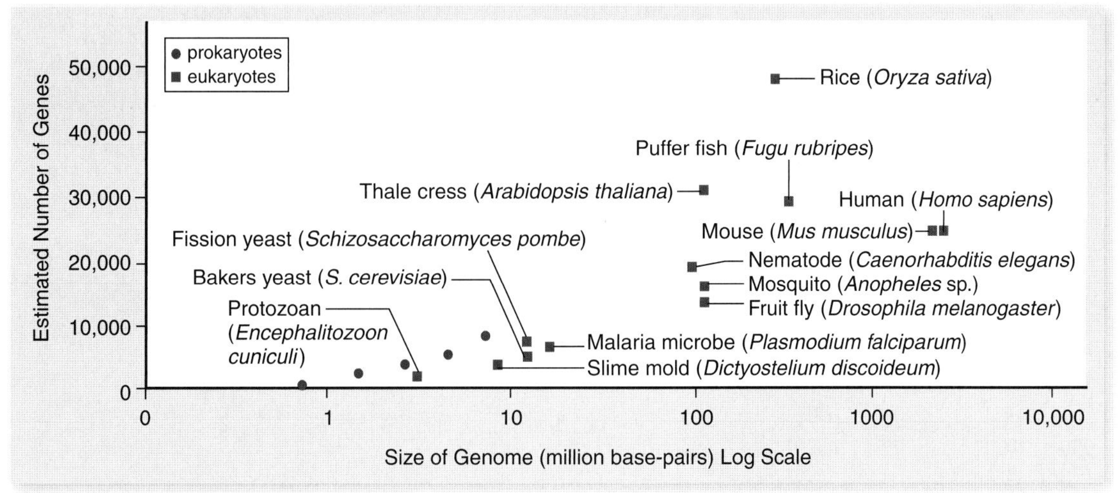

figure 18.6

SIZE AND COMPLEXITY OF GENOMES. In general, eukaryotic genomes are larger and have more genes than prokaryotic genomes, although the size of the organism is not the determining factor. The mouse genome is nearly as large as the human genome, and the rice genome contains more genes than the human genome.

also be found by comparing expressed sequences to genomic sequences. The analysis of expressed sequences is discussed later in this section.

Locating starts and stops

Information in the nucleotide sequence itself can also be used in the search for genes. A gene begins with a start codon, such as ATG, and it contains no stop codons (TAA, TGA, or TAG) for a distance long enough to encode a protein. This coding region is referred to as an **open reading frame (ORF).** Although these sequences between a start and a stop are likely to be genes, they may or may not actually be translated into a functional protein. Sequences for potential genes need to be tested experimentally to determine whether they have a function.

The addition of information to the basic sequence information, like identifying ORFs, is called sequence **annotation.** This process is what converts simple sequence data into something that we can recognize based on landmarks such as regions that are transcribed, and regions that are known or thought to encode proteins.

Inferring function across species: The BLAST algorithm

It is also possible to search genome databases for sequences that are homologous to known genes in other species. A researcher who has isolated a molecular clone for a gene of unknown function can search the database for similar sequences to infer function. The tool that makes this possible is a search algorithm called BLAST that is provided by the NCBI to search their databases. Using email and a networked computer, one can submit a sequence to the BLAST server and get back a reply with all possible similar sequences contained in the sequence database.

Using these techniques, sequences that are not part of ORFs have been identified that are conserved over millions of years of evolution. These sequences may be important for the regulation of the genes contained in the genome.

Using computer programs to search for genes, to compare genomes, and to assemble genomes are only a few of the new genomics approaches falling under the heading of **bioinformatics.**

Genomes contain both coding and noncoding DNA

When genome sequences are analyzed, regions that encode proteins and other regions that do not encode proteins are revealed. For many years investigators had known of the latter, but they did not know the extent and nature of the noncoding DNA. We first consider the types of coding DNA that have been found, then move on to look at types of noncoding DNA.

Protein-encoding DNA in eukaryotes

Four different classes of protein-encoding genes are found in eukaryotic genomes, differing largely in gene copy number.

Single-copy genes. Many genes exist as single copies on a particular chromosome. Most mutations in these genes result in recessive Mendelian inheritance.

Segmental duplications. Sometimes whole blocks of genes are copied from one chromosome to another, resulting in *segmental duplication.* Blocks of similar genes in the same order are found throughout the human genome. Chromosome 19 seems to have been the biggest borrower, sharing blocks of genes with 16 other chromosomes.

Multigene families. As more has been learned about eukaryotic genomes, many genes have been found to exist as parts of *multigene families,* groups of related but distinctly different genes that often occur together in clusters. These genes appear to have arisen from a single ancestral gene that duplicated during an uneven meiotic crossover in which genes were added to one chromosome and subtracted from the other. These multigene families may include silent copies called *pseudogenes,* inactivated by mutation.

Tandem clusters. Identical copies of genes can also be found in *tandem clusters.* These genes are all transcribed simultaneously, which increases the amount of mRNA available for protein production. Tandem clusters also include genes that do not encode proteins, such as rRNA genes that are typically present in clusters of several hundred copies.

Noncoding DNA in eukaryotes

Sequencing of several eukaryotic genomes has now been completed, and one of the most notable characteristics is the amount of noncoding DNA they possess. The Human Genome Project has revealed a particularly startling picture. Each of your cells has about 6 feet of DNA stuffed into it, but of that, less than 1 inch is devoted to genes! Nearly 99% of the DNA in your cells has little or nothing to do with the instructions that make you what you are.

True genes are scattered about the human genome in clumps among the much larger amount of noncoding DNA, like isolated hamlets in a desert. Six major sorts of noncoding human DNA have been described. (Table 18.1 shows the composition of the human genome, including noncoding DNA.)

Noncoding DNA within genes. As discussed in chapter 15, a human gene is not simply a stretch of DNA, like the letters of a word. Instead, a human gene is made up of numerous fragments of protein-encoding information (exons) embedded within a much larger matrix of noncoding DNA (introns). Together, introns make up about 24% of the human genome and exons less than 1.5%.

Structural DNA. Some regions of the chromosomes remain highly condensed, tightly coiled, and untranscribed throughout the cell cycle. Called *constitutive heterochromatin,* these portions tend to be localized around the centromere or located near the ends of the chromosome, at the telomeres.

TABLE 18.1 Classes of DNA Sequences Found in the Human Genome

Class	Frequency (%)	Description
Protein-encoding genes	1.5	Translated portions of the 25,000 genes scattered about the chromosomes
Introns	24	Noncoding DNA that constitutes the great majority of each human gene
Segmental duplications	5	Regions of the genome that have been duplicated
Pseudogenes (inactive genes)	2	Sequence that has characteristics of a gene but is not a functional gene
Structural DNA	20	Constitutive heterochromatin, localized near centromeres and telomeres
Simple sequence repeats	3	Stuttering repeats of a few nucleotides such as CGG, repeated thousands of times
Transposable elements	45	21%: Long interspersed elements (LINEs), which are active transposons 13%: Short interspersed elements (SINEs), which are active transposons 8%: Retrotransposons, which contain long terminal repeats (LTRs) at each end 3%: DNA transposon fossils

Simple sequence repeats. Scattered about chromosomes are **simple sequence repeats (SSRs).** An SSR is a one- to six-nucleotide sequence such as CA or CGG, repeated like a broken record thousands and thousands of times. SSRs can arise from DNA replication errors. SSRs make up about 3% of the human genome.

Segmental duplications. Blocks of genomic sequences composed of from 10,000 to 300,000 bp have duplicated and moved either within a chromosome or to a nonhomologous chromosome.

Pseudogenes. These are inactive genes that may have lost function because of mutation.

Transposable elements. Fully 45% of the human genome consists of mobile bits of DNA called *transposable elements.* Some of these elements code for proteins, but many do not. Because of the significance of these elements, we describe them more fully below.

Transposable elements: Mobile DNA

Discovered by Barbara McClintock in 1950, **transposable elements,** also termed *transposons* and *mobile genetic elements,* are bits of DNA that are able to move from one location on a chromosome to another. Barbara McClintock received the Nobel Prize in 1983 for discovery of these elements and their unexpected ability to change location.

Transposable elements move around in different ways. In some cases, the transposon is duplicated, and the duplicated DNA moves to a new place in the genome, so the number of copies of the transposon increases. Other types of transposons are excised without duplication and insert themselves elsewhere in the genome. The role of transposons in genome evolution is discussed in chapter 24.

Human chromosomes contain four sorts of transposable elements. Fully 21% of the genome consists of **long interspersed elements (LINEs).** These ancient and very successful elements are about 6000 bp long, and they contain all the equipment needed for transposition. LINEs encode a reverse transcriptase enzyme that can make a cDNA copy of the transcribed LINE RNA. The result is a double-stranded segment that can reinsert into the genome rather than undergo translation into a protein. Since these elements use an RNA intermediate, they are termed *retrotransposons.*

Short interspersed elements (SINEs) are similar to LINEs, but they cannot transpose without using the transposition machinery of LINEs. Nested within the genome's LINEs are over half a million copies of a SINE element called Alu (named for a restriction enzyme that cuts within the sequence). The Alu SINE represents 10% of the human genome. Like a flea on a dog, Alu moves with the LINE it resides within. Just as a flea sometimes jumps to a different dog, so Alu sometimes uses the enzymes of its LINE to move to a new chromosome location. Alu can jump right into genes, causing harmful mutations.

Two other sorts of transposable elements are also found in the human genome: 8% of the human genome is devoted to retrotransposons called **long terminal repeats (LTRs).** Although the transposition mechanism is a bit different from that of LINEs, LTRs also use reverse transcriptase to ensure that copies are double-stranded and can reintegrate into the genome.

Some 3% of the genome is devoted to dead transposons, elements that have lost the signals for replication and can no longer move.

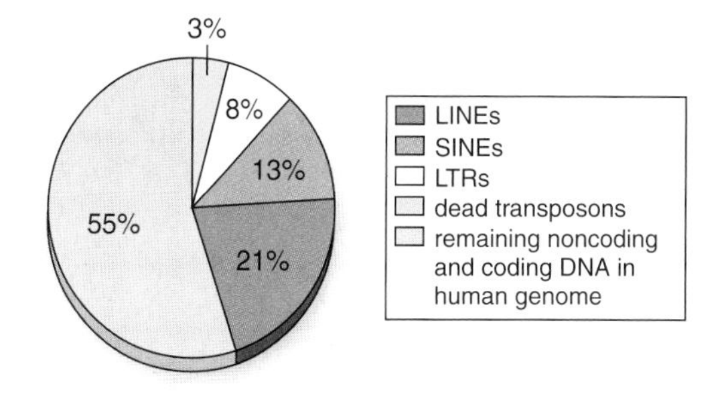

inquiry

How do you think these repetitive elements would affect the determination of gene order?

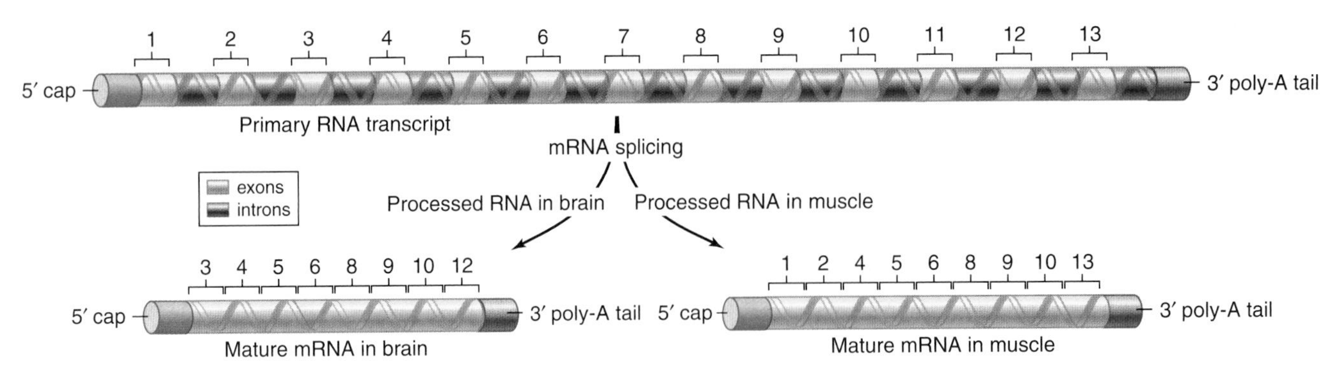

figure 18.7

ALTERNATIVE SPLICING CAN RESULT IN THE PRODUCTION OF DIFFERENT mRNAs FROM THE SAME CODING SEQUENCE. In some cells, exons can be excised along with neighboring introns, resulting in different proteins. Alternative splicing explains why 25,000 human genes can code for three to four times as many proteins.

Expressed sequence tags identify genes that are transcribed

Given the complexity of coding and noncoding DNA, it is important to be able to recognize regions of the genome that are actually expressed—that is, transcribed and then translated.

Because DNA is easier to work with than protein, one approach is to isolate mRNA, use this to make cDNA, then sequence one or both ends of as many cDNAs as possible. With automated sequencing, this task is not difficult, and these short sections of cDNA have been named **expressed sequence tags (ESTs).** An EST is another form of STS, and thus it can be included in physical maps. This technique does not tell us anything about the function of any particular EST, but it does provide one view, at the whole-genome level, of what genes are expressed, at least as mRNAs.

ESTs have been used to identify 87,000 cDNAs in different human tissues. About 80% of these cDNAs were previously unknown. You may wonder at this point how the estimated 25,000 genes of the human genome can result in 87,000 different cDNAs. The answer lies in the modularity of eukaryotic genes, which consist of exons interspersed with introns, as described in chapter 15.

Following transcription in eukaryotes, the introns are removed, and exons are spliced together. In some cells, some of the splice sites are skipped, and one or more exons is removed along with the introns. This process, called *alternative splicing* (figure 18.7), yields different proteins that can have different functions. Thus, the added complexity of proteins in the human genome comes not from additional genes, but from new ways to put existing parts of genes together.

SNPs are single-base differences between individuals

One fact becoming clear from analysis of the human genome is that a huge amount of genetic variation exists in our species. This information has practical use.

Single-nucleotide polymorphisms (SNPs) are sites where individuals differ by only a single nucleotide. To be classified as a polymorphism, an SNP must be present in at least 1% of the population. At present, the International SNP Map Working Group has identified 50,000 SNPs in coding regions of the genome and an additional 1.4 million in noncoding DNA. It is estimated that this represents about 10% of the variation available.

These SNPs are being used to look for associations between genes. We expect that the genetic recombination occurring during meiosis randomizes all but the most tightly linked genes. We call the tendency for genes *not* to be randomized **linkage disequilibrium.** This kind of association can be used to map genes.

The preliminary analysis of SNPs shows that many are in linkage disequilibrium. This unexpected result has led to the idea of genomic **haplotypes,** or regions of chromosomes that are not being exchanged by recombination. The existence of haplotypes allows the genetic characterization of genomic regions by describing a small number of SNPs (figure 18.8). If these haplotypes stand up to further analysis, they could greatly aid in mapping the genetic basis of disease. The Human Genome Project is now working on a haplotype map of the genome.

The human genome contains far fewer genes than expected: about 25,000. A significant amount of noncoding DNA is found in all eukaryotic genomes. Coding sequences can be single copy, repeated in clusters, part of segmental duplications, or part of a gene family. Among the human genome are a variety of transposable elements that are repeated many times. These elements are capable of movement in the genome and are found in all eukaryotic genomes. The number and location of expressed genes can be estimated by sequencing the ends of randomly selected cDNAs to produce expressed sequence tags (ESTs).

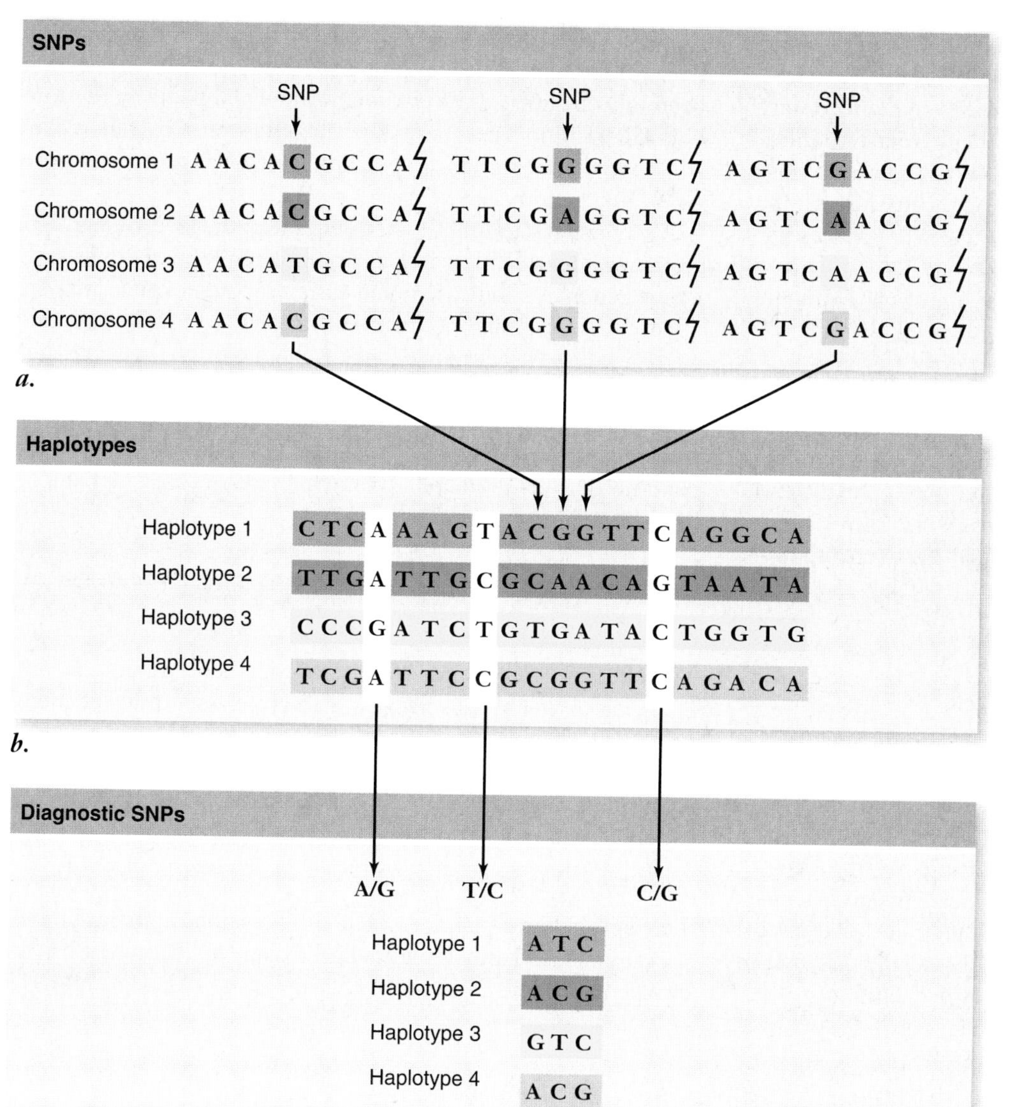

figure 18.8

CONSTRUCTION OF A HAPLOTYPE MAP. Single-nucleotide polymorphisms (SNPs) are single-base differences between individuals. Sections of DNA sequences from four individuals are shown in ***(a)***, with three SNPs indicated by arrows. ***b.*** These three SNPs are shown aligned along with 17 other SNPs from this chromosomal region. This represents a haplotype map for this region of the chromosome. Haplotypes are regions of the genome that are not exchanged by recombination during meiosis. ***c.*** Haplotypes can be identified using a small number of diagnostic SNPs that differ between the different haplotypes. In this case, 3 SNPs out of the 20 in this region are all that are needed to uniquely identify each haplotype. This greatly facilitates locating disease-causing genes, as haplotypes represent large regions of the genome that behave as a single site during meiosis.

18.4 Genomics and Proteomics

To fully understand how genes work, we need to characterize the proteins they produce. This information is essential in understanding cell biology, physiology, development, and evolution. In many ways, we continue to ask the same questions that Mendel asked, but at a much different level of organization.

Comparative genomics reveals conserved regions in genomes

With the large number of sequenced genomes, it is now possible to make comparisons at both the gene and genome level. One of the striking lessons learned from the sequence of the human genome is how very similar humans are to other organisms. More than half of the genes of *Drosophila* have human counterparts. Among mammals, the similarities are even greater. Humans have only 300 genes that have no counterpart in the mouse genome.

The flood of information from different genomes has given rise to a new field: **comparative genomics.** At this point, we have the complete sequences for close to 100 bacterial genomes. Among eukaryotes, we have the full genome sequences of both types of yeast used in genetics, *S. cerevisiae* and *S. pombe*, as well as the protist *Plasmodium*, the invertebrate animals *Drosophila* and *C. elegans*, and the vertebrates puffer fish (*Fugu* sp. and *Tetraodon* sp.), mouse, and human. In the plant kingdom, the genomes for *Arabidopsis* and rice have been completed. Most of these genomes, however, are draft sequences that include many gaps in regions of highly repetitive DNA.

The use of comparative genomics to ask evolutionary questions is also a field of great promise. The comparison of the many prokaryotic genomes already indicates a greater degree of lateral gene transfer than was previously suspected. The latest round of animal genomes sequenced has included the chimpanzee, our closest living relative. The draft sequence of the chimp (*Pan troglodytes*) genome has just been completed, and comparisons between the chimp and human genome may allow us to unravel what makes us uniquely human.

The early returns from this sequencing effort confirm that our genomes differ by only 1.23% in terms of nucleotide substitutions. At first glance, the largest difference between our genomes actually appears to be in transposable elements. In humans, the SINEs have been threefold more active than in the chimp, but the chimp has acquired two elements that are not found in the human genome. The differences due to insertion and deletion of bases are fewer than substitutions but account for about 1.5% of the euchromatic sequence being unique in each genome.

Synteny allows comparison of unsequenced genomes

Similarities and differences between highly conserved genes can be investigated on a gene-by-gene basis between species. Genome science allows for a much larger scale approach to comparing genomes by taking advantage of synteny.

Synteny refers to the conserved arrangements of segments of DNA in related genomes. Physical mapping techniques can be used to look for synteny in genomes that have not been sequenced. Comparisons with the sequenced, syntenous segment in another species can be very helpful.

To illustrate this, consider rice, already sequenced, and its grain relatives maize, barley, and wheat, none of which have been fully sequenced. Even though these plants diverged more than 50 million years ago, the chromosomes of rice, corn, wheat, and other grass crops show extensive synteny (figure 18.9). In a genomic sense, "rice is wheat."

By understanding the rice genome at the level of its DNA sequence, identification and isolation of genes from grains with larger genomes should be much easier. DNA sequence analysis of cereal grains could be valuable for identifying genes associated with disease resistance, crop yield, nutritional quality, and growth capacity.

The rice genome, as mentioned earlier, has more genes than the human genome. However, rice still has a much smaller genome than its grain relatives, which also represent a major food source for humans.

Organelle genomes have exchanged genes with the nuclear genome

Mitochondria and chloroplasts are considered to be descendants of ancient bacterial cells living in eukaryotes as a result of endosymbiosis (chapter 4). Their genomes have been sequenced in some species, and they are most like prokaryotic genomes. The chloroplast genome, having about 100 genes, is minute compared with the rice genome, with 32,000 to 55,000 genes.

The chloroplast genome

The chloroplast, a plant organelle that functions in photosynthesis, can independently replicate in the plant cell because it has its own genome. The DNA in the chloroplasts of all land plants have

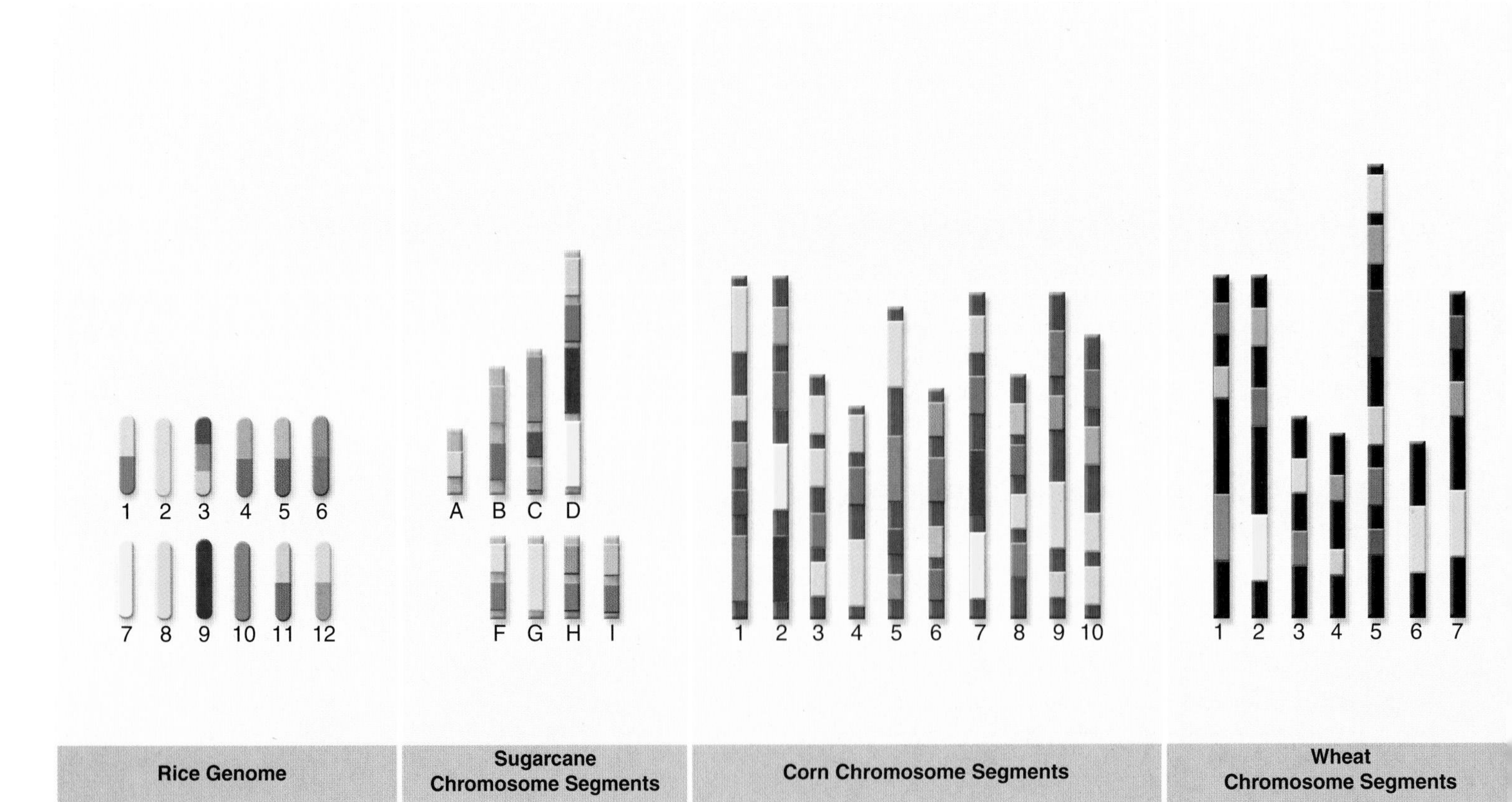

about the same number of genes, and they are present in about the same order. In contrast to the evolution of the DNA in the plant cell nucleus, chloroplast DNA has evolved at a more conservative pace and therefore shows a more easily interpretable evolutionary pattern when scientists study DNA sequence similarities. Chloroplast DNA is also not subject to modification caused by transposable elements or to mutations due to recombination.

Over time, some genetic exchange appears to have occurred between the nuclear and chloroplast genomes. For example, Rubisco, the key enzyme in the Calvin cycle of photosynthesis (chapter 8), consists of large and small subunits. The small subunit is encoded in the nuclear genome. The protein it encodes has a targeting sequence that allows it to enter the chloroplast and combine with large subunits, which are coded for and produced by the chloroplast.

The mitochondrial genome

Mitochondria are also constructed of components encoded by both the nuclear genome and the mitochondrial genome. For example, the electron transport chain (chapter 7) is made up of proteins that are encoded by both nuclear and mitochondrial genomes—and the pattern varies with different species. This observation implies a movement of genes from the mitochondria to the nuclear genome with some lineage-specific variation.

The evolutionary history of the localization of these genes is a puzzle. Comparative genomics and their evolutionary implications are explored in detail in chapter 24, after we have established the fundamentals of evolutionary theory.

Functional genomics reveals gene function at the genome level

Bioinformatics takes advantage of high-end computer technology to analyze the growing gene databases, look for relationships among genomes, and then hypothesize functions of genes based on sequence. Genomics is now shifting gears and moving back to hypothesis-driven science, to **functional genomics,** the study of the function of genes and their products.

Like sequencing whole genomes, finding how these genomes work requires the efforts of a large team. For example, an international community of researchers has come together with a plan to assign function to all of the 20,000–25,000 *Arabidopsis* genes by 2010 (Project 2010). One of the first steps is to determine when and where these genes are expressed. Each step beyond that will require additional enabling technology.

DNA microarrays

The earlier description of ESTs indicated that we could locate sequences that are transcribed on our DNA maps—but this tells us nothing about when and where these genes are turned on. To be able to analyze gene expression at the whole-genome level requires a representation of the genome that can be manipulated experimentally. This has led to the creation of **DNA microarrays** or "gene chips" (figure 18.10).

Preparation of a microarray To prepare a particular microarray, fragments of DNA are deposited on a microscope slide by a robot at indexed locations (i.e., an array). Silicon chips instead of slides can also be arrayed. These chips can then

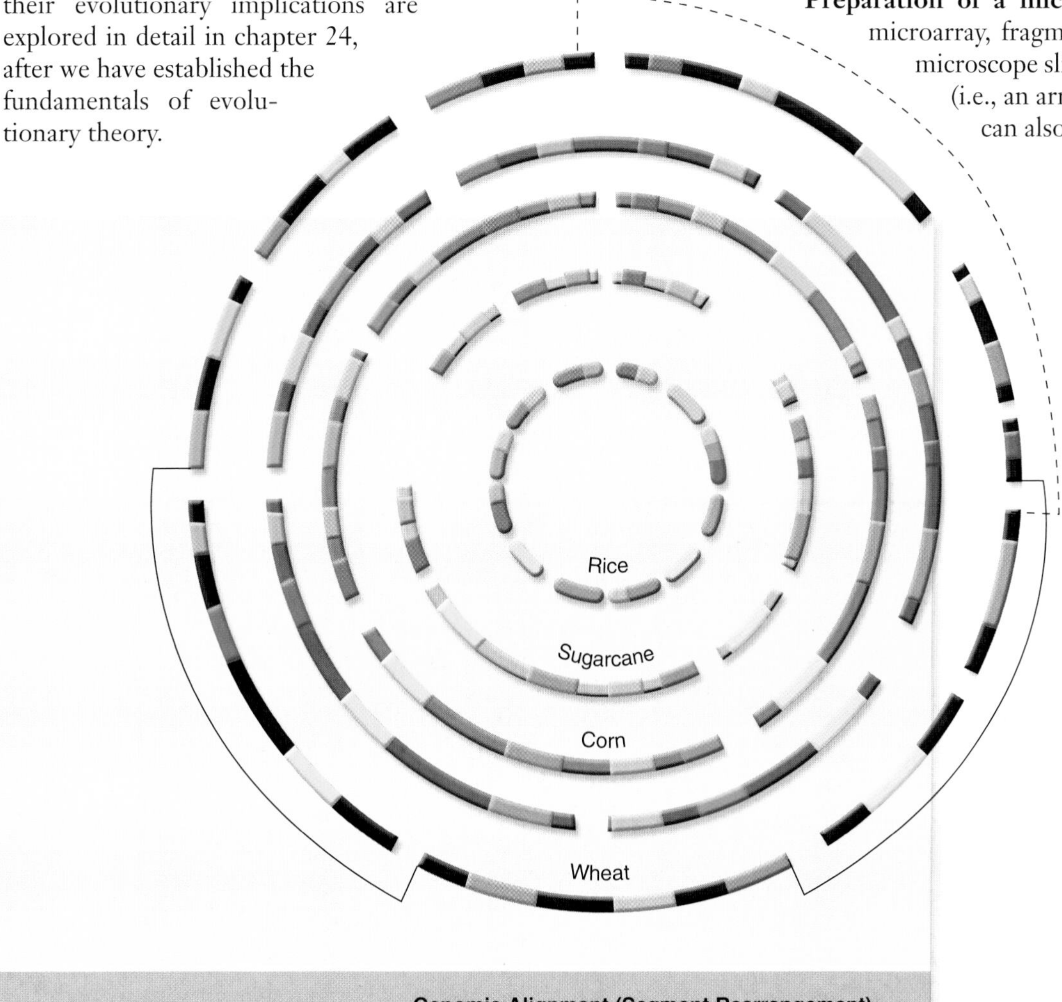

figure 18.9

GRAIN GENOMES ARE REARRANGEMENTS OF SIMILAR CHROMOSOME SEGMENTS. Shades of the same color represent pieces of DNA that are conserved among the different species but have been rearranged. By splitting the individual chromosomes of major grass species into segments and rearranging the segments, researchers have found that the genome components of rice, sugarcane, corn, and wheat are highly conserved. This implies that the order of the segments in the ancestral grass genome has been rearranged by recombination as the grasses have evolved.

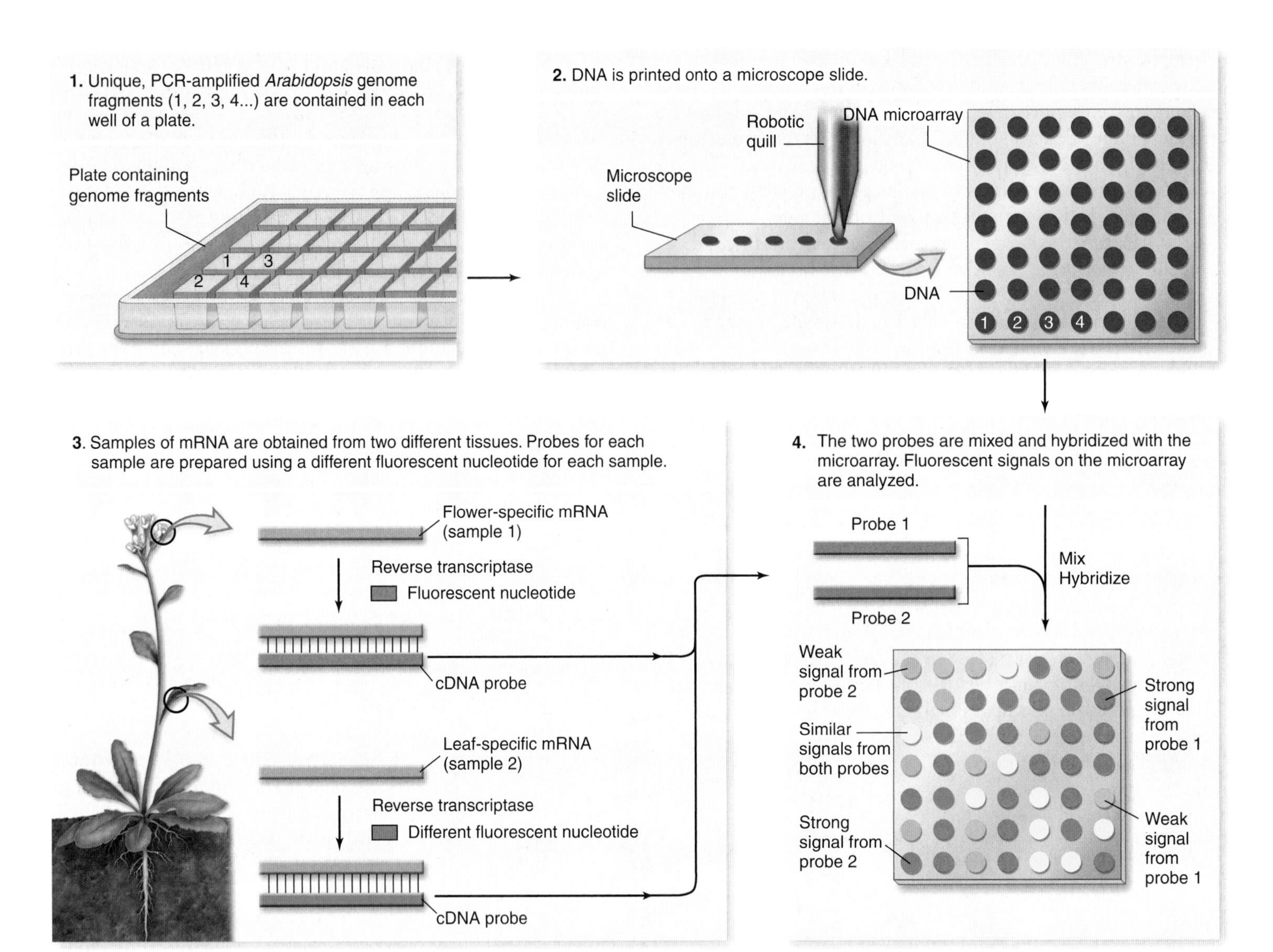

figure 18.10

MICROARRAYS. Microarrays are created by robotically placing DNA onto a microscope slide. The microarray can then be probed with RNA from tissues of interest to identify expressed DNA. The microarray with hybridized probes is analyzed and often displayed as a false-color image. If a gene is frequently expressed in one of the samples, the fluorescent signal will be strong (*red* or *green*) where the gene is located on the microarray. If a gene is rarely expressed in one of the samples, the signal will be weak (*pink* or *light green*). A yellow color indicates genes that are expressed at similar levels in each sample.

be used in hybridization experiments with labeled mRNA from different sources. This gives a high-level view of genes that are active and inactive in specific tissues.

Researchers are currently using a chip with 24,000 *Arabidopsis* genes on it to identify genes that are expressed developmentally in certain tissues or in response to environmental factors. RNA from these tissues can be isolated and used as a probe for these microarrays. Only those sequences that are expressed in the tissues will be present and will hybridize to the microarray.

Microarray analysis and cancer One of the most exciting uses of microarrays has been the profiling of gene expression patterns in human cancers. Microarray analysis has revealed that different cancers have different gene expression patterns. These findings are already being used to diagnose and design specific treatments for particular cancers.

From a large body of data, several patterns emerge:

1. Specific cancer types can be reliably distinguished from other cancer types and from normal tissue based on microarray data.
2. Subtypes of particular cancers often have different gene expression patterns in microarray data.
3. Gene expression patterns from microarray data can be used to predict disease recurrence, tendency to metastasize, and treatment response.

This represents an important step forward in both the diagnosis and treatment of human cancers.

Transgenics

How can we determine whether two genes from different species having similar sequences have the same function? And, how can we be sure that a gene identified by an annotation program actually functions as a gene in the organism? One way to address these questions is through **transgenics**—the creation of organisms containing genes from other species (transgenic organisms).

The technology for creating transgenic organisms was discussed in chapter 17; it is illustrated for plants in figure 18.11. For example, to test whether an *Arabidopsis* gene is homologous to a rice gene, the *Arabidopsis* gene can be inserted into cells of rice, which are then regenerated through cell culture into a rice plant (this type of cloning is discussed in chapter 42). Different markers can be incorporated into the gene so that its protein product can be visualized or isolated in the transgenic plant, demonstrating that the inserted gene is being transcribed. In some cases, the transgene (inserted foreign gene) may affect a visible phenotype. Of course, transgenics are but one of many ways to address questions about gene function.

Proteomics moves from genes to proteins

Proteins are much more difficult to study than DNA because of posttranslational modification and formation of protein complexes. And, as already mentioned, a single gene can code for multiple proteins using alternative splicing. Although all the DNA in a genome can be isolated from a single cell, only a portion of the proteome is expressed in a single cell or tissue.

Proteomics is the study of the **proteome**—all of the proteins encoded by the genome. Understanding the proteome for even a single cell will be a much more difficult task than determining the sequence of a genome. Because a single gene can produce more than one protein by alternative splicing, the first step is to characterize the **transcriptome**—all of the RNA that is present in a cell or tissue. Because of alternative splicing, both the transcriptome and the proteome are larger and more complex than the simple number of genes in the genome.

To make matters worse, a single protein can be modified posttranslationally to produce functionally different forms. The function of a protein can also depend on its association with other proteins. Nonetheless, since proteins perform most of the major functions of cells, understanding their diversity is essential.

inquiry

Why is the "proteome" likely to be different from simply the predicted protein products found in the complete genome sequence?

Predicting protein function

The use of new methods to quickly identify and characterize large numbers of proteins is the distinguishing feature between traditional protein biochemistry and proteomics. As with genomics, the challenge is one of scale.

Ideally, a researcher would like to be able to examine a nucleotide sequence and know what sort of functional

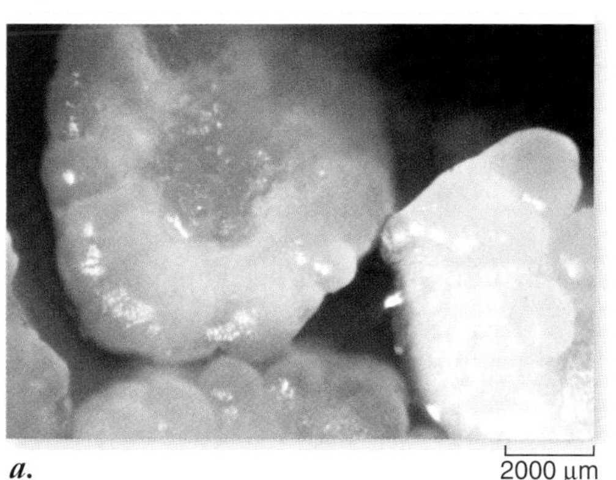

a.

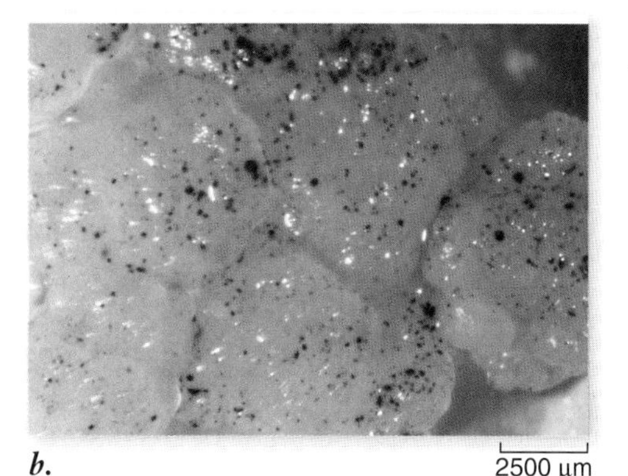

b.

c.

d.

figure 18.11

GROWTH OF A TRANSGENIC PLANT. DNA containing a gene for herbicide resistance was transferred into wheat (*Triticum aestivum*). The DNA also contains the *GUS* gene, which is used as a tag or label. The *GUS* gene produces an enzyme that catalyzes the conversion of a staining solution from clear to blue. ***a.*** Embryonic tissue just prior to insertion of foreign DNA. ***b.*** Following DNA transfer, callus cells containing the foreign DNA are indicated by color from the *GUS* gene (*blue spots*). ***c.*** Shoot formation in the transgenic plants growing on a selective medium. Here, the gene for herbicide resistance in the transgenic plants allows growth on the selective medium containing the herbicide. ***d.*** Comparison of growth on the selection medium for transgenic plants bearing the herbicide resistance gene (*left*) and a nontransgenic plant (*right*).

protein the sequence specifies. Databases of protein structures in different organisms can be searched to predict the structure and function of genes known only by sequence, as identified in genome projects. Analysis of these data provides a clearer picture of how gene sequence relates to protein structure and function. Having a greater number of DNA sequences available allows for more extensive comparisons as well as identification of common structural patterns as groups of proteins continue to emerge.

Fortunately, although there may be as many as a million different proteins, most are just variations on a handful of themes. The same shared structural motifs—barrels, helices, molecular zippers—are found in the proteins of plants, insects, and humans (figure 18.12; also see chapter 3 for more information on protein motifs). The maximum number of distinct motifs has been estimated to be fewer than 5000. About 1000 of these motifs have already been cataloged. Both publicly and privately financed efforts are now under way to detail the shapes of all the common motifs.

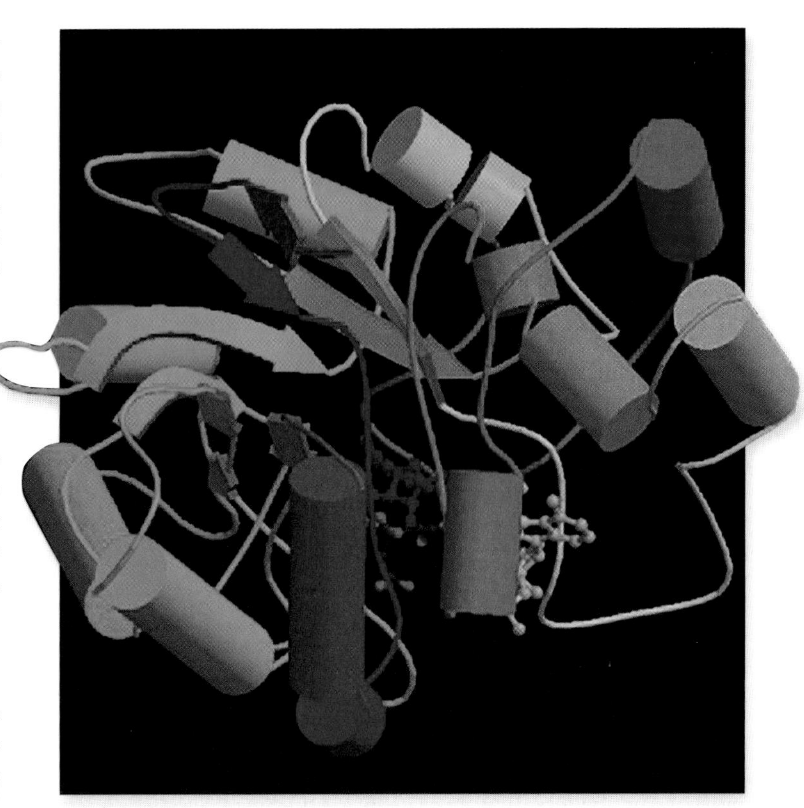

figure 18.12

COMPUTER-GENERATED MODEL OF AN ENZYME. Searchable databases contain known protein structures, including human aldose reductase shown here. Secondary structural motifs are shown in different colors.

inquiry

What is the relationship among genome, transcriptome, and proteome?

Protein microarrays

Protein microarrays, comparable to DNA microarrays, are being used to analyze large numbers of proteins simultaneously. Making a protein microarray starts with isolating the transcriptome of a cell or tissue. Then cDNAs are constructed and reproduced by cloning them into bacteria or viruses. Transcription and translation occur in the prokaryotic host, and micromolar quantities of protein are isolated and purified. These are then spotted onto glass slides.

Protein microarrays can be probed in at least three different ways. First, they can be screened with antibodies to specific proteins. Antibodies are labeled so that they can be detected, and the patterns on the protein array can be determined by computer analysis.

An array of proteins can also be screened with another protein to detect binding or other protein interactions. Thousands of interactions can be tested simultaneously. For example, calmodulin (which mediates Ca^{2+} function; see chapter 9) was labeled and used to probe a yeast proteome array with 5800 proteins. The screen revealed 39 proteins that bound calmodulin. Of those 39, 33 were previously unknown!

A third type of screen uses small molecules to assess whether they will bind to any of the proteins on the array. This approach shows promise for discovering new drugs that will inhibit proteins involved in disease.

Large-scale screens reveal protein–protein interactions

We often study proteins in isolation, compared with their normal cellular context. This approach is obviously artificial. One immediate goal of proteomics, therefore, is to map all the physical interaction between proteins in a cell. This is a daunting task that requires tools that can be automated, similarly to the way that genome sequencing was automated.

One approach is to use the yeast two-hybrid system discussed in the preceding chapter. This system can be automated once libraries of known cDNAs are available in each of the two vectors used.

The use of two-hybrid screens has been applied to budding yeast to generate a map of all possible interacting proteins. This method is difficult to apply to more complex multicellular organisms, but in a technical tour-de-force, it has been applied to *Drosophila melanogaster* as well.

For humans, mice, and other vertebrates, the two-hybrid system is being applied more selectively at present. Useful data can still be collected by concentrating on a biologically significant process, such as signal transduction. The technique can then be used to map all of the interacting proteins in a specific signaling pathway.

Comparative genomics uses comparisons of different genomes to infer structural, functional, and evolutionary relationships between genes and proteins. Functional genomics attempts to infer functions using genomic level information. Microarrays can be used to look at gene expression for many genes at once. Proteomics is moving this to the level of proteins. Protein microarrays allow the analysis of many proteins at once, and interaction screens using the yeast two-hybrid system can be scaled up to analyze large-scale interaction of proteins in cells.

18.5 Applications of Genomics

Space allows us to highlight only a few of the myriad applications of genomics to show the possibilities. The tools being developed truly represent a revolution in biology that will likely have a lasting impact on the way that we think about living systems.

Genomics can help to identify infectious diseases

The genomics revolution has yielded millions of new genes to be investigated. The potential of genomics to improve human health is enormous. Mutations in a single gene can explain some, but not most, hereditary diseases. With entire genomes to search, the probability of unraveling human, animal, and plant diseases is greatly improved.

Although proteomics will likely lead to new pharmaceuticals, the immediate impact of genomics is being seen in diagnostics. Both improved technology and gene discovery are enhancing diagnosis of genetic abnormalities.

Diagnostics are also being used to identify individuals. For example, STRs were among the forensic diagnostic tools used to identify remains of victims of the September 11, 2001, terrorist attack on the World Trade Center in New York City.

The September 11 attacks were followed by an increased awareness and concern about biological weapons. When cases of anthrax began appearing in the fall of 2001, genome sequencing allowed exploration of possible sources of the deadly bacteria and determination of whether they had been genetically engineered to increase their lethality.

In addition, substantial effort has been turned toward the use of genomic tools to distinguish between naturally occurring infections and intentional outbreaks of disease. The Centers for Disease Control and Prevention (CDC) have ranked bacteria and viruses that are likely targets for bioterrorism (table 18.2).

TABLE 18.2 High-Priority Pathogens for Genomic Research

Pathogen	Disease	Genome*
Variola major	Smallpox	Complete
Bacillus anthracis	Anthrax	Complete
Yersinia pestis	Plague	Complete
Clostridium botulinum	Botulism	In progress
Francisella tularensis	Tularemia	Complete
Filoviruses	Ebola and Marburg hemorrhagic fever	Both are complete
Arenaviruses	Lassa fever and Argentine hemorrhagic fever	Both are complete

*There are multiple strains of these viruses and bacteria. "Complete" indicates that at least one has been sequenced. For example, the Florida strain of anthrax was the first to be sequenced.

figure 18.13

RICE FIELD. Most of the rice grown globally is directly consumed by humans and is the dietary mainstay of 2 billion people.

Genomics can help improve agricultural crops

Globally speaking, nutrition is the greatest impediment to human health. Much of the excitement about the rice genome project is based on its potential for improving the yield and nutritional quality of rice and other cereals worldwide. The development of Golden Rice (chapter 17) is an example of improved nutrition through genetic approaches. About one-third of the world population obtains half its calories from rice (figure 18.13). In some regions, individuals consume up to 1.5 kg of rice daily. More than 500 million tons of rice is produced each year, but this may not be enough in the future.

Due in large part to scientific advances in crop breeding and farming techniques, in the last 50 years world grain production has more than doubled, with an increase in cropland of only 1%. The world now farms a total area the size of South America, but without the scientific advances of the past 50 years, an area equal to the entire western hemisphere would need to be farmed to produce enough food for the world.

Unfortunately, water usage for crops has tripled in that time period, and quality farmland is being lost to soil erosion. Scientists are also concerned about the effects of global climate change on agriculture worldwide. Increasing the yield and quality of crops, especially on more marginal farmland, will depend on many factors—but genetic engineering, built on the findings of genomics projects, can contribute significantly to the solution.

figure 18.14

CORN CROP PRODUCTIVITY WELL BELOW ITS GENETIC POTENTIAL DUE TO DROUGHT STRESS. Corn production can be limited by water deficiencies due to the drought that occurs during the growing season in dry climates. Global climate change may increase drought stress in areas where corn is the major crop.

inquiry

The corn genome has not been sequenced. How could you use information from the rice genome sequence to try to improve drought tolerance in corn?

Most crops grown in the United States produce less than half of their genetic potential because of environmental stresses (salt, water, and temperature), herbivores, and pathogens (figure 18.14). Identifying genes that can provide resistance to stress and pests is the focus of many current genomics research projects. Having access to entire genomic sequences will enhance the probability of identifying critical genes.

Genomics raises ethical issues over ownership of genomic information

Genome science is also a source of ethical challenges and dilemmas. One example is the issue of gene patents. Actually, it is the use of a gene, not the gene itself, that is patentable. For a gene-related patent, the product and its function must be known.

The public genome consortia, supported by federal funding, have been driven by the belief that the sequence of genomes should be freely available to all and should not be patented. Private companies patent gene functions, but they often make sequence data available with certain restrictions. The physical sciences have negotiated the landscape of public and for-profit research for decades, but this is relatively new territory for biologists.

Another ethical issue involves privacy. How sequence data are used is the focus of thoughtful and ongoing discussions. The Universal Declaration on the Human Genome and Human Rights states, "The human genome underlies the fundamental unity of all members of the human family, as well as the recognition of their inherent dignity and diversity. In a symbolic sense, it is the heritage of humanity."

Although we talk about "the" human genome, each of us has subtly different genomes that can be used to identify us. Genetic disorders such as cystic fibrosis and Huntington disease can already be identified by screening, but genomics will greatly increase the number of identifiable traits. What if employers or insurance companies gain access to your personal SNP profile? Could you be discriminated against because your genome indicates you have a genetic tendency toward chemical addiction or heart disease? What sort of legal protections should be in place to prevent this sort of discrimination?

On a more positive note, the U.S. Armed Forces require DNA samples from members for possible casualty identification, and DNA-based identification brought peace of mind to some families of the World Trade Center victims. Forensic use of DNA-based identification has already been described.

Behavioral genomics is an area that is also rich with possibilities and dilemmas. Very few behavioral traits can be accounted for by single genes. Two genes have been associated with fragile-X mental retardation, and three with early-onset Alzheimer disease. Comparisons of multiple genomes will likely lead to the identification of multiple genes controlling a range of behaviors. Will this change the way we view acceptable behavior?

In Iceland, the parliament has voted to have a private company create a database from pooled medical, genetic, and genealogical information about all Icelanders, a particularly fascinating population from a genetic perspective. Because minimal migration or immigration has occurred there over the last 800 years, the information that can be mined from the Icelandic database is phenomenal. Ultimately, the value of that information has to be weighed against any possible discrimination or stigmatization of individuals or groups.

Genomics has been used in the identification of remains of victims of disasters and can be also be used to identify bioweapon agents. Genomics is being used to improve agricultural crops and domesticated animals. This field has also created controversy over who owns genetic/genomic information.

concept review

18.1 Mapping Genomes

Maps provide landmarks to find your way around a genome.

- Genetic maps provide relative locations of genes on a chromosome based on recombination frequency and provide a link to phenotypes.
- Physical maps are based on landmarks in the actual DNA.
- There are several kinds of physical maps:
  - *Restriction maps are based on distances between restriction enzyme cleavage sites (figure 18.1).*
  - *Overlap between smaller fragments can be used to assemble them into contiguous segments called a contig.*
  - *Chromosomal banding using stains and hybridization results in low-resolution cytological maps.*
  - *Radiation hybrid maps are constructed by fusing an irradiated cell with another cell to identify overlapping fragments of chromosomes.*
  - *The ultimate physical map is the nucleotide sequence of a DNA.*
- Any physical site can be used as a sequence-tagged site (STS), based on a small stretch of a unique DNA sequence that allows unambiguous identification of a fragment.
- Physical and genetic maps can be correlated. Any gene that can be cloned can be placed within the genome sequence and mapped.

18.2 Whole-Genome Sequencing

Large-scale genome sequencing requires the use of high-throughput automated sequencing and computer analysis.

- Automated sequencing is accurate for DNA segments up to 800 bp long. Errors are possible, and 5–10 copies of the segments are sequenced and compared.
- Artificial chromosomes based on bacterial chromosomes allow the cloning of larger pieces of DNA.
- Clone-by-clone sequencing compares the overlapping of restriction maps using bacterial artificial chromosomes or identifying STS between clones (figure 18.3).
- Shotgun sequencing involves sequencing random clones, then using a computer to assemble the finished sequence.
- Whole-genome sequencing uses both clone-by-clone sequencing and shotgun sequencing (figure 18.5).

18.3 Characterizing Genomes

Genome sequences are descriptive and do not provide information on the organization of the genomes, their products, and how they are interrelated.

- Although eukaryotic genomes are larger and have more genes than prokaryotes, the size of the organism is not correlated with the size of the genome.
- Once a genome is sequenced, the genes are identified by looking for open-reading frames (ORF).
- An ORF begins with a start codon and contains no stop codon for a distance long enough to encode a protein.
- Bioinformatics uses computer programs to search for genes, compare genomes, and assemble genomes.
- Genomes contain both coding and noncoding DNA.
- In eukaryotes, protein-encoding DNA includes single-copy genes.
- Regions containing genes may be duplicated (segmental duplications).
- Protein-coding genes may be found as part of multigene families and tandem clusters.
- Noncoding DNA in eukaryotes makes up about 99% of DNA. Noncoding DNA can occur within genes (introns), be structural and untranscribable, or contain short repeated sequences.
- Protein-coding genes that are duplicated may accumulate mutations and become pseudogenes.
- Approximately 45% of the human genome is composed of mobile transposable elements that exist in many copies.
- The number and location of expressed genes can be estimated by sequencing the ends of randomly selected cDNAs to produce expressed sequence tags (ESTs).
- Alternative splicing of existing genes yields different proteins with different functions (figure 18.7).
- Genetic variation can exist as differences in a single nucleotide between individuals called single-nucleotide polymorphisms (SNPs).
- Genomic haplotypes are regions of chromosomes that are not exchanged by recombination. This is called linkage disequilibrium and can be used to map genes by association (figure 18.8).

18.4 Genomics and Proteomics

Proteomics characterizes the proteins produced by cells.

- The field of comparative genomics studies the conserved regions of genomes among organisms.
- The biggest difference between our genome and the chimpanzee genome is in transposable elements.
- Synteny refers to the conserved arrangements of segments of DNA in related genomes (figure 18.9).
- Chloroplasts and mitochondria contain components encoded by their own genomes and by the nuclear genome.
- Functional genomics, a hypothesis-driven science, studies gene function and gene products.
- DNA microarrays allow the expression of all of the genes in a cell to be monitored at once (figure 18.10).
- Proteomics characterizes all of the proteins produced by a cell. The transcriptome is all the mRNAs present in a cell at a specific time.
- Protein microarrays are used to identify and characterize large numbers of proteins.
- The yeast two-hybrid system is used to generate large-scale maps of interacting proteins.

18.5 Applications of Genomics

Genomics represents a revolution in biology that will have a lasting effect in the way we think about living systems.

- Genomics can help identify naturally occurring and intentional outbreaks of infectious diseases.
- Genomics can help improve domesticated animals, the nutritional value of crops, and their responses to environmental stresses.
- Genomics raises ethical issues over ownership of genomic information and personal privacy.

SELF TEST

1. A genetic map is based on—
   a. the sequence of the DNA.
   b. the relative position of genes on chromosomes.
   c. the location of sites of restriction enzyme cleavage.
   d. the banding pattern on a chromosome.
2. What is an STS?
   a. A unique sequence within the DNA that can be used for mapping
   b. A repeated sequence within the DNA that can be used for mapping
   c. An upstream element that allows for mapping of the 3′ region of a gene
   d. Both (b) and (c)
3. An artificial chromosome is useful because—
   a. it produces more consistent results than a natural chromosome.
   b. it allows for the isolation of larger DNA sequences.
   c. it provides a high copy number of a DNA sequence.
   d. it is linear.
4. Which of the following techniques relies on knowledge of overlapping sequences?
   a. Radiation hybrid mapping
   b. Shotgun method of genome sequencing
   c. FISH
   d. Clone-by-clone method of genome sequencing
5. Which number represents the total number of genes in the human genome?
   a. 2500
   b. 10,000
   c. 25,000
   d. 100,000
6. An open reading frame (ORF) is distinguished by the presence of—
   a. a stop codon.
   b. a start codon.
   c. a sequence of DNA long enough to encode a protein.
   d. All of the above.
7. What is a BLAST search?
   a. A mechanism for aligning consensus regions during whole-genome sequencing
   b. A search for similar gene sequences from other species
   c. A method of screening a DNA library
   d. A method for identifying ORFs
8. Which of the following is *not* an example of a protein-encoding gene?
   a. Single-copy gene
   b. Tandem clusters
   c. Pseudogene
   d. Multigene family
9. The duplication of a gene due to uneven meiotic crossing over is thought to lead to the production of a—
   a. segmental duplication.
   b. tandem duplication.
   c. simple sequence repeat.
   d. multigene family.
10. Which of the following is *not* an example of noncoding DNA?
   a. Promoter
   b. Intron
   c. Pseudogene
   d. Exon
11. Comparisons between genomes is made easier because of—
   a. synteny.
   b. haplotypes.
   c. transposons.
   d. expressed sequence tags.
12. What information can be obtained from a DNA microarray?
   a. The sequence of a particular gene
   b. The presence of genes within a specific tissue
   c. The pattern of gene expression
   d. Differences between genomes
13. Which of the following is true regarding microarray technology and cancer?
   a. A DNA microarray can determine the type of cancer.
   b. A DNA microarray can measure the response of a cancer to therapy.
   c. A DNA microarray can be used to predict whether the cancer will metastasize.
   d. All of the above.
14. What is a proteome?
   a. The collection of all genes encoding proteins
   b. The collection of all proteins encoded by the genome
   c. The collection of all proteins present in a cell
   d. The amino acid sequence of a protein
15. Which of the following techniques could be used to examine protein–protein interactions in a cell?
   a. Two-hybrid screens
   b. Protein structure databases
   c. Protein microarrays
   d. Both (a) and (c)

CHALLENGE QUESTIONS

1. You are in the early stages of a genome-sequencing project. You have isolated a number of clones from a BAC library and mapped the inserts in these clones using STSs. Use the STSs to align the clones into a contiguous sequence of the genome (a contig).

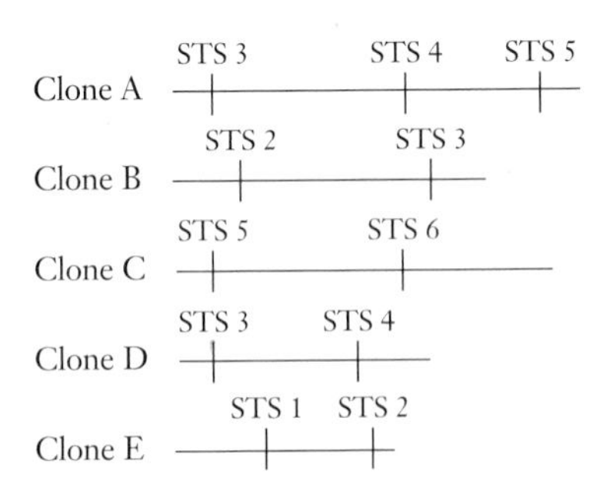

2. Genomic research can be used to determine if an outbreak of an infectious disease is natural or "intentional." Explain what a genomic researcher would be looking for in a suspected intentional outbreak of a disease like anthrax.

ARIS™ **Do you need additional review?** *Visit* www.ravenbiology.com *for practice quizzes, animations, videos, and activities designed to help you master the material in this chapter.*

chapter 19

Cellular Mechanisms of Development

introduction

COUNTLESS GENERATIONS OF YOUNG children have reveled in the discovery of frog tadpoles during summer visits to freshwater ponds, have watched with fascination as a chick pecks its way out of its shell, or have delighted as the first flowers of spring push their way up through the soil. For thousands of years, the wonder of such events has inspired a desire to understand how organisms arise, grow, change, and mature.

We have explored gene expression from the perspective of individual cells, examining the diverse mechanisms cells employ to control the transcription of particular genes. Now we broaden our perspective and look at the unique challenge posed by the development of a single cell, the fertilized egg, into a multicellular organism such as is occurring in these fish embryos. In the course of this developmental journey, a pattern of decisions about gene expression takes place that causes particular lines of cells to proceed along different paths, spinning an incredibly complex web of cause and effect. Yet, for all its complexity, this developmental program works with impressive precision. In this chapter, we explore the mechanisms of development in multicellular organisms.

concept outline

19.1 Overview of Development

19.2 Cell Division

- *Development begins with cell division*
- *Every cell division is known in the development of* C. elegans
- *Stem cells continue to divide and can form multiple kinds of tissue*
- *Plant growth occurs in specific areas called meristems*

19.3 Cell Differentiation

- *Cells become determined prior to differentiation*
- *Determination can be due to cytoplasmic determinants*
- *Induction can lead to cell differentiation*
- *Reversal of determination has allowed cloning*
- *Reproductive cloning has inherent problems*
- *Therapeutic cloning is a promising possibility*
- *Stem cell research has stimulated ethical debate*
- *Adult stem cells may offer an alternative to ES cells*

19.4 Pattern Formation

- Drosophila *embryogenesis produces a segmented larva*
- *Morphogen gradients form the basic body axes in* Drosophila
- *The body plan is produced by sequential activation of genes*
- *Segment identity arises from the action of homeotic genes*
- *Pattern formation in plants is also under genetic control*

19.5 Morphogenesis

- *Cell division during development may result in unequal cytokinesis*
- *Cells change shape and size as morphogenesis proceeds*
- *Programmed cell death is a necessary part of development*
- *Cell migration gets the right cells to the right places*
- *In seed plants, the plane of cell division determines morphogenesis*

19.6 Environmental Effects on Development

- *The environment affects normal development*
- *Endocrine disrupters can perturb development*

19.1 Overview of Development

Development can be defined as the process of systematic, gene-directed changes through which an organism forms the successive stages of its life cycle. Development is a continuum, and explorations of development can be focused on any point along this continuum. The study of development plays a central role in unifying the understanding of both the similarities and diversity of life on Earth.

We can divide the overall process of development into four subprocesses:

- **Growth (cell division).** A developing plant or animal begins as a fertilized egg, or zygote, that must undergo cell division to form the body mass of the new individual.
- **Differentiation.** As cells divide, orchestrated changes in gene expression result in differences between cells that ultimately result in cell specialization. In differentiated cells, certain genes are expressed at particular times, but other genes may not be expressed at all.
- **Pattern Formation.** Cells in a developing embryo must become oriented to the body plan of the organism the embryo will become. Pattern formation involves cells' abilities to detect positional information that guides their ultimate fate.
- **Morphogenesis.** As development proceeds, the form of the body is generated, namely its organs and anatomical features. Morphogenesis may involve cell death as well as cell division and differentiation.

Despite the overt differences between groups of plants and animals, most multicellular organisms develop according to molecular mechanisms that are fundamentally very similar. This observation suggests that these mechanisms evolved very early in the history of multicellular life.

19.2 Cell Division

When a frog tadpole hatches out of its protective coats, it is roughly the same overall mass as the fertilized egg from which it came. Instead of being made up of just one cell, however, the tadpole comprises a million or so cells, which are organized into tissues and organs with different functions. Thus, the very first process that must occur during embryogenesis is cell division.

Immediately following fertilization, the diploid zygote undergoes a period of rapid mitotic divisions that ultimately result in an early embryo comprising dozens to thousands of diploid cells. In animal embryos, the timing and number of these divisions are species-specific and are controlled by a set of molecules that we examined in chapter 10: the **cyclins** and **cyclin-dependent kinases (Cdks).** These molecules exert control over checkpoints in the cycle of mitosis.

Development begins with cell division

In animal embryos, the period of rapid cell division following fertilization is called **cleavage.** During cleavage, the enormous mass of the zygote is subdivided into a larger and larger number of smaller and smaller cells, called **blastomeres** (figure 19.1). Hence, cleavage is not accompanied by any increase in the overall size of the embryo. The G_1 and G_2 phases of the cell cycle, during which a cell increases its mass and size, are extremely shortened or eliminated during cleavage (figure 19.2).

Because of the absence of the two gap/growth phases, the rapid rate of mitotic divisions during cleavage is never again approached in the lifetime of any animal. For example, zebrafish blastomeres divide once every several minutes during cleavage, to create an embryo with a thousand cells in just under 3 hr! In contrast, cycling adult human intestinal epithelial cells divide on average only once every 19 hr.

When external sources of nutrients become available—for example, during larval feeding stages or after implantation of mammalian embryos—daughter cells can increase in size following cytokinesis, and an overall increase in the size of the organism occurs as more cells are produced.

Every cell division is known in the development of *C. elegans*

One of the most completely described models of development is the tiny nematode *Caenorhabditis elegans*. Only about 1 mm long, the adult worm consists of 959 somatic cells.

Because *C. elegans* is transparent, individual cells can be followed as they divide. By observing them, researchers have

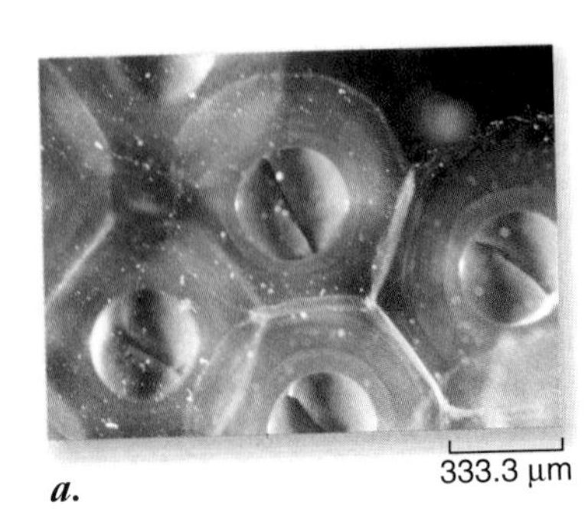

a.

b.

c.

figure 19.1

CLEAVAGE DIVISIONS IN A FROG EMBRYO. ***a.*** The first cleavage division divides the egg into two large blastomeres. ***b.*** After two more divisions, four small blastomeres sit on top of four large blastomeres, each of which continues to divide to produce (***c***) a compact mass of cells.

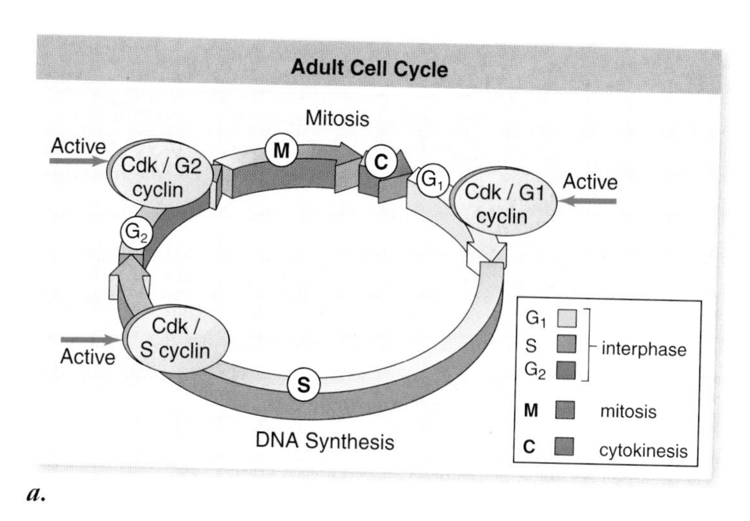

a.

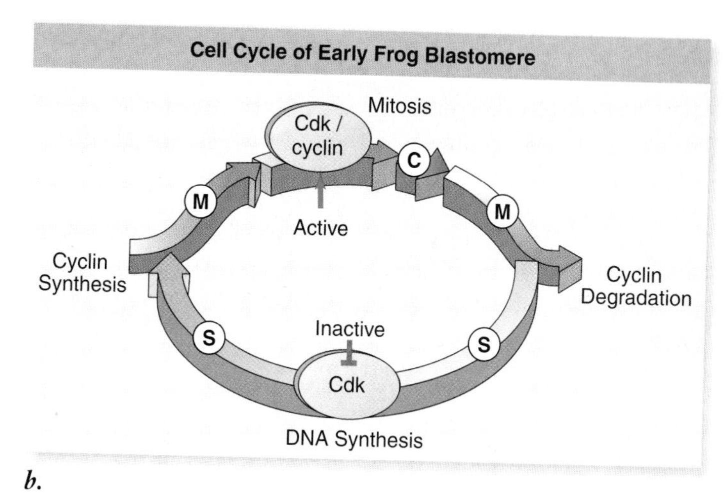

b.

figure 19.2

CELL CYCLE OF ADULT CELL AND EMBRYONIC CELL. In contrast to the cell cycle of adult somatic cells **(*a*)**, the dividing cells of early frog embryos lack G_1 and G_2 stages **(*b*)**, enabling the cleavage stage nuclei to rapidly cycle between DNA synthesis and mitosis. Large stores of cyclin mRNA are present in the unfertilized egg. Periodic translation of this message produces cyclin proteins. Cyclin degradation and Cdk inactivation allows the cell to complete mitosis and initiate the next round of DNA synthesis.

learned how each of the cells that make up the adult worm is derived from the fertilized egg. As shown on the lineage map in figure 19.3*a*, the egg divides into two cells, and these daughter cells continue to divide. Each horizontal line on the map represents one round of cell division. The length of each vertical line represents the time between cell divisions, and the end of each vertical line represents one fully differentiated cell. In figure 19.3*b*, the major organs of the worm are color-coded to match the colors of the corresponding groups of cells on the lineage map.

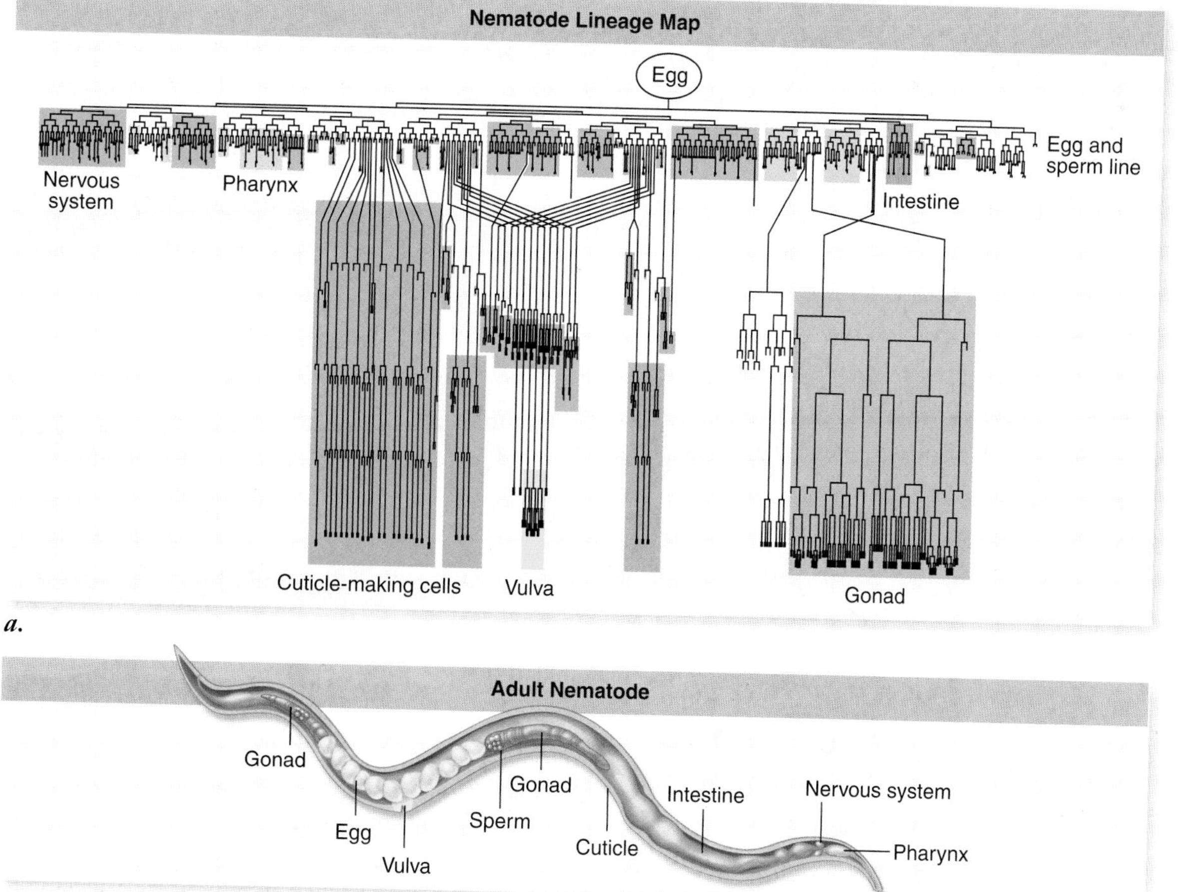

a.

b.

figure 19.3

STUDYING EMBRYONIC CELL DIVISION AND DEVELOPMENT IN THE NEMATODE. Development in *C. elegans* has been mapped out such that the fate of each cell from the single egg cell has been determined. ***a.*** The lineage map shows the number of cell divisions from the egg, and the color coding links their placement in **(*b*)** the adult organism.

M.E. Challinor illustration. From Howard Hughes Medical Institute © as published in *From Egg to Adult*, 1992. Reprinted by permission.

Some of these differentiated cells, such as some cells that generate the worm's external cuticle, are "born" after only 8 rounds of cell division; other cuticle cells require as many as 14 rounds. The cells that make up the worm's pharynx, or feeding organ, are born after 9 to 11 rounds of division, whereas cells in the gonads require up to 17 divisions.

Exactly 302 nerve cells are destined for the worm's nervous system. Exactly 131 cells are programmed to die, mostly within minutes of their "birth." The fate of each cell is the same in every *C. elegans* individual, except for the cells that will become eggs and sperm.

Stem cells continue to divide and can form multiple kinds of tissue

The blastomeres of cleavage stage mammalian embryos are nondifferentiated and can give rise to any tissue. As development proceeds, cells become limited in their fates, as discussed in the next section. Some cells, called **stem cells,** are set aside and will continue to divide while remaining undifferentiated. For example, one population of cells are set aside that will go on to form nerve cells, and another will go on to produce blood, and still others muscle. Each major tissue is represented by its own kind of **tissue-specific stem cell.** As development proceeds, these tissue-specific stem cells persist—even in adults.

Stem cells may give rise to a single cell type, such as muscle satellite cells that give rise to muscle cells, or give rise to multiple cell types, such as myeloid cells that give rise to different types of blood cells. Stem cells that form multiple cell types may be **totipotent,** meaning that they can become any cell type, or **pluripotent,** meaning that they can become multiple different cell types.

The cleavage stage in mammals continues for five or six days, producing a ball of cells called a **blastocyst.** This blastocyst consists of an outer layer that will become the placenta enclosing an inner cell mass that will go on to form the embryo. Stem cells can be isolated from this inner cell mass and grown in culture (figure 19.4); these cells are termed **embryonic stem cells (ES cells).** In mice, in whom these have been extensively studied, each ES cell is capable of developing into any tissue in the animal. If ES cells are removed from an early-stage embryo and are then placed back into a host embryo, interactions with the cells around them determine their fate.

Scientists can also stimulate differentiation of ES cells along different pathways in culture, by exposing the uncommitted ES cells to different chemical signals in the growth medium. The use of these cells is discussed later in the chapter.

Plant growth occurs in specific areas called meristems

A major difference between animals and plants is that most animals are mobile, at least in some phase of their life cycles, and therefore they can move away from unfavorable circumstances. Plants, in contrast, are anchored in position and must simply endure whatever environment they experience. Plants compensate for this restriction by allowing development to accommodate local circumstances.

Instead of creating a body in which every part is specified to have a fixed size and location, a plant assembles its body throughout its life span from a few types of modules, such as leaves, roots, branch nodes, and flowers. Each module has a rigidly controlled structure and organization, but how the modules are utilized is quite flexible—they can be adjusted to environmental conditions.

Plants develop by building their bodies outward, creating new parts from groups of stem cells that are contained in structures called **meristems.** As meristematic stem cells continually divide, they produce cells that can differentiate into the tissues of the plant.

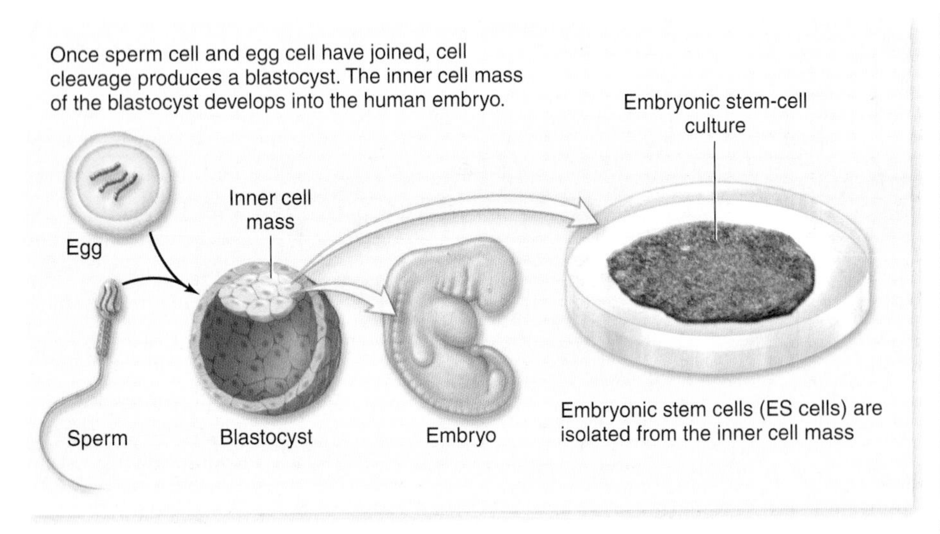

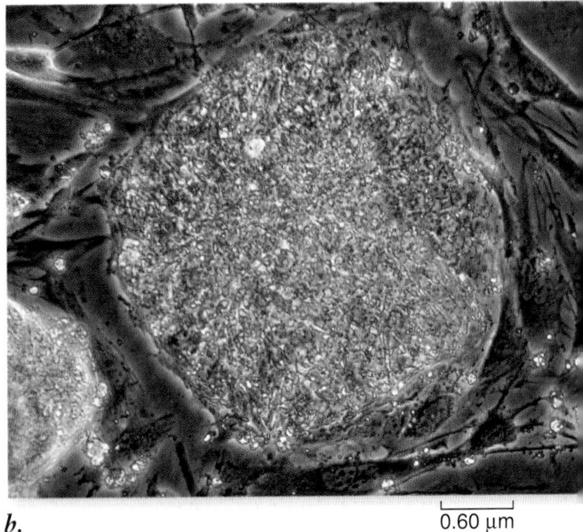

figure 19.4

ISOLATION OF EMBRYONIC STEM CELLS. ***a.*** Early cell divisions lead to the blastocyst stage that consists of an outer layer and an inner cell mass, which will go on to form the embryo. Embryonic stem cells (ES cells) can be isolated from this stage by disrupting the embryo and plating the cells. Stem cells removed from a six-day blastocyst can be established in culture and maintained indefinitely in an undifferentiated state. ***b.*** Human embryonic stem cells. This mass in the photograph is a colony of undifferentiated human embryonic stem cells surrounded by fibroblasts (elongated cells) that serve as a "feeder layer."

This simple scheme indicates a need to control the process of cell division. We know that cell-cycle control genes are present in both yeast (fungi) and animal cells, implying that these are a eukaryotic innovation—and in fact, the plant cell cycle is regulated by the same mechanisms, namely through cyclins and cyclin-dependent kinases. In one experiment, overexpression of a Cdk inhibitor in transgenic *Arabidopsis thaliana* plants resulted in strong inhibition of cell division in leaf meristems, leading to significant changes in leaf size and shape.

In animal embryos a series of rapid divisions convert the fertilized egg into many cells with no change in size. This is accomplished by eliminating G_1 and G_2 phases of mitosis. Every cell division that leads to the adult nematode *C. elegans* is known, and this pattern is invariant. Stem cells are nondifferentiated cells that have the potential to become a number of different tissues. In plants, growth is restricted to specific areas called meristems, where stem cells are retained.

19.3 Cell Differentiation

In chapter 16, we examined the mechanisms that control eukaryotic gene expression. These processes are critical for the development of multicellular organisms, in which life functions are carried out by different tissues and organs. In the course of development, cells become different from one another because of the differential expression of subsets of genes—not only at different times, but in different locations of the growing embryo. We now explore some of the mechanisms that lead to differential gene expression during development.

Cells become determined prior to differentiation

A human body contains more than 210 major types of differentiated cells. These differentiated cells are distinguishable from one another by the particular proteins that they synthesize, their morphologies, and their specific functions. A molecular decision to become a particular type of differentiated cell occurs prior to any overt changes in the cell. This molecular decision-making process is called **cell determination,** and it commits a cell to a particular developmental pathway.

Tracking determination

Determination is often not visible in the cell and can only be "seen" by experiment. The standard experiment to test whether a cell or group of cells is determined is to move the donor cell(s) to a different location in a host (recipient) embryo. If the cells of the transplant develop into the same type of cell as they would have if left undisturbed, then they are judged to be already determined (figure 19.5).

Determination has a time course; it depends on a series of intrinsic or extrinsic events, or both. For example, a cell in the prospective brain region of an amphibian embryo at the early gastrula stage (see chapter 53) has not yet been determined; if transplanted elsewhere in the embryo, it will develop according to the site of transplant. By the late gastrula stage, however, additional cell interactions have occurred, determination has taken place, and the cell will develop as neural tissue no matter where it is transplanted.

Determination often takes place in stages, with a cell first becoming partially committed, acquiring positional labels that reflect its location in the embryo. These labels can have a great influence on how the pattern of the body subsequently develops. In a chicken embryo, tissue at the base of the leg bud normally

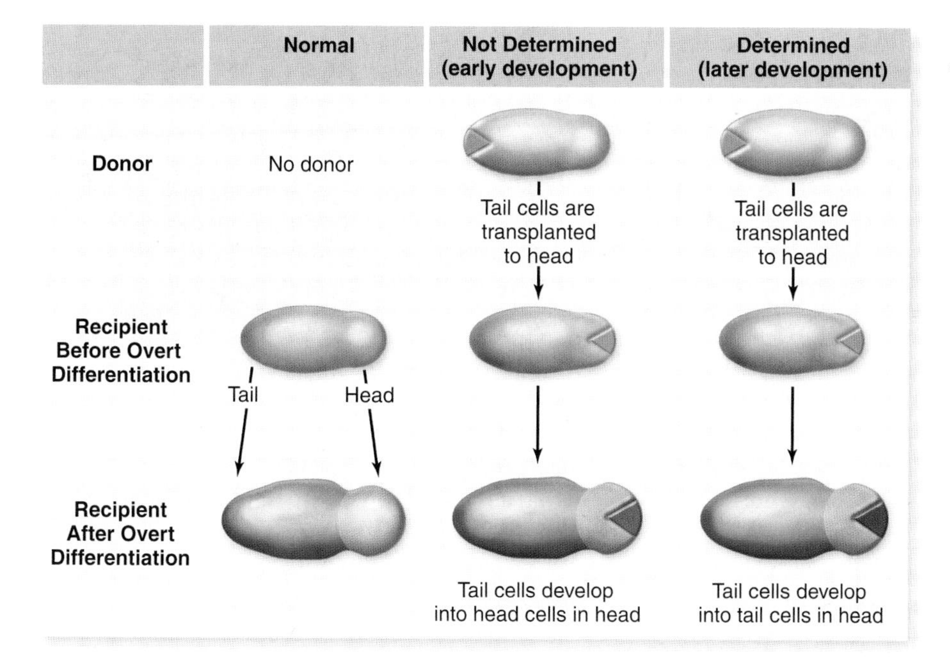

figure 19.5

THE STANDARD TEST FOR DETERMINATION. The gray ovals represent embryos at early stages of development. The cells to the right normally develop into head structures, whereas the cells to the left usually form tail structures. If prospective tail cells from an early embryo are transplanted to the opposite end of a host embryo, they develop according to their new position into head structures. These cells are not determined. At later stages of development, the tail cells are determined since they now develop into tail structures after transplantation into the opposite end of a host embryo!

gives rise to the thigh. If this tissue is transplanted to the tip of the identical-looking wing bud, which would normally give rise to the wing tip, the transplanted tissue will develop into a toe rather than a thigh. The tissue has already been determined as leg, but it is not yet committed to being a particular part of the leg. Therefore, it can be influenced by the positional signaling at the tip of the wing bud to form a tip (but in this case, a tip of leg).

The molecular basis of determination

Cells initiate developmental changes by using transcription factors to change patterns of gene expression. When genes encoding these transcription factors are activated, one of their effects is to reinforce their own activation. This reinforcement makes the developmental switch deterministic, initiating a chain of events that leads down a particular developmental pathway.

Cells in which a set of regulatory genes have been activated may not actually undergo differentiation until some time later, when other factors interact with the regulatory protein and cause it to activate still other genes. Nevertheless, once the initial "switch" is thrown, the cell is fully committed to its future developmental path.

Cells become committed to follow a particular developmental pathway in one of two ways:

(1) via the differential inheritance of cytoplasmic determinants, which are maternally produced and deposited into the egg during oogenesis; or
(2) via cell–cell interactions.

The first situation can be likened to a person's social status being determined by who his or her parents are and what he or she has inherited. In the second situation, the person's social standing is determined by interactions with his or her neighbors. Clearly both can be powerful factors in the development and maturation of that individual.

Determination can be due to cytoplasmic determinants

Many invertebrate embryos provide good visual examples of cell determination through the differential inheritance of cytoplasmic determinants. Tunicates are marine invertebrates (see chapter 35), and most adults have simple, saclike

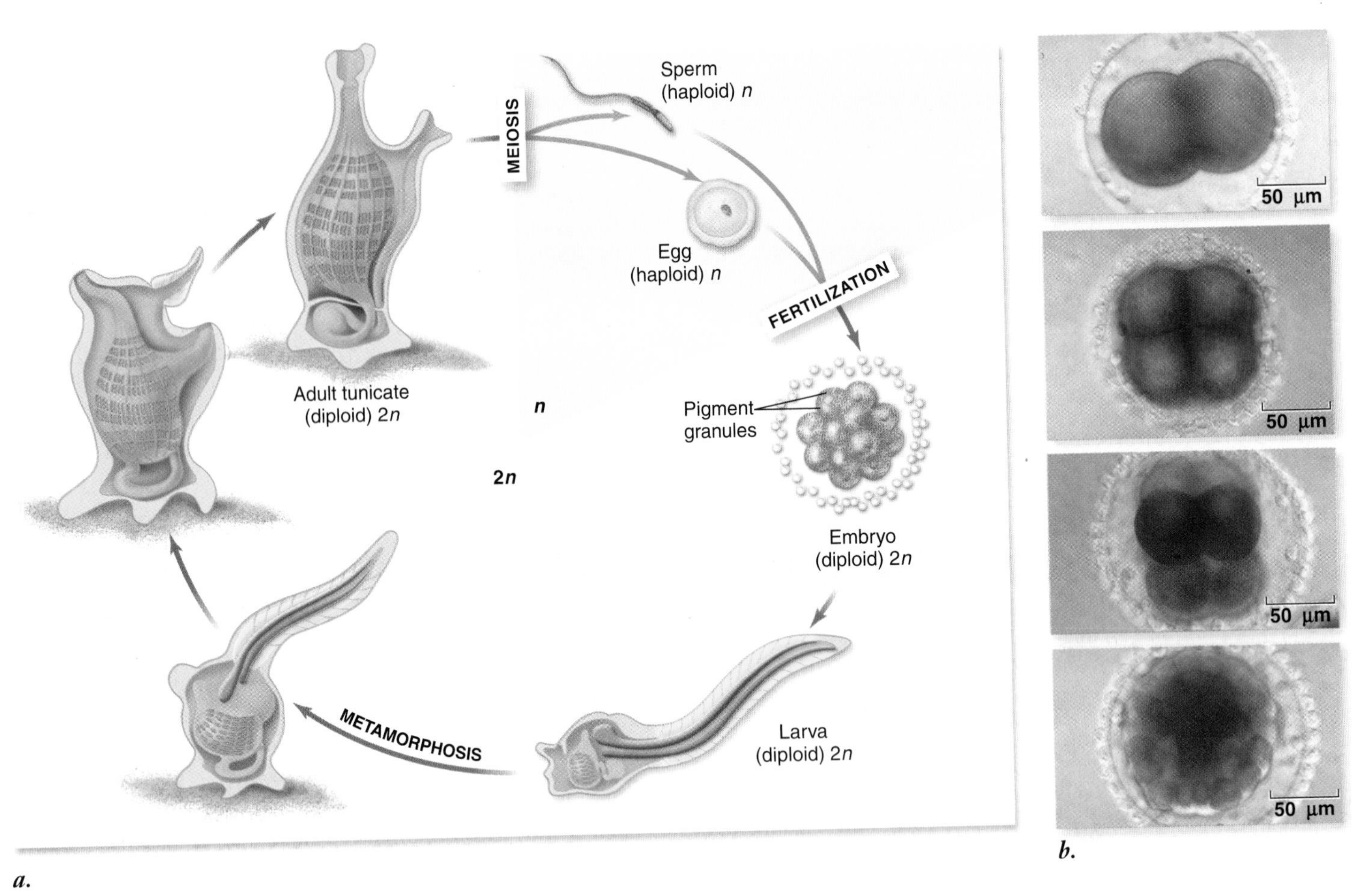

figure 19.6

MUSCLE DETERMINANTS IN TUNICATES. ***a.*** The life cycle of a solitary tunicate. Muscle cells that move the tail of the swimming tadpole are arranged on either side of the notochord and nerve cord. The tail is lost during metamorphosis into the sedentary adult. ***b.*** The egg of the tunicate *Styela* contains bright yellow pigment granules. These become asymmetrically localized in the egg following fertilization, and cells that inherit the yellow granules during cleavage will become the larval muscle cells. Embryos at the 2-cell, 4-cell, 8-cell, and 64-cell stages are shown. The tadpole tail will grow out from the lower region of the embryo in the bottom panel.

bodies that are attached to the underlying substratum. Tunicates are placed in the phylum Chordata, however, due to the characteristics of their swimming, tadpolelike larval stage, which has a dorsal nerve cord and notochord (figure 19.6*a*). The muscles that move the tail develop on either side of the notochord.

In many tunicate species, colored pigment granules become asymmetrically localized in the egg following fertilization and subsequently segregate to the tail muscle cell progenitors during cleavage (figure 19.6*b*). When these pigment granules are shifted experimentally into other cells that normally do not develop into muscle, their fate is changed and they become muscle cells. Thus, the molecules that flip the switch for muscle development appear to be associated with the pigment granules.

The next step is to determine the identity of the molecules involved. Experiments indicate that the female parent provides the egg with mRNA encoded by the *macho-1* gene. The elimination of *macho-1* function leads to a loss of tail muscle in the tadpole, and the misexpression of *macho-1* mRNA leads to the formation of additional (ectopic) muscle cells from nonmuscle lineage cells. The *macho-1* gene product has been shown to be a transcription factor that can activate the expression of several muscle-specific genes.

Induction can lead to cell differentiation

In chapter 9, we examined a variety of ways by which cells communicate with one another. We can demonstrate the importance of cell–cell interactions in development by separating the cells of an early frog embryo and allowing them to develop independently.

Under these conditions, blastomeres from one pole of the embryo (the "animal pole") develop features of ectoderm, and blastomeres from the opposite pole of the embryo (the "vegetal pole") develop features of endoderm. None of the two separated groups of cells ever develop features characteristic of mesoderm, the third main cell type. If animal-pole cells and vegetal-pole cells are placed next to each other, however, some of the animal-pole cells develop as mesoderm. The interaction between the two cell types triggers a switch in the developmental path of these cells. This change in cell fate due to interaction with an adjacent cell is called **induction.** Signaling molecules act to alter gene expression in the target cells, in this case, some of the animal-pole cells.

Another example of inductive cell interactions is the formation of the notochord and mesenchyme, a specific tissue, in tunicate embryos. Muscle, notochord, and mesenchyme all arise from mesodermal cells that form at the vegetal margin of the 32-cell stage embryo. These prospective mesodermal cells receive signals from the underlying endodermal precursor cells that lead to the formation of notochord and mesenchyme (figure 19.7).

The chemical signal is a member of the **fibroblast growth factor (FGF)** family of signaling molecules. It induces the overlying marginal zone cells to differentiate into either notochord (anterior) or mesenchyme (posterior). The FGF receptor on the marginal zone cells is a receptor tyrosine

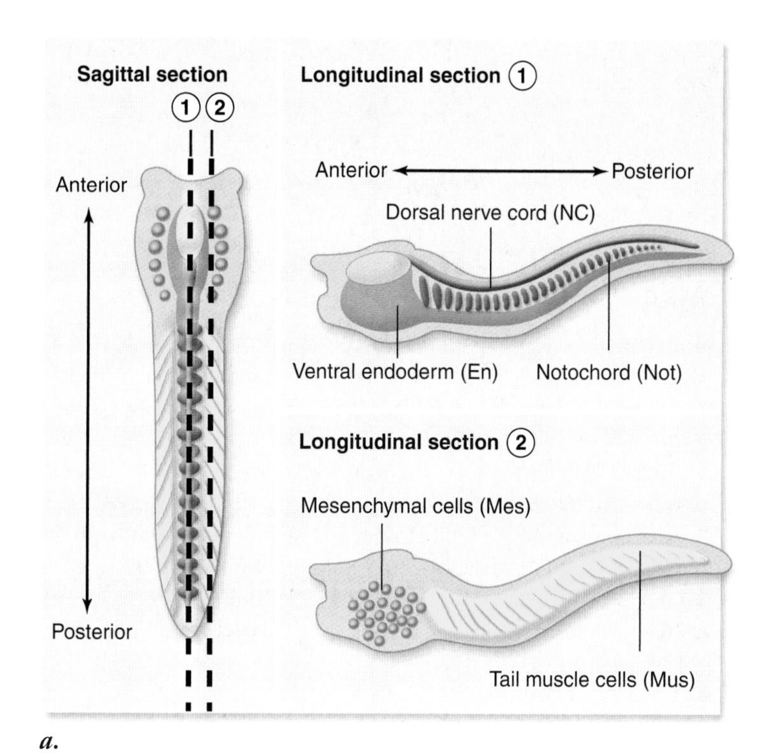

figure 19.7

INDUCTIVE INTERACTIONS CONTRIBUTE TO CELL FATE SPECIFICATION IN TUNICATE EMBRYOS.
a. Internal structures of a tunicate larva. To the left is a sagittal section through the larva with dotted lines indicating two longitudinal sections. Section 1, through the midline of a tadpole, shows the dorsal nerve cord (NC), the underlying notochord (NC) and the ventral endoderm cells (En). Section 2, a more lateral section, shows the mesenchymal cells (Mes) and the tail muscle cells (Mus). ***b.*** View of the 32-cell stage looking up at the endoderm precursor cells. FGF secreted by these cells is indicated with light-green arrows. Only the surfaces of the marginal cells that directly border the endoderm precursor cells bind FGF signal molecules. Note that the posterior vegetal blastomeres also contain the *macho-1* determinants (red and white stripes). ***c.*** Cell fates have been fixed by the 64-cell stage. Colors are as in (***a***) Cells on the anterior margin of the endoderm precursor cells become notochord and nerve cord, respectively, whereas cells that border the posterior margin of the endoderm cells become mesenchyme and muscle cells, respectively.

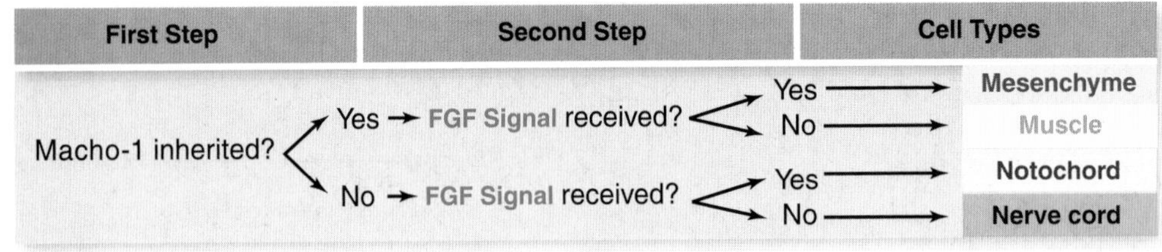

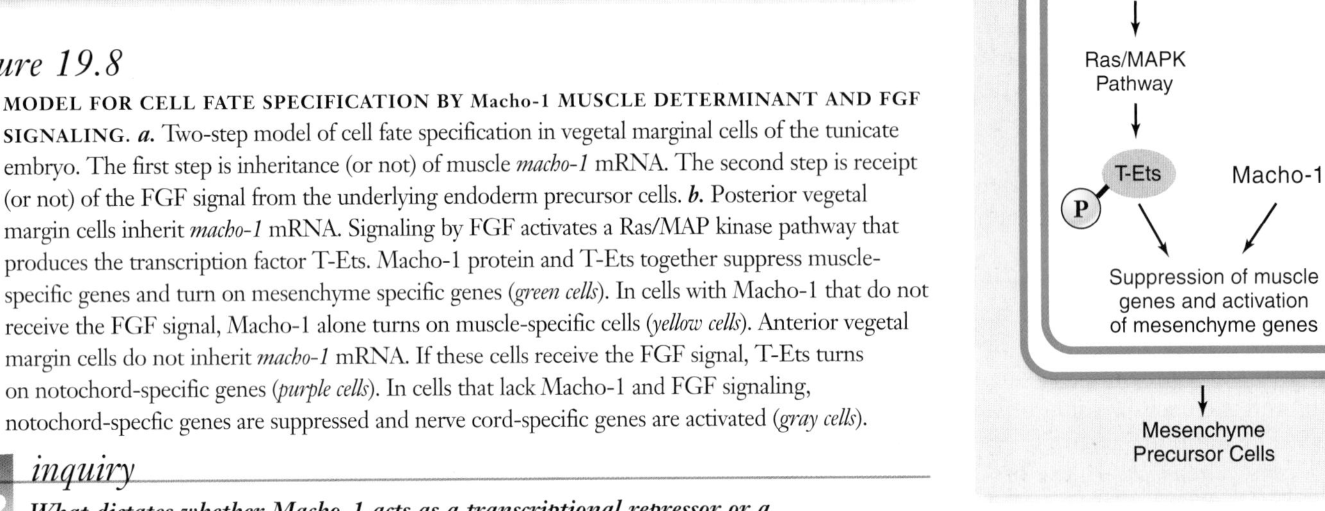

figure 19.8

MODEL FOR CELL FATE SPECIFICATION BY Macho-1 MUSCLE DETERMINANT AND FGF SIGNALING. ***a.*** Two-step model of cell fate specification in vegetal marginal cells of the tunicate embryo. The first step is inheritance (or not) of muscle *macho-1* mRNA. The second step is receipt (or not) of the FGF signal from the underlying endoderm precursor cells. ***b.*** Posterior vegetal margin cells inherit *macho-1* mRNA. Signaling by FGF activates a Ras/MAP kinase pathway that produces the transcription factor T-Ets. Macho-1 protein and T-Ets together suppress muscle-specific genes and turn on mesenchyme specific genes (*green cells*). In cells with Macho-1 that do not receive the FGF signal, Macho-1 alone turns on muscle-specific cells (*yellow cells*). Anterior vegetal margin cells do not inherit *macho-1* mRNA. If these cells receive the FGF signal, T-Ets turns on notochord-specific genes (*purple cells*). In cells that lack Macho-1 and FGF signaling, notochord-specfic genes are suppressed and nerve cord-specific genes are activated (*gray cells*).

inquiry

What dictates whether Macho-1 acts as a transcriptional repressor or a transcriptional activator?

kinase that signals through a MAP kinase cascade to activate a transcription factor that turns on gene expression resulting in differentiation (figure 19.8).

This example is also a case of two cells responding differently to the same signal. The presence or absence of the *macho-1* muscle determinant discussed earlier controls this difference in cell fate. In the presence of *macho-1*, cells differentiate into mesenchyme; in its absence, cells differentiate into notochord. Thus, the combination of *macho-1* and FGF signaling leads to four different cell types (see figure 19.8)

Reversal of determination has allowed cloning

Experiments carried out in the 1950s showed that single cells from fully differentiated tissue of an adult plant could develop into entire, mature plants. The cells of an early cleavage stage mammalian embryo are also totipotent. When mammalian embryos naturally split in two, identical twins result. If individual blastomeres are separated from one another, any one of them can produce a completely normal individual. In fact, this type of procedure has been used to produce sets of four or eight identical offspring in the commercial breeding of particularly valuable lines of cattle.

Early research in amphibians

Until very recently, biologists thought determination and cell differentiation were irreversible processes in animals. Experiments carried out in the 1950s by Briggs and King and in the 1960s and 1970s by John Gurdon and colleagues made what seemed a convincing case.

Using very fine pipettes (hollow glass tubes), these researchers sucked the nucleus out of a frog or toad egg and replaced the egg nucleus with a nucleus sucked out of a body

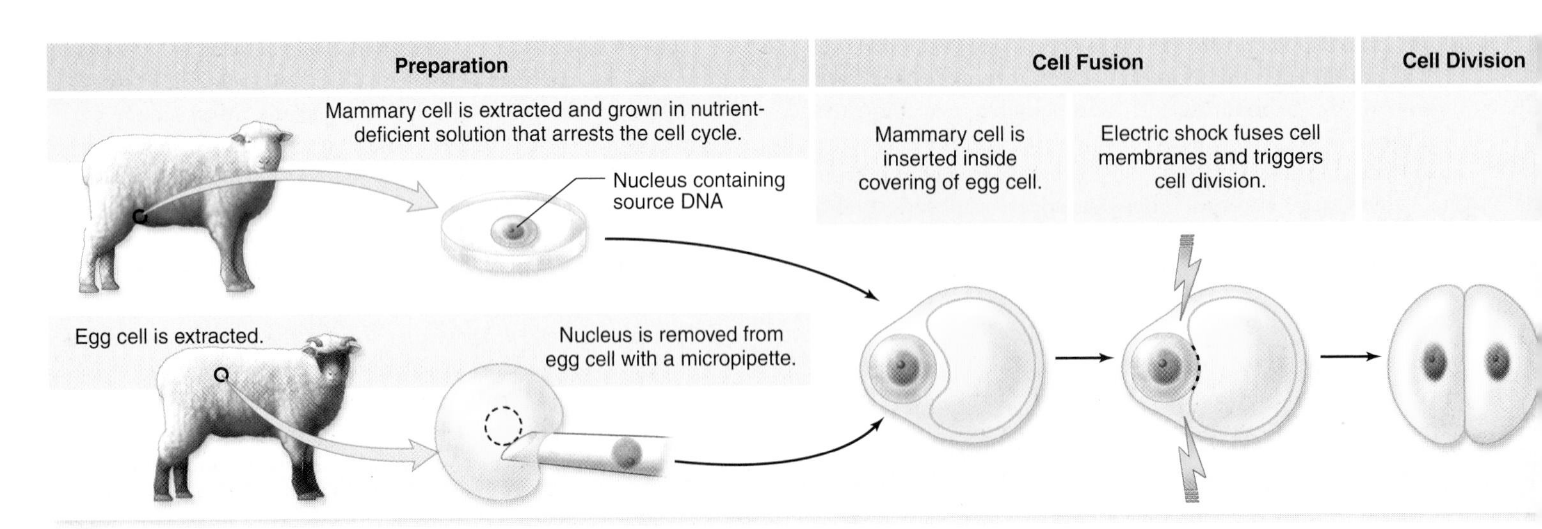

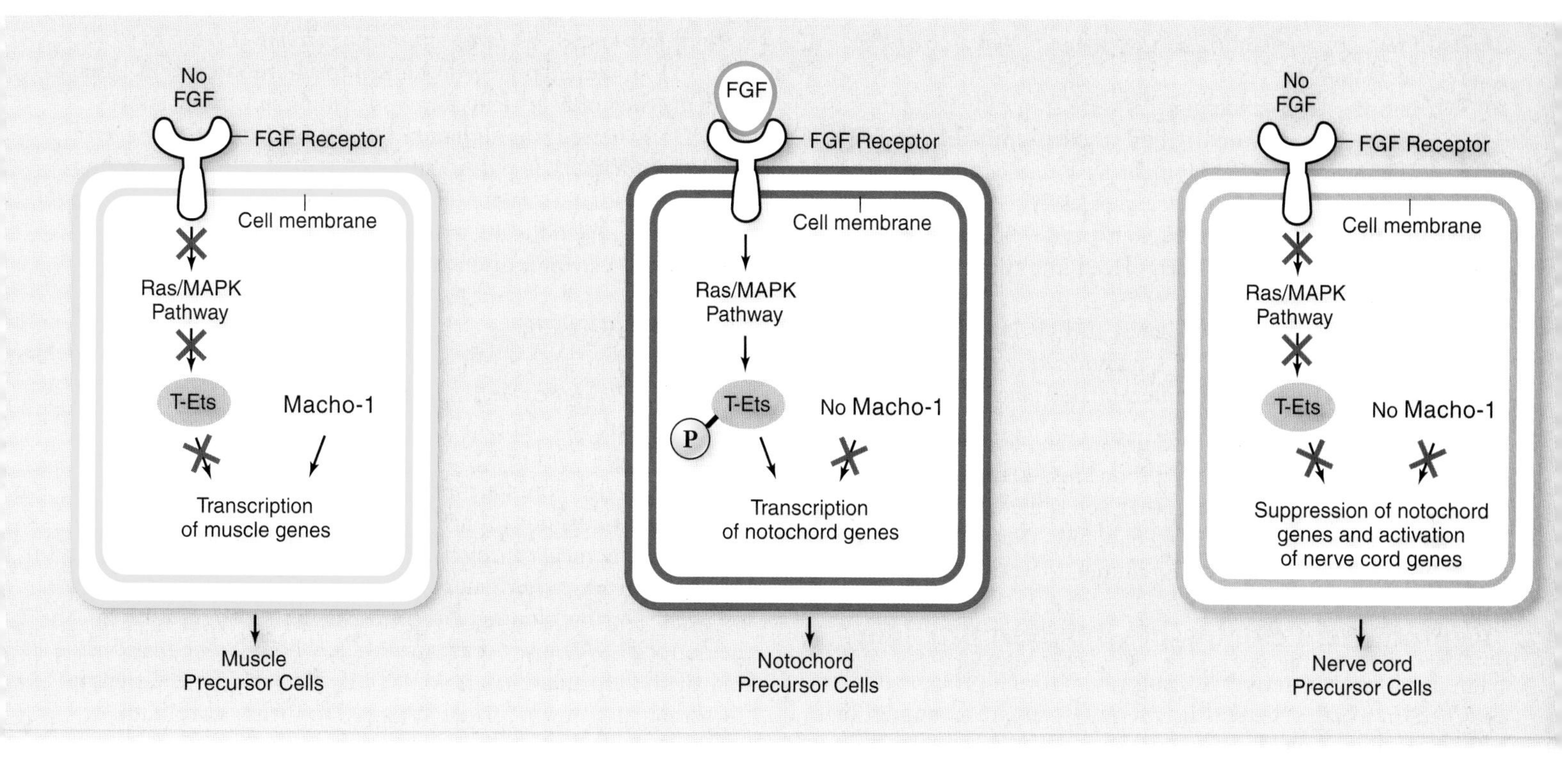

cell taken from another individual. If the transplanted nucleus was obtained from an advanced embryo, the egg went on to develop into a tadpole, but most died before becoming adult. Gurdon and colleagues also used donor nuclei from fully differentiated adult skin cells to direct tadpole development. However, all of these tadpoles died before the feeding stage.

Although these experiments showed that nuclei from adult cells have remarkable developmental potential, they did not provide evidence for the totipotency of these nuclei.

Successful nuclear transplant in mammals

Nuclear transplant experiments in mammals were attempted without success by many investigators, until finally, in 1984, a sheep was cloned using the nucleus from a cell of an early embryo. The key to this success was in picking a donor cell very early in development. This exciting result was soon replicated by others in a host of other organisms, including pigs and monkeys. Only early embryo cells seemed to work, however.

Geneticists at the Roslin Institute in Scotland reasoned that the egg and donated nucleus would need to be at the same stage of the cell cycle for successful development. To test this idea, they performed the following procedure (figure 19.9):

1. They removed differentiated mammary cells from the udder of a six-year-old sheep. The cells were grown in tissue culture, and then the concentration of serum nutrients was substantially reduced for five days, causing them to pause at the beginning of the cell cycle.

Development
Embryo begins to develop in vitro.
Implantation
Embryo is implanted into surrogate mother.
Birth of Clone
After a five-month pregnancy, a lamb genetically identical to the sheep from which the mammary cell was extracted is born.
Growth to Adulthood
Embryo

figure 19.9

PROOF THAT DETERMINATION IN ANIMALS IS REVERSIBLE. Scientists combined a nucleus from an adult mammary cell with an enucleated egg cell to successfully clone a sheep, named Dolly, who grew to be a normal adult and bore healthy offspring. This experiment, the first successful cloning of an adult animal, shows that a differentiated adult cell can be used to drive all of development.

2. In parallel preparation, eggs obtained from a ewe were enucleated.
3. Mammary cells and egg cells were surgically combined in a process called **somatic cell nuclear transfer (SCNT)** in January of 1996. Mammary cells and eggs were fused to introduce the mammary nucleus into egg.
4. Twenty-nine of 277 fused couplets developed into embryos, which were then placed into the reproductive tracts of surrogate mothers.
5. A little over five months later, on July 5, 1996, one sheep gave birth to a lamb named Dolly, the first clone generated from a fully differentiated animal cell.

Dolly matured into an adult ewe, and she was able to reproduce the old-fashioned way, producing six lambs. Thus, Dolly established beyond all dispute that determination in animals is reversible—that with the right techniques, the nucleus of a fully differentiated cell *can* be reprogrammed to be totipotent.

Reproductive cloning has inherent problems

The term **reproductive cloning** refers to the process just described, in which scientists use SCNT to create an animal that is genetically identical to another animal. Since Dolly's birth in 1997, scientists have successfully cloned one or more cats, rabbits, rats, mice, cattle, goats, pigs, and mules. All of these procedures used some form of adult cell.

Low success rate and age-associated diseases

The efficiency in all reproductive cloning is quite low—only 3–5% of adult nuclei transferred to donor eggs result in live births. In addition, many clones that are born usually die soon thereafter of liver failure or infections. Many become oversized, a condition known as *large offspring syndrome* (*LOS*). In 2003, three of four cloned piglets developed to adulthood, but all three suddenly died of heart failure at less than 6 months of age.

Dolly herself was euthanized at the relatively young age of six. Although she was put down because of virally induced lung cancer, she had been diagnosed with advanced-stage arthritis a year earlier. Thus, one difficulty in using genetic engineering and cloning to improve livestock is in producing enough healthy animals.

Lack of imprinting

The reason for these problems lies in a phenomenon discussed in chapter 13: **genomic imprinting.** Imprinted genes are expressed differently depending on parental origin—that is, they are turned off in either egg or sperm, and this "setting" continues through development into the adult. Normal mammalian development depends on precise genomic imprinting.

The chemical reprogramming of the DNA, which occurs in adult reproductive tissue, takes months for sperm and years for eggs. During cloning, by contrast, the reprogramming of the donor DNA must occur within a few hours. The organization of the chromatin in a somatic cell is also quite different from that in a newly fertilized egg. Significant chromatin remodeling of the transferred donor nucleus must also occur if the cloned embryo is to survive. Cloning fails because there is likely not enough time in these few hours to get the remodeling and reprogramming jobs done properly.

Therapeutic cloning is a promising possibility

One way to solve the problem of graft rejection, such as in skin grafts in severe burn cases, is to produce patient-specific lines of embryonic stem cells. Early in 2001, a

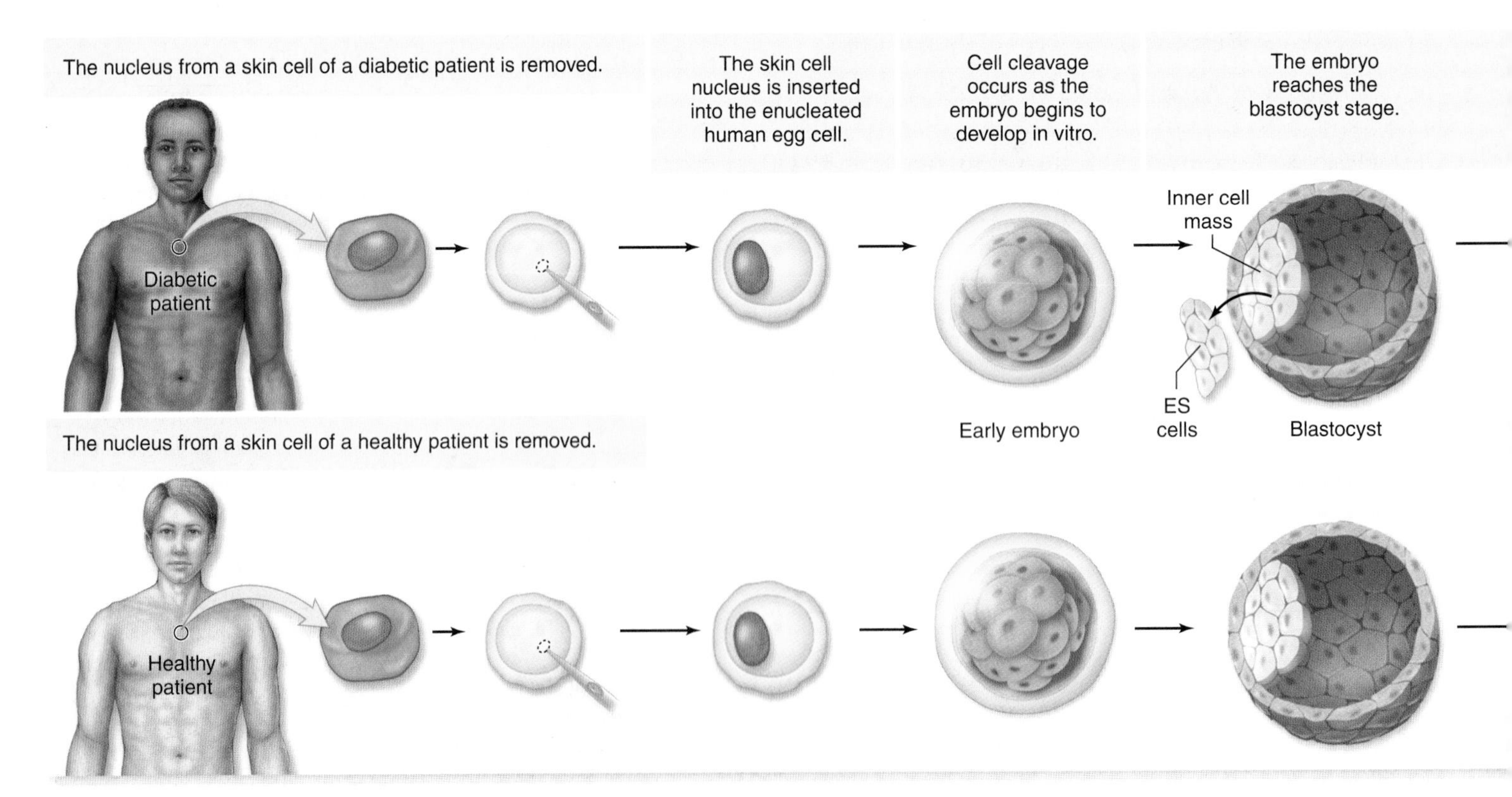

research team at Rockefeller University devised a way to accomplish this feat.

First, skin cells are isolated; then, using the same SCNT procedure that created Dolly, an embryo is assembled. After removing the nucleus from the skin cell, they insert it into an egg whose nucleus has already been removed. The egg with its skin cell nucleus is allowed to form a blastocyst stage embryo. This artificial embryo is then destroyed, and its cells are used as embryonic stem cells for transfer to injured tissue (figure 19.10).

Using this procedure, termed **therapeutic cloning,** the researchers succeeded in converting cells from the tail of a mouse into the dopamine-producing cells of the brain that are lost in Parkinson disease. Therapeutic cloning successfully addresses the key problem that must be solved before stem cells can be used to repair human tissues damaged by heart attack, nerve injury, diabetes, or Parkinson disease—the problem of immune acceptance. Since stem cells are cloned from a person's own tissues in therapeutic cloning, they pass the immune system's "self" identity check, and the body readily accepts them.

Stem cell research has stimulated ethical debate

Human embryonic stem cells have enormous promise for treating a wide range of diseases. ES cells are derived from blastocyst stage embryos, and preimplantation stage human embryos can be obtained from fertility clinics, which routinely produce excess embryos when helping infertile couples to have children by in vitro fertilization.

In therapeutic cloning, an early embryo must be taken apart to create human embryonic stem cells. For this reason, stem cell research has raised profound ethical issues. The timeless question of when human life begins cannot be avoided. In addition, the question of whether reproductive cloning should be undertaken in humans, as it was in sheep to produce Dolly, is highly controversial.

In Britain, reproductive cloning is banned, but stem cell research and therapeutic cloning to obtain clinically useful stem cells are both permitted. Careful ethical supervision of all research is provided by a variety of governmental oversight committees. Britain's Human Fertilization and Embryology Authority (HFEA), for example, is a panel of scientists and ethicists accountable to Parliament, which oversees government-funded research. Similar arrangements are being established in Japan and France.

China, Thailand, and South Korea have permissive policies toward human ES cell research and cloning. Germany and most Latin American countries, by contrast, discourage such research.

In the United States, the first human ES cell lines were created in private research labs using private funds. After a long debate, federal funds were made available in the summer of 2001 for research on the small number of already existing human embryonic stem cell lines. But the George W. Bush administration specifically prohibited the use of federal funds to create *new* lines of human ES cells (which requires the destruction of human embryos).

Because patient-specific therapeutic cloning requires the creation of new human ES cell lines, federally funded U.S. researchers are currently prohibited from doing this research. Some states, notably California, have passed legislation to allow human embryonic stem cell research, while at the same time banning reproductive cloning of humans.

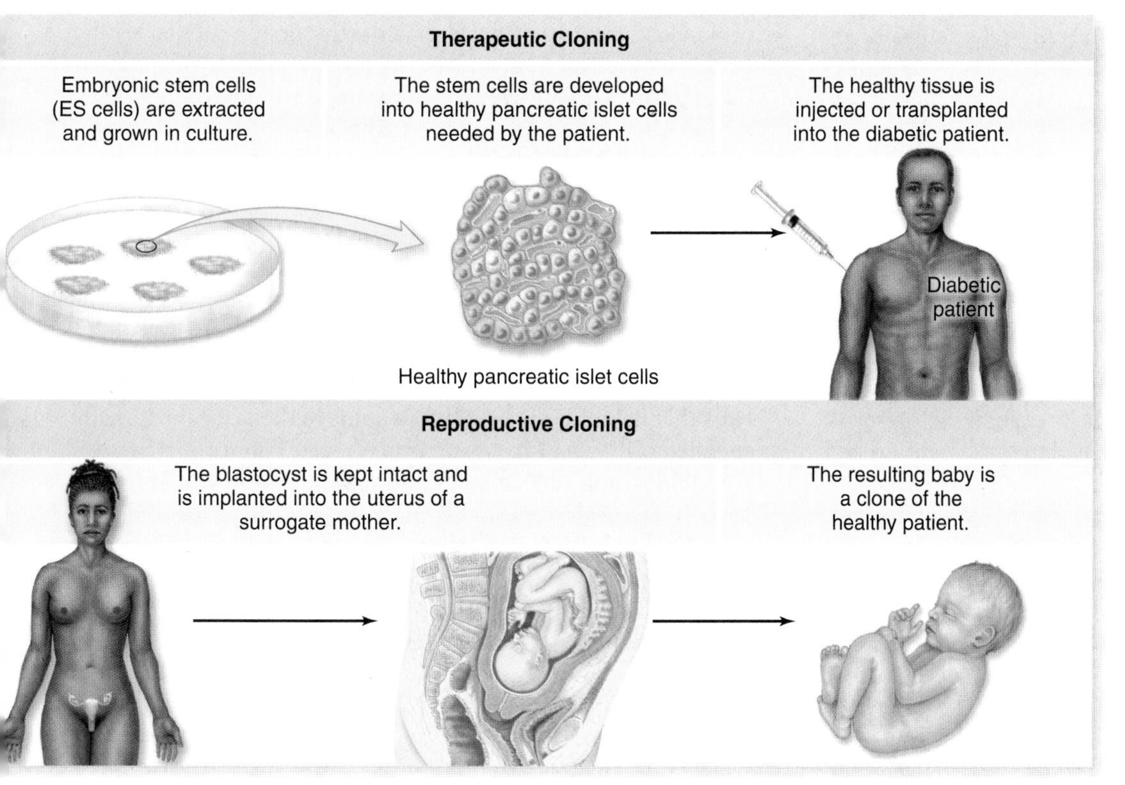

figure 19.10

HOW HUMAN EMBRYOS MIGHT BE USED FOR THERAPEUTIC CLONING. In therapeutic cloning, after initial stages to reproductive cloning, the embryo is broken apart and its embryonic stem cells are extracted. These are grown in culture and used to replace the diseased tissue of the individual who provided the DNA. This is useful only if the disease in question is not genetic as the stem cells are genetically identical to the patient. In reproductive cloning the intact embryo is implanted in the uterus of a surrogate mother, where it might grow to term. Because of health issues for both the mother and the cloned fetus, most scientists agree that reproductive cloning of humans should be banned.

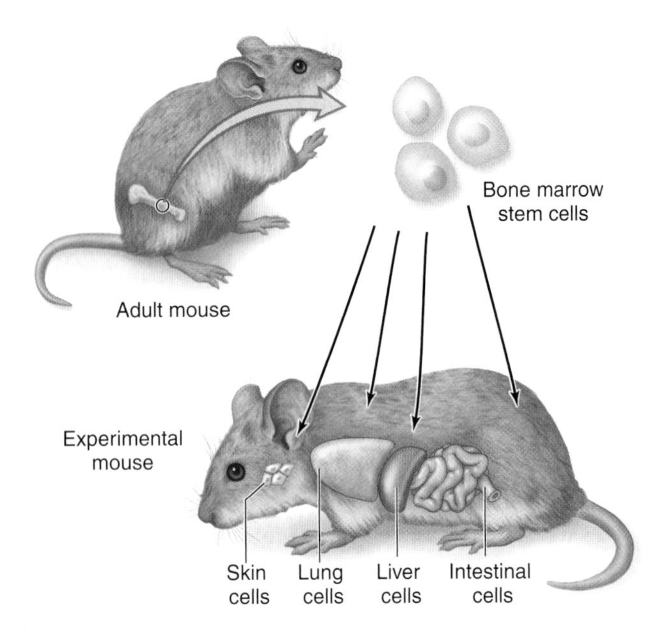

figure 19.11

MULTIPOTENT ADULT STEM CELLS. In May 2001, a single cell from the bone marrow of an adult mouse was claimed to have added functional cells to the lungs, liver, intestine, and skin of an experimental mouse. Cells isolated from adipose (fat) tissue may have similar capabilities. Since then these results have been challenged.

Adult stem cells may offer an alternative to ES cells

As discussed earlier, stem cells may be tissue specific, and they persist into adulthood in some tissues. Early reports on a variety of adult stem cells indicated that they may be reprogrammed to become other cell types than they normally would, that is, they are pluripotent (figure 19.11). These results have been challenged and the pluripotent status of adult stem cells is unclear. It also may be difficult to harvest these cells from the individual that you wish to treat. At this point the possible therapeutic use of both ES cells and adult stem cells is unclear.

Cell differentiation is preceded by determination when the cell is committed to a fate, but not yet differentiated. Differential inheritance of cytoplasmic factors can cause determination and differentiation, as can interactions between neighboring cells (induction). Inductive changes are mediated by signaling molecules that trigger signal transduction pathways. Determination may be reversible, as shown by reproductive cloning in some vertebrates. Cloned organisms, such as Dolly the sheep, have exhibited short life spans and early onset of disease, probably related to genomic imprinting. Human embryonic stem cells offer the possibility of replacing damaged or lost human tissues; however, the procedures are controversial and involve many ethical issues.

19.4 Pattern Formation

For cells in multicellular organisms to differentiate into appropriate cell types, they must gain information about their relative locations in the body. All multicellular organisms seem to use positional information to determine the basic pattern of body compartments and, thus, the overall architecture of the adult body. This positional information then leads to intrinsic changes in gene activity, so that cells ultimately adopt a fate appropriate for their location.

Pattern formation is an unfolding process. In the later stages, it may involve morphogenesis of organs (to be discussed later), but during the earliest events of development, the basic body plan is laid down, along with the establishment of the anterior–posterior (A/P, head-to-tail) axis and the dorsal–ventral (D/V, back-to-front) axis. Thus, pattern formation can be considered the process of taking a radially symmetrical cell and imposing two perpendicular axes to define the basic body plan, which in this way becomes bilaterally symmetrical. Developmental biologists use the term **polarity** to refer to the acquisition of axial differences in developing structures.

The fruit fly *Drosophila melanogaster* is the best understood animal in terms of the genetic control of early patterning. As described later, a hierarchy of gene expression that begins with maternally expressed genes controls the development of *Drosophila*. To understand the details of these gene interactions, we first need to briefly review the stages of *Drosophila* development.

Drosophila embryogenesis produces a segmented larva

Drosophila and many other insects produce two different kinds of bodies during their development: the first, a tubular eating machine called a **larva,** and the second, an adult flying sex machine with legs and wings. The passage from one body form to the other, called **metamorphosis,** involves a radical shift in development (figure 19.12). In this chapter, we concentrate on the process of going from a fertilized egg to a larva, which is termed *embryogenesis.*

Prefertilization maternal contribution

The development of an insect like *Drosophila* begins before fertilization, with the construction of the egg. Specialized **nurse cells** that help the egg grow move some of their own maternally encoded mRNAs into the maturing oocyte (figure 19.12*a*).

Following fertilization, the maternal mRNAs are transcribed into proteins, which initiate a cascade of sequential gene activations. Embryonic nuclei do not begin to function (that is, to direct new transcription of genes) until approximately 10 nuclear divisions have occurred. Therefore, the action of maternal, rather than zygotic, genes determines the initial course of *Drosophila* development.

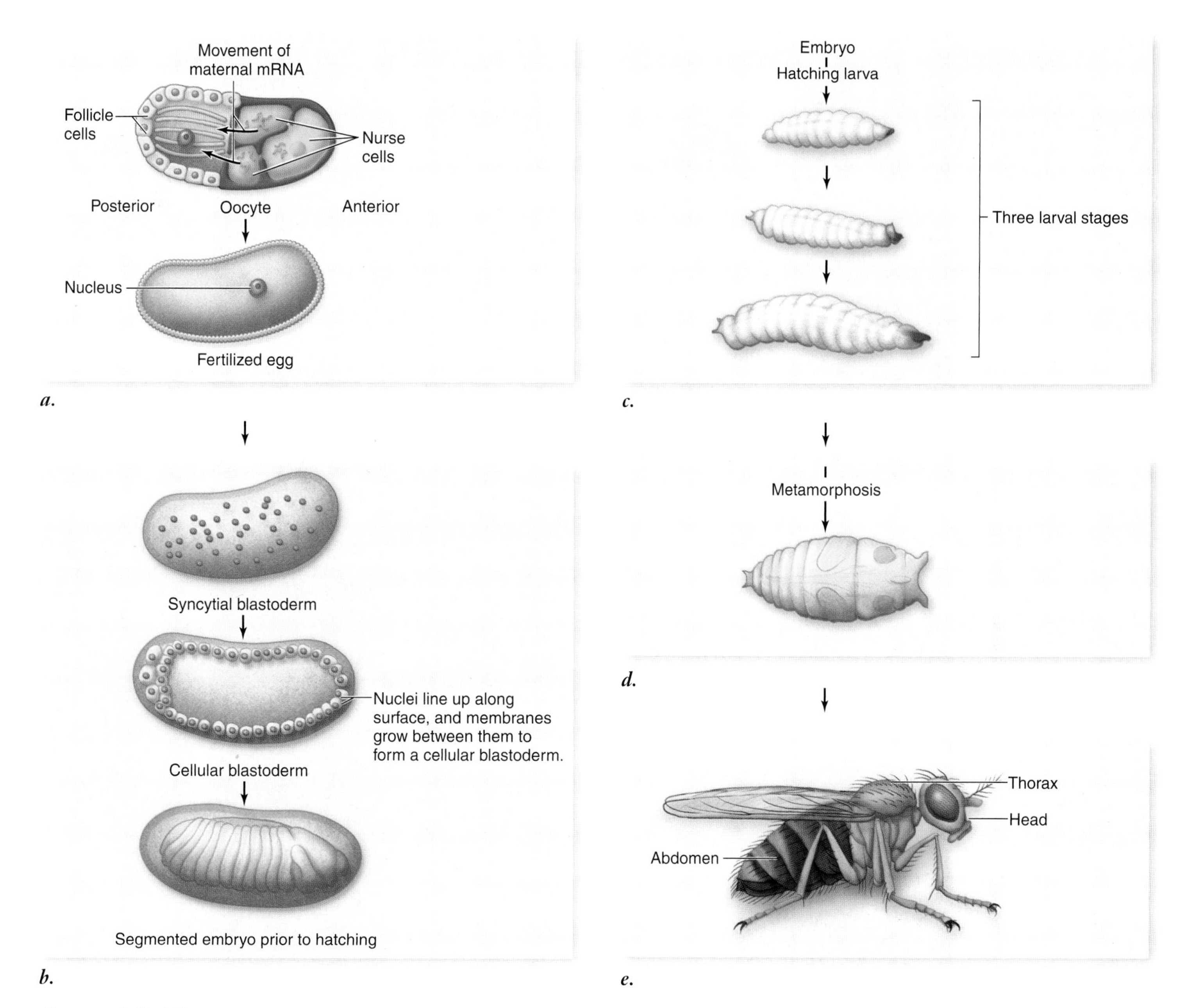

figure 19.12

THE PATH OF FRUIT FLY DEVELOPMENT. Major stages in the development of *Drosophila melanogaster* include formation of the **(*a*)** egg, **(*b*)** syncytial and cellular blastoderm, **(*c*)** larval instars, **(*d*)** pupa and metamorphosis into a **(*e*)** sexually mature adult.

Postfertilization events

After fertilization, 12 rounds of nuclear division without cytokinesis produce about 4000 nuclei, all within a single cytoplasm. All of the nuclei within this **syncytial blastoderm** (figure 19.12*b*) can freely communicate with one another, but nuclei located in different sectors of the egg encounter different maternal products.

Once the nuclei have spaced themselves evenly along the surface of the blastoderm, membranes grow between them to form the **cellular blastoderm.** Embryonic folding and primary tissue development soon follow, in a process fundamentally similar to that seen in vertebrate development. Within a day of fertilization, embryogenesis creates a segmented, tubular body—which is destined to hatch out of the protective coats of the egg as a larva.

Morphogen gradients form the basic body axes in *Drosophila*

Pattern formation in the early *Drosophila* embryo requires positional information encoded in labels that can be read by cells. The unraveling of this puzzle, work that earned the 1995 Nobel Prize

Establishing the Polarity of the Embryo

Fertilization of the egg triggers the production of bicoid protein from maternal RNA in the egg. The bicoid protein diffuses through the egg, forming a gradient. This gradient determines the polarity of the embryo, with the head and thorax developing in the zone of high concentration (*green* fluorescent dye in antibodies that bind bicoid protein allows visualization of the gradient).

Bicoid

500 μm

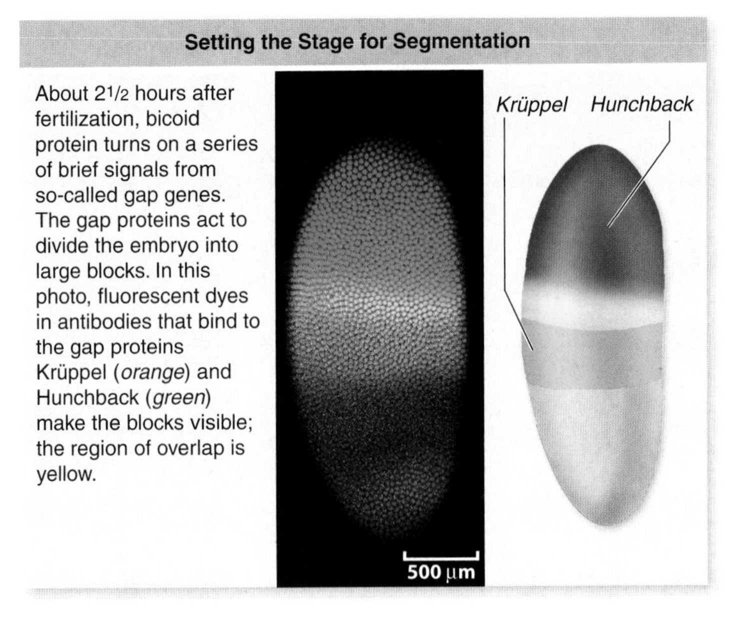

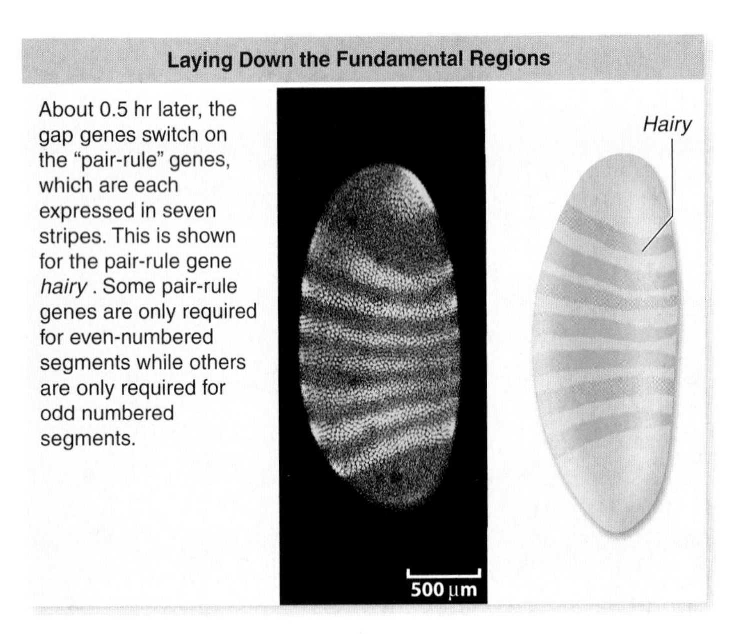

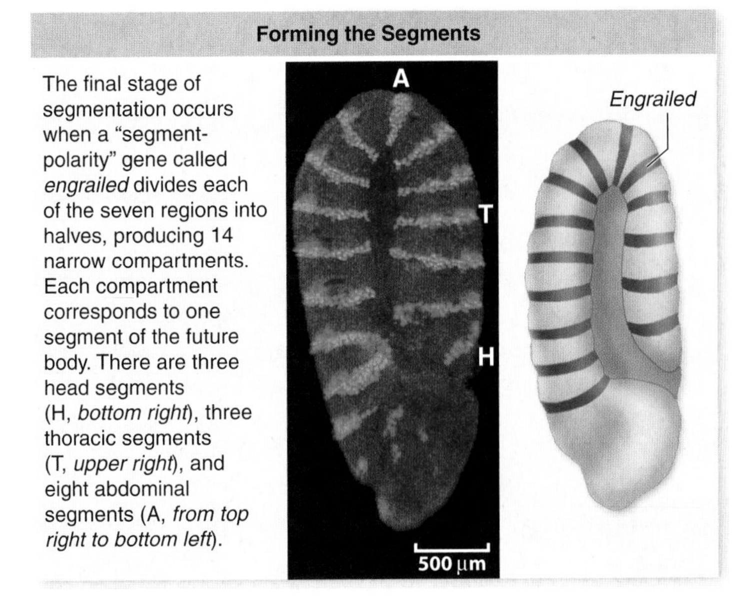

figure 19.13

BODY ORGANIZATION IN AN EARLY *DROSOPHILA* EMBRYO. In these fluorescent microscope images by 1995 Nobel laureate Christiane Nüsslein-Volhard and Sean Carroll, we watch a *Drosophila* egg pass through the early stages of development, in which the basic segmentation pattern of the embryo is established. The proteins in the photographs on the left were made visible by binding fluorescent antibodies to each specific protein. The drawings on the right help illustrate what is occurring in the photos.

for researchers Christiane Nüsslein-Volhard and Eric Wieschaus, is summarized in figure 19.13. We now know that two different genetic pathways control the establishment of A/P and D/V polarity in *Drosophila*.

Anterior–posterior axis

Formation of the A/P axis begins during maturation of the oocyte and is based on opposing gradients of two different proteins: **Bicoid** and **Nanos.** These protein gradients are established by an interesting mechanism.

Nurse cells in the ovary secrete maternally produced *bicoid* and *nanos* mRNAs into the maturing oocyte where they are differentially transported along microtubules to opposite poles of the oocyte (figure 19.14*a*). This differential transport comes about due to the use of different motor proteins to move the two mRNAs. The *bicoid* mRNA then becomes anchored in the cytoplasm at the end of the oocyte closest to the nurse cells, and this end will develop into the anterior end of the embryo. *Nanos* mRNA becomes anchored to the opposite end of the oocyte, which will become the posterior end of the embryo. Thus, by the end of oogenesis, the *bicoid* and *nanos* mRNAs are already set to function as cytoplasmic determinants in the fertilized egg (figure 19.14*b*).

Following fertilization, translation of the anchored mRNA and diffusion of the proteins away from their respec-

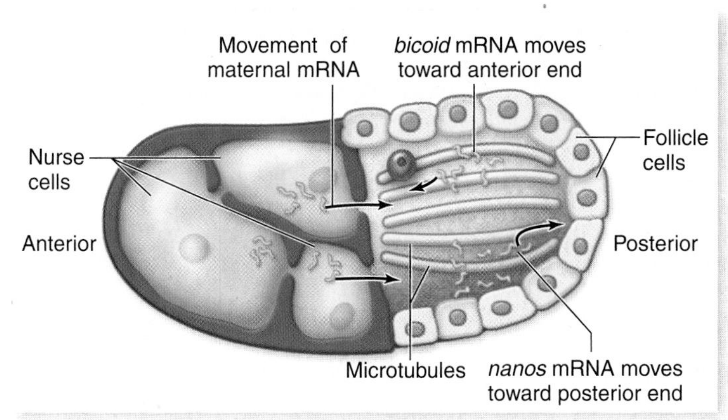

a.

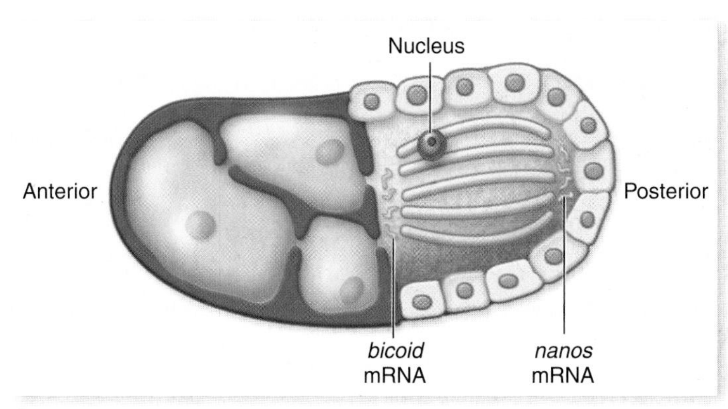

b.

figure 19.14

SPECIFYING THE A/P AXIS IN *DROSOPHILA* EMBRYOS I. ***a.*** In the ovary, nurse cells secrete maternal mRNAs into the cytoplasm of the oocyte. Clusters of microtubules direct oocyte growth and maturation. Motor proteins travel along the microtubules transporting molecules in two directions. *Bicoid* mRNAs are transported toward the anterior pole of the oocyte, *nanos* mRNA is transported toward the posterior pole of the oocyte. ***b.*** A mature oocyte, showing localization of *bicoid* mRNAs to the anterior pole and *nanos* mRNAs to the posterior pole.

tive sites of synthesis create opposing gradients of each protein: Highest levels of bicoid protein are at the anterior pole of the embryo (figure 19.15*a*), and highest levels of the nanos protein are at the posterior pole. Concentration gradients of soluble molecules can specify different cell fates along an axis, and proteins that act in this way, like Bicoid and Nanos, are called **morphogens.**

The Bicoid and Nanos proteins control the translation of two other maternal messages, *hunchback* and *caudal*, that encode transcription factors. **Hunchback** activates genes required for the formation of anterior structures, and **Caudal** activates genes required for the development of posterior (abdominal) structures. The *hunchback* and *caudal* mRNAs are evenly distributed across the egg (figure 19.15*b*), so how is it that proteins translated from these mRNAs become localized?

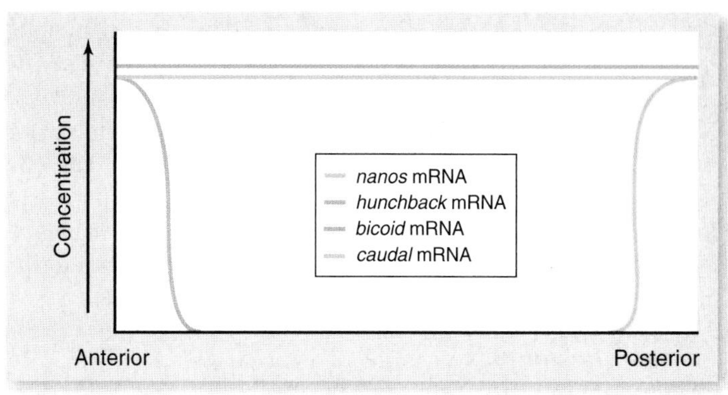

a. Oocyte mRNAs

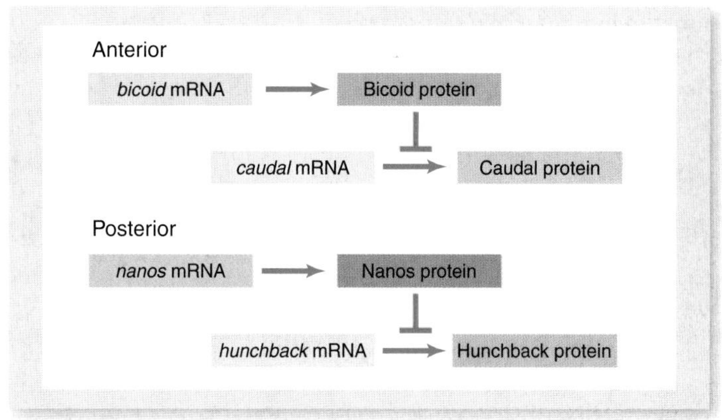

b. After fertilization

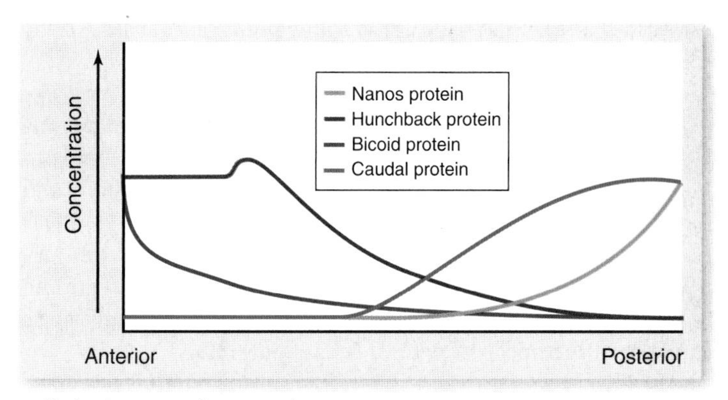

c. Early cleavage embryo proteins

figure 19.15

SPECIFYING THE A/P AXIS IN *DROSOPHILA* EMBRYOS II. ***a.*** Unlike *bicoid* and *nanos*, *hunchback* and *caudal* mRNAs are evenly distributed throughout the cytoplasm of the oocyte. ***b.*** Following fertilization, *bicoid* and *nanos* mRNAs are translated into protein, making opposing gradients of each protein. Bicoid binds to and represses translation of *caudal* mRNAs (in anterior regions of the egg). Nanos binds to and represses translation of *hunchback* mRNAs (in posterior regions of the egg). ***c.*** Translation of *hunchback* mRNAs in anterior regions of the egg will create a Hunchback gradient that mirrors the Bicoid gradient. Translation of *caudal* mRNAs in posterior regions of the embryo will create a Caudal gradient that mirrors the Nanos gradient.

The answer is that Bicoid protein binds to and inhibits translation of *caudal* mRNA. Therefore, *caudal* is only translated in the posterior regions of the egg where Bicoid is absent. Similarly, Nanos protein binds to and prevents translation of the *hunchback* mRNA. As a result, *hunchback* is only translated in the anterior regions of the egg (figure 19.15*c*). Thus, shortly after fertilization, four protein gradients exist in the embryo: anterior–posterior gradients of Bicoid and Hunchback proteins, and posterior–anterior gradients of Nanos and Caudal proteins (figure 19.15*d*).

Dorsal–ventral axis

The dorsal–ventral axis in *Drosophila* is established by actions of the *dorsal* gene product. Once again the process begins in the ovary, when maternal transcripts of the *dorsal* gene are put into the oocyte. However, unlike *bicoid* or *nanos*, the *dorsal* mRNA does not become asymmetrically localized. Instead, a series of steps are required for Dorsal to carry out its function.

First, the oocyte nucleus, which is located to one side of the oocyte, synthesizes *gurken* mRNA. The *gurken* mRNA then accumulates in a crescent between the nucleus and the membrane on that side of the oocyte (figure 19.16*a*). This will be the future dorsal side of the embryo.

The Gurken protein is a soluble cell-signaling molecule, and when it is translated and released from the oocyte, it binds to receptors in the membranes of the overlying follicle cells (figure 19.16*b*). These cells then differentiate into a dorsal morphology. Meanwhile, no Gurken signal is released from the other side of the oocyte, and the follicle cells on that side of the oocyte adopt a ventral fate.

Following fertilization, a signaling molecule is differentially activated on the ventral surface of the embryo in a complex sequence of steps. This signaling molecule then binds to a membrane receptor in the ventral cells of the embryo and activates a signal transduction pathway in those cells. Activation of this pathway results in the selected transport of the Dorsal protein (which is everywhere) into ventral nuclei, forming a gradient along the D/V axis. The Dorsal protein levels are highest in the nuclei of ventral cells (figure 19.16*c*).

The Dorsal protein is a transcription factor, and once it is transported into nuclei, it activates genes required for the proper development of ventral structures, simultaneously repressing genes that specify dorsal structures. Hence, the product of the *dorsal* gene ultimately directs the development of ventral structures.

(Note that many *Drosophila* genes are named for the mutant phenotype that results from a loss of function in that gene. A lack of *dorsal* function produces dorsalized embryos with no ventral structures.)

Although profoundly different mechanisms are involved, the unifying factor controlling the establishment of both A/P and D/V polarity in *Drosophila* is that *bicoid*, *nanos*, *gurken*, and *dorsal* are all maternally expressed genes. The polarity of the future embryo in both instances is therefore laid down in the oocyte using information coming from the maternal genome.

The preceding discussion simplifies events, but the outline is clear: Polarity is established by the creation of morpho-

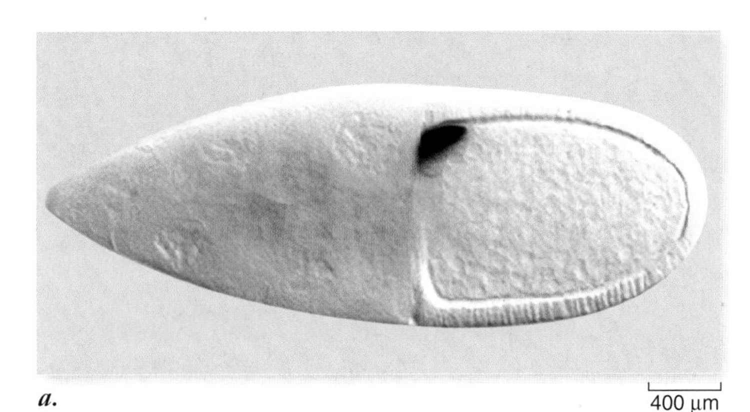

b. 400 μm

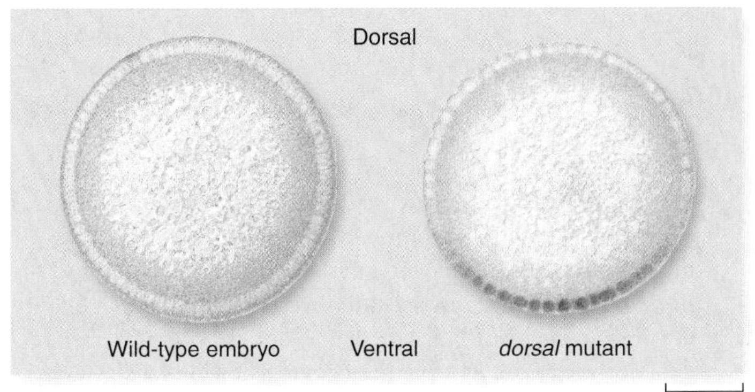

figure 19.16

SPECIFYING THE D/V AXIS IN *DROSOPHILA* EMBRYOS.
a. The *gurken* mRNA (*dark stain*) is concentrated between the oocyte nucleus (not visible) and the dorsal, anterior surface of the oocyte. ***b.*** In a more mature oocyte, Gurken protein (*yellow stain*) is secreted from the dorsal anterior surface of the oocyte, forming a gradient along the dorsal surface of the egg. Gurken then binds to membrane receptors in the overlying follicle cells. Double staining for actin (*red*) shows the cell boundaries of the oocyte, nurse cells, and follicle cells. ***c.*** For these images, cellular blastoderm stage embryos were cut in cross section, to visualize the nuclei of cells around the perimeter of the embryos. Dorsal protein (*dark stain*) is localized in nuclei on the ventral surface of the blastoderm in a wild-type embryo (*left*). The *dorsal* mutant on the right will not form ventral structures, and Dorsal is not present in ventral nuclei of this embryo.

gen gradients in the embryo based on maternal information in the egg. These gradients then drive the expression of the zygotic genes that will actually pattern the embryo. This reliance on a hierarchy of regulatory genes is a unifying theme for all of development.

The body plan is produced by sequential activation of genes

Let us now return to the process of pattern formation in *Drosophila* along the A/P axis. Determination of structures is accomplished by the sequential activation of three classes of **segmentation genes.** These genes create the hallmark segmented body plan of a fly, which consists of three fused head segments, three thoracic segments, and eight abdominal segments (see figure 19.12*e*).

To begin, Bicoid protein exerts its profound effect on the organization of the embryo by activating the translation and transcription of *hunchback* mRNA (which is the first mRNA to be transcribed after fertilization). *Hunchback* is a member of a group of nine genes called the **gap genes.** These genes map out the coarsest subdivision of the embryo along the A/P axis (see figure 19.13).

All of the gap genes encode transcription factors, which, in turn, activate the expression of eight or more **pair-rule genes.** Each of the pair-rule genes, such as *hairy*, produces seven distinct bands of protein, which appear as stripes when visualized with fluorescent reagents (see figure 19.13). These bands subdivide the broad gap regions and establish boundaries that divide the embryo into seven zones. When mutated, each of the pair-rule genes alters every other body segment.

All of the pair-rule genes also encode transcription factors, and they, in turn, regulate the expression of each other and of a group of nine or more **segment polarity genes.** The segment polarity genes are each expressed in 14 distinct bands of cells, which subdivide each of the seven zones specified by the pair-rule genes (see figure 19.13). The *engrailed* gene, for example, divides each of the seven zones established by *hairy* into anterior and posterior compartments. The segment polarity genes encode proteins that function in cell–cell signaling pathways. Thus, they function in inductive events—which occur *after* the syncytial blastoderm is divided into cells—to fix the anterior and posterior fates of cells within each segment.

In summary, within 3 hr after fertilization, a highly orchestrated cascade of segmentation gene activity transforms the broad gradients of the early embryo into a periodic, segmented structure with A/P and D/V polarity. The activation of the segmentation genes depends on the free diffusion of maternally encoded morphogens, which is only possible within the syncytial blastoderm of the early *Drosophila* embryo.

Segment identity arises from the action of homeotic genes

With the basic body plan laid down, the next step is to give identity to the segments of the embryo. A highly interesting class of *Drosophila* mutants has provided the starting point for understanding the creation of segment identity.

In these mutants, a particular segment seems to have changed its identity—that is, it has characteristics of a different segment. In wild-type flies, a pair of legs emerges from each of the three thoracic segments, but only the second thoracic segment has wings. Mutations in the *Ultrabithorax* gene cause a fly to grow an extra pair of wings, as though it has two second thoracic segments (figure 19.17). Even more bizarre are mutations in *Antennapedia*, which cause legs to grow out of the head in place of antennae!

Thus, mutations in these genes lead to the appearance of perfectly normal body parts in inappropriate places. Such mutants are termed *homeotic mutants* because the transformed body part looks similar (homeotic) to another. The genes in which such mutants occur are therefore called **homeotic genes.**

Homeotic gene complexes

In the early 1950s, geneticist and Nobel laureate Edward Lewis discovered that several homeotic genes, including *Ultrabithorax*, map together on the third chromosome of *Drosophila* in a tight cluster called the **bithorax complex.** Mutations in these genes all affect body parts of the thoracic and abdominal segments, and Lewis concluded that the genes of the bithorax complex control the development of body parts in the rear half of the thorax and all of the abdomen.

Interestingly, the order of the genes in the bithorax complex mirrors the order of the body parts they control, as though the genes are activated serially. Genes at the beginning of the cluster switch on development of the thorax; those in the middle control the anterior part of the abdomen; and those at the end affect the posterior tip of the abdomen.

A second cluster of homeotic genes, the **Antennapedia complex,** was discovered in 1980 by Thomas Kaufmann. The Antennapedia complex governs the anterior end of the fly, and

figure 19.17

MUTATIONS IN HOMEOTIC GENES. Three separate mutations in the Bithorax complex caused this fruit fly to develop an additional second thoracic segment, with accompanying wings.

the order of genes in this complex also corresponds to the order of segments they control (figure 19.18*a*).

The homeobox

An interesting relationship was discovered after the genes of the bithorax and Antennapedia complexes were cloned and sequenced. These genes all contain a conserved sequence of 180 nucleotides that codes for a 60-amino-acid, DNA-binding domain. Because this domain was found in all of the homeotic genes, it was named the *homeodomain*, and the DNA that encodes it is called the homeobox. Thus, the term ***Hox* gene** now refers to a homeobox-containing gene that specifies the identity of a body part. These genes function as transcription factors that bind DNA using their homeobox domain.

Clearly, the homeobox distinguishes portions of the genome that are devoted to pattern formation. How the *Hox* genes do this is the subject of much current research. Scientists believe that the ultimate targets of *Hox* gene function must be genes that control cell behaviors associated with organ morphogenesis.

Evolution of homeobox-containing genes

A large amount of research has been devoted to analyzing the clustered complexes of *Hox* genes in other organisms. These investigations have led to a fairly coherent view of homeotic gene evolution.

It is now clear that the *Drosophila* bithorax and Antennapedia complexes represent two parts of a single cluster of genes. In vertebrates, there are four copies of *Hox* gene clusters. As in *Drosophila*, the spatial domains of *Hox* gene expression correlate with the order of the genes on the chromosome (figure 19.18*b*). The existence of four *Hox* clusters in vertebrates is viewed by many as evidence that two duplication events of the entire genome have occurred in the vertebrate lineage.

This idea raises the issue of when the original cluster arose. To answer this question, researchers have turned to more primitive organisms, such as *Amphioxus* (now called *Branchiostoma*), a lancelet chordate (see chapter 35). The finding of only one cluster of *Hox* genes in *Amphioxus* implies that indeed there have been two duplications in the vertebrate lineage, at least of the *Hox* cluster. Given the single cluster in arthropods, this finding

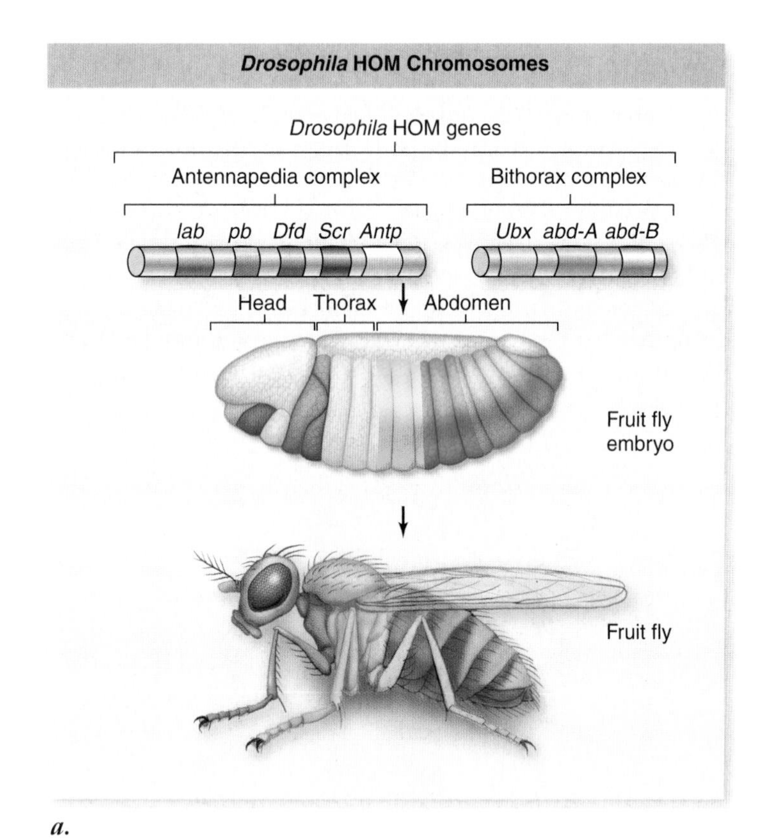

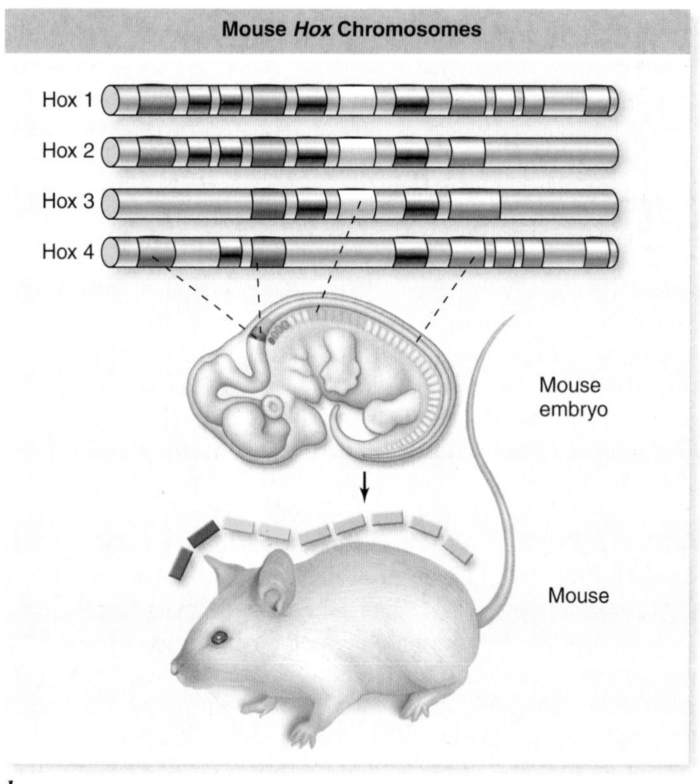

figure 19.18

A COMPARISON OF HOMEOTIC GENE CLUSTERS IN THE FRUIT FLY *Drosophila melanogaster* AND THE MOUSE *Mus musculus*. ***a.*** *Drosophila* homeotic genes. Called the homeotic gene complex, or HOM complex, the genes are grouped into two clusters: the Antennapedia complex (anterior) and the bithorax complex (posterior). ***b.*** The *Drosophila* HOM genes and the mouse *Hox* genes are related genes that control the regional differentiation of body parts in both animals. These genes are located on a single chromosome in the fly and on four separate chromosomes in mammals. In this illustration, the genes are color-coded to match the parts of the body along the A/P axis in which they are expressed. Note that the order of the genes along the chromosome(s) is mirrored by their pattern of expression in the embryo and in structures in the adult fly.

implies that the common ancestor to all animals with bilateral symmetry had a single *Hox* cluster as well.

The next logical step is to look at even more-primitive animals: the radially symmetrical cnidarians such as *Hydra* (see chapter 33). Thus far, *Hox* genes have been found in a number of cnidarian species, and recent sequence analyses suggest that cnidarian *Hox* genes are also arranged into clusters. Thus, the appearance of the ancestral *Hox* cluster likely preceded the divergence between radial and bilateral symmetries in animal evolution.

Pattern formation in plants is also under genetic control

The evolutionary split between plant and animal cell lineages occurred about 1.6 BYA, before the appearance of multicellular organisms with defined body plans. The implication is that multicellularity evolved independently in plants and animals. Because of the activity of meristems, additional modules can be added to plant bodies throughout their lifetimes. In addition, plant flowers and roots have a radial organization, in contrast to the bilateral symmetry of most animals. We may therefore expect that the genetic control of pattern formation in plants is fundamentally different from that of animals.

Although plants have homeobox-containing genes, they do not possess complexes of *Hox* genes similar to the ones that determine regional identity of developing structures in animals. Instead, the predominant homeotic gene family in plants appears to be the **MADS-box** genes.

MADS-box genes are a family of transcriptional regulators found in most eukaryotic organisms, including plants, animals, and fungi. The MADS-box is a conserved DNA-binding and dimerization domain, named after the first five genes to be discovered with this domain. Only a small number of MADS-box genes are found in animals, where their functions include the control of cell proliferation and tissue-specific gene expression in postmitotic muscle cells. They do not appear to play a role in the patterning of animal embryos.

In contrast, the number and functional diversity of MADS-box genes increased considerably during the evolution of land plants, and there are more than 100 MADS-box genes in the *Arabidopsis* genome. In flowering plants, the MADS-box genes dominate the control of development, regulating such processes as the transition from vegetative to reproductive growth, root development, and floral organ identity.

Although distinct from genes in the *Hox* clusters of animals, homeodomain-containing transcription factors in plants do have important developmental functions. One such example is the family of *knottedlike homeobox* (*knox*) genes, which are important regulators of shoot apical meristem development in both seed-bearing and nonseed-bearing plants. Mutations that affect expression of *knox* genes produce changes in leaf and petal shape, suggesting that these genes play an important role in generating leaf form.

Pattern formation in animals involves the coordinated expression of a hierarchy of genes. Gradients of morphogens in *Drosophila* specify A/P and D/V axes, then lead to sequential activation of segmentation genes that subdivide the embryo in progressively more defined segments. The action of homeotic genes acts to provide segment identity. Genes with a DNA-binding homeodomain sequence are called *Hox* genes (for *homeobox* genes), and they are organized into clusters. Plants also change gene expression to control development, but they use a different set of control genes called MADS-box genes.

19.5 Morphogenesis

At the end of cleavage, the *Drosophila* embryo still has a relatively simple structure: It comprises several thousand identical-looking cells, which are present in a single layer surrounding a central yolky region. The next step in embryonic development is **morphogenesis**—the generation of ordered form and structure.

Morphogenesis is the product of changes in cell structure and cell behavior. Animals regulate the following processes to achieve morphogenesis:

- The number, timing, and orientation of cell divisions;
- Cell growth and expansion;
- Changes in cell shape;
- Cell migration; and
- Cell death.

Plant and animal cells are fundamentally different in that animal cells have flexible surfaces and can move, but plant cells are immotile and encased within stiff cellulose walls. Each cell in a plant is fixed into position when it is created. Thus, animal cells use cell migration extensively during development while plants use the other four mechanisms but lack cell migration. We consider the morphogenetic changes in animals first, and then those that occur in plants.

Cell division during development may result in unequal cytokinesis

The orientation of the mitotic spindle determines the plane of cell division in eukaryotic cells. The coordinated function of microtubules and their motor proteins determines the respective position of the mitotic spindle within a cell (see chapter 10). If the spindle is centrally located in the dividing cell, two equal-sized daughter cells will result. If the spindle is off to one side, one large daughter cell and one small daughter cell will result.

The great diversity of cleavage patterns in animal embryos is determined by differences in spindle placement. In many cases, the fate of a cell is determined by its relative placement in the embryo during cleavage. For example, in preimplantation mammalian embryos, cells on the outside of the embryo usually differentiate into trophectoderm cells, which form only extraembryonic structures later in development (for example, a part of the placenta). In contrast, the embryo proper is derived from the inner cell mass, cells which, as the name implies, are in the interior of the embryo.

Cells change shape and size as morphogenesis proceeds

In animals, cell differentiation is often accompanied by profound changes in cell size and shape. For example, the large nerve cells that connect your spinal cord to the muscles in your big toe develop long processes called *axons* that span this entire distance. The cytoplasm of an axon contains microtubules, which are used for motor-driven transport of materials along the length of the axon.

As another example, muscle cells begin as *myoblasts*, undifferentiated muscle precursor cells. They eventually undergo conversion into the large, multinucleated *muscle fibers* that make up mammalian skeletal muscles. These changes begin with the expression of the *MyoD1* gene, which encodes a transcription factor that binds to the promoters of muscle-determining genes to initiate these changes.

Programmed cell death is a necessary part of development

Not every cell produced during development is destined to survive. For example, human embryos have webbed fingers and toes at an early stage of development. The cells that make up the webbing die in the normal course of morphogenesis. As another example, vertebrate embryos produce a very large number of neurons, ensuring that enough neurons are available to make the necessary synaptic connections, but over half of these neurons never make connections and die in an orderly way as the nervous system develops.

Unlike accidental cell deaths due to injury, these cell deaths are planned—and indeed required—for proper development and morphogenesis. Cells that die due to injury typically swell and burst, releasing their contents into the extracellular fluid. This form of cell death is called **necrosis.** In contrast, cells programmed to die shrivel and shrink in a process called **apoptosis,** which means "falling away," and their remains are taken up by surrounding cells.

Genetic control of apoptosis

Apoptosis occurs when a "death program" is activated. All animal cells appear to possess such programs. In *C. elegans*, the same 131 cells always die during development in a predictable and reproducible pattern.

Work on *C. elegans* showed that three genes are central to this process. Two (*ced-3* and *ced-4*) activate the death program itself; if either is mutant, those 131 cells do not die, and go on instead to form nervous tissue and other tissue. The third gene (*ced-9*) represses the death program encoded by the other two: All 1090 cells of the *C. elegans* embryo die in *ced-9* mutants. In *ced-9/ced-3* double mutants, all 1090 cells live, which suggests that *ced-9* inhibits cell death by functioning prior to *ced-3* in the apoptotic pathway (figure 19.19*a*).

The mechanism of apoptosis appears to have been highly conserved during the course of animal evolution. In human nerve cells, the *Apaf1* gene is similar to *ced-4* of *C. elegans* and activates the cell death program, and the human *bcl-2* gene acts similarly to *ced-9* to repress apoptosis. If a copy of the human *bcl-2* gene is transferred into a nematode with a defective *ced-9* gene, *bcl-2* suppresses the cell death program of *ced-3* and *ced-4*.

The mechanism of apoptosis

The product of the *C. elegans ced-4* gene is a protease that activates the product of the *ced-3* gene, which is also a protease. The human *Apaf1* gene is actually named for its role: *A*poptotic *p*rotease *a*ctivating *f*actor. It activates two proteases called caspases that have a role similar to the Ced-3 protease in *C. elegans* (figure 19.19*b*). When the final proteases are activated, they chew up proteins in important cellular structures such as the cytoskeleton and the nuclear lamina, leading to cell fragmentation.

The role of Ced-9/Bcl-2 is to inhibit this program. Specifically, it inhibits the activating protease, preventing the activation of the destructive proteases. The entire process is thus controlled by an inhibitor of the death program.

Both internal and external signals control the state of the Ced-9/Bcl-2 inhibitor. For example, in the human nervous system, neurons have a cytoplasmic inhibitor of Bcl-2 that allows the death program to proceed (figure 19.19*b*). In the presence of nerve growth factor, a signal transduction pathway leads to the cytoplasmic inhibitor being inactivated, allowing Bcl-2 to inhibit apoptosis and the nerve cell to survive.

Cell migration gets the right cells to the right places

The migration of cells is important during many stages of animal development. The movement of cells involves both adhesion and the loss of adhesion. Adhesion is necessary for cells to get "traction," but cells that are initially attached to others must lose this adhesion to be able to leave a site.

Cell movement also involves cell-to-substrate interactions, and the extracellular matrix may control the extent or route of cell migration. The central paradigm of morphogenetic cell movements in animals is a change in cell adhesiveness, which is mediated by changes in the composition of macromolecules in the plasma membranes of cells or in the extracellular matrix. Cell-to-cell interactions are often mediated through cadherins, but cell-to-substrate interactions often involve integrin-to-extracellular-matrix (ECM) interactions.

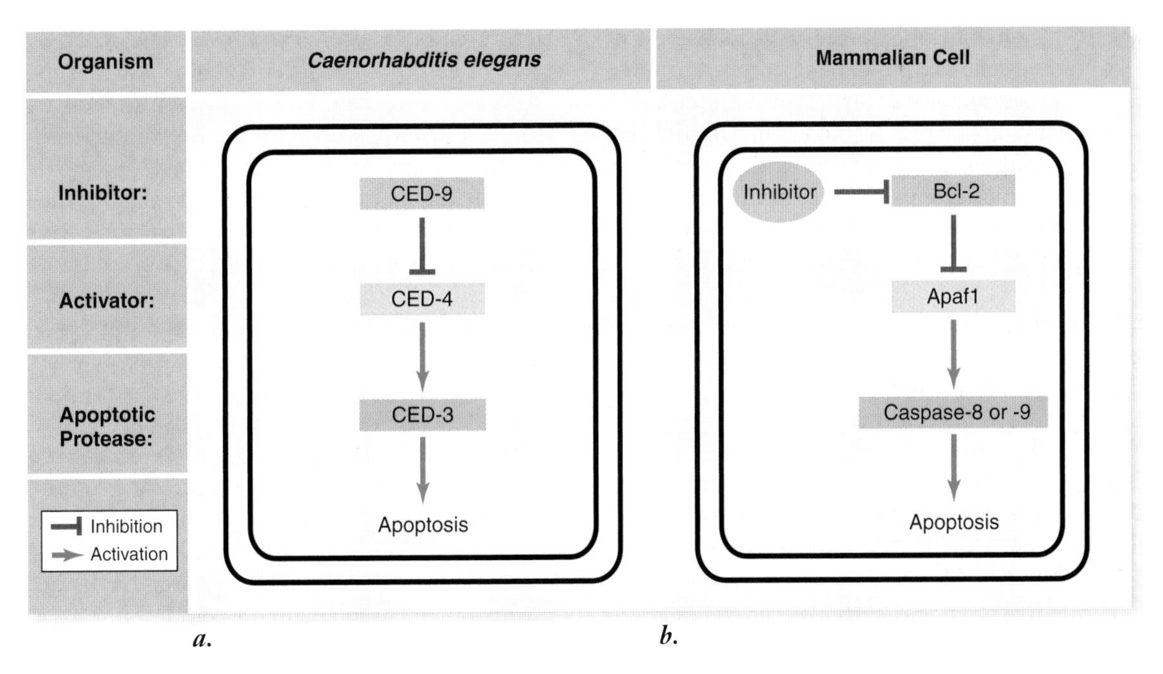

figure 19.19

PROGRAMMED CELL DEATH PATHWAY. Apoptosis, or programmed cell death, is necessary for the normal development of all animals. ***a.*** In the developing nematode, for example, two genes, *ced-3* and *ced-4*, code for proteins that cause the programmed cell death of 131 specific cells. In the other (surviving) cells of the developing nematode, the product of a third gene, *ced-9*, represses the death program encoded by *ced-3* and *ced-4*. ***b.*** The mammalian homologues of the apoptotic genes in *C. elegans* are *bcl-2* (*ced-9* homologue), *Apaf1* (*ced-4* homologue), and *caspase-8* or *-9* (*ced-3* homologues). In the absence of any cell survival factor, Bcl-2 is inhibited and apoptosis occurs. In the presence of nerve growth factor (NGF) and NGF receptor binding, Bcl-2 is activated, thereby inhibiting apoptosis.

Cadherins

Cadherins are a large gene family, with over 80 members identified in humans. In the genomes of *Drosophila*, *C. elegans*, and humans, the cadherins can be sorted into several subfamilies that exist in all three genomes.

The cadherin proteins are all transmembrane proteins that share a common motif, the *cadherin domain*, a 110-amino-acid domain in the extracellular portion of the protein that mediates Ca^{2+}-dependent binding between like cadherins (homophilic binding).

Experiments in which cells are allowed to sort in vitro illustrate the function of cadherins. Cells with the same cadherins adhere specifically to one another, while not adhering to other cells with different cadherins. If cell populations with different cadherins are dispersed and then allowed to reaggregate, they sort into two populations of cells based on the nature of the cadherins on their surface.

An example of the action of cadherins can be seen in the development of the vertebrate nervous system. All surface ectoderm cells of the embryo express E-cadherin. The formation of the nervous system begins when a central strip of cells on the dorsal surface of the embryo turns off E-cadherin expression and turns on N-cadherin expression. In the process of **neurulation,** the formation of the neural tube (see chapter 53), the central strip of N-cadherin-expressing cells folds up to form the tube. The neural tube pinches off from the overlying cells, which continue to express E-cadherin. The surface cells outside the tube differentiate into the epidermis of the skin, whereas the neural tube develops into the brain and spinal cord of the embryo.

Integrins

In some tissues, such as connective tissue, much of the volume of the tissue is taken up by the spaces *between* cells. These spaces are filled with a network of molecules secreted by surrounding cells, termed a *matrix.* In connective tissue such as cartilage, long polysaccharide chains are covalently linked to proteins (proteoglycans), within which are embedded strands of fibrous protein (collagen, elastin, and fibronectin). Migrating cells traverse this matrix by binding to it with cell surface proteins called **integrins.**

Integrins are attached to actin filaments of the cytoskeleton and protrude out from the cell surface in pairs, like two hands. The "hands" grasp a specific component of the matrix, such as collagen or fibronectin, thus linking the cytoskeleton to the fibers of the matrix. In addition to providing an anchor, this binding can initiate changes within the cell, alter the growth of the cytoskeleton, and activate gene expression and the production of new proteins.

The process of **gastrulation,** during which the hollow ball of animal embryonic cells folds in on itself to form a multilayered structure, depends on fibronectin–integrin interactions. For example, injection of antibodies against either fibronectin or integrins into salamander embryos blocks binding of cells to fibronectin in the ECM and inhibits gastrulation. The result is like a huge traffic jam following a major accident on a freeway: Cells (cars) keep coming, but they get backed up since they cannot get beyond the area of inhibition (accident site) (figure 19.20). Similarly, a targeted knockout of the fibronectin gene in mice resulted in gross defects in the migration, proliferation, and differentiation of embryonic mesoderm cells.

Thus, cell migration is largely a matter of changing patterns of cell adhesion. As a migrating cell travels, it continually extends projections that probe the nature of its environment. Tugged this way and that by different tentative attachments, the cell literally feels its way toward its ultimate target site.

In seed plants, the plane of cell division determines morphogenesis

The form of a plant body is largely determined by the plane in which cells divide. The first division of the fertilized egg in a flowering plant is off-center, so that one of the daughter cells is small, with dense cytoplasm (figure 19.21*a*). That cell, the future embryo, begins to divide repeatedly, forming a ball of cells. The other daughter cell also divides repeatedly, forming an elongated structure called a *suspensor*, which links the embryo to the nutrient tissue of the seed. The suspensor also provides a route for nutrients to reach the developing embryo.

Just as many animal embryos acquire their initial axis as a cell mass formed during cleavage divisions, so the plant embryo forms its root–shoot axis at this time. Cells near the suspensor are destined to form a root, whereas those at the other end of the axis ultimately become a shoot, the aboveground portion of the plant.

The relative position of cells within the plant embryo is also a primary determinant of cell differentiation. The outermost cells in a plant embryo become epidermal cells. The bulk of the embryonic interior consists of ground tissue cells that eventually function in food and water storage. Finally, cells at the core of the embryo are destined to form the future vascular tissue (figure 19.21*b*). (Plant tissues and development are described in detail in chapters 36 and 37.)

Soon after the three basic tissues form, a flowering plant embryo develops one or two seed leaves called *cotyledons.* At this point, development is arrested, and the embryo is either surrounded by nutritive tissue or has amassed stored food in its cotyledons (figure 19.21*c*). The resulting package, known as a *seed,* is resistant to drought and other unfavorable conditions.

A seed germinates in response to favorable changes in its environment. The embryo within the seed resumes development and grows rapidly, its roots extending downward and its leaf-bearing shoots extending upward (figure 19.21*d*). Plant development exhibits its great flexibility during the assembly of the modules that make up a plant body. Apical meristems at the root and shoot tips generate the large numbers of cells needed to form leaves, flowers, and all other components of the mature plant (figure 19.21*e*).

Growth within the developing flower is controlled by a cascade of transcription factors. A key member of this cascade is the *AINTEGUMENTA* (*ANT*) gene. Loss of ANT function reduces the number and size of floral organs, and inappropriate expression leads to larger floral organs.

Plant body form is also established by controlled changes in cell shape as cells expand osmotically after they form. Plant growth-regulating hormones and other factors influence the orientation of bundles of microtubules on the interior of the plasma membrane. These microtubules seem to guide cellulose deposition as the cell wall forms around the outside of a new cell. The orientation of the cellulose fibers, in turn, determines how the cell will elongate as it increases in volume due to osmosis, and so determines the cell's final shape.

Morphogenesis is the generation of ordered form and structure. Morphogenesis occurs by cell growth, cell shape change, cell death (apoptosis), and cell migration. Because plant cells cannot move, cell division and cell expansion are the primary morphogenetic processes in plants.

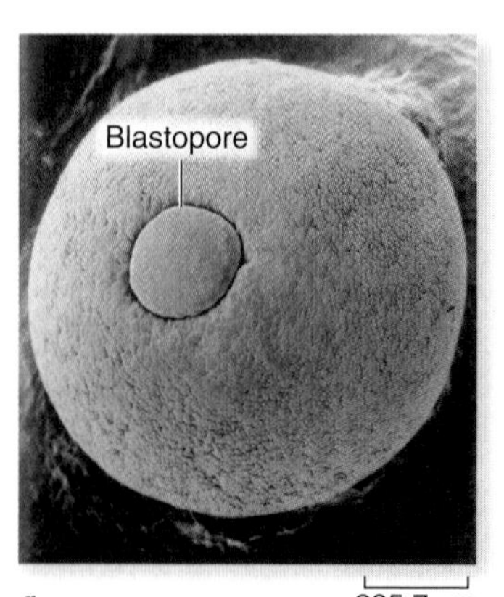

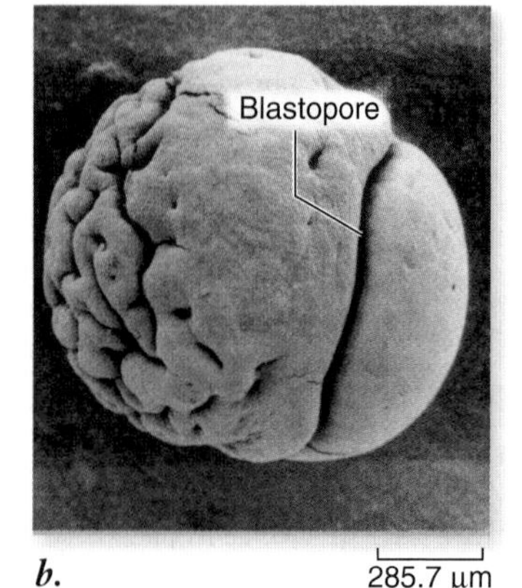

figure 19.20

REAGENTS THAT INTERFERE WITH CELL BINDING TO FIBRONECTIN INHIBIT GASTRULATION OF AMPHIBIAN EMBRYOS. ***a.*** Scanning electron micrograph of a normal salamander embryo during gastrulation. This embryo was injected with a control saline solution at the blastula stage. Cells have moved into the interior of the embryo around the circumference of the blastopore, allowing the outer cells to spread evenly over the surface of the embryo. ***b.*** Micrograph of a salamander embryo of the same age that was previously injected with fibronectin antibodies, which block binding of the migrating cells to the extracellular matrix. In this embryo, the cells lack traction to move into the interior of the embryo, and they pile up on the surface forming deep convolutions. Note also that the circumference of the blastopore has not decreased in these embryos.

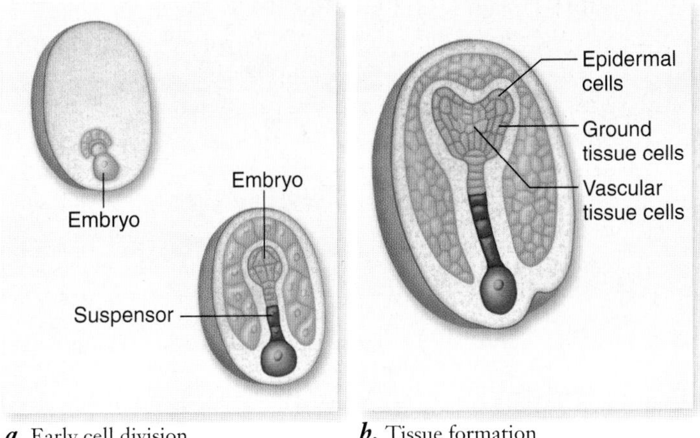

a. Early cell division

b. Tissue formation

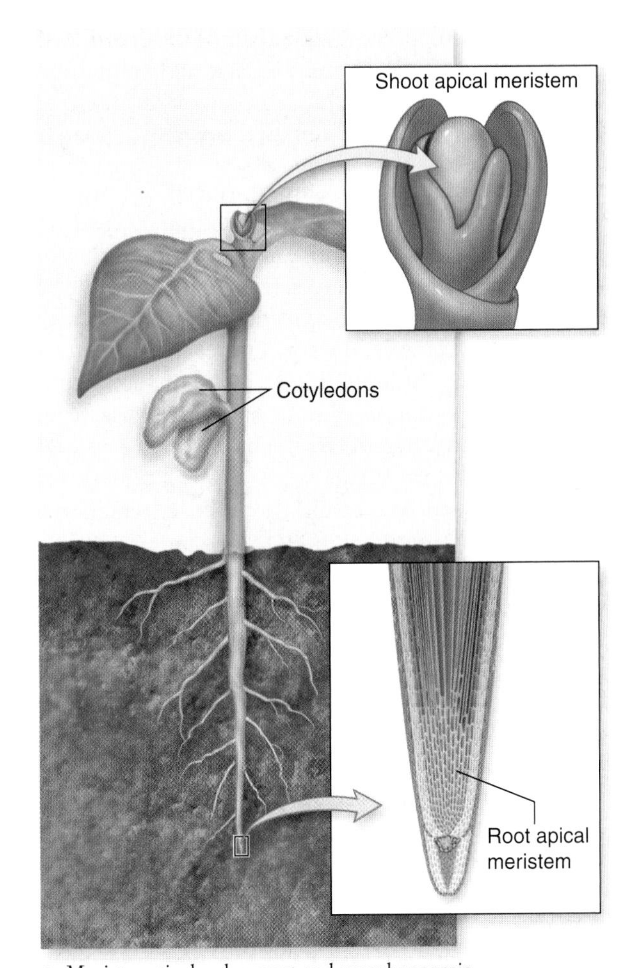

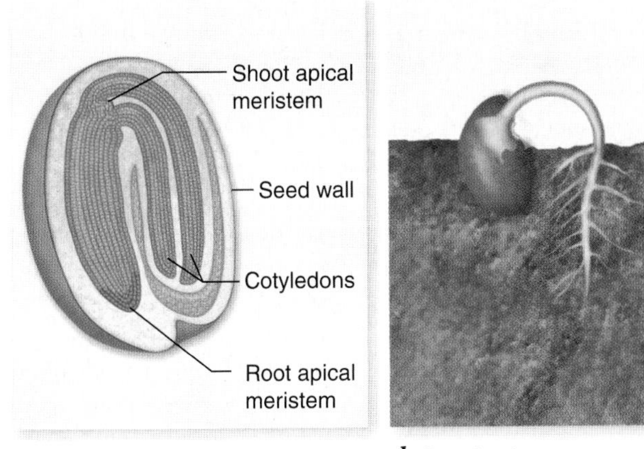

c. Seed formation

d. Germination

e. Meristematic development and morphogenesis

figure 19.21

THE PATH OF PLANT DEVELOPMENT. The developmental stages of *Arabidopsis thaliana* are (*a*) early embryonic cell division, (*b*) embryonic tissue formation, (*c*) seed formation, (*d*) germination, and (*e*) meristematic development and morphogenesis.

19.6 Environmental Effects on Development

In seed plants, embryogenesis is a brief stage in the life of the plant and results in the production of a seed. The environment affects essentially every subsequent step after embryogenesis, from seed dispersal to flower formation. For example, seed dispersal in jack pines can only occur following a fire because the heat of the fire causes the tightly closed cones to open up and release their seeds.

Germination of a dormant seed proceeds when soil conditions, temperature, and daylight hours are favorable. Similarly, a combination of environmental factors determines the timing of flower production in adult angiosperms.

Plant development is also influenced by interactions with other organisms. For example, the survival of a plant that is munched on by an animal depends on the quick regrowth afforded by meristems. Plant development is also guided by symbiotic relationships; the roots of legumes and a few other plant species develop nodules to house the nitrogen-fixing bacteria *Rhizobium*.

The effects of the environment on animal development are perhaps not so intuitive. Organisms such as *C. elegans* and *Drosophila* were selected as model systems for studying animal development because they develop so regularly under typical laboratory conditions. Organisms living in the wild, however, are subject to many different environmental changes, which may produce different phenotypes from a single genotype.

In mammals, embryonic and fetal development has a longer time course and is subject to the effects of blood-borne agents in the mother. The prescription of the sedative drug thalidomide to pregnant women in the 1950s and 1960s illustrated the profound effects that drugs can have on human development. Many of the women who took thalidomide gave birth to children with limb defects. Environmental agents such as lead compounds also affect the postnatal growth and maturation of children, leading to cognitive defects and mental retardation.

The environment affects normal development

The environment controls many aspects of normal animal development. The larvae of certain marine invertebrates, for example, do not metamorphose into adults until they encounter a specific surface on which to settle down. As with plants, animal development is often influenced by interactions with other organisms. For example, the small water flea *Daphnia* will change its morphology by doubling the size of the "helmet" on the top of its head after encountering a predatory fly larva (figure 19.22). Finally, when mice or zebrafish are reared under germ-free conditions, their guts are devoid of bacteria that normally colonize the gut. As a result, there are defects in intestinal differentiation and function in both species.

A particularly clear example of environmental effects on development is temperature-dependent sex determination (TSD). In many reptiles, the temperature of the soil in which the eggs incubate determines the sex of the hatchlings. In some species, one sex dominates at intermediate temperatures, whereas only the opposite sex develops at temperatures at either extreme of the natural thermal range. In other species, one end of the natural thermal range usually produces all males, the other end of the range produces all females, and intermediate temperatures are usually gender-neutral.

One possible hazard of TSD is that an increase in global temperatures may eventually skew sex ratios in wild populations of animals, leading to their demise. Some researchers have even speculated that the extinction of dinosaurs could be attributable to the effects of global temperature changes on sex determination and sex ratios in dinosaurs.

Endocrine disrupters can perturb development

The large family of **endocrine hormones** includes the androgens and estrogens that control sexual differentiation and function in animals. Endogenous endocrine activity is essential for the normal development and homeostasis of all complex animals. For example, hormones trigger metamorphosis in both frogs and flies, and if cells in the pituitary gland that produce human growth hormone malfunction during childhood, dwarfism or gigantism will result.

Although a number of endocrine disorders have a genetic basis, recent studies have shown that many environmental chemicals interfere with endocrine signaling. An **endocrine-disrupting chemical (EDC)** is any exogenous compound that interferes with the production, transport, or receptor binding of endogenous hormones.

Perhaps the best known endocrine disrupter is diethylstilbestrol (DES), which was prescribed to millions of women in the United States from 1938 to 1971 to prevent miscarriages and preterm births. The children of women who took DES while pregnant exhibited an increased incidence of abnormalities in differentiation of their reproductive organs, and DES daughters are at an increased risk of a rare kind of vaginal and cervical cancer.

figure 19.22

PREDATOR-INDUCED MORPHOLOGICAL CHANGES IN THE WATER FLEA *DAPHNIA*. These scanning electron micrographs show the difference between the morphology of *Daphnia* after encountering a predatory fly larva (left side) and the normal body shape (right side). *Daphnia* reproduces asexually, and these two individuals are genetic clones of one another. Thus, the environment can act on a *single* set of genes to induce the formation of *two different* body shapes.

Environmental EDCs are derived from three main sources: industrial wastes, agricultural practices, and the effluent of municipal sewage-treatment plants. Industrial EDCs include dioxins, heavy metals, and polychlorinated biphenyls (PCBs). Dioxins suppress immune system function in humans for long periods after exposure. PCBs, dioxins, and heavy metals impair spatial memory, learning, and other cognitive processes in primates and rodents.

Agricultural EDCs include the pesticides atrazine and DDT. The historic decline in bald eagle populations in North America was correlated with high levels of DDT in adults; females with high levels of DDT laid eggs with extremely weak shells, which broke easily. The subsequent banning of DDT use in the United States is thought to be the major reason for the rebounding of bald eagle populations from the verge of extinction.

A recent rise has been reported in the incidence of morphological defects in the urinary and reproductive structures of human males, which include abnormally placed urethral openings and undescended testes. A global decline in sperm count and quality and increased infertility has also been reported in men. All of these problems are exacerbated in areas with high EDC production.

A 2005 study in *Science* magazine showed that brief exposure of pregnant rats to two different EDCs—one a fungicide and the other a pesticide—caused decreased sperm number and viability and an increased incidence of infertility in adult male offspring of these mothers. Moreover, the reduced male fertility was passed on to nearly every male in future generations (examined F_1 to F_4). Thus, this startling set of experiments showed that the effects of EDCs can reach beyond affected individuals to affect multiple generations of offspring.

Both plant and animal development is affected by environmental influences. Sex determination in reptiles is controlled by temperature. Human development may be affected by environmental pollutants that mimic the effects of steroid hormones.

concept review

19.1 Overview of Development

- Development is the successive process of systematic, gene-directed changes throughout a life cycle.
- Development occurs in four subprocesses: growth, cell differentiation, pattern formation, and morphogenesis.

19.2 Cell Division

Early growth occurs by mitotic cell division, resulting in many undifferentiated cells.

- In animals, cleavage stage divisions divide the fertilized egg into numerous smaller cells called blastomeres.
- During cleavage the G_1 and G_2 phases of the cell cycle are shortened or eliminated (figure 19.2).
- The lineage of 959 adult somatic *Caenorhabditis elegans* cells is invariant.
- Stem cells can divide indefinitely and give rise to multiple cell types.
- Totipotent cells can give rise to any cell type; pluripotent cells can give rise to multiple cell types.
- Embryonic stem cells are derived from the inner cell mass of the blastocyst and are pluripotent (figure 19.4).
- Plant growth continues throughout the life span from meristematic stem cells that can differentiate into any plant tissue.

19.3 Cell Differentiation

During development, cells assume different fates due to both temporal and spatial differences in gene expression in the growing embryo.

- Cells that are committed to a particular developmental pathway are determined.
- Cells can become committed to a particular developmental pathway by the inheritance of cytoplasmic determinants or cell–cell interactions.
- Cytoplasmic determinants such as encoded maternal mRNA are produced during oogenesis and deposited into the egg.
- Induction occurs when one cell type produces signal molecules that induce gene expression in neighboring target cells.
- The nucleus of a fully differentiated cell can be reprogrammed to be totipotent (figure 19.9).
- Reproductive cloning suffers from a low success rate, and age-associated diseases.
- Therapeutic cloning uses stem cells from the recipient, solving the problem of tissue rejection in tissue and organ transplants.

19.4 Pattern Formation

For cells in multicellular organisms to differentiate into appropriate cell types, they must gain information about their relative locations in the body before their fate is determined.

- Pattern formation produces two perpendicular axes, or anterior–posterior and dorsal–ventral polarity, to a bilaterally symmetrical organism.
- Positional information leads to changes in gene activity so cells adopt a fate appropriate for their location.
- Fruit fly embryogenesis illustrates the genetic control of early patterning.
- Maternally encoded mRNAs are deposited in a maturing oocyte by nurse cells and they initiate a cascade of sequential gene activations.
- Formation of the anterior–posterior axis is based on opposing gradients of morphogens, Bicoid and Nanos, synthesized from maternal mRNA (figures 19.14, 19.15).
- The dorsal–ventral axis is established by a gradient of the Dorsal transcription factor.
- Gap genes encode transcription factors, which, in turn, activate the expression of pair-rule genes that divide the embryo into seven zones.
- Pair-rule genes regulate the expression of each other and the segment polarity genes that finish defining the embryonic segments.
- Homeotic genes give identity to segments of the embryo. They contain a DNA sequence called the homeobox and are called *Hox* genes.
- *Hox* genes are found in four clusters in vertebrates.
- Instead of *Hox* genes, plants have MADS box genes that control the transition from vegetative to reproductive growth, root development, and floral organ identity.

19.5 Morphogenesis

Morphogenesis is the product of changes in cell structure and cell behavior.

- Depending on the orientation of the mitotic spindle, cells of equal or different sizes can arise.
- Morphogenesis can arise by changes in cell shape, cell size, and by cell migration.
- Apoptosis, the programmed death of cells, is important in development to remove structures (figure 19.19).
- The migration of cells requires both the adhesion and loss of adhesion between cells and their substrate.
- Cell-to-cell interactions are often mediated by cadherin proteins, whereas cell-to-substrate interactions may involve integrin-to-extracellular-matrix interactions.
- Integrins bind to fibers found in the extracellular matrix. This can alter the cytoskeleton and activate gene expression.
- In plants, the primary morphogenetic processes are cell division, relative position of cells within the embryo, and changes in cell shape.
- Plant development stages begin with cell division and end with meristematic development and morphogenesis (figure 19.21).
- Relative position of cells in the plant embryo is the main determinant of cell differentiation.

19.6 Environmental Effects on Development

Both plant and animal development are affected by environmental factors.

- Seed dispersal, germination, and plant development are influenced by abiotic and biotic factors.
- In plants and animals, different phenotypic expressions of a single genotype can be influenced by the environment.
- In animals, blood-borne agents and environmental contaminants can affect embryonic development.
- The environment controls normal animal development by influencing attributes such as morphology and sex determination.
- Human development may be affected by exogenous compounds, called endocrine-disrupting chemicals, such as dioxin and PCBs, which interfere with the production, transport, or receptor binding of endogenous hormones.

review questions

SELF TEST

1. Which of the following developmental stages is associated with the generation of organs?
   a. Growth
   b. Pattern formation
   c. Differentiation
   d. Morphogenesis
2. Growth of the developing embryo involves rapid ________ cell divisions.
   a. mitotic
   b. meiotic
   c. binary fission
   d. sexual
3. The reduced size of a blastomere is the consequence of a shortened—
   a. M phase.
   b. S phase.
   c. G_1 and G_2 phases.
   d. All of the above.
4. A pluripotent cell is one that can—
   a. become any cell type.
   b. produce an indefinite supply of a single cell type.
   c. produce a limited amount of a specific cell type.
   d. produce multiple cell types.
5. Which of the following statements is *not* true regarding embryonic stem (ES) cells?
   a. They retain the ability to develop into any cell type.
   b. They are isolated from the inner cell mass of a developing embryo.
   c. They are tissue-specific.
   d. They are totipotent.
6. Plant meristems—
   a. are only present during development.
   b. contain stem cells.
   c. undergo meiosis.
   d. All of the above.
7. What is the common theme in cell determination by induction or cytoplasmic determinants?
   a. The activation of transcription factors
   b. The activation of cell signaling pathways
   c. A change in gene expression
   d. Both (a) and (c)
8. Which of the following is *not* a limitation to reproductive cloning?
   a. Efficiency of the process
   b. Ethical considerations
   c. Sources for donor DNA
   d. Genetic imprinting of the DNA
9. How do the products of therapeutic cloning differ from those of reproductive cloning?
   a. Therapeutic cloning provides a source of embryonic stem cells.
   b. Therapeutic cloning produces an embryo that can be implanted into a uterus.
   c. Therapeutic cloning produces whole tissue or organs.
   d. Therapeutic cloning provides a source of proteins.
10. The anterior–posterior axis of the fruit fly *Drosophila* is determined by—
    a. growth factors.
    b. zygotic RNA.
    c. morphogens.
    d. cellular blastoderm formation.
11. Which of the following best describes a morphogen?
    a. A cell that secretes a diffusible signal that specifies cell fate
    b. A diffusible signal that functions to determine cell fate
    c. A protein that helps mediate cell–cell interactions, altering cell fate
    d. A protein that allows a cell to become totipotent
12. Suppose that during a mutagenesis screen to isolate mutations in *Drosophila*, you come across a fly with legs growing out of its head. What gene cluster is likely affected?
    a. *bicoid*
    b. *hunchback*
    c. Bithorax
    d. Antennapedia
13. What would be the likely result of a mutation of the *bcl-2* gene on the level of apoptosis?
    a. No change
    b. A decrease in apoptosis
    c. An increase in apoptosis
    d. An initial increase, followed by a decrease in apoptosis
14. How is the body plan of a plant first determined?
    a. The activity of MADS box genes
    b. The first cell division following fertilization
    c. Gastrulation
    d. Both (a) and (b)
15. How does an endocrine-disrupting chemical (EDC) affect development?
    a. It alters the normal pathway of endocrine hormone activity.
    b. It induces mutations.
    c. It alters the sex determination of the developing embryo.
    d. Both (a) and (b).

CHALLENGE QUESTIONS

1. The fate map for *C. elegans* (refer to figure 19.3) diagrams development of a multicellular organism from a single cell. Use this fate map to determine the number of cell divisions required to establish the population of cells that will become (a) the nervous system and (b) the gonads.
2. Carefully examine the *C. elegans* fate map in figure 19.3. Notice that some of the branchpoints (daughter cells) do *not* go on to produce more cells. What is the cellular mechanism underlying this pattern?
3. You have generated a set of mutant embryonic mouse cells. Predict the developmental consequences for each of the following mutations.
   a. Knockout mutation for N-cadherin
   b. Knockout mutation for integrin
   c. Deletion of the cytoplasmic domain of integrin

Do you need additional review? *Visit* www.ravenbiology.com *for practice quizzes, animations, videos, and activities designed to help you master the material in this chapter.*

glossary

A

ABO blood group A set of four phenotypes produced by different combinations of three alleles at a single locus; blood types are A, B, AB, and O, depending on which alleles are expressed as antigens on the red blood cell surface.

abscission In vascular plants, the dropping of leaves, flowers, fruits, or stems at the end of the growing season, as the result of the formation of a layer of specialized cells (the abscission zone) and the action of a hormone (ethylene).

absorption spectrum The relationship of absorbance vs. wavelength for a pigment molecule. This indicates which wavelengths are absorbed maximally by a pigment. For example, chlorophyll *a* absorbs most strongly in the violet-blue and red regions of the visible light spectrum.

acceptor stem The 3′ end of a tRNA molecule; the portion that amino acids become attached to during the tRNA charging reaction.

accessory pigment A secondary light-absorbing pigment used in photosynthesis, including chlorophyll *b* and the carotenoids, that complement the absorption spectrum of chlorophyll *a*.

aceolomate An animal, such as a flatworm, having a body plan that has no body cavity; the space between mesoderm and endoderm is filled with cells and organic materials.

acetyl-CoA The product of the transition reaction between glycolysis and the Krebs cycle. Pyruvate is oxidized to acetyl-CoA by NAD^+, also producing CO_2, and NADH.

achiasmate segregation The lining up and subsequent separation of homologues during meiosis I without the formation of chiasmata between homologues; found in *Drosophila* males and some other species.

acid Any substance that dissociates in water to increase the hydrogen ion (H^+) concentration and thus lower the pH.

actin One of the two major proteins that make up vertebrate muscle; the other is myosin.

action potential A transient, all-or-none reversal of the electric potential across a membrane; in neurons, an action potential initiates transmission of a nerve impulse.

action spectrum A measure of the efficiency of different wavelengths of light for photosynthesis. In plants it corresponds to the absorption spectrum of chlorophylls.

activation energy The energy that must be processed by a molecule in order for it to undergo a specific chemical reaction.

active site The region of an enzyme surface to which a specific set of substrates binds, lowering the activation energy required for a particular chemical reaction and so facilitating it.

active transport The pumping of individual ions or other molecules across a cellular membrane from a region of lower concentration to one of higher concentration (i.e., against a concentration gradient); this transport process requires energy, which is typically supplied by the expenditure of ATP.

adaptation A peculiarity of structure, physiology, or behavior that promotes the likelihood of an organism's survival and reproduction in a particular environment.

adapter protein Any of a class of proteins that acts as a link between a receptor and other proteins to initiate signal transduction.

adaptive radiation The evolution of several divergent forms from a primitive and unspecialized ancestor.

adenosine triphosphate (ATP) A nucleotide consisting of adenine, ribose sugar, and three phosphate groups; ATP is the energy currency of cellular metabolism in all organisms.

adherins junction An anchoring junction that connects the actin filaments of one cell with those of adjacent cells or with the extracellular matrix.

ATP synthase The enzyme responsible for producing ATP in oxidative phosphorylation; it uses the energy from a proton gradient to catalyze the reaction $ADP + P_i \longrightarrow ATP$.

adenylyl cyclase An enzyme that produces large amounts of cAMP from ATP; the cAMP acts as a second messenger in a target cell.

adhesion The tendency of water to cling to other polar compounds due to hydrogen bonding.

adipose cells Fat cells, found in loose connective tissue, usually in large groups that form adipose tissue. Each adipose cell can store a droplet of fat (triacylglyceride).

adventitious Referring to a structure arising from an unusual place, such as stems from roots or roots from stems.

aerenchyma In plants, loose parenchymal tissue with large air spaces in it; often found in plants that grow in water.

aerobic Requiring free oxygen; any biological process that can occur in the presence of gaseous oxygen.

aerobic respiration The process that results in the complete oxidation of glucose using oxygen as the final electron acceptor. Oxygen acts as the final electron acceptor for an electron transport chain that produces a proton gradient for the chemiosmotic synthesis of ATP.

aleurone In plants, the outer layer of the endosperm in a seed; on germination, the aleurone produces α-amylase that breaks down the carbohydrates of the endosperm to nourish the embryo.

alga, pl. **algae** A unicellular or simple multicellular photosynthetic organism lacking multicellular sex organs.

allantois A membrane of the amniotic egg that functions in respiration and excretion in birds and reptiles and plays an important role in the development of the placenta in most mammals.

allele One of two or more alternative states of a gene.

allele frequency A measure of the occurrence of an allele in a population, expressed as proportion of the entire population, for example, an occurrence of 0.84 (84%).

allometric growth A pattern of growth in which different components grow at different rates.

allelopathy The release of a substance from the roots of one plant that block the germination of nearby seeds or inhibits the growth of a neighboring plant.

allopatric speciation The differentiation of geographically isolated populations into distinct species.

allopolyploid A polyploid organism that contains the genomes of two or more different species.

allosteric activator A substance that binds to an enzyme's allosteric site and keeps the enzyme in its active configuration.

allosteric inhibitor A noncompetitive inhibitor that binds to an enzyme's allosteric site and prevents the enzyme from changing to its active configuration.

allosteric site A part of an enzyme, away from its active site, that serves as an on/off switch for the function of the enzyme.

alpha (α) helix A form of secondary structure in proteins where the polypeptide chain is wound into a spiral due to interactions between amino and carboxyl groups in the peptide backbone.

alternation of generations A reproductive cycle in which a haploid (*n*) phase (the gametophyte), gives rise to gametes, which, after fusion to form a zygote, germinate to produce a diploid (2*n*) phase (the sporophyte). Spores produced by meiotic division from the sporophyte give rise to new gametophytes, completing the cycle.

alternative splicing In eukaryotes, the production of different mRNAs from a single primary transcript by including different sets of exons.

altruism Self-sacrifice for the benefit of others; in formal terms, the behavior that increases the fitness of the recipient while reducing the fitness of the altruistic individual.

alveolus, pl. **alveoli** One of many small, thin-walled air sacs within the lungs in which the bronchioles terminate.

amino acid The subunit structure from which proteins are produced, consisting of a central carbon atom with a carboxyl group (—COOH), an amino group ($—NH_2$), a hydrogen, and a side group (*R* group); only the side group differs from one amino acid to another.

aminoacyl-tRNA synthetase Any of a group of enzymes that attach specific amino acids to the correct tRNA during the tRNA-charging reaction. Each of the 20 amino acids has a corresponding enzyme.

amniocentesis Indirect examination of a fetus by tests on cell cultures grown from fetal cells obtained from a sample of the amniotic fluid or tests on the fluid itself.

amnion The innermost of the extraembryonic membranes; the amnion forms a fluid-filled sac around the embryo in amniotic eggs.

amniote A vertebrate that produces an egg surrounded by four membranes, one of which is the amnion; amniote groups are the reptiles, birds, and mammals.

amniotic egg An egg that is isolated and protected from the environment by a more or less impervious shell during the period of its development and that is completely self-sufficient, requiring only oxygen.

ampulla In echinoderms, a muscular sac at the base of a tube foot that contracts to extend the tube foot.

amyloplast A plant organelle called a plastid that specializes in storing starch.

anabolism The biosynthetic or constructive part of metabolism; those chemical reactions involved in biosynthesis.

anaerobic Any process that can occur without oxygen, such as anaerobic fermentation or H_2S photosynthesis.

anaerobic respiration The use of electron transport to generate a proton gradient for chemiosmotic synthesis of ATP using a final electron acceptor other than oxygen.

analogous Structures that are similar in function but different in evolutionary origin, such as the wing of a bat and the wing of a butterfly.

anaphase In mitosis and meiosis II, the stage initiated by the separation of sister chromatids, during which the daughter chromosomes move to opposite poles of the cell; in meiosis I, marked by separation of replicated homologous chromosomes.

anaphase-promoting complex (APC) A protein complex that triggers anaphase; it initiates a series of reactions that ultimately degrades cohesin, the protein complex that holds the sister chromatids together. The sister chromatids are then released and move toward opposite poles in the cell.

anchoring junction A type of cell junction that mechanically attaches the cytoskeleton of a cell to the cytoskeletons of adjacent cells or to the extracellular matrix.

androecium The floral whorl that comprises the stamens.

aneuploidy The condition in an organism whose cells have lost or gained a chromosome; Down syndrome, which results from an extra copy of human chromosome 21, is an example of aneuploidy in humans.

angiosperms The flowering plants, one of five phyla of seed plants. In angiosperms, the ovules at the time of pollination are completely enclosed by tissues.

animal pole In fish and other aquatic vertebrates with asymmetrical yolk distribution in their eggs, the hemisphere of the blastula comprising cells relatively poor in yolk.

anion A negatively charged ion.

annotation In genomics, the process of identifying and making note of "landmarks" in a DNA sequence to assist with recognition of coding and transcribed regions.

anonymous markers Genetic markers in a genome that do not cause a detectable phenotype, but that can be detected using molecular techniques.

antenna complex A complex of hundreds of pigment molecules in a photosystem that collects photons and feeds the light energy to a reaction center.

anther In angiosperm flowers, the pollen-bearing portion of a stamen.

antheridium, pl. **antheridia** A sperm-producing organ.

anthropoid Any member of the mammalian group consisting of monkeys, apes, and humans.

antibody A protein called immunoglobulin that is produced by lymphocytes in response to a foreign substance (antigen) and released into the bloodstream.

anticodon The three-nucleotide sequence at the end of a transfer RNA molecule that is complementary to, and base-pairs with, an amino-acid–specifying codon in messenger RNA.

antigen A foreign substance, usually a protein or polysaccharide, that stimulates an immune response.

antiporter A carrier protein in a cell's membrane that transports two molecules in opposite directions across the membrane.

anus The terminal opening of the gut; the solid residues of digestion are eliminated through the anus.

aorta (Gr. *aeirein*, to lift) The major artery of vertebrate systemic blood circulation; in mammals, carries oxygenated blood away from the heart to all regions of the body except the lungs.

apical meristem In vascular plants, the growing point at the tip of the root or stem.

apoplast route In plant roots, the pathway for movement of water and minerals that leads through cell walls and between cells.

apoptosis A process of programmed cell death, in which dying cells shrivel and shrink; used in all animal cell development to produce planned and orderly elimination of cells not destined to be present in the final tissue.

aposematic coloration An ecological strategy of some organisms that "advertise" their poisonous nature by the use of bright colors.

aquaporin A membrane channel that allows water to cross the membrane more easily than by diffusion through the membrane.

aquifers Permeable, saturated, underground layers of rock, sand, and gravel, which serve as reservoirs for groundwater.

archegonium, pl. **archegonia** The multicellular egg-producing organ in bryophytes and some vascular plants.

archenteron The principal cavity of a vertebrate embryo in the gastrula stage; lined with endoderm, it opens up to the outside and represents the future digestive cavity.

arteriole A smaller artery, leading from the arteries to the capillaries.

artificial selection Change in the genetic structure of populations due to selective breeding by humans. Many domestic animal breeds and crop varieties have been produced through artificial selection.

ascomycetes A large group comprising part of the "true fungi." They are characterized by separate hyphae, asexually produced conidiospores, and sexually produced ascospores within asci.

ascus, pl. **asci** A specialized cell, characteristic of the ascomycetes, in which two haploid nuclei fuse to produce a zygote that divides immediately by meiosis; at maturity, an ascus contains ascospores.

asexual reproduction The process by which an individual inherits all of its chromosomes from a single parent, thus being genetically identical to that parent; cell division is by mitosis only.

A site In a ribosome, the aminoacyl site, which binds to the tRNA carrying the next amino acid to be added to a polypeptide chain.

assembly The phase of a virus's reproductive cycle during which the newly made components are assembled into viral particles.

assortative mating A type of nonrandom mating in which phenotypically similar individuals mate more frequently.

aster In animal cell mitosis, a radial array of microtubules extending from the centrioles toward the plasma membrane, possibly serving to brace the centrioles for retraction of the spindle.

atom The smallest unit of an element that contains all the characteristics of that element. Atoms are the building blocks of matter.

atrial peptide Any of a group of small polypeptide hormones that may be useful in treatment of high blood pressure and kidney failure; produced by cells in the atria of the heart.

atrioventricular (AV) node A slender connection of cardiac muscle cells that receives the heartbeat impulses from the sinoatrial node and conducts them by way of the bundle of His.

atrium An antechamber; in the heart, a thin-walled chamber that receives venous blood and passes it on to the thick-walled ventricle; in the ear, the tympanic cavity.

autonomic nervous system The involuntary neurons and ganglia of the peripheral nervous system of vertebrates; regulates the heart, glands, visceral organs, and smooth muscle.

autopolyploid A polyploid organism that contains a duplicated genome of the same species; may result from a meiotic error.

autosome Any eukaryotic chromosome that is not a sex chromosome; autosomes are present in the same number and kind in both males and females of the species.

autotroph An organism able to build all the complex organic molecules that it requires as its own food source, using only simple inorganic compounds.

auxin (Gr. *auxein*, to increase) A plant hormone that controls cell elongation, among other effects.

auxotroph A mutation, or the organism that carries it, that affects a biochemical pathway causing a nutritional requirement.

avirulent pathogen Any type of normally pathogenic organism or virus that utilizes host resources but does not cause extensive damage or death.

axil In plants, the angle between a leaf's petiole and the stem to which it is attached.

axillary bud In plants, a bud found in the axil of a stem and leaf; an axillary bud may develop into a new shoot or may become a flower.

axon A process extending out from a neuron that conducts impulses away from the cell body.

B

***b6–f* complex** *See* cytochrome *b6–f* complex.
bacteriophage A virus that infects bacterial cells; also called a *phage*.
Barr body A deeply staining structure, seen in the interphase nucleus of a cell of an individual with more than one X chromosome, that is a condensed and inactivated X. Only one X remains active in each cell after early embryogenesis.
basal body A self-reproducing, cylindrical, cytoplasmic organelle composed of nine triplets of microtubules from which the flagella or cilia arise.
base Any substance that dissociates in water to absorb and therefore decrease the hydrogen ion (H^+) concentration and thus raise the pH.
base-pair A complementary pair of nucleotide bases, consisting of a purine and a pyrimidine.
basidium, pl. **basidia** A specialized reproductive cell of the basidiomycetes, often club-shaped, in which nuclear fusion and meiosis occur.
basophil A leukocyte containing granules that rupture and release chemicals that enhance the inflammatory response. Important in causing allergic responses.
Batesian mimicry A survival strategy in which a palatable or nontoxic organism resembles another kind of organism that is distasteful or toxic. Both species exhibit warning coloration.
B cell A type of lymphocyte that, when confronted with a suitable antigen, is capable of secreting a specific antibody protein.
behavioral ecology The study of how natural selection shapes behavior.
biennial A plant that normally requires two growing seasons to complete its life cycle. Biennials flower in the second year of their lives.
bilateral symmetry A single plane divides an organism into two structural halves that are mirror images of each other.
bile salts A solution of organic salts that is secreted by the vertebrate liver and temporarily stored in the gallbladder; emulsifies fats in the small intestine.
binary fission Asexual reproduction by division of one cell or body into two equal or nearly equal parts.
binomial distribution The distribution of phenotypes seen among the progeny of a cross in which there are only two alternative alleles.
binomial name The scientific name of a species that consists of two parts, the genus name and the specific species name, for example, *Apis mellifera*.
biochemical pathway A sequence of chemical reactions in which the product of one reaction becomes the substrate of the next reaction. The Krebs cycle is a biochemical pathway.
biodiversity The number of species and their range of behavioral, ecological, physiological, and other adaptations, in an area.
bioenergetics The analysis of how energy powers the activities of living systems.
biofilm A complex bacterial community comprising different species; plaque on teeth is a biofilm.
biogeography The study of the geographic distribution of species.
biological community All the populations of different species living together in one place; for example, all populations that inhabit a mountain meadow.
biological species concept (BSC) The concept that defines species as groups of populations that have the potential to interbreed and that are reproductively isolated from other groups.
biomass The total mass of all the living organisms in a given population, area, or other unit being measured.
biome One of the major terrestrial ecosystems, characterized by climatic and soil conditions; the largest ecological unit.
bipolar cell A specialized type of neuron connecting cone cells to ganglion cells in the visual system. Bipolar cells receive a hyperpolarized stimulus from the cone cell and then transmit a depolarization stimulus to the ganglion cell.
biramous Two-branched; describes the appendages of crustaceans.
blade The broad, expanded part of a leaf; also called the lamina.
blastocoel The central cavity of the blastula stage of vertebrate embryos.
blastodisc In the development of birds, a disclike area on the surface of a large, yolky egg that undergoes cleavage and gives rise to the embryo.
blastomere One of the cells of a blastula.
blastopore In vertebrate development, the opening that connects the archenteron cavity of a gastrula stage embryo with the outside.
blastula In vertebrates, an early embryonic stage consisting of a hollow, fluid-filled ball of cells one layer thick; a vertebrate embryo after cleavage and before gastrulation.
Bohr effect The release of oxygen by hemoglobin molecules in response to elevated ambient levels of CO_2.
bottleneck effect A loss of genetic variability that occurs when a population is reduced drastically in size.
Bowman's capsule In the vertebrate kidney, the bulbous unit of the nephron, which surrounds the glomerulus.
β-oxidation The oxygen-dependent reactions where 2-carbon units of fatty acids are cleaved and combined with CoA to produce acetyl-CoA, which then enters the Krebs cycle. This occurs cyclically until the entire fatty acid is oxidized.
β sheet A form of secondary structure in proteins where the polypeptide folds back on itself one or more times to form a planar structure stabilized by hydrogen bonding between amino and carboxyl groups in the peptide backbone. Also known as a β-pleated sheet.
book lung In some spiders, a unique respiratory system consisting of leaflike plates within a chamber over which gas exchange occurs.
bronchus, pl. **bronchi** One of a pair of respiratory tubes branching from the lower end of the trachea (windpipe) into either lung.
bud An asexually produced outgrowth that develops into a new individual. In plants, an embryonic shoot, often protected by young leaves; buds may give rise to branch shoots.
buffer A substance that resists changes in pH. It releases hydrogen ions (H^+) when a base is added and absorbs H^+ when an acid is added.

C

C_3 photosynthesis The main cycle of the dark reactions of photosynthesis, in which CO_2 binds to ribulose 1,5-bisphosphate (RuBP) to form two 3-carbon phosphoglycerate (PGA) molecules.
C_4 photosynthesis A process of CO_2 fixation in photosynthesis by which the first product is the 4-carbon oxaloacetate molecule.
cadherin One of a large group of transmembrane proteins that contain a Ca^{2+}-mediated binding between cells; these proteins are responsible for cell-to-cell adhesion between cells of the same type.
callus Undifferentiated tissue; a term used in tissue culture, grafting, and wound healing.
Calvin cycle The dark reactions of C_3 photosynthesis; also called the Calvin–Benson cycle.
calyx The sepals collectively; the outermost flower whorl.
Cambrian explosion The huge increase in animal diversity that occurred at the beginning of the Cambrian period.
cAMP response protein (CRP) *See* catabolite activator protein (CAP)
cancer The unrestrained growth and division of cells; it results from a failure of cell division control.
capillary The smallest of the blood vessels; the very thin walls of capillaries are permeable to many molecules, and exchanges between blood and the tissues occur across them; the vessels that connect arteries with veins.
capsid The outermost protein covering of a virus.
capsule In bacteria, a gelatinous layer surrounding the cell wall.
carapace (Fr. from Sp. *carapacho*, shell) Shieldlike plate covering the cephalothorax of decapod crustaceans; the dorsal part of the shell of a turtle.
carbohydrate An organic compound consisting of a chain or ring of carbon atoms to which hydrogen and oxygen atoms are attached in a ratio of approximately 2:1; having the generalized formula $(CH_2O)_n$; carbohydrates include sugars, starch, glycogen, and cellulose.
carbon fixation The conversion of CO_2 into organic compounds during photosynthesis; the first stage of the dark reactions of photosynthesis, in which carbon dioxide from the air is combined with ribulose 1,5-bisphosphate.
carotenoid Any of a group of accessory pigments found in plants; in addition to absorbing light energy, these pigments act as antioxidants, scavenging potentially damaging free radicals.
carpel A leaflike organ in angiosperms that encloses one or more ovules.
carrier protein A membrane protein that binds to a specific molecule that cannot cross the membrane and allows passage through the membrane.
carrying capacity The maximum population size that a habitat can support.
cartilage A connective tissue in skeletons of vertebrates. Cartilage forms much of the skeleton of embryos, very young vertebrates, and some adult vertebrates, such as sharks and their relatives.

Casparian strip In plants, a band that encircles the cell wall of root endodermal cells. Adjacent cells' strips connect, forming a layer through which water cannot pass; therefore, all water entering roots must pass through cell membranes and cytoplasm.

catabolism In a cell, those metabolic reactions that result in the breakdown of complex molecules into simpler compounds, often with the release of energy.

catabolite activator protein (CAP) A protein that, when bound to cAMP, can bind to DNA and activate transcription. The level of cAMP is inversely related to the level of glucose, and CAP/cAMP in *E. coli* activates the *lac* (lactose) operon. Also called *cAMP response protein* (*CRP*).

catalysis The process by which chemical subunits of larger organic molecules are held and positioned by enzymes that stress their chemical bonds, leading to the disassembly of the larger molecule into its subunits, often with the release of energy.

cation A positively charged ion.

cavitation In plants and animals, the blockage of a vessel by an air bubble that breaks the cohesion of the solution in the vessel; in animals more often called embolism.

$CD4^+$ cell A subtype of helper T cell that is identified by the presence of the CD4 protein on its surface. This cell type is targeted by the HIV virus that causes AIDS.

cecum In vertebrates, a blind pouch at the beginning of the large intestine.

cell cycle The repeating sequence of growth and division through which cells pass each generation.

cell determination The molecular "decision" process by which a cell becomes destined for a particular developmental pathway. This occurs before overt differentiation and can be a stepwise process.

cell plate The structure that forms at the equator of the spindle during early telophase in the dividing cells of plants and a few green algae.

cell-surface marker A glycoprotein or glycolipid on the outer surface of a cell's membrane that acts as an identifier; different cell types carry different markers.

cell-surface receptor A cell surface protein that binds a signal molecule and converts the extracellular signal into an intracellular one.

cellular blastoderm In insect embryonic development, the stage during which the nuclei of the syncitial blastoderm become separate cells through membrane formation.

cellular respiration The metabolic harvesting of energy by oxidation, ultimately dependent on molecular oxygen; carried out by the Krebs cycle and oxidative phosphorylation.

cellulose The chief constituent of the cell wall in all green plants, some algae, and a few other organisms; an insoluble complex carbohydrate formed of microfibrils of glucose molecules.

cell wall The rigid, outermost layer of the cells of plants, some protists, and most bacteria; the cell wall surrounds the plasma membrane.

central nervous system (CNS) That portion of the nervous system where most association occurs; in vertebrates, it is composed of the brain and spinal cord; in invertebrates, it usually consists of one or more cords of nervous tissue, together with their associated ganglia.

central vacuole A large, membrane-bounded sac found in plant cells that stores proteins, pigments, and waste materials, and is involved in water balance.

centriole A cytoplasmic organelle located outside the nuclear membrane, identical in structure to a basal body; found in animal cells and in the flagellated cells of other groups; divides and organizes spindle fibers during mitosis and meiosis.

centromere A visible point of constriction on a chromosome that contains repeated DNA sequences that bind specific proteins. These proteins make up the kinetochore to which microtubules attach during cell division.

cephalization The evolution of a head and brain area in the anterior end of animals; thought to be a consequence of bilateral symmetry.

cerebellum The hindbrain region of the vertebrate brain that lies above the medulla (brainstem) and behind the forebrain; it integrates information about body position and motion, coordinates muscular activities, and maintains equilibrium.

cerebral cortex The thin surface layer of neurons and glial cells covering the cerebrum; well developed only in mammals, and particularly prominent in humans. The cerebral cortex is the seat of conscious sensations and voluntary muscular activity.

cerebrum The portion of the vertebrate brain (the forebrain) that occupies the upper part of the skull, consisting of two cerebral hemispheres united by the corpus callosum. It is the primary association center of the brain. It coordinates and processes sensory input and coordinates motor responses.

chaetae Bristles of chitin on each body segment that help anchor annelid worms during locomotion.

channel protein A transmembrane protein with a hydrophilic interior that provides an aqueous channel allowing diffusion of species that cannot cross the membrane. Usually allows passage of specific ions such as K^+, Na^+, or Ca^{2+} across the membrane.

chaperone protein A class of enzymes that help proteins fold into the correct configuration and can refold proteins that have been misfolded or denatured.

character displacement A process in which natural selection favors individuals in a species that use resources not used by other species. This results in evolutionary change leading to species dissimilar in resource use.

character state In cladistics, one of two or more distinguishable forms of a character, such as the presence or absence of teeth in amniote vertebrates.

charging reaction The reaction by which an aminoacyl-tRNA synthetase attaches a specific amino acid to the correct tRNA using energy from ATP.

chelicera, pl. **chelicerae** The first pair of appendages in horseshoe crabs, sea spiders, and arachnids—the chelicerates, a group of arthropods. Chelicerae usually take the form of pincers or fangs.

chemical synapse A close association that allows chemical communication between neurons. A chemical signal (neurotransmitter) released by the first neuron binds to receptors in the membrane of the second neurons.

chemiosmosis The mechanism by which ATP is generated in mitochondria and chloroplasts; energetic electrons excited by light (in chloroplasts) or extracted by oxidation in the Krebs cycle (in mitochondria) are used to drive proton pumps, creating a proton concentration gradient; when protons subsequently flow back across the membrane, they pass through channels that couple their movement to the synthesis of ATP.

chiasma An X-shaped figure that can be seen in the light microscope during meiosis; evidence of crossing over, where two chromatids have exchanged parts; chiasmata move to the ends of the chromosome arms as the homologues separate.

chitin A tough, resistant, nitrogen-containing polysaccharide that forms the cell walls of certain fungi, the exoskeleton of arthropods, and the epidermal cuticle of other surface structures of certain other invertebrates.

chlorophyll The primary type of light-absorbing pigment in photosynthesis. Chlorophyll *a* absorbs light in the violet-blue and the red ranges of the visible light spectrum; chlorophyll *b* is an accessory pigment to chlorophyll *a*, absorbing light in the blue and red-orange ranges. Neither pigment absorbs light in the green range, 500–600 nm.

chloroplast A cell-like organelle present in algae and plants that contains chlorophyll (and usually other pigments) and carries out photosynthesis.

choanocyte A specialized flagellated cell found in sponges; choanocytes line the body interior.

chorion The outer member of the double membrane that surrounds the embryo of reptiles, birds, and mammals; in placental mammals, it contributes to the structure of the placenta.

chorionic villi sampling A technique in which fetal cells are sampled from the chorion of the placenta rather than from the amniotic fluid; this less invasive technique can be used earlier in pregnancy than amniocentesis.

chromatid One of the two daughter strands of a duplicated chromosome that is joined by a single centromere.

chromatin The complex of DNA and proteins of which eukaryotic chromosomes are composed; chromatin is highly uncoiled and diffuse in interphase nuclei, condensing to form the visible chromosomes in prophase.

chromatin-remodeling complex A large protein complex that has been found to modify histones and DNA and that can change the structure of chromatin, moving or transferring nucleosomes.

chromosomal mutation Any mutation that affects chromosome structure.

chromosome The vehicle by which hereditary information is physically transmitted from one generation to the next; in a bacterium, the chromosome consists of a single naked circle of DNA; in eukaryotes, each chromosome consists of a single linear DNA molecule and associated proteins.

chromosomal theory of inheritance The theory stating that hereditary traits are carried on chromosomes.

cilium A short cellular projection from the surface of a eukaryotic cell, having the same internal structure of microtubules in a 9 + 2 arrangement as seen in a flagellum.

circadian rhythm An endogenous cyclical rhythm that oscillates on a daily (24-hour) basis.

circulatory system A network of vessels in coelomate animals that carries fluids to and from different areas of the body.

cisterna A small collecting vessel that pinches off from the end of a Golgi body to form a transport vesicle that moves materials through the cytoplasm.

cisternal space The inner region of a membrane-bounded structure. Usually used to describe the interior of the endoplasmic reticulum; also called the *lumen*.

cladistics A taxonomic technique used for creating hierarchies of organisms that represent true phylogenetic relationship and descent.

class A taxonomic category between phyla and orders. A class contains one or more orders, and belongs to a particular phylum.

classical conditioning The repeated presentation of a stimulus in association with a response that causes the brain to form an association between the stimulus and the response, even if they have never been associated before.

clathrin A protein located just inside the plasma membrane in eukaryotic cells, in indentations called clathrin-coated pits.

cleavage In vertebrates, a rapid series of successive cell divisions of a fertilized egg, forming a hollow sphere of cells, the blastula.

cleavage furrow The constriction that forms during cytokinesis in animal cells that is responsible for dividing the cell into two daughter cells.

climax vegetation Vegetation encountered in a self-perpetuating community of plants that has proceeded through all the stages of succession and stabilized.

cloaca In some animals, the common exit chamber from the digestive, reproductive, and urinary system; in others, the cloaca may also serve as a respiratory duct.

clone-by-clone sequencing A method of genome sequencing in which a physical map is constructed first, followed by sequencing of fragments and identifying overlap regions.

cloning Producing a cell line or culture all of whose members contain identical copies of a particular nucleotide sequence; an essential element in genetic engineering.

closed circulatory system A circulatory system in which the blood is physically separated from other body fluids.

coacervate A spherical aggregation of lipid molecules in water, held together by hydrophobic forces.

coactivator A protein that functions to link transcriptional activators to the transcription complex consisting of RNA polymerase II and general transcription factors.

cochlea In terrestrial vertebrates, a tubular cavity of the inner ear containing the essential organs for hearing.

coding strand The strand of a DNA duplex that is the same as the RNA encoded by a gene. This strand is not used as a template in transcription, it is complementary to the template.

codominance Describes a case in which two or more alleles of a gene are each dominant to other alleles but not to each other. The phenotype of a heterozygote for codominant alleles exhibit characteristics of each of the homozygous forms. For example, in human blood types, a cross between an AA individual and a BB individual yields AB individuals.

codon The basic unit of the genetic code; a sequence of three adjacent nucleotides in DNA or mRNA that codes for one amino acid.

coelom In animals, a fluid-filled body cavity that develops entirely within the mesoderm.

coenzyme A nonprotein organic molecule such as NAD that plays an accessory role in enzyme-catalyzed processes, often by acting as a donor or acceptor of electrons.

coevolution The simultaneous development of adaptations in two or more populations, species, or other categories that interact so closely that each is a strong selective force on the other.

cofactor One or more nonprotein components required by enzymes in order to function; many cofactors are metal ions, others are organic coenzymes.

cohesin A protein complex that holds sister chromatids together during cell division. The loss of cohesins at the centromere allow the anaphase movement of chromosomes.

collenchyma cell In plants, the cells that form a supporting tissue called collenchyma; often found in regions of primary growth in stems and in some leaves.

colloblast A specialized type of cell found in members of the animal phylum Ctenophora (comb jellies) that bursts on contact with zooplankton, releasing an adhesive substance to help capture this prey.

colonial flagellate hypothesis The proposal first put forth by Haeckel that metazoans descended from colonial protists; supported by the similarity of sponges to choanoflagellate protists.

commensalism A relationship in which one individual lives close to or on another and benefits, and the host is unaffected; a kind of symbiosis.

community All of the species inhabiting a common environment and interacting with one another.

companion cell A specialized parenchyma cell that is associated with each sieve-tube member in the phloem of a plant.

competitive exclusion The hypothesis that two species with identical ecological requirements cannot exist in the same locality indefinitely, and that the more efficient of the two in utilizing the available scarce resources will exclude the other; also known as Gause's principle.

competitive inhibitor An inhibitor that binds to the same active site as an enzyme's substrate, thereby competing with the substrate.

complementary Describes genetic information in which each nucleotide base has a complementary partner with which it forms a base-pair.

complementary DNA (cDNA) A DNA copy of an mRNA transcript; produced by the action of the enzyme reverse transcriptase.

complement system The chemical defense of a vertebrate body that consists of a battery of proteins that become activated by the walls of bacteria and fungi.

complete digestive system A digestive system that has both a mouth and an anus, allowing unidirectional flow of ingested food.

compound eye An organ of sight in many arthropods composed of many independent visual units called ommatidia.

concentration gradient A difference in concentration of a substance from one location to another, often across a membrane.

condensin A protein complex involved in condensation of chromosomes during mitosis and meiosis.

cone (1) In plants, the reproductive structure of a conifer. (2) In vertebrates, a type of light-sensitive neuron in the retina concerned with the perception of color and with the most acute discrimination of detail.

conidia An asexually produced fungal spore.

conjugation Temporary union of two unicellular organisms, during which genetic material is transferred from one cell to the other; occurs in bacteria, protists, and certain algae and fungi.

consensus sequence In genome sequencing, the overall sequence that is consistent with the sequences of individual fragments; computer programs are used to compare sequences and generate a consensus sequence.

conservation of synteny The preservation over evolutionary time of arrangements of DNA segments in related species.

contig A contiguous segment of DNA assembled by analyzing sequence overlaps from smaller fragments.

continuous variation Variation in a trait that occurs along a continuum, such as the trait of height in human beings; often occurs when a trait is determined by more than one gene.

contractile vacuole In protists and some animals, a clear fluid-filled vacuole that takes up water from within the cell and then contracts, releasing it to the outside through a pore in a cyclical manner; functions primarily in osmoregulation and excretion.

conus arteriosus The anteriormost chamber of the embryonic heart in vertebrate animals.

convergent evolution The independent development of similar structures in organisms that are not directly related; often found in organisms living in similar environments.

cork cambium The lateral meristem that forms the periderm, producing cork (phellem) toward the surface (outside) of the plant and phelloderm toward the inside.

cornea The transparent outer layer of the vertebrate eye.

corolla The petals, collectively; usually the conspicuously colored flower whorl.

corpus callosum The band of nerve fibers that connects the two hemispheres of the cerebrum in humans and other primates.

corpus luteum A structure that develops from a ruptured follicle in the ovary after ovulation.

cortex The outer layer of a structure; in animals, the outer, as opposed to the inner, part of an organ; in vascular plants, the primary ground tissue of a stem or root.

cotyledon A seed leaf that generally stores food in dicots or absorbs it in monocots, providing nourishment used during seed germination.

crassulacean acid metabolism (CAM) A mode of carbon dioxide fixation by which CO_2 enters open leaf stomata at night and is used in photosynthesis during the day, when stomata are closed to prevent water loss.

crista A folded extension of the inner membrane of a mitochondrion. Mitochondria contain numerous cristae.

cross-current flow In bird lungs, the latticework of capillaries arranged across the air flow, at a 90° angle.

crossing over In meiosis, the exchange of corresponding chromatid segments between homologous chromosomes; responsible for genetic recombination between homologous chromosomes.

ctenidia Respiratory gills of mollusks; they consist of a system of filamentous projections of the mantle that are rich in blood vessels.

cuticle A waxy or fatty, noncellular layer (formed of a substance called cutin) on the outer wall of epidermal cells.

cutin In plants, a fatty layer produced by the epidermis that forms the cuticle on the outside surface.

cyanobacteria A group of photosynthetic bacteria, sometimes called the "blue-green algae," that contain the chlorophyll pigments most abundant in plants and algae, as well as other pigments.

cyclic AMP (cAMP) A form of adenosine monophosphate (AMP) in which the atoms of the phosphate group form a ring; found in almost all organisms, cAMP functions as an intracellular second messenger that regulates a diverse array of metabolic activities.

cyclic photophosphorylation Reactions that begin with the absorption of light by reaction center chlorophyll that excites an electron. The excited electron returns to the photosystem, generating ATP by chemiosmosis in the process. This is found in the single bacterial photosystem, and can occur in plants in photosystem I.

cyclin Any of a number of proteins that are produced in synchrony with the cell cycle and combine with certain protein kinases, the cyclin-dependent kinases, at certain points during cell division.

cyclin-dependent kinase (Cdk) Any of a group of protein kinase enzymes that control progress through the cell cycle. These enzymes are only active when complexed with cyclin. The cdc2 protein, produced by the *cdc2* gene, was the first Cdk enzyme discovered.

cytochrome Any of several iron-containing protein pigments that serve as electron carriers in transport chains of photosynthesis and cellular respiration.

cytochrome *b6–f* complex A proton pump found in the thylakoid membrane. This complex uses energy from excited electrons to pump protons from the stroma into the thylakoid compartment.

cytokinesis Division of the cytoplasm of a cell after nuclear division.

cytoplasm The material within a cell, excluding the nucleus; the protoplasm.

cytoskeleton A network of protein microfilaments and microtubules within the cytoplasm of a eukaryotic cell that maintains the shape of the cell, anchors its organelles, and is involved in animal cell motility.

cytosol The fluid portion of the cytoplasm; it contains dissolved organic molecules and ions.

cytotoxic T cell A special T cell activated during cell-mediated immune response that recognizes and destroys infected body cells.

D

deamination The removal of an amino group; part of the degradation of proteins into compounds that can enter the Krebs cycle.

deductive reasoning The logical application of general principles to predict a specific result. In science, deductive reasoning is used to test the validity of general ideas.

dehydration synthesis A type of chemical reaction in which two molecules join to form one larger molecule, simultaneously splitting out a molecule of water; one molecule is stripped of a hydrogen atom, and another is stripped of a hydroxyl group (—OH), resulting in the joining of the two molecules, while the H and —OH released may combine to form a water molecule.

dehydrogenation Chemical reaction involving the loss of a hydrogen atom. This is an oxidation that combines loss of an electron with loss of a proton.

deletion A mutation in which a portion of a chromosome is lost; if too much information is lost, the deletion can be fatal.

demography The properties of the rate of growth and the age structure of populations.

denaturation The loss of the native configuration of a protein or nucleic acid as a result of excessive heat, extremes of pH, chemical modification, or changes in solvent ionic strength or polarity that disrupt hydrophobic interactions; usually accompanied by loss of biological activity.

dendrite A process extending from the cell body of a neuron, typically branched, that conducts impulses toward the cell body.

deoxyribonucleic acid (DNA) The genetic material of all organisms; composed of two complementary chains of nucleotides wound in a double helix.

dephosphorylation The removal of a phosphate group, usually by a phosphatase enzyme. Many proteins can be activated or inactivated by dephosphorylation.

depolarization The movement of ions across a plasma membrane that locally wipes out an electrical potential difference.

derived character A characteristic used in taxonomic analysis representing a departure from the primitive form.

dermal tissue In multicellular organisms, a type of tissue that forms the outer layer of the body and is in contact with the environment; it has a protective function.

desmosome A type of anchoring junction that links adjacent cells by connecting their cytoskeletons with cadherin proteins.

derepression Seen in anabolic operons where the operon that encodes the enzymes for a biochemical pathway is repressed in the presence of the end product of the pathway and derepressed in the absence of the end product. This allows production of the enzymes only when they are necessary.

determinate development A type of development in animals in which each embryonic cell has a predetermined fate in terms of what kind of tissue it will form in the adult.

deuterostome Any member of a grouping of bilaterally symmetrical animals in which the anus develops first and the mouth second; echinoderms and vertebrates are deuterostome animals.

diacylglycerol (DAG) A second messenger that is released, along with inositol-1,4,5-trisphosphate (IP_3), when phospholipase C cleaves PIP_2. DAG can have a variety of cellular effects through activation of protein kinases.

diaphragm (1) In mammals, a sheet of muscle tissue that separates the abdominal and thoracic cavities and functions in breathing. (2) A contraceptive device used to block the entrance to the uterus temporarily and thus prevent sperm from entering during sexual intercourse.

diapsid Any of a group of reptiles that have two pairs of temporal openings in the skull, one lateral and one more dorsal; one lineage of this group gave rise to dinosaurs, modern reptiles, and birds.

diastolic pressure In the measurement of human blood pressure, the minimum pressure between heartbeats (repolarization of the ventricles). *Compare with* systolic pressure.

dicer An enzyme that generates small RNA molecules in a cell by chopping up double-stranded RNAs; dicer produces miRNAs and siRNAs.

dicot Short for dicotyledon; a class of flowering plants generally characterized as having two cotyledons, net-veined leaves, and flower parts usually in fours or fives.

dideoxynucleotide A nucleotide lacking —OH groups at both the 2′ and 3′ positions; used as a chain terminator in the enzymatic sequencing of DNA.

differentiation A developmental process by which a relatively unspecialized cell undergoes a progressive change to a more specialized form or function.

diffusion The net movement of dissolved molecules or other particles from a region where they are more concentrated to a region where they are less concentrated.

dihybrid An individual heterozygous at two different loci; for example *A/a B/b*.

dihybrid cross A single genetic cross involving two different traits, such as flower color and plant height.

dikaryotic In fungi, having pairs of nuclei within each cell.

dioecious Having the male and female elements on different individuals.

diploid Having two sets of chromosomes ($2n$); in animals, twice the number characteristic of gametes; in plants, the chromosome number characteristic of the sporophyte generation; in contrast to haploid (n).

directional selection A form of selection in which selection acts to eliminate one extreme from an array of phenotypes.

disaccharide A carbohydrate formed of two simple sugar molecules bonded covalently.

disruptive selection A form of selection in which selection acts to eliminate rather than favor the intermediate type.

dissociation In proteins, the reversible separation of protein subunits from a quaternary structure without altering their tertiary structure. Also refers to the dissolving of ionic compounds in water.

disassortative mating A type of nonrandom mating in which phenotypically different individuals mate more frequently.

diurnal Active during the day.

DNA-binding motif A region found in a regulatory protein that is capable of binding to a specific base sequence in DNA; a critical part of the protein's DNA-binding domain.

DNA fingerprinting An identification technique that makes use of a variety of molecular techniques to identify differences in the DNA of individuals.

DNA gyrase A topoisomerase involved in DNA replication; it relieves the torsional strain caused by unwinding the DNA strands.

DNA library A collection of DNAs in a vector (a plasmid, phage, or artificial chromosome) that taken together represent a complex mixture of DNAs, such as the entire genome, or the cDNAs made from all of the mRNA in a specific cell type.

DNA ligase The enzyme responsible for formation of phosphodiester bonds between adjacent nucleotides in DNA.

DNA microarray An array of DNA fragments on a microscope slide or silicon chip, used in hybridization experiments with labeled mRNA or DNA to identify active and inactive genes, or the presence or absence of particular sequences.

DNA polymerase A class of enzymes that all synthesize DNA from a preexisting template. All synthesize only in the 5′-to-3′ direction, and require a primer to extend.

DNA vaccine A type of vaccine that uses DNA from a virus or bacterium that stimulates the cellular immune response.

domain (1) A distinct modular region of a protein that serves a particular function in the action of the protein, such as a regulatory domain or a DNA-binding domain. (2) In taxonomy, the level higher than kingdom. The three domains currently recognized are Bacteria, Archaea, and Eukarya.

Domain Archaea In the three-domain system of taxonomy, the group that contains only the Archaea, a highly diverse group of unicellular prokaryotes.

Domain Bacteria In the three-domain system of taxonomy, the group that contains only the Bacteria, a vast group of unicellular prokaryotes.

Domain Eukarya In the three-domain system of taxonomy, the group that contains eukaryotic organisms including protists, fungi, plants, and animals.

dominant An allele that is expressed when present in either the heterozygous or the homozygous condition.

dosage compensation A phenomenon by which the expression of genes carried on sex chromosomes is kept the same in males and females, despite a different number of sex chromosomes. In mammals, inactivation of one of the X chromosomes in female cells accomplishes dosage compensation.

double fertilization The fusion of the egg and sperm (resulting in a $2n$ fertilized egg, the zygote) and the simultaneous fusion of the second male gamete with the polar nuclei (resulting in a primary endosperm nucleus, which is often triploid, $3n$); a unique characteristic of all angiosperms.

double helix The structure of DNA, in which two complementary polynucleotide strands coil around a common helical axis.

duodenum In vertebrates, the upper portion of the small intestine.

duplication A mutation in which a portion of a chromosome is duplicated; if the duplicated region does not lie within a gene, the duplication may have no effect.

E

ecdysis Shedding of outer, cuticular layer; molting, as in insects or crustaceans.

ecdysone Molting hormone of arthropods, which triggers when ecdysis occurs.

ecology The study of interactions of organisms with one another and with their physical environment.

ecosystem A major interacting system that includes organisms and their nonliving environment.

ecotype A locally adapted variant of an organism; differing genetically from other ecotypes.

ectoderm One of the three embryonic germ layers of early vertebrate embryos; ectoderm gives rise to the outer epithelium of the body (skin, hair, nails) and to the nerve tissue, including the sense organs, brain, and spinal cord.

ectomycorrhizae Externally developing mycorrhizae that do not penetrate the cells they surround.

ectotherms Animals such as reptiles, fish, or amphibians, whose body temperature is regulated by their behavior or by their surroundings.

electronegativity A property of atomic nuclei that refers to the affinity of the nuclei for valence electrons; a nucleus that is more electronegative has a greater pull on electrons than one that is less electronegative.

electron transport chain The passage of energetic electrons through a series of membrane-associated electron-carrier molecules to proton pumps embedded within mitochondrial or chloroplast membranes. *See* chemiosmosis.

elongation factor (Ef-Tu) In protein synthesis in *E. coli*, a factor that binds to GTP and to a charged tRNA to accomplish binding of the charged tRNA to the A site of the ribosome, so that elongation of the polypeptide chain can occur.

embryo A multicellular developmental stage that follows cell division of the zygote.

embryonic stem cell (ES cell) A stem cell derived from an early embryo that can develop into different adult tissues and give rise to an adult organism when injected into a blastocyst.

emergent properties Novel properties arising from the way in which components interact. Emergent properties often cannot be deduced solely from knowledge of the individual components.

emerging virus Any virus that originates in one organism but then passes to another; usually refers to transmission to humans.

endergonic Describes a chemical reaction in which the products contain more energy than the reactants, so that free energy must be put into the reaction from an outside source to allow it to proceed.

endocrine gland Ductless gland that secretes hormones into the extracellular spaces, from which they diffuse into the circulatory system.

endocytosis The uptake of material into cells by inclusion within an invagination of the plasma membrane; the uptake of solid material is phagocytosis, and that of dissolved material is pinocytosis.

endoderm One of the three embryonic germ layers of early vertebrate embryos, destined to give rise to the epithelium that lines internal structures and most of the digestive and respiratory tracts.

endodermis In vascular plants, a layer of cells forming the innermost layer of the cortex in roots and some stems.

endomembrane system A system of connected membranous compartments found in eukaryotic cells.

endometrium The lining of the uterus in mammals; thickens in response to secretion of estrogens and progesterone and is sloughed off in menstruation.

endomycorrhizae Mycorrhizae that develop within cells.

endonuclease An enzyme capable of cleaving phosphodiester bonds between nucleotides located internally in a DNA strand.

endoplasmic reticulum (ER) An internal membrane system that forms a netlike array of channels and interconnections of organelles within the cytoplasm of eukaryotic cells.

endorphin One of a group of small neuropeptides produced by the vertebrate brain; like morphine, endorphins modulate pain perception.

endosperm A storage tissue characteristic of the seeds of angiosperms, which develops from the union of a male nucleus and the polar nuclei of the embryo sac. The endosperm is digested by the growing sporophyte either before maturation of the seed or during its germination.

endospore A highly resistant, thick-walled bacterial spore that can survive harsh environmental stress, such as heat or dessication, and then germinate when conditions become favorable.

endosymbiosis Theory that proposes that eukaryotic cells evolved from a symbiosis between different species of prokaryotes.

endotherm An animal capable of maintaining a constant body temperature. *See* homeotherm.

energy level A discrete level, or quantum, of energy that an electron in an atom possesses. To change energy levels, an electron must absorb or release energy.

enhancer A site of regulatory protein binding on the DNA molecule distant from the promoter and start site for a gene's transcription.

enthalpy In a chemical reaction, the energy contained in the chemical bonds of the molecule, symbolized as H; in a cellular reaction, the free energy is equal to the enthalpy of the reactant molecules in the reaction.

entropy A measure of the randomness or disorder of a system; a measure of how much energy in a system has become so dispersed (usually as evenly distributed heat) that it is no longer available to do work.

enzyme A protein that is capable of speeding up specific chemical reactions by lowering the required activation energy.

enzyme–substrate complex The complex formed when an enzyme binds with its substrate. This complex often has an altered configuration compared with the nonbound enzyme.

epicotyl The region just above where the cotyledons are attached.

epidermal cell In plants, a cell that collectively forms the outermost layer of the primary plant body; includes specialized cells such as trichomes and guard cells.

epidermis The outermost layers of cells; in plants, the exterior primary tissue of leaves, young stems, and roots; in vertebrates, the nonvascular external layer of skin, of ectodermal origin; in invertebrates, a single layer of ectodermal epithelium.

epididymis A sperm storage vessel; a coiled part of the sperm duct that lies near the testis.

epistasis Interaction between two nonallelic genes in which one of them modifies the phenotypic expression of the other.

epithelium In animals, a type of tissue that covers an exposed surface or lines a tube or cavity.

equilibrium A stable condition; the point at which a chemical reaction proceeds as rapidly in the reverse direction as it does in the forward direction, so that there is no further net change in the concentrations of products or reactants. In ecology, a stable condition that resists change and fairly quickly returns to its original state if disturbed by humans or natural events.

erythrocyte Red blood cell, the carrier of hemoglobin.

erythropoiesis The manufacture of blood cells in the bone marrow.

E site In a ribosome, the exit site that binds to the tRNA that carried the previous amino acid added to the polypeptide chain.

estrus The period of maximum female sexual receptivity, associated with ovulation of the egg.

ethology The study of patterns of animal behavior in nature.

euchromatin That portion of a eukaryotic chromosome that is transcribed into mRNA; contains active genes that are not tightly condensed during interphase.

eukaryote A cell characterized by membrane-bounded organelles, most notably the nucleus, and one that possesses chromosomes whose DNA is associated with proteins; an organism composed of such cells.

eutherian A placental mammal.

eutrophic Refers to a lake in which an abundant supply of minerals and organic matter exists.

evolution Genetic change in a population of organisms; in general, evolution leads to progressive change from simple to complex.

excision repair A nonspecific mechanism to repair damage to DNA during synthesis. The damaged or mismatched region is excised, and DNA polymerase replaces the region removed.

exergonic Describes a chemical reaction in which the products contain less free energy than the reactants, so that free energy is released in the reaction.

exhalant siphon In bivalve mollusks, the siphon through which outgoing water leaves the body.

exocrine gland A type of gland that releases its secretion through a duct, such as a digestive gland or a sweat gland.

exocytosis A type of bulk transport out of cells in which a vacuole fuses with the plasma membrane, discharging the vacuole's contents to the outside.

exon A segment of DNA that is both transcribed into RNA and translated into protein. *See* intron.

exonuclease An enzyme capable of cutting phosphodiester bonds between nucleotides located at an end of a DNA strand. This allows sequential removal of nucleotides from the end of DNA.

exoskeleton An external skeleton, as in arthropods.

experiment A test of one or more hypotheses. Hypotheses make contrasting predictions that can be tested experimentally in control and test experiments where a single variable is altered.

expressed sequence tag (EST) A short sequence of a cDNA that unambiguously identifies the cDNA.

expression vector A type of vector (plasmid or phage) that contains the sequences necessary to drive expression of inserted DNA in a specific cell type.

exteroceptor A receptor that is excited by stimuli from the external world.

extremophile An archaean organism that lives in extreme environments; different archaean species may live in hot springs (thermophiles), highly saline environments (halophiles), highly acidic or basic environments, or under high pressure at the bottom of oceans.

F

5′ cap In eukaryotes, a structure added to the 5′ end of an mRNA consisting of methylated GTP attached by a 5′ to 5′ bond. The cap protects this end from degradation and is involved in the initiation of translation.

facilitated diffusion Carrier-assisted diffusion of molecules across a cellular membrane through specific channels from a region of higher concentration to one of lower concentration; the process is driven by the concentration gradient and does not require cellular energy from ATP.

family A taxonomic grouping of similar species above the level of genus.

fat A molecule composed of glycerol and three fatty acid molecules.

feedback inhibition Control mechanism whereby an increase in the concentration of some molecules inhibits the synthesis of that molecule.

fermentation The enzyme-catalyzed extraction of energy from organic compounds without the involvement of oxygen.

fertilization The fusion of two haploid gamete nuclei to form a diploid zygote nucleus.

fibroblast A flat, irregularly branching cell of connective tissue that secretes structurally strong proteins into the matrix between the cells.

first filial (F_1) generation The offspring resulting from a cross between a parental generation (P); in experimental crosses, these parents usually have different phenotypes.

First Law of Thermodynamics Energy cannot be created or destroyed, but can only undergo conversion from one form to another; thus, the amount of energy in the universe is unchangeable.

fitness The genetic contribution of an individual to succeeding generations. relative fitness refers to the fitness of an individual relative to other individuals in a population.

fixed action pattern A stereotyped animal behavior response, thought by ethologists to be based on programmed neural circuits.

flagellin The protein composing bacterial flagella, which allow a cell to move through an aqueous environment.

flagellum A long, threadlike structure protruding from the surface of a cell and used in locomotion.

flame cell A specialized cell found in the network of tubules inside flatworms that assists in water regulation and some waste excretion.

flavin adenine dinucleotide (FAD, $FADH_2$) A cofactor that acts as a soluble (not membrane-bound) electron carrier (can be reversibly oxidized and reduced).

fluorescent in situ hybridization (FISH) A cytological method used to find specific DNA sequences on chromosomes with a specific fluorescently labeled probe.

food security Having access to sufficient, safe food to avoid malnutrition and starvation; a global human issue.

foraging behavior A collective term for the many complex, evolved behaviors that influence what an animal eats and how the food is obtained.

founder effect The effect by which rare alleles and combinations of alleles may be enhanced in new populations.

fovea A small depression in the center of the retina with a high concentration of cones; the area of sharpest vision.

frameshift mutation A mutation in which a base is added or deleted from the DNA sequence. These changes alter the reading frame downstream of the mutation.

free energy Energy available to do work.

free radical An ionized atom with one or more unpaired electrons, resulting from electrons that have been energized by ionizing radiation being ejected from the atom; free radicals react violently with other molecules, such as DNA, causing damage by mutation.

frequency-dependent selection A type of selection that depends on how frequently or infrequently a phenotype occurs in a population.

fruit In angiosperms, a mature, ripened ovary (or group of ovaries), containing the seeds.

functional genomics The study of the function of genes and their products, beyond simply ascertaining gene sequences.

functional group A molecular group attached to a hydrocarbon that confers chemical properties or reactivities. Examples include hydroxyl (—OH), carboxylic acid (—COOH) and amino groups ($—NH_2$).

fundamental niche Also referred to as the hypothetical niche, this is the entire niche an organism could fill if there were no other interacting factors (such as competition or predation).

G

G_0 phase The stage of the cell cycle occupied by cells that are not actively dividing.

G_1 phase The phase of the cell cycle after cytokinesis and before DNA replication called the first "gap" phase. This phase is the primary growth phase of a cell.

G_1/S checkpoint The primary control point at which a cell "decides" whether or not to divide. Also called START and the restriction point.

G_2 phase The phase of the cell cycle between DNA replication and mitosis called the second "gap" phase. During this phase, the cell prepares for mitosis.

G_2/M checkpoint The second cell-division control point, at which division can be delayed if DNA has not been properly replicated or is damaged.

gametangium, pl. **gametangia** A cell or organ in which gametes are formed.

gamete A haploid reproductive cell.

gametocytes Cells in the malarial sporozoite life cycle capable of giving rise to gametes when in the correct host.

gametophyte In plants, the haploid (*n*), gamete-producing generation, which alternates with the diploid (2*n*) sporophyte.

ganglion, pl. **ganglia** An aggregation of nerve cell bodies; in invertebrates, ganglia are the integrative centers; in vertebrates, the term is restricted to aggregations of nerve cell bodies located outside the central nervous system.

gap gene Any of certain genes in *Drosophila* development that divide the embryo into large blocks in the process of segmentation; *hunchback* is a gap gene.

gap junction A junction between adjacent animal cells that allows the passage of materials between the cells.

gastrodermis In eumetazoan animals, the layer of digestive tissue that develops from the endoderm.

gastrula In vertebrates, the embryonic stage in which the blastula with its single layer of cells turns into a three-layered embryo made up of ectoderm, mesoderm, and endoderm.

gastrulation Developmental process that converts blastula into embryo with three embryonic germ layers: endoderm, mesoderm, and ectoderm. Involves massive cell migration to convert the hollow structure into a three-layered structure.

gene The basic unit of heredity; a sequence of DNA nucleotides on a chromosome that encodes a protein, tRNA, or rRNA molecule, or regulates the transcription of such a sequence.

gene conversion Alteration of one homologous chromosome by the cell's error-detection and repair system to make it resemble the sequence on the other homologue.

gene expression The conversion of the genotype into the phenotype; the process by which DNA is transcribed into RNA, which is then translated into a protein product.

gene pool All the alleles present in a species.

gene-for-gene hypothesis A plant defense mechanism in which a specific protein encoded by a viral, bacterial, or fungal pathogen binds to a protein encoded by a plant gene and triggers a defense response in the plant.

general transcription factor Any of a group of transcription factors that are required for formation of an initiation complex by RNA polymerase II at a promoter. This allows a basal level that can be increased by the action of specific factors.

generalized transduction A form of gene transfer in prokaryotes in which any gene can be transferred between cells. This uses a lytic bacteriophage as a carrier where the virion is accidentally packaged with host DNA.

genetic counseling The process of evaluating the risk of genetic defects occurring in offspring, testing for these defects in unborn children, and providing the parents with information about these risks and conditions.

genetic drift Random fluctuation in allele frequencies over time by chance.

genetic map An abstract map that places the relative location of genes on a chromosome based on recombination frequency.

genome The entire DNA sequence of an organism.

genomic imprinting Describes an exception to Mendelian genetics in some mammals in which the phenotype caused by an allele is exhibited when the allele comes from one parent, but not from the other.

genomic library A DNA library that contains a representation of the entire genome of an organism.

genomics The study of genomes as opposed to individual genes.

genotype The genetic constitution underlying a single trait or set of traits.

genotype frequency A measure of the occurrence of a genotype in a population, expressed as a proportion of the entire population, for example, an occurrence of 0.25 (25%) for a homozygous recessive genotype.

genus, pl. **genera** A taxonomic group that ranks below a family and above a species.

germination The resumption of growth and development by a spore or seed.

germ layers The three cell layers formed at gastrulation of the embryo that foreshadow the future organization of tissues; the layers, from the outside inward, are the ectoderm, the mesoderm, and the endoderm.

germ-line cells During zygote development, cells that are set aside from the somatic cells and that will eventually undergo meiosis to produce gametes.

gill (1) In aquatic animals, a respiratory organ, usually a thin-walled projection from some part of the external body surface, endowed with a rich capillary bed and having a large surface area. (2) In basidiomycete fungi, the plates on the underside of the cap.

globular protein Proteins with a compact tertiary structure with hydrophobic amino acids mainly in the interior.

glomerular filtrate The fluid that passes out of the capillaries of each glomerulus.

glomerulus A cluster of capillaries enclosed by Bowman's capsule.

glucagon A vertebrate hormone produced in the pancreas that acts to initiate the breakdown of glycogen to glucose subunits.

gluconeogenesis The synthesis of glucose from noncarbohydrates (such as proteins or fats).

glucose A common six-carbon sugar ($C_6H_{12}O_6$); the most common monosaccharide in most organisms.

glucose repression In *E. coli*, the preferential use of glucose even when other sugars are present; transcription of mRNA encoding the enzymes for utilizing the other sugars does not occur.

glycocalyx A "sugar coating" on the surface of a cell resulting from the presence of polysaccharides on glycolipids and glycoproteins embedded in the outer layer of the plasma membrane.

glycogen Animal starch; a complex branched polysaccharide that serves as a food reserve in animals, bacteria, and fungi.

glycolipid Lipid molecule modified within the Golgi complex by having a short sugar chain (polysaccharide) attached.

glycolysis The anaerobic breakdown of glucose; this enzyme-catalyzed process yields two molecules of pyruvate with a net of two molecules of ATP.

glycoprotein Protein molecule modified within the Golgi complex by having a short sugar chain (polysaccharide) attached.

glyoxysome A small cellular organelle or microbody containing enzymes necessary for conversion of fats into carbohydrates.

glyphosate A biodegradable herbicide that works by inhibiting EPSP synthetase, a plant enzyme that makes aromatic amino acids; genetic engineering has allowed crop species to be created that are resistant to glyphosate.

Golgi apparatus A collection of flattened stacks of membranes (each called a Golgi body) in the cytoplasm of eukaryotic cells; functions in collection, packaging, and distribution of molecules synthesized in the cell.

G protein A protein that binds guanosine triphosphate (GTP) and assists in the function of cell-surface receptors. When the receptor binds its signal molecule, the G protein binds GTP and is activated to start a chain of events within the cell.

G protein-coupled receptor (GPCR) A receptor that acts through a heterotrimeric (three component) G protein to activate effector proteins. The effector proteins then function as enzymes to produce second messengers such as cAMP or IP_3.

gradualism The view that species change very slowly in ways that may be imperceptible from one generation to the next but that accumulate and lead to major changes over thousands or millions of years.

Gram stain Staining technique that divides bacteria into gram-negative or gram-positive based on retention of a violet dye. Differences in staining are due to cell wall construction.

granum (pl. grana) A stacked column of flattened, interconnected disks (thylakoids) that are part of the thylakoid membrane system in chloroplasts.

gravitropism Growth response to gravity in plants; formerly called geotropism.

ground meristem The primary meristem, or meristematic tissue, that gives rise to the plant body (except for the epidermis and vascular tissues).

ground tissue In plants, a type of tissue that performs many functions, including support, storage, secretion, and photosynthesis; may consist of many cell types.

growth factor Any of a number of proteins that bind to membrane receptors and initiate intracellular signaling systems that result in cell growth and division.

guard cell In plants, one of a pair of sausage-shaped cells flanking a stoma; the guard cells open and close the stomata.

guttation The exudation of liquid water from leaves due to root pressure.

gymnosperm A seed plant with seeds not enclosed in an ovary; conifers are gymnosperms.

gynoecium The aggregate of carpels in the flower of a seed plant.

H

habitat The environment of an organism; the place where it is usually found.

habituation A form of learning; a diminishing response to a repeated stimulus.

halophyte A plant that is salt-tolerant.

haplodiploidy A phenomenon occurring in certain organisms such as wasps, wherein both haploid (male) and diploid (female) individuals are encountered.

haploid Having only one set of chromosomes (n), in contrast to diploid ($2n$).

haplotype A region of a chromosome that is usually inherited intact, that is, it does not undergo recombination. These are identified based on analysis of SNPs.

Hardy-Weinberg equilibrium A mathematical description of the fact that allele and genotype frequencies remain constant in a random-mating population in the absence of inbreeding, selection, or other evolutionary forces; usually stated: if the frequency of allele *a* is *p* and the frequency of allele *b* is *q*, then the genotype frequencies after one generation of random mating will always be $p_2 + 2pq + q_2 = 1$.

Haversian canal Narrow channels that run parallel to the length of a bone and contain blood vessels and nerve cells.

heat A measure of the random motion of molecules; the greater the heat, the greater the motion. Heat is one form of kinetic energy.

heat of vaporization The amount of energy required to change 1 g of a substance from a liquid to a gas.

heavy metal Any of the metallic elements with high atomic numbers, such as arsenic, cadmium, lead, etc. Many heavy metals are toxic to animals even in small amounts.

helicase Any of a group of enzymes that unwind the two DNA strands in the double helix to facilitate DNA replication.

helix-turn-helix motif A common DNA-binding motif found in regulatory proteins; it consists of two α-helices linked by a nonhelical segment (the "turn").

helper T cell A class of white blood cells that initiates both the cell-mediated immune response and the humoral immune response; helper T cells are the targets of the AIDS virus (HIV).

hemoglobin A globular protein in vertebrate red blood cells and in the plasma of many invertebrates that carries oxygen and carbon dioxide.

hemopoietic stem cell The cells in bone marrow where blood cells are formed.

hermaphroditism Condition in which an organism has both male and female functional reproductive organs.

heterochromatin The portion of a eukaryotic chromosome that is not transcribed into RNA; remains condensed in interphase and stains intensely in histological preparations.

heterochrony An alteration in the timing of developmental events due to a genetic change; for example, a mutation that delays flowering in plants.

heterokaryotic In fungi, having two or more genetically distinct types of nuclei within the same mycelium.

heterosporous In vascular plants, having spores of two kinds, namely, microspores and megaspores.

heterotroph An organism that cannot derive energy from photosynthesis or inorganic chemicals, and so must feed on other plants and animals, obtaining chemical energy by degrading their organic molecules.

heterozygote advantage The situation in which individuals heterozygous for a trait have a selective advantage over those who are homozygous; an example is sickle cell anemia.

heterozygous Having two different alleles of the same gene; the term is usually applied to one or more specific loci, as in "heterozygous with respect to the *W* locus" (that is, the genotype is *W/w*).

Hfr cell An *E. coli* cell that has a high frequency of recombination due to integration of an F plasmid into its genome.

histone One of a group of relatively small, very basic polypeptides, rich in arginine and lysine, forming the core of nucleosomes around which DNA is wrapped in the first stage of chromosome condensation.

histone protein Any of eight proteins with an overall positive charge that associate in a complex. The DNA duplex coils around a core of eight histone proteins, held by its negatively charged phosphate groups, forming a nucleosome.

holoblastic cleavage Process in vertebrate embryos in which the cleavage divisions all occur at the same rate, yielding a uniform cell size in the blastula.

homeobox A sequence of 180 nucleotides located in homeotic genes that produces a 60-amino-acid peptide sequence (the homeodomain) active in transcription factors.

homeodomain motif A special class of helix-turn-helix motifs found in regulatory proteins that control development in eukaryotes.

homeosis A change in the normal spatial pattern of gene expression that can result in homeotic mutants where a wild-type structure develops in the wrong place in or on the organism.

homeostasis The maintenance of a relatively stable internal physiological environment in an organism; usually involves some form of feedback self-regulation.

homeotherm An organism, such as a bird or mammal, capable of maintaining a stable body temperature independent of the environmental temperature. *See* endotherm.

homeotic gene One of a series of "master switch" genes that determine the form of segments developing in the embryo.

hominid Any primate in the human family, Hominidae. *Homo sapiens* is the only living representative.

hominoid Collectively, hominids and apes; the monkeys and hominoids constitute the anthropoid primates.

homokaryotic In fungi, having nuclei with the same genetic makeup within a mycelium.

homologue One of a pair of chromosomes of the same kind located in a diploid cell; one copy of each pair of homologues comes from each gamete that formed the zygote.

homologous (1) Refers to similar structures that have the same evolutionary origin. (2) Refers to a pair of the same kind of chromosome in a diploid cell.

homoplasy In cladistics, a shared character state that has not been inherited from a common ancestor exhibiting that state; may result from convergent evolution or evolutionary reversal. The wings of birds and of bats, which are convergent structures, are examples.

homosporous In some plants, production of only one type of spore rather than differentiated types. *Compare with* heterosporous.

homozygous Being a homozygote, having two identical alleles of the same gene; the term is usually applied to one or more specific loci, as in "homozygous with respect to the *W* locus" (i.e., the genotype is *W/W* or *w/w*).

horizontal gene transfer (HGT) The passing of genes laterally between species; more prevalent very early in the history of life.

hormone A molecule, usually a peptide or steroid, that is produced in one part of an organism and triggers a specific cellular reaction in target tissues and organs some distance away.

host range The range of organisms that can be infected by a particular virus.

***Hox* gene** A group of homeobox-containing genes that control developmental events, usually found organized into clusters of genes. These genes have been conserved in many different multicellular animals, both invertebrates and vertebrates, although the number of clusters changes in lineages, leading to four clusters in vertebrates.

humus Partly decayed organic material found in topsoil.

hybridization The mating of unlike parents.

hydration shell A "cloud" of water molecules surrounding a dissolved substance, such as sucrose or Na^+ and Cl^- ions.

hydrogen bond A weak association formed with hydrogen in polar covalent bonds. The partially positive hydrogen is attracted to partially negative atoms in polar covalent bonds. In water, oxygen and hydrogen in different water molecules form hydrogen bonds.

hydrolysis A reaction that breaks a bond by the addition of water. This is the reverse of dehydration, a reaction that joins molecules with the loss of water.

hydrophilic Literally translates as "water-loving" and describes substances that are soluble in water. These must be either polar or charged (ions).

hydrophobic Literally translates as "water-fearing" and describes nonpolar substances that are not soluble in water. Nonpolar molecules in water associate with each other and form droplets.

hydrophobic exclusion The tendency of nonpolar molecules to aggregate together when placed in water. Exclusion refers to the action of water in forcing these molecules together.

hydrostatic skeleton The skeleton of most soft-bodied invertebrates that have neither an internal nor an external skeleton. They use the relative incompressibility of the water within their bodies as a kind of skeleton.

hyperosmotic The condition in which a (hyperosmotic) solution has a higher osmotic concentration than that of a second solution. *Compare with* hypoosmotic.

hyperpolarization Above-normal negativity of a cell membrane during its resting potential.

hypersensitive response Plants respond to pathogens by selectively killing plant cells to block the spread of the pathogen.

hypertonic A solution with a higher concentration of solutes than the cell. A cell in a hypertonic solution tends to lose water by osmosis.

hypha, pl. **hyphae** A filament of a fungus or oomycete; collectively, the hyphae constitute the mycelium.

hypocotyl The region immediately below where the cotyledons are attached.

hypoosmotic The condition in which a (hypoosmotic) solution has a lower osmotic concentration than that of a second solution. *Compare with* hyperosmotic.

hypothalamus A region of the vertebrate brain just below the cerebral hemispheres, under the thalamus; a center of the autonomic nervous system, responsible for the integration and correlation of many neural and endocrine functions.

hypotonic A solution with a lower concentration of solutes than the cell. A cell in a hypotonic solution tends to take in water by osmosis.

I

icosahedron A structure consisting of 20 equilateral triangular facets; this is commonly seen in viruses and forms one kind of viral capsid.

imaginal disk One of about a dozen groups of cells set aside in the abdomen of a larval insect and committed to forming key parts of the adult insect's body.

immune response In vertebrates, a defensive reaction of the body to invasion by a foreign substance or organism. *See* antibody and B cell.

immunoglobulin An antibody.

in vitro mutagenesis The ability to create mutations at any site in a cloned gene to examine the mutations' effects on function.

inbreeding The breeding of genetically related plants or animals; inbreeding tends to increase homozygosity.

inclusive fitness Describes the sum of the number of genes directly passed on in an individual's offspring and those genes passed on indirectly by kin (other than offspring) whose existence results from the benefit of the individual's altruism.

incomplete dominance Describes a case in which two or more alleles of a gene do not display clear dominance. The phenotype of a heterozygote is intermediate between the homozygous forms. For example, crossing red-flowered with white-flowered four o'clocks yields pink heterozygotes.

independent assortment In a dihybrid cross, describes the random assortment of alleles for each of the genes. For genes on different chromosomes this results from the random orientations of different homologous pairs during metaphase I of meiosis. For genes on the same chromosome, this occurs when the two loci are far enough apart for roughly equal numbers of odd- and even-numbered multiple crossover events.

indeterminate development A type of development in animals in which the first few embryonic cells are identical daughter cells, any one of which could develop separately into a complete organism; their fate is indeterminate.

inducer exclusion Part of the mechanism of glucose repression in *E. coli* in which the presence of glucose prevents the entry of lactose such that the *lac* operon cannot be induced.

induction (1) Production of enzymes in response to a substrate; a mechanism by which binding of an inducer to a repressor allows transcription of an operon. This is seen in catabolic operons and results in production of enzymes to degrade a compound only when it is available. (2) In embryonic development, the process by which the development of a cell is influenced by interaction with an adjacent cell.

inductive reasoning The logical application of specific observations to make a generalization. In science, inductive reasoning is used to formulate testable hypotheses.

industrial melanism Phrase used to describe the evolutionary process in which initially light-colored organisms become dark as a result of natural selection.

inflammatory response A generalized nonspecific response to infection that acts to clear an infected area of infecting microbes and dead tissue cells so that tissue repair can begin.

inhalant siphon In bivalve mollusks, the siphon through which incoming water enters the body.

inheritance of acquired characteristics Also known as Lamarckism; the theory, now discounted, that individuals genetically pass on to their offspring physical and behavioral changes developed during the individuals' own lifetime.

inhibitor A substance that binds to an enzyme and decreases its activity.

initiation factor One of several proteins involved in the formation of an initiation complex in prokaryote polypeptide synthesis.

initiator tRNA A tRNA molecule involved in the beginning of translation. In prokaryotes, the initiator tRNA is charged with *N*-formylmethionine ($tRNA^{fMet}$); in eukaryotes, the tRNA is charged simply with methionine.

inorganic phosphate A phosphate molecule that is not a part of an organic molecule; inorganic phosphate groups are added and removed in the formation and breakdown of ATP and in many other cellular reactions.

insertional inactivation Destruction of a gene's function by the insertion of a transposon.

instar A larval developmental stage in insects.

integrin Any of a group of cell-surface proteins involved in adhesion of cells to substrates. Critical to migrating cells moving through the cell matrix in tissues such as connective tissue.

intercalary meristem A type of meristem that arises in stem internodes in some plants, such as corn and horsetails; responsible for elongation of the internodes.

interferon In vertebrates, a protein produced in virus-infected cells that inhibits viral multiplication.

intermembrane space The outer compartment of a mitochondrion that lies between the two membranes.

interneuron (association neuron) A nerve cell found only in the middle of the spinal cord that acts as a functional link between sensory neurons and motor neurons.

internode In plants, the region of a stem between two successive nodes.

interoceptor A receptor that senses information related to the body itself, its internal condition, and its position.

interphase The period between two mitotic or meiotic divisions in which a cell grows and its DNA replicates; includes G_1, S, and G_2 phases.

intracellular receptor A signal receptor that binds a ligand inside a cell, such as the receptors for NO, steroid hormones, vitamin D, and thyroid hormones.

intron Portion of mRNA as transcribed from eukaryotic DNA that is removed by enzymes before the mature mRNA is translated into protein. *See* exon.

inversion A reversal in order of a segment of a chromosome; also, to turn inside out, as in embryogenesis of sponges or discharge of a nematocyst.

ionizing radiation High-energy radiation that is highly mutagenic, producing free radicals that react with DNA; includes X-rays and γ-rays.

isomer One of a group of molecules identical in atomic composition but differing in structural arrangement; for example, glucose and fructose.

isosmotic The condition in which the osmotic concentrations of two solutions are equal, so that no net water movement occurs between them by osmosis.

isotonic A solution having the same concentration of solutes as the cell. A cell in an isotonic solution takes in and loses the same amount of water.

isotope Different forms of the same element with the same number of protons but different numbers of neutrons.

J

jasmonic acid An organic molecule that is part of a plant's wound response; it signals the production of a proteinase inhibitor.

K

karyotype The morphology of the chromosomes of an organism as viewed with a light microscope.

keratin A tough, fibrous protein formed in epidermal tissues and modified into skin, feathers, hair, and hard structures such as horns and nails.

key innovation A newly evolved trait in a species that allows members to use resources or other aspects of the environment that were previously inaccessible.

kidney In vertebrates, the organ that filters the blood to remove nitrogenous wastes and regulates the balance of water and solutes in blood plasma.

kilocalorie Unit describing the amount of heat required to raise the temperature of a kilogram of water by 1°C; sometimes called a Calorie, equivalent to 1000 calories.

kinase cascade A series of protein kinases that phosphorylate each other in succession; a kinase cascade can amplify signals during the signal transduction process.

kinesis Changes in activity level in an animal that are dependent on stimulus intensity. *See* kinetic energy.
kinetic energy The energy of motion.
kinetochore Disk-shaped protein structure within the centromere to which the spindle fibers attach during mitosis or meiosis. *See* centromere.
kingdom The second highest commonly used taxonomic category.
kin selection Selection favoring relatives; an increase in the frequency of related individuals (kin) in a population, leading to an increase in the relative frequency in the population of those alleles shared by members of the kin group.
knockout mice Mice in which a known gene is inactivated ("knocked out") using recombinant DNA and ES cells.
Krebs cycle Another name for the citric acid cycle; also called the tricarboxylic acid (TCA) cycle.

L

labrum The upper lip of insects and crustaceans situated above or in front of the mandibles.
***lac* operon** In *E. coli*, the operon containing genes that encode the enzymes to metabolize lactose.
lagging strand The DNA strand that must be synthesized discontinuously because of the 5′-to-3′ directionality of DNA polymerase during replication, and the antiparallel nature of DNA. Compare *leading strand.*
larva A developmental stage that is unlike the adult found in organisms that undergo metamorphosis. Embryos develop into larvae that produce the adult form by metamorphosis.
larynx The voice box; a cartilaginous organ that lies between the pharynx and trachea and is responsible for sound production in vertebrates.
lateral line system A sensory system encountered in fish, through which mechanoreceptors in a line down the side of the fish are sensitive to motion.
lateral meristems In vascular plants, the meristems that give rise to secondary tissue; the vascular cambium and cork cambium.
Law of Independent Assortment Mendel's second law of heredity, stating that genes located on nonhomologous chromosomes assort independently of one another.
Law of Segregation Mendel's first law of heredity, stating that alternative alleles for the same gene segregate from each other in production of gametes.
leading strand The DNA strand that can be synthesized continuously from the origin of replication. Compare *lagging strand.*
leaf primordium, pl. **primordia** A lateral outgrowth from the apical meristem that will eventually become a leaf.
lenticels Spongy areas in the cork surfaces of stem, roots, and other plant parts that allow interchange of gases between internal tissues and the atmosphere through the periderm.
leucine zipper motif A motif in regulatory proteins in which two different protein subunits associate to form a single DNA-binding site; the proteins are connected by an association between hydrophobic regions containing leucines (the "zipper").
leucoplast In plant cells, a colorless plastid in which starch grains are stored; usually found in cells not exposed to light.
leukocyte A white blood cell; a diverse array of nonhemoglobin-containing blood cells, including phagocytic macrophages and antibody-producing lymphocytes.
lichen Symbiotic association between a fungus and a photosynthetic organism such as a green alga or cyanobacterium.
ligand A signaling molecule that binds to a specific receptor protein, initiating signal transduction in cells.
light-dependent reactions In photosynthesis, the reactions in which light energy is captured and used in production of ATP and NADPH. In plants this involves the action of two linked photosystems.
light-independent reactions In photosynthesis, the reactions of the Calvin cycle in which ATP and NADPH from the light-dependent reactions are used to reduce CO_2 and produce organic compounds such as glucose. This involves the process of carbon fixation, or the conversion of inorganic carbon (CO_2) to organic carbon (ultimately carbohydrates).
lignin A highly branched polymer that makes plant cell walls more rigid; an important component of wood.
limbic system The hypothalamus, together with the network of neurons that link the hypothalamus to some areas of the cerebral cortex. Responsible for many of the most deep-seated drives and emotions of vertebrates, including pain, anger, sex, hunger, thirst, and pleasure.
linked genes Genes that are physically close together and therefore tend to segregate together; recombination occurring between linked genes can be used to produce a map of genetic distance for a chromosome.
lipase An enzyme that catalyzes the hydrolysis of fats.
lipid A nonpolar hydrophobic organic molecule that is insoluble in water (which is polar) but dissolves readily in nonpolar organic solvents; includes fats, oils, waxes, steroids, phospholipids, and carotenoids.
lipid bilayer The structure of a cellular membrane, in which two layers of phospholipids spontaneously align so that the hydrophilic head groups are exposed to water, while the hydrophobic fatty acid tails are pointed toward the center of the membrane.
lipopolysaccharide A lipid with a polysaccharide molecule attached; found in the outer membrane layer of gram-negative bacteria; the outer membrane layer protects the cell wall from antibiotic attack.
locus The position on a chromosome where a gene is located.
long interspersed element (LINE) Any of a type of large transposable element found in humans and other primates that contains all the biochemical machinery needed for transposition.
long terminal repeat (LTR) A particular type of retrotransposon that has repeated elements at its ends. These elements make up 8% of the human genome.
loop of Henle In the kidney of birds and mammals, a hairpin-shaped portion of the renal tubule in which water and salt are reabsorbed from the glomerular filtrate by diffusion.
lophophore A horseshoe-shaped crown of ciliated tentacles that surrounds the mouth of certain spiralian animals; seen in the phyla Brachiopoda and Bryozoa.
lumen A term for any bounded opening; for example, the cisternal space of the endoplasmic reticulum of eukaryotic cells, the passage through which blood flows inside a blood vessel, and the passage through which material moves inside the intestine during digestion.
luteal phase The second phase of the female reproductive cycle, during which the mature eggs are released into the fallopian tubes, a process called ovulation.
lymph In animals, a colorless fluid derived from blood by filtration through capillary walls in the tissues.
lymphatic system In animals, an open vascular system that reclaims water that has entered interstitial regions from the bloodstream (lymph); includes the lymph nodes, spleen, thymus, and tonsils.
lymphocyte A type of white blood cell. Lymphocytes are responsible for the immune response; there are two principal classes: B cells and T cells.
lymphokine A regulatory molecule that is secreted by lymphocytes. In the immune response, lymphokines secreted by helper T cells unleash the cell-mediated immune response.
lysis Disintegration of a cell by rupture of its plasma membrane.
lysogenic cycle A viral cycle in which the viral DNA becomes integrated into the host chromosome and is replicated during cell reproduction. Results in vertical rather than horizontal transmission.
lysosome A membrane-bounded vesicle containing digestive enzymes that is produced by the Golgi apparatus in eukaryotic cells.
lytic cycle A viral cycle in which the host cell is killed (lysed) by the virus after viral duplication to release viral particles.

M

macroevolution The creation of new species and the extinction of old ones.
macromolecule An extremely large biological molecule; refers specifically to proteins, nucleic acids, polysaccharides, lipids, and complexes of these.
macronutrients Inorganic chemical elements required in large amounts for plant growth, such as nitrogen, potassium, calcium, phosphorus, magnesium, and sulfur.
macrophage A large phagocytic cell that is able to engulf and digest cellular debris and invading bacteria.
madreporite A sievelike plate on the surface of echinoderms through which water enters the water–vascular system.
***MADS* box gene** Any of a family of genes identified by possessing shared motifs that are the predominant homeotic genes of plants; a small number of *MADS* box genes are also found in animals.
major groove The larger of the two grooves in a DNA helix, where the paired nucleotides' hydrogen bonds are accessible; regulatory

proteins can recognize and bind to regions in the major groove.

major histocompatibility complex (MHC) A set of protein cell-surface markers anchored in the plasma membrane, which the immune system uses to identify "self." All the cells of a given individual have the same "self" marker, called an MHC protein.

Malpighian tubules Blind tubules opening into the hindgut of terrestrial arthropods; they function as excretory organs.

mandibles In crustaceans, insects, and myriapods, the appendages immediately posterior to the antennae; used to seize, hold, bite, or chew food.

mantle The soft, outermost layer of the body wall in mollusks; the mantle secretes the shell.

map unit Each 1% of recombination frequency between two genetic loci; the unit is termed a centimorgan (cM) or simply a map unit (m.u.).

marsupial A mammal in which the young are born early in their development, sometimes as soon as eight days after fertilization, and are retained in a pouch.

mass extinction A relatively sudden, sharp decline in the number of species; for example, the extinction at the end of the Cretaceous period in which the dinosaurs and a variety of other organisms disappeared.

mass flow hypothesis The overall process by which materials move in the phloem of plants.

maternal inheritance A mode of uniparental inheritance from the female parent; for example, in humans mitochondria and their genomes are inherited from the mother.

matrix In mitochondria, the solution in the interior space surrounded by the cristae that contains the enzymes and other molecules involved in oxidative respiration; more generally, that part of a tissue within which an organ or process is embedded.

medusa A free-floating, often umbrella-shaped body form found in cnidarian animals, such as jellyfish.

megapascal (MPa) A unit of measure used for pressure in water potential.

megaphyll In plants, a leaf that has several to many veins connecting it to the vascular cylinder of the stem; most plants have megaphylls.

mesoglea A layer of gelatinous material found between the epidermis and gastrodermis of eumetazoans; it contains the muscles in most of these animals.

mesohyl A gelatinous, protein-rich matrix found between the choanocyte layer and the epithelial layer of the body of a sponge; various types of amoeboid cells may occur in the mesohyl.

metacercaria An encysted form of a larval liver fluke, found in muscle tissue of an infected animal; if the muscle is eaten, cysts dissolves in the digestive tract, releasing the flukes into the body of the new host.

methylation The addition of a methyl group to bases (primarily cytosine) in DNA. Cytosine methylation is correlated with DNA that is not expressed.

meiosis I The first round of cell division in meiosis; it is referred to as a "reduction division" because homologous chromosomes separate, and the daughter cells have only the haploid number of chromosomes.

meiosis II The second round of division in meiosis, during which the two haploid cells from meiosis I undergo a mitosis-like division without DNA replication to produce four haploid daughter cells.

membrane receptor A signal receptor present as an integral protein in the cell membrane, such as GPCRs, chemically gated ion channels in neurons, and RTKs.

Mendelian ratio The characteristic dominant-to-recessive phenotypic ratios that Mendel observed in his genetics experiments. For example, the F_2 generation in a monohybrid cross shows a ratio of 3:1; the F_2 generation in a dihybrid cross shows a ratio of 9:3:3:1.

menstruation Periodic sloughing off of the blood-enriched lining of the uterus when pregnancy does not occur.

meristem Undifferentiated plant tissue from which new cells arise.

meroblastic cleavage A type of cleavage in the eggs of reptiles, birds, and some fish. Occurs only on the blastodisc.

mesoderm One of the three embryonic germ layers that form in the gastrula; gives rise to muscle, bone and other connective tissue, the peritoneum, the circulatory system, and most of the excretory and reproductive systems.

mesophyll The photosynthetic parenchyma of a leaf, located within the epidermis.

messenger RNA (mRNA) The RNA transcribed from structural genes; RNA molecules complementary to a portion of one strand of DNA, which are translated by the ribosomes to form protein.

metabolism The sum of all chemical processes occurring within a living cell or organism.

metamorphosis Process in which a marked change in form takes place during postembryonic development as, for example, from tadpole to frog.

metaphase The stage of mitosis or meiosis during which microtubules become organized into a spindle and the chromosomes come to lie in the spindle's equatorial plane.

metastasis The process by which cancer cells move from their point of origin to other locations in the body; also, a population of cancer cells in a secondary location, the result of movement from the primary tumor.

methanogens Obligate, anaerobic archaebacteria that produce methane.

microarray DNA sequences are placed on a microscope slide or chip with a robot. The microarray can then be probed with RNA from specific tissues to identify expressed DNA.

microbody A cellular organelle bounded by a single membrane and containing a variety of enzymes; generally derived from endoplasmic reticulum; includes peroxisomes and glyoxysomes.

microevolution Refers to the evolutionary process itself. Evolution within a species. Also called adaptation.

micronutrient A mineral required in only minute amounts for plant growth, such as iron, chlorine, copper, manganese, zinc, molybdenum, and boron.

microphyll In plants, a leaf that has only one vein connecting it to the vascular cylinder of the stem; the club mosses in particular have microphylls.

micropyle In the ovules of seed plants, an opening in the integuments through which the pollen tube usually enters.

micro-RNA (miRNA) A class of RNAs that are very short and only recently could be detected. *See also* small interfering RNAs (siRNAs).

microtubule In eukaryotic cells, a long, hollow protein cylinder, composed of the protein tubulin; these influence cell shape, move the chromosomes in cell division, and provide the functional internal structure of cilia and flagella.

microvillus Cytoplasmic projection from epithelial cells; microvilli greatly increase the surface area of the small intestine.

middle lamella The layer of intercellular material, rich in pectic compounds, that cements together the primary walls of adjacent plant cells.

mimicry The resemblance in form, color, or behavior of certain organisms (mimics) to other more powerful or more protected ones (models).

miracidium The ciliated first-stage larva inside the egg of the liver fluke; eggs are passed in feces, and if they reach water they may be eaten by a host snail in which they continue their life cycle.

missense mutation A base substitution mutation that results in the alteration of a single amino acid.

mitogen-activated protein (MAP) kinase Any of a class of protein kinases that activate transcription factors to alter gene expression. A mitogen is any molecule that stimulates cell division. MAP kinases are activated by kinase cascades.

mitosis Somatic cell division; nuclear division in which the duplicated chromosomes separate to form two genetically identical daughter nuclei.

molar concentration Concentration expressed as moles of a substance in 1 L of pure water.

mole The weight of a substance in grams that corresponds to the atomic masses of all the component atoms in a molecule of that substance. One mole of a compound always contains 6.023×10^{23} molecules.

molecular clock method In evolutionary theory, the method in which the rate of evolution of a molecule is constant through time.

molecular cloning The isolation and amplification of a specific sequence of DNA.

monocot Short for monocotyledon; flowering plant in which the embryos have only one cotyledon, the floral parts are generally in threes, and the leaves typically are parallel-veined.

monocyte A type of leukocyte that becomes a phagocytic cell (macrophage) after moving into tissues.

monoecious A plant in which the staminate and pistillate flowers are separate, but borne on the same individual.

monomer The smallest chemical subunit of a polymer. The monosaccharide α-glucose is the monomer found in plant starch, a polysaccharide.

monophyletic In phylogenetic classification, a group that includes the most recent common ancestor of the group and all its descendants. A clade is a monophyletic group.

monosaccharide A simple sugar that cannot be decomposed into smaller sugar molecules.

monosomic Describes the condition in which a chromosome has been lost due to nondisjunction during meiosis, producing a diploid embryo with only one of these autosomes.

monotreme An egg-laying mammal.

morphogen A signal molecule produced by an embryonic organizer region that informs surrounding cells of their distance from the organizer, thus determining relative positions of cells during development.

morphogenesis The development of an organism's body form, namely its organs and anatomical features; it may involve apoptosis as well as cell division, differentiation, and changes in cell shape.

morphology The form and structure of an organism.

morula Solid ball of cells in the early stage of embryonic development.

mosaic development A pattern of embryonic development in which initial cells produced by cleavage divisions contain different developmental signals (determinants) from the egg, setting the individual cells on different developmental paths.

motif A substructure in proteins that confers function and can be found in multiple proteins. One example is the helix-turn-helix motif found in a number of proteins that is used to bind to DNA.

motor (efferent) neuron Neuron that transmits nerve impulses from the central nervous system to an effector, which is typically a muscle or gland.

M phase The phase of cell division during which chromosomes are separated. The spindle assembles, binds to the chromosomes, and moves the sister chromatids apart.

M phase-promoting factor (MPF) A Cdk enzyme active at the G_2/M checkpoint.

Müllerian mimicry A phenomenon in which two or more unrelated but protected species resemble one another, thus achieving a kind of group defense.

multidrug-resistant (MDR) strain Any bacterial strain that has become resistant to more than one antibiotic drug; MDR *Staphylococcus* strains, for example, are responsible for many infection deaths.

multienzyme complex An assembly consisting of several enzymes catalyzing different steps in a sequence of reactions. Close proximity of these related enzymes speeds the overall process, making it more efficient.

multigene family A collection of related genes on a single chromosome or on different chromosomes.

muscle fiber A long, cylindrical, multinucleated cell containing numerous myofibrils, which is capable of contraction when stimulated.

mutagen An agent that induces changes in DNA (mutations); includes physical agents that damage DNA and chemicals that alter DNA bases.

mutation A permanent change in a cell's DNA; includes changes in nucleotide sequence, alteration of gene position, gene loss or duplication, and insertion of foreign sequences.

mutualism A symbiotic association in which two (or more) organisms live together, and both members benefit.

mycelium, pl. **mycelia** In fungi, a mass of hyphae.

mycorrhiza, pl. **mycorrhizae** A symbiotic association between fungi and the roots of a plant.

myelin sheath A fatty layer surrounding the long axons of motor neurons in the peripheral nervous system of vertebrates.

myofilament A contractile microfilament, composed largely of actin and myosin, within muscle.

myosin One of the two protein components of microfilaments (the other is actin); a principal component of vertebrate muscle.

N

natural killer cell A cell that does not kill invading microbes, but rather, the cells infected by them.

natural selection The differential reproduction of genotypes; caused by factors in the environment; leads to evolutionary change.

nauplius A larval form characteristic of crustaceans.

negative control A type of control at the level of DNA transcription initiation in which the frequency of initiation is decreased; repressor proteins mediate negative control.

negative feedback A homeostatic control mechanism whereby an increase in some substance or activity inhibits the process leading to the increase; also known as feedback inhibition.

nematocyst A harpoonlike structure found in the cnidocytes of animals in the phylum Cnidaria, which includes the jellyfish among other groups; the nematocyst, when released, stings and helps capture prey.

nephridium, pl. **nephridia** In invertebrates, a tubular excretory structure.

nephrid organ A filtration system of many freshwater invertebrates in which water and waste pass from the body across the membrane into a collecting organ, from which they are expelled to the outside through a pore.

nephron Functional unit of the vertebrate kidney; one of numerous tubules involved in filtration and selective reabsorption of blood; each nephron consists of a Bowman's capsule, an enclosed glomerulus, and a long attached tubule; in humans, called a renal tubule.

nephrostome The funnel-shaped opening that leads to the nephridium, which is the excretory organ of mollusks.

nerve A group or bundle of nerve fibers (axons) with accompanying neurological cells, held together by connective tissue; located in the peripheral nervous system.

nerve cord One of the distinguishing features of chordates, running lengthwise just beneath the embryo's dorsal surface; in vertebrates, differentiates into the brain and spinal cord.

neural crest A special strip of cells that develops just before the neural groove closes over to form the neural tube in embryonic development.

neural groove The long groove formed along the long axis of the embryo by a layer of ectodermal cells.

neural tube The dorsal tube, formed from the neural plate, that differentiates into the brain and spinal cord.

neuroglia Nonconducting nerve cells that are intimately associated with neurons and appear to provide nutritional support.

neuromuscular junction The structure formed when the tips of axons contact (innervate) a muscle fiber.

neuron A nerve cell specialized for signal transmission; includes cell body, dendrites, and axon.

neurotransmitter A chemical released at the axon terminal of a neuron that travels across the synaptic cleft, binds a specific receptor on the far side, and depending on the nature of the receptor, depolarizes or hyperpolarizes a second neuron or a muscle or gland cell.

neurulation A process in early embryonic development by which a dorsal band of ectoderm thickens and rolls into the neural tube.

neutrophil An abundant type of granulocyte capable of engulfing microorganisms and other foreign particles; neutrophils comprise about 50–70% of the total number of white blood cells.

niche The role played by a particular species in its environment.

nicotinamide adenine dinucleotide (NAD) A molecule that becomes reduced (to NADH) as it carries high-energy electrons from oxidized molecules and delivers them to ATP-producing pathways in the cell.

NADH dehydrogenase An enzyme located on the inner mitochondrial membrane that catalyzes the oxidation by NAD^+ of pyruvate to acetyl-CoA. This reaction links glycolysis and the Krebs cycle.

nitrification The oxidization of ammonia or nitrite to produce nitrate, the form of nitrogen taken up by plants; some bacteria are capable of nitrification.

nociceptor A naked dendrite that acts as a receptor in response to a pain stimulus.

nocturnal Active primarily at night.

node The part of a plant stem where one or more leaves are attached. *See* internode.

node of Ranvier A gap formed at the point where two Schwann cells meet and where the axon is in direct contact with the surrounding intercellular fluid.

nodule In plants, a specialized tissue that surrounds and houses beneficial bacteria, such as root nodules of legumes that contain nitrogen-fixing bacteria.

nonassociative learning A learned behavior that does not require an animal to form an association between two stimuli, or between a stimulus and a response.

noncompetitive inhibitor An inhibitor that binds to a location other than the active site of an enzyme, changing the enzyme's shape so that it cannot bind the substrate.

noncyclic photophosphorylation The set of light-dependent reactions of the two plant photosystems, in which excited electrons are shuttled between the two photosystems, producing a proton gradient that is used for the chemiosmotic synthesis of ATP. The electrons are used to reduce NADP to NADPH. Lost electrons are replaced by the oxidation of water producing O_2.

nondisjunction The failure of homologues or sister chromatids to separate during mitosis or meiosis, resulting in an aneuploid cell or gamete.

nonextreme archaea Archaean groups that are not extremophiles, living in more moderate environments on Earth today.

nonpolar Said of a covalent bond that involves equal sharing of electrons. Can also refer to a compound held together by nonpolar covalent bonds.

nonsense codon One of three codons (UAA, UAG, and UGA) that are not recognized by tRNAs, thus serving as "stop" signals in the mRNA message and terminating translation.

nonsense mutation A base substitution in which a codon is changed into a stop codon. The protein is truncated because of premature termination.

Northern blot A blotting technique used to identify a specific mRNA sequence in a complex mixture. *See* Southern blot.

notochord In chordates, a dorsal rod of cartilage that runs the length of the body and forms the primitive axial skeleton in the embryos of all chordates.

nucellus Tissue composing the chief pair of young ovules, in which the embryo sac develops; equivalent to a megasporangium.

nuclear envelope The bounding structure of the eukaryotic nucleus. Composed of two phospholipid bilayers with the outer one connected to the endoplasmic reticulum.

nuclear pore One of a multitude of tiny but complex openings in the nuclear envelope that allow selective passage of proteins and nucleic acids into and out of the nucleus.

nuclear receptor Intracellular receptors are found in both the cytoplasm and the nucleus. The site of action of the hormone–receptor complex is in the nucleus where they modify gene expression.

nucleic acid A nucleotide polymer; chief types are deoxyribonucleic acid (DNA), which is double-stranded, and ribonucleic acid (RNA), which is typically single-stranded.

nucleoid The area of a prokaryotic cell, usually near the center, that contains the genome in the form of DNA compacted with protein.

nucleolus In eukaryotes, the site of rRNA synthesis; a spherical body composed chiefly of rRNA in the process of being transcribed from multiple copies of rRNA genes.

nucleosome A complex consisting of a DNA duplex wound around a core of eight histone proteins.

nucleotide A single unit of nucleic acid, composed of a phosphate, a five-carbon sugar (either ribose or deoxyribose), and a purine or a pyrimidine.

nucleus In atoms, the central core, containing positively charged protons and (in all but hydrogen) electrically neutral neutrons; in eukaryotic cells, the membranous organelle that houses the chromosomal DNA; in the central nervous system, a cluster of nerve cell bodies.

nutritional mutation A mutation affecting a synthetic pathway for a vital compound, such as an amino acid or vitamin; microorganisms with a nutritional mutation must be grown on medium that supplies the missing nutrient.

O

ocellus, pl. **ocelli** A simple light receptor common among invertebrates.

octet rule Rule to describe patterns of chemical bonding in main group elements that require a total of eight electrons to complete their outer electron shell.

Okazaki fragment A short segment of DNA produced by discontinuous replication elongating in the 5′-to-3′ direction away from the replication.

olfaction The function of smelling.

ommatidium, pl. **ommatidia** The visual unit in the compound eye of arthropods; contains light-sensitive cells and a lens able to form an image.

oncogene A mutant form of a growth-regulating gene that is inappropriately "on," causing unrestrained cell growth and division.

oocyst The zygote in a sporozoan life cycle. It is surrounded by a tough cyst to prevent dehydration or other damage.

open circulatory system A circulatory system in which the blood flows into sinuses in which it mixes with body fluid and then reenters the vessels in another location.

open reading frame (ORF) A region of DNA that encodes a sequence of amino acids with no stop codons in the reading frame.

operant conditioning A learning mechanism in which the reward follows only after the correct behavioral response.

operator A regulatory site on DNA to which a repressor can bind to prevent or decrease initiation of transcription.

operculum A flat, bony, external protective covering over the gill chamber in fish.

operon A cluster of adjacent structural genes transcribed as a unit into a single mRNA molecule.

opisthosoma The posterior portion of the body of an arachnid.

oral surface The surface on which the mouth is found; used as a reference when describing the body structure of echinoderms because of their adult radial symmetry.

orbital A region around the nucleus of an atom with a high probability of containing an electron. The position of electrons can only be described by these probability distributions.

order A category of classification above the level of family and below that of class.

organ A body structure composed of several different tissues grouped in a structural and functional unit.

organelle Specialized part of a cell; literally, a small cytoplasmic organ.

orthologues Genes that reflect the conservation of a single gene found in an ancestor.

oscillating selection The situation in which selection alternately favors one phenotype at one time, and a different phenotype at a another time, for example, during drought conditions versus during wet conditions.

osculum A specialized, larger pore in sponges through which filtered water is forced to the outside of the body.

osmoconformer An animal that maintains the osmotic concentration of its body fluids at about the same level as that of the medium in which it is living.

osmosis The diffusion of water across a selectively permeable membrane (a membrane that permits the free passage of water but prevents or retards the passage of a solute); in the absence of differences in pressure or volume, the net movement of water is from the side containing a lower concentration of solute to the side containing a higher concentration.

osmotic concentration The property of a solution that takes into account all dissolved solutes in the solution; if two solutions with different osmotic concentrations are separated by a water-permeable membrane, water will move from the solution with lower osmotic concentration to the solution with higher osmotic concentration.

osmotic pressure The potential pressure developed by a solution separated from pure water by a differentially permeable membrane. The higher the solute concentration, the greater the osmotic potential of the solution; also called *osmotic potential.*

ossicle Any of a number of movable or fixed calcium-rich plates that collectively make up the endoskeleton of echinoderms.

osteoblast A bone-forming cell.

osteocyte A mature osteoblast.

outcrossing Breeding with individuals other than oneself or one's close relatives.

ovary (1) In animals, the organ in which eggs are produced. (2) In flowering plants, the enlarged basal portion of a carpel that contains the ovule(s); the ovary matures to become the fruit.

oviduct In vertebrates, the passageway through which ova (eggs) travel from the ovary to the uterus.

oviparity Refers to a type of reproduction in which the eggs are developed after leaving the body of the mother, as in reptiles.

ovoviviparity Refers to a type of reproduction in which young hatch from eggs that are retained in the mother's uterus.

ovulation In animals, the release of an egg or eggs from the ovary.

ovum, pl. **ova** The egg cell; female gamete.

oxidation Loss of an electron by an atom or molecule; in metabolism, often associated with a gain of oxygen or a loss of hydrogen.

oxidation–reduction reaction A type of paired reaction in living systems in which electrons lost from one atom (oxidation) are gained by another atom (reduction). Termed a *redox reaction* for short.

oxidative respiration Process of cellular activity in which glucose or other molecules are broken down to water and carbon dioxide with the release of energy.

oxygen debt The amount of oxygen required to convert the lactic acid generated in the muscles during exercise back into glucose.

oxytocin A hormone of the posterior pituitary gland that affects uterine contractions during childbirth and stimulates lactation.

ozone O_3, a stratospheric layer of the Earth's atmosphere responsible for filtering out ultraviolet radiation supplied by the Sun.

P

***p53* gene** The gene that produces the p53 protein that monitors DNA integrity and halts cell division if DNA damage is detected. Many types of cancer are associated with a damaged or absent *p53* gene.

pacemaker A patch of excitatory tissue in the vertebrate heart that initiates the heartbeat.

pair-rule gene Any of certain genes in *Drosophila* development controlled by the gap genes that are expressed in stripes that subdivide the embryo in the process of segmentation.

paleopolyploid An ancient polyploid organism used in analysis of polyploidy events in the study of a species' genome evolution.
palisade parenchyma In plant leaves, the columnar, chloroplast-containing parenchyma cells of the mesophyll. Also called *palisade cells.*
panspermia The hypothesis that meteors or cosmic dust may have brought significant amounts of complex organic molecules to Earth, kicking off the evolution of life.
papilla A small projection of tissue.
paracrine A type of chemical signaling between cells in which the effects are local and short-lived.
paralogues Two genes within an organism that arose from the duplication of one gene in an ancestor.
paraphyletic In phylogenetic classification, a group that includes the most recent common ancestor of the group, but not all its descendants.
parapodia One of the paired lateral processes on each side of most segments in polychaete annelids.
parasexuality In certain fungi, the fusion and segregation of heterokaryotic haploid nuclei to produce recombinant nuclei.
parasitism A living arrangement in which an organism lives on or in an organism of a different species and derives nutrients from it.
parenchyma cell The most common type of plant cell; characterized by large vacuoles, thin walls, and functional nuclei.
parthenogenesis The development of an egg without fertilization, as in aphids, bees, ants, and some lizards.
partial diploid (merodiploid) Describes an *E. coli* cell that carries an F′ plasmid with host genes. This makes the cell diploid for the genes carried by the F′ plasmid.
partial pressure The components of each individual gas—such as nitrogen, oxygen, and carbon dioxide—that together constitute the total air pressure.
passive transport The movement of substances across a cell's membrane without the expenditure of energy.
pedigree A consistent graphic representation of matings and offspring over multiple generations for a particular genetic trait, such as albinism or hemophilia.
pedipalps A pair of specialized appendages found in arachnids; in male spiders, these are specialized as copulatory organs, whereas in scorpions they are large pincers.
pelagic Free-swimming, usually in open water.
pellicle A tough, flexible covering in ciliates and euglenoids.
pentaradial symmetry The five-part radial symmetry characteristic of adult echinoderms.
peptide bond The type of bond that links amino acids together in proteins through a dehydration reaction.
peptidoglycan A component of the cell wall of bacteria, consisting of carbohydrate polymers linked by protein cross-bridges.
peptidyl transferase In translation, the enzyme responsible for catalyzing the formation of a peptide bond between each new amino acid and the previous amino acid in a growing polypeptide chain.
perianth In flowering plants, the petals and sepals taken together.
pericycle In vascular plants, one or more cell layers surrounding the vascular tissues of the root, bounded externally by the endodermis and internally by the phloem.
periderm Outer protective tissue in vascular plants that is produced by the cork cambium and functionally replaces epidermis when it is destroyed during secondary growth; the periderm includes the cork, cork cambium, and phelloderm.
peristalsis In animals, a series of alternating contracting and relaxing muscle movements along the length of a tube such as the oviduct or alimentary canal that tend to force material such as an egg cell or food through the tube.
peroxisome A microbody that plays an important role in the breakdown of highly oxidative hydrogen peroxide by catalase.
petal A flower part, usually conspicuously colored; one of the units of the corolla.
petiole The stalk of a leaf.
phage conversion The phenomenon by which DNA from a virus, incorporated into a host cell's genome, alters the host cell's function in a significant way; for example, the conversion of *Vibrio cholerae* bacteria into a pathogenic form that releases cholera toxin.
phage lambda (λ) A well-known bacteriophage that has been widely used in genetic studies and is often a vector for DNA libraries.
phagocyte Any cell that engulfs and devours microorganisms or other particles.
phagocytosis Endocytosis of a solid particle; the plasma membrane folds inward around the particle (which may be another cell) and engulfs it to form a vacuole.
pharyngeal pouches In chordates, embryonic regions that become pharyngeal slits in aquatic and marine chordates and vertebrates, but do not develop openings to the outside in terrestrial vertebrates.
pharyngeal slits One of the distinguishing features of chordates; a group of openings on each side of the anterior region that form a passageway from the pharynx and esophagus to the external environment.
pharynx A muscular structure lying posterior to the mouth in many animals; aids in propelling food into the digestive tract.
phenotype The realized expression of the genotype; the physical appearance or functional expression of a trait.
pheromone Chemical substance released by one organism that influences the behavior or physiological processes of another organism of the same species. Pheromones serve as sex attractants, as trail markers, and as alarm signals.
phloem In vascular plants, a food-conducting tissue basically composed of sieve elements, various kinds of parenchyma cells, fibers, and sclereids.
phoronid Any of a group of lophophorate invertebrates, now classified in the phylum Brachiopoda, that burrows into soft underwater substrates and secretes a chitinous tube in which it lives out its life; it extends its lophophore tentacles to feed on drifting food particles.
phosphatase Any of a number of enzymes that removes a phosphate group from a protein, reversing the action of a kinase.
phosphodiester bond The linkage between two sugars in the backbone of a nucleic acid molecule; the phosphate group connects the pentose sugars through a pair of ester bonds.
phospholipid Similar in structure to a fat, but having only two fatty acids attached to the glycerol backbone, with the third space linked to a phosphorylated molecule; contains a polar hydrophilic "head" end (phosphate group) and a nonpolar hydrophobic "tail" end (fatty acids).
phospholipid bilayer The main component of cell membranes; phospholipids naturally associate in a bilayer with hydrophobic fatty acids oriented to the inside and hydrophilic phosphate groups facing outward on both sides.
phosphorylation Chemical reaction resulting in the addition of a phosphate group to an organic molecule. Phosphorylation of ADP yields ATP. Many proteins are also activated or inactivated by phosphorylation.
photoelectric effect The ability of a beam of light to excite electrons, creating an electrical current.
photon A particle of light having a discrete amount of energy. The wave concept of light explains the different colors of the spectrum, whereas the particle concept of light explains the energy transfers during photosynthesis.
photoperiodism The tendency of biological reactions to respond to the duration and timing of day and night; a mechanism for measuring seasonal time.
photoreceptor A light-sensitive sensory cell.
photorespiration Action of the enzyme rubisco, which catalyzes the oxidization of RuBP, releasing CO_2; this reverses carbon fixation and can reduce the yield of photosynthesis.
photosystem An organized complex of chlorophyll, other pigments, and proteins that traps light energy as excited electrons. Plants have two linked photosystems in the thylakoid membrane of chloroplasts. Photosystem II passes an excited electron through an electron transport chain to photosystem I to replace an excited electron passed to NADPH. The electron lost from photosystem II is replaced by the oxidation of water.
phototropism In plants, a growth response to a light stimulus.
pH scale A scale used to measure acidity and basicity. Defined as the negative log of H^+ concentration. Ranges from 0 to 14. A value of 7 is neutral; below 7 is acidic and above 7 is basic.
phycobiloprotein A type of accessory pigment found in cyanobacteria and some algae. Complexes of phycobiloprotein are able to absorb light energy in the green range.
phycologist A scientist who studies algae.
phyllotaxy In plants, a spiral pattern of leaf arrangement on a stem in which sequential leaves are at a 137.5° angle to one another, an angle related to the golden mean.
phylogenetic species concept (PSC) The concept that defines species on the basis of their phylogenetic relationships.
phylogenetic tree A pattern of descent generated by analysis of similarities and differences among organisms. Modern gene-sequencing techniques have produced phylogenetic trees showing the evolutionary history of individual genes.

phylogeny The evolutionary history of an organism, including which species are closely related and in what order related species evolved; often represented in the form of an evolutionary tree.

phylum, pl. **phyla** A major category, between kingdom and class, of taxonomic classifications.

physical map A map of the DNA sequence of a chromosome or genome based on actual landmarks within the DNA.

phytochrome A plant pigment that is associated with the absorption of light; photoreceptor for red to far-red light.

phytoestrogen One of a number of secondary metabolites in some plants that are structurally and functionally similar to the animal hormone estrogen.

phytoremediation The process that uses plants to remove contamination from soil or water.

pigment A molecule that absorbs light.

pilus, pl. **pili** Extensions of a bacterial cell enabling it to transfer genetic materials from one individual to another or to adhere to substrates.

pinocytosis The process of fluid uptake by endocytosis in a cell.

pistil Central organ of flowers, typically consisting of ovary, style, and stigma; a pistil may consist of one or more fused carpels and is more technically and better known as the gynoecium.

pith The ground tissue occupying the center of the stem or root within the vascular cylinder.

placenta, pl. **placentae** (1) In flowering plants, the part of the ovary wall to which the ovules or seeds are attached. (2) In mammals, a tissue formed in part from the inner lining of the uterus and in part from other membranes, through which the embryo (later the fetus) is nourished while in the uterus and through which wastes are carried away.

plankton Free-floating, mostly microscopic, aquatic organisms.

plant receptor kinase Any of a group of plant membrane receptors that, when activated by binding ligand, have kinase enzymatic activity. These receptors phosphorylate serine or threonine, unlike RTKs in animals that phosphorylate tyrosine.

planula A ciliated, free-swimming larva produced by the medusae of cnidarian animals.

plasma The fluid of vertebrate blood; contains dissolved salts, metabolic wastes, hormones, and a variety of proteins, including antibodies and albumin; blood minus the blood cells.

plasma cell An antibody-producing cell resulting from the multiplication and differentiation of a B lymphocyte that has interacted with an antigen.

plasma membrane The membrane surrounding the cytoplasm of a cell; consists of a single phospholipid bilayer with embedded proteins.

plasmid A small fragment of extrachromosomal DNA, usually circular, that replicates independently of the main chromosome, although it may have been derived from it.

plasmodesmata In plants, cytoplasmic connections between adjacent cells.

plasmodium Stage in the life cycle of myxomycetes (plasmodial slime molds); a multinucleate mass of protoplasm surrounded by a membrane.

plasmolysis The shrinking of a plant cell in a hypertonic solution such that it pulls away from the cell wall.

plastid An organelle in the cells of photosynthetic eukaryotes that is the site of photosynthesis and, in plants and green algae, of starch storage.

platelet In mammals, a fragment of a white blood cell that circulates in the blood and functions in the formation of blood clots at sites of injury.

pleiotropy Condition in which an individual allele has more than one effect on production of the phenotype.

plesiomorphy In cladistics, another term for an ancestral character state.

plumule The epicotyl of a plant with its two young leaves.

point mutation An alteration of one nucleotide in a chromosomal DNA molecule.

polar body Minute, nonfunctioning cell produced during the meiotic divisions leading to gamete formation in vertebrates.

polar covalent bond A covalent bond in which electrons are shared unequally due to differences in electronegativity of the atoms involved. One atom has a partial negative charge and the other a partial positive charge, even though the molecule is electrically neutral overall.

polarity (1) Refers to unequal charge distribution in a molecule such as water, which has a positive region and a negative region although it is neutral overall. (2) Refers to axial differences in a developing embryo that result in anterior–posterior and dorsal–ventral axes in a bilaterally symmetrical animal.

polarize In cladistics, to determine whether character states are ancestral or derived.

pollen tube A tube formed after germination of the pollen grain; carries the male gametes into the ovule.

pollination The transfer of pollen from an anther to a stigma.

polyandry The condition in which a female mates with more than one male.

polyclonal antibody An antibody response in which an antigen elicits many different antibodies, each fitting a different portion of the antigen surface.

polygenic inheritance Describes a mode of inheritance in which more than one gene affects a trait, such as height in human beings; polygenic inheritance may produce a continuous range of phenotypic values, rather than discrete either–or values.

polygyny A mating choice in which a male mates with more than one female.

polymer A molecule composed of many similar or identical molecular subunits; starch is a polymer of glucose.

polymerase chain reaction (PCR) A process by which DNA polymerase is used to copy a sequence of interest repeatedly, making millions of copies of the same DNA.

polymorphism The presence in a population of more than one allele of a gene at a frequency greater than that of newly arising mutations.

polyp A typically sessile, cylindrical body form found in cnidarian animals, such as hydras.

polypeptide A molecule consisting of many joined amino acids; not usually as complex as a protein.

polyphyletic In phylogenetic classification, a group that does not include the most recent common ancestor of all members of the group.

polyploidy Condition in which one or more entire sets of chromosomes is added to the diploid genome.

polysaccharide A carbohydrate composed of many monosaccharide sugar subunits linked together in a long chain; examples are glycogen, starch, and cellulose.

polyunsaturated fat A fat molecule having at least two double bonds between adjacent carbons in one or more of the fatty acid chains.

population Any group of individuals, usually of a single species, occupying a given area at the same time.

population genetics The study of the properties of genes in populations.

positive control A type of control at the level of DNA transcription initiation in which the frequency of initiation is increased; activator proteins mediate positive control.

posttranscriptional control A mechanism of control over gene expression that operates after the transcription of mRNA is complete.

postzygotic isolating mechanism A type of reproductive isolation in which zygotes are produced but are unable to develop into reproducing adults; these mechanisms may range from inviability of zygotes or embryos to adults that are sterile.

potential energy Energy that is not being used, but could be; energy in a potentially usable form; often called "energy of position."

precapillary sphincter A ring of muscle that guards each capillary loop and that, when closed, blocks flow through the capillary.

pre-mRNA splicing In eukaryotes, the process by which introns are removed from the primary transcript to produce mature mRNA; pre-mRNA splicing occurs in the nucleus.

pressure potential In plants, the turgor pressure resulting from pressure against the cell wall.

prezygotic isolating mechanism A type of reproductive isolation in which the formation of a zygote is prevented; these mechanisms may range from physical separation in different habitats to gametic in which gametes are incapable of fusing.

primary endosperm nucleus In flowering plants, the result of the fusion of a sperm nucleus and the (usually) two polar nuclei.

primary growth In vascular plants, growth originating in the apical meristems of shoots and roots; results in an increase in length.

primary immune response The first response of an immune system to a foreign antigen. If the system is challenged again with the same antigen, the memory cells created during the primary response will respond more quickly.

primary induction Inductions between the three primary tissue types: mesoderm and endoderm.

primary meristem Any of the three meristems produced by the apical meristem; primary meristems give rise to the dermal, vascular, and ground tissues.

primary nondisjunction Failure of chromosomes to separate properly at meiosis I.

primary phloem The cells involved in food conduction in plants.

primary plant body The part of a plant consisting of young, soft shoots and roots derived from apical meristem tissues.

primary productivity The amount of energy produced by photosynthetic organisms in a community.

primary structure The specific amino acid sequence of a protein.

primary tissues Tissues that make up the primary plant body.

primary transcript The initial mRNA molecule copied from a gene by RNA polymerase, containing a faithful copy of the entire gene, including introns as well as exons.

primary wall In plants, the wall layer deposited during the period of cell expansion.

primase The enzyme that synthesizes the RNA primers required by DNA polymerases.

primate Monkeys and apes (including humans).

primitive streak In the early embryos of birds, reptiles, and mammals, a dorsal, longitudinal strip of ectoderm and mesoderm that is equivalent to the blastopore in other forms.

primordium In plants, a bulge on the young shoot produced by the apical meristem; primordia can differentiate into leaves, other shoots, or flowers.

principle of parsimony Principle stating that scientists should favor the hypothesis that requires the fewest assumptions.

prions Infectious proteinaceous particles.

procambium In vascular plants, a primary meristematic tissue that gives rise to primary vascular tissues.

product rule *See* rule of multiplication.

proglottid A repeated body segment in tapeworms that contains both male and female reproductive organs; proglottids eventually form eggs and embryos, which leave the host's body in feces.

prokaryote A bacterium; a cell lacking a membrane-bounded nucleus or membrane-bounded organelles.

prometaphase The transitional phase between prophase and metaphase during which the spindle attaches to the kinetochores of sister chromatids.

promoter A DNA sequence that provides a recognition and attachment site for RNA polymerase to begin the process of gene transcription; it is located upstream from the transcription start site.

prophase The phase of cell division that begins when the condensed chromosomes become visible and ends when the nuclear envelope breaks down. The assembly of the spindle takes place during prophase.

proprioceptor In vertebrates, a sensory receptor that senses the body's position and movements.

prosimian Any member of the mammalian group that is a sister group to the anthropoids; prosimian means "before monkeys." Members include the lemurs, lorises, and tarsiers.

prosoma The anterior portion of the body of an arachnid, which bears all the appendages.

prostaglandins A group of modified fatty acids that function as chemical messengers.

prostate gland In male mammals, a mass of glandular tissue at the base of the urethra that secretes an alkaline fluid that has a stimulating effect on the sperm as they are released.

protease An enzyme that degrades proteins by breaking peptide bonds; in cells, proteases are often compartmentalized into vesicles such as lysosomes.

proteasome A large, cylindrical cellular organelle that degrades proteins marked with ubiquitin.

protein A chain of amino acids joined by peptide bonds.

protein kinase An enzyme that adds phosphate groups to proteins, changing their activity.

protein microarray An array of proteins on a microscope slide or silicon chip. The array may be used with a variety of probes, including antibodies, to analyze the presence or absence of specific proteins in a complex mixture.

proteome All the proteins coded for by a particular genome.

proteomics The study of the proteomes of organisms. This is related to functional genomics as the proteome is responsible for much of the function encoded by a genome.

protoderm The primary meristem that gives rise to the dermal tissue.

proton pump A protein channel in a membrane of the cell that expends energy to transport protons against a concentration gradient; involved in the chemiosmotic generation of ATP.

protooncogene A normal cellular gene that can act as an oncogene when mutated.

protostome Any member of a grouping of bilaterally symmetrical animals in which the mouth develops first and the anus second; flatworms, nematodes, mollusks, annelids, and arthropods are protostomes.

pseudocoel A body cavity located between the endoderm and mesoderm.

pseudogene A copy of a gene that is not transcribed.

pseudomurien A component of the cell wall of archaea; it is similar to peptidoglycan in structure and function but contains different components.

pseudopod A nonpermanent cytoplasmic extension of the cell body.

P site In a ribosome, the peptidyl site that binds to the tRNA attached to the growing polypeptide chain.

punctuated equilibrium A hypothesis about the mechanism of evolutionary change proposing that long periods of little or no change are punctuated by periods of rapid evolution.

Punnett square A diagrammatic way of showing the possible genotypes and phenotypes of genetic crosses.

pupa A developmental stage of some insects in which the organism is nonfeeding, immotile, and sometimes encapsulated or in a cocoon; the pupal stage occurs between the larval and adult phases.

purine The larger of the two general kinds of nucleotide base found in DNA and RNA; a nitrogenous base with a double-ring structure, such as adenine or guanine.

pyrimidine The smaller of two general kinds of nucleotide base found in DNA and RNA; a nitrogenous base with a single-ring structure, such as cytosine, thymine, or uracil.

pyruvate A three-carbon molecule that is the end product of glycolysis; each glucose molecule yields two pyruvate molecules.

Q

quantitative trait A trait that is determined by the effects of more than one gene; such a trait usually exhibits continuous variation rather than discrete either–or values.

quaternary structure The structural level of a protein composed of more than one polypeptide chain, each of which has its own tertiary structure; the individual chains are called subunits.

R

radial canal Any of five canals that connect to the ring canal of an echinoderm's water–vascular system.

radial cleavage The embryonic cleavage pattern of deuterostome animals in which cells divide parallel to and at right angles to the polar axis of the embryo.

radial symmetry A type of structural symmetry with a circular plan, such that dividing the body or structure through the midpoint in any direction yields two identical sections.

radicle The part of the plant embryo that develops into the root.

radioactive isotope An isotope that is unstable and undergoes radioactive decay, releasing energy.

radioactivity The emission of nuclear particles and rays by unstable atoms as they decay into more stable forms.

radula Rasping tongue found in most mollusks.

reaction center A transmembrane protein complex in a photosystem that receives energy from the antenna complex exciting an electron that is passed to an acceptor molecule.

reading frame The correct succession of nucleotides in triplet codons that specify amino acids on translation. The reading frame is established by the first codon in the sequence as there are no spaces in the genetic code.

realized niche The actual niche occupied by an organism when all biotic and abiotic interactions are taken into account.

receptor-mediated endocytosis Process by which specific macromolecules are transported into eukaryotic cells at clathrin-coated pits, after binding to specific cell-surface receptors.

receptor protein A highly specific cell-surface receptor embedded in a cell membrane that responds only to a specific messenger molecule.

receptor tyrosine kinase (RTK) A diverse group of membrane receptors that when activated have kinase enzymatic activity. Specifically, they phosphorylate proteins on tyrosine. Their activation can lead to diverse cellular responses.

recessive An allele that is only expressed when present in the homozygous condition, but being "hidden" by the expression of a dominant allele in the heterozygous condition.

redia A secondary, nonciliated larva produced in the sporocysts of liver flukes.

regulatory protein Any of a group of proteins that modulates the ability of RNA polymerase to bind to a promoter and begin DNA transcription.

replicon An origin of DNA replication and the DNA whose replication is controlled by this origin. In prokaryotic replication, the chromosome plus the origin consist of a single replicon; eukaryotic chromosomes consist of multiple replicons.

replisome The macromolecular assembly of enzymes involved in DNA replication; analogous to the ribosome in protein synthesis.

reciprocal altruism Performance of an altruistic act with the expectation that the favor will be returned. A key and very controversial assumption of many theories dealing with the evolution of social behavior. *See* altruism.

reciprocal cross A genetic cross involving a single trait in which the sex of the parents is reversed; for example, if pollen from a white-flowered plant is used to fertilize a purple-flowered plant, the reciprocal cross would be pollen from a purple-flowered plant used to fertilize a white-flowered plant.

reciprocal recombination A mechanism of genetic recombination that occurs only in eukaryotic organisms, in which two chromosomes trade segments; can occur between nonhomologous chromosomes as well as the more usual exchange between homologous chromosomes in meiosis.

recombinant DNA Fragments of DNA from two different species, such as a bacterium and a mammal, spliced together in the laboratory into a single molecule.

recombination frequency The value obtained by dividing the number of recombinant progeny by the total progeny in a genetic cross. This value is converted into a percentage, and each 1% is termed a map unit.

reduction The gain of an electron by an atom, often with an associated proton.

reflex In the nervous system, a motor response subject to little associative modification; a reflex is among the simplest neural pathways, involving only a sensory neuron, sometimes (but not always) an interneuron, and one or more motor neurons.

reflex arc The nerve path in the body that leads from stimulus to reflex action.

refractory period The recovery period after membrane depolarization during which the membrane is unable to respond to additional stimulation.

reinforcement In speciation, the process by which partial reproductive isolation between populations is increased by selection against mating between members of the two populations, eventually resulting in complete reproductive isolation.

replica plating A method of transferring bacterial colonies from one plate to another to make a copy of the original plate; an impression of colonies growing on a Petri plate is made on a velvet surface, which is then used to transfer the colonies to plates containing different media, such that auxotrophs can be identified.

replication fork The Y-shaped end of a growing replication bubble in a DNA molecule undergoing replication.

repolarization Return of the ions in a nerve to their resting potential distribution following depolarization.

repression In general, control of gene expression by preventing transcription. Specifically, in bacteria such as *E. coli* this is mediated by repressor proteins. In anabolic operons, repressors bind DNA in the absence of corepressors to repress an operon.

repressor A protein that regulates DNA transcription by preventing RNA polymerase from attaching to the promoter and transcribing the structural gene. *See* operator.

reproductive isolating mechanism Any barrier that prevents genetic exchange between species.

residual volume The amount of air remaining in the lungs after the maximum amount of air has been exhaled.

resting membrane potential The charge difference (difference in electric potential) that exists across a neuron at rest (about 70 mV).

restriction endonuclease An enzyme that cleaves a DNA duplex molecule at a particular base sequence, usually within or near a palindromic sequence; also called a restriction enzyme.

restriction fragment length polymorphism (RFLP) Restriction enzymes recognize very specific DNA sequences. Alleles of the same gene or surrounding sequences may have base-pair differences, so that DNA near one allele is cut into a different-length fragment than DNA near the other allele. These different fragments separate based on size on electrophoresis gels.

retina The photosensitive layer of the vertebrate eye; contains several layers of neurons and light receptors (rods and cones); receives the image formed by the lens and transmits it to the brain via the optic nerve.

retinoblastoma susceptibility gene (*Rb*) A gene that, when mutated, predisposes individuals to a rare form of cancer of the retina; one of the first tumor-suppressor genes discovered.

retrovirus An RNA virus. When a retrovirus enters a cell, a viral enzyme (reverse transcriptase) transcribes viral RNA into duplex DNA, which the cell's machinery then replicates and transcribes as if it were its own.

reverse genetics An approach by which a researcher uses a cloned gene of unknown function, creates a mutation, and introduces the mutant gene back into the organism to assess the effect of the mutation.

reverse transcriptase A viral enzyme found in retroviruses that is capable of converting their RNA genome into a DNA copy.

Rh blood group A set of cell-surface markers (antigens) on the surface of red blood cells in humans and rhesus monkeys (for which it is named); although there are several alleles, they are grouped into two main types: Rh-positive and Rh-negative.

rhizome In vascular plants, a more or less horizontal underground stem; may be enlarged for storage or may function in vegetative reproduction.

rhynchocoel A true coelomic cavity in ribbonworms that serves as a hydraulic power source for extending the proboscis.

ribonucleic acid (RNA) A class of nucleic acids characterized by the presence of the sugar ribose and the pyrimidine uracil; includes mRNA, tRNA, and rRNA.

ribosomal RNA (rRNA) A class of RNA molecules found, together with characteristic proteins, in ribosomes; transcribed from the DNA of the nucleolus.

ribosome The molecular machine that carries out protein synthesis; the most complicated aggregation of proteins in a cell, also containing three different rRNA molecules.

ribosome-binding sequence (RBS) In prokaryotes, a conserved sequence at the 5′ end of mRNA that is complementary to the 3′ end of a small subunit rRNA and helps to position the ribosome during initiation.

ribozyme An RNA molecule that can behave as an enzyme, sometimes catalyzing its own assembly; rRNA also acts as a ribozyme in the polymerization of amino acids to form protein.

ribulose 1,5-bisphosphate (RuBP) In the Calvin cycle, the five-carbon sugar to which CO_2 is attached, accomplishing carbon fixation. This reaction is catalyzed by the enzyme rubisco.

ribulose bisphosphate carboxylase/oxygenase (rubisco) The four-subunit enzyme in the chloroplast that catalyzes the carbon fixation reaction joining CO_2 to RuBP.

RNA interference A type of gene silencing in which the mRNA transcript is prevented from being translated; small interfering RNAs (siRNAs) have been found to bind to mRNA and target its degradation prior to its translation.

RNA polymerase An enzyme that catalyzes the assembly of an mRNA molecule, the sequence of which is complementary to a DNA molecule used as a template. *See* transcription.

RNA primer In DNA replication, a sequence of about 10 RNA nucleotides complementary to unwound DNA that attaches at a replication fork; the DNA polymerase uses the RNA primer as a starting point for addition of DNA nucleotides to form the new DNA strand; the RNA primer is later removed and replaced by DNA nucleotides.

RNA splicing A nuclear process by which intron sequences of a primary mRNA transcript are cut out and the exon sequences spliced together to give the correct linkages of genetic information that will be used in protein construction.

rod Light-sensitive nerve cell found in the vertebrate retina; sensitive to very dim light; responsible for "night vision."

root The usually descending axis of a plant, normally below ground, which anchors the plant and serves as the major point of entry for water and minerals.

root cap In plants, a tissue structure at the growing tips of roots that protects the root apical meristem as the root pushes through the soil; cells of the root cap are continually lost and replaced.

root hair In plants, a tubular extension from an epidermal cell located just behind the root tip; root hairs greatly increase the surface area for absorption.

root pressure In plants, pressure exerted by water in the roots in response to a solute potential in the absence of transpiration; often occurs at night. Root pressure can result in guttation, excretion of water from cells of leaves as dew.

root system In plants, the portion of the plant body that anchors the plant and absorbs ions and water.

R plasmid A resistance plasmid; a conjugative plasmid that picks up antibiotic resistance genes and can therefore transfer resistance from one bacterium to another.

rule of addition The rule stating that for two independent events, the probability of either event occurring is the sum of the individual probabilities.

rule of multiplication The rule stating that for two independent events, the probability of both events occurring is the product of the individual probabilities.

rumen An "extra stomach" in cows and related mammals wherein digestion of cellulose occurs and from which partially digested material can be ejected back into the mouth.

S

salicylic acid In plants, an organic molecule that is a long-distance signal in systemic acquired resistance.

saltatory conduction A very fast form of nerve impulse conduction in which the impulses leap from node to node over insulated portions.

saprobes Heterotrophic organisms that digest their food externally (e.g., most fungi).

sarcolemma The specialized cell membrane in a muscle cell.

sarcomere Fundamental unit of contraction in skeletal muscle; repeating bands of actin and myosin that appear between two Z lines.

sarcoplasmic reticulum The endoplasmic reticulum of a muscle cell. A sleeve of membrane that wraps around each myofilament.

satellite DNA A nontranscribed region of the chromosome with a distinctive base composition; a short nucleotide sequence repeated tandemly many thousands of times.

saturated fat A fat composed of fatty acids in which all the internal carbon atoms contain the maximum possible number of hydrogen atoms.

Schwann cells The supporting cells associated with projecting axons, along with all the other nerve cells that make up the peripheral nervous system.

sclereid In vascular plants, a sclerenchyma cell with a thick, lignified, secondary wall having many pits; not elongate like a fiber.

sclerenchyma cell Tough, thick-walled cells that strengthen plant tissues.

scolex The attachment organ at the anterior end of a tapeworm.

scrotum The pouch that contains the testes in most mammals.

scuttellum The modified cotyledon in cereal grains.

second filial (F_2) generation The offspring resulting from a cross between members of the first filial (F_1) generation.

secondary cell wall In plants, the innermost layer of the cell wall. Secondary walls have a highly organized microfibrillar structure and are often impregnated with lignin.

secondary growth In vascular plants, an increase in stem and root diameter made possible by cell division of the lateral meristems.

secondary immune response The swifter response of the body the second time it is invaded by the same pathogen because of the presence of memory cells, which quickly become antibody-producing plasma cells.

secondary induction An induction between tissues that have already differentiated.

secondary metabolite A molecule not directly involved in growth, development, or reproduction of an organism; in plants these molecules, which include nicotine, caffeine, tannins, and menthols, can discourage herbivores.

secondary plant body The part of a plant consisting of secondary tissues from lateral meristem tissues; the older trunk, branches, and roots of woody plants.

secondary structure In a protein, hydrogen-bonding interactions between —CO and —NH groups of the primary structure.

secondary tissue Any tissue formed from lateral meristems in trees and shrubs.

Second Law of Thermodynamics A statement concerning the transformation of potential energy into heat; it says that disorder (entropy) is continually increasing in the universe as energy changes occur, so disorder is more likely than order.

second messenger A small molecule or ion that carries the message from a receptor on the target cell surface into the cytoplasm.

seed bank Ungerminated seeds in the soil of an area. Regeneration of plants after events such as fire often depends on the presence of a seed bank.

seed coat In plants, the outer layers of the ovule, which become a relatively impermeable barrier to protect the dormant embryo and stored food.

segment polarity gene Any of certain genes in *Drosophila* development that are expressed in stripes that subdivide the stripes created by the pair-rule genes in the process of segmentation.

segmentation The division of the developing animal body into repeated units; segmentation allows for redundant systems and more efficient locomotion.

segmentation gene Any of the three classes of genes that control development of the segmented body plan of insects; includes the gap genes, pair-rule genes, and segment polarity genes.

segregation The process by which alternative forms of traits are expressed in offspring rather than blending each trait of the parents in the offspring.

selection The process by which some organisms leave more offspring than competing ones, and their genetic traits tend to appear in greater proportions among members of succeeding generations than the traits of those individuals that leave fewer offspring.

selectively permeable Condition in which a membrane is permeable to some substances but not to others.

self-fertilization The union of egg and sperm produced by a single hermaphroditic organism.

semen In reptiles and mammals, sperm-bearing fluid expelled from the penis during male orgasm.

semicircular canal Any of three fluid-filled canals in the inner ear that help to maintain balance.

semiconservative replication DNA replication in which each strand of the original duplex serves as the template for construction of a totally new complementary strand, so the original duplex is partially conserved in each of the two new DNA molecules.

senescent Aged, or in the process of aging.

sensory (afferent) neuron A neuron that transmits nerve impulses from a sensory receptor to the central nervous system or central ganglion.

sensory setae In insect, bristles attached to the nervous system that are sensitive mechanical and chemical stimulation; most abundant on antennae and legs.

sepal A member of the outermost floral whorl of a flowering plant.

septation In prokaryotic cell division, the formation of a septum where new cell membrane and cell wall is formed to separate the two daughter cells.

septum, pl. **septa** A wall between two cavities.

sequence-tagged site (STS) A small stretch of DNA that is unique in a genome, that is, it occurs only once; useful as a physical marker on genomic maps.

seta, pl. **setae** (L., bristle) In an annelid, bristles of chitin that help anchor the worm during locomotion or when it is in its burrow.

severe acute respiratory syndrome (SARS) A respiratory infection with an 8% mortality rate that is caused by a coronavirus.

sex chromosome A chromosome that is related to sex; in humans, the sex chromosomes are the X and Y chromosomes.

sex-linked A trait determined by a gene carried on the X chromosome and absent on the Y chromosome.

sexual reproduction The process of producing offspring through an alternation of fertilization (producing diploid cells) and meiotic reduction in chromosome number (producing haploid cells).

sexual selection A type of differential reproduction that results from variable success in obtaining mates.

shared derived character In cladistics, character states that are shared by species and that are different from the ancestral character state.

shoot In vascular plants, the aboveground portions, such as the stem and leaves.

short interspersed element (SINE) Any of a type of retrotransposon found in humans and other primates that does not contain the biochemical machinery needed for transposition; half a million copies of a SINE element called Alu is nested in the LINEs of the human genome.

shotgun sequencing The method of DNA sequencing in which the DNA is randomly cut into small fragments, and the fragments cloned and sequenced. A computer is then used to assemble a final sequence.

sieve cell In the phloem of vascular plants, a long, slender element with relatively unspecialized sieve areas and with tapering end walls that lack sieve plates.

signal recognition particle (SRP) In eukaryotes, a cytoplasmic complex of proteins that recognizes and binds to the signal sequence of a polypeptide, and then docks with a receptor that forms a channel in the ER membrane. In this way the polypeptide is released into the lumen of the ER.

signal transduction The events that occur within a cell on receipt of a signal, ligand binding to a receptor protein. Signal transduction pathways produce the cellular response to a signaling molecule.

simple sequence repeat (SSR) A one- to three-nucleotide sequence such as CA or CCG that is repeated thousands of times.

single-nucleotide polymorphism (SNP) A site present in at least 1% of the population at which individuals differ by a single nucleotide. These can be used as genetic markers to map unknown genes or traits.

sinus A cavity or space in tissues or in bone.

sister chromatid One of two identical copies of each chromosome, still linked at the centromere, produced as the chromosomes duplicate for mitotic division; similarly, one of two identical copies of each homologous chromosome present in a tetrad at meiosis.

small interfering RNAs (siRNAs) A class of micro-RNAs that appear to be involved in control of gene transcription and that play a role in protecting cells from viral attack.

small nuclear ribonucleoprotein particles (snRNP) In eukaryotes, a complex composed of snRNA and protein that clusters together with other snRNPs to form the spliceosome, which removes introns from the primary transcript.

small nuclear RNA (snRNA) In eukaryotes, a small RNA sequence that, as part of a small nuclear ribonucleoprotein complex, facilitates recognition and excision of introns by base-pairing with the 5′ end of an intron or at a branch site of the same intron.

sodium–potassium pump Transmembrane channels engaged in the active (ATP-driven) transport of Na^+, exchanging them for K^+, where both ions are being moved against their respective concentration gradients; maintains the resting membrane potential of neurons and other cells.

solute A molecule dissolved in some solution; as a general rule, solutes dissolve only in solutions of similar polarity; for example, glucose (polar) dissolves in (forms hydrogen bonds with) water (also polar), but not in vegetable oil (nonpolar).

solute potential The amount of osmotic pressure arising from the presence of a solute or solutes in water; measure by counterbalancing the pressure until osmosis stops.

solvent The medium in which one or more solutes is dissolved.

somatic cell Any of the cells of a multicellular organism except those that are destined to form gametes (germ-line cells).

somatic cell nuclear transfer (SCNT) The transfer of the nucleus of a somatic cell into an enucleated egg cell that then undergoes development. Can be used to make ES cells and to create cloned animals.

somatic mutation A change in genetic information (mutation) occurring in one of the somatic cells of a multicellular organism, not passed from one generation to the next.

somatic nervous system In vertebrates, the neurons of the peripheral nervous system that control skeletal muscle.

somite One of the blocks, or segments, of tissue into which the mesoderm is divided during differentiation of the vertebrate embryo.

Southern blot A technique in which DNA fragments are separated by gel electrophoresis, denatured into single-stranded DNA, and then "blotted" onto a sheet of filter paper; the filter is then incubated with a labeled probe to locate DNA sequences of interest.

S phase The phase of the cell cycle during which DNA replication occurs.

specialized transduction The transfer of only a few specific genes into a bacterium, using a lysogenic bacteriophage as a carrier.

speciation The process by which new species arise, either by transformation of one species into another, or by the splitting of one ancestral species into two descendant species.

species, pl. **species** A kind of organism; species are designated by binomial names written in italics.

specific heat The amount of heat that must be absorbed or lost by 1 g of a substance to raise or lower its temperature 1°C.

specific transcription factor Any of a great number of transcription factors that act in a time- or tissue-dependent manner to increase DNA transcription above the basal level.

spectrin A scaffold of proteins that links plasma membrane proteins to actin filaments in the cytoplasm of red blood cells, producing their characteristic biconcave shape.

spermatid In animals, each of four haploid (n) cells that result from the meiotic divisions of a spermatocyte; each spermatid differentiates into a sperm cell.

spermatozoa The male gamete, usually smaller than the female gamete, and usually motile.

sphincter In vertebrate animals, a ring-shaped muscle capable of closing a tubular opening by constriction (e.g., between stomach and small intestine or between anus and exterior).

spicule Any of a number of minute needles of silica or calcium carbonate made in the mesohyl by some kinds of sponges as a structural component.

spindle The structure composed of microtubules radiating from the poles of the dividing cell that will ultimately guide the sister chromatids to the two poles.

spindle apparatus The assembly that carries out the separation of chromosomes during cell division; composed of microtubules (spindle fibers) and assembled during prophase at the equator of the dividing cell.

spindle checkpoint The third cell-division checkpoint, at which all chromosomes must be attached to the spindle. Passage through this checkpoint commits the cell to anaphase.

spinnerets Organs at the posterior end of a spider's abdomen that secrete a fluid protein that becomes silk.

spiracle External opening of a trachea in arthropods.

spiral cleavage The embryonic cleavage pattern of some protostome animals in which cells divide at an angle oblique to the polar axis of the embryo; a line drawn through the sequence of dividing cells forms a spiral.

spiralian A member of a group of invertebrate animals; many groups exhibit spiral cleavage. Mollusks, annelids, and flatworms are examples of spiralians.

spliceosome In eukaryotes, a complex composed of multiple snRNPs and other associated proteins that is responsible for excision of introns and joining of exons to convert the primary transcript into the mature mRNA.

spongin A tough protein made by many kinds of sponges as a structural component within the mesohyl.

spongy parenchyma A leaf tissue composed of loosely arranged, chloroplast-bearing cells. *See* palisade parenchyma.

sporangium, pl. **sporangia** A structure in which spores are produced.

spore A haploid reproductive cell, usually unicellular, capable of developing into an adult without fusion with another cell.

sporophyte The spore-producing, diploid ($2n$) phase in the life cycle of a plant having alternation of generations.

stabilizing selection A form of selection in which selection acts to eliminate both extremes from a range of phenotypes.

stamen The organ of a flower that produces the pollen; usually consists of anther and filament; collectively, the stamens make up the androecium.

starch An insoluble polymer of glucose; the chief food storage substance of plants.

start codon The AUG triplet, which indicates the site of the beginning of mRNA translation; this codon also codes for the amino acid methionine.

stasis A period of time during which little evolutionary change occurs.

statocyst Sensory receptor sensitive to gravity and motion.

stele The central vascular cylinder of stems and roots.

stem cell A relatively undifferentiated cell in animal tissue that can divide to produce more differentiated tissue cells.

stereoscopic vision Ability to perceive a single, three-dimensional image from the simultaneous but slightly divergent two-dimensional images delivered to the brain by each eye.

stigma (1) In angiosperm flowers, the region of a carpel that serves as a receptive surface for pollen grains. (2) Light-sensitive eyespot of some algae.

stipules Leaflike appendages that occur at the base of some flowering plant leaves or stems.

stolon A stem that grows horizontally along the ground surface and may form adventitious roots, such as runners of the strawberry plant.

stoma, pl. **stomata** In plants, a minute opening bordered by guard cells in the epidermis of leaves and stems; water passes out of a plant mainly through the stomata.

stop codon Any of the three codons UAA, UAG, and UGA, that indicate the point at which mRNA translation is to be terminated.

stratify To hold plant seeds at a cold temperature for a certain period of time; seeds of many plants will not germinate without exposure to cold and subsequent warming.

stratum corneum The outer layer of the epidermis of the skin of the vertebrate body.

striated muscle Skeletal voluntary muscle and cardiac muscle.

stroma In chloroplasts, the semiliquid substance that surrounds the thylakoid system and that contains the enzymes needed to assemble organic molecules from CO_2.

stromatolite A fossilized mat of ancient bacteria formed as long as 2 BYA, in which the bacterial remains individually resemble some modern-day bacteria.

style In flowers, the slender column of tissue that arises from the top of the ovary and through which the pollen tube grows.

stylet A piercing organ, usually a mouthpart, in some species of invertebrates.

suberin In plants, a fatty acid chain that forms the impermeable barrier in the Casparian strip of root endoderm.

subspecies A geographically defined population or group of populations within a single species that has distinctive characteristics.

substrate (1) The foundation to which an organism is attached. (2) A molecule on which an enzyme acts.

subunit vaccine A type of vaccine created by using a subunit of a viral protein coat to elicit an immune response; may be useful in preventing viral diseases such as hepatitis B.

succession In ecology, the slow, orderly progression of changes in community composition that takes place through time.

summation Repetitive activation of the motor neuron resulting in maximum sustained contraction of a muscle.

supercoiling The coiling in space of double-stranded DNA molecules due to torsional strain, such as occurs when the helix is unwound.

surface tension A tautness of the surface of a liquid, caused by the cohesion of the molecules of liquid. Water has an extremely high surface tension.

surface area-to-volume ratio Relationship of the surface area of a structure, such as a cell, to the volume it contains.

suspensor In gymnosperms and angiosperms, the suspensor develops from one of the first two cells of a dividing zygote; the suspensor of an angiosperm is a nutrient conduit from maternal tissue to the embryo. In gymnosperms the suspensor positions the embryo closer to stored food reserves.

swim bladder An organ encountered only in the bony fish that helps the fish regulate its buoyancy by increasing or decreasing the amount of gas in the bladder via the esophagus or a specialized network of capillaries.

swimmerets In lobsters and crayfish, appendages that occur in lines along the ventral surface of the abdomen and are used in swimming and reproduction.

symbiosis The condition in which two or more dissimilar organisms live together in close association; includes parasitism (harmful to one of the organisms), commensalism (beneficial to one, of no significance to the other), and mutualism (advantageous to both).

sympatric speciation The differentiation of populations within a common geographic area into species.

symplast route In plant roots, the pathway for movement of water and minerals within the cell cytoplasm that leads through plasmodesmata that connect cells.

symplesiomorphy In cladistics, another term for a shared ancestral character state.

symporter A carrier protein in a cell's membrane that transports two molecules or ions in the same direction across the membrane.

synapomorphy In systematics, a derived character that is shared by clade members.

synapse A junction between a neuron and another neuron or muscle cell; the two cells do not touch, the gap being bridged by neurotransmitter molecules.

synapsid Any of an early group of reptiles that had a pair of temporal openings in the skull behind the eye sockets; jaw muscles attached to these openings. Early ancestors of mammals belonged to this group.

synapsis The point-by-point alignment (pairing) of homologous chromosomes that occurs before the first meiotic division; crossing over takes place during synapsis.

synaptic cleft The space between two adjacent neurons.

synaptic vesicle A vesicle of a neurotransmitter produced by the axon terminal of a nerve. The filled vesicle migrates to the presynaptic membrane, fuses with it, and releases the neurotransmitter into the synaptic cleft.

synaptonemal complex A protein lattice that forms between two homologous chromosomes in prophase I of meiosis, holding the replicated chromosomes in precise register with each other so that base-pairs can form between nonsister chromatids for crossing over that is usually exact within a gene sequence.

syncytial blastoderm A structure composed of a single large cytoplasm containing about 4000 nuclei in embryonic development of insects such as *Drosophila*.

syngamy The process by which two haploid cells (gametes) fuse to form a diploid zygote; fertilization.

synthetic polyploidy A polyploidy organism created by crossing organisms most closely related to an ancestral species and then manipulating the offspring.

systematics The reconstruction and study of evolutionary relationships.

systemic acquired resistance (SAR) In plants, a longer-term response to a pathogen or pest attack that can last days to weeks and allow the plant to respond quickly to later attacks by a range of pathogens.

systemin In plants, an 18-amino-acid peptide that is produced by damaged or injured leaves that leads to the wound response.

systolic pressure A measurement of how hard the heart is contracting. When measured during a blood pressure reading, ventricular systole (contraction) is what is being monitored.

T

3′ poly-A tail In eukaryotes, a series of 1–200 adenine residues added to the 3′ end of an mRNA; the tail appears to enhance the stability of the mRNA by protecting it from degradation.

T box A transcription factor protein domain that has been conserved, although with differing developmental effects, in invertebrates and chordates.

tagma, pl. tagmata A compound body section of an arthropod resulting from embryonic fusion of two or more segments; for example, head, thorax, abdomen.

Taq polymerase A DNA polymerase isolated from the thermophilic bacterium *Thermus aquaticus* (Taq); this polymerase is functional at higher temperatures, and is used in PCR amplification of DNA.

TATA box In eukaryotes, a sequence located upstream of the transcription start site. The TATA box is one element of eukaryotic core promoters for RNA polymerase II.

taxis, pl. taxes An orientation movement by a (usually) simple organism in response to an environmental stimulus.

taxonomy The science of classifying living things. By agreement among taxonomists, no two organisms can have the same name, and all names are expressed in Latin.

T cell A type of lymphocyte involved in cell-mediated immunity and interactions with B cells; the "T" refers to the fact that T cells are produced in the thymus.

telencephalon The most anterior portion of the brain, including the cerebrum and associated structures.

telomerase An enzyme that synthesizes telomeres on eukaryotic chromosomes using an internal RNA template.

telomere A specialized nontranscribed structure that caps each end of a chromosome.

telophase The phase of cell division during which the spindle breaks down, the nuclear envelope of each daughter cell forms, and the chromosomes uncoil and become diffuse.

telson The tail spine of lobsters and crayfish.

temperate (lysogenic) phage A virus that is capable of incorporating its DNA into the host cell's DNA, where it remains for an indeterminate length of time and is replicated as the cell's DNA replicates.

template strand The DNA strand that is used as a template in transcription. This strand is copied to produce a complementary mRNA transcript.

tendon (Gr. *tendon*, stretch) A strap of cartilage that attaches muscle to bone.

tensile strength A measure of the cohesiveness of a substance; its resistance to being broken apart. Water in narrow plant vessels has tensile strength that helps keep the water column continuous.

tertiary structure The folded shape of a protein, produced by hydrophobic interactions with water, ionic and covalent bonding between side chains of different amino acids, and van der Waal's forces; may be changed by denaturation so that the protein becomes inactive.

testcross A mating between a phenotypically dominant individual of unknown genotype and a homozygous "tester," done to determine whether the phenotypically dominant individual is homozygous or heterozygous for the relevant gene.

testis, pl. testes In mammals, the sperm-producing organ.

tetanus Sustained forceful muscle contraction with no relaxation.

thalamus That part of the vertebrate forebrain just posterior to the cerebrum; governs the flow of information from all other parts of the nervous system to the cerebrum.

therapeutic cloning The use of somatic cell nuclear transfer to create stem cells from a single individual that may be reimplanted in that individual to replace damaged cells, such as in a skin graft.

thermodynamics The study of transformations of energy, using heat as the most convenient form of measurement of energy.

thigmotropism In plants, unequal growth in some structure that comes about as a result of physical contact with an object.

threshold The minimum amount of stimulus required for a nerve to fire (depolarize).

thylakoid In chloroplasts, a complex, organized internal membrane composed of flattened disks, which contain the photosystems involved in the light-dependent reactions of photosynthesis.

Ti (tumor-inducing) plasmid A plasmid found in the plant bacterium *Agrobacterium tumefaciens* that has been extensively used to introduce recombinant DNA into broadleaf plants. Recent modifications have allowed its use with cereal grains as well.

tight junction Region of actual fusion of plasma membranes between two adjacent animal cells that prevents materials from leaking through the tissue.

tissue A group of similar cells organized into a structural and functional unit.

tissue plasminogen activator (TPA) A human protein that causes blood clots to dissolve; if used within 3 hours of an ischemic stroke, TPA may prevent disability.

tissue-specific stem cell A stem cell that is capable of developing into the cells of a certain tissue, such as muscle or epithelium; these cells persist even in adults.

tissue system In plants, any of the three types of tissue; called a system because the tissue extends throughout the roots and shoots.

tissue tropism The affinity of a virus for certain cells within a multicellular host; for example, hepatitis B virus targets liver cells.

tonoplast The membrane surrounding the central vacuole in plant cells that contains water channels; helps maintain the cell's osmotic balance.

topoisomerase Any of a class of enzymes that can change the topological state of DNA to relieve torsion caused by unwinding.

torsion The process in embryonic development of gastropods by which the mantle cavity and anus move from a posterior location to the front of the body, closer to the location of the mouth.

totipotent A cell that possesses the full genetic potential of the organism.

trachea, pl. **tracheae** A tube for breathing; in terrestrial vertebrates, the windpipe that carries air between the larynx and bronchi (which leads to the lungs); in insects and some other terrestrial arthropods, a system of chitin-lined air ducts.

tracheids In plant xylem, dead cells that taper at the ends and overlap one another.

tracheole The smallest branches of the respiratory system of terrestrial arthropods; tracheoles convey air from the tracheae, which connect to the outside of the body at spiracles.

trait In genetics, a characteristic that has alternative forms, such as purple or white flower color in pea plants or different blood type in humans.

transcription The enzyme-catalyzed assembly of an RNA molecule complementary to a strand of DNA.

transcription complex The complex of RNA polymerase II plus necessary activators, coactivators, transcription factors, and other factors that are engaged in actively transcribing DNA.

transcription factor One of a set of proteins required for RNA polymerase to bind to a eukaryotic promoter region, become stabilized, and begin the transcription process.

transcription bubble The region containing the RNA polymerase, the DNA template, and the RNA transcript, so called because of the locally unwound "bubble" of DNA.

transcription unit The region of DNA between a promoter and a terminator.

transcriptome All the RNA present in a cell or tissue at a given time.

transfection The transformation of eukaryotic cells in culture.

transfer RNA (tRNA) A class of small RNAs (about 80 nucleotides) with two functional sites; at one site, an "activating enzyme" adds a specific amino acid, while the other site carries the nucleotide triplet (anticodon) specific for that amino acid.

transformation The uptake of DNA directly from the environment; a natural process in some bacterial species.

transgenic organism An organism into which a gene has been introduced without conventional breeding, that is, through genetic engineering techniques.

translation The assembly of a protein on the ribosomes, using mRNA to specify the order of amino acids.

translation repressor protein One of a number of proteins that prevent translation of mRNA by binding to the beginning of the transcript and preventing its attachment to a ribosome.

translocation (1) In plants, the long-distance transport of soluble food molecules (mostly sucrose), which occurs primarily in the sieve tubes of phloem tissue. (2) In genetics, the interchange of chromosome segments between nonhomologous chromosomes.

transmembrane domain Hydrophobic region of a transmembrane protein that anchors it in the membrane. Often composed of α-helices, but sometimes utilizing β-pleated sheets to form a barrel-shaped pore.

transmembrane route In plant roots, the pathway for movement of water and minerals that crosses the cell membrane and also the membrane of vacuoles inside the cell.

transpiration The loss of water vapor by plant parts; most transpiration occurs through the stomata.

transposable elements Segments of DNA that are able to move from one location on a chromosome to another. Also termed *transposons* or *mobile genetic elements.*

transposition Type of genetic recombination in which transposable elements (transposons) move from one site in the DNA sequence to another, apparently randomly.

transposon DNA sequence capable of transposition.

trichome In plants, a hairlike outgrowth from an epidermal cell; glandular trichomes secrete oils or other substances that deter insects.

triglyceride (triacylglycerol) An individual fat molecule, composed of a glycerol and three fatty acids.

triploid Possessing three sets of chromosomes.

trisomic Describes the condition in which an additional chromosome has been gained due to nondisjunction during meiosis, and the diploid embryo therefore has three of these autosomes. In humans, trisomic individuals may survive if the autosome is small; Down syndrome individuals are trisomic for chromosome 21.

trochophore A specialized type of free-living larva found in lophotrochozoans.

trophic level A step in the movement of energy through an ecosystem.

trophoblast In vertebrate embryos, the outer ectodermal layer of the blastodermic vesicle; in mammals, it is part of the chorion and attaches to the uterine wall.

tropism Response to an external stimulus.

tropomyosin Low-molecular-weight protein surrounding the actin filaments of striated muscle.

troponin Complex of globular proteins positioned at intervals along the actin filament of skeletal muscle; thought to serve as a calcium-dependent "switch" in muscle contraction.

***trp* operon** In *E. coli*, the operon containing genes that code for enzymes that synthesize tryptophan.

true-breeding Said of a breed or variety of organism in which offspring are uniform and consistent from one generation to the next; for example. This is due to the genotypes that determine relevant traits being homozygous.

tube foot In echinoderms, a flexible, external extension of the water–vascular system that is capable of attaching to a surface through suction.

tubulin Globular protein subunit forming the hollow cylinder of microtubules.

tumor-suppressor gene A gene that normally functions to inhibit cell division; mutated forms can lead to the unrestrained cell division of cancer, but only when both copies of the gene are mutant.

turgor pressure The internal pressure inside a plant cell, resulting from osmotic intake of water, that presses its cell membrane tightly against the cell wall, making the cell rigid. Also known as *hydrostatic pressure.*

tympanum In some groups of insects, a thin membrane associated with the tracheal air sacs that functions as a sound receptor; paired on each side of the abdomen.

U

ubiquitin A 76-amino-acid protein that virtually all eukaryotic cells attach as a marker to proteins that are to be degraded.

unequal crossing over A process by which a crossover in a small region of misalignment at synapsis causes two homologous chromosomes to exchange segments of unequal length.

uniporter A carrier protein in a cell's membrane that transports only a single type of molecule or ion.

uniramous Single-branched; describes the appendages of insects.

unsaturated fat A fat molecule in which one or more of the fatty acids contain fewer than the maximum number of hydrogens attached to their carbons.

urea An organic molecule formed in the vertebrate liver; the principal form of disposal of nitrogenous wastes by mammals.

urethra The tube carrying urine from the bladder to the exterior of mammals.

uric acid Insoluble nitrogenous waste products produced largely by reptiles, birds, and insects.

urine The liquid waste filtered from the blood by the kidney and stored in the bladder pending elimination through the urethra.
uropod One of a group of flattened appendages at the end of the abdomen of lobsters and crayfish that collectively act as a tail for a rapid burst of speed.
uterus In mammals, a chamber in which the developing embryo is contained and nurtured during pregnancy.

V

vacuole A membrane-bounded sac in the cytoplasm of some cells, used for storage or digestion purposes in different kinds of cells; plant cells often contain a large central vacuole that stores water, proteins, and waste materials.
valence electron An electron in the outermost energy level of an atom.
variable A factor that influences a process, outcome, or observation. In experiments, scientists attempt to isolate variables to test hypotheses.
vascular cambium In vascular plants, a cylindrical sheath of meristematic cells, the division of which produces secondary phloem outwardly and secondary xylem inwardly; the activity of the vascular cambium increases stem or root diameter.
vascular tissue Containing or concerning vessels that conduct fluid.
vas deferens In mammals, the tube carrying sperm from the testes to the urethra.
vasopressin A posterior pituitary hormone that regulates the kidney's retention of water.
vector In molecular biology, a plasmid, phage or artificial chromosome that allows propagation of recombinant DNA in a host cell into which it is introduced.
vegetal pole The hemisphere of the zygote comprising cells rich in yolk.
vein (1) In plants, a vascular bundle forming a part of the framework of the conducting and supporting tissue of a stem or leaf. (2) In animals, a blood vessel carrying blood from the tissues to the heart.
veliger The second larval stage of mollusks following the trochophore stage, during which the beginning of a foot, shell, and mantle can be seen.
ventricle A muscular chamber of the heart that receives blood from an atrium and pumps blood out to either the lungs or the body tissues.
vertebrate A chordate with a spinal column; in vertebrates, the notochord develops into the vertebral column composed of a series of vertebrae that enclose and protect the dorsal nerve cord.
vertical gene transfer (VGT) The passing of genes from one generation to the next within a species.
vesicle A small intracellular, membrane-bounded sac in which various substances are transported or stored.
vessel element In vascular plants, a typically elongated cell, dead at maturity, which conducts water and solutes in the xylem.
vestibular apparatus The complicated sensory apparatus of the inner ear that provides for balance and orientation of the head in vertebrates.
vestigial structure A morphological feature that has no apparent current function and is thought to be an evolutionary relic; for example, the vestigial hip bones of boa constrictors.
villus, pl. villi In vertebrates, one of the minute, fingerlike projections lining the small intestine that serve to increase the absorptive surface area of the intestine.
virion A single virus particle.
viroid Any of a group of small, naked RNA molecules that are capable of causing plant diseases, presumably by disrupting chromosome integrity.
virus Any of a group of complex biochemical entities consisting of genetic material wrapped in protein; viruses can reproduce only within living host cells and are thus not considered organisms.
visceral mass Internal organs in the body cavity of an animal.
vitamin An organic substance that cannot be synthesized by a particular organism but is required in small amounts for normal metabolic function.
viviparity Refers to reproduction in which eggs develop within the mother's body and young are born free-living.
voltage-gated ion channel A transmembrane pathway for an ion that is opened or closed by a change in the voltage, or charge difference, across the plasma membrane.

W

water potential The potential energy of water molecules. Regardless of the reason (e.g., gravity, pressure, concentration of solute particles) for the water potential, water moves from a region where water potential is greater to a region where water potential is lower.
water–vascular system A fluid-filled hydraulic system found only in echinoderms that provides body support and a unique type of locomotion via extensions called tube feet.
Western blot A blotting technique used to identify specific protein sequences in a complex mixture. *See* Southern blot.
wild type In genetics, the phenotype or genotype that is characteristic of the majority of individuals of a species in a natural environment.
wobble pairing Refers to flexibility in the pairing between the base at the 5′ end of a tRNA anticodon and the base at the 3′ end of an mRNA codon. This flexibility allows a single tRNA to read more than one mRNA codon.
wound response In plants, a signaling pathway initiated by leaf damage, such as being chewed by a herbivore, and lead to the production of proteinase inhibitors that give herbivores indigestion.

X

X chromosome One of two sex chromosomes; in mammals and in *Drosophila*, female individuals have two X chromosomes.
xylem In vascular plants, a specialized tissue, composed primarily of elongate, thick-walled conducting cells, which transports water and solutes through the plant body.

Y

Y chromosome One of two sex chromosomes; in mammals and in *Drosophila*, male individuals have a Y chromosome and an X chromosome; the Y determines maleness.
yolk plug A plug occurring in the blastopore of amphibians during formation of the archenteron in embryological development.
yolk sac The membrane that surrounds the yolk of an egg and connects the yolk, a rich food supply, to the embryo via blood vessels.

Z

zinc finger motif A type of DNA-binding motif in regulatory proteins that incorporates zinc atoms in its structure.
zona pellucida An outer membrane that encases a mammalian egg.
zone of cell division In plants, the part of the young root that includes the root apical meristem and the cells just posterior to it; cells in this zone divide every 12–36 hr.
zone of elongation In plants, the part of the young root that lies just posterior to the zone of cell division; cells in this zone elongate, causing the root to lengthen.
zone of maturation In plants, the part of the root that lies posterior to the zone of elongation; cells in this zone differentiate into specific cell types.
zoospore A motile spore.
zooxanthellae Symbiotic photosynthetic protists in the tissues of corals.
zygomycetes A type of fungus whose chief characteristic is the production of sexual structures called zygosporangia, which result from the fusion of two of its simple reproductive organs.
zygote The diploid ($2n$) cell resulting from the fusion of male and female gametes (fertilization).

credits

Photographs

Chapter 1

Opener: © Soames Summerhays/Natural Visions; 1.1 (organelle): © S. Gschmeeissner/SPL/PUBLIPHOTO; (cell): © Lennart Nilsson/Albert Bonniers Förlag AB; (tissue): © Ed Reschke; (organism): © Russell Illig/Getty Images; (population): © Jeremy Woodhouse/Getty Images; (species (both)): © PhotoDisc/Volume 44/Getty Images; (community): © Steve Harper/Grant Heilman Photography; (ecosystem): © Robert & Jean Pollock; (biosphere): NASA; 1.5 (bottom left): © Huntington Library/Superstock; 1.11 (center right): © Dennis Kunkel/Phototake; (bottom right): © Karl E. Deckart/Phototake; 1.13 (1st row-left): © Alan L. Detrick/Photo Researchers Inc; (center): © David M. Dennis/Animals Animals-Earth Scenes; (right): Corbis/Volume 46; (2nd row-left): © Royalty-Free/CORBIS; (center): © Mediscan/CORBIS; (right): © PhotoDisc BS/Volume 15/Getty Images; (3rd row-left): © Royalty-Free/Corbis; (center): © Tom Brakefield/CORBIS; (right): © PhotoDisc/Volume 44/Getty Images; (4th row-left): Corbis/Volume 64; (center): © T.E. Adams/Visuals Unlimited; (right): © Douglas P. Wilson/Frank Lane Picture Agency/CORBIS; (5th row-left): © R. Robinson/Visuals Unlimited; (right): © Kari Lounatman/Photo Researchers Inc; (6th row-left): © Dwight R. Kuhn; (right): © Alfred Pasieka/Science Photo Library/Photo Researchers Inc.

Chapter 2

Opener: © IBM; 2.1: © Veeco Digital Instruments; 2.9a: © PhotoLink/Getty Images; 2.9b: © Glen Allison/Getty Images; 2.9c: © Jeff Vanuga/CORBIS; 2.12 (bottom left): © Hermann Eisenbeiss/National Audubon Society Collection/Photo Researchers Inc.

Chapter 3

Opener: © Jacob Halaska/IndexStock; 3.9 (center right): © Asa Thoresen/Photo Researchers Inc; 3.9 (bottom right): © J. Carson/Custom Medical Stock Photo; 3.10 (center): © J.D. Litvay/Visuals Unlimited; 3.11 (bottom right): © Scott Johnson/Animals Animals/Earth Scenes; 3.12a (bottom left): © Driscoll: Youngquist & Baldeschwieler, Caltech/SPL/Photo Researchers Inc; 3.12b (bottom center): © M. Freeman/PhotoLink/Getty Images; 3.18 (top left): © Dr. Tim Evans/Photo Researchers Inc; 3.18 (top right): © PhotoDisc/Volume 6/Getty Images; 3.18 (left 2nd from top): © C.W. SCHWARTZ; 3.18 (right 2nd from top): © Royalty-Free/CORBIS; 3.18 (left bottom): © Chad Baker/Getty Images; 3.18 (right 2nd from bottom): © Lon C. Diehl/Photo Edit; 3.18 (right bottom): © DPA/NVM/The Image Works.

Chapter 4

Opener: © Dr. Gopal Murti/Photo Researchers Inc; 4.1 (1st from top): © David M. Phillips/Visuals Unlimited; (2nd from top): © Mike Abbey/Visuals Unlimited; (3rd from top): © David M. Phillips/Visuals Unlimited; (4th from top): © Mike Abbey/Visuals Unlimited; (5th from top): Dr. Torsten Wittmann/Photo Researchers Inc; (6th from top): © Med. Mic. Sciences: Cardiff Uni./Wellcome Photo Library; (7th from top): © Microworks/Phototake; (8th from top): © Stanley Flegler/Visuals Unlimited; 4.2 (top left): © Dr. Don W. Fawcett/Visuals Unlimited; 4.3 (bottom right): © Phototake; 4.4: Courtesy of E.H. Newcomb & T.D. Pugh: University of Wisconsin; 4.5 (top left): © Eye of Science/Photo Researchers Inc; 4.8 (center left): © Dr. Richard Kessel & Dr. Gene Shih/Visuals Unlimited; 4.8 (center right): © John T. Hansen, Ph.D/Phototake; 4.8d (bottom right): Courtesy of Ueli Aebi, M.E. Mueller Institute for Structural Biology, Biozentrum, University of Basel, Switzerland.Picture adapted from, Aebi, U., Cohn, J., Buhle, E.L. and Gerace, L. (1986). The Nuclear Lamina is a Meshwork of Intermediate-type Filaments. Nature (Lond.) 323: 560–564; 4.9 (bottom left): © Ed Reschke; 4.11 (bottom right): © R. Bolender & D. Fawcett/Visuals Unlimited; 4.12 (center left): © Dennis Kunkel/Phototake; 4.15 (center left): Courtesy of E.H. Newcomb & S.E. Frederick University of Wisconsin. Reprinted with permission from Science, Vol 163, 1353–1355 © 1969, American Association for the Advancement of Science; 4.16 (bottom right): © Dr. Henry Aldrich/Visuals Unlimited; 4.17 (bottom right): © Dr. Donald Fawcett & Dr. Porter/Visuals Unlimited; 4.18 (top right): © Dr. Jeremy Burgess/Photo Researchers Inc; 4.23 (both): © William Dentler, University of Kansas; 4.24 (both): © SPL/Photo Researchers Inc; 4.25 (bottom right): © BioPhoto Associates/Photo Researchers Inc.

Chapter 5

Opener: © Dr. Gopal Murti/SPL/Photo Researchers Inc; 5.1 (pg 88): © Don W. Fawcett/Photo Researchers Inc; 5.3 (top right): © Dr. Don W. Fawcett/Visuals Unlimited; 5.13 (all): © Dr. David M. Phillips/Visuals Unlimited; 5.14: © Wim van Egmond/Visuals Unlimited; 5.17a: Micrograph Courtesy of the CDC/Dr. Edwin P. Ewing, Jr; 5.17b: BCC Microimaging, Inc. Reproduced with permission; 5.17d: © The Company of Biologists Limited; 5.18b: Dr. Brigit Satir.

Chapter 6

Opener: © Robert A. Caputo/Aurora & Quanta Productions Inc; 6.3 (both): © Spencer Grant/Photo Edit; 6.10 (bottom right): Professor Emeritus Lester J. Reed, University of Texas at Austin.

Chapter 7

Opener: © Creatas/PunchStock; 7.9 (center right): Royalty-Free/CORBIS; 7.18 (bottom left): © Wolfgang Baumeister/Photo Researchers Inc; 7.18 (bottom right): National Park Service.

Chapter 8

Opener: © Royalty-Free/Corbis; 8.1 (center right): Courtesy Dr. Kenneth Miller, Brown University; 8.7 (both): © Eric Soder/Tom Stack & Associates; 8.18 (bottom right): © Dr. Jeremy Burgess/Photo Researchers Inc; 8.20 (top center): © John Shaw/Photo Researchers Inc; 8.20 (bottom center): © Joseph Nettis/National Audubon Society Collection/Photo Researchers Inc; 8.22 (top right): © Eric Soder/Tom Stack & Associates.

Chapter 9

Opener: © RMF/Scientifica/Visuals Unlimited; 9.17 (top left): Courtesy of Daniel Goodenough; 9.17 (center left): © Dr. Donald Fawcett/Visuals Unlimited; 9.17 (bottom left): © Dr. Donald Fawcett/D. Albertini/Visuals Unlimited.

Chapter 10

Opener: © Stem Jems/Photo Researchers Inc; 10.2 (both): Courtesy of William Margolin; 10.4 (bottom right): Biophoto Associates/Photo Researchers Inc; 10.6 (bottom left): CNRI/Photo Researchers Inc; 10.10 (top right): Image courtesy of S. Hauf and J-M. Peters, IMP, Vienna, Austria; 10.11 (all), 10.12; © Andrew S. Bauer, University of Oregon; 10.13 (both): © Dr. Jeremy Pickett-Heaps; 10.14 (bottom left): © Dr. David M. Phillips/Visuals Unlimited; 10.14 (bottom center): Guenter Albrecht-Buehler, Northwestern University, Chicago; 10.15 (top right): © B.A. Palevits & E.H. Newcomb/BPS/Tom Stack & Associates.

Chapter 11

Opener: © Science VU/L. Maziarski; 11.4b: Reprinted, with permission, from the Annual Review of Genetics, Volume 6 © 1972 by Annual Reviews, www.annualreviews.org; 11.8 (all): © Clare A. Hasenkampf/Biological Photo Service.

Chapter 12

Opener: © Corbis; 12.1: © Norbert Schaefer/Corbis; 12.2: © David Sieren/Visuals Unlimited; 12.3: © Leslie Holzer/Photo Researchers Inc; 12.5 (both) © Wally Eberhart/Visuals Unlimited; 12.12: From Albert & Blakeslee Corn and Man Journal of Heredity, Vol. 5, pg 511, 1914, Oxford University Press; 12.15: © DK Limited/Corbis.

Chapter 13

Opener: © Adrian T. Sumner/Photo Researchers Inc; 13.1 (both): © Cabisco/Phototake; 13.2: © Biophoto Associates/Photo Researchers Inc; 13.3: © Bettmann/Corbis; p. 241 (left): PNAS 2004, Chadwick and Willard; 13.4 (top right): © Kenneth Mason; 13.11: © Jackie Lewin, Royal Free Hospital/Photo Researchers Inc; 13.13: © R. Hutchings/Photo Researchers Inc.

Chapter 14

Opener: © Vol. 29/PhotoDisc/Getty Images; 14.6 (both): From "The Double Helix," by J.D. Watson,

Atheneum Press, NY, 1968; 14.7: © Barrington Brown/Photo Researchers Inc; 14.12 (bottom right): From M. Meselson and F.W. Stahl/ Proceedings of the Nat. Acad. of Sci. 44 (1958): 671; 14.17 (both): From Biochemistry 4e by Stryer © 1995 by Lupert Stryer. Used with permission of W.H. Freeman and Company; 14.21: © Dr. Don W. Fawcett/Visuals Unlimited; 14.22: Courtesy of Dr. David Wolstenholme.

Chapter 15

Opener: © Dr. Gopal Murti/Visuals Unlimited; 15.3: © University of Missouri, Extension and Agriculture Information; 15.4: From R.C. Williams, Proc. Nat. Acad. of Sci. 74 (1977): 2313; 15.8 (top right): © Dr. Oscar Miller; 15.11b: Courtesy of Dr. Bert O'Malley, Baylor College of Medicine; 15.13 (space-filled model): Created by John Beaver using ProteinWorkshop, a product of the RCSB PDB, and built using the Molecular Biology Toolkit developed by John Moreland and Apostol Gramada (mbt.sdsc.edu). The MBT is financed by grant GM63208; 15.16: © P. Nissen, N. Ban, P.B. Moore, and T.A. Steitz (from Science 289: 920-930 [2000].

Chapter 16

Opener: © Dr. Claus Pelling; 16.10 (both): Courtesy of Dr. Harrison Echols; 16.20: Reprinted with permission from the Annual Review of Biochemistry, Volume 68 © 1999 by Annual Reviews www.annualreviews.org.

Chapter 17

Opener: © Prof. Stanley Cohen/Photo Researchers Inc; 17.2d: Courtesy of Biorad Laboratories; 17.7 (bottom right): © SSPL/The Image Works; 17.9: Courtesy of Lifecodes Corp, Stamford CT; 17.10: © Matt Meadows/Peter Arnold Inc; 17.15: R.L. Brinster, U. of Pennsylvania Sch. of Vet. Med.; 17.18: © Rob Horsch, Monsanto Company.

Chapter 18

Opener: © William C. Ray, Ohio State University; 18.2a: © B. Trask and Colleagues, Fred Hutchinson Cancer Research Center: *Nature* 409, Feb. 15, 2001, pg 953; 18.2b: © Dr. Cynthia Morton: Nature 409, Feb. 15, 2001, pg 953; 18.4: Courtesy of Celera Genomics; 18.11a,b: Photographs provided by Indra K. Vasil; 18.11c-d: With permission from Altpeter et al, Plant Cell Reports 16: 12-17, 1996, photos provided by Indra Vasil; 18.12: Courtesy of Research Collaboratory for Structural Bioinformatics; 18.13: © Royalty Free/Corbis; 18.14: © Grant Heilman/ Grant Heilman Photography.

Chapter 19

Opener: © Dwight Kuhn; 19.1 (all): © Carolina Biological Supply Company/Phototake; 19.4b: © University of Wisconsin-Madison News & Public Affairs; 19.6b: © J. Richard Whittaker, used by permission; 19.9 (bottom right): AP/WideWorld Photos; 19.13 (top left): © Steve Paddock and Sean Carroll; 19.13 (top right, center left, center right): © Jim Langeland, Steve Paddock and Sean Carroll; 19.16a: © Dr. Daniel St. Johnston/Wellcome Photo Library; 19.16b: © Schupbach, T. and van Buskirk, C.; 19.16c: from Rotha et al., 1989, courtesy of Siegfried Roth; 19.17: Courtesy of E.B. Lewis; 19.20 (both): from Boucaut et al., 1984, courtesy of J-C Boucaut; 19.22: © Christian Laforsch/Photo Researchers Inc.

Chapter 20

Opener: © Cathy & Gordon ILLG; 20.2: © Royalty-Free/Corbis.

Chapter 21

Opener: © PhotoDisc/Getty Images; 21.3 (both): © Breck P. Kent/Animals Animals/Earth Scenes; 21.8 (both): © Courtesy of Lyudmilla N. Trut, Institute of Cytology & Genetics, Siberian Dept. of the Russian Academy of Sciences; 21.11: © Kevin Schafer/Peter Arnold Inc.

Chapter 22

Opener: © Chris Johns/National Geographic/Getty Images; 22.2: © Porterfield/Chickering/Photo Researchers Inc; 22.3: © Barbara Gerlach/Visuals Unlimited; 22.5 (1): © John Shaw/Tom Stack & Associates; (2): © Rob & Ann Simpson/Visuals Unlimited; (3): © Suzanne L. Collins & Joseph T. Collins/National Audubon Society Collection/Photo Researchers; (4): © Phil A. Dotson/National Audubon Society Collection/Photo Researchers; 22.7 (left to right): © Jonathan Losos; © Chas. McRae/ Visuals Unlimited; © Jonathan Losos; 22.13a: © Jeffrey Taylor; 22.13b: © Kenneth Y. Kaneshiro, Center for Conservation Research & Training, University of Hawaii; 22.16 (left to right): © Photo New Zealand/Nick Groves; © Jim Harding/ firstlight; © Colin Harris/Light Touch Images/ Alamy; © Focus New Zealand Photo Library; 22.16: © Focus New Zealand Photo Library.

Chapter 23

Opener: © G. Mermet/Peter Arnold Inc; 23.1a: by permission of the Syndics of Cambridge University Library; 23.8a: image #5789, photo by D. Finnin/ American Museum of Natural History; 23.8b: © Roger De La Harpe/Animals Animals; 23.10 (bottom left): © Lee W. Wilcox; (bottom right): © Dr. Richard Kessel & Dr. Gene Shih/Visuals Unlimited.

Chapter 25

Opener: © Michael & Patricia Fogden/Minden Pictures; 25.4a: © Michael Persson; 25.4b: © E.R. Degginger/Photo Researchers Inc; 25.5: Dr. Anna Di Gregorio, Weill Cornell Medical College; 25.10 (bottom left): © Chuck Pefley/Getty Images; (bottom center left): © Darwin Dale/Photo Researchers Inc; (bottom center right): © Aldo Brando Peter Arnold Inc; (bottom right): © Tom E. Adams/Peter Arnold Inc; 25.11 (all): Courtesy of Walter Gehring, reprinted with permission from Induction of Ectopic Eyes by Targeted Expression of the Eyeless Gene in Drosophila, G. Halder, P. Callaerts, Walter J. Gehring, Science Vol. 267 © 24 March 1995 American Association for the Advancement of Science; 25.12 (both): © Dr. William Jeffrey.

Chapter 26

Opener: © Jeff Hunter/The Image Bank/Getty Images; 26.1: © T.E. Adams/Visuals Unlimited; pg 504 (bottom right): © NASA/Photo Researchers Inc; 26.2: © NASA/JPL-Caltech; 26.5 (top left): © Tom Walker/Riser/Getty Images; (top right): © Corbis/Volume 8; (bottom left): © Corbis/Volume 102; (bottom right): © PhotoDisc Volume 1/Getty Images; 26.14: © Sean W. Graham, UBC Botanical Garden & Centre for Plant Research, University of British Columbia.

Chapter 27

Opener: © Dr. Gopal Murti/Visuals Unlimited; 27.2: From J. H. Hogle et al, "Three Dimensional Structure of Poliovirus @ 2.9 A Resolution," Science 229: 1360, Sept. 27, 1985. © 1985 by the AAAS; 27.3a: Dept. of Biology, Biozentrum/SPL/Photo Researchers; 27.8: © Corbis/Volume 40.

Chapter 28

Opener: © Dr. David Phillips/Visuals Unlimited; 28.6, pg 544 (left to right): © SPL/Photo Researchers Inc; © Dr. R. Rachel and Prof. Dr. K. O. Stetten, University of Regensburg, Lehrstuhl fuer Mikrobiologie, Regensburg, Germany; © Andrew Syred/SPL/Photo Researchers, Inc.; © Microfield Scientific Ltd/SPL/Photo Researchers Inc; © Alfred Paseika/SPL/Photo Researchers, Inc.; pg 545 (left to right): © Science VU/S. Watson/Visuals Unlimited; © Dennis Kunkel Microscopy Inc; Prof. Dr. Hans Reichenbach, Helmholtz Centre for Infection Research, Braunschweig; 28.3 (center left to right): Dr. Gary Gaugler/Science Photo Library/Photo Researchers, Inc.; © CNRI/Photo Researchers, Inc.; © Dr. Richard Kessel & Dr. Gene Shih/Visuals Unlimited; 28.7b: © Jack Bostrack/Visuals Unlimited; 28.9b: © Julius Adler; 28.10a: © Science VU/S. W. Watson/Visuals Unlimited; 28.10b: © Norma J. Lang/Biological Photo Service; 28.11a: © Dr. Dennis Kunkel/Visuals Unlimited; 28.16: © CNRI/Photo Researchers, Inc.; 28.18: © Science/Visuals Unlimited.

Chapter 29

Opener: © Wim van Egmond/Visuals Unlimited; 29.2: © Andrew H. Knoll, Harvard University; 29.6, 29.7: © Science VU/E. White/Visuals Unlimited; 29.8a: © Andrew Syred/Photo Researchers; 29.9a: © Manfred Kage/Peter Arnold Inc; 29.9b: © Edward S. Ross; 29.10: Michael Delannoy (John Hopkins University School of Medicine Microscope Facility) and Vern B. Carruthers (University of Michigan School of Medicine); 29.12: © Dr. David M. Phillips/Visuals Unlimited; 29.14: © Michael Abbey/Visuals Unlimited; 29.15: © Prof. David J.P. Ferguson, Oxford University; 29.17: © Brian Parker/Tom Stack & Associates; 29.18: © Michele Bahr and D. J. Patterson, used under license to MBL; 29.19: © Randy Morse/Earth Scenes; 29.21: © Dennis Kunkel/Phototake; 29.22: © Andrew Syred/Photo Researchers, Inc.; 29.23 (center left): © Manfred Kage/Peter Arnold Inc; (center right): © Wim van Egmond/Visuals Unlimited; (bottom right): Runk/ Schoenberger/Grant Heilman Photography; 29.24: © William Bourland, image used under license to MBL; 29.25: © Eye of Science/Photo Researchers Inc; 29.26: Phil A. Harrington/Peter Arnold Inc; 29.27: © Manfred Kage/Peter Arnold Inc; 29.28: © Ric Ergenbright/CORBIS; 29.29: © Peter Arnold Inc/ Alamy; 29.30: © John Shaw/Tom Stack & Associates; 29.31: © Mark J. Grimson and Richard L. Blanton, Biological Sciences Electron Microscopy Laboratory, Texas Tech University.

Chapter 30

Opener: © S.J. Krasemann/Peter Arnold; 30.3: © Dr. Richard Kessel & Dr. Gene Shih/Visuals Unlimited; 30.4: © Wim van Egmond/Visuals Unlimited; 30.5: © Marevision/agefotostock; 30.6 (bottom left): © Robert Calentine/Visuals Unlimited; (bottom right): © Wim van Egmond/Visuals Unlimited; 30.7: © David Sieren/Visuals Unlimited; 30.8: © Lee Wilcox; 30.9: © Edward S. Ross; 30.11: Courtesy of

Hans Steur, The Netherlands; 30.12: © Ed Reschke/ Peter Arnold; 30.13: © Kingsley R. Stern; 30.14: © Stephen P. Parker/Photo Researchers Inc; 30.15: © Mark Bowler/NHPA; 30.16 (left): © Ed Reschke; (right): © Mike Zensa/CORBIS; p. 593: © Biology Media/Photo Researchers Inc; 30.18: © Patti Murray/ Earth Scenes; 30.20a: © Jim Strawser/Grant Heilman Photography; 30.20b: © Nancy Hoyt Belcher/Grant Heilman Photography; 30.20c: © Robert Gustafson/ Visuals Unlimited; 30.21: David Dilcher and Ge Sun; 30.22: Courtesy of Sandra Floyd.

Chapter 31

Opener: © Ullstein-Joker/Peter Arnold Inc; 31.1a: © Dean A. Glawe/Biological Photo Service; 31.1b: © Carolina Biological Supply Company/Phototake; 31.1c: Yolande Dalpé, Agriculture and Agri-Food Canada, Spores of Glomus Intradices Schenck-Smith Glomeronmycota, In-vitro Collection–GINCO/ DAOM 197198; 31.1d: © Michael & Patricia Fogden; 31.2: © Dr. Garry T. Cole/Biological Photo Service; 31.3: © Michael & Patricia Fogden/CORBIS; (inset): © Micro Discovery/CORBIS; 31.4: © Microfield Scientific Ltd/Photo Researchers Inc; 31.5a: © Carolina Biological Supply Company/Phototake; 31.5b: © L. West/Photo Researchers, Inc.; 31.7a, 31.8a: © Carolina Biological Supply Company/ Phototake; 31.10: © David Scharf/Photo Researchers Inc; 31.9a: © Richard Kolar/Earth Scenes; 31.9b: © Ed Reschke/Peter Arnold Inc.; 31.11a: © Alexandra Lowry/The National Audubon Society Collection/ Photo Researchers, Inc.; 31.12a: © Dr. Fred Hossler/ Visuals Unlimited; 31.12b: © Eye of Science/Photo Researchers Inc; 31.13a (left): © Holt Studios International Ltd./Alamy; 31.13b (right): © B. Borrell Casal/Frank Lane Picture Agency/CORBIS; 31.14a: © Ken Wagner/Phototake; 31.14b: © Robert & Jean Pollack/Visuals Unlimited; 31.14c: © Robert Lee/Photo Researchers, Inc.; 31.15: © Ed Reschke; 31.16a: © Eye of Science/Photo Researchers, Inc.; 31.16b: © Dr. Gerald Van Dyke/Visuals Unlimited; 31.17: © Scott Camazine/Photo Researchers, Inc.; 31.18a: © Ralph Williams/USDA Forest Service; 31.18b: © Chris Mattison/Superstock; 31.18c: © USDA Forest Service Archives, USDA Forest Service, www.forestryimages.org; 31.19a: © Dayton Wild/Visuals Unlimited; 31.19b: © Manfred Kage/Peter Arnold Inc; 31.20: Courtesy of Zoology Dept/University of Canterbury, New Zealand; 31.20b (inset): Courtesy of Dr. Peter Dazak.

Chapter 32

Opener: © Corbis/Volume 53; 32.1, pg 422 (top): © Corbis/Volume 86, (2nd from top): © Corbis/Volume 65, (3rd from top): © David M. Phillips/Visuals Unlimited, (4th from top): © Royalty-Free/Corbis, pg 323 (top): © Edward S. Ross, (2nd from top): © Corbis, (3rd from top): © Cleveland P. Hickman, (4th from top): © Cabisco/Phototake, (5th from top): © Ed Reschke; 32.4: Courtesy of Dr. Igor Eeckhaut.

Chapter 33

Opener: © Denise Tackett/Tom Stack & Associates; 33.3a: © Andrew J. Martinez/Photo Researchers Inc; 33.5: © Roland Birke/Phototake; 33.6: © Ed Reschke; 33.7: © Amos Nachoum/CORBIS; 33.10: © David Wrobel/Visuals Unlimited; 33.8: © Kelvin Aitken/Peter Arnold Inc; 33.9: © Neil G. McDaniel/ Photo Researchers; 33.11 (top): © Tom Adams/ Visuals Unlimited; 33.12 (bottom left): © Dwight Kuhn; 33.13 (top left): © Dennis Kunkel/Phototake; 33.14 © L. Newman & A. Flowers/Photo Researchers; 33.15: © Kjell Sandved/Butterfly Alphabet; 33.16 (top left): © Peter Funch, University of Aarhus; 33.17: © Gary D. Gaugler/Photo Researchers; 33.18: © Educational Images Ltd, Elmira NY, USA; 33.19: © T.E. Adams/Visuals Unlimited.

Chapter 34

Opener: © James H. Robinson/Animals Animals/ Earth Scenes; 34.1a: © Marty Snyderman/Visuals Unlimited; 34.1b: © Alex Kerstitch/Visuals Unlimited; 34.1c: © Douglas Faulkner/Photo Researchers Inc; 34.1d: © age fotostock/SuperStock; 34.2: © A. Flowers & L. Newman/Photo Researchers Inc; 34.4: © Eye of Science/Photo Researchers Inc; 34.5b: © Kjell Sandved/Butterfly Alphabet; 34.6: © Kelvin Aitken/Peter Arnold Inc; 34.7: © Milton Rand/Tom Stack & Associates; 34.8: © PhotoDisc/ Getty Images; 34.10: © AFP/Getty Images; 34.11: © Jeff Rotman/Photo Researchers Inc; 34.12: © Ken Lucas/Visuals Unlimited; 34.14: © Ronald L. Shimek; 34.15: © Fred Grassle, Woods Hole Oceanographic Institution; 34.16: © David M. Dennis/Animals Animals Enterprises; 34.17: © Pascal Goetgheluck/Photo Researchers Inc; 34.18b: © Robert Brons/BiologicalPhoto Service; 34.19: © Fred Bavendam/Peter Arnold Inc; 34.26a: © National Geographic/Getty Images; 34.26b: © S. Camazine/K. Visscher/Photo Researchers Inc; 34.27a: © Alex Kerstich/Visuals Unlimited; 34.27b: © Edward S. Ross; 34.29: © T.E. Adams/Visuals Unlimited; 34.31: © Kjell Sandved/Butterfly Alphabet; 34.32a: © Cleveland P. Hickman; 34.32b: © Valorie Hodgson/Visuals Unlimited; 34.32c: © Gyorgy Csoka, Hungary Forest Research Institute www.forestryimages.org; 34.32d: © Kjell Sandved/ Butterfly Alphabet; 34.32e: © Greg Johnston/Lonely Planet Images/Getty Images; 34.32f: © Nature's Images/Photo Researchers Inc; 34.34: © Dwight Kuhn; 34.35, 34.36: © Kjell Sandved/Butterfly Alphabet; 34.37: © Wim van Egmond/Visuals Unlimited; 34.38a: © Alex Kerstitch/Visuals Unlimited; 34.38b: © Randy Morse/Tom Stack & Associates; 34.38c: © Daniel W. Gotshall/Visuals Unlimited; 34.38d: © Reinhard Dirscherl/Visuals Unlimited; 34.38e: © Jeff Rotman/Photo Researchers Inc.

Chapter 35

Opener: © Phone Ferrero J.P./Labat J.M./Peter Arnold Inc; 35.2: © Eric N. Olson, PhD/The University of Texas MD Anderson Cancer Center; 35.4a: © Rick Harbo/Marine Images; 35.5: © Heather Angel; 35.9 (top): © agefotostock/ Superstock; (left): © Royalty-Free/CORBIS; (right): © Brandon Cole/www.brandoncole.com/Visuals Unlimited; 35.11: Corbis/Volume 33; 35.13a: © Federico Cabello/Superstock; 35.13b: © Raymond Tercafs-Production Services/Bruce Coleman Inc; 35.16a: © John Shaw/Tom Stack & Associates; 35.16b: © Suzanne L. Collins & Joseph T. Collins/ Photo Researchers Inc; 35.16c: © Jany Sauvanet/ Photo Researchers Inc; 35.22: © Paul Sareno, University of Chicago; 35.24a (left): © William J. Weber/Visuals Unlimited; (right): © Frans Lemmens/Getty Images; 35.24b: © Jonathan Losos; 35.24c (right): © Rod Planck/Tom Stack & Associates; 35.25 (right): © Rod Planck/Tom Stack & Associates; 35.25d (left): © Corbis/Volume 6; (right): © Zig Leszczynski/Animals Animals; 35.28: © Layne Kennedy/CORBIS; 35.29a: © Corbis; 35.29b: Arthur C. Smith III/Grant Heilman Photography; 35.29c: © David Boyle/Animals Animals; 35.29d: © John Cancalosi/Peter Arnold; 35.32: © Stephen Dalton/ National Audubon Society Collection/Photo Researchers; 35.33a (left): © B.J Alcock/Visuals Unlimited; (right): © Dave Watts/Alamy; 35.33b (left): © CORBIS Volume 6; (right): © W. Perry Conway/CORBIS; 35.33c (left): © Stephen J. Krasemann/DRK Photo; (right): © J. & C. Sohns/ Animals Animals Enterprises; 35.34: © Alan Nelson/ Animals Animals/Earth Scenes; 35.35a: © Peter Arnold Inc/Alamy; 35.35b: © Martin Harvey/Peter Arnold; 35.35c (left): © Joe McDonald/Visuals Unlimited; 35.35c (right): © Dynamic Graphics Group/IT Stock Free/Alamy; 35.38: National Museums of Kenya, Nairobi © 1985 David L. Brill; 35.40: © AP/Wide World Photos.

Chapter 36

Opener: © Susan Singer; 36.4 (left side): Dr. Robert Lyndon; (right side): © Biodisc/Visuals Unlimited; 36.6a: © Brian Sullivan/Visuals Unlimited; 36.6b, 36.6c: © EM Unit, Royal Holloway, University of London, Egham, Surrey; 36.7: © Jessica Lucas & Fred Sack; 36.8: © Andrew Syred/Science Photo Library/Photo Researchers Inc; 36.10: © Runk/ Shoenberger/Grant Heilman Photography; 36.9a (both): Courtesy of Allan Lloyd; 36.11a: © Lee Wilcox; 36.11b: © George Wilder/Visuals Unlimited; 36.11c: © Lee Wilcox; 36.12 (upper): © NC Brown Center for Ultrastructure Studies, SUNY, College of Environmental Science and Forestry, Syracuse, NY; (lower): Tom Kuster/USDA; 36.13b: © Dr. Richard Kessel & Dr. Gene Shih/Visuals Unlimited; 36.14: © Biodisc/Visuals Unlimited; 36.15b: © John Schiefelbein, University of Michigan; 36.16b: © Courtesy of Dr. Philip Benfey, from Wysocka-Diller, J.W., Helariutta, Y. Fukaki, H., Malamy, J.E. and P.N. Benfey (2000) Molecular analysis of SCARECROW function reveals a radial patterning mechanism common to root and shoot development, Cell 127, 595–603; 36.17 (bottom left upper): © Carolina Biological Supply Company/Phototake; (bottom right upper): 36:17 (bottom right; upper/lower): © George Ellmore, Tufts University; (bottom left, lower): © Lee Wilcox; 36.19a: © E.R. Degginger/Photo Researchers, Inc.; 36.19b: © Peter Frischmuth/Peter Arnold Inc; 36.19c: © Walter H. Hodge/Peter Arnold Inc; 36.19d: © Gerald & Buff Corsi/Visuals Unlimited; 36.19e: © Kingsley R. Stern; 36.20: Courtesy of J.H. Troughton and L. Donaldson/ Industrial Research Ltd; 36.23a,b: © Ed Reschke; 36.26: © Ed Reschke/Peter Arnold Inc; 36.27a: © Ed Reschke; 36.27b: © Biodisc/Visuals Unlimited; 36.28a: © Jerome Wexler/Visuals Unlimited; 36.28b: © Lee Wilcox; 36.28c: © Runk/Shoenberger/Grant Heilman Photography; 36.28d: © Chase Studio/ Photo Researchers Inc.; 36.28e: © Charles D. Winters/Photo Researchers, Inc.; 36.28f: © Lee Wilcox; 36.29 (both): © Scott Poethig, University of Pennsylvania; 36.30a (upper): © Kjell Sandved/ Butterfly Alphabet; 36.30b (lower): © Pat Anderson/ Visuals Unlimited; 36.31a: © Gusto/Photo Researchers; 36.31b: © Glenn M. Oliver/Visuals Unlimited; 36.31c: © Joel Arrington/Visuals Unlimited; 36.33: © Ed Reschke.

Chapter 37

Opener: © Norm Thomas/The National Audubon Society Collection/Photo Researchers Inc; 37.4: Edward

Yeung (University of Calgary) and David Meinke (Oklahoma State University); 37.6 (all): Kindly provided by Prof. Chun-ming Liu, Institute of Botany, Chinese Academy of Sciences; 37.7: Max Planck Institute for Developmental Biology; 37.8c: © Ben Scheres, University of Utrecht; 37.8d, 37.8e: Courtesy of George Stamatiou and Thomas Berleth; 37.11 (corn): © photocuisine/Corbis; 37.11 (bean): © Barry L. Runk/ Grant Heilman Photography; 37.13a: © Ed Reschke/ Peter Arnold Inc; 37.13b: © David Sieren/Visuals Unlimited; 37.15 (top left), (top center), (bottom center): © Kingsley R. Stern; 37.15 (bottom left): © James Richardson/Visuals Unlimited; 37.15 (top right): Courtesy of Robert A. Schisling; 37.15 (bottom right): © Charles D. Winters/Photo Researchers Inc; 37.16a: © Edward S. Ross; 37.16b, c, d: © James Castner; 37.18a: © PHONE Thiriet Claudius/Peter Arnold, Inc; 37.18b: © Holt Studios International Ltd/Alamy.

Chapter 38

Opener: © Richard Rowan's Collection Inc/Photo Researchers Inc; 38.4 (both): © Jim Strawser/Grant Heilman Photography; 38.7: © Ken Wagner/ Phototake; 38.11a: (closed): © Dr. Ryder/Jason Borns/ Phototake; 38.11b (open): © Dr. Ryder/Jason Borns/ Phototake; 38.14: © Herve Conge/ISM/Phototake; 38.15a: © Ed Reschke; 38.15b: © Jon Bertsch/Visuals Unlimited; 38.16: © Mark Boulton/Photo Researchers Inc; 38.17a: © Andrew Syred/Photo Researchers Inc; 38.17b: © Bruce Iverson Photomicrography.

Chapter 39

Opener: © PhotoDisc/Getty Images; 39.4a: © Hulton Archive/Getty Images; 39.4b: © M.L. Bonsirven-Fontana/UNESCO; 39.5 (all): International Plant Nutrition Institute (IPNI), Norcross, GA U.S.A.; 39.7: © George Bernard/ Animals Animals/Earth Scenes; 39.8 (bottom left): © Ken Wagner/Phototake; (bottom center): © Bruce Iverson; 39.10a: © Kjell Sandved/Butterfly Alphabet; 39.10b: © Runk/Schoenberger/Grant Heilman Photography; 39.10c: © Perennov Nuridsany/Photo Researchers, Inc.; 39.10d: © Barry Rice; 39.12: © Don Albert; 39.15 (both): Courtesy of Nicholas School of the Environment and Earth Sciences, Duke University; 39.17: © Greg Harvey USAF; 39.18b: © REUTERS; 39.18c: © AP/Wide World Photo.

Chapter 40

Opener: © fstop/Getty Images; 40.1: © Richard la Val/Animals Animals; 40.2, 40.3a: USDA/Agricultural Research Service; 40.3b: © USDA/Agricultural Research Service; 40.5: © Allan Morgan/Peter Arnold Inc; 40.6: © Adam Jones/Photo Researchers Inc; p. 793 (1-3): © Inga Spence/Visuals Unlimited; (4): © Heather Angel/Natural Visions; (5): © Pallava Bagla/ Corbis; 40.7a: © Gilbert S. Grant/Photo Researchers Inc; 40.7b: © Lee Wilcox; 40.8: © Michael J. Doolittle/Peter Arnold Inc; 40.12: © Courtesy R.X. Latin. Reprinted with permission from Compendium of Cucurbit Diseases, 1996, American Phytopathological Society, St. Paul, MN.

Chapter 41

Opener: © Alan G. Nelson/Animals Animals; 41.3a (all):f 41.8: © Ray Evert; 41.10 (all): Jee Jung & Philip Benfey; 41.11: © Lee Wilcox; 41.13: © Frank Krahmar/Zefa/Corbis; 41.16: © Don Grall/Index Stock Imagery; 41.26: © Prof. Malcolm B. Wilkins, Botany Dept, Glasgow University; 41.28: © Robert Calentine/Visuals Unlimited; 41.29: © Runk/ Schoenberger/Grant Heilman Photography; 41.31: Amnon Lichter, The Volcani Center; 41.34a: © John Solden/Visuals Unlimited; 41.34b: Courtesy of Donald R. McCarty, from "Molecular Analysis of viviparous-1: An Abscisic Acid-Insensitive Mutant of Maize: The Plant Cell. v. 1, 523–532 © 1989 American Society of Plant Physiologists; 41.34c: © ISM/Phototake.

Chapter 42

Opener: © Heather Angel/Natural Visions; 42.3a: © Fred Habegger/Grant Heilman Photography; 42.3b: © Pat Breen, Oregon State University; 42.4: © Richard La Val/Animals Animals; 42.5a: © Mack Henley/Visuals Unlimited; 42.5a (inset): © Michael Gadomski; 42.5b: Max Planck Institute; 42.5b (inset): © Henrik Bohlenius; 42.7: © Jim Strawser/Grant Heilman Photography; 42.12 (all): © John Bowman; 42.15: © John Bishop/Visuals Unlimited; 42.16: © Paul Gier/Visuals Unlimited; 42.17 (both): Courtesy of Enrico Coen; 42.19 (both): © L. DeVos—Free University of Brussels; 42.20: © Kingsley R. Stern; 42.21: © David Cappaert, www.forestryimages.org; 42.22: © Topham/The Image Works; 42.23: © Michael Fogden; 42.24 (both): © Thomas Eisner; 42.25: © John D. Cunningham/Visuals Unlimited; 42.26: © Edward S. Ross; 42.27a: © David Sieren/ Visuals Unlimited; 42.27b: © Barbara Gerlach/Visuals Unlimited; 42.30: © Jerome Wexler/Photo Researchers Inc; 42.31a: © Sinclair Stammers/Photo Researchers; 42.31 (b-d): Courtesy of Dr. Hans Ulrich Koop, from Plant Cell Reports, 17:601–604eports, 17: 601-604; 42.32a: © Lee Wilcox; 42.32b: © David Lazenby/Animals Animals/Earth Scenes.

Chapter 43

Opener: © Dr. Roger C. Wagner, Professor Emeritus of Biological Sciences, University of Delaware; Table 43.1, pg 855 (top to bottom): © Ed Reschke; © Arthur Siegelman/Visuals Unlimited; © Ed Reschke; © Gladden Willis, M.D./Visuals Unlimited; © Ed Reschke; 43.3: © J. Gross, Biozentrum/Photo Researchers Inc; 43.4: © Biophoto Associates/Photo Researchers Inc; Table 43.2, pg 856 (top to bottom): © Ed Reschke; © Gladden Willis, M.D./Visuals Unlimited; © Chuck Brown/Photo Researchers Inc; © Ed Reschke; © Kenneth Eward/Photo Researchers Inc; 43.3 (all): © Ed Reschke.

Chapter 44

Opener: Courtesy of David I. Vaney, University of Queensland Australia; 44.3: © Enrico Mugnaini/ Visuals Unlimited; 44.12: © John Heuser, Washington University School of Medicine, St. Louis, MO; 44.14: © Ed Reschke; 44.25: © Dr. Marcus E. Rachle, Washington University, McDonnell Center for High Brain Function; 44.26: © Lennart Nilsson/Albert Bonniers Förlag AB; 44.29: © E.R. Lewis/Biological Photo Service.

Chapter 45

Opener: © Omikron/Photo Researchers Inc; 45.22: © Leonard L. Rue, III.

Chapter 46

Opener: © Nature's Images/Photo Researchers Inc; 46.10: © John Paul Kay/Peter Arnold Inc; 46.11: © Bettmann/Corbis; 46.16: © Robert & Linda Mitchell.

Chapter 47

Opener: © Hal Beral/Grant Heilman Photography; 47.4 (top): © Ed Reschke/Peter Arnold; (2nd row-left): © David Scharf/Peter Arnold; (center): © Dr. Holger Jastrow; (right): © CNRI/Photo Researchers Inc; (3rd row-left): © Dr. Kessel & Dr. Kardon/Tissue & Organs/Visuals Unlimited/Getty Images; (center) © Ed Reschke/Peter Arnold; (right): © Ed Reschke; (bottom right): © Ed Reschke/Peter Arnold; 47.11 (both): © Dr. H.E. Huxley; 47.22: © Treat Davidson/ Photo Researchers Inc.

Chapter 48

Opener: © John Gerlach, Animals Animals/Earth Scenes; 48.11: © Ron Boardman/Stone/Getty Images; 48.19: from O.T. Avery. C.M. McLeod & M. McCarty, "Studies on the chemical nature of the substance inducing transformation of pneumococcal types" reproduced from *The Journal of Experimental Medicine* 79 (1944): 137–158, fig 1 by copyright permission of the Rockefeller University press, reproduced by permission. Photograph made by Mr. Joseph B. Haulenbeck.

Chapter 49

Figure 49.12: © Tom Drysdale; 49.13a,b: © Ed Reschke; 49.13c: © Dr. Gladden Willis/Visuals Unlimited; 49.18: © Bruce Watkins/Animals Animals; 49.30a: © Clark Overton/Phototake; 49.30b: © Martin Rotker/Phototake; 49.31: © Kenneth Eward/BioGrafx/Photo Researchers Inc.

Chapter 50

Opener: © Ann & Steve Toon/Robert Harding World Imagery/Getty Images.

Chapter 51

Opener: AP/Wide World Photos; 51.1: © Manfred Kage/Peter Arnold Inc; 51.10 (both): © Dr. Andrejs Liepins/Photo Researchers, Inc; 51.21: © CDC/ Science Source; 51.4: © Wellcome Library London.

Chapter 52

Opener: © Michael Fogden/DRK Photo; 52.1: © Dennis Kunkel Microscopy Inc; 52.2a: © Chuck Wise/Animals Animals/Earth Scenes; 52.2b: © Fred McConnaughey/The National Audubon Society Collection/Photo Researchers Inc; 52.4: © David Doubilet; 52.5: © Hans Pfletschinger/Peter Arnold Inc; 52.7: © Cleveland P. Hickman; 52.8: © Frans Lanting/Minden Pictures; 52.9a: © Jean Phllippe Varin/Jacana/Photo Researchers Inc; 52.9b: © Tom McHugh/The National Audubon Society Collection/Photo Researchers Inc; 52.9c: © Corbis Volume 86; 52.12a: © David M. Phillips/Photo Researchers Inc; 52.17: © Ed Reschke.

Chapter 53

Opener: © Lennart Nilsson/Albert Bonniers Förlag AB, A Child Is Born, Dell Publishing Company; 53.1d: © David M. Philips/Visuals Unlimited; 53.3 (all): Dr. Mathias Hafner (Mannheim University of Applied Sciences, Institute for Molecular Biology, Mannheim, Germany) and Dr. Gerald Schatten (Pittsburgh Development Center Deputy Director, Magee-Women's Research Institute Professor and Vice-Chair of Obstetrics, Gynecology & Reproductive Sciences and Professor of Cell Biology & Physiology Director, Division of Developmental and Regenerative Medicine University of Pittsburgh School of Medicine Pittsburgh, PA 15213); 53.7: © David M. Philips/Visuals Unlimited; 53.8a: © Cabisco/Phototake; 53.9: © David M. Philips/ Visuals Unlimited; 53.11 (all): From "An Atlas of the Development of the Sea Urchin *Lytechinus variegatus*." Provided by Dr. John B. Merrill (left to right) Plate 20, pg 62, #I; Plate 33, pg 93, #C; Plate 38, pg 105, #6; 53.17 (both): © Courtesy of Manfred Frasch; 53.20 (both): © Roger Fleischman, University of Kentucky; 53.27b, 53.28 (all): © Lennart Nilsson/Albert Bonniers Förlag AB, A Child Is Born, Dell Publishing Company.

Chapter 54

Opener: © K. Ammann/Bruce Coleman Inc; 54.4 (both): from J.R. Brown et al, "A defect in nurturing mice lacking . . . gene for fosB" Cell v. 86, 1996 pp. 297–308, © Cell Press; 54.5 (both): © Larry Young Emory University—Yerkes Research Center; 54.6 (all): © Lee Boltin Picture Library; 54.7: © William Grenfell/Visuals Unlimited; 54.8: © Thomas McAvoy, Life Magazine/Time, Inc.; 54.9: Harlow Primate Laboratory—University of Wisconsin-Madison; 54.11: © Roger Wilmhurst/The National Audubon Society Collection/Photo Researchers Inc; 54.12a: © Linda Koebner/Bruce Coleman; 54.12b: © Jeff Foott/Tom Stack & Associates; 54.13: © Superstock; 54.14: Courtesy of Bernd Heinrich; 54.15b: © Fred Breunner/Peter Arnold Inc; 54.15c: © George Lepp/Getty Images; 54.18: © Dwight Kuhn; 54.19: © PhotoLink/Getty Images; 54.20: © Tom Leeson; 54.21b: © Scott Camazine/Photo Researchers, Inc; 54.22a: © Gerald Cubitt; 54.23: © Corbis Volume 53; 54.24: © Nina Leen, Life Magazine/Time, Inc; 54.27a: © Peter Steyn/Getty Images; 54.27b: © Gerald C. Kelley/Photo Researchers Inc; 54.28a: © Bruce Beehler/Photo Researchers Inc; 54.28b: © B. Chudleigh/Vireo; 54.30: © Cathy & Gordon ILLG; 54.32a: Courtesy of T.A. Burke, Reprinted by permission from Nature, "Parental care and mating behavior of polyandrous dunnocks," 338: 247–251, 1989; 54.33: © Merlin B. Tuttle/Bat Conservation International; 54.35: © Heinrich Van DEN Berf/Peter Arnold Inc; 54.36: © Edward S. Ross; 54.38: © Mark Moffett/Minden Pictures; 54.39: © Nigel Dennis/National Audubon Society Collection/Photo Researchers Inc.

Chapter 55

Opener: © PhotoDisc/Volume 44/Getty Images; 55.1: © Michael Fogden/Animals Animals; 55.4: © Tom Baugh; 55.14: © Christian Kerihuel; 55.16: Courtesy of Barry Sinervo; 55.22 (bottom left): © Juan Medina/Reuters/Corbis; (bottom center): © Oxford Scientific/Photolibrary.

Chapter 56

Opener: © Corbis; 56.1: © Daryl & Sharna Balfour/Okopia/Photo Researchers Inc; 56.6 (all): © J.B. Losos; 56.10a: © Edward S. Ross; 56.10b: © Raymond Mendez/Animals Animals; 56.11 (both): © Lincoln P. Brower; 56.12: © Michael & Patricia Fogden/Corbis; 56.13: © James L. Castner; 56.15: © Merlin D. Tuttle/Bat Conservation International; 56.16: © Eastcott/Momatiuk/The Image Works; 56.17: © PhotoDisc/Volume 44/Getty Images; 56.18: © Michael Fogden/DRK Photo; 56.19: © David Moorhead; 56.21a: © F. Stuart Westmorland/Photo Researchers Inc; 56.21b: © Ann Weirtheim/Animals Animals/Earth Scenes; 56.23: © David Hosking/National Audubon Society Collection/Photo Researchers Inc; 56.24 (all): © Tom Bean; 56.25 (both): © Dani/Jeske/Earth Scenes; 56.26: © Educational Images Ltd., Elmira, NY, USA. Used by Permission.

Chapter 57

Opener: © PhotoDisc/Getty Images; 57.3: © Worldwide Picture Library/Alamy; 57.7a: © U.S. Forest Service; 57.19a: © Lane Kennedy/Corbis.

Chapter 58

Opener: NASA; 58.10, pg 1218, 1st row, (left to right): © age fotostock/SuperStock; © Luiz C. Marigo/Peter Arnold; © age fotostock/SuperStock; 2nd row, (left to right): © Mitsuaki Iwago/Minden Pictures; © Konrad Wothe/Minden Pictures; © age fotostock/SuperStock; 3rd row, (left to right): © Adam Jones/Getty Images; © age fotostock/SuperStock; 4th row, (left to right): © Jon Arnold Images/Alamy; © Jim Brandenburg/Minden Pictures; © Eastcott Momatiuk/Getty Images; p. 1219 (1st row-left to right): © SuperStock, Inc./SuperStock; © Dwight Kuhn; © Claudia Adams/Dembinsky Photo Associates; (2nd row-left to right): © Jim Lundgren/Alamy; © Craig Tuttle/Corbis; © Dennis M. Dennis/Animals Animals; (3rd row-left to right): © Stephen J. Krasemann/Photo Researchers Inc; © Joe McDonald/Corbis; © imagebroker/Alamy; (4th row-left to right): © Eastcott Momatiuk/Getty Images; © Science Faction/Getty Images; © Michio Hoshino/Minden Pictures; 58.13a: © Art Wolfe/Photo Researchers Inc; 58.13b: © Bill Banaszowski/Visuals Unlimited; 58.15: Provided by the SeaWiFS Project, NASA/Goddard Space Flight Center, and ORBIMAGE; 58.16: © Digital Vision/Picture Quest; 58.18a: © Jim Church; 58.18b: © Ralph White/Corbis; 58.21a: © Peter May/Peter Arnold Inc; 58.21b: © Frans Lanting/Minden Pictures; 58.22: © Gilbert S. Grant/National Audubon Society Collection/Photo Researchers, Inc; 58.24a: NASA; 58.25: NASA/Goddard Institute for Space Studies, New York; 58.27 (bottom left upper): © Lonnie G. Thompson, The Ohio State University; 58.27b (bottom left lower): © Dr. Bruno Messerli.

Chapter 59

Opener: Cornell Lab of Ornithology; 59.2: © John Elk; 59.3 (left to right): © Frank Krahmer/Masterfile; © Michael & Patricia Fogden/Minden Pictures; © Heather Angel/Natural Visions; © NHPA/Martin Harvey; 59.5a: © Edward S. Ross; 59.5b: © Inga Spence/Getty Images; 59.6a: © Jean-Leo Dugast/Peter Arnold Inc; 59.6b: © Oxford Scientific/PhotoLibrary; 59.8: © Michael Fogden/DRK Photo; 59.9 (left to right): © Brian Rogers/Natural Visions; © David M. Dennis/Animals Animals; © Michael Turco, 2006; © David A. Northcott/Corbis; 59.13 (right): © Randall Hyman; (left): © Dr. Morley Read/Photo Researchers Inc; 59.14: © John Gerlach/Animals Animals; 59.16: © Peter Yates/Science Photo Library; 59.17 (left): © Jack Jeffrey; (right): © Jack Jeffrey/Photo Reseource Hawaii; 59.18: © Tom McHugh/Photo Researchers Inc; 59.20: © Merlin B. Tuttle/Bat Conservation International; 59.21: U.S. Fish and Wildlife Service; 59.21a: ANSP © Steven Holt/stockpix.com; 59.23: Wm. J. Weber/Visuals Unlimited; 59.24 (both): University of Wisconsin, Madison Arboretum; 59.26b: © Dani/Jeske/Earth Scenes.

Line Art and Text

Chapter 8

Figure 8.5: From Raven et al., *Biology of Plants*, 5e. Reprinted by permission of W.H. Freeman and Company/Worth Publishers. Figure 8.12: From Lincoln Taiz and Eduardo Zeiger, *Plant Physiology*, 1991 Benjamin-Cummings Publishing. Reprinted with permission of the authors.

Chapter 17

TA17.1: CALVIN AND HOBBES © 1995 Watterson. Dist. By UNIVERSAL PRESS SYNDICATE. Reprinted with permission. All rights reserved.

Chapter 18

Figure 18.8: After *Nature*, page 790, December 2003. Figure 18.9: G. More, K.M. Devos, Z. Wang, and M.D. Gale: "Grass, line up and form a circle," *Current Biology*, 1995, vol. 5, pp. 737–739.

Chapter 19

Figure 19.5: After Alberts, et al., *Molecular Biology of the Cell*, 3e, 1994, Garland Publishing, Inc.

Chapter 20

Figure 20.7: Data from P.A. Powers, et al., "A Multidisciplinary Approach to the Selectionist/Neutralist Controversy." *Oxford Surveys in Evolutionary Biology*. Oxford University Press, 1993. Figure 20.9: From R.F. Preziosi and D.J. Fairbairn, "Sexual Size Dimorphism and Selection in the Wild in the Waterstrider *Aquarius remigis:* Lifetime Fecundity Selection on Female Total Length and Its Components," *Evolution, International Journal of Organic Evolution* 51:467–474, 1997. Figure 20.10: Data from M.R. MacNair in J.M. Bishops & L.M. Cook, *Genetic Consequences of Man-Made Change*, Academic Press, 1981, pp. 177–207. Figure 20.11: Adapted from Clark, B. "Balanced Polymorphism and the Diversity of Sympatric Species," Syst. Assoc. Publ., Vol. 4, 1962.

Chapter 21

Figure 21.2a: Data from Grant, "Natural Selection and Darwin's Finches" in *Scientific American*, October 1991. Figure 21.2b: Data from Grant, "Natural Selection and Darwin's Finches" in *Scientific American*, October 1991. Figure 21.4: Data from Grant, et al., "Parallel Rise and Fall of Melanic Peppered Moths" in *Journal of Heredity*, vol. 87, 1996, Oxford University Press. Figure 21.5: Data from G. Dayton and A. Roberson, *Journal of Genetics*, Vol. 55, p. 154, 1957.

Chapter 22

Figure 22.1: Data from R. Conant & J.T. Collins, *Reptiles & Amphibians of Eastern/Central North America*, 3rd edition, 1991. Houghton Mifflin Company.

Chapter 23

Figure 23.11: Adapted from Duda and Palumbi, *Evolution*, August 1999, Proceedings for the National Academy of Science. Figure 23.14: Adapted from Collin, *Evolution*, Proceedings for the National Academy of Science. Figure 23.16: Data from Hahn et al., *Science*, 1980.

Chapter 24

Figure 24.4: Data from Adams and Wendell, *Current Opinion in Plant Biology*, 2005. Figure 24.6: Data from Blane and Wolfe, *The Plant Cell*, 2004; Adams and Wendell, *Current Opinion in Plant Biology*, 2005.

Chapter 28

Figure 28.18: Data from U.S. Centers for Disease Control and Prevention, Atlanta, GA.

Chapter 37

Figure 37.3: From Ralph Quantrano, Washington University. Figure 37.9d & e: Data from *The EMBO Journal*, vol. 17, 1998.

Chapter 38

Figure 38.11: Data from Raven, Evert, and Eichhorn, *Biology of Plants*, 7e, W. H. Freeman and Co. Figure 38.13: Data from Raven, Evert, and Eichhorn, *Biology of Plants*, 7e, W. H. Freeman and Co.

Chapter 41

Figure 41.9: Based on data from The Arabidopsis Book. Figure 41.10: Based on data from The Arabidopsis Book.

Chapter 52

Table 52.2: Data from American College of Obstetricians and Gynecologists: Contraception, Patient Education Pamphlet No. AP005. ACOG, Washington, D.C., 1990.

Chapter 54

Figure 54.4c & d: Data from J.R. Brown et al., "A Defect in Nurturing in Mice Lacking the Immediate Early Gene for fosB" *Cell*, 1996. Figure 54.10: Reprinted from *Animal Behavior*, 51, Brood Parasite Figure by M.D. Beecher et al., Copyright © 1996 with permission from Elsevier. Figure 54.22: Reprinted from *Animal Behavior*, 36(2), Private Semantics Figure by John Alcock, p. 484, Copyright © 1988, with permission from Elsevier. Figure 54.28c: Data from M. Petrie, et al. "Peahens Prefer Peacocks with Elaborate Trains," *Animal Behavior*, 1991. Figure 54.32: Data from H.L. Gibbs et al., "Realized Reproductive Excess of Polygynous Red-Winged Blackbirds Revealed by NDA Markers," *Science*, 1990.

Chapter 55

Figure 55.5: Data from Brown & Lomolino, *Biogeography*, 3rd edition, 1998, Sinauer Associates, Inc. Figure 55.6: Data from Brown & Lomolino, *Biogeography*, 3rd edition, 1998, Sinauer Associates, Inc. After A.T. Smith, *Ecology*, 1974. Figure 55.9: Data from *Patch Occupation and Population Size of the Glanville Fritillary in the Anland Islands*, Metapopulation Research Group, Helsinki, Finland. Figure 55.19b: Data from C.E. Goulden, L.L. Henry, and A.J. Tessier, *Ecology*, 1982. Figure 55.23: From "Population Changes in German Forest Pests," by G.C. Varley in *Journal of Animal Ecology*, 18. Copyright © 1949 British Ecology Society. Reprinted by permission.

Chapter 56

Figure 56.2: Abundance of tree species along a moisture gradient in the Santa Catalina Mountains of southeastern Arizona, from *The Economy of Nature*, 4e, by Robert E. Ricklets. Copyright 1973, 1979 by Chiron Press, Inc. © 1990, 2000 by W.H. Freeman and Company. Reprinted by permission. Figure 56.3: Change in community composition across an ecotone, from *The Economy of Nature*, 4e, by Robert E. Ricklets. Copyright 1973, 1979 by Chiron Press, Inc. © 1990, 2000 by W.H. Freeman and Company. Reprinted by permission. Figure 56.5: Data from Begon et al., *Ecology*, 1996. After: W.B. Clapham, *Natural Ecosystems*, Clover, Macmillan. Figure 56.7: Data from E.J. Heske, et al., *Ecology*, 1994. Figure 56.8: Data from E.J. Heske, et al., *Ecology*, 1994.

Chapter 57

Figure 57.1: Data in: Begon, M., J.L. Harper, and C. R. Townsend, *Ecology*, 3/e, Blackwell Science 1996, page 715. Original Source: Whittaker, R. H. *Communities and Ecosystems*, 2/e Macmillan, London, 1975. Figure 57.11: Data in: Begon, M., J.L. Harper, and C.R. Townsend, *Ecology*, Blackwell Science 1996, page 715. Original Source: Whittaker, R.H. *Communities and Ecosystems*, 2/e Macmillan, London, 1975. Figure 57.14: Data from Flecker, A.S. and Townsend, C.R., "Community-Wide Consequences of Trout Introduction in New Zealand Streams." In *Ecosystem Management: Selected Readings*, F.B. Samson and F.L. Knopf eds., Springer-Verlag, New York, 1996. Figure 57.18: Data from J.T. Wooten and M.E. Power, "Productivity, Consumers & the Structure of a River Food Chain," *Proceedings National Academic Science*, 1993. Figure 57.20: Data from F. Morrin, *Community Ecology*, Blackwell, 1999.

Chapter 59

Figure 59.4: From *Nature's Place: Human Population and the Future of Biological Diversity*, by Richard P. Cincotta and Robert Engelman, 2000. Reprinted by permission. Figure 59.11: Data assembled by Pima, 1991. Figure 59.14: Data from Marra, Hobson, Holmes, "Linking Winter & Summer Events," in *Science*, Dec. 1998. Figure 59.15: Data from UNEP, *Environmental Data Report*, 1993, 1994. Figure 59.22: Data from H.L. Billington, "Effects of Population Size on a Genetic Variation in a Dioecious Conifer" in *Conservation Geology*, Blackwell Scientific Publication, Inc., 1999. Figure 59.25: Data from The Peregrine Fund.

Electronic Publishing Services Inc. Illustration Team

Art Director: Kim E. Moss
Lead Illustrator: Erin Daniel
Lead Illustrator: Gwen DeCelles
Lead Illustrator: Matthew McAdams
Art Coordinator: Donna Hallera
Art Coordinator: Haydee Martinez

Art Development

Jen Christiansen: Chapters 2–7, 24, 27–29, 35, 51, 57, 58.

Martin Huber: Chapters 8–11, 14, 19, 25, 30, 43–50, 52, 53.

Eliza Jewett: Chapters 36–42.

Kim Moss: Chapters 1, 12, 13, 15–18, 20–23, 26, 31–34, 54–56, 59.

Illustrations

Philip Ashley: Figures: 32.01, 32.06, 33.19, 34.29, 58.19, 58.28

Christopher Burke: Figures: 17.07, 17.12

Raychel Ciemma: Figures: 1.07, ta4.02, 4.03, 4.05, 4.06, 4.12, 4.13, 4.14, 4.17, 4.18, 4.19, 4.20, 4.23, 4.25, ta5.02, ta5.03, ta5.04, 5.05, 5.06, 5.07, 5.09, 5.10, 5.11, 5.12, 5.17, 7.05, 7.13, 7.14, 7.15, 8.02, 8.06, 8.09, 8.14, 8.15, 9.01, 9.02, 9.03, 9.05, 9.06, 9.07, 9.10, 9.11, 9.13, 9.14, 9.15, 9.16, 9.17, 9.18, 9.19, 10.05, 11.04, 12.16, 14.09, 14.11, 14.16, 14.18, 14.19, 14.20, 14.23, 14.24, 14.25, 14.26, ta14.01, 17.13, 21.01, 21.09, 21.12, 22.09, 22.14, 36.01, 36.02, 36.09, 37.09, 37.15, 38.08, 38.11, 39.10, 42.13, 54.17, 54.18, 54.33, 55.03, 55.07, 55.19, 55.23, ta55.01, 59.19, 55.25

Erin Daniel: Figures: 21.04, 21.05, 21.06, 21.07, 21.10, 21.13, 21.14, 21.17, 21.18, 21.19, 21.20, 22.16

Laurie Decopain: Figure ta34.06

Mica Duran: Figures: 17.19, 19.11

Eli Ensor: Figures: 35.20, 35.22, 35.27

Shawn Gould: Figures: 29.01, 29.11, 29.20, 47.02, 47.21, 47.23, 53.01, 53.08, 53.10–53.14, 53.20, 53.22, 53.24, 53.25

Stephen Halker: Figure ta53.03

Joyce Lavery Hall: Figures: 20.03, 34.22, 34.25, 46.01, 46.03, 47.03b, 50.03, 50.07, 50.09, 50.10, 50.11, 50.15, 50.16, 50.17, 50.18, 50.20, 52.06, 52.11, 52.20

Jonathan Higgins: Figures: 20.09, 20.14, 20.17, 20.20, 21.14, 57.01, 57.02, 57.04, 57.05, 59.01

Kellie Holoski: Figures: 22.08, 44.11, 44.17, 44.18, 44.20, 44.21, 50.19

Martin Huber: Figures: 14.11, 15.05, 15.06, 15.07, 15.09, 15.13, 15.14, 15.15, 15.20, 15.21, 19.02, 19.08, 31.02, 31.07, 31.09, 31.16, 46.12, 46.17, 49.01, 49.03, 49.06, 49.09, 49.11, 49.17, 49.19, 49.20, 49.22, 49.28, 49.29, 49.34, 50.02, 51.02, 51.03, ta52.01, 51.05, 51.06, ta51.01, 51.08, 51.09, 51.11–51.15, 51.17, 51.18, 51.19, 51.20, 53.16, 53.26, 57.9

Sara Krause: Figures: 19.09, 24.01, 24.05, 25.03, 33.13, 34.02, 34.19, 34.20, 34.21, 35.12, 35.18, 37.17, 42.08

Tiana Litwak: Figures: 19.12, 19.14, 23.05, 23.11, 23.13, 34.34a, 35.19, 35.21, 35.22, 40.09

Jacqueline Mahannah: Figures: ta32.02, 35.14b, 35.15, 35.39, 46.08, 47.09, 47.18

Gwen DeCelles: Figures: 16.19, 16.21, 18.09, 19.05, 19.12, 19.14, 19.21, 23.11, 25.02, 25.07, 26.12, 28.11, 28.13, 29.11, 29.13, 33.12, 34.09, 35.04, 35.13, 37.08, 39.01, 41.06, 41.20, 41.29, 42.02, 48.15, 50.03, 53.06, 59.01

Matthew McAdams: Figures: 20.05, 20.06, 20.12, 22.05, 22.08, 22.15, 55.06, 55.08, 55.15

Thomas Moorman: Figures: 1.01, 6.21, 20.05, ta22.01, 54.01, 54.02, 54.18, 54.29, 55.12

Fiona Morris: Figures: 3.16, 20.05, 21.20, 54.25, 56.03, 56.04, 56.06, 56.08, 56.14, 56.20

Kim Moss: Figures: 1.12, 3.15, 3.22, 3.29, 10.05, 13.12, 15.17, 15.18, 15.19, 15.22, 16.02, 16.09, 16.12, 16.14, 17.02, 17.11, 18.03, 23.13, ta30.01, 30.23, 43.02, 45.02, 45.20, 46.08, 47.05, 47.10, 48.06, 48.15, 48.16, 49.11, 57.15, 57.16

Evelyn Pence: Figures: 4.05a, 4.06, 4.07, 4.08a, 4.10, 4.12, 4.14, 4.17, 4.18, 4.19, 4.20, 4.22, 4.26, 5.02, 8.19, 8.21, 19.06, 19.07, 22.15, CH 27, CH 28, 36.32, 37.01, 37.02, 37.03, 37.05, 41.12, 41.18, 41.19, 41.20, 41.21

Curtis Perone: Figures: 16.02, 16.07, 20.07

Michael Rothman: Figure 37.15

Tara Russo: Figures: 3.02, 3.03, 3.21, 3.24, 3.25, 3.27, 3.30, 5.13, 6.01, 6.09, 6.10, 6.12, 6.13, 6.14, 22.19

Alison Schroer: Figure: 43.02

Cameron Slayden: Figures: 35.14, 3515, 35.39

Rick Simonson: Figures: 19.13, 23.09, 25.01, 25.04, 26.16, 36.21, 42.27

Zeke Smith: Figures: ta35.04, 47.06b, 47.08, 50.12, 53.31

Tami Tolpa: Figures: 15.12, 17.02

Brook Wainwright: Figures: 16.03, 16.04, 16.05, 16.06, 20.01, 20.11, 26.08, 28.08, 28.11, 28.12

Travis Vermilye: Figures: ta43.02, ta43.03, 45.11, 45.13, 45.22

index

Boldface page numbers correspond with **boldface terms** in the text. Page numbers followed by an "f" indicate figures; page numbers followed by a "t" indicate tabular material.

C

D

E

F

G

J

K

L

M

O

P

S

T

U

V